Optical Fiber Communications

ONE WEEK LOAN

We work with leading authors to develop the
strongest educational materials in engineering,
bringing cutting-edge thinking and best
learning practice to a global market.

Under a range of well-known imprints, including
Prentice Hall, we craft high quality print and
electronic publications which help readers to
understand and apply their content,
whether studying or at work.

To find out more about the complete range of our
publishing, please visit us on the World Wide Web at:
www.pearsoned.co.uk

Optical Fiber Communications Principles and Practice

Third edition

John M. Senior

assisted by

M. Yousif Jamro

 **FT** Prentice Hall

FINANCIAL TIMES

An imprint of **Pearson Education**

Harlow, England • London • New York • Boston • San Francisco • Toronto • Sydney • Singapore • Hong Kong
Tokyo • Seoul • Taipei • New Delhi • Cape Town • Madrid • Mexico City • Amsterdam • Munich • Paris • Milan

Pearson Education Limited
Edinburgh Gate
Harlow
Essex CM20 2JE
England

and Associated Companies throughout the world

Visit us on the World Wide Web at:
www.pearsoned.co.uk

First published 1985
Second edition 1992
Third edition published 2009

© Prentice Hall Europe 1985, 1992
© Pearson Education Limited 2009

ISBN: 978-0-13-032681-2

British Library Cataloguing-in-Publication Data
A catalogue record for this book is available from the British Library

Library of Congress Cataloging-in-Publication Data
Senior, John M., 1951–
    Optical fiber communications : principles and practice / John M. Senior, assisted by
M. Yousif Jamro. — 3rd ed.
        p.   cm.
    Includes bibliographical references and index.
    ISBN-13: 978-0-13-032681-2 (alk. paper)   1. Optical communications.   2. Fiber optics.
I. Jamro, M. Yousif.   II. Title.
TK5103.59.S46 2008
621.382′75—dc22
                                                                            2008018133

10   9   8   7   6   5   4   3   2   1
12   11   10   09   08

Typeset in 10/12 Times by 35
Printed and bound by Ashford Colour Press Ltd, Gosport

To Judy and my mother Joan, and in memory of my father Ken

Contents

Chapter 3: Transmission characteristics of optical fibers   86

Contents ix

Chapter 4: Optical fibers and cables   169

Chapter 9: Direct detection receiver performance considerations   502

Chapter 12: Optical fiber systems 1: intensity modulation/direct detection   673

Supporting resources

Visit **www.pearsoned.co.uk/senior-optical** to find valuable online resources

For instructors
- An Instructor's Manual that provides full solutions to all the numerical problems, which are provided at the end of each chapter in the book.

For more information please contact your local Pearson Education sales representative or visit **www.pearsoned.co.uk/senior-optical**

Preface

The preface to the second edition drew attention to the relentless onslaught in the development of optical fiber communications technology identified in the first edition in the context of the 1980s. Indeed, although optical fiber communications could now, nearly two decades after that period finished, be defined as mature, this statement fails to signal the continuing rapid and extensive developments that have subsequently taken place. Furthermore the pace of innovation and deployment fuelled, in particular, by the Internet is set to continue with developments in the next decade likely to match or even exceed those which have occurred in the last decade. Hence this third edition seeks to record and explain the improvements in both the technology and its utilization within what is largely an optical fiber global communications network.

Major advances which have occurred while the second edition has been in print include: those associated with low-water-peak and high-performance single-mode fibers; the development of photonic crystal fibers; a new generation of multimode graded index plastic optical fibers; quantum-dot fabrication for optical sources and detectors; improvements in optical amplifier technology and, in particular, all-optical regeneration; the realization of photonic integrated circuits to provide ultrafast optical signal processing together with silicon photonics; developments in digital signal processing to mitigate fiber transmission impairments and the application of forward error correction strategies. In addition, there have been substantial enhancements in transmission and multiplexing techniques such as the use of duobinary-encoded transmission, orthogonal frequency division multiplexing and coarse/dense wavelength division multiplexing, while, more recently, there has been a resurgence of activity concerned with coherent and, especially, phase-modulated transmission. Finally, optical networking techniques and optical networks have become established employing both specific reference models for the optical transport network together with developments originating from local area networks based on Ethernet to provide for the future optical Internet (i.e. 100 Gigabit Ethernet for carrier-class transport networks). Moreover, driven by similar broadband considerations, activity has significantly increased in relation to optical fiber solutions for the telecommunication access network.

Although a long period has elapsed since the publication of the second edition in 1992, it has continued to be used extensively in both academia and industry. Furthermore, as delays associated with my ability to devote the necessary time to writing the updates for this edition became apparent, it has been most gratifying that interest from the extensive user community of the second edition has encouraged me to find ways to pursue the necessary revision and enhancement of the book. A major strategy to enable this process has been the support provided by my former student and now colleague, Dr M. Yousif Jamro, working with me, undertaking primary literature searches and producing update drafts for many chapters which formed the first stage of the development for the new edition. An extensive series of iterations, modifications and further additions then ensued to craft the final text.

In common with the other editions, this edition relies upon source material from the numerous research and other publications in the field including, most recently, the Proceedings of the 33rd European Conference on Optical Communications (ECOC'07) which took place in Berlin, Germany, in September 2007. Furthermore, it also draws upon the research activities of the research group focused on optical systems and networks that I established at the University of Hertfordshire when I took up the post as Dean of Faculty in 1998, having moved from Manchester Metropolitan University. Although the book remains a comprehensive introductory text for use by both undergraduate and postgraduate engineers and scientists to provide them with a firm grounding in all significant aspects of the technology, it now also encompasses a substantial chapter devoted to optical networks and networking concepts as this area, in totality, constitutes the most important and extensive range of developments in the field to have taken place since the publication of the second edition.

In keeping with a substantial revision and updating of the content, then, the practical nature of the coverage combined with the inclusion of the relevant up-to-date standardization developments has been retained to ensure that this third edition can continue to be widely employed as a reference text for practicing engineers and scientists. Following very positive feedback from reviewers in relation to its primary intended use as a teaching/learning text, the number of worked examples interspersed throughout the book has been increased to over 120, while a total of 372 problems are now provided at the end of relevant chapters to enable testing of the reader's understanding and to assist tutorial work. Furthermore, in a number of cases they are designed to extend the learning experience facilitated by the book. Answers to the numerical problems are provided at the end of the relevant sections in the book and the full solutions can be accessed on the publisher's website using an appropriate password.

Although the third edition has grown into a larger book, its status as an introductory text ensures that the fundamentals are included where necessary, while there has been no attempt to cover the entire field in full mathematical rigor. Selected proofs are developed, however, in important areas throughout the text. It is assumed that the reader is conversant with differential and integral calculus and differential equations. In addition, the reader will find it useful to have a grounding in optics as well as a reasonable familiarity with the fundamentals of solid-state physics.

This third edition is structured into 15 chapters to facilitate a logical progression of material and to enable straightforward access to topics by providing the appropriate background and theoretical support. Chapter 1 gives a short introduction to optical fiber communications by considering the historical development, the general system and the major advantages provided by this technology. In Chapter 2 the concept of the optical fiber as a transmission medium is introduced using the simple ray theory approach. This is followed by discussion of electromagnetic wave theory applied to optical fibers prior to consideration of lightwave transmission within the various fiber types. In particular, single-mode fiber, together with a more recent class of microstructured optical fiber, referred to as photonic crystal fiber, are covered in further detail. The major transmission characteristics of optical fibers are then dealt with in Chapter 3. Again there is a specific focus on the properties and characteristics of single-mode fibers including, in this third edition, enhanced discussion of single-mode fiber types, polarization mode dispersion, nonlinear effects and, in particular, soliton propagation.

Chapters 4 and 5 deal with the more practical aspects of optical fiber communications and therefore could be omitted from an initial teaching program. A number of these areas, however, are of crucial importance and thus should not be lightly overlooked. Chapter 4 deals with the manufacturing and cabling of the various fiber types, while in Chapter 5 the different techniques to provide optical fiber connection are described. In this latter chapter both fiber-to-fiber joints (i.e. connectors and splices) are discussed as well as fiber branching devices, or couplers, which provide versatility within the configuration of optical fiber systems and networks. Furthermore, a new section incorporating coverage of optical isolators and circulators which are utilized for the manipulation of signals within optical networks has been included.

Chapters 6 and 7 describe the light sources employed in optical fiber communications. In Chapter 6 the fundamental physical principles of photoemission and laser action are discussed prior to consideration of the various types of semiconductor and nonsemiconductor laser currently in use, or under investigation, for optical fiber communications. The other important semiconductor optical source, namely the light-emitting diode, is dealt with in Chapter 7.

The next two chapters are devoted to the detection of the optical signal and the amplification of the electrical signal obtained. Chapter 8 discusses the basic principles of optical detection in semiconductors; this is followed by a description of the various types of photodetector currently employed. The optical fiber direct detection receiver is then considered in Chapter 9, with particular emphasis on its performance characteristics.

Enhanced coverage of optical amplifiers and amplification is provided in Chapter 10, which also incorporates major new sections concerned with wavelength conversion processes and optical regeneration. Both of these areas are of key importance for current and future global optical networks. Chapter 11 then focuses on the fundamentals and ongoing developments in integrated optics and photonics providing descriptions of device technology, optoelectronic integration and photonic integrated circuits. In addition, the chapter includes a discussion of optical bistability and digital optics which leads into an overview of optical computation.

Chapter 12 draws together the preceding material in a detailed discussion of the major current implementations of optical fiber communication systems (i.e. those using intensity modulation and the direct detection process) in order to give an insight into the design criteria and practices for all the main aspects of both digital and analog fiber systems. Two new sections have been incorporated into this third edition dealing with the crucial topic of dispersion management and describing the research activities into the performance attributes and realization of optical soliton systems.

Over the initial period since the publication of the second edition, research interest and activities concerned with coherent optical fiber communications ceased as a result of the improved performance which could be achieved using optical amplification with conventional intensity modulation–direct detection optical fiber systems. Hence no significant progress in this area was made for around a decade until a renewed focus on coherent optical systems was initiated in 2002 following experimental demonstrations using phase-modulated transmission. Coherent and phase-modulated optical systems are therefore dealt with in some detail in Chapter 13 which covers both the fundamentals and the initial period of research and development associated with coherent transmission prior to 1992, together with the important recent experimental system and field trial demonstrations primarily focused on phase-modulated transmission that have taken place since 2002. In particular a major new section describing differential phase shift keying systems together with new sections on polarization multiplexing and high-capacity transmission have been incorporated into this third edition.

Chapter 14 provides a general treatment of the major measurements which may be undertaken on optical fibers in both the laboratory and the field. The chapter is incorporated at this stage in the book to enable the reader to obtain a more complete understanding of optical fiber subsystems and systems prior to consideration of these issues. It continues to include the measurements required to be taken on single-mode fibers and it addresses the measurement techniques which have been adopted as national and international standards.

Finally, Chapter 15 on optical networks comprises an almost entirely new chapter for the third edition which provides both a detailed overview of this expanding field and a discussion of all the major aspects and technological solutions currently being explored. In particular, important implementations of wavelength routing and optical switching networks are described prior to consideration of the various optical network deployments that have occurred or are under active investigation. The chapter finishes with a section which addresses optical network protection and survivability.

The book is also referenced throughout to extensive end-of-chapter references which provide a guide for further reading and also indicate a source for those equations that have been quoted without derivation. A complete list of symbols, together with a list of common abbreviations in the text, is also provided. SI units are used throughout the book.

I must extend my gratitude for the many useful comments and suggestions provided by the diligent reviewers that have both encouraged and stimulated improvements to the text. Many thanks are also given to the authors of the multitude of journal and conference papers, articles and books that have been consulted and referenced in the preparation of this third edition and especially to those authors, publishers and companies who have kindly granted permission for the reproduction of diagrams and photographs. I would also like to thank the many readers of the second edition for their constructive and courteous feedback which has enabled me to make the substantial improvements that now comprise this third edition. Furthermore, I remain extremely grateful to my family and friends who have continued to be supportive and express interest over the long period of the revision for this edition of the book. In particular, my very special thanks go to Judy for her continued patience and unwavering support which enabled me to finally complete the task, albeit at the expense of evenings and weekends which could have been spent more frequently together.

John M. Senior

Acknowledgements

We are grateful to the following for permission to reproduce copyright material:

Figures 2.17 and 2.18 from Weakly guiding fibers in *Applied Optics*, 10, p. 2552, OSA (Gloge, D. 1971), with permission from The Optical Society of America; Figure 2.30 from Fiber manufacture at AT&T with the MCVD process in *Journal of Lightwave Technology*, LT-4(8), pp. 1016–1019, OSA (Jablonowski, D. P. 1986), with permission from The Optical Society of America; Figure 2.35 from Gaussian approximation of the fundamental modes of graded-index fibers in *Journal of the Optical Society of America*, 68, p. 103, OSA (Marcuse, D. 1978), with permission from The Optical Society of America; Figure 2.36 from *Applied Optics*, 19, p. 3151, OSA (Matsumura, H. and Suganuma, T. 1980), with permission from The Optical Society of America; Figures 3.1 and 3.3 from Ultimate low-loss single-mode fibre at 1.55 mum in *Electronic Letters*, 15(4), pp. 106–108, Institution of Engineering and Technology (T Miya, T., Teramuna, Y., Hosaka, Y. and Miyashita, T. 1979), with permission from IET; Figure 3.2 from *Applied Physics Letters*, 22, 307, Copyright 1973, American Institute of Physics (Keck, D. B., Maurer, R. D. and Schultz, P. C. 1973), reproduced with permission; Figure 3.10 from *Electronic Letters*, 11, p. 176, Institution of Engineering and Technology (Payne, D. N. and Gambling, W. A. 1975), with permission from IET; Figures 3.15 and 3.17 from *The Radio and Electronic Engineer*, 51, p. 313, Institution of Engineering and Technology (Gambling, W. A., Hartog, A. H. and Ragdale, C. M. 1981), with permission from IET; Figure 3.18 from High-speed optical pulse transmission at 1.29 mum wavelength using low-loss single-mode fibers in *IEEE Journal of Quantum Electronics*, QE-14, p. 791, IEEE (Yamada, J. I., Saruwatari, M., Asatani, K., Tsuchiya, H., Kawana, A., Sugiyama, K. and Kumara, T. 1978), © IEEE 1978, reproduced with permission; Figure 3.30 from Polarization-maintaining fibers and their applications in *Journal of Lightwave Technology*, LT-4(8), pp. 1071–1089, OSA (Noda, J., Okamoto, K. and Susaki, Y. 1986), with permission from The Optical Society of America; Figure 3.34 from Nonlinear phenomena in optical fibers in *IEEE Communications Magazine*, 26, p. 36, IEEE (Tomlinson, W. J. and Stolen, R. H. 1988), © IEEE 1988, reproduced with permission; Figures 4.1 and 4.4 from Preparation of sodium borosilicate glass fibers for optical communication in *Proceedings of IEE*, 123, pp. 591–595, Institution of Engineering and Technology (Beales, K. J., Day, C. R., Duncan, W. J., Midwinter, J. E. and Newns, G. R. 1976), with permission from IET; Figure 4.5 from A review of glass fibers for optical communications in *Phys. Chem. Glasses*, 21(1), p. 5, Society of Glass Technology (Beales, K. J. and Day, C. R. 1980), reproduced with permission; Figure 4.7 Reprinted from *Optics Communication*, 25, pp. 43–48, D. B. Keck and R. Bouilile, Measurements on high-bandwidth optical waveguides, copyright 1978, with permission from Elsevier; Figure 4.8 from Low-OH-content optical fiber fabricated by vapor-phase axial-deposition method in *Electronic Letters*, 14(17), pp. 534–535,

Institution of Engineering and Technology (Sudo, S., Kawachi, M., Edahiro, M., Izawa, T., Shoida, T. and Gotoh, H. 1978), with permission from IET; Figure 4.20 from Optical fibre cables in *Radio and Electronic Engineer* (IERE J.), 51(7/8), p. 327, Institution of Engineering and Technology (Reeve, M. H. 1981), with permission from IET; Figure 4.21 from Power loss, modal noise and distortion due to microbending of optical fibres in *Applied Optics*, 24, pp. 2323, OSA (Das, S., Englefield, C. G. and Goud, P. A. 1985), with permission from The Optical Society of America; Figure 4.22 from Hydrogen induced loss in MCVD fibers, Optical Fiber Communication Conference, OFC 1985, USA, TUII, February 1985, OFC/NFOEC (Lemaire, P. J. and Tomita, A. 1985), with permission from The Optical Society of America; Figures 5.2 (a) and 5.16 (a) from Connectors for optical fibre systems in *Radio and Electronic Engineer* (*J. IERE*), 51(7/8), p. 333, Institution of Engineering and Technology (Mossman, P. 1981), with permission from IET; Figure 5.5 (b) from Jointing loss in single-loss fibres in *Electronic Letters*, 14(3), pp. 54–55, Institution of Engineering and Technology (Gambling, W. A., Matsumura, H. and Cowley, A. G. 1978), with permission from IET; Figure 5.7 (a) from Figure 1, page 1, *Optical Fiber Arc Fusion Splicer FSM-45F*, No.: B- 06F0013Cm, 13 February 2007, http://www.fujikura.co.jp/00/splicer/front-page/pdf/e_fsm-45f.pdf; with permission from Fujikura Limited; Figure 5.7 (b) from Figure 4, page 1, Arc Fusion Splicer, SpliceMate, *SpliceMate Brochure*, http://www.fujikura.co.jp/00/splicer/front-page/pdf/splicemate_brochure.pdf, with permission from Fujikura Limited; Figure 5.8 (a) from Optical communications research and technology in *Proceedings of the IEEE*, 66(7), pp. 744–780, IEEE (Giallorenzi, T. G. 1978), © IEEE 1978, reproduced with permission; Figure 5.13 from Simple high-performance mechanical splice for single mode fibers in *Proceedings of the Optical Fiber Communication Conference*, OFC 1985, USA, paper M12, OFC/NFOEC (Miller, C. M., DeVeau, G. F. and Smith, M. Y. 1985), with permission from The Optical Society of America; Figure 5.15 from Rapid ribbon splice for multimode fiber splicing in *Proceedings of the Optical Fiber Communication Conference*, OFC1985, USA, paper TUQ27, OFC/NFOEC (Hardwick, N. E. and Davies, S. T. 1985), with permission from The Optical Society of America; Figure 5.21 (a) from Demountable multiple connector with precise V-grooved silicon in *Electronic Letters*, 15(14), pp. 424–425, Institution of Engineering and Technology (Fujii, Y., Minowa, J. and Suzuki, N. 1979), with permission from IET; Figure 5.21 (b) from Very small single-mode ten-fiber connector in *Journal of Lightwave Technology*, 6(2), pp. 269–272, OSA (Sakake, T., Kashima, N. and Oki, M. 1988), with permission from The Optical Society of America; Figure 5.20 from High-coupling-efficiency optical interconnection using a 90-degree bent fiber array connector in optical printed circuit boards in *IEEE Photonics Technology Letters*, 17(3), pp. 690–692, IEEE (Cho, M. H., Hwang, S. H., Cho, H. S. and Park, H. H. 2005), © IEEE 2005, reproduced with permission; Figure 5.22 (a) from Practical low-loss lens connector for optical fibers in *Electronic Letters*, 14(16), pp. 511–512, Institution of Engineering and Technology (Nicia, A. 1978), with permission from IET; Figure 5.23 from Assembly technology for multi-fiber optical connectivity solutions in *Proceedings of IEEE/LEOS Workshop on Fibres and Optical Passive Components*, 22–24 June 2005, Mondello, Italy, IEEE (Bauknecht, R. Kunde, J., Krahenbuhl, R., Grossman, S. and Bosshard, C. 2005), © IEEE 2005, reproduced with permission; Figure 5.31 from Polarization-independent optical circulator consisting of two fiber-optic polarizing beamsplitters and two YIG spherical lenses in *Electronic Letters*, 22, pp. 370–372, Institution of Engineering and

Technology (Yokohama, I., Okamoto, K. and Noda, J. 1985), with permission from IET; Figures 5.36 and 5.38 (a) from Optical demultiplexer using a silicon echette grating in *IEEE Journal of Quantum Electronics*, QE-16, pp. 165–169, IEEE (Fujii, Y., Aoyama, K. and Minowa, J. 1980), © IEEE 1980, reproduced with permission; Figure 5.44 from Filterless 'add' multiplexer based on novel complex gratings assisted coupler in *IEEE Photonics Technology Letters*, 17(7), pp. 1450–1452, IEEE (Greenberg, M. and Orenstein, M. 2005), © IEEE 2005, reproduced with permission; Figure 6.33 from Low threshold operation of 1.5 μm DFB laser diodes in *Journal of Lightwave Technology*, LT-5, p. 822, IEEE (Tsuji, S., Ohishi, A., Nakamura, H., Hirao, M., Chinone, N. and Matsumura, H. 1987), © IEEE 1987, reproduced with permission; Figure 6.37 adapted from *Vertical-Cavity Surface-Emitting Lasers: Design, Fabrication, Characterization, and Applications,* Cambridge University Press (Wilmsen, C. W., Temkin, H. and Coldren, L. A. 2001), reproduced with permission; Figure 6.38 from Semiconductor laser sources for optical communication in *Radio and Electronic Engineer*, 51, p. 362, Institution of Engineering and Technology (Kirby, P. A. 1981), with permission from IET; Figure 6.47 from Optical amplification in an erbium-doped fluorozirconate fibre between 1480 nm and 1600 nm in *IEE Conference Publication* 292, Pt 1, p. 66, Institution of Engineering and Technology (Millar, C. A., Brierley, M. C. and France, P. W. 1988), with permission from IET; Figure 6.48 (a) from High efficiency Nd-doped fibre lasers using direct-coated dielectric mirrors in *Electronic Letters*, 23, p. 768, Institution of Engineering and Technology (Shimtzu, M., Suda, H. and Horiguchi, M. 1987), with permission from IET; Figure 6.48 (b) from Rare-earth-doped fibre lasers and amplifiers in *IEE Conference Publication*, 292. Pt 1, p. 49, Institution of Engineering and Technology (Payne, D. N. and Reekie, L. 1988), with permission from IET; Figure 6.53 from Wavelength-tunable and single-frequency semiconductor lasers for photonic communications networks in *IEEE Communications Magazine*, October, p. 42, IEEE (Lee, T. P. and Zah, C. E. 1989), © IEEE 1989, reproduced with permission; Figure 6.55 from Single longitudinal-mode operation on an Nd3+-doped fibre laser in *Electronic Letters*, 24, pp. 24–26, IEEE (Jauncey, I. M., Reekie, L., Townsend, K. E. and Payne, D. N. 1988), © IEEE 1988, reproduced with permission; Figure 6.56 from Tunable single-mode fiber lasers in *Journal of Lightwave Technology*, LT-4, p. 956, IEEE (Reekie, L., Mears, R. J., Poole, S. B. and Payne, D. N. 1986), © IEEE 1986, reproduced with permission; Figure 6.58 reprinted from *Semiconductors and Semimetals: Lightwave communication technology*, 22C, Y. Horikoshi, 'Semiconductor lasers with wavelengths exceeding 2 μm', pp. 93–151, 1985, edited by W. T. Tsang (volume editor), copyright 1985, with permission from Elsevier; Figure 6.59 from PbEuTe lasers with 4–6 μm wavelength mode with hot-well epitaxy in *IEEE Journal of Quantum Electronics*, 25(6), pp. 1381–1384, IEEE (Ebe, H., Nishijima, Y. and Shinohara, K. 1989), © IEEE 1989, reproduced with permission; Figure 7.5 reprinted from *Optical Communications*, 4, C. A. Burrus and B. I. Miller, Small-area double heterostructure aluminum-gallium arsenide electroluminsecent diode sources for optical fiber transmission lines, pp. 307–369, 1971, copyright 1971, with permission from Elsevier; Figure 7.6 from High-power single-mode optical-fiber coupling to InGaAsP 1.3 μm mesa-structure surface-emitting LEDs in *Electronic Letters*, 21(10), pp. 418–419, Institution of Engineering and Technology (Uji, T. and Hayashi, J. 1985), with permission from IET; Figure 7.8 from Sources and detectors for optical fiber communications applications: the first 20 years in *IEE Proceedings on Optoelectronics*, 133(3), pp. 213–228,

Institution of Engineering and Technology (Newman, D. H. and Ritchie, S. 1986), with permission from IET; Figure 7.9 (a) from 2 Gbit/s and 600 Mbit/s single-mode fibre-transmission experiments using a high-speed Zn-doped 1.3 μm edge-emitting LED in *Electronic Letters*, 13(12), pp. 636–637, Institution of Engineering and Technology (Fujita, S., Hayashi, J., Isoda, Y., Uji, T. and Shikada, M. 1987), with permission from IET; Figure 7.9 (b) from Gigabit single-mode fiber transmission using 1.3 μm edge-emitting LEDs for broadband subscriber loops in *Journal of Lightwave Technology*, LT-5(10) pp. 1534–1541, OSA (Ohtsuka, T., Fujimoto, N., Yamaguchi, K., Taniguchi, A., Naitou, N. and Nabeshima, Y. 1987), with permission from The Optical Society of America; Figure 7.10 (a) from A stripe-geometry double-heterostructure amplified-spontaneous-emission (superluminescent) diode in IEEE Journal of Quantum Electronics QE-9, p. 820 (Lee, T. P., Burrus, C. A. and Miller, B. I. 1973), with permission from IET; Figure 7.10 (b) from High output power GaInAsP/InP superluminescent diode at 1.3 μm in *Electronic Letters*, 24(24) pp. 1507–1508, Institution of Engineering and Technology (Kashima, Y., Kobayashi, M. and Takano, T. 1988), with permission from IET; Figure 7.14 from Highly efficient long lived GaAlAs LEDs for fiber-optical communications in *IEEE Trans. Electron Devices*, ED-24(7) pp. 990–994, Institution of Engineering and Technology (Abe, M., Umebu, I., Hasegawa, O., Yamakoshi, S., Yamaoka, T., Kotani, T., Okada, H., and Takamashi, H. 1977), with permission from IET; Figure 7.15 from CaInAsP/InP fast, high radiance, 1.05–1.3 μm wavelength LEDs with efficient lens coupling to small numerical aperture silica optical fibers in *IEEE Trans Electron. Devices*, ED-26(8), pp. 1215–1220, Institution of Engineering and Technology (Goodfellow, R. C., Carter, A. C., Griffith, I. and Bradley, R. R. 1979), with permission from IET; Figures 7.19 and 7.23 were published in *Optical Fiber Telecommunications II*, T. P. Lee, C. A. Burrus Jr and R. H. Saul, Light-emitting diodes for telecommunications, pp. 467–507, edited by S. E. Miller and I. P. Kaminow, 1988, Copyright Elsevier 1988; Figure 7.20 from Lateral confinement InGaAsP superluminescent diode at 1.3 μm in *IEEE Journal of Quantum Electronics*, QE19, p. 79, IEEE (Kaminow, I. P., Eisenstein, G., Stulz, L. W. and Dentai, A. G. 1983), © IEEE 1983, reproduced with permission; Figure 7.21 adapted from Figure 6, page 121 of AlGaInN resonant-cavity LED devices studied by electromodulated reflectance and carrier lifetime techniques in *IEE Proceedings on Optoelectronics*, vol. 152, no. 2, pp. 118–124, 8 April 2005, Institution of Engineering and Technology (Blume, G., Hosea, T. J. C., Sweeney, S. J., de Mierry, P., Lancefield, D. 2005), with permission from IET; Figure 7.22 (b) from Light-emitting diodes for optical fibre systems in *Radio and Electronic Engineer* (J. IERE), 51(7/8), p. 41, Institution of Engineering and Technology (Carter, A. C. 1981), with permission from IET; Figure 8.3 from *Optical Communications Essentials (Telecommunications)*, McGraw-Hill Companies (Keiser, G. 2003), with permission of the McGraw-Hill Companies; Figure 8.19 (a) from Improved germanium avalanche photodiodes in *IEEE Journal of Quantum Electronics*, QE-16(9), pp. 1002–1007 (Mikami, O., Ando, H., Kanbe, H., Mikawa, T., Kaneda, T. and Toyama, Y. 1980), © IEEE 1980, reproduced with permission; Figure 8.19 (b) from High-sensitivity Hi-Lo germanium avalanche photodiode for 1.5 μm wavelength optical communication in *Electronic Letters*, 20(13), pp. 552–553, Institution of Engineering and Technology (Niwa, M., Tashiro, Y., Minemura, K. and Iwasaki, H. 1984), with permission from IET; Figure 8.24 from Impact ionisation in multi-layer heterojunction structures in *Electronic Letters*, 16(12), pp. 467–468, Institution of Engineering and Technology (Chin, R.,

Holonyak, N., Stillman, G. E., Tang, J. Y. and Hess, K. 1980), with permission from IET; Figure 8.25 Reused with permission from Federico Capasso, *Journal of Vacuum Science & Technology* B, 1, 457 (1983). Copyright 1983, AVS The Science & Technology Society; Figure 8.29 Reused with permission from P. D. Wright, R. J. Nelson, and T. Cella, *Applied Physics Letters*, 37, 192 (1980). Copyright 1980, American Institute of Physics; Figure 8.32 from MSM-based integrated CMOS wavelength-tunable optical receiver in *IEEE Photonics Technology Letters*, 17(6) pp. 1271–1273 (Chen, R., Chin, H., Miller, D. A. B., Ma, K. and Harris Jr., J. S. 2005); © IEEE 2005, reproduced with permission; Figure 9.5 from Receivers for optical fibre communications in *Electronic and Radio Engineer*, 51(7/8), p. 349, Institution of Engineering and Technology (Garrett, I. 1981), with permission from IET; Figure 9.7 from Photoreceiver architectures beyond 40 Gbit/s, IEEE Symposium on Compound Semiconductor Integrated circuits, Monterey, California, USA, pp. 85–88, October (Ito, H. 2004), © IEEE 2004, reproduced with permission; Figure 9.14 from GaAs FET tranimpedance front-end design for a wideband optical receiver in *Electronic Letters*, 15(20), pp. 650–652, Institution of Engineering and Technology (Ogawa, K. and Chinnock, E. L. 1979), with permission from IET; Figure 9.15 published in *Optical Fiber Telecommunications II*, B. L. Kaspar, Receiver design, p. 689, edited by S. E. Miller and I. P. Kaminow, 1988, Copyright Elsevier 1988; Figure 9.17 from An APD/FET optical receiver operating at 8 Gbit/s in *Journal of Lightwave Technology*, LT-5(3) pp. 344–347, OSA (Kaspar, B. L., Campbell, J. C., Talman, J. R., Gnauck, A. H., Bowers, J. E. and Holden, W. S. 1987), with permission from The Optical Society of America; Figure 9.23 Reprinted from *Optical Fiber Telecommunications IV A: Components*, B. L. Kaspar, O. Mizuhara and Y. K. Chen, High bit-rates receivers, transmiters and electronics, pp. 784–852, Figure 1.13, page 807, edited by I. P. Kaminow and T. Li, Copyright 2002, with permission from Elsevier; Figure 10.3 from Semiconductor laser optical amplifiers for use in future fiber systems in *Journal of Lightwave Technology* 6(4), p. 53, OSA (O'Mahony, M. J. 1988), with permission from The Optical Society of America; Figure 10.8 from Noise performance of semiconductor optical amplifiers, International Conference on Trends in Communication, EUROCON, 2001, Bratislava, Slovakia, 1, pp. 161–163, July (Udvary, E. 2001), © IEEE 2001, reproduced with permission; Figure 10.17 from Properties of fiber Raman amplifiers and their applicability to digital optical communication systems in *Journal of Lightwave Technology*, 6(7), p. 1225, IEEE (Aoki, Y. 1988), © IEEE 1988, reproduced with permission; Figure 10.18 (a) from Semiconductor Raman amplifier for terahertz bandwidth optical communication in *Journal of Lightwave Technology*, 20(4), pp. 705–711, IEEE (Suto, K., Saito, T., Kimura, T., Nishizawa, J. I. and Tanube, T. 2002), © IEEE 2002, reproduced with permission; Figure 11.2 from Scaling rules for thin-film optical waveguides, *Applied Optics*, 13(8), p. 1857, OSA (Kogelnik, H. and Ramaswamy, V. 1974), with permission from the Optical Society of America; Figure 11.7 Reused with permission from M. Papuchon, Y. Combemale, X. Mathieu, D. B. Ostrowsky, L. Reiber, A. M. Roy, B. Sejourne, and M. Werner, *Applied Physics Letters*, 27, 289 (1975). Copyright 1975, American Institute of Physics; Figure 11.13 from Beam-steering micromirrors for large optical cross-connects in *Journal of Lightwave Technology*, 21(3), pp. 634–642, OSA (Aksyuk, V. A. *et al.* 2003), with permission from The Optical Society of America; Figure 11.23 from 5 Git/s modulation characteristics of optical intensity modulator monolithically integrated with DFB laser in *Electronic Letters*, 25(5), pp. 1285–1287, Institution of Engineering and

Technology (Soda, H., Furutsa, M., Sato, K., Matsuda, M. and Ishikawa, H. 1989), with permission from IET; Figure 11.24 from Widely tunable EAM-integrated SGDBR laser transmitter for analog applications in *IEEE Photonics Technology Letters*, 15(9), pp. 1285–1297, IEEE (Johansson, L. A., Alkulova, Y. A., Fish, G. A. and Coldren, L. A. 2003), © IEEE 2003, reproduced with permission; Figure 11.25 from 80-Gb/s InP-based waveguide-integrated photoreceiver in *IEEE Journal of Sel. Top. Quantum Electronics*, 11(2), pp. 356–360, IEEE (Mekonne, G. G., Bach, H. G., Beling, A., Kunkel, R., Schmidt, D. and Schlaak, W. 2005), © IEEE 2005, reproduced with permission; Figure 11.27 from Wafer-scale replication of optical components on VCSEL wafers in *Proceedings of Optical Fiber Communication*, OFC 2004, Los Angeles, USA, vol. 1, 23–27 February, © IEEE 2004, reproduced with permission; Figure 11.28 from Terabus: terabit/second-class card-level optical interconnect technologies in *IEEE Journal of Sel. Top. Quantum Electronics*, 12(5), pp. 1032–1044, IEEE (Schares, L., Kash, J. A., Doany, F. E., Schow, C. L., Schuster, C., Kuchta, D. M., Pepeljugoski, P. K., Trewhella, J. M., Baks, C. W. and John, R. A. 2006), © IEEE 2006, reproduced with permission; Figure 11.29 from Figure 2, http://www.fujitsu.com/global/news/pr/archives/month/2007/20070119-01.html, courtesy of Fujitsu Limited; Figure 11.33 (b) from Large-scale InP photonic integrated circuits: enabling efficient scaling of optical transport networks, *IEEE Journal of Se. Top. Quantum Electronics*, 13(1) pp. 22–31, IEEE (Welch, D. F. *et al.* 2007), © IEEE 2007, reproduced with permission; Figure 11.33 (c) from Monolithically integrated 100-channel WDM channel selector employing low-crosstalk AWG in *IEEE Photonics Technology Letters*, 16(11), pp. 2481–2483, IEEE (Kikuchi, N., Shibata, Y., Okamoto, H., Kawaguchi, Y., Oku, S., Kondo, Y. and Tohmori, Y. 2004), © IEEE 2004, reproduced with permission; Figure 11.35 Reused with permission from P. W. Smith, I. P. Kaminow, P. J. Maloney, and L. W. Stulz, *Applied Physics Letters*, 33, 24 (1978). Copyright 1978, American Institute of Physics; Figure 11.38 from All-optical flip-flop multimode interference bistable laser diode in *IEEE Photonics Technology Letters*, 17(5), pp. 968–970, IEEE (Takenaka, M., Raburn, M. and Nakano, Y. 2005), © IEEE 2005, reproduced with permission; Figure 11.42 from Optical bistability, phonomic logic and optical computation in *Applied Optics*, 25, pp. 1550–1564, OSA (Smith, S. D. 1986), with permission from The Optical Society of America; Figure 12.4 from Non-linear phase distortion and its compensation in LED direct modulation in Electronic Letters, 13(6), pp. 162–163, Institution of Engineering and Technology (Asatani, K. and Kimura, T. 1977), with permission from IET; Figures 12.6 and 12.7 from Springer-Verlag, Topics in Applied Physics, vol. 39, 1982, pp. 161–200, Lightwave transmitters, P. W. Schumate Jr. and M. DiDomenico Jr., in H. Kressel, ed., *Semiconductor Devices for Optical Communications*, with kind permission from Springer Science and Business Media; Figure 12.12 from Electronic circuits for high bit rate digital fiber optic communication systems in *IEEE Trans. Communications*, COM-26(7), pp. 1088–1098, IEEE (Gruber, J., Marten, P., Petschacher, R. and Russer, P. 1978), © IEEE 1978, reproduced with permission; Figure 12.13 from Design and stability analysis of a CMOS feedback laser driver in *IEEE Trans. Instrum. Meas.*, 53(1), pp. 102–108, IEEE (Zivojinovic, P., Lescure, M. and Tap-Beteille, H. 2004), © IEEE 2004, reproduced with permission; Figure 12.14 from Laser automatic level control for optical communications systems in Third European Conference on Optical Communications, Munich, Germany, 14–16 September (S. R. Salter, S. R., Smith, D. R., White, B. R. and Webb, R. P. 1977), with permission from VDE-Verlag GMBH;

Figure 12.15 from Electronic circuits for high bit rate digital fiber optic communication systems in *IEEE Trans. Communications*, COM-26(7), pp. 1088–1098, IEEE (Gruber, J., Marten, P., Petschacher, R. and Russer, P. 1978), © IEEE 1978, reproduced with permission; Figure 12.35 from NRZ versus RZ in 10–14-Gb/s dispersion-managed WDM transmission systems in *IEEE Photonics Technology Letters*, 11(8), pp. 991–993, IEEE (Hayee, M. I., Willner, A. E., Syst, T. S. and Eacontown, N. J. 1999), © IEEE 1999, reproduced with permission; Figure 12.36 from Dispersion-tolerant optical transmission system using duobinary transmitter and binary receiver in *Journal of Lightwave Technology*, 15(8), pp. 1530–1537, OSA (Yonenaga, K. and Kuwano, S. 1997), with permission from The Optical Society of America; Figure 12.44 Copyright BAE systems Plc. Reproduced with permission from Fibre optic systems for analogue transmission, *Marconi Review*, XLIV(221), pp. 78–100 (Windus, G. G. 1981); Figure 12.53 from Performance of optical OFDM in ultralong-haul WDM lightwave systems in *Journal of Lightwave Technology*, 25(1), pp. 131–138, OSA (Lowery, A. J., Du, L. B. and Armstrong, J. 2007), with permission from The Optical Society of America; Figure 12.58 from 110 channels × 2.35 Gb/s from a single femtosecond laser in *IEEE Photonics Technology Letters*, 11(4), pp. 466–468, IEEE (Boivin, L., Wegmueller, M., Nuss, M. C. and Knox, W. H. 1999), © IEEE 1999, reproduced with permission; Figure 12.64 from 10 000-hop cascaded in-line all-optical 3R regeneration to achieve 1 250 000-km 10-Gb/s transmission in *IEEE Photonics Technology Letters*, 18(5), pp. 718–720, IEEE (Zuqing, Z., Funabashi, M., Zhong, P., Paraschis, L. and Yoo, S. J. B. 2006), © IEEE 2006, reproduced with permission; Figure 12.65 from An experimental analysis of performance fluctuations in high-capacity repeaterless WDM systems in Proceedings of OFC/Fiber Optics Engineering Conference (NFOEC) 2006, Anaheim, CA, USA, p. 3, 5–10 March, OSA (Bakhshi, B., Richardson, L., Golovchenko, E. A., Mohs, G. and Manna, M. 2006), with permission from The Optical Society of America; Figure 12.73 from Springer-Verlag, *Massive WDM and TDM Soliton Transmission Systems* 2002, A. Hasegawa, © 2002 Springer, with kind permission from Springer Science and Business Media; Figure 13.7 from Techniques for multigigabit coherent optical transmission in *Journal of Lightwave Technology*, LT-5, p. 1466, IEEE (Smith, D. W. 1987), © IEEE 1987, reproduced with permission; Figure 13.15 from Costas loop experiments for a 10.6 μm communications receiver in *IEEE Tras. Communications*, COM-31(8), pp. 1000–1002, IEEE (Phillip, H. K., Scholtz, A. L., Bonekand, E. and Leeb, W. 1983), © IEEE 1983, reproduced with permission; Figure 13.18 from Semiconductor laser homodyne optical phase lock loop in *Electronic Letters*, 22, pp. 421–422, Institution of Engineering and Technology (Malyon, D. J., Smith, D. W. and Wyatt, R. 1986), with permission from IET; Figure 13.32 from A consideration of factors affecting future coherent lightwave communication systems in *Journal of Lightwave Technology*, 6, p. 686, OSA (Nosu, K. and Iwashita, K. 1988), with permission from The Optical Society of America; Figure 13.34 from RZ-DPSK field trail over 13 100 km of installed non-slope matched submarine fibers in *Journal of Lightwave Technology*, 23(1) pp. 95–103, OSA (Cai, J. X. *et al.* 2005), with permission from The Optical Society of America; Figure 13.35 from Polarization-multiplexed 2.8 Gbit/s synchronous QPSK transmission with real-time polarization tracking in Proceedings of the 33rd European Conference on Optical Communications, Berlin, Germany, pp. 263–264, 3 September (Pfau, T. *et al.* 2007), with permission from VDE Verlag GMBH; Figure 13.36 from Hybrid 107-Gb/s polarization-multiplexed DQPSK and 42.7-Gb/s DQPSK transmission at 1.4 bits/s/Hz spectral efficiency over 1280 km of SSMF and 4 bandwidth-managed ROADMs in Proceedings of the 33rd European Conference on Optical Communications, Berlin, Germany, PD1.9, September, with permission from VDE Verlag GMBH; Figure 13.37 from Coherent optical orthogonal frequency division multiplexing in *Electronic Letters*, 42(10), pp. 587–588, Institution of Engineering and Technology (Shieh, W. and Athaudage, C. 2006), with permission from IET; Figure 14.1 (a) from Mode scrambler for optical fibres in *Applied Optics*, 16(4), pp. 1045–1049, OSA (Ikeda, M., Murakami, Y. and Kitayama, C. 1977), with permission from The Optical Society of America; Figure 14.1 (b) from Measurement of baseband frequency reponse of multimode fibre by using a new type of mode scrambler in *Electronic Letters*, 13(5), pp. 146–147, Institution of Engineering and Technology (Seikai, S., Tokuda, M., Yoshida, K. and Uchida, N. 1977), with permission from IET; Figure 14.6 from An improved technique for the measurement of low optical absorption losses in bulk glass in *Opto-electronics*, 5, p. 323, Institution of Engineering and Technology (White, K. I. and Midwinter, J. E. 1973); with permission from IET; Figure 14.8 from Self pulsing GaAs laser for fiber dispersion measurement in *IEEE Journal of Quantum Electronics*, QE-8, pp. 844–846, IEEE (Gloge, D., Chinnock, E. L. and Lee, T. P. 1972), © IEEE 1972, reproduced with permission; Figure 14.12, image of 86038B Optical Dispersion Analyzer, © Agilent Technologies, Inc. 2005, Reproduced with Permission, Courtesy of Agilent Technologies, Inc.; Figure 14.13 (a) from Refractive index profile measurements of diffused optical waveguides in *Applied Optics*, 13(9), pp. 2112–2116, OSA (Martin, W. E. 1974), with permission from The Optical Society of America; Figures 14.13 (b) and 14.14 Reused with permission from L. G. Cohen, P. Kaiser, J. B. Mac Chesney, P. B. O'Connor, and H. M. Presby, *Applied Physics Letters*, 26, 472 (1975). Copyright 1975, American Institute of Physics; Figures 14.17 and 14.18 (a) Reused with permission from F. M. E. Sladen, D. N. Payne, and M. J. Adams, *Applied Physics Letters*, 28, 255 (1976). Copyright 1976, American Institute of Physics; Figure 14.19 from *An Introduction to Optical Fibers*, McGraw-Hill Companies (Cherin, A. H. 1983), with permission of the McGraw-Hill Companies; Figure 14.26 Reused with permission from L. G. Cohen and P. Glynn, *Review of Scientific Instruments*, 44, 1749 (1973). Copyright 1973, American Institute of Physics; Figure 14.31 (a) from EXFO http://documents.exfo.com/appnotes/anote044-ang.pdf, accessed 21 September 2007, with permission from EXFO; Figure 14.31 (b) from http://www.afltelecommunications.com, Afltelecommunications Inc., accessed 21 September 2007, with permission from Fujikura Limited; Figure 14.35 from EXFO, OTDR FTB-7000B http://documents.exfo.com/appnotes/anote087-ang.pdf, accessed 21 September 2007, with permission from EXFO; Figure 14.36 from EXFO P-OTDR, accessed 21 September 2007, with permission from EXFO; Figure 15.1 from Future optical networks in *Journal of Lightwave Technology*, 24(12), pp. 4684–4696 (O'Mahony, M. J., Politi, C., Klonidis, D., Nejabati, R. and Simeonidou, D. 2006), with permission from The Optical Society of America; Figures 15.14 (a) and (b) from ITU-T Recommendation G.709/Y.1331(03/03) Interfaces for the Optical Transport Network (TON), 2003, reproduced with kind permission from ITU; Figures 15.25 and 15.26 from Enabling technologies for next-generation optical packet-switching networks in *Proceedings of IEEE* 94(5), pp. 892–910 (Gee-Kung, C., Jianjun, Y., Yong-Kee, Y., Chowdhury, A. and Zhensheng, J. 2006), © IEEE 2006, reproduced with permission; Figures 15.30, 15.31 and 15.32 from www.telegeography.com, accessed 17 October 2007, reproduced with permission; Figure 15.34 from Transparent optical protection ring architectures and applications in *Journal of Lightwave Technology*, 23(10), pp. 3388–3403 (Ming-Jun, L., Soulliere, M. J., Tebben, D. J., Nderlof, L., Vaughn, M. D. and R. E. Wagner, R. E. 2005), with permission from The Optical Society of America; Figure 15.46 from Hybrid DWDM-TDM long-reach PON for next-generation optical access in *Journal of Lightwave Technology*, 24(7), pp. 2827–2834 (Talli, G. and Townsend, P. D. 2006), with permission from The Optical Society of America; Figure 15.52 from IEEE 802.3 CSMA/CD (ETHERNET), accessed 17 October 2007, reproduced with permission; Figure 15.53 (a) from ITU-T Recommendation G.985 (03/2003) 100 Mbit/s point-to-pint Ethernet based optical access system, accessed 22 October 2007, reproduced with kind permission from ITU; Figure 15.53 (b) from ITU-T Recommendation Q.838.1 (10/2004) Requirements and analysis for the management interface of Ethernet passive optical networks (EPON), accessed 19 October 2007, reproduced with kind permission from ITU; Table 15.4 from Deployment of submarine optical fiber bacle and communication systems since 2001, www.atlantic-cable.com/Cables/CableTimeLine/index2001.htm, reproduced with permission.

In some instances we have been unable to trace the owners of copyright material, and we would appreciate any information that would enable us to do so.

List of symbols and abbreviations

A	constant, area (cross-section, emission), far-field pattern size, mode amplitude, wave amplitude (A_0)
A_{21}	Einstein coefficient of spontaneous emission
A_c	peak amplitude of the subcarrier waveform (analog transmission)
a	fiber core radius, parameter which defines the asymmetry of a planar guide (given by Eq. (10.21)), baseband message signal ($a(t)$)
$a_b(\lambda)$	effective fiber core radius
a_{eff}	bend attenuation fiber
a_k	integer 1 or 0
$a_m(\lambda)$	relative attenuation between optical powers launched into multimode and single-mode fibers
B	constant, electrical bandwidth (post-detection), magnetic flux density, mode amplitude, wave amplitude (B_0)
B_{12}, B_{21}	Einstein coefficients of absorption, stimulated emission
B_F	modal birefringence
B_{fib}	fiber bandwidth
B_{FPA}	mode bandwidth (Fabry–Pérot amplifier)
B_m	bandwidth of an intensity-modulated optical signal $m(t)$, maximum 3 dB bandwidth (photodiode)
B_{opt}	optical bandwidth
B_r	recombination coefficient for electrons and holes
B_T	bit rate, when the system becomes dispersion limited ($B_T(DL)$)
b	normalized propagation constant for a fiber, ratio of luminance to composite video, linewidth broadening factor (injection laser)
C	constant, capacitance, crack depth (fiber), wave coupling coefficient per unit length, coefficient incorporating Einstein coefficients
C_a	effective input capacitance of an optical fiber receiver amplifier
C_d	optical detector capacitance
C_f	capacitance associated with the feedback resistor of a transimpedance optical fiber receiver amplifier
C_j	junction capacitance (photodiode)
C_L	total optical fiber channel loss in decibels, including the dispersion–equalization penalty (C_{LD})
C_0	wave amplitude
C_T	total capacitance
CT	polarization crosstalk

c	velocity of light in a vacuum, constant (c_1, c_2)
c_i	tap coefficients for a transversal equalizer
D	amplitude coefficient, electric flux density, distance, diffusion coefficient, corrugation period, decision threshold in digital optical fiber transmission, fiber dispersion parameters: material (D_M); profile (D_P); total first order (D_T); waveguide (D_W), detectivity (photodiode), specific detectivity (D^*)
D_c	minority carrier diffusion coefficient
D_f	frequency deviation ratio (subcarrier FM)
D_L	dispersion–equalization penalty in decibels
D_P	frequency deviation ratio (subcarrier PM)
D_T	total chromatic dispersion (fibers)
d	fiber core diameter, hole diameter, distance, width of the absorption region (photodetector), thickness of recombination region (optical source), pin diameter (mode scrambler)
$d_{f\text{-}}$	far-field mode-field diameter (single-mode fiber)
d_n	near-field mode-field diameter (single-mode fiber)
d_o	fiber outer (cladding) diameter
E	electric field, energy, Young's modulus, expected value of a random variable, elcctron energy
E_a	activation energy of homogeneous degradation for an LED
E_F	Fermi level (energy), quasi-Fermi level located in the conduction band (E_{Fc}), valence band (E_{Fv}) of a semiconductor
E_g	separation energy between the valence and conduction bands in a semiconductor (bandgap energy)
$E_m(t)$	subcarrier electric field (analog transmission)
E_o	optical energy
E_q	separation energy of the quasi-Fermi levels
e	electronic charge, base for natural logarithms
F	probability of failure, transmission factor of a semiconductor–external interface, excess avalanche noise factor ($F(M)$), optical amplifier noise figure
$\mathscr{F}$	Fourier transformation
F_n	noise figure (electronic amplifier)
F_{to}	total noise figure for system of cascaded optical amplifiers
f	frequency
f_D	peak-to-peak frequency deviation (PFM–IM)
f_d	peak frequency deviation (subcarrier FM and PM)
f_o	Fabry–Pérot resonant frequency (optical amplifier), pulse rate (PFM–IM)
G	open loop gain of an optical fiber receiver amplifier, photoconductive gain, cavity gain of a semiconductor laser amplifier
$G_i(r)$	amplitude function in the WKB method
G_o	optical gain (phototransistor)
G_p	parametric gain (fiber amplifier)
G_R	Raman gain (fiber amplifier)
G_s	single-pass gain of a semiconductor laser amplifier
Gsn	Gaussian (distribution)
g	degeneracy parameter

$\bar{g}$	gain coefficient per unit length (laser cavity)
g_m	transconductance of a field effect transistor, material gain coefficient
g_0	unsaturated material gain coefficient
g_R	power Raman gain coefficient
$\bar{g}_{th}$	threshold gain per unit length (laser cavity)
H	magnetic field
$H(\omega)$	optical power transfer function (fiber), circuit transfer function
$H_A(\omega)$	optical fiber receiver amplifier frequency response (including any equalization)
$H_{CL}(\omega)$	closed loop current to voltage transfer function (receiver amplifier)
$H_{eq}(\omega)$	equalizer transfer function (frequency response)
$H_{OL}(\omega)$	open loop current to voltage transfer function (receiver amplifier)
$H_{out}(\omega)$	output pulse spectrum from an optical fiber receiver
h	Planck's constant, thickness of a planar waveguide, power impulse response for optical fiber ($h(t)$), mode coupling parameter (PM fiber)
$h_A(t)$	optical fiber receiver amplifier impulse response (including any equalization)
h_{eff}	effective thickness of a planar waveguide
h_{FE}	common emitter current gain for a bipolar transistor
$h_f(t)$	optical fiber impulse response
$h_{out}(t)$	output pulse shape from an optical fiber receiver
$h_p(t)$	input pulse shape to an optical fiber receiver
$h_t(t)$	transmitted pulse shape on an optical fiber link
I	electric current, optical intensity
I_b	background-radiation-induced photocurrent (optical receiver)
I_{bias}	bias current for an optical detector
I_c	collector current (phototransistor)
I_d	dark current (optical detector)
I_o	maximum optical intensity
I_p	photocurrent generated in an optical detector
I_S	output current from photodetector resulting from intermediate frequency in coherent receiver
I_{th}	threshold current (injection laser)
i	electric current
i_a	optical receiver preamplifier shunt noise current
i_{amp}	optical receiver, preamplifier total noise current
i_D	decision threshold current (digital transmission)
i_d	photodiode dark noise current
i_{det}	output current from an optical detector
i_f	noise current generated in the feedback resistor of an optical fiber receiver transimpedance preamplifier
i_N	total noise current at a digital optical fiber receiver
i_n	multiplied shot noise current at the output of an APD excluding dark noise current
i_s	shot noise current on the photocurrent for a photodiode
i_{SA}	multiplied shot noise current at the output of an APD including the noise current

i_{sig}	signal current obtained in an optical fiber receiver
i_t	thermal noise current generated in a resistor
i_{TS}	total shot noise current for a photodiode without internal gain
J	Bessel function, current density
J_{th}	threshold current density (injection laser)
j	$\sqrt{-1}$
K	Boltzmann's constant, constant, modified Bessel function
K_I	stress intensity factor, for an elliptical crack (K_{IC})
k	wave propagation constant in a vacuum (free space wave number), wave vector for an electron in a crystal, ratio of ionization rates for holes and electrons, integer, coupling coefficient for two interacting waveguide modes, constant
k_f	angular frequency deviation (subcarrier FM)
k_p	phase deviation constant (subcarrier PM)
L	length (fiber), distance between mirrors (laser), coupling length (waveguide modes)
L_a	length of amplifier (asymmetric twin-waveguide)
L_A	amplifying space (soliton transmission)
L_{ac}	insertion loss of access coupler in distribution system
L_B	beat length in a monomode optical fiber
L_{bc}	coherence length in a monomode optical fiber
L_c	characteristic length (fiber)
L_D	diffusion length of charge carriers (LED), fiber dispersion length
L_{ex}	star coupler excess loss in distribution system
L_{map}	dispersion management map period
L_0	constant with dimensions of length
L_t	lateral misalignment loss at an optical fiber joint
L_{tr}	tap ratio loss in distribution system
$\mathscr{L}$	transmission loss factor (transmissivity) of an optical fiber
l	azimuthal mode number, distance, length
l_a	atomic spacing (bond distance)
l_0	wave coupling length
M	avalanche multiplication factor, material dispersion parameter, total number of guided modes or mode volume; for a multimode step index fiber (M_s); for multimode graded index fiber (M_g), mean value (M_1) and mean square value (M_2) of a random variable
M_a	safety margin in an optical power budget
M_{op}	optimum avalanche multiplication factor
M^x	excess avalanche noise factor (also denoted as $F(M)$)
m	radial mode number, Weibull distribution parameter, intensity-modulated optical signal ($m(t)$), mean value of a random variable, integer, optical modulation index (subcarrier amplitude modulation)
m_a	modulation index
N	integer, density of atoms in a particular energy level (e.g. N_1, N_2, N_3), minority carrier concentration in n-type semiconductor material, number of input/output ports on a fiber star coupler, number of nodes on distribution

	system, noise current, dimensionless combination of pulse and fiber parameters (soliton)
NA	numerical aperture of an optical fiber
NEP	noise equivalent power
N_g	group index of an optical waveguide
N_{ge}	effective group index or group index of a single-mode waveguide
N_0	defined by Eq. (11.80)
N_p	number of photons per bit (coherent transmission)
n	refractive index (e.g. n_1, n_2, n_3), stress corrosion susceptibility, negative-type semiconductor material, electron density, number of chips (OCDM)
n_e	effective refractive index of a planar waveguide
n_{eff}	effective refractive index of a single-mode fiber
n_0	refractive index of air
n_{sp}	spontaneous emission factor (injection laser)
P	electric power, minority carrier concentration in p-type semiconductor material, probability of error ($P(e)$), of detecting a zero level ($P(0)$), of detecting a one level ($P(1)$), of detecting z photons in a particular time period ($P(z)$), conditional probability of detecting a zero when a one is transmitted ($P(0|1)$), of detecting a one when a zero is transmitted ($P(1|0)$), optical power (P_1, P_2, etc.)
P_a	total power in a baseband message signal $a(t)$
P_B	threshold optical power for Brillouin scattering
P_b	backward traveling signal power (semiconductor laser amplifier), power transmitted through fiber sample
P_c	optical power coupled into a step index fiber, optical power level
P_D	optical power density
P_{dc}	d.c. optical output power
P_e	optical power emitted from an optical source
P_G	optical power in a guided mode
P_i	mean input (transmitted) optical power launched into a fiber
P_{in}	input signal power (semiconductor laser amplifier)
P_{int}	internally generated optical power (optical source)
P_L	optical power of local oscillator signal (coherent system)
P_m	total power in an intensity-modulated optical signal $m(t)$
P_o	mean output (received) optical power from a fiber
P_{opt}	mean optical power traveling in a fiber
P_{out}	initial output optical (prior to degradation) power from an optical source
P_p	optical pump power (fiber amplifier)
P_{po}	peak received optical power
P_r	reference optical power level, optical power level
P_R	threshold optical power for Raman scattering
$P_{Ra}(t)$	backscattered optical power (Rayleigh) within a fiber
P_S	optical power of incoming signal (coherent system)
P_s	total power transmitted through a fiber sample
P_{sc}	optical power scattered from a fiber
P_t	optical transmitter power, launch power (P_{tx})

p	crystal momentum, average photoelastic coefficient, positive-type semiconductor material, probability density function ($p(x)$)
q	integer, fringe shift
q_0	dimensionless parameter (soliton transmission)
R	photodiode responsivity, radius of curvature of a fiber bend, electrical resistance (e.g. R_{in}, R_{out}); facet reflectivity (R_1, R_2)
R_{12}	upward transition rate for electrons from energy level 1 to level 2
R_{21}	downward transition rate for electrons from energy level 2 to level 1
R_a	effective input resistance of an optical fiber receiver preamplifier
R_b	bias resistance, for optical fiber receiver preamplifier (R_{ba})
R_c	critical radius of an optical fiber
R_D	radiance of an optical source
RE_{dB}	ratio of electrical output power to electrical input power in decibels for an optical fiber system
R_f	feedback resistance in an optical fiber receiver transimpedance prcamplifier
R_L	load resistance associated with an optical fiber detector
RO_{dB}	ratio of optical output power to optical input power in decibels for an optical fiber system
R_t	total carrier recombination rate (semiconductor optical source)
R_{TL}	total load resistance within an optical fiber receiver
r	radial distance from the fiber axis, Fresnel reflection coefficient, mirror reflectivity, electro-optic coefficient.
r_e	generated electron rate in an optical detector
r_{ER}, r_{ET}	reflection and transmission coefficients, respectively, for the electric field at a planar, guide–cladding interface
r_{HR}, r_{HT}	reflection and transmission coefficients respectively for the magnetic field at a planar, guide–cladding interface
r_{nr}	nonradiative carrier recombination rate per unit volume
r_p	incident photon rate at an optical detector
r_r	radiative carrier recombination rate per unit volume
r_t	total carrier recombination rate per unit volume
S	fraction of captured optical power, macroscopic stress, dispersion slope (fiber), power spectral density $S(\omega)$
S_f	fracture stress
$S_i(r)$	phase function in the WKB method
$S_m(\psi)$	spectral density of the intensity-modulated optical signal $m(t)$
S/N	peak signal power to rms noise power ratio, with peak-to-peak signal power $[(S/N)_{p-p}]$ with rms signal power $[(S/N)_{rms}]$
S_0	scale parameter; zero-dispersion slope (fiber)
S_t	theoretical cohesive strength
s	pin spacing (mode scrambler)
T	temperature, time, arbitrary parameter representing soliton pulse duration
T_a	insertion loss resulting from an angular offset between jointed optical fibers
T_c	10 to 90% rise time arising from chromatic dispersion on an optical fiber link
T_D	10 to 90% rise time for an optical detector

T_F	fictive temperature
T_l	insertion loss resulting from a lateral offset between jointed optical fibers
T_n	10 to 90% rise time arising from intermodal dispersion on an optical fiber link
T_0	threshold temperature (injection laser), nominal pulse period (PFM–IM)
T_R	10 to 90% rise time at the regenerator circuit input (PFM–IM)
T_S	10 to 90% rise time for an optical source
T_{syst}	total 10 to 90% rise time for an optical fiber system
T_T	total insertion loss at an optical fiber joint
T_t	temperature rise at time t
T_∞	maximum temperature rise
t	time, carrier transit time, slow(t_s), fast (t_f)
t_c	time constant
t_d	switch-on delay (laser)
t_e	$1/e$ pulse width from the center
t_r	10 to 90% rise time
U	eigenvalue of the fiber core
V	electrical voltage, normalized frequency for an optical fiber or planar waveguide
V_{bias}	bias voltage for a photodiode
V_c	cutoff value of normalized frequency (fiber)
V_{CC}	collector supply voltage
V_{CE}	collector–emitter voltage (bipolar transistor)
V_{EE}	emitter supply voltage
V_{eff}	effective normalized frequency (fiber)
V_{opt}	voltage reading corresponding to the total optical power in a fiber
V_{sc}	voltage reading corresponding to the scattered optical power in a fiber
v	electrical voltage
v_a	amplifier series noise voltage
$v_A(t)$	receiver amplifier output voltage
v_c	crack velocity
v_d	drift velocity of carriers (photodiode)
v_g	group velocity
$v_{out}(t)$	output voltage from an RC filter circuit
v_p	phase velocity
W	eigenvalue of the fiber cladding, random variable
W_e	electric pulse width
W_o	optical pulse width
w	depletion layer width (photodiode)
X	random variable
x	coordinate, distance, constant, evanescent field penetration depth, slab thickness, grating line spacing
Y	constant, shunt admittance, random variable
y	coordinate, lateral offset at a fiber joint
Z	random variable, constant
Z_0	electrical impedance

z	coordinate, number of photons
z_m	average or mean number of photons arriving at a detector in a time period τ
z_{md}	average number of photons detected in a time period τ
α	characteristic refractive index profile for fiber (profile parameter), optimum profile parameter (α_{op}), linewidth enhancement factor (injection laser), optical link loss
$\bar{\alpha}$	loss coefficient per unit length (laser cavity)
α_{cr}	connector loss at transmitter and receiver in decibels
α_{dB}	signal attenuation in decibels per unit length
α_{fc}	fiber cable loss in decibels per kilometer
α_i	internal wavelength loss per unit length (injection laser)
α_j	fiber joint loss in decibels per kilometer
α_m	mirror loss per unit length (injection laser)
α_N	signal attenuation in nepers
α_0	absorption coefficient
α_p	fiber transmission loss at the pump wavelength (fiber amplifier)
α_r	radiation attenuation coefficient
β	wave propagation constant
$\bar{\beta}$	gain factor (injection laser cavity)
β_c	isothermal compressibility
β_0	proportionality constant
β_2	second-order dispersion coefficient
β_r	degradation rate
Γ	optical confinement factor (semiconductor laser amplifier)
γ	angle, attenuation coefficient per unit length for a fiber, nonlinear coefficient resulting from the Kerr effect
γ_p	surface energy of a material
γ_R	Rayleigh scattering coefficient for a fiber
Δ	relative refractive index difference between the fiber core and cladding
Δf	linewidth of single-frequency injection laser
ΔG	peak–trough ratio of the passband ripple (semiconductor laser amplifier)
Δn	index difference between fiber core and cladding ($\Delta n/n_1$ fractional index difference)
δ_E	phase shift associated with transverse electric waves
δf	uncorrelated source frequency widths
δ_H	phase shift associated with transverse magnetic waves
$\delta\lambda$	optical source spectral width (linewidth), mode spacing (laser)
δT	intermodal dispersion time in an optical fiber
δT_g	delay difference between an extreme meridional ray and an axial ray for a graded index fiber
δT_s	delay difference between an extreme meridional ray and an axial ray for a step index fiber, with mode coupling (δT_{sc})
δT_g	polarization mode dispersion in fiber
ε	electric permittivity, of free space (ε_0), relative (ε_r), semiconductor (ε_s), extinction ratio (optical transmitter)

ζ	solid acceptance angle
η	quantum efficiency (optical detector)
η_{ang}	angular coupling efficiency (fiber joint)
η_c	coupling efficiency (optical source to fiber)
η_D	differential external quantum efficiency (optical source)
η_{ep}	external power efficiency (optical source)
η_{ext}	external quantum efficiency (light-emitting devices)
η_i	internal quantum efficiency injection laser
η_{int}	internal quantum efficiency (LED)
η_{lat}	lateral coupling efficiency (fiber joint)
η_{pc}	overall power conversion efficiency (optical source)
η_T	total external quantum efficiency (optical source)
θ	angle, fiber acceptance angle (θ_a)
θ_B	Bragg diffraction angle, blaze angle diffraction grating
Λ	acoustic wavelength, period for perturbations in a fiber, optical grating period, spacing between holes and pitch (photonic crystal fiber)
Λ_c	cutoff period for perturbations in a fiber
λ	optical wavelength
λ_B	Bragg wavelength (DFB laser)
λ_c	long-wavelength cutoff (photodiode), cutoff wavelength for single-mode fiber, effective cutoff wavelength (λ_{ce})
λ_0	wavelength at which first-order dispersion is zero
μ	magnetic permeability, relative permeability, (μ_r), permeability of free space (μ_0)
v	optical source bandwidth in gigahertz
ρ	polarization rotation in a single-mode optical fiber
ρ_f	spectral density of the radiation energy at a transition frequency f
σ	standard deviation (rms pulse width), variance (σ^2)
σ_c	rms pulse broadening resulting from chromatic dispersion in a fiber
σ_m	rms pulse broadening resulting from material dispersion in a fiber
σ_n	rms pulse broadening resulting from intermodal dispersion in a graded index fiber (σ_g), in a step index fiber (σ_s)
σ_T	total rms pulse broadening in a fiber or fiber link
τ	time period, bit period, signaling interval, pulse duration, 3 dB pulse width ($\tau(3\ dB)$), retarded time
τ_{21}	spontaneous transition lifetime between energy levels 2 and 1
τ_E	time delay in a transversal equalizer
τ_e	$1/e$ full width pulse broadening due to dispersion on an optical fiber link
τ_g	group delay
τ_i	injected (minority) carrier lifetime
τ_{ph}	photon lifetime (semiconductor laser)
τ_r	radiative minority carrier lifetime
τ_{sp}	spontaneous emission lifetime (equivalent to τ_{21})
Φ	linear retardation
ϕ	angle, critical angle (ϕ_c), photon density, phase shift
ψ	scalar quantity representing $\mathbf{E}$ or $\mathbf{H}$ field

ω	angular frequency, of the subcarrier waveform in analog transmission (ω_c), of the modulating signal in analog transmission (ω_m), pump frequency (ω_p), Stokes component (ω_s), anti-Stokes component (ω_a), intermediate frequency of coherent heterodyne receiver (ω_{IF}), normalized spot size of the fundamental mode
ω_0	spot size of the fundamental mode
∇	vector operator, Laplacian operator (∇^2)

A–D	analog to digital
a.c.	alternating current
ADCCP	advanced data communications control procedure (optical networks)
AFC	automatic frequency control
AGC	automatic gain control
AM	amplitude modulation
AMI	alternate mark inversion (line code)
ANSI	American National Standards Institute
AOWC	all-optical wavelength converter
APD	avalanche photodiode
AR	antireflection (surface, coating)
ARROW	antiresonant reflecting optical waveguide
ASE	amplified spontaneous emission (optical amplifier)
ASK	amplitude shift keying
ASON	automatic switched optical network
ATM	alternative test method (fiber), asynchronous transfer mode (transmission)
AWG	arrayed-waveguide grating
BCH	Bose Chowdhry Hocquenghem (line codes)
BER	bit-error-rate
BERTS	bit-error-rate test set
BGP	Border Gateway Protocol
BH	buried heterostructure (injection laser)
BHC	burst header cell (optical switch)
BHP	burst header packet (optical switch)
BLSR	bi-directional line-switched ring (optical networks)
BOD	bistable optical device
BPSK	binary phase shift keying
BXC	waveband cross-connect (optical networks)
CAPEX	capital expenditure
CATV	common antenna television
CCTV	closed circuit television
CDH	constricted double heterojunction (injection laser)
CMI	coded mark inversion
CMOS	complementary metal oxide silicon
CNR	carrier to noise ratio
CO	central office (telephone switching center)
CPFSK	continuous phase frequency shift keying
CPU	central processing unit

CRZ	chirped return to zero
CSMA/CD	Carrier Sense Multiple Access with Collision Detection
CSP	channelled substrate planar (injection laser)
CSRZ	carrier-suppressed return to zero
CW	continuous wave or operation
CWDM	coarse wavelength division multiplexing
D–A	digital to analog
DB	duobinary (line code)
dB	decibel
DBPSK	differential binary phase shift keying
DBR	distributed Bragg reflector (laser)
D–IM	direct intensity modulation
DC	depressed cladding (fiber design)
d.c.	direct current
DCC	data control channel (optical networks)
DCF	dispersion-compensating fiber
DDF	dispersion-decreasing fiber
DF	dispersion flattened (single-mode fiber)
DFB	distributed feedback (injection laser)
DFF	dispersion-flattened fiber
DFG	difference frequency generation (nonlinear effect)
DGD	differential group delay
DGE	dynamic gain equalizer
DH	double heterostructure or heterojunction (injection laser or LED)
DI	delay interferometer
DLD	dark line defect (semiconductor optical source)
DMS	dispersion-managed soliton
DOP	degree of polarization
DPSK	differential phase shift keying
DQPSK	differential quadrature phase shift keying
DS	dispersion shifted (single-mode fiber)
DSB	double sideband (amplitude modulation)
DSD	dark spot defect (laser)
DSF	dispersion-shifted fiber
DSL	digital subscriber line, asymmetrical (ADSL), very high speed (VDSL)
DSP	digital signal processing
DSTM	dynamic synchronous transfer mode
DUT	device under test (fiber measurement)
DWDM	dense wavelength division multiplexing
DWELL	dots-in-well (photodiode)
DXC	digital cross-connect
E/O	electrical (or electronic) to optical conversion
EAM	electro-absorption modulator
ECL	emitter-coupler logic
EDFA	erbium-doped fiber amplifier
EDWA	erbium-doped waveguide amplifier

EH	traditional mode designation
EIA	Electronics Industries Association
ELED	edge-emitting light-emitting diode
ELH	extended long haul
EMFA	erbium micro-fiber amplifier
EMI	electromagnetic interference
EMP	electromagnetic pulse
EPON	Ethernet passive optical network
erf	error function
erfc	complementary error function
ESI	equivalent step index (fiber)
ESTI	European Telecommunications Standards Institute
ETDM	electrical time division multiplexing
EYDFA	erbium–ytterbium-doped fiber amplifier
FAST	field assembly simple technique (optical connector)
FBG	fiber Bragg grating
FBT	fused biconical taper (fiber coupler)
FC	fiber connector, ferrule connector
FDDI	Fiber Distributed Data Interface
FDM	frequency division multiplexing
FEC	forward error correction
FET	field effect transistor, junction (JFET)
FFT	fast Fourier transform
FM	frequency modulation
FOTP	Fiber Optic Test Procedure
FPA	Fabry–Pérot amplifier
FSAN	Full Service Access Network
FSK	frequency shift keying
FTTB	fiber-to-the-building
FTTC	fiber-to-the-curb
FTTCab	fiber-to-the-cabinet
FTTH	fiber-to-the-home
FWHP	full width half power
FWHM	full width half maximum (equivalent to FWHP)
FWM	four-wave mixing (nonlinear effect)
FXC	fiber cross-connect
GbE	gigabit Ethernet
GC-SOA	gain clamped-semiconductor optical amplifier
GCSR	grating-assisted codirectional coupler with sampled reflector (laser)
GEM	gigabit passive optical network encapsulation method
GEPON	gigabit Ethernet passive optical network
GFF	gain flattening filter
GFP	generic frame procedure (network protocols)
GI	guard time insertion
GMPLS	Generalized Multiprotocol Label Switching
GPON	gigabit passive optical network

GRIN	graded index (rod lens)
GTC	gigabit passive optical network transmission convergence
GVD	group velocity delay (fiber)
HB	high birefringence (fiber)
HBT	heterojunction bipolar transmitter
HDB	high-density bipolar
HDLC	High-level Data Link Control (network protocol)
HDTV	high-definition television
HE	traditional mode designation
HEMT	high electron mobility transistor
He–Ne	helium–neon (laser)
HF	high frequency
HFC	hybrid fiber coaxial
HV	high voltage
IF	intermediate frequency
ILD	injection laser diode
IM	intensity modulation, with direct detection (IM/DD)
I3O	ion-implanted integrated optics
IEC	International Electrotechnical Commission
IEEE	Institute of Electrical and Electronics Engineers
IET	Institution of Engineering and Technology
IFFT	inverse fast Fourier transform
IGA	induced-grating autocorrelation (fiber measurement)
ILM	integrated laser modulator
INCITS	International Committee for Information Technology Standards (ANSI)
IO	integrated optics
I/O	input/output
I & Q	inphase and quadrature (coherent receiver)
IP	integrated photonics, Internet Protocol
ISDN	integrated services digital network, broadband (BISDN)
ISI	intersymbol interference
IS–IS	Intermediate-System-to-Intermediate-System (optical network protocol)
ISO	International Organization for Standardization
ITU-T	International Telecommunication Union – Telecom sector
JET	just-enough-time (optical network protocol)
JIT	just-in-time (optical network protocol)
LAN	local area network
LB	low birefringence (fiber)
LC	Lucent connector, local connector
LDPC	low-density parity check codes
LEC	long external cavity (laser)
LED	light-emitting diode
LH	long haul
LLC	logical link control (LAN)
LO	local oscillator
LOA	linear optical amplifier

LOC	large optical cavity (injection laser)
LP	linearly polarized (mode notation)
LPE	liquid-phase epitaxy
LR-PON	long-reach passive optical network
LSP	label-switched path (optical networks)
LSR	label-switching router (optical networks)
LWPF	low-water-peak fiber
MAC	Medium Access Control (LAN), isochronous (I-MAC)
MAN	metropolitan area network
MBE	molecular beam epitaxy
MC	matched cladding (fiber design)
MCVD	modified chemical vapor deposition
MEMS	micro-electro-mechanical systems, optical (OMEMS)
MESFET	metal Schottky field effect transistor
MFD	mode-field diameter (single-mode fiber)
MFSK	multilevel frequency shift keying
MG-OXC	multi-granular optical cross-connect (optical networks)
MI	Michelson interferometer
MISFET	metal integrated-semiconductor field effect transistor
MMF	multimode fiber
MMI	multimode interference (optical coupler)
MOSFET	metal oxide semiconductor field effect transistor
MOVPE	metal oxide vapor-phase epitaxy
MPLS	Multiple Protocol Label Switching
MPO	multi-fiber push-on (fiber connector)
MQW	multiquantum well
MSM	metal–semiconductor–metal (photodetector)
MTP	multi fiber termination push-on (fiber connector)
MT-RJ	mechanical transfer registered jack (fiber connector)
MUSE	multiple sub-Nyquist sampling encoding
MZI	Mach–Zehnder interferometer
MZM	Mach–Zehnder modulator
NdDFA	neodymium-doped fiber amplifier
NDF	negative dispersion fiber
Nd : YAG	neodymium-doped yttrium–aluminum–garnet (laser)
NIU	networking interface unit
NODG	nonlinear optical digital gate
NOLM	nonlinear optical loop mirror
NRZ	nonreturn to zero
NT	network termination
NZDF	nonzero-dispersion fiber
NZ-DSF	nonzero-dispersion-shifted fiber
O/E	optical to electrical (or electronic) conversion
OADM	optical add/drop multiplexer
OBPF	optical bandpass filter
OBS	optical burst switch(ed)

ROADM	reconfigurable optical add/drop multiplexer
RS	Reed–Solomon (line codes)
RTM	reference test method (fiber)
RWA	routing and wavelength assignment (optical networks)
Rx	receiver
RZ	return-to-zero
SACM	separate absorption, charge and multiplication (avalanche photodiode)
SAGCM	separate absorption, grading, charge and multiplication (avalanche photodiode)
SAM	separate absorption and multiplication (avalanche photodiode)
SAT	South Atlantic (optical fiber cable)
SAW	surface acoustic wave
SBS	stimulated Brillouin scattering
SC	subscriber connector (fiber)
SCM	subcarrier multiplexing
SDH	synchronous digital hierarchy
SDM	space division multiplexing
SEED	self-electro-optic device, symmetric (S-SEED)
SFF	small form factor (fiber connector)
SFP	small form pluggable (fiber connector)
SG-DBR	sampled-grating distributed Bragg reflector (laser)
SHF	super high frequency
SIU	subscriber interface unit
SLA	semiconductor laser amplifier
SLD	superluminescent diode
SLED	surface emitter light-emitting diode
SMA	subminiature A (fiber connector)
SMC	subminiature C (fiber connector)
SMF	single-mode fiber
SML	separated multiclad layer (injection laser)
SMT	station management (FDDI)
SNI	service network interface
SNR	signal-to-noise ratio
SONET	synchronous optical network
SOA	semiconductor optical amplifier
SOI	silicon-on-insulator (integrated optics)
SOL	silicon-on-liquid (integrated optics)
SOP	state of polarization
SOS	silica-on-silicon (integrated optics)
SPAD	single-photon-counting avalanche photodetector
SPE	synchronous payload envelope (SONET)
SPM	self-phase modulation (nonlinear effect)
SQW	single quantum well
SRS	stimulated Raman scattering
SSG-DBR	superstructure grating distributed Bragg reflector (laser)
SSMF	standard single-mode fiber
ST	straight tip (fiber connector)

STM	synchronous transport module (SDH)
STS	synchronous transport signal (SONET)
TAG	tell-and-go (optical network protocol)
TAT	transatlantic (optical fiber cables)
TAW	tell-and-wait (optical network protocol)
TCP	Transmission Control Protocol
TDFA	thulium-doped fiber amplifier
TDS	time domain sampling
TDM	time division multiplexing
TDMA	time division multiple access
TE	transverse electric
Te-EDFA	tellurium–erbium-doped fluoride fiber amplifier
TEM	transverse electromagnetic
ThDFA	thorium-doped fiber amplifier
TIA	Telecommunication Industry Association
TJS	transverse junction stripe (injection laser)
TM	transverse magnetic
TOAD	terahertz optical asymmetrical demultiplexer
TRC	time-resolved chirp (lasers)
TTL	transistor–transistor logic
TWA	traveling wave amplifier
Tx	transmitter
UDP	User Datagram Protocol
UHF	ultra high frequency
ULH	ultra long haul
UNI	user network interface
UTC	unitraveling carrier (photodiode)
VAD	vapor axial deposition
VC	virtual concatenation
VCO	voltage-controlled oscillator
VCSEL	vertical cavity surface-emitting laser
VHF	very high frequency
VIFO	variable input–fixed output (wavelength conversion)
VOA	variable optical attenuator
VPE	vapor-phase epitaxy
VPN	virtual private network
VSB	vestigial sideband (modulation)
VT	virtual tributary (SONET)
WADD	wavelength add/drop device
WAN	wide area network
WBC	waveband cross-connect (optical networks)
WBS	waveband switching (optical networks)
WC	wavelength converter
WCB	wavelength converter bank
WDM	wavelength division multiplexing
WIXC	wavelength exchange cross-connect (optical networks)

WKB	Wentzel, Kramers, Brillouin (analysis technique for graded fiber)
WRA	wavelength routing assignment (optical networks)
WSS	wavelength selective switch
WXC	wavelength cross-connecting (optical networks)
XAM	cross-absorption modulation (nonlinear effect)
XGM	cross-gain modulation (nonlinear effect)
XPM	cross-phase modulation (nonlinear effect)
ZD	Zener diode
ZMD	zero material dispersion (fiber)
ZWP	zero water peak (fiber)

Introduction

Communication may be broadly defined as the transfer of information from one point to another. When the information is to be conveyed over any distance a communication system is usually required. Within a communication system the information transfer is frequently achieved by superimposing or modulating the information onto an electromagnetic wave which acts as a carrier for the information signal. This modulated carrier is then transmitted to the required destination where it is received and the original information signal is obtained by demodulation. Sophisticated techniques have been developed for this process using electromagnetic carrier waves operating at radio frequencies as well as microwave and millimeter wave frequencies. However, 'communication' may also be achieved using an electromagnetic carrier which is selected from the optical range of frequencies.

1.1 Historical development

The use of visible optical carrier waves or light for communication has been common for many years. Simple systems such as signal fires, reflecting mirrors and, more recently, signaling lamps have provided successful, if limited, information transfer. Moreover, as early as 1880 Alexander Graham Bell reported the transmission of speech using a light beam [Ref. 1]. The photophone proposed by Bell just four years after the invention of the telephone modulated sunlight with a diaphragm giving speech transmission over a distance of 200 m. However, although some investigation of optical communication continued in the early part of the twentieth century [Refs 2 and 3] its use was limited to mobile, low-capacity communication links. This was due to both the lack of suitable light sources and the problem that light transmission in the atmosphere is restricted to line of sight and is severely affected by disturbances such as rain, snow, fog, dust and atmospheric turbulence. Nevertheless lower frequency and hence longer wavelength electromagnetic waves* (i.e. radio and microwave) proved suitable carriers for information transfer in the atmosphere, being far less affected by these atmospheric conditions. Depending on their wavelengths, these electromagnetic carriers can be transmitted over considerable distances but are limited in the amount of information they can convey by their frequencies (i.e. the information-carrying capacity is directly related to the bandwidth or frequency extent of the modulated carrier, which is generally limited to a fixed fraction of the carrier frequency). In theory, the greater the carrier frequency, the larger the available transmission bandwidth and thus the information-carrying capacity of the communication system. For this reason radio communication was developed to higher frequencies (i.e. VHF and UHF) leading to the introduction of the even higher frequency microwave and, latterly, millimeter wave transmission. The relative frequencies and wavelengths of these types of electromagnetic wave can be observed from the electromagnetic spectrum shown in Figure 1.1. In this context it may also be noted that communication at optical frequencies offers an increase in the potential usable bandwidth by a factor of around 10^4 over high-frequency microwave transmission. An additional benefit of the use of high carrier frequencies is the general ability of the communication system to concentrate the available power within the transmitted electromagnetic wave, thus giving an improved system performance [Ref. 4].

A renewed interest in optical communication was stimulated in the early 1960s with the invention of the laser [Ref. 5]. This device provided a powerful coherent light source, together with the possibility of modulation at high frequency. In addition the low beam divergence of the laser made enhanced free space optical transmission a practical possibility. However, the previously mentioned constraints of light transmission in the atmosphere tended to restrict these systems to short-distance applications. Nevertheless, despite the problems some modest free space optical communication links have been implemented for applications such as the linking of a television camera to a base vehicle and for data links of a few hundred meters between buildings. There is also some interest in optical communication between satellites in outer space using similar techniques [Ref. 6].

Although the use of the laser for free space optical communication proved somewhat limited, the invention of the laser instigated a tremendous research effort into the study of optical components to achieve reliable information transfer using a lightwave carrier. The proposals for optical communication via dielectric waveguides or optical fibers fabricated from glass to avoid degradation of the optical signal by the atmosphere were made almost simultaneously in 1966 by Kao and Hockham [Ref. 7] and Werts [Ref. 8]. Such systems were viewed as a replacement for coaxial cable or carrier transmission systems. Initially the optical fibers exhibited very high attenuation (i.e. 1000 dB km^{-1}) and were therefore not comparable with the coaxial cables they were to replace (i.e. 5 to 10 dB km^{-1}). There were also serious problems involved in jointing the fiber cables in a satisfactory manner to achieve low loss and to enable the process to be performed relatively easily and repeatedly

---

* For the propagation of electromagnetic waves in free space, the wavelength λ equals the velocity of light in a vacuum c times the reciprocal of the frequency f in hertz or $\lambda = c/f$.

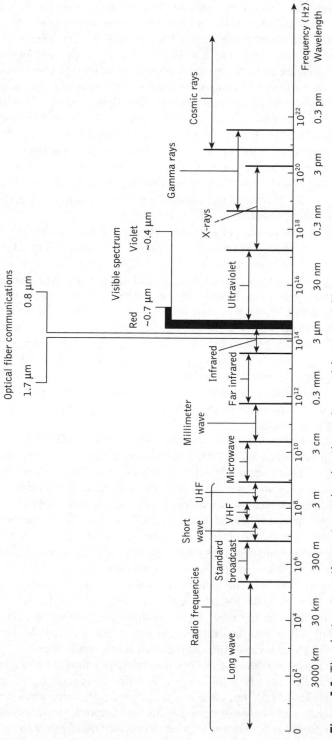

Figure 1.1 The electromagnetic spectrum showing the region used for optical fiber communications

in the field. Nevertheless, within the space of 10 years optical fiber losses were reduced to below 5 dB km^{-1} and suitable low-loss jointing techniques were perfected.

In parallel with the development of the fiber waveguide, attention was also focused on the other optical components which would constitute the optical fiber communication system. Since optical frequencies are accompanied by extremely small wavelengths, the development of all these optical components essentially required a new technology. Thus semiconductor optical sources (i.e. injection lasers and light-emitting diodes) and detectors (i.e. photodiodes and to a lesser extent phototransistors) compatible in size with optical fibers were designed and fabricated to enable successful implementation of the optical fiber system. Initially the semiconductor lasers exhibited very short lifetimes of at best a few hours, but significant advances in the device structure enabled lifetimes greater than 1000 h [Ref. 9] and 7000 h [Ref. 10] to be obtained by 1973 and 1977 respectively. These devices were originally fabricated from alloys of gallium arsenide (AlGaAs) which emitted in the near infrared between 0.8 and 0.9 μm.

Subsequently the above wavelength range was extended to include the 1.1 to 1.6 μm region by the use of other semiconductor alloys (see Section 6.3.6) to take advantage of the enhanced performance characteristics displayed by optical fibers over this range. In particular for this longer wavelength region around 1.3 μm and 1.55 μm, semiconductor lasers and also the simpler structured light-emitting diodes based on the quaternary alloy InGaAsP-grown lattice matched to an InP substrate have been available since the late 1980s with projected median lifetimes in excess of 25 years (when operated at 10 °C) for the former and 100 years (when operated at 70 °C) for the latter device types [Ref. 11]. Hence the materials growth and fabrication technology has been developed specifically for telecommunication applications and it is now mature [Ref. 12]. Moreover, for telecommunication applications such lasers are often provided with a thermoelectric cooler together with a monitoring photodiode in the device package in order to facilitate current and thus temperature control.

Direct modulation of commercial semiconductor lasers at 2.5 Gbit s^{-1} over single-mode fiber transmission distances up to 200 km at a wavelength of 1.55 μm can be achieved and this may be extended up to 10 Gbit s^{-1} over shorter unrepeated fiber links [Ref. 13]. Indeed, more recent research and development has focused on 40 Gbit s^{-1} transmission where external laser modulation is required using, for example, a Mach–Zehnder or an electroabsorption modulator (see Section 11.4.2) [Refs 14, 15]. This aspect also proves useful in the first longer wavelength window region around 1.3 μm where fiber intramodal dispersion is minimized and hence the transmission bandwidth is maximized, particularly for single-mode fibers. It is also noteworthy that this fiber type quickly came to dominate system applications within telecommunications since its initial field trial demonstration in 1982 [Ref. 16]. Moreover, the lowest silica glass fiber losses to date of 0.1484 dB km^{-1} were reported in 2002 for the other longer wavelength window at 1.57 μm [Ref. 17] but, unfortunately, chromatic dispersion is greater at this wavelength, thus limiting the maximum bandwidth achievable with conventional single-mode fiber.

To obtain low loss over the entire fiber transmission longer wavelength region from 1.3 to 1.6 μm, or alternatively, very low loss and low dispersion at the same operating wavelength of typically 1.55 μm, advanced single-mode fiber structures have been commercially realized: namely, low-water-peak fiber and nonzero dispersion-shifted fiber. Although developments in fiber technology have continued rapidly over recent years, certain previously favored areas of interest such as the application of fluoride fibers for even longer wavelength operation in the mid-infrared (2 to 5 μm) and far-infrared (8 to 12 μm) regions have declined due to their failure to demonstrate practically the theoretically predicted, extremely low fiber losses combined with the emergence of optical amplifiers suitable for use with silica-based fibers.

An important development, however, concerns the discovery of the phenomenon of photonic bandgaps which can be created in structures which propagate light, such as crystals or optical fibers. One particular form of photonic crystal fiber, for example, comprises a microstructured regular lattice of air holes running along its length (see Section 2.6). Such 'holey' fibers have the unusual property that they only transmit a single mode of light and hence form an entirely new single-mode fiber type which can carry more optical power than a conventional one. A further class of photonic bandgap fiber is defined by a large hollow core in which the light is guided. Such air guiding or hollow-core optical fibers could find application in photonic bandgap devices to provide dispersion compensation on long-haul fiber links or for high-resolution, tunable spectral filters [Ref. 18]. Nevertheless, even without the commercial availability of photonic bandgap devices, the implementation of a wide range of conventional fiber components (splices, connectors, couplers, etc.) and active optoelectronic devices (sources, detectors, amplifiers, etc.) has also moved to a stage of maturity. High-performance, reliable optical fiber communication systems and networks are therefore now widely deployed within the worldwide telecommunication network and in many more localized communication application areas.

1.2 The general system

An optical fiber communication system is similar in basic concept to any type of communication system. A block schematic of a general communication system is shown in Figure 1.2(a), the function of which is to convey the signal from the information source over the transmission medium to the destination. The communication system therefore consists of a transmitter or modulator linked to the information source, the transmission medium, and a receiver or demodulator at the destination point. In electrical communications the information source provides an electrical signal, usually derived from a message signal which is not electrical (e.g. sound), to a transmitter comprising electrical and electronic components which converts the signal into a suitable form for propagation over the transmission medium. This is often achieved by modulating a carrier, which, as mentioned previously, may be an electromagnetic wave. The transmission medium can consist of a pair of wires, a coaxial cable or a radio link through free space down which the signal is transmitted to the receiver, where it is transformed into the original electrical information signal (demodulated) before being passed to the destination. However, it must be noted that in any transmission medium the signal is attenuated, or suffers loss, and is subject to degradations due to contamination by random signals and noise, as well as possible distortions imposed by mechanisms within the medium itself. Therefore, in any communication system there is a maximum permitted distance between the transmitter and the receiver beyond which the system effectively ceases to give intelligible communication. For long-haul applications these factors necessitate the installation of repeaters or line amplifiers

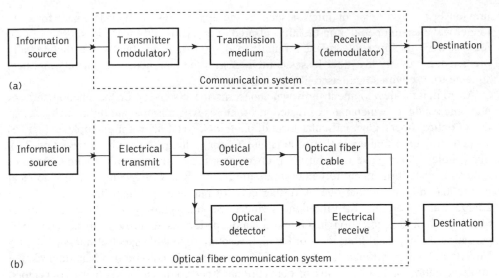

Figure 1.2 (a) The general communication system. (b) The optical fiber communication system

(see Sections 12.4 and 12.10) at intervals, both to remove signal distortion and to increase signal level before transmission is continued down the link.

For optical fiber communications the system shown in Figure 1.2(a) may be considered in slightly greater detail, as given in Figure 1.2(b). In this case the information source provides an electrical signal to a transmitter comprising an electrical stage which drives an optical source to give modulation of the lightwave carrier. The optical source which provides the electrical–optical conversion may be either a semiconductor laser or light-emitting diode (LED). The transmission medium consists of an optical fiber cable and the receiver consists of an optical detector which drives a further electrical stage and hence provides demodulation of the optical carrier. Photodiodes (p–n, p–i–n or avalanche) and, in some instances, phototransistors and photoconductors are utilized for the detection of the optical signal and the optical–electrical conversion. Thus there is a requirement for electrical interfacing at either end of the optical link and at present the signal processing is usually performed electrically.*

The optical carrier may be modulated using either an analog or digital information signal. In the system shown in Figure 1.2(b) analog modulation involves the variation of the light emitted from the optical source in a continuous manner. With digital modulation, however, discrete changes in the light intensity are obtained (i.e. on–off pulses). Although often simpler to implement, analog modulation with an optical fiber communication system is less efficient, requiring a far higher signal-to-noise ratio at the receiver than digital modulation. Also, the linearity needed for analog modulation is not always provided by semiconductor optical sources, especially at high modulation frequencies. For these reasons, analog optical fiber communication links are generally limited to shorter distances and lower bandwidth operation than digital links.

* Significant developments have taken place in devices for optical signal processing which are starting to alter this situation (see Chapter 11).

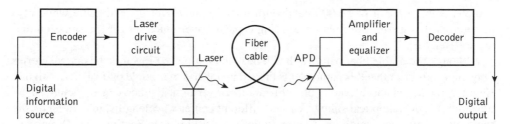

Figure 1.3 A digital optical fiber link using a semiconductor laser source and an avalanche photodiode (APD) detector

Figure 1.3 shows a block schematic of a typical digital optical fiber link. Initially, the input digital signal from the information source is suitably encoded for optical transmission. The laser drive circuit directly modulates the intensity of the semiconductor laser with the encoded digital signal. Hence a digital optical signal is launched into the optical fiber cable. The avalanche photodiode (APD) detector is followed by a front-end amplifier and equalizer or filter to provide gain as well as linear signal processing and noise bandwidth reduction. Finally, the signal obtained is decoded to give the original digital information. The various elements of this and alternative optical fiber system configurations are discussed in detail in the following chapters. However, at this stage it is instructive to consider the advantages provided by lightwave communication via optical fibers in comparison with other forms of line and radio communication which have brought about the extensive use of such systems in many areas throughout the world.

1.3 Advantages of optical fiber communication

Communication using an optical carrier wave guided along a glass fiber has a number of extremely attractive features, several of which were apparent when the technique was originally conceived. Furthermore, the advances in the technology to date have surpassed even the most optimistic predictions, creating additional advantages. Hence it is useful to consider the merits and special features offered by optical fiber communications over more conventional electrical communications. In this context we commence with the originally foreseen advantages and then consider additional features which have become apparent as the technology has been developed.

(a) Enormous potential bandwidth. The optical carrier frequency in the range 10^{13} to 10^{16} Hz (generally in the near infrared around 10^{14} Hz or 10^5 GHz) yields a far greater potential transmission bandwidth than metallic cable systems (i.e. coaxial cable bandwidth typically around 20 MHz over distances up to a maximum of 10 km) or even millimeter wave radio systems (i.e. systems currently operating with modulation bandwidths of 700 MHz over a few hundreds of meters). Indeed, by the year 2000 the typical bandwidth multiplied by length product for an optical fiber link incorporating fiber amplifiers (see Section 10.4) was 5000 GHz km in comparison with the typical bandwidth–length product for coaxial cable of around 100 MHz km. Hence at this time optical fiber was already

demonstrating a factor of 50 000 bandwidth improvement over coaxial cable while also providing this superior information-carrying capacity over much longer transmission distances [Ref. 16].

Although the usable fiber bandwidth will be extended further towards the optical carrier frequency, it is clear that this parameter is limited by the use of a single optical carrier signal. Hence a much enhanced bandwidth utilization for an optical fiber can be achieved by transmitting several optical signals, each at different center wavelengths, in parallel on the same fiber. This wavelength division multiplexed operation (see Section 12.9.4), particularly with dense packing of the optical wavelengths (or, essentially, fine frequency spacing), offers the potential for a fiber information-carrying capacity that is many orders of magnitude in excess of that obtained using copper cables or a wideband radio system.

(b) Small size and weight. Optical fibers have very small diameters which are often no greater than the diameter of a human hair. Hence, even when such fibers are covered with protective coatings they are far smaller and much lighter than corresponding copper cables. This is a tremendous boon towards the alleviation of duct congestion in cities, as well as allowing for an expansion of signal transmission within mobiles such as aircraft, satellites and even ships.

(c) Electrical isolation. Optical fibers which are fabricated from glass, or sometimes a plastic polymer, are electrical insulators and therefore, unlike their metallic counterparts, they do not exhibit earth loop and interface problems. Furthermore, this property makes optical fiber transmission ideally suited for communication in electrically hazardous environments as the fibers create no arcing or spark hazard at abrasions or short circuits.

(d) Immunity to interference and crosstalk. Optical fibers form a dielectric waveguide and are therefore free from electromagnetic interference (EMI), radio-frequency interference (RFI), or switching transients giving electromagnetic pulses (EMPs). Hence the operation of an optical fiber communication system is unaffected by transmission through an electrically noisy environment and the fiber cable requires no shielding from EMI. The fiber cable is also not susceptible to lightning strikes if used overhead rather than underground. Moreover, it is fairly easy to ensure that there is no optical interference between fibers and hence, unlike communication using electrical conductors, crosstalk is negligible, even when many fibers are cabled together.

(e) Signal security. The light from optical fibers does not radiate significantly and therefore they provide a high degree of signal security. Unlike the situation with copper cables, a transmitted optical signal cannot be obtained from a fiber in a noninvasive manner (i.e. without drawing optical power from the fiber). Therefore, in theory, any attempt to acquire a message signal transmitted optically may be detected. This feature is obviously attractive for military, banking and general data transmission (i.e. computer network) applications.

(f) Low transmission loss. The development of optical fibers over the last 20 years has resulted in the production of optical fiber cables which exhibit very low attenuation or transmission loss in comparison with the best copper conductors. Fibers have been

fabricated with losses as low as 0.15 dB km^{-1} (see Section 3.3.2) and this feature has become a major advantage of optical fiber communications. It facilitates the implementation of communication links with extremely wide optical repeater or amplifier spacings, thus reducing both system cost and complexity. Together with the already proven modulation bandwidth capability of fiber cables, this property has provided a totally compelling case for the adoption of optical fiber communications in the majority of long-haul telecommunication applications, replacing not only copper cables, but also satellite communications, as a consequence of the very noticeable delay incurred for voice transmission when using this latter approach.

(g) Ruggedness and flexibility. Although protective coatings are essential, optical fibers may be manufactured with very high tensile strengths (see Section 4.6). Perhaps surprisingly for a glassy substance, the fibers may also be bent to quite small radii or twisted without damage. Furthermore, cable structures have been developed (see Section 4.8.4) which have proved flexible, compact and extremely rugged. Taking the size and weight advantage into account, these optical fiber cables are generally superior in terms of storage, transportation, handling and installation to corresponding copper cables, while exhibiting at least comparable strength and durability.

(h) System reliability and ease of maintenance. These features primarily stem from the low-loss property of optical fiber cables which reduces the requirement for intermediate repeaters or line amplifiers to boost the transmitted signal strength. Hence with fewer optical repeaters or amplifiers, system reliability is generally enhanced in comparison with conventional electrical conductor systems. Furthermore, the reliability of the optical components is no longer a problem with predicted lifetimes of 20 to 30 years being quite common. Both these factors also tend to reduce maintenance time and costs.

(i) Potential low cost. The glass which generally provides the optical fiber transmission medium is made from sand – not a scarce resource. So, in comparison with copper conductors, optical fibers offer the potential for low-cost line communication. Although over recent years this potential has largely been realized in the costs of the optical fiber transmission medium which for bulk purchases has become competitive with copper wires (i.e. twisted pairs), it has not yet been achieved in all the other component areas associated with optical fiber communications. For example, the costs of high-performance semiconductor lasers and detector photodiodes are still relatively high, as well as some of those concerned with the connection technology (demountable connectors, couplers, etc.).

Overall system costs when utilizing optical fiber communication on long-haul links, however, are substantially less than those for equivalent electrical line systems because of the low-loss and wideband properties of the optical transmission medium. As indicated in (f), the requirement for intermediate repeaters and the associated electronics is reduced, giving a substantial cost advantage. Although this cost benefit gives a net gain for long-haul links, it is not always the case in short-haul applications where the additional cost incurred, due to the electrical–optical conversion (and vice versa), may be a deciding factor. Nevertheless, there are other possible cost advantages in relation to shipping, handling, installation and maintenance, as well as the features indicated in (c) and (d) which may prove significant in the system choice.

The reducing costs of optical fiber communications has provided strong competition not only with electrical line transmission systems, but also for microwave and millimeter wave radio transmission systems. Although these systems are reasonably wideband, the relatively short-span 'line of sight' transmission necessitates expensive aerial towers at intervals no greater than a few tens of kilometers. Hence, with the exception of the telecommunication access network (see Section 15.6.3) due primarily to current first installed cost constraints, optical fiber has become the dominant transmission medium within the major industrialized societies.

Many advantages are therefore provided by the use of a lightwave carrier within a transmission medium consisting of an optical fiber. The fundamental principles giving rise to these enhanced performance characteristics, together with their practical realization, are described in the following chapters. However, a general understanding of the basic nature and properties of light is assumed. If this is lacking, the reader is directed to the many excellent texts encompassing the topic, a few of which are indicated in Refs 19 to 23.

References

[1] A. G. Bell, 'Selenium and the photophone', *The Electrician*, pp. 214, 215, 220, 221, 1880.
[2] W. S. Huxford and J. R. Platt, 'Survey of near infra-red communication systems', *J. Opt. Soc. Am.*, **38**, pp. 253–268, 1948.
[3] N. C. Beese, 'Light sources for optical communication', *Infrared Phys.*, **1**, pp. 5–16, 1961.
[4] R. M. Gagliardi and S. Karp, *Optical Communications*, Wiley, 1976.
[5] T. H. Maiman, 'Stimulated optical radiation in ruby', *Nature*, **187**, pp. 493–494, 1960.
[6] A. R. Kraemer, 'Free-space optical communications', *Signal*, pp. 26–32, 1977.
[7] K. C. Kao and G. A. Hockham, 'Dielectric fiber surface waveguides for optical frequencies', *Proc. IEE*, **113**(7), pp. 1151–1158, 1966.
[8] A. Werts, 'Propagation de la lumière cohérente dans les fibres optiques', *L'Onde Electrique*, **46**, pp. 967–980, 1966.
[9] R. L. Hartman, J. C. Dyment, C. J. Hwang and H. Kuhn, 'Continuous operation of GaAs–Ga$_x$Al$_{1-x}$As, double heterostructure lasers with 330 °C half lives exceeding 1000 h', *Appl. Phys. Lett.*, **23**(4), pp. 181–183, 1973.
[10] A. R. Goodwin, J. F. Peters, M. Pion and W. O. Bourne, 'GaAs lasers with consistently low degradation rates at room temperature', *Appl. Phys. Lett.*, **30**(2), pp. 110–113, 1977.
[11] S. E. Miller, 'Overview and summary of progress', in S. E. Miller and I. P. Kaminow (Eds), *Optical Fiber Telecommunications II*, pp. 1–27, Academic Press, 1988.
[12] E. Garmire, 'Sources, modulators, and detectors for fiber-optic communication systems', in M. Bass and Eric W. Van Stryland (Eds), *Fiber Optics Handbook*, pp. 4.1–4.80, McGraw-Hill, 2002.
[13] D. A. Ackerman, J. E. Johnson, L. J. P. Ketelsen, L. E. Eng, P. A. Kiely and T. G. B. Mason, 'Telecommunication lasers', in I. P. Kaminow and T. Li (Eds), *Optical Fiber Telecommunications IVA*, pp. 587–665, Academic Press, 2002.
[14] R. DeSalvo *et al.*, 'Advanced components and sub-system solutions for 40 Gb/s transmission', *J. Lightwave Technol.*, **20**(12), pp. 2154–2181, 2002.
[15] A. Belahlou *et al.*, 'Fiber design considerations for 40 Gb/s systems', *J. Lightwave Technol.*, **20**(12), pp. 2290–2305, 2002.

[16] W. A. Gambling, 'The rise and rise of optical fibers', *IEEE J. Sel. Top. Quantum Elecbron*, **6**(6), pp. 1084–1093, 2000.

[17] K. Nayayama, M. Kakui, M. Matsui, T. Saitoh and Y. Chigusa, 'Ultra-low-loss (0.1484 dB/km) pure silica core fibre and extension of transmission distance', *Electron. Lett.*, **38**(20), pp. 1168–1169, 2002.

[18] K. Oh, S. Choi, Y. Jung and J. W. Lee, 'Novel hollow optical fibers and their applications in photonic devices for optical communications', *J. Lightwave Technol.*, **23**(2), pp. 524–532, 2005.

[19] M. Born and E. Wolf, *Principles of Optics* (7th edn), Cambridge University Press, 1999.

[20] W. J. Smith, *Modern Optical Engineering* (3rd edn), McGraw-Hill, 2000.

[21] E. Hecht and A. Zajac, *Optics* (4th edn), Addison-Wesley, 2003.

[22] F. L. Pedrotti, L. S. Pedrotti and L. M. Pedrotti, *Introduction to Optics* (3rd edn), Prentice Hall, 2006.

[23] F. Graham Smith, T. A. King and D. Wilkins, *Optics and Photonics: An Introduction* (2nd edn), Wiley, 2007.

CHAPTER 2

Optical fiber waveguides

2.1 Introduction

The transmission of light via a dielectric waveguide structure was first proposed and investigated at the beginning of the twentieth century. In 1910 Hondros and Debye [Ref. 1] conducted a theoretical study, and experimental work was reported by Schriever in 1920 [Ref. 2]. However, a transparent dielectric rod, typically of silica glass with a refractive index of around 1.5, surrounded by air, proved to be an impractical waveguide due to its unsupported structure (especially when very thin waveguides were considered in order to limit the number of optical modes propagated) and the excessive losses at any discontinuities of the glass–air interface. Nevertheless, interest in the application of dielectric optical waveguides in such areas as optical imaging and medical diagnosis (e.g. endoscopes) led to proposals [Refs 3, 4] for a clad dielectric rod in the mid-1950s in order to overcome these problems. This structure is illustrated in Figure 2.1, which shows a transparent core with a refractive index n_1 surrounded by a transparent cladding of slightly lower refractive index n_2. The cladding supports the waveguide structure while also, when

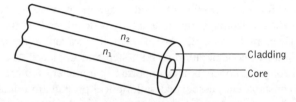

Figure 2.1 Optical fiber waveguide showing the core of refractive index n_1, surrounded by the cladding of slightly lower refractive index n_2

sufficiently thick, substantially reducing the radiation loss into the surrounding air. In essence, the light energy travels in both the core and the cladding allowing the associated fields to decay to a negligible value at the cladding–air interface.

The invention of the clad waveguide structure led to the first serious proposals by Kao and Hockham [Ref. 5] and Werts [Ref. 6], in 1966, to utilize optical fibers as a communications medium, even though they had losses in excess of 1000 dB km^{-1}. These proposals stimulated tremendous efforts to reduce the attenuation by purification of the materials. This has resulted in improved conventional glass refining techniques giving fibers with losses of around 4.2 dB km^{-1} [Ref. 7]. Also, progress in glass refining processes such as depositing vapor-phase reagents to form silica [Ref. 8] allowed fibers with losses below 1 dB km^{-1} to be fabricated.

Most of this work was focused on the 0.8 to 0.9 μm wavelength band because the first generation of optical sources fabricated from gallium aluminum arsenide alloys operated in this region. However, as silica fibers were studied in further detail it became apparent that transmission at longer wavelengths (1.1 to 1.6 μm) would result in lower losses and reduced signal dispersion. This produced a shift in optical fiber source and detector technology in order to provide operation at these longer wavelengths. Hence at longer wavelengths, especially around 1.55 μm, typical high-performance fibers have losses of 0.2 dB km^{-1} [Ref. 9].

As such losses are very close to the theoretical lower limit for silicate glass fiber, there is interest in glass-forming systems which can provide low-loss transmission in the mid-infrared (2 to 5 μm) optical wavelength regions. Although a system based on fluoride glass offers the potential for ultra-low-loss transmission of 0.01 dB km^{-1} at a wavelength of 2.55 μm, such fibers still exhibit losses of at least 0.65 dB km^{-1} and they also cannot yet be produced with the robust mechanical properties of silica fibers [Ref. 10].

In order to appreciate the transmission mechanism of optical fibers with dimensions approximating to those of a human hair, it is necessary to consider the optical waveguiding of a cylindrical glass fiber. Such a fiber acts as an open optical waveguide, which may be analyzed utilizing simple ray theory. However, the concepts of geometric optics are not sufficient when considering all types of optical fiber, and electromagnetic mode theory must be used to give a complete picture. The following sections will therefore outline the transmission of light in optical fibers prior to a more detailed discussion of the various types of fiber.

In Section 2.2 we continue the discussion of light propagation in optical fibers using the ray theory approach in order to develop some of the fundamental parameters associated with optical fiber transmission (acceptance angle, numerical aperture, etc.). Furthermore, this provides a basis for the discussion of electromagnetic wave propagation presented in Section 2.3, where the electromagnetic mode theory is developed for the planar (rectangular) waveguide. Then, in Section 2.4, we discuss the waveguiding mechanism within cylindrical fibers prior to consideration of both step and graded index fibers. Finally, in Section 2.5 the theoretical concepts and important parameters (cutoff wavelength, spot size, propagation constant, etc.) associated with optical propagation in single-mode fibers are introduced and approximate techniques to obtain values for these parameters are described.

All consideration in the above sections is concerned with what can be referred to as conventional optical fiber in the context that it comprises both solid-core and cladding regions as depicted in Figure 2.1. In the mid-1990s, however, a new class of microstructured optical fiber, termed photonic crystal fiber, was experimentally demonstrated [Ref. 11] which has subsequently exhibited the potential to deliver applications ranging from light transmission over distance to optical device implementations (e.g. power splitters, amplifiers, bistable switches, wavelength converters). The significant physical feature of this microstructured optical fiber is that it typically contains an array of air holes running along the longitudinal axis rather than consisting of a solid silica rod structure. Moreover, the presence of these holes provides an additional dimension to fiber design which has already resulted in new developments for both guiding and controlling light. Hence the major photonic crystal fiber structures and their guidance mechanisms are outlined and discussed in Section 2.6 in order to give an insight into the fundamental developments of this increasingly important fiber class.

2.2 Ray theory transmission

2.2.1 Total internal reflection

To consider the propagation of light within an optical fiber utilizing the ray theory model it is necessary to take account of the refractive index of the dielectric medium. The refractive index of a medium is defined as the ratio of the velocity of light in a vacuum to the velocity of light in the medium. A ray of light travels more slowly in an optically dense medium than in one that is less dense, and the refractive index gives a measure of this effect. When a ray is incident on the interface between two dielectrics of differing refractive indices (e.g. glass–air), refraction occurs, as illustrated in Figure 2.2(a). It may be observed that the ray approaching the interface is propagating in a dielectric of refractive index n_1 and is at an angle ϕ_1 to the normal at the surface of the interface. If the dielectric on the other side of the interface has a refractive index n_2 which is less than n_1, then the refraction is such that the ray path in this lower index medium is at an angle ϕ_2 to the normal, where ϕ_2 is greater than ϕ_1. The angles of incidence ϕ_1 and refraction ϕ_2 are related to each other and to the refractive indices of the dielectrics by Snell's law of refraction [Ref. 12], which states that:

$$n_1 \sin \phi_1 = n_2 \sin \phi_2$$

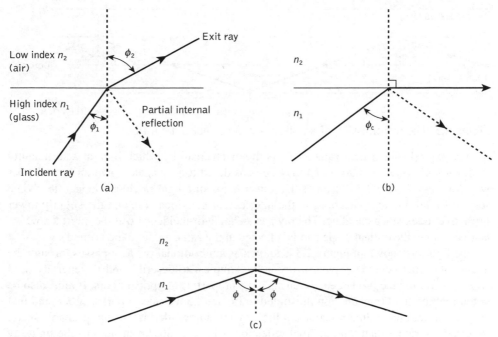

Figure 2.2 Light rays incident on a high to low refractive index interface (e.g. glass–air): (a) refraction; (b) the limiting case of refraction showing the critical ray at an angle ϕ_c; (c) total internal reflection where $\phi > \phi_c$

or:

$$\frac{\sin \phi_1}{\sin \phi_2} = \frac{n_2}{n_1} \tag{2.1}$$

It may also be observed in Figure 2.2(a) that a small amount of light is reflected back into the originating dielectric medium (partial internal reflection). As n_1 is greater than n_2, the angle of refraction is always greater than the angle of incidence. Thus when the angle of refraction is 90° and the refracted ray emerges parallel to the interface between the dielectrics, the angle of incidence must be less than 90°. This is the limiting case of refraction and the angle of incidence is now known as the critical angle ϕ_c, as shown in Figure 2.2(b). From Eq. (2.1) the value of the critical angle is given by:

$$\sin \phi_c = \frac{n_2}{n_1} \tag{2.2}$$

At angles of incidence greater than the critical angle the light is reflected back into the originating dielectric medium (total internal reflection) with high efficiency (around 99.9%). Hence, it may be observed in Figure 2.2(c) that total internal reflection occurs at the interface between two dielectrics of differing refractive indices when light is incident on the dielectric of lower index from the dielectric of higher index, and the angle of incidence of

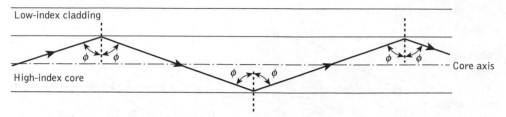

Figure 2.3 The transmission of a light ray in a perfect optical fiber

the ray exceeds the critical value. This is the mechanism by which light at a sufficiently shallow angle (less than $90° − \phi_c$) may be considered to propagate down an optical fiber with low loss. Figure 2.3 illustrates the transmission of a light ray in an optical fiber via a series of total internal reflections at the interface of the silica core and the slightly lower refractive index silica cladding. The ray has an angle of incidence ϕ at the interface which is greater than the critical angle and is reflected at the same angle to the normal.

The light ray shown in Figure 2.3 is known as a meridional ray as it passes through the axis of the fiber core. This type of ray is the simplest to describe and is generally used when illustrating the fundamental transmission properties of optical fibers. It must also be noted that the light transmission illustrated in Figure 2.3 assumes a perfect fiber, and that any discontinuities or imperfections at the core–cladding interface would probably result in refraction rather than total internal reflection, with the subsequent loss of the light ray into the cladding.

2.2.2 Acceptance angle

Having considered the propagation of light in an optical fiber through total internal reflection at the core–cladding interface, it is useful to enlarge upon the geometric optics approach with reference to light rays entering the fiber. Since only rays with a sufficiently shallow grazing angle (i.e. with an angle to the normal greater than ϕ_c) at the core–cladding interface are transmitted by total internal reflection, it is clear that not all rays entering the fiber core will continue to be propagated down its length.

The geometry concerned with launching a light ray into an optical fiber is shown in Figure 2.4, which illustrates a meridional ray A at the critical angle ϕ_c within the fiber at the core–cladding interface. It may be observed that this ray enters the fiber core at an angle θ_a to the fiber axis and is refracted at the air–core interface before transmission to the core–cladding interface at the critical angle. Hence, any rays which are incident into the fiber core at an angle greater than θ_a will be transmitted to the core–cladding interface at an angle less than ϕ_c, and will not be totally internally reflected. This situation is also illustrated in Figure 2.4, where the incident ray B at an angle greater than θ_a is refracted into the cladding and eventually lost by radiation. Thus for rays to be transmitted by total internal reflection within the fiber core they must be incident on the fiber core within an acceptance cone defined by the conical half angle θ_a. Hence θ_a is the maximum angle to the axis at which light may enter the fiber in order to be propagated, and is often referred to as the acceptance angle* for the fiber.

* θ_a is sometimes referred to as the maximum or total acceptance angle.

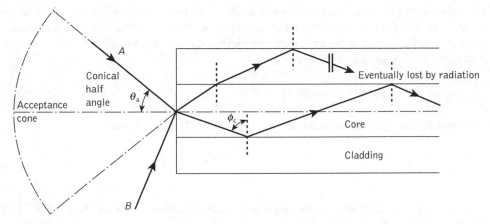

Figure 2.4 The acceptance angle θ_a when launching light into an optical fiber

If the fiber has a regular cross-section (i.e. the core–cladding interfaces are parallel and there are no discontinuities) an incident meridional ray at greater than the critical angle will continue to be reflected and will be transmitted through the fiber. From symmetry considerations it may be noted that the output angle to the axis will be equal to the input angle for the ray, assuming the ray emerges into a medium of the same refractive index from which it was input.

2.2.3 Numerical aperture

The acceptance angle for an optical fiber was defined in the preceding section. However, it is possible to continue the ray theory analysis to obtain a relationship between the acceptance angle and the refractive indices of the three media involved, namely the core, cladding and air. This leads to the definition of a more generally used term, the numerical aperture of the fiber. It must be noted that within this analysis, as with the preceding discussion of acceptance angle, we are concerned with meridional rays within the fiber.

Figure 2.5 shows a light ray incident on the fiber core at an angle θ_1 to the fiber axis which is less than the acceptance angle for the fiber θ_a. The ray enters the fiber from a

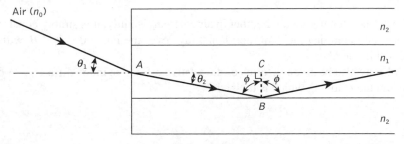

Figure 2.5 The ray path for a meridional ray launched into an optical fiber in air at an input angle less than the acceptance angle for the fiber

medium (air) of refractive index n_0, and the fiber core has a refractive index n_1, which is slightly greater than the cladding refractive index n_2. Assuming the entrance face at the fiber core to be normal to the axis, then considering the refraction at the air–core interface and using Snell's law given by Eq. (2.1):

$$n_0 \sin \theta_1 = n_1 \sin \theta_2 \qquad (2.3)$$

Considering the right-angled triangle ABC indicated in Figure 2.5, then:

$$\phi = \frac{\pi}{2} - \theta_2 \qquad (2.4)$$

where ϕ is greater than the critical angle at the core–cladding interface. Hence Eq. (2.3) becomes:

$$n_0 \sin \theta_1 = n_1 \cos \phi \qquad (2.5)$$

Using the trigonometrical relationship $\sin^2 \phi + \cos^2 \phi = 1$, Eq. (2.5) may be written in the form:

$$n_0 \sin \theta_1 = n_1(1 - \sin^2 \phi)^{\frac{1}{2}} \qquad (2.6)$$

When the limiting case for total internal reflection is considered, ϕ becomes equal to the critical angle for the core–cladding interface and is given by Eq. (2.2). Also in this limiting case θ_1 becomes the acceptance angle for the fiber θ_a. Combining these limiting cases into Eq. (2.6) gives:

$$n_0 \sin \theta_a = (n_1^2 - n_2^2)^{\frac{1}{2}} \qquad (2.7)$$

Equation (2.7), apart from relating the acceptance angle to the refractive indices, serves as the basis for the definition of the important optical fiber parameter, the numerical aperture (NA). Hence the NA is defined as:

$$NA = n_0 \sin \theta_a = (n_1^2 - n_2^2)^{\frac{1}{2}} \qquad (2.8)$$

Since the NA is often used with the fiber in air where n_0 is unity, it is simply equal to $\sin \theta_a$. It may also be noted that incident meridional rays over the range $0 \le \theta_1 \le \theta_a$ will be propagated within the fiber.

The NA may also be given in terms of the relative refractive index difference Δ between the core and the cladding which is defined as:*

* Sometimes another parameter $\Delta n = n_1 - n_2$ is referred to as the index difference and $\Delta n/n_1$ as the fractional index difference. Hence Δ also approximates to the fractional index difference.

$$\Delta = \frac{n_1^2 - n_2^2}{2n_1^2}$$

$$\simeq \frac{n_1 - n_2}{n_1} \qquad \text{for } \Delta \ll 1 \tag{2.9}$$

Hence combining Eq. (2.8) with Eq. (2.9) we can write:

$$NA = n_1(2\Delta)^{\frac{1}{2}} \tag{2.10}$$

The relationships given in Eqs (2.8) and (2.10) for the numerical aperture are a very useful measure of the light-collecting ability of a fiber. They are independent of the fiber core diameter and will hold for diameters as small as 8 μm. However, for smaller diameters they break down as the geometric optics approach is invalid. This is because the ray theory model is only a partial description of the character of light. It describes the direction a plane wave component takes in the fiber but does not take into account interference between such components. When interference phenomena are considered it is found that only rays with certain discrete characteristics propagate in the fiber core. Thus the fiber will only support a discrete number of guided modes. This becomes critical in small-core-diameter fibers which only support one or a few modes. Hence electromagnetic mode theory must be applied in these cases (see Section 2.3).

Example 2.1

A silica optical fiber with a core diameter large enough to be considered by ray theory analysis has a core refractive index of 1.50 and a cladding refractive index of 1.47.

Determine: (a) the critical angle at the core–cladding interface; (b) the *NA* for the fiber; (c) the acceptance angle in air for the fiber.

Solution: (a) The critical angle ϕ_c at the core–cladding interface is given by Eq. (2.2) where:

$$\phi_c = \sin^{-1}\frac{n_2}{n_1} = \sin^{-1}\frac{1.47}{1.50}$$

$$= 78.5°$$

(b) From Eq. (2.8) the *NA* is:

$$NA = (n_1^2 - n_2^2)^{\frac{1}{2}} = (1.50^2 - 1.47^2)^{\frac{1}{2}}$$
$$= (2.25 - 2.16)^{\frac{1}{2}}$$
$$= 0.30$$

(c) Considering Eq. (2.8) the acceptance angle in air θ_a is given by:

$$\theta_a = \sin^{-1} NA = \sin^{-1} 0.30$$
$$= 17.4°$$

Example 2.2

A typical relative refractive index difference for an optical fiber designed for long-distance transmission is 1%. Estimate the *NA* and the solid acceptance angle in air for the fiber when the core index is 1.46. Further, calculate the critical angle at the core–cladding interface within the fiber. It may be assumed that the concepts of geometric optics hold for the fiber.

Solution: Using Eq. (2.10) with $\Delta = 0.01$ gives the *NA* as:

$$NA = n_1(2\Delta)^{\frac{1}{2}} = 1.46(0.02)^{\frac{1}{2}}$$
$$= 0.21$$

For small angles the solid acceptance angle in air ζ is given by:

$$\zeta \simeq \pi\theta_a^2 = \pi \sin^2 \theta_a$$

Hence from Eq. (2.8):

$$\zeta \simeq \pi(NA)^2 = \pi \times 0.04$$
$$= 0.13 \text{ rad}$$

Using Eq. (2.9) for the relative refractive index difference Δ gives:

$$\Delta \simeq \frac{n_1 - n_2}{n_1} = 1 - \frac{n_2}{n_1}$$

Hence

$$\frac{n_2}{n_1} = 1 - \Delta = 1 - 0.01$$
$$= 0.99$$

From Eq. (2.2) the critical angle at the core–cladding interface is:

$$\phi_c = \sin^{-1} \frac{n_2}{n_1} = \sin^{-1} 0.99$$
$$= 81.9°$$

2.2.4 Skew rays

In the preceding sections we have considered the propagation of meridional rays in the optical waveguide. However, another category of ray exists which is transmitted without passing through the fiber axis. These rays, which greatly outnumber the meridional rays,

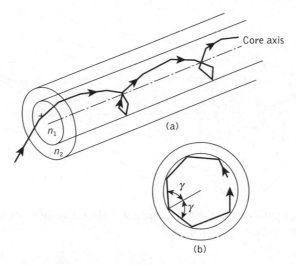

Figure 2.6 The helical path taken by a skew ray in an optical fiber: (a) skew ray path down the fiber; (b) cross-sectional view of the fiber

follow a helical path through the fiber, as illustrated in Figure 2.6, and are called skew rays. It is not easy to visualize the skew ray paths in two dimensions, but it may be observed from Figure 2.6(b) that the helical path traced through the fiber gives a change in direction of 2γ at each reflection, where γ is the angle between the projection of the ray in two dimensions and the radius of the fiber core at the point of reflection. Hence, unlike meridional rays, the point of emergence of skew rays from the fiber in air will depend upon the number of reflections they undergo rather than the input conditions to the fiber. When the light input to the fiber is nonuniform, skew rays will therefore tend to have a smoothing effect on the distribution of the light as it is transmitted, giving a more uniform output. The amount of smoothing is dependent on the number of reflections encountered by the skew rays.

A further possible advantage of the transmission of skew rays becomes apparent when their acceptance conditions are considered. In order to calculate the acceptance angle for a skew ray it is necessary to define the direction of the ray in two perpendicular planes. The geometry of the situation is illustrated in Figure 2.7 where a skew ray is shown incident on the fiber core at the point A, at an angle θ_s to the normal at the fiber end face. The ray is refracted at the air–core interface before traveling to the point B in the same plane. The angles of incidence and reflection at the point B are ϕ, which is greater than the critical angle for the core–cladding interface.

When considering the ray between A and B it is necessary to resolve the direction of the ray path AB to the core radius at the point B. As the incident and reflected rays at the point B are in the same plane, this is simply $\cos \phi$. However, if the two perpendicular planes through which the ray path AB traverses are considered, then γ is the angle between the core radius and the projection of the ray onto a plane BRS normal to the core axis, and θ is the angle between the ray and a line AT drawn parallel to the core axis. Thus to resolve the ray path AB relative to the radius BR in these two perpendicular planes requires multiplication by $\cos \gamma$ and $\sin \theta$.

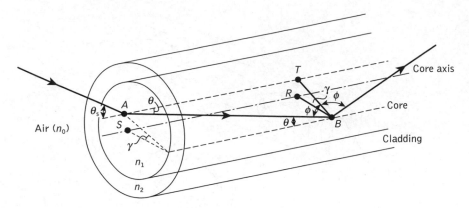

Figure 2.7 The ray path within the fiber core for a skew ray incident at an angle θ_s to the normal at the air–core interface

Hence, the reflection at point B at an angle ϕ may be given by:

$$\cos \gamma \sin \theta = \cos \phi \qquad (2.11)$$

Using the trigonometrical relationship $\sin^2 \phi + \cos^2 \phi = 1$, Eq. (2.11) becomes:

$$\cos \gamma \sin \theta = \cos \phi = (1 - \sin^2 \phi)^{\frac{1}{2}} \qquad (2.12)$$

If the limiting case for total internal reflection is now considered, then ϕ becomes equal to the critical angle ϕ_c for the core–cladding interface and, following Eq. (2.2), is given by $\sin \phi_c = n_2/n_1$. Hence, Eq. (2.12) may be written as:

$$\cos \gamma \sin \theta \leq \cos \phi_c = \left(1 - \frac{n_2^2}{n_1^2}\right)^{\frac{1}{2}} \qquad (2.13)$$

Furthermore, using Snell's law at the point A, following Eq. (2.1) we can write:

$$n_0 \sin \theta_a = n_1 \sin \theta \qquad (2.14)$$

where θ_a represents the maximum input axial angle for meridional rays, as expressed in Section 2.2.2, and θ is the internal axial angle. Hence substituting for $\sin \theta$ from Eq. (2.13) into Eq. (2.14) gives:

$$\sin \theta_{as} = \frac{n_1 \cos \phi_c}{n_0 \cos \gamma} = \frac{n_1}{n_0 \cos \gamma} \left(1 - \frac{n_2^2}{n_1^2}\right)^{\frac{1}{2}} \qquad (2.15)$$

where θ_{as} now represents the maximum input angle or acceptance angle for skew rays. It may be noted that the inequality shown in Eq. (2.13) is no longer necessary as all the terms in Eq. (2.15) are specified for the limiting case. Thus the acceptance conditions for skew rays are:

$$n_0 \sin \theta_{as} \cos \gamma = (n_1^2 - n_2^2)^{\frac{1}{2}} = NA \qquad (2.16)$$

and in the case of the fiber in air ($n_0 = 1$):

$$\sin \theta_{as} \cos \gamma = NA \qquad (2.17)$$

Therefore by comparison with Eq. (2.8) derived for meridional rays, it may be noted that skew rays are accepted at larger axial angles in a given fiber than meridional rays, depending upon the value of cos γ. In fact, for meridional rays cos γ is equal to unity and θ_{as} becomes equal to θ_a. Thus although θ_a is the maximum conical half angle for the acceptance of meridional rays, it defines the minimum input angle for skew rays. Hence, as may be observed from Figure 2.6, skew rays tend to propagate only in the annular region near the outer surface of the core, and do not fully utilize the core as a transmission medium. However, they are complementary to meridional rays and increase the light-gathering capacity of the fiber. This increased light-gathering ability may be significant for large NA fibers, but for most communication design purposes the expressions given in Eqs (2.8) and (2.10) for meridional rays are considered adequate.

Example 2.3

An optical fiber in air has an NA of 0.4. Compare the acceptance angle for meridional rays with that for skew rays which change direction by 100° at each reflection.

Solution: The acceptance angle for meridional rays is given by Eq. (2.8) with $n_0 = 1$ as:

$$\theta_a = \sin^{-1} NA = \sin^{-1} 0.4$$
$$= 23.6°$$

The skew rays change direction by 100° at each reflection, therefore $\gamma = 50°$. Hence using Eq. (2.17) the acceptance angle for skew rays is:

$$\theta_{as} = \sin^{-1} \left(\frac{NA}{\cos \gamma} \right) = \sin^{-1} \left(\frac{0.4}{\cos 50°} \right)$$
$$= 38.5°$$

In this example, the acceptance angle for the skew rays is about 15° greater than the corresponding angle for meridional rays. However, it must be noted that we have only compared the acceptance angle of one particular skew ray path. When the light input to the fiber is at an angle to the fiber axis, it is possible that γ will vary from zero for meridional rays to 90° for rays which enter the fiber at the core–cladding interface giving acceptance of skew rays over a conical half angle of $\pi/2$ radians.

2.3 Electromagnetic mode theory for optical propagation

2.3.1 Electromagnetic waves

In order to obtain an improved model for the propagation of light in an optical fiber, electromagnetic wave theory must be considered. The basis for the study of electromagnetic wave propagation is provided by Maxwell's equations [Ref. 13]. For a medium with zero conductivity these vector relationships may be written in terms of the electric field **E**, magnetic field **H**, electric flux density **D** and magnetic flux density **B** as the curl equations:

$$\nabla \times \mathbf{E} = -\frac{\partial \mathbf{B}}{\partial t} \tag{2.18}$$

$$\nabla \times \mathbf{H} = \frac{\partial \mathbf{D}}{\partial t} \tag{2.19}$$

and the divergence conditions:

$$\nabla \cdot \mathbf{D} = 0 \qquad \text{(no free charges)} \tag{2.20}$$

$$\nabla \cdot \mathbf{B} = 0 \qquad \text{(no free poles)} \tag{2.21}$$

where ∇ is a vector operator.

The four field vectors are related by the relations:

$$\mathbf{D} = \varepsilon \mathbf{E} \tag{2.22}$$
$$\mathbf{B} = \mu \mathbf{H}$$

where ε is the dielectric permittivity and μ is the magnetic permeability of the medium.

Substituting for **D** and **B** and taking the curl of Eqs (2.18) and (2.19) gives:

$$\nabla \times (\nabla \times \mathbf{E}) = -\mu \varepsilon \frac{\partial^2 \mathbf{E}}{\partial t^2} \tag{2.23}$$

$$\nabla \times (\nabla \times \mathbf{H}) = -\mu \varepsilon \frac{\partial^2 \mathbf{H}}{\partial t^2} \tag{2.24}$$

Then using the divergence conditions of Eqs (2.20) and (2.21) with the vector identity:

$$\nabla \times (\nabla \times \mathbf{Y}) = \nabla(\nabla \cdot \mathbf{Y}) - \nabla^2(\mathbf{Y})$$

we obtain the nondispersive wave equations:

$$\nabla^2 \mathbf{E} = \mu \varepsilon \frac{\partial^2 \mathbf{E}}{\partial t^2} \tag{2.25}$$

and:

$$\nabla^2 \mathbf{H} = \mu \varepsilon \frac{\partial^2 \mathbf{H}}{\partial t^2} \tag{2.26}$$

where ∇^2 is the Laplacian operator. For rectangular Cartesian and cylindrical polar coordinates the above wave equations hold for each component of the field vector, every component satisfying the scalar wave equation:

$$\nabla^2 \psi = \frac{1}{v_p^2} \frac{\partial^2 \psi}{\partial t^2} \tag{2.27}$$

where ψ may represent a component of the $\mathbf{E}$ or $\mathbf{H}$ field and v_p is the phase velocity (velocity of propagation of a point of constant phase in the wave) in the dielectric medium. It follows that:

$$v_p = \frac{1}{(\mu \varepsilon)^{\frac{1}{2}}} = \frac{1}{(\mu_r \mu_0 \varepsilon_r \varepsilon_0)^{\frac{1}{2}}} \tag{2.28}$$

where μ_r and ε_r are the relative permeability and permittivity for the dielectric medium and μ_0 and ε_0 are the permeability and permittivity of free space. The velocity of light in free space c is therefore:

$$c = \frac{1}{(\mu_0 \varepsilon_0)^{\frac{1}{2}}} \tag{2.29}$$

If planar waveguides, described by rectangular Cartesian coordinates (x, y, z), or circular fibers, described by cylindrical polar coordinates (r, ϕ, z), are considered, then the Laplacian operator takes the form:

$$\nabla^2 \psi = \frac{\partial^2 \psi}{\partial x^2} + \frac{\partial^2 \psi}{\partial y^2} + \frac{\partial^2 \psi}{\partial z^2} \tag{2.30}$$

or:

$$\nabla^2 \psi = \frac{\partial^2 \psi}{\partial r^2} + \frac{1}{r} \frac{\partial \psi}{\partial r} + \frac{1}{r^2} \frac{\partial^2 \psi}{\partial \phi^2} + \frac{\partial^2 \psi}{\partial z^2} \tag{2.31}$$

respectively. It is necessary to consider both these forms for a complete treatment of optical propagation in the fiber, although many of the properties of interest may be dealt with using Cartesian coordinates.

The basic solution of the wave equation is a sinusoidal wave, the most important form of which is a uniform plane wave given by:

$$\psi = \psi_0 \exp[j(\omega t - \mathbf{k} \cdot \mathbf{r})] \tag{2.32}$$

where ω is the angular frequency of the field, t is the time, $\mathbf{k}$ is the propagation vector which gives the direction of propagation and the rate of change of phase with distance, while the components of $\mathbf{r}$ specify the coordinate point at which the field is observed. When λ is the optical wavelength in a vacuum, the magnitude of the propagation vector or the vacuum phase propagation constant k (where $k = |\mathbf{k}|$) is given by:

$$k = \frac{2\pi}{\lambda} \tag{2.33}$$

It should be noted that in this case k is also referred to as the free space wave number.

2.3.2 Modes in a planar guide

The planar guide is the simplest form of optical waveguide. We may assume it consists of a slab of dielectric with refractive index n_1 sandwiched between two regions of lower refractive index n_2. In order to obtain an improved model for optical propagation it is useful to consider the interference of plane wave components within this dielectric waveguide.

The conceptual transition from ray to wave theory may be aided by consideration of a plane monochromatic wave propagating in the direction of the ray path within the guide (see Figure 2.8(a)). As the refractive index within the guide is n_1, the optical wavelength in this region is reduced to λ/n_1, while the vacuum propagation constant is increased to $n_1 k$. When θ is the angle between the wave propagation vector or the equivalent ray and the

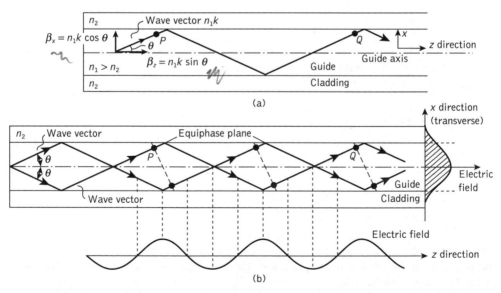

Figure 2.8 The formation of a mode in a planar dielectric guide: (a) a plane wave propagating in the guide shown by its wave vector or equivalent ray – the wave vector is resolved into components in the z and x directions; (b) the interference of plane waves in the guide forming the lowest order mode ($m = 0$)

guide axis, the plane wave can be resolved into two component plane waves propagating in the z and x directions, as shown in Figure 2.8(a). The component of the phase propagation constant in the z direction β_z is given by:

$$\beta_z = n_1 k \cos \theta \tag{2.34}$$

The component of the phase propagation constant in the x direction β_x is:

$$\beta_x = n_1 k \sin \theta \tag{2.35}$$

The component of the plane wave in the x direction is reflected at the interface between the higher and lower refractive index media. When the total phase change* after two successive reflections at the upper and lower interfaces (between the points P and Q) is equal to $2m\pi$ radians, where m is an integer, then constructive interference occurs and a standing wave is obtained in the x direction. This situation is illustrated in Figure 2.8(b), where the interference of two plane waves is shown. In this illustration it is assumed that the interference forms the lowest order (where $m = 0$) standing wave, where the electric field is a maximum at the center of the guide decaying towards zero at the boundary between the guide and cladding. However, it may be observed from Figure 2.8(b) that the electric field penetrates some distance into the cladding, a phenomenon which is discussed in Section 2.3.4.

Nevertheless, the optical wave is effectively confined within the guide and the electric field distribution in the x direction does not change as the wave propagates in the z direction. The sinusoidally varying electric field in the z direction is also shown in Figure 2.8(b). The stable field distribution in the x direction with only a periodic z dependence is known as a mode. A specific mode is obtained only when the angle between the propagation vectors or the rays and the interface have a particular value, as indicated in Figure 2.8(b). In effect, Eqs (2.34) and (2.35) define a group or congruence of rays which in the case described represents the lowest order mode. Hence the light propagating within the guide is formed into discrete modes, each typified by a distinct value of θ. These modes have a periodic z dependence of the form $\exp(- j\beta_z z)$ where β_z becomes the propagation constant for the mode as the modal field pattern is invariant except for a periodic z dependence. Hence, for notational simplicity, and in common with accepted practice, we denote the mode propagation constant by β, where $\beta = \beta_z$. If we now assume a time dependence for the monochromatic electromagnetic light field with angular frequency ω of $\exp(j\omega t)$, then the combined factor $\exp[j(\omega t - \beta z)]$ describes a mode propagating in the z direction.

To visualize the dominant modes propagating in the z direction we may consider plane waves corresponding to rays at different specific angles in the planar guide. These plane waves give constructive interference to form standing wave patterns across the guide following a sine or cosine formula. Figure 2.9 shows examples of such rays for $m = 1, 2, 3$, together with the electric field distributions in the x direction. It may be observed that m

---

* It should be noted that there is a phase shift on reflection of the plane wave at the interface as well as a phase change with distance traveled. The phase shift on reflection at a dielectric interface is dealt with in Section 2.3.4.

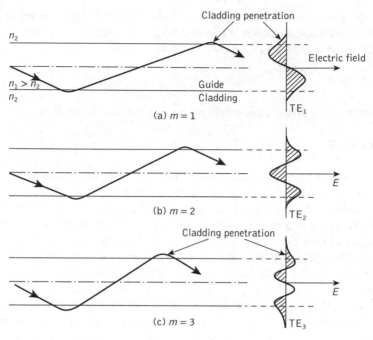

Figure 2.9 Physical model showing the ray propagation and the corresponding transverse electric (TE) field patterns of three lower order models ($m = 1, 2, 3$) in the planar dielectric guide

denotes the number of zeros in this transverse field pattern. In this way m signifies the order of the mode and is known as the mode number.

When light is described as an electromagnetic wave it consists of a periodically varying electric field **E** and magnetic field **H** which are orientated at right angles to each other. The transverse modes shown in Figure 2.9 illustrate the case when the electric field is perpendicular to the direction of propagation and hence $E_z = 0$, but a corresponding component of the magnetic field **H** is in the direction of propagation. In this instance the modes are said to be transverse electric (TE). Alternatively, when a component of the **E** field is in the direction of propagation, but $H_z = 0$, the modes formed are called transverse magnetic (TM). The mode numbers are incorporated into this nomenclature by referring to the TE_m and TM_m modes, as illustrated for the transverse electric modes shown in Figure 2.9. When the total field lies in the transverse plane, transverse electromagnetic (TEM) waves exist where both E_z and H_z are zero. However, although TEM waves occur in metallic conductors (e.g. coaxial cables) they are seldom found in optical waveguides.

2.3.3 Phase and group velocity

Within all electromagnetic waves, whether plane or otherwise, there are points of constant phase. For plane waves these constant phase points form a surface which is referred to as a wavefront. As a monochromatic lightwave propagates along a waveguide in the z direction these points of constant phase travel at a phase velocity v_p given by:

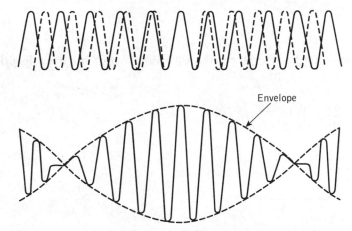

Figure 2.10 The formation of a wave packet from the combination of two waves with nearly equal frequencies. The envelope of the wave package or group of waves travels at a group velocity v_g

$$v_p = \frac{\omega}{\beta} \qquad (2.36)$$

where ω is the angular frequency of the wave. However, it is impossible in practice to produce perfectly monochromatic lightwaves, and light energy is generally composed of a sum of plane wave components of different frequencies. Often the situation exists where a group of waves with closely similar frequencies propagate so that their resultant forms a packet of waves. The formation of such a wave packet resulting from the combination of two waves of slightly different frequency propagating together is illustrated in Figure 2.10. This wave packet does not travel at the phase velocity of the individual waves but is observed to move at a group velocity v_g given by:

$$v_g = \frac{\delta\omega}{\delta\beta} \qquad (2.37)$$

The group velocity is of greatest importance in the study of the transmission characteristics of optical fibers as it relates to the propagation characteristics of observable wave groups or packets of light.

If propagation in an infinite medium of refractive index n_1 is considered, then the propagation constant may be written as:

$$\beta = n_1 \frac{2\pi}{\lambda} = \frac{n_1 \omega}{c} \qquad (2.38)$$

where c is the velocity of light in free space. Equation (2.38) follows from Eqs (2.33) and (2.34) where we assume propagation in the z direction only and hence cos θ is equal to unity. Using Eq. (2.36) we obtain the following relationship for the phase velocity:

$$v_{\mathrm{p}} = \frac{c}{n_1} \tag{2.39}$$

Similarly, employing Eq. (2.37), where in the limit $\delta\omega/\delta\beta$ becomes $\mathrm{d}\omega/\mathrm{d}\beta$, the group velocity:

$$
\begin{aligned}
v_{\mathrm{g}} &= \frac{\mathrm{d}\lambda}{\mathrm{d}\beta} \cdot \frac{\mathrm{d}\omega}{\mathrm{d}\lambda} = \frac{\mathrm{d}}{\mathrm{d}\lambda}\left(n_1 \frac{2\pi}{\lambda}\right)^{-1}\left(\frac{-\omega}{\lambda}\right) \\
&= \frac{-\omega}{2\pi\lambda}\left(\frac{1}{\lambda}\frac{\mathrm{d}n_1}{\mathrm{d}\lambda} - \frac{n_1}{\lambda^2}\right)^{-1} \\
&= \frac{c}{\left(n_1 - \lambda\dfrac{\mathrm{d}n_1}{\mathrm{d}\lambda}\right)} = \frac{c}{N_{\mathrm{g}}}
\end{aligned}
\tag{2.40}
$$

The parameter N_{g} is known as the group index of the guide.

2.3.4 Phase shift with total internal reflection and the evanescent field

The discussion of electromagnetic wave propagation in the planar waveguide given in Section 2.3.2 drew attention to certain phenomena that occur at the guide–cladding interface which are not apparent from ray theory considerations of optical propagation. In order to appreciate these phenomena it is necessary to use the wave theory model for total internal reflection at a planar interface. This is illustrated in Figure 2.11, where the arrowed lines represent wave propagation vectors and a component of the wave energy is

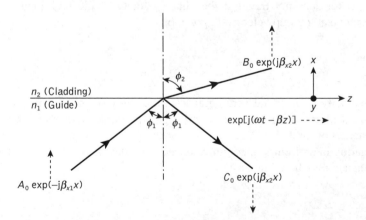

Figure 2.11 A wave incident on the guide–cladding interface of a planar dielectric waveguide. The wave vectors of the incident, transmitted and reflected waves are indicated (solid arrowed lines) together with their components in the z and x directions (dashed arrowed lines)

shown to be transmitted through the interface into the cladding. The wave equation in Cartesian coordinates for the electric field in a lossless medium is:

$$\nabla^2 \mathbf{E} = \mu\varepsilon \frac{\partial^2 \mathbf{E}}{\partial t^2} = \frac{\partial^2 \mathbf{E}}{\partial x^2} + \frac{\partial^2 \mathbf{E}}{\partial y^2} + \frac{\partial^2 \mathbf{E}}{\partial z^2} \tag{2.41}$$

As the guide–cladding interface lies in the y–z plane and the wave is incident in the x–z plane onto the interface, then $\partial/\partial y$ may be assumed to be zero. Since the phase fronts must match all points along the interface in the z direction, the three waves shown in Figure 2.11 will have the same propagation constant β in this direction. Therefore from the discussion of Section 2.3.2 the wave propagation in the z direction may be described by $\exp[\,j(\omega t - \beta z)]$. In addition, there will also be propagation in the x direction. When the components are resolved in this plane:

$$\beta_{x1} = n_1 k \cos \phi_1 \tag{2.42}$$

$$\beta_{x2} = n_2 k \cos \phi_2 \tag{2.43}$$

where β_{x1} and β_{x2} are propagation constants in the x direction for the guide and cladding respectively. Thus the three waves in the waveguide indicated in Figure 2.11, the incident, the transmitted and the reflected, with amplitudes A, B and C, respectively, will have the forms:

$$A = A_0 \exp[-(j\beta_{x1}x)] \exp[\,j(\omega t - \beta z)] \tag{2.44}$$

$$B = B_0 \exp[-(j\beta_{x2}x)] \exp[\,j(\omega t - \beta z)] \tag{2.45}$$

$$C = C_0 \exp[(j\beta_{x1}x)] \exp[\,j(\omega t - \beta z)] \tag{2.46}$$

Using the simple trigonometrical relationship $\cos^2 \phi + \sin^2 \phi = 1$:

$$\beta_{x1}^2 = (n_1^2 k^2 - \beta^2) = -\xi_1^2 \tag{2.47}$$

and:

$$\beta_{x2}^2 = (n_2^2 k^2 - \beta^2) = -\xi_2^2 \tag{2.48}$$

When an electromagnetic wave is incident upon an interface between two dielectric media, Maxwell's equations require that both the tangential components of $\mathbf{E}$ and $\mathbf{H}$ and the normal components of $\mathbf{D}$ ($= \varepsilon\mathbf{E}$) and $\mathbf{B}$ ($= \mu\mathbf{H}$) are continuous across the boundary. If the boundary is defined at $x = 0$ we may consider the cases of the transverse electric (TE) and transverse magnetic (TM) modes.

Initially, let us consider the TE field at the boundary. When Eqs (2.44) and (2.46) are used to represent the electric field components in the y direction E_y and the boundary conditions are applied, then the normal components of the $\mathbf{E}$ and $\mathbf{H}$ fields at the interface may be equated giving:

$$A_0 + C_0 = B_0 \tag{2.49}$$

Furthermore, it can be shown (see Appendix A) that an electric field component in the y direction is related to the tangential magnetic field component H_z following:

$$H_z = \frac{j}{\mu_r \mu_0 \omega} \frac{\partial E_y}{\partial x} \tag{2.50}$$

Applying the tangential boundary conditions and equating H_z by differentiating E_y gives:

$$-\beta_{x1} A_0 + \beta_{x2} C_0 = -\beta_{x2} B_0 \tag{2.51}$$

Algebraic manipulation of Eqs (2.49) and (2.51) provides the following results:

$$C_0 = A_0 \left(\frac{\beta_{x1} - \beta_{x2}}{\beta_{x1} + \beta_{x2}} \right) = A_0 r_{ER} \tag{2.52}$$

$$B_0 = A_0 \left(\frac{2\beta_{x1}}{\beta_{x1} + \beta_{x2}} \right) = A_0 r_{ET} \tag{2.53}$$

where r_{ER} and r_{ET} are the reflection and transmission coefficients for the **E** field at the interface respectively. The expressions obtained in Eqs (2.52) and (2.53) correspond to the Fresnel relationships [Ref. 12] for radiation polarized perpendicular to the interface (**E** polarization).

When both β_{x1} and β_{x2} are real it is clear that the reflected wave C is in phase with the incident wave A. This corresponds to partial reflection of the incident beam. However, as ϕ_1 is increased the component β_z (i.e. β) increases and, following Eqs (2.47) and (2.48), the components β_{x1} and β_{x2} decrease. Continuation of this process results in β_{x2} passing through zero, a point which is signified by ϕ_1 reaching the critical angle for total internal reflection. If ϕ_1 is further increased the component β_{x2} becomes imaginary and we may write it in the form $-j\xi_2$. During this process β_{x1} remains real because we have assumed that $n_1 > n_2$. Under the conditions of total internal reflection Eq. (2.52) may therefore be written as:

$$C_0 = A_0 \left(\frac{\beta_{x1} + j\xi_2}{\beta_{x2} - j\xi_2} \right) = A_0 \exp(2j\delta_E) \tag{2.54}$$

where we observe there is a phase shift of the reflected wave relative to the incident wave. This is signified by δ_E which is given by:

$$\tan \delta_E = \frac{\xi_2}{\beta_{x1}} \tag{2.55}$$

Furthermore, the modulus of the reflected wave is identical to the modulus of the incident wave ($|C_0| = |A_0|$). The curves of the amplitude reflection coefficient $|r_{ER}|$ and phase shift on reflection, against angle of incidence ϕ_1, for TE waves incident on a glass–air interface are displayed in Figure 2.12 [Ref. 14]. These curves illustrate the above results,

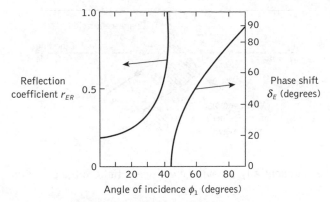

Figure 2.12 Curves showing the reflection coefficient and phase shift on reflection for transverse electric waves against the angle of incidence for a glass–air interface ($n_1 = 1.5$, $n_2 = 1.0$). From J. E. Midwinter, *Optical Fibers for Transmission*, John Wiley & Sons Inc., 1979

where under conditions of total internal reflection the reflected wave has an equal amplitude to the incident wave, but undergoes a phase shift corresponding to δ_E degrees.

A similar analysis may be applied to the TM modes at the interface, which leads to expressions for reflection and transmission of the form [Ref. 14]:

$$C_0 = A_0 \left(\frac{\beta_{x1} n_2^2 - \beta_{x2} n_1^2}{\beta_{x1} n_2^2 + \beta_{x2} n_1^2} \right) = A_0 r_{HR} \tag{2.56}$$

and:

$$B_0 = A_0 \left(\frac{2 \beta_{x1} n_2^2}{\beta_{x1} n_2^2 + \beta_{x2} n_1^2} \right) = A_0 r_{HT} \tag{2.57}$$

where r_{HR} and r_{HT} are, respectively, the reflection and transmission coefficients for the **H** field at the interface. Again, the expressions given in Eqs (2.56) and (2.57) correspond to Fresnel relationships [Ref. 12], but in this case they apply to radiation polarized parallel to the interface (**H** polarization). Furthermore, considerations of an increasing angle of incidence ϕ_1, such that β_{x2} goes to zero and then becomes imaginary, again results in a phase shift when total internal reflection occurs. However, in this case a different phase shift is obtained corresponding to:

$$C_0 = A_0 \exp(2j\delta_H) \tag{2.58}$$

where:

$$\tan \delta_H = \left(\frac{n_1}{n_2} \right)^2 \tan \delta_E \tag{2.59}$$

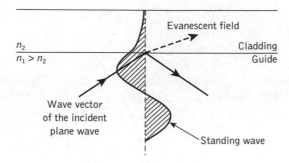

Figure 2.13 The exponentially decaying evanescent field in the cladding of the optical waveguide

Thus the phase shift obtained on total internal reflection is dependent upon both the angle of incidence and the polarization (either TE or TM) of the radiation.

The second phenomenon of interest under conditions of total internal reflection is the form of the electric field in the cladding of the guide. Before the critical angle for total internal reflection is reached, and hence when there is only partial reflection, the field in the cladding is of the form given by Eq. (2.45). However, as indicated previously, when total internal reflection occurs, β_{x2} becomes imaginary and may be written as $-j\xi_2$. Substituting for β_{x2} in Eq. (2.45) gives the transmitted wave in the cladding as:

$$B = B_0 \exp(-\xi_2 x)\, \exp[\,j(\omega t - \beta z)] \tag{2.60}$$

Thus the amplitude of the field in the cladding is observed to decay exponentially* in the x direction. Such a field, exhibiting an exponentially decaying amplitude, is often referred to as an evanescent field. Figure 2.13 shows a diagrammatic representation of the evanescent field. A field of this type stores energy and transports it in the direction of propagation (z) but does not transport energy in the transverse direction (x). Nevertheless, the existence of an evanescent field beyond the plane of reflection in the lower index medium indicates that optical energy is transmitted into the cladding.

The penetration of energy into the cladding underlines the importance of the choice of cladding material. It gives rise to the following requirements:

1. The cladding should be transparent to light at the wavelengths over which the guide is to operate.

2. Ideally, the cladding should consist of a solid material in order to avoid both damage to the guide and the accumulation of foreign matter on the guide walls. These effects degrade the reflection process by interaction with the evanescent field. This in part explains the poor performance (high losses) of early optical waveguides with air cladding.

3. The cladding thickness must be sufficient to allow the evanescent field to decay to a low value or losses from the penetrating energy may be encountered. In many

* It should be noted that we have chosen the sign of ξ_2 so that the exponential field decays rather than grows with distance into the cladding. In this case a growing exponential field is a physically improbable solution.

cases, however, the magnitude of the field falls off rapidly with distance from the guide–cladding interface. This may occur within distances equivalent to a few wavelengths of the transmitted light.

Therefore, the most widely used optical fibers consist of a core and cladding, both made of glass. The cladding refractive index is thus higher than would be the case with liquid or gaseous cladding giving a lower numerical aperture for the fiber, but it provides a far more practical solution.

2.3.5 Goos–Haenchen shift

The phase change incurred with the total internal reflection of a light beam on a planar dielectric interface may be understood from physical observation. Careful examination shows that the reflected beam is shifted laterally from the trajectory predicted by simple ray theory analysis, as illustrated in Figure 2.14. This lateral displacement is known as the Goos–Haenchen shift, after its first observers.

The geometric reflection appears to take place at a virtual reflecting plane which is parallel to the dielectric interface in the lower index medium, as indicated in Figure 2.14. Utilizing wave theory it is possible to determine this lateral shift [Ref. 14] although it is very small ($d \simeq 0.06$ to $0.10\ \mu$m for a silvered glass interface at a wavelength of $0.55\ \mu$m) and difficult to observe. However, this concept provides an important insight into the guidance mechanism of dielectric optical waveguides.

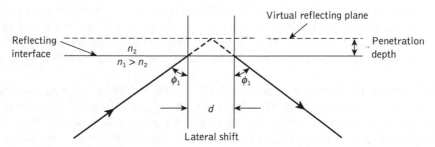

Figure 2.14 The lateral displacement of a light beam on reflection at a dielectric interface (Goos–Haenchen shift)

2.4 Cylindrical fiber

2.4.1 Modes

The exact solution of Maxwell's equations for a cylindrical homogeneous core dielectric waveguide* involves much algebra and yields a complex result [Ref. 15]. Although the

---

* This type of optical waveguide with a constant refractive index core is known as a step index fiber (see Section 2.4.3).

presentation of this mathematics is beyond the scope of this text, it is useful to consider the resulting modal fields. In common with the planar guide (Section 2.3.2), TE (where $E_z = 0$) and TM (where $H_z = 0$) modes are obtained within the dielectric cylinder. The cylindrical waveguide, however, is bounded in two dimensions rather than one. Thus two integers, l and m, are necessary in order to specify the modes, in contrast to the single integer (m) required for the planar guide. For the cylindrical waveguide we therefore refer to TE_{lm} and TM_{lm} modes. These modes correspond to meridional rays (see Section 2.2.1) traveling within the fiber. However, hybrid modes where E_z and H_z are nonzero also occur within the cylindrical waveguide. These modes, which result from skew ray propagation (see Section 2.2.4) within the fiber, are designated HE_{lm} and EH_{lm} depending upon whether the components of **H** or **E** make the larger contribution to the transverse (to the fiber axis) field. Thus an exact description of the modal fields in a step index fiber proves somewhat complicated.

Fortunately, the analysis may be simplified when considering optical fibers for communication purposes. These fibers satisfy the weakly guiding approximation [Ref. 16] where the relative index difference $\Delta \ll 1$. This corresponds to small grazing angles θ in Eq. (2.34). In fact Δ is usually less than 0.03 (3%) for optical communications fibers. For weakly guiding structures with dominant forward propagation, mode theory gives dominant transverse field components. Hence approximate solutions for the full set of HE, EH, TE and TM modes may be given by two linearly polarized components [Ref. 16]. These linearly polarized (LP) modes are not exact modes of the fiber except for the fundamental (lowest order) mode. However, as Δ in weakly guiding fibers is very small, then HE–EH mode pairs occur which have almost identical propagation constants. Such modes are said to be degenerate. The superpositions of these degenerating modes characterized by a common propagation constant correspond to particular LP modes regardless of their HE, EH, TE or TM field configurations. This linear combination of degenerate modes obtained from the exact solution produces a useful simplification in the analysis of weakly guiding fibers.

The relationship between the traditional HE, EH, TE and TM mode designations and the LP_{lm} mode designations is shown in Table 2.1. The mode subscripts l and m are related to the electric field intensity profile for a particular LP mode (see Figure 2.15(d)). There are in general $2l$ field maxima around the circumference of the fiber core and m field

Table 2.1 Correspondence between the lower order in linearly polarized modes and the traditional exact modes from which they are formed

Linearly polarized	Exact
LP_{01}	HE_{11}
LP_{11}	HE_{21}, TE_{01}, TM_{01}
LP_{21}	HE_{31}, EH_{11}
LP_{02}	HE_{12}
LP_{31}	HE_{41}, EH_{21}
LP_{12}	HE_{22}, TE_{02}, TM_{02}
LP_{lm}	HE_{2m}, TE_{0m}, TM_{0m}
LP_{lm} ($l \neq 0$ or 1)	$HE_{l+1.m}$, $EH_{l-1.m}$

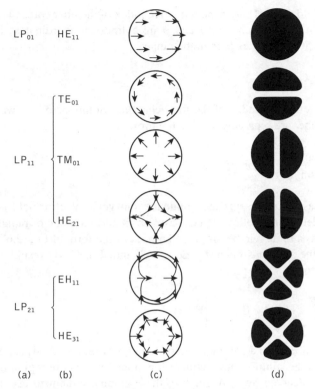

Figure 2.15 The electric field configurations for the three lowest LP modes illustrated in terms of their constituent exact modes: (a) LP mode designations; (b) exact mode designations; (c) electric field distribution of the exact modes; (d) intensity distribution of E_x for the exact modes indicating the electric field intensity profile for the corresponding LP modes

maxima along a radius vector. Furthermore, it may be observed from Table 2.1 that the notation for labeling the HE and EH modes has changed from that specified for the exact solution in the cylindrical waveguide mentioned previously. The subscript l in the LP notation now corresponds to HE and EH modes with labels $l + 1$ and $l - 1$ respectively.

The electric field intensity profiles for the lowest three LP modes, together with the electric field distribution of their constituent exact modes, are shown in Figure 2.15. It may be observed from the field configurations of the exact modes that the field strength in the transverse direction (E_x or E_y) is identical for the modes which belong to the same LP mode. Hence the origin of the term 'linearly polarized'.

Using Eq. (2.31) for the cylindrical homogeneous core waveguide under the weak guidance conditions outlined above, the scalar wave equation can be written in the form [Ref. 17]:

$$\frac{d^2\psi}{dr^2} + \frac{1}{r}\frac{d\psi}{dr} + \frac{1}{r^2}\frac{d^2\psi}{d\phi^2} + (n_1^2 k^2 - \beta^2)\psi = 0 \tag{2.61}$$

where ψ is the field ($\mathbf{E}$ or $\mathbf{H}$), n_1 is the refractive index of the fiber core, k is the propagation constant for light in a vacuum, and r and ϕ are cylindrical coordinates. The propagation constants of the guided modes β lie in the range:

$$n_2 k < \beta < n_1 k \tag{2.62}$$

where n_2 is the refractive index of the fiber cladding. Solutions of the wave equation for the cylindrical fiber are separable, having the form:

$$\psi = E(r)\left[\begin{array}{c}\cos l\phi \\ \sin l\phi\end{array} \exp(\omega t - \beta z)\right] \tag{2.63}$$

where in this case ψ represents the dominant transverse electric field component. The periodic dependence on ϕ following $\cos l\phi$ or $\sin l\phi$ gives a mode of radial order l. Hence the fiber supports a finite number of guided modes of the form of Eq. (2.63).

Introducing the solutions given by Eq. (2.63) into Eq. (2.61) results in a differential equation of the form:

$$\frac{d^2 \mathbf{E}}{dr^2} + \frac{1}{r}\frac{d\mathbf{E}}{dr} + \left[(n_1 k^2 - \beta^2) - \frac{l^2}{r^2}\right]\mathbf{E} = 0 \tag{2.64}$$

For a step index fiber with a constant refractive index core, Eq. (2.64) is a Bessel differential equation and the solutions are cylinder functions. In the core region the solutions are Bessel functions denoted by J_l. A graph of these gradually damped oscillatory functions (with respect to r) is shown in Figure 2.16(a). It may be noted that the field is finite at $r = 0$ and may be represented by the zero-order Bessel function J_0. However, the field vanishes as r goes to infinity and the solutions in the cladding are therefore modified Bessel functions denoted by K_l. These modified functions decay exponentially with respect to r, as illustrated in Figure 2.16(b). The electric field may therefore be given by:

$$\mathbf{E}(r) = G J_l(UR) \qquad \text{for } R < 1 \text{ (core)}$$

$$= G J_l(U)\frac{K_l(WR)}{K_l(W)} \qquad \text{for } R > 1 \text{ (cladding)} \tag{2.65}$$

where G is the amplitude coefficient and $R = r/a$ is the normalized radial coordinate when a is the radius of the fiber core; U and W, which are the eigenvalues in the core and cladding respectively,* are defined as [Ref. 17]:

$$U = a(n_1^2 k^2 - \beta^2)^{\frac{1}{2}} \tag{2.66}$$

$$W = a(\beta^2 - n_2^2 k^2)^{\frac{1}{2}} \tag{2.67}$$

_____

* U is also referred to as the radial phase parameter or the radial propagation constant, whereas W is known as the cladding decay parameter [Ref. 19].

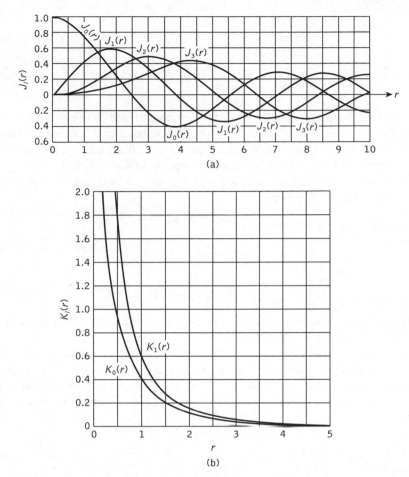

Figure 2.16 (a) Variation of the Bessel function $J_l(r)$ for $l = 0, 1, 2, 3$ (first four orders), plotted against r. (b) Graph of the modified Bessel function $K_l(r)$ against r for $l = 0, 1$

The sum of the squares of U and W defines a very useful quantity [Ref. 18] which is usually referred to as the normalized frequency* V where:

$$V = (U^2 + W^2)^{\frac{1}{2}} = ka(n_1^2 - n_2^2)^{\frac{1}{2}} \tag{2.68}$$

It may be observed that the commonly used symbol for this parameter is the same as that normally adopted for voltage. However, within this chapter there should be no confusion over this point. Furthermore, using Eqs (2.8) and (2.10) the normalized frequency may be expressed in terms of the numerical aperture NA and the relative refractive index difference Δ, respectively, as:

---

* When used in the context of the planar waveguide, V is sometimes known as the normalized film thickness as it relates to the thickness of the guide layer (see Section 10.5.1).

$$V = \frac{2\pi}{\lambda} a(NA) \tag{2.69}$$

$$V = \frac{2\pi}{\lambda} an_1(2\Delta)^{\frac{1}{2}} \tag{2.70}$$

The normalized frequency is a dimensionless parameter and hence is also sometimes simply called the *V* number or value of the fiber. It combines in a very useful manner the information about three important design variables for the fiber: namely, the core radius *a*, the relative refractive index difference Δ and the operating wavelength λ.

It is also possible to define the normalized propagation constant *b* for a fiber in terms of the parameters of Eq. (2.68) so that:

$$b = 1 - \frac{U^2}{V^2} = \frac{(\beta/k)^2 - n_2^2}{n_1^2 - n_2^2}$$

$$= \frac{(\beta/k)^2 - n_2^2}{2n_1^2\Delta} \tag{2.71}$$

Referring to the expression for the guided modes given in Eq. (2.62), the limits of β are $n_2 k$ and $n_1 k$, hence *b* must lie between 0 and 1.

In the weak guidance approximation the field matching conditions at the boundary require continuity of the transverse and tangential electric field components at the core–cladding interface (at *r* = *a*). Therefore, using the Bessel function relations outlined previously, an eigenvalue equation for the LP modes may be written in the following form [Ref. 20]:

$$U \frac{J_{l\pm1}(U)}{J_l(U)} = \pm W \frac{K_{l\pm1}(W)}{K_l(W)} \tag{2.72}$$

Solving Eq. (2.72) with Eqs (2.66) and (2.67) allows the eigenvalue *U* and hence β to be calculated as a function of the normalized frequency. In this way the propagation characteristics of the various modes, and their dependence on the optical wavelength and the fiber parameters, may be determined.

Considering the limit of mode propagation when $\beta = n_2 k$, then the mode phase velocity is equal to the velocity of light in the cladding and the mode is no longer properly guided. In this case the mode is said to be cut off and the eigenvalue $W = 0$ (Eq. 2.67). Unguided or radiation modes have frequencies below cutoff where $\beta < kn_2$, and hence *W* is imaginary. Nevertheless, wave propagation does not cease abruptly below cutoff. Modes exist where $\beta < kn_2$ but the difference is very small, such that some of the energy loss due to radiation is prevented by an angular momentum barrier [Ref. 21] formed near the core–cladding interface. Solutions of the wave equation giving these states are called leaky modes, and often behave as very lossy guided modes rather than radiation modes. Alternatively, as β is increased above $n_2 k$, less power is propagated in the cladding until at $\beta = n_1 k$ all the power is confined to the fiber core. As indicated previously, this range of values for β signifies the guided modes of the fiber.

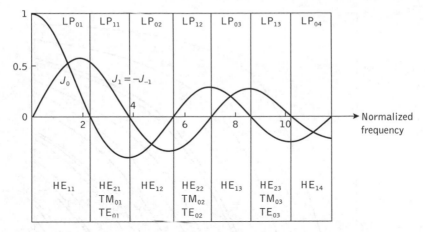

Figure 2.17 The allowed regions for the LP modes of order $l = 0, 1$ against normalized frequency (V) for a circular optical waveguide with a constant refractive index core (step index fiber). Reproduced with permission from D. Gloge. *Appl. Opt.*, **10**, p. 2552, 1971

The lower order modes obtained in a cylindrical homogeneous core waveguide are shown in Figure 2.17 [Ref. 16]. Both the LP notation and the corresponding traditional HE, EH, TE and TM mode notations are indicated. In addition, the Bessel functions J_0 and J_1 are plotted against the normalized frequency and where they cross the zero gives the cutoff point for the various modes. Hence, the cutoff point for a particular mode corresponds to a distinctive value of the normalized frequency (where $V = V_c$) for the fiber. It may be observed from Figure 2.17 that the value of V_c is different for different modes. For example, the first zero crossing J_1 occurs when the normalized frequency is 0 and this corresponds to the cutoff for the LP_{01} mode. However, the first zero crossing for J_0 is when the normalized frequency is 2.405, giving a cutoff value V_c of 2.405 for the LP_{11} mode. Similarly, the second zero of J_1 corresponds to a normalized frequency of 3.83, giving a cutoff value V_c for the LP_{02} mode of 3.83. It is therefore apparent that fibers may be produced with particular values of normalized frequency which allow only certain modes to propagate. This is further illustrated in Figure 2.18 [Ref. 16] which shows the normalized propagation constant b for a number of LP modes as a function of V. It may be observed that the cutoff value of normalized frequency V_c which occurs when $\beta = n_2 k$ corresponds to $b = 0$.

The propagation of particular modes within a fiber may also be confirmed through visual analysis. The electric field distribution of different modes gives similar distributions of light intensity within the fiber core. These waveguide patterns (often called mode patterns) may give an indication of the predominant modes propagating in the fiber. The field intensity distributions for the three lower order LP modes were shown in Figure 2.15. In Figure 2.19 we illustrate the mode patterns for two higher order LP modes. However, unless the fiber is designed for the propagation of a particular mode it is likely that the superposition of many modes will result in no distinctive pattern.

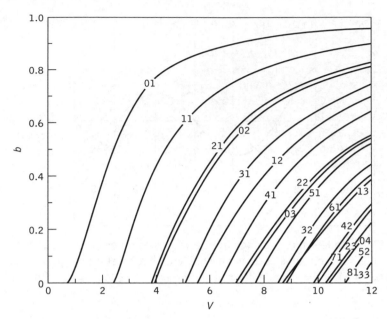

Figure 2.18 The normalized propagation constant b as a function of normalized frequency V for a number of LP modes. Reproduced with permission from D. Gloge. *Appl. Opt.*, **10**, p. 2552, 1971

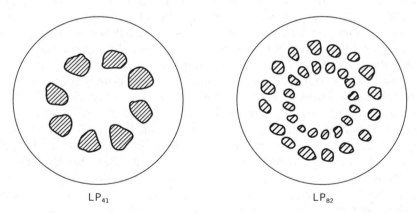

Figure 2.19 Sketches of fiber cross-sections illustrating the distinctive light intensity distributions (mode patterns) generated by propagation of individual linearly polarized modes

2.4.2 Mode coupling

We have thus far considered the propagation aspects of perfect dielectric waveguides. However, waveguide perturbations such as deviations of the fiber axis from straightness, variations in the core diameter, irregularities at the core–cladding interface and refractive index variations may change the propagation characteristics of the fiber. These will have the effect of coupling energy traveling in one mode to another depending on the specific perturbation.

Ray theory aids the understanding of this phenomenon, as shown in Figure 2.20, which illustrates two types of perturbation. It may be observed that in both cases the ray no longer maintains the same angle with the axis. In electromagnetic wave theory this corresponds to a change in the propagating mode for the light. Thus individual modes do not normally propagate throughout the length of the fiber without large energy transfers to adjacent modes, even when the fiber is exceptionally good quality and is not strained or bent by its surroundings. This mode conversion is known as mode coupling or mixing. It is usually analyzed using coupled mode equations which can be obtained directly from Maxwell's equations. However, the theory is beyond the scope of this text and the reader is directed to Ref. 17 for a comprehensive treatment. Mode coupling affects the transmission properties of fibers in several important ways, a major one being in relation to the dispersive properties of fibers over long distances. This is pursued further in Sections 3.8 to 3.11.

2.4.3 Step index fibers

The optical fiber considered in the preceding sections with a core of constant refractive index n_1 and a cladding of a slightly lower refractive index n_2 is known as step index fiber. This is because the refractive index profile for this type of fiber makes a step change at the

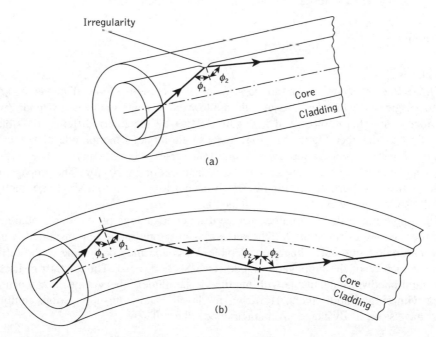

Figure 2.20 Ray theory illustrations showing two of the possible fiber perturbations which give mode coupling: (a) irregularity at the core–cladding interface; (b) fiber bend

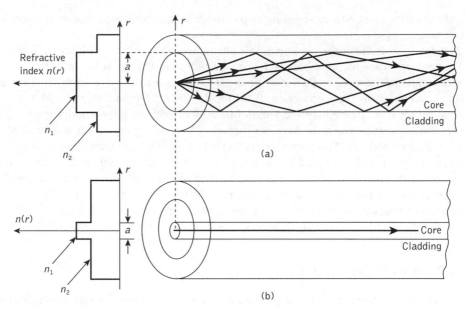

Figure 2.21 The refractive index profile and ray transmission in step index fibers:
(a) multimode step index fiber; (b) single-mode step index fiber

core–cladding interface, as indicated in Figure 2.21, which illustrates the two major types
of step index fiber. The refractive index profile may be defined as:

$$n(r) = \begin{cases} n_1 & r < a \quad \text{(core)} \\ n_2 & r \geq a \quad \text{(cladding)} \end{cases} \tag{2.73}$$

in both cases.

Figure 2.21(a) shows a multimode step index fiber with a core diameter of around
50 µm or greater, which is large enough to allow the propagation of many modes within
the fiber core. This is illustrated in Figure 2.21(a) by the many different possible ray
paths through the fiber. Figure 2.21(b) shows a single-mode or monomode step index fiber
which allows the propagation of only one transverse electromagnetic mode (typically
HE_{11}), and hence the core diameter must be of the order of 2 to 10 µm. The propagation of
a single mode is illustrated in Figure 2.21(b) as corresponding to a single ray path only
(usually shown as the axial ray) through the fiber.

The single-mode step index fiber has the distinct advantage of low intermodal dispersion (broadening of transmitted light pulses), as only one mode is transmitted, whereas
with multimode step index fiber considerable dispersion may occur due to the differing
group velocities of the propagating modes (see Section 3.10). This in turn restricts the
maximum bandwidth attainable with multimode step index fibers, especially when compared with single-mode fibers. However, for lower bandwidth applications multimode
fibers have several advantages over single-mode fibers. These are:

(a) the use of spatially incoherent optical sources (e.g. most light-emitting diodes)
which cannot be efficiently coupled to single-mode fibers;

(b) larger numerical apertures, as well as core diameters, facilitating easier coupling to optical sources;

(c) lower tolerance requirements on fiber connectors.

Multimode step index fibers allow the propagation of a finite number of guided modes along the channel. The number of guided modes is dependent upon the physical parameters (i.e. relative refractive index difference, core radius) of the fiber and the wavelengths of the transmitted light which are included in the normalized frequency V for the fiber. It was indicated in Section 2.4.1 that there is a cutoff value of normalized frequency V_c for guided modes below which they cannot exist. However, mode propagation does not entirely cease below cutoff. Modes may propagate as unguided or leaky modes which can travel considerable distances along the fiber. Nevertheless, it is the guided modes which are of paramount importance in optical fiber communications as these are confined to the fiber over its full length. It can be shown [Ref. 16] that the total number of guided modes or mode volume M_s for a step index fiber is related to the V value for the fiber by the approximate expression:

$$M_s \simeq \frac{V^2}{2} \tag{2.74}$$

which allows an estimate of the number of guided modes propagating in a particular multimode step index fiber.

Example 2.4

A multimode step index fiber with a core diameter of 80 μm and a relative index difference of 1.5% is operating at a wavelength of 0.85 μm. If the core refractive index is 1.48, estimate: (a) the normalized frequency for the fiber; (b) the number of guided modes.

Solution: (a) The normalized frequency may be obtained from Eq. (2.70) where:

$$V \simeq \frac{2\pi}{\lambda} \, an_1(2\Delta)^{\frac{1}{2}} = \frac{2\pi \times 40 \times 10^{-6} \times 1.48}{0.85 \times 10^{-6}} (2 \times 0.015)^{\frac{1}{2}} = 75.8$$

(b) The total number of guided modes is given by Eq. (2.74) as:

$$M_s \simeq \frac{V^2}{2} = \frac{5745.6}{2}$$

$$= 2873$$

Hence this fiber has a V number of approximately 76, giving nearly 3000 guided modes.

Therefore, as illustrated in Example 2.4, the optical power is launched into a large number of guided modes, each having different spatial field distributions, propagation constants, etc. In an ideal multimode step index fiber with properties (i.e. relative index difference, core diameter) which are independent of distance, there is no mode coupling, and the optical power launched into a particular mode remains in that mode and travels independently of the power launched into the other guided modes. Also, the majority of these guided modes operate far from cutoff, and are well confined to the fiber core [Ref. 16]. Thus most of the optical power is carried in the core region and not in the cladding. The properties of the cladding (e.g. thickness) do not therefore significantly affect the propagation of these modes.

2.4.4 Graded index fibers

Graded index fibers do not have a constant refractive index in the core* but a decreasing core index $n(r)$ with radial distance from a maximum value of n_1 at the axis to a constant value n_2 beyond the core radius a in the cladding. This index variation may be represented as:

$$n(r) = \begin{cases} n_1(1 - 2\Delta(r/a)^{\alpha})^{\frac{1}{2}} & r < a \quad \text{(core)} \\ n_1(1 - 2\Delta)^{\frac{1}{2}} = n_2 & r \geq a \quad \text{(cladding)} \end{cases} \tag{2.75}$$

where Δ is the relative refractive index difference and α is the profile parameter which gives the characteristic refractive index profile of the fiber core. Equation (2.75) which is a convenient method of expressing the refractive index profile of the fiber core as a variation of α, allows representation of the step index profile when $\alpha = \infty$, a parabolic profile when $\alpha = 2$ and a triangular profile when $\alpha = 1$. This range of refractive index profiles is illustrated in Figure 2.22.

The graded index profiles which at present produce the best results for multimode optical propagation have a near parabolic refractive index profile core with $\alpha \approx 2$. Fibers

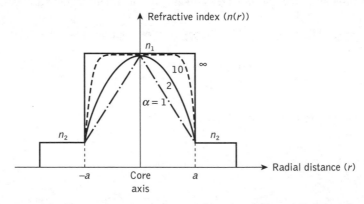

Figure 2.22 Possible fiber refractive index profiles for different values of α (given in Eq. (2.75))

* Graded index fibers are therefore sometimes referred to as inhomogeneous core fibers.

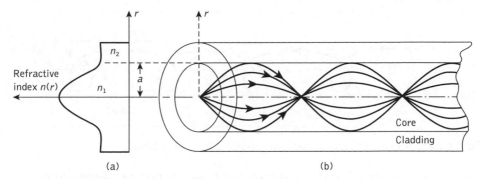

Figure 2.23 The refractive index profile and ray transmission in a multimode graded index fiber

with such core index profiles are well established and consequently when the term 'graded index' is used without qualification it usually refers to a fiber with this profile. For this reason in this section we consider the waveguiding properties of graded index fiber with a parabolic refractive index profile core.

A multimode graded index fiber with a parabolic index profile core is illustrated in Figure 2.23. It may be observed that the meridional rays shown appear to follow curved paths through the fiber core. Using the concepts of geometric optics, the gradual decrease in refractive index from the center of the core creates many refractions of the rays as they are effectively incident on a large number or high to low index interfaces. This mechanism is illustrated in Figure 2.24 where a ray is shown to be gradually curved, with an ever-increasing angle of incidence, until the conditions for total internal reflection are met, and the ray travels back towards the core axis, again being continuously refracted.

Multimode graded index fibers exhibit far less intermodal dispersion (see Section 3.10.2) than multimode step index fibers due to their refractive index profile. Although many different modes are excited in the graded index fiber, the different group velocities of the modes tend to be normalized by the index grading. Again considering ray theory, the rays traveling close to the fiber axis have shorter paths when compared with rays which travel

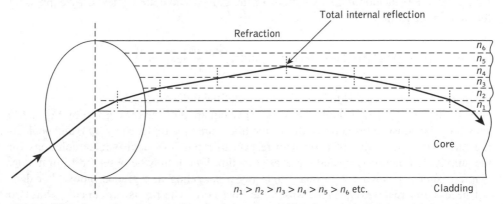

$n_1 > n_2 > n_3 > n_4 > n_5 > n_6$ etc.

Figure 2.24 An expanded ray diagram showing refraction at the various high to low index interfaces within a graded index fiber, giving an overall curved ray path

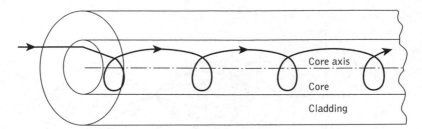

Figure 2.25 A helical skew ray path within a graded index fiber

into the outer regions of the core. However, the near axial rays are transmitted through a
region of higher refractive index and therefore travel with a lower velocity than the more
extreme rays. This compensates for the shorter path lengths and reduces dispersion in the
fiber. A similar situation exists for skew rays which follow longer helical paths, as illustrated in Figure 2.25. These travel for the most part in the lower index region at greater
speeds, thus giving the same mechanism of mode transit time equalization. Hence, multimode graded index fibers with parabolic or near-parabolic index profile cores have transmission bandwidths which may be orders of magnitude greater than multimode step index
fiber bandwidths. Consequently, although they are not capable of the bandwidths attainable with single-mode fibers, such multimode graded index fibers have the advantage of
large core diameters (greater than 30 μm) coupled with bandwidths suitable for long-
distance communication.

The parameters defined for step index fibers (i.e. NA, Δ, V) may be applied to graded
index fibers and give a comparison between the two fiber types. However, it must be noted
that for graded index fibers the situation is more complicated since the numerical aperture
is a function of the radial distance from the fiber axis. Graded index fibers, therefore,
accept less light than corresponding step index fibers with the same relative refractive
index difference.

Electromagnetic mode theory may also be utilized with the graded profiles. Approximate field solutions of the same order as geometric optics are often obtained employing the
WKB method from quantum mechanics after Wentzel, Kramers and Brillouin [Ref. 22].
Using the WKB method modal solutions of the guided wave are achieved by expressing
the field in the form:

$$E_x = \tfrac{1}{2}\{G_1(r)\,\exp[jS(r)] + G_2(r)\,\exp[-jS(r)]\}\left(\frac{\cos l\phi}{\sin l\phi}\right)\exp(j\beta z) \qquad (2.76)$$

where G and S are assumed to be real functions of the radial distance r.

Substitution of Eq. (2.76) into the scalar wave equation of the form given by Eq. (2.61)
(in which the constant refractive index of the fiber core n_1 is replaced by $n(r)$) and neglecting the second derivative of $G_i(r)$ with respect to r provides approximate solutions for
the amplitude function $G_i(r)$ and the phase function $S(r)$. It may be observed from the ray
diagram shown in Figure 2.23 that a light ray propagating in a graded index fiber does
not necessarily reach every point within the fiber core. The ray is contained within two
cylindrical caustic surfaces and for most rays a caustic does not coincide with the core–
cladding interface. Hence the caustics define the classical turning points of the light ray

within the graded fiber core. These turning points defined by the two caustics may be designated as occurring at $r = r_1$ and $r = r_2$.

The result of the WKB approximation yields an oscillatory field in the region $r_1 < r < r_2$ between the caustics where:

$$G_1(r) = G_2(r) = D/[(n^2(r)k^2 - \beta^2)r^2 - l^2]^{\frac{1}{4}} \tag{2.77}$$

(where D is an amplitude coefficient) and:

$$S(r) = \int_{r_1}^{r_2} [(n^2(r)k^2 - \beta^2)r^2 - l^2]^{\frac{1}{2}} \frac{dr}{r} - \frac{\pi}{4} \tag{2.78}$$

Outside the interval $r_1 < r < r_2$ the field solution must have an evanescent form. In the region inside the inner caustic defined by $r < r_1$ and assuming r_1 is not too close to $r = 0$, the field decays towards the fiber axis giving:

$$G_1(r) = D \exp(jmx)/[l^2 - (n^2(r)k^2 - \beta^2)r^2]^{\frac{1}{4}} \tag{2.79}$$

$$G_2(r) = 0 \tag{2.80}$$

where the integer m is the radial mode number and:

$$S(r) = j \int_r^{r_1} [l^2 - (n^2(r)k^2 - \beta^2)r^2]^{\frac{1}{2}} \frac{dr}{r} \tag{2.81}$$

Also outside the outer caustic in the region $r > r_2$, the field decays away from the fiber axis and is described by the equations:

$$G_1(r) = D \exp(jmx)/[l^2 - (n^2(r)k^2 - \beta^2)r^2]^{\frac{1}{4}} \tag{2.82}$$

$$G_2(r) = 0 \tag{2.83}$$

$$S(r) = j \int_{r_2}^r [l^2 - (n^2(r)k^2 - \beta^2)r^2]^{\frac{1}{2}} \frac{dr}{r} \tag{2.84}$$

The WKB method does not initially provide valid solutions of the wave equation in the vicinity of the turning points. Fortunately, this may be amended by replacing the actual refractive index profile by a linear approximation at the location of the caustics. The solutions at the turning points can then be expressed in terms of Hankel functions of the first and second kind of order $\frac{1}{3}$ [Ref. 23]. This facilitates the joining together of the two separate solutions described previously for inside and outside the interval $r_1 < r < r_2$. Thus the WKB theory provides an approximate eigenvalue equation for the propagation constant β of the guided modes which cannot be determined using ray theory. The WKB eigenvalue equation of which β is a solution is given by [Ref. 23]:

$$\int_{r_1}^{r_2} [(n^2(r)k^2 - \beta^2)r^2 - l^2]^{\frac{1}{2}} \frac{dr}{r} = (2m - 1) \frac{\pi}{2} \tag{2.85}$$

where the radial mode number $m = 1, 2, 3 \ldots$ and determines the number of maxima of
the oscillatory field in the radial direction. This eigenvalue equation can only be solved in
a closed analytical form for a few simple refractive index profiles. Hence, in most cases it
must be solved approximately or with the use of numerical techniques [Refs 24, 25].

Finally the amplitude coefficient D may be expressed in terms of the total optical power
P_G within the guided mode. Considering the power carried between the turning points r_1
and r_2 gives a geometric optics approximation of [Ref. 26]:

$$D = \frac{4(\mu_0/\varepsilon_0)^{\frac{1}{2}}P_G^{\frac{1}{2}}}{n_1\pi a^2 I} \tag{2.86}$$

where:

$$I = \int_{r_1/a}^{r_2/a} \frac{x\,dx}{[(n^2(ax)k^2 - \beta^2)a^2x^2 - l^2]^{\frac{1}{2}}} \tag{2.87}$$

The properties of the WKB solution may by observed from a graphical representation
of the integrand given in Eq. (2.78). This is shown in Figure 2.26, together with the corresponding WKB solution. Figure 2.26 illustrates the functions $(n^2(r)k^2 - \beta^2)$ and (l^2/r^2). The

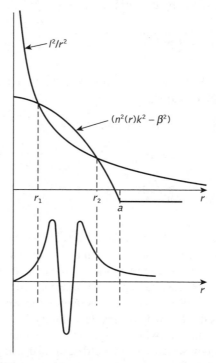

Figure 2.26 Graphical representation of the functions $(n^2(r)k^2 - \beta^2)$ and (l^2/r^2) that
are important in the WKB solution and which define the turning points r_1 and r_2. Also
shown is an example of the corresponding WKB solution for a guided mode where an
oscillatory wave exists in the region between the turning points

two curves intersect at the turning points $r = r_1$ and $r = r_2$. The oscillatory nature of the WKB solution between the turning points (i.e. when $l^2/r^2 < n^2(r)k^2 - \beta^2$) which changes into a decaying exponential (evanescent) form outside the interval $r_1 < r < r_2$ (i.e. when $l^2/r^2 > n^2(r)k^2 - \beta^2$) can also be clearly seen.

It may be noted that as the azimuthal mode number l increases, the curve (l^2/r^2) moves higher and the region between the two turning points becomes narrower. In addition, even when l is fixed the curve $(n^2(r)k^2 - \beta^2)$ is shifted up and down with alterations in the value of the propagation constant β. Therefore, modes far from cutoff which have large values of β exhibit more closely spaced turning points. As the value of β decreases below n_2k, $(n^2(r)k^2 - \beta^2)$ is no longer negative for large values of r and the guided mode situation depicted in Figure 2.26 changes to one corresponding to Figure 2.27. In this case a third turning point $r = r_3$ is created when at $r = a$ the curve $(n^2(r)k^2 - \beta^2)$ becomes constant, thus allowing the curve (l^2/r^2) to drop below it. Now the field displays an evanescent, exponentially decaying form in the region $r_2 < r < r_3$, as shown in Figure 2.27. Moreover, for $r > r_3$ the field resumes an oscillatory behavior and therefore carries power away from the fiber core. Unless mode cutoff occurs at $\beta = n_2k$, the guided mode is no longer fully contained within the fiber core but loses power through leakage or tunneling into the cladding. This situation corresponds to the leaky modes mentioned previously in Section 2.4.1.

The WKB method may be used to calculate the propagation constants for the modes in a parabolic refractive index profile core fiber where, following Eq. (2.75):

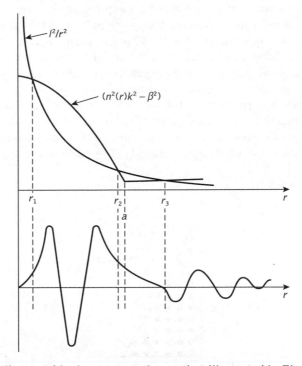

Figure 2.27 Similar graphical representation as that illustrated in Figure 2.26. Here the curve $(n^2(r)k^2 - \beta^2)$ no longer goes negative and a third turning point r_3 occurs. This corresponds to leaky mode solutions in the WKB method

$$n^2(r) = n_1^2 \left[1 - 2\left(\frac{r}{a}\right)^2 \Delta \right] \quad \text{for } r < a \tag{2.88}$$

Substitution of Eq. (2.88) into Eq. (2.85) gives:

$$\int_{r_1}^{r_2} \left[n_1^2 k^2 - \beta^2 - 2n_1^2 k^2 \left(\frac{r}{a}\right)^2 \Delta - \frac{l^2}{r^2} \right]^{\frac{1}{2}} dr = \left(m + \frac{1}{2} \right)\pi \tag{2.89}$$

The integral shown in Eq. (2.89) can be evaluated using a change of variable from r to $u = r^2$. The integral obtained may be found in a standard table of indefinite integrals [Ref. 27]. As the square root term in the resulting expression goes to zero at the turning points (i.e. $r = r_1$ and $r = r_2$), then we can write:

$$\left[\frac{a(n_1 k^2 - \beta^2)}{4n_1 k \sqrt{(2\Delta)}} - \frac{l}{2} \right]\pi = \left(m + \frac{1}{2} \right)\pi \tag{2.90}$$

Solving Eq. (2.90) for β^2 gives:

$$\beta^2 = n_1^2 k^2 \left[\frac{1 - 2\sqrt{(2\Delta)}}{n_1 ka}(2m + l + 1) \right] \tag{2.91}$$

It is interesting to note that the solution for the propagation constant for the various modes in a parabolic refractive index core fiber given in Eq. (2.91) is exact even though it was derived from the approximate WKB eigenvalue equation (Eq. (2.85)). However, although Eq. (2.91) is an exact solution of the scalar wave equation for an infinitely extended parabolic profile medium, the wave equation is only an approximate representation of Maxwell's equation. Furthermore, practical parabolic refractive index profile core fibers exhibit a truncated parabolic distribution which merges into a constant refractive index at the cladding. Hence Eq. (2.91) is not exact for real fibers.

Equation (2.91) does, however, allow us to consider the mode number plane spanned by the radial and azimuthal mode numbers m and l. This plane is displayed in Figure 2.28,

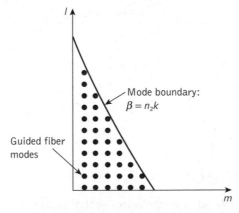

Figure 2.28 The mode number plane illustrating the mode boundary and the guided fiber modes

where each mode of the fiber described by a pair of mode numbers is represented as a point in the plane. The mode number plane contains guided, leaky and radiation modes. The mode boundary which separates the guided modes from the leaky and radiation modes is indicated by the solid line in Figure 2.28. It depicts a constant value of β following Eq. (2.91) and occurs when $\beta = n_2 k$. Therefore, all the points in the mode number plane lying below the line $\beta = n_2 k$ are associated with guided modes, whereas the region above the line is occupied by leaky and radiation modes. The concept of the mode plane allows us to count the total number of guided modes within the fiber. For each pair of mode numbers m and l the corresponding mode field can have azimuthal mode dependence $\cos l\phi$ or $\sin l\phi$ and can exist in two possible polarizations (see Section 3.13). Hence the modes are said to be fourfold degenerate.* If we define the mode boundary as the function $m = f(l)$, then the total number of guided modes M is given by:

$$M = 4 \int_0^{l_{max}} f(l)\, \mathrm{d}l \tag{2.92}$$

as each representation point corresponding to four modes occupies an element of unit area in the mode plane. Equation (2.92) allows the derivation of the total number of guided modes or mode volume M_g supported by the graded index fiber. It can be shown [Ref. 23] that:

$$M_g = \left(\frac{\alpha}{\alpha+2}\right)(n_1 ka)^2 \Delta \tag{2.93}$$

Furthermore, utilizing Eq. (2.70), the normalized frequency V for the fiber when $\Delta \ll 1$ is approximately given by:

$$V = n_1 ka(2\Delta)^{\frac{1}{2}} \tag{2.94}$$

Substituting Eq. (2.94) into Eq. (2.93), we have:

$$M_g \simeq \left(\frac{\alpha}{\alpha+2}\right)\left(\frac{V^2}{2}\right) \tag{2.95}$$

Hence for a parabolic refractive index profile core fiber ($\alpha = 2$), $M_g \approx V^2/4$, which is half the number supported by a step index fiber ($\alpha = \infty$) with the same V value.

* An exception to this are the modes that occur when $l = 0$ which are only doubly degenerate as $\cos l\phi$ becomes unity and $\sin l\phi$ vanishes. However, these modes represent only a small minority and therefore may be neglected.

Example 2.5

A graded index fiber has a core with a parabolic refractive index profile which has a diameter of 50 μm. The fiber has a numerical aperture of 0.2. Estimate the total number of guided modes propagating in the fiber when it is operating at a wavelength of 1 μm.

Solution: Using Eq. (2.69), the normalized frequency for the fiber is:

$$V = \frac{2\pi}{\lambda} a(NA) = \frac{2\pi \times 25 \times 10^{-6} \times 0.2}{1 \times 10^{-6}}$$

$$= 31.4$$

The mode volume may be obtained from Eq. (2.95) where for a parabolic profile:

$$M_g \simeq \frac{V^2}{4} = \frac{986}{4} = 247$$

Hence the fiber supports approximately 247 guided modes.

2.5 Single-mode fibers

The advantage of the propagation of a single mode within an optical fiber is that the signal dispersion caused by the delay differences between different modes in a multimode fiber may be avoided (see Section 3.10). Multimode step index fibers do not lend themselves to the propagation of a single mode due to the difficulties of maintaining single-mode operation within the fiber when mode conversion (i.e. coupling) to other guided modes takes place at both input mismatches and fiber imperfections. Hence, for the transmission of a single mode the fiber must be designed to allow propagation of only one mode, while all other modes are attenuated by leakage or absorption [Refs 28–34].

Following the preceding discussion of multimode fibers, this may be achieved through choice of a suitable normalized frequency for the fiber. For single-mode operation, only the fundamental LP_{01} mode can exist. Hence the limit of single-mode operation depends on the lower limit of guided propagation for the LP_{11} mode. The cutoff normalized frequency for the LP_{11} mode in step index fibers occurs at $V_c = 2.405$ (see Section 2.4.1). Thus single-mode propagation of the LP_{01} mode in step index fibers is possible over the range:

$$0 \leq V < 2.405 \tag{2.96}$$

as there is no cutoff for the fundamental mode. It must be noted that there are in fact two modes with orthogonal polarization over this range, and the term single-mode applies to propagation of light of a particular polarization. Also, it is apparent that the normalized frequency for the fiber may be adjusted to within the range given in Eq. (2.96) by reduction of the core radius, and possibly the relative refractive index difference following Eq. (2.70), which, for single-mode fibers, is usually less than 1%.

Example 2.6

Estimate the maximum core diameter for an optical fiber with the same relative refractive index difference (1.5%) and core refractive index (1.48) as the fiber given in Example 2.4 in order that it may be suitable for single-mode operation. It may be assumed that the fiber is operating at the same wavelength (0.85 µm). Further, estimate the new maximum core diameter for single-mode operation when the relative refractive index difference is reduced by a factor of 10.

Solution: Considering the relationship given in Eq. (2.96), the maximum V value for a fiber which gives single-mode operation is 2.4. Hence, from Eq. (2.70) the core radius a is:

$$a = \frac{V\lambda}{2\pi n_1 (2\Delta)^{\frac{1}{2}}} = \frac{2.4 \times 0.85 \times 10^{-6}}{2\pi \times 1.48 \times (0.03)^{\frac{1}{2}}}$$

$$= 1.3 \ \mu m$$

Therefore the maximum core diameter for single-mode operation is approximately 2.6 µm.

Reducing the relative refractive index difference by a factor of 10 and again using Eq. (2.70) gives:

$$a = \frac{2.4 \times 0.85 \times 10^{-6}}{2\pi \times 1.48 \times (0.003)^{\frac{1}{2}}} = 4.0 \ \mu m$$

Hence the maximum core diameter for single-mode operation is now approximately 8 µm.

It is clear from Example 2.6 that in order to obtain single-mode operation with a maximum V number of 2.4, the single-mode fiber must have a much smaller core diameter than the equivalent multimode step index fiber (in this case by a factor of 32). However, it is possible to achieve single-mode operation with a slightly larger core diameter, albeit still much less than the diameter of multimode step index fiber, by reducing the relative refractive index difference of the fiber.* Both these factors create difficulties with single-mode fibers. The small core diameters pose problems with launching light into the fiber and with field jointing, and the reduced relative refractive index difference presents difficulties in the fiber fabrication process.

---

* Practical values for single-mode step index fiber designed for operation at a wavelength of 1.3 µm are $\Delta = 0.3\%$, giving $2a = 8.5$ µm.

Graded index fibers may also be designed for single-mode operation and some specialist fiber designs do adopt such non step index profiles (see Section 3.12). However, it may be shown [Ref. 35] that the cutoff value of normalized frequency V_c to support a single mode in a graded index fiber is given by:

$$V_c = 2.405(1 + 2/\alpha)^{\frac{1}{2}} \tag{2.97}$$

Therefore, as in the step index case, it is possible to determine the fiber parameters which give single-mode operation.

Example 2.7

A graded index fiber with a parabolic refractive index profile core has a refractive index at the core axis of 1.5 and a relative index difference of 1%. Estimate the maximum possible core diameter which allows single-mode operation at a wavelength of 1.3 μm.

Solution: Using Eq. (2.97) the maximum value of normalized frequency for single-mode operation is:

$$V = 2.4(1 + 2/\alpha)^{\frac{1}{2}} = 2.4(1 + 2/2)^{\frac{1}{2}}$$
$$= 2.4\sqrt{2}$$

The maximum core radius may be obtained from Eq. (2.70) where:

$$a = \frac{V\lambda}{2\pi n_1(2\Delta)^{\frac{1}{2}}} = \frac{2.4\sqrt{2} \times 1.3 \times 10^{-6}}{2\pi \times 1.5 \times (0.02)^{\frac{1}{2}}}$$
$$= 3.3 \ \mu m$$

Hence the maximum core diameter which allows single-mode operation is approximately 6.6 μm.

It may be noted that the critical value of normalized frequency for the parabolic profile graded index fiber is increased by a factor of $\sqrt{2}$ on the step index case. This gives a core diameter increased by a similar factor for the graded index fiber over a step index fiber with the equivalent core refractive index (equivalent to the core axis index) and the same relative refractive index difference.

The maximum V number which permits single-mode operation can be increased still further when a graded index fiber with a triangular profile is employed. It is apparent from Eq. (2.97) that the increase in this case is by a factor of $\sqrt{3}$ over a comparable step index fiber. Hence, significantly larger core diameter single-mode fibers may be produced utilizing this index profile. Such advanced refractive index profiles, which came under serious investigation in the early 1980s [Ref. 36], have now been adopted, particularly in the area of dispersion modified fiber design (see Section 3.12).

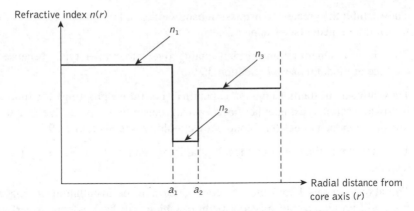

Figure 2.29 The refractive index profile for a single-mode W fiber

A further problem with single-mode fibers with low relative refractive index differences and low V values is that the electromagnetic field associated with the LP_{10} mode extends appreciably into the cladding. For instance, with V values less than 1.4, over half the modal power propagates in the cladding [Ref. 21]. Thus the exponentially decaying evanescent field may extend significant distances into the cladding. It is therefore essential that the cladding is of a suitable thickness, and has low absorption and scattering losses in order to reduce attenuation of the mode. Estimates [Ref. 37] show that the necessary cladding thickness is of the order of 50 μm to avoid prohibitive losses (greater than 1 dB km^{-1}) in single-mode fibers, especially when additional losses resulting from microbending (see Section 4.7.1) are taken into account. Therefore, the total fiber cross-section for single-mode fibers is of a comparable size to multimode fibers [Ref. 38].

Another approach to single-mode fiber design which allows the V value to be increased above 2.405 is the W fiber [Ref. 39]. The refractive index profile for this fiber is illustrated in Figure 2.29 where two cladding regions may be observed. Use of such two-step cladding allows the loss threshold between the desirable and undesirable modes to be substantially increased. The fundamental mode will be fully supported with small cladding loss when its propagation constant lies in the range $kn_3 < \beta < kn_1$.

If the undesirable higher order modes are excited or converted to have values of propagation constant $\beta < kn_3$, they will leak through the barrier layer between a_1 and a_2 (Figure 2.29) into the outer cladding region n_3. Consequently these modes will lose power by radiation into the lossy surroundings. This design can provide single-mode fibers with larger core diameters than can the conventional single-cladding approach which proves useful for easing jointing difficulties; W fibers also tend to give reduced losses at bends in comparison with conventional single-mode fibers.

Following the emergence of single-mode fibers as a viable communication medium in 1983, they quickly became the dominant and the most widely used fiber type within telecommunications.* Major reasons for this situation are as follows:

---

* Multimode fibers are still finding significant use within more localized communications (e.g. for short data links and on-board automobile/aircraft applications).

1. They exhibit the greatest transmission bandwidths and the lowest losses of the fiber transmission media (see Chapter 3).

2. They have a superior transmission quality over other fiber types because of the absence of modal noise (see Section 3.10.3).

3. They offer a substantial upgrade capability (i.e. future proofing) for future wide-bandwidth services using either faster optical transmitters and receivers or advanced transmission techniques (e.g. coherent technology, see Section 13.9.2).

4. They are compatible with the developing integrated optics technology (see Chapter 11).

5. The above reasons 1 to 4 provide confidence that the installation of single-mode fiber will provide a transmission medium which will have adequate performance such that it will not require replacement over its anticipated lifetime of more than 20 years.

Widely deployed single-mode fibers employ a step index (or near step index) profile design and are dispersion optimized (referred to as standard single-mode fibers, see Section 3.11.2) for operation in the 1.3 μm wavelength region. These fibers are either of a matched-cladding (MC) or a depressed-cladding (DC) design, as illustrated in Figure 2.30. In the conventional MC fibers, the region external to the core has a constant uniform refractive index which is slightly lower than the core region, typically consisting of pure silica.

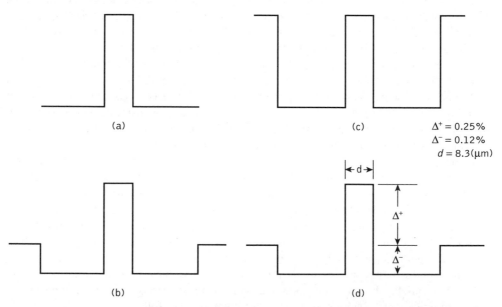

Figure 2.30 Single-mode fiber step index profiles optimized for operation at a wavelength of 1.3 μm: (a) conventional matched-cladding design; (b) segmented core matched-cladding design; (c) depressed-cladding design; (d) profile specifications of a depressed-cladding fiber [Ref. 42]

Alternatively, when the core region comprises pure silica then the lower index cladding is obtained through fluorine doping. A mode-field diameter (MFD) (see Section 2.5.2) of 10 μm is typical for MC fibers with relative refractive index differences of around 0.3%. However, improved bend loss performance (see Section 3.6) has been achieved in the 1.55 μm wavelength region with reduced MFDs of about 9.5 μm and relative refractive index differences of 0.37% [Ref. 40].

An alternative MC fiber design employs a segmented core as shown in Figure 2.30(b) [Ref. 41]. Such a structure provides standard single-mode dispersion-optimized performance at wavelengths around 1.3 μm but is multimoded with a few modes (two or three) in the shorter wavelength region around 0.8 μm. The multimode operating region is intended to help relax both the tight tolerances involved when coupling LEDs to such single-mode fibers (see Section 7.3.7) and their connectorization. Thus segmented core fiber of this type provides for applications which require an inexpensive initial solution but upgradeability to standard single-mode fiber performance at the 1.3 μm wavelength in the future.

In the DC fibers shown in Figure 2.30 the cladding region immediately adjacent to the core is of a lower refractive index than that of an outer cladding region. A typical MFD (see Section 2.5.2) of a DC fiber is 9 μm with positive and negative relative refractive index differences of 0.25% and 0.12% (see Figure 2.30(d)) [Ref. 42].

2.5.1 Cutoff wavelength

It may be noted by rearrangement of Eq. (2.70) that single-mode operation only occurs above a theoretical cutoff wavelength λ_c given by:

$$\lambda_c = \frac{2\pi a n_1}{V_c} (2\Delta)^{\frac{1}{2}} \qquad (2.98)$$

where V_c is the cutoff normalized frequency. Hence λ_c is the wavelength above which a particular fiber becomes single-moded. Dividing Eq. (2.98) by Eq. (2.70) for the same fiber we obtain the inverse relationship:

$$\frac{\lambda_c}{\lambda} = \frac{V}{V_c} \qquad (2.99)$$

Thus for step index fiber where $V_c = 2.405$, the cutoff wavelength is given by [Ref. 43]:

$$\lambda_c = \frac{V\lambda}{2.405} \qquad (2.100)$$

An effective cutoff wavelength has been defined by the ITU-T [Ref. 44] which is obtained from a 2 m length of fiber containing a single 14 cm radius loop. This definition was produced because the first higher order LP_{11} mode is strongly affected by fiber length and curvature near cutoff. Recommended cutoff wavelength values for primary coated fiber range from 1.1 to 1.28 μm for single-mode fiber designed for operation in the 1.3 μm wavelength region in order to avoid modal noise and dispersion problems. Moreover, practical transmission systems are generally operated close to the effective cutoff wavelength in order to enhance the fundamental mode confinement, but sufficiently distant from cutoff so that no power is transmitted in the second-order LP_{11} mode.

Example 2.8

Determine the cutoff wavelength for a step index fiber to exhibit single-mode operation when the core refractive index and radius are 1.46 and 4.5 μm, respectively, with the relative index difference being 0.25%.

Solution: Using Eq. (2.98) with $V_c = 2.405$ gives:

$$\lambda_c = \frac{2\pi a n_1 (2\Delta)^{\frac{1}{2}}}{2.405} = \frac{2\pi 4.5 \times 1.46(0.005)^{\frac{1}{2}}}{2.405} \ \mu m$$

$$= 1.214 \ \mu m$$
$$= 1214 \ nm$$

Hence the fiber is single-moded to a wavelength of 1214 nm.

2.5.2 Mode-field diameter and spot size

Many properties of the fundamental mode are determined by the radial extent of its electromagnetic field including losses at launching and jointing, microbend losses, waveguide dispersion and the width of the radiation pattern. Therefore, the MFD is an important parameter for characterizing single-mode fiber properties which takes into account the wavelength-dependent field penetration into the fiber cladding. In this context it is a better measure of the functional properties of single-mode fiber than the core diameter. For step index and graded (near parabolic profile) single-mode fibers operating near the cutoff wavelength λ_c, the field is well approximated by a Gaussian distribution (see Section 2.5.5). In this case the MFD is generally taken as the distance between the opposite $1/e = 0.37$ field amplitude points and the power $1/e^2 = 0.135$ points in relation to the corresponding values on the fiber axis, as shown in Figure 2.31.

Another parameter which is directly related to the MFD of a single-mode fiber is the spot size (or mode-field radius) ω_0. Hence MFD $= 2\omega_0$, where ω_0 is the nominal half width of the input excitation (see Figure 2.31). The MFD can therefore be regarded as the single-mode analog of the fiber core diameter in multimode fibers [Ref. 45]. However, for many refractive index profiles and at typical operating wavelengths the MFD is slightly larger than the single-mode fiber core diameter.

Often, for real fibers and those with arbitrary refractive index profiles, the radial field distribution is not strictly Gaussian and hence alternative techniques have been proposed. However, the problem of defining the MFD and spot size for non-Gaussian field distributions is a difficult one and at least eight definitions exist [Ref. 19]. Nevertheless, a more general definition based on the second moment of the far field and known as the

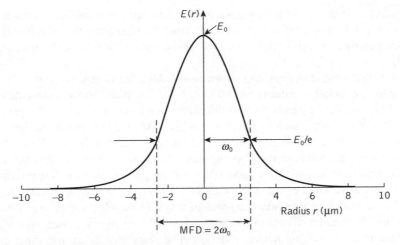

Figure 2.31 Field amplitude distribution $E(r)$ of the fundamental mode in a single-mode fiber illustrating the mode-field diameter (MFD) and spot size (ω_0)

Petermann II definition [Ref. 46] is recommended by the ITU-T. Moreover, good agreement has been obtained using this definition for the MFD using different measurement techniques on arbitrary index fibers [Ref. 47].

2.5.3 Effective refractive index

The rate of change of phase of the fundamental LP_{01} mode propagating along a straight fiber is determined by the phase propagation constant β (see Section 2.3.2). It is directly related to the wavelength of the LP_{01} mode λ_{01} by the factor 2π, since β gives the increase in phase angle per unit length. Hence:

$$\beta\lambda_{01} = 2\pi \quad \text{or} \quad \lambda_{01} = \frac{2\pi}{\beta} \qquad (2.101)$$

Moreover, it is convenient to define an effective refractive index for single-mode fiber, sometimes referred to as a phase index or normalized phase change coefficient [Ref. 48] n_{eff}, by the ratio of the propagation constant of the fundamental mode to that of the vacuum propagation constant:

$$n_{\text{eff}} = \frac{\beta}{k} \qquad (2.102)$$

Hence, the wavelength of the fundamental mode λ_{01} is smaller than the vacuum wavelength λ by the factor $1/n_{\text{eff}}$ where:

$$\lambda_{01} = \frac{\lambda}{n_{\text{eff}}} \qquad (2.103)$$

It should be noted that the fundamental mode propagates in a medium with a refractive index $n(r)$ which is dependent on the distance r from the fiber axis. The effective refractive index can therefore be considered as an average over the refractive index of this medium [Ref. 19].

Within a normally clad fiber, not depressed-cladded fibers (see Section 2.5), at long wavelengths (i.e. small V values) the MFD is large compared to the core diameter and hence the electric field extends far into the cladding region. In this case the propagation constant β will be approximately equal to n_2k (i.e. the cladding wave number) and the effective index will be similar to the refractive index of the cladding n_2. Physically, most of the power is transmitted in the cladding material. At short wavelengths, however, the field is concentrated in the core region and the propagation constant β approximates to the maximum wave number n_1k. Following this discussion, and as indicated previously in Eq. (2.62), then the propagation constant in single-mode fiber varies over the interval $n_2k < \beta < n_1k$. Hence, the effective refractive index will vary over the range $n_2 < n_{\text{eff}} < n_1$.

In addition, a relationship between the effective refractive index and the normalized propagation constant b defined in Eq. (2.71) as:

$$b = \frac{(\beta/k)^2 - n_2^2}{n_1^2 - n_2^2} = \frac{\beta^2 - n_2^2k^2}{n_1^2k^2 - n_2^2k^2} \tag{2.104}$$

may be obtained. Making use of the mathematical relation $A^2 - B^2 = (A + B)(A - B)$, Eq. (2.104) can be written in the form:

$$b = \frac{(\beta + n_2k)(\beta - n_2k)}{(n_1k + n_2k)(n_1k - n_2k)} \tag{2.105}$$

However, taking regard of the fact that $\beta \simeq n_1k$, then Eq. (2.105) becomes:

$$b \simeq \frac{\beta - n_2k}{n_1k - n_2k} = \frac{\beta/k - n_2}{n_1 - n_2}$$

Finally, in Eq. (2.102) n_{eff} is equal to β/k, therefore:

$$b \simeq \frac{n_{\text{eff}} - n_2}{n_1 - n_2} \tag{2.106}$$

The dimensionless parameter b which varies between 0 and 1 is particularly useful in the theory of single-mode fibers because the relative refractive index difference is very small, giving only a small range for β. Moreover, it allows a simple graphical representation of results to be presented as illustrated by the characteristic shown in Figure 2.32 of the normalized phase constant of β as a function of normalized frequency V in a step index fiber.* It should also be noted that $b(V)$ is a universal function which does not depend explicitly on other fiber parameters [Ref. 49].

* For step index fibers the eigenvalue U, which determines the radial field distribution in the core, can be obtained from the plot of b against V because, from Eq. (2.71), $U^2 = V^2(1 - b)$.

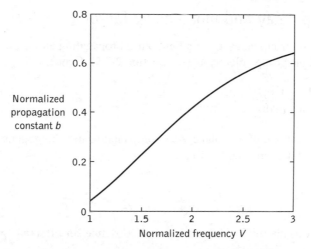

Figure 2.32 The normalized propagation constant (b) of the fundamental mode in a step index fiber shown as a function of the normalized frequency (V)

Example 2.9

Given that a useful approximation for the eigenvalue of the single-mode step index fiber cladding W is [Ref. 43]:

$$W(V) \simeq 1.1428V - 0.9960$$

deduce an approximation for the normalized propagation constant $b(V)$.

Solution: Substituting from Eq. (2.68) into Eq. (2.71), the normalized propagation constant is given by:

$$b(V) = 1 - \frac{(V^2 - W^2)}{V^2} = \frac{W^2}{V^2}$$

Then substitution of the approximation above gives:

$$b(V) \simeq \frac{(1.1428V - 0.9960)^2}{V^2}$$

$$= \left(1.1428 - \frac{0.9960}{V}\right)^2$$

The relative error on this approximation for $b(V)$ is less than 0.2% for $1.5 \leq V \leq 2.5$ and less than 2% for $1 \leq V \leq 3$ [Ref. 43].

2.5.4 Group delay and mode delay factor

The transit time or group delay τ_g for a light pulse propagating along a unit length of fiber is the inverse of the group velocity v_g (see Section 2.3.3). Hence:

$$\tau_g = \frac{1}{v_g} = \frac{d\beta}{d\omega} = \frac{1}{c}\frac{d\beta}{dk} \tag{2.107}$$

The group index of a uniform plane wave propagating in a homogeneous medium has been determined following Eq. (2.40) as:

$$N_g = \frac{c}{v_g}$$

However, for a single-mode fiber, it is usual to define an effective group index* N_{ge} [Ref. 48] by:

$$N_{ge} = \frac{c}{v_g} \tag{2.108}$$

where v_g is considered to be the group velocity of the fundamental fiber mode. Hence, the specific group delay of the fundamental fiber mode becomes:

$$\tau_g = \frac{N_{ge}}{c} \tag{2.109}$$

Moreover, the effective group index may be written in terms of the effective refractive index n_{eff} defined in Eq. (2.102) as:

$$N_{ge} = n_{eff} - \lambda\frac{dn_{eff}}{d\lambda} \tag{2.110}$$

It may be noted that Eq. (2.110) is of the same form as the denominator of Eq. (2.40) which gives the relationship between the group index and the refractive index in a transparent medium (planar guide).

Rearranging Eq. (2.71), β may be expressed in terms of the relative index difference Δ and the normalized propagation constant b by the following approximate expression:

$$\beta = k[(n_1^2 - n_2^2)b + n_2^2] \simeq kn_2[1 + b\Delta] \tag{2.111}$$

Furthermore, approximating the relative refractive index difference as $(n_1 - n_2)/n_2$, for a weakly guiding fiber where $\Delta \ll 1$, we can use the approximation [Ref. 16]:

$$\frac{n_1 - n_2}{n_2} \simeq \frac{N_{g1} - N_{g2}}{N_{g2}} \tag{2.112}$$

---

* N_{ge} may also be referred to as the group index of the single-mode waveguide.

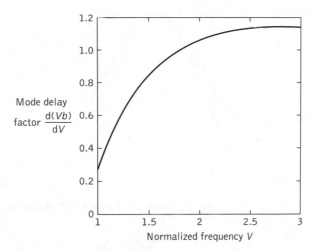

Figure 2.33 The mode delay factor (d(Vb)/dV) for the fundamental mode in a step index fiber shown as a function of normalized frequency (V)

where N_{g1} and N_{g2} are the group indices for the fiber core and cladding regions respectively. Substituting Eq. (2.111) for β into Eq. (2.107) and using the approximate expression given in Eq. (2.112), we obtain the group delay per unit distance as:

$$\tau_{g} = \frac{1}{c}\left[N_{g2} + (N_{g1} - N_{g2})\frac{d(Vb)}{dV}\right] \tag{2.113}$$

The dispersive properties of the fiber core and the cladding are often about the same and therefore the wavelength dependence of Δ can be ignored [Ref. 19]. Hence the group delay can be written as:

$$\tau_{g} = \frac{1}{c}\left[N_{g2} + n_{2}\Delta\frac{d(Vb)}{dV}\right] \tag{2.114}$$

The initial term in Eq. (2.114) gives the dependence of the group delay on wavelength caused when a uniform plane wave is propagating in an infinitely extended medium with a refractive index which is equivalent to that of the fiber cladding. However, the second term results from the waveguiding properties of the fiber only and is determined by the mode delay factor d(Vb)/dV, which describes the change in group delay caused by the changes in power distribution between the fiber core and cladding. The mode delay factor [Ref. 50] is a further universal parameter which plays a major part in the theory of single-mode fibers. Its variation with normalized frequency for the fundamental mode in a step index fiber is shown in Figure 2.33.

2.5.5 The Gaussian approximation

The field shape of the fundamental guided mode within a single-mode step index fiber for two values of normalized frequency is displayed in Figure 2.34. As may be expected,

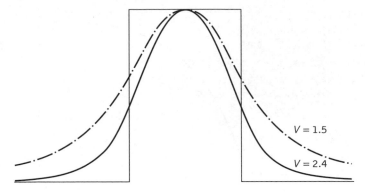

Figure 2.34 Field shape of the fundamental mode for normalized frequencies, $V = 1.5$ and $V = 2.4$

considering the discussion in Section 2.4.1, it has the form of a Bessel function ($J_0(r)$) in the core region matched to a modified Bessel function ($K_0(r)$) in the cladding. Depending on the value of the normalized frequency, a significant proportion of the modal power is propagated in the cladding region, as mentioned earlier. Hence, even at the cutoff value (i.e. V_c) only about 80% of the power propagates within the fiber core.

It may be observed from Figure 2.34 that the shape of the fundamental LP_{01} mode is similar to a Gaussian shape, which allows an approximation of the exact field distribution by a Gaussian function.* The approximation may be investigated by writing the scalar wave equation Eq. (2.27) in the form:

$$\nabla^2 \psi + n^2 k^2 \psi = 0 \tag{2.115}$$

where k is the propagation vector defined in Eq. (2.33) and $n(x, y)$ is the refractive index of the fiber, which does not generally depend on z, the coordinate along the fiber axis. It should be noted that the time dependence $\exp(j\omega t)$ has been omitted from the scalar wave equation to give the reduced wave equation† in Eq. (2.115) [Ref. 23]. This representation is valid since the guided modes of a fiber with a small refractive index difference (i.e. $\Delta \ll 1$) have one predominant transverse field component, for example E_y. By contrast E_x and the longitudinal component are very much smaller [Ref. 23].

The field of the fundamental guided mode may therefore be considered as a scalar quantity and need not be described by the full set of Maxwell's equations. Hence Eq. (2.115) may be written as:

$$\nabla^2 \phi + n^2 k^2 \phi = 0 \tag{2.116}$$

where ϕ represents the dominant transverse electric field component.

---

* However, it should be noted that $K_0(r)$ decays as $\exp(-r)$ which is much slower than a true Gaussian.
† Eq. (2.115) is also known as the Helmholtz equation.

The near-Gaussian shape of the predominant transverse field component of the fundamental mode has been demonstrated [Ref. 51] for fibers with a wide range of refractive index distributions. This proves to be the case not only for the LP_{01} mode of the step index fiber, but also for the modes with fibers displaying arbitrary graded refractive index distributions. Therefore, the predominant electric field component of the single guided mode may be written as the Gaussian function [Ref. 23]:

$$\phi = \left(\frac{2}{\pi}\right)^{\frac{1}{2}} \frac{1}{\omega_0} \exp(-r^2/\omega_0^2) \exp(-j\beta z) \tag{2.117}$$

where the radius parameter $r^2 = x^2 + y^2$, ω_0 is a width parameter which is often called the spot size or radius of the fundamental mode (see Section 2.5.2) and β is the propagation constant of the guided mode field.

The factor preceding the exponential function is arbitrary and is chosen for normalization purposes. If it is accepted that Eq. (2.117) is to a good approximation the correct shape [Ref. 26], then the parameters β and ω_0 may be obtained either by substitution [Ref. 52] or by using a variational principle [Ref. 26]. Using the latter technique, solutions of the wave equation, Eq. (2.116), are claimed to be functions of the minimum integral:

$$J = \int_V [(\nabla\phi) \cdot (\nabla\phi^*) - n^2 k^2 \phi\phi^*] \, dV = \min \tag{2.118}$$

where the asterisk indicates complex conjugation. The integration range in Eq. (2.118) extends over a large cylinder with the fiber at its axis. Moreover, the length of the cylinder L is arbitrary and its radius is assumed to tend towards infinity.

Use of variational calculus [Ref. 53] indicates that the wave equation Eq. (2.116) is the Euler equation of the variational expression given in Eq. (2.118). Hence, the functions that minimize J satisfy the wave equation. Firstly, it can be shown [Ref. 23] that the minimum value of J is zero if ϕ is a legitimate guided mode field. We do this by performing a partial integration of Eq. (2.118) which can be written as:

$$J = \int_s \phi^*(\nabla\phi) \, ds - \int_V [\nabla^2\phi + n^2 k^2 \phi]\phi^* \, dV \tag{2.119}$$

where the surface element ds represents a vector in a direction normal to the outside of the cylinder. However, the function ϕ for a guided mode disappears on the curved cylindrical surface with infinite radius. In this case the guided mode field may be expressed as:

$$\phi = \hat{\phi}(x, y) \exp(-j\beta z) \tag{2.120}$$

It may be observed from Eq. (2.120) that the z dependence is limited to the exponential function and therefore the integrand of the surface integral in Eq. (2.119) is independent of z. This indicates that the contributions to the surface integral from the two end faces of the cylinder are equal in value, opposite in sign and independent of the cylinder length. Thus the entire surface integral goes to zero. Moreover, when the function ϕ is a solution of the wave equation, the volume integral in Eq. (2.119) is zero and hence J is also equal to zero.

The variational expression given in Eq. (2.118) can now be altered by substituting Eq. (2.120). In this case the volume integral becomes an integral over the infinite cross-section of the cylinder (i.e. the fiber) which may be integrated over the length coordinate z. Integration over z effectively multiplies the remaining integral over the cross-section by the cylinder length L because the integrand is independent of z. Hence dividing by L we can write:

$$\frac{J}{L} = \int_{-\infty}^{\infty} \int_{-\infty}^{\infty} \{(\nabla_t \hat{\phi})(\nabla_t \hat{\phi}^*) - [n^2(x,y)k^2 - \beta^2]\hat{\phi}\hat{\phi}^*\}\, dx\, dy \tag{2.121}$$

where the operator ∇_t indicates the transverse part (i.e. the x and y derivatives) of ∇.

We have now obtained in Eq. (2.121) the required variational expression that will facilitate the determination of spot size and propagation constant for the guided mode field. The latter parameter may be obtained by solving Eq. (2.121) for β^2 with $J = 0$, as has been proven to be the case for solutions of the wave equation. Thus:

$$\beta^2 = \frac{\displaystyle\int_{-\infty}^{\infty}\int_{-\infty}^{\infty}[n^2k^2\,\hat{\phi}\hat{\phi}^* - (\nabla_t\hat{\phi})(\nabla_t\hat{\phi}^*)]\, dx\, dy}{\displaystyle\int_{-\infty}^{\infty}\int_{-\infty}^{\infty}\hat{\phi}\hat{\phi}^*\, dx\, dy} \tag{2.122}$$

Equation (2.122) allows calculation of the propagation constant of the fundamental mode if the function ϕ is known. However, the integral expression in Eq. (2.122) exhibits a stationary value such that it remains unchanged to the first order when the exact mode function $\hat{\phi}$ is substituted by a slightly perturbed function. Hence a good approximation to the propagation constant can be obtained using a function that only reasonably approximates to the exact function. The Gaussian approximation given in Eq. (2.117) can therefore be substituted into Eq. (2.122) to obtain:

$$\beta^2 = \left[\frac{4k^2}{\omega_0^2}\int_0^{\infty} rn^2(r)\exp(-2r^2/\omega_0^2)\, dr\right] - \frac{2}{\omega_0^2} \tag{2.123}$$

Two points should be noted in relation to Eq. (2.123). Firstly, following Marcuse [Ref. 23] the normalization was picked to bring the denominator of Eq. (2.122) to unity. Secondly, the stationary expression of Eq. (2.123) was obtained from Eq. (2.122) by assuming that the refractive index was dependent only upon the radial coordinate r. This condition is, however, satisfied by most common optical fiber types.

Finally, to derive an expression for the spot size ω_0 we again make use of the stationary property of Eqs (2.122) and (2.123). Hence, if the Gaussian function of Eq. (2.117) is the correct mode function to give a value for ω_0, then β^2 will not alter if ω_0 is changed slightly. This indicates that the derivative of β^2 with respect to ω_0 becomes zero (i.e. $d\beta^2/d\omega_0 = 0$). Therefore, differentiation of Eq. (2.123) and setting the result to zero yields:

$$1 + 2k^2\int_0^{\infty} r\left(\frac{2r^2}{\omega_0^2} - 1\right)n^2(r)\exp(-2r^2/\omega_0^2)\, dr = 0 \tag{2.124}$$

Equation (2.124) allows the Gaussian approximation for the fundamental mode within single-mode fiber to be obtained by providing a value for the spot size ω_0. This value may be utilized in Eq. (2.123) to determine the propagation constant β.

For step index profiles it can be shown [Ref. 52] that an optimum value of the spot size ω_0 divided by the core radius is only a function of the normalized frequency V. The optimum values of ω_0/a can be approximated to better than 1% accuracy by the empirical formula [Ref. 52]:

$$\frac{\omega_0}{a} = 0.65 + 1.619V^{-\frac{3}{2}} + 2.879V^{-6} \tag{2.125}$$

$$= 0.65 + 1.619\left(2.40\,\frac{\lambda_c}{\lambda}\right)^{-\frac{3}{2}} + 2.879\left(2.405\,\frac{\lambda_c}{\lambda}\right)^{-6}$$

$$\omega_0 = a\left[0.65 + 0.434\left(\frac{\lambda}{\lambda_c}\right)^{\frac{3}{2}} + 0.0149\left(\frac{\lambda}{\lambda_c}\right)^{6}\right] \tag{2.126}$$

The approximate expression for spot size given in Eq. (2.126) is frequently used to determine the parameter for step index fibers over the usual range of λ/λ_c (i.e. 0.8 to 1.9) [Ref. 43].

Example 2.10

Estimate the fiber core diameter for a single-mode step index fiber which has an MFD of 11.6 μm when the normalized frequency is 2.2.

Solution: Using the Gaussian approximation, from Eq. (2.125) the fiber core radius is:

$$a = \frac{\omega_0}{0.65 + 1.619(V)^{-\frac{3}{2}} + 2.879(V)^{-6}}$$

$$= \frac{5.8 \times 10^{-6}}{0.65 + 1.619(2.2)^{-\frac{3}{2}} + 2.879(2.2)^{-6}}$$

$$= 4.95 \text{ μm}$$

Hence the fiber core diameter is 9.9 μm.

The accuracy of the Gaussian approximation has also been demonstrated for graded index fibers [Ref. 54], having a refractive index profile given by Eq. (2.76) (i.e. power-law profiles in the core region). When the near-parabolic refractive index profile is considered (i.e. $\alpha = 2$) and the square-law medium is assumed to extend to infinity rather than to the cladding where $n(r) = n_2$, for $r \geq a$ (Eq. (2.76)), then the Gaussian spot size given in Eq. (2.124) reduces to:

$$\omega_0^2 = \frac{a}{n_1 k}\left(\frac{2}{\Delta}\right)^{\frac{1}{2}} \tag{2.127}$$

Furthermore, the propagation constant becomes:

$$\beta^2 = n_1^2 k^2 \left[1 - \frac{2(2\Delta)^{\frac{1}{2}}}{n_1 ka} \right]$$ (2.128)

It is interesting to note that the above relationships for ω_0 and β in this case are identical to the solutions obtained from exact analysis of the square-law medium [Ref. 26].

Numerical solutions of Eqs (2.123) and (2.124) are shown in Figure 2.35 (dashed lines) for values of α of 6 and ∞ for profiles with constant refractive indices in the cladding region [Ref. 51]. In this case Eqs (2.123) and (2.124) cannot be solved analytically and computer solutions must be obtained. The solid lines in Figure 2.35 show the corresponding solutions of the wave equation, also obtained by a direct numerical technique. These results for the spot size and propagation constant are provided for comparison as they are not influenced by the prior assumption of Gaussian shape.

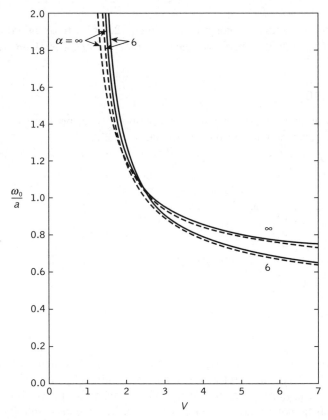

Figure 2.35 Comparison of ω_0/a approximation obtained from Eqs (2.123) and (2.124) (dashed lines) with values obtained from numerical integration of the wave equation and subsequent optimization of its width (solid lines). Reproduced with permission from D. Marcuse, 'Gaussian approximation of the fundamental modes of graded-index fibers', *J. Opt. Soc. Am.*, **68**, p. 103, 1978

The Gaussian approximation for the transverse field distribution is very much simpler than the exact solution and is very useful for calculations involving both launching efficiency at the single-mode fiber input as well as coupling losses at splices or connectors. In this context it describes very well the field inside the fiber core and provides good approximate values for the guided mode propagation constant. It is a particularly good approximation for fibers operated near the cutoff wavelength of the second-order mode [Ref. 26] but when the wavelength increases, the approximation becomes less accurate. In addition, for single-mode fibers with homogeneous cladding, the true field distribution is never exactly Gaussian since the evanescent field in the cladding tends to a more exponential function for which the Gaussian provides an underestimate.

However, for the calculations involving cladding absorption, bend losses, crosstalk between fibers and the properties of directional couplers, then the Gaussian approximation should not be utilized [Ref. 26]. Better approximations for the field profile in these cases can, however, be employed, such as the exponential function [Ref. 55], or the modified Hankel function of zero order [Ref. 56], giving the Gaussian–exponential and the Gaussian–Hankel approximations respectively. Unfortunately, these approximations lose the major simplicity of the Gaussian approximation, in which essentially one parameter (the spot size) defines the radial amplitude distribution, because they necessitate two parameters to characterize the same distribution.

2.5.6 Equivalent step index methods

Another strategy to obtain approximate values for the cutoff wavelength and spot size in graded index single-mode fibers (or arbitrary refractive index profile fibers) is to define an equivalent step index (ESI) fiber on which to model the fiber to be investigated. Various methods have been proposed in the literature [e.g. Refs 57–62] which commence from the observation that the fields in the core regions of graded index fibers often appear similar to the fields within step index fibers. Hence, as step index fiber characteristics are well known, it is convenient to replace the exact methods for graded index single-mode fibers [Refs 63, 64] by approximate techniques based on step index fibers. In addition, such ESI methods allow the propagation characteristics of single-mode fibers to be represented by a few parameters.

Several different suggestions have been advanced for the choice of the core radius a_{ESI}, and the relative index difference Δ_{ESI}, of the ESI fiber which lead to good approximations for the spot size (and hence joint and bend losses) for the actual graded index fiber. They are all conceptually related to the Gaussian approximation (see Section 2.5.5) in that they utilize the close resemblance of the field distribution of the LP_{01} mode to the Gaussian distribution within single-mode fiber. An early proposal for the ESI method [Ref. 58] involved transformation of the basic fiber parameters following:

$$a_{\mathrm{s}} = Xa \qquad V_{\mathrm{s}} = YV \qquad NA_{\mathrm{s}} = (Y/X)NA \qquad\qquad (2.129)$$

where the subscript s is for the ESI fiber and X, Y are constants which must be determined. However, these ESI fiber representations are only valid for a particular value of normalized frequency V and hence there is a different X, Y pair for each wavelength. The transformation can be carried out on the basis of either compared radii or relative refractive index

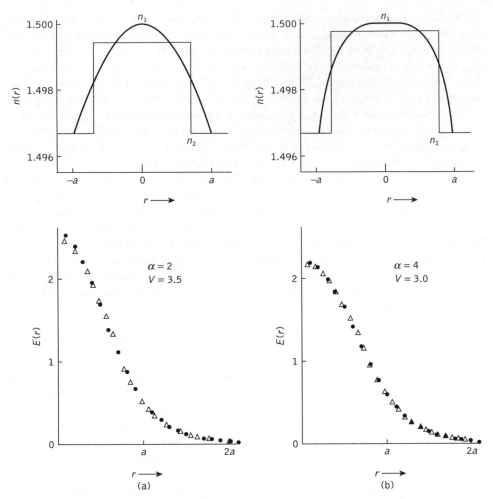

Figure 2.36 Refractive index distributions $n(r)$ and electric field distributions $E(r)$ for graded index fibers and their ESI fibers for: (a) $\alpha = 2$, $V = 3.5$; (b) $\alpha = 4$, $V = 3.0$. The field distributions for the graded index and corresponding ESI profiles are shown by solid circles and open triangles respectively. Reproduced with permission from H. Matsumura and T. Suganuma, *Appl. Opt.*, **19**, p. 3151, 1980

differences. Figure 2.36 compares the refractive index profiles and the electric field distributions for two graded index fibres ($\alpha = 2$, 4) and their ESI fibers. It may be observed that their fields differ slightly only near the axis.

An alternative ESI technique is to normalize the spot size ω_0 with respect to an optimum effective fiber core radius a_{eff} [Ref. 61]. This latter quantity is obtained from the experimental measurement of the first minimum (angle θ_{min}) in the diffraction pattern using transverse illumination of the fiber immersed in an index-matching fluid. Hence:

$$a_{\text{eff}} = 3.832/k \sin \theta_{\text{min}} \tag{2.130}$$

where $k = 2\pi/\lambda$. In order to obtain the full comparison with single-mode step index fiber, the results may be expressed in terms of an effective normalized frequency V_{eff} which relates the cutoff frequencies/wavelengths for the two fibers:

$$V_{eff} = 2.405(V/V_c) = 2.405(\lambda_c/\lambda) \tag{2.131}$$

The technique provides a dependence of ω_0/a_{eff} on V_{eff} which is almost identical for a reasonably wide range of profiles which are of interest for minimizing dispersion (i.e. $1.5 < V_{eff} < 2.4$).

A good analytical approximation for this dependence is given by [Ref. 61]:

$$\frac{\omega_0}{a_{eff}} = 0.6043 + 1.755V_{eff}^{-\frac{3}{2}} + 2.78V_{eff}^{-6} \tag{2.132}$$

Refractive index profile-dependent deviations from the relationship shown in Eq. (2.132) are within ±2% for general power-law graded index profiles.

Example 2.11

A parabolic profile graded index single-mode fiber designed for operation at a wavelength of 1.30 μm has a cutoff wavelength of 1.08 μm. From experimental measurement it is established that the first minimum in the diffraction pattern occurs at an angle of 12°. Using an ESI technique, determine the spot size at the operating wavelength.

Solution: Using Eq. (2.130), the effective core radius is:

$$a_{eff} = \frac{3.832\lambda}{2\pi \sin \theta_{min}} = \frac{3.832 \times 1.30 \times 10^{-6}}{2\pi \sin 12°}$$

$$= 3.81 \ \mu m$$

The effective normalized frequency can be obtained from Eq. (2.131) as:

$$V_{eff} = 2.405 \frac{\lambda_c}{\lambda} = 2.405 \frac{1.08}{1.30} = 2.00$$

Hence the spot size is given by Eq. (2.132) as:

$$\omega_0 = 3.81 \times 10^{-6}[0.6043 + 1.755(2.00)^{-\frac{3}{2}} + 2.78(2.00)^{-6}]$$

$$= 4.83 \ \mu m$$

Other ESI methods involve the determination of the equivalent parameters from experimental curves of spot size against wavelength [Ref. 62]. All require an empirical formula, relating spot size to the normalized frequency for a step index fiber, to be fitted by some means to the data. The usual empirical formula employed is that derived by Marcuse for the Gaussian approximation and given in Eq. (2.125). An alternative formula which is close to Eq. (2.125) is provided by Snyder [Ref. 65] as:

$$\omega_0 = a(\ln V)^{-\frac{3}{2}} \tag{2.133}$$

However, it is suggested [Ref. 62] that the expression given in Eq. (2.133) is probably less accurate than that provided by Eq. (2.125).

A cutoff method can also be utilized to obtain the ESI parameters [Ref. 66]. In this case the cutoff wavelength λ_c and spot size ω_0 are known. Therefore, substituting $V = 2.405$ into Eq. (2.125) gives:

$$\omega_0 = 1.099 a_{ESI} \quad \text{or} \quad 2a_{ESI} = 1.820\omega_0 \tag{2.134}$$

Then using Eq. (2.70) the ESI relative index difference is:

$$\Delta_{ESI} = (0.293/n_1^2)(\lambda_c/2a_{ESI})^2 \tag{2.135}$$

where n_1 is the maximum refractive index of the fiber core.

Example 2.12

Obtain the ESI relative refractive index difference for a graded index fiber which has a cutoff wavelength and spot size of 1.190 μm and 5.2 μm respectively. The maximum refractive index of the fiber core is 1.485.

Solution: The ESI core radius may be obtained from Eq. (2.134) where:

$$2a_{ESI} = 1.820 \times 5.2 \times 10^{-6} = 9.464 \text{ μm}$$

Using Eq. (2.135), the ESI relative index difference is given by:

$$\Delta_{ESI} = (0.293/1.485^2)\,(1.190/9.464)^2$$
$$= 2.101 \times 10^{-3} \text{ or } 0.21\%$$

Alternatively, performing a least squares fit on Eq. (2.125) provides 'best values' for the ESI diameter ($2a_{ESI}$) and relative index difference (Δ_{ESI}) [Ref. 62]. It must be noted, however, that these best values are dependent on the application and the least squares method appears most useful in estimating losses at fiber joints [Ref. 67]. In addition, some work [Ref. 68] has attempted to provide a more consistent relationship between the ESI parameters and the fiber MFD. Overall, the concept of the ESI fiber has been relatively useful in the specification of standard MC and DC fibers by their equivalent a_{ESI} and Δ_{ESI} values. Unfortunately, ESI methods are unable accurately to predict MFDs and waveguide dispersion in dispersion-shifted and dispersion-flattened (see Section 3.12) fibers [Ref. 19].

2.6 Photonic crystal fibers

The previous discussion in this chapter has concentrated on optical fibers comprising solid silica core and cladding regions in which the light is guided by a small increase in refractive index in the core facilitated through doping the silicon with germanium. More recently, however, a new class of microstructured optical fiber containing a fine array of air holes running longitudinally down the fiber cladding [Ref. 69] has been developed. Since the microstructure within the fiber is often highly periodic due to the fabrication process, these fibers are usually referred to as photonic crystal fibers (PCFs), or sometimes just as holey fibers [Ref. 70]. Whereas in conventional optical fibers electromagnetic modes are guided by total internal reflection in the core region, which has a slightly raised refractive index, in PCFs two distinct guidance mechanisms arise.

Although the guided modes can be trapped in a fiber core which exhibits a higher average index than the cladding containing the air holes by an effect similar to total internal reflection, alternatively they may be trapped in a core of either higher, or indeed lower, average index by a photonic bandgap effect. In the former case the effect is often termed modified total internal reflection and the fibers are referred to as index guided, while in the latter they are called photonic bandgap fibers. Furthermore, the existence of two different guidance mechanisms makes PCFs versatile in their range of potential applications. For example, PCFs have been used to realize various optical components and devices including long period gratings [Ref. 71], multimode interference power splitters [Ref. 72], tunable coupled cavity fiber lasers [Ref. 73], fiber amplifiers [Ref. 74], multichannel add/drop filters [Ref. 75], wavelength converters [Ref. 76] and wavelength demultiplexers [Ref. 77]. As with conventional optical fibers, however, a crucial issue with PCFs has been the reduction in overall transmission losses which were initially several hundred decibels per kilometer even with the most straightforward designs. Increased control over the homogeneity of the fiber structures together with the use of highly purified silicon as the base material has now lowered these losses to a level of a very few decibels per kilometer for most PCF types, with a loss of just 0.3 dB km^{-1} at 1.55 μm for a 100 km span being recently reported [Ref. 78].

2.6.1 Index-guided microstructures

Although the principles of guidance and the characteristics of index-guided PCFs are similar to those of conventional fiber, there is greater index contrast since the cladding contains air holes with a refractive index of 1 in comparison with the normal silica cladding index of 1.457 which is close to the germanium-doped core index of 1.462. A fundamental physical difference, however, between index-guided PCFs and conventional fibers arises from the manner in which the guided mode interacts with the cladding region. Whereas in a conventional fiber this interaction is largely first order and independent of wavelength, the large index contrast combined with the small structure dimensions cause the effective cladding index to be a strong function of wavelength. For short wavelengths the effective cladding index is only slightly lower than the core index and hence they remain tightly confined to the core. At longer wavelengths, however, the mode samples more of the cladding and the effective index contrast is larger. This wavelength dependence results in

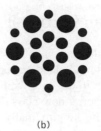

(a)

(b)

Figure 2.37 Two index-guided photonic crystal fiber structures. The dark areas are air holes while the white areas are silica

a large number of unusual optical properties which can be tailored. For example, the high index contrast enables the PCF core to be reduced from around 8 μm in conventional fiber to less than 1 μm, which increases the intensity of the light in the core and enhances the nonlinear effects.

Two common index-guided PCF designs are shown diagrammatically in Figure 2.37. In both cases a solid-core region is surrounded by a cladding region containing air holes. The cladding region in Figure 2.37(a) comprises a hexagonal array of air holes while in Figure 2.37(b) the cladding air holes are not uniform in size and do not extend too far from the core. It should be noted that the hole diameter d and hole to hole spacing or pitch Λ are critical design parameters used to specify the structure of the PCF. For example, in a silica PCF with the structure depicted in Figure 2.37(a) when the air fill fraction is low (i.e. $d/\Lambda < 0.4$), then the fiber can be single-moded at all wavelengths [Ref. 79]. This property, which cannot be attained in conventional fibers, is particularly significant for broadband applications such as wavelength division multiplexed transmission [Ref. 80].

As PCFs have a wider range of optical properties in comparison with standard optical fibers, they provide for the possibility of new and technologically important fiber devices. When the holey region covers more than 20% of the fiber cross-section, for instance, index-guided PCFs display an interesting range of dispersive properties which could find application as dispersion-compensating or dispersion-controlling fiber components [Ref. 81]. In such fibers it is possible to produce very high optical nonlinearity per unit length in which modest light intensities can induce substantial nonlinear effects. For example, while several kilometers of conventional fiber are normally required to achieve 2R data re-generation (see Section 10.6), it was obtained with just 3.3 m of large air-filling fraction PCF [Ref. 82]. In addition, filling the cladding holes with polymers or liquid crystals allows external fields to be used to dynamically vary the fiber properties. The temperature sensitivity of a polymer within the cladding holes may be employed to tune a Bragg grating written into the core [Ref. 83]. By contrast, index-guided PCFs with small holes and large hole spacings provide very large mode area (and hence low optical nonlinearities) and have potential applications in high-power delivery (e.g. laser welding and machining) as well as high-power fiber lasers and amplifiers [Ref. 74]. Furthermore, the large index contrast between silica and air enables production of such PCFs with large multimoded cores which also have very high numerical aperture values (greater than 0.7). Hence these fibers are useful for the collection and transmission of high optical powers in situations where signal distortion is not an issue. Finally, it is apparent that PCFs can be readily spliced to conventional fibers, thus enabling their integration with existing components and subsystems.

2.6.2 Photonic bandgap fibers

Photonic bandgap (PBG) fibers are a class of microstructured fiber in which a periodic arrangement of air holes is required to ensure guidance. This periodic arrangement of cladding air holes provides for the formation of a photonic bandgap in the transverse plane of the fiber. As a PBG fiber exhibits a two-dimensional bandgap, then wavelengths within this bandgap cannot propagate perpendicular to the fiber axis (i.e. in the cladding) and they can therefore be confined to propagate within a region in which the refractive index is lower than the surrounding material. Hence utilizing the photonic bandgap effect light can, for example, be guided within a low-index, air-filled core region creating fiber properties quite different from those obtained without the bandgap. Although, as with index-guided PCFs, PBG fibers can also guide light in regions with higher refractive index, it is the lower index region guidance feature which is of particular interest. In addition, a further distinctive feature is that while index-guiding fibers usually have a guided mode at all wavelengths, PBG fibers only guide in certain wavelength bands, and furthermore it is possible to have wavelengths at which higher order modes are guided while the fundamental mode is not.

Two important PBG fiber structures are displayed in Figure 2.38. The honeycomb fiber design shown in Figure 2.38(a) was the first PBG fiber to be experimentally realized in 1998 [Ref. 84] and adaptations of this structure continue to be pursued [Ref. 85]. A triangular array of air holes of sufficient size as displayed in Figure 2.38(b), however, provides for the possibility, unique to PBG fibers, of guiding electromagnetic modes in air. In this case a large hollow core has been defined by removing the silica around seven air holes in the center of the structure. These fibers, which are termed air-guiding or hollow-core PBG fibers, enable more than 98% of the guided mode field energy to propagate in the air regions [Ref. 81]. Such air-guiding fibers have attracted attention because they potentially provide an environment in which optical propagation can take place with little attenuation as the localization of light in the air core removes the limitations caused by material absorption losses. The fabrication of hollow-core fiber with low propagation losses, however, has proved to be quite difficult, with losses of the order of 13 dB km^{-1} [Ref. 86]. Moreover, the fibers tend to be highly dispersive with narrow transmission windows and

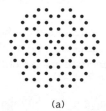

(a)   (b)

Figure 2.38 Photonic bandgap (PBG) fiber structures in which the dark areas are air (lower refractive index) and the lighter area is the higher refractive index: (a) honeycomb PBG fiber; (b) air-guiding PBG fiber

while single-mode operation is possible, it is not as straightforward to achieve in comparison with index-guiding PCFs.

More recently, the fabrication and characterization of a new type of solid silica-based photonic crystal fiber which guides light using the PBG mechanism has been reported [Refs 87, 88]. This fiber employed a two-dimensional periodic array of germanium-doped rods in the core region. It was therefore referred to as a nanostructure core fiber and exhibited a minimum attenuation of 2.6 dB km^{-1} at a wavelength of 1.59 μm [Ref. 87]. Furthermore, the fiber displayed greater bending sensitivity than conventional single-mode fiber as a result of the much smaller index difference between the core and the leaky modes which could provide for potential applications in the optical sensing of curvature and stress. In addition, it is indicated that the all-solid silica structure would facilitate fiber fabrication using existing technology (see Sections 4.2 to 4.4), and birefringence (see Section 3.13.1) of the order of 10^{-4} is easily achievable with a large mode field diameter up to 10 μm, thus enabling its use within fiber lasers (see Section 6.10.3) and gyroscope applications [Ref. 88].

Problems

2.1 Using simple ray theory, describe the mechanism for the transmission of light within an optical fiber. Briefly discuss with the aid of a suitable diagram what is meant by the acceptance angle for an optical fiber. Show how this is related to the fiber numerical aperture and the refractive indices for the fiber core and cladding.

An optical fiber has a numerical aperture of 0.20 and a cladding refractive index of 1.59. Determine:
(a) the acceptance angle for the fiber in water which has a refractive index of 1.33;
(b) the critical angle at the core–cladding interface.
Comment on any assumptions made about the fiber.

2.2 The velocity of light in the core of a step index fiber is 2.01×10^8 m s^{-1}, and the critical angle at the core–cladding interface is 80°. Determine the numerical aperture and the acceptance angle for the fiber in air, assuming it has a core diameter suitable for consideration by ray analysis. The velocity of light in a vacuum is 2.998×10^3 m s^{-1}.

2.3 Define the relative refractive index difference for an optical fiber and show how it may be related to the numerical aperture.

A step index fiber with a large core diameter compared with the wavelength of the transmitted light has an acceptance angle in air of 22° and a relative refractive index difference of 3%. Estimate the numerical aperture and the critical angle at the core–cladding interface for the fiber.

2.4 A step index fiber has a solid acceptance angle in air of 0.115 radians and a relative refractive index difference of 0.9%. Estimate the speed of light in the fiber core.

2.5 Briefly indicate with the aid of suitable diagrams the difference between meridional and skew ray paths in step index fibers.

Derive an expression for the acceptance angle for a skew ray which changes direction by an angle 2γ at each reflection in a step index fiber in terms of the fiber NA and γ. It may be assumed that ray theory holds for the fiber.

A step index fiber with a suitably large core diameter for ray theory considerations has core and cladding refractive indices of 1.44 and 1.42 respectively. Calculate the acceptance angle in air for skew rays which change direction by 150° at each reflection.

2.6 Skew rays are accepted into a large core diameter (compared with the wavelength of the transmitted light) step index fiber in air at a maximum axial angle of 42°. Within the fiber they change direction by 90° at each reflection. Determine the acceptance angle for meridional rays for the fiber in air.

2.7 Explain the concept of electromagnetic modes in relation to a planar optical waveguide.

Discuss the modifications that may be made to electromagnetic mode theory in a planar waveguide in order to describe optical propagation in a cylindrical fiber.

2.8 Briefly discuss, with the aid of suitable diagrams, the following concepts in optical fiber transmission:
(a) the evanescent field;
(b) Goos–Haenchen shift;
(c) mode coupling.
Describe the effects of these phenomena on the propagation of light in optical fibers.

2.9 Define the normalized frequency for an optical fiber and explain its use in the determination of the number of guided modes propagating within a step index fiber.

A step index fiber in air has a numerical aperture of 0.16, a core refractive index of 1.45 and a core diameter of 60 μm. Determine the normalized frequency for the fiber when light at a wavelength of 0.9 μm is transmitted. Further, estimate the number of guided modes propagating in the fiber.

2.10 Describe with the aid of simple ray diagrams:
(a) the multimode step index fiber;
(b) the single-mode step index fiber.
Compare the advantages and disadvantages of these two types of fiber for use as an optical channel.

2.11 A multimode step index fiber has a relative refractive index difference of 1% and a core refractive index of 1.5. The number of modes propagating at a wavelength of 1.3 μm is 1100. Estimate the diameter of the fiber core.

2.12 Explain what is meant by a graded index optical fiber, giving an expression for the possible refractive index profile. Using simple ray theory concepts, discuss the transmission of light through the fiber. Indicate the major advantage of this type of fiber with regard to multimode propagation.

2.13 The relative refractive index difference between the core axis and the cladding of a graded index fiber is 0.7% when the refractive index at the core axis is 1.45. Estimate values for the numerical aperture of the fiber when:

(a) the index profile is not taken into account; and
(b) the index profile is assumed to be triangular.
Comment on the results.

2.14 A multimode graded index fiber has an acceptance angle in air of 8°. Estimate the relative refractive index difference between the core axis and the cladding when the refractive index at the core axis is 1.52.

2.15 The WKB value for the propagation constant β given in Eq. (2.91) in a parabolic refractive index core fiber assumes an infinitely extended parabolic profile medium. When in a practical fiber the parabolic index profile is truncated, show that the mode numbers m and l are limited by the following condition:

$$2(2m + l + 1) \leq ka(n_1^2 - n_2^2)^{\frac{1}{2}}$$

2.16 A graded index fiber with a parabolic index profile supports the propagation of 742 guided modes. The fiber has a numerical aperture in air of 0.3 and a core diameter of 70 μm. Determine the wavelength of the light propagating in the fiber.

Further estimate the maximum diameter of the fiber which gives single-mode operation at the same wavelength.

2.17 A graded index fiber with a core axis refractive index of 1.5 has a characteristic index profile (α) of 1.90, a relative refractive index difference of 1.3% and a core diameter of 40 μm. Estimate the number of guided modes propagating in the fiber when the transmitted light has a wavelength of 1.55 μm, and determine the cutoff value of the normalized frequency for single-mode transmission in the fiber.

2.18 A single-mode step index fiber has a core diameter of 7 μm and a core refractive index of 1.49. Estimate the shortest wavelength of light which allows single-mode operation when the relative refractive index difference for the fiber is 1%.

2.19 In Problem 2.18, it is required to increase the fiber core diameter to 10 μm while maintaining single-mode operation at the same wavelength. Estimate the maximum possible relative refractive index difference for the fiber.

2.20 Show that the maximum value of a/λ is approximately 1.4 times larger for a parabolic refractive index profile single-mode fiber than for a single-mode step index fiber. Hence, sketch the relationship between the maximum core diameter and the propagating optical wavelength which will facilitate single-mode transmission in the parabolic profile fiber.

2.21 A single-mode step index fiber which is designed for operation at a wavelength of 1.3 μm has core and cladding refractive indices of 1.447 and 1.442 respectively. When the core diameter is 7.2 μm, confirm that the fiber will permit single-mode transmission and estimate the range of wavelengths over which this will occur.

2.22 A single-mode step index fiber has core and cladding refractive indices of 1.498 and 1.495 respectively. Determine the core diameter required for the fiber to permit its operation over the wavelength range 1.48 to 1.60 μm. Calculate the new fiber core diameter to enable single-mode transmission at a wavelength of 1.30 μm.

2.23 A single-mode fiber has a core refractive index of 1.47. Sketch a design characteristic of relative refractive index difference Δ against core radius for the fiber to operate at a wavelength of 1.30 μm. Determine whether the fiber remains single-mode at a transmission wavelength of 0.85 μm when its core radius is 4.5 μm.

2.24 Convert the approximation for the normalized propagation constant of a single-mode step index fiber given in Example 2.9 into a relationship involving the normalized wavelength λ/λ_c in place of the normalized frequency. Hence, determine the range of values of this parameter over which the relative error in the approximation is between 0.2% and 2%.

2.25 Given that the Gaussian function for the electric field distribution of the fundamental mode in a single-mode fiber of Eq. (2.117) takes the form:

$$E(r) = E_0 \exp(-r^2/\omega_0^2)$$

where $E(r)$ and E_0 are shown in Figure 2.31, use the approximation of Eq. (2.125) to evaluate and sketch $E(r)/E_0$ against r/a over the range 0 to 3 for values of normalized frequency $V = 1.0, 1.5, 2.0, 2.5, 3.0$.

2.26 The approximate expression provided in Eq. (2.125) is valid over the range of normalized frequency $1.2 < V < 2.4$. Sketch ω_0/a against V over this range for the fundamental mode in a step index fiber. Comment on the magnitude of ω_0/a as the normalized frequency is reduced significantly below 2.4 and suggest what this indicates about the distribution of the light within the fiber.

2.27 The spot size in a parabolic profile graded index single-mode fiber is 11.0 μm at a transmission wavelength of 1.55 μm. In addition, the cutoff wavelength for the fiber is 1.22 μm. Using an ESI technique, determine the fiber effective core radius and hence estimate the angle at which the first minimum in the diffraction pattern from the fiber would occur.

2.28 The cutoff method is employed to obtain the ESI parameters for a graded index single-mode fiber. If the ESI relative index difference was found to be 0.30% when the spot size and cutoff wavelength were 4.6 μm and 1.29 μm, respectively, calculate the maximum refractive index of the fiber core.

2.29 Describe what is implied by the term photonic crystal fiber (PCF) and explain the guidance mechanisms for electromagnetic modes in such optical fibers.

2.30 Compare and contrast the performance attributes, potential drawbacks and possible applications of index-guided PCFs and photonic bandgap fibers.

Answers to numerical problems

2.1	(a) 8.6°; (b) 83.6°	**2.6**	28.2°
2.2	0.263, 15.2°	**2.9**	33.5, 561
2.3	0.375, 75.9°	**2.11**	92 μm
2.4	2.11×10^8 m s^{-1}	**2.13**	(a) 0.172; (b) 0.171
2.5	34.6°	**2.14**	0.42%

2.16	1.2 μm, 4.4 μm	**2.22**	12.0 μm, 10.5 μm
2.17	94, 3.45	**2.24**	$0.8 \le \lambda/\lambda_c \le 1.0$ and $1.6 \le \lambda/\lambda_c \le 2.4$
2.18	1.36 μm	**2.27**	3.0 μm, 18.4°
2.19	0.24%	**2.28**	1.523
2.21	down to 1139 nm		

References

[1] D. Hondros and P. Debye, 'Electromagnetic waves along long cylinders of dielectric', *Ann. Phys.*, **32**(3), pp. 465–476, 1910.

[2] O. Schriever, 'Electromagnetic waves in dielectric wires', *Ann. Phys.*, **63**(7), pp. 645–673, 1920.

[3] A. C. S. van Heel, 'A new method of transporting optical images without aberrations', *Nature*, **173**, p. 39, 1954.

[4] H. H. Hopkins and N. S. Kapany, 'A flexible fibrescope, using static scanning', *Nature*, **113**, pp. 39–41, 1954.

[5] K. C. Kao and G. A. Hockham, 'Dielectric-fibre surface waveguides for optical frequencies', *Proc. IEE*, **113**, pp. 1151–1158, 1966.

[6] A. Werts, 'Propagation de la lumière coherente dans les fibres optiques', *L'Onde Electr.*, **46**, pp. 967–980, 1966.

[7] S. Takahashi and T. Kawashima, 'Preparation of low loss multi-component glass fiber', *Tech. Dig. Int. Conf. on Integrated Optics and Optical Fiber Communication*, p. 621, 1977.

[8] J. B. MacChesney, P. B. O'Connor, F. W. DiMarcello, J. R. Simpson and P. D. Lazay, 'Preparation of low-loss optical fibres using simultaneous vapour phase deposition and fusion', *Proc. 10th Int. Conf. on Glass*, paper 6–40, 1974.

[9] T. Miya, Y. Terunuma, T. Hosaka and T. Miyashita, 'Ultimate low-loss single-mode fibre at 1.55 μm', *Electron. Lett.*, **15**(4), pp. 106–108, 1979.

[10] J. A. Harrington, 'Infrared fibers', in M. Bass and E. W. Van Stryland (Eds), *Fiber Optics Handbook*, pp. 14.1–14.16, McGraw-Hill, 2002.

[11] J. C. Knight, T. A. Birks, P. S. J. Russell and D. M. Atkin, 'All-silica single-mode optical fiber with photonic crystal cladding', *Opt. Lett.*, **21**, pp. 1547–1549, 1996.

[12] M. Born and E. Wolf, *Principles of Optics* (7th edn), Cambridge University Press, 1999.

[13] R. P. Feyman, *The Feyman Lectures on Physics*, Vol. 2, Addison-Wesley, 1969.

[14] J. E. Midwinter, *Optical Fibers for Transmission*, Wiley, 1979.

[15] E. Snitzer, 'Cylindrical dielectric waveguide modes', *J. Opt. Soc. Am.*, **51**, pp. 491–498, 1961.

[16] D. Gloge, 'Weakly guiding fibers', *Appl. Opt.*, **10**, pp. 2252–2258, 1971.

[17] D. Marcuse, *Theory of Dielectric Optical Waveguides*, Academic Press, 1974.

[18] A. W. Snyder, 'Asymptotic expressions for eigenfunctions and eigenvalues of a dielectric or optical waveguide', *IEEE Trans. Microw. Theory Tech.*, **MTT-17**, pp. 1130–1138, 1969.

[19] E. G. Neumann, *Single-Mode Fibers: Fundamentals*, Springer-Verlag, 1988.

[20] D. Gloge, 'Optical power flow in multimode fibers', *Bell Syst. Tech. J.*, **51**, pp. 1767–1783, 1972.

[21] R. Olshansky, 'Propagation in glass optical waveguides', *Rev. Mod. Phys.*, **51**(2), pp. 341–366, 1979.

[22] P. M. Morse and H. Fesbach, *Methods of Theoretical Physics*, Vol. II, McGraw-Hill, 1953.

[23] D. Marcuse, *Light Transmission Optics* (2nd edn), Van Nostrand Reinhold, 1982.

[24] A. Ghatak and K. Thyagarajan, 'Graded index optical waveguides', in E. Wolf (Ed.), *Progress in Optics Vol. XVIII*, pp. 3–128, North-Holland, 1980.

[25] K. Okamoto, *Fundamentals of Optical Waveguides* (2nd edn), Academic Press, 2006.

[26] D. Marcuse, D. Gloge, E. A. J. Marcatili, 'Guiding properties of fibers', in S. E. Miller and A. G. Chynoweth (Eds), *Optical Fiber Telecommunications*, pp. 37–100, Academic Press, 1979.

[27] I. S. Gradshteyn and I. M. Ryzhik, *Tables of Integrals, Series and Products* (4th edn), Academic Press, 1965.

[28] C. W. Yeh, 'Optical waveguide theory', *IEEE Trans. Circuits Syst.*, **CAS-26**(12), pp. 1011–1019, 1979.

[29] C. Pask and R. A. Sammut, 'Developments in the theory of fibre optics', *Proc. IREE Aust.*, **40**(3), pp. 89–101, 1979.

[30] W. A. Gambling, A. H. Hartog and C. M. Ragdale, 'Optical fibre transmission lines', *Radio Electron. Eng.*, **51**(7/8), pp. 313–325, 1981.

[31] H. G. Unger, *Planar Optical Waveguides and Fibres*, Clarendon Press, 1977.

[32] M. J. Adams, *An Introduction to Optical Waveguides*, Wiley, 1981.

[33] Y. Suematsu and K.-I. Iga, *Introduction to Optical Fibre Communications*, Wiley, 1982.

[34] T. Okoshi, *Optical Fibers*, Academic Press, 1982.

[35] K. Okamoto and T. Okoshi, 'Analysis of wave propagation in optical fibers having core with α-power refractive-index distribution and uniform cladding', *IEEE Trans. Microw. Theory Tech.*, **MTT-24**, pp. 416–421, 1976.

[36] M. A. Saifi, 'Triangular index monomode fibres', *Proc. SPIE*, **374**, pp. 13–15, 1983.

[37] D. Gloge, 'The optical fibre as a transmission medium', *Rep. Prog. Phys.*, **42**, pp. 1777–1824, 1979.

[38] M. M. Ramsey and G. A. Hockham, 'Propagation in optical fibre waveguides', in C. P. Sandbank (Ed.), *Optical Fibre Communication Systems*, pp. 25–41, Wiley, 1980.

[39] S. Kawakami and S. Nishida, 'Characteristics of a doubly clad optical fiber with a low index cladding', *IEEE J. Quantum Electron*, **QE-10**, pp. 879–887, 1974.

[40] H. Kanamori, H. Yokota, G. Tanaka, M. Watanabe, Y. Ishiguro, I. Yoshida, T. Kakii, S. Itoh, Y. Asano and S. Takana, 'Transmission characteristics and reliability of silica-core single-mode fibers', *J. Lightwave Technol.*, **LT-4**(8), pp. 1144–1150, 1986.

[41] V. A. Bhagavatula, J. C. Lapp, A. J. Morrow and J. E. Ritter, 'Segmented-core fiber for long haul and local-area-network applications', *J. Lightwave Technol.*, **6**(10), pp. 1466–1469, 1988.

[42] D. P. Jablonowski, 'Fiber manufacture at AT&T with the MCVD Process', *J. Lightwave Technol.*, **LT-4**(8), pp. 1016–1019, 1986.

[43] L. B. Jeunhomme, *Single-Mode Fiber-Optics*, Marcel Dekker, 1983.

[44] ITU-T Recommendation G.652, 'Characteristics of single-mode optical fiber cable', October 2000.

[45] K. I. White, 'Methods of measurements of optical fiber properties', *J. Phys. E: Sci. Instrum.*, **18**, pp. 813–821, 1985.

[46] K. Petermann, 'Constraints for the fundamental-mode spot size for broadband dispersion-compensated single-mode fibres', *Electron. Lett.*, **19**, pp. 712–714, 1983.

[47] W. T. Anderson, V. Shah, L. Curtis, A. J. Johnson and J. P. Kilmer, 'Mode-field diameter measurements for single-mode fibers with non-Gaussian field profiles', *J. Lightwave Technol.*, **LT-5**, pp. 211–217, 1987.

[48] H. Kogelnik and H. P. Weber, 'Rays, stored energy, and power flow in dielectric waveguides', *J. Opt. Soc. Am.*, **64**, pp. 174–185, 1974.

[49] A. W. Snyder and J. D. Love, *Optical Waveguide Theory*, Chapman and Hall, 1983.

[50] H. G. Unger, *Planar Optical Waveguides and Fibres*, Clarendon Press, 1977.

[51] D. Marcuse, 'Gaussian approximation of the fundamental modes of graded-index fibers', *J. Opt. Soc. Am.*, **68**(1), pp. 103–109, 1978.

[52] D. Marcuse, 'Loss analysis of single-mode fiber splices', *Bell Syst. Tech. J.*, **56**(5), pp. 703–718, 1977.

[53] G. A. Bliss, *Lectures on the Calculus of Variations*, University of Chicago Press, 1946.

[54] D. Marcuse, 'Excitation of the dominant mode of a round fiber by a Gaussian beam', *Bell. Syst. Tech. J.*, **49**, pp. 1695–1703, 1970.

[55] A. K. Ghatak, R. Srivastava, I. F. Faria, K. Thyagaranjan and R. Tiwari, 'Accurate method for characterizing single-mode fibers: theory and experiment', *Electron. Lett.*, **19**, pp. 97–99, 1983.

[56] E. K. Sharma and R. Tewari, 'Accurate estimation of single-mode fiber characteristics from near-field measurements', *Electron. Lett.*, **20**, pp. 805–806, 1984.

[57] A. W. Snyder and R. A. Sammut, 'Fundamental (HE) modes of graded optical fibers', *J. Opt. Soc. Am.*, **69**, pp. 1663–1671, 1979.

[58] H. Matsumura and T. Suganama, 'Normalization of single-mode fibers having arbitrary index profile', *Appl. Opt.*, **19**, pp. 3151–3158, 1980.

[59] R. A. Sammut and A. W. Snyder, 'Graded monomode fibres and planar waveguides', *Electron. Lett.*, **16**, pp. 32–34, 1980.

[60] C. Pask and R. A. Sammut, 'Experimental characterisation of graded-index single-mode fibres', *Electron. Lett.*, **16**, pp. 310–311, 1980.

[61] J. Streckert and E. Brinkmeyer, 'Characteristic parameters of monomode fibers', *Appl. Opt.*, **21**, pp. 1910–1915, 1982.

[62] M. Fox, 'Calculation of equivalent step-index parameters for single-mode fibres', *Opt. Quantum Electron.*, **15**, pp. 451–455, 1983.

[63] W. A. Gambling and H. Matsumura, 'Propagation in radially-inhomogeneous single-mode fibre', *Opt. Quantum Electron.*, **10**, pp. 31–40, 1978.

[64] W. A. Gambling, H. Matsumura and C. M. Ragdale, 'Wave propagation in a single-mode fibre with dip in the refractive index', *Opt. Quantum Electron.*, **10**, pp. 301–309, 1978.

[65] A. W. Snyder, 'Understanding monomode optical fibers', *Proc. IEEE*, **69**(1), pp. 6–13, 1981.

[66] V. A. Bhagavatula, 'Estimation of single-mode waveguide dispersion using an equivalent-step-index approach', *Electron. Lett.*, **18**(8), pp. 319–320, 1982.

[67] D. Davidson, 'Single-mode wave propagation in cylindrical optical fibers', in E. E. Basch (Ed.), *Optical-Fiber Transmission*, pp. 27–64, H. W. Sams & Co., 1987.

[68] F. Martinez and C. D. Hussey, '(E) ESI determination from mode-field diameter and refractive index profile measurements on single-mode fibers', *IEE Proc., Optoelectron.*, **135**(3), pp. 202–210, 1988.

[69] D. J. DiGiovanni, S. K. Das, L. L. Blyler, W. White and R. Boncek, 'Design of optical fibers for communication systems', in I. P. Kaminow and T. Li (Eds), *Optical Fiber Telecommunications IVA*, pp. 17–79, Academic Press, 2002.

[70] D. J. Richardson, T. M. Monro, W. Belardi and K. Furusawa, 'Holey fibers: new possibilities for guiding and manipulating light', *Proc. IEEE/LEOS Workshop on Fiber and Optical Passive Components*, pp. 169–175, June 2002.

[71] K. Morishita and Y. Miyake, 'Fabrication and resonance wavelengths of long-period gratings written in a pure-silica photonic crystal fiber by the glass structure change', *J. Lightwave Technol.*, **22**(2), pp. 625–630, 2004.

[72] L. Tao, A. R. Zakharian, M. Fallahi, J. V. Moloney and M. Mansuripur, 'Multimode interference-based photonic crystal waveguide power splitter', *J. Lightwave Technol.*, **22**(12), pp. 2842–2846, 2004.

[73] S. Mahnkopf, R. Marz, M. Kamp, H. D. Guang, F. Lelarge and A. Forchel, 'Tunable photonic crystal coupled-cavity laser', *IEEE J. Quantum Electron*, **40**(9), pp. 1306–1314, 2004.

[74] A. Cucinotta, F. Poli and S. Selleri, 'Design of erbium-doped triangular photonic-crystal-fiber-based amplifiers', *IEEE Photonics Technol. Lett.*, **16**(9), pp. 2027–2029, 2004.

[75] S. Bong-Shik, T. Asano, Y. Akahane, Y. Tanaka and S. Noda, 'Multichannel add/drop filter based on in-plane hetero photonic crystals', *J. Lightwave Technol.*, **23**(3), pp. 1449–1455, 2005.

[76] K. K. Chow, C. Shu, L. Chinlon and A. Bjarklev, 'Polarization-insensitive widely tunable wavelength converter based on four-wave mixing in a dispersion-flattened nonlinear photonic crystal fiber', *IEEE Photonics Technol. Lett.*, **17**(3), pp. 624–626, 2005.

[77] T. Niemi, L. H. Frandsen, K. K. Hede, A. Harpoth, P. I. Borel and M. Kristensen, 'Wavelength division demultiplexing using photonic crystal waveguides', *IEEE Photonics Lett.*, **18**(1), pp. 226–228, 2006.

[78] K. Kurokawa, K. Tajima, K. Tsujikawa, K. Nakajima, T. Matsui, I. Sankawa and T. Haibara, 'Penalty-free dispersion-managed soliton transmission over a 100-km low-loss PCF', *J. Lightwave Technol.*, **24**(1), pp. 32–37, 2006.

[79] T. A. Birks, J. C. Knight and P. St J. Russell, 'Endlessly singlemode photonic crystal fibers', *Opt. Lett.*, **22**, pp. 961–963, 1997.

[80] K. Tajima, J. Zhon, K. Nakajima and K. Sato, 'Ultralow loss and long length photonic crystal fiber', *J. Lightwave Technol.*, **22**(1), pp. 7–10, 2004.

[81] J. Laegsgaard, K. P. Hansen, M. D. Nielsen, T. P. Hansen, J. Riishede, K. Hougaard, T. Sørenen, T. T. Larsen, N. A. Mortensen, J. Broeng, J. B. Jensen and A. Bjarkeev, 'Photonic crystal fibers', *Proc. SBMO/IEEE MTT-S IMOC*, pp. 259–264, 2003.

[82] L. P. Shen, W. P. Huang and S. S. Jian, 'Design of photonic crystal fibers for dispersion-related applications', *J. Lightwave Technol.*, **21**(7), pp. 1644–1651, 2003.

[83] A. A. Abramov, A. Hale, R. S. Windeler and T. A. Strasser, 'Widely tunable long-period gratings', *Electron. Lett.*, **35**(1), pp. 1–2, 1999.

[84] J. C. Knight, J. Broeng, T. A. Birks and P. S. J. Russell, 'Photonic band gap guidance in optical fibers', *Science*, **282**, pp. 1476–1478, 1998.

[85] Y. Li, C. Y. Wang, M. Hu, B. Liu, X. Sun and L. Chai, 'Photonic bandgap fibers based on a composite honeycomb lattice', *IEEE Photonics Technol. Lett.*, **18**(1), pp. 262–264, 2006.

[86] T. P. Hansen, J. Broeng, C. Jakobsen, G. Vienna, H. R. Simonsen, M. D. Nielsen, P. M. W. Skovgaard, J. R. Folkenberg and A. Bjarklev, 'Air-guiding photonic bandgap fibers: spectral properties, macrobending loss and practical handling', *J. Lightwave Technol.*, **22**(1), pp. 11–15, 2004.

[87] X. Pu, P. Shum, N. Q. Ngo, W. J. Tong, J. Luo, G. B. Ren, Y. D. Gong and J. Q. Zhou, 'Silica-based nanostructure core fiber', *IEEE Photonics Technol. Lett.*, **20**(2), pp. 162–164, 2008.

[88] X. Pu, P. Shum, M. Yan and G. B. Ren, 'Silica-based birefringent large-mode-area fiber with a nanostructure core', *IEEE Photonics Technol. Lett.*, **20**(4), pp. 246–248, 2008.

CHAPTER 3

Transmission characteristics of optical fibers

3.1 Introduction

The basic transmission mechanisms of the various types of optical fiber waveguide have been discussed in Chapter 2. However, the factors which affect the performance of optical fibers as a transmission medium were not dealt with in detail. These transmission characteristics are of utmost importance when the suitability of optical fibers for communication purposes is investigated. The transmission characteristics of most interest are those of attenuation (or loss) and bandwidth.

The huge potential bandwidth of optical communications helped stimulate the birth of the idea that a dielectric waveguide made of glass could be used to carry wideband telecommunication signals. This occurred, as indicated in Section 2.1 in the celebrated papers by Kao and Hockham, and Werts, in 1966. However, at the time the idea may have seemed somewhat ludicrous as a typical block of glass could support optical transmission for at best a few tens of meters before it was attenuated to an unacceptable level. Nevertheless, careful investigation of the attenuation showed that it was largely due to absorption in the glass, caused by impurities such as iron, copper, manganese and other transition metals which occur in the third row of the periodic table. Hence, research was stimulated towards a new generation of 'pure' glasses for use in optical fiber communications.

A major breakthrough came in 1970 when the first fiber with an attenuation below 20 dB km^{-1} was reported [Ref. 1]. This level of attenuation was seen as the absolute minimum that had to be achieved before an optical fiber system could in any way compete economically with existing communication systems. Since 1970 tremendous improvements have been made, leading to silica-based glass fibers with losses of less than 0.2 dB km^{-1} in the laboratory by the late 1980s [Ref. 2]. Hence, comparatively low-loss fibers have been incorporated into optical communication systems throughout the world. Although the fundamental lower limits for attenuation in silicate glass fibers were largely achieved by 1990, continuing significant progress has been made in relation to the removal of the water impurity peak within the operational wavelength range [Ref. 3]. The investigation of other material systems which can exhibit substantially lower losses when operated at longer wavelengths [Ref. 2] has, however, slowed down in relation to telecommunication transmission due to difficulties in the production of fiber with both optical and mechanical properties that will compete with silica. In particular, such mid-infrared (and possibly far-infrared) transmitting fibers continue to exhibit both relatively high losses and low strength [Ref. 4].

The other characteristic of primary importance is the bandwidth of the fiber. This is limited by the signal dispersion within the fiber, which determines the number of bits of information transmitted in a given time period. Therefore, once the attenuation was reduced to acceptable levels, attention was directed towards the dispersive properties of fibers Again, this has led to substantial improvements, giving wideband fiber bandwidths of many tens of gigahertz over a number of kilometers.

In order to appreciate these advances and possible future developments, the optical transmission characteristics of fibers must be considered in greater depth. Therefore, in this chapter we discuss the mechanisms within optical fibers which give rise to the major transmission characteristics mentioned previously (attenuation and dispersion), while also considering other, perhaps less obvious, effects when light is propagating down an optical fiber (modal noise, polarization and nonlinear phenomena).

We begin the discussion of attenuation in Section 3.2 with calculation of the total losses incurred in optical fibers. The various attenuation mechanisms (material absorption, linear scattering, nonlinear scattering, fiber bends) are then considered in detail in Sections 3.3 to 3.6. The primary focus within these sections is on silica-based glass fibers. However, in Section 3.7 consideration is given to other material systems which are employed for mid-infrared and far-infrared optical transmission. Dispersion in optical fibers is described in Section 3.8, together with the associated limitations on fiber bandwidth. Sections 3.9 and 3.10 deal with chromatic (intramodal) and intermodal dispersion mechanisms and included in the latter section is a discussion of the modal noise phenomenon associated with intermodal dispersion. Overall signal dispersion in both multimode and single-mode fibers is then considered in Section 3.11. This is followed in Section 3.12 by a review of the modification of the dispersion characteristics within single-mode fibers in order to obtain dispersion-shifted, dispersion-flattened and nonzero-dispersion-shifted fibers. Section 3.13 presents an account of the polarization within single-mode fibers which includes discussion of both polarization mode dispersion and the salient features of polarization-maintaining fibers. Nonlinear optical effects, which can occur at relatively high optical power levels within single-mode fibers, are then dealt with in Section 3.14 prior to a final Section 3.15 describing the special case of nonlinear pulse propagation referred to as soliton propagation.

3.2 Attenuation

The attenuation or transmission loss of optical fibers has proved to be one of the most important factors in bringing about their wide acceptance in telecommunications. As channel attenuation largely determined the maximum transmission distance prior to signal restoration, optical fiber communications became especially attractive when the transmission losses of fibers were reduced below those of the competing metallic conductors (less than 5 dB km^{-1}).

Signal attenuation within optical fibers, as with metallic conductors, is usually expressed in the logarithmic unit of the decibel. The decibel, which is used for comparing two power levels, may be defined for a particular optical wavelength as the ratio of the input (transmitted) optical power P_i into a fiber to the output (received) optical power P_o from the fiber as:

$$\text{Number of decibels (dB)} = 10 \log_{10} \frac{P_i}{P_o} \tag{3.1}$$

This logarithmic unit has the advantage that the operations of multiplication and division reduce to addition and subtraction, while powers and roots reduce to multiplication and division. However, addition and subtraction require a conversion to numerical values which may be obtained using the relationship:

$$\frac{P_i}{P_o} = 10^{(dB/10)} \tag{3.2}$$

In optical fiber communications the attenuation is usually expressed in decibels per unit length (i.e. dB km^{-1}) following:

$$\alpha_{dB}L = 10 \log_{10} \frac{P_i}{P_o}$$ (3.3)

where α_{dB} is the signal attenuation per unit length in decibels which is also referred to as the fiber loss parameter and L is the fiber length.

Example 3.1

When the mean optical power launched into an 8 km length of fiber is 120 μW, the mean optical power at the fiber output is 3 μW.
Determine:

(a) the overall signal attenuation or loss in decibels through the fiber assuming there are no connectors or splices;

(b) the signal attenuation per kilometer for the fiber.

(c) the overall signal attenuation for a 10 km optical link using the same fiber with splices at 1 km intervals, each giving an attenuation of 1 dB;

(d) the numerical input/output power ratio in (c).

Solution: (a) Using Eq. (3.1), the overall signal attenuation in decibels through the fiber is:

$$\text{Signal attenuation} = 10 \log_{10} \frac{P_i}{P_o} = 10 \log_{10} \frac{120 \times 10^{-6}}{3 \times 10^{-6}}$$

$$= 10 \log_{10} 40 = 16.0 \text{ dB}$$

(b) The signal attenuation per kilometer for the fiber may be simply obtained by dividing the result in (a) by the fiber length which corresponds to it using Eq. (3.3) where:

$$\alpha_{dB}L = 16.0 \text{ dB}$$

hence:

$$\alpha_{dB} = \frac{16.0}{8}$$

$$= 2.0 \text{ dB km}^{-1}$$

▶

(c) As $\alpha_{dB} = 2$ dB km^{-1}, the loss incurred along 10 km of the fiber is given by:

$$\alpha_{dB}L = 2 \times 10 = 20 \text{ dB}$$

However, the link also has nine splices (at 1 km intervals) each with an attenuation of 1 dB. Therefore, the loss due to the splices is 9 dB.

Hence, the overall signal attenuation for the link is:

$$\text{Signal attenuation} = 20 + 9$$
$$= 29 \text{ dB}$$

(d) To obtain a numerical value for the input/output power ratio, Eq. (3.2) may be used where:

$$\frac{P_i}{P_o} = 10^{29/10} = 794.3$$

A number of mechanisms are responsible for the signal attenuation within optical fibers. These mechanisms are influenced by the material composition, the preparation and purification technique, and the waveguide structure. They may be categorized within several major areas which include material absorption, material scattering (linear and nonlinear scattering), curve and microbending losses, mode coupling radiation losses and losses due to leaky modes. There are also losses at connectors and splices, as illustrated in Example 3.1. However, in this chapter we are interested solely in the characteristics of the fiber; connector and splice losses are dealt with in Section 5.2. It is instructive to consider in some detail the loss mechanisms within optical fibers in order to obtain an understanding of the problems associated with the design and fabrication of low-loss waveguides.

3.3 Material absorption losses in silica glass fibers

Material absorption is a loss mechanism related to the material composition and the fabrication process for the fiber, which results in the dissipation of some of the transmitted optical power as heat in the waveguide. The absorption of the light may be intrinsic (caused by the interaction with one or more of the major components of the glass) or extrinsic (caused by impurities within the glass).

3.3.1 Intrinsic absorption

An absolutely pure silicate glass has little intrinsic absorption due to its basic material structure in the near-infrared region. However, it does have two major intrinsic absorption

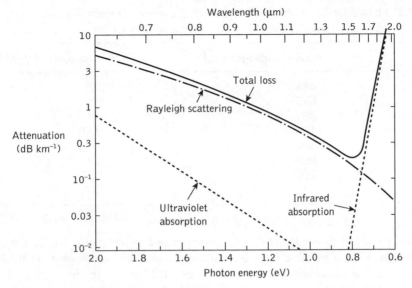

Figure 3.1 The attenuation spectra for the intrinsic loss mechanisms in pure GeO_2–SiO_2 glass [Ref. 5]

mechanisms at optical wavelengths which leave a low intrinsic absorption window over the 0.8 to 1.7 μm wavelength range, as illustrated in Figure 3.1, which shows a possible optical attenuation against wavelength characteristic for absolutely pure glass [Ref. 5]. It may be observed that there is a fundamental absorption edge, the peaks of which are centered in the ultraviolet wavelength region. This is due to the stimulation of electron transitions within the glass by higher energy excitations. The tail of this peak may extend into the window region at the shorter wavelengths, as illustrated in Figure 3.1. Also in the infrared and far infrared, normally at wavelengths above 7 μm, fundamentals of absorption bands from the interaction of photons with molecular vibrations within the glass occur. These give absorption peaks which again extend into the window region. The strong absorption bands occur due to oscillations of structural units such as Si–O (9.2 μm), P–O (8.1 μm), B–O (7.2 μm) and Ge–O (11.0 μm) within the glass. Hence, above 1.5 μm the tails of these largely far-infrared absorption peaks tend to cause most of the pure glass losses.

However, the effects of both these processes may be minimized by suitable choice of both core and cladding compositions. For instance, in some nonoxide glasses such as fluorides and chlorides, the infrared absorption peaks occur at much longer wavelengths which are well into the far infrared (up to 50 μm), giving less attenuation to longer wavelength transmission compared with oxide glasses.

3.3.2 Extrinsic absorption

In practical optical fibers prepared by conventional melting techniques (see Section 4.3), a major source of signal attenuation is extrinsic absorption from transition metal element impurities. Some of the more common metallic impurities found in glasses are shown in

Table 3.1 Absorption losses caused by some of the more common metallic ion impurities in glasses, together with the absorption peak wavelength

	Peak wavelength (nm)	One part in 10^9 (dB km^{-1})
Cr^{3+}	625	1.6
C^{2+}	685	0.1
Cu^{2+}	850	1.1
Fe^{2+}	1100	0.68
Fe^{3+}	400	0.15
Ni^{2+}	650	0.1
Mn^{3+}	460	0.2
V^{4+}	725	2.7

the Table 3.1, together with the absorption losses caused by one part in 10^9 [Ref. 6]. It may be noted that certain of these impurities, namely chromium and copper, in their worst valence state can cause attenuation in excess of 1 dB km^{-1} in the near-infrared region. Transition element contamination may be reduced to acceptable levels (i.e. one part in 10^{10}) by glass refining techniques such as vapor-phase oxidation [Ref. 7] (see Section 4.4), which largely eliminates the effects of these metallic impurities.

However, another major extrinsic loss mechanism is caused by absorption due to water (as the hydroxyl or OH ion) dissolved in the glass. These hydroxyl groups are bonded into the glass structure and have fundamental stretching vibrations which occur at wavelengths between 2.7 and 4.2 μm depending on group position in the glass network. The fundamental vibrations give rise to overtones appearing almost harmonically at 1.38, 0.95 and 0.72 μm, as illustrated in Figure 3.2. This shows the absorption spectrum for the hydroxyl

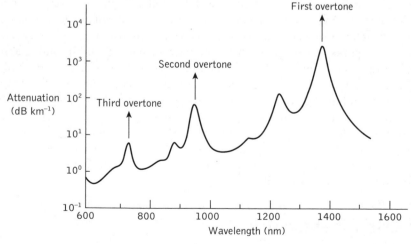

Figure 3.2 The absorption spectrum for the hydroxyl (OH) group in silica. Reproduced with permission from D. B. Keck, R. D. Maurer and P. C. Schultz, *Appl. Phys. Lett.*, **22**, p. 307, 1973. Copyright © 1973, American Institute of Physics

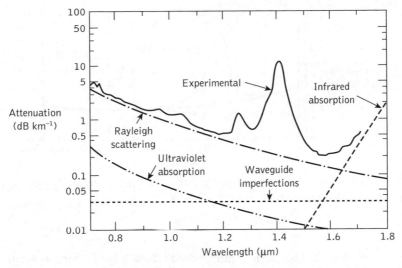

Figure 3.3 The measured attenuation spectrum for an ultra-low-loss single-mode fiber (solid line) with the calculated attenuation spectra for some of the loss mechanisms contributing to the overall fiber attenuation (dashed and dotted lines) [Ref. 5]

group in silica. Furthermore, combinations between the overtones and the fundamental SiO_2 vibration occur at 1.24, 1.13 and 0.88 μm, completing the absorption spectrum shown in Figure 3.2.

It may also be observed in Figure 3.2 that the only significant absorption band in the region below a wavelength of 1 μm is the second overtone at 0.95 μm which causes attenuation of about 1 dB km^{-1} for one part per million (ppm) of hydroxyl. At longer wavelengths the first overtone at 1.383 μm and its sideband at 1.24 μm are strong absorbers giving attenuation of about 2 dB km^{-1} ppm and 4 dB km^{-1} ppm respectively. Since most resonances are sharply peaked, narrow windows exist in the longer wavelength region around 1.31 and 1.55 μm which are essentially unaffected by OH absorption once the impurity level has been reduced below one part in 10^7. This situation is illustrated in Figure 3.3, which shows the attenuation spectrum of a low-loss single-mode fiber produced in 1979 [Ref. 5]. It may be observed that the lowest attenuation for this fiber occurs at a wavelength of 1.55 μm and is 0.2 dB km^{-1}. Despite this value approaching the minimum possible attenuation of around 0.18 dB km^{-1} at the 1.55 μm wavelength [Ref. 8], it should be noted that the transmission loss of an ultra-low-loss pure silica core fiber was more recently measured as 0.1484 dB km^{-1} at the slightly longer wavelength of 1.57 μm [Ref. 9].

Although in standard, modern single-mode fibers the loss caused by the primary OH peak at 1.383 μm has been reduced below 1 dB km^{-1}, it still limits operation over significant distances to the lower loss windows at 1.31 and 1.55 μm. A more recent major advance, however, has enabled the production of a revolutionary fiber type* in which the

---

* An example is the Lucent AllWave fiber which has typical losses of 0.32, 0.28 and 0.19 dB at wavelengths of 1.310, 1.383 and 1.550 μm, respectively. This fiber is referred to as exhibiting a zero water peak (ZWP) in the Lucent specification literature.

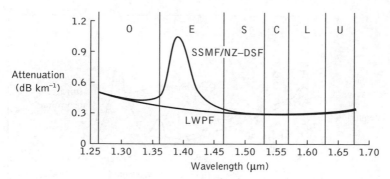

Figure 3.4 Fiber attenuation spectra: low-water-peak fiber compared with standard single-mode and nonzero-dispersion-shifted fibers

1.383 µm water peak has been permanently reduced to such levels that it is virtually eliminated [Ref. 10]. The attenuation spectrum for this low-water-peak fiber (LWPF), or dry fiber, is shown in Figure 3.4 where it is compared with standard single-mode fiber (SSMF) [Ref. 3].

The LWPF permits the transmission of optical signals over the full 1.260 to 1.675 µm wavelength range with losses less than 0.4 dB km^{-1} and therefore better facilitates wavelength division multiplexing (see Section 12.9.3). It may also be seen that the optical transmission wavelength band designations are also identified on the wavelength axis of Figure 3.4. These International Telecommunications Union (ITU) spectral band designations for both intermediate-range and long-distance optical fiber communications are indicated by the letters O, E, S, C, L and U, which are defined in Table 3.2 and are in common use in the field. It should be noted that long-haul transmission first took place in the O- and C-bands, subsequently followed by the L-band region. In addition, it is apparent that LWPF has enabled the use of the 1.460 to 1.530 µm window or S-band which is affected by the water peak in SSMF.

Table 3.2 ITU spectral band definitions

Name	ITU band	Wavelength range (µm)
Original band	O-band	1.260 to 1.360
Extended band	E-band	1.360 to 1.460
Short band	S-band	1.460 to 1.530
Conventional band	C-band	1.530 to 1.565
Long band	L-band	1.565 to 1.625
Ultralong band	U-band	1.625 to 1.675

3.4 Linear scattering losses

Linear scattering mechanisms cause the transfer of some or all of the optical power contained within one propagating mode to be transferred linearly (proportionally to the mode power) into a different mode. This process tends to result in attenuation of the transmitted light as the transfer may be to a leaky or radiation mode which does not continue to propagate within the fiber core, but is radiated from the fiber. It must be noted that as with all linear processes, there is no change of frequency on scattering.

Linear scattering may be categorized into two major types: Rayleigh and Mie scattering. Both result from the nonideal physical properties of the manufactured fiber which are difficult and, in certain cases, impossible to eradicate at present.

3.4.1 Rayleigh scattering

Rayleigh scattering is the dominant intrinsic loss mechanism in the low-absorption window between the ultraviolet and infrared absorption tails. It results from inhomogeneities of a random nature occurring on a small scale compared with the wavelength of the light. These inhomogeneities manifest themselves as refractive index fluctuations and arise from density and compositional variations which are frozen into the glass lattice on cooling. The compositional variations may be reduced by improved fabrication, but the index fluctuations caused by the freezing-in of density inhomogeneities are fundamental and cannot be avoided. The subsequent scattering due to the density fluctuations, which is in almost all directions, produces an attenuation proportional to $1/\lambda^4$ following the Rayleigh scattering formula [Ref. 11]. For a single-component glass this is given by:

$$\gamma_R = \frac{8\pi^3}{3\lambda^4} n^8 p^2 \beta_c K T_F \tag{3.4}$$

where γ_R is the Rayleigh scattering coefficient, λ is the optical wavelength, n is the refractive index of the medium, p is the average photoelastic coefficient, β_c is the isothermal compressibility at a fictive temperature T_F, and K is Boltzmann's constant. The fictive temperature is defined as the temperature at which the glass can reach a state of thermal equilibrium and is closely related to the anneal temperature. Furthermore, the Rayleigh scattering coefficient is related to the transmission loss factor (transmissivity) of the fiber $\mathcal{L}$ following the relation [Ref. 12]:

$$\mathcal{L} = \exp(-\gamma_R L) \tag{3.5}$$

where L is the length of the fiber. It is apparent from Eq. (3.4) that the fundamental component of Rayleigh scattering is strongly reduced by operating at the longest possible wavelength. This point is illustrated in Example 3.2.

Example 3.2

Silica has an estimated fictive temperature of 1400 K with an isothermal compressibility of 7×10^{-11} m^2 N^{-1} [Ref. 13]. The refractive index and the photoelastic coefficient for silica are 1.46 and 0.286 respectively [Ref. 13]. Determine the theoretical attenuation in decibels per kilometer due to the fundamental Rayleigh scattering in silica at optical wavelengths of 0.63, 1.00 and 1.30 μm. Boltzmann's constant is 1.381×10^{-21} J K^{-1}.

Solution: The Rayleigh scattering coefficient may be obtained from Eq. (3.4) for each wavelength. However, the only variable in each case is the wavelength, and therefore the constant of proportionality of Eq. (3.4) applies in all cases. Hence:

$$\gamma_R = \frac{8\pi^3 n^8 p^2 \beta_c K T_F}{3\lambda^4}$$

$$= \frac{248.15 \times 20.65 \times 0.082 \times 7 \times 10^{-11} \times 1.381 \times 10^{-23} \times 1400}{3 \times \lambda^4}$$

$$= \frac{1.895 \times 10^{-28}}{\lambda^4} \ \text{m}^{-1}$$

At a wavelength of 0.63 μm:

$$\gamma_R = \frac{1.895 \times 10^{-28}}{0.158 \times 10^{-24}} = 1.199 \times 10^{-3} \ \text{m}^{-1}$$

The transmission loss factor for 1 kilometer of fiber may be obtained using Eq. (3.5):

$$\mathcal{L}_{km} = \exp(-\gamma_R L) = \exp(-1.199 \times 10^{-3} \times 10^3)$$
$$= 0.301$$

The attenuation due to Rayleigh scattering in decibels per kilometer may be obtained from Eq. (3.1) where:

$$\text{Attenuation} = 10 \log_{10}(1/\mathcal{L}_{km}) = 10 \log_{10} 3.322$$
$$= 5.2 \ \text{dB km}^{-1}$$

At a wavelength of 1.0 μm:

$$\gamma_R = \frac{1.895 \times 10^{-28}}{10^{-24}} = 1.895 \times 10^{-4} \ \text{m}^{-1}$$

Using Eq. (3.5):

$$\mathcal{L}_{km} = \exp(-1.895 \times 10^{-4} \times 10^3) = \exp(-0.1895)$$
$$= 0.827$$

and Eq. (3.1):

Attenuation $= 10 \log_{10} 1.209 = 0.8$ dB km^{-1}

At a wavelength of 1.30 μm:

$$\gamma_R = \frac{1.895 \times 10^{-28}}{2.856 \times 10^{-24}} = 0.664 \times 10^{-4}$$

Using Eq. (3.5):

$$\mathscr{L}_{km} = \exp(-0.664 \times 10^{-4} \times 10^3) = 0.936$$

and Eq. (3.1):

Attenuation $= 10 \log_{10} 1.069 = 0.3$ dB km^{-1}

The theoretical attenuation due to Rayleigh scattering in silica at wavelengths of 0.63, 1.00 and 1.30 μm, from Example 3.2, is 5.2, 0.8 and 0.3 dB km^{-1} respectively. These theoretical results are in reasonable agreement with experimental work. For instance, a low reported value for Rayleigh scattering in silica at a wavelength of 0.6328 μm is 3.9 dB km^{-1} [Ref. 13]. However, values of 4.8 dB km^{-1} [Ref. 14] and 5.4 dB km^{-1} [Ref. 15] have also been reported. The predicted attenuation due to Rayleigh scattering against wavelength is indicated by a dashed line on the attenuation characteristics shown in Figures 3.1 and 3.3.

3.4.2 Mie scattering

Linear scattering may also occur at inhomogeneities which are comparable in size with the guided wavelength. These result from the nonperfect cylindrical structure of the waveguide and may be caused by fiber imperfections such as irregularities in the core–cladding interface, core–cladding refractive index differences along the fiber length, diameter fluctuations, strains and bubbles. When the scattering inhomogeneity size is greater than $\lambda/10$, the scattered intensity which has an angular dependence can be very large.

The scattering created by such inhomogeneities is mainly in the forward direction and is called Mie scattering. Depending upon the fiber material, design and manufacture, Mie scattering can cause significant losses. The inhomogeneities may be reduced by:

(a) removing imperfections due to the glass manufacturing process;

(b) carefully controlled extrusion and coating of the fiber;

(c) increasing the fiber guidance by increasing the relative refractive index difference.

By these means it is possible to reduce Mie scattering to insignificant levels.

3.5 Nonlinear scattering losses

Optical waveguides do not always behave as completely linear channels whose increase in output optical power is directly proportional to the input optical power. Several nonlinear effects occur, which in the case of scattering cause disproportionate attenuation, usually at high optical power levels. This nonlinear scattering causes the optical power from one mode to be transferred in either the forward or backward direction to the same, or other modes, at a different frequency. It depends critically upon the optical power density within the fiber and hence only becomes significant above threshold power levels.

The most important types of nonlinear scattering within optical fibers are stimulated Brillouin and Raman scattering, both of which are usually only observed at high optical power densities in long single-mode fibers. These scattering mechanisms in fact give optical gain but with a shift in frequency, thus contributing to attenuation for light transmission at a specific wavelength. However, it may be noted that such nonlinear phenomena can also be used to give optical amplification in the context of integrated optical techniques (see Section 11.7). In addition, these nonlinear processes are explored in further detail both following and in Section 3.14.

3.5.1 Stimulated Brillouin scattering

Stimulated Brillouin scattering (SBS) may be regarded as the modulation of light through thermal molecular vibrations within the fiber. The scattered light appears as upper and lower sidebands which are separated from the incident light by the modulation frequency. The incident photon in this scattering process produces a phonon* of acoustic frequency as well as a scattered photon. This produces an optical frequency shift which varies with the scattering angle because the frequency of the sound wave varies with acoustic wavelength. The frequency shift is a maximum in the backward direction, reducing to zero in the forward direction, making SBS a mainly backward process.

As indicated previously, Brillouin scattering is only significant above a threshold power density. Assuming that the polarization state of the transmitted light is not maintained (see Section 3.12), it may be shown [Ref. 16] that the threshold power P_B is given by:

$$P_B = 4.4 \times 10^{-3} d^2 \lambda^2 \alpha_{dB} v \text{ watts} \tag{3.6}$$

where d and λ are the fiber core diameter and the operating wavelength, respectively, both measured in micrometers, α_{dB} is the fiber attenuation in decibels per kilometer and v is the source bandwidth (i.e. injection laser) in gigahertz. The expression given in Eq. (3.6) allows the determination of the threshold optical power which must be launched into a single-mode optical fiber before SBS occurs (see Example 3.3).

---

* The phonon is a quantum of an elastic wave in a crystal lattice. When the elastic wave has a frequency f, the quantized unit of the phonon has energy hf joules, where h is Planck's constant.

3.5.2 Stimulated Raman scattering

Stimulated Raman scattering (SRS) is similar to SBS except that a high-frequency optical phonon rather than an acoustic phonon is generated in the scattering process. Also, SRS can occur in both the forward and backward directions in an optical fiber, and may have an optical power threshold of up to three orders of magnitude higher than the Brillouin threshold in a particular fiber.

Using the same criteria as those specified for the Brillouin scattering threshold given in Eq. (3.6), it may be shown [Ref. 16] that the threshold optical power for SRS P_R in a long single-mode fiber is given by:

$$P_R = 5.9 \times 10^{-2} d^2 \lambda \alpha_{dB} \text{ watts} \tag{3.7}$$

where d, λ and α_{dB} are as specified for Eq. (3.6).

Example 3.3

A long single-mode optical fiber has an attenuation of 0.5 dB km^{-1} when operating at a wavelength of 1.3 μm. The fiber core diameter is 6 μm and the laser source bandwidth is 600 MHz. Compare the threshold optical powers for stimulated Brillouin and Raman scattering within the fiber at the wavelength specified.

Solution: The threshold optical power for SBS is given by Eq. (3.6) as:

$$
\begin{aligned}
P_B &= 4.4 \times 10^{-3} d^2 \lambda^2 \alpha_{dB} \nu \\
&= 4.4 \times 10^{-3} \times 6^2 \times 1.3^2 \times 0.5 \times 0.6 \\
&= 80.3 \text{ mW}
\end{aligned}
$$

The threshold optical power for SRS may be obtained from Eq. (3.7), where:

$$
\begin{aligned}
P_R &= 5.9 \times 10^{-2} d^2 \lambda \alpha_{dB} \\
&= 5.9 \times 10^{-2} \times 6^2 \times 1.3 \times 0.5 \\
&= 1.38 \text{ W}
\end{aligned}
$$

In Example 3.3, the Brillouin threshold occurs at an optical power level of around 80 mW while the Raman threshold is approximately 17 times larger. It is therefore apparent that the losses introduced by nonlinear scattering may be avoided by use of a suitable optical signal level (i.e. working below the threshold optical powers). However, it must be noted that the Brillouin threshold has been reported [Ref. 17] as occurring at optical powers as low as 10 mW in single-mode fibers. Nevertheless, this is still a high power level for optical communications and may be easily avoided. SBS and SRS are not usually observed in multimode fibers because their relatively large core diameters make the threshold optical power levels extremely high. Moreover, it should be noted that the threshold optical powers for both these scattering mechanisms may be increased by suitable adjustment of the other parameters in Eqs (3.6) and (3.7). In this context, operation at the longest possible wavelength is advantageous although this may be offset by the reduced fiber attenuation (from Rayleigh scattering and material absorption) normally obtained.

3.6 Fiber bend loss

Optical fibers suffer radiation losses at bends or curves on their paths. This is due to the energy in the evanescent field at the bend exceeding the velocity of light in the cladding and hence the guidance mechanism is inhibited, which causes light energy to be radiated from the fiber. An illustration of this situation is shown in Figure 3.5. The part of the mode which is on the outside of the bend is required to travel faster than that on the inside so that a wavefront perpendicular to the direction of propagation is maintained. Hence, part of the mode in the cladding needs to travel faster than the velocity of light in that medium. As this is not possible, the energy associated with this part of the mode is lost through radiation. The loss can generally be represented by a radiation attenuation coefficient which has the form [Ref. 18]:

$$\alpha_r = c_1 \exp(-c_2 R)$$

where R is the radius of curvature of the fiber bend and c_1, c_2 are constants which are independent of R. Furthermore, large bending losses tend to occur in multimode fibers at a critical radius of curvature R_c which may be estimated from [Ref. 19]:

$$R_c \simeq \frac{3n_1^2 \lambda}{4\pi(n_1^2 - n_2^2)^{\frac{3}{2}}} \qquad (3.8)$$

It may be observed from the expression given in Eq. (3.8) that potential macrobending losses may be reduced by:

(a) designing fibers with large relative refractive index differences;

(b) operating at the shortest wavelength possible.

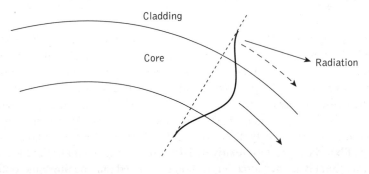

Figure 3.5 An illustration of the radiation loss at a fiber bend. The part of the mode in the cladding outside the dashed arrowed line may be required to travel faster than the velocity of light in order to maintain a plane wavefront. Since it cannot do this, the energy contained in this part of the mode is radiated away

The above criteria for the reduction of bend losses also apply to single-mode fibers. One theory [Ref. 20], based on the concept of a single quasi-guided mode, provides an expression from which the critical radius of curvature for a single-mode fiber R_{cs} can be estimated as:

$$R_{cs} \simeq \frac{20\lambda}{(n_1 - n_2)^{\frac{3}{2}}}\left(2.748 - 0.996\frac{\lambda}{\lambda_c}\right)^{-3} \tag{3.9}$$

where λ_c is the cutoff wavelength for the single-mode fiber. Hence again, for a specific single-mode fiber (i.e. a fixed relative index difference and cutoff wavelength), the critical wavelength of the radiated light becomes progressively shorter as the bend radius is decreased. The effect of this factor and that of the relative refractive index difference on the critical bending radius is demonstrated in the following example.

Example 3.4

Two step index fibers exhibit the following parameters:

(a) a multimode fiber with a core refractive index of 1.500, a relative refractive index difference of 3% and an operating wavelength of 0.82 μm;

(b) an 8 μm core diameter single-mode fiber with a core refractive index the same as (a), a relative refractive index difference of 0.3% and an operating wavelength of 1.55 μm.

Estimate the critical radius of curvature at which large bending losses occur in both cases.

Solution: (a) The relative refractive index difference is given by Eq. (2.9) as:

$$\Delta = \frac{n_1^2 - n_2^2}{2n_1^2}$$

Hence:

$$n_2^2 = n_1^2 - 2\Delta n_1^2 = 2.250 - 0.06 \times 2.250$$
$$= 2.115$$

Using Eq. (3.8) for the multimode fiber critical radius of curvature:

$$R_c \simeq \frac{3n_1^2\lambda}{4\pi(n_1^2 - n_2^2)^{\frac{1}{2}}} = \frac{3 \times 2.250 \times 0.82 \times 10^{-6}}{4\pi \times (0.135)^{\frac{1}{2}}}$$
$$= 9 \text{ μm}$$

(b) Again, from Eq. (2.9):

▶

$$n_2^2 = n_1^2 - 2\Delta n_1^2 = 2.250 - (0.006 \times 2.250)$$
$$= 2.237$$

The cutoff wavelength for the single-mode fiber is given by Eq. (2.98) as:

$$\lambda_c = \frac{2\pi a n_1 (2\Delta)^{\frac{1}{2}}}{2.405}$$

$$= \frac{2\pi \times 4 \times 10^{-6} \times 1.500\ (0.06)^{\frac{1}{2}}}{2.405}$$

$$= 1.214\ \mu m$$

Substituting into Eq. (3.9) for the critical radius of curvature for the single-mode fiber gives:

$$R_{cs} \simeq \frac{20 \times 1.55 \times 10^{-6}}{(0.043)^{\frac{3}{2}}} \left(2.748 - \frac{0.996 \times 1.55 \times 10^{-6}}{1.214 \times 10^{-6}}\right)^{-3}$$

$$= 34\ mm$$

Example 3.4 shows that the critical radius of curvature for guided modes can be made extremely small (e.g. 9 µm), although this may be in conflict with the preferred design and operational characteristics. Nevertheless, for most practical purposes, the critical radius of curvature is relatively small (even when considering the case of a long-wavelength single-mode fiber, it was found to be around 34 mm) to avoid severe attenuation of the guided mode(s) at fiber bends. However, modes propagating close to cutoff, which are no longer fully guided within the fiber core, may radiate at substantially larger radii of curvature. Thus it is essential that sharp bends, with a radius of curvature approaching the critical radius, are avoided when optical fiber cables are installed. Finally, it is important that microscopic bends with radii of curvature approximating to the fiber radius are not produced in the fiber cabling process. These so-called microbends, which can cause significant losses from cabled fiber, are discussed further in Section 4.7.1.

3.7 Mid-infrared and far-infrared transmission

In the near-infrared region of the optical spectrum, fundamental silica fiber attenuation is dominated by Rayleigh scattering and multiphonon absorption from the infrared absorption edge (see Figure 3.2). Therefore, the total loss decreases as the operational transmission wavelength increases until a crossover point is reached around a wavelength of 1.55 µm where the total fiber loss again increases because at longer wavelengths the loss is dominated by the phonon absorption edge. Since the near fundamental attenuation limits for near-infrared silicate class fibers have been achieved, more recently researchers have turned their attention to the mid-infrared (2 to 5 µm) and the far-infrared (8 to 12 µm) optical wavelengths.

In order to obtain lower loss fibers it is necessary to produce glasses exhibiting longer infrared cutoff wavelengths. Potentially, much lower losses can be achieved if the transmission window of the material can be extended further into the infrared by utilizing constituent atoms of higher atomic mass and if it can be drawn into fiber exhibiting suitable strength and chemical durability. The reason for this possible loss reduction is due to Rayleigh scattering which displays a λ^{-4} dependence and hence becomes much reduced as the wavelength is increased. For example, the scattering loss is reduced by a factor of 16 when the optical wavelength is doubled. Thus it may be possible to obtain losses of the order of 0.01 dB km^{-1} at a wavelength of 2.55 μm, with even lower losses at wavelengths of between 3 and 5 μm [Ref. 21].

Candidate glass-forming systems for mid-infrared transmission are fluoride, fluoride–chloride, chalcogenide and oxide. In particular, oxide glasses such as Al_2O_3 (i.e. sapphire) offer a near equivalent transmittance range to many of the fluoride glasses and have benefits of high melting points, chemical inertness, and the ability to be readily melted and grown in air. Chalcogenide glasses, which generally comprise one or more elements Ge, Si, As and Sb, are capable of optical transmission in both the mid-infrared and far-infrared regions. A typical chalcogenide fiber glass is therefore arsenide trisulfide (As_2S_3). However, research activities into far-infrared transmission using chalcogenide glasses, halide glasses, polycrystalline halide fibers (e.g. silver and thallium) and hollow glass waveguides are primarily concerned with radiometry, infrared imaging, optical wireless, optical sensing and optical power transmission rather than telecommunications [Refs 22, 23].

Research activities into ultra-low-loss fibers for long-haul repeaterless communications in the 1980s and early 1990s centered on the fluorozirconates, with zirconium fluoride (ZrF_4) as the major constituent, and fluorides of barium, lanthanum, aluminum, gadolinium, sodium, lithium and occasionally lead added as modifiers and stabilizers [Ref. 24]. Such alkali additives improve the glass stability and working characteristics. Moreover, the two most popular heavy metal fluoride glasses for fabrication into fiber are fluorozirconate and fluoroaluminate glasses [Ref. 4]. Extensive work has been undertaken on a common fluorozirconate system comprising ZrF_4–BaF_2–LaF_3–AlF_3–NaF which forms ZBLAN, while an important fluoroaluminate comprises AlF_3–ZrF_4–BaF_2–CaF_3–YF_3. Although ZBLAN can theoretically provide for the lowest transmission losses over the mid-infrared wavelength region, it has a significantly lower glass transition (melting) temperature than the fluoroaluminate glass and is therefore less durable when subject to both thermal and mechanical perturbations.

The fabrication of low-loss, long-length fluoride fibers presents a basic problem with reducing the extrinsic losses which remains to be resolved [Refs 4, 25]. In practice, however, the most critical and difficult problems are associated with the minimization of the scattering losses resulting from extrinsic factors such as defects, waveguide imperfections and radiation caused by mechanical deformation. The estimated losses of around 0.01 dB km^{-1} at a wavelength of 2.55 μm for ZrF_4-based fibers are derived from an extrapolation of the intrinsic losses due to ultraviolet and infrared absorptions together with Rayleigh scattering [Ref. 21]. Moreover, refinements of scattering loss have increased this loss value slightly to 0.024 dB km^{-1} which is still around eight times lower than that of a silica fiber [Ref. 26]. Nevertheless, practical fiber losses remain much higher, as may be observed from the attenuation spectra for the common mid- and far-infrared fibers shown in Figure 3.6 in which the fluoride fiber (ZBLAN) is exhibiting a loss of several decibels per kilometer [Ref. 4].

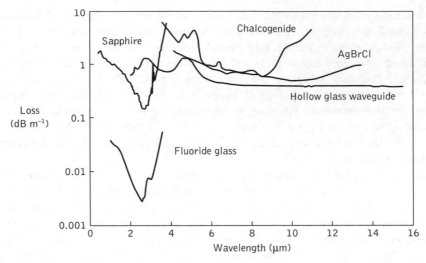

Figure 3.6 Attenuation spectra for some common mid- and far-infrared fibers [Ref. 4]

The loss spectrum for a single-crystal sapphire fiber which also transmits in the mid-infrared is also shown in Figure 3.6. Although they have robust physical properties, including a Young's modulus six times greater as well as a thermal expansion some ten times higher than that of silica, these fibers lend themselves to optical power delivery applications [Ref. 27], not specifically optical communications. Chalcogenide glasses which have their lowest losses over both the mid- and far-infrared ranges are very stable, durable and insensitive to moisture. Arsenic trisulfide fiber, being one of the simplest, has a spectral range from 0.7 to around 6 µm. Hence it has a cut off at long wavelength significantly before the chalcogenide fibers containing heavier elements such as Te, Ge and Se, an attenuation spectrum for the latter being incorporated in Figure 3.6. In general, chalcogenide glass fibers have proved to be useful in areas such as optical sensing, infrared imaging and for the production of fiber infrared lasers and amplifiers.

The loss spectrum for the polycrystalline fiber AgBrCl is also displayed in Figure 3.6. Although these fibers are transmissive over the entire far-infrared wavelength region and they were initially considered to hold significant potential as ultra-low-loss fibers because their intrinsic losses were estimated to be around 10^{-3} dB m^{-1} [Ref. 4], they are mechanically weak in comparison with silica fibers. In addition, the estimated low losses are far from being achieved, with experimental loss values being not even close to the predicted minimum as can be observed in Figure 3.6. Furthermore, polycrystalline fibers plastically deform resulting in increased transmission loss well before they fracture.

Finally, a hollow glass waveguide spectral characteristic is also shown in Figure 3.6. This hollow glass tube with a 530 µm bore was designed for optimum response at a transmission wavelength of 10 µm [Ref. 4]. Such hollow glass waveguides have been successfully employed for infrared laser power delivery at both 2.94 µm (Er:YAG laser) and 10.6 µm (CO_2 laser) [Ref. 28]. In summary, the remaining limitations of high loss (in comparison with theory) and low strength have inhibited the prospect of long-distance mid- or far-infrared transmission for communications for even the most promising fluoride fibers, while a range of alternative nontelecommunications applications for the various fiber and waveguide types have been developed.

3.8 Dispersion

Dispersion of the transmitted optical signal causes distortion for both digital and analog transmission along optical fibers. When considering the major implementation of optical fiber transmission which involves some form of digital modulation, then dispersion mechanisms within the fiber cause broadening of the transmitted light pulses as they travel along the channel. The phenomenon is illustrated in Figure 3.7, where it may be observed that each pulse broadens and overlaps with its neighbors, eventually becoming indistinguishable at the receiver input. The effect is known as intersymbol interference (ISI). Thus an increasing number of errors may be encountered on the digital optical channel as the ISI becomes more pronounced. The error rate is also a function of the signal attenuation on the link and the subsequent signal-to-noise ratio (SNR) at the receiver. This factor is not pursued further here but is considered in detail in Section 12.6.3. However, signal dispersion alone limits the maximum possible bandwidth attainable with a particular optical fiber to the point where individual symbols can no longer be distinguished.

For no overlapping of light pulses down on an optical fiber link the digital bit rate B_T must be less than the reciprocal of the broadened (through dispersion) pulse duration (2τ). Hence:

$$B_T \leq \frac{1}{2\tau} \tag{3.10}$$

This assumes that the pulse broadening due to dispersion on the channel is τ which dictates the input pulse duration which is also τ. Hence Eq. (3.10) gives a conservative estimate of the maximum bit rate that may be obtained on an optical fiber link as $1/2\tau$.

Another more accurate estimate of the maximum bit rate for an optical channel with dispersion may be obtained by considering the light pulses at the output to have a Gaussian shape with an rms width of σ. Unlike the relationship given in Eq. (3.10), this analysis allows for the existence of a certain amount of signal overlap on the channel, while avoiding any SNR penalty which occurs when ISI becomes pronounced. The maximum bit rate is given approximately by (see Appendix B):

$$B_T(\text{max}) \simeq \frac{0.2}{\sigma} \text{ bit s}^{-1} \tag{3.11}$$

It must be noted that certain sources [Refs 29, 30] give the constant term in the numerator of Eq. (3.11) as 0.25. However, we take the slightly more conservative estimate given, following Olshansky [Ref. 11] and Gambling et al. [Ref. 31]. Equation (3.11) gives a reasonably good approximation for other pulse shapes which may occur on the channel resulting from the various dispersive mechanisms within the fiber. Also, σ may be

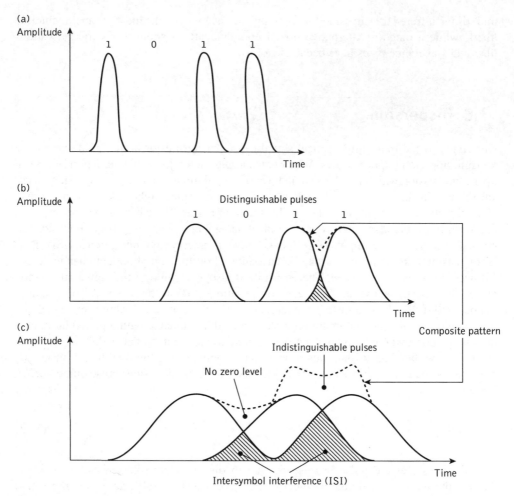

Figure 3.7 An illustration using the digital bit pattern 1011 of the broadening of light pulses as they are transmitted along a fiber: (a) fiber input; (b) fiber output at a distance L_1; (c) fiber output at a distance $L_2 > L_1$

assumed to represent the rms impulse response for the channel, as discussed further in Section 3.10.1.

The conversion of bit rate to bandwidth in hertz depends on the digital coding format used. For metallic conductors when a nonreturn-to-zero code is employed, the binary 1 level is held for the whole bit period τ. In this case there are two bit periods in one wavelength (i.e. 2 bits per second per hertz), as illustrated in Figure 3.8(a). Hence the maximum bandwidth B is one-half the maximum data rate or:

$$B_T(\text{max}) = 2B \tag{3.12}$$

However, when a return-to-zero code is considered, as shown in Figure 3.8(b), the binary 1 level is held for only part (usually half) of the bit period. For this signaling scheme the

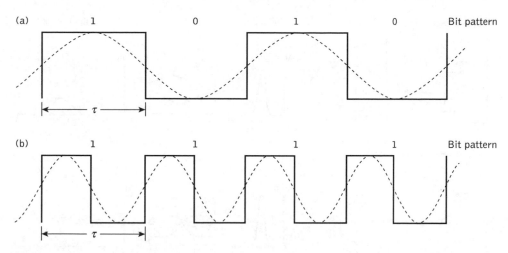

(a) ... 1 ... 0 ... 1 ... 0 ... Bit pattern

τ

(b) ... 1 ... 1 ... 1 ... 1 ... Bit pattern

τ

Figure 3.8 Schematic illustration of the relationships of the bit rate to wavelength for digital codes: (a) nonreturn-to-zero (NRZ); (b) return-to-zero (RZ)

data rate is equal to the bandwidth in hertz (i.e. 1 bit per second per hertz) and thus $B_T = B$. The bandwidth B for metallic conductors is also usually defined by the electrical 3 dB points (i.e. the frequencies at which the electric power has dropped to one-half of its constant maximum value). However, when the 3 dB optical bandwidth of a fiber is considered it is significantly larger than the corresponding 3 dB electrical bandwidth for the reasons discussed in Section 7.4.3. Hence, when the limitations in the bandwidth of a fiber due to dispersion are stated (i.e. optical bandwidth B_{opt}), it is usually with regard to a return to zero code where the bandwidth in hertz is considered equal to the digital bit rate. Within the context of dispersion the bandwidths expressed in this chapter will follow this general criterion unless otherwise stated. However, as is made clear in Section 7.4.3, when electro-optic devices and optical fiber systems are considered it is more usual to state the electrical 3 dB bandwidth, this being the more useful measurement when interfacing an optical fiber link to electrical terminal equipment. Unfortunately, the terms of bandwidth measurement are not always made clear and the reader must be warned that this omission may lead to some confusion when specifying components and materials for optical fiber communication systems.

Figure 3.9 shows the three common optical fiber structures, namely multimode step index, multimode graded index and single-mode step index, while diagrammatically illustrating the respective pulse broadening associated with each fiber type. It may be observed that the multimode step index fiber exhibits the greatest dispersion of a transmitted light pulse and the multimode graded index fiber gives a considerably improved performance. Finally, the single-mode fiber gives the minimum pulse broadening and thus is capable of the greatest transmission bandwidths which are currently in the gigahertz range, whereas transmission via multimode step index fiber is usually limited to bandwidths of a few tens of megahertz. However, the amount of pulse broadening is dependent upon the distance the pulse travels within the fiber, and hence for a given optical fiber link the restriction on usable bandwidth is dictated by the distance between regenerative repeaters (i.e. the

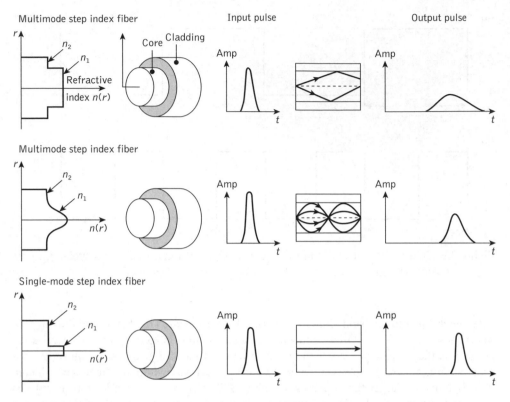

Figure 3.9 Schematic diagram showing a multimode step index fiber, multimode graded index fiber and single-mode step index fiber, and illustrating the pulse broadening due to intermodal dispersion in each fiber type

distance the light pulse travels before it is reconstituted). Thus the measurement of the dispersive properties of a particular fiber is usually stated as the pulse broadening in time over a unit length of the fiber (i.e. ns km^{-1}).

Hence, the number of optical signal pulses which may be transmitted in a given period, and therefore the information-carrying capacity of the fiber, is restricted by the amount of pulse dispersion per unit length. In the absence of mode coupling or filtering, the pulse broadening increases linearly with fiber length and thus the bandwidth is inversely proportional to distance. This leads to the adoption of a more useful parameter for the information-carrying capacity of an optical fiber which is known as the bandwidth–length product (i.e. $B_{opt} \times L$). The typical best bandwidth–length products for the three fibers shown in Figure 3.9 are 20 MHz km, 1 GHz km and 100 GHz km for multimode step index, multimode graded index and single-mode step index fibers respectively.

In order to appreciate the reasons for the different amounts of pulse broadening within the various types of optical fiber, it is necessary to consider the dispersive mechanisms involved. These include material dispersion, waveguide dispersion, intermodal dispersion and profile dispersion which are considered in the following sections.

Example 3.5

A multimode graded index fiber exhibits total pulse broadening of 0.1 μs over a distance of 15 km. Estimate:

(a) the maximum possible bandwidth on the link assuming no intersymbol interference;

(b) the pulse dispersion per unit length;

(c) the bandwidth–length product for the fiber.

Solution: (a) The maximum possible optical bandwidth which is equivalent to the maximum possible bit rate (for return to zero pulses) assuming no ISI may be obtained from Eq. (3.10), where:

$$B_{opt} = B_T = \frac{1}{2\tau} = \frac{1}{0.2 \times 10^{-6}} = 5 \text{ MHz}$$

(b) The dispersion per unit length may be acquired simply by dividing the total dispersion by the total length of the fiber:

$$\text{Dispersion} = \frac{0.1 \times 10^{-6}}{15} = 6.67 \text{ ns km}^{-1}$$

(c) The bandwidth–length product may be obtained in two ways. Firstly by simply multiplying the maximum bandwidth for the fiber link by its length. Hence:

$$B_{opt}L = 5 \text{ MHz} \times 15 \text{ km} = 75 \text{ MHz km}$$

Alternatively, it may be obtained from the dispersion per unit length using Eq. (3.10) where:

$$B_{opt}L = \frac{1}{2 \times 6.67 \times 10^{-6}} = 75 \text{ MHz km}$$

3.9 Chromatic dispersion

Chromatic or intramodal dispersion may occur in all types of optical fiber and results from the finite spectral linewidth of the optical source. Since optical sources do not emit just a single frequency but a band of frequencies (in the case of the injection laser corresponding to only a fraction of a percent of the center frequency, whereas for the LED it is likely to be a significant percentage), then there may be propagation delay differences between the different spectral components of the transmitted signal. This causes broadening of each transmitted mode and hence intramodal dispersion. The delay differences may be caused by the dispersive properties of the waveguide material (material dispersion) and also guidance effects within the fiber structure (waveguide dispersion).

3.9.1 Material dispersion

Pulse broadening due to material dispersion results from the different group velocities of the various spectral components launched into the fiber from the optical source. It occurs when the phase velocity of a plane wave propagating in the dielectric medium varies non-linearly with wavelength, and a material is said to exhibit material dispersion when the second differential of the refractive index with respect to wavelength is not zero (i.e. $d^2n/d\lambda^2 \neq 0$). The pulse spread due to material dispersion may be obtained by considering the group delay τ_g in the optical fiber which is the reciprocal of the group velocity v_g defined by Eqs (2.37) and (2.40). Hence the group delay is given by:

$$\tau_g = \frac{d\beta}{d\omega} = \frac{1}{c}\left(n_1 - \lambda\frac{dn_1}{d\lambda}\right) \tag{3.13}$$

where n_1 is the refractive index of the core material. The pulse delay τ_m due to material dispersion in a fiber of length L is therefore:

$$\tau_m = \frac{L}{c}\left(n_1 - \lambda\frac{dn_1}{d\lambda}\right) \tag{3.14}$$

For a source with rms spectral width σ_λ and a mean wavelength λ, the rms pulse broadening due to material dispersion σ_m may be obtained from the expansion of Eq. (3.14) in a Taylor series about λ where:

$$\sigma_m = \sigma_\lambda\frac{d\tau_m}{d\lambda} + \sigma_\lambda\frac{2d^2\tau_m}{d\lambda^2} + \ldots \tag{3.15}$$

As the first term in Eq. (3.15) usually dominates, especially for sources operating over the 0.8 to 0.9 μm wavelength range, then:

$$\sigma_m \simeq \sigma_\lambda\frac{d\tau_m}{d\lambda} \tag{3.16}$$

Hence the pulse spread may be evaluated by considering the dependence of τ_m on λ, where from Eq. (3.14):

$$\frac{d\tau_m}{d\lambda} = \frac{L\lambda}{c}\left[\frac{dn_1}{d\lambda} - \frac{d^2n_1}{d\lambda^2} - \frac{dn_1}{d\lambda}\right]$$

$$= \frac{-L\lambda}{c}\frac{d^2n_1}{d\lambda^2} \tag{3.17}$$

Therefore, substituting the expression obtained in Eq. (3.17) into Eq. (3.16), the rms pulse broadening due to material dispersion is given by:

$$\sigma_m \simeq \frac{\sigma_\lambda L}{c} \left| \lambda \frac{d^2 n_1}{d\lambda^2} \right| \tag{3.18}$$

The material dispersion for optical fibers is sometimes quoted as a value for $|\lambda^2(d^2 n_1/d\lambda^2)|$ or simply $|d^2 n_1/d\lambda^2|$.

However, it may be given in terms of a material dispersion parameter M which is defined as:

$$M = \frac{1}{L} \frac{d\tau_m}{d\lambda} = \frac{\lambda}{c} \left| \frac{d^2 n_1}{d\lambda^2} \right| \tag{3.19}$$

and which is often expressed in units of ps nm^{-1} km^{-1}.

Example 3.6

A glass fiber exhibits material dispersion given by $|\lambda^2(d^2 n_1/d\lambda^2)|$ of 0.025. Determine the material dispersion parameter at a wavelength of 0.85 μm, and estimate the rms pulse broadening per kilometer for a good LED source with an rms spectral width of 20 nm at this wavelength.

Solution: The material dispersion parameter may be obtained from Eq. (3.19):

$$M = \frac{\lambda}{c} \left| \frac{d^2 n_1}{d\lambda^2} \right| = \frac{1}{c\lambda} \left| \lambda^2 \frac{d^2 n_1}{d\lambda^2} \right|$$

$$= \frac{0.025}{2.998 \times 10^5 \times 850} \text{ s nm}^{-1} \text{ km}^{-1}$$

$$= 98.1 \text{ ps nm}^{-1} \text{ km}^{-1}$$

The rms pulse broadening is given by Eq. (3.18) as:

$$\sigma_m \simeq \frac{\sigma_\lambda L}{c} \left| \lambda \frac{d^2 n_1}{d\lambda^2} \right|$$

Therefore in terms of the material dispersion parameter M defined by Eq. (3.19):

$$\sigma_m \simeq \sigma_\lambda L M$$

Hence, the rms pulse broadening per kilometer due to material dispersion:

$$\sigma_m(1 \text{ km}) = 20 \times 1 \times 98.1 \times 10^{-12} = 1.96 \text{ ns km}^{-1}$$

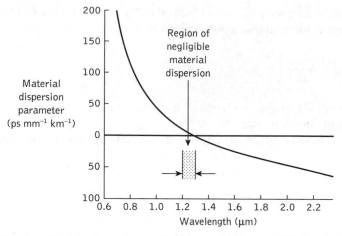

Figure 3.10 The material dispersion parameter for silica as a function of wavelength. Reproduced with permission from D. N. Payne and W. A. Gambling, *Electron. Lett.*, 11, p. 176, 1975

Figure 3.10 shows the variation of the material dispersion parameter M with wavelength for pure silica [Ref. 32]. It may be observed that the material dispersion tends to zero in the longer wavelength region around 1.3 μm (for pure silica). This provides an additional incentive (other than low attenuation) for operation at longer wavelengths where the material dispersion may be minimized. Also, the use of an injection laser with a narrow spectral width rather than an LED as the optical source leads to a substantial reduction in the pulse broadening due to material dispersion, even in the shorter wavelength region.

Example 3.7

Estimate the rms pulse broadening per kilometer for the fiber in Example 3.6 when the optical source used is an injection laser with a relative spectral width σ_λ/λ of 0.0012 at a wavelength of 0.85 μm.

Solution: The rms spectral width may be obtained from the relative spectral width by:

$$\sigma_\lambda = 0.0012\lambda = 0.0012 \times 0.85 \times 10^{-6}$$
$$= 1.02 \text{ nm}$$

The rms pulse broadening in terms of the material dispersion parameter following Example 3.6 is given by:

$$\sigma_m \simeq \sigma_\lambda L M$$

Therefore, the rms pulse broadening per kilometer due to material dispersion is:

$$\sigma_m \simeq 1.02 \times 1 \times 98.1 \times 10^{-12} = 0.10 \text{ ns km}^{-1}$$

Hence, in this example the rms pulse broadening is reduced by a factor of around 20 (i.e. equivalent to the reduced rms spectral width of the injection laser source) compared with that obtained with the LED source of Example 3.6.

3.9.2 Waveguide dispersion

The waveguiding of the fiber may also create chromatic dispersion. This results from the variation in group velocity with wavelength for a particular mode. Considering the ray theory approach, it is equivalent to the angle between the ray and the fiber axis varying with wavelength which subsequently leads to a variation in the transmission times for the rays, and hence dispersion. For a single mode whose propagation constant is β, the fiber exhibits waveguide dispersion when $d^2\beta/d\lambda^2 \neq 0$. Multimode fibers, where the majority of modes propagate far from cutoff, are almost free of waveguide dispersion and it is generally negligible compared with material dispersion (≈ 0.1 to 0.2 ns km^{-1}) [Ref. 32]. However, with single-mode fibers where the effects of the different dispersion mechanisms are not easy to separate, waveguide dispersion may be significant (see Section 3.11.2).

3.10 Intermodal dispersion

Pulse broadening due to intermodal dispersion (sometimes referred to simply as modal or mode dispersion) results from the propagation delay differences between modes within a multimode fiber. As the different modes which constitute a pulse in a multimode fiber travel along the channel at different group velocities, the pulse width at the output is dependent upon the transmission times of the slowest and fastest modes. This dispersion mechanism creates the fundamental difference in the overall dispersion for the three types of fiber shown in Figure 3.9. Thus multimode step index fibers exhibit a large amount of intermodal dispersion which gives the greatest pulse broadening. However, intermodal dispersion in multimode fibers may be reduced by adoption of an optimum refractive index profile which is provided by the near-parabolic profile of most graded index fibers. Hence, the overall pulse broadening in multimode graded index fibers is far less than that obtained in multimode step index fibers (typically by a factor of 100). Thus graded index fibers used with a multimode source give a tremendous bandwidth advantage over multimode step index fibers.

Under purely single-mode operation there is no intermodal dispersion and therefore pulse broadening is solely due to the intramodal dispersion mechanisms. In theory, this is the case with single-mode step index fibers where only a single mode is allowed to propagate. Hence they exhibit the least pulse broadening and have the greatest possible bandwidths, but in general are only usefully operated with single-mode sources.

In order to obtain a simple comparison for intermodal pulse broadening between multi-mode step index and multimode graded index fibers, it is useful to consider the geometric optics picture for the two types of fiber.

3.10.1 Multimode step index fiber

Using the ray theory model, the fastest and slowest modes propagating in the step index fiber may be represented by the axial ray and the extreme meridional ray (which is incident at the core–cladding interface at the critical angle ϕ_c) respectively. The paths taken by these two rays in a perfectly structured step index fiber are shown in Figure 3.11. The delay difference between these two rays when traveling in the fiber core allows estimation of the pulse broadening resulting from intermodal dispersion within the fiber. As both rays are traveling at the same velocity within the constant refractive index fiber core, then the delay difference is directly related to their respective path lengths within the fiber. Hence the time taken for the axial ray to travel along a fiber of length L gives the minimum delay time T_{Min} and:

$$T_{\text{Min}} = \frac{\text{distance}}{\text{velocity}} = \frac{L}{(c/n_1)} = \frac{Ln_1}{c} \qquad (3.20)$$

where n_1 is the refractive index of the core and c is the velocity of light in a vacuum.

The extreme meridional ray exhibits the maximum delay time T_{Max} where:

$$T_{\text{Max}} = \frac{L/\cos\theta}{c/n_1} = \frac{Ln_1}{c\cos\theta} \qquad (3.21)$$

Using Snell's law of refraction at the core–cladding interface following Eq. (2.2):

$$\sin\phi_c = \frac{n_2}{n_1} = \cos\theta \qquad (3.22)$$

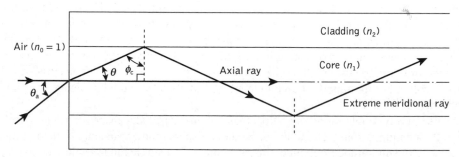

Figure 3.11 The paths taken by the axial and an extreme meridional ray in a perfect multimode step index fiber

where n_2 is the refractive index of the cladding. Furthermore, substituting into Eq. (3.21) for cos θ gives:

$$T_{\text{Max}} = \frac{Ln_1^2}{cn_2} \qquad (3.23)$$

The delay difference δT_s between the extreme meridional ray and the axial ray may be obtained by subtracting Eq. (3.20) from Eq. (3.23). Hence:

$$\delta T_s = T_{\text{Max}} - T_{\text{Min}} = \frac{Ln_1^2}{cn_2} - \frac{Ln_1}{c}$$

$$= \frac{Ln_1^2}{cn_2}\left(\frac{n_1 - n_2}{n_1}\right) \qquad (3.24)$$

$$\simeq \frac{Ln_1^2\Delta}{cn_2} \quad \text{when } \Delta \ll 1 \qquad (3.25)$$

where Δ is the relative refractive index difference. However, when $\Delta \ll 1$, then from the definition given by Eq. (2.9), the relative refractive index difference may also be given approximately by:

$$\Delta \simeq \frac{n_1 - n_2}{n_2} \qquad (3.26)$$

Hence rearranging Eq. (3.24):

$$\delta T_s = \frac{Ln_1}{c}\left(\frac{n_1 - n_2}{n_2}\right) \simeq \frac{Ln_1\Delta}{c} \qquad (3.27)$$

Also substituting for Δ from Eq. (2.10) gives:

$$\delta T_s \simeq \frac{L(NA)^2}{2n_1c} \qquad (3.28)$$

where NA is the numerical aperture for the fiber. The approximate expressions for the delay difference given in Eqs (3.27) and (3.28) are usually employed to estimate the maximum pulse broadening in time due to intermodal dispersion in multimode step index fibers. It must be noted that this simple analysis only considers pulse broadening due to meridional rays and totally ignores skew rays with acceptance angles $\theta_{as} > \theta_a$ (see Section 2.2.4).

Again considering the perfect step index fiber, another useful quantity with regard to intermodal dispersion on an optical fiber link is the rms pulse broadening resulting from this dispersion mechanism along the fiber. When the optical input to the fiber is a pulse $p_i(t)$ of unit area, as illustrated in Figure 3.12, then [Ref. 33]:

$$\int_{-\infty}^{\infty} p_i(t)\,dt = 1 \qquad (3.29)$$

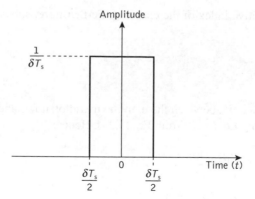

Figure 3.12 An illustration of the light input to the multimode step index fiber consisting of an ideal pulse or rectangular function with unit area

It may be noted that $p_i(t)$ has a constant amplitude of $1/\delta T_s$ over the range:

$$\frac{-\delta T_s}{2} \le p(t) \le \frac{\delta T_s}{2}$$

The rms pulse broadening at the fiber output due to intermodal dispersion for the multimode step index fiber σ_s (i.e. the standard deviation) may be given in terms of the variance σ_s^2 as (see Appendix C):

$$\sigma_s^2 = M_2 - M_1^2 \tag{3.30}$$

where M_1 is the first temporal moment which is equivalent to the mean value of the pulse and M_2, the second temporal moment, is equivalent to the mean square value of the pulse. Hence:

$$M_1 = \int_{-\infty}^{\infty} t p_i(t)\, \mathrm{d}t \tag{3.31}$$

and:

$$M_2 = \int_{-\infty}^{\infty} t^2 p_i(t)\, \mathrm{d}t \tag{3.32}$$

The mean value M_1 for the unit input pulse of Figure 3.12 is zero, and assuming this is maintained for the output pulse, then from Eqs (3.30) and (3.32):

$$\sigma_s^2 = M_2 = \int_{-\infty}^{\infty} t^2 p_i(t)\, \mathrm{d}t \tag{3.33}$$

Integrating over the limits of the input pulse (Figure 3.12) and substituting for $p_i(t)$ in Eq. (3.33) over this range gives:

$$\sigma_s^2 = \int_{-\delta T_s/2}^{\delta T_s/2} \frac{1}{\delta T_s} t^2 \, dt$$

$$= \frac{1}{\delta T_s} \left[\frac{t^3}{3} \right]_{-\delta T_s/2}^{\delta T_s/2} = \frac{1}{3} \left(\frac{\delta T_s}{2} \right)^2 \qquad (3.34)$$

Hence substituting from Eq. (3.27) for δT_s gives:

$$\sigma_s \simeq \frac{L n_1 \Delta}{2\sqrt{3}c} \simeq \frac{L(NA)^2}{4\sqrt{3}n_1 c} \qquad (3.35)$$

Equation (3.35) allows estimation of the rms impulse response of a multimode step index fiber if it is assumed that intermodal dispersion dominates and there is a uniform distribution of light rays over the range $0 \le \theta \le \theta_a$. The pulse broadening is directly proportional to the relative refractive index difference Δ and the length of the fiber L. The latter emphasizes the bandwidth–length trade-off that exists, especially with multimode step index fibers, and which inhibits their use for wideband long-haul (between repeaters) systems. Furthermore, the pulse broadening is reduced by reduction of the relative refractive index difference Δ for the fiber. This suggests that weakly guiding fibers (see Section 2.4.1) with small Δ are best for low-dispersion transmission. However, as may be seen from Eq. (3.35) this is also subject to a trade-off as a reduction in Δ reduces the acceptance angle θ_a and the NA, thus worsening the launch conditions.

Example 3.8

A 6 km optical link consists of multimode step index fiber with a core refractive index of 1.5 and a relative refractive index difference of 1%. Estimate:

(a) the delay difference between the slowest and fastest modes at the fiber output;

(b) the rms pulse broadening due to intermodal dispersion on the link;

(c) the maximum bit rate that may be obtained without substantial errors on the link assuming only intermodal dispersion;

(d) the bandwidth–length product corresponding to (c).

Solution: (a) The delay difference is given by Eq. (3.27) as:

$$\delta T_s \simeq \frac{L n_1 \Delta}{c} = \frac{6 \times 10^3 \times 1.5 \times 0.01}{2.998 \times 10^8}$$

$$= 300 \text{ ns}$$

▶

(b) The rms pulse broadening due to intermodal dispersion may be obtained from Eq. (3.35) where:

$$\sigma_s = \frac{Ln_1\Delta}{2\sqrt{3}c} = \frac{1}{2\sqrt{3}} \frac{6 \times 10^3 \times 1.5 \times 0.01}{2.998 \times 10^8}$$

$$= 86.7 \text{ ns}$$

(c) The maximum bit rate may be estimated in two ways. Firstly, to get an idea of the maximum bit rate when assuming no pulse overlap, Eq. (3.10) may be used where:

$$B_T(\text{max}) = \frac{1}{2\tau} = \frac{1}{2\delta T_s} = \frac{1}{600 \times 10^{-9}}$$

$$= 1.7 \text{ Mbit s}^{-1}$$

Alternatively an improved estimate may be obtained using the calculated rms pulse broadening in Eq. (3.11) where:

$$B_T(\text{max}) = \frac{0.2}{\sigma_s} = \frac{0.2}{86.7 \times 10^{-9}}$$

$$= 2.3 \text{ Mbit s}^{-1}$$

(d) Using the most accurate estimate of the maximum bit rate from (c), and assuming return to zero pulses, the bandwidth–length product is:

$$B_{\text{opt}} \times L = 2.3 \text{ MHz} \times 6 \text{ km} = 13.8 \text{ MHz km}$$

Intermodal dispersion may be reduced by propagation mechanisms within practical fibers. For instance, there is differential attenuation of the various modes in a step index fiber. This is due to the greater field penetration of the higher order modes into the cladding of the waveguide. These slower modes therefore exhibit larger losses at any core–cladding irregularities, which tends to concentrate the transmitted optical power into the faster lower order modes. Thus the differential attenuation of modes reduces intermodal pulse broadening on a multimode optical link.

Another mechanism which reduces intermodal pulse broadening in nonperfect (i.e. practical) multimode fibers is the mode coupling or mixing discussed in Section 2.4.2. The coupling between guided modes transfers optical power from the slower to the faster modes, and vice versa. Hence, with strong coupling the optical power tends to be transmitted at an average speed, which is the mean of the various propagating modes. This reduces the intermodal dispersion on the link and makes it advantageous to encourage mode coupling within multimode fibers.

The expression for delay difference given in Eq. (3.27) for a perfect step index fiber may be modified for the fiber with mode coupling among all guided modes to [Ref. 34]:

$$\delta T_{sc} \simeq \frac{n_1 \Delta}{c} (LL_c)^{\frac{1}{2}} \tag{3.36}$$

where L_c is a characteristic length for the fiber which is inversely proportional to the coupling strength. Hence, the delay difference increases at a slower rate proportional to $(LL_c)^{\frac{1}{2}}$ instead of the direct proportionality to L given in Eq. (3.27). However, the most successful technique for reducing intermodal dispersion in multimode fibers is by grading the core refractive index to follow a near-parabolic profile. This has the effect of equalizing the transmission times of the various modes as discussed in the following section.

3.10.2 Multimode graded index fiber

Intermodal dispersion in multimode fibers is minimized with the use of graded index fibers. Hence, multimode graded index fibers show substantial bandwidth improvement over multimode step index fibers. The reason for the improved performance of graded index fibers may be observed by considering the ray diagram for a graded index fiber shown in Figure 3.13. The fiber shown has a parabolic index profile with a maximum at the core axis, as illustrated in Figure 3.13(a). Analytically, the index profile is given by Eq. (2.75) with $\alpha = 2$ as:

$$n(r) = n_1[1 - 2\Delta(r/a)^2]^{\frac{1}{2}} \qquad r < a \text{ (core)} \tag{3.37}$$
$$= n_1(1 - 2\Delta)^{\frac{1}{2}} = n_2 \qquad r \geq a \text{ (cladding)}$$

Figure 3.13(b) shows several meridional ray paths within the fiber core. It may be observed that apart from the axial ray, the meridional rays follow sinusoidal trajectories of different path lengths which result from the index grading, as was discussed in Section 2.4.4. However, following Eq. (2.40) the local group velocity is inversely proportional to

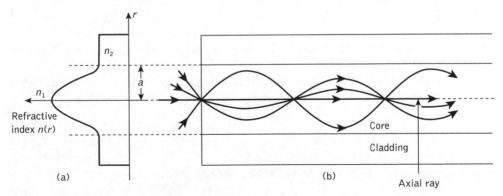

Figure 3.13 A multimode graded index fiber: (a) parabolic refractive index profile; (b) meridional ray paths within the fiber core

the local refractive index and therefore the longer sinusoidal paths are compensated for by higher speeds in the lower index medium away from the axis. Hence there is an equalization of the transmission times of the various trajectories towards the transmission time of the axial ray which travels exclusively in the high-index region at the core axis, and at the slowest speed. As these various ray paths may be considered to represent the different modes propagating in the fiber, then the graded profile reduces the disparity in the mode transit times.

The dramatic improvement in multimode fiber bandwidth achieved with a parabolic or near-parabolic refractive index profile is highlighted by consideration of the reduced delay difference between the fastest and slowest modes for this graded index fiber δT_g. Using a ray theory approach the delay difference is given by [Ref. 35]:

$$\delta T_g \simeq \frac{Ln_1\Delta^2}{2c} \simeq \frac{(NA)^4}{8n_1^3c} \tag{3.38}$$

As in the step index case, Eq. (2.10) is used for conversion between the two expressions shown.

However, a more rigorous analysis using electromagnetic mode theory gives an absolute temporal width at the fiber output of [Refs 36, 37]:

$$\delta T_g = \frac{Ln_1\Delta^2}{8c} \tag{3.39}$$

which corresponds to an increase in transmission time for the slowest mode of $\Delta^2/8$ over the fastest mode. The expression given in Eq. (3.39) does not restrict the bandwidth to pulses with time slots corresponding to δT_g as 70% of the optical power is concentrated in the first half of the interval. Hence the rms pulse broadening is a useful parameter for assessment of intermodal dispersion in multimode graded index fibers. It may be shown [Ref. 37] that the rms pulse broadening of a near-parabolic index profile graded index fiber σ_g is reduced compared with similar broadening for the corresponding step index fiber σ_s (i.e. with the same relative refractive index difference) following:

$$\sigma_g = \frac{\Delta}{D}\sigma_s \tag{3.40}$$

where D is a constant between 4 and 10 depending on the precise evaluation and the exact optimum profile chosen.

The best minimum theoretical intermodal rms pulse broadening for a graded index fiber with an optimum characteristic refractive index profile for the core α_{op} of [Refs 37, 38]:

$$\alpha_{op} = 2 - \frac{12\Delta}{5} \tag{3.41}$$

is given by combining Eqs (3.27) and (3.40) as [Refs 31, 38]:

$$\sigma_g = \frac{Ln_1\Delta^2}{20\sqrt{3}c} \tag{3.42}$$

Example 3.9

Compare the rms pulse broadening per kilometer due to intermodal dispersion for the multimode step index fiber of Example 3.8 with the corresponding rms pulse broadening for an optimum near-parabolic profile graded index fiber with the same core axis refractive index and relative refractive index difference.

Solution: In Example 3.8, σ_s over 6 km of fiber is 86.7 ns. Hence the rms pulse broadening per kilometer for the multimode step index fiber is:

$$\frac{\sigma_s(1 \text{ km})}{L} = \frac{86.7}{6} = 14.4 \text{ ns km}^{-1}$$

Using Eq. (3.42), the rms pulse broadening per kilometer for the corresponding graded index fiber is:

$$\sigma_g(1 \text{ km}) = \frac{Ln_1\Delta^2}{20\sqrt{3}c} = \frac{10^3 \times 1.5 \times (0.01)^2}{20\sqrt{3} \times 2.998 \times 10^8}$$

$$= 14.4 \text{ ps km}^{-1}$$

Hence, from Example 3.9, the theoretical improvement factor of the graded index fiber in relation to intermodal rms pulse broadening is 1000. However, this level of improvement is not usually achieved in practice due to difficulties in controlling the refractive index profile radially over long lengths of fiber. Any deviation in the refractive index profile from the optimum results in increased intermodal pulse broadening. This may be observed from the curve shown in Figure 3.14, which gives the variation in intermodal pulse broadening (δT_g) as a function of the characteristic refractive index profile α for typical graded

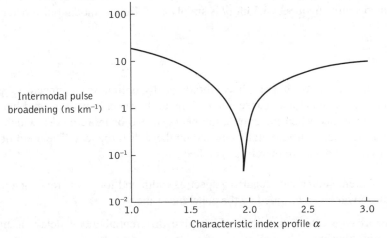

Figure 3.14 The intermodal pulse broadening δT_g for graded index fibers having $\Delta = 1\%$, against the characteristic refractive index profile α

index fibers (where $\Delta = 1\%$). The curve displays a sharp minimum at a characteristic refractive index profile slightly less than 2 ($\alpha = 1.98$). This corresponds to the optimum value of α in order to minimize intermodal dispersion. Furthermore, the extreme sensitivity of the intermodal pulse broadening to slight variations in α from this optimum value is evident. Thus at present improvement factors for practical graded index fibers over corresponding step index fibers with regard to intermodal dispersion are around 100 [Ref. 36].

Another important factor in the determination of the optimum refractive index profile for a graded index fiber is the dispersion incurred due to the difference in refractive index between the fiber core and cladding. It results from a variation in the refractive index profile with optical wavelength in the graded fiber and is often given by a profile dispersion parameter $d\Delta/d\lambda$. Thus the optimized profile at a given wavelength is not necessarily optimized at another wavelength. As all optical fiber sources (e.g. injection lasers and LEDs) have a finite spectral width, the profile shape must be altered to compensate for this dispersion mechanism. Moreover, the minimum overall dispersion for graded index fiber is also limited by the other intramodal dispersion mechanisms (i.e. material and waveguide dispersion). These give temporal pulse broadening of around 0.08 and 1 ns km^{-1} with injection lasers and LEDs respectively. Therefore, practical pulse broadening values for graded index fibers lie in the range 0.2 to 1 ns km^{-1}. This gives bandwidth–length products of between 0.5 and 2.5 GHz km when using lasers and optimum profile fiber.

3.10.3 Modal noise

The intermodal dispersion properties of multimode optical fibers (see Sections 3.10.1 and 3.10.2) create another phenomenon which affects the transmitted signals on the optical channel. It is exhibited within the speckle patterns observed in multimode fiber as fluctuations which have characteristic times longer than the resolution time of the detector, and is known as modal or speckle noise. The speckle patterns are formed by the interference of the modes from a coherent source when the coherence time of the source is greater than the intermodal dispersion time δT within the fiber. The coherence time for a source with uncorrelated source frequency width δf is simply $1/\delta f$. Hence, modal noise occurs when:

$$\delta f \gg \frac{1}{\delta T} \qquad\qquad (3.43)$$

Disturbances along the fiber such as vibrations, discontinuities, connectors, splices and source/detector coupling may cause fluctuations in the speckle patterns and hence modal noise. It is generated when the correlation between two or more modes which gives the original interference is differentially delayed by these disturbances. The conditions which give rise to modal noise are therefore specified as:

(a) a coherent source with a narrow spectral width and long coherence length (propagation velocity multiplied by the coherence time);

(b) disturbances along the fiber which give differential mode delay or modal and spatial filtering;

(c) phase correlation between the modes.

Measurements [Ref. 39] of rms signal to modal noise ratio using good narrow-linewidth injection lasers show large signal-to-noise ratio penalties under the previously mentioned conditions. The measurements were carried out by misaligning connectors to create disturbances. They gave carrier to noise ratios reduced by around 10 dB when the attenuation at each connector was 20 dB due to substantial axial misalignment.

Modal noise may be avoided by removing one of the conditions (they must all be present) which give rise to this degradation. Hence modal-noise-free transmission may be obtained by the following:

1. The use of a broad spectrum source in order to eliminate the modal interference effects. This may be achieved by either (a) increasing the width of the single longitudinal mode and hence decreasing its coherence time or (b) by increasing the number of longitudinal modes and averaging out of the interference patterns [Ref. 40].

2. In conjunction with 1(b) it is found that fibers with large numerical apertures support the transmission of a large number of modes giving a greater number of speckles, and hence reduce the modal noise generating effect of individual speckles [Ref. 41].

3. The use of single-mode fiber which does not support the transmission of different modes and thus there is no intermodal interference.

4. The removal of disturbances along the fiber. This has been investigated with regard to connector design [Ref. 42] in order to reduce the shift in speckle pattern induced by mechanical vibration and fiber misalignment.

Hence, modal noise may be prevented on an optical fiber link through suitable choice of the system components. However, this may not always be possible and then certain levels of modal noise must be tolerated. This tends to be the case on high-quality analog optical fiber links where multimode injection lasers are frequently used. Analog transmission is also more susceptible to modal noise due to the higher optical power levels required at the receiver when quantum noise effects are considered (see Section 10.2.5). Therefore, it is important that modal noise is taken into account within the design considerations for these systems.

Modal noise, however, can be present in single-mode fiber links when propagation of the two fundamental modes with orthogonal polarization is allowed or, alternatively, when the second-order modes* are not sufficiently attenuated. The former modal noise type, which is known as polarization modal noise, is outlined in Section 3.13.1. For the latter type, it is apparent that at shorter wavelengths, a nominally single-mode fiber can also guide four second-order LP modes (see Section 2.4.1). Modal noise can therefore be introduced into single-mode fiber systems by time-varying interference between the LP_{01} and the LP_{11} modes when the fiber is operated at a wavelength which is smaller than the cutoff wavelength of the second-order modes. The effect has been observed in overmoded single-mode fibers [Ref. 43] and may be caused by a number of conditions. In particular

---

* In addition to the two orthogonal LP_{01} modes, at shorter wavelengths 'single-mode' fiber can propagate four LP_{11} modes.

the insertion of a short jumper cable or repair section, with a lateral offset, in a long single-mode fiber can excite the second-order LP_{11} mode [Rcf. 44]. Moreover, such a repair section can also attenuate the fundamental LP_{01} mode if its operating wavelength is near the cutoff wavelength for this mode. Hence, to reduce modal noise, repair sections should use special fibers with a lower value of cutoff wavelength than that in the long single-mode fiber link; also offsets at joints should be minimized.

3.11 Overall fiber dispersion

3.11.1 Multimode fibers

The overall dispersion in multimode fibers comprises both chromatic and intermodal terms. The total rms pulse broadening σ_T is given (see Appendix D) by:

$$\sigma_T = (\sigma_c^2 + \sigma_n^2)^{\frac{1}{2}} \tag{3.44}$$

where σ_c is the intramodal or chromatic broadening and σ_n is the intermodal broadening caused by delay differences between the modes (i.e. σ_s for multimode step index fiber and σ_g for multimode graded index fiber). The chromatic term σ_c consists of pulse broadening due to both material and waveguide dispersion. However, since waveguide dispersion is generally negligible compared with material dispersion in multimode fibers, then $\sigma_c \simeq \sigma_m$.

Example 3.10

A multimode step index fiber has a numerical aperture of 0.3 and a core refractive index of 1.45. The material dispersion parameter for the fiber is 250 ps nm^{-1} km^{-1} which makes material dispersion the totally dominating chromatic dispersion mechanism. Estimate (a) the total rms pulse broadening per kilometer when the fiber is used with an LED source of rms spectral width 50 nm and (b) the corresponding bandwidth–length product for the fiber.

Solution: (a) The rms pulse broadening per kilometer due to material dispersion may be obtained from Eq. (3.18), where:

$$\sigma_m(1 \text{ km}) \simeq \frac{\sigma_\lambda L \lambda}{c} \left| \frac{d^2 n_1}{d\lambda^2} \right| = \sigma_\lambda L M = 50 \times 1 \times 250 \text{ ps km}^{-1}$$
$$= 12.5 \text{ ns km}^{-1}$$

The rms pulse broadening per kilometer due to intermodal dispersion for the step index fiber is given by Eq. (3.35) as:

$$\sigma_s(1 \text{ km}) \simeq \frac{L(NA)^2}{4\sqrt{3}n_1 c} = \frac{10^3 \times 0.09}{4\sqrt{3} \times 1.45 \times 2.998 \times 10^8}$$

$$= 29.9 \text{ ns km}^{-1}$$

The total rms pulse broadening per kilometer may be obtained using Eq. (3.43), where $\sigma_c \approx \sigma_m$ as the waveguide dispersion is negligible and $\sigma_n = \sigma_s$ for the multimode step index fiber. Hence:

$$\sigma_T = (\sigma_m^2 + \sigma_s^2)^{\frac{1}{2}} = (12.5^2 + 29.9^2)^{\frac{1}{2}}$$

$$= 32.4 \text{ ns km}^{-1}$$

(b) The bandwidth–length product may be estimated from the relationship given in Eq. (3.11) where:

$$B_{opt} \times L = \frac{0.2}{\sigma_T} = \frac{0.2}{32.4 \times 10^{-9}}$$

$$= 6.2 \text{ MHz km}$$

3.11.2 Single-mode fibers

The pulse broadening in single-mode fibers results almost entirely from chromatic or intramodal dispersion as only a single-mode is allowed to propagate.* Hence the bandwidth is limited by the finite spectral width of the source. Unlike the situation in multimode fibers, the mechanisms giving chromatic dispersion in single-mode fibers tend to be interrelated in a complex manner. The transit time or specific group delay τ_g for a light pulse propagating along a unit length of single-mode fiber may be given, following Eq. (2.107), as:

$$\tau_g = \frac{1}{c}\frac{d\beta}{dk} \tag{3.45}$$

where c is the velocity of light in a vacuum, β is the propagation constant for a mode within the fiber core of refractive index n_1 and k is the propagation constant for the mode in a vacuum.

The total first-order dispersion parameter or the chromatic dispersion of a single-mode fiber, D_T, is given by the derivative of the specific group delay with respect to the vacuum wavelength λ as:

$$D_T = \frac{d\tau_g}{d\lambda} \tag{3.46}$$

* Polarization mode dispersion can, however, occur in single-mode fibers (see Section 3.13.2).

In common with the material dispersion parameter it is usually expressed in units of ps nm^{-1} km^{-1}. When the variable λ is replaced by ω, then the total dispersion parameter becomes:

$$D_\mathrm{T} = -\frac{\omega}{\lambda}\frac{\mathrm{d}\tau_\mathrm{g}}{\mathrm{d}\omega} = -\frac{\omega}{\lambda}\frac{\mathrm{d}^2\beta}{\mathrm{d}\omega^2} \qquad (3.47)$$

The fiber exhibits intramodal dispersion when β varies nonlinearly with wavelength. From Eq. (2.71) β may be expressed in terms of the relative refractive index difference Δ and the normalized propagation constant b as:

$$\beta = kn_1[1 - 2\Delta(1-b)]^{\frac{1}{2}} \qquad (3.48)$$

The rms pulse broadening caused by chromatic dispersion down a fiber of length L is given by the derivative of the group delay with respect to wavelength as [Ref. 45]:

$$\text{Total rms pulse broadening} = \sigma_\lambda L \left| \frac{\mathrm{d}\tau_\mathrm{g}}{\mathrm{d}\lambda} \right|$$

$$= \frac{\sigma_\lambda L 2\pi}{c\lambda^2}\frac{\mathrm{d}^2\beta}{\mathrm{d}k^2} \qquad (3.49)$$

where σ_λ is the source rms spectral linewidth centered at a wavelength λ.

When Eq. (3.44) is substituted into Eq. (3.45), detailed calculation of the first and second derivatives with respect to k gives the dependence of the pulse broadening on the fiber material's properties and the normalized propagation constant b. This gives rise to three interrelated effects which involve complicated cross-product terms. However, the final expression may be separated into three composite dispersion components in such a way that one of the effects dominates each term [Ref. 46]. The dominating effects are as follows:

1. The material dispersion parameter D_M defined by $\lambda/c \mid \mathrm{d}^2n/\mathrm{d}\lambda^2 \mid$ where $n = n_1$ or n_2 for the core or cladding respectively.

2. The waveguide dispersion parameter D_W, which may be obtained from Eq. (3.47) by substitution from Eq. (2.114) for τ_g, is defined as:*

$$D_\mathrm{W} = -\left(\frac{n_1 - n_2}{\lambda c}\right) V \frac{\mathrm{d}^2(Vb)}{\mathrm{d}V^2} \qquad (3.50)$$

where V is the normalized frequency for the fiber. Since the normalized propagation constant b for a specific fiber is only dependent on V, then the normalized

---

* Equation (3.50) does not provide the composite waveguide dispersion term (i.e. taking into account both the fiber core and the cladding) from which it differs by a factor near unity which contains $\mathrm{d}n_2/\mathrm{d}\lambda$ [Ref. 47].

waveguide dispersion coefficient $V \mathrm{d}^2(Vb)/\mathrm{d}V^2$ also depends on V. This latter function is another universal parameter which plays a central role in the theory of single-mode fibers.

3. A profile dispersion parameter D_P which is proportional to $\mathrm{d}\Delta/\mathrm{d}\lambda$.

This situation is different from multimode fibers where the majority of modes propagate far from cutoff and hence most of the power is transmitted in the fiber core. In the multimode case the composite dispersion components may be simplified and separated into two chromatic terms which depend on either material or waveguide dispersion, as was discussed in Section 3.9. Also, especially when considering step index multimode fibers, the effect of profile dispersion is negligible. Although material and waveguide dispersion tend to be dominant in single-mode fibers, the composite profile should not be ignored. However, the profile dispersion parameter D_P can be quite small (e.g. less than $0.5 \text{ ps nm}^{-1} \text{ km}^{-1}$), especially at long wavelengths, and hence is often neglected in rough estimates of total dispersion within single-mode fibers.

Strictly speaking, in single-mode fiber with a power-law refractive index profile the composite dispersion terms should be employed [Ref. 47]. Nevertheless, it is useful to consider the total first-order dispersion D_T in a practical single-mode fiber as comprising:

$$D_T = D_M + D_W + D_P \qquad (\text{ps nm}^{-1} \text{ km}^{-1}) \tag{3.51}$$

which is simply the addition of the material dispersion D_M, the waveguide dispersion D_W and the profile dispersion D_P components. However, in standard single-mode fibers the total dispersion tends to be dominated by the material dispersion of fused silica. This parameter is shown plotted against wavelength in Figure 3.10. It may be observed that the characteristic goes through zero at a wavelength of 1.27 μm. This zero material dispersion (ZMD) point can be shifted anywhere in the wavelength range 1.2 to 1.4 μm by the addition of suitable dopants. For instance, the ZMD point shifts from 1.27 μm to approximately 1.37 μm as the GeO_2 dopant concentration is increased from 0 to 15%. However, the ZMD point alone does not represent a point of zero pulse broadening since the pulse dispersion is influenced by both waveguide and profile dispersion.

With ZMD the pulse spreading is dictated by the waveguide dispersion coefficient $V \mathrm{d}^2(Vb)/\mathrm{d}V^2$, which is illustrated in Figure 3.15 as a function of normalized frequency for the LP_{01} mode. It may be seen that in the single-mode region where the normalized frequency is less than 2.405 (see Section 2.5) the waveguide dispersion is always positive and has a maximum at $V = 1.15$. In this case the waveguide dispersion goes to zero outside the true single-mode region at $V = 3.0$. However, a change in the fiber parameters (such as core radius) or in the operating wavelength alters the normalized frequency and therefore the waveguide dispersion.

The total fiber dispersion, which depends on both the fiber material composition and dimensions, may be minimized by trading off material and waveguide dispersion while limiting the profile dispersion (i.e. restricting the variation in refractive index with wavelength). For wavelengths longer than the ZMD point, the material dispersion parameter is positive whereas the waveguide dispersion parameter is negative, as shown in Figure 3.16. However, the total dispersion D_T is approximately equal to the sum of the material dispersion D_M and the waveguide dispersion D_W following Eq. (3.50). Hence for a particular

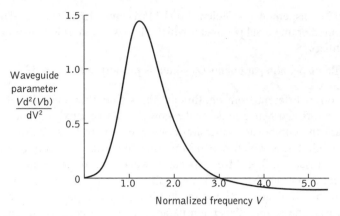

Figure 3.15 The waveguide parameter $V \, d^2(Vb)/dV^2$ as a function of the normalized frequency V for the LP_{01} mode. Reproduced with permission from W. A. Gambling. A. H. Hartog and C. M. Ragdale, *The Radio Electron. Eng.*, **51**, p. 313, 1981

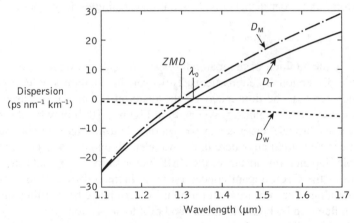

Figure 3.16 The material dispersion parameter (D_M), the waveguide dispersion parameter (D_W) and the total dispersion parameter (D_T) as functions of wavelength for a conventional single-mode fiber

wavelength, designated λ_0, which is slightly larger than the ZMD point wavelength, the waveguide dispersion compensates for the material dispersion and the total first-order dispersion parameter D_T becomes zero (see Figure 3.16). The wavelength at which the first-order dispersion is zero, λ_0, may be selected in the range 1.3 to 2 μm by careful control of the fiber core diameter and profile [Ref. 46]. This point is illustrated in Figure 3.17 where the total first-order dispersion as a function of wavelength is shown for three single-mode fibers with core diameters of 4, 5 and 6 μm.

The effect of the interaction of material and waveguide dispersion on λ_0 is also demonstrated in the dispersion against wavelength characteristics for a single-mode silica core fiber shown in Figure 3.18. It may be noted that the ZMD point occurs at a wavelength of

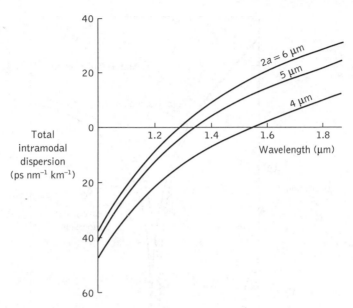

Figure 3.17 The total first order intramodal dispersion as a function of wavelength for single-mode fibers with core diameters of 4, 5, and 6 μm. Reproduced with permission from W. A. Gambling. A. H. Hartog, and C. M. Ragdale, *The Radio and Electron. Eng.*, **51**, p. 313, 1981

1.276 μm for pure silica, but that the influence of waveguide dispersion shifts the total dispersion minimum towards the longer wavelength giving a λ_0 of 1.32 μm.

The wavelength at which the first-order dispersion is zero λ_0 may be extended to wavelengths of 1.55 μm and beyond by a combination of three techniques. These are:

(a) lowering the normalized frequency (*V* value) for the fiber;

(b) increasing the relative refractive index difference Δ for the fiber;

(c) suitable doping of the silica with germanium.

This allows bandwidth–length products for such single-mode fibers to be in excess of 100 GHz km^{-1} [Ref. 48] at the slight disadvantage of increased attenuation due to Rayleigh scattering within the doped silica.

For standard single-mode fibers optimized for operation at a wavelength of 1.31 μm, their performance characteristics are specified by the International Telecommunications Union ITU-T Recommendation G.652 [Ref. 49]. A typical example exhibits C-band (1.530 to 1.565 μm) total first-order chromatic dispersion D_T in the range 16 to 19 ps nm^{-1} km^{-1} with a zero-dispersion wavelength λ_0 between 1.302 and 1.322 μm. However, although the wavelength of zero first-order chromatic dispersion (i.e. $D_T = 0$) is often called the zero-dispersion wavelength, it is more correct to refer to it as the wavelength of minimum dispersion because of the significant second-order effects.

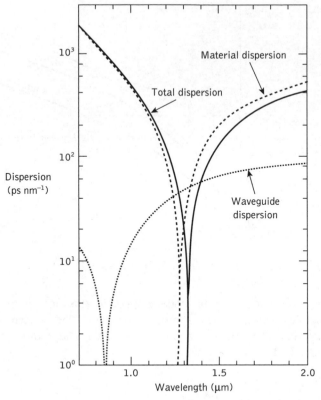

Figure 3.18 The pulse dispersion as a function of wavelength in 11 km single-mode fiber showing the major contributing dispersion mechanisms (dashed and dotted curves) and the overall dispersion (solid curve). Reprinted with permission from J. I. Yamada, M. Saruwatari, K. Asatani, H. Tsuchiya, A. Kawana, K. Sugiyama and T. Kumara, 'High speed optical pulse transmission at 1.29 μm wavelength using low-loss single-mode fibers', *IEEE J. Quantum Electron.*, **QE-14**, p. 791, 1978. Copyright © 1980, IEEE

The variation of the chromatic dispersion with wavelength is usually characterized by the second-order dispersion parameter or dispersion slope S which may be written as [Ref. 50]:

$$S = \frac{\mathrm{d}D_\mathrm{T}}{\mathrm{d}\lambda} = \frac{\mathrm{d}^2\tau_\mathrm{g}}{\mathrm{d}\lambda^2} \tag{3.52}$$

Whereas the first-order dispersion parameter D_T may be seen to be related only to the second derivative of the propagation constant β with respect to angular frequency in Eq. (3.47), the dispersion slope can be shown to be related to both the second and third derivatives [Ref. 47] following:

$$S = \frac{(2\pi c)^3}{\lambda^4}\frac{\mathrm{d}^3\beta}{\mathrm{d}\omega^3} + \frac{4\pi c}{\lambda^3}\frac{\mathrm{d}^2\beta}{\mathrm{d}\omega^2} \tag{3.53}$$

It should be noted that although there is zero first-order dispersion at λ_0, these higher order chromatic effects impose limitations on the possible bandwidths that may be achieved with single-mode fibers. For example, a fundamental lower limit to pulse spreading in silica-based fibers of around 2.50×10^{-2} ps nm^{-1} km^{-1} is suggested at a wavelength of 1.273 μm [Ref. 51]. These secondary effects, such as birefringence arising from ellipticity or mechanical stress in the fiber core, are considered further in Section 3.13. However, they may cause dispersion, especially in the case of mechanical stress of between 2 and 40 ps km^{-1}. If mechanical stress is avoided, pulse dispersion around the lower limit may be obtained in the longer wavelength region (i.e. 1.3 to 1.7 μm). By contrast the minimum pulse spread at a wavelength of 0.85 μm is around 100 ps nm^{-1} km^{-1} [Ref. 35].

An important value of the dispersion slope $S(\lambda)$ is obtained at the wavelength of minimum chromatic dispersion λ_0 such that:

$$S_0 = S(\lambda_0) \qquad (3.54)$$

where S_0 is called the zero-dispersion slope which, from Eqs (3.46) and (3.52), is determined only by the third derivative of β. Typical values for the dispersion slope for standard single-mode fiber at λ_0 are in the region 0.085 to 0.095 ps nm^{-1} km^{-1}. The total chromatic dispersion at an arbitrary wavelength can be estimated when the two parameters λ_0 and S_0 are specified according to [Ref. 52]:

$$D_T(\lambda) = \frac{\lambda S_0}{4}\left[1 - \left(\frac{\lambda_0}{\lambda}\right)^4\right] \qquad (3.55)$$

Example 3.11

A typical single-mode fiber has a zero-dispersion wavelength of 1.31 μm with a dispersion slope of 0.09 ps nm^{-2} km^{-1}. Compare the total first-order dispersion for the fiber at the wavelengths of 1.28 μm and 1.55 μm. When the material dispersion and profile dispersion at the latter wavelength are 13.5 ps nm^{-1} km^{-1} and 0.4 ps nm^{-1} km^{-1}, respectively, determine the waveguide dispersion at this wavelength.

Solution: The total first-order dispersion for the fiber at the two wavelengths may be obtained from Eq. (3.55). Hence:

$$D_T(1280 \text{ nm}) = \frac{\lambda S_0}{4}\left[1 - \left(\frac{\lambda_0}{\lambda}\right)^4\right]$$

$$= \frac{1280 \times 0.09 \times 10^{-12}}{4}\left[1 - \left(\frac{1310}{1280}\right)^4\right]$$

$$= -2.8 \text{ ps nm}^{-1} \text{ km}^{-1}$$

▶

and:

$$D_T(1550 \text{ nm}) = \frac{1550 \times 0.09 \times 10^{-12}}{4} \left[1 - \left(\frac{1310}{1550} \right)^4 \right]$$

$$= 17.1 \text{ ps nm}^{-1} \text{ km}^{-1}$$

The total dispersion at the 1.28 μm wavelength exhibits a negative sign due to the influence of the waveguide dispersion. Furthermore, as anticipated the total dispersion at the longer wavelength (1.55 μm) is considerably greater than that obtained near the zero-dispersion wavelength.

The waveguide dispersion for the fiber at a wavelength of 1.55 μm is given by Eq. (3.51) where:

$$D_W = D_T - (D_M + D_P)$$
$$= 17.1 - (13.5 + 0.4)$$
$$= 3.2 \text{ ps nm}^{-1} \text{ km}^{-1}$$

3.12 Dispersion-modified single-mode fibers

It was suggested in Section 3.11.2 that it is possible to modify the dispersion characteristics of single-mode fibers by the tailoring of specific fiber parameters. However, the major trade-off which occurs in this process between material dispersion (Eq. 3.19) and waveguide dispersion (Eq. 3.50) may be expressed as:

$$D_T = D_M + D_W = \underbrace{\frac{\lambda}{c} \left| \frac{d^2 n_1}{d\lambda^2} \right|}_{\text{material dispersion}} - \underbrace{\left[\frac{n_1 - n_2}{\lambda c} \right] \frac{V d^2 (Vb)}{dV^2}}_{\text{waveguide dispersion}} \qquad (3.56)$$

At wavelengths longer than the ZMD point in most common fiber designs, the D_M and D_W components are of opposite sign and can therefore be made to cancel at some longer wavelength. Hence the wavelength of zero first-order chromatic dispersion can be shifted to the lowest loss wavelength for silicate glass fibers at 1.55 μm to provide both low-dispersion and low-loss fiber. This may be achieved by such mechanisms as a reduction in the fiber core diameter with an accompanying increase in the relative or fractional index difference to create so-called dispersion-shifted single-mode fibers (DSFs). However, the design flexibility required to obtain particular dispersion, attenuation, mode-field diameter and bend loss characteristics has resulted in specific, different refractive index profiles for these dispersion-modified fibers [Ref. 53].

An alternative modification of the dispersion characteristics of single-mode fibers involves the achievement of a low-dispersion window over the low-loss wavelength region between 1.3 and 1.6 μm. Such fibers, which relax the spectral requirements for

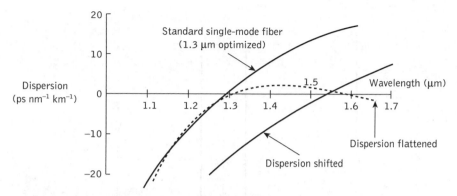

Figure 3.19 Total dispersion characteristics for the various types of single-mode fiber

optical sources and allow flexible wavelength division multiplexing (see Section 12.9.4) are known as dispersion-flattened single-mode fibers (DFFs). In order to obtain DFFs multilayer index profiles are fabricated with increased waveguide dispersion which is tailored to provide overall dispersion (e.g. less than 2 ps nm^{-1} km^{-1}) over the entire wavelength range 1.3 to 1.6 μm [Ref. 54]. In effect these fibers exhibit two wavelengths of zero total chromatic dispersion. This factor may be observed in Figure 3.19 which shows the overall dispersion characteristics as a function of optical wavelength for standard single-mode fiber (SSMF) optimized for operation at 1.3 μm in comparison with both DSF and DFF [Ref. 55]. Furthermore, the low-water-peak fiber discussed in Section 3.3.2 currently exhibits overall dispersion characteristics that are the same as those for the SSMF shown in Figure 3.19. It should also be noted that although DFF is characterized by low dispersion over a large wavelength range, it has been superseded by nonzero dispersion-shifted fiber (see Section 3.12.3) and therefore it does not currently find use in practical applications.

3.12.1 Dispersion-shifted fibers

A wide variety of single-mode fiber refractive index profiles are capable of modification in order to tune the zero-dispersion wavelength point λ_0 to a specific wavelength within a region adjacent to the ZMD point. In the simplest case, the step index profile illustrated in Figure 3.20 gives a shift to longer wavelength by reducing the core diameter and increasing the fractional index difference. Typical values for the two parameters are 4.4 μm and 0.012 respectively [Ref. 56]. For comparison, the standard nonshifted design is shown dashed in Figure 3.20.

It was indicated in Section 3.11.2 that λ_0 could be shifted to longer wavelength by altering the material composition of the single-mode fiber. For suitable power confinement of the fundamental mode, the normalized frequency V should be maintained in the range 1.5 to 2.4 μm and the fractional index difference must be increased as a square function while the core diameter is linearly reduced to keep V constant. This is normally achieved by substantially increasing the level of germanium doping in the fiber core. Figure 3.21 [Ref. 56] displays typical material and waveguide dispersion characteristics for single-mode step

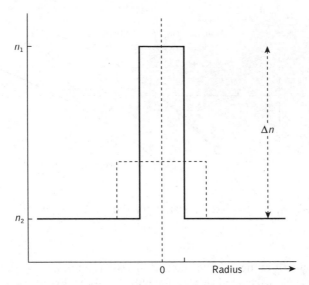

Figure 3.20 Refractive index profile of a step index dispersion-shifted fiber (solid) with a conventional nonshifted profile design (dashed)

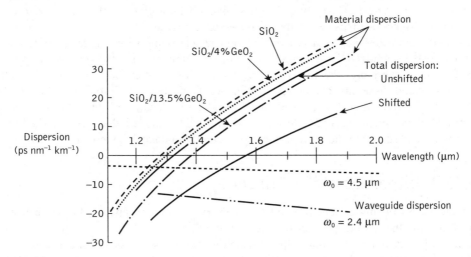

Figure 3.21 Material, waveguide and total dispersion characteristics for conventional and dispersion-shifted step index single-mode fibers showing variation with composition and spot size (ω_0)

index fibers with various compositions and core radii. It may be observed that higher concentrations of the dopant cause a shift to longer wavelength which, when coupled with a reduction in the mode-field diameter (MFD), giving a larger value (negative of waveguide dispersion), leads to the shifted fiber characteristic shown in Figure 3.21.

A problem that arises with the simple step index approach to dispersion shifting displayed in Figure 3.20 is that the fibers produced exhibit relatively high dopant-dependent

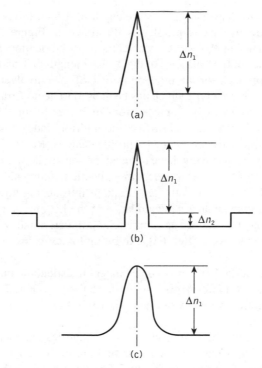

Figure 3.22 Refractive index profiles for graded index dispersion-shifted fibers:
(a) triangular profile; (b) depressed-cladding triangular profile; (c) Gaussian profile

losses at operation wavelengths around 1.55 μm. This excess optical loss, which may be of the order of 2 dB km^{-1} [Ref. 56], could be caused by stress-induced defects which occur in the region of the core–cladding interface [Ref. 57]. Alternatively, it may result from refractive index inhomogeneities associated with waveguide variations at the core–cladding interface [Ref. 58]. A logical assumption is that any stress occurring across the core–cladding interface might be reduced by grading the material composition and therefore an investigation of graded index single-mode fiber designs was undertaken.

Several of the graded refractive index profile DSF types are illustrated in Figure 3.22. The triangular profile shown in Figure 3.22(a) is the simplest and was the first to exhibit the same low loss (i.e. 0.24 dB km^{-1}) at a wavelength of 1.56 μm (i.e. λ_0) as conventional nonshifted single-mode fiber [Ref. 59]. Furthermore, such fiber designs also provide an increased MFD over equivalent step index structures which assists with fiber splicing [Ref. 56]. However, in the basic triangular profile design the optimum parameters giving low loss together with zero dispersion at a wavelength of 1.55 μm cause the LP_{11} mode to cut off in the wavelength region 0.85 to 0.9 μm. Thus the fiber must be operated far from cutoff, which produces sensitivity to bend-induced losses (in particular microbending) at the 1.55 μm wavelength [Ref. 60]. One method to overcome this drawback is to employ a triangular index profile combined with a depressed cladding index, as shown in Figure 3.22(b) [Ref. 61]. In this case the susceptibility to microbending losses is reduced through a shift of the LP_{11} cutoff wavelength to around 1.1 μm with an MFD of 7 μm at 1.55 μm.

Low losses and zero dispersion at a wavelength of 1.55 μm have also been obtained with a Gaussian refractive index profile, as illustrated in Figure 3.22(c). This profile, which was achieved using the vapor axial deposition fabrication process (see Section 4.4.2), produced losses of 0.21 dB km^{-1} at the λ_0 wavelength of 1.55 μm [Ref. 62].

The alternative approach for the production of DSF has involved the use of multiple index designs. One such fiber type which has been used to demonstrate dispersion shifting but which has been more widely employed for DFFs (see Section 3.12.2) is the doubly clad or W fiber (see Section 2.5). However, the multiple index triangular profile fibers [Ref. 63] and the segmented-core triangular profile designs [Ref. 64], which are shown in Figure 3.23(a) and (b), respectively, have reduced the sensitivity to microbending by shifting the LP_{11} mode cutoff to longer wavelength while maintaining an MFD of around 9 μm at a wavelength of 1.55 μm. The latter technique of introducing a ring of elevated index around the triangular core enhances the guidance of the LP_{11} mode towards longer wavelength. Such fibers may be obtained as commercial products and have been utilized within the telecommunication network [Ref. 65], exhibiting losses as low as 0.17 dB at 1.55 μm [Ref. 66].

Dual-shaped core DSFs have also come under investigation in order to provide an improvement in bend loss performance over the 1.55 μm wavelength region [Refs 67, 68]. A dual-shaped core refractive index profile is shown in Figure 3.23(c), which illustrates a step index fiber design.

DSF has more recently been subject to the standardization recommendation ITU-T G.653 [Ref. 69]. Typical values for its attenuation are in the range 0.22 to 0.24 dB km^{-1} while it exhibits dispersion between 0 and 2.7 ps nm^{-1} km^{-1}, both at a wavelength of 1.55 μm. Although DSF has been installed to provide for high-speed transmission of a single 1.55 μm wavelength channel, it presents dispersion problems for wavelength division multiplexed operation when many wavelength channels are packed into one or more of the ITU spectral bands (see Section 3.3.2). In particular, as all the transmitted channels need to be grouped around the 1.55 μm wavelength to reduce dispersion, they then travel at the same speed creating four-wave mixing interaction and the associated high crosstalk levels (see Section 3.14.2). Hence DSF is no longer recommended for deployment.

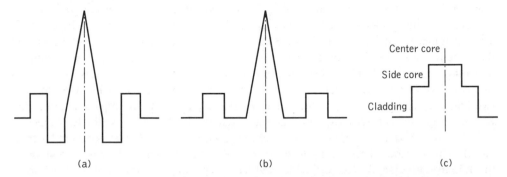

Figure 3.23 Advanced refractive index profiles for dispersion-shifted fibers: (a) triangular profile multiple index design; (b) segmented-core triangular profile design; (c) dual-shaped core design

3.12.2 Dispersion-flattened fibers

The original W fiber structure mentioned in Section 3.12.1 was initially employed to modify the dispersion characteristics of single-mode fibers in order to give two wavelengths of zero dispersion, as illustrated in Figure 3.19. A typical W fiber index profile (double clad) is shown in Figure 3.24(a). The first practical demonstration of dispersion flattening using the W structure was reported in 1981 [Ref. 70]. However, drawbacks with the W structural design included the requirement for a high degree of dimensional control so as to make reproducible DFF [Ref. 71], comparatively high overall fiber losses (around 0.3 dB km^{-1}), as well as a very high sensitivity to fiber bend losses. The last factor results from operation very close to the cutoff (or leakage) of the fundamental mode in the long-wavelength window in order to obtain a flat dispersion characteristic.

To reduce the sensitivity to bend losses associated with the W fiber structure, the light which penetrates into the outer cladding area can be retrapped by introducing a further region of raised index into the structure. This approach has resulted in the triple clad (TC) and quadruple clad (QC) structures shown in Figure 3.24(b) and (c) [Refs 72, 73]. An independent but similar program produced segmented-core DFF designs [Ref. 66]. Although reports of low attenuation of 0.19 dB km^{-1} for DFF at a wavelength of 1.55 μm [Ref. 74] with significantly reduced bending losses [Ref. 75] have been made, it has proved difficult to balance the performance attributes of this fiber type.

More recent efforts have focused on DFFs that exhibit low-dispersion slopes in the C-band while also providing acceptably large effective core areas in order to reduce fiber nonlinear effects [Ref. 76]. It is particularly difficult, however, to realize both near-zero-dispersion slopes and large effective core areas while, in addition, reducing their current bend loss sensitivity. Furthermore, there remain some fundamental problems associated with DFF fabrication as the chromatic dispersion characteristics are highly dependent on changes in the fiber structural parameters including the core diameter and refractive index difference. It is therefore still a concern that commercial production of DFFs will require both substantial stability and controllability of the fabrication process [Ref. 77]. Hence such fibers have yet to find widespread deployment in telecommunication networks.

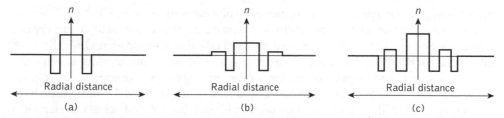

Figure 3.24 Dispersion-flattened fiber refractive index profiles: (a) double clad fiber (W fiber); (b) triple clad fiber; (c) quadruple clad fiber

3.12.3 Nonzero-dispersion-shifted fibers

Nonzero-dispersion-shifted fiber (NZ-DSF) is sometimes simply called nonzero-dispersion fiber (NZDF) and a variant of this fiber type is negative-dispersion fiber (NDF)

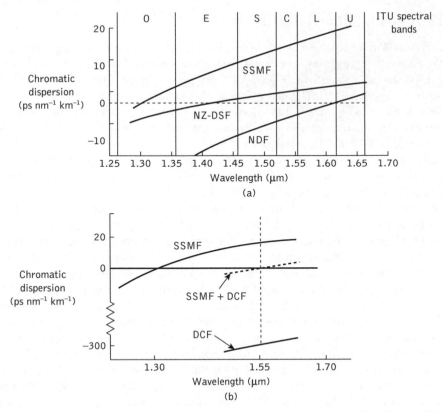

Figure 3.25 Single-mode fiber dispersion characteristics: (a) comparing the profiles for nonzero-dispersion-shifted fiber (NZ-DSF) and negative-dispersion fiber (NDF) with standard single-mode fiber (SSMF); (b) dispersion compensation by using negative-dispersion, dispersion-compensating fiber (DCF)

which can also be referred to as dispersion compensating fiber (DCF). NZ-DSF was introduced in the mid-1990s to better provide for wavelength division multiplexing applications (see Section 12.9.4). It is specified in the ITU-T Recommendation G.655 [Ref. 78] in which, unlike DSF, its principal attribute is that it has a nonzero-dispersion value over the entire C-band as may be observed from the dispersion characteristic profile displayed in Figure 3.25(a). In comparison, however, with the SSMF profile also shown in Figure 3.25(a), the chromatic dispersion exhibited by NZ-DSF at a wavelength of 1.55 μm is much lower, being in the range 2 to 4 ps nm^{-1} km^{-1} rather than the typical 17 ps nm^{-1} km^{-1} for SSMF.

It should also be noted that the dispersion profile for NZ-DSF shown in Figure 3.25(a) is also referred to as extended band or bandwidth nonzero-dispersion-shifted fiber which was introduced in 2000 to enable wavelength division multiplexed applications to be extended into the S-band. Hence it can be seen that it exhibits nonzero dispersion over the full C- and S-bands, whereas with the original NZ-DSF the dispersion varies from negative values through zero to positive values in the S-band. In addition, Figure 3.25(a) also

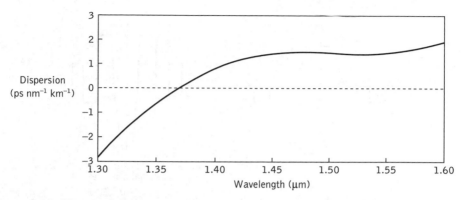

Figure 3.26 Flattened dispersion spectrum for an extended band NZ-DSF

displays the dispersion profile for NDF,* which exhibits negative dispersion over the entire 1.30 to 1.60 μm wavelength range (E-, S- and C-bands) as its dispersion zero is shifted to around 1.62 μm. Such fiber facilitates dispersion compensation when used together with either SSMF or NZ-DSF, both of which have positive dispersion over the majority of the aforementioned spectral range. The dispersion compensation attribute of negative-dispersion fiber identified as dispersion compensating fiber (DCF) is illustrated in the dispersion characteristic shown in Figure 32.5(b) where the total chromatic dispersion accumulated in the SSMF is cancelled by inserting an appropriate length of DCF (often exhibiting large absolute dispersion around -300 ps nm^{-1} km^{-1}) which provides the equivalent total negative dispersion, thus giving zero overall dispersion at the 1.55 μm wavelength on the transmission link.

The dispersion characteristic for an extended band NZ-DSF can also be flattened, as indicated in Figure 3.26, which then provides low, but noticeable, dispersion in S- and C-bands, further reducing the nonlinear crosstalk problem associated with four-wave mixing. This reduction in the dispersion slope or flattening is an important issue in the design of extended band NZ-DSFs together with the enlargement of the effective core area. The former adjusts the chromatic dispersion to appropriate levels across a wide wavelength range while the latter is intended to suppress the nonlinear effects in the fiber (see Section 3.14). Pursuing both these objectives produces a trade-off in the fiber design process which has resulted in a number of refractive index profiles being explored to improve the performance of NZ-DSF.

The refractive index profiles shown in Figure 3.23(b) and (c) for DSF have also been utilized to provide NZ-DSF. Indeed, the segmented core triangular profile displayed in Figure 3.23(b) is employed by the commercially produced Corning LEAF fiber which exhibits a large effective core area at the expense of a high-dispersion slope. A more typical NZ-DSF multiple index profile is shown in Figure 3.27(a) whereas a NDF refractive index profile is illustrated in Figure 3.27(b).

Although NZ-DSF has now effectively replaced both DSF and DFF for use in long-haul optical fiber communications in order to remove four-wave mixing crosstalk with multichannel operation, it is still subject to greater problems with nonlinearities in

---

* An example is the Corning MetroCor fiber.

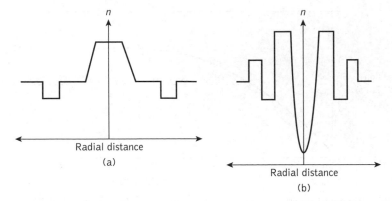

Figure 3.27 Typical refractive index profiles: (a) nonzero-dispersion-shifted fiber (NZ-DSF); (b) negative-dispersion fiber (NDF)

Table 3.3 Typical performance parameters for common nonzero-dispersion-shifted fibers (NZ-DSFs) all at 1.55 μm, excepting the zero-dispersion wavelength [Ref. 77]

Parameter	Reduced slope NZ-DSF	Large effective core NZ-DSF	Extended band NZ-DSF
Zero-dispersion wavelength (μm)	1.46	1.50	1.42
Dispersion (ps nm^{-1} km^{-1})	4	4	8
Dispersion slope (ps nm^{-2} km^{-1})	0.045	0.085	0.058
Effective core area (μm^2)	50	70	63

comparison with SSMF. This situation occurs because in general NZ-DSF has a smaller cross-sectional area and therefore the nonlinear threshold power levels can be reached at lower values of transmitted power. To keep the ratio of power to effective area low, however, a so-called large effective area NZ-DSF can be used to lower the power density below that of a conventional NZ-DSF [Ref. 79]. A further advantage of the large effective area approach is that it provides improved splicing capability to SSMF.

Typical parameters for some common NZ-DSFs are summarized in Table 3.3 [Ref. 77]. It may be observed that NZ-DSF with a reduced dispersion slope exhibits a significantly smaller effective core area in comparison with the large effective core NZ-DSF and also the extended band NZ-DSF. Examples of a reduced dispersion slope and an extended band NZ-DSFs are the Lucent TrueWave-RS and the Alcatel TeraLight fibers respectively.

3.13 Polarization

Cylindrical optical fibers do not generally maintain the polarization state of the light input for more than a few meters, and hence for many applications involving optical fiber transmission some form of intensity modulation (see Section 7.5) of the optical source is utilized. The optical signal is thus detected by a photodiode which is insensitive to optical polarization or phase of the lightwave within the fiber. Nevertheless, systems and applications have been investigated [Ref. 80] (see Section 13.1) which could require the polarization states of the input light to be maintained over significant distances, and fibers have been designed for this purpose. These fibers are single mode and the maintenance of the polarization state is described in terms of a phenomenon known as fiber birefringence.

3.13.1 Fiber birefringence

Single-mode fibers with nominal circular symmetry about the core axis allow the propagation of two nearly degenerate modes with orthogonal polarizations. They are therefore bimodal supporting HE_{11}^x and HE_{11}^y modes where the principal axes x and y are determined by the symmetry elements of the fiber cross section. Hence in an optical fiber with an ideal optically circularly symmetric core both polarization modes propagate with identical velocities. Manufactured optical fibers, however, exhibit some birefringence resulting from differences in the core geometry (i.e. ellipticity) resulting from variations in the internal and external stresses, and fiber bending. The fiber therefore behaves as a birefringent medium due to the difference in the effective refractive indices, and hence phase velocities, for these two orthogonally polarized modes. The modes therefore have different propagation constants β_x and β_y which are dictated by the anisotropy of the fiber cross section. In this case β_x and β_y are the propagation constants for the slow mode and the fast mode respectively. When the fiber cross-section is independent of the fiber length L in the z direction, then the modal birefringence B_F for the fiber is given by [Ref. 81]:

$$B_F = \frac{(\beta_x - \beta_y)}{(2\pi/\lambda)} \tag{3.57}$$

where λ is the optical wavelength. Light polarized along one of the principal axes will retain its polarization for all L.

The difference in phase velocities causes the fiber to exhibit a linear retardation $\Phi(z)$ which depends on the fiber length L in the z direction and is given by [Ref. 81]:

$$\Phi(z) = (\beta_x - \beta_y)L \tag{3.58}$$

assuming that the phase coherence of the two mode components is maintained. The phase coherence of the two mode components is achieved when the delay between the two transit times is less than the coherence time of the source. As indicated in Section 3.11, the coherence time for the source is equal to the reciprocal of the uncorrelated source frequency width ($1/\delta f$).

It may be shown [Ref. 82] that birefringent coherence is maintained over a length of fiber L_{bc} (i.e. coherence length) when:

$$L_{bc} \simeq \frac{c}{B_F \delta f} = \frac{\lambda^2}{B_F \delta \lambda} \tag{3.59}$$

where c is the velocity of light in a vacuum and $\delta\lambda$ is the source linewidth.

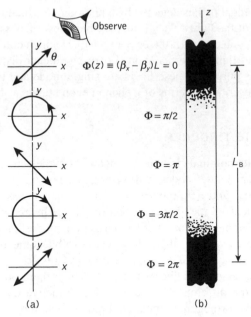

Figure 3.28 An illustration of the beat length in a single-mode optical fiber [Ref. 81]: (a) the polarization states against $\Phi(z)$; (b) the light intensity distribution over the beat length within the fiber

However, when phase coherence is maintained (i.e. over the coherence length) Eq. (3.58) leads to a polarization state which is generally elliptical but which varies periodically along the fiber. This situation is illustrated in Figure 3.28(a) [Ref. 81] where the incident linear polarization which is at 45° with respect to the x axis becomes circular polarization at $\Phi = \pi/2$ and linear again at $\Phi = \pi$. The process continues through another circular polarization at $\Phi = 3\pi/2$ before returning to the initial linear polarization at $\Phi = 2\pi$. The characteristic length L_B for this process corresponding to the propagation distance for which a 2π phase difference accumulates between the two modes is known as the beat length. It is given by:

$$L_B = \frac{\lambda}{B_F} \tag{3.60}$$

Substituting for B_F from Eq. (3.47) gives:

$$L_B = \frac{2\pi}{(\beta_x - \beta_y)} \tag{3.61}$$

It may be noted that Eq. (3.61) may be obtained directly from Eq. (3.58) where:

$$\Phi(L_B) = (\beta_x - \beta_y)L_B = 2\pi \tag{3.62}$$

Typical single-mode fibers are found to have beat lengths of a few centimeters [Ref. 83], and the effect may be observed directly within a fiber via Rayleigh scattering with use of a suitable visible source (e.g. He–Ne laser) [Ref. 84]. It appears as a series of bright and dark bands with a period corresponding to the beat length, as shown in Figure 3.28.(b). The modal birefringence B_F may be determined from these observations of beat length.

Example 3.12

The beat length in a single-mode optical fiber is 9 cm when light from an injection laser with a spectral linewidth of 1 nm and a peak wavelength of 0.9 μm is launched into it. Determine the modal birefringence and estimate the coherence length in this situation. In addition calculate the difference between the propagation constants for the two orthogonal modes and check the result.

Solution: To find the modal birefringence Eq. (3.60) may be used where:

$$B_F = \frac{\lambda}{L_B} = \frac{0.9 \times 10^{-6}}{0.09} = 1 \times 10^{-5}$$

Knowing B_F, Eq. (3.59) may be used to obtain the coherence length:

$$L_{bc} \simeq \frac{\lambda^2}{B_F \delta\lambda} = \frac{0.81 \times 10^{-12}}{10^{-5} \times 10^{-9}} = 81 \text{ m}$$

The difference between the propagation constant for the two orthogonal modes may be obtained from Eq. (3.61) where:

$$\beta_x - \beta_y = \frac{2\pi}{L_B} = \frac{2\pi}{0.09} = 69.8$$

The result may be checked by using Eq. (3.57) where:

$$\beta_x - \beta_y = \frac{2\pi B_F}{\lambda} = \frac{2\pi \times 10^{-5}}{0.9 \times 10^{-6}}$$

$$= 69.8$$

In a nonperfect fiber various perturbations along the fiber length such as strain or variations in the fiber geometry and composition lead to coupling of energy from one polarization to the other. These perturbations are difficult to eradicate as they may easily occur in the fiber manufacture and cabling. The energy transfer is at a maximum when the perturbations have a period Λ, corresponding to the beat length, and defined by [Ref. 80]:

$$\Lambda = \frac{\lambda}{B_F} \tag{3.63}$$

However, the cross-polarizing effect may be minimized when the period of the perturbations is less than a cutoff period Λ_c (around 1 mm). Hence polarization-maintaining fibers may be designed by either:

(a) high (large) birefringence: the maximization of the fiber birefringence, which, following Eq. (3.60), may be achieved by reducing the beat length L_B to around 1 mm or less; or

(b) low (small) birefringence: the minimization of the polarization coupling perturbations with a period of Λ. This may be achieved by increasing Λ_c giving a large beat length of around 50 m or more.

Example 3.13

Two polarization-maintaining fibers operating at a wavelength of 1.3 μm have beat lengths of 0.7 mm and 80 m. Determine the fiber birefringence in each case and comment on the results.

Solution: Using Eq. (3.60), the modal birefringence is given by:

$$B_F = \frac{\lambda}{L_B}$$

Hence, for a beat length of 0.7 mm:

$$B_F = \frac{1.3 \times 10^{-6}}{0.7 \times 10^{-3}} = 1.86 \times 10^{-3}$$

This typifies a high birefringence fiber.
  For a beat length of 80 m:

$$B_F = \frac{1.3 \times 10^{-6}}{80} = 1.63 \times 10^{-8}$$

which indicates a low birefringence fiber.

3.13.2 Polarization mode dispersion

Polarization mode dispersion (PMD) is a source of pulse broadening which results from fiber birefringence and it can become a limiting factor for optical fiber communications at high transmission rates. It is a random effect due to both intrinsic (caused by non-circular fiber core geometry and residual stresses in the glass material near the core region) and extrinsic (caused by stress from mechanical loading, bending or twisting of the fiber) factors which in actual manufactured fibers result in group velocity variation with polarization state.

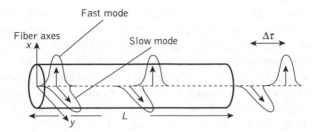

Figure 3.29 Time domain effect of polarization mode dispersion in a short fiber length with a pulse being launched with equal power on the two birefringent axes, *x* and *y*, becoming two pulses at the output separated by the differential group delay

When considering a short section of single-mode fiber within a long fiber span, as shown in the time domain illustration of Figure 3.29, it can be assumed that any perturbations acting on it are constant over its entire length rather than varying along it. In this case the fiber becomes bimodal due to a loss of degeneracy for the two HE_{11} modes. As these two modes have different phase propagation constants β_x and β_y they exhibit different specific group delays (see Section 2.5.4.). In the time domain for a short section of fiber, the differential group delay (DGD), $\Delta\tau = \delta\tau_g L$, is defined as the group delay difference between the slow and the fast modes over the fiber lengths as indicated in Figure 3.29. The DGD can be obtained from the frequency derivative of the difference in the phase propagation constants from Eq. (2.107) as:

$$\delta\tau_g = \frac{\Delta\tau}{L} = \frac{d}{d\omega}(\beta_x - \beta_y) = \frac{d}{d\omega}\left(\frac{\omega n_x}{c} - \frac{\omega n_y}{c}\right)$$

$$= \frac{d}{d\omega}\frac{\omega}{c}\Delta n_{\text{eff}} = \frac{\Delta n_{\text{eff}}}{c} - \frac{\omega}{c}\frac{d}{d\omega}\Delta n_{\text{eff}} \tag{3.64}$$

where $\delta\tau_g$, the differential group delay per unit length, is referred to as the polarization mode dispersion (PMD) of the fiber and is usually expressed in units of picoseconds per kilometer of fiber such that the DGD is the time domain manifestation of PMD as shown in Figure 3.29. This linear relationship to fiber length, however, applies only to short fiber lengths or intrinsic PMD in which the birefringence can be assumed to be uniform. Hence for polarization-maintaining fibers $\delta\tau_g$ can be quite large at around 1 ns km^{-1} when the two components are equally exited at the fiber input.

For conventional single-mode fibers, however, the axis of the birefringence (and its magnitude) varies randomly along the fiber. This phenomenon causes polarization mode coupling such that the fast and slow polarization modes of one segment of a long fiber migrate into both fast and slow modes in the next span. The polarization mode coupling which results from localized stress during cabling, from splices and connectors, and from variations in the fiber drawing process therefore tends to reduce the overall dispersion because the PMD effects do not accumulate linearly in very long fiber spans. In this context low-birefringence fibers which exhibit low PMD can also be fabricated by spinning the fiber in the drawing process (see Section 3.3.2.). Hence, as a result of mode coupling in long fiber lengths, the birefringence of each segment adds to, or subtracts from, the total birefringence so that the DGD does not accumulate linearly. Indeed, in long fiber spans it has been shown that the PMD increases on average with the square root of the length [Ref. 85]. To determine a more precise value for the PMD of a specific fiber link, however, requires a statistical approach [Ref. 86]. Using this method it is possible to categorize fiber into a short- or a long-length regime on the basis of a parameter called the correlation length, defined as the length over which the two polarization modes remain correlated.

The statistical approach to the theory of PMD [Refs 87, 88] provides an expression for the mean square DGD in terms of the fiber polarization beat length L_B and the correlation length L_c defined as the length over which the two polarization modes remain correlated such that:

$$<\Delta\tau^2> = 2\left(\Delta\tau_B \frac{L_c}{L_B}\right)^2\left[\exp\left(-\frac{L}{L_c}\right) + \frac{L}{L_c} - 1\right] \tag{3.65}$$

where $\Delta\tau_B$ is the DGD corresponding to the beat length and L is the fiber length. The correlation length can vary over quite a wide range, from 1 m to 1 km, depending on the specific fiber type, with typical values around 10 m. Moreover, the correlation length can be seen to define two distinct PMD regimes as follows. For $L \ll L_c$, Eq. (3.65) simplifies to:

$$(<\delta\tau^2>)^{\frac{1}{2}} = \delta\tau_{rms} = \delta\tau_B \frac{L_c}{L_B} \tag{3.66}$$

indicating the linear relationship. When $L \gg L_c$, however, then Eq. (3.65) becomes:

$$\delta\tau_{rms} = \frac{\delta\tau_B}{L_B} (2\,LL_c)^{\frac{1}{2}} \tag{3.67}$$

demonstrating the dependence on the square root of the length L. Since optical fiber transmission systems usually operate in the long-length regime, then fiber PMD is often specified using a PMD coefficient having units of ps km^{-1} [Ref. 86]. Although legacy fibers from the 1980s can exhibit mean PMD coefficients greater than 0.8 ps km^{-1}, recently manufactured fibers usually have mean PMD coefficients lower than 0.1 ps km^{-1}. Hence, as a result of the $L^{\frac{1}{2}}$ dependence, PMD-induced pulse broadening is often small in comparison with the combined material and waveguide dispersion effects, with $\Delta\tau_{rms}$ only around 1 ps for 100 km fiber lengths. Nevertheless, PMD can become a limiting factor in optical communication systems operating over long distances at high transmission rates [Ref. 89] and hence PMD compensation techniques and devices have been developed for these situations [Refs 85, 90, 91]. In addition, it should be noted that the relationship in Eq. (3.65) takes no account of other elements in the fiber link which may exhibit polarization-dependent gain or loss, the latter of which can cause additional broadening. The effects of second- and higher order PMD are also important at higher transmission rates of 40 Gbit s^{-1} and above, particularly for a system where the first-order effects have been removed using a compensator device [Ref. 92].

Although certain single-mode fibers can be fabricated to propagate only one polarization mode (see Section 3.13.3), fibers which transmit two orthogonally polarized fundamental modes can exhibit interference between the modes which may cause polarization modal noise. This phenomenon occurs when the fiber is slightly birefringent and there is a component with polarization-dependent loss. Hence, when the fiber link contains an element whose insertion loss is dependent on the state of polarization, then the transmitted optical power will depend on the phase difference between the normal modes and it will fluctuate if the transmitted wavelength or the birefringence alters. Any polarization-sensitive loss will therefore result in modal noise within single-mode fiber [Ref. 93].

Polarization modal noise is generally of larger amplitude than modal noise obtained within multimode fibers (see Section 3.10.3). It can therefore significantly degrade the performance of a communication system such that high-quality analog transmission may prove impossible [Ref. 47]. Moreover, with digital transmission it is usually necessary to increase the system channel loss margin (see Section 12.6.4). It is therefore important to minimize the use of elements with polarization-dependent insertion losses (e.g. beam splitters, polarization-selective power dividers, couplers to single-polarization optical components, bends in high-birefringence fibers) on single-mode optical fiber links. However, other types of fiber perturbation such as bends in low-birefringence fibers, splices and directional couplers do not appear to introduce significant polarization sensitive losses [Ref. 80].

Techniques have been developed to produce both high- and low-birefringence fibers, initially to facilitate coherent optical communication systems. Birefringence occurs when the circular symmetry in single-mode fibers is broken, which can result from the effect of geometrical shape or stress. Alternatively, to design low-birefringence fibers it is necessary to reduce the possible perturbations within the fiber manufacture. These fiber types are discussed in the following section.

3.13.3 Polarization-maintaining fibers

Although the polarization state of the light arriving at a conventional photodetector is not distinguished and hence of little concern, it is of considerable importance in coherent lightwave systems in which the incident signal is superimposed on the field of a local oscillator (see Section 13.3). Moreover, interference and delay differences between the orthogonally polarized modes in birefringent fibers may cause polarization modal noise and PMD respectively (see Section 3.13.2). Finally, polarization is also of concern when a single-mode fiber is coupled to a modulator or other waveguide device (see Section 11.4.2) that can require the light to be linearly polarized for efficient operation. Hence, there are several reasons why it may be desirable to use fibers that will permit light to pass through while retaining its state of polarization. Such polarization-maintaining (PM) fibers can be classified into two major groups: namely, high-birefringence (HB) and low-birefringence (LB) fibers.

The birefringence of conventional single-mode fibers is in the range $B_F = 10^{-6}$ to 10^{-5} [Ref. 94]. An HB fiber requires $B_F > 10^{-5}$ and a value better than 10^{-4} is a minimum for polarization maintenance [Ref. 95]. HB fibers can be separated into two types which are generally referred to as two-polarization fibers and single-polarization fibers. In the latter case, in order to allow only one polarization mode to propagate through the fiber, a cutoff condition is imposed on the other mode by utilizing the difference in bending loss between the two polarization modes.

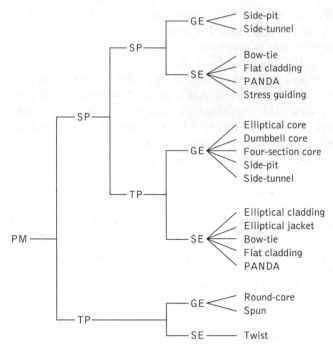

Figure 3.30 Polarization-maintaining fiber types classified from a linear polarization maintenance viewpoint. PM: polarization-maintaining, HB: high-birefringence, LB: low-birefringence, SP: single-polarization, TP: two-polarization, GE: geometrical effect, SE: stress effect [Ref. 96]

The various types of PM fiber, classified in terms of their linear polarization maintenance, are shown in Figure 3.30 [Ref. 96]. In addition, a selection of the most common structures is illustrated in Figure 3.31. The fiber types illustrated in Figure 3.31(a) and (b) employ geometrical shape birefringence, while Figure 3.31(c) to (g) utilize various stress effects. Geometrical birefringence is a somewhat weak effect and a large relative refractive index difference between the fiber core and cladding is required to produce high birefringence. Therefore, the elliptical core fiber of Figure 3.31(a) generally has high doping levels which tend to increase the optical losses as well as the polarization cross-coupling [Ref. 97]. Alternatively, deep low refractive index side-pits can be employed to produce HB fibers, as depicted in Figure 3.31(b).

Stress birefringence may be induced using an elliptical cladding (Figure 3.31(c)) with a high thermal expansion coefficient. For example, borosilicate glass with some added germanium or phosphorus to provide index compensation can be utilized [Ref. 98]. The HB fibers shown in Figure 3.31(d) and (e) employ two distinct stress regions and are often referred to as the bow-tie [Ref. 99] and PANDA* [Ref. 100] fibers because of the shape of these regions. Alternatively, the flat cladding fiber design illustrated in Figure 3.31(f) has

* The mnemonic PANDA, however, represents polarization-maintaining and absorption-reducing.

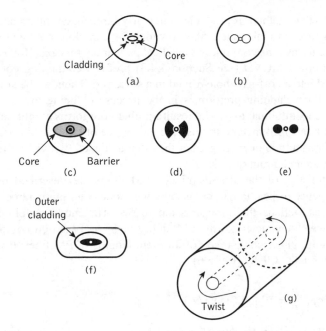

Figure 3.31 Polarization-maintaining fiber structure: (a) elliptical core; (b) side-pit fiber; (c) elliptical stress cladding; (d) bow-tie stress regions; (e) circular stress regions (PANDA fiber); (f) flat fiber; (g) twisted fiber

the outer edge of its elliptical cladding touching the fiber core which therefore divides the stressed cladding into two separate regions [Ref. 101].

In order to produce LB fibers attempts have been made to fabricate near-perfect, round-shaped core fibers. Ellipticity of less than 0.1% and modal birefringence of 4.5×10^{-9} have been achieved using the MCVD (see Section 4.4.3) fabricational technique [Refs 102, 103]. Moreover, the residual birefringence within conventional single-mode fibers can be compensated for by twisting the fiber after manufacture, as shown in Figure 3.31(g). A twist rate of around five turns per meter is sufficient to reduce crosstalk significantly between the polarization modes [Ref. 96].

The reduction occurs because a high degree of circular birefringence is created by the twisting process. Hence, it is found that the propagation constants of the modes polarized in the left hand and right hand circular directions are different. This has the effect of averaging out the linear birefringence and thus produces an LB fiber. Unfortunately, the method has limitations as the fiber tends to break when beat lengths are reduced to around 10 cm [Ref. 104].

An alternative method of compensation for the residual birefringence in conventional circularly symmetric single-mode fibers is to rotate the glass preform during the fiber drawing process to produce spun fiber [Ref. 105]. This geometric effect also decreases the residual linear birefringence on average by introducing circular birefringence, but without introducing shear stress. The technique has produced fibers with modal birefringence as low as 4.3×10^{-9} [Ref. 96].

Another effective method of producing circularly birefringent fibers and thus reducing linear birefringence is to fabricate a fiber in which the core does not lie along the longitudinal fiber axis; instead the core follows a helical path about this axis [Ref. 106]. To obtain such fibers a normal MCVD (see Section 4.4.3) preform containing core and cladding glass is inserted into an off-axis hole drilled in a silica rod. Then, as the silica rod containing the offset core–cladding preform is in the process of being drawn into fiber, it is rotated about its longitudinal axis. The resulting fiber core forms a tight helix which has a pitch length of a few millimeters. In this case the degree of circular birefringence tends to be an order of magnitude or more greater than that achieved by twisting the fiber, giving beat lengths of around 5 mm or less.

The characteristics of the aforementioned PM fibers are described not only by the modal birefringence or beat length but also by the mode coupling parameters or polarization crosstalk as well as their transmission losses. The mode coupling parameter or coefficient h, which characterizes the PM ability of fibers based on random mode coupling, proves useful in the comparison of different lengths of PM fiber. It is related to the polarization crosstalk* CT by [Ref. 96]:

$$CT = 10 \log_{10} \frac{P_y}{P_x} = 10 \log_{10} \tanh(hL) \qquad (3.68)$$

where P_x and P_y represent the optical power in the excited (i.e. unwanted) mode and the coupled (i.e. launch) mode, respectively, in an ensemble of fiber length L. However, it should be noted that the expression given in Eq. (3.68) applies with greater accuracy to two-polarization fibers because the crosstalk in a single-polarization fiber becomes almost constant around −30 dB and is independent of the fiber length beyond 200 m [Ref. 107].

Example 3.14

A 3.5 km length, of two-polarization mode PM fiber has a polarization crosstalk of −27 dB at its output end. Determine the mode coupling parameter for the fiber.

Solution: Using Eq. (3.68) relating the mode coupling parameter h to the polarization crosstalk CT:

$$\log_{10} \tanh(hL) = \frac{CT}{10} = -2.7$$

Thus $\tanh(hL) = 2 \times 10^{-3}$ and $hL \simeq 2 \times 10^{-3}$. Hence:

$$h = \frac{2 \times 10^{-3}}{3.5 \times 10^{-3}} = 5.7 \times 10^{-7} \text{ m}^{-1}$$

* The crosstalk is also referred to as the extinction ratio at the fiber output between the unwanted mode and the launch mode.

The generally higher transmission losses exhibited by PM fibers over conventional single-mode fibers is a major consideration in their possible utilization within coherent optical fiber communication systems. This factor is, however, less important when dealing with the short fiber lengths employed in fiber devices (see Section 5.6). Nevertheless, care is required in the determination of the cutoff wavelength or the measurement of fiber loss at longer wavelengths than cutoff because the transmission losses of the HE_{11}^x and HE_{11}^y modes in HB fibers exhibit different wavelength dependencies. PM fibers with losses approaching those of conventional single-mode fiber have been fabricated. For example, optical losses of around 0.23 dB km^{-1} at a wavelength of 1.55 μm with polarization cross-talk of −36 dB km^{-1} have been obtained [Refs 108, 109]. Such PM fibers could therefore eventually find application within long-haul coherent optical fiber transmission systems.

3.14 Nonlinear effects

Usually lightwaves or photons transmitted through a fiber have little interaction with each other, and are not changed by their passage through the fiber (except for absorption and scattering). There are exceptions, however, arising from the interactions between light-waves and the material transmitting them, which can affect optical signals. These processes are normally referred to as nonlinear effects or phenomena because their strength typically depends on the square (or some higher power) of the optical intensity. Hence nonlinear effects are weak at low powers but they can become much stronger at high optical intensities. This situation can result either when the power is increased, or when it is concentrated in a small area such as the core of a single-mode optical fiber.

Although the nonlinear effects in optical fibers are small, they accumulate as light passes through many kilometers of single-mode fiber. The small core diameters, together with the long transmission distances that may be obtained with these fibers, have enabled the occurrence of nonlinear phenomena at power levels of a few milliwatts which are well within the capability of semiconductor lasers. Furthermore, the optical power levels become much larger when wavelength division multiplexing (see Section 12.9.4) packs many signal channels into one single-mode fiber such that the overall power level is the summation of the individual channel optical powers [Ref. 110].

There are two broad categories of nonlinear effects that can be separated based on their characteristics: namely, scattering and Kerr effects. These fiber nonlinearities are identified in Figure 3.32 where the fiber attenuation associated with nonlinear scattering was discussed in Section 3.5, but both these and the Kerr effects shown may also be employed in important applications for single-mode fibers including distributed in-fiber amplification, wavelength conversion (see Section 10.5.4), multiplexing and demultiplexing, pulse regeneration, optical monitoring and optical switching [Ref. 111].

3.14.1 Scattering effects

It was indicated in Section 3.5 that when an optical wave is within a fiber medium incident photons may be scattered, producing a phonon emitted at acoustic frequencies by exciting molecular vibrations, together with another photon at a shifted frequency. In quantum

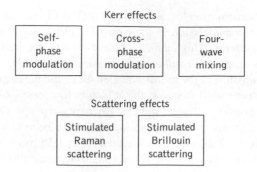

Figure 3.32 Block schematic showing the fiber nonlinear effects

mechanical terms this process can be described as the molecule absorbing the photon at the original frequency while emitting a photon at the shifted frequency and simultaneously making a transition between vibrational states. The scattered photon therefore emerges at a frequency shifted below or above the incident photon frequency with the energy difference between the two photons being deposited or extracted from the scattering medium. An upshifted photon frequency is only possible if the material gives up quantum energy equal to the energy difference between the incident and scattered photon. The material must therefore be in a thermally excited state before the incident photon arrives, and at room temperature (i.e. 300 K) the upshifted scattering intensity is much weaker than the downshifted one. The former scattered wave is known as the Stokes component whereas the latter is referred to as the anti-Stokes component. In contrast to linear scattering (i.e. Rayleigh), which is said to be elastic because the scattered wave has the same frequency as the incident wave, these nonlinear scattering processes are clearly inelastic. A schematic of the spectrum obtained from these inelastic scattering processes is shown in Figure 3.33. It should be noted that the schematic depicts the spontaneous scattering spectrum rather than the stimulated one.

The frequency shifts associated with inelastic scattering can be small (less than 1 cm^{-1}), which typifies Brillouin scattering with an acoustic frequency phonon. Larger frequency shifts (greater than 100 cm^{-1}) characterize the Raman regime where the photon is scattered by local molecular vibrations or by optical frequency phonons. An interesting feature of these inelastic scattering processes is that they not only result in a frequency shift but for sufficiently high incident intensity also provide optical gain at the shifted frequency. The incident optical frequency is also known as the pump frequency ω_p, which gives the Stokes (ω_s) and anti-Stokes (ω_a) components of the scattered radiation (see Figure 3.33). For a typical fiber, a pump power of around 1 watt in 100 m of fiber results in a Raman gain of about a factor of 2 [Ref. 112]. By contrast, the peak Brillouin gain is more than two orders of magnitude greater than the Raman gain, but the Brillouin frequency shift and gain bandwidth are much smaller. Furthermore, Brillouin gain only exists for light propagation in the opposite direction to the pump light while Raman amplification will occur for light propagating in either direction.

Raman gain also extends over a substantial bandwidth, as may be observed in Figure 3.34 [Ref. 113]. Hence, with a suitable pump source, a fiber can function as a relatively high-gain, broad-bandwidth, bidirectional optical amplifier (see Section 10.4.2). Although

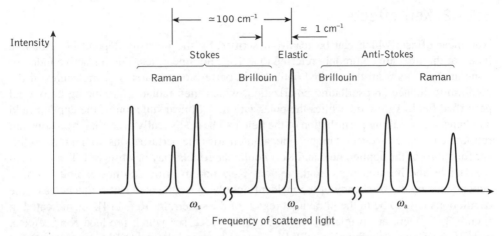

Figure 3.33 Spectrum of scattered light showing the inelastic scattering processes. Not drawn to scale as the intensities of the anti-Stokes Raman lines are far less than those of the Stokes Raman lines

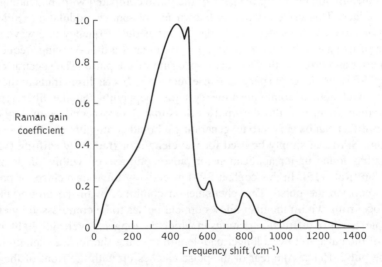

Figure 3.34 Raman gain spectrum for a silica core single-mode fiber. The peak gain occurred at 440 cm^{-1} with a pump wavelength of 0.532 μm. Reprinted with permission from W. J. Tomlinson and R. H. Stolen, 'Nonlinear phenomena in optical fibers', *IEEE Commun. Mag.*, **26**, p. 36, 1988. Copyright © 1988 IEEE

given its much greater peak gain, it might be expected that Brillouin amplification would dominate over Raman amplification. At present this is not usually the case because of the narrow bandwidth associated with the Brillouin process which is often in the range 20 to 80 MHz. Pulsed semiconductor laser sources generally have much broader bandwidths and therefore prove inefficient pumps for such a narrow-gain spectrum.

3.14.2 Kerr effects

Nonlinear effects which can be readily described by the intensity-dependent refractive index of the fiber are commonly referred to as Kerr nonlinearities. The refractive index of a medium results from the applied optical field perturbing the atoms or molecules of the medium to induce an oscillating polarization, which then radiates, producing an overall perturbed field. At low intensities the polarization is a linear function of the applied field and hence the resulting perturbation of the field can be realistically described by a constant refractive index. However, at higher optical intensities the perturbations do not remain linear functions of the applied field and Kerr nonlinear effects may be observed. Typically, in the visible and infrared wavelength regions Kerr nonlinearities do not exhibit a strong dependence on the frequency of the incident light because the resonant frequencies of the oscillations tend to be in the ultraviolet region of the spectrum [Ref. 114]. As indicated in Figure 3.32, there are primarily, however, three processes which produce Kerr effects: namely, self-phase modulation (SPM), cross-phase modulation (XPM) and four-wave mixing (FWM).

The intensity-dependent refractive index causes an intensity-dependent phase shift in the fiber. Hence, for a light pulse propagating in the fiber, Kerr nonlinearities result in a different transmission phase for the peak of the pulse compared with the leading and trailing pulse edges. This effect, which is known as self-phase modulation (SPM), causes modifications to the pulse spectrum. As the instantaneous frequency of a wave is the time derivative of its phase, then a time-varying phase creates a time-varying frequency. Thus SPM can alter and broaden the frequency spectrum of the pulse. The spectral broadening caused by SPM produces dispersion-like effects which can limit transmission rates in some long-haul optical communication systems, depending on the fiber type and its chromatic dispersion. For ultrashort pulses (less than 1 picosecond) with very high peak powers, its effect can be very strong, generating a broad continuum of wavelengths.

Although SPM can simply be used for wavelength or frequency shifting (see Section 11.4.4), it has found major application for pulse compression within single-mode fiber transmission [Ref. 115]. In this context SPM effectively imposes a chirp, or positive frequency sweep, on the pulse. This phenomenon combined with the group velocity dispersion* occurring within the fiber allows optical pulses to be compressed by employing, for example, a pair of diffraction gratings in which the longer wavelength light traveling at the front of the pulse follows a longer path length than the shorter wavelength light at the rear of the pulse. Hence, the rear of the pulse catches up with the front of the pulse and compression occurs. Furthermore, for critical pulse shapes and at high optical power levels, such pulse compression can be obtained in the fiber itself which forms the basis of so-called soliton propagation (see Section 3.15).

Cross-phase modulation (XPM) is a similar effect to SPM except that overlapping but distinguishable pulses, possessing, for example, different wavelengths or polarizations, are involved. In this case variations in intensity of one pulse will modulate the refractive

---

* It is usual to describe the group velocity dispersion resulting from the frequency dependence of the group velocity (i.e. the different spectral components within a pulse exhibit a different group delay τ_g thus causing pulse spread) in terms of the chromatic or intramodal dispersion which for a unit length of fiber is defined by $d\tau_g/d\lambda$.

index of the fiber which causes phase modulation of the overlapping pulse(s). As with SPM, this phase modulation translates into frequency modulation which broadens the pulse spectrum. Thus XPM is exhibited as a crosstalk mechanism between channels when either intensity modulation is used in dispersive optical fiber transmission or, alternatively, when phase encoding is employed [Ref. 116]. Moreover, the strength of XPM increases with the number of channels and it also becomes stronger as the channel spacing is made smaller. There is no energy transfer, however, between channels, which distinguishes the effect from other crosstalk processes in which the increase in signal power in a channel takes place only by a reduction in power in another one. Although the overall strength of XPM is twice that of SPM because the total intensity is the square of the sum of two electric field amplitudes, the effect is weakened as pulses with different wavelengths or polarizations are usually not group velocity matched and therefore the overlap is not maintained [Ref. 111].

The beating between light at different frequencies or wavelengths in multichannel fiber transmission causes phase modulation of the channels and hence the generation of modulation sidebands at new frequencies which are termed four-wave mixing (FWM). When three wave components copropagate at angular frequencies ω_1, ω_2 and ω_3, then a new wave is generated at frequency ω_4 where $\omega_4 = \omega_1 + \omega_2 - \omega_3$. This frequency combination can be problematic for multichannel optical communications as they can become phase matched if the channel wavelengths are close to the zero-dispersion wavelength. FWM is therefore one of a broad class of harmonic mixing or harmonic generation processes in which two or more waves combine to generate waves at a different frequency that is the sum (or difference) of the signals that are mixed. Such second-harmonic generation or frequency doubling is common in optics; it combines two waves at the same frequency to generate a wave at twice the frequency (or, equivalently, at half the wavelength). This phenomenon can occur in optical fibers, but the first harmonic of the 1.55 μm wavelength is at 0.775 μm which is not quite in the optical communications band and thus it does not interfere with the signal transmission wavelength.

Although FWM is a weak effect, it can accumulate when multichannel signals remain in phase with each other over long transmission distances, which is typically when the fiber chromatic dispersion is very close to zero. Hence pulses transmitted over different optical channels at different wavelengths stay in the same relative positions along the fiber length because the signals experience near-zero dispersion. In this case the effect of FWM is amplified and a noise signal builds up which interferes with other channels on the system. Hence one method to minimize crosstalk resulting from FWM in wavelength division multiplexed systems based on low-dispersion fiber is to employ unequal channel spacing so that the FWM noise components are not generated at frequencies which correspond to the channel frequencies [Ref. 117].

3.15 Soliton propagation

Soliton propagation results from a special case of nonlinear dispersion compensation in which the nonlinear chirp caused by SPM balances, and hence postpones, the temporal broadening induced by group velocity delay (GVD). Although both of these phenomena limit the propagation distance that can be achieved when acting independently, if balanced at the necessary critical pulse intensity they enable the pulse to propagate without any distortion (i.e. its shape is self-maintaining) as a soliton. In essence a soliton has two distinctive features which are potentially important for the provision of high-speed optical fiber communications: it propagates without changing shape; and the shape is unaffected, that of a soliton, after a collision with another soliton. Hence the former soliton property overcomes the dispersion limitation and avoids intersymbol interference while the collision invariance potentially provides for efficient wavelength division multiplexing (see Section 12.9.4).

Optical soliton propagation was initially determined theoretically where, in an ideal optical fiber exhibiting only second-order dispersion and the Kerr nonlinearity, the wave envelope of the light follows the nonlinear Schrödinger equation written as [Ref. 118]:

$$j\frac{\partial U}{\partial z} - \frac{\beta_2}{2}\frac{\partial^2 U}{\partial \tau^2} + \gamma|U|^2 U = 0 \tag{3.69}$$

where U represents the complex wave envelope, z is the distance along the fiber, β_2 is the second-order dispersion coefficient, and γ is the nonlinear coefficient resulting from the Kerr effect. Furthermore, the quantity τ, referred to as the retarded time, is equal to $t - z/v_g$ where t is the physical time with v_g being the group velocity. It can therefore be shown that when $\beta_2 < 0$, Eq. (3.69) has a soliton solution* for the input pulse to the fiber which is:

$$U = \left(\frac{|\beta_2|}{\gamma T^2}\right)^{\frac{1}{2}} \operatorname{sech}\left(\frac{\tau}{T}\right) \exp\left(j\frac{|\beta_2|}{2T^2}z\right) \tag{3.70}$$

where T is an arbitrary parameter representing the soliton pulse duration. The variable term $(|\beta_2|/\gamma T^2)^{\frac{1}{2}}$ in Eq. (3.70), sometimes referred to as the parameter N, comprises a dimensionless combination of pulse and fiber parameters. When the input pulse represented by Eq. (3.70) is launched into the fiber its shape remains unchanged for $N = 1$ which is known as the fundamental soliton, while higher order solitons exist for integer values of N greater than 1. The pulse intensity over distance along the fiber is shown in Figure 3.35 for both the fundamental ($N = 1$) and the third-order ($N = 3$) solitons over a soliton period. It may be noted that the shape of the fundamental soliton remains unchanged over the period whereas although the shape of the third-order soliton changes, it does so in a periodic manner recovering its original shape after a distance equal to the soliton period.†

Although a single soliton pulse in an ideal lossless fiber is completely stable, in an actual optical fiber the fiber attenuation must be taken into account. This fiber loss may be readily compensated using optical amplifiers (see Chapter 10) which leads to a periodic evolution of the pulse amplitude but also to the accumulation of other optical impairments along the fiber link (e.g. amplified spontaneous emission noise). Hence if the soliton energy

---

* For the condition $\beta_2 < 0$, the solutions are pulse-like solitons also referred to as a bright solitons, whereas in the case of normal dispersion when $\beta_2 > 0$, the solutions which exhibit a dip in a constant intensity background are referred to as dark solitons; these are not currently being pursued for practical communication applications as they exist in the spectral region where fibers have high losses.
† The soliton period is, in fact, defined as the distance over which higher order solitons recover their original shape.

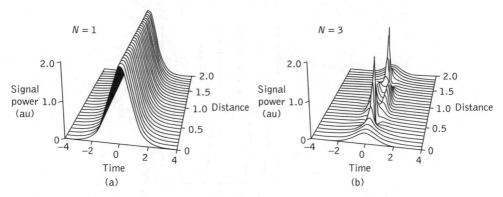

Figure 3.35 Pulse intensity over one soliton period: (a) fundamental soliton;
(b) third-order soliton

is maintained within the fiber through periodic optical amplification, the SPM compensates
for the GVD and the solitons maintain their widths and are not destroyed. When the
amplifier spacing on the fiber link is determined in order to maintain such soliton propagation, then the solitons are referred to as being loss-managed.

In the above case it has been assumed that the GVD is constant along the fiber link,
whereas in practice it may vary (or be varied through dispersion management techniques
[Ref. 118]). Nevertheless it turns out that solitons can both form and propagate when the
GVD (and hence β_2) varies along the fiber link [Ref. 119]. Indeed for such a dispersion-
managed soliton propagation regime a new single-mode fiber type has been developed
called dispersion-decreasing fiber [Ref. 120]. This fiber type is designed so that the
reduced SPM experienced by solitons which have been attenuated through the fiber losses
is counteracted by a decreasing GVD. Furthermore, when the GVD is designed to
decrease exponentially in a fiber section between two optical amplifiers, then the spacing
of the amplifiers is independent of the fiber losses. Hence in these circumstances the
fundamental soliton maintains its shape and width even in a high-loss fiber. A practical
technique for fabricating dispersion-decreasing single-mode fibers consists of reducing
the fiber core diameter along the fiber length during the fiber drawing process (see
Section 4.3.1) and in this way fibers with a nearly exponential GVD profile have been
obtained [Ref. 121].

A further soliton propagation issue results from the amplified spontaneous emission
noise on an optically amplified fiber link which acts to produce random variations of the
solitons' central frequencies. Chromatic dispersion within the single-mode fiber then converts these variations in frequency to a jitter in pulse arrival times which is known as the
Gordon–Haus effect [Ref. 122]. This timing jitter (see Section 12.6.1.) can cause some of
the pulses to move out of their correct bit time slots which will then create errors in the
soliton transmission. These bit time slots are depicted by the dashed lines in Figure 3.36
which illustrates the nonreturn-to-zero (NRZ) transmission format often used in intensity-
modulated direct detection fiber systems (see Section 12.6.7) together with the soliton
pulse format. It may be noted that a binary 1 in the NRZ format fills the bit time slot
almost uniformly with optical energy, whereas the binary 0 puts as little energy as possible
in the time slot. Although the optical energy is concentrated in the center of the bit time

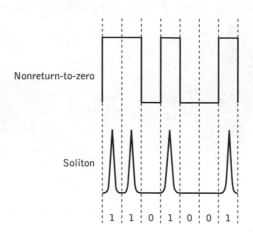

Figure 3.36 Comparison of the nonreturn-to-zero and soliton transmission formats

slot with the soliton format, timing jitter resulting from the Gordon–Haus effect can shift soliton pulses into adjacent bit time slots, creating errors in the received bit pattern.

Problems

3.1   The mean optical power launched into an optical fiber link is 1.5 mW and the fiber has an attenuation of 0.5 dB km^{-1}. Determine the maximum possible link length without repeaters (assuming lossless connectors) when the minimum mean optical power level required at the detector is 2 µW.

3.2   The numerical input/output mean optical power ratio in a 1 km, length of optical fiber is found to be 2.5. Calculate the received mean optical power when a mean optical power of 1 mW is launched into a 5 km length of the fiber (assuming no joints or connectors).

3.3   A 15 km optical fiber link uses fiber with a loss of 1.5 dB km^{-1}. The fiber is jointed every kilometer with connectors which give an attenuation of 0.8 dB each. Determine the minimum mean optical power which must be launched into the fiber in order to maintain a mean optical power level of 0.3 µW at the detector.

3.4   Discuss absorption losses in optical fibers, comparing and contrasting the intrinsic and extrinsic absorption mechanisms.

3.5   Briefly describe linear scattering losses in optical fibers with regard to:
(a) Rayleigh scattering;
(b) Mie scattering.
The photoelastic coefficient and the refractive index for silica are 0.286 and 1.46 respectively. Silica has an isothermal compressibility of 7×10^{-11} m^2 N^{-1} and an estimated fictive temperature of 1400 K. Determine the theoretical attenuation in

decibels per kilometer due to the fundamental Rayleigh scattering in silica at optical wavelengths of 0.85 and 1.55 μm. Boltzmann's constant is 1.381×10^{-23} J K^{-1}.

3.6 A K_2O–SiO_2 glass core optical fiber has an attenuation resulting from Rayleigh scattering of 0.46 dB km^{-1} at a wavelength of 1 μm. The glass has an estimated fictive temperature of 758 K, isothermal compressibility of 8.4×10^{-11} m^2 N^{-1}, and a photoelastic coefficient of 0.245. Determine from theoretical considerations the refractive index of the glass.

3.7 Compare stimulated Brillouin and stimulated Raman scattering in optical fibers, and indicate the way in which they may be avoided in optical fiber communications.

The threshold optical powers for stimulated Brillouin and Raman scattering in a single-mode fiber with a long 8 μm core diameter are found to be 190 mW and 1.70 W, respectively, when using an injection laser source with a bandwidth of 1 GHz. Calculate the operating wavelength of the laser and the attenuation in decibels per kilometer of the fiber at this wavelength.

3.8 The threshold optical power for stimulated Brillouin scattering at a wavelength of 0.85 μm in a long single-mode fiber using an injection laser source with a bandwidth of 800 MHz is 127 mW. The fiber has an attenuation of 2 dB km^{-1} at this wavelength. Determine the threshold optical power for stimulated Raman scattering within the fiber at a wavelength of 0.9 μm assuming the fiber attenuation is reduced to 1.8 dB km^{-1} at this wavelength.

3.9 Explain what is meant by the critical bending radius for an optical fiber.

A multimode graded index fiber has a refractive index at the core axis of 1.46 with a cladding refractive index of 1.45. The critical radius of curvature which allows large bending losses to occur is 84 μm when the fiber is transmitting light of a particular wavelength. Determine the wavelength of the transmitted light.

3.10 A single-mode step index fiber with a core refractive index of 1.49 has a critical bending radius of 10.4 mm when illuminated with light at a wavelength of 1.30 μm. If the cutoff wavelength for the fiber is 1.15 μm calculate its relative refractive index difference.

3.11 (a) A multimode step index fiber gives a total pulse broadening of 95 ns over a 5 km length. Estimate the bandwidth–length product for the fiber when a nonreturn to zero digital code is used.

(b) A single-mode step index fiber has a bandwidth–length product of 10 GHz km. Estimate the rms pulse broadening over a 40 km digital optical link without repeaters consisting of the fiber, and using a return to zero code.

3.12 An 8 km optical fiber link without repeaters uses multimode graded index fiber which has a bandwidth–length product of 400 MHz km. Estimate:
(a) the total pulse broadening on the link;
(b) the rms pulse broadening on the link.
It may be assumed that a return to zero code is used.

3.13 Briefly explain the reasons for pulse broadening due to material dispersion in optical fibers.

The group delay τ_g in an optical fiber is given by:

$$\tau_g = \frac{1}{c}\left(n_1 - \frac{\lambda \mathrm{d}n_1}{\mathrm{d}\lambda}\right)$$

where c is the velocity of light in a vacuum, n_1 is the core refractive index and λ is the wavelength of the transmitted light. Derive an expression for the rms pulse broadening due to material dispersion in an optical fiber and define the material dispersion parameter.

The material dispersion parameter for a glass fiber is $20\ \mathrm{ps\ nm^{-1}\ km^{-1}}$ at a wavelength of 1.5 μm. Estimate the pulse broadening due to material dispersion within the fiber when light is launched from an injection laser source with a peak wavelength of 1.5 μm and an rms spectral width of 2 nm into a 30 km length of the fiber.

3.14   The material dispersion in an optical fiber defined by $|\ \mathrm{d}^2 n_1/\mathrm{d}\lambda^2\ |$ is $4.0 \times 10^{-2}\ \mathrm{μm^{-2}}$. Estimate the pulse broadening per kilometer due to material dispersion within the fiber when it is illuminated with an LED source with a peak wavelength of 0.9 μm and an rms spectral width of 45 nm.

3.15   Describe the mechanism of intermodal dispersion in a multimode step index fiber.

Show that the total broadening of a light pulse δT_s due to intermodal dispersion in a multimode step index fiber may be given by:

$$\delta T_s \simeq \frac{L(NA)^2}{2n_1 c}$$

where L is the fiber length, NA is the numerical aperture of the fiber, n_1 is the core refractive index and c is the velocity of light in a vacuum.

A multimode step index fiber has a numerical aperture of 0.2 and a core refractive index of 1.47. Estimate the bandwidth–length product for the fiber assuming only intermodal dispersion and a return to zero code when:
(a) there is no mode coupling between the guided modes;
(b) mode coupling between the guided modes gives a characteristic length equivalent to 0.6 of the actual fiber length.

3.16   Using the relation for δT_s given in Problem 3.15, derive an expression for the rms pulse broadening due to intermodal dispersion in a multimode step index fiber. Compare this expression with a similar expression which may be obtained for an optimum near-parabolic profile graded index fiber.

Estimate the bandwidth–length product for the step index fiber specified in Problem 3.15 considering the rms pulse broadening due to intermodal dispersion within the fiber and comment on the result. Indicate the possible improvement in the bandwidth–length product when an optimum near-parabolic profile graded index fiber with the same relative refractive index difference and core axis refractive index is used. In both cases assume only intermodal dispersion within the fiber and the use of a return to zero code.

3.17 An 11 km optical fiber link consisting of optimum near-parabolic profile graded index fiber exhibits rms intermodal pulse broadening of 346 ps over its length. If the fiber has a relative refractive index difference of 1.5%, estimate the core axis refractive index. Hence determine the numerical aperture for the fiber.

3.18 A multimode, optimum, near-parabolic profile graded index fiber has a material dispersion parameter of 30 ps nm^{-1} km^{-1} when used with a good LED source of rms spectral width 25 nm. The fiber has a numerical aperture of 0.4 and a core axis refractive index of 1.48. Estimate the total rms pulse broadening per kilometer within the fiber assuming waveguide dispersion to be negligible. Hence, estimate the bandwidth–length product for the fiber.

3.19 A multimode step index fiber has a relative refractive index difference of 1% and a core refractive index of 1.46. The maximum optical bandwidth that may be obtained with a particular source on a 4.5 km link is 3.1 MHz.
   (a) Determine the rms pulse broadening per kilometer resulting from chromatic dispersion mechanisms.
   (b) Assuming waveguide dispersion may be ignored, estimate the rms spectral width of the source used, if the material dispersion parameter for the fiber at the operating wavelength is 90 ps nm^{-1} km^{-1}.

3.20 Describe the phenomenon of modal noise in optical fibers and suggest how it may be avoided.

3.21 Discuss dispersion mechanisms with regard to single-mode fibers indicating the dominating effects. Hence, describe how intramodal dispersion may be minimized within the single-mode region.

3.22 An approximation for the normalized propagation constant in a single-mode step index fiber shown in Example 2.9 is:

$$b(V) \simeq \left(1.1428 - \frac{0.9960}{V}\right)^2$$

Obtain a corresponding approximation for the waveguide parameter $V\, d^2(Vb)/dV^2$ and hence write down an expression for the waveguide dispersion in the fiber.

   Estimate the waveguide dispersion in a single-mode step index fiber at a wavelength of 1.34 μm when the fiber core radius and refractive index are 4.4 μm and 1.48 respectively.

3.23 A single-mode step index fiber exhibits material dispersion of 7 ps nm^{-1} km^{-1} at an operating wavelength of 1.55 μm. Using the approximation obtained in Problem 3.22, estimate the fiber core diameter which will enable the waveguide dispersion to cancel the material dispersion so that zero intramodal dispersion is obtained at this wavelength. The refractive index of the fiber core is 1.45.

3.24 A single-mode step index fiber has a zero-dispersion wavelength of 1.29 μm and exhibits total first-order dispersion of 3.5 ps nm^{-1} km^{-1} at a wavelength of 1.32 μm. Determine the total first-order dispersion in the fiber at a wavelength of 1.54 μm.

3.25 Describe the techniques employed and the fiber structures utilized to provide:
(a) dispersion-shifted single-mode fibers;
(b) dispersion-flattened single-mode fibers.
(c) nonzero-dispersion shifted single-mode fibers.

3.26 Explain what is meant by:
(a) fiber birefringence;
(b) the beat length;
in single-mode fibers.

   The difference between the propagation constants for the two orthogonal modes in a single-mode fiber is 250. It is illuminated with light of peak wavelength 1.55 μm from an injection laser source with a spectral linewidth of 0.8 nm. Estimate the coherence length within the fiber.

3.27 The difference in the effective refractive indices $(n_x - n_y)$ for the two orthogonally polarized modes in conventional single-mode fibers is in the range $9.3 \times 10^{-7} < n_x - n_y < 1.1 \times 10^{-5}$. Determine the corresponding range for the beat lengths of the fibers when they are operating at a transmission wavelength of 1.3 μm. Hence obtain the range of the modal birefringence for the fibers.

3.28 A single-mode fiber maintains birefringent coherence over a length of 100 km when it is illuminated with an injection laser source with a spectral linewidth of 1.5 nm and a peak wavelength of 1.32 μm. Estimate the beat length within the fiber and comment on the result.

3.29 Provide a definition for polarization mode dispersion (PMD) in single-mode optical fibers.

   Discuss the statistical approach to the theory of PMD explaining the relationship between the fiber polarization beat length and the correlation length. Hence describe the two distinct PMD regions as determined by the correlation length, suggesting the reason why with recently deployed single-mode fiber links PMD is not necessarily a major limitation on link performance.

3.30 Describe, with the aid of sketches, the techniques that can be employed to produce both high- and low-birefringence PM fibers.

   A two-polarization mode PM fiber has a mode coupling parameter of 2.3×10^{-5} m^{-1} when operating at a wavelength of 1.55 μm. Estimate the polarization crosstalk for the fiber at this wavelength.

3.31 Explain what is meant by self-phase modulation.

   Identify and discuss a major application area for this nonlinear phenomenon.

3.32 Describe the two distinctive features of an optical soliton pulse and indicate how loss-managed solitons are produced and maintained on an actual single-mode fiber link.

3.33 Explain the phenomenon referred to as the Gordon–Haus effect and identify the major problem this may cause in optical soliton communications.

Answers to numerical problems

3.1	57.5 km	**3.15**	(a) 11.0 MHz km; (b) 14.2 MHz km
3.2	10.0 μW	**3.16**	15.3 MHz km; improvement to 10.9 GHz km
3.3	703 μW	**3.17**	1.45, 0.25
3.5	1.57 dB km^{-1}, 0.14 dB km^{-1}	**3.18**	774 ps km^{-1}, 258 MHz km
3.6	1.49	**3.19**	(a) 2.82 ns km^{-1}; (b) 31 nm
3.7	1.50 μm, 0.30 dB km^{-1}	**3.22**	−3.92 ps nm^{-1} km^{-1}
3.8	2.4 W	**3.23**	7.2 μm
3.9	0.86 μm	**3.24**	23.6 ps nm^{-1} km^{-1}
3.10	0.47%	**3.26**	48.6 m
3.11	(a) 13.2 MHz km; (b) 800 ps	**3.27**	12 cm $< L_B <$ 1.4 m; 9.3 × 10^{-7} $< B_F <$ 1.1 × 10^{-5}
3.12	(a) 10 ns; (b) 4 ns		
3.13	1.2 ns	**3.28**	113.6 m
3.14	5.4 ns km^{-1}	**3.30**	−16.4 dB km^{-1}

References

[1] F. P. Kapron, D. B. Keck and R. D. Maurer, 'Radiation losses in optical waveguides', *Appl. Phys. Lett.*, **10**, pp. 423–425, 1970.

[2] S. R. Nagel, 'Optical fiber – the expanding medium', *IEEE Commun. Mag.*, **25**(4), pp. 33–43, 1987.

[3] D. J. DiGiovanni, S. K. Das, L. L. Blyler, W. White, R. K. Boncek and S. E. Golowich, 'Design of optical fibers for communication systems', in I. P. Kaminow and T. Li (Eds), *Optical Fiber Telecommunications IVA*, pp. 17–79, Academic Press, 2002.

[4] J. A. Harrington, 'Infrared fibers', in M. Bass and E. W. Van Stryland (Eds), *Fiber Optics Handbook*, pp. 14.1–14.16, McGraw-Hill, 2002.

[5] T. Miya, Y. Teramuna, Y. Hosaka and T. Miyashita, 'Ultimate low-loss single-mode fibre at 1.55 μm', *Electron. Lett.*, **15**(4), pp. 106–108, 1979.

[6] P. C. Schultz. 'Preparation of very low loss optical waveguides', *J. Am. Ceram. Soc.*, **52**(4), pp. 383–385, 1973.

[7] H. Osanai, T. Shioda, T. Morivama, S. Araki, M. Horiguchi, T. Izawa and H. Takata, 'Effect of dopants on transmission loss of low OH-content optical fibres', *Electron. Lett.*, **12**(21), pp. 549–550, 1976.

[8] K. J. Beales and C. R. Day, 'A review of glass fibres for optical communications', *Phys. Chem. Glasses*, **21**(1), pp. 5–21, 1980.

[9] K. Nagayama, M. Kakui, M. Matsui, T. Siatoh and Y. Chigusa, 'Ultra-low-loss (0.1484 dB/km) pure silica core fiber and extension of transmission distance', *Electron. Lett.*, **38**(20), pp. 1168–1169, 2002.

[10] G. A. Thomas, B. L. Shraiman, P. F. Glodis and M. J. Stephan, 'Towards the clarity limit in optical fibre', *Nature*, **404**, pp. 262–264, 2000.

[11] R. Olshansky, 'Propagation in glass optical waveguides', *Rev. Mod. Phys.*, **51**(2), pp. 341–367, 1979.

[12] R. M. Gagliardi and S. Karp, *Optical Communications*, Wiley, 1976.

[13] J. Schroeder, R. Mohr, P. B. Macedo and C. J. Montrose, 'Rayleigh and Brillouin scattering in K$_2$O–SiO$_2$ glasses', *J. Am. Ceram. Soc.*, **56**, pp. 510–514, 1973.

[14] R. D. Maurer, 'Glass fibers for optical communications', *Proc. IEEE*, **61**, pp. 452–462, 1973.

[15] D. A. Pinnow, T. C. Rich, F. W. Ostermayer Jr and M. DiDomenico Jr, 'Fundamental optical attenuation limits in the liquid and glassy state with application to fiber optical waveguide materials', *Appl. Phys. Lett.*, **22**, pp. 527–529, 1973.

[16] R. H. Stolen, 'Nonlinearity in fiber transmission', *Proc. IEEE*, **68**(10), pp. 1232–1236, 1980.

[17] R. H. Stolen, 'Nonlinear properties of optical fibers', in S. E. Miller and A. G. Chynoweth (Eds), *Optical Fiber Telecommunications*, pp. 125–150, Academic Press, 1979.

[18] M. M. Ramsay and G. A. Hockham, 'Propagation in optical fibre waveguides', in C. P. Sandbank (Ed.), *Optical Fibre Communication Systems*, pp. 25–41, Wiley, 1980.

[19] H. F. Wolf, 'Optical waveguides', in H. F. Wolf (Ed.), *Handbook of Fiber Optics: Theory and Applications*, pp. 43–152, Granada, 1979.

[20] W. A. Gambling, H. Matsumura and C. M. Ragdale, 'Curvature and microbending losses in single mode optical fibres', *Opt. Quantum Electron.*, **11**, pp. 43–59, 1979.

[21] J. A. Savage, 'Materials for infrared fibre optics', *Mater. Sci. Rep.*, **2**, pp. 99–138, 1987.

[22] M. Saito, M. Takizawa and M. Miyagi, 'Optical and mechanical properties of infrared fibers', *J. Lightwave Technol.*, **6**(2), pp. 233–239, 1988.

[23] M. Saito and K. K. Kikuchi, 'Infrared optical fiber sensors', *Opt. Rev.*, **4**, pp. 527–538, 1997.

[24] S. R. Nagel, 'Fiber materials and fabrication methods', in S. E. Miller and I. P. Kaminow (Eds), *Optical Fiber Telecommunications II*, pp. 121–215, Academic Press, 1988.

[25] S. Sakaguchi and S. Takahashi, 'Low-loss fluoride optical fibers for midinfrared optical communication', *J. Lightwave Technol.*, **LT-5**(9), pp. 1219–1228, 1987.

[26] S. F. Carter, M. W. Moore, D. Szebesta, D. Ransom and P. W. France, 'Low loss fluoride fiber by reduced pressure casting', *Electron. Lett.*, **26**, pp. 2115–2117, 1990.

[27] R. Nubling and J. A. Harrington, 'Single-crystal LHPG sapphire fibers for Er:YAG laser power delivery', *Appl. Opt.*, **37**, pp. 4777–4781, 1998.

[28] R. Nubling and J. A. Harrington, 'Hollow-waveguide delivery systems for high-power industrial CO_2 lasers', *Appl. Opt.*, **34**, pp. 372–380, 1996.

[29] I. P. Kaminow, D. Marcuse and H. M. Presby, 'Multimode fiber bandwidth: theory and practice', *Proc. IEEE*, **68**(10), pp. 1209–1213, 1980.

[30] M. J. Adams, D. N. Payne, F. M. Sladen and A. H. Hartog, 'Optimum operating wavelength for chromatic equalisation in multimode optical fibres', *Electron. Lett.*, **14**(3), pp. 64–66, 1978.

[31] W. A. Gambling, A. H. Hartog and C. M. Ragdale, 'Optical fibre transmission lines', *Radio Electron. Eng. J. IERE*, **51**(7/8), pp. 313–325, 1981.

[32] D. N. Payne and W. A. Gambling, 'Zero material dispersion in optical fibres', *Electron. Lett.*, **11**(8), pp. 176–178, 1975.

[33] F. G. Stremler, *Introduction in Communication Systems* (2nd edn), Addison-Wesley, 1982.

[34] D. Botez and G. J. Herskowitz, 'Components for optical communication systems: a review', *Proc. IEEE*, **68**(6), pp. 689–730, 1980.

[35] A. Ghatak and K. Thyagarajan, 'Graded index optical waveguides: a review', in E. Wolf (Ed.), *Progress in Optics*, pp. 1–109, North-Holland, 1980.

[36] D. Gloge and E. A. Marcatili, 'Multimode theory of graded-core fibers', *Bell Syst. Tech. J.*, **52**, pp. 1563–1578, 1973.

[37] J. E. Midwinter, *Optical Fibers for Transmission*, Wiley, 1979.

[38] R. Olshansky and D. B. Keck, 'Pulse broadening in graded-index optical fibers', *Appl. Opt.*, **15**(12), pp. 483–491, 1976.

[39] R. E. Epworth, 'The phenomenon of modal noise in analogue and digital optical fibre systems', in *Proc. 4th Eur. Conf. on Optical Communications*, Italy, pp. 492–501, 1978.

[40] A. R. Godwin, A. W. Davis, P. A. Kirkby, R. E. Epworth and R. G. Plumb, 'Narrow stripe semiconductor laser for improved performance of optical communication systems', *Proc. 5th Eur. Conf. on Optical Communications*, The Netherlands, paper 4–3, 1979.

[41] K. Sato and K. Asatani, 'Analogue baseband TV transmission experiments using semiconductor laser diodes', *Electron. Lett.*, **15**(24), pp. 794–795, 1979.

[42] B. Culshaw, 'Minimisation of modal noise in optical-fibre connectors', *Electron. Lett.*, **15**(17), pp. 529–531, 1979.

[43] N. K. Cheung, A. Tomita and P. F. Glodis, 'Observation of modal noise in single-mode fiber transmission systems', *Electron. Lett.*, **21**, pp. 5–7, 1985.

[44] F. M. Sears, I. A. White, R. B. Kummer and F. T. Stone, 'Probability of modal noise in single-mode lightguide systems', *J. Lightwave Technol.*, **LT-4**, pp. 652–655, 1986.

[45] D. Gloge, 'Dispersion in weakly guiding fibers', *Appl. Opt.*, **10**(11), pp. 2442–2445, 1971.

[46] W. A. Gambling, H. Matsumura and C. M. Ragdale, 'Mode dispersion, material dispersion and profile dispersion in graded index single-mode fibers', *IEE J. Microw. Opt. Acoust.*, **3**(6), pp. 239–246, 1979.

[47] E. G. Neumann, *Single-Mode Fibers: Fundamentals*, Springer-Verlag, 1988.

[48] J. I. Yamada, M. Saruwatari, K. Asatani, H. Tsuchiya, A. Kawana, K. Sugiyama and T. Kimura, 'High speed optical pulse transmission at 1.29 μm wavelength using low-loss single-mode fibers', *IEEE J. Quantum Electron.*, **QE-14**, pp. 791–800, 1978.

[49] ITU-T Recommendation G.652, 'Characteristics of single-mode optical fiber cable', October 2000.

[50] F. P. Kapron, 'Dispersion-slope parameter for monomode fiber bandwidth', *Conf. on Optical Fiber Communication, OFC'84*, USA, pp. 90–92, January 1984.

[51] F. P. Kapron, 'Maximum information capacity of fibre-optic waveguides', *Electron. Lett.*, **13**(4), pp. 96–97, 1977.

[52] F. P. Kapron, 'Chromatic dispersion format for single-mode and multimode fibers', *Conf. Dig. of Optical Fiber Communication, OFC'87*, USA, paper TUQ2, January 1987.

[53] L. Osterberg, 'Optical fiber, cable and connectors', in C. DeCusatis, E. Maass, D. P. Clement and R. C. Lasky (Eds), *Handbook of Fiber Optic Data Communication*, Academic Press, 1998.

[54] V. A. Bhagavatula, J. C. Lapp, A. J. Morrow and J. E. Ritter, 'Segmented-core fiber for long-haul and local-area-network applications', *J. Lightwave Technol.*, **6**(10), pp. 1466–1469, 1988.

[55] L. G. Cohen, 'Comparison of single mode fiber dispersion measurement techniques', *J. Lightwave Technol.*, **LT-3**, pp. 958–966, 1985.

[56] B. J. Ainslie and C. R. Day, 'A review of single-mode fibers with modified dispersion characteristics', *J. Lightwave Technol.*, **LT-4**(8), pp. 967–979, 1986.

[57] B. J. Ainslie, K. J. Beales, C. R. Day and J. D. Rush, 'Interplay of design parameters and fabrication conditions on the performance of monomode fibers made by MCVD', *IEEE. J. Quantum Electron.*, **QE-17**, pp. 854–857, 1981.

[58] M. A. Saifi, 'Triangular index monomode fibres', *Proc. SPIE*, **374**, pp. 13–15, 1983.

[59] W. A. Gambling, H. Matsumura and C. M. Ragdale, 'Zero total dispersion in graded-index single mode fibres', *Electron. Lett.*, **15**, pp. 474–476, 1979.

[60] B. J. Ainslie, K. J. Beales, D. M. Cooper and C. R. Day, 'Monomode optical fibres with graded-index cores for low dispersion at 1.55 μm', *Br. Telecom Technol. J.*, **2**(2), pp. 25–34, 1984.

[61] H.-T. Shang, T. A. Lenahan, P. F. Glodis and D. Kalish, 'Design and fabrication of dispersion-shifted depressed-clad triangular-profile (DDT) single-mode fibre', *Electron. Lett.*, **21**, pp. 484–486, 1982.

[62] M. Miyamoto, T. Abiru, T. Ohashi, R. Yamauchi and O. Fukuda, 'Gaussian profile dispersion-shifted fibers made by VAD method', *Proc. IOOC-ECOC'85*, Venice, Italy, pp. 193–196, 1985.

[63] D. M. Cooper, S. P. Craig, C. R. Day and B. J. Ainslie, 'Multiple index structures for dispersion shifted single mode fibers using multiple index structures', *Br. Telecom Technol. J.*, **3**, pp. 52–58, 1985.

[64] V. A. Bhagavatula and P. E. Blaszyk, 'Single mode fiber with segmented core', *Conf. Dig. of Optical Fiber Communication, OFC'83*, New Orleans, USA, Paper MF5, 1983.

[65] A. R. Hunwicks, P. A. Rosher, L. Bickers and D. Stanley, 'Installation of dispersion-shifted fibre in the British Telecom trunk network', *Electron. Lett.*, **24**(9), pp. 536–537, 1988.

[66] V. Bhagavatula, M. S. Spotz, W. F. Love and D. B. Keck, 'Segmented core single mode fibre with low loss and low dispersion', *Electron. Lett.*, **19**(9), pp. 317–318, 1983.

[67] N. Kuwaki, M. Ohashi, C. Tanaka, N. Uesugi, S. Seikai and Y. Negishi, 'Characteristics of dispersion-shifted dual shape core single-mode fiber', *J. Lightwave Technol.*, **LT-5**(6), pp. 792–797, 1987.

[68] K. Nishide, D. Tanaka, M. Miyamoto, R. Yamauchi and K. Inada, 'Long-length and high-strength dual-shaped core dispersion-shifted fibers made by a fully synthesized VAD method', *Conf. Dig. of Optical Fiber Communication, OFC'88*, New Orleans, USA, paper WI2, 1988.

[69] ITU-T Recommendation G.653, 'Characteristics of dispersion shifted optical fiber cable', October 2000.

[70] T. Miya, K. Okamoto, Y. Ohmori and Y. Sasaki, 'Fabrication of low dispersion single mode fibers over a wide spectral range', *IEEE J. Quantum Electron.*, **QE-17**, pp. 858–861, 1981.

[71] J. J. Bernard, C. Brehm, P. H. Dupont, G. M. Gabriagues, C. Le Sergeant, M. Liegois, P. L. Francois, M. Monerie and P. Sansonetti, 'Investigation of the properties of depressed inner cladding single-mode fibres', *Proc. 8th Eur. Conf. on Optical Communication*, Cannes, France, pp. 133–138, 1982.

[72] L. G. Cohen, W. L. Mammel and S. J. Jang, 'Low-loss quadruple-clad single-mode light-guides with dispersion below 2 ps/km nm over the 1.28 μm–1.65 μm wavelength range', *Electron. Lett.*, **18**, pp. 1023–1024, 1982.

[73] S. J. Jang, J. Sanchez, K. D. Pohl and L. D. L'Esperance, 'Graded-index single-mode fibers with multiple claddings', *Proc. IOOC'83*, Tokyo, Japan, pp. 396–397, 1983.

[74] V. A. Bhagavatula, 'Dispersion-modified fibers', *Conf. Dig. of Optical Fiber Communication, OFC'88*, New Orleans, USA, paper WI1, 1988.

[75] P. K. Backmann, D. Leers, H. Wehr, D. U. Wiechert, J. A. Steenwijk, D. L. A. Tjaden and E. R. Wehrhatim, 'Dispersion-flattened single-mode fibers prepared with PCVD: performance, limitations, design optimization', *J. Lightwave Technol.*, **LT-4**, pp. 858–863, 1986.

[76] N. Kumano, K. Mukasa, M. Sakano, H. Moridaira, T. Yagi and K. Kokura, 'Novel NZ-DSF with ultra-low dispersion slope lower than 0.020 ps/nm²/km', *Proc. Eur. Conf. on Optical Communication, ECOC2001*, **6**, paper PD.A.1.5, pp. 54–55, 2001.

[77] M. Nishimura, 'Optical fibers and fiber dispersion compensators for high-speed optical communication', *Opt. Fiber Commun. Rep.*, **2**, pp. 115–139, 2005.

[78] ITU-T Recommendation G.655, 'Characteristics of a nonzero dispersion shifted single-mode optical fiber and fiber cable', October 2000.

[79] G. Mahlke and P. Gossing, *Fiber optic cables* (4th edn), Publicis MCD Corporate Publishing, 2001.

[80] I. P. Kaminow, 'Polarization in fibers', *Laser Focus*, **16**(6), pp. 80–84, 1980.

[81] I. P. Kaminow, 'Polarization in optical fibers', *IEEE J. Quantum Electron.*, **QE-17**(1), pp. 15–22, 1981.

[82] S. C. Rashleigh and R. Ulrich, 'Polarization mode dispersion in single-mode fibers', *Opt. Lett.*, **3**, pp. 60–62, 1978.

[83] V. Ramaswamy, R. D. Standley, D. Sze and W. G. French, 'Polarisation effects in short length, single mode fibres', *Bell Syst. Tech. J.*, **57**, pp. 635–651, 1978.

[84] A. Papp and H. Harms, 'Polarization optics of index-gradient optical waveguide fibers', *Appl. Opt.*, **14**, pp. 2406–2411, 1975.

[85] M. F. Ferreira, A. N. Pinto, P. S. Andre, N. J. Muga, J. E. Machado, R. N. Nogueira, S. V. Latas, M. H. Sousa and J. F. Rocha, 'Polarization mode dispersion in high speed optical communication systems', *Fiber Integr. Opt*, **24**, pp. 261–285, 2005.

[86] H. Kogelnik, R. M. Jopson and L. E. Nelson, 'Polarization mode dispersion', in I. P. Kaminow and T. Li (Eds), *Optical Fiber Telecommunications IVB*, pp. 725–861, Academic Press, 2002.

[87] G. J. Foschini and C. D. Poole, 'Statistical theory of polarization dispersion in single mode fibers', *J. Lightwave Technol.*, **9**(11), pp. 1439–1456, 1991.

[88] P. K. A. Wai and C. R. Menyuk, 'Polarization mode dispersion, decorrelation and diffusion in optical fibers with randomly varying birefringence', *J. Lightwave Technol.*, **14**(2), pp. 148–157, 1996.

[89] D. A. Nolan, X. Chen and M. J. Li, 'Fibers with low polarization-mode dispersion', *J. Lightwave Technol.*, **22**(4), pp. 1066–1077, 2004.

[90] S. Lanne and E. Corbel, 'Practical considerations for optical polarization-mode dispersion compensators', *J. Lightwave Technol.*, **22**(4), pp. 1033–1040, 2004.

[91] S. Kiekbusch, S. Ferber, H. Rosenfeldt, R. Ludwig, C. Boerner, A. Ehrhardt, E. Brinkmeyer and H. G. Weber, 'Automatic PMD compensator in a 160-Gb/s OTDM transmission over deployed fiber using RZ-DPSK modulation format', *J. Lightwave Technol.*, **23**(1), pp. 165–171, 2005.

[92] J. M. Fini and H. A. Haus, 'Accumulation of polarization-mode dispersion in cascades of compensated optical fibers', *IEEE Photonics Technol. Lett.*, **13**(2), pp. 124–126, 2001.

[93] S. Heckmann, 'Modal noise in single-mode fibers', *Opt. Lett.*, 6, pp. 201–203, 1981.

[94] D. N. Payne, A. J. Barlow and J. J. Ramskov Hansen, 'Development of low-and-high-birefringence optical fibers', *IEEE J. Quantum Electron.*, **QE-18**(4) pp. 477–487, 1982.

[95] R. H. Stolen and R. P. De Paula, 'Single-mode fiber components', *Proc. IEEE*, **75**(11), pp. 1498–1511, 1987.

[96] J. Noda, K. Okamoto and Y. Sasaki, 'Polarization-maintaining fibers and their applications', *J. Lightwave Technol.*, **LT-4**(8), pp. 1071–1089, 1986.

[97] W. Eickhoff and E. Brinkmeyer, 'Scattering loss vs. polarization holding ability of single-mode fibers', *Appl. Opt.*, **23**, pp. 1131–1132, 1984.

[98] I. P. Kaminow and V. Ramaswamy, 'Single-polarization optical fibers: slab model', *Appl. Phys. Lett.*, **34**, pp. 268–70, 1979.

[99] R. D. Birch, D. N. Payne and M. P. Varnham, 'Fabrication of polarization-maintaining fibers using gas-phase etching', *Electron. Lett.*, **18**, pp. 1036–1038, 1982.

[100] T. Hosaka, Y. Sasaki, J. Noda and M. Horiguchi, 'Low-loss and low-crosstalk polarization-maintaining optical fibers', *Electron. Lett.*, **21**, pp. 920–921, 1985.

[101] R. H. Stolen, W. Pleibel and J. R. Simpson, 'High-birefringence optical fibers by preform deformation', *J. Lightwave Technol.*, **LT-2**, pp. 639–641, 1985.

[102] H. Schneider, H. Harms, A. Rapp and H. Aulich, 'Low birefringence single-mode fibers: preparation and polarization characteristics', *Appl. Opt.*, **17**(19), pp. 3035–3037, 1978.

[103] S. R. Norman, D. N. Payne, M. J. Adams and A. M. Smith, 'Fabrication of single-mode fibers exhibiting extremely low polarization birefringence', *Electron. Lett.*, **15** (11), pp. 309–311, 1979.

[104] D. N. Payne, A. J. Barlow and J. J. Ramskov Hansen, 'Development of low and high birefringent optical fibers', *IEEE J. Quantum Electron.*, **QE-18**(4), pp. 477–487, 1982.

[105] W. A. Gambling and S. B. Poole, 'Optical fibers for sensors', in J. P. Dakin and B. Culshaw (Eds), *Optical Fiber Sensors: Principles and components*, Artech House, pp. 249–276, 1988.

[106] R. D. Birch, 'Fabrication and characterisation of circularly-birefringent helical fibres', *Electron Lett.*, 23, pp. 50–52, 1987.

[107] T. Hosaka, Y. Sasaki and K. Okamoto, '3-km long single-polarization single-mode fiber', *Electron. Lett.*, **21**(22), pp. 1023–1024, 1985.

[108] H. Kajioka, Y. Takuma, K. Yamada and T. Tokunaga, 'Low-loss polarization-maintaining single-mode fibers for 1.55 μm operation', in *Tech. Dig. of Optical Fiber Communication Conference, OFC'88*, New Orleans, USA, paper WA5, 1988.

[109] K. Tajima and Y. Sasaki, 'Transmission loss of a 125 μm diameter PANDA fiber with circular stress-applying parts', *J. Lightwave Technol.*, **7**(4), pp. 674–679, 1989.

[110] P. Bayvel and R. Killey, 'Nonlinear effects in WDM transmission', in I. P. Kaminow and T. Li (Eds), *Optical Fiber Telecommunications IVB*, Academic Press, pp. 611–641, 2002.

[111] J. Toulouse, 'Optical nonlinearities in fibers: review, recent examples, and system applications', *J. Lightwave Technol.*, **23**(11), pp. 3625–3641, 2005.

[112] L. B. Jeunhomme, *Single-Mode Fiber Optics*, Marcel Dekker, 1983.

[113] W. J. Tomlinson and R. H. Stolen, 'Nonlinear phenomena in optical fibers', *IEEE Commun. Mag.*, **26**(4), pp. 36–44, 1988.

[114] R. H. Stolen, C. Lee and R. K. Jain, 'Development of the stimulated Raman spectrum in single-mode silica fibers', *J. Opt. Soc. Am.* **B1**, pp. 652–657, 1984.

[115] E. H. Lee, K. H. Kim and H. K. Lee, 'Nonlinear effects in optical fiber: advantages and disadvantages for high capacity all-optical communication application', *Opt. Quantum Electron.*, **34**(12), pp. 1167–1174, 2002.

[116] J. A. Buck, 'Nonlinear effects in optical fibers', in M. Bass and E. W. Van Stryland (Eds), *Fiber Optics Handbook*, pp. 3.1–3.14, McGraw-Hill, 2002.

[117] F. Forghieri, R. W. Tkach and A. R. Chraplyvy, 'WDM systems with unequally spaced channels', *J. Lightwave Technol.*, **13**(5), pp. 889–897, 1995.

[118] C. R. Menyuk, G. M. Carter, W. L. Kath and R. M. Mu, 'Dispersion managed solitons and chirped return to zero: what is the difference?', in I. P. Kaminow and T. Li (Eds), *Optical Fiber Telecommunications IVB*, Academic Press, pp. 305–328, 2002.

[119] P. V. Mamyshev, 'Solitons for optical fiber communication systems', in M. Bass and E. W. Van Stryland (Eds), *Fiber Optics Handbook*, pp. 7.1–7.20, McGraw-Hill, 2002.

[120] C. F. Wehmann, L. M. Fernandes, C. S. Sobrinho, J. L. S. Lima, M. G. da Silva, E. F. Almeida, J. A. M. Neto and A. S. B. Sombra, 'Analysis of the four wave mixing effect (FWM) in dispersion decreasing fiber (DDF) for a WDM system', *Opt. Fiber Technol.: Mater., Devices Syst.*, **11**(3), pp. 306–318, 2005.

[121] V. A. Bogatyrjov, M. M. Bubnov, E. M. Dianov and A. A. Sysoliatin, 'Advanced fibres for soliton systems', *Pure Appl. Opt.*, **4**, pp. 345–347, 1994.

[122] J. P. Gordon and H. A. Haus, 'Random walk of coherently amplified solitons in optical fiber transmission', *Opt. Lett.*, **11**, pp. 665–667, 1986.

CHAPTER | 4

Optical fibers and cables

4.1 Introduction

Optical fiber waveguides and their transmission characteristics have been considered in some detail in Chapters 2 and 3. However, we have yet to discuss the practical considerations and problems associated with the production, application and installation of optical fibers within a line transmission system. These factors are of paramount importance if optical fiber communication systems are to be considered as viable replacements for conventional metallic line communication systems. Optical fiber communication is of little use if the many advantages of optical fiber transmission lines outlined in the preceding chapters may not be applied in practice in the telecommunications network without severe degradation of the lines' performance.

It is therefore essential that:

1. Optical fibers may be produced with good stable transmission characteristics in long lengths at a minimum cost and with maximum reproducibility.

2. A range of optical fiber types with regard to size, refractive indices and index profiles, operating wavelengths, materials, etc., be available in order to fulfill many different system applications.

3. The fibers may be converted into practical cables which can be handled in a similar manner to conventional electrical transmission cables without problems associated with the degradation of their characteristics or damage.

4. The fibers and fiber cables may be terminated and connected together (jointed) without excessive practical difficulties and in ways which limit the effect of this process on the fiber transmission characteristics to keep them within acceptable operating levels. It is important that these jointing techniques may be applied with ease in the field locations where cable connection takes place.

In this chapter, we therefore consider the first three of the above practical elements associated with optical fiber communications. The final element, however, concerned with fiber termination and jointing is discussed immediately following, in Chapter 5. The various methods of preparation for silica-based optical fibers (both liquid and vapor phase) with characteristics suitable for telecommunications applications are dealt with in Sections 4.2 to 4.4. This is followed in Section 4.5 by consideration of the major commercially available fibers describing in general terms both the types and their characteristics. In particular, an outline of the range of single-mode silica optical fibers specified by standards together with the developments in the area of plastic or polymeric fibers for use in lower bandwidth, shorter distance applications is provided. The requirements for optical fiber cabling in relation to fiber protection are then discussed in Section 4.6 prior to discussion in Section 4.7 of the factors which cause modification to the cabled fiber transmission characteristics in a practical operating environment (i.e. microbending, hydrogen absorption, nuclear radiation exposure). Finally, cable design strategies and their influence upon typical examples of optical fiber cable constructions are dealt with in Section 4.8.

4.2 Preparation of optical fibers

From the considerations of optical waveguiding of Chapter 2 it is clear that a variation of refractive index inside the optical fiber (i.e. between the core and the cladding) is a fundamental necessity in the fabrication of fibers for light transmission. Hence at least two different materials which are transparent to light over the major operating wavelength range (0.8 to 1.7 μm) are required. In practice these materials must exhibit relatively low optical attenuation and they must therefore have low intrinsic absorption and scattering losses. A number of organic and inorganic insulating substances meet these conditions in the visible and near-infrared regions of the spectrum.

However, in order to avoid scattering losses in excess of the fundamental intrinsic losses, scattering centers such as bubbles, strains and grain boundaries must be eradicated.

This tends to limit the choice of suitable materials for the fabrication of optical fibers to either glasses (or glass-like materials) and monocrystalline structures (certain plastics).

It is also useful, and in the case of graded index fibers essential, that the refractive index of the material may be varied by suitable doping with another compatible material. Hence these two materials should have mutual solubility over a relatively wide range of concentrations. This is only achieved in glasses or glass-like materials, and therefore monocrystalline materials are unsuitable for the fabrication of graded index fibers, but may be used for step index fibers. However, it is apparent that glasses exhibit the best overall material characteristics for use in the fabrication of low-loss optical fibers. They are therefore used almost exclusively in the preparation of fibers for telecommunications applications. Plastic-clad [Ref. 1] and all plastic fibers find some use in short-haul, low-bandwidth applications.

In this chapter the discussion will therefore be confined to the preparation of glass fibers. This is a two-stage process in which initially the pure glass is produced and converted into a form (rod or preform) suitable for making the fiber. A drawing or pulling technique is then employed to acquire the end product. The methods of preparing the extremely pure optical glasses generally fall into two major categories which are:

(a) conventional glass refining techniques in which the glass is processed in the molten state (melting methods) producing a multicomponent glass structure;

(b) vapor-phase deposition methods producing silica-rich glasses which have melting temperatures that are too high to allow the conventional melt process.

Although the vapor-phase methods are the ones that are now used to produce silica-based fibers with very low attenuation, both processes, with their respective drawing techniques, are described in the following sections.

4.3 Liquid-phase (melting) techniques

The first stage in this process is the preparation of ultrapure material powders which are usually oxides or carbonates of the required constituents. These include oxides such as SiO_2, GeO_2, B_2O_2 and A_2O_3, and carbonates such as Na_2CO_3, K_2CO_3, $CaCO_3$ and $BaCO_3$ which will decompose into oxides during the glass melting. Very high initial purity is essential and purification accounts for a large proportion of the material cost; nevertheless these compounds are commercially available with total transition metal contents below 20 parts in 10^9 and below 1 part in 10^9 for some specific impurities [Ref. 2]. The purification may therefore involve combined techniques of fine filtration and coprecipitation, followed by solvent extraction before recrystallization and final drying in a vacuum to remove any residual OH ions [Ref. 3].

The next stage is to melt these high-purity, powdered, low-melting-point glass materials to form a homogeneous, bubble-free multicomponent glass. A refractive index variation may be achieved by either a change in the composition of the various constituents or by ion exchange when the materials are in the molten phase. The melting of these multicomponent glass systems occurs at relatively low temperatures between 900 and 1300 °C

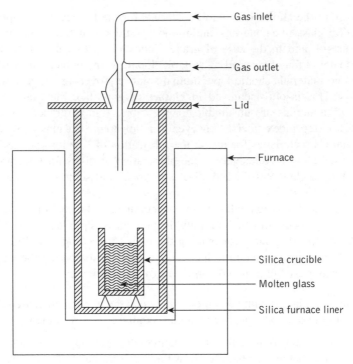

Figure 4.1 Glassmaking furnace for the production of high-purity glasses [Ref. 4]

and may take place in a silica crucible as shown in Figure 4.1 [Ref. 4]. However, contamination can arise during melting from several sources including the furnace environment and the crucible. Both fused silica and platinum crucibles have been used with some success, although an increase in impurity content was observed when the melt was held in a platinum crucible at high temperatures over long periods [Ref. 5].

Silica crucibles can give dissolution into the melt which may introduce inhomogeneities into the glass, especially at high melting temperatures. A technique for avoiding this involves melting the glass directly into a radio-frequency (RF approximately 5 MHz) induction furnace while cooling the silica by gas or water flow, as shown in Figure 4.2 [Refs 6–8]. The materials are preheated to around 1000°C where they exhibit sufficient ionic conductivity to enable coupling between the melt and the RF field. The melt is also protected from any impurities in the crucible by a thin layer of solidified pure glass which forms due to the temperature difference between the melt and the cooled silica crucible.

In both techniques the glass is homogenized and dried by bubbling pure gases through the melt, while protecting against any airborne dust particles either originating in the melt furnace or present as atmospheric contamination. After the melt has been suitably processed, it is cooled and formed into long rods (cane) of multicomponent glass.

4.3.1 Fiber drawing

An original technique for producing fine optical fiber waveguides was to make a preform using the rod in tube process. A rod of core glass was inserted into a tube of cladding glass

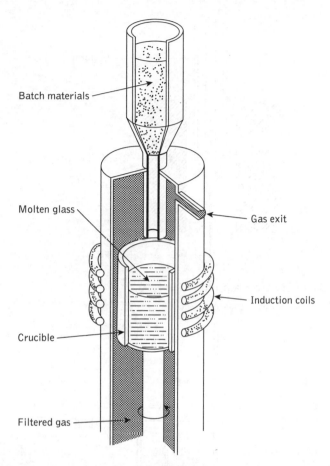

Batch materials

Molten glass

Gas exit

Induction coils

Crucible

Filtered gas

Figure 4.2 High-purity melting using a radio-frequency induction furnace [Refs 6–8]

and the preform was drawn in a vertical muffle furnace, as illustrated in Figure 4.3 [Ref. 9]. This technique was useful for the production of step index fibers with large core and cladding diameters where the achievement of low attenuation was not critical as there was a danger of including bubbles and particulate matter at the core–cladding interface. Indeed, these minute perturbations and impurities can result in very high losses of between 500 and 1000 dB km^{-1} after the fiber is drawn [Ref. 10].

Subsequent development in the drawing of optical fibers (especially graded index) produced by liquid-phase techniques has concentrated on the double-crucible method. In this method the core and cladding glass in the form of separate rods is fed into two concentric platinum crucibles, as illustrated in Figure 4.4 [Ref. 4]. The assembly is usually located in a muffle furnace capable of heating the crucible contents to a temperature of between 800 and 1200°C. The crucibles have nozzles in their bases from which the clad fiber is drawn directly from the melt, as shown in Figure 4.4. Index grading may be achieved through the diffusion of mobile ions across the core–cladding interface within the molten glass. It is possible to achieve a reasonable refractive index profile via this diffusion process, although due to lack of precise control it is not possible to obtain the

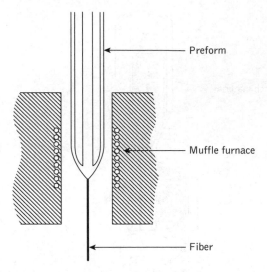

Figure 4.3 Optical fiber from a preform [Ref. 9]

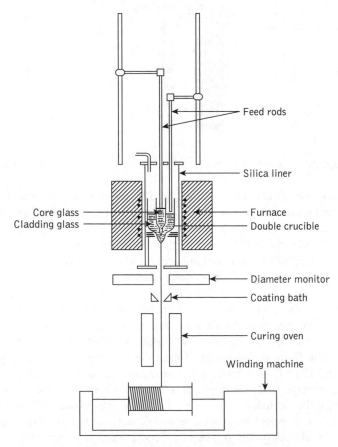

Figure 4.4 The double-crucible method for fiber drawing [Ref. 4]

Table 4.1 Material systems used in the fabrication of multicomponent glass fibers by the double crucible technique

Step index

Core glass	Cladding glass
Na_2–B_2O_3–SiO_2	Na_2O–B_2O_3–SiO_2
Na_2–LiO–CaO–SiO_2	Na_2O–Li_2O–CaO–SiO_2
Na_2–CaO–GeO_2	Na_2O–CaO–SiO_2
Tl_2O–Na_2O–B_2O_3–GeO_2–BaO–CaO–SiO_2	Na_2O–B_2O_3–SiO_2
Na_2O–BaO–GeO_2–B_2O_3–SiO_2	Na_2O–B_2O_5–SiO_2
P_2O_5–Ga_2O_3–GeO_2	P_2O_5–Ga_2O_3–SiO_2

Graded index

Base glass	Diffusion mechanism
R_2O–GeO_2–CaO–SiO_2	$Na^+ \rightleftharpoons K^+$
R_2O–B_2O_3–SiO_2	$Tl^+ \rightleftharpoons Na^+$
Na_2O–B_2O_3–SiO_2	Na_2O diffusion
Na_2O–B_2O_3–SiO_2	CaO, BaO, diffusion

optimum near-parabolic profile which yields the minimum pulse dispersion (see Section 3.10.2). Hence graded index fibers produced by this technique are subsequently less dispersive than step index fibers, but do not have the bandwidth–length products of optimum profile fibers. Pulse dispersion of 1 to 6 ns km^{-1} [Refs 11, 12] is quite typical, depending on the material system used.

Some of the material systems used in the fabrication of multicomponent glass step index and graded index fibers are given in Table 4.1.

Using very high-purity melting techniques and the double-crucible drawing method, step index and graded index fibers with attenuations as low as 3.4 dB km^{-1} [Ref. 13] and 1.1 dB km^{-1} [Ref. 2], respectively, have been produced. However, such low losses cannot be consistently obtained using liquid-phase techniques and typical losses for multicomponent glass fibers prepared continuously by these methods remain in the range 5 to 20 dB km^{-1} at a wavelength of 0.85 μm [Ref. 10]. Hence the method is particularly used for the production of fibers with a large core diameter of 200 μm and above which now rarely find application in mainstream communications. Nevertheless, a benefit of these techniques is their potential for continuous production (both melting and drawing) of optical fibers.

4.4 Vapor-phase deposition techniques

Vapor-phase deposition techniques are used to produce silica-rich glasses of the highest transparency and with the optimal optical properties. The starting materials are volatile compounds such as $SiCl_4$, $GeCl_4$, SiF_4, BCl_3, O_2, BBr_3 and $POCl_3$ which may be distilled to reduce the concentration of most transition metal impurities to below one part in 10^9, giving negligible absorption losses from these elements. Refractive index modification is achieved through the formation of dopants from the nonsilica starting materials. These vapor-phase dopants include TiO_2, GeO_2, P_2O_5, Al_2O_3, B_2O_3 and F, the effects of which on

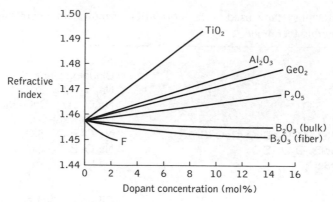

Figure 4.5 The variation in the refractive index of silica using various dopants. Reproduced with permission from the publishers, Society of Glass Technology, *Phys. Chem. Glasses*, **21**, p. 5, 1980

the refractive index of silica are shown in Figure 4.5 [Ref. 2]. Gaseous mixtures of the silica-containing compound, the doping material and oxygen are combined in a vapor-phase oxidation reaction where the deposition of oxides occurs. The deposition is usually onto a substrate or within a hollow tube and is built up as a stack of successive layers. Hence the dopant concentration may be varied gradually to produce a graded index profile or maintained to give a step index profile. In the case of the substrate this directly results in a solid rod or preform whereas the hollow tube must be collapsed to give a solid preform from which the fiber may be drawn.

There are a number of variations of vapor-phase deposition which have been successfully utilized to produce low-loss fibers. These methods are currently used for mass production of preforms in the manufacture of multimode, and particularly single-mode, fibers with extremely low attenuation (i.e. 0.18 dB km^{-1} at 1.55 μm) and low dispersion (i.e. less than 3.5 ps nm^{-1} km^{-1} between 1.285 and 1.330 μm) [Ref. 10]. The major techniques are illustrated in Figure 4.6, which also indicates the plane (horizontal or vertical) in which the deposition takes place as well as the formation of the preform. These vapor-phase deposition techniques fall into two broad categories: flame hydrolysis and chemical vapor deposition (CVD) methods. The individual techniques are considered in the following sections.

4.4.1 Outside vapor-phase oxidation process

This process which uses flame hydrolysis stems from work on 'soot' processes originally developed by Hyde [Ref. 14] which were used to produce the first fiber with losses of less than 20 dB km^{-1} [Ref. 15]. The best known technique of this type is often referred to as the outside vapor-phase oxidation (OVPO) or the outside vapor-phase deposition (OVD) process. In this process the required glass composition is deposited laterally from a 'soot' generated by hydrolyzing the halide vapors in an oxygen–hydrogen flame. Oxygen is passed through the appropriate silicon compound (i.e. SiCl$_4$) which is vaporized, removing any impurities. Dopants such as GeCl$_4$ or TiCl$_4$ are added and the mixture is blown through the oxygen–hydrogen flame giving the following reactions:

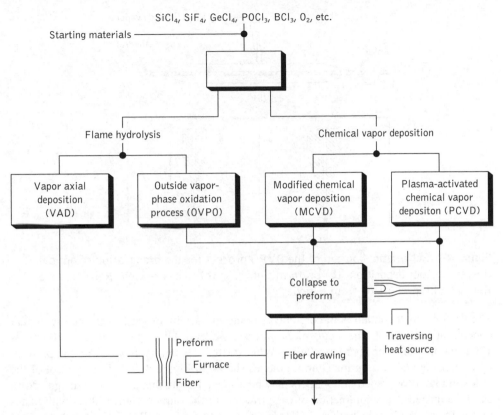

$SiCl_4$, SiF_4, $GeCl_4$, $POCl_3$, BCl_3, O_2, etc.

Starting materials

Flame hydrolysis

Chemical vapor deposition

Vapor axial deposition (VAD)

Outside vapor-phase oxidation process (OVPO)

Modified chemical vapor deposition (MCVD)

Plasma-activated chemical vapor depositon (PCVD)

Collapse to preform

Traversing heat source

Preform

Furnace

Fiber

Fiber drawing

Figure 4.6 Schematic illustration of the vapor-phase deposition techniques used in the preparation of low-loss optical fibers

$$SiCl_4 + 2H_2O \xrightarrow{\text{heat}} SiO_2 + 4HCl \qquad (4.1)$$
$$\text{(vapor)} \quad \text{(vapor)} \qquad \text{(solid)} \quad \text{(gas)}$$

and:

$$SiCL_4 + O_2 \xrightarrow{\text{heat}} SiO_2 + 2Cl_2 \qquad (4.2)$$
$$\text{(vapor)} \quad \text{(gas)} \qquad \text{(solid)} \quad \text{(gas)}$$

$$GeCl_4 + O_2 \xrightarrow{\text{heat}} GeO_2 + 2Cl_2 \qquad (4.3)$$
$$\text{(vapor)} \quad \text{(gas)} \qquad \text{(solid)} \quad \text{(gas)}$$

or:

$$TiCl_4 + O_2 \xrightarrow{\text{heat}} TiO_2 + 2Cl_2 \qquad (4.4)$$
$$\text{(vapor)} \quad \text{(gas)} \qquad \text{(solid)} \quad \text{(gas)}$$

The silica is generated as a fine soot which is deposited on a cool rotating mandrel, as illustrated in Figure 4.7(a) [Ref. 16]. The flame of the burner is reversed back and forth over the length of the mandrel until a sufficient number of layers of silica (approximately 200) are deposited on it. When this process is completed the mandrel is removed and the porous mass of silica soot is sintered (to form a glass body), as illustrated in Figure 4.7(b).

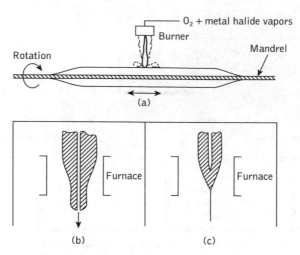

Figure 4.7 Schematic diagram of the OVPO process for the preparation of optical fibers: (a) soot deposition: (b) preform sintering; (c) fiber drawing. Reprinted from Ref. 17 with permission from Elsevier

The preform may contain both core and cladding glasses by properly varying the dopant concentrations during the deposition process. Several kilometers (around 10 km of 120 μm core diameter fiber have been produced [Ref. 2]) can be drawn from the preform by collapsing and closing the central hole, as shown in Figure 4.7(c). Fine control of the index gradient for graded index fibers may be achieved using this process as the gas flows can be adjusted at the completion of each traverse of the burner. Hence fibers with band-width–length products as high as 3 GHz km have been reported [Ref. 17] through accurate index grading with this process.

The purity of the glass fiber depends on the purity of the feeding materials and also upon the amount of OH impurity from the exposure of the silica to water vapor in the flame following the reactions given in Eqs (4.1) to (4.4). Typically, the OH content is between 50 and 200 parts per million and this contributes to the fiber attenuation. It is possible to reduce the OH impurity content by employing gaseous chlorine as a drying agent during sintering. [Ref. 18].

Other problems stem from the use of the mandrel which can create some difficulties in the formation of the fiber preform. Cracks may form due to stress concentration on the surface of the inside wall when the mandrel is removed. Also the refractive index profile has a central depression due to the collapsed hole when the fiber is drawn. Therefore, although the OVPO process is a useful fiber preparation technique, it has several drawbacks. Furthermore, it is a batch process, which limits its use for the volume production of optical fibers. Nevertheless, a number of proprietary approaches to scaling up the process have provided preforms capable of producing 250 km of fiber [Ref. 19].

4.4.2 Vapor axial deposition (VAD)

This process was developed by Izawa *et al.* [Ref. 20] in the search for a continuous (rather than batch) technique for the production of low-loss optical fibers. The VAD technique

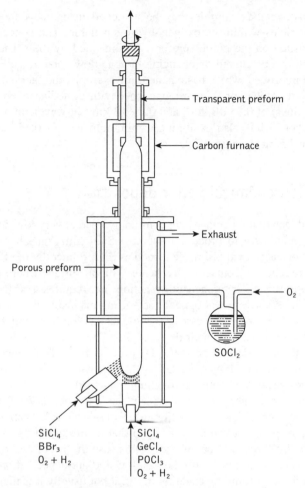

Figure 4.8 The VAD process [Ref. 21]

uses an end-on deposition onto a rotating fused silica target, as illustrated in Figure 4.8 [Ref. 21]. The vaporized constituents are injected from burners and react to form silica soot by flame hydrolysis. This is deposited on the end of the starting target in the axial direction forming a solid porous glass preform in the shape of a boule. The preform which is growing in the axial direction is pulled upwards at a rate which corresponds to the growth rate. It is initially dehydrated by heating with $SOCl_2$ using the reaction:

$$\underset{\text{(vapor)}}{H_2O} + \underset{\text{(vapor)}}{SOCl_2} \xrightarrow{\text{heat}} \underset{\text{(gas)}}{2HCl} + \underset{\text{(gas)}}{SO_2} \tag{4.5}$$

and is then sintered into a solid preform in a graphite resistance furnace at an elevated temperature of around 1500 °C. Therefore, in principle this process may be adapted to draw fiber continuously, although at present it tends to be operated as a batch process partly because the resultant preforms can yield more than 100 km of fiber [Ref. 19].

A spatial refractive index profile may be achieved using the deposition properties of SiO_2–GeO_2 particles within the oxygen–hydrogen flame. The concentration of these constituents deposited on the porous preform is controlled by the substrate temperature distribution which can be altered by changing the gas flow conditions. Finally, the VAD process has been improved, which has enabled, for example, the fabrication of extremely low-attenuation pure silica core single-mode fiber with a median attenuation (for more than 2000 km of fiber) of 0.35 dB km^{-1} and 0.21 dB km^{-1} at wavelengths of 1.30 μm and 1.55 μm respectively while also exhibiting a minimum loss of 0.154 dB km^{-1} over the wavelength range 1.55 to 1.56 μm [Ref. 22].

4.4.3 Modified chemical vapor deposition

Chemical vapor deposition techniques are commonly used at very low deposition rates in the semiconductor industry to produce protective SiO_2 films on silicon semiconductor devices. Usually an easily oxidized reagent such as SiH_4 diluted by inert gases and mixed with oxygen is brought into contact with a heated silicon surface where it forms a glassy transparent silica film. This heterogeneous reaction (i.e. requires a surface to take place) was pioneered for the fabrication of optical fibers using the inside surface of a fused quartz tube [Ref. 23]. However, these processes gave low deposition rates and were prone to OH contamination due to the use of hydride reactants. This led to the development of the modified chemical vapor deposition (MCVD) process by Bell Telephone Laboratories [Ref. 24] and Southampton University, UK [Ref. 25], which overcomes these problems and has found widespread application throughout the world.

The MCVD process is also an inside vapor-phase oxidation (IVPO) technique taking place inside a silica tube, as shown in Figure 4.9. However, the vapor-phase reactants (halide and oxygen) pass through a hot zone so that a substantial part of the reaction is homogeneous (i.e. involves only one phase; in this case the vapor phase). Glass particles formed during this reaction travel with the gas flow and are deposited on the walls of the silica tube. The tube may form the cladding material but usually it is merely a supporting structure which is heated on the outside by an oxygen–hydrogen flame to temperatures between 1400 and 1600 °C. Thus a hot zone is created which encourages high-temperature oxidation reactions such as those given in Eqs (4.2) and (4.3) or (4.4) (not Eq. (4.1)). These reactions reduce the OH impurity concentration to levels below those found in fibers prepared by hydride oxidation or flame hydrolysis.

The hot zone is moved back and forth along the tube allowing the particles to be deposited on a layer-by-layer basis giving a sintered transparent silica film on the walls of the tube. The film may be up to 10 μm in thickness and uniformity is maintained by rotating the tube. A graded refractive index profile can be created by changing the composition of the layers as the glass is deposited. Usually, when sufficient thickness has been formed by successive traverses of the burner for the cladding, vaporized chlorides of germanium ($GeCl_4$) or phosphorus ($POCl_3$) are added to the gas flow. The core glass is then formed by the deposition of successive layers of germanosilicate or phosphosilicate glass. The cladding layer is important as it acts as a barrier which suppresses OH absorption losses due to the diffusion of OH ions from the silica tube into the core glass as it is deposited. After the deposition is completed the temperature is increased to between 1700 and

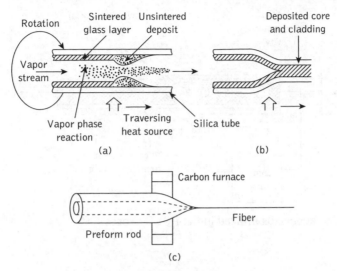

Figure 4.9 Schematic diagram showing the MCVD method for the preparation of optical fibers: (a) deposition; (b) collapse to produce a preform; (c) fiber drawing

1900 °C. The tube is then collapsed to give a solid preform which may then be drawn into fiber at temperatures of 2000 to 2200 °C as illustrated in Figure 4.9.

This technique is the most widely used at present as it allows the fabrication of fiber with the lowest losses. Apart from the reduced OH impurity contamination the MCVD process has the advantage that deposition occurs within an enclosed reactor which ensures a very clean environment. Hence, gaseous and particulate impurities may be avoided during both the layer deposition and the preform collapse phases. The process also allows the use of a variety of materials and glass compositions. It has produced GeO_2-doped silica single-mode fiber with minimum losses of only 0.2 dB km^{-1} at a wavelength of 1.55 μm [Ref. 26]. More generally, the GeO_2–B_2O_3–SiO_2 system (B_2O_3 is added to reduce the viscosity and assist fining) has shown minimum losses of 0.34 dB km^{-1} with multimode fiber at a wavelength of 1.55 μm [Ref. 27]. Also, graded index germanium phosphosilicate fibers have exhibited losses near the intrinsic level for their composition of 2.8, 0.45 and 0.35 dB km^{-1} at wavelengths of 0.82, 1.3 and 1.5 μm respectively [Ref. 28]. Although it is not a continuous process, the MCVD technique has proved suitable for the widespread mass production of high-performance optical fibers [Ref. 29]. Moreover, it can be scaled up to produce preforms which provide 100 to 200 km of fiber [Ref. 19].

4.4.4 Plasma-activated chemical vapor deposition (PCVD)

A variation on the MCVD technique is the use of various types of plasma to supply energy for the vapor-phase oxidation of halides. This method, first developed by Kuppers and Koenings [Ref. 30], involves plasma-induced chemical vapor deposition inside a silica tube, as shown in Figure 4.10. The essential difference between this technique and the

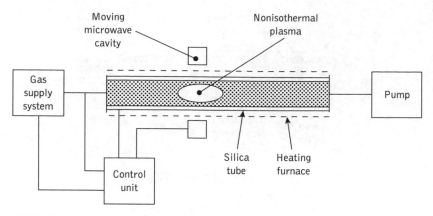

Figure 4.10 The apparatus utilized in the PCVD process

MCVD process is the stimulation of oxide formation by means of a nonisothermal plasma maintained at low pressure in a microwave cavity (2.45 GHz) which surrounds the tube. Volatile reactants are introduced into the tube where they react heterogeneously within the microwave cavity, and no particulate matter is formed in the vapor phase.

The reaction zone is moved backwards and forwards along the tube by control of the microwave cavity and a circularly symmetric layer growth is formed. Rotation of the tube is unnecessary and the deposition is virtually 100% efficient. Film deposition can occur at temperatures as low as 500 °C but a high chlorine content may cause expansivity and cracking of the film. Hence the tube is heated to around 1000 °C during deposition using a stationary furnace.

The high deposition efficiency allows the composition of the layers to be accurately varied by control of the vapor-phase reactants. Also, when the plasma zone is moved rapidly backwards and forwards along the tube, very thin layer deposition may be achieved, giving the formation of up to 2000 individual layers. This enables very good graded index profiles to be realized which are a close approximation to the optimum near-parabolic profile. Thus low-pulse dispersion of less than 0.8 ns km^{-1}, for fibers with attenuations of between 3 and 4 dB km^{-1}, at a wavelength of 0.85 μm has been reported [Ref. 2]. Finally, the PCVD method also lends itself to large-scale production of optical fibers with preform sizes that would allow the preparation of over 200 km of fiber [Ref. 31].

4.4.5 Summary of vapor-phase deposition techniques

The salient features of the four major vapor-phase deposition techniques are summarized in Table 4.2 [Ref. 32]. These techniques have all demonstrated relatively similar performance for the fabrication of both multimode and single-mode fiber of standard step and graded index designs [Ref. 19]. For the production of polarization-maintaining fiber (see Section 3.13.3), however, the MCVD and VAD processes have been employed, together with a hybrid MCVD–VAD technique.

Table 4.2 Summary of vapor-phase deposition techniques used in the preparation of low-loss optical fibers

Reaction type	
Flame hydrolysis	OVPO, VAD
High-temperature oxidation	MCVD
Low-temperature oxidation	PCVD
Depositional direction	
Outside layer deposition	OVPO
Inside layer deposition	MCVD, PCVD
Axial layer deposition	VAD
Refractive index profile formation	
Layer approximation	OVPO, MCVD, PCVD
Simultaneous formation	VAD
Process	
Batch	OVPO, MCVD, PCVD
Continuous	VAD

4.5 Optical fibers

In order to plan the use of optical fibers in a variety of line communication applications it is necessary to consider the various optical fibers currently available. The following is a summary of the dominant optical fiber types with an indication of their general characteristics. The performance characteristics of the various fiber types discussed vary considerably depending upon the materials used in the fabrication process and the preparation technique involved. The values quoted are based upon both manufacturers' and suppliers' data, and practical descriptions [Refs 33–40] for commercially available fibers, presented in a general form rather than for specific fibers. Hence in some cases the fibers may appear to have somewhat poorer performance characteristics than those stated for the equivalent fiber types produced by the best possible techniques and in the best possible conditions which were indicated in Chapter 3. It is interesting to note, however, that although the high-performance values quoted in Chapter 3 were generally for fibers produced and tested in the laboratory, the performance characteristics of commercially available fibers in many cases are now quite close to these values. This factor is indicative of the improvements made over recent years in the fiber materials preparation and fabrication technologies.

This section therefore reflects the maturity of the technology associated with the production of both multicomponent and silica glass fibers, and also plastic optical fibers. In particular, a variety of high-performance silica-based single-mode fibers for operation over the 1.260 to 1.625 μm wavelength range (O to L spectral bands; see Section 3.3.2) are now widely commercially available. A number of these fibers have found substantial application within the telecommunications network while the more specialized polarization-maintaining fibers (see Section 3.13.3) are also commercially available, but do not at present find widespread application, and therefore these fibers are not dealt with in this section. Moreover, fibers developed for both mid- and far-infrared transmission can also be obtained commercially but they continue to exhibit limitations in relation to relatively high losses and low strength (see Section 3.7) which negates their consideration in this section.

Finally, it should be noted that the bandwidths quoted are specified over a 1 km length of fiber (i.e. $B_{opt} \times L$). These are generally obtained from manufacturers' data which does not always indicate whether the electrical or the optical bandwidth has been measured. It is likely that these are in fact optical bandwidths which are significantly greater than their electrical equivalents (see Section 7.4.3).

4.5.1 Multimode step index fibers

Multimode step index fibers may be fabricated from either multicomponent glass compounds or doped silica. These fibers can have reasonably large core diameters and large numerical apertures to facilitate efficient coupling to incoherent light sources such as LEDs. The performance characteristics of this fiber type may vary considerably depending on the materials used and the method of preparation; the doped silica fibers exhibit the best performance. Multicomponent glass and doped silica fibers are often referred to as multicomponent glass/glass (glass-clad glass) and silica/silica (silica-clad silica), respectively, although the glass-clad glass terminology is sometimes used somewhat vaguely to denote both types. A typical structure for a glass multimode step index fiber is shown in Figure 4.11.

Structure
Core diameter: 100 to 300 μm
Cladding diameter: 140 to 400 μm
Buffer jacket diameter: 400 to 1000 μm
Numerical aperture: 0.16 to 0.5.

Performance characteristics
Attenuation: 2.6 to 50 dB km^{-1} at a wavelength of 0.85 μm, limited by absorption or scattering. The wide variation in attenuation is due to the large differences both within and between the two overall preparation methods (melting and deposition). To illustrate this point Figure 4.12 shows the attenuation spectra for a multicomponent glass fiber (glass-clad glass) and a doped silica fiber (silica-clad silica). It may be observed that the multicomponent glass fiber has an attenuation of around 40 dB km^{-1} at a wavelength of 0.85 μm, whereas the doped silica fiber has an attenuation of less than 5 dB km^{-1} at a similar wavelength. Furthermore, at a wavelength of 1.31 μm losses reduced to around 0.4 dB km^{-1} can be obtained [Ref. 36].

Bandwidth: 6 to 50 MHz km.

Applications: These fibers are best suited for short-haul, limited bandwidth and relatively low-cost applications.

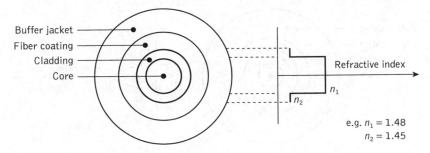

Figure 4.11 Typical structure for a glass multimode step index fiber

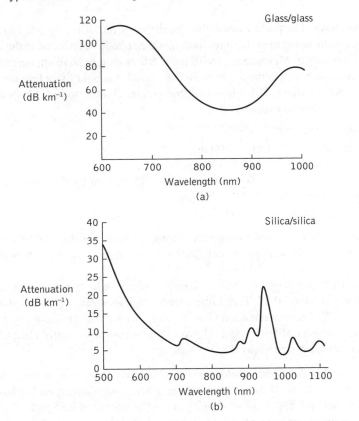

Figure 4.12 Attenuation spectra for multimode step index fibers: (a) multicomponent glass fiber; (b) doped silica fiber

4.5.2 Multimode graded index fibers

These multimode fibers which have a graded index profile may also be fabricated using multicomponent glasses or doped silica. However, they tend to be manufactured from materials with higher purity than the majority of multimode step index fibers in order to

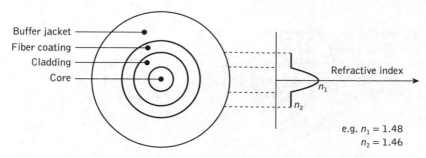

Figure 4.13 Typical structure for a glass multimode graded index fiber

reduce fiber losses. The performance characteristics of multimode graded index fibers are therefore generally better than those for multimode step index fibers due to the index grading and lower attenuation. Multimode graded index fibers tend to have smaller core diameters than multimode step index fibers, although the overall diameter including the buffer jacket is usually about the same. This gives the fiber greater rigidity to resist bending. A typical structure is illustrated in Figure 4.13.

Structure

Core diameter:	50 to 100 μm
Cladding diameter:	125 to 150 μm
Coating diameter:	200 to 300 μm (e.g. 245 ± 5 μm for Corning fibers)
Buffer jacket diameter:	400 to 1000 μm
Numerical aperture:	0.2 to 0.3.

Although the above general parameters encompass most of the currently available multimode graded index fibers, in particular the following major groups are now in use:

1. 50 μm/125 μm (core–cladding) diameter fibers with typical numerical apertures between 0.20 and 0.24. These fibers were originally developed and standardized by the ITU-T (Recommendation G. 651) [Ref. 38] for telecommunication applications at wavelengths of 0.85 and 1.31 μm but now they are mainly utilized within data links and local area networks (LANs).

2. 62.5 μm/125 μm (core–cladding) diameter fibers with typical numerical apertures between 0.26 and 0.29. Although these fibers were developed for longer distance access network applications at operating wavelengths of 0.85 and 1.31 μm, they are now mainly used within LANs (see Section 15.6.4).

3. 85 μm/125 μm (core/cladding) diameter fibers with typical numerical apertures of between 0.26 and 0.30. These fibers were developed for operation at wavelengths of 0.85 and 1.31 μm in short-haul systems and LANs.

4. 100 μm/140 μm (core–cladding) diameter fibers with a numerical aperture of 0.29. These fibers were developed to provide high coupling efficiency to LEDs at a wavelength of 0.85 μm in low-cost, short-distance applications. They can, however, be utilized at the 1.31 μm operating wavelength and have therefore also found application within LANs.

Performance characteristics

Attenuation: 2 to 10 dB km^{-1} at a wavelength of 0.85 μm with generally a scattering limit. Average losses of around 0.4 and 0.25 dB km^{-1} can be obtained at wavelengths of 1.31 and 1.55 μm respectively [Refs 10, 33]

Bandwidth: 200 MHz km to 3 GHz km.

Applications: Although these fibers were initially used for medium haul, they are now best suited to short-haul and medium- to high-bandwidth applications using either incoherent or coherent multimode sources (i.e. LEDs or injection laser diodes respectively).

It is useful to note that quasi-step index or partially graded index fibers are also commercially available. These fibers generally exhibit slightly better performance characteristics than corresponding multimode step index fibers but are somewhat inferior to the fully graded index fibers described above.

4.5.3 Single-mode fibers

Single-mode fibers can have either a step index or graded index profile. The benefits of using a graded index profile are to provide dispersion-modified single-mode fibers (see Section 3.12). The more sophisticated single-mode fiber structures used to produce polarization-maintaining fibers (see Section 3.13.3) make these fibers quite expensive at present and thus they are not generally utilized within optical fiber communication systems. Therefore currently commercially available single-mode fibers are designed to conform with the appropriate ITU-T recommendations, being fabricated from doped silica (silica-clad silica) to produce high-quality, both medium- and long-haul, wideband transmission fibers suitable for the full range of telecommunication applications.

Although single-mode fibers have small core diameters to allow single-mode propagation, the cladding diameter must be at least 10 times the core diameter to avoid losses from the evanescent field. Hence with a coating and buffer jacket to provide protection and strength, single-mode fibers have similar overall diameters to multimode fibers.

4.5.3.1 Standard single-mode fiber

A typical example of the standard single-mode fiber (SSMF) which usually comprises a step index profile and is specified in the ITU-T Recommendation G.652.A [Ref. 40] is shown in Figure 4.14. Such fiber is also referred to as nondispersion shifted as it has

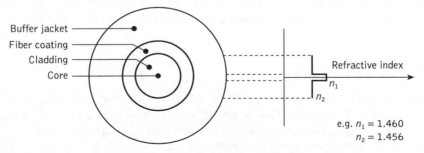

Figure 4.14 Typical structure for a standard single-mode step index fiber

a zero-dispersion wavelength at 1.31 μm and is therefore particularly suited to single-wavelength transmission in the O-band. Although SSMF can be utilized for operation at a wavelength of 1.55 μm, it is not optimized for operation in the C- and L-bands where it exhibits high dispersion in the range 16 to 20 ps nm^{-1} km^{-1}. A commercially available example of this fiber type is the Corning SMF-28.

Structure

Mode-field diameter	7 to 11 μm, typically between 9 and 10 μm at the 1.31 μm wavelength
Cladding diameter:	generally 125 μm
Coating diameter:	200 to 300 μm (e.g. 245 ± 5 μm for Corning fibers)
Buffer jacket diameter:	500 to 1000 μm
Numerical aperture:	0.08 to 0.15, usually around 0.10.

Performance characteristics

	2 to 5 dB km^{-1} with a scattering limit of around 1 dB km^{-1} at a wavelength of 0.85 μm. In addition, maximum losses around 0.35 and 0.20 dB km^{-1} at wavelengths of 1.31 and 1.55 μm can be obtained in a manufacturing environment.
Bandwidth:	Greater than 500 MHz km. In theory the bandwidth is limited by waveguide and material dispersion to approximately 40 GHz km at a wavelength of 0.85 μm. However, practical bandwidths in excess of 10 GHz km are obtained at a wavelength of 1.31 μm.
Applications:	These fibers are ideally suited for high-bandwidth and medium- and long-haul applications using single-mode injection laser sources.

4.5.3.2 Low-water-peak nondispersion-shifted fiber

The concept of low-water-peak fiber (LWPF) was introduced in Section 3.3.2 and it has now become a major single-mode fiber type specified by ITU-T G.652.C [Ref. 40]. As the OH absorption around the wavelength of 1.383 μm SSMF has been removed to create LWPF, this provides a fiber which can be employed for transmission across a wide wavelength range from 1.260 to 1.625 μm. In this context LWPF is also called extended band single-mode fiber as it exhibits low attenuation across the O-, E-, S- and L-bands, although its dispersion performance is the same as SSMF and hence this parameter is not optimized for operation outside the O-band. It should be noted that a low polarization mode dispersion variant of the LWPF is also commercially available, meeting ITU-T G.652.B/D in which the dispersion parameter is reduced from 0.5 ps km^{-1} in ITU-T G.652.A/C to less than 0.2 ps km^{-1}.

4.5.3.3 Loss-minimized fiber

A fiber optimized for operation around the 1.55 μm wavelength is specified in ITU-T G.654 [Ref. 41]. The fiber, which is generally fabricated with a pure silica core, has its cutoff wavelength shifted to typically 1.50 μm providing operation restricted to the 1.55 μm wavelength region where it exhibits a typical loss of only 0.19 dB km^{-1}. Such loss-minimized fiber also has high chromatic dispersion at 1.55 μm with the zero-dispersion wavelength being in the range 1.30 to 1.33 μm and it has been designed for very long-haul undersea applications.

4.5.3.4 Nonzero-dispersion-shifted fiber

The limitations in relation to nonlinear effects associated with four-wave mixing (FWM) have meant that the variants of nonzero-dispersion-shifted fiber (NZ-DSF) have now superseded both dispersion-shifted (see Section 3.12.1) and dispersion-flattened (see Section 3.12.2) single-mode fibers. As indicated in Section 3.12.3, the initial NZ-DSF subsequently specified in ITU-T G.655.A [Ref. 42] was first deployed in the late 1990s providing low but nonzero dispersion around the 1.55 μm wavelength to reduce the non-linear effects such as FWM, self- and cross-phase modulation (see Section 3.14) which cause problems particularly with wavelength division multiplexed optical communication systems.

The zero-dispersion wavelength in NZ-DSF is translated outside the 1.55 μm operating window creating two fiber families termed nonzero dispersion, NZD+ and NZD−, with their zero-dispersion wavelengths typically being around 1.51 μm and 1.58 μm respectively (i.e. falling before and after 1.55 μm). Significant nonlinear effects, however, are still exhibited by the original NZ-DSF (i.e. G.655.A fiber) as it is specified with a very small lower dispersion limit of less than 0.1 ps nm^{-1} km^{-1} and it also exhibits a small effective core area. Hence improved NZ-DSF is now commercially available, compliant with ITU-T G.655.B/C in which the lower dispersion limit is increased to 1 ps nm^{-1} km^{-1} and the effective core area is increased even though it still remains smaller than that provided by SSMF. In addition, the G.655.C fiber also displays reduced polarization mode dispersion at 0.2 ps km^{-1} in comparison with the 0.5 ps km^{-1} value obtained with the G.655.A/B fiber.

Although the G.655.B/C NZ-DSF provides for more efficient suppression of FWM in comparison with the original G.655.A fiber, the range of the dispersion coefficient within the recommendation is from 1 to 10 ps nm^{-1} km^{-1} over only the C-band (1.530 to 1.565 μm). Enhanced suppression of nonlinear effects over the S-, C- and L-bands (1.460 to 1.625 μm) is provided, however, with a more recent development of ultra-broadband NZ-DSF specified in ITU-T G.656 [Ref. 43] where a low value for the chromatic dispersion coefficient in the range 2 to 14 ps nm^{-1} km^{-1} is maintained over the three aforementioned bands. In all other respects the characteristics of the ultra-broadband NZ-DSF matches that of the G.655.C fiber including the improved polarization mode dispersion performance.

NZ-DSF compliant with both ITU-T G.655 and G.656 but also with a suppressed water peak (i.e. exhibiting a low water peak) is also commercially available. An example is the Lucent Truewave Reach LWPF which is optimized for long-haul, high-capacity applications. This fiber typically exhibits attenuations of 0.35 dB km^{-1} and 0.2 dB km^{-1} at wavelengths of 1.383 μm and 1.550 μm respectively, while displaying chromatic dispersion in the range 2.0 to 11.4 ps nm^{-1} km^{-1} over the S-, C- and L-bands. Furthermore, it has a zero-dispersion wavelength of 1.405 μm with a typical effective core area of 55 μm^2 at a wavelength of 1.55 μm.

Finally, the more recent ITU-T Recommendation G.657 [Ref. 44] addresses the issue of reducing single-mode fiber bend losses (see Section 3.6) so that such fibers can be deployed with smaller bend radii without incurring the optical attenuation usually associated with this condition. The imperative for this standardization development is the perceived urgent requirement for a range of single-mode fibers providing improved macrobending perform- ance for deployment in optical access networks which operate from the local exchange/ office to the business or home user (see Section 15.6.3). For example, a commercially available single-mode fiber exceeding the bend performance requirements of ITU-T G.657.A and B which is compatible with SSMF is the Corning Clearcurve fiber [Ref. 45]. This fiber is designed for tight bend deployments (down to a 5 mm bend radius) common in high- rise apartment buildings and multidwelling units where it suffers a loss of no more than 0.1 dB per full turn at a 5 mm bend radius and an operational wavelength of 1.55 μm.

4.5.4 Plastic-clad fibers

Plastic-clad fibers are multimode and have either a step index or a graded index profile. They have a plastic cladding (often a silicone rubber) and a glass core which is frequently silica (i.e. plastic-clad silica (PCS) fibers). The PCS fibers exhibit lower radiation-induced losses than silica-clad silica fibers and, therefore, have an improved performance in certain environments. PCS is the original plastic-clad type where the silicone cladding is quite easy to strip from the silica core that has installation benefits but also causes reliability problems. Hard-clad silica fibers with a tougher plastic cladding are also commercially available which provide for increased durability. Plastic-clad fibers are generally slightly cheaper than the corresponding glass fibers, but usually have more limited performance characteristics. A typical structure for a step index plastic-clad or hard-clad fiber is shown in Figure 4.15.

Structure

Core diameter:	Step index	100 to 1000 μm
	Graded index	50 to 100 μm
Cladding diameter:	Step index	300 to 1400 μm
	Graded index	125 to 150 μm
Buffer jacket diameter:	Step index	500 to 1600 μm
	Graded index	250 to 1000 μm
Numerical aperture:	Step index	0.2 to 0.5
	Graded index	0.2 to 0.3.

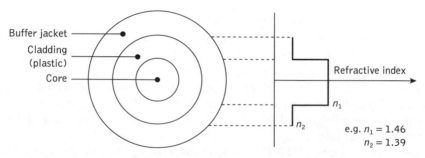

Figure 4.15 Typical structure for a plastic-clad silica multimode step index fiber

Performance characteristics
Attenuation: Step index   5 to 50 dB km^{-1}
                Graded index   4 to 15 dB km^{-1}

4.5.5 Plastic optical fibers

Plastic or polymeric optical fibers (POFs) are fabricated from organic polymers for both the core and cladding regions exhibiting large core and cladding diameters. Hence there is a reduced requirement for a buffer jacket for fiber protection and strengthening. These fibers are usually cheaper to produce and easier to handle than the corresponding silica-based glass variety. However, their performance (especially for optical transmission in the infrared) is restricted, giving them limited use in communication applications. POFs which are multimode with either a step or graded index profile have large numerical apertures as a consequence of the core–cladding refractive index difference which allow easier coupling of light into the fiber from a multimode source.

Early plastic fibers fabricated with a polymethyl methacrylate (PMMA) and a fluorinated acrylic cladding exhibited losses around 500 dB km^{-1}. Subsequently, a continuous casting process was developed for PMMA and losses as low as 110 dB km^{-1} were achieved in the visible wavelength region. The loss mechanisms in PMMA, polystyrene and polycarbonate core fibers are similar to those in glass fibers. These fibers exhibit both intrinsic and extrinsic loss mechanisms including absorption and Rayleigh scattering which results from density fluctuations and the anisotropic structure of the polymers. Significant absorption occurs due to the long-wavelength tail caused by the carbon–hydrogen bonds in these polymers and in particular, strong optical absorption in PMMA resulting from the overtones of the carbon–hydrogen stretching vibration at 3.2 μm which restricts transmission to a single window around 0.65 μm. Moreover, extrinsic absorption results from transition metal and organic contaminants as well as overtone bands from the OH ion.

Structure
Core diameter: 125 to 1880 μm
Cladding diameter: 1250 to 2000 μm
Numerical aperture: 0.3 to 0.6.

Performance characteristics
Attenuation: 50 to 1000 dB km^{-1} at a wavelength of 0.65 μm.
Bandwidth: up to 10 MHz km
Applications: These fibers can only be used for very short-haul (i.e. 'in-house') low cost links. However, fiber coupling and termination are relatively easy and do not require sophisticated techniques.

Although substantial progress in the fabrication of PMMA core fibers has been made, the typical losses for commercially produced fibers have remained in the range 70 to 100 dB km^{-1} and 125 to 150 dB km^{-1} at wavelengths of 0.57 μm and 0.65 μm respectively [Ref. 37]. Moreover, polycarbonate core fibers exhibit far higher losses usually in the range 600 to 700 dB km^{-1} while the lower attenuation provided by polystyrene, typically in the range 70 to 90 dB km^{-1}, is offset by the brittle nature of the material which restricts its application. Hence these POF types have limited maximum link transmission distances

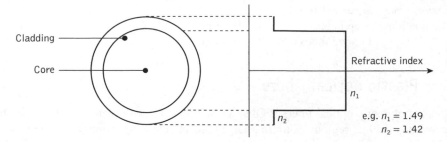

Figure 4.16 Typical structure for a PMMA plastic fiber

up to around 50 m and, in addition, for step index profile fiber large intermodal dispersion also severely reduces the bandwidth capability [Ref. 34]. The structure of a typical step index PMMA core fiber is illustrated in Figure 4.16. A common dimension for this fiber type is 980 μm core diameter plus a carbon polymer cladding with an additional thickness of only 10 μm giving an overall cladding diameter of 1000 μm [Ref. 37]. It also exhibits an attenuation of 150 dB km^{-1} at a wavelength of 0.65 μm with a bandwidth–distance product of 4 MHz km. Other common PMMA fibers have core/cladding diameters of 480/500 μm and 735/750 μm, each with a numerical aperture of 0.5 ± 0.15 [Ref. 37].

Although multimode step index PMMA fiber has been commercially available for many years with a bandwidth limited to a few megahertz kilometres, it is only more recently that an improvement in bandwidth has been obtained by grading the refractive index profile. Hence graded index PMMA POF has demonstrated a bandwidth–distance product of 0.5 GHz km at a wavelength of 0.65 μm [Ref. 46]. Nevertheless, the large fiber attenuation at this wavelength still proves a restriction and also there are very few suitable high-speed light sources available.

Reduction of transmission loss for POF was achieved, however, in 1996 by employing amorphous perfluorinated polymer for the core material. Hence a graded index POF using poly perfluoro-butenylvinyl ether or PFBVE provided for both lower attenuation and potentially high capacity [Refs 47, 48]. This new type of POF, which has been named perfluorinated (PF) plastic optical fiber (PF-POF) produced by Asahi Glass Co. (the perfluorinated material is also called CYTOP®), has been commercially available since June 2000 [Ref. 49].

The attenuation spectrum for a graded index PF-POF is shown in Figure 4.17 where it is compared with the spectra of a graded index PMMA and an LWP silica core fiber [Refs 38, 46]. It may be observed that the PF-POF has a low-loss (for POF) wavelength region from 0.65 μm to 1.31 μm with attenuation of around 40 dB km^{-1} at the latter wavelength. The good near-infrared transparency of PFBVE results from the material's lack of carbon–hydrogen bonds, thus removing the major mechanism causing losses in PMMA fibers. Nevertheless, PF-POF is still subject to significant attenuation due to both intrinsic and extrinsic scattering which causes high losses in it in comparison with silica core fiber. A spectral characteristic displaying the theoretical lower limit of PF-POF attenuation is also displayed in Figure 4.17, the losses in this case being caused by intrinsic scattering resulting from thermodynamic fluctuations of density and of orientational order within the chemical structure of the polymer, the former phenomenon being greater when a dopant is utilized to provide a graded index profile. Although the estimated intrinsic losses of

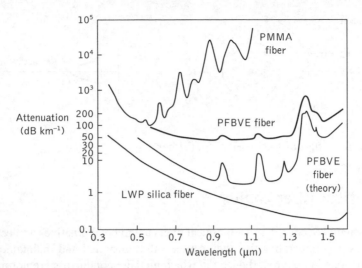

Figure 4.17 Spectral attenuation for plastic optical fibers. Adapted from Refs [34] and [46]

PF-POF are 9.9 dB km^{-1} and 1.8 dB km^{-1} at the wavelengths of 0.85 μm and 1.31 μm respectively [Ref. 38], the attenuation spectrum for the practical graded index PF-POF is dominated by extrinsic scattering induced by processing defects such as impurities and geometric perturbations. These loss mechanisms have been reduced, however, by recent material and processing improvements, such that attenuation below 40 dB km^{-1} over the 0.65 μm to 1.31 μm wavelength region has been obtained, with losses as low as 10 dB km^{-1} at a wavelength of 1.21 μm also being achieved [Ref. 45].

Combined with the lower attenuation facilitated by the PFBVE polymer, graded index PF-POFs exhibit low material dispersion. Indeed, while PMMA fibers have higher material dispersion than silica fibers, PF-POFs have substantially lower material dispersion. The other major dispersive mechanism in the latter fibers is intermodal dispersion which is largely dependent on the shape of the core refractive index profile together with the degree of mode coupling and differential mode attenuation that may occur. Nevertheless, it is predicted that a graded index PF-POF with an index which is optimized to minimize dispersion should be able to achieve a bandwidth–distance product in the order of 10 GHz km [Ref. 50].

The structure of the Lucina® PF-POF is shown in Figure 4.18. This commercially available CYTOP fiber has a graded index profile and transmission at 1.25 Gbit s^{-1} over 1 km of the fiber has been demonstrated [Refs 50, 51].

Structure
Core diameter: 120 μm
Cladding diameter: 230 μm
Buffer diameter: 500 μm
Numerical aperture: 0.170 to 0.195.

Performance characteristics
Attenuation: 20 dB km^{-1} at a wavelength of 0.85 μm
Bandwidth: 2.5 GHz over 100 m.

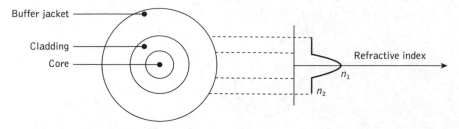

Figure 4.18 Structure of Lucina® PFBVE fiber

4.6 Optical fiber cables

It was indicated in Section 4.1 that if optical fibers are to be alternatives to electrical transmission lines it is imperative that they can be safely installed and maintained in all the environments (e.g. underground ducts) in which metallic conductors are normally placed. Therefore, when optical fibers are to be installed in a working environment their mechanical properties are of prime importance. In this respect the unprotected optical fiber has several disadvantages with regard to its strength and durability. Bare glass fibers are brittle and have small cross-sectional areas which make them very susceptible to damage when employing normal transmission line handling procedures. It is therefore necessary to cover the fibers to improve their tensile strength and to protect them against external influences. This is usually achieved by surrounding the fiber with a series of protective layers, which is referred to as coating and cabling. The initial coating of plastic with high elastic modulus is applied directly to the fiber cladding, as illustrated in Section 4.5. It is then necessary to incorporate the coated and buffered fiber into an optical cable to increase its resistance to mechanical strain and stress as well as adverse environmental conditions.

The functions of the optical cable may be summarized into four main areas. These are as follows:

1. Fiber protection. The major function of the optical cable is to protect against fiber damage and breakage both during installation and throughout the life of the fiber.

2. Stability of the fiber transmission characteristics. The cabled fiber must have good stable transmission characteristics which are comparable with the uncabled fiber. Increases in optical attenuation due to cabling are quite usual and must be minimized within the cable design.

3. Cable strength. Optical cables must have similar mechanical properties to electrical transmission cables in order that they may be handled in the same manner. These mechanical properties include tension, torsion, compression, bending, squeezing and vibration. Hence the cable strength may be improved by incorporating a suitable strength member and by giving the cable a properly designed thick outer sheath.

4. Identification and jointing of the fibers within the cable. This is especially important for cables including a large number of optical fibers. If the fibers are arranged in a suitable geometry it may be possible to use multiple jointing techniques rather than jointing each fiber individually.

In order to consider the cabling requirements for fibers with regard to areas 1 and 2, it is necessary to discuss the fiber strength and durability as well as any possible sources of degradation of the fiber transmission characteristics which are likely to occur due to cabling.

4.6.1 Fiber strength and durability

Optical fibers for telecommunications usage are almost exclusively fabricated from silica or a compound of glass (multicomponent glass). These materials are brittle and exhibit almost perfect elasticity until their breaking point is reached. The bulk material strength of flawless glass is quite high and may be estimated for individual materials using the relationship [Ref. 33]:

$$S_t = \left(\frac{\gamma_p E}{4 l_a}\right)^{\frac{1}{2}} \tag{4.6}$$

where S_t is the theoretical cohesive strength, γ_p is the surface energy of the material, E is Young's modulus for the material (stress/strain), and l_a is the atomic spacing or bond distance. However, the bulk material strength may be drastically reduced by the presence of surface flaws within the material.

In order to treat surface flaws in glass analytically, the Griffith theory [Ref. 52] is normally used. This theory assumes that the surface flaws are narrow cracks with small radii of curvature at their tips, as illustrated in Figure 4.19. It postulates that the stress is concentrated at the tip of the crack, which leads to crack growth and eventually catastrophic failure. Figure 4.19 shows the concentration of stress lines at the crack tip which indicates that deeper cracks have higher stress at their tips. The Griffith theory gives a stress intensity factor K_I as:

$$K_I = SYC^{\frac{1}{2}} \tag{4.7}$$

where S is the macroscopic stress on the fiber, Y is a constant dictated by the shape of the crack (e.g. $Y = \pi^{\frac{1}{2}}$ for an elliptical crack, as illustrated in Figure 4.19) and C is the depth of the crack (this is the semimajor axis length for an elliptical crack).

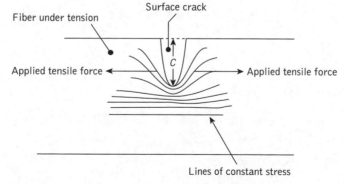

Figure 4.19 An elliptical surface crack in a tensioned optical fiber

Further, the Griffith theory gives an expression for the critical stress intensity factor K_{IC} where fracture occurs as:

$$K_{\mathrm{IC}} = (2E\gamma_{\mathrm{p}})^{\frac{1}{2}} \tag{4.8}$$

Combining Eqs (4.7) and (4.8) gives the Griffith equation for fracture stress of a crack S_{f} as:

$$S_{\mathrm{f}} = \left(\frac{2E\gamma_{\mathrm{p}}}{Y^2 C}\right)^{\frac{1}{2}} \tag{4.9}$$

It is interesting to note that S_{f} is proportional to $C^{-\frac{1}{2}}$. Therefore, S_{f} decreases by a factor of 2 for a fourfold increase in the crack depth C.

Example 4.1

The Si–O bond has a theoretical cohesive strength of 2.6×10^6 psi which corresponds to a bond distance of 0.16 nm. A silica optical fiber has an elliptical crack of depth 10 nm at a point along its length. Estimate

(a) the fracture stress in psi for the fiber if it is dependent upon this crack:

(b) the percentage strain at the break.

Young's modulus for silica is approximately 9×10^{10} N m^{-2} and 1 psi $\equiv$ 6894.76 N m^{-2}.
Solution: (a) Using Eq. (4.6), the theoretical cohesive strength for the Si–O bond is:

$$S_{\mathrm{t}} = \left(\frac{\gamma_{\mathrm{p}}E}{4l_{\mathrm{a}}}\right)^{\frac{1}{2}}$$

Hence:

$$\gamma_{\mathrm{p}} = \frac{4l_{\mathrm{a}}S_{\mathrm{t}}^2}{E} = \frac{4 \times 0.16 \times 10^{-9}(2.6 \times 10^6 \times 6894.76)^2}{9 \times 10^{-10}}$$

$$= 2.29 \text{ J}$$

The fracture stress for the silica fiber may be obtained from Eq. (4.9) where:

$$S_{\mathrm{f}} = \left(\frac{2E\gamma_{\mathrm{p}}}{Y^2 C}\right)^{\frac{1}{2}}$$

For an elliptical crack:

$$S_{\mathrm{f}} = \left(\frac{2E\gamma_{\mathrm{p}}}{\pi C}\right)^{\frac{1}{2}} = \left(\frac{2 \times 9 \times 10^{10} \times 2.29}{\pi \times 10^{-8}}\right)^{\frac{1}{2}}$$

$$= 3.62 \times 10^9 \text{ N m}^{-1}$$
$$= 5.25 \times 10^5 \text{ psi}$$

It may be noted that the fracture stress is reduced from the theoretical value for flawless silica of 2.6×10^6 psi by a factor of approximately 5.

(b)   Young's modulus is defined as:

$$E = \frac{\text{stress}}{\text{strain}}$$

Therefore:

$$\text{Strain} = \frac{\text{stress}}{E} = \frac{S_f}{E} = \frac{3.62 \times 10^9}{9 \times 10^{10}} = 0.04$$

Hence the strain at the break is 4%, which corresponds to the change in length over the original length for the fiber.

In Example 4.1 we considered only a single crack when predicting the fiber fracture. However, when a fiber surface is exposed to the environment and is handled, many flaws may develop. The fracture stress of a length of fiber is then dependent upon the dominant crack (i.e. the deepest) which will give a fiber fracture at the lowest strain. Hence, the fiber surface must be protected from abrasion in order to ensure high fiber strength. A primary protective plastic coating is usually applied to the fiber at the end of the initial production process so that mechanically induced flaws may be minimized. Flaws also occur due to chemical and structural causes. These flaws are generally smaller than the mechanically induced flaws and may be minimized within the fiber fabrication process.

There is another effect which reduces the fiber fracture stress below that predicted by the Griffith equation. It is due to the slow growth of flaws under the action of stress and water and is known as stress corrosion. Stress corrosion occurs because the molecular bonds at the tip of the crack are attacked by water when they are under stress. This causes the flaw to grow until breakage eventually occurs. Hence stress corrosion must be taken into account when designing and testing optical fiber cables. It is usual for optical fiber cables to have some form of water-protective barrier, as is the case for most electrical cable designs.

In order to predict the life of practical optical fibers under particular stresses it is necessary to use a technique which takes into account the many flaws a fiber may possess, rather than just the single surface flaw considered in Example 4.1. This is approached using statistical methods because of the nature of the problem which involves many flaws of varying depths over different lengths of fiber.

Calculations of strengths of optical fibers are usually conducted using Weibull statistics [Ref. 53] which describe the strength behavior of a system that is dependent on the weakest link within the system. In the case of optical fibers this reflects fiber breakage due to the dominant or deepest crack. The empirical relationship established by Weibull and applied to optical fibers indicates that the probability of failure F at a stress S is given by:

$$F = 1 - \exp\left[-\left(\frac{S}{S_0}\right)^m \left(\frac{L}{L_0}\right)\right] \tag{4.10}$$

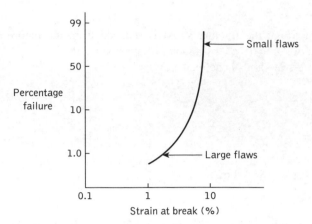

Figure 4.20 A schematic representation of a Weibull plot. Reproduced with permission from M. H. Reeve, *Radio Electron. Eng.*, **51**, p. 327, 1981

where m is the Weibull distribution parameter, S_0 is a scale parameter, L is the fiber length and L_0 is a constant with dimensions of length.

The expression given in Eq. (4.10) may be plotted for a fiber under test by breaking a large number of 10 to 20 m fiber lengths and measuring the strain at the break. The various strains are plotted against the cumulative probability of their occurrence to give the Weibull plot as illustrated in Figure 4.20 [Ref. 54]. It may be observed from Figure 4.20 that most of the fiber tested breaks at strain due to the prevalence of many shallow surface flaws. However, some of the fiber tested contains deeper flaws (possibly due to external damage) giving the failure at lower strain depicted by the tail of the plot. This reduced strength region is of greatest interest when determining the fiber's lifetime under stress.

Finally, the additional problem of stress corrosion must be added to the information on the fiber under stress gained from the Weibull plot. The stress corrosion is usually predicted using an empirical relationship for the crack velocity v_c in terms of the applied stress intensity factor K_I, where [Ref. 54]:

$$v_c = AK_I^n \tag{4.11}$$

The constant n is called the stress corrosion susceptibility (typically in the range 15 to 50 for glass), and A is also a constant for the fiber material. Equation (4.11) allows estimation of the time to failure of a fiber under stress corrosion conditions. Therefore, from a combination of fiber testing (Weibull plot) and stress corrosion, information estimates of the maximum allowable fiber strain can be made available to the cable designer. These estimates may be confirmed by straining the fiber up to a specified level (proof testing) such as 1% strain. Fiber which survives this test can be accepted. However, proof testing presents further problems, as it may cause fiber damage. Also, it is necessary to derate the maximum allowable fiber strain from the proof test value to increase confidence in fiber survival under stress conditions. It is suggested [Ref. 54] that a reasonable derating for use by the cable designer for fiber which has survived a 1% strain proof test is around 0.3% in order that the fiber has a reasonable chance of surviving with a continuous strain for 20 years.

4.7 Stability of the fiber transmission characteristics

Optical fiber cables must be designed so that the transmission characteristics of the fiber are maintained after the cabling process and cable installation. Therefore, potential increases in the optical attenuation and reduction in the bandwidth of the cabled fiber should be avoided.

Certain problems can occur either within the cabling process or subsequently which can significantly affect the fiber transmission characteristics. In particular, a problem which often occurs in the cabling of optical fiber is the meandering of the fiber core axis on a microscopic scale within the cable form. This phenomenon, known as microbending, results from small lateral forces exerted on the fiber during the cabling process and it causes losses due to radiation in both multimode and single-mode fibers.

In addition to microbending losses caused by fiber stress and deformation on a micron scale, macrobending losses occur when the fiber cable is subjected to a significant amount of bending above a critical value of curvature. Such fiber bend losses are discussed in Section 3.6. However, additional optical losses can occur when fiber cables are *in situ*. These losses may result from hydrogen absorption by the fiber material or from exposure of the fiber cable to ionizing radiation. The above phenomena are discussed in this section in order to provide an insight into the problems associated with the stability of the cabled fiber transmission characteristics.

4.7.1 Microbending

Microscopic meandering of the fiber core axis, known as microbending, can be generated at any stage during the manufacturing process, the cable installation process or during service. This is due to environmental effects, particularly temperature variations causing differential expansion or contraction [Ref. 55]. Microbending introduces slight surface imperfections which can cause mode coupling between adjacent modes, which in turn creates a radiative loss which is dependent on the amount of applied fiber deformation, the length of fiber, and the exact distribution of power among the different modes.

It has become accepted to consider, in particular, two forms of modal power distribution. The first form occurs when a fiber is excited by a diffuse Lambertian source, launching all possible modes, and is referred to as a uniform or fully filled mode distribution. The second form occurs when, due to a significant amount of mode coupling and mode attenuation, the distribution of optical power becomes essentially invariant with the distance of propagation along the fiber. This second distribution is generally referred to as a steady-state or equilibrium mode distribution, which typically occurs after transmission over approximately 1 km of fiber (see Section 14.1).

Since microbending losses are mode dependent and from Eq. (2.69) the number of modes is an inverse function of the wavelength of the transmitted light within a particular fiber, it is to be expected that microbending losses will be wavelength dependent. This effect is demonstrated for multimode fiber in Figure 4.21 [Ref. 56], which illustrates the theoretical microbending loss for both the uniform and the steady-state mode distributions as a function of applied linear pressure (i.e. simulated microbending), for a normalized frequency $V = 39$, corresponding to a wavelength of 0. 82 µm, and $V = 21$, corresponding

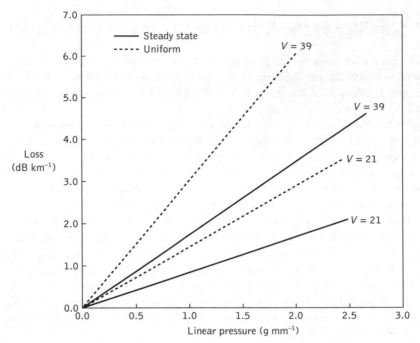

Figure 4.21 Theoretical microbending loss against linear pressure for graded index multimode fibers ($NA = 0.2$, core diameter of 50 μm). Reproduced with permission from S. Das, C. G. Englefield and P. A. Goud, *Appl. Opt.*, **24**, p. 2323, 1985

to a wavelength of 1.55 μm. It may be observed from Figure 4.21 that the microbending loss decreases at longer wavelengths, and that it is also dependent on the modal power distribution present within the fiber; microbending losses corresponding to a uniform power distribution are approximately 1.75 times greater than those obtained with a steady-state distribution. In addition it has been predicted [Ref. 57] that microbending losses for single-mode fiber follow an approximately exponential form, with increasing losses at longer wavelengths. Minimal losses were predicted at operating wavelengths below 1.3 μm, with a rapid rise in attenuation at wavelengths above 1.5 μm. Experimental measurements have confirmed these predictions.

It is clear that excessive microbending can create additional fiber losses to an unacceptable level. To avoid deterioration in the optical fiber transmission characteristics resulting from mode-coupling-induced microbending, it is important that the fiber is free from irregular external pressure within the cable. Carefully controlled coating and cabling of the fiber is therefore essential in order to minimize the cabled fiber attenuation. Furthermore, the fiber cabling must be capable of maintaining this situation under all the strain and environmental conditions envisaged in its lifetime.

4.7.2 Hydrogen absorption

The diffusion of hydrogen into optical fiber has been shown to affect the spectral attenuation characteristic [Ref. 58]. There are two fundamental mechanisms by which hydrogen

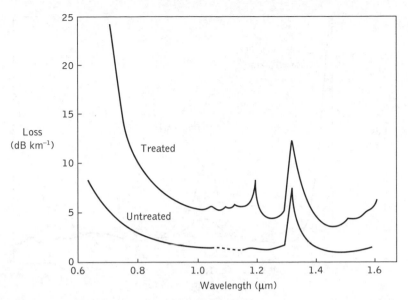

Figure 4.22 Attenuation spectra for multimode fiber before (untreated) and after (treated) hydrogen diffusion [Ref. 59]

absorption causes an increase in optical fiber losses [Ref. 55]. The first is where hydrogen diffuses into interstitial spaces in the glass, thereby altering the spectral loss characteristics through the formation of new absorption peaks. This phenomenon has been found to affect all silica-based glass fibers, both multimode and single mode. However, the extra losses obtained can be reversed if the hydrogen source is removed. Typically, it causes losses in the range 0.2 to 0.3 dB km^{-1} atm^{-1} at an optical wavelength of 1.3 μm and a temperature of 25°C with 500 h exposure [Ref. 58]. At higher temperatures these additional losses may be substantially increased, as can be observed from Figure 4.22, which displays the change in spectral attenuation obtained for a fiber with 68 h hydrogen exposure at a temperature of 150°C [Ref. 59].

The second mechanism occurs when hydrogen reacts with the fiber deposits to give P–OH, Ge–OH or Si–OH absorption. These losses are permanent and can be greater than 25 dB km^{-1} [Ref. 55]. Studies suggest that hydrogen can be generated by either chemical decomposition of the fiber coating materials or through metal–electrolytic action (i.e. moisture affecting the metal sheathing of the fiber cable). These effects can be minimized by careful selection of the cable, the prevention of immersion of the cable in water, or by pressurizing the cable to prevent water ingress. Alternatively, the fiber cable may be periodically purged using an inert gas.

4.7.3 Nuclear radiation exposure

The optical transmission characteristics of fiber cables can be seriously degraded by exposure to nuclear radiation. Such radiation forms color centers in the fiber core which can cause spectral attenuation [Ref. 60]. The precise nature of this attenuation depends upon a number of factors including: fiber parameters such as structure; core and cladding material

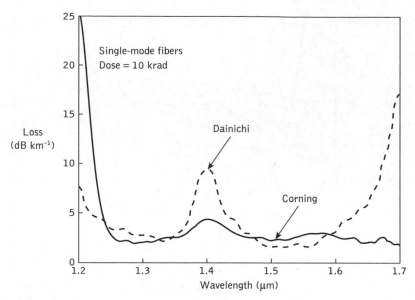

Figure 4.23 Effect of nuclear radiation on the spectral attenuation characteristics of two single-mode fibers. From E. J. Friebele, K. J. Long, C. G. Askins, M. E. Gingerich, M. J. Marrone and D. L. Griscom, 'Overview of radiation effects in fiber optics', *Proc. SPIE, Int. Soc. Opt. Eng., Radiation Effects in Optical Materials*, **541**, p. 70, 1985

composition; system parameters such as optical intensity and wavelength as well as temperature; and radiation parameters such as total dose, dose rate and energy levels, together with the length of recovery time allowed. The radiation-induced attenuation comprises a permanent component which is irreversible, and a metastable component which is reversible and contains both a transient (with decay time less than 1 s) and steady-state (with decay time less than 10 s) constituents. The nature of both the permanent and decaying components of the attenuation is dependent on the fiber composition.

Typical measured spectral loss characteristics for two single-mode fibers, following exposure to a 10 krad dose of steady-state radiation for 1 h, are shown in Figure 4.23 [Ref. 61]. The Corning fiber under test had a Ge-doped silica core and a pure silica cladding, whereas the Dainichi fiber had a pure silica core and an F–P-doped silica cladding. Figure 4.23 displays the spectral loss characteristics over the wavelength range 1.2 to 1.7 μm, but it should also be noted that pure silica core fibers exhibit considerable radiation-induced losses at wavelengths around 0.85 μm. These losses initially increase linearly with increasing radiation dose and can become hundreds of decibels per kilometer [Ref. 60].

Radiation-resistant fibers have been developed which are less sensitive to the effects of nuclear radiation. For example, hydrogen treatment of a pure silica core fiber, or use of boron–fluoride-doped silica cladding fiber, has been found to reduce gamma-ray-induced attenuation in the visible wavelength region [Ref. 61]. It has also been reported that radiation-induced attenuation can be reduced through photobleaching [Ref. 62]. In general, however, the only fiber structures likely to have an acceptable performance over a wide wavelength range when exposed to ionizing radiation are those having pure undoped silica cores, or those with core dopants of germanium and germanium with small amounts of fluorine–phosphorus [Ref. 63].

Clearly, radiation exposure can induce a considerable amount of attenuation in optical fibers, although the number of possible variable parameters, relating to both fiber structure and the nature of the radiation, make it difficult to generalize on the precise spectral effects. Nevertheless, more specific details relating to these effects can be found in the literature [Refs 60–64].

4.8 Cable design

The design of optical fiber cables must take account of the constraints discussed in Section 4.6. In particular, the cable must be designed so that the strain on the fiber in the cable does not exceed 0.2% [Ref. 36]. Alternatively, it is suggested that the permanent strain on the fiber should be less than 0.1% [Ref. 34]. In practice, these constraints may be overcome in various ways which are, to some extent, dependent upon the cable's application. Nevertheless, cable design may generally be separated into a number of major considerations. These can be summarized into the categories of fiber buffering, cable structural and strength members, and cable sheath and water barrier.

4.8.1 Fiber buffering

It was indicated in Section 4.6 that the fiber is given a primary coating during production (typically 5 to 10 µm of Teflon) in order to prevent abrasion of the glass surface and subsequent flaws in the material. The primary coated fiber is then given a secondary or buffer coating (jacket) to provide protection against external mechanical and environmental influences. This buffer jacket is designed to protect the fiber from microbending losses and may take several different forms. These generally fall into one of three distinct types which are illustrated in Figure 4.24 [Refs 10, 39, 65]. A tight buffer jacket is shown in Figure 4.24(a), which usually consists of a hard plastic (e.g. nylon, Hytrel, Tefzel) and is in direct contact with the primary coated fiber which has a typical diameter of 250 µm. This thick buffer coating (usually 900 µm in diameter) provides stiffening for the fiber

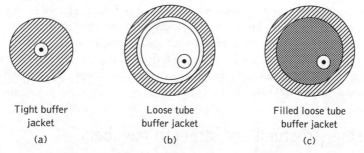

Tight buffer jacket   Loose tube buffer jacket   Filled loose tube buffer jacket

(a)   (b)   (c)

Figure 4.24 Techniques for buffering of optical fibers [Ref. 65]: (a) tight buffer jacket; (b) loose tube buffer jacket; (c) filled loose tube buffer jacket

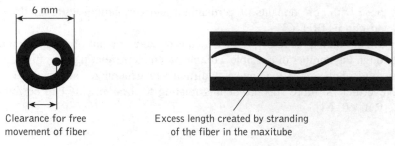

6 mm

Clearance for free
movement of fiber

Excess length created by stranding
of the fiber in the maxitube

Figure 4.25 Maxitube loose buffer design showing fiber excess length

against outside microbending influences, but it must be applied in such a manner as not to cause microbending losses itself.

An alternative and now common approach, which is shown in Figure 4.24(b), is the use of a loose tube buffer jacket. This produces an over sized cavity in which the fiber is placed and which mechanically isolates the fiber from external forces. Loose tube buffering is generally achieved by using a hard, smooth, flexible material, or combination of materials (e.g. polyester and polyamide), in the form of an extruded tube with an outer diameter of typically 1.4 mm. As the buffer tube is smooth inside, it exhibits a low resistance to movement of the fiber. In addition, it provides the benefit that it can be easily stripped for jointing or fiber termination.

A variation on the loose tube buffering in which the over sized cavity is filled with a moisture-resistant compound is depicted in Figure 4.24(c). This technique, which combines the advantages of the previous methods, also provides a water barrier in the immediate vicinity of the fiber. The filling material must be soft, self-healing and stable over a wide range of temperatures (i.e. it should not drip out or freeze between −30 and +70°C), and therefore usually consists of specially blended petroleum or silicon-based compounds.

Although single-fiber loose tube buffers have proved useful within cable constructions, the buffer size limits these cables to low fiber counts (e.g. 16 fibers). In order to reduce the cable complexity, size and overall weight while increasing the number of fibers, several single-mode or multimode fibers (typically between 2 and 12) can be inserted into a larger loose tube buffer jacket with an outside diameter of 1.8 to 3.5 mm [Ref. 10] to create a multifiber loose tube buffer. As with the single-fiber approach, the hollow cavity space is often filled with a moisture-resistant compound.

A further development of the multifiber loose tube buffer is referred to as the maxitube [Ref. 10]. This comprises a larger buffer jacket with an outside diameter of typically 6 mm, inside which a greater number of either individual fibers or fibers in the form of bundles or ribbon structures can be placed. Optical fibers located inside the multifiber maxitube are stranded and through this process an excess length of approximately 0.5% can be obtained, as illustrated in Figure 4.25.

4.8.2 Cable structural and strength members

One or more structural members are usually included in the optical fiber cable to serve as a cable core foundation around which the buffered fibers may be wrapped, or into which

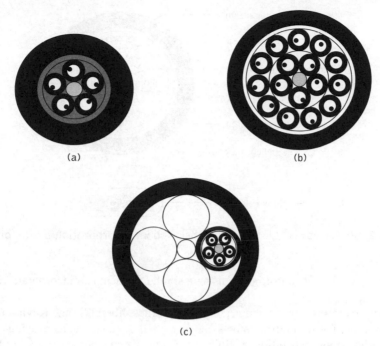

Figure 4.26 Optical fiber cable structures: (a) one-layer cable incorporating single-fiber loose tube buffers; (b) layer cable incorporating single-fiber loose tube buffers in two layers; (c) unit cable construction

they may be slotted. This approach, which is referred to as stranding, is illustrated in Figure 4.26. It may be observed that the cable elements in Figure 4.26(a) and (b) are stranded in one, two or indeed several layers around the central structural member. When the stranding is composed of individual elements (e.g. single-fiber or multifiber loose tube buffer, or a maxitube) then the cable is described as a layer cable. If, however, the cable core consists of stranding elements each of which comprises a unit of stranded elements, then this is termed an optical unit cable. Such a unit cable construction is shown in Figure 4.26(c), which typically enables increased fiber packing density.

The structural member may also be a strength or tensile member if it consists of suitable material such as solid or stranded steel wire, dielectric aramid yarns, often simply referred to as a dielectric strength member (e.g. Kevlar (DuPont Ltd)) and/or glass elements. This situation is indicated in Figure 4.26(a), (b) and (c) where the central steel member acts as both a structural and strength member. In these cases the central steel member is the primary load-bearing element providing strain relief.

Figure 4.27 shows a slotted core cable which comprises a special design of layer cable with a single layer incorporating an extruded plastic structural member around a central steel wire strength member. In this case optical fiber ribbons lie in helical grooves or slots formed in the surface of the structural member rather than the fibers being inside buffer jackets and stranded in layers. Moreover, the primary function of the structural member is therefore not load bearing, but to provide suitable accommodation for the fiber ribbons within the cable. It should be noted that although the overall fiber count provided by this

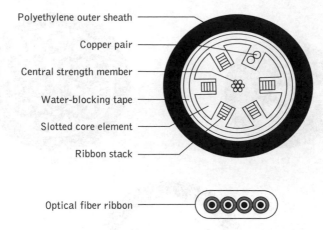

Polyethylene outer sheath

Copper pair

Central strength member

Water-blocking tape

Slotted core element

Ribbon stack

Optical fiber ribbon

Figure 4.27 Slotted core cable with four fiber ribbons incorporating a total of 100 fibers

cable is 100, a unit cable construction using a similar element can incorporate 1000 fibers [Ref. 10].

Structural members may be nonmetallic with plastics, fiberglass and Kevlar often being used. However, for strength members the preferred features include a high Young's modulus, high strain capability, flexibility and low weight per unit length. Therefore, although similar materials are frequently utilized for both strength and structural members, the requirement for additional tensile strength of the strength member must be considered within the cable design.

Flexibility in strength members formed of materials with high Young's modulii may be improved by using a stranded or bunched assembly of smaller units, as in the case of steel wire. Similar techniques are also employed with other materials used for strength members which include plastic monofilaments (i.e. specially processed polyester), textile fiber (nylon, Terylene, Dacron and the widely used Kevlar) and carbon and glass fibers. These materials provide a variety of tensile strengths for different cable applications. However, it is worth noting that Kevlar, an aromatic polyester, has a very high Young's modulus (up to 13×10^{10} N m^{-2}) which gives it a strength to weight ratio advantage four times that of steel.

It is usual when utilizing a stranded strength member to cover it with a coating of extruded plastic or helically applied tape. This is to provide the strength member with a smooth (cushioned) surface which is especially important for the prevention of microbending losses when the member is in contact with the buffered optical fibers.

4.8.3 Cable sheath, water barrier and cable core

The cable is normally covered with a substantial outer plastic sheath in order to reduce abrasion and to provide the cable with extra protection against external mechanical effects such as crushing. The cable sheath is said to contain the cable core and may vary in complexity from a single extruded plastic jacket to a multilayer structure comprising two or more jackets with intermediate armouring. The common and well-proven polyethylene

(PE) sheath material is most often utilized, while polyvinyl chloride (PVC) finds frequent application for indoor cables [Ref. 10]. Hence, an additional water barrier is usually incorporated for outdoor cables. This may take the form of an axially laid aluminum foil/polyethylene laminated film immediately inside the sheath.

Alternatively, the ingress of water may be prevented by filling the spaces in the cable with moisture-resistant compounds. Specially formulated silicone rubber or petroleum-based compounds are often used, which do not cause difficulties in identification and handling of individual optical fibers within the cable form. These filling compounds are also easily removed from the cable and provide protection from corrosion for any metallic strength members within the fiber. Also, the filling compounds must not cause degradation of the other materials within the cable and must remain stable under pressure and temperature variations. Finally, if filled cables are required to have a metal-free sheath construction, then a barrier plastic layer comprising, for example, polyimide–thermoplastic adhesive can be incorporated between the cable sheath and the yarns or filling compound. This approach also prevents migration of the filling compound into the sheath from the cable core.

4.8.4 Examples of fiber cables

A number of different cable designs have emerged and been adopted by various organizations throughout the world. In this section we consider some of the more common designs used in optical fiber cable construction to provide the reader with an insight into the developments in this important field.

Leading cable designs in the late 1970s included the use of loose buffer tubes or, alternatively, fiber ribbons [Ref. 66]. In the former design the fibers are enclosed in tubes which are stranded around a central strength (see Figure 4.26(a)) prior to the application of a polymeric sheath, thus providing large fiber strain relief. In the latter case high fiber packing density as well as ease of connectorization can be obtained. More recently, other designs have also found widespread application, in particular the slotted core design, an example of which is shown in Figure 4.27, and the loose fiber bundle design in which the fibers are packaged into bundles before being enclosed in a single, loose-fitting tube [Ref. 65]. All of the aforementioned multifiber cable designs are available with both single-mode and multimode fibers.

Figure 4.28 [Refs 35, 39] shows two examples of cable construction for single fibers. In Figure 4.28(a) the normally 250 μm diameter fiber protective coating is followed by a tight buffer jacket (usually 900 μm diameter) which is surrounded by a layer of Kevlar for strengthening. Finally, an outer cable jacket with a typical diameter of 2.4 mm is provided. In this construction the optical fiber itself acts as a central strength member. The cable construction illustrated in Figure 4.28(b) uses a loose tube buffer around the central optical fiber. This is surrounded by a Kevlar strength member which is protected by an inner sheath or jacket before the outer sheath layer. The strength members of single optical fiber cables are not usually incorporated at the center of the cable (unless the fiber is acting as a strength member), but are placed in the surrounding cable form, as illustrated in Figure 4.28(b).

Although the single-fiber cables shown in Figure 4.28 can be utilized for indoor applications, alternative two- and multiple-fiber designs have been produced for such areas. A

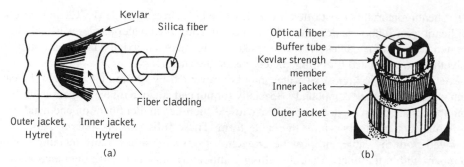

Figure 4.28 Single-fiber cables [Refs 35, 39]: (a) tight buffer jacket design; (b) loose buffer jacket design

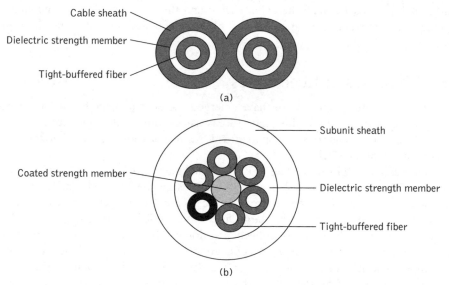

Figure 4.29 Indoor cables: (a) interconnect cable incorporating two optical fibers; (b) 6-fiber subunit of a 48-fiber cable

duplex tight-buffered fiber interconnect cable is shown in Figure 4.29(a). Each of the buffered fibers is surrounded by a dielectric (aramid yarn) strength member and then they are both located within the cable sheath. Figure 4.29(b) depicts a 6-fiber subunit for a multifiber indoor cable which can incorporate a total of eight such subunits creating a 48-fiber cable [Ref. 10]. The tight-buffered fibers in the subunit are placed around a coated central strength member and then surrounded by a further aramid yarn strength member prior to a subunit sheath creating a structure with a diameter of 2.4 mm for standard duty applications. This dimension can be reduced to 2.0 mm for light-duty or increased to 2.7 mm for heavy-duty tasks [Ref. 10].

Multifiber cables for outside plant applications, as mentioned previously, exhibit several different design methodologies. The use of a central strength member, as illustrated in Figure 4.26(a), is a common technique for the incorporation of either loose buffer or tight

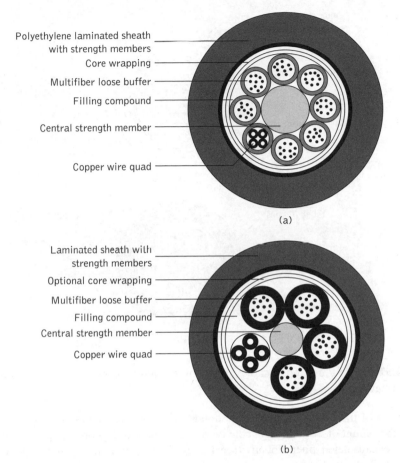

Polyethylene laminated sheath
with strength members

Core wrapping

Multifiber loose buffer

Filling compound

Central strength member

Copper wire quad

(a)

Laminated sheath with
strength members

Optional core wrapping

Multifiber loose buffer

Filling compound

Central strength member

Copper wire quad

(b)

Figure 4.30 Multifiber cables for outdoor applications: (a) multimode fiber loose buffer tube cable; (b) single-mode fiber loose buffer tube cable

buffer jacketed fibers. Such a structural member is utilized in the loose tube cable structure for multimode fibers shown in Figure 4.30(a) in which filled loose tubes are extruded over fiber bundles with typically 10 fibers per tube [Refs 10, 66]. The tubes are then stranded around a central strength member, forming the cable core which is completed with a polyethylene sheath. A similar multifiber cable construction is illustrated in Figure 4.30(b). This type of layered cable design with loose tube buffers incorporating between 2 and 12 single-mode fibers per tube predominates in the core telecommunications network. A number (e.g. four) of such cable cores can also be incorporated into a unit cable structure to provide for a much higher fiber count (e.g. 240 fibers).

Although the slotted core structural member design depicted in Figure 4.27 is a useful approach, an alternative for incorporating fiber ribbons in order to produce high-fiber-count cables is illustrated in Figure 4.31(a) [Ref. 67]. In this common modular ribbon cable design the ribbons are located in a central tube which is surrounded by an appropriate water-retardant filling compound. It can be observed that in this case the strength members are, by necessity, placed in the cable sheath which also provides for an element

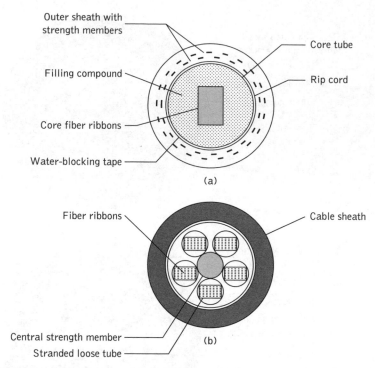

Outer sheath with
strength members

Core tube

Filling compound

Rip cord

Core fiber ribbons

Water-blocking tape

(a)

Fiber ribbons

Cable sheath

Central strength member

(b)

Stranded loose tube

Figure 4.31 Loose tube ribbon cables: (a) central tube construction; (b) multiple loose tube unit cable

of armoring of the cable. Eighteen edge-bonded ribbons each containing 16 fibers give a typical fiber count of 288 for this cable type, while the slotted core design which usually employs encapsulated fiber ribbons (see Figure 4.27) can similarly incorporate a large number of single-mode fibers.

Furthermore, as with the slotted core design, a unit cable approach using the central tube can scale up the fiber count by stranding multiple tubes around a central strength member, as shown in Figure 4.31(b). In addition, it should be noted that the use of fiber ribbons provides the most efficient technique for simultaneously locating, handling, splicing and connectorizing a large number of fibers while also maximizing fiber packing density and facilitating mechanical robustness [Ref. 68]. As indicated above, one of three basic ribbon structures may be distinguished based on the way in which the fibers are joined to form the ribbon. Individual fibers are positioned in parallel and equally spaced from each other in a single layer. They are then sandwiched, edge bonded or encapsulated. In the first structure they are bonded together in one layer between two polyester adhesive foils. By contrast, in the edge-bonded structure glue is applied without the foils in the interstices between the fibers and in the encapsulated structure the bonding matrix extends well beyond the outer boundary of the fibers providing complete encapsulation in a plastic coating [Ref. 10].

Finally, two significant cable designs incorporating standard loose tube buffered fibers are depicted in Figure 4.32. A 264-fiber indoor–outdoor cable is shown in Figure 4.32(a) which is configured in two layers. As this cable finds use either indoors or outside, or both,

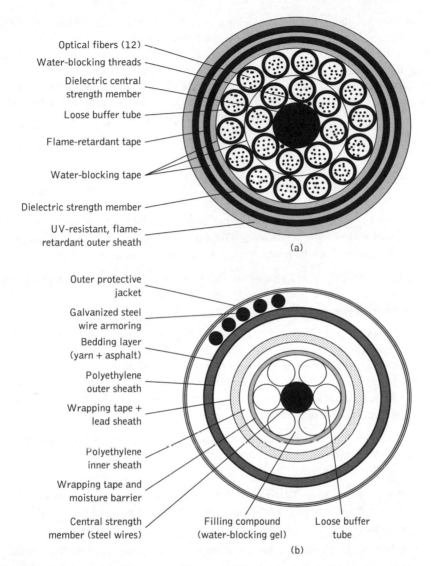

Optical fibers (12)
Water-blocking threads
Dielectric central
strength member
Loose buffer tube
Flame-retardant tape
Water-blocking tape
Dielectric strength member
UV-resistant, flame-
retardant outer sheath

(a)

Outer protective
jacket
Galvanized steel
wire armoring
Bedding layer
(yarn + asphalt)
Polyethylene
outer sheath
Wrapping tape +
lead sheath
Polyethylene
inner sheath
Wrapping tape and
moisture barrier
Central strength
member (steel wires)

Filling compound
(water-blocking gel)

Loose buffer
tube

(b)

Figure 4.32 High-performance loose buffer tube fiber cables: (a) indoor–outdoor cable incorporating 264 fibers; (b) optical submarine cable

eliminating the need for transitional splicing or termination between buildings, it is fully water blocked. The loose buffer tubes containing 12 fibers each, together with filling compound, are stranded around a dielectric (i.e. aramid yarn) central member and wrapped in water-blocking tapes. Although the cable core is not filled, it contains further water-blocking threads. In addition, a dielectric strength member is also incorporated beneath the cable outer sheath.

Figure 4.32(b) displays an optical submarine cable design incorporating fibers in loose tube buffer structures stranded around a steel wire central strength member. In this case the cable core is filled with water-blocking gel and a further water barrier is contained together with the wrapping tape around the inner core prior to a polyethylene inner sheath. The inner core is surrounded by a lead sheath before further layers of polyethylene (outer sheath) and aramid yarn. A heavy armor jacket containing galvanized steel wire provides substantial armoring underneath the final outer protection layer. Such cables also typically contain copper wires rather than fibers within one of the loose buffer tubes to enable the provision of electric power to submerged optical amplifiers or regenerators.

Problems

4.1   Describe in general terms liquid-phase techniques for the preparation of multicomponent glasses for optical fibers. Discuss with the aid of a suitable diagram one melting method for the preparation of multicomponent glass.

4.2   Indicate the major advantages of vapor-phase deposition in the preparation of glasses for optical fibers. Briefly describe the various vapor-phase techniques currently in use.

4.3   (a) Compare and contrast, using suitable diagrams, the outside vapor-phase oxidation (OVPO) process and the modified chemical vapor deposition (MCVD) technique for the preparation of low-loss optical fibers.
(b) Briefly describe the salient features of vapor axial deposition (VAD) and the plasma-activated chemical vapor deposition (PCVD) when applied to the preparation of optical fibers.

4.4   Discuss the drawing of optical fibers from prepared glasses with regard to:
(a) multicomponent glass fibers;
(b) silica-rich fibers.

4.5   List the various silica-based optical fiber types currently on the market indicating their important features. Hence, briefly describe the general areas of application for each type.

4.6   Outline the developments that have taken place in relation to plastic optical fibers since 1996, with particular reference to contrasting the performance attributes of PF-POF with PMMA POF.

4.7   Briefly describe the major reasons for the cabling of optical fibers which are to be placed in a field environment. Thus state the functions of the optical fiber cable.

4.8   Explain how the Griffith theory is developed in order to predict the fracture stress of an optical fiber with an elliptical crack.
  Silica has a Young's modulus of 9×10^{10} N m^{-2} and a surface energy of 2.29 J. Estimate the fracture stress in psi for a silica optical fiber with a dominant elliptical crack of depth 0.5 μm. Also, determine the strain at the break for the fiber (1 psi ≡ 6894.76 N m^{-2}).

4.9 Another length of the optical fiber described in Problem 4.8 is found to break at 1% strain. The failure is due to a single dominant elliptical crack. Estimate the depth of this crack.

4.10 Describe the effects of stress corrosion on optical fiber strength and durability.

It is found that a 20 m length of fused silica optical fiber may be extended to 24 m at liquid nitrogen temperatures (i.e. little stress corrosion) before failure occurs. Estimate the fracture stress in psi for the fiber under these conditions. Young's modulus for silica is 9×10^{10} N m^{-2} and 1 psi $\equiv$ 6894.76 N m^{-2}.

4.11 Outline the phenomena that can affect the stability of the transmission characteristics in optical fiber cables and describe any techniques by which these problems may be avoided.

4.12 Discuss optical fiber cable design with regard to:
(a) fiber buffering;
(b) cable strength and structural members;
(c) layered cable construction;
(d) cable sheath and water barrier.
Further, compare and contrast possible cable designs for multifiber cables, making particular reference to unit cables.

Answers to numerical problems

4.8 7.43×10^4 psi, 0.6%
4.9 0.2 μm
4.10 2.61×10^6 psi

References

[1] S. Tanaka, K. Inada, T. Akimoko and M. Kozima, 'Silicone-clad fused-silica-core fiber', *Electron. Lett.*, **11**(7), pp. 153–154, 1975.
[2] K. J. Beales and C. R. Day, 'A review of glass fibers for optical communications', *Phys. Chem. Glass*, **21**(1), pp. 5–21, 1980.
[3] T. Yamazuki and M. Yoshiyagawa, 'Fabrication of low-loss, multicomponent glass fibers with graded index and pseudo-step-index borosilicate compound glass fibers', *Dig. Int. Conf. on Integrated Optics and Optical Fiber Communications*, Osaka, pp. 617–620, 1977.
[4] K. J. Beales, C. R. Day, W. J. Duncan, J. E. Midwinter and G. R. Newns, 'Preparation of sodium borosilicate glass fibers for optical communication', *Proc. IEE*, **123**, pp. 591–595, 1976.
[5] G. R. Newns, P. Pantelis, J. L. Wilson, R. W. J. Uffen and R. Worthington, 'Absorption losses in glasses and glass fiber waveguides', *Opto-Electronics*, **5**, pp. 289–296, 1973.
[6] B. Scott and H. Rawson, 'Techniques for producing low loss glasses for optical fibre communication systems', *Glass Technol.*, **14**(5), pp. 115–124, 1973.
[7] C. E. E. Stewart, D. Tyldesley, B. Scott, H. Rawson and G. R. Newns, 'High-purity glasses for optical-fibre communication', *Electron. Lett.*, **9**(21), pp. 482–483, 1973.

[8] B. Scott and H. Rawson, 'Preparation of low loss glasses for optical fiber communication', *Opto-Electronics*, **5**(4), pp. 285–288, 1973.

[9] N. S. Kapany, *Fiber Optics*, Academic Press, 1967.

[10] G. Mahlke and P. Gossing, *Fiber Optic Cables* (4th edn), Publicis MCD Corporate Publishing, 2001.

[11] G. R. Newns, 'Compound glass optical fibres', *2nd Eur. Conf. on Optical Fiber Communication*, Paris, pp. 21–26, 1976.

[12] K. J. Beales, C. R. Day, W. J. Duncan, A. G. Dunn, P. L. Dunn, G. R. Newns and J. V. Wright, 'Low loss graded index fiber by the double crucible technique', *5th Eur. Conf. on Optical Fiber Communications*, Amsterdam, paper 3.2, 1979.

[13] K. J. Beales, C. R. Day, W. J. Duncan and G. R. Newns, 'Low-loss compound-glass optical fibre', *Electron. Lett.*, **13**(24), pp. 755–756, 1977.

[14] J. F. Hyde, US Patent 2 272 342, 1942.

[15] F. P. Kapron, D. B. Keck and R. D. Maurer, 'Radiation losses in optical waveguides', *Appl. Phys. Lett.*, **10**, pp. 423–425, 1970.

[16] B. Bendow and S. S. Mitra, *Fiber Optics*, Plenum Press, 1979.

[17] D. B. Keck and R. Bouillie, 'Measurements on high-bandwidth optical waveguides', *Opt. Commun.*, **25**, pp. 43–48, 1978.

[18] B. S. Aronson, D. R. Powers and R. Sommer, 'Chloride drying of doped deposited silica pre-form simultaneous to consolidation', *Tech. Dig. Top. Meet. on Optical Fiber Communication*, Washington, DC, p. 42, 1979.

[19] S. R. Nagel, 'Fiber materials and fabrication methods', in S. E. Miller and I. P. Kaminow (Eds), *Optical Fiber Telecommunications II*, pp. 121–215, Academic Press, 1988.

[20] T. Izawa, T. Miyashita and F. Hanawa, US Patent 4 062 665, 1977.

[21] S. Sudo, M. Kawachi, M. Edahiro, T. Izawa, T. Shoida and H. Gotoh, 'Low-OH-content optical fiber fabricated by vapor-phase axial-deposition method', *Electron. Lett.*, **14**(17), pp. 534–535, 1978.

[22] H. Kanamori, H. Yokota, G. Tanaka, M. Watanabe, Y. Ishiguro, I. Yoshida, T. Kakii, S. Itoh, Y. Asano and S. Tanaka, 'Transmission characteristics and reliability of pure-silica-core single-mode-fibers', *J. Lightwave Technol.*, **LT-4**(8), pp. 1144–1150, 1986.

[23] D. B. Keck and P. C. Schultz, US Patent 3 711 262, 1973.

[24] W. G. French, J. B. MacChesney, P. B. O'Conner and G. W. Tasker, 'Optical waveguides with very low losses', *Bell Syst. Tech. J.*, **53**, pp. 951–954, 1974.

[25] D. N. Payne and W. A. Gambling, 'New silica-based low-loss optical fibres', *Electron. Lett.*, **10**(15), pp. 289–90, 1974.

[26] T. Miya, Y. Terunuma, T. Mosaka and T. Miyashita, 'Ultimate low-loss single-mode fibre at 1.55 μm', *Electron. Lett.*, **15**(4), pp. 106–108, 1979.

[27] D. Gloge, 'The optical fibre as a transmission medium', *Rep. Prog. Phys.*, **42**, pp. 1778–1824, 1979.

[28] S. R. Nagel, J. B. MacChesney and K. L. Walker, 'An overview of the modified chemical vapor deposition (MCVD) process and performance', *IEEE J. Quantum Electron.*, **QE-18**(4), pp. 459–477, 1982.

[29] W. A. Gambling, 'The rise and rise of optical fibers', *IEEE J. Sel. Top. Quantum Electron.*, **6**(6), pp. 1084–1093, 2000.

[30] D. Kuppers and J. Koenings, 'Preform fabrication by deposition of thousands of layers with the aid of plasma activated CVD', *2nd Eur. Conf. on Optical Fiber Communications*, Paris, p. 49, 1976.

[31] H. Lydtin, 'PCVD: a technique suitable for large-scale fabrication of optical fibers', *J. Lightwave Technol.*, **LT-4**(8), pp. 1034–1038, 1986.

[32] N. Nobukazu, 'Recent progress in glass fibers for optical communication', *Jpn. J. Appl. Phys.*, **20**(8), pp. 1347–1360, 1981.

[33] H. Murata, *Handbook of Optical Fibers and Cables* (2nd edn), Marcel Dekker, 1996.

[34] D. J. DiGiovanni, S. K. Das, L. L. Blyler, W. White and R. K. Boncek, 'Design of optical fibers for communication systems', in I. P. Kaminow and T. Li (Eds), *Optical Fiber Telecommunications IVA*, Academic Press, pp. 17–79, 2002.

[35] G. Keiser, *Optical Communications Essentials*, McGraw-Hill, 2003.

[36] M. Nishimura, 'Optical fibers and fiber dispersion compensators for high-speed optical communication', *J. Opt. Fiber Commun. Rep.*, **2**, pp. 115–139, 2005.

[37] A. Weinert, *Plastic Optical Fibers*, Publicis MCD Verlag, 1999.

[38] ITU-T Recommendation G.651, 'Characteristics of a 50/125 μm multimode graded index optical fiber', February 1998.

[39] R. Neat, 'New and developing standards and specifications relating to multimode optical fiber', *Proc. 12th Eur. Conf. on Networks and Optical Communications*, Acreo, Kista, Sweden, pp. 63–67, June 2007.

[40] ITU-T Recommendation G.652, 'Characteristics of a single-mode optical fiber and cable', April 1997.

[41] ITU-T Recommendation G.654, 'Characteristics of a cut-off shifted single-mode optical fiber and cable', April 1997.

[42] ITU-T Recommendation G.655, 'Characteristics of a non-zero dispersion-shifted single-mode optical fiber and cable', October 1996.

[43] ITU-T Recommendation G.656, 'Characteristics of a fiber and cable with non-zero dispersion for wideband optical transport', June 2004.

[44] ITU-T Recommendation G.657, 'Characteristics of a bending loss insensitive single mode optical fiber and cable for the access network', December 2006.

[45] Corning Clearcurve solutions, http://www.corning.com/clearcurve/solutions.html, 21 February 2008.

[46] I. T. Monroy, H. P. A. van den Boom, A. M. J. Koonen, G. D. Khoe, Y. Watanabe, Y. Koike and T. Ishigure, 'Data transmission over polymer optical fibers', *Opt. Fiber Technol.*, **9**, pp. 159–171, 2003.

[47] Y. Koike, 'Progress of plastic optical fiber technology', *Proc. Eur. Conf. on Optical Communications (ECOC'96)*, MoB.3.1, **41**, 1996.

[48] H. P. A. van den Boom, W. Li, P. K. van Bennekom, I. Tafur Monroy and Giok-Djan Khoe, 'High capacity transmission over polymer optical fiber', *IEEE J. Sel. Top. Quantum Electron.*, **7(3)**, pp. 461–470, 2001.

[49] Asahi Glass Company, 'Lucina: graded index-Cytop optical fiber', *Technical Note T009E*, July 2000.

[50] Y. Koike, 'POF technology for the 21st century', *Proc. 10th Plastic Fibers Conf.*, Amsterdam, pp. 5–8, September 2001.

[51] W. Daum, J. Kraurer, P. E. Zamzow and O. Ziemann, *POF-Polymer Optical Fibers for Data Communication*, Springer-Verlag, 2002.

[52] A. A. Griffith, 'Phenomena of rupture and flow in solids', *Philos. Trans. R. Soc. Ser. A*, **221**, pp. 163–168, 1920.

[53] W. Weibull, 'A statistical theory of the strength of materials', *Proc. R. Swedish Inst. Res.*, No. 151, publication no. 4, 1939.

[54] M. H. Reeve, 'Optical fibre cables', *Radio Electron. Eng. (IERE J.)*, **51**(7/8), pp. 327–332, 1981.

[55] B. Wiltshire and M. H. Reeve, 'A review of the environmental factors affecting optical cable design', *J. Lightwave Technol.*, **LT-6**, pp. 179–185, 1988.

[56] S. Das, G. S. Englefield and P. A. Goud, 'Power loss, modal noise and distortion due to microbending of optical fibres', *Appl. Opt.*, **24**, pp. 2323–2333, 1985.

[57] P. Danielsen, 'Simple power spectrum of microbending in single mode fibers', *Electron. Lett.*, **19**, p. 318, 1983.

[58] R. S. Ashpole and R. J. W. Powell, 'Hydrogen in optical cables', *IEE Proc., Optoelectron.*, **132**, pp. 162–168, 1985.

[59] P. J. Lemaire and A. Tomita, 'Hydrogen induced loss in MCVD fibres', *Optical Fiber Communication Conf., OFC'85*, USA, TUII, February 1985.

[60] E. J. Friebele, K. J. Long, C. G. Askins, M. E. Gingerich, M. J. Marrone and D. L. Griscom, 'Overview of radiation effects in fiber optics', *Proc. SPIE*, **541**, pp. 70–88, 1985.

[61] A. Iino and J. Tamura, 'Radiation resistivity in silica optical fibers', *J. Lightwave Technol.*, **LT-6**, pp. 145–149, 1988.

[62] E. J. Friebele and M. E. Gingerich, 'Photobleaching effects in optical fibre waveguides', *Appl. Opt.*, **20**, pp. 3448–3452, 1981.

[63] R. H. West, 'A local view of radiation effects in fiber optics', *J. Lightwave Technol.*, **LT-6**, pp. 155–164, 1988.

[64] E. J. Friebele, E. W. Waylor, G. T. De Beauregard, J. A. Wall and C. E. Barnes, 'Interlaboratory comparison of radiation-induced attenuation in optical fibers. Part 1: steady state exposure', *J. Lightwave Technol.*, **LT-6**, pp. 165–178, 1988.

E. J. Friebele *et al.*, 'Interlaboratory comparison of radiation-induced attenuation in optical fibers'. Part 2, *J. Lightwave Technol.*, **LT-8**, pp. 967–988, 1990.

[65] P. R. Bank and D. O. Lawrence, 'Emerging standards in fiber optic telecommunications cable'. *Proc. SPIE*, **224**, pp. 149–158, 1980.

[66] C. H. Gartside III, P. D. Patel and M. R. Santana, 'Optical fiber cables', in S. E. Miller and I. P. Kaminow (Eds), *Optical Fiber Telecommunications II*, Academic Press, 1988.

[67] K. W. Jackson, T. D. Mathis, P. D. Patel, M. R. Santana and P. M. Thomas, 'Advances in cable design', in I. P. Kaminow and T. L. Koch (Eds), *Optical Fiber Telecommunications IIIA*, Academic Press, pp. 92–113, 1997.

[68] V. Alwayn, *Optical Network Design and Implementation*, Cisco Press, 2004.

CHAPTER 5

Optical fiber connection: joints, couplers and isolators

5.1 Introduction

Optical fiber links, in common with any line communication system, have a requirement for both jointing and termination of the transmission medium. 'The number of intermediate fiber connections or joints is dependent upon the link length (between repeaters), the continuous length of fiber cable that may be produced by the preparation methods outlined in Sections 4.2 to 4.4, and the length of the fiber cable that may be practically or conveniently installed as a continuous section on the link. Although scaling up of the preparation processes now provides the capability to produce very large preforms allowing continuous single-mode fiber lengths of around 200 km, such fiber spans cannot be readily installed [Ref. 1]. However, continuous cable lengths of tens of kilometers have already been deployed, in particular within submarine systems where continuous cable laying presents fewer problems [Ref. 2].

Repeater spacing on optical fiber telecommunication links is a continuously increasing parameter with currently installed digital systems operating over spacings in the range 40 to 60 km at transmission rates of between 2.5 Gbit s^{-1} and 10 Gbit s^{-1}. For example, a transatlantic optical fiber system operating over a distance of 6000 km employs 120 optical repeaters with a 50 km spacing and is capable of carrying 32 wavelength division multiplexed channels each at a transmission rate of 10 Gbit s^{-1} [Ref. 3]. An experimental optical fiber link covering a distance of 4800 km with optoelectronic repeater spacing of 120 km has also been demonstrated [Ref. 4]. It carried 64 wavelength multiplexed channels each operating at a transmission rate of 10 Gbit s^{-1}. In addition, more advanced optical fiber communication links utilize optical amplifiers (see Section 10.12) and optical regenerators (see Section 10.6) [Ref. 5]. A field trial using a combination of both erbium-doped fiber and Raman amplifiers has produced high-capacity wavelength division multiplexed transmission of 1.28 Tbit s^{-1} with 32 channels each operating at a transmission rate of 43 Gbit s^{-1} [Ref. 6]. Also an eight-channel wavelength division multiplexed system with each channel operating at a rate of 170 Gbit s^{-1} utilizing a Raman fiber amplifier per channel has demonstrated successful transmission over a 185 km span of standard single-mode fiber [Ref. 7].

It is therefore apparent that fiber-to-fiber connection with low loss and minimum distortion (i.e. modal noise) remains an important aspect of optical fiber communication systems (fiber, sources, detectors, etc.). In addition, it also serves to increase the number of terminal connections permissible within the developing optical fiber communication networks (see Chapter 15). Although fiber jointing techniques appeared to lag behind the technologies associated with the other components required in optical fiber communication systems (fiber sources, detectors, etc.) it is clear that, in recent years, significant developments have continued to be made. Therefore, in this and the sections immediately following we review the theoretical and practical aspects of fiber–fiber connection with regard to both multimode and single-mode systems. Fiber termination to sources and detectors is not considered since the important aspects of these topics are discussed in the chapters covering sources and detectors (Chapters 6, 7 and 8). Nevertheless, the discussion on fiber jointing is relevant to both source and detector coupling, as many manufacturers supply these electro-optic devices already terminated to a fiber optic pigtail in order to facilitate direct fiber–fiber connection to an optical fiber link.

Before we consider fiber–fiber connection in further detail it is necessary to indicate the two major categories of fiber joint currently in both use and development. These are as follows:

1. Fiber splices. These are semipermanent or permanent joints which find major use in most optical fiber telecommunication systems (analogous to electrical soldered joints).

2. Demountable fiber connectors or simple connectors. These are removable joints which allow easy, fast, manual coupling and uncoupling of fibers (analogous to electrical plugs and sockets).

The above fiber–fiber joints are designed ideally to couple all the light propagating in one fiber into the adjoining fiber. By contrast fiber couplers are branching devices that split all the light from a main fiber into two or more fibers or, alternatively, couple a proportion of the light propagating in the main fiber into a branch fiber. Moreover, these devices are often bidirectional, providing for the combining of light from one or more branch fibers into a main fiber. The importance and variety of these fiber couplers have increased substantially over recent years in order to facilitate the widespread deployment of optical fiber within communication networks. Although the requirement for such devices was less in earlier point-to-point fiber links, the growing demand for more sophisticated fiber network configurations (see Chapter 15) has made them essential components within optical fiber communications.

In this chapter we therefore consider the basic techniques and technology associated with both fiber joints and couplers. A crucial aspect of fiber jointing concerns the optical loss associated with the connection. This joint loss is critically dependent upon the alignment of the two fibers. Hence, in Section 5.2 the mechanisms which cause optical losses at fiber joints are outlined, with particular attention being paid to the fiber alignment. This discussion provides a grounding for consideration of the techniques employed for jointing optical fibers. Permanent fiber joints (i.e. splices) are then dealt with in Section 5.3 prior to discussion of the two generic types of demountable connector in Sections 5.4 and 5.5. Then, in Section 5.6, the basic construction and performance characteristics of the various fiber directional coupler types are described. Finally, the function and implementation of optical isolators is dealt with in Section 5.7 including their interconnection to provide an optical circulator which can also be employed to facilitate add/drop wavelength multiplexing.

5.2 Fiber alignment and joint loss

A major consideration with all types of fiber–fiber connection is the optical loss encountered at the interface. Even when the two jointed fiber ends are smooth and perpendicular to the fiber axes, and the two fiber axes are perfectly aligned, a small proportion of the light may be reflected back into the transmitting fiber causing attenuation at the joint. This phenomenon, known as Fresnel reflection, is associated with the step changes in refractive index at the jointed interface (i.e. glass–air–glass). The magnitude of this partial reflection of the light transmitted through the interface may be estimated using the classical Fresnel formula for light of normal incidence and is given by [Ref. 8]:

$$r = \left(\frac{n_1 - n}{n_1 + n}\right)^2 \tag{5.1}$$

where r is the fraction of the light reflected at a single interface, n_1 is the refractive index of the fiber core and n is the refractive index of the medium between the two jointed fibers (i.e. for air $n = 1$). However, in order to determine the amount of light reflected at a fiber joint, Fresnel reflection at both fiber interfaces must be taken into account. The loss in decibels due to Fresnel reflection at a single interface is given by:

$$\text{Loss}_{\text{Fres}} = -10 \log_{10}(1 - r) \tag{5.2}$$

Hence, using the relationships given in Eqs (5.1) and (5.2) it is possible to determine the optical attenuation due to Fresnel reflection at a fiber–fiber joint.

It is apparent that Fresnel reflection may give a significant loss at a fiber joint even when all other aspects of the connection are ideal. However, the effect of Fresnel reflection at a fiber–fiber connection can be reduced to a very low level through the use of an index-matching fluid in the gap between the jointed fibers. When the index-matching fluid has the same refractive index as the fiber core, losses due to Fresnel reflection are in theory eradicated.

Unfortunately, Fresnel reflection is only one possible source of optical loss at a fiber joint. A potentially greater source of loss at a fiber–fiber connection is caused by misalignment of the two jointed fibers. In order to appreciate the development and relative success of various connection techniques it is useful to discuss fiber alignment in greater detail.

Example 5.1

An optical fiber has a core refractive index of 1.5. Two lengths of the fiber with smooth and perpendicular (to the core axes) end faces are butted together. Assuming the fiber axes are perfectly aligned, calculate the optical loss in decibels at the joint (due to Fresnel reflection) when there is a small air gap between the fiber end faces.

Solution: The magnitude of the Fresnel reflection at the fiber–air interface is given by Eq. (5.1) where:

$$r = \left(\frac{n_1 - n}{n_1 + n}\right)^2 = \left(\frac{1.5 - 1.0}{1.5 + 1.0}\right)^2$$

$$= \left(\frac{0.5}{2.5}\right)^2$$

$$= 0.04$$

The value obtained for r corresponds to a reflection of 4% of the transmitted light at the single interface. Further, the optical loss in decibels at the single interface may be obtained using Eq. (5.2) where:

$$\text{Loss}_{\text{Fres}} = -10 \log_{10}(1 - r) = -10 \log_{10} 0.96$$
$$= 0.18 \text{ dB}$$

A similar calculation may be performed for the other interface (air–fiber). However, from considerations of symmetry it is clear that the optical loss at the second interface is also. 0.18 dB.

Hence the total loss due to Fresnel reflection at the fiber joint is approximately 0.36 dB.

Any deviations in the geometrical and optical parameters of the two optical fibers which are jointed will affect the optical attenuation (insertion loss) through the connection. It is not possible within any particular connection technique to allow for all these variations. Hence, there are inherent connection problems when jointing fibers with, for instance:

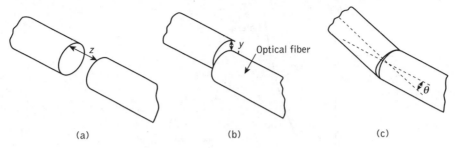

Figure 5.1 The three possible types of misalignment which may occur when jointing compatible optical fibers [Ref. 9]: (a) longitudinal misalignment; (b) lateral misalignment; (c) angular misalignment

(a) different core and/or cladding diameters;

(b) different numerical apertures and/or relative refractive index differences;

(c) different refractive index profiles;

(d) fiber faults (core ellipticity, core concentricity, etc.).

The losses caused by the above factors together with those of Fresnel reflection are usually referred to as intrinsic joint losses.

The best results are therefore achieved with compatible (same) fibers which are manufactured to the lowest tolerance. In this case there is still the problem of the quality of the fiber alignment provided by the jointing mechanism. Examples of possible misalignment between coupled compatible optical fibers are illustrated in Figure 5.1 [Ref. 9]. It is apparent that misalignment may occur in three dimensions: the separation between the fibers (longitudinal misalignment), the offset perpendicular to the fiber core axes (lateral/radial/axial misalignment) and the angle between the core axes (angular misalignment).

Optical losses resulting from these three types of misalignment depend upon the fiber type, core diameter and the distribution of the optical power between the propagating modes. Examples of the measured optical losses due to the various types of misalignment are shown in Figure 5.2. Figure 5.2(a) [Ref. 9] shows the attenuation characteristic for both longitudinal and lateral misalignment of a graded index fiber of 50 µm core diameter. It may be observed that the lateral misalignment gives significantly greater losses per unit displacement than the longitudinal misalignment. For instance, in this case a lateral displacement of 10 µm gives about 1 dB insertion loss whereas a similar longitudinal displacement gives an insertion loss of around 0.1 dB. Figure 5.2(b) [Ref. 10] shows the attenuation characteristic for the angular misalignment of two multimode step index fibers with numerical apertures of 0.22 and 0.3. An insertion loss of around 1 dB is obtained with angular misalignment of 4° and 5° for the $NA = 0.22$ and $NA = 0.3$ fibers respectively. It may also be observed in Figure 5.2(b) that the effect of an index-matching fluid in the fiber gap causes increased losses with angular misalignment. Therefore, it is clear that relatively small levels of lateral and/or angular misalignment can cause significant attenuation at a fiber joint. This is especially the case for fibers of small core diameter (less than 150 µm) which are currently employed for most telecommunication purposes.

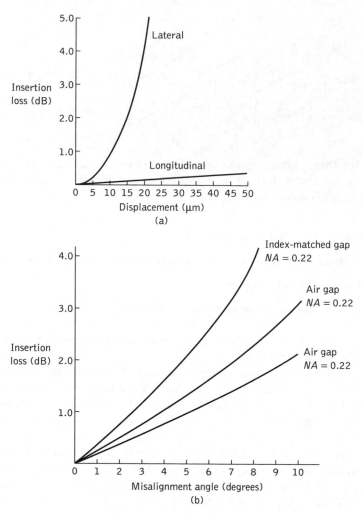

Figure 5.2 Insertion loss characteristics for jointed optical fibers with various types of misalignment: (a) insertion loss due to lateral and longitudinal misalignment for a graded index fiber of 50 μm core diameter. Reproduced with permission from P. Mossman, *Radio Electron. Eng.*, **51**, p. 333. 1981; (b) insertion loss due to angular misalignment for joints in two multimode step index fibers with numerical apertures of 0.22 and 0.3. From C. P. Sandback (Ed.), *Optical Fiber Communication Systems*, John Wiley & Sons, 1980

5.2.1 Multimode fiber joints

Theoretical and experimental studies of fiber misalignment in optical fiber connections [Refs 11–19] allow approximate determination of the losses encountered with the various misalignments of different fiber types. We consider here some of the expressions used to calculate losses due to lateral and angular misalignment of optical fiber joints. Longitudinal misalignment is not discussed in detail as it tends to be the least important effect and may be largely avoided in fiber connection. Also there is some disagreement over the magnitude of the losses due to longitudinal misalignment when it is calculated theoretically between Miyazaki *et al.* [Ref. 12] and Tsuchiya *et al.* [Ref. 13]. Both groups of workers claim good agreement with experimental results, which is perhaps understandable when considering the number of variables involved in the measurement. However, it is worth noting that the lower losses predicted by Tsuchiya *et al.* agree more closely with a third group of researchers [Ref. 14]. Also, all groups predict higher losses for fibers with larger numerical apertures, which is consistent with intuitive considerations (i.e. the larger the numerical aperture, the greater the spread of the output light and the higher the optical loss at a longitudinally misaligned joint).

Theoretical expressions for the determination of lateral and angular misalignment losses are by no means definitive, although in all cases they claim reasonable agreement with experimental results. However, experimental results from different sources tend to vary (especially for angular misalignment losses) due to difficulties of measurement. It is therefore not implied that the expressions given in the text are necessarily the most accurate, as at present the choice appears somewhat arbitrary.

Lateral misalignment reduces the overlap region between the two fiber cores. Assuming uniform excitation of all the optical modes in a multimode step index fiber, the overlapped area between both fiber cores approximately gives the lateral coupling efficiency η_{lat}. Hence, the lateral coupling efficiency for two similar step index fibers may be written as [Ref. 13]:

$$\eta_{lat} \simeq \frac{16(n_1/n)^2}{[1+(n_1/n)]^4} \frac{1}{\pi} \left\{ 2\cos^{-1}\left(\frac{y}{2a}\right) - \left(\frac{y}{a}\right)\left[1-\left(\frac{y}{2a}\right)^2\right]^{\frac{1}{2}} \right\} \tag{5.3}$$

where n_1 is the core refractive index, n is the refractive index of the medium between the fibers, y is the lateral offset of the fiber core axes, and a is the fiber core radius. The lateral misalignment loss in decibels may be determined using:

$$Loss_{lat} = -10\log_{10}\eta_{lat}\ dB \tag{5.4}$$

The predicted losses obtained using the formulas given in Eqs (5.3) and (5.4) are generally slightly higher than the measured values due to the assumption that all modes are equally excited. This assumption is only correct for certain cases of optical fiber transmission. Also, certain authors [Refs 12, 18] assume index matching and hence no Fresnel reflection, which makes the first term in Eq. (5.3) equal to unity (as $n_1/n = 1$). This may be valid if the two fiber ends are assumed to be in close contact (i.e. no air gap in between) and gives lower predicted losses. Nevertheless, bearing in mind these possible inconsistencies, useful estimates for the attenuation due to lateral misalignment of multimode step index fibers may be obtained.

Lateral misalignment loss in multimode graded index fibers assuming a uniform distribution of optical power throughout all guided modes was calculated by Gloge [Ref. 16]. He estimated that the lateral misalignment loss was dependent on the refractive index gradient α for small lateral offset and may be obtained from:

$$L_t = \frac{2}{\pi}\left(\frac{y}{a}\right)\left(\frac{\alpha+2}{\alpha+1}\right) \quad \text{for } 0 \leq y \leq 0.2a \tag{5.5}$$

where the lateral coupling efficiency was given by:

$$\eta_{\text{lat}} = 1 - L_t \tag{5.6}$$

Hence Eq. (5.6) may be utilized to obtain the lateral misalignment loss in decibels. With a parabolic refractive index profile where $\alpha = 2$, Eq. (5.5) gives:

$$L_t = \frac{8}{3\pi}\left(\frac{y}{a}\right) = 0.85\left(\frac{y}{a}\right) \tag{5.7}$$

A further estimate including the leaky modes gave a revised expression for the lateral misalignment loss given in Eq. (5.6) of $0.75(y/a)$. This analysis was also extended to step index fibers (where $\alpha = \infty$) and gave lateral misalignment losses of $0.64(y/a)$ and $0.5(y/a)$ for the cases of guided modes only and both guided plus leaky modes respectively.

Example 5.2

A step index fiber has a core refractive index of 1.5 and a core diameter of 50 μm. The fiber is jointed with a lateral misalignment between the core axes of 5 μm. Estimate the insertion loss at the joint due to the lateral misalignment assuming a uniform distribution of power between all guided modes when:

(a) there is a small air gap at the joint;

(b) the joint is considered index matched.

Solution: (a) The coupling efficiency for a multimode step index fiber with uniform illumination of all propagating modes is given by Eq. (5.3) as:

$$\eta_{\text{lat}} \simeq \frac{16(n_1/n)^2}{[1+(n_1/n)]^4}\frac{1}{\pi}\left\{2\cos^{-1}\left(\frac{y}{2a}\right) - \left(\frac{y}{a}\right)\left[1-\left(\frac{y}{2a}\right)^2\right]^{\frac{1}{2}}\right\}$$

$$= \frac{16(1.5)^2}{[1+1.5]^4}\frac{1}{\pi}\left\{2\cos^{-1}\left(\frac{5}{50}\right) - \left(\frac{5}{25}\right)\left[1-\left(\frac{5}{50}\right)^2\right]^{\frac{1}{2}}\right\}$$

$$= 0.293\{2(1.471) - 0.2[0.99]^{\frac{1}{2}}\}$$

$$= 0.804$$

The insertion loss due to lateral misalignment is given by Eq. (5.4) where:

$$\text{Loss}_{\text{lat}} = -10\log_{10}\eta_{\text{lat}} = -10\log_{10}0.804$$
$$= 0.95 \text{ dB}$$

Hence, assuming a small air gap at the joint, the insertion loss is approximately 1 dB when the lateral offset is 10% of the fiber diameter.

(b) When the joint is considered index matched (i.e. no air gap) the coupling efficiency may again be obtained from Eq. (5.3) where:

$$\eta_{\text{lat}} \simeq \frac{1}{\pi}\left\{2\cos^{-1}\left(\frac{5}{50}\right) - \left(\frac{5}{25}\right)\left[1 - \left(\frac{5}{50}\right)^2\right]^{\frac{1}{2}}\right\}$$

$$= 0.318\{2(1.471) - 0.2[0.99]^{\frac{1}{2}}\}$$

$$= 0.872$$

Therefore the insertion loss is:

$$\text{Loss}_{\text{lat}} = -10\log_{10}0.872 = 0.59\text{ dB}$$

With index matching, the insertion loss at the joint in Example 5.2 is reduced to approximately 0.36 dB. It may be noted that the difference between the losses obtained in parts (a) and (b) corresponds to the optical loss due to Fresnel reflection at the similar fiber–air–fiber interface determined in Example 5.1.

The result may be checked using the formulas derived by Gloge for a multimode step index fiber where the lateral misalignment loss assuming uniform illumination of all guided modes is obtained using:

$$L_{\text{t}} = 0.64\left(\frac{y}{a}\right) = 0.64\left(\frac{5}{25}\right) = 0.128$$

Hence the lateral coupling efficiency is given by Eq. (5.6) as:

$$\eta_{\text{lat}} = 1 - 0.128 = 0.872$$

Again using Eq. (5.4), the insertion loss due to the lateral misalignment assuming index matching is:

$$\text{Loss}_{\text{lat}} = -10\log_{10}0.872 = 0.59\text{ dB}$$

Hence using the expression derived by Gloge we obtain the same value of approximately 0.6 dB for the insertion loss with the inherent assumption that there is no change in refractive index at the joint interface. Although this estimate of insertion loss may be shown to agree with certain experimental results [Ref. 12], a value of around 1 dB insertion loss for a 10% lateral displacement with regard to the core diameter (as estimated in Example 5.2(a)) is more usually found to be the case with multimode step index fibers [Refs 8, 19–21]. Further, it is generally accepted that the lateral offset must be kept below 5% of the fiber core diameter in order to reduce insertion loss at a joint to below 0.5 dB [Ref. 19].

Example 5.3

A graded index fiber has a parabolic refractive index profile ($\alpha = 2$) and a core diameter of 50 μm. Estimate the insertion loss due to a 3 μm lateral misalignment at a fiber joint when there is index matching and assuming:

(a) there is uniform illumination of all guided modes only;

(b) there is uniform illumination of all guided and leaky modes.

Solution: (a) Assuming uniform illumination of guided modes only, the misalignment loss may be obtained using Eq. (5.7), where:

$$L_t = 0.85\left(\frac{y}{a}\right) = 0.85\left(\frac{3}{25}\right) = 0.102$$

The coupling efficiency is given by Eq. (5.6) as:

$$\eta_{lat} = 1 - L_t = 1 - 0.102 = 0.898$$

Hence the insertion loss due to the lateral misalignment is given by Eq. (5.4), where:

$$\text{Loss}_{lat} = -10 \log_{10} 0.898 = 0.47 \text{ dB}$$

(b) When assuming the uniform illumination of both guided and leaky modes Gloge's formula becomes:

$$L_t = 0.75\left(\frac{y}{a}\right) = 0.75\left(\frac{3}{25}\right) = 0.090$$

Therefore the coupling efficiency is:

$$\eta_{lat} = 1 - 0.090 = 0.910$$

and the insertion loss due to lateral misalignment is:

$$\text{Loss}_{lat} = -10 \log_{10} 0.910 = 0.41 \text{ dB}$$

It may be noted by observing Figure 5.2(a), which shows the measured lateral misalignment loss for a 50 μm diameter graded index fiber, that the losses predicted above are very pessimistic (the loss for 3 μm offset shown in Figure 5.2(a) is less than 0.2 dB). A model which is found to predict insertion loss due to lateral misalignment in graded index fibers with greater accuracy was proposed by Miller and Mettler [Ref. 17]. In this model they assumed the power distribution at the fiber output to be of Gaussian form. Unfortunately, the analysis is too detailed for this text as it involves integration using numerical techniques.

We therefore limit estimates of insertion losses due to lateral misalignment in multimode graded index fibers to the use of Gloge's formula.

Angular misalignment losses at joints in multimode step index fibers may be predicted with reasonable accuracy using an expression for the angular coupling efficiency η_{ang} given by [Ref. 13]:

$$\eta_{ang} \simeq \frac{16(n_1/n)^2}{[1 + (n_1/n)]^4} \left[1 - \frac{n\theta}{\pi n_1 (2\Delta)^{\frac{1}{2}}}\right] \tag{5.8}$$

where θ is the angular displacement in radians and Δ is the relative refractive index difference for the fiber. The insertion loss due to angular misalignment may be obtained from the angular coupling efficiency in the same manner as the lateral misalignment loss following:

$$\text{Loss}_{ang} = -10 \log_{10} \eta_{ang} \tag{5.9}$$

The formulas given in Eqs (5.8) and (5.9) predict that the smaller the values of Δ, the larger the insertion loss due to angular misalignment. This appears intuitively correct as small values of Δ imply small numerical aperture fibers, which will be more affected by angular misalignment. It is confirmed by the measurements shown in Figure 5.2(b) and demonstrated in Example 5.4.

Example 5.4

Two multimode step index fibers have numerical apertures of 0.2 and 0.4, respectively, and both have the same core refractive index of 1.48. Estimate the insertion loss at a joint in each fiber caused by a 5° angular misalignment of the fiber core axes. It may be assumed that the medium between the fibers is air.

Solution: The angular coupling efficiency is given by Eq. (5.8) as:

$$\eta_{ang} \simeq \frac{16(n_1/n)^2}{[1 + (n_1/n)]^4} \left[1 - \frac{n\theta}{\pi n_1 (2\Delta)^{\frac{1}{2}}}\right]$$

The numerical aperture is related to the relative refractive index difference following Eq. (2.10) where:

$$NA \simeq n_1 (2\Delta)^{\frac{1}{2}}$$

Hence:

$$\eta_{ang} \simeq \frac{16(n_1/n)^2}{[1 + (n_1/n)]^4} \left[1 - \frac{n\theta}{\pi NA}\right]$$

For the $NA = 0.2$ fiber:

$$\eta_{ang} \simeq \frac{16(1.48)^2}{[1 + 1.48]^4} \left[1 - \frac{5\pi/180}{\pi 0.2}\right]$$

$$= 0.797$$

▶

The insertion loss due to the angular misalignment may be obtained from Eq. (5.9), where:

$$\text{Loss}_{\text{ang}} = -10 \log_{10} \eta_{\text{ang}} = -10 \log_{10} 0.797$$
$$= 0.98 \text{ dB}$$

For the $NA = 0.4$ fiber:

$$\eta_{\text{ang}} \simeq 0.926 \left[1 - \frac{5\pi/180}{\pi 0.4} \right]$$

$$\simeq$$

0.862

The insertion loss due to the angular misalignment is therefore:

$$Loss_{\text{ang}} = -10 \log_{10} 0.862$$

Hence it may be noted from Example 5.4 that the insertion loss due to angular misalignment is reduced by using fibers with large numerical apertures. This is the opposite trend to the increasing insertion loss with numerical aperture for fiber longitudinal misalignment at a joint.

Factors causing fiber–fiber intrinsic losses were listed in Section 5.2; the major ones comprising a mismatch in the fiber core diameters, a mismatch in the fiber numerical apertures and differing fiber refractive index profiles are illustrated in Figure 5.3. Connections between multimode fibers with certain of these parameters being different can be quite common, particularly when a pigtailed optical source is used, the fiber pigtail of which has different characteristics from the main transmission fiber. Moreover, as indicated previously, diameter variations can occur with the same fiber type.

Assuming all the modes are equally excited in a multimode step or graded index fiber, and that the numerical apertures and index profiles are the same, then the loss resulting from a mismatch of core diameters (see Figure 5.3(a)) is given by [Refs 11, 22]:

$$\text{Loss}_{\text{CD}} = \begin{cases} -10 \log_{10} \left(\dfrac{a_2}{a_1} \right)^2 & \text{(dB)} \quad a_2 < a_1 \\ 0 & \text{(dB)} \quad a_2 \geq a_1 \end{cases} \tag{5.10}$$

where a_1 and a_2 are the core radii of the transmitting and receiving fibers respectively. It may be observed from Eq. (5.10) that no loss is incurred if the receiving fiber has a larger core diameter than the transmitting one. In addition, only a relatively small loss (0.09 dB) is obtained when the receiving fiber core diameter is 1% smaller than that of the transmitting fiber.

When the transmitting fiber has a higher numerical aperture than the receiving fiber, then some of the emitted light rays will fall outside the acceptance angle of the receiving

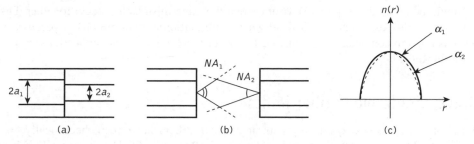

Figure 5.3 Some intrinsic coupling losses at fiber joints: (a) core diameter mismatch; (b) numerical aperture mismatch; (c) refractive index profile difference

fiber and they will therefore not be coupled through the joint. Again assuming a uniform modal power distribution, and fibers with equivalent refractive index profiles and core diameters, then the loss caused by a mismatch of numerical apertures (see Figure 5.3(b)) can be obtained from [Refs 18, 22]:

$$
\text{Loss}_{\text{NA}} = \begin{cases} -10 \log_{10} \left(\dfrac{NA_2}{NA_1} \right)^2 \ (\text{dB}) & NA_2 < NA_1 \\ 0 & (\text{dB}) \quad NA_2 \geq NA_1 \end{cases}
\tag{5.11}
$$

where NA_1 and NA_2 are the numerical apertures for the transmitting and receiving fibers respectively. Equation (5.11) is valid for both step and graded index* fibers and in common with Eq. (5.10) it demonstrates that no losses occur when the receiving parameter (i.e. numerical aperture) is larger than the transmitting one.

Finally, a mismatch in refractive index profiles (see Figure 5.3(a)) results in a loss which can be shown to be [Ref. 22]:

$$
\text{Loss}_{\text{RI}} = \begin{cases} -10 \log_{10} \dfrac{\alpha_2(\alpha_1 + 2)}{\alpha_1(\alpha_2 + 2)} \ (\text{dB}) & \alpha_2 < \alpha_1 \\ 0 & (\text{dB}) \quad \alpha_2 \geq \alpha_1 \end{cases}
\tag{5.12}
$$

where α_1 and α_2 are the profile parameters for the transmitting and receiving fibers respectively (see Section 2.4.4). When connecting from a step index fiber with $\alpha_1 = \infty$ to a parabolic profile graded index fiber with $\alpha_2 = 2$, both having the same core diameter and axial numerical aperture, then a loss of 3 dB is produced. The reverse connection, however, does not incur a loss due to refractive index profile mismatch.

The intrinsic losses obtained at multimode fiber–fiber joints provided by Eqs (5.10) to (5.12) can be combined into a single expression as follows:

$$
\text{Loss}_{\text{int}} = \begin{cases} -10 \log_{10} \dfrac{(a_2 NA_2)^2(\alpha_1 + 2)\alpha_2}{(a_1 NA_1)^2(\alpha_2 + 2)\alpha_1} \ (\text{dB}) & a_2 > a_1, NA_2 > NA_1, \alpha_2 > \alpha_1 \\ 0 & (\text{dB}) \quad a_2 \leq a_1, NA_2 \leq NA_1, \alpha_2 \leq \alpha_1 \end{cases}
\tag{5.13}
$$

---

* In the case of graded index fibers the numerical aperture on the fiber core axis must be used.

It should be noted that Eq. (5.13) assumes that the three mismatches occur together. Distributions of losses which are obtained when, with particular distributions of parameters, various random combinations of mismatches occur in a long series of connections are provided in Ref. 11.

5.2.2 Single-mode fiber joints

Misalignment losses at connections in single-mode fibers have been theoretically considered by Marcuse [Ref. 23] and Gambling *et al.* [Refs 24, 25]. The theoretical analysis which was instigated by Marcuse is based upon the Gaussian or near-Gaussian shape of the modes propagating in single-mode fibers regardless of the fiber type (i.e. step index or graded index). Further development of this theory by Gambling *et al.* [Ref. 25] gave simplified formulas for both the lateral and angular misalignment losses at joints in single-mode fibers. In the absence of angular misalignment Gambling *et al.* calculated that the loss T_1 due to lateral offset y was given by:

$$T_1 = 2.17 \left(\frac{y}{\omega}\right)^2 dB \tag{5.14}$$

where ω is the normalized spot size of the fundamental mode.* However, the normalized spot size for the LP_{01} mode (which corresponds to the HE mode) may be obtained from the empirical formula [Refs 19, 24]:

$$\omega = a \frac{(0.65 + 1.62V^{-\frac{3}{2}} + 2.88V^{-6})}{2^{\frac{1}{2}}} \tag{5.15}$$

where ω is the spot size in μm, a is the fiber core radius and V is the normalized frequency for the fiber. Alternatively, the insertion loss T_a caused by an angular misalignment θ (in radians) at a joint in a single-mode fiber may be given by:

$$T_a = 2.17 \left(\frac{\theta \omega n_1 V}{aNA}\right)^2 dB \tag{5.16}$$

where n_1 is the fiber core refractive index and NA is the numerical aperture of the fiber. It must be noted that the formulas given in Eqs (5.15) and (5.16) assume that the spot sizes of the modes in the two coupled fibers are the same. Gambling *et al.* [Ref. 25] also derived a somewhat complicated formula which gave a good approximation for the combined losses due to both lateral and angular misalignment at a fiber joint. However, they indicate that for small total losses (less than 0.75 dB) a reasonable approximation is obtained by simply combining Eqs (5.14) and (5.16).

---

* The spot size for single-mode fibers is discussed in Section 2.5.2. It should be noted, however, that the normalization factor for the spot size causes it to differ in Eq. (5.15) by a factor of $2^{\frac{1}{2}}$ from that provided in Eq. (2.125).

Example 5.5

A single-mode fiber has the following parameters:

normalized frequency $(V) = 2.40$
core refractive index (n_1) $= 1.46$
core diameter $(2a)$ $= 8$ μm
numerical aperture (NA) $= 0.1$

Estimate the total insertion loss of a fiber joint with a lateral misalignment of 1 μm and an angular misalignment of 1°.

Solution: Initially it is necessary to determine the normalized spot size in the fiber. This may be obtained from Eq. (5.15) where:

$$\omega = a\frac{(0.65 + 1.62V^{-\frac{3}{2}} + 2.88V^{-6})}{2^{\frac{1}{2}}}$$

$$= 4\frac{(0.65 + 1.62(2.4)^{-1.5} + 2.88(2.4)^{-6})}{2^{\frac{1}{2}}}$$

$$= 3.12 \text{ μm}$$

The loss due to the lateral offset is given by Eq. (5.14) as:

$$T_1 = 2.17\left(\frac{y}{\omega}\right)^2 = 2.17\left(\frac{1}{3.12}\right)^2$$

$$= 0.22 \text{ dB}$$

The loss due to angular misalignment may be obtained from Eq. (5.16) where:

$$T_a = 2.17\left(\frac{\theta\omega n_1 V}{aNA}\right)^2$$

$$= 2.17\left(\frac{(\pi/180) \times 3.12 \times 1.46 \times 2.4}{4 \times 0.1}\right)$$

$$= 0.49 \text{ dB}$$

Hence, the total insertion loss is:

$$T_T \simeq T_1 + T_a = 0.22 + 0.49$$

$$= 0.71 \text{ dB}$$

In this example the loss due to angular misalignment is significantly larger than that due to lateral misalignment. However, aside from the actual magnitudes of the respective misalignments, the insertion losses incurred are also strongly dependent upon the normalized frequency of the fiber. This is especially the case with angular misalignment at a single-mode fiber joint where insertion losses of less than 0.3 dB may be obtained when the angular misalignment is 1° with fibers of appropriate V value. Nevertheless, for low-loss single-mode fiber joints it is important that angular alignment is better than 1°.

The theoretical model developed by Marcuse [Ref. 23] has been utilized by Nemota and Makimoto [Ref. 26] in a derivation of a general equation for determining the coupling loss between single-mode fibers. Their full expression takes account of all the extrinsic factors (lateral, angular and longitudinal misalignments, and Fresnel reflection), as well as the intrinsic factor associated with the connection of fibers with unequal mode-field diameters. Moreover, good agreement with various experimental investigations has been obtained using this generalized equation [Ref. 27]. Although consideration of the full expression [Ref. 28] is beyond the scope of this text, a reduced equation, to allow calculation of the intrinsic factor which quite commonly occurs in the interconnection of single-mode fibers, may be employed. Hence, assuming that no losses are present due to the extrinsic factors, the intrinsic coupling loss is given by [Ref. 27]:

$$\text{Loss}_{\text{int}} = -10 \log_{10} \left[4 \left(\frac{\omega_{02}}{\omega_{01}} + \frac{\omega_{01}}{\omega_{02}} \right)^{-2} \right] \text{(dB)} \qquad (5.17)$$

where ω_{01} and ω_{02} are the spot sizes of the transmitting and receiving fibers respectively. Equation (5.17) therefore enables the additional coupling loss resulting from mode-field diameter mismatch between two single-mode fibers to be calculated.

Example 5.6

Two single-mode fibers with mode-field diameters of 9.2 μm and 8.4 μm are to be connected together. Assuming no extrinsic losses, determine the loss at the connection due to the mode-field diameter mismatch.

Solution: The intrinsic loss is obtained using Eq. (5.17) where:

$$\text{Loss}_{\text{int}} = -10 \log_{10} \left[4 \left(\frac{\omega_{02}}{\omega_{01}} + \frac{\omega_{01}}{\omega_{02}} \right)^{-2} \right]$$

$$= -10 \log_{10} \left[4 \left(\frac{4.2}{5.6} + \frac{5.6}{4.2} \right)^{-2} \right]$$

$$= -10 \log_{10} 0.922$$

$$= 0.35 \text{ dB}$$

It should be noted from Example 5.6 that the same result is obtained irrespective of which fiber is transmitting or receiving through the connection. Hence, by contrast to the situation with multimode fibers (see Section 5.2.1), the intrinsic loss through a single-mode fiber joint is independent of the direction of propagation.

We have considered in some detail the optical attenuation at fiber–fiber connections. However, we have not yet discussed the possible distortion of the transmitted signal at a fiber joint. Although work in this area is in its infancy, increased interest has been generated with the use of highly coherent sources (injection lasers) and very low dispersion fibers. It is apparent that fiber connections strongly affect the signal transmission causing modal noise (see Section 3.10.3) and nonlinear distortion [Ref. 29] when a coherent light source is utilized with a multimode fiber. Also, it has been reported [Ref. 30] that the transmission loss of a connection in a coherent multimode system is extremely wavelength dependent, exhibiting a possible 10% change in the transmitted optical wavelength for a very small change (0.001 nm) in the laser emission wavelength. Although it has been found that these problems may be reduced by the use of single-mode optical fiber [Ref. 29], a theoretical model for the wavelength dependence of joint losses in single-mode fiber has been obtained [Ref. 31]. This model predicts that as the wavelength increases then the width of the fundamental mode field increases and hence for a given lateral offset or angular tilt the joint loss decreases. For example, the lateral offset loss at a wavelength of 1.5 μm was calculated to be only around 80% of the loss at a wavelength of 1.3 μm.

Furthermore, the above modal effects become negligible when an incoherent source (LED) is used with multimode fiber. However, in this instance there is often some mode conversion at the fiber joint which can make the connection effectively act as a mode mixer or filter [Ref. 32]. Indications are that this phenomenon, which has been invest-igated [Ref. 33] with regard to fiber splices, is more pronounced with fusion splices than with mechanical splices, both of which are described in Section 5.3.

5.3 Fiber splices

A permanent joint formed between two individual optical fibers in the field or factory is known as a fiber splice. Fiber splicing is frequently used to establish long-haul optical fiber links where smaller fiber lengths need to be joined, and there is no requirement for repeated connection and disconnection. Splices may be divided into two broad categories depending upon the splicing technique utilized. These are fusion splicing or welding and mechanical splicing.

Fusion splicing is accomplished by applying localized heating (e.g. by a flame or an electric arc) at the interface between two butted, prealigned fiber ends causing them to soften and fuse. Mechanical splicing, in which the fibers are held in alignment by some mechanical means, may be achieved by various methods including the use of tubes around the fiber ends (tube splices) or V-grooves into which the butted fibers are placed (groove splices). All these techniques seek to optimize the splice performance (i.e. reduce the insertion loss at the joint) through both fiber end preparation and alignment of the two joint fibers. Typical average splice insertion losses for multimode fibers are in the range 0.1 to 0.2 dB [Ref. 34] which is generally a better performance than that exhibited by demountable connections (see Sections 5.4 and 5.5). It may be noted that the insertion losses of fiber splices are generally much less than the possible Fresnel reflection loss at a butted fiber–fiber joint. This is because there is no large step change in refractive index with the fusion splice as it forms a continuous fiber connection, and some method of index matching (e.g. a fluid) tends to be utilized with mechanical splices. Although fiber splicing (especially fusion splicing) can be a somewhat difficult process to perform in a field environment, these problems have been overcome through the development of field-usable equipment.

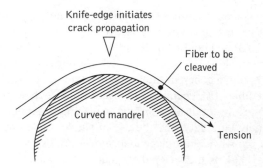

Figure 5.4 Optical fiber end preparation: the principle of scribe and break cutting [Ref. 35]

A requirement with fibers intended for splicing is that they have smooth and square end faces. In general this end preparation may be achieved using a suitable tool which cleaves the fiber as illustrated in Figure 5.4 [Ref. 35]. This process is often referred to as scribe and break or score and break as it involves the scoring of the fiber surface under tension with a cutting tool (e.g. sapphire, diamond, tungsten carbide blade). The surface scoring creates failure as the fiber is tensioned and a clean, reasonably square fiber end can be produced. Figure 5.4 illustrates this process with the fiber tensioned around a curved mandrel. However, straight pull, scribe and break tools are also utilized, which arguably give better results [Ref. 36]. An alternative technique involves circumferential scoring which provides a controlled method of lightly scoring around the fiber circumference [Ref. 31]. In this case the score can be made smooth and uniform and large-diameter fibers may be prepared by a simple straight pull with end angles less than 1°.

5.3.1 Fusion splices

The fusion splicing of single fibers involves the heating of the two prepared fiber ends to their fusing point with the application of sufficient axial pressure between the two optical fibers. It is therefore essential that the stripped (of cabling and buffer coating) fiber ends are adequately positioned and aligned in order to achieve good continuity of the transmission medium at the junction point. Hence the fibers are usually positioned and clamped with the aid of an inspection microscope.

Flame heating sources such as microplasma torches (argon and hydrogen) and oxhydric microburners (oxygen, hydrogen and alcohol vapor) have been utilized with some success [Ref. 37]. However, the most widely used heating source is an electric arc. This technique offers advantages of consistent, easily controlled heat with adaptability for use under field conditions. A schematic diagram of the basic arc fusion method is given in Figure 5.5(a) [Refs 34, 35] illustrating how the two fibers are welded together. Figure 5.5(b) [Ref. 24] shows a development of the basic arc fusion process which involves the rounding of the fiber ends with a low-energy discharge before pressing the fibers together and fusing with a stronger arc. This technique, known as prefusion, removes the requirement for fiber end preparation which has a distinct advantage in the field environment. It has been utilized with multimode fibers giving average splice losses of 0.09 dB [Ref. 39].

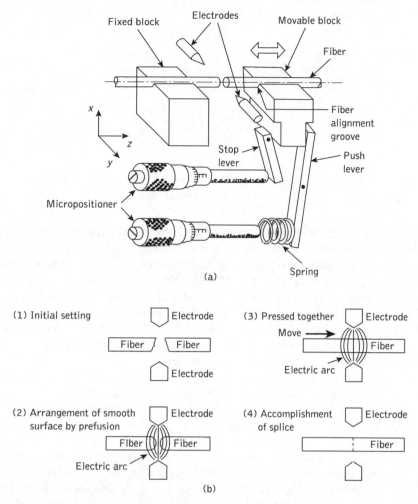

Figure 5.5 Electric arc fusion splicing: (a) an example of fusion splicing apparatus [Refs 34, 38]; (b) schematic illustration of the prefusion method for accurately splicing optical fibers [Ref. 24]

Fusion splicing of single-mode fibers with typical core diameters between 5 and 10 μm presents problems of more critical fiber alignment (i.e. lateral offsets of less than 1 μm are required for low-loss joints). However, splice insertion losses below 0.3 dB may be achieved due to a self-alignment phenomenon which partially compensates for any lateral offset.

Self-alignment, illustrated in Figure 5.6 [Refs 38, 40, 41], is caused by surface tension effects between the two fiber ends during fusing. An early field trial of single-mode fiber fusion splicing over a 31.6 km link gave mean splice insertion losses of 0.18 and 0.12 dB at wavelengths of 1.3 and 1.55 μm respectively [Ref. 42]. Mean splice losses of only 0.06 dB have also been obtained with a fully automatic single-mode fiber fusion splicing machine [Ref. 43].

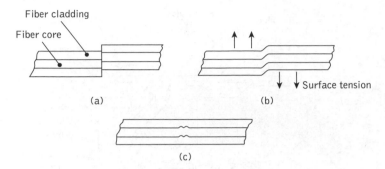

Figure 5.6 Self-alignment phenomenon which takes place during fusion splicing: (a) before fusion; (b) during fusion; (c) after fusion [Refs 38, 40, 41]

A possible drawback with fusion splicing is that the heat necessary to fuse the fibers may weaken the fiber in the vicinity of the splice. It has been found that even with careful handling, the tensile strength of the fused fiber may be as low as 30% of that of the uncoated fiber before fusion [Ref. 44]. The fiber fracture generally occurs in the heat-affected zone adjacent to the fused joint. The reduced tensile strength is attributed [Refs 44, 45] to the combined effects of surface damage caused by handling, surface defect growth during heating and induced residential stresses due to changes in chemical composition. It is therefore necessary that the completed splice is packaged so as to reduce tensile loading upon the fiber in the vicinity of the splice.

Commercial fusion splicers are produced in various sizes from handheld to laptop or tabletop depending on the requirements of the optical network [Refs 46, 47]. Two fusion splicers are shown in Figure 5.7 where a tabletop model is depicted in Figure 5.7(a) and a smaller handheld size device is presented in Figure 5.7(b). Both these instruments are capable of splicing various fiber configurations including the ribbon fiber with average splicing loss of 0.01 dB and 0.02 dB for multimode and single-mode fiber, respectively [Ref. 47]. Furthermore, the tabletop device can splice a wide range of fiber types including erbium-doped, PANDA and bow-tie fibers (see Section 3.13). Moreover, digital interface and control options are provided on these instruments which facilitate data analysis both during and after the splicing process.

5.3.2 Mechanical splices

A number of mechanical techniques for splicing individual optical fibers have been developed. A common method involves the use of an accurately produced rigid alignment tube into which the prepared fiber ends are permanently bonded. This snug tube splice is illustrated in Figure 5.8(a) [Ref. 48] and may utilize a glass or ceramic capillary with an inner diameter just large enough to accept the optical fibers. Transparent adhesive (e.g. epoxy resin) is injected through a transverse bore in the capillary to give mechanical sealing and index matching of the splice. Average insertion losses as low as 0.1 dB have been obtained [Ref. 47] with multimode graded index and single-mode fibers using ceramic capillaries. However, in general, snug tube splices exhibit problems with capillary tolerance requirements. Hence as a commercial product they may exhibit losses of up to 0.5 dB [Ref. 49].

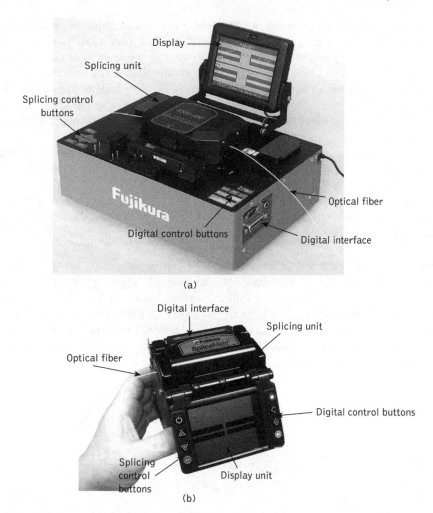

Figure 5.7 Fujikura optical fiber fusion splicers: (a) tabletop model; (b) handheld device [Ref. 46]

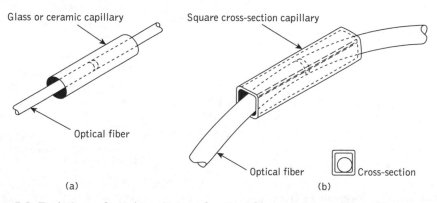

Figure 5.8 Techniques for tube splicing of optical fibers: (a) snug tube splice. Reprinted with permission from Ref. 48 © IEEE 1978; (b) loose tube splice utilizing square cross-section capillary [Ref. 50]

A mechanical splicing technique which avoids the critical tolerance requirements of the snug tube splice is shown in Figure 5.8(b) [Ref. 50]. This loose tube splice uses an over-sized square-section metal tube which easily accepts the prepared fiber ends. Transparent adhesive is first inserted into the tube followed by the fibers. The splice is self-aligning when the fibers are curved in the same plane, forcing the fiber ends simultaneously into the same corner of the tube, as indicated in Figure 5.8(b). Mean splice insertion losses of 0.073 dB have been achieved [Refs 41, 51] using multimode graded index fibers with the loose tube approach.

Other common mechanical splicing techniques involve the use of grooves to secure the fibers to be jointed. A simple method utilizes a V-groove into which the two prepared fiber ends are pressed. The V-groove splice which is illustrated in Figure 5.9(a) [Ref. 52] gives alignment of the prepared fiber ends through insertion in the groove. The splice is made permanent by securing the fibers in the V-groove with epoxy resin. Jigs for producing V-groove splices have proved quite successful, giving joint insertion losses of around 0.1 dB [Ref. 35].

V-groove splices formed by sandwiching the butted fiber ends between a V-groove glass substrate and a flat glass retainer plate, as shown in Figure 5.9(b), have also proved very successful in the laboratory. Splice insertion losses of less than 0.01 dB when coupling single-mode fibers have been reported [Ref. 53] using this technique. However, reservations are expressed regarding the field implementation of these splices with respect to manufactured fiber geometry, and housing of the splice in order to avoid additional losses due to local fiber bending.

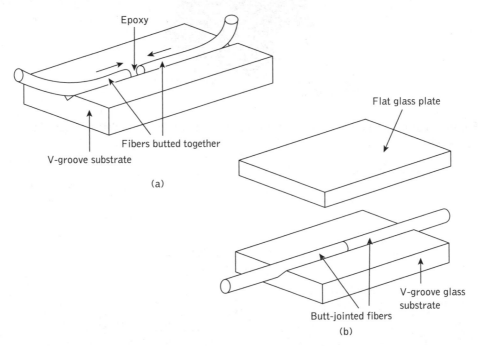

Figure 5.9 V-groove splices [Ref. 52]

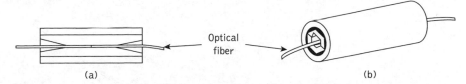

Figure 5.10 The elastomeric splice [Ref. 54]: (a) cross-section; (b) assembly

A further variant on the V-groove technique is the elastic tube or elastomeric splice shown in Figure 5.10 [Ref. 54]. The device comprises two elastomeric internal parts, one of which contains a V-groove. An outer sleeve holds the two elastic parts in compression to ensure alignment of the fibers in the V-groove, and fibers with different diameters tend to be centered and hence may be successfully spliced. Although originally intended for multimode fiber connection, the device has become a widely used commercial product [Ref. 49] which is employed with single-mode fibres, albeit often as a temporary splice for laboratory investigations. The splice loss for the elastic tube device was originally reported as 0.12 dB or less [Ref. 54] but is generally specified as around 0.25 dB for the commercial product [Ref. 49]. In addition, index-matching gel is normally employed within the device to improve its performance.

A slightly more complex groove splice known as the Springroove® splice utilized a bracket containing two cylindrical pins which serve as an alignment guide for the two prepared fiber ends. The cylindrical pin diameter was chosen to allow the fibers to protrude above the cylinders, as shown in Figure 5.11(a) [Ref. 55]. An elastic element (a spring) was used to press the fibers into a groove and maintain the fiber end alignment, as illustrated in Figure 5.11(b). The complete assembly was secured using a drop of epoxy resin. Mean splice insertion losses of 0.05 dB [Ref. 41] were obtained using multimode graded index fibers with the Springroove® splice. This device found practical use in Italy.

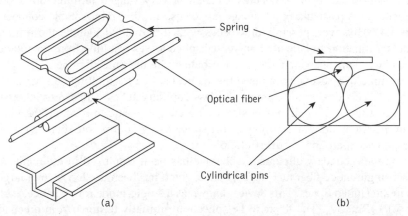

Figure 5.11 The Springroove® splice [Ref. 55]: (a) expanded overview of the splice; (b) schematic cross-section of the splice

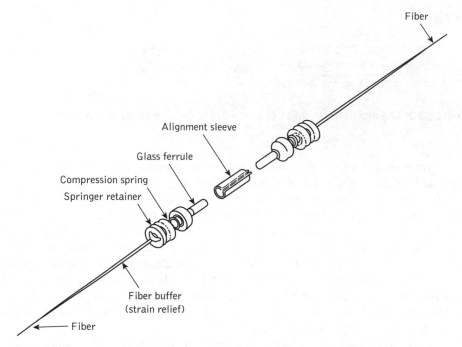

Figure 5.12 Multimode fiber mechanical splice using glass capillary tubes [Ref. 56]

The aforementioned mechanical splicing methods employ alignment of the bare fibers, whereas subsequently alignment of secondary elements around the bare fibers is a technique which has gained favor [Ref. 31]. Secondary alignment generally gives increased ruggedness and provides a structure that can be ground and polished for fiber end preparation. Furthermore, with a good design the fiber coating can be terminated within the secondary element leaving only the fiber end face exposed. Hence when the fiber end face is polished flat to the secondary element, a very rugged termination is produced. This technique is particularly advantageous for use in fiber remountable connectors (see Section 5.4). However, possible drawbacks with this method include the time taken to make the termination and the often increased splice losses resulting from the tolerances on the secondary elements which tend to contribute to the fiber misalignment.

An example of a secondary aligned mechanical splice for multimode fiber is shown in Figure 5.12. This device uses precision glass capillary tubes called ferrules as the secondary elements with an alignment sleeve of metal or plastic into which the glass tubed fibers are inserted. Normal assembly of the splice using 50 μm core diameter fiber yields an average loss of around 0.2 dB [Ref. 56].

Finally, the secondary alignment technique has been employed in the realization of a low-loss, single-mode fiber mechanical splice which has been used in several large installations in the United States. This device, known as a single-mode rotary splice, is shown in Figure 5.13 [Ref. 57]. The fibers to be spliced are initially terminated in precision glass capillary tubes which are designed to make use of the small eccentricity that is present, as illustrated in Figure 5.13(a). An ultraviolet curable adhesive is used to cement the fibers in

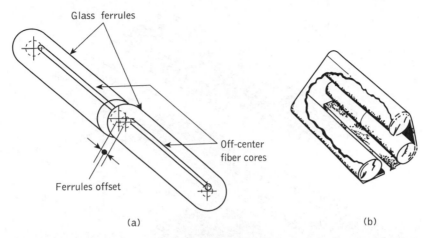

Glass ferrules

Off-center
fiber cores

Ferrules offset

(a)

(b)

Figure 5.13 Rotary splice for single-mode fibers [Ref. 57]: (a) alignment technique using glass ferrules; (b) glass rod alignment sleeve

the glass tubes and the fiber terminations are prepared with a simple grinding and polishing operation.

Alignment accuracies of the order of 0.05 μm are obtained using the three glass rod alignment sleeve shown in Figure 5.13(b). Such alignment accuracies are necessary to obtain low losses as the mode-field diameter for single-mode fiber is generally in the range 8 to 10 μm. The sleeve has a built-in offset such that when each ferrule is rotated within it, the two circular paths of the center of each fiber core cross each other. Excellent alignment is obtained utilizing a simple algorithm, and strong metal springs provide positive alignment retention. Using index-matching gel such splices have demonstrated mean losses of 0.03 dB with a standard deviation of 0.018 dB [Ref. 31]. Moreover, these results were obtained in the field, suggesting that the rotary splicing technique was not affected by the skill level of the splicer in that harsh environment.

5.3.3 Multiple splices

Multiple simultaneous fusion splicing of an array of fibers in a ribbon cable has been demonstrated for both multimode [Ref. 58] and single-mode [Ref. 59] fibers. In both cases a 12-fiber ribbon was prepared by scoring and breaking prior to pressing the fiber ends onto a contact plate to avoid difficulties with varying gaps between the fibers to be fused. An electric are fusing device was then employed to provide simultaneous fusion. Such a device is now commercially available to allow the splicing of 12 fibers simultaneously in a time of around 6 minutes, which requires only 30 seconds per splice [Refs 46, 47]. Splice losses using this device with multimode graded index fiber range from an average of 0.04 dB to a maximum of 0.12 dB, whereas for single-mode fiber the average loss is 0.04 dB with a 0.4 dB maximum.

A simple technique employed for multiple simultaneous splicing involves mechanical splicing of an array of fibers, usually in a ribbon cable. The V-groove multiple-splice secondary element comprising etched silicon chips has been used extensively in the

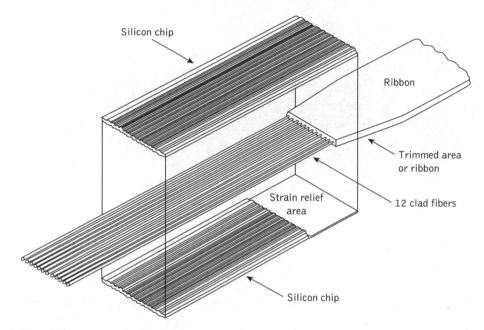

Figure 5.14 Multiple-fiber splicing using a silicon chip array

United States [Ref. 31] for splicing multimode fibers. In this technique a 12-fiber splice is prepared by stripping the ribbon and coating material from the fibers. Then the 12 fibers are laid into the trapezoidal* grooves of a silicon chip using a comb structure, as shown in Figure 5.14. The top silicon chip is then positioned prior to applying epoxy to the chip–ribbon interface. Finally, after curing, the front end face is ground and polished.

The process is normally carried out in the factory and the arrays are clipped together in the field, putting index-matching silica gel between the fiber ends. The average splice loss obtained with this technique in the field is 0.12 dB, with the majority of the loss resulting from intrinsic fiber mismatch. Major advantages of this method are the substantial reduction in splicing time (by more than a factor of 10) per fiber and the increased robustness of the final connection. Although early array splicing investigations using silicon chips [Refs 60, 61] demonstrated the feasibility of connecting 12 × 12 fiber arrays, in practice only single 12-fiber ribbons have been spliced at one time due to concerns in relation to splice tolerance and the large number of telecommunication channels which would be present in the two-dimensional array [Ref. 31].

An alternative V-groove flat chip molded from a glass-filled polymer resin has been employed in France [Ref. 62]. Moreover, direct mass splicing of 12-fiber ribbons has also been accomplished [Ref. 63]. In this technique simultaneous end preparation of all 24 fibers was achieved using a ribbon grinding and polishing procedure. The ribbons were then laid in guides and all 12 fibers were positioned in grooves in the glass-filled plastic

---

* A natural consequence of etching.

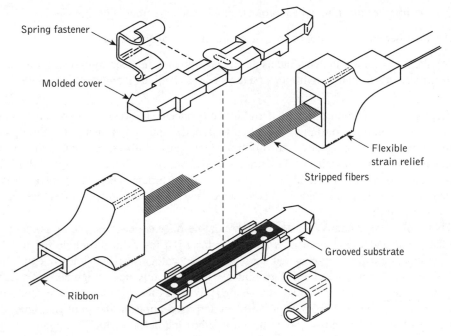

Figure 5.15 Splicing of V-groove polymer resin ribbon fiber [Ref. 63]

substrate shown in Figure 5.15. A vacuum technique was used to hold the fibers in position while the cover plate was applied, and spring clips were used to hold the assembly together. Index-matching gel was applied through a hole in the cover plate giving average splice losses of 0.18 dB with multimode fiber.

5.4 Fiber connectors

Demountable fiber connectors are more difficult to achieve than optical fiber splices. This is because they must maintain similar tolerance requirements to splices in order to couple light between fibers efficiently, but they must accomplish it in a removable fashion. Also, the connector design must allow for repeated connection and disconnection without problems of fiber alignment, which may lead to degradation in the performance of the transmission line at the joint. Hence to operate satisfactorily the demountable connector must provide reproducible accurate alignment of the optical fibers.

In order to maintain an optimum performance the connection must also protect the fiber ends from damage which may occur due to handling (connection and disconnection), must be insensitive to environmental factors (e.g. moisture and dust) and must cope with tensile load on the cable. Additionally, the connector should ideally be a low-cost component which can be fitted with relative ease. Hence optical fiber connectors may be considered in three major areas, which are:

(a) the fiber termination, which protects and locates the fiber ends;

(b) the fiber end alignment to provide optimum optical coupling;

(c) the outer shell, which maintains the connection and the fiber alignment, protects the fiber ends from the environment and provides adequate strength at the joint.

The use of an index-matching material in the connector between the two jointed fibers can assist the connector design in two ways. It increases the light transmission through the connection while keeping dust and dirt from between the fibers. However, this design aspect is not always practical with demountable connectors, especially where fluids are concerned. Apart from problems of sealing and replacement when the joint is disconnected and reconnected, liquids in this instance may have a detrimental effect, attracting dust and dirt to the connection.

There are a large number of demountable single-fiber connectors, both commercially available and under development, which have insertion losses in the range 0.2 to 3 dB. Fiber connectors may be separated into two broad categories: butt-jointed connectors and expanded beam connectors. Butt-jointed connectors rely upon alignment of the two prepared fiber ends in close proximity (butted) to each other so that the fiber core axes coincide. Expanded beam connectors utilize interposed optics at the joint (i.e. lenses) in order to expand the beam from the transmitting fiber end before reducing it again to a size compatible with the receiving fiber end.

Butt-jointed connectors are the most widely used connector type and a substantial number have been reported. In this section we review some of the more common butt-jointed connector designs which have been developed for use with both multimode and single-mode fibers. In Section 5.5, following, expanded beam connectors are discussed.

5.4.1 Cylindrical ferrule connectors

The basic ferrule connector (sometimes referred to as a concentric sleeve connector), which is perhaps the simplest optical fiber connector design, is illustrated in Figure 5.16(a) [Ref. 9]. The two fibers to be connected are permanently bonded (with epoxy resin) in metal plugs known as ferrules which have an accurately drilled central hole in their end faces where the stripped (of buffer coating) fiber is located. Within the connector the two ferrules are placed in an alignment sleeve which, using accurately machined components, allows the fiber ends to be butt jointed. The ferrules are held in place via a retaining mechanism which, in the example shown in Figure 5.16(a), is a spring.

It is essential with this type of connector that the fiber end faces are smooth and square (i.e. perpendicular to the fiber axis). This may be achieved with varying success by:

(a) cleaving the fiber before insertion into the ferrule;

(b) inserting and bonding before cleaving the fiber close to the ferrule end face;

(c) using either (a) or (b) and polishing the fiber end face until it is flush with the end of the ferrule.

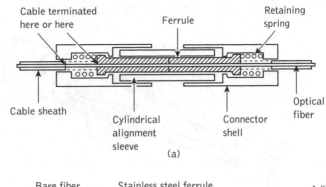

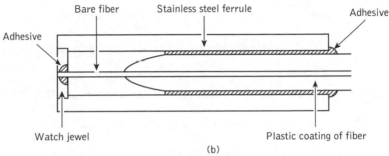

Figure 5.16 Ferrule connectors: (a) structure of a basic ferrule connector [Ref. 9]; (b) structure of a watch jewel connector ferrule [Ref. 10]

Polishing the fiber end face after insertion and bonding provides the best results but it tends to be time consuming and inconvenient, especially in the field.

The fiber alignment accuracy of the basic ferrule connector is largely dependent upon the ferrule hole into which the fiber is inserted. Hence, some ferrule connectors have incorporated a watch jewel in the ferrule end face (jeweled ferrule connector), as illustrated in Figure 5.16(b) [Ref. 10]. In this case the fiber is centered with respect to the ferrule through the watch jewel hole. The use of the watch jewel allows the close diameter and tolerance requirements of the ferrule end face hole to be obtained more easily than simply through drilling of the metallic ferrule end face alone. Nevertheless, typical concentricity errors between the fiber core and the outside diameter of the jeweled ferrule are in the range 2 to 6 μm giving insertion losses in the range 1 to 2 dB with multimode step index fibers.

Subsequently, capillary ferrules manufactured from ceramic materials (e.g. alumina porcelain) found widespread application within precision ferrule connectors. Such capillary ferrules have a precision bore which is accurately centered in the ferrule. Final assembly of the connector includes the fixture of the fiber within the ferrule, using adhesive prior to the grinding and polishing for end preparation. The ceramic materials possess outstanding thermal, mechanical and chemical resistance characteristics in comparison with metals and plastics [Ref. 64]. In addition, unlike metal and plastic components, the ceramic ferrule material is harder than the optical fiber and is therefore unaffected by the grinding and polishing process, a factor which assists in the production of low-loss fiber

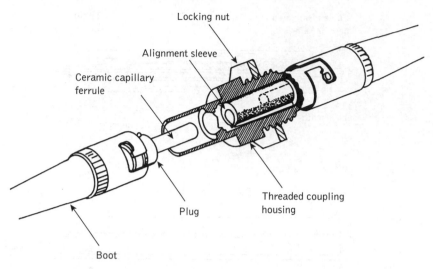

Figure 5.17 ST series multimode fiber connector using ceramic capillary ferrules

connectors. Typical average losses for multimode graded index fiber (i.e. core/cladding: 50/125 μm) and single-mode fiber (i.e. core/cladding: 9/125 μm) with the precision ceramic ferrule connector are 0.2 and 0.3 dB respectively [Ref. 60]. For example, an early ferrule-type connector widely used as part of jumper cable in a variety of applications in the United States was the biconical plug connector [Refs 34, 65]. The plugs were either transfer molded directly onto the fiber or cast around the fiber employing a silica-loaded epoxy resin, and when modified for single-mode fiber to ensure core eccentricity to 0.33 μm or less gave an average connector loss of 0.28 dB [Ref. 60].

Numerous cylindrical sleeve ferrule connectors are commercially available for both multimode and single-mode fiber termination. The most common design types are the straight tip (ST), the subminiature assembly (SMA), the fiber connector (FC), the miniature unit (MU), the subscriber connector (SC) and the D4 connector [Refs 27, 31, 60, 65]. An example of an ST series multimode fiber connector is shown in Figure 5.17, which exhibits an optimized cylindrical sleeve with a cross-section designed to expand uniformly when the ferrules are inserted. Hence, the constant circumferential pressure provides accurate alignment, even when the ferrule diameters differ slightly. In addition, the straight ceramic ferrule may be observed in Figure 5.17 which contrasts with the stepped ferrule (i.e. a ferrule with a single step which reduces the diameter midway along its length) provided in the SMA connector design. The average loss obtained using this connector with multimode graded index fiber (i.e. core/cladding: 62.5/125 μm) was 0.22 dB with less than 0.1 dB change in loss after 1000 reconnections [Ref. 66].

More recently, an improved version of the ST connector known as the field assembly simple technique (FAST) connector has been produced which exhibits an average insertion loss of 0.20 dB for single-mode fiber while giving an average insertion loss of only 0.03 dB for 62.5/125 μm multimode graded index fiber [Ref. 67]. A summary depicting some of the available connector types is provided in Section 5.4.3.

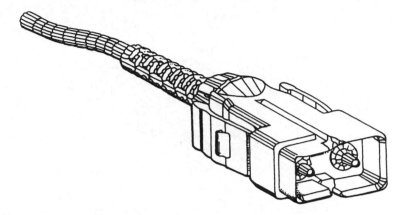

Figure 5.18 An example media interface plug for a duplex fiber connector

5.4.2 Duplex and multiple-fiber connectors

A number of duplex fiber connector designs were developed in order to provide two-way communication, but few have found widespread use [Ref. 27]. For example, AT&T produced a duplex version of the ST single-fiber connector (see Section 5.4.1). Moreover, the media interface connector plug shown in Figure 5.18 was part of a duplex fiber connector which was developed to meet the American National Standards Institute (ANSI) specification for use within optical fiber LANs [Ref. 68]. This connector plug will mate directly with connectorized optical LAN components (i.e. transmitters and receivers). A duplex fiber connector for use with the Fiber Distributed Data Interface also subsequently became commercially available. It comprised two ST ferrules housed in a protective molded shroud and exhibits a typical insertion loss of 0.6 dB. Hence, such duplex connectors were preferred for their simplicity.

Multiple-fiber connection is obviously advantageous when interconnecting a large number of fibers. Both cylindrical and biconical ferrule connectors (see Sections 5.4.1 and 5.4.2) can be assembled in housings to form multiple-fiber configurations [Ref. 31]. Single-ferrule connectors generally allow the alignment sleeve to float within the housing, thus removing any requirement for high tolerance on ferrule positioning within multiple-ferrule versions. However, the force needed to insert multiple cylindrical ferrules can be large when many ferrules are involved. In this case multiple biconical ferrule connectors prove advantageous due to the low insertion force of the biconic configuration.

In addition to assembling a number of single-fiber connectors to form a multiple-fiber connector, other examples of multiple-fiber connectors have been explored. Silicon chip arrays were suggested for the jointing of fiber ribbon cable for many years [Ref. 69]. However, difficulties were experienced in the design of an appropriate coupler for the two arrays. These problems were then largely overcome by the multiple-connector design shown in Figure 5.19(a) which utilizes V-grooved silicon chips [Ref. 70]. In this connector, ribbon fibers were mounted and bonded into the V-grooves in order to form a plug together with precision metal guiding rods and coil springs. The fiber connections were

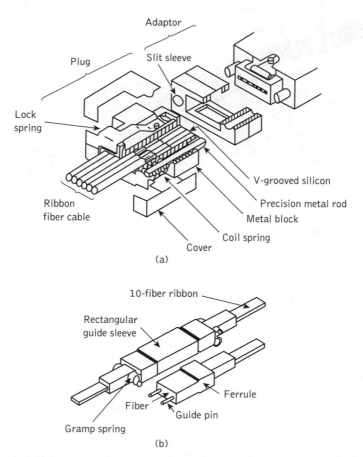

Figure 5.19 Multiple-fiber connectors: (a) fiber ribbon connector using V-groove silicon chips [Ref. 70]; (b) single-mode 10-fiber connector [Ref. 73]

then accomplished by butt jointing the two pairs of guiding rods in the slitted sleeves located in the adaptor, also illustrated in Figure 5.19(b). This multiple-fiber connector exhibited average insertion losses of 0.8 dB which were reduced to 0.4 dB by the use of index-matching fluid. Improved loss characteristics were obtained with another five-fiber molded connector, also used with fiber ribbons [Ref. 71]. In this case the mean loss and standard deviation without index matching were only 0.45 dB and 0.12 dB, respectively, when terminating 50 μm core multimode fibers [Ref. 72].

The structure of a small plastic molded single-mode 10-fiber connector is shown in Figure 5.19(b) [Ref. 73]. It comprised two molded ferrules with 10-fiber ribbon cables which are accurately aligned by guide pins, then held stable with a rectangular guide sleeve and a cramp spring. This compact multifiber connector which has dimensions of only 6 × 4 mm exhibited an average connection loss of 0.43 dB when used with single-mode fibers having a spot size (ω_0) of 5 μm.

5.4.3 Fiber connector-type summary

Table 5.1 provides a summary listing of the common fiber connector types used for both multimode and single-mode fiber systems. Multimode optical connectors are generally used within data communications (i.e. LANs), transport (i.e. automobiles and aircraft) and with specific test instruments, while single-mode fiber connectors are employed extensively in optical fiber telecommunication systems.

It should be noted that the majority of the commercially available fiber connectors are designed to specifications determined by international standards bodies such as the Telecommunication Industry Association (TIA), International Electrotechnical Commission (IEC) and American National Standards Institute (ANSI). Furthermore, reduced size and low weight are significant features of small form factor (SFF) and small form pluggable (SFP) connectors enabling the production of cost-effective components. Hence SFF and SFP connectors can be used to provide high-density interconnections where several fiber connectors of the same (or different) type are combined to form an array using bulkhead

Table 5.1 Fiber connector types

Type	Shape	Insertion loss (dB)	Features and applications
SMA		1.00–1.50	A slotted screw-on connector; preferred in multimode fiber, data communication, multimedia and instrumentation connections
FDDI		0.20–0.70	A push-on, pull-off type of dual connector primarily used with multimode fiber in LANs.
D4		0.30–1.00	A slotted screw-on type of multimode and single-mode fiber connector; used for data communications, instrumentation connections and telecommunication applications
ST		0.20–0.50	A slotted bayonet (push-in, twist-out) type of metallic multimode or single-mode fiber connector with a ceramic ferrule; widely used in inter/intra building, data communication and also telecommunication applications
SC (simplex and duplex)		0.20–0.45	A push-on, pull-off type of multimode or single-mode connector with a ceramic ferrule and an SFF design in a simplex or a duplex plastic housing; often used for LANs and data communication

Table 5.1 (*continued*)

Type	Shape	Insertion loss (dB)	Features and applications
FC		0.25–1.00	A screw-on metallic connector with a ceramic ferrule; widely used with single-mode fiber, for active device termination and in high-vibration environments
MU		0.10–0.30	A push-on, pull-off type of multimode or single-mode fiber connector with plastic housing and ceramic ferrule; SFF and SFP designs with packaging density that is greater than the SC connector and can be simplex or duplex; useful for board-mounted applications and high-density interconnections
LC (simplex and duplex)		0.10–0.50	A push-on, pull-off, multimode or single-mode fiber type of connector containing a standard RJ 45 telephone plug housing with a ceramic ferrule in a simplex or duplex plastic housing; SFF and SFP designs are suitable for high-density interconnection and also useful for instrumentation and test equipment interconnections
E2000		0.12–0.30	A push-on, pull-off type of connector, mostly preferred for single-mode fiber operation; SFF and SFP designs similar to the SC type but also contains an eye protection safety cover built into the end face
MT-RJ (single or multiple)		0.25–0.75	A push-on, pull-off type of connector with two (or more) fibers in a single plastic ferrule where the housing uses the standard RJ 45 latch mechanism; it can connect up to 72 fibers, and it is suitable for both local and metropolitan area networks, particularly with high-density interconnections
MTP/ MPO		0.25–1.00	A push-on, pull-off type of connector for multi fiber ribbon cable (4 to 72 fibers) based on multiple MT–RJ connectors in a plastic housing and used for high-density interconnections

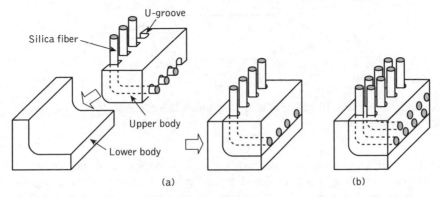

Figure 5.20 A 90° bent fiber connector: (a) different sections of the optical fiber connector for single-layer connection; (b) multilayer connection. Reprinted with permission from Ref. 74 © IEEE 2005

adaptors. For example, a commercially available multiple MT–RJ connector can combine 6 cables each with 12 fibers to form an MMC® connector accommodating a total of 72 optical fibers. Moreover, several other fiber connectors are produced with SFF and SFP features which include the variants of LC-, MU- and MTP-type optical connectors. The use of screw-on connectors (i.e. SMA and D4), however, is declining as they are not compliant with the SFF and SFP requirements for optical networking.

Finally, the coupling of signals into optical fiber with a 90° bend is important when fiber is required to be housed in a small space. Figure 5.20(a) displays the structure for optical interconnection using a 90° bent fiber connector which comprises two parts: namely, a lower body and an upper body to provide U-shaped grooves which support the optical fibers [Ref. 74]. The connector facilitates four fiber connections using a single-layer structure. To achieve more interconnections it is possible to produce a multilayer connector employing a similar approach. A multilayered structure providing eight interconnections is illustrated in Figure 5.20(b). Such single-layered or multilayered fiber connectors oriented at 90° can also be very useful for implementing optical printed circuit boards and in this function they enable surface- or bottom-emitting/receiving devices to be interfaced with the circuit board. These connectors have displayed a total interconnection loss of around 1.3 dB between a transmitter and a receiver while also exhibiting relatively low optical crosstalk of 53 dB between neighboring channels when operating at transmission rates of 2.5 Gbit s^{-1} [Ref. 74].

5.5 Expanded beam connectors

An alternative to connection via direct butt joints between optical fibers is offered by the principle of the expanded beam. Fiber connection utilizing this principle is illustrated in Figure 5.21, which shows a connector consisting of two lenses for collimating and refocusing the light from one fiber into the other. The use of these interposed optics makes the achievement of lateral alignment much less critical than with a butt-jointed fiber connector.

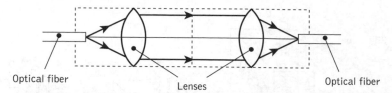

Optical fiber   Lenses   Optical fiber

Figure 5.21 Schematic illustration of an expanded beam connector showing the principle of operation

Also, the longitudinal separation between the two mated halves of the connector ceases to be critical. However, this is achieved at the expense of more stringent angular alignment. Nevertheless, expanded beam connectors are useful for multifiber connection and edge connection for printed circuit boards where lateral and longitudinal alignment are frequently difficult to achieve.

Two examples of lens-coupled expanded beam connectors are illustrated in Figure 5.22. The connector shown in Figure 5.22(a) [Ref. 75] utilized spherical microlenses for beam expansion and reduction. It exhibited average losses of 1 dB which were reduced to 0.7 dB with the application of an antireflection coating on the lenses and the use of graded index fiber of 50 μm core diameter. A similar configuration has been used for single-mode fiber connection in which the lenses have a 2.5 mm diameter [Ref. 76]. Again with antireflection-coated lenses, average losses around 0.7 dB were obtained using single-mode

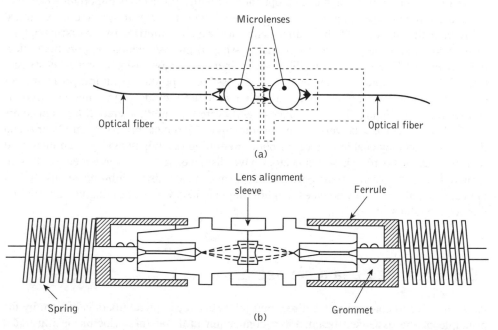

Figure 5.22 Lens-coupled expanded beam connectors: (a) schematic diagram of a connector with two microlenses making a 1:1 image of the emitting fiber upon the receiving one [Ref. 75]; (b) molded plastic lens connector assembly [Ref. 77]

fibers of 8 μm core diameter. Furthermore, successful single-mode fiber connection has been achieved with a much smaller (250 μm diameter) sapphire ball lens expanded beam design [Ref. 31]. In this case losses in the range 0.4 to 0.7 dB were demonstrated over 1000 connections.

Figure 5.22(b) shows an expanded beam connector which employs a molded spherical lens [Ref. 77]. The fiber is positioned approximately at the focal length of the lens in order to obtain a collimated beam and hence minimize lens-to-lens longitudinal misalignment effects. A lens alignment sleeve is used to minimize the effects of angular misalignment which, together with a ferrule, grommet, spring and external housing, provides the complete connector structure. The repeatability of this relatively straightforward lens design was found to be good, incurring losses of around 0.7 dB.

More recently, an array of microlenses has been used to connect several fibers simultaneously. Figure 5.23 shows an assembly where two arrays of microlenses are employed to interconnect two arrays of fibers. It displays a multifiber connector assembly in which the fibers are placed onto a tray of V-grooves inside the adaptor by a mechanical fixture which provides a permanent bond. This multifiber connector can be inserted into an adaptor containing two arrays of microlenses also indicated in Figure 5.23. The microlens arrays translate the divergent beams from the optical fibers into collimated beams and vice versa. Furthermore, optical coupling losses (per lens transition) remain under 1 dB for multimode fiber and around 0.5 dB for single-mode fiber arrays [Ref. 78].

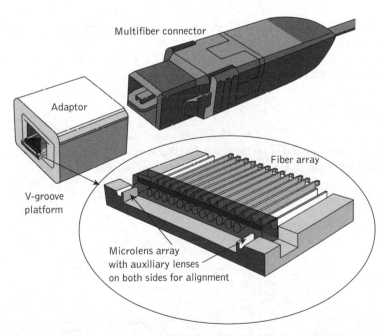

Figure 5.23 Multifiber connectivity using microlenses showing an assembly for a multifiber connector and an adaptor with a V-groove platform for the fiber array which also contains microlens arrays. Reprinted with permission from Ref. 78 © IEEE 2005

5.5.1 GRIN-rod lenses

An alternative lens geometry to facilitate efficient beam expansion and collimation within expanded beam connectors is that of the graded index (GRIN) rod lens [Ref. 79]. In addition the focusing properties of such microlens devices have enabled them to find application within both fiber couplers (see Section 5.6) and source-to-fiber coupling (see Section 6.8).

The GRIN-rod lens, which arose from developments on graded index fiber waveguides [Ref. 80], comprises a cylindrical glass rod typically 0.2 to 2 mm [Ref. 81] in diameter which exhibits a parabolic refractive index profile with a maximum at the axis similar to graded index fiber. Light propagation through the lens is determined by the lens dimensions and, because refractive index is a wavelength-dependent parameter, by the wavelength of the light. The GRIN-rod lens can produce a collimated output beam with a divergent angle α of between 1° and 5° from a light source situated on, or near to, the opposite lens face, as illustrated in Figure 5.24. Conversely, it can focus an incoming light beam onto a small area located at the center of the opposite lens face. Typically, light launched from a 50 μm diameter fiber core using a GRIN-rod lens results in a collimated output beam of between 0.5 and 1 mm.

Ray propagation through the GRIN-rod lens medium is approximately governed by the paraxial ray equation:

$$\frac{\mathrm{d}^2 r}{\mathrm{d}z^2} = \frac{1}{n}\frac{\mathrm{d}n}{\mathrm{d}r} \tag{5.18}$$

where r is the radial coordinate, z is the distance along the optical axis and n is the refractive index at a point.

Furthermore, the refractive index at r following Eq. (2.75), distance r from the optical axis in a gradient index medium, may be expressed as [Ref. 82]:

$$n(r) = n_1\left(1 - \frac{Ar^2}{2}\right) \tag{5.19}$$

where n_1 is the refractive index on the optical axis and A is a positive constant.

Using Eqs (5.18) and (5.19), the position r of the ray is given by:

$$\frac{\mathrm{d}^2 r}{\mathrm{d}z^2} = -Ar \tag{5.20}$$

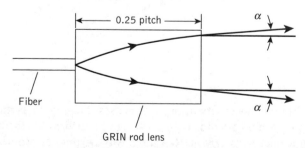

Figure 5.24 Formation of a collimated output beam from a GRIN-rod lens

Following Miller [Ref. 83], the general solution of Eq. (5.20) becomes:

$$r = K_1 \cos A^{\frac{1}{2}}r + K_2 \sin A^{\frac{1}{2}}r \qquad (5.21)$$

where K_1 and K_2 are constants.

The refractive index variation with radius therefore causes all the input rays to follow a sinusoidal path through the lens medium. The traversion of one sinusoidal period is termed one full pitch and GRIN-rod lenses are manufactured with several pitch lengths. Three major pitch lengths are as follows:

1. The quarter pitch (0.25 pitch) lens, which produces a perfectly collimated output beam when the input light emanates from a point source on the opposite lens face. Conversely, the lens focuses an incoming light beam to a point at the center of the opposite lens face (Figure 5.25(a)). Thus the focal point of the quarter pitch GRIN-rod

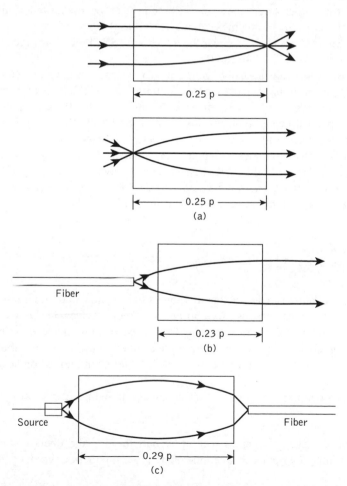

Figure 5.25 Operation of various GRIN-rod lenses: (a) the quarter pitch lens; (b) the 0.23 pitch lens; (c) the 0.29 pitch lens

lens is coincident with the lens faces, thus providing efficient direct butted connection to optical fiber.

2. The 0.23 pitch lens is designed such that its focal point lies outside the lens when a collimated beam is projected on the opposite lens face. It is often employed to convert the diverging beam from a fiber or laser diode into a collimated beam, as illustrated in Figure 5.25(b) [Ref. 84].

3. The 0.29 pitch lens is designed such that both focal points lie just outside the lens end faces. It is frequently used to convert a diverging beam from a laser diode into a converging beam. Hence, it proves useful for coupling the output from a laser diode into an optical fiber (Figure 5.25(c)), or alternatively for coupling the output from an optical fiber into a photodetector.

The majority of GRIN-rod lenses which have diameters in the range 0.5 and 2 mm may be employed with either single-mode or multimode (step or graded index) fiber. Various fractional pitch lenses, including those above as well as 0.5 p and 0.75 p, may be obtained from Nippon Sheet Glass Co. Ltd under the trade name SELFOC. They are available with numerical apertures of 0.37, 0.46 and 0.6.

A number of factors can cause divergence of the collimated beam from a GRIN rod lens. These include errors in the lens cut length, the finite size of the fiber core and chromatic aberration. As indicated previously, divergence angles as small as 1° may be obtained which yield expanded beam connector losses of around 1 dB [Ref. 31]. Furthermore, in contrast to butt-jointed multimode fiber connectors, GRIN-rod lens connectors have demonstrated loss characteristics which are independent of the modal power distribution in the fiber [Ref. 85]. In addition, GRIN-rod lenses have been employed to efficiently connect microstructured fiber (see Section 2.6) demonstrating low transmission loss of 0.4 dB, a value which could not be otherwise achieved [Ref. 81].

5.6 Fiber couplers

An optical fiber coupler is a device that distributes light from a main fiber into one or more branch fibers.* The latter case is more normal and such devices are known as multiport fiber couplers. Requirements are increasing for the use of these devices to divide or combine optical signals for application within optical fiber information distribution systems including data buses, LANs, computer networks and telecommunication access networks (see Chapter 15).

Optical fiber couplers are often passive devices in which the power transfer takes place either:

(a) through the fiber core cross-section by butt jointing the fibers or by using some form of imaging optics between the fibers (core interaction type); or

* Devices of this type are also referred to as directional couplers.

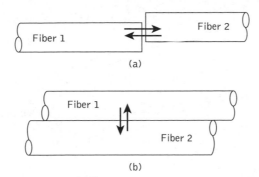

Figure 5.26 Classification of optical fiber couplers: (a) core interaction type; (b) surface interaction type

(b) through the fiber surface and normal to its axis by converting the guided core modes to both cladding and refracted modes which then enable the power-sharing mechanism (surface interaction type).

The mechanisms associated with these two broad categories are illustrated in Figure 5.26. Active waveguide directional couplers are also available which are realized using integrated optical fabrication techniques. Such device types, however, are dealt with in Section 11.4.1 and thus in this section the discussion is restricted to the above passive coupling strategies.

Multiport optical fiber couplers can also be subdivided into the following three main groups [Ref. 86], as illustrated in Figure 5.27.

1. Three- and four-port* couplers, which are used for signal splitting, distribution and combining.

2. Star couplers, which are generally used for distributing a single input signal to multiple outputs.

3. Wavelength division multiplexing (WDM) devices, which are a specialized form of coupler designed to permit a number of different peak wavelength optical signals to be transmitted in parallel on a single fiber (see Section 12.9.4). In this context WDM couplers either combine the different wavelength optical signal onto the fiber (i.e. multiplex) or separate the different wavelength optical signals output from the fiber (i.e. demultiplex).

Ideal fiber couplers should distribute light among the branch fibers with no scattering loss[†] or the generation of noise, and they should function with complete insensitivity to factors including the distribution of light between the fiber modes, as well as the state of polarization of the light. Unfortunately, in practice passive fiber couplers do not display all of the above properties and hence the characteristics of the devices affect the performance of

* Four-port couplers may also be referred to as 2 × 2 star couplers.
† The scattering loss through the coupler is often referred to as the excess loss.

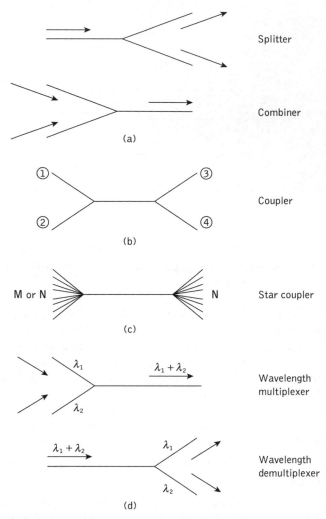

Figure 5.27 Optical fiber coupler types and functions: (a) three-port couplers; (b) four-port coupler; (c) star coupler; (d) wavelength division multiplexing and demultiplexing couplers

optical fiber networks. In particular, the finite scattering loss at the coupler limits the number of terminals that can be connected, or alternatively the span of the network, whereas the generation of noise and modal effects can cause problems in the specification of the network performance. Hence, couplers in a network cannot usually be treated as individual components with known parameters, a factor which necessitates certain compromises in their application. In this section, therefore, a selection of the more common fiber coupler types is described in relation to the coupling mechanisms, their performance and limitations.

5.6.1 Three- and four-port couplers

Several methods are employed to fabricate three- and four-port optical fiber couplers [Refs 86–89]. The lateral offset method, illustrated in Figure 5.28(a), relies on the overlapping of the fiber end faces. Light from the input fiber is coupled to the output fibers according to the degree of overlap. Hence the input power can be distributed in a well-defined proportion by appropriate control of the amount of lateral offset between the fibers. This technique, which can provide a bidirectional coupling capability, is well suited for use with multimode step index fibers but may incur higher excess losses than other methods as all the input light cannot be coupled into the output fibers.

Another coupling technique is to incorporate a beam splitter element between the fibers. The semitransparent mirror method provides an ingenious way to accomplish such a fiber coupler, as shown in Figure 5.28(b). A partially reflecting surface can be applied directly to the fiber end face cut at an angle of 45° to form a thin-film beam splitter. The input power may be split in any desired ratio between the reflected and transmitted beams depending upon the properties of the intervening mirror, and typical excess losses for the device lie in the range 1 to 2 dB. Using this technology both three- and four-port couplers with both multimode and single-mode fibers have been fabricated [Ref. 88]. In addition, with suitable wavelength-selective interference coatings this coupler type can form a WDM device (see Section 5.6.3).

A fast-growing category of optical fiber coupler is based on the use of micro-optic components. In particular, a complete range of couplers has been developed which utilize the beam expansion and collimation properties of the GRIN-rod lens (see Section 5.5.1)

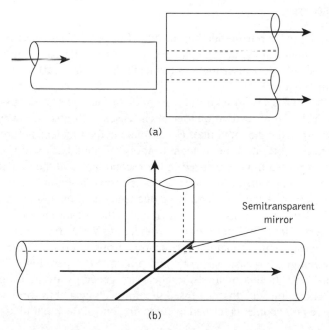

(a)

(b)

Figure 5.28 Fabrication techniques for three-port fiber couplers: (a) the lateral offset method; (b) the semitransparent mirror method

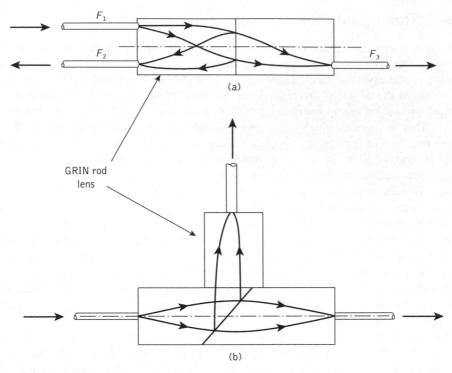

Figure 5.29 GRIN-rod lens micro-optic fiber couplers: (a) parallel surface type; (b) slant surface type

combined with spherical retro-reflecting mirrors [Ref. 88]. These devices, two of which are displayed in Figure 5.29, are miniature optical assemblies of compact construction which generally exhibit low insertion loss (typically less than 1 dB) and are insensitive to modal power distribution.

Figure 5.29(a) shows the structure of a parallel surface type of GRIN-rod lens three-port coupler which comprises two quarter pitch lenses with a semitransparent mirror in between. Light rays from the input fiber F_1 collimate in the first lens before they are incident on the mirror. A portion of the incident beam is reflected back and is coupled to fiber F_2, while the transmitted light is focused in the second lens and then coupled to fiber F_3. The slant surface version of the similar coupler is shown in Figure 5.29(b). The parallel surface type, however, is the most attractive due to its ease of fabrication, compactness, simplicity and relatively low insertion loss. Finally, the substitution of the mirror by an interference filter* offers application of these devices to WDM (see Section 5.6.3).

Perhaps the most common method for manufacturing couplers is the fused biconical taper (FBT) technique, the basic structure and principle of operation of which are illustrated in Figure 5.30. In this method the fibers are generally twisted together and then spot fused under tension such that the fused section is elongated to form a biconical taper structure. A three-port coupler is formed by removing one of the input fibers. Optical power

---

* Such a dichroic device transmits only a certain wavelength band and reflects all other shorter or longer wavelengths.

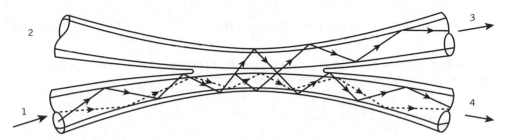

Figure 5.30 Structure and principle of operation for the fiber fused biconical taper coupler

launched into the input fiber propagates in the form of guided core modes. The higher order modes, however, leave the fiber core because of its reduced size in the tapered-down region and are therefore guided as cladding modes. These modes transfer back to guided core modes in the tapered-up region of the output fiber with an approximately even distribution between the two fibers.

Often only a portion of the total power is coupled between the two fibers because only the higher order modes take part in the process, the lower order modes generally remaining within the main fiber. In this case a mode-dependent (and therefore wavelength-dependent) coupling ratio is obtained. However, when the waist of the taper is made sufficiently narrow, then the entire mode volume can be encouraged to participate in the coupling process and a larger proportion of input power can be shared between the output fibers. This strategy gives an improvement in both the power and modal uniformity of the coupler.

The various loss parameters associated with four-port couplers may be written down with reference to Figure 5.30. Hence, the excess loss which is defined as the ratio of power input to power output is given by:

$$\text{Excess loss (four-port coupler)} = 10 \log_{10} \frac{P_1}{(P_3 + P_4)} \text{ (dB)} \tag{5.22}$$

The insertion loss, however, is generally defined as the loss obtained for a particular port-to-port optical path.* Therefore, considering Figure 5.32:

$$\text{Insertion loss (ports 1 to 4)} = 10 \log_{10} \frac{P_1}{P_4} \text{ (dB)} \tag{5.23}$$

The crosstalk which provides a measure of the directional isolation[†] achieved by the device is the ratio of the backscattered power received at the second input port to the input power which may be written as:

---

* It should be noted that there is some confusion in the literature between coupler insertion loss and excess loss. Insertion loss is sometimes referred to when the value quoted is actually the excess loss. However, the author has not noticed the opposite where excess loss is used in place of insertion loss.
† The directional isolation and the crosstalk associated with a coupler are the same value in decibels but the former parameter is normally given as a positive value whereas the latter is a negative value. Sometimes the directional isolation is referred to as the insertion loss between the two particular ports of the coupler which would be ports 1 to 2 in Figure 5.30.

$$\text{Crosstalk (four-port coupler)} = 10 \log_{10} \frac{P_2}{P_1} \text{ (dB)} \tag{5.24}$$

Finally, the splitting or coupling ratio indicates the percentage division of optical power between the output ports. Again referring to Figure 5.30:

$$\text{Split ratio} = \left[\frac{P_3}{(P_3 + P_4)} \right] \times 100\% \tag{5.25}$$

$$= \left[1 - \frac{P_4}{(P_3 + P_4)} \right] \times 100\% \tag{5.26}$$

Example 5.7

A four-port multimode fiber FBT coupler has 60 μW optical power launched into port 1. The measured output powers at ports 2, 3 and 4 are 0.004, 26.0 and 27.5 μW respectively. Determine the excess loss, the insertion losses between the input and output ports, the crosstalk and the split ratio for the device.

Solution: The excess loss for the coupler may be obtained from Eq. (5.22) where:

$$\text{Excess loss} = 10 \log_{10} \frac{P_1}{(P_3 + P_4)} = 10 \log_{10} \frac{60}{53.5}$$

$$= 0.5 \text{ dB}$$

The insertion loss is provided by Eq. (5.23) as:

$$\text{Insertion loss (ports 1 to 3)} = 10 \log_{10} \frac{P_1}{P_3} = 10 \log_{10} \frac{60}{26}$$

$$= 3.63 \text{ dB}$$

$$\text{Insertion loss (ports 1 to 4)} = 10 \log_{10} \frac{60}{27.5} = 3.39 \text{ dB}$$

Crosstalk is given by Eq. (5.24) where:

$$\text{Crosstalk} = 10 \log_{10} \frac{P_2}{P_1} = 10 \log_{10} \frac{0.004}{60}$$

$$= -41.8 \text{ dB}$$

Finally, the split ratio can be obtained from Eq. (5.25) as:

$$\text{Split ratio} = \left[\frac{P_3}{(P_3 + P_4)} \times 100 \right] = \frac{26}{53.5} \times 100$$

$$= 48.6\%$$

The split ratio for the FBT coupler is determined by the difference in the relative cross-sections of the fibers, and the mode coupling mechanism is observed in both multimode and single-mode fibers [Refs 87, 89]. An advantage of the FBT structure is its relatively low excess loss which is typically less than 0.5 dB,* with low crosstalk being usually better than −50 dB. A further advantage is the capability to fabricate FBT couplers with almost any fiber and geometry. Hence, they can be tailored to meet the specific requirements of a system or network. A major disadvantage, however, concerns the modal basis of the coupling action. The mode-dependent splitting can result in differing losses through the coupler, a wavelength-dependent performance, as well as the generation of modal noise when coherent light sources are employed [Ref. 90].

The precise spectral behavior of FBT couplers is quite complex. It depends upon the dimensions and the geometry of the fused cross-section, and on whether the fusing process produces a coupling region where the two cores are close (strongly fused) or relatively far apart (weakly fused) [Ref. 91]. It can also depend upon the refractive index of the surrounding medium [Ref. 92] and, in coherent systems, on the state of polarization of the optical field. Theoretical considerations [Ref. 91] show that for a single-mode FBT coupler, a minimum wavelength dependence on the splitting ratio is achieved for small cladding radii and strong fusing (i.e. the fiber cores placed close together). In order to obtain such performance it is necessary to taper the fibers down to a radius of around 15 μm or less, and to ensure that the rate of taper is such that the major proportion of the coupling occurs in the neck region. The wavelength-dependent behavior associated with single-mode FBT couplers follows an approximately sinusoidal pattern over the wavelength range 0.8 to 1.5 μm as a result of the single-mode coupling length between the two fibers [Ref. 93]. This mechanism has been used in the manufacture of WDM multiplexer/demultiplexer couplers (see Section 5.6.3).

Single-mode fiber couplers have also been fabricated from polarization-maintaining fiber (so-called hi-birefringence couplers) which preserve the polarization of the input signals (see Section 3.13.3). Moreover, using polarization-maintaining fiber, it is possible to fabricate polarization-sensitive couplers, which effectively function as polarizing beam splitters [Ref. 94].

An alternative technology to fiber joint couplers, micro-optic lensed devices or fused fiber couplers is the optical waveguide coupler. Corning has demonstrated [Ref. 95] the way in which such passive optical waveguide coupling components compatible with both multimode and single-mode fiber can be fabricated. The production involved two basic processes. Firstly, a mask of the desired branching function was deposited onto a glass substrate using a photolithographic process. The substrate was then subjected to a two-stage ion exchange [Ref. 96], which created virtually circular waveguides embedded within the surface of the substrate on which the mask was deposited. An example of a three-port integrated waveguide coupler fabricated using the above process is shown schematically in Figure 5.31. Multimode fibers are bonded to the structure using etched V-grooves. Excess losses were measured at 0.5 dB for the three-port coupler and at 0.8 dB for the 1 × 8 star coupler [Ref. 95]. Clearly, this type of waveguide coupler is attractive because

---

* Environmentally stable single-mode fused fiber couplers with excess losses less than 0.1 dB are commercially available.

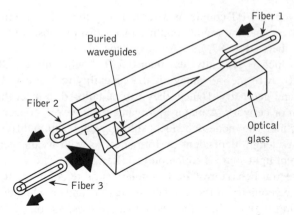

Figure 5.31 The Corning™ multimode fiber integrated waveguide three-port coupler [Ref. 94].

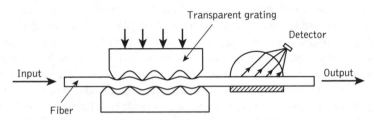

Figure 5.32 Schematic diagram of a microbend-type coupler

of the flexibility it allows at the masking stage. Furthermore, the same technique has been employed to fabricate WDM multiplexing demultiplexing devices (see Section 5.6.3).

Finally, directional couplers have been produced which use the mode coupling that takes place between the guided and radiation modes when a periodic deformation is applied to the fiber. The principle of operation for this microbend* type of coupler is illustrated in Figure 5.32 [Ref. 88]. Mode coupling between the guided and radiation modes may be obtained by pressing the fiber in close contact with a transparent mechanical grating. The radiated optical power can be collected by a lens or a shaped, curved glass plate. Interesting features of such devices are their variable coupling ratios which may be controlled over a wide range by altering the pressure on the fiber. In this context low light levels can be extracted from the fiber with very little excess loss (e.g. estimated at 0.05 dB [Ref. 97].

5.6.2 Star couplers

Star couplers distribute an optical signal from a single-input fiber to multiple-output fibers, as may be observed in Figure 5.27. The two principal manufacturing techniques for

---

* This coupler operates in a similar manner to a microbend optical fiber sensor.

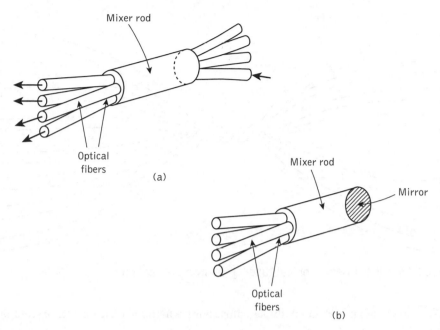

Figure 5.33 Fiber star couplers using the mixer-rod technique: (a) transmissive star coupler; (b) reflective star coupler

producing multimode fiber star couplers are the mixer-rod and the FBT methods. In the mixer-rod method illustrated in Figure 5.33 a thin platelet of glass is employed, which effectively mixes the light from one fiber, dividing it among the outgoing fibers. This method can be used to produce a transmissive star coupler or a reflective star coupler, as displayed in Figure 5.33. The typical insertion loss for an 8×8 mixer-rod transmissive star coupler with fiber pigtails is 12.5 dB with port-to-port uniformity of ±0.7 dB [Ref. 86].

The manufacturing process for the FBT star coupler is similar to that discussed in Section 5.6.1 for the three- and four-port FBT coupler. Thus the fibers which constitute the star coupler are bundled, twisted, heated and pulled, to form the device illustrated in Figure. 5.34. With multimode fiber this method relies upon the coupling of higher order modes between the different fibers. It is therefore highly mode dependent, which results in a relatively wide port-to-port output variation in comparison with star couplers based on the mixer-rod technique [Ref. 86].

In an ideal star coupler the optical power from any input fiber is evenly distributed among the output fibers. The total loss associated with the star coupler comprises its theoretical splitting loss together with the excess loss. The splitting loss is related to the number of output ports N following:

$$\text{Splitting loss (star coupler)} = 10 \log_{10} N \text{ (dB)} \tag{5.27}$$

It should be noted that for a reflective star coupler N is equal to the total number of ports (both input and output combined).

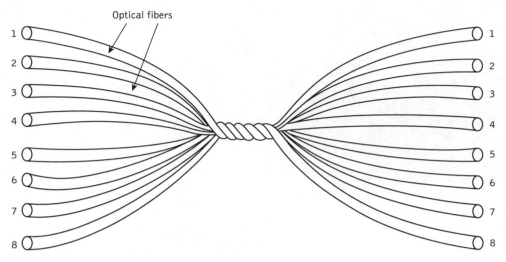

Figure 5.34 Fiber fused biconical taper 8 × 8 port star coupler

For a single input port and multiple output ports where $j = 1, N$, then the excess loss is given by:

$$\text{Excess loss (star coupler)} = 10 \log_{10}\left(P_i \bigg/ \sum_{1}^{N} P_j \right) \text{(dB)} \tag{5.28}$$

The insertion loss between any two ports on the star coupler may be obtained in a similar manner to the four-port coupler using Eq. (5.23). Similarly, the crosstalk between any two input ports is given by Eq. (5.24).

Example 5.8

A 32 × 32 port multimode fiber transmissive star coupler has 1 mW of optical power launched into a single input port. The average measured optical power at each output port is 14 μW. Calculate the total loss incurred by the star coupler and the average insertion loss through the device.

Solution: The total loss incurred by the star coupler comprises the splitting loss and the excess loss through the device. The splitting loss is given by Eq. (5.27) as.

$$\begin{aligned}
\text{Splitting loss} &= 10 \log_{10} N = 10 \log_{10} 32 \\
&= 15.05 \text{ dB}
\end{aligned}$$

The excess loss may be obtained from Eq. (5.28) where:

$$\text{Excess loss} = 10 \log_{10}\left(P_i \bigg/ \sum_{1}^{N} P_j \right) = 10 \log_{10}(10^3/32 \times 14) = 3.49 \text{ dB}$$

Hence the total loss for the star coupler:

$$\text{Total loss} = \text{splitting loss} + \text{excess loss} = 15.05 + 3.49$$
$$= 18.54 \text{ dB}$$

The average insertion loss from the input port to an output port is provided by Eq. (5.23) as:

$$\text{Insertion loss} = 10 \log_{10} \frac{10^3}{14} = 18.54 \text{ dB}$$

Therefore, as may have been anticipated, the total loss incurred by the star coupler is equivalent to the average insertion loss through the device. This result occurs because the total loss is the loss incurred on a single (average) optical path through the coupler which effectively defines the average insertion loss for the device.

An alternative strategy for the realization of a star coupler is to construct a ladder coupler, as illustrated in Figure 5.35. The ladder coupler generally comprises a number of cascaded stages, each incorporating three- or four-port FBT couplers in order to obtain a multiport output. Hence, the example shown in Figure 5.35 consists of three stages, which gives eight output ports. It must be noted, however, that when three-port couplers are used

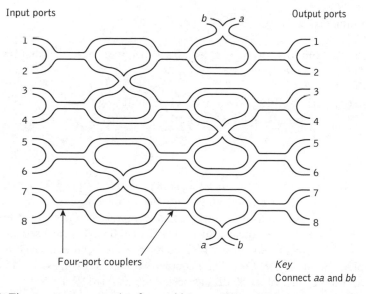

Figure 5.35 The 8 × 8 star coupler formed by cascading 12 four-port couplers (ladder coupler). This strategy is often used to produce low-loss single-mode fiber star or tree couplers

such devices do not form symmetrical star couplers* in that they provide a $1 \times N$ rather than $N \times N$ configuration. Nevertheless, the ladder coupler presents a useful device to achieve a multiport output with relatively low insertion loss. Furthermore, when four-port couplers are employed, then a true $N \times N$ star coupler may be obtained. It may be deduced from Figure 5.37 that the number of output ports N obtained with an M-stage ladder coupler is 2^M. These devices have found relatively widespread application for the production of single-mode fiber star couplers.

Example 5.9

A number of three-port single-mode fiber couplers are utilized in the fabrication of a tree (ladder) coupler with 16 output ports. The three-port couplers each have an excess loss of 0.2 dB with a split ratio of 50%. In addition, there is a splice loss of 0.1 dB at the interconnection of each stage. Determine the insertion loss associated with one optical path through the device.

Solution: The number of stages M within the ladder design is given by $2^M = 16$. Hence $M = 4$. Thus the excess loss through four stages of the coupler with three splices is:

$$\text{Excess loss} = (4 \times 0.2) + (3 \times 0.1) = 1.1 \text{ dB}$$

Assuming a 50% split ratio at each stage, the splitting loss for the coupler may be obtained using Eq. (5.27) as:

$$\text{Splitting loss} = 10 \log_{10} 16 = 12.04 \text{ dB}$$

Hence the insertion loss for the coupler which is equivalent to the total loss for one optical path though the device is:

$$\begin{aligned} \text{Insertion loss} &= \text{splitting loss} + \text{excess loss (four stages)} \\ &= 12.04 + 1.1 = 13.14 \text{ dB} \end{aligned}$$

Significantly lower excess losses than that indicated in Example 5.9 have been achieved with single-mode fiber ladder couplers. In particular, a mean excess loss of only 0.13 dB for an 8×8 star coupler constructed using this technique has been reported [Ref. 98]. Four-port FBT couplers with mean excess losses of 0.05 dB were used in this device. Alternatively, 3×3 single-mode fiber FBT couplers have been employed as a basis for ladder couplers. For example, a 9×9 star coupler with an excess loss of 1.46 dB and output port power uniformity of ± 1.50 dB has been demonstrated [Ref. 99].

---

* Such devices are sometimes referred to as tree couplers.

5.6.3 Wavelength division multiplexing couplers

It was indicated in Section 5.6 that WDM devices are a specialized coupler type which enable light from two or more optical sources of differing nominal peak optical wavelength to be launched in parallel into a single optical fiber. Hence such couplers perform as either wavelength multiplexers or wavelength demultiplexers (see Section 12.9.4). The spectral performance characteristic for a typical five-channel WDM device is shown in Figure 5.36. The important optical parameters associated with the WDM coupler are the attenuation of the light over a particular wavelength band, the interband isolation and the wavelength band or channel separation. Ideally, the device should have a low-loss transmission window for each wavelength band, giving a low insertion loss.* In addition, the device should exhibit high interband isolation, thus minimizing crosstalk. However, in practice, high inter-channel isolation is only required at the receiver (demultiplexer) end of the link or at both ends in a bidirectional system. Finally, the channel separation should be as small as may be permitted by light source availability and stability together with crosstalk considerations.

Numerous techniques have been developed for the implementation of WDM couplers. Passive devices, however, may be classified into three major categories [Ref. 101], two of which are core interaction types: namely, angularly dispersive (usually diffraction grating) and filter, and a surface interaction type which may be employed with single-mode fiber in the form of a directional coupler. Any other implementations tend to be hybrid combinations of the two core interaction types.

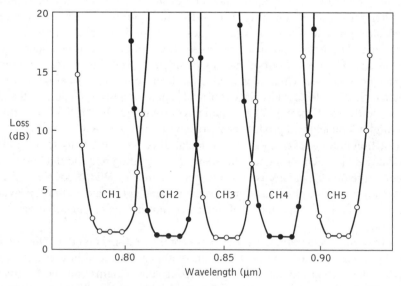

Figure 5.36 Typical flat passband spectral output characteristic for a WDM demultiplexer device (diffraction grating type). Reprinted with permission from Ref. 100 © IEEE 1980

* In the case of the WDM coupler the device loss is specified by the insertion loss associated with a particular wavelength band. The use of excess loss as in the case of other fiber couplers is inappropriate because the optical signals are separated into different wavelength bands.

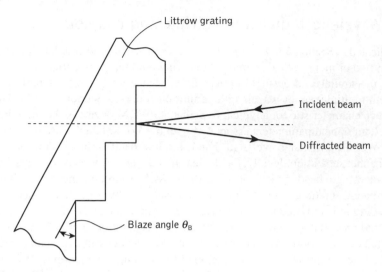

Figure 5.37 Littrow mounted diffraction grating

Although a glass prism may be utilized as an angularly dispersive element to facilitate wavelength multiplexing and demultiplexing, the principal angularly dispersive element used in this context is the diffraction grating. Any arrangement which is equivalent in its action to a number of parallel equidistant slits of the same width may be referred to as a diffraction grating. A common form of diffraction grating comprises an epoxy layer deposited on a glass substrate, on which lines are blazed. There are two main types of blazed grating. The first is produced by conventional mechanical techniques, while the other is fabricated by the anisotropic etching of single-crystal silicon [Ref. 100] and hence is called a silicon grating. The silicon grating has been found to be superior to the conventional mechanically ruled device, since it provides greater design freedom in the choice of blazing angle θ_B (see Figure 5.37) and grating constant (number of lines per unit length). It is also highly efficient and produces a more environmentally stable surface.

A diffraction grating reflects light in particular directions according to the grating constant, the angle at which the light is incident on the grating and the optical wavelength. Two main structural types are used in the manufacture of WDM couplers: the Littrow device, which employs a single lens and a separate plane grating; and the concave grating, which does not utilize a lens since both focusing and diffraction functions are performed by the grating.

In a Littrow mounted grating, the blaze angle of the grating is such that the incident and reflected light beams follow virtually the same path, as illustrated in Figure 5.37, thereby maximizing the grating efficiency and minimizing lens astigmatism. For a given center wavelength λ, the blaze angle is set such that [Ref. 102]:

$$\theta_B = \sin^{-1}\left(\frac{\lambda}{2\Lambda}\right) \tag{5.29}$$

where Λ is the line spacing on the grating (i.e. grating period). Schematic diagrams of Littrow-type grating demultiplexers employing a conventional lens [Ref. 100] and a

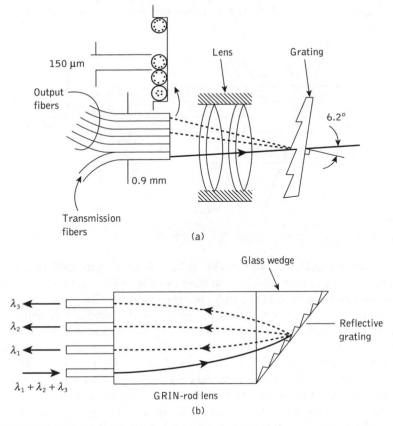

Figure 5.38 Littrow-type grating demultiplexers: (a) using a conventional lens. Reprinted with permission from Ref. 100 © IEEE 1980; (b) using a GRIN-rod lens [Ref. 103]

GRIN-rod lens [Ref. 103] are shown in Figure 5.38. The use of a spherical ball microlens has also been reported [Ref. 104]. Although all the lens-type devices exhibit similar operating mechanisms and hence performance, the GRIN-rod lens configuration proves advantageous for its compactness and ease of alignment. Therefore the operation of a GRIN-rod lens type of demultiplexer is considered in greater detail.

Referring to Figure 5.38(b), the single input fiber and multiple output fibers are arranged on the focal plane of the lens, which, for a quarter pitch GRIN-rod lens, is coincident with the fiber end face (see Section 5.5.1). The input wavelength multiplexed optical beam is collimated by the lens and hence transmitted to the diffraction grating, which is offset at the blaze angle so that the incoming light is incident virtually normal to the groove faces. The required offset angle can be produced by interposing a prism (glass wedge) between the lens and the grating, as illustrated in Figure 5.38(b) or, alternatively, by cutting and polishing the GRIN-rod lens and by mounting the grating on its end face. The former method gives superior performance since the optical properties of the GRIN-rod lens are not altered [Ref. 104]. On reflection from the grating, the diffraction process causes the light to be angularly dispersed according to the optical wavelength. Finally, the different

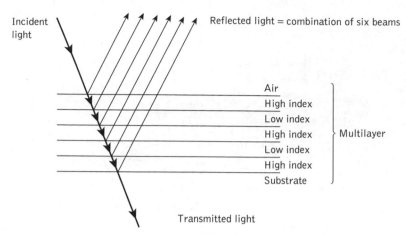

Figure 5.39 Multilayer interference filter structure

optical wavelengths pass through the lens and are focused onto the different collecting output fibers. Devices of this type have demonstrated channel insertion losses of less than 2 dB and channel spacings of 18 nm with low crosstalk [Ref. 105].

In addition, single-mode wavelength multiplexer and demultiplexer pairs based on a planar diffraction grating and a lithium niobate strip waveguide structure have also been reported [Ref. 106]. Six wavelength multiplexed channels were demonstrated, three over the wavelength region from 1275 to 1335 nm and three over the wavelength range from 1510 to 1570 nm. Crosstalk levels were less than −25 dB, with insertion losses for the multiplexer and demultiplexer of 5 to 8 dB and 1 to 2.2 dB respectively.

The other major core interaction types of WDM devices employ optical filter technology. Optical spectral filters fall into two main categories: namely, interference filters and absorption filters. Dielectric thin-film (DTF) interference filters can be constructed from alternate layers of high refractive index (e.g. zinc sulfide) and low refractive index (e.g. magnesium fluoride) materials, each of which is one-quarter wavelength thick [Ref. 107]. In this structure, shown schematically in Figure 5.39, light which is reflected within the high-index layers does not suffer any phase shift on reflection, while those optical beams reflected within the low-index layers undergo a phase shift of 180°. Thus the successive reflected beams recombine constructively at the filter front face, producing a high reflectance over a limited wavelength region which is dependent upon the ratio between the high and low refractive indices. Outside this high-reflectance region, the reflectance changes abruptly to a low value. Consequently, the quarter-wave stack can be used as a high-pass filter, a low-pass filter or a high-reflectance coating.

Absorption filters comprise a thin film of material (e.g. germanium) which exhibits an absorption edge at a specific wavelength. Absorption filters usually display very high rejection in the cutoff region. However, as their operation is dependent upon the fundamental optical properties of the material structure, they tend to be inflexible because the edge positions are fixed. Nevertheless, by fabricating interference filters onto an absorption layer substrate, a filter can be obtained which combines the sharp rejection of the absorption filter together with the flexibility of the interference filter. Such combined structures can be used as high-performance edge filters.

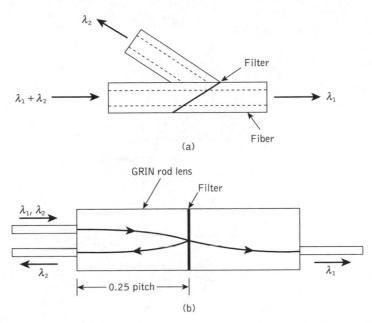

Figure 5.40 Two wavelength interference filter demultiplexers: (a) fiber end device; (b) GRIN-rod lens device

Specific filter WDM coupler designs are now considered in further detail. Firstly, edge filters are generally used in devices which require the separation of two wavelengths (generally reasonably widely separated by 10% or more of median wavelength). A configuration which has been adopted [Ref. 108] is one in which the fiber is cleaved at a specific angle and then an edge filter is interposed between the two fiber ends, as illustrated in Figure 5.40(a). In a demultiplexing structure light at one wavelength is reflected by the filter and collected by a suitably positioned receive fiber, while the other optical wavelength is transmitted through the filter and then propagates down the cleaved fiber. Such a device, which has been tested with LED sources emitting at center wavelengths of 755 and 825 nm, exhibited insertion losses of 2 to 3 dB with crosstalk levels less than −60 dB [Ref. 108]. An alternative two-wavelength WDM device employing a cascaded BPF sandwiched between two GRIN-rod lenses is shown in Figure 5.40(b). A practical two-channel (operating at wavelengths of 1.2 and 1.3 μm) multiplex/demultiplex system which is capable of operation in both directions using this WDM design has been reported [Ref. 107] to exhibit low insertion losses of around 1.5 dB with crosstalk levels less than −58 dB. This device also displayed acceptable environmental stability with insertion loss variations of less than 0.3 dB throughout a range of tests (i.e. vibration, temperature cycling and damp/heat tests).

Multiple wavelength multiplexer/demultiplexer devices employing DTF interference filters may be constructed from a suitably aligned series of bandpass filters with different passband wavelength regions, cascaded in such a way that each filter transmits a particular wavelength, but reflects all others. Such a multiple-reflection demultiplexing device is illustrated in Figure 5.41. This structure has the disadvantage that the insertion losses increase linearly with the number of multiplexed channels since losses are incurred at each

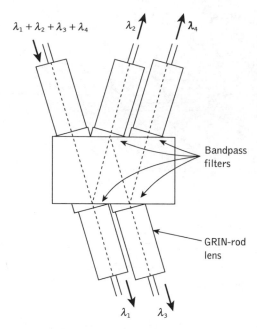

Figure 5.41 GRIN-rod lensed bandpass demultiplexer

successive reflection due to filter imperfections and the difficulties of maintaining good alignment [Ref. 101].

A two-channel slab waveguide version of a filter WDM device has been introduced by Corning, which is based on the same technology as its optical waveguide coupler (see Section 5.6.1). The wavelength separation is accomplished within the waveguide using a dichroic filter which intersects the path of the incoming light beam. Longer wavelengths are transmitted and shorter wavelengths reflected. The multiplexer/demultiplexer device reported [Ref. 109] is compatible with both 50/125 μm and 85/125 μm graded index fibers. It combines/separates optical wavelength regions between 0.8 to 0.9 μm and 1.2 to 1.4 μm with an insertion loss lower than 1.5 dB and crosstalk levels less than −25 dB.

The wavelength-dependent characteristics of single-mode fiber directional couplers were mentioned in Section 5.6.1. Both single-mode ground fiber and FBT fiber couplers can be fabricated to provide the complete transfer of optical power between the two fibers. However, since the optical power coupling characteristic of such single-mode fiber couplers is highly wavelength dependent, they can be used to fabricate WDM devices.

Optical power transfer within multimode fiber couplers is a mode-dependent phenomenon which, in general, takes place between the higher order modes propagating in the outer reaches of the fiber cores as well as in the cladding regions. These higher order modes couple more freely when the fibers are in close proximity. Consequently, the spectral dependence of light transfer within multimode fiber couplers is far less pronounced and predictable than that exhibited by single-mode fiber structures. Therefore, multimode fiber WDM devices cannot readily be fabricated using the FBT or ground fiber techniques.

Optical power is coupled between two single-mode fibers by bringing the fiber cores close together over a region known as the interaction length. One or two methods are generally

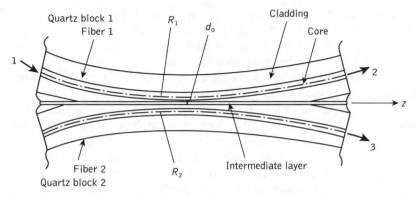

Figure 5.42 Schematic diagram showing ground (polished) single-mode fiber coupler

used to perform this function. The first technique [Ref. 110] necessitates bending and fixing the two fibers into two blocks of suitable material (e.g. quartz) prior to grinding the two blocks down so that a proportion of each of the fiber cladding regions is worn away. Finally, the two blocks are brought together, as shown in Figure 5.42.

Parallel single-mode fiber waveguides exchange energy with a spatial period (coupling length) $L = 2\pi/k$, where k is the coupling coefficient (units of inverse length) for the two interacting waveguide modes [Ref. 111]. This result can be extended to curved regions where spacing between the waveguides over the interaction length is no longer fixed [Ref. 112]. Thus, for a pair of fibers curved against each other, the coupling coefficient k is a nonlinear function of the interaction length (which in turn is proportional to the square root of the radius of curvature R), the minimum spacing between the fiber cores, the refractive index of the intervening material, the fiber parameters and the wavelength of the light. The wavelength-dependent properties of a single mode ground fiber coupler can therefore be altered by adjusting several different parameters.

An early demonstration of such a two-channel ground fiber directional coupler was made from two identical single-mode fibers of 2 μm core diameter [Ref. 113]. This device, with a radius of curvature $R_1 = R_2 = 70$ cm and a minimum core separation of 4.5 μm, gave a measured coupling ratio which followed the typical sinusoidal pattern, with approximately two periods over the 0.45 to 0.9 μm wavelength region. By offsetting the cores laterally (i.e. in the direction z indicated in Figure 5.42), and effectively altering d_0, the spectral characteristics of the coupler were altered. The sinusoidal response curve was shifted by around 400 nm with a lateral offset of 5 μm. Interchannel wavelength spacings (wavelength separation between minimum and maximum on the sinusoidal pattern) were about 140 nm for this structure, which exhibited insertion losses as low as 0.1 dB using suitable index matching.

Wavelength-selective ground fiber directional couplers constructed from single-mode fibers of different core diameters and refractive indices exhibit propagation constants which are matched at only one wavelength and hence can be used to produce true bandpass filters. The center wavelength and spectral bandwidth of these couplers are essentially determined by the fiber parameters [Ref. 114]. Measured insertion losses for such devices fabricated for operation in the longer wavelength region were between 0.5 and 0.6 dB with crosstalk levels less than −22 dB [Ref. 110].

The second method of fabricating a single-mode fiber WDM coupler is the FBT technique [Refs 87, 89]. Carefully fabricated fused couplers display very low insertion losses and provide a high degree of environmental stability. The manufacturing process requires the single-mode fibers to be fused together at around 1500 °C before being pulled while heat is still applied. The pulling process decreases the fiber core size causing the evanescent field of the transmitted optical signal to spread out further from the fiber core, which enables light to couple into the adjacent fiber. In practice this manufacturing process necessitates the monitoring of the optical power output from the two fibers, the process being halted when the required coupling ratio is reached [Ref. 115].

In common with the ground fiber coupler constructed by using similar fibers, the optical power transferred between the two fibers (or the coupling ratio) in an FBT coupler as a function of wavelength is sinusoidal with a period dependent on the dimensions and the geometry of the fused cross-section, and on the refractive index of the surrounding medium [Ref. 92]. It can also depend upon whether the fusing process produces a coupling region where the two fiber cores are close (strongly fused) or relatively far apart (weakly fused) [Ref. 91]. The most popular method of varying the periodic coupling function in such fused WDM couplers is to extend the interaction length by continuing the stretching process during fusing. An increase in the interaction length has the effect of increasing the coupling ratio period. Such two-channel devices have displayed insertion losses of 0.25 and 0.37 dB with crosstalk levels less than −22 dB [Ref. 116]. It should be noted, however, that a limitation with these WDM couplers is that they are not well suited for the provision of closely spaced or multiple channels.

More recently, fiber Bragg grating (FBG) assisted devices to couple optical signals into fiber have been demonstrated [Refs 117, 118]. Such an approach reduces the need for optical filters and lenses when coupling a signal from one fiber into another fiber. The FBG operates by facilitating reflections where an optical signal at the Bragg wavelength propagating through alternating regions of different refractive indices has a portion of signal power reflected back at each interface between the regions. If the spacing between regions is such that all the partial reflections are constructively in phase then the total reflection can grow to nearly 100%. Figure 5.43 illustrates this situation showing a fiber core comprising four FBG sections transmitting and reflecting an optical signal. The output signal is therefore transmitted through these Bragg gratings while the reflected signal due to the back reflections from each grating appears at the input to the fiber core. To form an FBG the variations in refractive index can be incorporated by exposing the core of the fiber to an intense ultraviolet optical interference pattern that has a periodicity equal to the periodicity of the grating to be formed. This process of altering the refractive index of the core

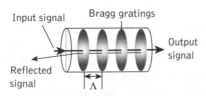

Figure 5.43 Schematic diagram of an optical fiber core containing four fiber Bragg gratings

through exposure to high-intensity radiation is referred to as photosensitivity [Ref. 119]. The reflections are dependent on the Bragg wavelength, λ_B, given by:

$$\lambda_B = 2n\Lambda \tag{5.30}$$

where n is the refractive index of the material and Λ is the grating period.

Example 5.10

An FBG is developed within a fiber core which has a refractive index of 1.46. Find the grating period for it to reflect an optical signal with a wavelength of 1.55 μm.

Solution: The grating period of the FBG can be obtained by rearranging Eq. (5.30) as:

$$\Lambda = \frac{\lambda_B}{2n} = \frac{1.55 \times 10^{-6}}{2 \times 1.46} = 0.53 \text{ μm}$$

The grating period of the FBG is therefore 0.53 μm in order to reflect an optical signal at a wavelength of 1.55 μm.

Equation (5.30) implies that any variation in refractive index of the material or the grating period produces a different Bragg wavelength, and therefore it is possible to construct FBGs capable of reflecting back or transmitting through an optical signal at any desired wavelength. When there is a uniform period between all the Bragg gratings then the FBG reflects an optical signal at a particular wavelength. However, when the period between each Bragg grating is linearly varied along the length of the fiber core, then the FGB is referred to as being chirped with each grating element reflecting a different optical wavelength. Although in an ideal case the optical signal power that is not transmitted through the FBG should be reflected within the fiber core, depending upon the angle of reflection (see Section 2.2), it can be reflected into the cladding. It should also be noted that the refractive index of an optical fiber can vary with changes in temperature and therefore the spectral response of FBGs is also temperature dependent. The variation in spectral response, however, remains within the range of ±50 pm of the Bragg wavelength when operating over a temperature range of 0 to 65 °C [Ref. 119].

FBGs are also useful devices to perform wavelength division multiplexing (WDM) where optical signals at desired wavelengths can be multiplexed or demultiplexed using gratings combined with simple optical couplers. For example, a grating-assisted three-port optical fiber coupler which functions as an add-multiplexer for WDM transmission is displayed in Figure 5.44. The coupler comprises two active parallel waveguides of InGaAs material separated by 1 μm which are buried in an indium phosphide medium. Core and cladding refractive indices for the waveguides are 3.60 and 3.41 respectively, while the lower waveguide shown in Figure 5.44 incorporates a unidirectional complex FBG structure. It may be noted that the device operates without any interception of through-traffic WDM channels when another signal at wavelength λ_4 is combined with the three transmission channels λ_1, λ_2 and λ_3. Finally, the device shown in Figure 5.44, which has a

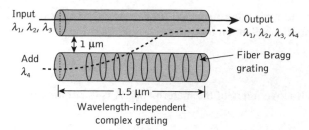

Figure 5.44 Grating-assisted three-port coupler add-multiplexer showing the operation of adding a further wavelength channel to a WDM signal. Reprinted with permission from Ref. 118 © IEEE 2005

length of just 1.5 μm, exhibits a 3 dB loss when operating over the wavelength range from 1.5 to 1.6 μm [Ref. 118].

Another category of passive optical multiplexer and demultiplexer coupler using the diffraction grating mechanism is the arrayed waveguide grating (AWG) [Refs 120–124]. These devices can potentially replace FGB-based devices which support only a limited number of wavelengths as, in particular, the several times smaller size AWG can perform multiplexing/demultiplexing functions in dense WDM networks with narrow channel spacing. An AWG essentially comprises a number of waveguides with different lengths (i.e. a waveguide array) converging at the same point(s). Optical signals passing through each of these waveguides interfere with the signals passing through their neighbouring waveguides at the convergence points. Depending upon the phase difference of interfering signals (i.e. constructive or destructive) an optical signal at a desired wavelength can be obtained at the device output. The AWG can therefore be used as a wavelength selective filter or a wavelength switch thus providing an add/drop multiplexer function in optical networks (see Section 15.2.2).

An AWG primarily comprises five elements as illustrated in Figure 5.45(a) which include input/output waveguides, arrayed waveguides of different lengths (i.e. shown by ΔL for topmost waveguide channel of order M) and two focusing slab waveguides, each one to split and combine at the input and the output ports of the device, respectively. The basic operation of wavelength multiplexing/demultiplexing is carried out in these two focusing slab waveguides each of which act as multimode interference coupler or a free space propagation region. When a wavelength division multiplexed signal is coupled to the input waveguide the multiplexed signal propagates through the input waveguide slab region where it illuminates the grating by splitting the optical signal into each arrayed waveguide (often more than 64 waveguides) with a Gaussian distribution. Although curved arrayed waveguides are preferred to produce waveguide channels over a suitable distance, dispersive elements can also be incorporated into the arrayed waveguide structure to modify the refractive index and hence propagation time thus avoiding the need for the curved shape of the arrayed waveguide structure [Ref. 121]. The optical signals then travel down this waveguide array to the other waveguide slab. Since each arrayed waveguide exhibits a different path length then the optical wavefronts reach the input ports of second slab out-of-phase with one another.

As the output waveguide slab performs as a combiner the overall AWG therefore becomes a wavelength demultiplexer. Each output signal from arrayed waveguide interferes

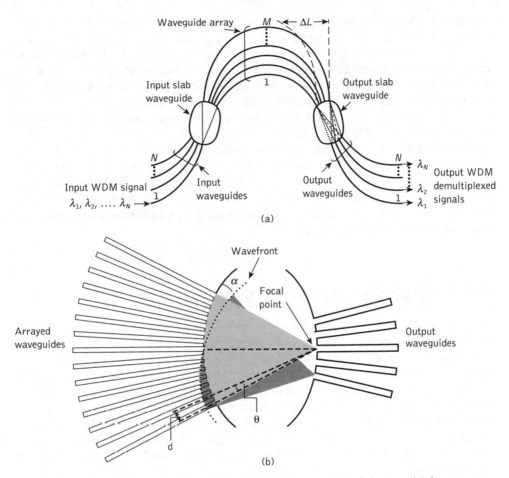

Figure 5.45 Arrayed waveguide grating: (a) basic operation of device; (b) free space propagation through output waveguide slab

with all the others within the second slab waveguide. As a consequence of constructive interference each single wavelength signal present in the original WDM signal will be coupled into exactly one of the output waveguides as illustrated in Figure 5.45(b). The optical signal propagating at the central wavelength λ_c output from the array converges in the output waveguide slab and is focused into the central output channel in the image plane. If the wavelength is shifted to $\lambda_c + \Delta\lambda$, there will be a phase change in the individual waveguides that increases linearly from the lower to the upper channel. As a result the phase front at the output aperture of the array will be slightly tilted as identified by angle α in Figure 5.45(b) so the beam is focused on a different position in the image plane (e.g. the last output waveguide). The angle θ in Figure 5.45(b) describes the divergence angle between any two array channels in the array aperture that is obtained by moving distance d equivalent to the distance between two channels subtended by the focal length of the array (i.e. distance to the focal point) as shown in Figure 5.45(b).

It should be noted that due to the optically passive nature of the device, the AWG operates as a multiplexer when operating in the opposite direction. The design of an AWG relies mainly on determining the geometry of the arrayed and slab waveguides in order to set the correct path length differences and conditions for wavelength-selective constructive interference. Several combinations of input/output ports for an AWG (i.e. $1 \times N$, $N \times M$ where N and M are the positive integers) and concatenations of two AWGs or their use with a Mach–Zehnder interferometer can be arranged to achieve different WDM functions [Refs 120, 125, 126]. Channel spacings of 100, 50, 25 GHz are common in commercial AWG devices while a narrower channel spacing of 6.25 GHz has also been achieved enabling the transmission of 1024 WDM channels (i.e. 8×128) with each channel operating at transmission rate of 2.67 Gbit s^{-1} [Ref. 124]. In this case the operating wavelength range covered both the C- and the L-bands (i.e. from 1.53 to 1.6 μm). Moreover, the optical input signal power per channel was −15 dBm with incurred an adjacent-channel crosstalk of −21 dB.

5.7 Optical isolators and circulators

An optical isolator is essentially a passive device which allows the flow of optical signal power (for a particular wavelength or a wavelength band) in only one direction preventing reflections in the backward direction. Ideally, an optical isolator should transmit all the signal power in the desired forward direction. Material imperfections in the isolator medium, however, do generate backward reflections. Additionally, both the insertion loss and isolation determine the limitations for the device to transmit optical power from one terminal to another. Figure 5.46(a) illustrates the basic function of an isolator where an incident optical signal is shown to be transmitted through the device and then it appears at the output terminal. Furthermore, a small amount of optical signal power is reflected back to the input port.

Optical isolators can be implemented by using FBGs. These devices permit the optical signal to pass through the isolator and propagate to the output terminal, or, alternatively, they reflect it backwards. Since FBGs are wavelength dependent then optical isolators can be designed to allow or block the optical signal at a particular (or a range of) wavelength(s). Furthermore, the wavelength blocking feature makes the optical isolator a very attractive device for use with optical amplifiers in order to protect them from backward reflections. In addition, magneto-optic devices can be used to function as isolators [Ref. 127]. Magneto-optic devices utilize the principle of Faraday rotation which relates the TM mode characteristic and polarization state of an optical signal with its direction of propagation, according to which the rotation of the plane of polarization is proportional to the intensity of the component of the magnetic field in the direction of the optical signal. Therefore it is possible to block or divert an optical signal as desired using the magneto-optic properties of the material [Ref. 128]. Magnetic oxide materials can be used in the fabrication of optical waveguides to construct optical isolators, in particular by using photonic crystal waveguides (see Section 2.6) [Ref. 129]. Furthermore, it is also possible to develop optical waveguide isolators using either the TE or TM modes for the propagation of an optical signal.

More recently, the use of semiconductor optical amplifiers (SOAs) (see Section 10.3) to construct such optical waveguide isolators based on either TE or TM modes has also been demonstrated [Refs 130, 131]. In this approach an SOA incorporating a ferromagnetic metal contact very close to the active region functioned as an optical waveguide isolator. Based on the TM mode and using InGaAlAs deposited onto an indium phosphide substrate, the SOA exhibited an optical isolation ratio (i.e. an optical signal power ratio between forward and backward directions) of 11.4 dB when operating at wavelength of 1.30 µm [Ref. 130]. Another device using an SOA fabricated from InGaAsP on an indium phosphide substrate operating on the TE mode demonstrated an optical isolation ratio of 14.7 dB at a signal wavelength of 1.55 µm [Ref. 131]. The main advantage of this type of isolator, however, is its ability to facilitate monolithic integration with other photonic integrated devices (see Section 11.5).

Isolators can also be connected together to form multiport devices where, depending upon their isolation characteristics, an optical signal can leave the device at an end terminal or it can continue to flow towards the next connected isolator. The resulting device is generally known as a circulator, taking its name from the path of the optical signal which follows a closed loop or a circle. Such a device is shown in Figure 5.46(b) where three isolators are interconnected to form a three-port device which does not discard the backward reflections but directs them to another isolator. Therefore the signal continues to travel from isolator 1 to isolator 2 and finally it terminates at the end terminal of isolator 3. In order to prevent the signal going back to the input port 1, no connection is usually permitted between port 3 and port 1. When a signal is transmitted from port 1 to port 2, however, the device simultaneously allows another optical signal to travel from port 2 to port 3.

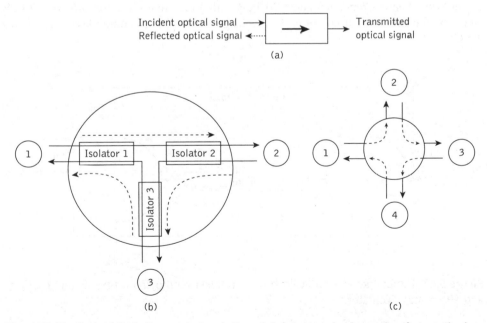

Figure 5.46 Optical isolation and circulation: (a) functional schematic of an optical fiber isolator; (b) three-port optical circulator; (c) four-port optical circulator

A four-port optical circulator which operates in a similar manner to the three-port device but incorporates an additional isolator is displayed in Figure 5.46(c). Although it is also possible to produce a circulator with a larger number of ports, the device complexity increases with increasing number of ports and therefore in practice only three- or four-port circulators have proved useful for optical interconnection [Ref. 132]. Commercially available optical circulators exhibit insertion losses around 1 dB and high isolation in the range of 40 to 50 dB centered at signal wavelengths of 1.3 and 1.5 μm [Ref. 133].

Two optical circulators may be incorporated with an FBG where the latter device is used to enable a specific wavelength channel to exit at a particular terminal or to allow it to continue to flow to the next terminal. Such combined devices can therefore be employed to perform an all-optical add/drop wavelength multiplexer function where, for instance, an optical circulator is incorporated at both the input and the output ports as shown in Figure 5.47 [Refs 134, 135]. It may be observed that the FBG is placed in between the two circulators to transmit the selected wavelength channels from λ_2 to λ_N while reflecting back the optical channel at λ_1. Since the signal at wavelength λ_1 is removed/dropped at circulator 1, then another optical signal at this wavelength may be added at optical circulator 2 as indicated in Figure 5.47.

A combination of an FBG and optical circulators can also be used to produce non blocking $N \times M$ optical wavelength division add/drop multiplexers where N and M represent the number of wavelength channels in a wavelength multiplexed signal and the intended add/drop channels, respectively. For example, such a 4×4 FBG-based optical wavelength multiplexer has facilitated the add/drop function for four wavelength channels simultaneously, operating over a range of wavelengths from 1548.8 to 1551.2 nm with a channel spacing of 100 GHz [Ref. 134]. In this case the maximum crosstalk level between optical wavelength channels remained under 20.4 dB with a maximum insertion loss of 2.14 dB when the transmission bit rate of each wavelength channel in the multiplexed signal was 2.5 Gbit s^{-1}.

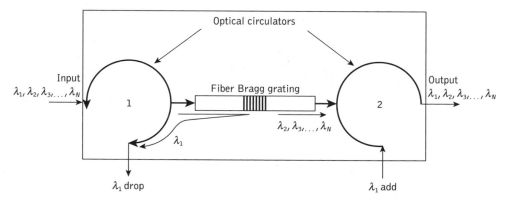

Figure 5.47 Optical add/drop wavelength multiplexer employing a fiber Bragg grating and all-optical circulators

Problems

5.1   State the two major categories of fiber–fiber joint, indicating the differences between them. Briefly discuss the problem of Fresnel reflection at all types of optical fiber joint, and indicate how it may be avoided.

A silica multimode step index fiber has a core refractive index of 1.46. Determine the optical loss in decibels due to Fresnel reflection at a fiber joint with:
(a) a small air gap;
(b) an index-matching epoxy which has a refractive index of 1.40.
It may be assumed that the fiber axes and end faces are perfectly aligned at the joint.

5.2   The Fresnel reflection at a butt joint with an air gap in a multimode step index fiber is 0.46 dB. Determine the refractive index of the fiber core.

5.3   Describe the three types of fiber misalignment which may contribute to insertion loss at an optical fiber joint.

A step index fiber with a 200 μm core diameter is butt jointed. The joint which is index matched has a lateral offset of 10 μm but no longitudinal or angular misalignment. Using two methods, estimate the insertion loss at the joint assuming the uniform illumination of all guided modes.

5.4   A graded index fiber has a characteristic refractive index profile (α) of 1.85 and a core diameter of 60 μm. Estimate the insertion loss due to a 5 μm lateral offset at an index-matched fiber joint assuming the uniform illumination of all guided modes.

5.5   A graded index fiber with a parabolic refractive index profile ($\alpha = 2$) has a core diameter of 40 μm. Determine the difference in the estimated insertion losses at an index-matched fiber joint with a lateral offset of 1 μm (no longitudinal or angular misalignment). When performing the calculation assume (a) the uniform illumination of only the guided modes and (b) the uniform illumination of both guided and leaky modes.

5.6   A graded index fiber with a 50 μm core diameter has a characteristic refractive index profile (α) of 2.25. The fiber is jointed with index matching and the connection exhibits an optical loss of 0.62 dB. This is found to be solely due to a lateral offset of the fiber ends. Estimate the magnitude of the lateral offset assuming the uniform illumination of all guided modes in the fiber core.

5.7   A step index fiber has a core refractive index of 1.47, a relative refractive index difference of 2% and a core diameter of 80 μm. The fiber is jointed with a lateral offset of 2 μm, an angular misalignment of the core axes of 3° and a small air gap (no longitudinal misalignment). Estimate the total insertion loss at the joint which may be assumed to comprise the sum of the misalignment losses.

5.8   Briefly outline the factors which cause intrinsic losses of fiber–fiber joints.
(a) Plot the loss resulting from a mismatch in multimode fiber core diameters or numerical apertures over a mismatch range 0 to 50%.

(b) An optical source is packaged with a fiber pigtail comprising 62.5/125 μm graded index fiber with a numerical aperture of 0.28 and a profile parameter of 2.1. The fiber pigtail is spliced to a main transmission fiber which is 50/125 μm graded index fiber with a numerical aperture of 0.22 and a profile parameter of 1.9. When the fiber axes are aligned without a gap, radial or angular misalignment, calculate the insertion loss at the splice.

5.9 Describe what is meant by the fusion splicing of optical fibers. Discuss the advantages and drawbacks of this jointing technique.

A multimode step index fiber with a core refractive index of 1.52 is fusion spliced. The splice exhibits an insertion loss of 0.8 dB. This insertion loss is found to be entirely due to the angular misalignment of the fiber core axes which is 7°. Determine the numerical aperture of the fiber.

5.10 Describe, with the aid of suitable diagrams, three common techniques used for the mechanical splicing of optical fibers.

A mechanical splice in a multimode step index fiber has a lateral offset of 16% of the fiber core radius. The fiber core has a refractive index of 1.49, and an index-matching fluid with a refractive index of 1.45 is inserted in the splice between the butt-jointed fiber ends. Assuming no longitudinal or angular misalignment, estimate the insertion loss of the splice.

5.11 Discuss the principles of operation of the two major categories of demountable optical fiber connector. Describe in detail a common technique for achieving a butt-jointed fiber connector.

A butt-jointed fiber connector used on a multimode step index fiber with a core refractive index of 1.42 and a relative refractive index difference of 1% has an angular misalignment of 9°. There is no longitudinal or lateral misalignment but there is a small air gap between the fibers in the connector. Estimate the insertion loss of the connector.

5.12 Briefly describe the types of demountable connector that may be used with single-mode fibers. Further, indicate the problems involved with the connection of single-mode fibers.

A single-mode fiber connector is used with a silica (refractive index 1.46) step index fiber of 6 μm core diameter which has a normalized frequency of 2.2 and a numerical aperture of 0.9. The connector has a lateral offset of 0.7 μm and an angular misalignment of 0.8°. Estimate the total insertion loss of the connector assuming that the joint is index matched and that there is no longitudinal misalignment.

5.13 A single-mode fiber of 10 μm core diameter has a normalized frequency of 2.0. A fusion splice at a point along its length exhibits an insertion loss of 0.15 dB. Assuming only lateral misalignment contributes to the splice insertion loss, estimate the magnitude of the lateral misalignment.

5.14 A single-mode step index fiber of 5 μm core diameter has a normalized frequency of 1.7, a core refractive index of 1.48 and a numerical aperture of 0.14. The loss in decibels due to angular misalignment at a fusion splice with a lateral offset of

0.4 μm is twice that due to the lateral offset. Estimate the magnitude in degrees of the angular misalignment.

5.15 Given the following parameters for a single-mode step index fiber with a fusion splice, estimate (a) the fiber core diameter and (b) the numerical aperture for the fiber:

Fiber normalized frequency = 1.9
Fiber core refractive index = 1.46
Splice lateral offset = 0.5 μm
Splice lateral offset loss = 0.05 dB
Splice angular misalignment = 0.3°
Splice angular misalignment loss = 0.04 dB

5.16 Two single-mode fibers have mode-field diameters of 9 μm and 11 μm. Assuming that there are no extrinsic losses, calculate the coupling loss between the fibers as a result of the mode-field diameter mismatch. Comment on the result in relation to the direction of transmission of the optical signal between the two fibers.

Determine the loss if the mode-field diameter mismatch between the fibers is increased to 30%.

5.17 With the aid of simple sketches outline the major categories of multiport optical fiber coupler.

Describe two common methods used in the fabrication of three- and four-port fiber couplers.

5.18 A four-port FBT coupler is shown in Figure 5.32. In addition a section of a tapered multimode step index fiber from such a coupler may be observed in Figure 5.48. A meridional ray propagating along the taper (characterized by the taper angle γ) is shown to undergo an increase in its propagation angle (i.e. the angle formed with the fiber axis). However, as long as the angle of incidence remains larger than the critical angle, then the ray is still guided and it emerges from the taper region forming an angle θ_o with the fiber axis. When the taper is smooth and the number of reflections is high, then, in Figure 5.49, $\sin \theta_o = R_1/R_2 \sin \theta_i$, where R_1 and R_2 are the core radii before and after the taper respectively. Show that the numerical aperture for the tapered fiber NA_T is given by:

$$NA_T = \frac{R_2}{R_1} (n_1^2 - n_2^2)^{\frac{1}{2}}$$

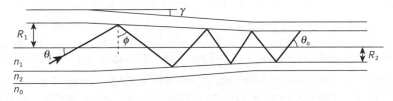

Figure 5.48 Section of a tapered multimode step index fiber for Problem 5.18

where n_1 and n_2 are the refractive indices of the fiber core and cladding respectively. Comment on this result when considering the modes of the light launched into the coupler.

5.19   The measured optical output powers from ports 3 and 4 of a multimode fiber FBT coupler are 47.0 μW and 52.0 μW respectively. If the excess loss specified for the device is 0.7 dB, calculate the amount of optical power that is launched into port 1 in order to obtain these output power levels. Hence, determine the insertion losses between the input and two output ports, as well as the split ratio for the device.

When the specified crosstalk for the coupler is −45 dB, calculate the optical output power level that would be measured at port 2 when the above input power level is maintained.

5.20   Indicate the distinction between fiber star and tree couplers.

Discuss the major techniques used in the fabrication of multimode fiber star couplers and describe how this differs from the strategy that tends to be adopted to produce single-mode fiber star couplers.

5.21   A 64 × 64 port transmissive star coupler has 1.6 mW of optical power launched into a single input port. If the device exhibits an excess loss of 3.90 dB, determine the total loss through the device and the average optical power level that would be expected at each output port.

5.22   An 8 × 8 port multimode fiber reflective star coupler has −8.0 dBm of optical power launched into a single port. The average measured optical power at each output port is −22.8 dBm. Obtain the excess loss for the device and hence the total loss experienced by an optical signal in transmission through the coupler. Check the result.

5.23   A number of four-port single-mode fiber couplers are employed in the fabrication of a 32 × 32 port star coupler. Each four-port coupler has a split ratio of 50% and when an optical input power level of −6 dBm is launched into port 1, the output power level from port 3 is found to be 122 μW. Furthermore, there is a splice loss of 0.06 dB at the interconnection of each stage within the ladder design. Calculate the optical power emitted from each of the output ports when the −6 dBm power level is launched into any one of the input ports. Check the result.

5.24   Outline the three major categories of passive wavelength division multiplexing coupler. Describe in detail one implementation of each category. Comment on the relative merits and drawbacks associated with each of the WDM devices you have described.

5.25   Describe the structure of the fiber Bragg grating assisted coupler and explain how it can effectively block a specific optical signal at a particular wavelength.

5.26   A fiber Bragg grating assisted coupler is designed to block an incoming optical signal present at the input port of the device. When the fiber core refractive index is 1.6 and the grating period is 0.42 μm, determine the wavelength of the blocked signal.

5.27   Explain the operation of both optical isolators and optical circulators. Discuss the use of these devices in wavelength division multiplexing systems as three- and four-ports devices.

Answers to numerical problems

5.1	(a) 0.31 dB; (b) 3.8×10^{-4} dB
5.2	1.59
5.3	0.29 dB
5.4	0.67 dB
5.5	(a) 0.19 dB; (b) 0.17 dB; difference 0.02 dB
5.6	4.0 µm
5.7	0.71 dB
5.8	4.25 dB
5.9	0.35
5.10	0.47 dB

5.11	1.51 dB
5.12	0.54 dB
5.13	1.2 µm
5.14	0.65°
5.15	(a) 7.0 µm; (b) 0.10
5.16	0.17 dB, 0.54 dB
5.19	116.3 µm, 3.93 dB, 3.50 dB, 47.5%, 3.7 nW
5.21	21.96 dB, 10.18 µW
5.22	5.77 dB, 14.80 dB
5.23	6.40 µW
5.26	1.34 µm

References

[1] E. Desurvire, 'Capacity demand and technology challenges for lightwave systems in the next two decades', *J. Lightwave Technol.*, **24**(12), pp. 4697–4710, 2006.

[2] ITU-T Recommendation, Series G Supplement 41, 'Design guidelines for optical fiber submarine cable systems', May 2005.

[3] J. Zyskind, R. Bary, G. Pendock, M. Cahill and J. Ranka, 'High-capacity ultra-long-haul networks', in I. P. Kaminow and T. Li (Eds), *Optical Fiber Telecommunications IVB*, pp. 189–231, Academic Press, 2002.

[4] M. Vaa, W. Anderson, L. Rahman, S. Jiang, D. I. Kovsh, E. A. Golovchenko, A. Pilipetskii and S. M. Abbott, 'Transmission capacity study using cost effective undersea system technology with 120 km repeater spacing', *Proc. of Optical Fiber Communication OFC'04*, Los Angeles, USA, p. 3, 2004.

[5] J. M. Simmons, 'On determining the optimal optical reach for a long-haul network', *J. Lightwave Technol.*, **23**(3), pp. 1039–1048, 2005.

[6] H. Masuda, H. Kawakami, S. Kuwahara, A. Hirano, K. Sato and Y. Miyamoto, '1.28 Tbit/s (32 × 43 Gbit/s) field trial over 528 km (6 × 88 km) DSF using L-band remotely-pumped EDF/distributed Raman hybrid inline amplifiers', *Electron. Lett.*, **39**(23), pp. 1668–1670, 2003.

[7] M. Schneiders, S. Vorbeck, R. Leppla, E. Lach, M. Schmidt, S. B. Papernyi and K. Sanapi, 'Field transmission of 8 × 170 Gb/s over high-loss SSMF link using third-order distributed Raman amplification', *J. Lightwave Technol.*, **24**(1), pp. 175–182, 2006.

[8] M. Born and W. Wolf, *Principles of Optics* (7th edn), Cambridge University Press, 1999.

[9] P. Mossman, 'Connectors for optical fibre systems', *Radio Electron. Eng. (J. IERE)*, **51**(7/8), pp. 333–340, 1981.

[10] J. S. Leach, M. A. Matthews and E. Dalgoutte, 'Optical fibre cable connections', in C. P. Sandbank (Ed.), *Optical Fibre Communication Systems*, pp. 86–105, Wiley, 1980.

[11] F. L. Thiel and R. M. Hawk, 'Optical waveguide cable connection', *Appl. Opt.*, **15**(11), pp. 2785–2791, 1976.

[12] K. Miyazaki *et al.*, 'Theoretical and experimental considerations of optical fiber connector', *OSA Top. Meet. on Optical Fiber Transmission*, Williamsburg, VA, USA, paper WA 4-1, 1975.

[13] H. Tsuchiya, H. Nakagome, N. Shimizu and S. Ohara, 'Double eccentric connectors for optical fibers', *Appl. Opt.*, **16**(5), pp. 1323–1331, 1977.

[14] K. J. Fenton and R. L. McCartney, 'Connecting the thread of light', *Electronic Connector Study Group Symposium, 9th Annu. Symp. Proc.*, Cherry Hill, NJ, USA, p. 63, 1976.

[15] C. M. Miller, 'Transmission vs transverse offset for parabolic-profile fiber splices with unequal core diameters', *Bell Syst. Tech. J.*, **55**(7), pp. 917–927, 1976.

[16] D. Gloge, 'Offset and tilt loss in optical fiber splices', *Bell Syst. Tech. J.*, **55**(7), pp. 905–916, 1976.

[17] C. M. Miller and S. C. Mettler, 'A loss model for parabolic-profile fiber splices', *Bell Syst. Tech. J.*, **57**(9), pp. 3167–3180, 1978.

[18] J. J. Esposito, 'Optical connectors, couplers and switches', in H. F. Wolf (Ed.), *Handbook of Fiber Optics, Theory and Applications*, pp. 241–303, Granada, 1979.

[19] J. F. Dalgleish, 'Connections', *Electronics*, pp. 96–98, 5 August 1976.

[20] D. Botez and G. J. Herskowitz, 'Components for optical communications systems: a review', *Proc. IEEE*, **68**(6), pp. 689–731, 1980.

[21] G. Coppa and P. Di Vita, 'Length dependence of joint losses in multimode optical fibres', *Electron. Lett.*, **18**(2), pp. 84–85, 1982.

[22] W. van Etten and J. van Der Platts, *Fundamentals of Optical Fiber Communications*, Prentice Hall International, 1991.

[23] D. Marcuse, 'Loss analysis of single-mode fiber splices', *Bell Syst. Tech. J.*, **56**(5), pp. 703–718, 1977.

[24] W. A. Gambling, H. Matsumura and A. G. Cowley, 'Jointing loss in single-mode fibres', *Electron. Lett.*, **14**(3), pp. 54–55, 1978.

[25] W. A. Gambling, H. Matsumura and C. M. Ragdale, 'Joint loss in single-mode fibres', *Electron. Lett.*, **14**(15), pp. 491–493, 1978.

[26] S. Nemoto and T. Makimoto, 'Analysis of splice loss in single-mode fibers using a Gaussian field approximation', *Opt. Quantum Electron.*, **11**, pp. 447–457, 1979.

[27] W. C. Young and D. R. Frey, 'Fiber connectors', in S. E. Miller and I. P. Kaminow (Eds), *Optical Fiber Telecommunications II*, pp. 301–326, Academic Press, 1988.

[28] Y. Ushui, T. Ohshima, Y. Toda, Y. Kato and M. Tateda, 'Exact splice loss prediction for single-mode fiber', *IEEE J. Quantum Electron.*, **QE-18**(4), pp. 755–757, 1982.

[29] K. Petermann, 'Nonlinear distortions due to fibre connectors', *Proc. 6th Eur. Conf. on Optical Communications*, UK, pp. 80–83, 1980.

[30] K. Petermann, 'Wavelength-dependent transmission at fibre connectors', *Electron. Lett.*, **15**(22), pp. 706–708, 1979.

[31] C. M. Miller, S. C. Mettler and I. A. White, *Optical Fiber Splices and Connectors: Theory and methods*, Marcel Dekker, 1986.

[32] M. Ikeda, Y. Murakami and K. Kitayama, 'Mode scrambler for optical fibers', *Appl. Opt.*, **16**(4), pp. 1045–1049, 1977.

[33] N. Nashima and N. Uchida, 'Relation between splice loss and mode conversion in a graded-index optical fibre', *Electron. Lett.*, **15**(12), pp. 336–338, 1979.

[34] A. H. Cherin and J. F. Dalgleish, 'Splices and connectors for optical fibre communications', *Telecommun. J. (Eng. Ed.) Switzerland*, **48**(11), pp. 657–665, 1981.

[35] J. E. Midwinter, *Optical Fibers for Transmission*, Wiley, 1979.

[36] E. A. Lacy, *Fiber Optics*, Prentice Hall, 1982.

[37] R. Jocteur and A. Tardy, 'Optical fiber splicing with plasma torch and oxyhydric microburner', *2nd Eur. Conf. on Optical Fibre Communications*, Paris, 1976.

[38] I. Hatakeyama and H. Tsuchiya, 'Fusion splices for single-mode optical fibers', *IEEE J. Quantum Electron.*, **QE-14**(8), pp. 614–619, 1978.

[39] M. Hirai and N. Uchida, 'Melt splice of multimode optical fibre with an electric arc', *Electron. Lett.*, **13**(5), pp. 123–125, 1977.

[40] M. Tsuchiya and I. Hatakeyama, 'Fusion splices for single-mode optical fibres', *Optical Fiber Transmission II*, Williamsburg, VA, USA, pp. 1–4, February 1977.

[41] F. Esposto and E. Vezzoni, 'Connecting and splicing techniques', *Optical Fibre Communication*, by Technical Staff of CSELT, pp. 541–643, McGraw-Hill, 1981.

[42] D. B. Payne, D. J. McCartney and P. Healey, 'Fusion splicing of a 31.6 km monomode optical fibre system', *Electron. Lett.*, **18**(2), pp. 82–84, 1982.

[43] O. Kawata, K. Hoshino, Y. Miyajima, M. Ohnishi and K. Ishihara, 'A splicing end inspection technique for single-mode fibers using direct core monitoring', *J. Lightwave Technol.*, **LT-2**, pp. 185–190, 1984.

[44] I. Hatakeyama, M. Tachikura and H. Tsuchiya, 'Mechanical strength of fusion-spliced optical fibres', *Electron. Lett.*, **14**(19), pp. 613–614, 1978.

[45] C. K. Pacey and J. F. Dalgleish, 'Fusion splicing of optical fibres', *Electron. Lett.*, **15**(1), pp. 32–34, 1978.

[46] A. D. Yablon, *Optical Fiber Fusion Splicing*, Springer-Verlag, 2005.

[47] Fujikura®, 'Arc fusion splicers and test equipments', http://www.fujikura.co.jp/splicer/frontpage/front-page.html, 11 September 2007.

[48] T. G. Giallorenzi, 'Optical communications research and technology', *Proc. IEEE*, **66**(7), pp. 744–780, 1978.

[49] J. G. Woods, 'Fiber optic splices', *Proc. SPIE*, **512**, pp. 44–56, 1984.

[50] C. M. Miller, 'Loose tube splice for optical fibres', *Bell Syst. Tech. J.*, **54**(7), pp. 1215–1225, 1975.

[51] D. Gloge, A. H. Cherin, C. M. Miller and P. W. Smith, 'Fiber splicing', in S. E. Miller (Ed.), *Optical Fiber Telecommunications*, pp. 455–482, Academic Press, 1979.

[52] P. Hensel, J. C. North and J. H. Stewart, 'Connecting optical fibers', *Electron. Power*, **23**(2), pp. 133–135, 1977.

[53] A. R. Tynes and R. M. Derosier, 'Low-loss splices for single-mode fibres', *Electron. Lett.*, **13**(22), pp. 673–674, 1977.

[54] D. N. Knecht, W, J. Carlsen and P. Melman, 'Fiber optic field splice', *Proc. SPIE*, pp. 44–50, 1982.

[55] G. Cocito, B. Costa, S. Longoni, L. Michetti, L. Silvestri, D. Tribone and F. Tosco, 'COS 2 experiment in Turin: field test on an optical cable in ducts', *IEEE Trans. Commun.*, **COM-26**(7), pp. 1028–1036, 1978.

[56] J. A. Aberson and K. M. Yasinski, 'Multimode mechanical splices', *Proc. Tenth ECOC*, Germany, p. 182, 1984.

[57] C. M. Miller, G. F. DeVeau and M. Y. Smith, 'Simple high-performance mechanical splice for single mode fibers', *Proc. Optical Fiber Communication Conf., OFC'85*, USA, paper MI2, 1985.

[58] M. Kawase, M. Tachikura, F. Nihei and H. Murata, 'Mass fusion splices for high density optical fiber units', *Proc. Eighth ECOC*, France, paper AX-5, 1982.

[59] Y. Katsuyama, S. Hatano, K. Hogari, T. Matsumoto and T. Kokubun, 'Single mode optical fibre ribbon cable', *Electron. Lett.*, **21**, pp. 134–135, 1985.

[60] H. Murata, *Handbook of Optical Fibers and Cables*, Marcel Dekker, 1996.

[61] E. L. Chinnock, D. Gloge, D. L. Bisbee and P. W. Smith, 'Preparation of optical fiber ends for low-loss tape splices', *Bell Syst. Tech. J.*, **54**, pp. 471–477, 1975.

[62] R. Delebecque, E. Chazelas and D. Boscher, 'Flat mass splicing process for cylindrical V-groved cables', *Proc. IWCS '82*, Cherry Hill, NJ, USA, pp. 184–187, 1982.

[63] N. E. Hardwick and S. T. Davies, 'Rapid ribbon splice for multimode fiber splicing', *Proc. Optical Fiber Communication Conf., OFC'85*, USA, paper TUQ27, 1985.

[64] T. W. Tamulevich, 'Fiber optic ceramic capillary connectors', *Photonics Spectra*, pp. 65–70, October, 1984.

[65] W. C. Young, P. Kaiser, N. K. Cheung, L. Curtis, R. E. Wagner and D. M. Folkes, 'A transfer molded biconic connector with insertion losses below 0.3 dB without index match', *Proc. 6th Eur. Conf. on Optical Communications*, pp. 310–313, 1980.

[66] G. Keiser, *Optical Communications Essentials (Telecommunications)*, McGraw-Hill Professional, 2003.

[67] Fujikura®, 'Fast-ST-connectors', http://www.fujikura.co.uk/pdf/fibre_optics_connectors_fast_st.pdf, 11 September 2007.

[68] T. King, 'Fibre optic components for the fibre distributed data interface (FDDI) 100 Mbit/s local area network', *Proc. SPIE*, **949**, pp. 2–13, 1988.

[69] P. W. Smith, D. L. Bisbee, D. Gloge and E. L. Chinnock, 'A moulded-plastic technique for connecting and splicing optical fiber tapes and cables', *Bell Syst. Tech. J.*, **54**(6), pp. 971–984, 1975.

[70] Y. Fujii, J. Minowa and N. Suzuki, 'Demountable multiple connector with precise V-grooved silicon', *Electron. Lett.*, **15**(14), pp. 424–425, 1979.

[71] M. Oda, M. Ogai, A. Ohtake, S. Tachigami, S. Ohkubo, F. Nihei and N. Kashima, 'Nylon extruded fiber ribbon and its connection', *Proc. Optical Fiber Communication Conf., OFC'82*, USA, p. 46, 1982.

[72] S. Tachigami, A. Ohtake, T. Hayashi, T. Iso and T. Shirasawa, 'Fabrication and evaluation of high density multi-fiber plastic connector', *Proc. IWCS*, Cherry Hill, NJ, USA, pp. 70–75, 1983.

[73] T. Sakake, N. Kashima and M. Oki, 'Very small single-mode ten-fiber connector', *J. Lightwave Technol.*, **6**(2), pp. 269–272, 1988.

[74] M. H. Cho, S. H. Hwang, H. S. Cho and H. H. Park, 'High-coupling-efficiency optical interconnection using a 90° bent fiber array connector in optical printed circuit boards', *IEEE Photonics Technol. Lett.*, **17**(3), pp. 690–692, 2005.

[75] A. Nicia, 'Practical low-loss lens connector for optical fibers', *Electron. Lett.*, **14**(16), pp. 511–512, 1978.

[76] A. Nicia and A. Tholen, 'High efficiency ball-lens connector and related functional devices for single-mode fibers', *Proc. Seventh ECOC*, Denmark, paper 7.5, 1981.

[77] D. M. Knecht and W. J. Carlsen, 'Expanded beam fiber optic connectors', *Proc. SPIE*, pp. 44–50, 1983.

[78] R. Bauknecht, J. Kunde, R. Krahenbuhl, S. Grossman and C. Bosshard, 'Assembly technology for multi-fiber optical connectivity solutions', *Proc. IEEE/LEOS Workshop on Fibres and Optical Passive Components '05*, Mondello, Italy, pp. 92–97, 22–24 June 2005.

[79] W. J. Tomlinson, 'Applications of GRIN rod lenses in optical fiber communication systems', *Appl. Opt.*, **19**, pp. 1127–1138, 1980.

[80] T. Uchida, M. Furukawa, I. Kitano, K. Koizumi and H. Matsomura, 'Optical characteristics of a light focusing guide and its application', *IEEE J. Quantum Electron.*, **QE-6**, pp. 606–612, 1970.

[81] A. D. Yablon and R. T. Bise, 'Low-loss high-strength microstructured fiber fusion splices using GRIN fiber lenses', *IEEE Photonics Technol. Lett.*, **17**(1), pp. 118–120, 2005.

[82] D. Marcuse and S. E. Miller, 'Analysis of a tubular gas lens', *Bell Syst. Tech. J.*, **43**, pp. 1159–1782, 1965.

[83] S. E. Miller, 'Light propagation in generalized lenslike media', *Bell Syst. Tech. J.*, **44**, pp. 2017–2064, 1965.

[84] K. Sono, 'Graded index rod lenses', *Laser Focus*, **17**, pp. 70–74, 1981.

[85] J. M. Senior, S. D. Cusworth, N. G. Burrow and A. D. Muirhead, 'Misalignment losses at multimode graded-index fiber splices and GRIN rod lens couplers', *Appl. Opt.*, **24**, pp. 977–982, 1985.

[86] S. van Dorn, 'Fiber optic couplers', *Proc. SPIE*, **574**, pp. 2–8, 1985.

[87] K. O. Hill, D. C. Johnson and R. G. Lamont, 'Optical fiber directional couplers: biconical taper technology and device applications', *Proc. SPIE*, **574**, pp. 92–99, 1985.

[88] A. K. Agarwal, 'Review of optical fiber couplers', *Fiber Integr. Opt.*, **6**(1), pp. 27–53, 1987.

[89] J. P. Goure and I. Verrier, *Optical Fibre Devices*, CRC Press, 2002.

[90] B. S. Kawasaki, K. O. Hill and Y. Tremblay, 'Modal-noise generation in biconical taper couplers', *Opt. Lett.*, **6**, p. 499, 1981.

[91] J. , V. Wright, 'Wavelength dependence of fused couplers', *Electron. Lett.*, **22**, pp. 329–331, 1986.

[92] F. P. Payne, 'Dependence of fused taper couplers on external refractive index', *Electron. Lett.*, **22**, pp. 1207–1208, 1986.

[93] D. T. Cassidy, D. C. Johnson and K. O. Hill, 'Wavelength dependent transmission of monomode optical fiber tapers', *Appl. Opt.*, **24**, pp. 945–950, 1985.

[94] I. Yokohama, K. Okamoto and J. Noda, 'Polarization-independent optical circulator consisting of two fiber-optic polarizing beamsplitters and two YIG spherical lenses', *Electron. Lett.*, **22**, pp. 370–372, 1985.

[95] E. Paillard, 'Recent developments in integrated optics', *Proc. SPIE*, **734**, pp. 131–136, 1987.

[96] T. Findalky, 'Glass waveguides by ion exchange', *Opt. Eng.*, **24**, pp. 244–250, 1985.

[97] J. P. Dakin, M. G. Holliday and S. W. Hickling, 'Non invasive optical bus for video distribution', *Proc. SPIE*, **949**, pp. 36–40, 1988.

[98] G. D. Khoe and H. Lydtin, 'European optical fibers and passive components: status and trends', *IEEE J. Sel. Areas Commun.*, **SAC-4**(4), pp. 457–471, 1986.

[99] C. C. Wang, W. K. Burns and C. A. Villaruel, '9 × 9 single-mode fiber optic star couplers', *Opt. Lett.*, **10**(1), pp. 49–51, 1985.

[100] Y. Fujii, K. Aoyama and J. Minowa, 'Optical demultiplexer using a silicon echette grating', *IEEE J. Quantum Electron.*, **QE-16**, pp. 165–169, 1980.

[101] J. M. Senior and S. D. Cusworth, 'Devices for wavelength multiplexing and demultiplexing', *IEE Proc., Pt J*, **136**(3), pp. 183–202, 1989.

[102] F. L. Pedrotti, L. S. Pedrotti and L. M. Pedrotti, *Introduction to Optics* (3rd edn), Prentice Hall.

[103] R. Erdmann, 'Prism gratings for fiber optic multiplexing', *Proc. SPIE*, **417**, pp. 12–17, 1983.

[104] A. Nicia, 'Wavelength multiplexing and demultiplexing systems for single mode and multimode fibers', *Seventh Eur. Conf. on Optical Communications (ECOC'81)*, pp. 8.1–7, September 1981.

[105] J. Lipson, C. A. Young, P. D. Yeates, J. C. Masland, S. A. Wartonick, G. T. Harvey and P. H. Read, 'A four channel lightwave subsystem using wavelength multiplexing', *J. Lightwave Technol.*, **LT-3**, pp. 16–20, 1985.

[106] J. Lispon, W. J. Minford, E. J. Murphy, T. C. Rice, R. A. Linke and G. T. Harvey, 'A six-channel wavelength multiplexer and demultiplexer for single mode systems', *J. Lightwave Technol.*, **LT-3**, pp. 1159–1161, 1985.

[107] Y. Fujii, J. Minowa and H. Tanada, 'Practical two-wavelength multiplexer and demultiplexer: design and performance', *Appl. Opt.*, **22**, pp. 3090–3097, 1983.

[108] G. Winzer, H. F. Mahlein and A. Reichelt, 'Single-mode and multimode all-fiber directional couplers for WDM', *Appl. Opt.*, **20**, pp. 3128–3135, 1981.

[109] M. McCourt and J. L. Malinge, 'Application of ion exchange techniques to the fabrication of multimode wavelength division multiplexers', *Proc. SPIE*, **949**, pp. 131–137, 1988.

[110] R. Zengerle and O. G. Leminger, 'Wavelength-selective directional coupler made of non-identical single-mode fibers', *J. Lightwave Technol.*, **LT-4**, pp. 823–826, 1986.

[111] D. Marcuse, 'Coupling of degenerative modes in two parallel dielectric waveguides', *Bell Syst. Tech. J.*, **50**, pp. 1791–1816, 1971.

[112] B. S. Kawasaki and K. O. Hill, 'Low loss access coupler for multimode optical fiber distribution networks', *Appl. Opt.*, **16**, pp. 327–328, 1977.

[113] M. J. F. Digonnet and H. J. Shaw, 'Wavelength multiplexing in single mode fiber couplers', *Appl. Opt.*, **22**, pp. 484–492, 1983.

[114] O. Leminger and R. Zengerle, 'Bandwidth of directional-coupler wavelength filters made of dissimilar optical fibres', *Electron. Lett.*, **23**, pp. 241–242, 1987.

[115] R. Zengerle and O. Leminger, 'Narrow band wavelength selective directional-coupler made of dissimilar optical fibres', *J. Lightwave Technol.*, **LT-5**, pp. 1196–1198, 1987.

[116] H. A. Roberts, 'Single-mode fused wavelength division multiplexer', *Proc. SPIE*, **574**, pp. 100–104, 1985.

[117] A. Vivek, *Optical Network Design and Implementation*, Cisco Press, 2004.

[118] M. Greenberg and M. Orenstein, 'Filterless "add" multiplexer based on novel complex gratings assisted coupler', *IEEE Photonics Technol. Lett.*, **17**(7), pp. 1450–1452, 2005.

[119] E. Shafir, G. Berkovic, Y. Sadi, S. Rotter and S. Gali, 'Practical strain isolation in embedded fiber Bragg gratings', *J. Smart Mater. Struct.*, **14**(4), pp. N26–N28, 2005.

[120] C. R. Daerr and K. Okamoto, 'Planar lightwave circuits in fiber-optic communications', in I. P. Kaminow, T. Li and A. E. Willner (Eds), *Optical Fiber Telecommunications VA*, pp. 269–341, Elsevier/Academic Press, 2008.

[121] O. M. Matos, M. L. Calvo, P. Cheben, S. Janz, J. A. Rodrigo, D. X. Xu and A. Delâge 'Arrayed waveguide grating based on group-index modification', *J. Lightwave Technol.*, **24**(3), pp. 1551–1557, 2006.

[122] H. C. Lu and W. S. Wang, 'Cyclic arrayed waveguide grating devices with flat-top passband and uniform spectral response', *IEEE Photon. Technol. Lett.*, **20**(1), pp. 3–5, 2008.

[123] P. Muñoz, D. Pastor and J. Capmany, 'Modeling and design of arrayed waveguide gratings', *J. Lightwave Technol.*, **20**(4), pp. 661–674, 2002.

[124] T. Ohara, H. Takara, T. Yamamoto, H. Masuda, T. Morioka, M. Abe and H. Takahashi, 'Over-1000-channel ultradense WDM transmission with supercontinuum multicarrier source', *J. Lightwave Technol.*, **24**(6), pp. 2311–2317, 2006.

[125] S. Kakehashi, H. Hasegawa, K. Sato, O. Moriwaki, S. Kamei, Y. Jinnouchi and M. Okuno, 'Waveband MUX/DEMUX using concatenated AWGs-formulation of waveguide connection and fabrication', *Proc. Optical Fiber Communication and National Fiber Optic Engineers Conference (OFC/NFOEC '07)*, Anaheim, California, USA, JThA94, pp. 1–3, March 2007.

[126] I. S. Joe and O. Solgaard, 'Scalable optical switches with large port count based on a waveguide grating router and passive couplers', *IEEE Photon. Technol. Lett.*, **20**(7), pp. 508–510, 2008.

[127] C. Koerdt, G. L. J. A. Rikken and E. P. Petrov, 'Faraday effect of photonic crystals', *Appl. Phys. Lett.*, **82**(10), pp. 1538–1540, 2003.

[128] R. A. Chilton and R. Lee, 'Chirping unit cell length to increase frozen-mode bandwidth in nonreciprocal MPCs', *IEEE Trans. Microw. Theory Technol.*, **54**(1), pp. 473–480, 2006.

[129] H. Dotsch, N. Bahlmann, O. Zhuromskyy, M. Hammer, L. Wilkens, R. Gerhardt and P. Hertel, 'Applications of magneto-optical waveguides in integrated optics: review', *J. Opt. Soc. Am. B.*, **22**(1), pp. 240–253, 2005.

[130] W. Van Parys, D. Van Thourhout, R. Baets, B. Dagens, J. Decobert, O. Le Gouezigou, D. Make, R. Vanheertum, L. Lagae, '11.4 dB Isolation on an amplifying AlGaInAs/InP optical waveguide isolator', *Proc. Optical Fiber Communication, OFC'06*, Anaheim, USA, pp. OFA2, 1–3, 5–10 March 2006.

[131] H. Shimizu and Y. Nakano, 'Fabrication and characterization of an InGaAsP/InP active waveguide optical isolator with 14.7 dB/mm TE mode nonreciprocal attenuation', *J. Lightwave Technol.*, **24**(1), pp. 38–43, 2006.

[132] N. A. Riza and N. Madamopoulos, 'Compact switched-retroreflection-based 2 × 2 optical switching fabric for WDM applications', *J. Lightwave Technol.*, **23**(1), pp. 247–259, 2005.

[133] oeMarket.com, 'Optical circulators', http://www.oemarket.com/index.php?cPath=22_24, 17 September 2007.

[134] O. Frazão, I. Terroso, J. P. Carvalho and H. M. Salgado, 'Optical cross-connect based on tuneable FBG-OC with full scalability and bidirectionality', *Opt. Commun.*, **220**(1–3), pp. 105–109, 2003.

[135] S. P Majumder and S. Dey, 'Performance limitations due to crosstalk in a WDM system using optical cross-connect based on tunable fiber Bragg gratings and optical circulators', *IEEE/IFIP Conf. on Wireless and Optical Communications Network, WOCN 2005*, Piscataway, NJ, USA, pp. 14–17, 6–8 March 2005.

Optical sources 1: the laser

6.1 Introduction

The optical source is often considered to be the active component in an optical fiber communication system. Its fundamental function is to convert electrical energy in the form of a current into optical energy (light) in an efficient manner which allows the light output to be effectively launched or coupled into the optical fiber. Three main types of optical light source are available. These are:

(a) wideband 'continuous spectra' sources (incandescent lamps);

(b) monochromatic incoherent sources (light-emitting diodes, LEDs);

(c) monochromatic coherent sources (lasers).

To aid consideration of the sources currently in major use, the historical aspect must be mentioned. In the early stages of optical fiber communications the most powerful narrow-band coherent light sources were necessary due to severe attenuation and dispersion in the fibers. Therefore, gas lasers (helium-neon) were utilized initially. However, the development of the semiconductor injection laser and the LED, together with the substantial improvement in the properties of optical fibers, has given prominence to these two specific sources.

To a large extent these two sources fulfill the major requirements for an optical fiber emitter which are outlined below:

1. A size and configuration compatible with launching light into an optical fiber. Ideally, the light output should be highly directional.

2. Must accurately track the electrical input signal to minimize distortion and noise. Ideally, the source should be linear.

3. Should emit light at wavelengths where the fiber has low losses and low dispersion and where the detectors are efficient.

4. Preferably capable of simple signal modulation (i.e. direct – see Section 7.5) over a wide bandwidth extending from audio frequencies to beyond the gigahertz range.

5. Must couple sufficient optical power to overcome attenuation in the fiber plus additional connector losses and leave adequate power to drive the detector.

6. Should have a very narrow spectral bandwidth (linewidth) in order to minimize dispersion in the fiber.

7. Must be capable of maintaining a stable optical output which is largely unaffected by changes in ambient conditions (e.g. temperature).

8. It is essential that the source is comparatively cheap and highly reliable in order to compete with conventional transmission techniques.

In order to form some comparison between these two types of light source, the historical aspect must be enlarged upon. The first-generation optical communication sources were designed to operate between 0.8 and 0.9 μm (ideally around 0.85 μm) because initially the properties of the semiconductor materials used lent themselves to emission at this wavelength. Also, as suggested in requirement 3 this wavelength avoided the loss incurred in many fibers near 0.9 μm due to the OH ion (see Section 3.3.2). These early systems utilized multimode step index fibers which required the superior performance of semiconductor lasers for links of reasonable bandwidth (tens of megahertz) and distances (several kilometers). The LED (being a lower power source generally exhibiting little spatial or temporal coherence) was not suitable for long-distance wideband transmission, although it found use in more moderate distance applications.

However, the role of the LED as a source for optical fiber communications was enhanced following the development of multimode graded index fiber. The substantial reduction in intermodal dispersion provided by this fiber type over multimode step index fiber allowed incoherent LEDs emitting in the 0.8 to 0.9 μm wavelength band to be utilized for applications requiring wider bandwidths. This position was further consolidated with the development of second-generation optical fiber sources operating at wavelengths between 1.1 and 1.6 μm where both material losses and dispersion are greatly reduced. In this wavelength region, wideband graded index fiber systems utilizing LED sources may be operated over long distances without the need for intermediate repeaters. Furthermore, LEDs offer the advantages of relatively simple construction and operation with the inherent effects of these factors on cost and extended, trouble-free life.

In parallel with these later developments in multimode optical propagation came advances in single-mode fiber construction. This has stimulated the development of single-mode laser sources to take advantage of the extremely low dispersion offered by single-mode fibers. These systems are ideally suited to extra wideband, very long-haul applications and became under intensive investigation for long-distance telecommunications. On the other hand, light is usually emitted from the LED in many spatial modes which cannot be as efficiently focused and coupled into single-mode fiber. Nevertheless, more recently advanced LED sources were developed that allowed moderate optical power levels to be launched into single-mode fiber (see Chapter 7). However, to date the LED has been utilized primarily as a multimode source giving acceptable coupling efficiencies into multimode fiber. Moreover, in this capacity the LED remains the major multimode source which is extensively used for increasingly wider bandwidth, longer haul applications. Therefore at present the LED is chosen for many applications using multimode fibers and the injection laser diode (ILD) tends to find more use as a single-mode device in single-mode fiber systems. Although other laser types (e.g. Nd : YAG and glass fiber lasers, Section 6.9), as well as the injection laser, may eventually find significant use in optical fiber communications, this chapter and the following one will deal primarily with major structures and configurations of semiconductor sources (ILD and LED), taking into account recent developments and possible future advances.

We begin by describing in Section 6.2 the basic principles of laser operation which may be applied to all laser types. Immediately following, in Section 6.3, is a discussion of optical emission from semiconductors in which we concentrate on the fundamental operating principles, the structure and the materials for the semiconductor laser. Aspects concerning practical semiconductor lasers are then considered in Section 6.4 prior to a more specific discussion of the structure and operation of some common injection laser types including both quantum-well and quantum dot devices in Section 6.5. Then, in Section 6.6, the major single-frequency injection laser structures which provide single-mode operation, primarily in the longer wavelength region (1.1 to 1.6 μm), are then described. In Section 6.7 we consider the operating characteristics which are common to all injection laser types, before a short discussion of injection laser to optical fiber coupling is presented in Section 6.8. Major nonsemiconductor laser devices which have found use in optical fiber communications (the neodymium-doped yttrium–aluminium–garnet (Nd : YAG) laser and the glass fiber laser) are then outlined in Section 6.9. This is followed in Section 6.10 with a discussion of advanced linewidth-narrowed and wavelength tunable laser types. Finally, in Section 6.11, developments in laser sources for transmission in the mid-infrared and far-infrared wavelength regions (2 to 12 µm) are considered to give an insight into this potentially important area. In particular, the structure and operation of the quantum cascade laser for use over this wavelength range are discussed.

6.2 Basic concepts

To gain an understanding of the light-generating mechanisms within the major optical sources used in optical fiber communications it is necessary to consider both the fundamental atomic concepts and the device structure. In this context the requirements for the laser source are far more stringent than those for the LED. Unlike the LED, strictly speaking, the laser is a device which amplifies light – hence the derivation of the term LASER as an acronym for Light Amplification by Stimulated Emission of Radiation. Lasers, however, are seldom used as amplifiers since there are practical difficulties in relation to the achievement of high gain while avoiding oscillation from the required energy feedback. Thus the practical realization of the laser is as an optical oscillator. The operation of the device may be described by the formation of an electromagnetic standing wave within a cavity (or optical resonator) which provides an output of monochromatic, highly coherent radiation. By contrast the LED provides optical emission without an inherent gain mechanism. This results in incoherent light output.

In this section we elaborate on the basic principles which govern the operation of both these optical sources. It is clear, however, that the operation of the laser must be discussed in some detail in order to provide an appreciation of the way it functions as an optical source. Hence we concentrate first on the general principles of laser action.

6.2.1 Absorption and emission of radiation

The interaction of light with matter takes place in discrete packets of energy or quanta, called photons. Furthermore, the quantum theory suggests that atoms exist only in certain discrete energy states such that absorption and emission of light causes them to make a transition from one discrete energy state to another. The frequency of the absorbed or emitted radiation f is related to the difference in energy E between the higher energy state E_2 and the lower energy state E_1 by the expression:

$$E = E_2 - E_1 = hf \tag{6.1}$$

where $h = 6.626 \times 10^{-34}$ J s is Planck's constant. These discrete energy states for the atom may be considered to correspond to electrons occurring in particular energy levels relative to the nucleus. Hence, different energy states for the atom correspond to different electron configurations, and a single electron transition between two energy levels within the atom will provide a change in energy suitable for the absorption or emission of a photon. It must be noted, however, that modern quantum theory [Ref. 1] gives a probabilistic description which specifies the energy levels in which electrons are most likely to be found. Nevertheless, the concept of stable atomic energy states and electron transitions between energy levels is still valid.

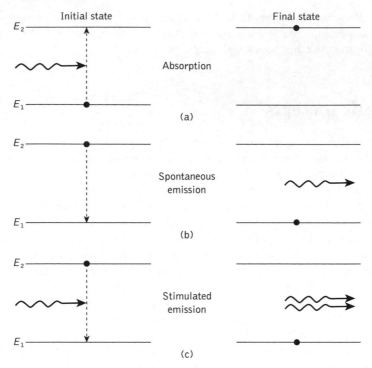

Figure 6.1 Energy state diagram showing: (a) absorption; (b) spontaneous emission; (c) stimulated emission. The black dot indicates the state of the atom before and after a transition takes place

Figure 6.1(a) illustrates a two energy state or level atomic system where an atom is initially in the lower energy state E_1. When a photon with energy $(E_2 - E_1)$ is incident on the atom it may be excited into the higher energy state E_2 through absorption of the photon. This process is sometimes referred to as stimulated absorption. Alternatively, when the atom is initially in the higher energy state E_2 it can make a transition to the lower energy state E_1 providing the emission of a photon at a frequency corresponding to Eq. (6.1). This emission process can occur in two ways:

(a) by spontaneous emission in which the atom returns to the lower energy state in an entirely random manner;

(b) by stimulated emission when a photon having an energy equal to the energy difference between the two states $(E_2 - E_1)$ interacts with the atom in the upper energy state causing it to return to the lower state with the creation of a second photon.

These two emission processes are illustrated in Figure 6.1(b) and (c) respectively. The random nature of the spontaneous emission process where light is emitted by electronic transitions from a large number of atoms gives incoherent radiation. A similar emission process in semiconductors provides the basic mechanism for light generation within the LED (see Section 6.3.2).

It is the stimulated emission process, however, which gives the laser its special properties as an optical source. Firstly, the photon produced by stimulated emission is generally* of an identical energy to the one which caused it and hence the light associated with them is of the same frequency. Secondly, the light associated with the stimulating and stimulated photon is in phase and has the same polarization. Therefore, in contrast to spontaneous emission, coherent radiation is obtained. Furthermore, this means that when an atom is stimulated to emit light energy by an incident wave, the liberated energy can add to the wave in a constructive manner, providing amplification.

6.2.2 The Einstein relations

Prior to a discussion of laser action in semiconductors it is useful to consider optical amplification in the two-level atomic system shown in Figure 6.1. In 1917 Einstein [Ref. 2] demonstrated that the rates of the three transition processes of absorption, spontaneous emission and stimulated emission were related mathematically. He achieved this by considering the atomic system to be in thermal equilibrium such that the rate of the upward transitions must equal the rate of the downward transitions. The population of the two energy levels of such a system is described by Boltzmann statistics which give:

$$\frac{N_1}{N_2} = \frac{g_1 \exp(-E_1/KT)}{g_2 \exp(-E_2/KT)} = \frac{g_1}{g_2} \exp(E_2 - E_1/KT)$$

$$= \frac{g_1}{g_2} \exp(hf/KT) \tag{6.2}$$

where N_1 and N_2 represent the density of atoms in energy levels E_1 and E_2, respectively, with g_1 and g_2 being the corresponding degeneracies[†] of the levels, K is Boltzmann's constant and T is the absolute temperature.

As the density of atoms in the lower or ground energy state E_1 is N_1, the rate of upward transition or absorption is proportional to both N_1 and the spectral density ρ_f of the radiation energy at the transition frequency f. Hence, the upward transition rate R_{12} (indicating an electron transition from level 1 to level 2) may be written as:

$$R_{12} = N_1 \rho_f B_{12} \tag{6.3}$$

where the constant of proportionality B_{12} is known as the Einstein coefficient of absorption.

By contrast, atoms in the higher or excited energy state can undergo electron transitions from level 2 to level 1 either spontaneously or through stimulation by the radiation field.

---

* A photon with energy hf will not necessarily always stimulate another photon with energy hf. Photons may be stimulated over a small range of energies around hf providing an emission which has a finite frequency or wavelength spread (linewidth).

† In many cases the atom has several sublevels of equal energy within an energy level which is then said to be degenerate. The degeneracy parameters g_1 and g_2 indicate the number of sublevels within the energy levels E_1 and E_2 respectively. If the system is not degenerate, then g_1 and g_2 may be set to unity [Ref. 1].

For spontaneous emission the average time that an electron exists in the excited state before a transition occurs is known as the spontaneous lifetime τ_{21}. If the density of atoms within the system with energy E_2 is N_2, then the spontaneous emission rate is given by the product of N_2 and $1/\tau_2$. This may be written as $N_2 A_{21}$ where A_{21}, the Einstein coefficient of spontaneous emission, is equal to the reciprocal of the spontaneous lifetime.

The rate of stimulated downward transition of an electron from level 2 to level 1 may be obtained in a similar manner to the rate of stimulated upward transition. Hence the rate of stimulated emission is given by $N_2 \rho_f B_{21}$, where B_{21} is the Einstein coefficient of stimulated emission. The total transition rate from level 2 to level 1, R_{21}, is the sum of the spontaneous and stimulated contributions. Hence:

$$R_{21} = N_2 A_{21} + N_2 \rho_f B_{21} \tag{6.4}$$

For a system in thermal equilibrium, the upward and downward transition rates must be equal and therefore $R_{12} = R_{21}$, or:

$$N_1 \rho_f B_{12} = N_2 A_{21} + N_2 \rho_f B_{21} \tag{6.5}$$

It follows that:

$$\rho_f = \frac{N_2 A_{21}}{N_1 B_{12} - N_2 B_{21}}$$

and:

$$\rho_f = \frac{A_{21}/B_{21}}{(B_{12} N_1 / B_{21} N_2) - 1} \tag{6.6}$$

Substituting Eq. (6.2) into Eq. (6.6) gives:

$$\rho_f = \frac{A_{21}/B_{21}}{[(g_1 B_{12}/g_2 B_{21}) \exp(hf/KT)] - 1} \tag{6.7}$$

However, since the atomic system under consideration is in thermal equilibrium it produces a radiation density which is identical to black body radiation. Planck showed that the radiation spectral density for a black body radiating within a frequency range f to $f + df$ is given by [Ref. 3]:

$$\rho_f = \frac{8\pi h f^3}{c^3} \left[\frac{1}{\exp(hf/KT) - 1} \right] \tag{6.8}$$

Comparing Eq. (6.8) with Eq. (6.7) we obtain the Einstein relations:

$$B_{12} = \left(\frac{g_2}{g_1} \right) B_{21} \tag{6.9}$$

and:

$$\frac{A_{21}}{B_{21}} = \frac{8\pi h f^3}{c^3} \tag{6.10}$$

It may be observed from Eq. (6.9) that when the degeneracies of the two levels are equal $(g_1 = g_2)$, then the probabilities of absorption and stimulated emission are equal. Furthermore, the ratio of the stimulated emission rate to the spontaneous emission rate is given by:

$$\frac{\text{Stimulated emission rate}}{\text{Spontaneous emission rate}} = \frac{B_{21}\rho_f}{A_{21}} = \frac{1}{\exp(hf/KT) - 1} \tag{6.11}$$

Example 6.1

Calculate the ratio of the stimulated emission rate to the spontaneous emission rate for an incandescent lamp operating at a temperature of 1000 K. It may be assumed that the average operating wavelength is 0.5 μm.

Solution: The average operating frequency is given by:

$$f = \frac{c}{\lambda} = \frac{2.998 \times 10^8}{0.5 \times 10^{-6}} \simeq 6.0 \times 10^{14} \text{ Hz}$$

Using Eq. (6.11) the ratio is:

$$\frac{\text{Stimulated emission rate}}{\text{Spontaneous emission rate}} = \frac{1}{\exp\left(\dfrac{6.626 \times 10^{-34} \times 6 \times 10^{14}}{1.381 \times 10^{-23} \times 1000}\right)}$$

$$= \exp(-28.8)$$
$$= 3.1 \times 10^{-13}$$

The result obtained in Example 6.1 indicates that for systems in thermal equilibrium spontaneous emission is by far the dominant mechanism. Furthermore, it illustrates that the radiation emitted from ordinary optical sources in the visible spectrum occurs in a random manner, proving that these sources are incoherent.

It is apparent that in order to produce a coherent optical source and amplification of a light beam the rate of stimulated emission must be increased far above the level indicated by Example 6.1. From consideration of Eq. (6.5) it may be noted that for stimulated emission to dominate over absorption and spontaneous emission in a two-level system, both the radiation density and the population density of the upper energy level N_2 must be increased in relation to the population density of the lower energy level N_1.

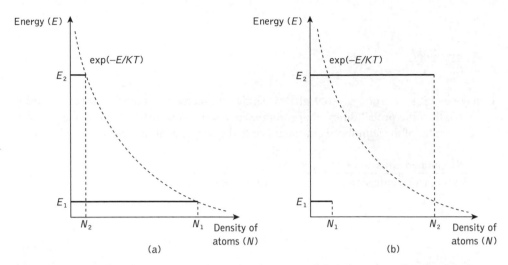

Figure 6.2 Populations in a two-energy-level system: (a) Boltzmann distribution for a system in thermal equilibrium; (b) a nonequilibrium distribution showing population inversion

6.2.3 Population inversion

Under the conditions of thermal equilibrium given by the Boltzmann distribution (Eq. (6.2)) the lower energy level E_1 of the two-level atomic system contains more atoms than the upper energy level E_2. This situation, which is normal for structures at room temperature, is illustrated in Figure 6.2(a). However, to achieve optical amplification it is necessary to create a nonequilibrium distribution of atoms such that the population of the upper energy level is greater than that of the lower energy level (i.e. $N_2 > N_1$). This condition, which is known as population inversion, is illustrated in Figure 6.2(b).

In order to achieve population inversion it is necessary to excite atoms into the upper energy level E_2 and hence obtain a nonequilibrium distribution. This process is achieved using an external energy source and is referred to as 'pumping'. A common method used for pumping involves the application of intense radiation (e.g. from an optical flash tube or high-frequency radio field). In the former case atoms are excited into the higher energy state through stimulated absorption. However, the two-level system discussed above does not lend itself to suitable population inversion. Referring to Eq. (6.9), when the two levels are equally degenerate (or not degenerate), then $B_{12} = B_{21}$. Thus the probabilities of absorption and stimulated emission are equal, providing at best equal populations in the two levels.

Population inversion, however, may be obtained in systems with three or four energy levels. The energy-level diagrams for two such systems, which correspond to two non-semiconductor lasers, are illustrated in Figure 6.3. To aid attainment of population inversion both systems display a central metastable state in which the atoms spend an unusually long time. It is from this metastable level that the stimulated emission or lasing takes place. The three-level system (Figure 6.3(a)) consists of a ground level E_0, a metastable level E_1 and a third level above the metastable level E_2. Initially, the atomic distribution

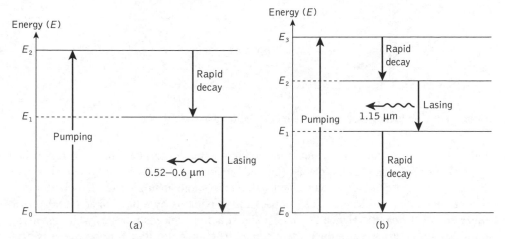

Figure 6.3 Energy-level diagrams showing population inversion and lasing for two nonsemiconductor lasers: (a) three-level system – ruby (crystal) laser; (b) four-level system – He–Ne (gas) laser

will follow Boltzmann's law. However, with suitable pumping the electrons in some of the atoms may be excited from the ground state into the higher level E_2. Since E_2 is a normal level the electrons will rapidly decay by nonradiative processes to either E_1 or directly to E_0. Hence empty states will always be provided in E_2. The metastable level E_1 exhibits a much longer lifetime than E_2 which allows a large number of atoms to accumulate at E_1. Over a period the density of atoms in the metastable state N_1 increases above those in the ground state N_0 and a population inversion is obtained between these two levels. Stimulated emission and hence lasing can then occur, creating radiative electron transitions between levels E_1 and E_0. A drawback with the three-level system such as the ruby laser is that it generally requires very high pump powers because the terminal state of the laser transition is the ground state. Hence more than half the ground state atoms must be pumped into the metastable state to achieve population inversion.

By contrast, a four-level system such as the He–Ne laser illustrated in Figure 6.3(b) is characterized by much lower pumping requirements. In this case the pumping excites the atoms from the ground state into energy level E_3 and they decay rapidly to the metastable level E_2. However, since the populations of E_3 and E_1 remain essentially unchanged, a small increase in the number of atoms in energy level E_2 creates population inversion, and lasing takes place between this level and level E_1.

6.2.4 Optical feedback and laser oscillation

Light amplification in the laser occurs when a photon colliding with an atom in the excited energy state causes the stimulated emission of a second photon and then both these photons release two more. Continuation of this process effectively creates avalanche multiplication, and when the electromagnetic waves associated with these photons are in phase, amplified coherent emission is obtained. To achieve this laser action it is necessary to contain photons

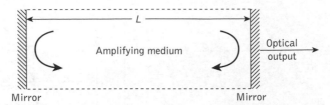

Figure 6.4 The basic laser structure incorporating plane mirrors

within the laser medium and maintain the conditions for coherence. This is accomplished by placing or forming mirrors (plane or curved) at either end of the amplifying medium, as illustrated in Figure 6.4. The optical cavity formed is more analogous to an oscillator than an amplifier as it provides positive feedback of the photons by reflection at the mirrors at either end of the cavity. Hence the optical signal is fed back many times while receiving amplification as it passes through the medium. The structure therefore acts as a Fabry–Pérot resonator. Although the amplification of the signal from a single pass through the medium is quite small, after multiple passes the net gain can be large. Furthermore, if one mirror is made partially transmitting, useful radiation may escape from the cavity.

A stable output is obtained at saturation when the optical gain is exactly matched by the losses incurred in the amplifying medium. The major losses result from factors such as absorption and scattering in the amplifying medium, absorption, scattering and diffraction at the mirrors and nonuseful transmission through the mirrors.

Oscillations occur in the laser cavity over a small range of frequencies where the cavity gain is sufficient to overcome the above losses. Hence the device is not a perfectly monochromatic source but emits over a narrow spectral band. The central frequency of this spectral band is determined by the mean energy-level difference of the stimulated emission transition. Other oscillation frequencies within the spectral band result from frequency variations due to the thermal motion of atoms within the amplifying medium (known as Doppler broadening)* and by atomic collisions.† Hence the amplification within the laser medium results in a broadened laser transition or gain curve over a finite spectral width, as illustrated in Figure 6.5. The spectral emission from the device therefore lies within the frequency range dictated by this gain curve.

Since the structure forms a resonant cavity, when sufficient population inversion exists in the amplifying medium the radiation builds up and becomes established as standing waves between the mirrors. These standing waves exist only at frequencies for which the distance between the mirrors is an integral number of half wavelengths. Thus when the optical spacing between the mirrors is L, the resonance condition along the axis of the cavity is given by [Ref. 4]:

$$L = \frac{\lambda q}{2n} \tag{6.12}$$

* Doppler broadening is referred to as an inhomogeneous broadening mechanism since individual groups of atoms in the collection have different apparent resonance frequencies.

† Atomic collisions provide homogeneous broadening as every atom in the collection has the same resonant frequency and spectral spread.

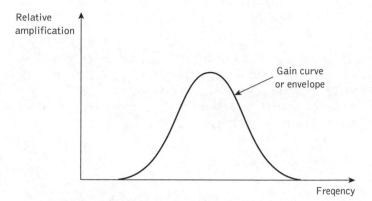

Figure 6.5 The relative amplification in the laser amplifying medium showing the broadened laser transition line or gain curve

where λ is the emission wavelength, n is the refractive index of the amplifying medium and q is an integer. Alternatively, discrete emission frequencies f are defined by:

$$f = \frac{qc}{2nL} \tag{6.13}$$

where c is the velocity of light. The different frequencies of oscillation within the laser cavity are determined by the various integer values of q and each constitutes a resonance or mode. Since Eqs (6.12) and (6.13) apply for the case when L is along the longitudinal axis of the structure (Figure 6.4) the frequencies given by Eq. (6.13) are known as the longitudinal or axial modes. Furthermore, from Eq. (6.13) it may be observed that these modes are separated by a frequency interval δf where:

$$\delta f = \frac{c}{2nL} \tag{6.14}$$

The mode separation in terms of the free space wavelength, assuming $\delta f \ll f$ and as $f = c/\lambda$, is given by:

$$\delta \lambda = \frac{\lambda \delta f}{f} = \frac{\lambda^2}{c} \, \delta f \tag{6.15}$$

Hence substituting for δf from Eq. (6.14) gives:

$$\delta \lambda = \frac{\lambda^2}{2nL} \tag{6.16}$$

In addition it should be noted that Eq. (6.15) can be used to determine the device spectral linewidth as a function of wavelength when it is quoted in hertz, or vice versa (see Problem 6.4).

Example 6.2

A ruby laser contains a crystal of length 4 cm with a refractive index of 1.78. The peak emission wavelength from the device is 0.55 μm. Determine the number of longitudinal modes and their frequency separation.

Solution: The number of longitudinal modes supported within the structure may be obtained from Eq. (6.12) where:

$$q = \frac{2nL}{\lambda} = \frac{2 \times 1.78 \times 0.04}{0.55 \times 10^{-6}} = 2.6 \times 10^5$$

Using Eq. (6.14) the frequency separation of the modes is:

$$\delta f = \frac{2.998 \times 10^8}{2 \times 1.78 \times 0.04} = 2.1 \text{ GHz}$$

Although the result of Example 6.2 indicates that a large number of modes may be generated within the laser cavity, the spectral output from the device is defined by the gain curve. Hence the laser emission will only include the longitudinal modes contained within the spectral width of the gain curve. This situation is illustrated in Figure 6.6 where several modes are shown to be present in the laser output. Such a device is said to be multimode.

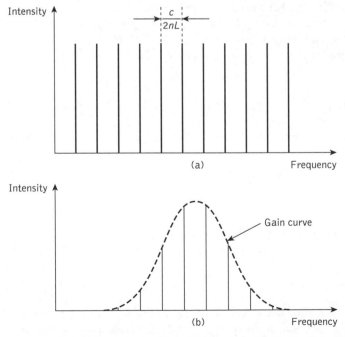

Figure 6.6 (a) The modes in the laser cavity. (b) The longitudinal modes in the laser output

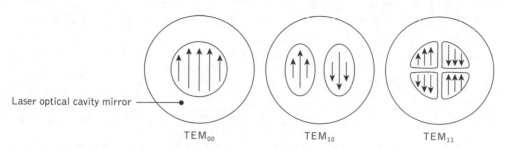

Laser optical cavity mirror

TEM_{00} TEM_{10} TEM_{11}

Figure 6.7 The lower order transverse modes of a laser

    Laser oscillation may also occur in a direction which is transverse to the axis of the cavity. This gives rise to resonant modes which are transverse to the direction of propagation. These transverse electromagnetic modes are designated in a similar manner to transverse modes in waveguides (Section 2.3.2) by TEM_{lm} where the integers l and m indicate the number of transverse modes (see Figure 6.7). Unlike the longitudinal modes which contribute only a single spot of light to the laser output, transverse modes may give rise to a pattern of spots at the output. This may be observed from the low-order transverse mode patterns shown in Figure 6.7 on which the direction of the electric field is also indicated. In the case of the TEM_{00} mode all parts of the propagating wavefront are in phase. This is not so, however, with higher order modes (TEM_{10}, TEM_{11}, etc.) where phase reversals produce the various mode patterns. Thus the greatest degree of coherence, together with the highest level of spectral purity, may be obtained from a laser which operates in only the TEM_{00} mode. Higher order transverse modes only occur when the width of the cavity is sufficient for them to oscillate. Consequently, they may be eliminated by suitable narrowing of the laser cavity.

6.2.5 Threshold condition for laser oscillation

It has been indicated that steady-state conditions for laser oscillation are achieved when the gain in the amplifying medium exactly balances the total losses.* Hence, although population inversion between the energy levels providing the laser transition is necessary for oscillation to be established, it is not alone sufficient for lasing to occur. In addition a minimum or threshold gain within the amplifying medium must be attained such that laser oscillations are initiated and sustained. This threshold gain may be determined by considering the change in energy of a light beam as it passes through the amplifying medium. For simplicity, all the losses except those due to transmission through the mirrors may be included in a single loss coefficient per unit length, $\bar{\alpha}$ cm^{-1}. Again we assume the amplifying medium occupies a length L completely filling the region between the two mirrors which have reflectivities r_1 and r_2. On each round trip the beam passes through the medium twice. Hence the fractional loss incurred by the light beam is:

$$\text{Fractional loss} = r_1 r_2 \exp(-2\bar{\alpha}L) \qquad (6.17)$$

* This applies to a CW laser which gives a continuous output, rather than pulsed devices for which slightly different conditions exist. For oscillation to commence, the fractional gain and loss must be matched.

Furthermore, it is found that the increase in beam intensity resulting from stimulated emission is exponential [Ref. 4]. Therefore if the gain coefficient per unit length produced by stimulated emission is $\bar{g}$ cm^{-1}, the fractional round trip gain is given by:

$$\text{Fractional gain} = \exp(2\bar{g}L) \tag{6.18}$$

Hence:

$$\exp(2\bar{g}L) \times r_1 r_2 \exp(-2\bar{\alpha}L) = 1$$

and:

$$r_1 r_2 \exp[2(\bar{g} - \bar{\alpha})L] = 1 \tag{6.19}$$

The threshold gain per unit length may be obtained by rearranging the above expression to give:

$$\bar{g}_{th} = \bar{\alpha} + \frac{1}{2L} \ln \frac{1}{r_1 r_2} \tag{6.20}$$

The second term on the right hand side of Eq. (6.20) represents the transmission loss through the mirrors.*

For laser action to be easily achieved it is clear that a high threshold gain per unit length is required in order to balance the losses from the cavity. However, it must be noted that the parameters displayed in Eq. (6.20) are totally dependent on the laser type.

Example 6.3

An injection laser has an active cavity with losses of 30 cm^{-1} and the reflectivity of the each cleaved laser facet is 30%. Determine the laser gain coefficient for the cavity when it has a length of 600 µm.

Solution: The threshold gain per unit length where $r_1 = r_2 = r$ is given by Eq. (6.20) as:

$$\bar{g}_{th} = \bar{\alpha} + \frac{1}{L} \ln \frac{1}{r}$$

$$= 30 + \frac{1}{0.06} + \ln \frac{1}{0.3}$$

$$= 50 \text{ cm}^{-1}$$

The threshold gain per unit length is equivalent to the laser gain coefficient for the active cavity, which is 50 cm^{-1}.

* This term is sometimes expressed in the form 1/L ln 1/r, where r, the reflectivity of the mirrored ends, is equal to $\sqrt{(r_1 r_2)}$.

6.3 Optical emission from semiconductors

6.3.1 The *p–n* junction

To allow consideration of semiconductor optical sources it is necessary to review some of the properties of semiconductor materials, especially with regard to the *p–n* junction. A perfect semiconductor crystal containing no impurities or lattice defects is said to be intrinsic. The energy band structure [Ref. 1] of an intrinsic semiconductor is illustrated in Figure 6.8(a) which shows the valence and conduction bands separated by a forbidden energy gap or bandgap E_g, the width of which varies for different semiconductor materials.

Figure 6.8(a) shows the situation in the semiconductor at a temperature above absolute zero where thermal excitation raises some electrons from the valence band into the conduction band, leaving empty hole states in the valence band. These thermally excited electrons in the conduction band and the holes left in the valence band allow conduction through the material, and are called carriers.

For a semiconductor in thermal equilibrium the energy-level occupation is described by the Fermi–Dirac distribution function (rather than the Boltzmann). Consequently, the probability $P(E)$ that an electron gains sufficient thermal energy at an absolute temperature T, such that it will be found occupying a particular energy level E, is given by the Fermi–Dirac distribution [Ref. 1]:

$$P(E) = \frac{1}{1 + \exp(E - E_F)/KT} \tag{6.21}$$

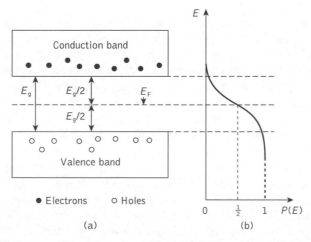

Figure 6.8 (a) The energy band structure of an intrinsic semiconductor at a temperature above absolute zero, showing an equal number of electrons and holes in the conduction band and the valence band respectively. (b) The Fermi–Dirac probability distribution corresponding to (a)

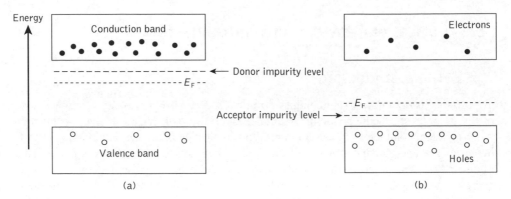

Figure 6.9 Energy band diagrams: (a) *n*-type semiconductor; (b) *p*-type semiconductor

where K is Boltzmann's constant and E_F is known as the Fermi energy or Fermi level. The Fermi level is only a mathematical parameter but it gives an indication of the distribution of carriers within the material. This is shown in Figure 6.8(b) for the intrinsic semiconductor where the Fermi level is at the center of the bandgap, indicating that there is a small probability of electrons occupying energy levels at the bottom of the conduction band and a corresponding number of holes occupying energy levels at the top of the valence band.

To create an extrinsic semiconductor the material is doped with impurity atoms which create either more free electrons (donor impurity) or holes (acceptor impurity). These two situations are shown in Figure 6.9 where the donor impurities form energy levels just below the conduction band while acceptor impurities form energy levels just above the valence band.

When donor impurities are added, thermally excited electrons from the donor levels are raised into the conduction band to create an excess of negative charge carriers and the semiconductor is said to be *n*-type, with the majority carriers being electrons. The Fermi level corresponding to this carrier distribution is raised to a position above the center of the bandgap, as illustrated in Figure 6.9(a). When acceptor impurities are added, as shown in Figure 6.9(b), thermally excited electrons are raised from the valence band to the acceptor impurity levels leaving an excess of positive charge carriers in the valence band and creating a *p*-type semiconductor where the majority carriers are holes. In this case Fermi level is lowered below the center of the bandgap.

The *p–n* junction diode is formed by creating adjoining *p*- and *n*-type semiconductor layers in a single crystal, as shown in Figure 6.10(a). A thin depletion region or layer is formed at the junction through carrier recombination which effectively leaves it free of mobile charge carriers (both electrons and holes). This establishes a potential barrier between the *p*- and *n*-type regions which restricts the interdiffusion of majority carriers from their respective regions, as illustrated in Figure 6.10(b). In the absence of an externally applied voltage no current flows as the potential barrier prevents the net flow of carriers from one region to another. When the junction is in this equilibrium state the Fermi level for the *p*- and *n*-type semiconductor is the same as shown Figure 6.10(b).

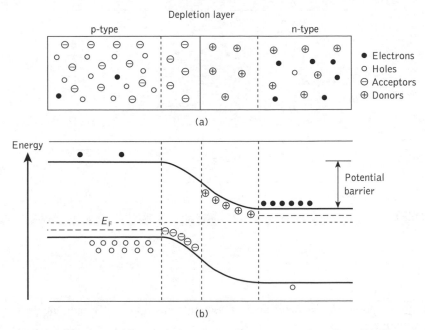

Figure 6.10 (a) The impurities and charge carriers at a *p–n* junction. (b) The energy band diagram corresponding to (a)

The width of the depletion region and thus the magnitude of the potential barrier is dependent upon the carrier concentrations (doping) in the *p*- and *n*-type regions and any external applied voltage. When an external positive voltage is applied to the *p*-type region with respect to the *n*-type, both the depletion region width and the resulting potential barrier are reduced and the diode is said to be forward biased. Electrons from the *n*-type region and holes from the *p*-type region can flow more readily across the junction into the opposite type region. These minority carriers are effectively injected across the junction by the application of the external voltage and form a current flow through the device as they continuously diffuse away from the interface. However, this situation in suitable semiconductor materials allows carrier recombination with the emission of light.

6.3.2 Spontaneous emission

The increased concentration of minority carriers in the opposite type region in the forward-biased *p–n* diode leads to the recombination of carriers across the bandgap. This process is shown in Figure 6.11 for a direct bandgap (see Section 6.3.3) semiconductor material where the normally empty electron states in the conduction band of the *p*-type material and the normally empty hole states in the valence band of the *n*-type material are populated by injected carriers which recombine across the bandgap. The energy released by this electron–hole recombination is approximately equal to the bandgap energy E_{g}.

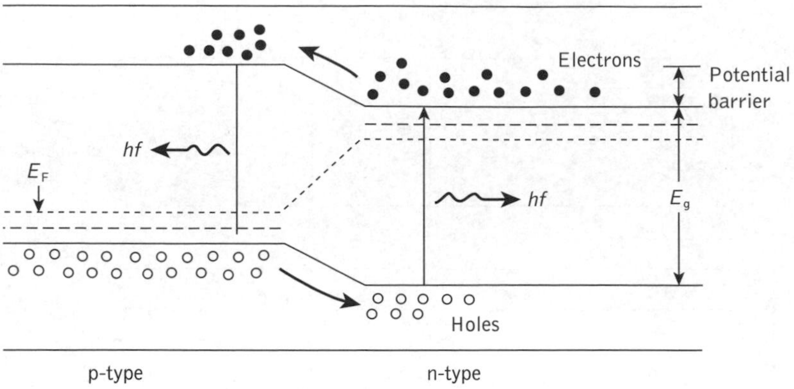

Figure 6.11 The p–n junction with forward bias giving spontaneous emission of photons

Excess carrier population is therefore decreased by recombination which may be radiative or nonradiative.

In nonradiative recombination the energy released is dissipated in the form of lattice vibrations and thus heat. However, in band-to-band radiative recombination the energy is released with the creation of a photon (see Figure 6.11) with a frequency following Eq. (6.1) where the energy is approximately equal to the bandgap energy E_g and therefore:

$$E_g = hf = \frac{hc}{\lambda} \tag{6.22}$$

where c is the velocity of light in a vacuum and λ is the optical wavelength. Substituting the appropriate values for h and c in Eq. (6.22) and rearranging gives:

$$\lambda = \frac{1.24}{E_g} \tag{6.23}$$

where λ is written in μm and E_g in eV.

This spontaneous emission of light from within the diode structure is known as electroluminescence.* The light is emitted at the site of carrier recombination which is primarily close to the junction, although recombination may take place through the hole diode structure as carriers diffuse away from the junction region (see Figure 6.12). However, the amount of radiative, recombination and the emission area within the structure is dependent upon the semiconductor materials used and the fabrication of the device.

---

* The term electroluminescence is used when the optical emission results from the application of an electric field.

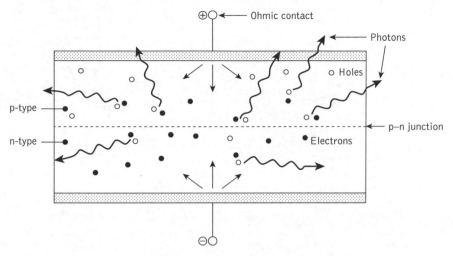

Figure 6.12 An illustration of carrier recombination giving spontaneous emission of light in a *p–n* junction diode

6.3.3 Carrier recombination

6.3.3.1 Direct and indirect bandgap semiconductors

In order to encourage electroluminescence it is necessary to select an appropriate semiconductor material. The most useful materials for this purpose are direct bandgap semiconductors in which electrons and holes on either side of the forbidden energy gap have the same value of crystal momentum and thus direct recombination is possible. This process is illustrated in Figure 6.13(a) with an energy–momentum diagram for a direct bandgap semiconductor. It may be observed that the energy maximum of the valence band occurs at the same (or very nearly the same) value of electron crystal momentum* as the energy minimum of the conduction band. Hence when electron–hole recombination occurs the momentum of the electron remains virtually constant and the energy released, which corresponds to the bandgap energy E_g, may be emitted as light. This direct transition of an electron across the energy gap provides an efficient mechanism for photon emission and the average time that the minority carrier remains in a free state before recombination (the minority carrier lifetime) is short (10^{-8} to 10^{-10} s). Some commonly used direct bandgap semiconductor materials are shown in Table 6.1 [Refs 3, 5].

In indirect bandgap semiconductors, however, the maximum and minimum energies occur at different values of crystal momentum (Figure 6.13(b)). For electron–hole recombination to take place it is essential that the electron loses momentum such that it has a value of momentum corresponding to the maximum energy of the valence band. The conservation of momentum requires the emission or absorption of a third particle, a phonon.

---

* The crystal momentum p is related to the wave vector k for an electron in a crystal by $p = 2\pi hk$, where h is Planck's constant [Ref. 1]. Hence the abscissa of Figure 6.13 is often shown as the electron wave vector rather than momentum.

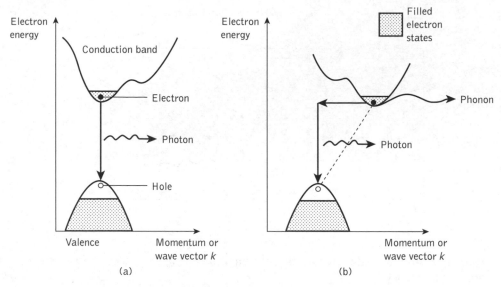

Figure 6.13 Energy–momentum diagrams showing the types of transition: (a) direct bandgap semiconductor; (b) indirect bandgap semiconductor

Table 6.1 Some direct and indirect bandgap semiconductors with calculated recombination coefficients

Semiconductor material	Energy bandgap (eV)	Recombination coefficient B_r (cm^3 s^{-1})
GaAs	Direct: 1.43	7.21×10^{-10}
CaSb	Direct: 0.73	2.39×10^{-10}
InAs	Direct: 0.35	8.5×10^{-11}
InSb	Direct: 0.18	4.58×10^{-11}
Si	Indirect: 1.12	1.79×10^{-15}
Ge	Indirect: 0.67	5.25×10^{-14}
GaP	Indirect: 2.26	5.37×10^{-14}

This three-particle recombination process is far less probable than the two-particle process exhibited by direct bandgap semiconductors. Hence, the recombination in indirect bandgap semiconductors is relatively slow (10^{-2} to 10^{-4} s). This is reflected by a much longer minority carrier lifetime, together with a greater probability of nonradiative transitions. The competing nonradiative recombination processes which involve lattice defects and impurities (e.g. precipitates of commonly used dopants) become more likely as they allow carrier recombination in a relatively short time in most materials. Thus the indirect bandgap emitters such as silicon and germanium shown in Table 6.1 give insignificant levels of electroluminescence. This disparity is further illustrated in Table 6.1 by the values of the recombination coefficient B_r given for both the direct and indirect bandgap recombination semiconductors shown.

The recombination coefficient is obtained from the measured absorption coefficient of the semiconductor, and for low injected minority carrier density relative to the majority carriers it is related approximately to the radiative minority carrier lifetime* τ_r by [Ref. 4]:

$$\tau_r = [B_r(N + P)]^{-1} \tag{6.24}$$

where N and P are the respective majority carrier concentrations in the n- and p-type regions. The significant difference between the recombination coefficients for the direct and indirect bandgap semiconductors shown underlines the importance of the use of direct bandgap materials for electroluminescent sources. Direct bandgap semiconductor devices in general have a much higher internal quantum efficiency. This is the ratio of the number of radiative recombinations (photons produced within the structure) to the number of injected carriers which is often expressed as a percentage.

Example 6.4

Compare the approximate radiative minority carrier lifetimes in gallium arsenide and silicon when the minority carriers are electrons injected into the p-type region which has a hole concentration of 10^{18} cm^{-3}. The injected electron density is small compared with the majority carrier density.

Solution: Equation (6.24) gives the radiative minority carrier lifetime τ_r as:

$$\tau_r \simeq [B_r(N + P)]^{-1}$$

In the p-type region the hole concentration determines the radiative carrier lifetime as $P \gg N$. Hence:

$$\tau_r \simeq [B_r N]^{-1}$$

Thus for gallium arsenide:

$$\tau_r \simeq [7.21 \times 10^{-10} \times 10^{18}]^{-1}$$
$$= 1.39 \times 10^{-9}$$
$$= 1.39 \text{ ns}$$

For silicon:

$$\tau_r \simeq [1.79 \times 10^{-15} \times 10^{18}]^{-1}$$
$$= 5.58 \times 10^{-4}$$
$$= 0.56 \text{ ms}$$

Thus the direct bandgap gallium arsenide has a radiative carrier lifetime factor of around 2.5×10^{-6} less than the indirect bandgap silicon.

---

* The radiative minority carrier lifetime is defined as the average time that a minority carrier can exist in a free state before radiative recombination takes place.

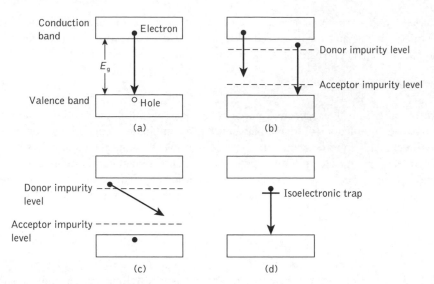

Figure 6.14 Major radiative recombination processes at 300 K: (a) conduction to valence band (band-to-band) transition: (b) conduction band to acceptor impurity, and donor impurity to valence band transition; (c) donor impurity to acceptor impurity transition; (d) recombination from an isoelectronic impurity to the valence band

6.3.3.2 Other radiative recombination processes

In the preceding sections, only full bandgap transitions have been considered to give radiative recombination. However, energy levels may be introduced into the bandgap by impurities or lattice defects within the material structure which may greatly increase the electron–hole recombination (effectively reduce the carrier lifetime). The recombination process through such impurity or defect centers may be either radiative or nonradiative. Major radiative recombination processes at 300 K other than band-to-band transitions are shown in Figure 6.14. These are band to impurity center or impurity center to band, donor level to acceptor level and recombination involving isoelectronic impurities.

Hence, an indirect bandgap semiconductor may be made into a more useful electroluminescent material by the addition of impurity centers which will effectively convert it into a direct bandgap material. An example of this is the introduction of nitrogen as an impurity into gallium phosphide. In this case the nitrogen forms an isoelectronic impurity as it has the same number of valence (outer shell) electrons as phosphorus but with a different covalent radius and higher electronegativity [Ref. 1]. The nitrogen impurity center thus captures an electron and acts as an isoelectronic trap which has a large spread of momentum. This trap then attracts the oppositely charged carrier (a hole) and a direct transition takes place between the impurity center and the valence band. Hence gallium phosphide may become an efficient light emitter when nitrogen is incorporated. However, such conversion of indirect to direct bandgap transitions is only readily achieved in materials where the direct and indirect bandgaps have a small energy difference. This is the case with gallium phosphide but not with silicon or germanium.

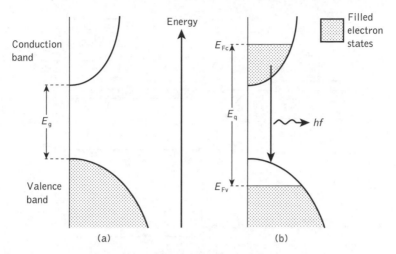

Figure 6.15 The filled electron states for an intrinsic direct bandgap semiconductor at absolute zero [Ref. 6]: (a) in equilibrium; (b) with high carrier injection

6.3.4 Stimulated emission and lasing

The general concept of stimulated emission via population inversion was indicated in Section 6.2.3. Carrier population inversion is achieved in an intrinsic (undoped) semiconductor by the injection of electrons into the conduction band of the material. This is illustrated in Figure 6.15 where the electron energy and the corresponding filled states are shown. Figure 6.15(a) shows the situation at absolute zero when the conduction band contains no electrons. Electrons injected into the material fill the lower energy states in the conduction band up to the injection energy or the quasi-Fermi level for electrons. Since charge neutrality is conserved within the material, an equal density of holes is created in the top of the valence band by the absence of electrons, as shown in Figure 6.15(b) [Ref. 6].

Incident photons with energy E_g but less than the separation energy of the quasi-Fermi levels $E_q = E_{Fc} - E_{Fv}$ cannot be absorbed because the necessary conduction band states are occupied. However, these photons can induce a downward transition of an electron from the filled conduction band states into the empty valence band states, thus stimulating the emission of another photon. The basic condition for stimulated emission is therefore dependent on the quasi-Fermi level separation energy as well as the bandgap energy and may be defined as:

$$E_{Fc} - E_{Fv} > hf > E_g \tag{6.25}$$

However, it must be noted that we have described an ideal situation whereas at normal operating temperatures the distribution of electrons and holes is less well defined but the condition for stimulated emission is largely maintained.

Population inversion may be obtained at a p–n junction by heavy doping (degenerative doping) of both the p- and n-type material. Heavy p-type doping with acceptor impurities causes a lowering of the Fermi level or boundary between the filled and empty states into

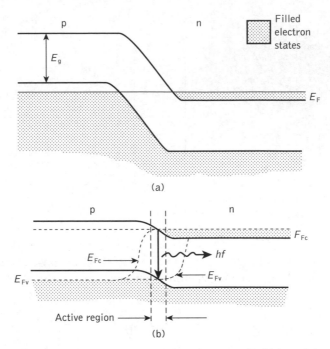

Figure 6.16 The degenerate p–n junction: (a) with no applied bias; (b) with strong forward bias such that the separation of the quasi-Fermi levels is higher than the electron–hole recombination energy hf in the narrow active region. Hence stimulated emission is obtained in this region

the valence band. Similarly, degenerative n-type doping causes the Fermi level to enter the conduction band of the material. Energy band diagrams of a degenerate p–n junction are shown in Figure 6.16. The position of the Fermi level and the electron occupation (shading) with no applied bias are shown in Figure 6.16(a). Since in this case the junction is in thermal equilibrium, the Fermi energy has the same value throughout the material. Figure 6.16(b) shows the p–n junction when a forward bias nearly equal to the bandgap voltage is applied and hence there is direct conduction. At high injection carrier density* in such a junction there exists an active region near the depletion layer that contains simultaneously degenerate populations of electrons and holes (sometimes termed doubly degenerate). For this region the condition for stimulated emission of Eq. (6.22) is satisfied for electromagnetic radiation of frequency $E_g/h < f < (E_{Fc} - E_{Fv})/h$. Therefore, any radiation of this frequency which is confined to the active region will be amplified. In general, the degenerative doping distinguishes a p–n junction which provides stimulated emission from one which gives only spontaneous emission as in the case of the LED.

Finally, it must be noted that high impurity concentration within a semiconductor causes differences in the energy bands in comparison with an intrinsic semiconductor. These differences are particularly apparent in the degeneratively doped p–n junctions used

---

* This may be largely considered to be electrons injected into the p–n region because of their greater mobility.

for semiconductor lasers. For instance, at high donor-level concentrations in gallium arsenide, the donor impurity levels form a band that merges with the conduction band. These energy states, sometimes referred to as 'bandtail' states [Ref. 7], extend into the forbidden energy gap. The laser transition may take place from one of these states. Furthermore, the transitions may terminate on acceptor states which because of their high concentration also extend as a band into the energy gap. In this way the lasing transitions may occur at energies less than the bandgap energy E_g. When transitions of this type dominate, the lasing peak energy is less than the bandgap energy. Hence the effective lasing wavelength can be varied within the electroluminescent semiconductor used to fabricate the junction laser through variation of the impurity concentration. For example, the lasing wavelength of gallium arsenide may be varied between 0.85 and 0.95 μm, although the best performance is usually achieved in the 0.88 to 0.91 μm band (see Problem 6.6).

However, a further requirement of the junction diode is necessary to establish lasing. This involves the provision of optical feedback to give laser oscillation. It may be achieved by the formation of an optical cavity (Fabry–Pérot cavity, see Section 6.2.4) within the structure by polishing the end faces of the junction diode to act as mirrors. Each end of the junction is polished or cleaved and the sides are roughened to prevent any unwanted light emission and hence wasted population inversion.

The behavior of the semiconductor laser can be described by rate equations for electron and photon density in the active layer of the device. These equations assist in providing an understanding of the laser electrical and optical performance characteristics under direct current modulation as well as its potential limitations. The problems associated with high-speed direct current modulation are unique to the semiconductor laser whose major application area is that of a source within optical fiber communications.

Although the rate equations may be approached with some rigor [Ref. 8] we adopt a simplified analysis which is valid within certain constraints [Ref. 9]. In particular, the equations represent an average behavior for the active medium within the laser cavity and they are not applicable when the time period is short compared with the transit time of the optical wave in the laser cavity.* The two rate equations for electron density n, and photon density ϕ, are:

$$\frac{dn}{dt} = \frac{J}{ed} - \frac{n}{\tau_{sp}} - Cn\phi \qquad (m^{-3}\ s^{-1}) \tag{6.26}$$

and:

$$\frac{d\phi}{dt} = Cn\phi + \delta\frac{n}{\tau_{sp}} - \frac{\phi}{\tau_{ph}} \qquad (m^{-3}\ s^{-1}) \tag{6.27}$$

where J is the current density, in amperes per square meter, e is the charge on an electron, d is the thickness of the recombination region, τ_{sp} is the spontaneous emission lifetime which is equivalent to τ_{21} in Section 6.2.2, C is a coefficient which incorporates the B coefficients in Section 6.2.2, δ is a small fractional value and τ_{ph} is the photon lifetime.

---

* Thus performance characteristics derived from these rate equations become questionable when the time scale is less than 10 ps or the modulation bandwidth is greater than 100 GHz.

The rate equations given in Eqs (6.26) and (6.27) may be balanced by taking into account all the factors which affect the numbers of electrons and holes in the laser structure. Hence, in Eq. (6.26), the first term indicates the increase in the electron concentration in the conduction band as the current flows into the junction diode. The electrons lost from the conduction band by spontaneous and stimulated transitions are provided by the second and third terms respectively. In Eq. (6.27) the first term depicts the stimulated emission as a source of photons. The fraction of photons produced by spontaneous emission which combine to the energy in the lasing mode is given by the second term. This term is often neglected, however, as δ is small. The final term represents the decay in the number of photons resulting from losses in the optical cavity.

Although these rate equations may be used to study both the transient and steady-state behavior of the semiconductor laser, we are particularly concerned with the steady-state solutions. The steady state is characterized by the left hand side of Eqs (6.26) and (6.27) being equal to zero, when n and ϕ have nonzero values. In addition, the fields in the optical cavity which are represented by ϕ must build up from small initial values, and hence $d\phi/dt$ must be positive when ϕ is small. Therefore, setting δ equal to zero in Eq. (6.27), it is clear that for any value of ϕ, $d\phi/dt$ will only be positive when:

$$ Cn - \frac{1}{\tau_{ph}} \geq 0 \tag{6.28} $$

There is therefore a threshold value of n which satisfies the equality of Eq. (6.28). If n is larger than this threshold value, then ϕ can increase; however, when n is smaller it cannot. From Eq. (6.28) the threshold value for the electron density n_{th} is:

$$ n_{th} = \frac{1}{C\tau_{ph}} \quad (m^{-3}) \tag{6.29} $$

The threshold current written in terms of its current density J_{th}, required to maintain $n = n_{th}$ in the steady state when $\phi = 0$, may be obtained from Eq. (6.26) as:

$$ \frac{J_{th}}{ed} = \frac{n_{th}}{\tau_{sp}} \quad (m^{-3}\,s^{-1}) \tag{6.30} $$

Hence Eq. (6.30) defines the current required to sustain an excess electron density in the laser when spontaneous emission provides the only decay mechanism. The steady-state photon density ϕ_s is provided by substituting Eq. (6.30) in Eq. (6.26) giving:

$$ 0 = \frac{(J - J_{th})}{ed} - Cn_{th}\phi_s $$

Rearranging we obtain:

$$ \phi_s = \frac{1}{Cn_{th}} \frac{(J - J_{th})}{ed} \quad (m^{-3}) \tag{6.31} $$

Substituting for Cn_{th} from Eq. (6.29) we can write Eq. (6.31) in the form:

$$\phi_s = \frac{\tau_{ph}}{ed}(J - J_{th}) \quad (\text{m}^{-3}) \tag{6.32}$$

The photon density ϕ_s cannot be a negative quantity as this is meaningless, and for ϕ_s to be greater than zero the current must exceed its threshold value. Moreover, ϕ_s is proportional to the amount by which J exceeds its threshold value. As each photon has energy hf it is possible to determine the optical power density in W m^{-2} by assuming that half the photons are traveling in each of two directions.

An idealized optical output power against current characteristic (also called light output against current characteristic) for a semiconductor laser is illustrated in Figure 6.17. The solid line represents the laser characteristic, whereas the dashed line is a plot of Eq. (6.32) showing the current threshold. It may be observed that the device gives little light output in the region below the threshold current which corresponds to spontaneous emission only within the structure. However, after the threshold current density is reached, the light output increases substantially for small increases in current through the device. This corresponds to the region of stimulated emission when the laser is acting as an amplifier of light.

In common with all other laser types a requirement for the initiation and maintenance of laser oscillation is that the optical gain matches the optical losses within the cavity (see Section 6.2.5). For the p–n junction or semiconductor laser this occurs at a particular photon energy within the spectrum of spontaneous emission (usually near the peak wavelength of spontaneous emission). Thus when extremely high currents are passed through the device (i.e. injection levels of around 10^{18} carriers cm^{-3}), spontaneous emission with a wide spectrum (linewidth) becomes lasing (when a current threshold is passed) and the linewidth subsequently narrows.

For strongly confined structures the threshold current density for stimulated emission J_{th} is to a fair approximation [Refs 4, 10] related to the threshold gain coefficient $\bar{g}_{th}$ for the laser cavity through:

$$\bar{g}_{th} = \bar{\beta}J_{th} \tag{6.33}$$

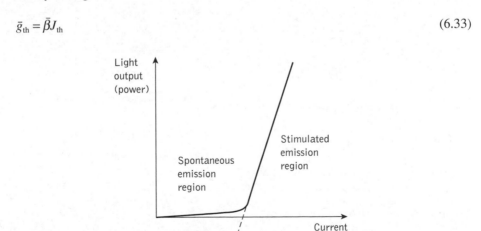

Figure 6.17 The ideal light output against current characteristic for an injection laser

where the gain factor $\bar{\beta}$ is a constant appropriate to specific devices. Detailed discussion of the more exact relationship is given in Ref. 4.

Substituting for $\bar{g}_{th}$ from Eq. (6.18) and rearranging we obtain:

$$J_{th} = \frac{1}{\bar{\beta}}\left[\bar{\alpha} + \frac{1}{2L}\ln\frac{1}{r_1 r_2}\right] \tag{6.34}$$

Since for the semiconductor laser the mirrors are formed by a dielectric plane and are often uncoated, the mirror reflectivities r_1 and r_2 may be calculated using the Fresnel reflection relationship of Eq. (5.1).

Example 6.5

A GaAs injection laser has an optical cavity of length 250 μm and width 100 μm. At normal operating temperature the gain factor $\bar{\beta}$ is 21×10^{-3} A cm^{-3} and the loss coefficient $\bar{\alpha}$ per cm is 10. Determine the threshold current density and hence the threshold current for the device. It may be assumed that the cleaved mirrors are uncoated and that the current is restricted to the optical cavity. The refractive index of GaAs may be taken as 3.6.

Solution: The reflectivity for normal incidence of a plane wave on the GaAs–air interface may be obtained from Eq. (5.1) where:

$$r_1 = r_2 = r = \left(\frac{n-1}{n+1}\right)^2$$

$$= \left(\frac{3.6-1}{3.6+1}\right)^2 \simeq 0.32$$

The threshold current density may be obtained from Eq. (6.34) where:

$$J_{th} = \frac{1}{\bar{\beta}}\left[\bar{\alpha}\frac{1}{L}\ln\frac{1}{r}\right]$$

$$= \frac{1}{21 \times 10^{-3}}\left[10 + \frac{1}{250 \times 10^{-4}}\ln\frac{1}{0.32}\right]$$

$$= 2.65 \times 10^3 \text{ A cm}^{-2}$$

The threshold current I_{th} is given by:

$$I_{th} = J_{th} \times \text{area of the optical cavity}$$
$$= 2.65 \times 10^3 \times 250 \times 100 \times 10^{-8}$$
$$\simeq 663 \text{ mA}$$

Therefore the threshold current for this device is 663 mA if the current flow is restricted to the optical cavity.

As the stimulated emission minority carrier lifetime is much shorter (typically 10^{-11} s) than that due to spontaneous emission, further increases in input current above the threshold will result almost entirely in stimulated emission, giving a high internal quantum efficiency (50 to 100%). Also, whereas incoherent spontaneous emission has a linewidth of tens of nanometers, stimulated coherent emission has a linewidth of a nanometer or less.

6.3.5 Heterojunctions

The preceding sections have considered the photoemissive properties of a single *p–n* junction fabricated from a single-crystal semiconductor material. This is known as a homojunction. However, the radiative properties of a junction diode may be improved by the use of heterojunctions. A heterojunction is an interface between two adjoining single-crystal semiconductors with different bandgap energies. Devices which are fabricated with heterojunctions are said to have heterostructure.

Heterojunctions are classified into either an isotype (n–n or p–p) or an anisotype (*p–p*). The isotype heterojunction provides a potential barrier within the structure which is useful for the confinement of minority carriers to a small active region (carrier confinement). It effectively reduces the carrier diffusion length and thus the volume within the structure where radiative recombination may take place. This technique is widely used for the fabrication of injection lasers and high-radiance LEDs. Isotype heterojunctions are also extensively used in LEDs to provide a transparent layer close to the active region which substantially reduces the absorption of light emitted from the structure.

Alternatively, anisotype heterojunctions with sufficiently large bandgap differences improve the injection efficiency of either electrons or holes. Both types of heterojunction provide a dielectric step due to the different refractive indices at either side of the junction. This may be used to provide radiation confinement to the active region (i.e. the walls of an optical waveguide). The efficiency of the containment depends upon the magnitude of the step which is dictated by the difference in bandgap energies and the wavelength of the radiation.

It is useful to consider the application of heterojunctions in the fabrication of a particular device. They were first used to provide potential barriers in injection lasers. When a double-heterojunction (DH) structure was implemented, the resulting carrier and optical confinement reduced the threshold currents necessary for lasing by a factor of around 100. Thus stimulated emission was obtained with relatively small threshold currents (50 to 200 mA). The layer structure and an energy band diagram for a DH injection laser are illustrated in Figure 6.18. A heterojunction is shown either side of the active layer for laser oscillation. The forward bias is supplied by connecting a positive electrode of a supply to the p side of the structure and a negative electrode to the n side. When a voltage which corresponds to the bandgap energy of the active layer is applied, a large number of electrons (or holes) are injected into the active layer and laser oscillation commences. These carriers are confined to the active layer by the energy barriers provided by the heterojunctions which are placed within the diffusion length of the injected carriers. It may also be observed from Figure 6.18(c) that a refractive index step (usually a difference of 5 to 10%) at the heterojunctions provides radiation containment to the active layer. In effect

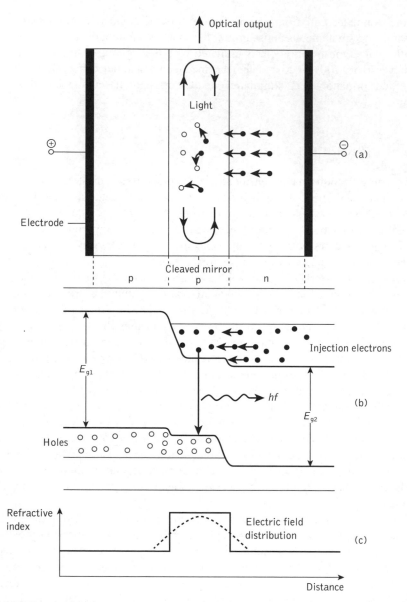

Figure 6.18 The double-heterojunction injection laser: (a) the layer structure, shown with an applied forward bias; (b) energy band diagram indicating a p–p heterojunction on the left and a *p–n* heterojunction on the right; (c) the corresponding refractive index diagram and electric field distribution

the active layer forms the center of a dielectric waveguide which strongly confines the electroluminescence within this region, as illustrated in Figure 6.18(c). The refractive index step shown is the same for each heterojunction, which is desirable in order to prevent losses due to lack of waveguiding which can occur if the structure is not symmetrical.

Careful fabrication of the heterojunctions is also important in order to reduce defects at the interfaces such as misfit dislocations or inclusions which cause nonradiative recombination and thus reduce the internal quantum efficiency. Lattice matching is therefore an important criterion for the materials used to form the interface. Ideally, heterojunctions should have a very small lattice parameter mismatch of no greater than 0.1%. However, it is often not possible to obtain such good lattice parameter matching with the semiconductor materials required to give emission at the desired wavelength and therefore much higher lattice parameter mismatch is often tolerated ($\approx$0.6%).

6.3.6 Semiconductor materials

The semiconductor materials used for optical sources must broadly fulfill several criteria. These are as follows:

1. *p–n* junction formation. The materials must lend themselves to the formation of *p–n* junctions with suitable characteristics for carrier injection.

2. Efficient electroluminescence. The devices fabricated must have a high probability of radiative transitions and therefore a high internal quantum efficiency. Hence the materials utilized must be either direct bandgap semiconductors or indirect bandgap semiconductors with appropriate impurity centers.

3. Useful emission wavelength. The materials must emit light at a suitable wavelength to be utilized with current optical fibers and detectors (0.8 to 1.7 μm). Ideally, they should allow bandgap variation with appropriate doping and fabrication in order that emission at a desired specific wavelength may be achieved.

Initial investigation of electroluminescent materials for LEDs in the early 1960s centered around the direct bandgap III–V alloy semiconductors including the binary compounds gallium arsenide (GaAs) and gallium phosphide (GaP) and the ternary gallium arsenide phosphide (GaAs$_x$P$_{1-x}$). Gallium arsenide gives efficient electroluminescence over an appropriate wavelength band (0.88 to 0.91 μm) and for the first-generation optical fiber communication systems was the first material to be fabricated into homojunction semiconductor lasers operating at low temperature. It was quickly realized that improved devices could be fabricated with heterojunction structures which through carrier and radiation confinement would give enhanced light output for drastically reduced device currents. These heterostructure devices were first fabricated using liquid-phase epitaxy (LPE) to produce GaAs/Al$_x$Ga$_{1-x}$As single-heterojunction lasers. This process involves the precipitation of material from a cooling solution onto an underlying substrate. When the substrate consists of a single crystal and the lattice constant or parameter of the precipitating material is the same or very similar to that of the substrate (i.e. the unit cells within the two crystalline structures are of a similar dimension), the precipitating material forms an epitaxial layer on the substrate surface. Subsequently, the same technique was used to produce double heterojunctions consisting of Al$_x$Ga$_{1-x}$As/GaAs/Al$_x$Ga$_{1-x}$As epitaxial layers, which gave continuous wave (CW) operation at room temperature [Refs 11–13]. Some of the common material systems now utilized for DH device fabrication, together with their useful wavelength ranges, are shown in Table 6.2.

Table 6.2 Some common material systems used in the fabrication of electroluminescent sources for optical fiber communications

Material systems active layer/confining layers	Useful wavelength range (µm)	Substrate
$GaAs/Al_xGa_{1-x}As$	0.8–0.9	GaAs
$GaAs/In_xGa_{1-x}P$	0.9	GaAs
$Al_yGa_{1-y}As/Al_xGa_{1-x}As$	0.65–0.9	GaAs
$In_yGa_{1-y}As/In_xGa_{1-x}P$	0.85–1.1	GaAs
$GaAs_{1-x}Sb_x/Ga_{1-y}Al_yAs_{1-x}Sb_x$	0.9–1.1	GaAs
$Ga_{1-y}Al_yAs_{1-x}Sb_x/GaSb$	1.0–1.7	GaSb
$In_{1-x}Ga_xAs_yP_{1-y}/InP$	0.92–1.7	InP
$In_xGa_{1-x}As/InGaAlAs$	1.3	InGaAs
$In_{1-x}GaN_yAs_{1-y}/GaNAs$	1.3–1.55	GaAs
$In_{1-x}Ga_xN_{1-y}As_ySb/Ga_{1-x}Al_xAs$	1.31	GaAs

The GaAs/AlGaAs DH system is the best developed and is used for fabricating both lasers and LEDs for the shorter wavelength region. The bandgap in this material may be 'tailored' to span the entire 0.8 to 0.9 µm wavelength band by changing the AlGa composition. Also there is very little lattice mismatch (0.017%) between the AlGaAs epitaxial layer and the GaAs substrate which gives good internal quantum efficiency. In the longer wavelength region (1.1 to 1.6 µm) a number of III–V alloys have been utilized which are compatible with GaAs, InP and GaSb substrates. These include ternary alloys such as $GaAs_{1-x}Sb_x$ and $In_xGa_{1-x}As$ grown on GaAs.

However, although the ternary alloys allow bandgap tailoring they have a fixed lattice parameter. Therefore, quaternary alloys which allow both bandgap tailoring and control of the lattice parameter (i.e. a range of lattice parameters is available for each bandgap) appear to be of more use for the longer wavelength region. The most advanced are $In_{1-x}Ga_xAs_yP_{1-y}$ lattice matched to InP and $Ga_{1-y}Al_yAs_{1-x}Sb_x$ lattice matched to GaSb. Both these material systems allow emission over the entire 1.0 to 1.7 µm wavelength band. The InGaAsP/InP material system remains the most favorable for both long-wavelength light sources and detectors. This is due to the ease of fabrication with lattice matching on InP which is also a suitable material for the active region with a bandgap energy of 1.35 eV at 300 K. Hence, InP/InGaAsP (active/confining) devices may be fabricated. Conversely, GaSb is a low-bandgap material (0.78 eV at 300 K) and the quaternary alloy must be used for the active region in the GaAlAsSb/GaSb system. Thus compositional control must be maintained for three layers in this system in order to minimize lattice mismatch in the active region, whereas it is only necessary for one layer in the InP/InGaAsP system.

A ternary substrate based on InGaAs has a lattice constant between that of GaAs and InP, and therefore this substrate is considered useful to provide for devices operating at the 1.30 µm wavelength exhibiting small size and low threshold current density [Ref. 13]. Furthermore ternary-nitride-based alloys also have an advantage over GaAs or InP-based alloys because of their better index match with silica optical fibers [Ref. 14]. These alloys can be produced by introducing nitrogen, which reduces the bandgap sufficiently to create a new material that contains properties of GaAs [Ref. 15]. For example, gallium nitride (GaN), which is used in the construction of blue lasers [Ref. 16] and hence normally operates at short wavelengths, can be extended to the 1.30 to 1.60 μm wavelength range when indium arsenide is added [Refs 17, 18]. The bandgap of the resulting alloy (i.e. InGaAsN) can be tailored to emit at wavelengths in the range from 1.30 to 1.55 μm. Recently, a low-threshold CW laser using InGaNAsSb/GaNAs functioning as the active/confinement layers grown on a GaAs substrate has demonstrated emission at the wavelength around 1.50 μm [Refs 15, 19, 20]. Moreover, an AlGaN/GaN-based heterostructure enables the realization of both sources and detectors for operation at 1.55 μm [Ref. 21]. In addition, erbium-doped ternary nitride waveguide-based optical amplifiers [Ref. 22] may provide integration with lasers and other functional optical devices, such as optical switches, wavelength routers and detectors. Further developments, however, in relation to optical propagation loss, temperature sensitivity, polarization effects in heterostructures, carrier-induced change of the index of refraction, modulation efficiency and modulation-induced polarization effects, are required to enable the full exploitation of the ternary-nitride-based alloys [Ref. 22].

6.4 The semiconductor injection laser

The electroluminescent properties of the forward-biased p–n junction diode have been considered in the preceding sections. Stimulated emission by the recombination of the injected carriers is encouraged in the semiconductor injection laser (also called the injection laser diode (ILD) or simply the injection laser) by the provision of an optical cavity in the crystal structure in order to provide the feedback of photons. This gives the injection laser several major advantages over other semiconductor sources (e.g. LEDs) that may be used for optical communications. These are as follows:

1. High radiance due to the amplifying effect of stimulated emission. Injection lasers will generally supply milliwatts of optical output power.

2. Narrow linewidth on the order of 1 nm (10 Å) or less which is useful in minimizing the effects of material dispersion.

3. Modulation capabilities which at present extend up into the gigahertz range and will undoubtedly be improved upon.

4. Relative temporal coherence which is considered essential to allow heterodyne (coherent) detection in high-capacity systems, but at present is primarily of use in single-mode systems.

5. Good spatial coherence which allows the output to be focused by a lens into a spot which has a greater intensity than the dispersed unfocused emission. This permits efficient coupling of the optical output power into the fiber even for fibers with low numerical aperture. The spatial fold matching to the optical fiber which may be obtained with the laser source is not possible with an incoherent emitter and, consequently, coupling efficiencies are much reduced.

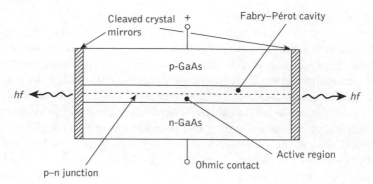

Figure 6.19 Schematic diagram of a GaAs homojunction injection laser with a Fabry–Pérot cavity

These advantages, together with the compatibility of the injection laser with optical fibers (e.g. size), led to the early developments of the device in the 1960s. Early injection lasers had the form of a Fabry–Pérot cavity often fabricated in gallium arsenide which was the major III–V compound semiconductor with electroluminescent properties at the appropriate wavelength for first-generation systems. The basic structure of this homojunction device is shown in Figure 6.19, where the cleaved ends of the crystal act as partial mirrors in order to encourage stimulated emission in the cavity when electrons are injected into the *p*-type region. However, as mentioned previously these devices had a high threshold current density (greater than 10^4 A cm^{-2}) due to their lack of carrier containment and proved inefficient light sources. The high current densities required dictated that these devices when operated at 300 K were largely utilized in a pulsed mode in order to minimize the junction temperature and thus avert damage.

Improved carrier containment and thus lower threshold current densities (around 10^3 A cm^{-2}) were achieved using heterojunction structures (see Section 6.3.5). The DH injection laser fabricated from lattice-matched III–V alloys provided both carrier and optical confinement on both sides of the *p–n* junction, giving the injection laser a greatly enhanced performance. This enabled these devices with the appropriate heat sinking to be operated in a CW mode at 300 K with obvious advantages for optical communications (e.g. analog transmission). However, in order to provide reliable CW operation of the DH injection laser it was necessary to provide further carrier and optical confinement which led to the introduction of stripe geometry DH laser configurations. Prior to discussion of this structure, however, it is useful to consider the efficiency of the semiconductor injection laser as an optical source.

6.4.1 Efficiency

There are a number of ways in which the operational efficiency of the semiconductor laser may be defined. A useful definition is that of the differential external quantum efficiency η_D which is the ratio of the increase in photon output rate for a given increase in the number of injected electrons. If P_e is the optical power emitted from the device, I is the current, e is the charge on an electron and hf is the photon energy, then:

$$\eta_D = \frac{dP_e/hf}{dI/e} \simeq \frac{dP_e}{dI(E_g)} \tag{6.35}$$

where E_g is the bandgap energy expressed in eV. It may be noted that η_D gives a measure of the rate of change of the optical output power with current and hence defines the slope of the output characteristic (Figure 6.17) in the lasing region for a particular device. Hence η_D is sometimes referred to as the slope quantum efficiency. For a CW semiconductor laser it usually has values in the range 40 to 60%. Alternatively, the internal quantum efficiency of the semiconductor laser η_i, which was defined in Section 6.3.3.1 as:

$$\eta_i = \frac{\text{number of photons produced in the laser cavity}}{\text{number of injected electrons}} \tag{6.36}$$

may be quite high with values usually in the range 50 to 100%. It is related to the differential external quantum efficiency by the expression [Ref. 4]:

$$\eta_D = \eta_i \left[\frac{1}{1 + (2\bar{\alpha}L/\ln(1/r_1 r_2))} \right] \tag{6.37}$$

where $\bar{\alpha}$ is the loss coefficient of the laser cavity, L is the length of the laser cavity and r_1, r_2 are the cleaved mirror reflectivities.

Another parameter is the total efficiency (external quantum efficiency) η_T which is efficiency defined as:

$$\eta_T = \frac{\text{total number of output photons}}{\text{total number of injected electrons}} \tag{6.38}$$

$$= \frac{P_e/hf}{I/e} \simeq \frac{P_e}{IE_g} \tag{6.39}$$

As the power emitted P_e changes linearly when the injection current I is greater than the threshold current I_{th}, then:

$$\eta_T \simeq \eta_D \left(1 - \frac{I_{th}}{I} \right) \tag{6.40}$$

For high injection current (e.g. $I = 5I_{th}$) then $\eta_T \simeq \eta_D$, whereas for lower currents ($I \simeq 2I_{th}$) the total efficiency is lower and around 15 to 25%.

The external power efficiency of the device (or device efficiency) η_{ep} in converting electrical input to optical output is given by:

$$\eta_{ep} = \frac{P_e}{P} \times 100 = \frac{P_e}{IV} \times 100\% \tag{6.41}$$

where $P = IV$ is the d.c. electrical input power.

Using Eq. (6.39) for the total efficiency we find:

$$\eta_{ep} = \eta_T \left(\frac{E_g}{V} \right) \times 100\% \tag{6.42}$$

Example 6.6

The total efficiency of an injection laser with a GaAs active region is 18%. The voltage applied to the device is 2.5 V and the bandgap energy for GaAs is 1.43 eV. Calculate the external power efficiency of the device.

Solution: Using Eq. (6.42), the external power efficiency is given by:

$$\eta_{ep} = 0.18 \left(\frac{1.43}{2.5} \right) \times 100 \simeq 10\%$$

This result indicates the possibility of achieving high overall power efficiencies from semiconductor lasers which are much larger than for other laser types.

6.4.2 Stripe geometry

The DH laser structure provides optical confinement in the vertical direction through the refractive index step at the heterojunction interfaces, but lasing takes place across the whole width of the device. This situation is illustrated in Figure 6.20 which shows the broad-area DH laser where the sides of the cavity are simply formed by roughening the

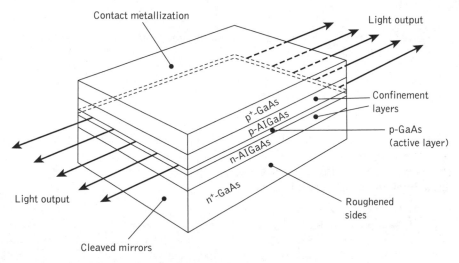

Figure 6.20 A broad-area GaAs/AlGaAs DH injection laser

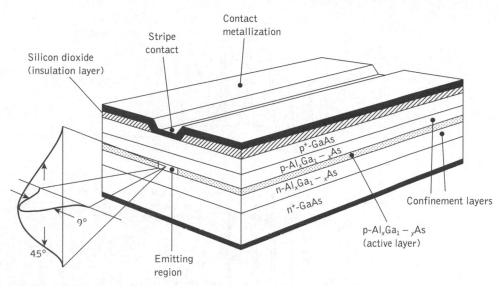

Figure 6.21 Schematic representation of an oxide stripe AlGaAs DH injection laser

edges of the device in order to reduce unwanted emission in these directions and limit the number of horizontal transverse modes. However, the broad emission area creates several problems including difficult heat sinking, lasing from multiple filaments in the relatively wide active area and unsuitable light output geometry for efficient coupling to the cylindrical fibers.

To overcome these problems while also reducing the required threshold current, laser structures in which the active region does not extend to the edges of the device were developed. A common technique involved the introduction of stripe geometry to the structure to provide optical containment in the horizontal plane. The structure of a DH stripe contact laser is shown in Figure 6.21 where the major current flow through the device and hence the active region is within the stripe. Generally, the stripe is formed by the creation of high-resistance areas on either side by techniques such as proton bombardment [Ref. 11] or oxide isolation [Ref. 12]. The stripe therefore acts as a guiding mechanism which overcomes the major problems of the broad-area device. However, although the active area width is reduced the light output is still not particularly well collimated due to isotropic emission from a small active region and diffraction within the structure. The optical output and far-field emission pattern are also illustrated in Figure 6.21. The output beam divergence is typically 45° perpendicular to the plane of the junction and 9° parallel to it. Nevertheless, this is a substantial improvement on the broad-area laser.

The stripe contact device also gives, with the correct balance of guiding, single transverse (in a direction parallel to the junction plane) mode operation, whereas the broad-area device tends to allow multimode operation in this horizontal plane. Numerous stripe geometry laser structures have been investigated with stripe widths ranging from 2 to 65 μm, and the DH stripe geometry structure has been widely utilized for optical fiber communications. Such structures have active regions which are planar and continuous.

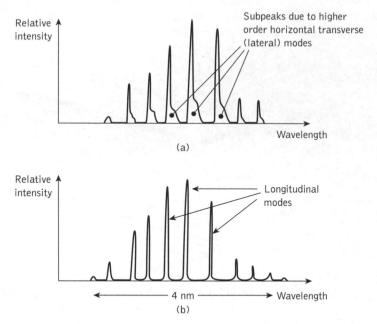

Figure 6.22 Output spectra for multimode injection lasers: (a) broad-area device with multitransverse modes; (b) stripe geomety device with single transverse mode

Hence the stimulated emission characteristics of these injection lasers are determined by the carrier distribution (which provides optical gain) along the junction plane. The optical mode distribution along the junction plane is, however, decided by the optical gain and therefore these devices are said to be gain-guided laser structures (see Section 6.5.1). In addition, ridge waveguide laser diodes (see Section 6.5.2) with stripe widths of 5, 10 and 20 μm and lengths ranging from 400 to 1500 μm have also been reported [Refs 19, 23].

6.4.3 Laser modes

The typical output spectrum for a broad-area injection laser is shown in Figure 6.22(a). It does not consist of a single wavelength output but a series of wavelength peaks corresponding to different longitudinal (in the plane of the junction, along the optical cavity) modes within the structure. As indicated in Section 6.2.4, the spacing of these modes is dependent on the optical cavity length as each one corresponds to an integral number of lengths. They are generally separated by a few tenths of a nanometer, and the laser is said to be a multimode device. However, Figure 6.22(a) also indicates some broadening of the longitudinal mode peaks due to subpeaks caused by higher order horizontal transverse modes.* These higher order lateral modes may exist in the broad-area device due to the

---

* Tranverse modes in the plane of the junction are often called lateral modes, transverse modes being reserved for modes perpendicular to the junction plane.

unrestricted width of the active region. The correct stripe geometry inhibits the occurrence of the higher order lateral modes by limiting the width of the optical cavity, leaving only a single lateral mode which gives the output spectrum shown in Figure 6.22(b) where only the longitudinal modes may be observed. This represents the typical output spectrum for a good multimode injection laser.

6.4.4 Single-mode operation

For single-mode operation, the optical output from a laser must contain only a single longitudinal and single transverse mode. Hence the spectral width of the emission from the single-mode device is far smaller than the broadened transition linewidth discussed in Section 6.2.4. It was indicated that an inhomogeneously broadened laser can support a number of longitudinal and transverse modes simultaneously, giving a multimode output. Single transverse mode operation, however, may be obtained by reducing the aperture of the resonant cavity such that only the TEM_{00} mode is supported. To obtain single-mode operation it is then necessary to eliminate all but one of the longitudinal modes.

One method of achieving single longitudinal mode operation is to reduce the length L of the cavity until the frequency separation of the adjacent modes given by Eq. (6.14) as $\delta f = c/2nL$ is larger than the laser transition linewidth or gain curve. Then only the single mode which falls within the transition linewidth can oscillate within the laser cavity. However, it is clear that rigid control of the cavity parameters is essential to provide the mode stabilization necessary to achieve and maintain this single-mode operation.

The structures required to give mode stability are discussed with regard to the multimode injection laser in Section 6.5 and similar techniques can be employed to produce a laser emitting a single longitudinal and transverse mode. For example, the correct DH structure will restrict the vertical width of the waveguiding region to less than 0.4 μm allowing only the fundamental transverse mode to be supported and removing any interference of the higher order transverse modes on the emitted longitudinal modes.

The lateral modes (in the plane of the junction) may be confined by the restrictions on the current flow provided by the stripe geometry. In general, only the lower order modes are excited, which appear as satellites to each of the longitudinal modes. However, as will be discussed in Section 6.5.1, stripe contact devices often have instabilities and strong nonlinearities (e.g. kinks) in their light output against current characteristics. Tight current confinement as well as good waveguiding are therefore essential in order to achieve only the required longitudinal modes which form between the mirror facets in the plane of the junction. Finally, as indicated above, single-mode operation may be obtained through control of the optical cavity length such that only a single longitudinal mode falls within the gain bandwidth of the device. Figure 6.23 shows a typical output spectrum for a single-mode device.

However, injection lasers with short cavity lengths (around 50 μm) are difficult to handle and have not been particularly successful. Nevertheless, such devices, together with the major alternative structures which provide single-mode operation, are dealt with in Section 6.6 under the title of single-frequency injection lasers.

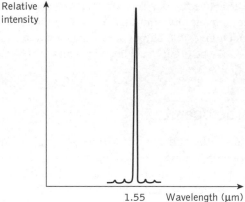

Figure 6.23 Typical single longitudinal mode output spectrum from a single-mode injection laser

6.5 Some injection laser structures

6.5.1 Gain-guided lasers

Fabrication of multimode injection lasers with a single or small number of lateral modes is achieved by the use of stripe geometry. These devices are often called gain-guided lasers as indicated in Section 6.4.2. The constriction of the current flow to the stripe is realized in the structure either by implanting the regions outside the stripe with protons (proton-isolated stripe) to make them highly resistive, or by oxide or $p–n$ junction isolation. The structure for an aluminum gallium arsenide oxide isolated stripe DH laser was shown in Figure 6.21. It has an active region of gallium arsenide bounded on both sides by aluminum gallium arsenide regions. This technique has been widely applied, especially for multimode laser structures used in the shorter wavelength region. The current is confined by etching a narrow stripe in a silicon dioxide film.

Two other basic techniques for the fabrication of gain-guided laser structures are illustrated in Figure 6.24(a) and (b) which show the proton-isolated stripe and the $p–n$ junction isolated stripe structures respectively. In Figure 6.24(a) the resistive region formed by the proton bombardment gives better current confinement than the simple oxide stripe and has superior thermal properties due to the absence of the silicon dioxide layer; $p–n$ junction isolation involves a selective diffusion through the n-type surface region in order to reach the p-type layers, as illustrated in Figure 6.24(b). None of these structures confines all the radiation and current to the stripe region and spreading occurs on both sides of the stripe. With stripe widths of 10 μm or less, such planar stripe lasers provide highly efficient coupling into multimode fibers, but significantly lower coupling efficiency is achieved into small-core-diameter single-mode fibers.

The optical output power against current characteristic for the ideal semiconductor laser was illustrated in Figure 6.17. However, with certain practical laser diodes the

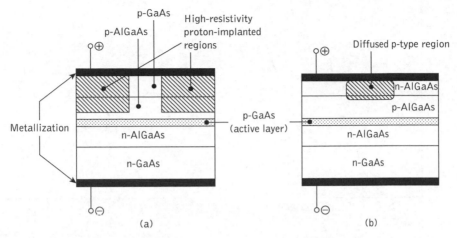

Figure 6.24 Schematic representation of structures for stripe geometry injection lasers: (a) proton-isolated stripe GaAs/AlGaAs laser; (b) *p–n* junction isolated (diffused planar stripe) GaAs/AlGaAs laser

characteristic is not linear in the simulated emission region, but exhibits kinks. This phenomenon is particularly prevalent with gain-guided injection laser devices. The kinks may be classified into two broad categories.

The first type of kink results from changes in the dominant lateral mode of the laser as the current is changed. The output characteristic for laser *A* in Figure 6.25(a) illustrates this type of kink where lasing from the device changes from the fundamental lateral mode to a higher order lateral mode (second order) in a current region corresponding to a change in slope. The second type of kink involves a 'spike', as observed for laser *B* of Figure 6.25(a). These spikes have been shown to be associated with filamentary behavior within the active region of the device [Ref. 4]. The filaments result from defects within the crystal structure.

Both these mechanisms affect the near- and far-field intensity distributions (patterns) obtained from the laser. A typical near-field intensity distribution corresponding to a single optical output power level in the plane of the junction is shown in Figure 6.25(b). As this distribution is in the lateral direction, it is determined by the nature of the lateral waveguide. The single intensity maximum shown indicates that the fundamental lateral mode is dominant. To maintain such a near-field pattern the stripe geometry of the device is important. In general, relatively narrow stripe devices (< 10 μm) formed by a planar process allow the fundamental lateral mode to dominate. This is especially the case at low power levels where near-field patterns similar to Figure 6.25(b) may be obtained.

Although gain-guided lasers are commercially available for operation in both the shorter wavelength range (using GaAs active regions) and the longer wavelength range (using InGaAsP active regions) they exhibit several undesirable characteristics. Apart from the nonlinearities in the light output versus current characteristics discussed above, gain-guided injection lasers have relatively high threshold currents (100 to 150 mA) as well as low differential quantum efficiency [Ref. 24]. These effects are primarily caused

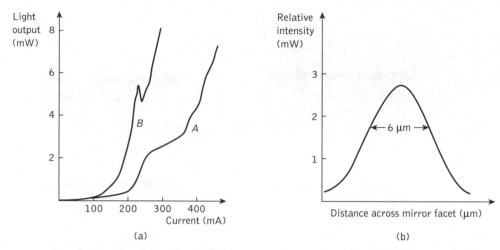

Figure 6.25 (a) The light output against current characteristic for an injection laser with nonlinearities or a kink in the stimulated emission region. (b) A typical near-field intensity distribution (pattern) in the plane of the junction for an injection laser

by the small carrier-induced refractive index reduction within the devices which results in the movement of the optical mode along the junction plane. The problems can be greatly reduced by introducing some real refractive index variation into the lateral structure of the laser such that the optical mode along the junction plane is essentially determined by the device structure.

6.5.2 Index-guided lasers

The drawbacks associated with the gain-guided laser structures were largely overcome through the development of index-guided injection lasers. In some such structures with weak index guiding, the active region waveguide thickness is varied by growing it over a channel or ridge in the substrate. A ridge is produced above the active region and the surrounding areas are etched close to it (i.e. within 0.2 to 0.3 µm). Insulating coatings on these surrounding areas confine the current flow through the ridge and active stripe while the edges of the ridge reflect light, guiding it within the active layer, and thus forming a waveguide. Hence in the ridge waveguide laser shown in Figure 6.26(a), the ridge not only provides the location for the weak index guiding but also acts as the narrow current-confining stripe [Ref. 25]. These devices have been fabricated to operate at various wavelengths with a single lateral mode, and room temperature CW threshold currents as low as 18 mA with output powers of 25 mW have been reported [Ref. 26]. More typically, the threshold currents for such weakly index-guided structures are in the range 40 to 60 mA, as illustrated in Figure 6.26(b) which compares a light output versus current characteristic for a ridge waveguide laser with that of an oxide stripe gain-guided device.

Alternatively, the application of a uniformly thick, planar active waveguide can be achieved through lateral variations in the confinement layer thickness or the refractive

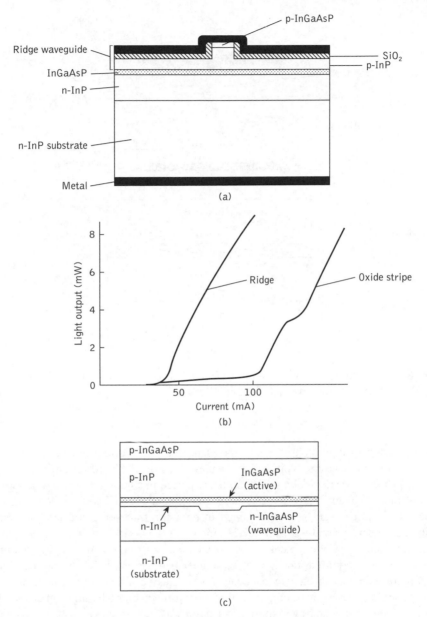

Figure 6.26 Index-guided lasers: (a) ridge waveguide injection laser structures; (b) light output versus current characteristic for (a) compared with that of an oxide stripe (gain-guided) laser; (c) rib (plano-convex) waveguide injection laser structure

index. The inverted-rib waveguide device (sometimes called plano-convex waveguide) illustrated in Figure 6.26(c) is an example of this structure. However, room temperature CW threshold currents are between 70 and 90 mA with output powers of around 20 mW for InGaAsP devices operating at a wavelength of 1.3 μm [Ref. 27].

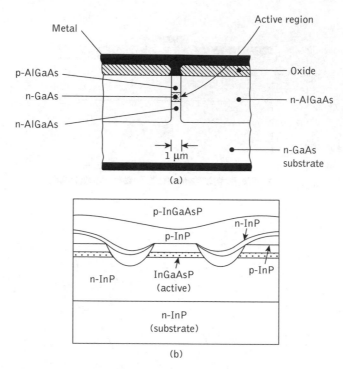

Figure 6.27 Buried heterostructure laser structures: (a) GaAs/AlGaAs BH device; (b) InGaAsP/InP double-channel planar BH device

Strong index guiding along the junction plane can provide improved transverse mode control in injection lasers. This can be achieved using a buried heterostructure (BH) device in which the active volume is completely buried in a material of wider bandgap and lower refractive index [Refs 28, 29]. The structure of a BH laser is shown in Figure 6.27(a). The optical field is well confined in both the transverse and lateral directions within these lasers, providing strong index guiding of the optical mode together with good carrier confinement. Confinement of the injected current to the active region is obtained through the reverse-biased junctions of the higher bandgap material. It may be observed from Figure 6.27 that the higher bandgap, low refractive index confinement material is AlGaAs for GaAs lasers operating in the 0.8 to 0.9 μm wavelength range, whereas it is InP in InGaAsP devices operating in the 1.1 to 1.6 μm wavelength range.

A wide variety of BH laser configurations are commercially available offering both multimode and single-mode operation. In general, the lateral current confinement provided by these devices leads to lower threshold currents (10 to 20 mA) than may be obtained with either weakly index-guided or gain-guided structures. A more complex structure called the double-channel planar buried heterostructure (DCPBH) laser is illustrated in Figure 6.27(b) [Ref. 30]. This device, which has a planar InGaAsP active region, provides very high-power operation with CW output powers up to 40 mW in the longer wavelength region. Room temperature threshold currents are in the range 15 to 20 mA for both 1.3 μm and 1.55 μm emitting devices [Ref. 31]. Lateral mode control may be achieved by reducing the dimension of the active region, with a cross-sectional area of 0.3 μm^2 being required for fundamental mode operation [Ref. 24].

Parasitic capacitances resulting from the use of the reverse-biased current confinement layers can reduce the high-speed modulation capabilities of BH lasers. However, this problem has been overcome through either the regrowth of semi-insulating material [Ref. 21] or the deposition of a dielectric material [Ref. 33]. Using these techniques, modulation speeds in excess of 20 GHz have been achieved which are limited by the active region rather than the parasitic capacitances [Ref. 29].

6.5.3 Quantum-well lasers

DH lasers have also been fabricated with very thin active layer thicknesses of around 10 nm instead of the typical range for conventional DH structures of 0.1 to 0.3 μm. The carrier motion normal to the active layer in these devices is restricted, resulting in a quantization of the kinetic energy into discrete energy levels for the carriers moving in that direction. This effect is similar to the well-known quantum mechanical problem of a one-dimensional potential well [Ref. 24] and therefore these devices are known as quantum-well lasers. In this structure the thin active layer causes drastic changes to the electronic and optical properties in comparison with a conventional DH laser. These changes are due to the quantized nature of the discrete energy levels with a step-like density of states which differs from the continuum normally obtained. Hence, quantum-well lasers exhibit an inherent advantage over conventional DH devices in that they allow high gain at low carrier density, thus providing the possibility of significantly lower threshold currents.

Both single-quantum-well (SQW), corresponding to a single active region, and multiquantum-well (MQW), corresponding to multiple active regions, lasers are utilized [Ref. 24]. In the latter structure, the layers separating the active regions are called barrier layers. Energy band diagrams for the active regions of these structures are displayed in Figure 6.28. It may be observed in Figure 6.28(c) that when the bandgap energy of the barrier layer differs from the cladding layer in an MQW device, it is usually referred to as a modified multiquantum-well laser [Ref. 34].

Better confinement of the optical mode is obtained in MQW lasers in comparison with SQW lasers, resulting in a lower threshold current density for these devices. A substantial amount of experimental work has been carried out on MQW lasers using the AlGaAs/GaAs material system. It has demonstrated the superior characteristics of MQW devices over conventional DH lasers in relation to lower threshold currents, narrower linewidths, higher modulation speeds, lower frequency chirp and less temperature dependence (see Section 6.7.1) [Ref. 29].

6.5.4 Quantum-dot lasers

More recently, quantum-well lasers have been developed in which the device contains a single discrete atomic structure or so-called quantum dot (QD) [Ref. 35]. Quantum dots are small elements that contain a tiny droplet of free electrons forming a quantum-well structure. Hence a QD laser is also referred to as a dot-in-a-well device [Ref. 36]. They are

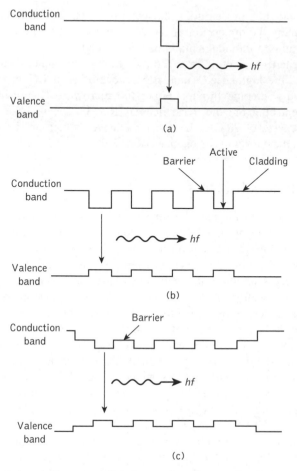

Figure 6.28 Energy band diagrams showing various types of quantum-well structure: (a) single quantum well; (b) multiquantum well; (c) modified multiquantum well

fabricated using semiconductor crystalline materials and have typical dimensions between nanometers and a few microns. The size and shape of these structures and therefore the number of electrons they contain may be precisely controlled such that a QD can have anything from a single electron to several thousand electrons. Theoretical treatment of QDs indicates that they do not suffer from thermal broadening and their threshold current is also temperature insensitive [Ref. 37]. If the conventional injection laser diode is regarded as three dimensional and a quantum well (i.e. an SQW where an array of SQWs forms an MQW structure) is confined to two dimensions, then the QD structure can be considered to be zero dimensional. It should be noted, however, that the single dimensional structure forms a quantum wire or dash.

The above hierarchy is illustrated in Figure 6.29 which identifies four different possible structures for the semiconductor laser with their corresponding energy responses with respect to carrier densities shown underneath. The three-dimensional structure of the

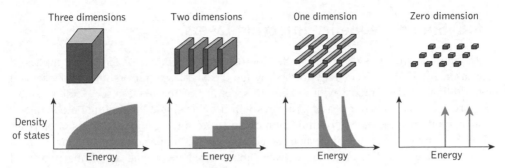

Figure 6.29 Schematic illustration and density states for semiconductor lasers. From left to right: conventional injection laser diode; multiple quantum wells; array of quantum wires; an array of quantum dots. Shown underneath each of these illustrations is the corresponding density of states for each type of laser structure

conventional injection laser diode on the left displays an exponential variation in the density of states for the charge carriers. An SQW structure exhibits two dimensions (i.e. length and height) where the corresponding energy representation is shown in Figure 6.29 by a staircase response in the carrier density of states. However, when this structure is reduced to one dimension (i.e. length only) it displays a sharp rise and an exponential fall in the carrier density variation. Since this one-dimensional quantum-well structure is confined to only the device length, then, in general, it appears as a long wire and hence it is known as a quantum wire. The zero-dimensional (i.e. single-point) structure shown on the right of Figure 6.29, however, corresponds to a single QD which results in an impulse response for the variation in the charge carrier density with increasing number of carriers.

The size and shape of the structure for a QD laser can be altered as required during the fabrication process [Ref. 38]. For example, in fabrication arrays of QDs can be formed on a GaAs substrate with different shapes being produced. Shapes such as the cube, circular disk, cylinder, pyramid or truncated pyramid can be created from self-organized crystalline growth of InGaAs material on the GaAs substrate [Ref. 39]. Each of these crystal shapes possesses different material characteristics (i.e. elasticity, stress, strain distribution, etc.) and therefore their different shapes and sizes produce a varying impact on the operation of the QD laser (i.e. emission wavelength, polarization and operating temperature) [Ref. 40]. By contrast, regularity of size and shape in an array improves the control of the QD device lasing frequency and intensity.

One of the important features of the QD laser is its very low-threshold current density. For example, low-threshold current densities between 6 and 20 A cm^{-2} have been obtained with InAs/InGaAs QD lasers emitting at the wavelengths of 1.3 μm and 1.5 μm [Refs 37, 41, 42]. These low values of threshold current density make it possible to create stacked or cascaded QD structures thus providing high optical gain suitable for the short-cavity transmitters and vertical cavity surface-emitting lasers (see Section 6.6.2). Despite the potential benefits of QD technology, issues remain in relation to materials technology and in the design and fabrication techniques to facilitate the large-scale production of QD devices [Refs 38, 43].

6.6 Single-frequency injection lasers

Although the structures described in Section 6.5 provide control of the lateral modes of the laser, the Fabry–Pérot cavity formed by the cleaved laser mirrors may allow several longitudinal modes to exist within the gain spectrum of the device (see Section 6.4.3). Nevertheless, such injection laser structures will provide single longitudinal mode operation, even though the mode discrimination obtained from the gain spectrum is often poor. However, improved longitudinal mode selectivity can be achieved using structures which give adequate loss discrimination between the desired mode and all of the unwanted modes of the laser resonator. It was indicated in Section 6.4.4 that such mode discrimination could be obtained by shortening the laser cavity. This technique has met with only limited success [Ref. 29] and therefore alternative structures have been developed to give the necessary electrical and optical containment to allow stable single longitudinal mode operation.

Longitudinal mode selectivity may be improved through the use of frequency-selective feedback so that the cavity loss is different for the various longitudinal modes. Devices which employ this technique to provide single longitudinal mode operation are often referred to as single-frequency or dynamic single-mode (DSM) lasers [Refs 44, 45]. Such lasers are of increasing interest not only to reduce fiber intramodal dispersion within high-speed systems, but also for the provision of suitable sources for coherent optical transmission (see Chapter 13). Strategies which have proved successful in relation to single-frequency operation are the use of short-cavity resonators, coupled cavity resonators and distributed feedback.

6.6.1 Short- and coupled-cavity lasers

It was suggested in Section 6.4.4 that a straightforward method for increasing the longitudinal mode discrimination of an injection laser is to shorten the cavity length; shortening from, say, 250 to 25 μm will have the effect of increasing the mode spacing from 1 to 10 nm. The peak of the gain curve can then be adjusted to provide the desired single-mode operation. Conventional cleaved mirror structures are, however, difficult to fabricate with cavity lengths below 50 μm and therefore configurations employing resonators, either microcleaved [Ref. 46] or etched [Ref. 47], have been utilized. Such resonators form a short cavity of length 10 to 20 μm in a direction normal to the active region providing stable single-frequency operation.

Multiple-element resonators or resonators with distributed reflectors also give a loss mechanism with a frequency dependence which is strong enough to provide single-frequency oscillation under most operating conditions. Mode selectivity in such a coupled-cavity laser is obtained when the longitudinal modes of each Fabry–Pérot cavity coincide and therefore constitute the longitudinal modes of the coupled system for which both cavities are in resonance. One example of a three-mirror resonator shown in Figure 6.30(a) uses a graded index (GRIN) rod lens (see Section 5.5.1) to enhance the coupling to an external mirror [Ref. 48].

An alternative approach is illustrated in Figure 6.30(b) in which two active laser sections are separated by a gap of approximately a single wavelength. When the gap is

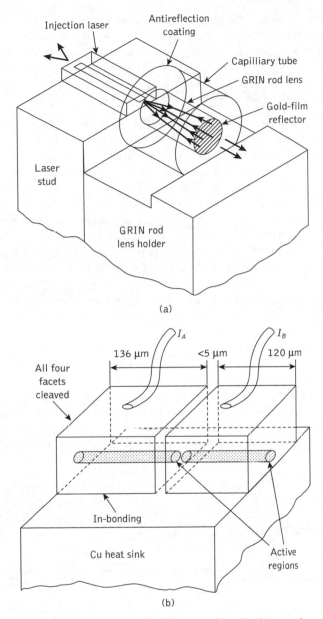

Figure 6.30 Coupled-cavity lasers: (a) short external cavity laser using GRIN-rod lens; (b) cleaved-coupled-cavity laser

obtained by recleaving a finished laser chip into two partially attached segments it yields the cleaved-coupled-cavity (C^3) laser [Ref. 49]. This four-mirror resonator device has provided dynamic single-mode operation with side mode suppression ratios of several thousand being achieved through control of the magnitudes and the relative phases of the two injection currents, as well as the temperature [Ref. 50]. Another attribute of the C^3 device is that its single-frequency emission can be tuned discretely over a range of some 26 nm by varying the current through one section [Ref. 24]. This tunability, which occurs through mode jumps of around 2 nm each, is discussed further in Section 6.10.

6.6.2 Distributed feedback lasers

An elegant approach to single-frequency operation which has found widespread application involves the use of distributed resonators, fabricated into the laser structure to give integrated wavelength selectivity. The structure which is employed is the distributed Bragg diffraction grating which provides periodic variation in refractive index in the laser heterostructure waveguide along the direction of wave propagation so that feedback of optical energy is obtained through Bragg reflection (see Section 11.4.3) rather than by the usual cleaved mirrors. Hence the corrugated grating structure shown in Figure 6.31(a) determines the wavelength of the longitudinal mode emission instead of the Fabry–Pérot gain curve shown in Figure 6.31(b). When the period of the corrugation is equal to $l\lambda_B/2n_e$, where l is the integer order of the grating, λ_B is the Bragg wavelength and n_e is the effective refractive index of the waveguide, then only the mode near the Bragg wavelength λ_B is reflected constructively (i.e. Bragg reflection). Therefore, as may be observed in Figure 6.31(a), this particular mode will lase while the other modes exhibiting higher losses are suppressed from oscillation.

It should be noted that first-order gratings (i.e. $l = 1$) provide the strongest coupling within the device. Nevertheless, second-order gratings are sometimes used as their larger spatial period eases fabrication.

From the viewpoint of device operation, semiconductor lasers employing the distributed feedback mechanism can be classified into two broad categories, referred to as the distributed feedback (DFB) laser [Ref. 51] and the distributed Bragg reflector (DBR) laser [Ref. 52]. These two device structures are shown schematically in Figure 6.32. In the DFB

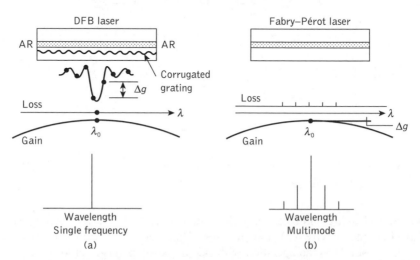

Figure 6.31 Illustration showing the single-frequency operation of (a) the distributed feedback (DFB) laser in comparison with (b) the Fabry–Pérot laser

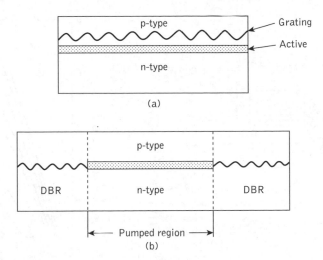

Figure 6.32 Schematic of distributed feedback lasers: (a) DFB laser; (b) DBR laser

laser the optical grating is usually applied over the entire active region which is pumped, whereas in the DBR laser the grating is etched only near the cavity ends and hence distributed feedback does not occur in the central active region. The unpumped corrugated end regions effectively act as mirrors whose reflectivity results from the distributed feedback mechanism which is therefore dependent on wavelength. In addition, this latter device displays the advantage of separating the perturbed regions from the active region but proves somewhat lossy due to optical absorption in the unpumped distributed reflectors. It should be noted that in Figure 6.32 the grating is shown in a passive waveguide layer adjacent to the active gain region for both device structures. This structure has evolved as a result of the performance deterioration with earlier devices (at temperatures above 80 K) in which the corrugations were applied directly to the active layer [Ref. 53].

Both DBR and DFB lasers are used to provide single-frequency semiconductor optical sources. Any of the semiconductor laser structures discussed in Section 6.5 can be employed to fabricate a DFB laser after etching a grating into an appropriate cladding layer adjacent to the active layer. The grating period is determined by the desired emission frequency from the structure following the Bragg condition (see Section 11.4.3). In particular, DFB BH lasers have been developed in many laboratories, which exhibit low threshold currents (10 to 20 mA), high modulation speeds (several Gbit s^{-1}) and output powers comparable with Fabry–Pérot devices with similar BH geometries [Refs 24, 29, 54]. The structure and the light output against current characteristic for a low-threshold-current DFB BH laser operating at a wavelength of 1.55 μm is displayed in Figure 6.33 [Ref. 55]. A substantial change in the output characteristic with increasing temperature may be observed for this separate confinement DH device (see Section 6.7.1).

In theory, when considering a DFB laser with both end facets antireflection (AR) coated (see Figure 6.31(a)), the two modes located symmetrically on either side of the Bragg wavelength will experience the same lowest threshold gain within an ideal symmetrical structure and will therefore lase simultaneously. However, in practice, the randomness associated with the cleaving process creates different end phases, thus removing the

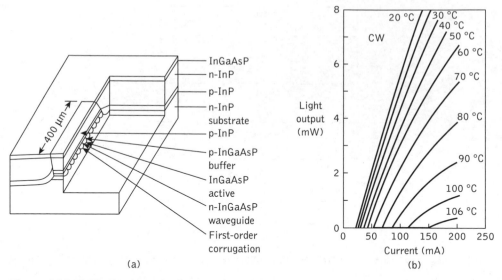

(a)   (b)

Figure 6.33 DFB BH laser with a window structure: (a) structure; (b) light output against current characteristics for various temperatures under CW operation. Reprinted with permission from S. Tsuji, A. Ohishi, H. Nakamura, M. Hirao, N. Chinone and H. Matsumura, 'Low threshold operation of 1.5 μm DFB laser diodes', *J. Lightwave Technol.*, **LT-5**, p. 822, 1987. Copyright © 1987 IEEE

degeneracy of the modal gain and providing only single-mode operation. Moreover, facet asymmetry can be increased by placing a high-reflection coating on one end facet and a low-reflection coating on the other (known as the hi–lo structure) in order to improve the power output for single-frequency operation [Ref. 29].

Another technique to improve the performance of the DFB laser is to modify the grating at a central point to introduce an additional optical phase shift, typically a quarter wavelength or less [Refs 56, 57]. Such a device is shown in Figure 6.34 which illustrates the structure of an InGaAsP/InP double-channel planar DFB BH laser with a quarter-wavelength-shifted first-order grating [Ref. 58]. This structure, which provides excellent,

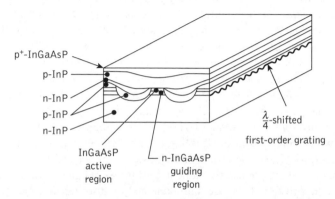

Figure 6.34 Quarter-wavelength-shifted double-channel planar DFB BH laser

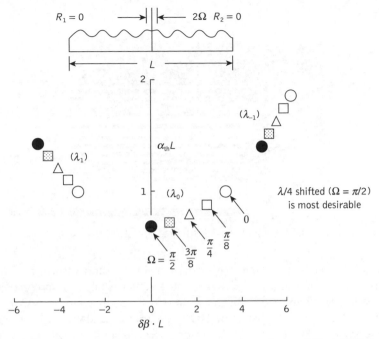

Figure 6.35 The threshold gain and mode frequency in a phase-shifted DFB laser

stable, single-frequency operation, incorporates a $\pi/2$ phase shift (equivalent to one-quarter wavelength) in the corrugation at the center of the laser cavity with both end facets AR coated. The threshold gain and the mode frequency (relative to the Bragg wavelength) for the device as the phase shift is varied from 0 to $\pi/2$ is shown in Figure 6.35. It may be observed that the lowest threshold gain for the central mode (at λ_0) is obtained precisely at the Bragg wavelength when the phase shift is $\pi/2$. Furthermore, the gain difference between the central mode and the nearest side mode (at λ_1) has the largest value at this phase shift.

The performance of the quarter-wavelength-shifted DFB laser is superior to that of the conventional DFB structure because the large gain difference between the central mode and the side modes gives improved dynamic single-mode stability with negligible mode partition noise (see Section 6.7.4) at multigigabit per second modulation speeds [Ref. 58]. In addition, narrow linewidths of around 3 MHz ($\approx 2 \times 10^{-5}$ nm) have been obtained under CW operation [Ref. 29], which is substantially less than the typical 100 MHz ($\approx 6 \times 10^{-4}$ nm) linewidth associated with the Fabry–Pérot injection laser. Linewidth narrowing is achieved within such DFB lasers by detuning the lasing wavelength towards the shorter wavelength side of the gain peak (i.e. towards λ_{-1} in Figure 6.35) in order to increase the differential gain between the central mode and the nearest side mode (λ_1 in Figure 6.35). This strategy is sometimes referred to as Bragg wavelength detuning.

6.6.3 Vertical cavity surface-emitting lasers

The vertical cavity surface-emitting laser (VCSEL, pronounced 'vixel') emits a coherent optical signal perpendicular to the device substrate. In comparison with edge-emitting

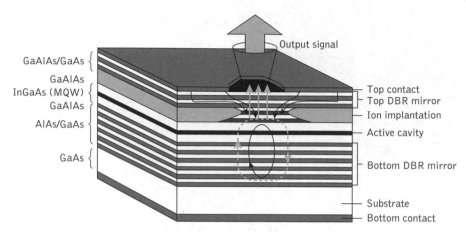

GaAlAs/GaAs {
GaAlAs
InGaAs (MQW) {
GaAlAs {
AlAs/GaAs {

GaAs {

Output signal

Top contact
Top DBR mirror
Ion implantation
Active cavity

Bottom DBR mirror

Substrate
Bottom contact

Figure 6.36 Structure of a vertical cavity surface-emitting laser

lasers, the VCSEL structure is somewhat different since a short vertical cavity is formed by the surfaces of epitaxial layers and the optical output is taken from one of the mirror surfaces [Ref. 59]. Figure 6.36 illustrates the structure of a typical VCSEL where a Fabry–Pérot active cavity consisting of MQW material is sandwiched between two mirrors each formed by multilayered DBR mirrors. The top surface DBR mirror comprising p-type material possesses low facet reflectivity as compared with the n-type DBR mirror at the bottom of the device. The number of Bragg gratings determines the amount of facet reflectivity and it generally requires between 10 and 30 Bragg grating periods to develop satisfactory facet reflectivity for the top or bottom DBR mirrors where the particular grating number depends upon the specific semiconductor material composition.

The top and bottom surface DBR mirrors in Figure 6.36 form a p–n junction laser diode where the arrows emerging from the top contact into the active cavity region represent the flow of electric current from the top to bottom surfaces. Therefore an optical signal at the resonant wavelength can be emitted from this p–n junction to the surface of the VCSEL while the active cavity region shown by the circular dotted arrowed line in Figure 6.36 provides for further buildup of optical power in a similar manner to a conventional semiconductor laser. An advantage of the DBR mirror design is that any light reflected back towards the laser from any other part of the structure cannot re-enter the resonator and hence the VCSEL is effectively isolated against any other optical signals entering it. Moreover, the output signal pattern emitted from the surface of the VCSEL is circular instead of the elliptical pattern produced by edge-emitting lasers [Ref. 60].

A short cavity for the VCSEL may be obtained by reducing the cross-sectional area in which gain occurs. Current confinement can be realized by ion implantation* while also effectively providing for the short cavity. Ions are implanted into a selected area of the semiconductor material to make it non-conducting. Since the introduction of ions into a semiconductor produces permanent damage to the crystalline structure of the implanted

---

* Ion implantation is a technique used to render the material nonconducting around the active cavity by producing permanent defects in the implanted area (see Section 6.5.1). The approach therefore concentrates the injection current in the active region [Ref. 61].

area, it is therefore important that implanted ions do not reach the active layer. Although protons are most often used, ions including O[+], N[+] and H[+] have also been employed to provide effective ion implantation in VCSELs [Ref. 62].

Another method referred to as 'selectively oxidized' utilizes the formation of an insulating aperture of aluminum oxide between the cavity layer and one of the DBR mirrors [Refs 59, 63]. In this technique an oxidization layer is created, and since the oxide layer has a low refractive index compared with the semiconductor material, it acts as a waveguide. The output of such a VCSEL is limited, however, by the oxide aperture which confines both the electric current and also the optical modes. Hence the aperture must be kept small (i.e. less than 7 μm) to ensure fundamental mode operation which consequently reduces the possible output signal power. Alternatively, photonic crystals can be used to introduce permanent defects in order to confine the electric current [Ref. 64].

Due to epitaxial complexity in the fabrication of VCSELs their development is dependent on progress in compound materials for semiconductor growth as well as the processing technologies. Figure 6.37 compares the lattice mismatch of the different materials on gallium arsenide and indium phosphide substrates used to produce VCSELs emitting optical signals over the fiber communication wavelength range [Ref. 61]. The right hand axis shows the wavelengths related to different material types. For example, materials such as InGaAlP/AlGaAs can be used to fabricate VCSELs to emit within the wavelength range from 0.63 to 0.68 μm which is suitable for plastic optical fibers (see Section 4.5.5). On the other hand, the InGaAsP/InAs material system can emit optical signals in the wavelength range from 1.30 to 1.55 μm compatible with standard silica multimode or single-mode fiber. In the latter case, however, particularly at the 1.55 μm wavelength, the resultant

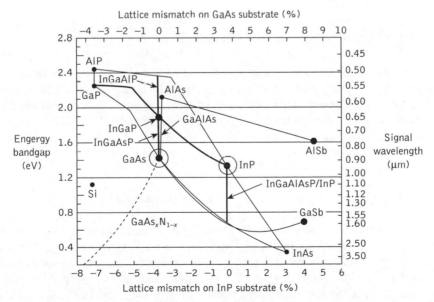

Figure 6.37 Bandgap energies and lattice mismatch on GaAs and InP substrates for use in the fabrication of vertical cavity surface-emitting lasers [Ref. 61]

compound materials do not necessarily yield sufficient stimulated gain [Ref. 65]. In addition a CW VCSEL fabricated from InGaAs/InAs and using quantum dot technology has successfully demonstrated emission at a signal wavelength of 1.30 μm [Ref. 66]. The device delivered an output power of 0.33 mW at room temperature when the threshold current was only 1.7 mA. Furthermore, it displayed a modulation bandwidth of 2 GHz which could be extended to 4 GHz when the losses due to absorption and thermal noise were reduced [Ref. 67].

VCSELs can also be fabricated to provide wavelength tunability in either multiple longitudinal or single-mode operation. Various tuning mechanisms can be adopted to select a wavelength within the gain band of the active region. Since Bragg gratings are temperature-dependent structures where the temperature variation changes the refractive index of the DBR material, then VCSELs incorporating DBRs can be readily tuned using temperature variation to change the transmission wavelength of the lasers [Refs 68, 69].

Micro-electromechanical systems (see Section 11.4.3) have also been used to adjust the emission wavelengths of VCSELs [Refs 70, 71]. In this approach the top mirror is divided into two sections with the lower section remaining in position within the active cavity, whereas the upper section is made movable to adjust its distance from the active cavity medium. Hence the upper section of the DBR can move vertically up or down producing a variable gap between it and the top DBR layer, and therefore consequently this variable DBR mirror provides tuning for the VCSEL. At room temperature a continuous tuning range of 40 nm with a gain peak signal power of 100 μW at a wavelength of 1.55 μm has been obtained [Ref. 72]. Finally, since the short active cavity requires very low current thresholds, and also because they can be arranged in an array formation, VCSELs are considered important optical sources to facilitate photonic integration (see Section 11.2).

6.7 Injection laser characteristics

When considering the use of the injection laser for optical fiber communications it is necessary to be aware of certain of its characteristics which may affect its efficient operation. The following sections outline the major operating characteristics of the device (the ones which have not been dealt with in detail previously) which generally apply to all the various materials and structures previously discussed, although there is substantial variation in behavior between them.

6.7.1 Threshold current temperature dependence

Figure 6.38 shows the variation in threshold current with temperature for two gain-guided (oxide insulated stripe) injection lasers [Ref. 73]. Both devices had stripe widths of approximately 20 μm but were fabricated from different material systems for emission at wavelengths of 0.85 μm and 1.55 μm (AlGaAs and InGaAsP devices respectively).

In general terms the threshold current tends to increase with temperature, the temperature dependence of the threshold current density J_{th} being approximately exponential [Ref. 4] for most common structures. It is given by:

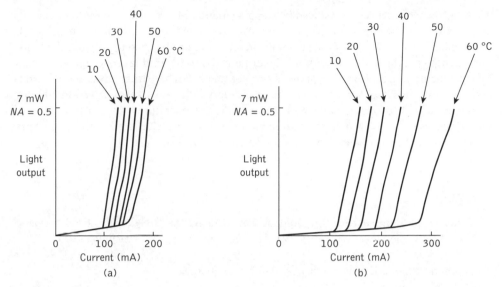

Figure 6.38 Variation in threshold current with temperature for gain-guided injection lasers: (a) AlGaAs device; (b) InGaAsP device. Reproduced with permission from P. A. Kirby, 'Semiconductor laser sources for optical communication', *Radio Electron. Eng.*, **51**, p. 362, 1981

$$J_{th} \propto \exp \frac{T}{T_0} \tag{6.43}$$

where T is the device absolute temperature and T_0 is the threshold temperature coefficient which is a characteristic temperature describing the quality of the material, but which is also affected by the structure of the device. For AlGaAs devices, T_0 is usually in the range 120 to 190 K, whereas for InGaAsP devices it is between 40 and 75 K [Ref. 74]. This emphasizes the stronger temperature dependence of InGaAsP structures which is illustrated in Figure 6.38 and Example 6.7. The increase in threshold current with temperature for AlGaAs devices can be accounted for with reasonable accuracy by consideration of the increasing energy spread of electrons and holes injected into the conduction and valence bands. It appears that the intrinsic physical properties of the InGaAsP material system may cause its higher temperature sensitivity; these include Auger recombination, intervalence band absorption and carrier leakage effects over the heterojunctions [Ref. 75].

Auger recombination is a process where the energy released during the recombination of an electron–hole event is transferred to another carrier (i.e. an electron or hole). During this process, when a carrier is excited to a higher energy level, it loses its surplus energy by emitting a phonon in order to maintain thermal equilibrium. Auger recombination is not a single process but consists of many different processes (i.e. more than 80), each of which may involve at least three particles (i.e. two electrons and one hole, or one electron and two holes, etc.). Although Auger recombination is not the main loss mechanism at room temperature, it dominates, however, at elevated threshold current densities [Ref. 20].

These adverse effects can be reduced by using a strained MQW structure for the laser. The strain, which can be either compressive or tensile (i.e. bending or stretching), modifies the valence band energy levels of the material and therefore can be used to enhance the transition strength (i.e. increase energy). It is incorporated into the thin layers of quantum wells by introducing small differences in the lattice constants. Higher strain, however, should be avoided as it can cause damage in the thin quantum-well layers [Refs 76, 77]. In addition, carrier leakage also contributes significantly at high temperatures since it represents all those processes that prevent carriers from recombination. Carrier leakage therefore raises the lasing threshold and hence reduces device efficiency [Refs 78, 79].

Example 6.7

Compare the ratio of the threshold current densities at 20 °C and 80 °C for an AlGaAs injection laser with $T_0 = 160$ K and the similar ratio for an InGaAsP device with $T_0 = 55$ K.

Solution: From Eq. (6.43) the threshold current density:

$$J_{th} \propto \exp \frac{T}{T_0}$$

For the AlGaAs device:

$$J_{th} (20\,°C) \propto \exp \frac{293}{160} = 6.24$$

$$J_{th} (80\,°C) \propto \exp \frac{353}{160} = 9.08$$

Hence the ratio of the current densities:

$$\frac{J_{th} (80\,°C)}{J_{th} (20\,°C)} = \frac{9.08}{6.24} = 1.46$$

For the InGaAsP device:

$$J_{th} (20\,°C) \propto \exp \frac{293}{55} = 205.88$$

$$J_{th} (80\,°C) \propto \exp \frac{353}{55} = 612.89$$

Hence the ratio of the current densities:

$$\frac{J_{th} (80\,°C)}{J_{th} (20\,°C)} = \frac{612.89}{205.88} = 2.98$$

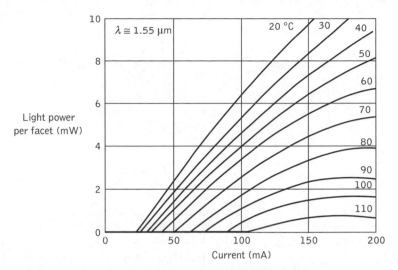

Figure 6.39 Light output against current characteristics at various temperatures for an InGaAsP double-channel planar BH laser emitting at a wavelength of 1.55 μm. From N. K. Dutta, 'Optical sources for lightwave system applications', in E. E. Basch (Ed.), *Optical-Fiber Transmission*, H. W. Sams & Co., p. 265, 1987

Thus in Example 6.7 the threshold current density for the AlGaAs device increases by a factor of 1.5 over the temperature range, whereas the threshold current density for the InGaAsP device increases by a factor of 3. Hence the stronger dependence of threshold current on temperature for InGaAsP structures is shown in this comparison of two average devices. It may also be noted that it is important to obtain high values of T_0 for the devices in order to minimize temperature dependence.

The increased temperature dependence for the InGaAsP/InP material system is also displayed by the more advanced, mode-stabilized device structures. Figure 6.39 provides the light output against current characteristic at various device temperatures for a strongly index-guided DCPBH injection laser (see Section 6.5.2) emitting at a wavelength of 1.55 μm [Ref. 24]. Moreover, the similar characteristic for a DFB BH laser was shown in Figure 6.33(b). It is therefore necessary to pay substantial attention to thermal dissipation in order to provide efficient heat-sinking arrangements (e.g. thermoelectric cooling etc.) to achieve low operating currents. In addition, the need to minimize or eliminate the thermal resistance degradation associated with the solder bond on such devices (an effect which could, to a certain extent, be tolerated with GaAs injection lasers) has also become critically important [Ref. 80]. In all cases, however, adequate heat sinking along with consideration of the working environment are essential so that devices operate reliably over the anticipated current range.

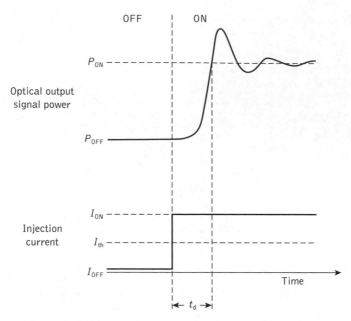

Figure 6.40 The dynamic behavior of an injection laser and the corresponding injection current of the device showing relaxation oscillations and the switch-on delay

6.7.2 Dynamic response

The dynamic behavior of the injection laser is critical, especially when it is used in high bit rate (wideband) optical fiber communication systems. The application of a current step to the device results in a switch-on delay, often followed by high-frequency (one to tens of gigahertz) damped oscillations known as relaxation oscillations (ROs). These transient phenomena occur while the electron and photon populations within the structure come into equilibrium and are illustrated in Figure 6.40. In addition, when a current pulse reaches a laser which has significant parasitic capacitance after the initial delay time, the pulse will be broadened because the capacitance provides a source of current over the period when the photon density is high. Consequently, the injection laser output can comprise several pulses as the electron density is repetitively built up and quickly reduced, thus causing ROs. The switch-on delay t_d may last for 0.5 ns and the RO for perhaps twice that period. At data rates above 100 Mbit s^{-1} this behavior can produce a serious deterioration in the pulse shape. Hence, reducing t_d and damping the relaxation oscillations is highly desirable.

The switch-on or turn-on delay is caused by the initial build up of photon density resulting from stimulated emission. It is related to the minority carrier lifetime and the current through the device [Ref. 7]. The current term, and hence the switch-on delay, may be reduced by biasing the laser near threshold (prebiasing). However, damping of the ROs is less straightforward. They are basic laser phenomena which vary with device structure and operating conditions; however, RO damping has been observed, and is believed to be due to several mechanisms including lateral carrier diffusion [Refs 81, 82], the feeding of the spontaneous emission into the lasing mode [Ref. 83] and gain nonlinearities [Ref. 84].

Narrow stripe geometry DH lasers and all the mode-stabilized devices (see Sections 6.5 and 6.6) give RO damping, but it tends to coincide with a relatively slow increase in output power. This is thought to be the result of lateral carrier diffusion due to lack of lateral carrier confinement. However, it appears that RO damping and fast response may be obtained in BH structures with stripe widths less than the carrier diffusion length (i.e. less than 3 µm) [Ref. 85]. Moreover, (this phenomenon has been employed within a digital transmission system by biasing the laser near threshold and then by using a single RO as a 'one' bit [Ref. 86].

6.7.3 Frequency chirp

The d.c. modulation of a single longitudinal mode semiconductor laser can cause a dynamic shift of the peak wavelength emitted from the device [Ref. 87]. This phenomenon, which results in dynamic linewidth broadening under the direct modulation of the injection current, is referred to as frequency chirping. It arises from gain-induced variations in the laser refractive index due to the strong coupling between the free carrier density and the index of refraction which is present in any semiconductor structure. Hence, even small changes in carrier density, apart from producing relaxation oscillations in the device output, will also result in a phase shift of the optical field, giving an associated change in the resonance frequency within both Fabry–Pérot and DFB laser structures.

The laser linewidth broadening or chirping combined with the chromatic dispersion characteristics of single-mode fibers (see Section 3.9) can cause a significant performance degradation within high transmission rate systems [Ref. 88]. In particular, it may result in a shift in operating wavelength from the zero-dispersion wavelength of the fiber, which can ultimately limit the achievable system performance. For example, theoretical predictions [Ref. 89] of the wavelength shift that may occur with an InGaAsP laser under modulation of a few gigabits per second are around 0.05 nm (6.4 GHz frequency shift).

A number of techniques can be employed to reduce frequency chirp. One approach is to bias the laser sufficiently above threshold so that the modulation current does not drive the device below the threshold where the rate of change of optical output power varies rapidly with time. Unfortunately, this strategy gives an extinction ratio penalty (see Section 12.2.1.6) of the order of several decibels at the receiver. Another method involves the damping of the relaxation oscillations that can occur at turn-on and turn-off which result in large power fluctuations. This has been achieved, for instance, by shaping the electrical drive pulses [Ref. 90].

Certain device structures also prove advantageous for chirp reduction. In particular, quantum-well lasers (see Section 6.5.3), Bragg wavelength detuned DFB lasers (see Section 6.6.2) and multielectrode DFB lasers (see Section 6.10.2) provide improved performance under d.c. modulation in relation to frequency chirping. Such lasers, however, require complex fabricational processes. An alternative technique which has proved effective in minimizing the effects of chirp is to allow the laser to emit continuously and to impress the data onto the optical carrier using an external modulator [Ref. 89]. Such devices, which may be separate lithium-niobate-based components or can be monolithically integrated with the laser [Ref. 91], are described in Chapter 11.

In semiconductor lasers a direct approach can be used to obtain reduced chirp operation by ensuring that the devices have small values of the linewidth enhancement factor which is also known as the α-parameter (see Section 6.10). This parameter determines the variation of refractive index due to coupling of spontaneous emission into the lasing mode. Typical values of the α-parameter for semiconductor lasers range from 2 to 8, whereas for the QD and electroabsorption modulator lasers (see Section 11.4.2) smaller values less than 1 or even negative values may be obtained [Ref. 92]. Moreover, the chirp is said to be negative when a negative value of the α-parameter is employed, whereas an optical signal with zero chirp is referred to as a chirpless signal. Chirpless or negative chirp optical source properties are generally required to achieve very high transmission rates (i.e. 40 Gbit s^{-1} and above) when using standard single-mode fiber operating at a wavelength of 1.55 µm [Refs 93–95].

It is important to accurately measure values for the frequency chirp in order to determine efficient dispersion control for the optical signal. Useful methods that can be employed to measure the value of the α-parameter utilize either interferometric or optical feedback approaches [Refs 96–98]. In such techniques frequency modulation is converted into amplitude modulation to determine the chirp on the optical signal. Based on this principle, several commercial simulation software tools have been developed for optoelectronic equipment such as optical spectrum analyzers [Refs 99, 100] to predict the accurate values for the chirp on the output signal. These methods are known as time-resolved chirp (TRC) measurement techniques and they provide a bit-by-bit measurement of the pattern-dependent instantaneous laser chirp.

The measurement process to acquire TRC is relatively straightforward in that it utilizes the conversion of phase modulation to amplitude modulation using a fiber-based Mach–Zehnder interferometer. Such TRC measurement techniques provide moderate accuracy for the chirp using different modulation schemes and therefore can be used for both directly modulated and externally modulated lasers. It is thus possible to predict chirp on the signal if it is required to introduce chirp before it is launched into the optical fiber (i.e. pre-chirping) or after the detection of the signal (i.e. post-chirping) [Refs 101, 102]. In the case of WDM (see Section 12.9.4) pre-chirping techniques may not be effective since the short pulse duration used at high transmission rates requires large frequency spacing amongst the channels and therefore wide bandwidth is required. Furthermore, when a wide-bandwidth signal travels through an optical amplifier it increases the amplified spontaneous emission noise (see Section 10.3). Alternatively, pre-chirping techniques, when used with positive transmitter chirp, substantially increase the maximum reach of the optical signal by reducing the dispersion effects in standard single-mode fiber at a wavelength of 1.55 µm [Refs 102, 103].

6.7.4 Noise

Another important characteristic of injection laser operation involves the noise behavior of the device. This is especially the case when considering analog transmission. The sources of noise are:

(a) phase or frequency noise;

(b) instabilities in operation such as kinks in the light output against current characteristic (see Section 6.5.1) and self-pulsation;

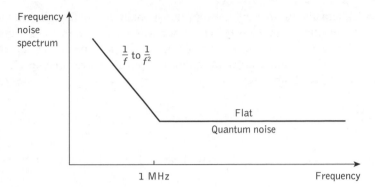

Figure 6.41 Spectral characteristic showing injection laser phase noise

(c) reflection of light back into the device;

(d) mode partition noise.

It is possible to reduce, if not remove, (b), (c) and (d) by using mode-stabilized devices and optical isolators. Phase noise, however, is an intrinsic property of all laser types. It results from the discrete and random spontaneous or simulated transitions which cause intensity fluctuations in the optical emission and are an inevitable aspect of laser operation. Each event causes a sudden jump (of random magnitude and sign) in the phase of the electromagnetic field generated by the device. It has been observed that the spectral density of this phase or frequency noise has a characteristic represented by $1/f$ to $1/f^2$ up to a frequency (f) of around 1 MHz, as illustrated in Figure 6.41 [Ref. 104]

At frequencies above 1 MHz the noise spectrum is flat or white and is associated with quantum fluctuations (sometimes referred to as quantum noise, see Figure 6.41) which are a principal cause of linewidth broadening within semiconductor lasers [Ref. 105]. Although the low-frequency components can easily be tracked and therefore are not a significant problem within optical fiber communications, this is not the case for the white noise component where, as time elapses, the phase executes a random walk away from the value it would have had in the absence of spontaneous emission.

For injection lasers operating at frequencies less than 100 MHz quantum noise levels are usually low (signal-to-noise ratios less than −80 dB) unless the device is biased within 100% of threshold. Over this region the noise spectrum is flat. However, for wideband systems when the laser is operating above threshold, quantum noise becomes more pronounced. This is especially the case with multimode devices (signal-to-noise ratios of around −60 dB). The higher noise level would appear to result from a peak in the noise spectrum due to a relaxation resonance which typically occurs between 200 MHz and 1 GHz [Ref. 7]. Single-mode lasers have demonstrated greater noise immunity by as much as 30 dB when the current is raised above threshold [Ref. 106]. Nevertheless, the wandering of the phase determines both the laser linewidth and the coherence time which are both major considerations, particularly within coherent optical fiber communications [Ref. 107].

Fluctuations in the amplitude or intensity of the output from semiconductor injection lasers also lead to optical intensity noise. These fluctuations may be caused by temperature variations or, alternatively, they result from the spontaneous emission contained in the laser output, as mentioned previously. The random intensity fluctuations create a noise source referred to as relative intensity noise (RIN), which may be defined in terms of the mean square power fluctuation $\overline{\delta P_e^2}$ and the mean optical power squared $(\bar{P}_e)^2$ which is emitted from the device following:

$$
\text{RIN} = \frac{\overline{\delta P_1^2}}{\overline{(P_e)^2}}
\tag{6.44}
$$

The above definition allows the RIN to be measured in dB Hz^{-1} where the power fluctuation is written as:

$$
\overline{\delta P_e^2}(t) = \int_0^\infty S_{\text{RIN}}(f)\,\mathrm{d}f
\tag{6.45}
$$

where $S_{\text{RIN}}(f)$ is related to the power spectral density of the relative intensity noise $S_{\text{RIN}}(\omega)$ by:

$$
S_{\text{RIN}}(f) = 2\pi S_{\text{RIN}}(\omega)
\tag{6.46}
$$

where $\omega = 2\pi f$.

Hence from Eq. (6.44), the RIN as a relative power fluctuation over a bandwidth B which is defined as 1 Hz:

$$
\text{RIN} = \frac{S_{\text{RIN}}(f)\,B(=1\ \text{Hz})}{\overline{(P_e)^2}}
\tag{6.47}
$$

It should be noted that the RIN spectrum is not flat and hence it cannot be considered as a white noise source. To simplify link budget analyses, however, the RIN value is assumed to remain constant only when the bandwidth B is limited to 1 Hz as identified in Eq. (6.47). Typical values for the RIN decrease exponentially from -130 to -160 dB Hz^{-1} when the injection current remains within the range from 30 to 40 mA without optical feedback [Ref. 108]. A VCSEL can display a RIN value between -140 and -145 dB Hz^{-1} while improved performance for DFB lasers with CW operation can be obtained at slightly higher values of RIN in the range -150 to -160 dB Hz^{-1} [Refs 53, 59, 109]. However, when a semiconductor laser is biased near threshold with optical feedback, low levels of RIN are expected and in this case Fabry–Pérot devices typically exhibit values between -125 and -130 dB Hz^{-1} [Ref. 108]. Finally, it should also be noted that the relative intensity noise decreases as the injection current level I increases following the relation:

$$
\text{RIN} \propto \left(\frac{I_{-1}}{I_{\text{th}}}\right)^{-3}
\tag{6.48}
$$

where I_{th} is the laser threshold current.

From the discussion of optical detectors following in Section 8.6 it is clear that when an optical field at a frequency f is incident with power $P_o(t)$ on a photodetector whose quantum efficiency (electrons per photon) is η, the output photocurrent $I_p(t)$ is:

$$I_p(t) = \frac{\eta e P_o(t)}{hf} \tag{6.49}$$

where e is the charge on an electron and h is Planck's constant. Therefore an optical power fluctuation $\delta P_o(t)$ will cause a fluctuating current component $\delta I_p(t) = \eta e \delta P_o(t)/hf$ which exhibits a mean square value:

$$\overline{i^2}(t) = \overline{\delta I_p^2}(t) = \frac{\eta^2 e^2}{(hf)^2} \overline{\delta P_o^2}(t) \tag{6.50}$$

Now, considering the fluctuation in the incident optical power at the detector to result from the RIN in the laser emission, using Eqs (6.44) and (6.47), and transposing P_e for P_o, the mean square noise current in the output of the detector i_{RIN}^2 due to these fluctuations is:

$$\overline{i_{RIN}^2} = \frac{\eta^2 e^2}{(hf)^2} (RIN)(\overline{P_e})^2 B \tag{6.51}$$

Example 6.8

The output from a single-mode semiconductor laser with a RIN value of 10^{-15} Hz^{-1} is incident directly on an optical detector which has a bandwidth of 100 MHz. The device is emitting at a wavelength of 1.55 μm, at which the detector has a quantum efficiency of 60%. If the mean optical power incident on the detector is 2 mW, determine: (a) the rms value of the power fluctuation and (b) the rms noise current at the output of the detector.

Solution: (a) The relative mean square fluctuation in the detected current is equal to $\overline{\delta P_e^2}/(\overline{P_e})^2$, which, using Eqs (6.44) and (6.47), can be written as:

$$\frac{\overline{\delta P_e^2}}{(\overline{P_e})^2} = \frac{S_{RIN}(f)}{(\overline{P_e})^2} B = 10^{-15} \times 100 \times 10^6 = 10^{-7}$$

Hence the rms value of this power fluctuation is:

$$\frac{(\overline{\delta P_e^2})^{\frac{1}{2}}}{\overline{P_e}} = 3.16 \times 10^{-4} \text{ W}$$

(b) The rms noise current at the detector output may be obtained from Eq. (6.51) as:

$$(\overline{i_{RIN}^2})^{\frac{1}{2}} = \frac{e\eta}{hf} (RIN)^{\frac{1}{2}} \overline{P_e} B^{\frac{1}{2}} = \frac{e\eta\lambda}{hc} (RIN)^{\frac{1}{2}} \overline{P_e} B^{\frac{1}{2}}$$

$$= \frac{1.602 \times 10^{-19} \times 0.6 \times 1.55 \times 10^{-6} \times 3.16 \times 10^{-8} \times 2 \times 10^{-3} \times 10^4}{6.626 \times 10^{-34} \times 2.998 \times 10^8}$$

$$= 4.74 \times 10^{-7} \text{ A}$$

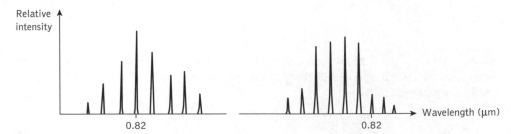

Figure 6.42 The effect of partition noise in a multimode injection laser. It is displayed as a variation in the distribution of the various longitudinal modes emitted from the device

Optical feedback from unwanted external reflections can also affect the intensity and frequency stability of semiconductor lasers [Ref. 110]. With multimode lasers, however, this effect is reduced because the reflections are distributed among many fiber modes and therefore they are only weakly coupled back into the laser mode [Ref. 111]. The stronger fiber-to-laser coupling in single-mode systems, particularly those operating at 1.55 μm, can result in reflection-induced frequency hops and linewidth broadening [Ref. 89]. In these cases an optical isolator (see Section 5.7), which is a nonreciprocal device that allows light to pass in the forward direction but strongly attenuates it in the reverse direction, may be required to provide reliable single-mode operation.

Mode partition noise is a phenomenon which occurs in multimode semiconductor lasers when the modes are not well stabilized [Ref. 112]. Even when the total output power from a laser is maintained nearly constant, temperature changes can cause the relative intensities of the various longitudinal modes in the laser's output spectrum to vary considerably from one pulse to the next, as illustrated in Figure 6.42. These spectral fluctuations combined with the fiber dispersion produce random distortion of received pulses on a digital channel, causing an increase in bit-error-rate. As mode partition noise is a function of laser spectral fluctuations then a reduced number of modes results in less pulse-width spreading thus providing low values of intermodal dispersion in the fiber [Ref. 113]. Hence, as a rule of thumb, reducing the number of modes in the multimode fiber decreases the mode partition noise [Ref. 114].

Mode partition noise can also occur in single-mode devices as a result of the residual side modes in the laser output spectrum. The effect varies between lasers emitting at 1.3 μm and those operating at 1.55 μm but, overall, a degree of side mode suppression is required in both cases in order to avoid additional errors at the receiver [Ref. 89]. Finally, various methods have been identified to reduce the mode partition noise including the use of injection mode-locked lasers [Ref. 115] and semiconductor optical amplifiers [Refs 116, 117].

6.7.5 Mode hopping

The single longitudinal mode output spectrum of a single-mode laser is illustrated in Figure 6.43(a). Mode hopping to a longer wavelength as the current is increased above threshold is demonstrated by comparison with the output spectrum shown in Figure 6.43(b).

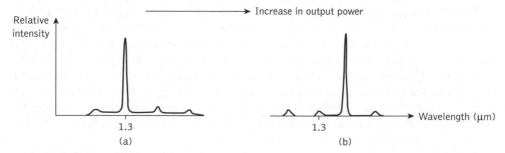

Figure 6.43 Mode hopping in a single-mode injection laser: (a) single longitudinal mode optical output: (b) mode hop to a longer peak emission wavelength at an increased optical output power

This behavior occurs in all single-mode injection lasers and is a consequence of increases in temperature of the device junction. The transition (hopping) from one mode to another is not a continuous function of the drive current but occurs suddenly over only 1 to 2 mA. Mode hopping alters the light output against current characteristics of the laser, and is responsible for the kinks observed in the characteristics of many single-mode devices.

Between hops the mode tends to shift slightly with temperature in the range 0.05 to 0.08 nm K^{-1}. Stabilization against mode hopping and mode shift may be obtained with adequate heat sinking or thermoelectric cooling. However, at constant heat sink temperature, shifts due to thermal increases can only be fully controlled by the use of feedback from external or internal grating structures (see Section 12.2.3). More recently, fiber Bragg gratings have been proposed to provide such control [Refs 118, 119] and an integrated external cavity laser device incorporating this structure has demonstrated the suppression of mode hopping when operating over the temperature range 16 to 56 °C [Ref. 120].

6.7.6 Reliability

Device reliability has been a major problem with injection lasers and although it has been extensively studied, not all aspects of the failure mechanisms are fully understood [Ref. 24]. Nevertheless, much progress has been made since the early days when device lifetimes were very short (a few hours).

The degradation behavior may be separated into two major processes known as 'catastrophic' and 'gradual' degradation. Catastrophic degradation is the result of mechanical damage of the mirror facets and leads to partial or complete laser failure. It is caused by the average optical flux density within the structure at the facet and therefore may be limited by using the device in a pulsed mode. However, its occurrence may severely restrict the operation (to low optical power levels) and lifetime of CW devices.

Gradual degradation mechanisms can be separated into two categories which are: (a) defect formation in the active region; and (b) degradation of the current-confining junctions. These degradations are normally characterized by an increase in the threshold current for the laser which is often accompanied by a decrease in its external quantum efficiency [Ref. 121].

Defect formation in the active region can be promoted by the high density of recombining holes within the device [Ref. 122]. Internal damage may be caused by the energy released, resulting in the possible presence of strain and thermal gradients by these nonradiative carrier recombination processes. Hence if nonradiative electron–hole recombination occurs, for instance at the damaged surface of a laser where it has been roughened, this accelerates the diffusion of the point defects into the active region of the device. The emission characteristics of the active region therefore gradually deteriorate through the accumulation of point defects until the device is no longer useful. These defect structures are generally observed as dark spot defects (DSDs).

Mobile impurities formed by the precipitation process, such as oxygen, copper or interstitial beryllium or zinc atoms, may also be displaced into the active region of the laser. These atoms tend to cluster around existing dislocations encouraging high local absorption of photons. This causes dark lines in the output spectrum of the device which are a major problem associated with gradual degradation. Such defect structures are normally referred to as dark line defects (DLDs). Both DLDs and DSDs have been observed in aging AlGaAs lasers as well as in InGaAsP lasers [Ref. 122].

Degradation of the current-confining junctions occurs in many index-guided laser structures (see Section 6.5.2) which utilize current restriction layers so that most of the injected current will flow through the active region. For example, the current flowing outside the active region in BH lasers is known as leakage current. Hence a mode of degradation that is associated with this laser structure is an increase in the leakage current which increases the device threshold and decreases the external differential quantum efficiency with aging.

Over recent years techniques have evolved to reduce, if not eliminate, the introduction of defects, particularly into the injection laser active region. These include the use of substrates with low dislocation densities (i.e. less than 10^{-3} cm^{-2}), passivating the mirror facets to avoid surface-related effects and mounting with soft solders to avoid external strain. Together with improvements in crystal growth, device fabrication and material selection, this has led to CW injection lasers with reported mean lifetimes in excess of 10^6 hours, or more than 100 years. These projections have been reported [Ref. 123] for a variety of GaAs/AlGaAs laser structures. In the longer wavelength region where techniques were not as well advanced, earlier reported extrapolated lifetimes for CW InGaAsP lasers were around 10^5 hours [Ref. 124]. Subsequently, however, InGaAsP and InGaAlAs lasers emitting at 1.3 μm have been tested which display statistically estimated mean lifetimes in excess of 10^6 hours at operating temperatures of 85 °C [Ref. 125]. In addition DFB lasers emitting at 1.55 μm subject to accelerated aging at a temperature of 60 °C have demonstrated stable aging characteristics for more than 2000 hours of operating time.

6.8 Injection laser to fiber coupling

One of the major difficulties with using semiconductor lasers within optical fiber communication systems concerns the problems associated with the efficient coupling of light between the laser and the optical fiber (particularly single-mode fiber with its small core diameter and low numerical aperture). Although injection lasers are relatively directional

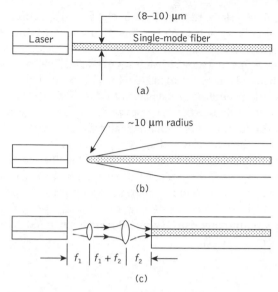

Figure 6.44 Techniques for coupling injection lasers to optical fiber, illustrated using single-mode fiber: (a) butt coupling; (b) tapered hemispherical fiber coupling; (c) confocal lens system

they have diverging output fields which do not correspond to the narrow acceptance angles of single-mode fibers. Thus butt coupling (see Figure 6.44(a)) efficiency from the laser to the fiber is often low at around 10%, even with good alignment and the use of a fiber with a well-cleaved end [Ref. 126]. In this case the optimum coupling efficiency is obtained by positioning the fiber end very close to the laser facet. Unfortunately, this technique allows back reflections from the fiber to couple strongly into the laser which produce noise at the device output that can cause performance degradations in high-speed systems [Ref. 127].

The coupling efficiency can be substantially improved when the output field from the laser is matched to the output field of the fiber. Such matching is usually achieved using a lens (or lens system) positioned between the laser and the fiber. A simple and popular technique is to employ a hemispherical lens formed on the end of a tapered optical fiber,* as illustrated in Figure 6.44(b) [Refs 128, 129]. The numbers of piece parts are therefore minimized and only one alignment step is required. Measured coupling efficiencies up to 65% have been obtained using this method [Ref. 130]. Alternative strategies for micro-lensed fiber coupling include the use of an etched fiber end with lens [Ref. 131] and a high-index lens on the end of a fiber taper [Ref. 132]. Coupling efficiencies of 60% and 55%, respectively, have been achieved using these techniques.

Injection laser coupling using designs based on discrete lenses have also proved fruit-ful. In particular, such lens systems provide for a relaxation in the alignment tolerances normally required to achieve efficient microlensed fiber coupling. For example, the confocal lens system shown in Figure 6.44(c) allows a relaxation in the 1 dB tolerance by about a factor of 4 in comparison with an 8 μm radius microlensed fiber [Ref. 130]. The combination

---

* Such techniques are sometimes referred to as microlensed fibers [Ref. 130].

of the sphere lens and the GRIN-rod lens (see Section 5.5.1) is common within such systems because of the simplicity of the components. Coupling efficiencies of 40% have been obtained with the sphere and GRIN-rod lens in a confocal design. Furthermore, slightly higher efficiencies have been achieved using a GRIN-rod lens with one convex surface (49%) and with a silicon plano-convex lens (55%). The use of a silicon lens within a confocal system has provided coupling efficiencies of up to 70% [Ref. 130].

More recently, a new technique has been developed in order to effectively couple laser power into standard single-mode fiber without using lenses. The technique involves fabricating a polymer tip on the end of the fiber. A single polymer tip measuring between 15 and 150 μm in length may be viewed as an extension of the fiber core. Using this approach a coupling efficiency up to 70% with 1.5 dB coupling loss has been demonstrated [Ref. 133]. Moreover, it is also possible to produce multipeaked tips on multimode fiber by applying mechanical strain to the fiber during the tip growth process, where the multipeaked tip essentially comprises a three-dimensional mold of the intensity distribution within the fiber [Ref. 134].

6.9 Nonsemiconductor lasers

Although at present injection lasers are the major lasing source for optical fiber communications, certain nonsemiconductor sources are of increasing interest for application within this field. Both crystalline and glass waveguiding structures doped with rare earth ions (e.g. neodymium) show potential for use as optical communication sources. In particular, the latter devices in which the short waveguiding structures are glass optical fibers have formed an area of significant development only since 1985 [Ref. 135]. Prior to consideration of these rare-earth-doped fiber lasers, however, this section briefly discusses the most advanced of the crystalline solid-state lasers which could find use within optical fiber communications: the Nd : YAG laser.

6.9.1 The Nd : YAG laser

The crystalline waveguiding material which forms the active medium for this laser is yttrium–aluminum–garnet ($Y_3Al_5O_{12}$) doped with the rare earth metal ion neodymium (Nd^{3+}) to form the Nd : YAG structure. The energy levels for both the lasing transitions and the pumping are provided by the neodymium ions which are randomly distributed as substitutional impurities on lattice sites normally occupied by yttrium ions within the crystal structure. However, the maximum possible doping level is around 1.5%. This laser, which is currently utilized in a variety of areas [Ref. 136], has the following several important properties that may enable its use as an optical fiber communication source:

1. Single-mode operation near 1.064 and 1.32 μm, making it a suitable source for single-mode systems.

2. A narrow linewidth (<0.01 nm) which is useful for reducing dispersion on optical links.

3. A potentially long lifetime, although comparatively little data is available.

4. The possibility that the dimensions of the laser may be reduced to match those of the single-mode fiber.

However, the Nd : YAG laser also has the following drawbacks which are common to all neodymium-doped solid-state devices:

1. The device must be optically pumped. However, long-lifetime AlGaAs LEDs may be utilized which improve the overall lifetime of the laser.

2. A long fluorescence lifetime of the order of 10^{-4} seconds which only allows direct modulation (see Section 7.5) of the device at very low bandwidths. Thus an external optical modulator is necessary if the laser is to be usefully utilized in optical fiber communications.

3. The device cannot take advantage of the well-developed technology associated with semiconductors and integrated circuits.

4. The above requirements (i.e. pumping and modulation) tend to give a cost disadvantage in comparison with semiconductor lasers.

An illustration of a typical end-pumped Nd : YAG laser is shown in Figure 6.45. It comprises an Nd : YAG rod with its ends ground flat and then silvered. One mirror is made fully reflecting while the other is about 10% transmitting to give the output.

The Nd : YAG laser is a four-level system (see Section 6.2.3) with a number of pumping bands and fluorescent transitions. The strongest pumping bands are at wavelengths of 0.75 and 0.81 μm, giving major useful lasing transitions at 1.064 and 1.32 μm. Single-mode emission is obtained at these wavelengths with devices which are usually only around 1 cm in length [Ref. 136]. Although the Nd : YAG laser has the specific advantages and drawbacks noted above, it also has a cost disadvantage in comparison with rare-earth-doped glass fiber lasers (see next section) in that it is far easier and less expensive to fabricate glass fiber than it is to grow YAG crystals.

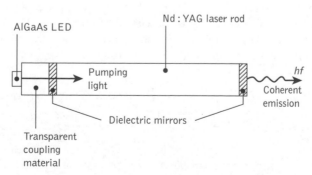

Figure 6.45 Schematic diagram of an end-pumped Nd : YAG laser

6.9.2 Glass fiber lasers

The basic structure of a glass fiber laser is shown in Figure 6.46. An optical fiber, the core of which is doped with rare earth ions, is positioned between two mirrors adjacent to its end faces which form the laser cavity. Light from a pumping laser source is launched through one mirror into the fiber core which is a waveguiding resonant structure forming a Fabry–Pérot cavity. The optical output from the device is coupled through the mirror on the other fiber end face, as illustrated in Figure 6.46. Thus the fiber laser is effectively an optical wavelength converter in which the photons at the pumping wavelength are absorbed to produce the required population inversion and stimulated emission; this provides a lasing output at a wavelength which is characterized by the dopant in the fiber.

The rare earth elements, or lanthanides, number 15 and occupy the penultimate row of the periodic table. They range from lanthanum (La), with an atomic number of 57, to lutetium, which has an atomic number of 71. Ionization of the rare earths normally takes place to form a trivalent state and the two major dopants currently employed for fiber lasers are neodymium (Nd^{3+}) and erbium (Er^{3+}). In common with the Nd : YAG laser (see Section 6.9.1) the former element provides a four-level scheme with significant lasing outputs at wavelengths of 0.90, 1.06 and 1.32 μm. The latter element gives a three-level scheme (see Section 6.2.3) with major useful lasing transitions at 0.80, 0.98 and 1.55 μm [Ref. 135]. One consequence of the number of levels involved in the laser action that is of particular significance to fiber lasers is the length dependence of the threshold power. Provided that the imperfection losses are low, then in a four-level system the threshold power decreases inversely with the length of the fiber gain medium. In a three-level system, however, there is an optimum length that gives the minimum threshold power which is independent of the value of the imperfection losses [Ref. 135].

The glasses which form the host materials for the rare-earth-doped fiber lasers mainly comprise covalently bonded molecules in the form of a disordered matrix with a wide range of bond lengths and bond angles [Ref. 137]. The rare earth ions which are impurities either act as network modifiers or are interstitially located within the glass network. To date, silica-based glasses have provided the major host material, although fluorozirconate fibers (see Section 3.7) doped with both neodymium and erbium ions have produced lasers emitting at wavelengths of 1.05 and 1.35 μm, and 1.55 μm respectively. In addition, fluoride glasses with other dopants give lasing outputs in the mid-infrared wavelength range (see Section 6.11).

Both neodymium- and erbium-doped silica fiber lasers employ codopants such as phosphorus pentoxide (P_2O_5), germania (e.g. GeO_2, $GeCl_4$) or alumina (Al_2O_3). Dopant levels are generally low (at 400 parts per million) in order to avoid concentration quenching which causes a reduction in the population of the upper lasing levels as well as crystallization

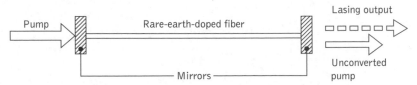

Figure 6.46 Schematic diagram showing the structure of a fiber laser

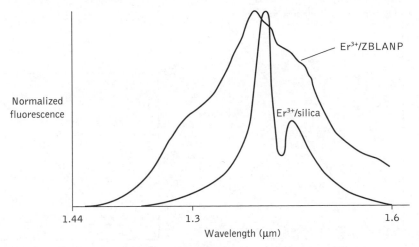

Figure 6.47 Normalized fluorescence from erbium-doped silica and ZBLANP fibers. Reproduced with permission from C. A. Millar, M. C. Brierley and P. W. France, 'Optical amplification in an erbium-doped fluorozirconate fibre between 1480 nm and 1600 nm', *IEE Conf. Publ.*, **292**, Pt 1, p. 66, 1988

within the glass matrix [Ref. 135]. In addition, certain properties of the glass host materials lead to significant spectral broadening of the laser outputs through several mechanisms [Ref. 138] in contrast to what occurs with the Nd : YAG gain medium (see Section 6.9.1). For example, the different fluorescence spectra for an erbium-doped silica fiber and a similarly doped fluorozirconate fiber (ZBLANP)* may be observed in Figure 6.47 [Ref. 139].

The light output against absorbed pump power characteristics for two fiber lasers are displayed in Figure 6.48. The characteristic shown in Figure 6.48(a) corresponds to a neodymium-doped silica fiber laser in which every effort was made to optimize the optical components in the cavity [Ref. 140]. This device in which the mirrors were dielectric coatings deposited directly onto the fiber end faces emitted at a wavelength of 1.06 µm. It may be observed from Figure 6.48(a) that the fiber laser provided a CW output power in excess of 4 mW with a threshold power of 1.51 mW. In addition, the characteristic is linear above threshold with a slope efficiency of 55%. Figure 6.48(b) corresponds to an erbium/ ytterbium with alumina codoped silica fiber laser emitting at a wavelength of 1.56 µm [Ref. 141]. The device, which could be injection laser pumped without the need for stringent pump laser wavelength selection, gave 1 mW of CW output power with a threshold power of 2 mW.

The basic Fabry–Pérot cavity fiber laser shown in Figure 6.45 can be easily constructed from standard optical components but it has several limitations. In particular, the launching of light from the pump laser through one of the mirrored fiber ends can cause damage to the mirror coating as well as a substantial reduction in the launch efficiency. Furthermore, as mentioned previously, the gain spectrum of most rare earth ions extends over a wavelength range of some 50 nm. Unless the dielectric coatings on the mirrors are specially designed for broadband performance, however, the lasing output will be

* ZBLANP fiber has lead fluoride added to the core glass to raise the relative refractive index.

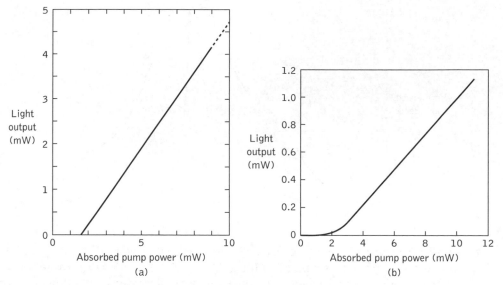

Figure 6.48 Light output against absorbed pump power characteristics for fiber lasers. (a) Neodymium-doped silica fiber. Reproduced with permission from M. Shimitzu, H. Suda and M. Horiguchi, 'High efficiency Nd-doped fibre lasers using direct-coated dielectric mirrors', *Electron. Lett.*, **23**, p. 768, 1987 (IEE). (b) Erbium/ytterbium with alumina codoped silica fiber. Reproduced with permission from D. N. Payne and L. Reekie, 'Rare-earth-doped fibre lasers and amplifers', *IEE Conf. Publ.*, **292**, Pt 1, p. 49, 1988

restricted to between 5 and 10 nm. Such a linewidth is too narrow for the provision of a broadband optical source but too wide to be used in single-frequency laser applications such as coherent transmission. A number of alternative fiber laser structures have therefore been fabricated which do not require dielectric or metallic mirrors. Two of these structures, which are illustrated in Figure 6.49, are the fiber ring resonator [Ref. 142] and the fiber loop reflector made from a series concatenation of distributed reflectors using loops of fiber [Refs 143, 144].

The fiber ring resonator may employ the coherent beam splitting properties of the single-mode fiber fused directional coupler (see Section 5.6.1). In this case two of the arms of the coupler are spliced together as shown in Figure 6.49(a) to form a circulating pathway in which light can travel. Hence an optical cavity without mirrors is formed, the finesse* of which is determined by the splitting ratio of the coupler. When the splitting ratio is low, the finesse is high and the energy storage on resonance is high, thus lowering the laser threshold. However, as with the Fabry–Pérot laser, the lower threshold is obtained at the expense of a reduction in the slope efficiency. Alternatively, high-performance fiber ring resonators can be fabricated by forming the two halves from a single fiber length. This technique has the effect of both reducing losses and of producing a higher finesse.

---

* The finesse of the Fabry–Pérot cavity provides a measure of its filtering properties and can be defined as the free spectral range divided by the full width half maximum permitted by the cavity.

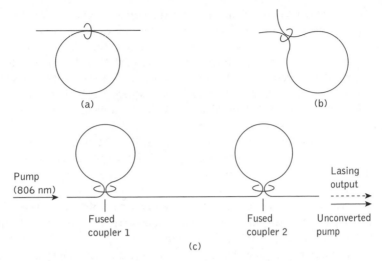

Figure 6.49 Fiber laser structures: (a) fiber ring resonator; (b) fiber loop reflector; (c) all-fiber laser made from two loops in series

The structure of a fiber loop reflector which may also be based on a directional coupler is illustrated in Figure 6.49(b). However, in contrast to the fiber ring where there is energy storage within the resonant structure, the fiber loop is a nonresonant interferometer (it constitutes a Sagnac interferometer, see Section 10.5.2). Light entering the loop through the input fiber end forms forward and backward (reflected) waves which are counterpropagating, providing a coherent superposition of the clockwise and counterclockwise propagating fields. Hence the single fiber loop performs as a distributed all-fiber reflector which may be used to form a fiber laser [Ref. 143]. In addition, when two such loops are joined together in series a resonator is obtained, as shown in Figure 6.49(c). This two-loop structure provided the all-fiber laser which was fabricated from a single length of neodymium-doped fiber without a splice [Ref. 144]. The excess loss of the couplers was only 0.04 dB, giving efficient laser action when the device was pumped with an AlGaAs injection laser at a wavelength of 0.806 µm and with a launch power of 470 µW. Lasing output from the device was obtained at a wavelength of 1.064 µm and was combined with the unconverted pump emission.

Narrow-linewidth and frequency-tunable rare-earth-doped fiber lasers have also been implemented and these devices are discussed in the following section.

6.10 Narrow-linewidth and wavelength-tunable lasers

The single-frequency injection lasers described in Section 6.6 have been developed to minimize the transmission limitations resulting from fiber dispersion in high-speed digital systems. For systems employing intensity modulation with direct detection of the optical signal, however, the laser linewidth and its absolute stability are of secondary importance.

This is not the case with coherent optical fiber transmission where laser linewidth and stability are critical factors affecting the system performance (see Section 13.4.1). Laser linewidths in the range of 1 MHz and below are required for such system applications which are around two orders of magnitude smaller than the 100 MHz linewidths obtained with 250 μm long Fabry–Pérot or DFB devices which emit a few milliwatts without special linewidth control. In addition, wavelength- or frequency-tunable devices are considered to be key components for the provision of both the transmitter and local oscillator optical sources [Refs 145–147].

Injection laser linewidth broadening occurs as a result of the change in lasing frequency with gain [Ref. 148]. It is a fundamental consequence of the spontaneous emission process which is directly related to fluctuations in the phase of the optical field. These phase fluctuations arise from the phase noise directly associated with the spontaneous emission process as well as the conversion of spontaneous emission amplitude noise to phase noise through a coupling mechanism between the photon and carrier densities. In the latter case, because the refractive index is strongly dependent on the carrier density which produces the gain, the fluctuations of gain due to spontaneous emission produce a substantial change in the refractive index which therefore increases the frequency/phase noise in the laser emission. The relationship for the linewidth Δf of an injection laser in terms of the emitted power P_e is given by [Ref. 148].

$$\Delta f = \frac{V_g^2 E n_{sp} \alpha_m}{8 \pi P_e} \left(\alpha_i + \alpha_m\right) \left(1 + \alpha^2\right) \tag{6.52}$$

where V_g is the group velocity, E is the carrier (electron) energy, n_{sp} (in the range 2 to 3) is the spontaneous emission factor, α_i is the internal waveguide loss per unit length,* α_m is the mirror loss per unit length and α is called the linewidth enhancement factor. This last parameter is defined as the ratio of the refractive index change with electron density to the differential gain change with electron density and is a measure of the amplitude to phase fluctuation conversion caused by the spontaneous emission. It can take up values between 2 and 10 depending upon the device material composition, structure and operating wavelength. The term $(1 + \alpha^2)$ in Eq. (6.52) results from the contributions to the linewidth of the two phase fluctuation effects.

It is clear that as the laser power increases, the spontaneous emission becomes relatively less important at the higher photon densities and hence the device linewidth decreases. However, as the output power of the laser cannot be made arbitrarily large, then a more effective method to reduce the linewidth is to make the cavity longer. The linewidth is decreased by increasing the laser length because the effective mirror loss α_m per unit length in Eq. (6.52) is decreased. Two techniques which can be utilized to increase the injection laser cavity length are either to use a long laser chip or to extend the cavity with a passive medium such as air, an optical fiber or an appropriate semiconductor integrated passive waveguide [Refs 94, 149–151]. The latter external cavity devices also provide wavelength/ frequency tunability.

---

* α_i is the injection laser equivalent of the laser loss coefficient per unit length $\bar{\alpha}$ defined in Section 6.2.5.

6.10.1 Long external cavity lasers

Extension of the laser cavity length by the introduction of external feedback can be achieved by using an external cavity with a wavelength dispersive element as part of the cavity. Such devices are often referred to as long external cavity (LEC) lasers. A wavelength dispersive element is required because the long resonator structure has very closely spaced longitudinal modes which necessitates additional wavelength selectivity. A common technique for laboratory use is illustrated in Figure 6.50 where a diffraction grating is employed as an external mirror in order to filter the lasing emission from the wide gain spectrum of a laser chip giving a narrow linewidth at a desired wavelength [Refs 152, 153]. Spectral linewidths as narrow as 10 kHz have been reported with such devices [Refs 145, 152]. Furthermore, wavelength tuning of the output may be achieved by mechanical rotation of the grating such that the lasing wavelength moves with mode hops from one longitudinal mode to the next. In general, coarse spectral adjustment is obtained by rotation of the grating, while fine tuning can be achieved by lateral translation of the grating, as shown in Figure 6.50. Coarse tuning of a single-mode 1.5 μm laser over 90 nm through rotation of the external grating with fine tuning of the same device over approximately 1 GHz by lateral translation of the grating has been demonstrated [Ref. 154].

Another long external cavity method which has been proposed [Ref. 155] employs an external prism grating and GRIN-rod lens (see Section 5.5.1) combination. This technique enabled coarse wavelength adjustment of a BH single-mode device over a range of 40 nm through the lateral displacement of the GRIN-rod lens relative to the laser chip. Fine tuning of around 6 GHz μm^{-1} could be achieved by slight variations in the separation between the laser chip and the GRIN-rod lens end face. The principal disadvantage with these mechanically tuned devices is their relatively low switching speeds. However, by using electro-optic [Ref. 156], acousto-optic [Ref. 157] devices to modulate the external cavity, much higher switching speeds can be achieved. Wavelength selection can then be produced by altering the electro-optic or acousto-optic drive frequency. For example, an acousto-optic filter and modulator pair has been used to select wavelengths over a range of 35 nm for a 0.85 μm laser, with switching speeds of 10 ns [Ref. 158].

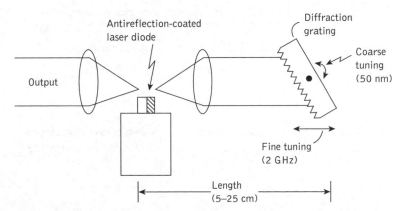

Figure 6.50 Wavelength tuning of an ILD using an external reflective diffraction grating (long external cavity technique)

6.10.2 Integrated external cavity lasers

An alternative technique for the provision of the external cavity is the integrated wave-guide approach. Such monolithic integrated devices often utilize the DFB or the DBR structure. An example of an integrated external cavity DBR laser providing narrow-linewidth dynamic single-mode (DSM) operation at a wavelength of 1.51 μm is shown in Figure 6.51 [Ref. 159]. This device, which had a cavity length of 4.5 mm, exhibited a spectral linewidth of 2 MHz with some 6 mW of optical output power.

Monolithic integrated DSM lasers also offer the potential for wavelength tuning. There are, in principle, two techniques which can be employed to tune these devices. One method is to use the mode selectivity of a coupled-cavity structure such as a C^3 laser (see Section 6.6.1) [Ref. 49]. In this case the effective gain peak wavelength is controlled by the multicavity structure together with multisegment electrodes, as illustrated in Figure 6.52(a). Hence the lasing wavelength can be varied within the effective gain width which is a range in excess of 15 nm for a 1.5 μm InGaAsP laser [Ref. 160]. The device wavelength changes, however, with mode jumps and thus this technique does not provide continuous wavelength tunability.

The other wavelength tuning method for monolithic integrated lasers is to use a refractive index change in the device cavity provided by current injection or the application of an electric field. Typically, this is achieved by employing a multiple-electrode DFB or DBR structure [Ref. 57]. For example, with a single-electrode DFB laser operated above threshold, the high injected carrier density (10^{18} cm^{-3}) reduces the effective refractive index in the corrugation region (Bragg region), thereby decreasing the lasing wavelength. Most of the injected carriers recombine, however, to produce photons, which results in a very small increase in the carrier density leading to only a very small change in the lasing wavelength. The two-electrode DFB laser shown in Figure 6.52(b) allows the wavelength tuning range to be improved by the application of a large current to one electrode and a small current to the other.

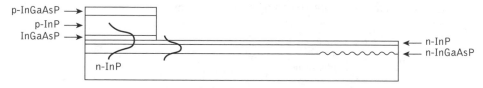

Figure 6.51 Structure of an integrated external cavity DBR laser

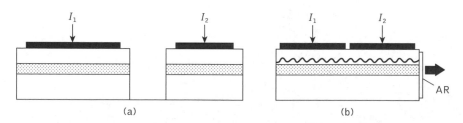

Figure 6.52 Monolithic integrated dynamic single-mode lasers: (a) cleaved-coupled-cavity (C^3) laser; (b) double-sectioned DFB laser

With the asymmetric DFB laser structure of Figure 6.52(b), the optical field is higher in the region near the output port where the facet is nonreflecting (antireflection (AR) coated, as shown in diagram), and the device operating wavelength is primarily determined by the effective refractive index in this region. When the aforementioned section is pumped at current densities at or slightly below the threshold density (under uniform pumping) simply to overcome the absorption losses, then it acts as a Bragg reflector. Furthermore, the injected carriers do not contribute significantly to the generation of photons because of the low pumping level. This factor results in a large change of refractive index which gives wavelength tuning. It should be noted that the gain is provided by another section (not shown in Figure 6.52(b)) which is pumped well above threshold. A maximum continuous tuning range of 3.3 nm with 1 mW output power has been obtained with such a device [Ref. 161]. The spectral linewidth of this laser was 15 MHz and the tuning range reduced to 2 nm at an output power of 5 mW.

Three-electrode DFB lasers have also demonstrated good tunability. A $\lambda/4$-shifted device (see Section 6.6.2) in which the two outer electrodes were electrically connected to a common current supply while the central electrode was supplied with a different current has given a continuous tuning range of 2 nm by varying the two currents [Ref. 162]. In addition, the device displayed a spectral linewidth of only 500 kHz. Although such tunable DFB lasers have a limited tuning range in comparison with coupled-cavity devices, they exhibit advantages of ease of fabrication as well as providing continuous tuning rather than discrete jumps.

Multiple-electrode DBR laser structures have also been developed to allow wavelength tuning [Ref. 160]. In particular, wider wavelength tuning ranges have been obtained not only by separating the Bragg region in the passive waveguide (a large bandgap material) from the active region (a small bandgap material) inside the laser cavity, but also by introducing a phase region within the waveguide. The structure of such a three-sectioned DBR laser is illustrated in Figure 6.53(a). The wavelength of this device can simply be electronically tuned by current injection into the DBR section. This region exhibits a high reflectance within a certain wavelength band (the stop band) which is nominally between 2 and 4 nm wide. The mechanism which results from a refractive index change in the passive waveguide layer is known as Bragg wavelength control. A continuous tuning range, however, is limited to the resonant mode spacing which is defined from the effective cavity length of the laser. It is the mode which is nearest to the center of the stop band and which simultaneously satisfies the 2π round trip phase condition that lases. Therefore, the introduction of the phase region in the waveguide (Figure 6.53(a)) which is independently controlled by the injection current allows the lasing wavelength to be tuned around each Bragg wavelength. Such a region provides phase control which again occurs through refractive index changes in the passive waveguide.

The combination of the two types of tuning (Bragg wavelength and phase tuning) provides a significantly larger tuning range because the lasing wavelength deviation from the Bragg wavelength can be compensated by phase control. With good design and the independent adjustment of the three currents in the active Bragg and phase regions, quasi-continuous tuning ranges between 8 and 10 nm have been obtained [Refs 58, 163]. In addition, a continuous tuning range of 6.2 nm has been achieved with a similar device [Ref. 164]. Alternatively, for continuous wavelength tuning, one control current has been divided in a prescribed proportion into the Bragg and phase sections as illustrated in

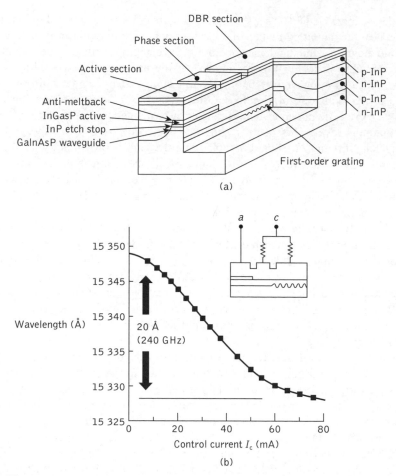

Figure 6.53 Three-sectioned DBR laser: (a) structure; (b) range of continuous wavelength tuning. Reprinted with permission from T. P. Lee and C. E. Zah, 'Wavelength-tunable and single-frequency semiconductor lasers for photonic communications networks', *IEEE Commun. Mag.*, p. 42, October, 1989. Copyright © 1989 IEEE

Figure 6.53(b). Continuous tuning ranges of between 2 and 4 nm have been reported using this method [Ref. 58].

An increased tuning range of between 40 and 60 nm can be achieved, however, by adding two diffraction Bragg grating sections, a gain, a phase and an amplifier section, as displayed in Figure 6.54(a) in which all five sections are longitudinally integrated together on a semiconductor substrate. The structure is called a sampled-grating distributed Bragg reflector (SG-DBR) laser [Ref. 165] or sometimes it is referred to as a superstructure grating distributed Bragg reflector (SSG-DBR) laser [Ref. 166]. It should be noted that a sampled grating is a modification of a continuous grating in which the grating teeth have been periodically removed along its length. The front and back mirrors of the laser are sampled

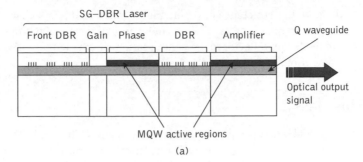

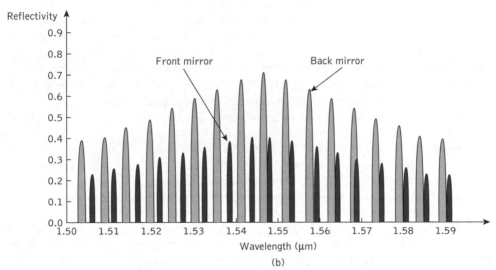

Figure 6.54 Sampled-grating distributed Bragg reflector (SG–DBR) laser: (a) device structure; (b) reflectivity of the mirror against wavelength, showing front and back mirror reflections

in this way with different periods such that only one of their multiple reflection peaks can coincide at a time, as shown in Figure 6.54(b). The desired channel can be selected by tuning the two mirrors when the closest reflection peak of each mirror is aligned to the desired channel and at that point lasing occurs. This tuning process is referred to as a vernier effect [Ref. 167] because the tuning scheme resembles the measurement technique employing a vernier scale used to determine the length of physical quantities. Both grating sections have a slightly different pitch and therefore the wavelength of the output signal is tuned by varying the current to the grating sections which changes the refractive index of each section to produce the required output signal wavelength. Although the output power is generally limited to about 2 mW, this can be increased up to 10 mW by using the optical amplifier section also shown in Figure 6.54(a) [Ref. 168].

Another type of DBR laser capable of a tuning range greater than 40 nm is the grating-assisted codirectional coupler with sampled reflector (GCSR) device [Ref. 169]. The GCSR laser can be produced when a coupler section is introduced between the amplifier and phase sections of the structure shown in Figure 6.54(a). In this case the current-controlled waveguide coupler acts as a coarse tuner to deliver a narrow range of signal wavelengths from the Bragg reflector to the phase section which then provides fine tuning, thus ensuring lasing of only one cavity mode.

6.10.3 Fiber lasers

Techniques are also under investigation to obtain narrow-linewidth output from glass fiber lasers [Ref. 135]. The rare-earth-doped fiber lasers described in Section 6.9.2 have spectral linewidths typically in the range 0.1 to 1 nm which are too broad for high-speed transmission. One method to achieve narrower spectral linewidths employed polished silica blocks with surface gratings, as illustrated in Figure 6.55 [Ref. 170]. In this case the holographic gratings acted as distributed feedback reflectors which reflected only a narrow band of wavelengths. The reflector (Figure 6.55) through which the pump beam was launched was a dielectric mirror butted against the fiber end. Moreover, the fiber in the coupler block was undoped and one end was butt jointed to an erbium-doped fiber. An output spectral linewidth of 0.04 nm (5 GHz) was obtained which is indicative of the relative state of development of fiber lasers in comparison with semiconductor devices.

Substantially narrower spectral linewidths have, however, been obtained with fiber lasers using a fiber Fox–Smith resonator design [Ref. 135]. This device, which employs a fused coupler fabricated from erbium-doped fiber, has demonstrated a lasing linewidth of less than 1 MHz which compares favorably with the linewidths obtained from conventional semiconductor DFB lasers but not external cavity lasers.

Finally, wavelength tuning has also been obtained with fiber lasers. In particular, the use of silica as the laser medium provides good power handling characteristics and broadens the rare earth transitions, enabling tunable devices. An investigation of wavelength tuning in a neodymium (Nd^{3+}) doped single-mode fiber laser employed the experimental configuration shown in Figure 6.56(a) [Ref. 171]. Tuning was accomplished by changing the angle of the diffraction grating, which was mounted on a sine-bar-driven turntable. A tuning range of 80 nm was obtained, as may be observed from the characteristic (including the fluorescence spectrum of Nd^{3+} ions in silica) displayed in Figure 6.56(b). Furthermore, the wavelength tuning of an erbium (Er^{3+}) doped single-mode fiber laser was also reported [Ref. 171] to provide a tuning range of 25 nm around the 1.54 µm wavelength region using a similar experimental configuration. In addition, a wider tuning range greater than 100 nm around the 1.55 µm wavelength using an erbium-doped photonic crystal fiber (see Section 2.6) has also been obtained [Ref. 172].

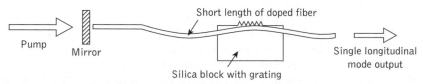

Figure 6.55 Fiber laser with a cavity incorporating a polished silica block and grating reflector. Reprinted with permission from Ref. 170 © IEEE 1988

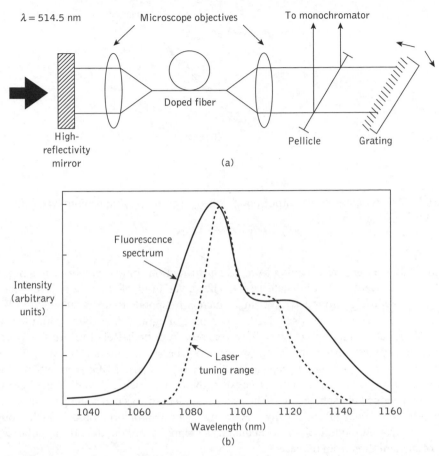

Figure 6.56 Tunable neodymium-doped single mode fiber laser: (a) configuration; (b) fluorescence spectrum for doped fiber and the laser tuning range. Reprinted with permission from L. Reekie, R. J. Mears, S. B. Poole and D. N. Payne, 'Tunable single-mode fiber lasers', *J. Lightwave Technol.*, **LT-4**, p. 956, 1986. Copyright © 1986 IEEE

An alternative method for wavelength tuning of fiber lasers employed the loop reflector discussed in Section 6.9.2. In this case a temperature shift was used to adjust the coupling ratio through the directional coupler which had a direct effect on the output optical wavelength. A 60 °C variation in temperature provided a tuning range of around 33 nm [Ref. 173].

More recently, the generation of short optical pulses using a cavity with frequency-shifted feedback based on a Q-passively mode-locked fiber laser has been demonstrated [Ref. 174]. This device uses an acousto-optic modulator inside the cavity to achieve the frequency shifting. The fiber laser generated stable mode-locked pulses of 2 ps deviation when emitting at a wavelength of 1.55 μm [Ref. 174]. Moreover, the device could be tuned over the entire erbium fiber gain bandwidth (i.e. 1.52 to 1.62 μm).

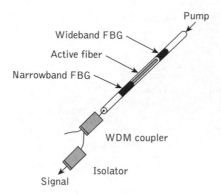

Figure 6.57 FBG-based high-output-power fiber laser employing erbium–ytterbium codoped phosphate material

Fiber-based lasers are capable of providing diffraction-limited power at much higher levels than conventional solid-state lasers. The lower limit of the power level that can damage pure silica is around 10^{10} W cm^{-2} which corresponds to approximately 5 kW for an 8 μm core diameter fiber [Ref. 175]. A compact integrated fiber laser with more than 200 mW of stable output power has, however, been reported [Ref. 150]. In this case the laser cavity was established by using two passive fiber Bragg gratings (FBGs), one narrowband and one wideband, as depicted in Figure 6.57. These FBGs were attached with a very short piece of active material between them comprising heavily doped erbium–ytterbium codoped phosphate glass. The lasing wavelength was determined by the spectral overlap between both FBGs. Finally, the arrangement also incorporated a WDM coupler and a polarization mode isolator to maintain the signal wavelength and the polarization state of the emission from the laser.

6.11 Mid-infrared and far-infrared lasers

Laser sources for transmission at wavelengths beyond 2 μm, in particular gas and solid-state lasers as well as low-temperature injection lasers, have been utilized in nontelecommunication applications such as high-resolution spectroscopy, materials processing and remote monitoring. Development of the potentially ultra-low-loss fibers for mid-infrared transmission (see Section 3.7) operating over the wavelength range 2 to 5 μm has, however, encouraged greater activity in the pursuit of longer wavelength optical sources. For practical communication systems in the mid-infrared wavelength region the requirement is for semiconductor or fiber lasers which are capable of operating at, or close to, room temperature.

Semiconductor materials with direct bandgaps which encompass both the mid-infrared and far-infrared (8 to 12 μm) wavelength range include many of the III–V, II–VI and IV–VI alloys. Injection lasers operating in this longer wavelength region, however, are subject to increased carrier losses over devices emitting at wavelengths up to 1.6 μm which result from nonradiative recombination via the Auger interaction [Ref. 176]. The recombination energy of the injected carriers is dissipated as thermal energy to the remaining free carriers by this process. Moreover, the probability of the occurrence of such a process increases as the bandgap of the semiconductor is reduced. In addition, optical losses due to free carrier absorption are also greater because of their dependence on the square of the wavelength. Both of these effects present more problems in the mid-infrared wavelength range and they exhibit increased importance at higher temperatures as a result of the higher concentration of free carriers. They therefore play a major role in the determination of the injection laser threshold current and efficiency, as well as providing a limit to the maximum operating temperature of the device.

The total current required to provide the injection laser threshold is greater than the amount attributable only to radiative recombination by the addition of an Auger current. Although the Auger current depends upon the precise electronic band structure of the material, and often consists of contributions from a number of different Auger transitions, it is generally large for materials with bandgaps which provide longer wavelength emission. In this context the results of calculations for threshold current and internal quantum efficiency for several long-wavelength semiconductor alloys are displayed in Figure 6.58 [Ref. 177]. A comparison of the highest predicted oscillation temperatures of pulsed DH lasers fabricated from various compounds as a function of wavelength, based on estimates of the temperature at which the device internal quantum efficiency at current threshold falls to 2.5%, is shown. In addition, experimental observations are depicted as data points in the figure. It may be observed from Figure 6.58 that this data indicates an overall limit to room temperature laser action at wavelengths slightly above 2 μm for any of the semiconductor alloys investigated.

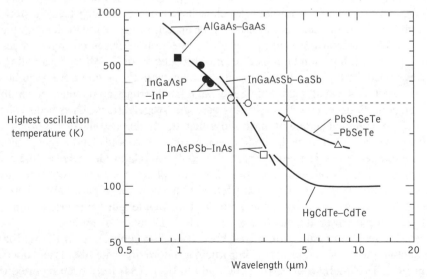

Figure 6.58 Characteristics showing maximum temperature of pulsed operation for DH lasers against wavelength for several material systems. Reprinted from Ref. 177 with permission from Elsevier

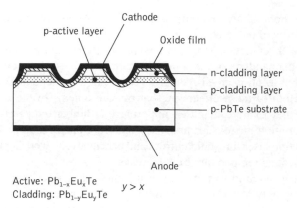

Figure 6.59 Structure of PbEuTe/PbTe DH laser. Reprinted with permission from Ref. 181 © IEEE 1989

Room temperature operation of III–V alloy semiconductor lasers fabricated from InGaAsSb, and GaAlAsSb lattice matched to either GaSb or InAs, has been obtained in the wavelength range 2.2 to 2.3 μm [Refs 178, 179]. Low-threshold-current density of 1.7 kA cm^{-2} at room temperature has also been reported [Ref. 180] but although laser oscillation is predicted to occur up to a wavelength of 4.4 μm, it is at a temperature of only 77 K due to the presence of the Auger current [Ref. 29]. In addition, the InAsPSb lattice matched to InAs offers the potential for operation over the 2 to 3.5 μm wavelength region but calculations indicate a similar dependence of the maximum operating temperature on wavelength to GaInAsSb (see Figure 6.59).

An example of a II–VI alloy semiconductor is the HgCdTe material system, also shown in Figure 6.59, from which infrared detectors have been fabricated (see Section 8.10). Although LEDs and optically pumped lasers for operation over the wavelength range 2 to 4 μm have been demonstrated using this alloy, injection laser sources have as yet to be reported [Ref. 29]. Injection lasers, however, fabricated from IV–VI lead–salt alloys have been developed for high-resolution spectroscopic as well as gas monitoring applications. Devices based on the quaternary PbSnSeTe and related ternary compounds generally emit at wavelengths longer than 4 μm. In this case the Auger effects have been calculated [Ref. 177] to be less in certain of these alloys than those obtained in III–V semiconductor materials, which could provide both lower current thresholds and higher maximum operating temperatures. The replacement of Sn with Eu, Cd or Ge increases the bandgap to provide shorter wavelength operation. For example, the structure of some reported ternary alloy PbEuTe/PbTe DH lasers [Ref. 181] is shown in Figure 6.59. These mesa-stripe devices which emitted over the 3.5 to 6.5 μm wavelength range provided in excess of 200 μW output power at temperatures up to 210 K in pulsed operation.

The investigation of rare-earth-doped fiber lasers for application in the mid-infrared wavelength region is also under way. In particular, fluorozirconate fiber lasers doped with erbium [Ref. 182], holium [Ref. 183] and thulium [Ref. 184] have been reported to provide emissions in the 2 to 3 μm wavelength range. The 2.702 μm transition in erbium which had only previously been obtained in bulk fluorozirconate glass samples [Ref. 135] was demonstrated in a CW fiber laser pumped at twice threshold [Ref. 182]. Lasing was obtained when 191 mW of pump light at a wavelength of 0.477 µm was launched into the doped fluorozirconate fiber.

The holium-doped fluorozirconate fiber was made to lase with a CW output at wavelengths of 1.38 µm and 2.08 µm [Ref. 183]. In both cases, pumping was obtained from an argon ion source at a wavelength of 0.488 µm and the 2.08 µm emission was the first report of the operation of a fiber laser at wavelengths beyond 1.55 µm [Ref. 135]. Finally, the thulium-doped fiber laser emitted at a wavelength of 2.3 µm when pumped with the pulsed output from an alexandrite laser at 0.786 µm [Ref. 184]. Unlike the longer wavelength holium emission which originates from a three-level system, the thulium system at 2.3 µm is four level in which the pump band is also the upper lasing level.

6.11.1 Quantum cascade lasers

A fiber laser produced for operation around the 3 µm wavelength point is the diode-cladding-pumped erbium praseodymium–doped fluoride device [Refs 185, 186]. Although this laser is capable of producing very high output power of more than 1 W, it comprises expensive double-clad fluoride fiber which is difficult to cleave with consistently high optical quality over the entire cross-section of the pump fiber cladding. More recently, however, a new technique based on intersubband transitions has resulted in a device known as a quantum cascade (QC) laser which has been successfully demonstrated for mid-infrared emission [Refs 187, 188]. In principle this technique provides for emission of an optical signal across the full wavelength range of the mid- and far-infrared regions (i.e. 2 to 12 µm) since the emitted wavelength is determined by quantum mechanical band structure engineering. The QC laser is a layered semiconductor device comprising a series of coupled quantum wells grown on GaAs or InP substrates [Refs 189, 190].

A basic energy-level structure for the QC laser is provided in Figure 6.60 which shows two cascaded quantum-well stages (i.e. stages A and B). Each stage is divided into two sections which act as the injector and active region, respectively. Furthermore, the injector section can be further divided into two parts for the injection and collection of electrons, the latter part being situated adjacent to the active stage. These parts are sometimes also referred to individually as the injector and the collector.

The operation of the QC laser can be likened to an electronic waterfall which functions by pumping up the energy level of electrons which then instead of dropping back in a single step lose their energy in a controlled manner so that they give up some energy each time over several steps. Since the QC laser structure contains a series of energy levels, the same electron can therefore emit a number of photons as it cascades down through each energy level. Furthermore, when the size of the layers (i.e. quantum wells) is reduced to a size comparable with the emission wavelength, then the motion of the electron becomes perpendicular to the plane of the layer. This effect, which is referred to as the quantum confinement,* causes the electrons to only jump from one state to the other in discrete steps, each time emitting a photon of light. It should be noted that the three energy levels identified as 1, 2 and 3 in Figure 6.60 represent only the conduction band since the QC

---

* Quantum confinement is the trapping of electrons or holes (i.e. charge carriers) in a small area and typically occurs in quantum wells at the nanometer scale.

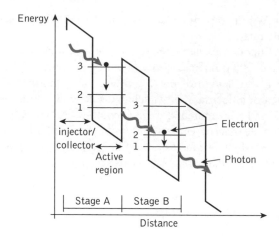

Figure 6.60 Energy-level diagram for a quantum cascade laser

laser uses only *n*-type charge carriers and the holes play no part in the device operation, such that it is sometimes simply referred to as the unipolar laser. Each quantum-well stage produces a single photon when an electron falls from a higher energy level (identified as 3) to the lower energy level (identified as 2) and then to the lowest level dropping from 2 to 1. Therefore a single electron can produce several photons depending on the number of cascaded stages, unlike the conventional semiconductor injection laser where one electron generates only a single photon. The energy given up by the electron at each cascade stage determines the wavelength of the radiation which does not depend on the properties of the material but on the thickness of layer. Hence by selecting layers of different thicknesses it is possible to obtain different output signal wavelengths.

Cascading several stages to form a QC laser makes the overall structure complex but nevertheless the device still remains small (typically 1.5 to 3.0 μm) in comparison with a conventional injection laser [Ref. 191]. In addition, it is also useful to create several injector/collector and active regions in a single stage where each region contains a single quantum well. Such a structure (i.e. MQW cascade laser) allows more injection/collection of current and thus produces a greater number of photons. The formation of many (typically 25 to 75 [Ref. 192]) alternating injector/collectors and a number of active regions can be achieved through the precise control of several hundred layers of material, each one only a few nanometers thick.

The QC laser can exhibit improved performance characteristics in comparison with the conventional injection laser as it provides an increased output signal power (greater than 1000) at the same wavelength due to the large number of cascaded stages and its ability to carry large currents [Ref. 192]. Also, since the QC laser employs a larger energy bandgap it can transmit at any desired signal wavelength within the infrared region. Hence QC lasers emitting at signal wavelengths in both the mid- and far-infrared regions using different material systems based on InP and also photonic crystal bandgap materials (see Section 2.6.2) have been demonstrated [Refs 193, 194]. In addition, QC lasers can be designed to emit a signal vertically in the same manner as the VCSEL [Ref. 195]. Moreover, as a result of their larger energy bandgaps QC laser materials are much easier to process, less prone to defect formation and more reliable than low-bandgap semiconductors such as lead salts, indium-arsenide- and indium-antimonide-based alloys used for conventional mid-infrared laser diodes. Heat generation in QC lasers, however, needs to be carefully controlled since the electrons in the active regions, influenced by lattice vibrations, create phonons that cause large heat generation [Refs 196, 197]. These phonons also cause the electrons to drift into lower energy levels (i.e. subband) instead of reaching their anticipated higher energy level [Ref. 197]. For this reason QC lasers operate better in pulsed mode at room temperature [Ref. 198]. Device performance for CW mode is, however, improved when laser cooling techniques are incorporated to reduce these adverse heating effects [Ref. 199]. Finally, it should be noted that mid- and far-infrared fibers are at present not being seriously considered for long-haul communications (see Section 3.7) and therefore QC lasers are finding utilization in nontelecommunication applications.

Problems

6.1 Briefly outline the general requirements for a source in optical fiber communications.
Discuss the areas in which the injection laser fulfills these requirements, and comment on any drawbacks of using this device as an optical fiber communication source.

6.2 Briefly describe the two processes by which light can be emitted from an atom. Discuss the requirement for population inversion in order that stimulated emission may dominate over spontaneous emission. Illustrate your answer with an energy-level diagram of a common nonsemiconductor laser.

6.3 Discuss the mechanism of optical feedback to provide oscillation and hence amplification within the laser. Indicate how this provides a distinctive spectral output from the device.
The longitudinal modes of a gallium arsenide injection laser emitting at a wavelength of 0.87 μm are separated in frequency by 278 GHz. Determine the length of the optical cavity and the number of longitudinal modes emitted. The refractive index of gallium arsenide is 3.6.

6.4 An injection laser has a GaAs active region with a bandgap energy of 1.43 eV. Estimate the wavelength of optical emission from the device and determine its linewidth in hertz when the measured spectral width is 0.1 nm.

6.5 The refractive index of the InGaAsP active region of an injection laser at a wavelength of 1.5 μm is 3.5 and the device has an active cavity length of 400 μm. For laser operation at a wavelength of 1.5 μm determine: (a) the laser emission mode index; (b) the eligible number of wavelengths inside the cavity; (c) the frequency separation of the modes in the active cavity in order to produce constructive interference.

6.6 When GaSb is used in the fabrication of an electroluminescent source, estimate the necessary hole concentration in the p-type region in order that the radiative minority carrier lifetime is 1 ns.

6.7   The energy bandgap for lightly doped gallium arsenide at room temperature is 1.43 eV. When the material is heavily doped (degenerative) it is found that the lasing transitions involve 'bandtail' states which effectively reduce the bandgap transition by 8%. Determine the difference in the emission wavelength of the light between the lightly doped and this heavily doped case.

6.8   With the aid of suitable diagrams, discuss the principles of operation of the injection laser.
   Outline the semiconductor materials used for emission over the wavelength range 0.8 to 1.7 μm and give reasons for their choice.

6.9   Determine the range of bandgap energies for:
   (a) $Al_yGa_{1-y}As/Al_xGa_{1-x}As$;
   (b) $In_{1-x}Ga_xAs_yP_{1-y}/InP$.

6.10 A DH injection laser has an optical cavity of length 50 μm and width 15 μm. At normal operating temperature the loss coefficient is 10 cm^{-1} and the current threshold is 50 mA. When the mirror reflectivity at each end of the optical cavity is 0.3, estimate the gain factor $\bar{\beta}$ for the device. It may be assumed that the current is confined to the optical cavity.

6.11 The coated mirror reflectivity at either end of the 350 μm long optical cavity of an injection laser is 0.5 and 0.65. At normal operating temperature the threshold current density for the device is 2×10^3 A cm^{-2} and the gain factor β is 22×10^{-3} cm A^{-1}. Estimate the loss coefficient in the optical cavity.

6.12 Describe the techniques used to give both electrical and optical confinement in multimode injection lasers. Contrast these techniques when used in gain-guided and index-guided lasers.

6.13 A gallium arsenide injection laser with a cavity of length 500 μm has a loss coefficient of 20 cm^{-1}. The measured differential external quantum efficiency of the device is 45%. Calculate the internal quantum efficiency of the laser. The refractive index of gallium arsenide is 3.6.

6.14 Compare the ideal light output against current characteristic for the injection laser with one from a typical gain-guided device. Describe the points of significance on the characteristics and suggest why the two differ.

6.15 Describe, with the aid of suitable diagrams, the major strategies and structures utilized in the fabrication of single-frequency injection lasers. Indicate the reasons for the great interest in such devices.

6.16 The threshold current density for a stripe geometry AlGaAs laser is 3000 A cm^{-1} at a temperature of 15 °C. Estimate the required threshold current at a temperature of 60 °C when the threshold temperature coefficient T_0 for the device is 180 K, and the contact stripe is 20×100 μm.

6.17 Briefly describe what is meant by the following terms when they are used in relation to injection lasers:
   (a) relaxation oscillations;
   (b) frequency chirp;

(c) partition noise;

(d) mode hopping.

6.18 Explain the concept of quantum-dot and quantum wire lasers and describe their operation in comparison with conventional injection laser diodes.

6.19 Discuss the operation of a vertical cavity surface-emitting laser (VCSEL). Briefly indicate the three methods to provide wavelength tuning for a VCSEL.

6.20 The rms value of the power fluctuation on the output from a single-mode semiconductor laser is 2×10^{-4} W when the relative intensity noise (RIN) is -160 dB Hz^{-1}. The emission, which is at a wavelength of 1.30 μm, is directly incident on an optical detector with a quantum efficiency of 70% at this wavelength. If the rms noise current at the detector output is 0.53 μA, and assuming that the RIN is the dominant noise source, calculate the mean optical power incident on the photodetector.

6.21 A single-mode injection laser launches light with a 3 dB linewidth Δf into a fiber link which has two connectors exhibiting reflectivities r_1 and r_2. It is known that the worst case relative intensity noise (RIN) occurs when the direct and doubly reflected optical fields interfere in quadrature [Ref. 200] following:

$$\mathrm{RIN}(f) = \frac{4r_1 r_2}{\pi} \frac{\Delta f}{f^2 + \Delta f^2} [1 + \exp(-4\pi \,\Delta f\, \tau) - 2 \exp(-2\pi \,\Delta f\, \tau) \cos(2\pi f \tau)]$$

Demonstrate that the above expression reduces to:

$$\mathrm{RIN}(f) = \frac{16 r_1 r_2}{\pi} \Delta f \tau^2 \qquad \text{for } \Delta f \cdot \tau \ll 1$$

and:

$$\mathrm{RIN}(f) = \frac{4 r_1 r_2}{\pi} \frac{\Delta f}{f^2 + \Delta f^2} \qquad \text{for } f \cdot \tau \gg 1$$

6.22 A DFB laser has a 3 dB linewidth Δf of 50 MHz. It is connected to a short optical jumper cable such that $\Delta f \cdot \tau$ is 0.1. Using the relationship given in Problem 6.21 when the frequency f is also 50 MHz, obtain the average reflectivity for each of the connectors so that the RIN is reduced below a level of -130 dB Hz^{-1}.

6.23 Discuss degradation mechanisms in injection lasers. Comment on these with regard to the CW lifetime of the devices.

6.24 Describe the structure and operation of a glass fiber laser. Comment on the glass compounds currently employed together with their fluorescence spectra.

6.25 Discuss linewidth narrowing and wavelength tunability associated with single-frequency injection lasers. Outline the major techniques which are being adopted to facilitate these characteristics.

6.26 Describe the structure and explain the operation of a sample-grating DBR laser. Using a diagram, briefly discuss the function of the vernier effect in this laser type.

6.27 Describe the energy-level structure and operation of a quantum cascade (QC) laser for mid-infrared transmission. Indicate the benefits of the QC approach in comparison with a conventional semiconductor injection laser.

Answers to numerical problems

6.3 150 μm, 1241
6.4 0.87 μm, 39.6 GHz
6.5 (a) 1866
    (b) 933
    (c) 107 Hz
6.6 4.2×10^{18} cm^{-3}
6.7 0.07 μm

6.9 (a) 1.38 to 1.91 eV
    (b) 0.73 to 1.35 eV
6.10 3.76×10^{-2} cm A^{-1}
6.11 28 cm^{-1}
6.13 84.5%
6.16 77.0 mA
6.20 3.6 mW
6.22 −25.2 dB

References

[1] C. Kittel, *Introduction to Solid State Physics* (8th edn), Wiley, 2004.
[2] E. S. Yang, *Fundamentals of Semiconductor Devices*, McGraw-Hill, 1978.
[3] Y. P. Varshni, 'Band to band radiative recombination in groups IV, VI and III–V semiconductor I', *Phys. Status Solidi*, **19**(2), pp. 459–514, 1967.
[4] H. Kressel and J. K. Butler, *Semiconductor Lasers and Heterojunction LEDs*, Academic Press, 1977.
[5] T. Landsberg, *Recombination in Semiconductors*, Cambridge University Press, 2003.
[6] H. Kressel, 'Electroluminescent sources for fiber systems', in M. K. Barnoski (Ed.), *Fundamentals of Optical Fiber Communications*, pp. 109–141, Academic Press, 1976.
[7] H. C. Casey and M. B. Parish, *Heterostructure Lasers: Part A and B*, Academic Press, 1978.
[8] A. Yariv, *Optical Electronics* (4th edn), Holt, Rinehart and Winston, 1991.
[9] K. Y. Lau and A. Yariv, 'High-frequency current modulation of semiconductor injection lasers', in *Semiconductors and Semimetals*, 22, Pt B, pp. 70–152, Academic Press, 1985.
[10] R. G. Hunsperger, *Integrated Optics*, (5th edn), Springer-Verlag, 2002.
[11] J. C. Dyment, L. A. D'Asaro, J. C. Norht, B. I. Miller and J. E. Ripper, 'Proton-bombardment formation of stripe-geometry heterostructure lasers for 300 K CW operation', *Proc. IEEE*, **60**, pp. 726–728, 1982.
[12] H. Kressel and M. Ettenburg, 'Low-threshold double heterojunction AlGaAs/GaAs laser diodes: theory and experiment', *J. Appl. Phys.*, **47**(8), pp. 3533–3537, 1976.
[13] H. Shoji, K. Otsubo, T. Fujii and H. Ishikawa, 'Calculated performances of 1.3 μm vertical-cavity surface-emitting lasers on InGaAs ternary substrates', *IEEE J. Quantum Electron.*, **33**(2), pp. 238–245, 1997.
[14] K. Otsubo, Y. Nishijima and H. Ishikawa, 'Long-wavelength semiconductor lasers on InGaAs ternary substrates with excellent temperature characteristics', *J. Fujiisu Sci. Technol.*, **34**(2), pp. 212–222, 1998.
[15] D. Gollub, M. Fischer and A. Forchel, 'Towards high performance GaInAsN/GaAsN laser diodes in 1.5 μm range', *Electron. Lett.*, **38**(20), pp. 1183–1184, 2002.
[16] S. Nakamura, G. Fasol and S. J. Pearton, *The Blue Laser Diode: The Complete Story*, Springer-Verlag, 1997.

[17] S. Nakamura, 'InGaN-based violet laser diodes: topical review', *J. Semicond. Sci. Technol.*, **14**, R27–R40, 1999.

[18] K. B. Nam, J. Li, M. L. Nakarmi, J. Y. Lin and H. X. Jiang, 'Deep ultraviolet picosecond time-resolved photoluminescence studies of AlN epilayers', *Appl. Phys. Lett.*, **82**(11), pp. 1694–1696, 2003.

[19] W. Ha, V. Gambin, S. Bank, M. Wistey, H. Yuen, S. Kim and J. S. Harris Jr, 'Long-wavelength GaInNAs(Sb) lasers on GaAs', *IEEE J. Quantum Electron.*, **38**(9), pp. 1260–1267, 2002.

[20] S. R. Bank, M. A. Wistey, L. L. Goddard, H. B. Yuen, V. Lordi and J. S. Harris Jr, 'Low-threshold continuous-wave 1.5 μm GaInNAsSb lasers grown on GaAs', *IEEE J. Quantum Electron.*, **40**(6), pp. 656–664, 2004.

[21] X. Hu, J. Deng, N. Pala, R. Gaska, M. S. Shur, C. Q. Chen, J. Yang, G. Simin, M. A. Khan, J. C. Rojo and L. J. Schowalter, 'AlGaN/GaN heterostructure field-effect transistors on single-crystal bulk AlN', *Appl. Phys. Lett.*, **82**(8), pp. 1299–1301, 2003.

[22] R. Hui, Y. Wan, J. Li, S. Jin, J. Lin and H. Jiang, 'III-nitride-based planar lightwave circuits for long wavelength optical communications', *IEEE J. Quantum Electron.*, **41**(1), pp. 100–110, 2005.

[23] M. Pessa, C. S. Peng, T. Jouhti, E.-M. Pavelescu, W. Li, S. Karirinne, H. Liu and O. Okhotnikov, 'Towards high-performance nitride lasers at 1.3 μm and beyond', *IEE Proc., Optoelecron.*, **150**(1), pp. 12–21, 2003.

[24] N. K. Dutta, 'Optical sources for lightwave systems applications', in E. E. Basch (Ed.), *Optical-Fiber Transmission*, H. W. Sams & Co., 1987.

[25] I. P. Kaminow, L. W. Stulz, J. S. Ko, A. G. Dentai, R. E. Nahory, J. C. DeWinter and R. L. Hartman, 'Low-threshold InGaAsP ridge waveguide lasers at 1.3 μm', *IEEE J. Quantum Electron.*, **QE-19**, pp. 1312–1319, 1983.

[26] C. J. Armistead, S. A. Wheeler, R. G. Plumb and R. W. Musk, 'Low threshold ridge waveguide lasers at $\lambda = 1.5$ μm', *Electron. Lett.*, **22**, pp. 1145–1147, 1986.

[27] S. E. H. Turley, G. D. Henshall, P. D. Greene, V. P. Knight, D. M. Moule and S. A. Wheeler, 'Properties of inverted rib-waveguide lasers operating at 1.3 μm wavelength', *Electron. Lett.*, **17**, pp. 868–870, 1981.

[28] K. Saito and R. Ito, 'Buried-heterostructure AlGaAs lasers', *IEEE J. Quantum Electron.*, **QE-16**(2), pp. 205–215, 1980.

[29] J. E. Bowers and M. A. Pollack, 'Semiconductor lasers for telecommunications', in S. E. Miller and I. P. Kaminow (Eds), *Optical Fiber Telecommunications II*, pp. 509–568, Academic Press, 1988.

[30] I. Mito, M. Kitamura, K. Kobayashi, S. Murata, M. Seki, Y. Odagiri, H. Nishimoto, M. Yamaguchi and K. Kobayashi, 'InGaAsP double-channel planar buried-heterostructure laser diode (DCPBH LD) with effective current confinement', *IEEE J. Lightwave Technol.*, **LT-1**, pp. 195–202, 1983.

[31] N. K. Dutta, R. B. Wilson, D. P. Wilt, P. Besomi, R. L. Brown, R. J. Nelson and R. W. Dixon, 'Performance comparison of InGaAsP lasers emitting at 1.3 and 1.55 μm for lightwave system applications', *AT&T Tech. J.*, **64**, pp. 1857–1884, 1985.

[32] S. E. Miller, 'Integrated low-noise lasers', *Electron., Lett.* **22**, pp. 256–257, 1986.

[33] J. E. Bowers, B. R. Hemenway, A. H. Gnauck, T. J. Bridges and E. G. Burkhardt, 'High-frequency constricted mesa lasers', *Appl. Phys. Lett.*, **47**, pp. 78–80, 1985.

[34] W. T. Tsang, 'Extremely low threshold AlGaAs modified multiquantum-well heterostructure lasers grown by MBE', *Appl. Phys. Lett.*, **39**, p. 786, 1981.

[35] P. Michler, *Single Quantum Dots*, Springer-Verlag, 2004.

[36] X. Huang, A. Stintz, H. Li, J. Cheng and K. J. Malloy, 'Modeling of long wavelength quantum-dot lasers with dots-in-a-well structure', *Conf. Proc. CLEO'02. Tech. Dig.*, **1**, p. 551, 2002.

[37] G. T. Liu, A. Stintz, H. Li, T. C. Newell, A. L. Gray, P. M. Varangis, K. J. Malloy and L. F. Lester, 'The influence of quantum-well composition on the performance of quantum dot

lasers using InAs-InGaAs dots-in-a-well (DWELL) structures', *IEEE J. Quantum Electron.*, **36**(11), pp. 1272–1279, 2000.

[38] V. M. Ustinov, A. E. Zhukov, A. Y. Egorov and N. A. Maleev, *Quantum Dot Lasers*, Oxford University Press, 2003.

[39] F.-Y. Chang, C.-S. Lee, G.-H. Liao and H.-H. Lin, 'InAs/InGaAs quantum dot laser with high ground-state modal gain grown by solid-source molecular-beam epitaxy', *16th Int. Conf. IPRM'04*, Kagoshima, Japan, pp. 461–464, 2004.

[40] A. D. Andreev, J. R. Downes, D. A. Faux and E. P. O'Reilly, 'Strain distributions in quantum dots of arbitrary shape', *J. Appl. Phys.*, **86**(1), pp. 297–305, 1999.

[41] N. N. Ledentsov, 'Long-wavelength quantum-dot lasers on GaAs substrates: from media to device concepts', *IEEE J. Sel. Top. Quantum Electron.*, **8**(5), pp. 1015–1024, 2002.

[42] S. R. Jin, S. J. Sweeney, S. Tomic, A. R. Adams and H. Riechert, 'High-pressure studies of recombination mechanisms in 1.3 μm GaInNAs quantum-well lasers', *IEEE J. Sel. Top. Quantum Electron.*, **9**(5), pp. 1196–1201, 2003.

[43] M. Rossetti, A. Markus, A. Fiore, L. Occhi and C. Velez, 'Quantum dot superluminescent diodes emitting at 1.3 μm', *IEEE Photonics Technol. Lett.*, **17**(3), pp. 540–542, 2005.

[44] T. E. Bell, 'Single-frequency semiconductor lasers', *IEEE Spectrum*, **20**, p. 38, 1983.

[45] T. Nakagami and T. Sakurai, 'Optical and optoelectronic devices for optical fiber transmission systems', *IEEE Commun. Mag.*, **26**(1), pp. 28–33, 1988.

[46] H. Blauvelt, N. Bar-Chaim, D. Fekete, S. Margalet and A. Yariv, 'AlGaAs lasers with micro-cleaved mirrors suitable for monolithic integration', *Appl. Phys. Lett.*, **40**, pp. 289–290, 1982.

[47] L. A. Coldren, K. Furuya, B. I. Miller and J. A. Rentschler, 'Etched mirror and groove-coupled GaInAsP/InP laser devices for integrated optics', *IEEE J. Quantum Electron.*, **QE-18**, pp. 1679–1688, 1982.

[48] K.-Y. Liou, C. A. Burrus, R. A. Linke, I. P. Kaminow, S. W. Granlund, C. B. Swan and P. Besomi, 'Single longitudinal-mode stabilized graded-index-rod external coupled-cavity laser', *Appl. Phys. Lett.*, **45**, p. 729, 1984.

[49] W. T. Tsang, N. A. Olsson and R. A. Logan, 'High-speed direct single-frequency modulation with large tuning rate in cleaved-coupled-cavity lasers', *Appl. Phys. Lett.*, **42**(8), pp. 650–651, 1983.

[50] L. A. Coldren, G. D. Boyd, J. E. Bowers and C. A. Burrus, 'Reduced dynamic linewidth in three-terminal two-section diode lasers', *Appl. Phys. Lett.*, **46**, pp. 125–127, 1985.

[51] H. S. Ghafouri, *Distributed Feedback Laser Diodes and Optical Tunable Filters* (2nd edn), Wiley, 2003.

[52] W. T. Silfvast, *Laser Fundamentals*, Cambridge University Press, 2004.

[53] C. J. Chang-Hasnain, 'VCSEL for metro communications', in I. P. Kaminow and T. Li (Eds), *Optical Fiber Telecommunications IVA: Components*, pp. 666–698, Academic Press, 2002.

[54] T. Suhara and S. Suhara, *Semiconductor Laser Fundamentals*, Marcel Dekker, 2004.

[55] S. Tsuji, A. Ohishi, H. Nakamura, M. Hirao, N. Chinone and H. Matsumura, 'Low threshold operation of 1.5 μm DFB laser diodes', *J. Lightwave Technol.*, **LT-5**(6), pp. 822–826, 1987.

[56] K. Utaka, S. Akiba, K. Sakai and Y. Matsushima, 'λ/14 shifted InGaAsP/InP DFB lasers', *IEEE J. Quantum Electron.*, **QE-22**, pp. 1042–1051, 1986.

[57] S. Akiba, M. Usami and K. Utaka, '1.5 μm λ/4 shifted InGaAsP/InP DFB lasers', *J. Lightwave Technol.*, **LT-5**(11), pp. 1564–1573, 1987.

[58] T.-P. Lee and C. Zah, 'Wavelength-tunable and single-frequency semiconductor lasers for photonic communications networks', *IEEE Commun. Mag.*, pp. 42–51, October 1989.

[59] S. F. Yu, *Analysis and Design of Vertical Cavity Surface Emitting Lasers*, Wiley–IEEE, 2003.

[60] H. Li, *Vertical-Cavity Surface-Emitting Laser Devices*, Springer-Verlag, 2002.

[61] A. V. P. Coelho, H. Boudinov, T. Lippen, H. H. Tan and C. Jagadish, 'Implant isolation of AlGaAs multilayer DBR', *Nucl. Instrum. Methods Phys. Res. B*, **218**(6), pp. 381–385, 2004.

[62] H. P. D. Yang, F. I. Lai, Y. H. Chang, H. C. Yu, C. P. Sung, H. C. Kuo, S. C. Wang, S. Y. Lin and J. Y. Chi, 'Single mode (SMSR>40 dB) proton-implanted photonic crystal vertical-cavity surface-emitting lasers', *Electron. Lett.*, **41**(6), pp. 326–328, 2005.

[63] L.-H. Laih, H. C. Kuo, G.-R. Lin, L.-W. Laih and S. C. Wang, 'As$^+$ implanted AlGaAs oxide-confined VCSEL with enhanced oxidation rate and high performance uniformity', *IEEE Photonics Technol. Lett.*, **16**(6), pp. 1423–1425, 2004.

[64] P.-T. Lee, J. R. Cao, S.-J. Choi, Z.-J. Wei, J. D. O'Brien and P. D. Dapkus, 'Room-temperature operation of VCSEL-pumped photonic crystal lasers', *IEEE Photonics Technol. Lett.*, **14**(4), pp. 435–437, 2002.

[65] C. W. Wilmsen, H. Temkin and L. A. Coldren, *Vertical-Cavity Surface-Emitting Lasers: Design, Fabrication, Characterization, and Applications*, Cambridge University Press, 2001.

[66] H. C. Yu, J. S. Wang, Y. K. Su, S. J. Chang, F. I. Lai, Y. H. Chang, H. C. Kuo, C. P. Sung, H. P. D. Yang, K. F. Lin, J. M. Wang, J. Y. Chi, R. S. Hsiao and S. Mikhrin, '1.3 μm InAs-InGaAs quantum-dot vertical-cavity surface-emitting laser with fully doped DBRs grown by MBE', *IEEE Photonics Technol. Lett.*, **18**(2), pp. 418–420, 2006.

[67] E. S. Bjorlin, J. Geske, M. Mehta, J. Piprek and J. E. Bowers, 'Temperature dependence of the relaxation resonance frequency of long-wavelength vertical-cavity lasers', *IEEE Photonics Technol. Lett.*, **17**(5), pp. 944–946, 2005.

[68] C. J. Chang-Hasnain, 'Tunable VCSEL', *IEEE J. Sel. Top. Quantum Electron.*, **6**(6), pp. 978–987, 2000.

[69] S. Mogg, N. Chitica, U. Christiansson, R. Schatz, P. Sundgren, C. Asplund and M. Hammar, 'Temperature sensitivity of the threshold current of long-wavelength InGaAs-GaAs VCSELs with large gain-cavity detuning', *IEEE J. Quantum Electron.*, **40**(5), pp. 453–462, 2004.

[70] C. Prott, F. Romer, E. O. Ataro, J. Daleiden, S. Irmer, A. Tarraf and H. Hillmer, 'Modeling of ultrawidely tunable vertical cavity air-gap filters and VCSELs', *IEEE J. Sel. Top. Quantum Electron.*, **9**(3), pp. 918–928, 2003.

[71] A. Tarraf, F. Riemenschneider, M. Strassner, J. Daleiden, S. Irmer, H. Halbritter, H. Hillmer and P. Meissner, 'Continuously tunable 1.55 μm VCSEL implemented by precisely curved dielectric top DBR involving tailored stress', *IEEE Photonics Technol. Lett.*, **16**(3), pp. 720–722, 2004.

[72] F. Riemenschneider, M. Maute, H. Halbritter, G. Boehm, M.-C. Amann and P. Meissner, 'Continuously tunable long-wavelength MEMS-VCSEL with over 40-nm tuning range', *IEEE Photonics Technol. Lett.*, **16**(10), pp. 2212–2214, 2004.

[73] P. A. Kirby, 'Semiconductor laser sources for optical communications', *Radio Electron. Eng., J. IERE*, **51**(7/8), pp. 363–376, 1981.

[74] D. Botez and G. J. Herskowitz, 'Components for optical communications systems: a review', *Proc. IEEE*, **68**(6), pp. 689–730, 1980.

[75] H. C. Casey, 'Temperature dependence of the threshold current density in InP-Ga$_{0.28}$In$_{0.72}$As$_{0.6}$P$_{0.4}$ ($\lambda = 1.3$ μm) double heterostructure lasers', *J. Appl. Phys.*, **56**, p. 1959, 1984.

[76] J. Wang, P. von Allmen, J.-P. Leburton and K. J. Linden, 'Auger recombination in long-wavelength strained-layer quantum-well structures', *IEEE J. Quantum Electron.*, **31**(5), pp. 864–875, 1995.

[77] I. P. Marko, A. D. Andreev, A. R. Adams, R. Krebs, J. P. Reithmaier and A. Forchel, 'The role of Auger recombination in InAs 1.3 μm quantum-dot lasers investigated using high hydrostatic pressure', *IEEE J. Sel. Top. Quantum Electron.*, **9**(5), pp. 1300–1307, 2003.

[78] P. Bhattacharya, S. Ghosh, S. Pradhan, J. Singh, Z.-K. Wu, J. Urayama, K. Kim and T. B. Norris, 'Carrier dynamics and high-speed modulation properties of tunnel injection InGaAs-GaAs quantum-dot lasers', *IEEE J. Quantum Electron.*, **39**(8), pp. 952–962, 2003.

[79] T. J. Houle, K. A. Williams, B. Murray, J. M. Rorison, I. H. White, A. J. Springthorpe, K. White, P. Paddon, P. A. Crump and M. Silver, 'A detailed comparison of the temperature

sensitivity of threshold of InGaAsP/InP, AlGaAs/GaAs, and AlInGaAs/InP lasers', *Conf. Proc. CLEO'01*, Baltimore, Maryland, USA, pp. 206–207, 2001.

[80] D. H. Newman and S. Ritchie, 'Sources and detectors for optical fibre communications applications: the first 20 years', *IEE Proc.*, Optoelectron. **133**(3), pp. 213–229, 1986.

[81] T. Ikegami, 'Spectrum broadening and tailing effect in direct-modulated injection lasers', *Proc. 1st Eur. Conf. on Optical Fiber Communications*, London, UK, p. 111, 1975.

[82] K. Furuya, Y. Suematsu and T. Hong, 'Reduction of resonance like peak in direct modulation due to carrier diffusion in injection laser', *Appl. Opt.*, **17**(12), pp. 1949–1952, 1978.

[83] P. M. Boers, M. T. Vlaardingerbroek and M. Danielson, 'Dynamic behaviour of semiconductor lasers', *Electron. Lett.*, **11**(10), pp. 206–208, 1975.

[84] D. J. Channin, 'Effect of gain saturation on injection laser switching', *J. Appl. Phys.*, **50**(6), pp. 3858–3860, 1979.

[85] N. Chinane, K. Aiki, M. Nakamura and R. Ito, 'Effects of lateral mode and carrier density profile on dynamic behaviour of semiconductor lasers', *IEEE J. Quantum Electron.*, **QE-14**, pp. 625–631, 1977.

[86] R. S. Tucker, A. H. Gnauck, J. M. Wiesenfield and J. E. Bowers, '8 Gb/s return to zero modulation of a semiconductor laser by gain switching', *Int. Conf. on Integrated Optics and Optical Fiber Communications, Tech. Dig. Ser. 1987 (OSA)*, **3**, p. 178, 1987.

[87] R. A. Linke, 'Modulation induced transient chirping in single frequency lasers', *IEEE J. Quantum Electron.*, **QE-21**, pp. 593–597, 1985.

[88] J. C. Cartledge and G. S. Burley, 'The effect of laser chirping on lightwave system performance', *J. Lightwave Technol.*, **7**(3), pp. 568–573, 1989.

[89] P. S. Henry, R. A. Linke and A. H. Gnauck, 'Introduction to lightwave systems', in S. E. Miller and I. P. Kaminow (Eds), *Optical Fiber Telecommunications II*, pp. 781–831, Academic Press, 1988.

[90] L. Bickers and L. P. Westbrook, 'Reduction in laser chirp in 1.5 μm DFB lasers by modulation pulse shaping', *Electron. Lett.*, **21**, pp. 103–104, 1985.

[91] R. G. Hunsperger, *Integrated Optics: Theory and Technology*, Springer-Verlag, 2002.

[92] J. Muszalski, J. Houlihan, G. Huyet and B. Corbett, 'Measurement of linewidth enhancement factor in self-assembled quantum dot semiconductor lasers emitting at 1310 nm', *Electron. Lett.*, **40**(7), pp. 428–430, 2004.

[93] L. Billia, J. Zhu, T. Ranganath, D. P. Bour, S. W. Corzine and G. E. Hofler, '40-gb/s EA modulators with wide temperature operation and negative chirp', *IEEE Photonics Technol. Lett.*, **17**(1), pp. 49–51, 2005.

[94] K. Sato, S. Kuwahara and Y. Miyamoto, 'Chirp characteristics of 40-gb/s directly modulated distributed-feedback laser diodes', *J. Lightwave Technol.*, **23**(11), pp. 3790–3797, 2005.

[95] E. Tangdiongga, Y. Liu, H. deWaardt, G. D. Khoe and H. J. S. Dorren, '320-to-40-Gb/s demultiplexing using a single SOA assisted by an optical filter', *IEEE Photonics Technol. Lett.*, **18**(8), pp. 908–910, 2006.

[96] S. Oikawa, T. Kawanishi and M. Izutsu, 'Measurement of chirp parameters and halfwave voltages of Mach-Zehnder-type optical modulators by using a small signal operation', *IEEE Photonics Technol. Lett.*, **15**(5), pp. 682–684, 2003.

[97] K.-G. Gan and J. E. Bowers, 'Measurement of gain, group index, group velocity dispersion, and linewidth enhancement factor of an InGaN multiple quantum-well laser diode', *IEEE Photonics Technol. Lett.*, **16**(5), pp. 1256–1258, 2004.

[98] Y. Yu, G. Giuliani and S. Donati, 'Measurement of the linewidth enhancement factor of semiconductor lasers based on the optical feedback self-mixing effect', *IEEE Photonics Technol. Lett.*, **16**(4), pp. 990–992, 2004.

[99] Agilent 86146B Optical Spectrum Analyzer: Technical Specifications, http://cp.literature. agilent.com/litweb/pdf/5989-4403EN.pdf, 29 February 2008.

[100] 'Optical complex spectrum analyzer', http://www.apex-t.com/pdf/ap2440a.pdf, http://www.apex-t.com/apex_instruments.htm#complex, 12 November 2007.

[101] J. Hansryd, J. van Howe and C. Xu, 'Nonlinear crosstalk and compensation in QDPASK optical communication systems', *IEEE Photonics Technol. Lett.*, **16**(8), pp. 1975–1977, 2004.

[102] A. Hodzic, B. Konrad and K. Petermann, 'Prechirp in NRZ-based 40-Gb/s single-channel and WDM transmission systems', *IEEE Photonics Technol. Lett.*, **14**(2), pp. 152–154, 2002.

[103] I. Neokosmidis, T. Kamalakis, A. Chipouras and T. Sphicopoulos, 'New techniques for the suppression of the four-wave mixing-induced distortion in nonzero dispersion fiber WDM systems', *J. Lightwave Technol.*, **23**(3), pp. 1137–1144, 2005.

[104] J. Saltz, 'Modulation and detection for coherent lightwave communications', *IEEE Commun. Mag.*, **24**(6), pp. 38–49, 1986.

[105] F. G. Walther and J. E. Kaufmann, 'Characterization of GaAlAs laser diode frequency noise', *Sixth Top. Mtg. on Optical Fiber Communications*, USA, paper TUJ5, 1983.

[106] Y. Suematsu and T. Hong, 'Suppression of relaxation oscillations in light output of injection lasers by electrical resonance circuit', *IEEE J. Quantum Electron.*, **QE-13**(9), pp. 756–762, 1977.

[107] C. H. Henry, 'Phase noise in semiconductor lasers', *J. Lightwave Technol.*, **LT-4**(3), pp. 298–310, 1986.

[108] Y. Hong, S. Bandyopadhyay, S. Sivaprakasam, P. S. Spencer and K. A. Shore, 'Noise characteristics of a single-mode laser diode subject to strong optical feedback', *J. Lightwave Technol.*, **20**(10), pp. 1847–1850, 2002.

[109] I. Darwazeh, 'Electronics for optics: introduction to MMICS', in A. Vilcot, B. Cabon and J. Chazelas (Eds), *Microwave Photonics: From Components to Applications and Systems*, Springer-Verlag, 2003.

[110] C. H. Henry and R. F. Kazarinov, 'Instability of semiconductor lasers due to optical feedback from distant reflectors', *IEEE J. Quantum Electron.*, **QE-22**, pp. 294–301, 1986.

[111] S. D. Personick, *Fiber Optic Technology and Applications*, Plenum Press, 1985.

[112] K. Ogawa, 'Analysis of mode partition noise in laser transmission systems', *IEEE J. Quantum Electron.*, **QE-18**, pp. 849–855, 1982.

[113] M. Bass, *Handbook of Optics and Nonlinear Optics* (2nd edn), Vol. IV, McGraw-Hill Professional, 2000.

[114] W. B. Leigh, *Devices for Optoelectronics*, Marcel Dekker, 1996.

[115] M. Mielke, P. J. Delfyett and G. A. Alphonse, 'Suppression of mode partition noise in a multiwavelength semiconductor laser through hybrid mode locking', *Opt. Lett.*, **27**(12), pp. 1064–1066, 2002.

[116] K. Sato and H. Toba, 'Reduction of mode partition noise by using semiconductor optical amplifiers', *IEEE J. Sel. Top. Quantum Electron.*, **7**(2), pp. 328–333, 2001.

[117] H.-C. Kwon, Y.-Y. Won and S.-K. Han, 'Noise suppressed Fabry-Perot laser diode with gain-saturated semiconductor optical amplifier for hybrid WDM/SCM-PON link', *IEEE Photonics Technol. Lett.*, **18**(4), pp. 640–642, 2006.

[118] H. Ishii, F. Kano, Y. Yoshikuni and H. Yasaka, 'Mode stabilization method for superstructure-grating DBR lasers', *J. Lightwave Technol.*, **16**(3), pp. 433–442, 1998.

[119] N. Hashizume and H. Nasu, 'Mode hopping control and lasing wavelength stabilization of fiber grating lasers', *Furukawa Rev.*, **20**, pp. 7–10, 2001.

[120] T. Tanaka, Y. Hibino, T. Hashimoto, R. Kasahara, M. Abe and Y. Tohmori, 'Hybrid-integrated external-cavity laser without temperature-dependent mode hopping', *J. Lightwave Technol.*, **20**(9), pp. 1730–1739, 2002.

[121] F. R. Nash, W. J. Sundberg, R. L. Hartman, J. R. Pawlik, D. A. Ackerman, N. K. Dutta and R. W. Dixon, 'Implementation of the proposed reliability assurance strategy for an InGaAsP/InP planar mesa BH laser for use in a submarine cable', *AT&T Tech. J.*, **64**, p. 809, 1985.

[122] N. K. Dutta and C. L. Zipfel, 'Reliability of lasers and LEDs', in S. E. Miller and I. P. Kaminow (Eds), *Optical Fiber Telecommunications II*, pp. 671–687, Academic Press, 1988.

[123] H. Kogelnik, 'Devices for optical communications', *Solid State Devices Research Conf. (ESSDERC) and 4th Symp. on Solid Device Technology*, Munich, W. Germany, **53**, pp. 1–19, 1980.

[124] T. Yamamoto, K. Sakai and S. Akiba, '10000-h continuous CW operation of $In_{1-x}Ga_xAs_yP_{1-y}$/ InP DH lasers at room temperature', *IEEE J. Quantum Electron.*, **QE-15**(8), pp. 684–687, 1979.

[125] V. Vilokkinen, P. Savolainen and P. Sipila, 'Reliability analysis of AlGaInAs lasers at 1.3 μm', *Electron. Lett.*, **40**(23), pp. 1489–1490, 2004.

[126] I. W. Marshall, 'Low loss coupling between semiconductor lasers and single-mode fibre using tapered lensed fibres', *Br. Telecom Technol. J.*, **4**(2), pp. 114–121, 1986.

[127] H. Kawahara, Y. Onada, M. Goto and T. Nakagami, 'Reflected light in the coupling of semi-conductor lasers with tapered hemispherical end fibres', *Appl. Opt.*, **22**, pp. 2732–2738, 1983.

[128] H. Kawahara, M. Sasaki and N. Tokoyo, 'Efficient coupling from semiconductor lasers into single-mode fibres with tapered hemispherical ends', Appl. *Opt.*, **19**, pp. 2578–2583, 1980.

[129] T. Schwander, B. Schwaderer and H. Storm, 'Coupling of lasers to single-mode fibres with high efficiency and low optical feedback', *Electron. Lett.*, **21**, pp. 287–289, 1985.

[130] J. Lipson, R. T. Ku and R. E. Scotti, 'Opto-mechanical considerations for laser-fiber coupling and packaging', *Proc. SPIE*, **5543**, pp. 308–312, 1985.

[131] R. T. Ku, 'Progress in efficient/reliable semiconductor laser-to-single mode fiber coupler development', *Conf. on Optical Fiber Communication*, Washington, DC, USA, *Tech. Dig.*, pp. 4–6, January 1984.

[132] G. D. Khoe, H. G. Kock, D. Kuppers, J. H. F. M. Poulissen and H. M. DeVrieze, 'Progress in monomode optical fiber interconnection devices', *J. Lightwave Technol.*, **LT-2**(3), pp. 217–227, 1984.

[133] R. Bachelot, A. Fares, R. Fikri, D. Barchiesi, G. Lerondel and P. Royer, 'Coupling semicon-ductor lasers into single-mode optical fibers by use of tips grown by photopolymerization', *Opt. Lett.*, **29**(17), pp. 1971–1973, 2004.

[134] M. Hocine, R. Bachelot, C. Ecoffet, N. Fressengeas, P. Royer and G. Kugel, 'End-of-fiber polymer tip: manufacturing and modeling', *Synth. Met.*, **17**(1), pp. 313–318, 2002.

[135] P. Urquhart, 'Review of rare earth doped fiber lasers and amplifiers', *IEE Proc., Optoelectron.*, **135**(6), pp. 385–407, 1988.

[136] J. Wilson and J. F. B. Hawkes, *Lasers: Principles and Applications*, Prentice Hall, 1987.

[137] K. Patek, *Glass Lasers*, Butterworth, 1970.

[138] R. M. MacFarlane and R. M. Shelby, 'Coherent transient and holeburning spectroscopy of rare earth ions in solids', in R. M. MacFarlane and A. A. Kaplyanskii (Eds), *Spectroscopy of Solids Containing Rare Earth Ions*, pp. 51–184, North-Holland, 1987.

[139] C. A. Millar, M. C. Brierley and P. W. France, 'Optical amplification in an erbium-doped fluorozirconate fibre between 1480 nm and 1600 nm', *Fourteenth Eur. Conf. on Optical Communications, ECOC'8 (UK), IEE Conf. Publ.*, **292**, Pt 1, pp. 66–69, 1988.

[140] M. Shimitzu, H. Suda and M. Horiguchi, 'High efficiency Nd-doped fibre lasers using direct-coated dielectric mirrors', *Electron. Lett.*, **23**, pp. 768–769, 1987.

[141] D. N. Payne and L. Reekie, 'Rare-earth-doped fibre lasers and amplifiers', *Fourteenth Eur. Conf. on Optical Communications, ECOC'88, (UK), IEE Conf. Publ.*, **292**, Pt 1, pp. 49–51, 1988.

[142] L. F. Stokes, M. Chodorow and H. J. Shaw, 'All single-mode fibre resonator', *Opt. Lett.*, **7**, p. 288, 1982.

[143] I. D. Miller, D. B. Mortimore, P. Urquhart, B. J. Ainslie, S. P. Craig, C. A. Millar and D. B. Payne, 'A Nd^{3+}-doped CW fibre laser using all-fibre reflectors', *Appl. Opt.*, **26**, pp. 2197–2201, 1987.

[144] I. D. Miller, D. B. Mortimore, B. J. Ainslie, P. Urquhart, S. P. Craig, C. A. Millar and D. B. Payne, 'New all-fiber laser', *Optical Fiber Communication Conf., OFC'87*, USA, January 1987.

[145] K. T. Kim, S. Hwangbo, J. P. Mah and K. R. Sohn, 'Widely tunable filter based on coupling between a side-polished fiber and a tapered planar waveguide', *IEEE Photonics Technol. Lett.*, **17**(1), pp. 142–144, 2005.

[146] M. L. Masanovic, V. Lal, J. A. Summers, J. S. Barton, E. J. Skogen, L. G. Rau, L. A. Coldren and D. J. Blumenthal, 'Widely tunable monolithically integrated all-optical wavelength converters in InP', *J. Lightwave Technol.*, **23**(3), pp. 1350–1362, 2005.

[147] J. Buus and E. J. Murphy, 'Tunable lasers in optical networks', *J. Lightwave Technol.*, **24**(1), pp. 5–11, 2006.

[148] C. H. Henry, 'Theory of the linewidth of semiconductor lasers', *IEEE J. Quantum Electron.*, **QE-18**, pp. 259–264, 1982.

[149] T. Kimoto, T. Kurobe, K. Muranushi, T. Mukaihara and A. Kasukawa, 'Reduction of spectral-linewidth in high power SOA integrated wavelength selectable laser', *IEEE J. Sel. Top. Quantum Electron.*, **11**(5), pp. 919–923, 2005.

[150] C. Spiegelberg, J. Geng, Y. Hu, Y. Kaneda, S. Jiang and N. Peyghambarian, 'Low-noise narrow-linewidth fiber laser at 1550 nm', *J. Lightwave Technol.*, **22**(1), pp. 57–62, 2004.

[151] J. Xi, Y. Yu, J. F. Chicharo and T. Bosch, 'Estimating the parameters of semiconductor lasers based on weak optical feedback self-mixing interferometry', *IEEE J. Quantum Electron.*, **41**(8), pp. 1058–1064, 2005.

[152] R. Wyatt and W. J. Devlin, '10-kHz linewidth 1.5-μm InGaAsP external cavity laser with 55-nm tuning range', *Electron. Lett.*, **19**, pp. 110–112, 1983.

[153] N. A. Olsson and J. P. van der Ziel, 'Performance characteristics of 1.5 micron external cavity semiconductor lasers for coherent optical communication', *J. Lightwave Technol.*, **LT-5**, pp. 510–515, 1987.

[154] W. V. Severin and H. J. Shaw, 'A single-mode fiber evanescent grating reflector', *J. Lightwave Technol.*, **LT-3**, pp. 1041–1048, 1985.

[155] J. Wittmann and G. Gaukel, 'Narrow-linewidth laser with a prism grating: GRINrod lens combination serving an external cavity', *Electron. Lett.*, **23**, pp. 524–525, 1987.

[156] F. Heismann, R. C. Alferness, L. L. Buhl, G. Eisenstein, S. K. Korotky, J. J. Veselka, L. W. Stulz and C. A. Burrus, 'Narrow-linewidth, electro-optically tunable InGaAsP–Ti: LiNbO₃ extended cavity laser', *Appl. Phys. Lett.*, **51**, pp. 164–165, 1987.

[157] G. Coquin, K. W. Cheung and M. M. Choy, 'Single and multiple wavelength operation of acousto-optically tuned semiconductor lasers at 1.3 microns', *Proc. 11th IEEE Int. Semiconductor Laser Conf.*, USA, pp. 130–131, 1988.

[158] G. Coquin and K. W. Cheung, 'An electronically tunable external cavity semiconductor laser', *Electron. Lett.*, **24**, pp. 599–600, 1988.

[159] N. K. Dutta, T. Cella, A. B. Piccirilli and R. L. Brown, 'Integrated external cavity lasers', *Conf. on Lasers and Electrooptics, CLEO'87*, USA, MF2, April 1987.

[160] Y. Suematsu and S. Arai, 'Integrated optics approach for advanced semiconductor lasers', *Proc. IEEE*, **75**(11), pp. 1472–1487, 1987.

[161] M. Okai, S. Sakano and N. Chinone, 'Wide-range continuous tunable double-sectioned distributed feedback lasers', *Fifteenth Eur. Conf. on Optical Communications*, Sweden, pp. 122–125, September 1989.

[162] H. Imai, 'Tuning results of 3-sectioned DFB lasers', *Semiconductor Laser Workshop, Conf. on Lasers and Electrooptics, CLEO'89*, USA, 1989.

[163] S. Murata, I. Mito and K. Kobayashi, 'Tuning ranges for 1.5 μm wavelength tunable DBR lasers', *Electron. Lett.*, **24**, pp. 577–579, 1988.

[164] Y. Kotaki, M. Matsuda, H. Ishikawa and H. Imni, 'Tunable DBR laser with wide tuning range', *Electron. Lett.*, **24**, pp. 503–505, 1988.

[165] L. A. Johansson, J. T. Getty, Y. A. Akulova, G. A. Fish and L. A. Coldren, 'Sampled-grating DBR laser-based analog optical transmitters', *J. Lightwave Technol.*, **21**(12), pp. 2968–2976, 2003.

[166] G. Sarlet, G. Morthier and R. Baets, 'Control of widely tunable SSG-DBR lasers for dense wavelength division multiplexing', *J. Lightwave Technol.*, **18**(8), pp. 1128–1138, 2000.

[167] A. Bergonzo, J. Jacquet, D. De Gaudemaris, J. Landreau, A. Plais, A. Vuong, H. Sillard, T. Fillion, O. Durand, H. Krol, A. Accard and I. Riant, 'Widely vernier tunable external cavity laser including a sampled fiber Bragg grating with digital wavelength selection', *IEEE Photonics Technol. Lett.*, **15**(8), pp. 1144–1146, 2003.

[168] Y. A. Akulova *et al.*, 'Widely tunable electroabsorption-modulated sampled-grating DBR laser transmitter', *IEEE J. Sel. Top. Quantum Electron.*, **8**(6), pp. 1349–1357, 2002.

[169] E. Buimovich and D. Sadot, 'Physical limitation of tuning time and system considerations in implementing fast tuning of GCSR lasers', *J. Lightwave Technol.*, **22**(2), pp. 582–588, 2004.

[170] I. M. Jauncey, L. Reekie, J. E. Townsend and D. N. Payne, 'Single longitudinal-mode operation of an Nd^{3+}-doped fibre laser', *Electron. Lett.*, **24**, pp. 24–26, 1988.

[171] L. Reekie, R. J. Mears, S. B. Poole and D. N. Payne, 'Tunable single-mode fibre lasers', *J. Lightwave Technol.*, **LT-4**(7), pp. 956–960, 1986.

[172] K. Furusawa, T. Kogure, J. K. Sahu, J. H. Lee, T. M. Monro and D. J. Richardson, 'Efficient low-threshold lasers based on an erbium-doped holey fiber', *IEEE Photonics Technol. Lett.*, **17**(1), pp. 25–27, 2005.

[173] C. A. Millar, I. D. Miller, D. B. Mortimore, B. J. Ainslie and P. Urquhart, 'Fibre laser with adjustable fibre reflector for wavelength tuning and variable output coupling', *IEE Proc., Optoelectron.*, **135**, pp. 303–304, 1988.

[174] S. U. Alam and A. B. Grudinin, 'Tunable picosecond frequency-shifted feedback fiber laser at 1550 nm', *IEEE Photonics Technol. Lett.*, **16**(9), pp. 2012–2014, 2004.

[175] J. P. Goure and I. Verrier, *Optical Fiber Devices*, IOP Publishing, 2002.

[176] G. H. B. Thompson, *Physics of Semiconductor Laser Devices*, Wiley, 1980.

[177] Y. Horikoshi, 'Semiconductor lasers with wavelengths exceeding 2 µm', in W. T. Tsang (Vol. Ed.), *Semiconductors and Semimetals: Lightwave Communication Technology*, 22C, pp. 93–151, Academic Press, 1985.

[178] C. Caneau, A. K. Srivastava, A. G. Dentai, J. L. Zyskind and M. A. Pollack, 'Room temperature GaInAsSb/AlGaAsSb DH injection lasers at 2.2 µm', *Electron. Lett.*, **21**, pp. 815–817, 1985.

[179] A. E. Bockarev, L. M. Dolginov, A. E. Drakin, L. V. Druzhinina, P. G. Eliseev and B. N. Sverdlov, 'Injection InGaAsSb lasers emitting radiation of wavelengths 1.2–2.3 µm at room temperature', *Sov. J. Quantum Electron.*, **15**, pp. 869–870, 1985.

[180] C. Caneau, J. L. Zyskind, J. W. Sulhoff, T. E. Glover, J. Centanni, C. A. Burrus, A. G. Dentai and M. A. Pollack, '2.2 µm GaInAsSb/AlGaAsSb injection lasers with low threshold current density', *Appl. Phys. Lett.*, **51**, pp. 764–766, 1987.

[181] H. Ebe, Y. Nishijima and K. Shinohara, 'PbEuTe lasers with 4–6 µm wavelength mode with hot-well epitaxy', *IEEE J. Quantum Electron.*, **25**(6), pp. 1381–1384, 1989.

[182] M. C. Brierley and P. W. France, 'Continuous wave lasing at 2.7 µm in an erbium-doped fluorozirconate fibre', *Electron. Lett.*, **24**, pp. 935–937, 1988.

[183] M. C. Brierley, P. W. France and C. A. Millar, 'Lasing at 2.08 µm in a holmium-doped fluorozirconate fibre laser', *Electron. Lett.*, **24**, pp. 539–540, 1988.

[184] L. Esterowitz, R. Allen and I. Aggarwal, 'Pulsed laser emission at 2.3 µm in a thulium-doped fluorozirconate fibre', *Electron. Lett.*, **24**, p. 1104, 1988.

[185] S. D. Jackson, T. King and M. Pollnau, 'Diode-pumped 1.7 W erbium 3 µm fiber laser', *Opt. Lett.*, **24**(16), pp. 1133–1135, 1999.

[186] S. D. Jackson, 'Singly Ho^{3+} doped flouride fibre laser operating at 2.92 µm', *Electron. Lett.*, **40**(22), pp. 1400–1401, 2004.

[187] K. Ohtani, K. Fujita and H. Ohno, 'Mid-infrared InAs/AlGaSb superlattice quantum-cascade lasers', *Appl. Phy. Lett.*, **87**(21), pp. 211113, 2005.

[188] S. Banerjee, K. A. Shore, C. J. Mitchell, J. L. Sly and M. Missous, 'Current-voltage and light-current characteristics in highly strained InGaAs/InAlAs quantum cascade laser structures', *Proc. IEE Circuit Dev. Syst.*, **152**(5), pp. 497–501, 2005.

[189] J. Faist, F. Capasso, D. L. Sivco, C. Sirtori, A. L. Hutchinson and A. Y. Cho, 'Quantum cascade laser', *Science*, **264**, pp, 553–556, 1994.

[190] R. Q. Yang, J. L. Bradshaw, J. D. Bruno, J. T. Pham and D. E. Wortman, 'Mid-infrared type-II interband cascade lasers', *IEEE J. Quantum Electron.*, **38**(6), pp. 559–568, 2002.

[191] G. Luo, C. Peng, H. Q. Le, S.-S. Pei, H. Lee, W.-Y. Hwang, B. Ishaug and J. Zheng, 'Broadly wavelength-tunable external cavity, mid-infrared quantum cascade lasers', *IEEE J. Quantum Electron.*, **38**(5), pp. 486–494, 2002.

[192] C. Gmachl, A. Straub, R. Colombelli, F. Capasso, D. L. Sivco, A. M. Sergent and A. Y. Cho, 'Single-mode, tunable distributed-feedback and multiple-wavelength quantum cascade lasers', *IEEE J. Quantum Electron.*, **38**(6), pp. 569–581, 2002.

[193] I. Vurgaftman and J. R. Meyer, 'Photonic-crystal distributed-feedback quantum cascade lasers', *IEEE J. Quantum Electron.*, **38**(6), pp. 592–602, 2002.

[194] C. L. Walker, C. D. Farmer, C. R. Stanley and C. N. Ironside, 'Progress towards photonic crystal quantum cascade laser', *IEE Proc., Optoelectron.*, **151**(6), pp. 502–507, 2004.

[195] C. Pflugl, M. Austerer, W. Schrenk, S. Golka, G. Strasser, R. P. Green, L. R. Wilson, J. W. Cockburn, A. B. Krysa and J. S. Roberts, 'Single-mode surface-emitting quantum-cascade lasers', *Appl. Phy. Lett.*, **86**(21), pp. 211102:1–3, 2005.

[196] C. A. Evans, V. D. Jovanovic, D. Indjin, Z. Ikonic and P. Harrison, 'Investigation of thermal effects in quantum-cascade lasers', *IEEE J. Quantum Electron.*, **42**(9), pp. 859–867, 2006.

[197] A. Soibel, K. Mansour, D. L. Sivco and A. Y. Cho, 'Evaluation of thermal crosstalk in quantum cascade laser arrays', *IEEE Photonics Technol. Lett.*, **19**(6), pp. 375–377, 2007.

[198] J. Faist, 'Recent progress in long wavelength quantum cascade lasers', *IEEE Int. Semiconductor Laser Conf.*, Hawaii, pp. 3–4, September 2006.

[199] J. Z. Chen, Z. Liu, Y. S. Rumala, D. L. Sivco and C. F. Gmachl, 'Direct liquid cooling of room-temperature operated quantum cascade lasers', *Electron. Lett.*, **42**(9), pp. 534–535, 2006.

[200] R. W. Tkach and A. R. Chaplyry, 'Phase noise and linewidth in an InGaAsP laser', *J. Lightwave Technol.*, **LT-4**, pp. 1711–1716, 1986.

CHAPTER 7

Optical sources 2: the light-emitting diode

7.1 Introduction

Spontaneous emission of radiation in the visible and infrared regions of the spectrum from a forward-biased p–n junction was discussed in Section 6.3.2. The normally empty conduction band of the semiconductor is populated by electrons injected into it by the forward current through the junction, and light is generated when these electrons recombine with holes in the valence band to emit a photon. This is the mechanism by which light is emitted from an LED, but stimulated emission is not encouraged, as it is in the injection laser, by the addition of an optical cavity and mirror facets to provide feedback of photons.

The LED can therefore operate at lower current densities than the injection laser, but the emitted photons have random phases and the device is an incoherent optical source. Also, the energy of the emitted photons is only roughly equal to the bandgap energy of the semiconductor material, which gives a much wider spectral linewidth (possibly by a factor of 100) than the injection laser. The linewidth for an LED corresponds to a range of photon energy between 1 and $3.5KT$, where K is Boltzmann's constant and T is the absolute temperature. This gives linewidths of 30 to 40 nm for GaAs-based devices operating at room temperature. Thus the LED supports many optical modes within its structure and is therefore often used as a multimode source, although the coupling of LEDs to single-mode fibers has been pursued with success, particularly when advanced structures are employed. Also, LEDs have several further drawbacks in comparison with injection lasers. These include:

(a) generally lower optical power coupled into a fiber (microwatts);

(b) usually lower modulation bandwidth;

(c) harmonic distortion.

However, although these problems may initially appear to make the LED a less attractive optical source than the injection laser, the device has a number of distinct advantages which have given it a prominent place in optical fiber communications:

1. *Simpler fabrication.* There are no mirror facets and in some structures no striped geometry.

2. *Cost.* The simpler construction of the LED leads to much reduced cost which is always likely to be maintained.

3. *Reliability.* The LED does not exhibit catastrophic degradation and has proved far less sensitive to gradual degradation than the injection laser. It is also immune to self-pulsation and modal noise problems.

4. *Generally less temperature dependence.* The light output against current characteristic is less affected by temperature than the corresponding characteristic for the injection laser. Furthermore, the LED is not a threshold device and therefore raising the temperature does not increase the threshold current above the operating point and hence halt operation.

5. *Simpler drive circuitry.* This is due to the generally lower drive currents and reduced temperature dependence which makes temperature compensation circuits unnecessary.

6. *Linearity.* Ideally, the LED has a linear light output against current characteristic (see Section 7.4.1), unlike the injection laser. This can prove advantageous where analog modulation is concerned.

These advantages combined with the development of high-radiance, relatively high-bandwidth devices have ensured that the LED remains an extensively used source for optical fiber communications.

Structures fabricated using the GaAs/AlGaAs material system are well tried for operation in the shorter wavelength region. In addition, there have been substantial advances in devices based on the InGaAsP/InP material structure for use in the longer wavelength region especially around 1.3 μm. At this wavelength, the material dispersion in silica glass fibers goes through zero and hence the wider linewidth of the LED imposes a far slighter limitation on link length than does intermodal dispersion within multimode fiber. Furthermore, the reduced fiber attenuation at this operating wavelength can allow longer haul LED systems.

Although longer wavelength LED systems using multimode graded index fiber have been developed, particularly for nontelecommunication applications (see Section 15.6.4), activity has also been concerned with both high-speed operation and with the coupling of these InGaAsP LEDs to single-mode fiber. A major impetus for these strategies has been the potential deployment of such single-mode LED systems in the telecommunication access network or subscriber loop (see Section 15.6.3) In this context, theoretical studies of both LED coupling [Ref. 1] and transmission [Ref. 2] with single-mode fiber have been undertaken, as well as numerous practical investigations, some of which are outlined in the following sections. It is therefore apparent that LEDs are likely to remain a significant optical fiber communication source for many system applications including operation over shorter distances with single-mode fiber at transmission rates that may exceed 1 Gbit s^{-1}.

Having dealt with the basic operating principles for the LED in Section 6.3.2, we continue in Section 7.2 with a discussion of LED power and efficiency in relation to the launching of light into optical fibers. Moreover, at the end of this section we include a brief account of the operation of an efficient LED which employs a double heterostructure. This leads into a discussion in Section 7.3 of the major practical LED structures where again we have regard to their light coupling efficiency. Also included in this section are the more advanced device structures such as the superluminescent, resonant cavity and quantum-dot LED. The various operating characteristics and limitations on LED performance are then described in Section 7.4. Finally, in Section 7.5, we include a brief discussion on the possible modulation techniques for semiconductor optical sources.

7.2 LED power and efficiency

The absence of optical amplification through stimulated emission in the LED tends to limit the internal quantum efficiency (ratio of photons generated to injected electrons) of the device. Reliance on spontaneous emission allows nonradiative recombination to take place within the structure due to crystalline imperfections and impurities giving, at best, an internal quantum efficiency of 50% for simple homojunction devices. However, as with injection lasers, double-heterojunction (DH) structures have been implemented which recombination lifetime measurements suggest [Ref. 3] give internal quantum efficiencies of 60 to 80%.

The power generated internally by an LED may be determined by consideration of the excess electrons and holes in the p- and n-type material respectively (i.e. the minority carriers) when it is forward biased and carrier injection takes place at the device contacts (see Section 6.3.2). The excess density of electrons Δn and holes Δp is equal since the injected carriers are created and recombined in pairs such that charge neutrality is maintained within the structure. In extrinsic materials one carrier type will have a much higher concentration than the other and hence in the p-type region, for example, the hole concentration will be much greater than the electron concentration. Generally, the excess minority carrier density decays exponentially with time t [Ref. 4] according to the relation:

$$\Delta n = \Delta n(0) \exp(-t/\tau) \qquad (7.1)$$

where $\Delta n(0)$ is the initial injected excess electron density and τ represents the total carrier recombination lifetime. In most cases, however, Δn is only a small fraction of the majority carriers and comprises all of the minority carriers. Therefore, in these cases, the carrier recombination lifetime becomes the minority or injected carrier lifetime τ_i.

When there is a constant current flow into the junction diode, an equilibrium condition is established. In this case, the total rate at which carriers are generated will be the sum of the externally supplied and the thermal generation rates. The current density J in amperes per square meter may be written as J/ed in electrons per cubic meter per second, where e is the charge on an electron and d is the thickness of the recombination region. Hence a rate equation for carrier recombination in the LED can be expressed in the form [Ref. 4]:

$$\frac{d(\Delta n)}{dt} = \frac{J}{ed} - \frac{\Delta n}{\tau} \quad (m^{-3}\,s^{-1}) \tag{7.2}$$

The condition for equilibrium is obtained by setting the derivative in Eq. (7.2) to zero. Hence:

$$\Delta n = \frac{J\tau}{ed} \quad (m^{-3}) \tag{7.3}$$

Equation (7.3) therefore gives the steady-state electron density when a constant current is flowing into the junction region.

It is also apparent from Eq. (7.2) that in the steady state the total number of carrier recombinations per second or the recombination rate r_t will be:

$$r_t = \frac{J}{ed} \quad (m^{-3}) \tag{7.4}$$

$$= r_r + r_{nr} \quad (m^{-3}) \tag{7.5}$$

where r_r is the radiative recombination rate per unit volume and r_{nr} is the nonradiative recombination rate per unit volume. Moreover, when the forward-biased current into the device is i, then from Eq. (7.4) the total number of recombinations per second R_t becomes:

$$R_t = \frac{i}{e} \tag{7.6}$$

It was indicated in Section 6.3.3.1 that excess carriers can recombine either radiatively or nonradiatively. While in the former case a photon is generated, in the latter case the energy is released in the form of heat (i.e. lattice vibrations). Moreover, for a DH device with a thin active region (a few microns), the nonradiative recombination tends to be dominated by surface recombination at the heterojunction interfaces.

The LED internal quantum efficiency* η_{int}, which can be defined as the ratio of the radiative recombination rate to the total recombination rate, following Eq. (7.5) may be written as [Ref. 5]:

---

* The internal quantum efficiency for the LED is obtained only from the spontaneous radiation and hence is written as η_{int}. By contrast, the internal quantum efficiency for the injection laser combined the internal quantum efficiencies for both spontaneous and simulated radiation. It was therefore denoted as η_i (see Section 6.4.1).

$$\eta_{int} = \frac{r_r}{r_t} = \frac{r_r}{r_r + r_{nr}} \tag{7.7}$$

$$= \frac{R_r}{R_t} \tag{7.8}$$

where R_r is the total number of *radiative* recombinations per second. Rearranging Eq. (7.8) and substituting from Eq. (7.6) gives:

$$R_r = \eta_{int} \frac{i}{e} \tag{7.9}$$

Since R_r is also equivalent to the total number of photons generated per second and from Eq. (6.1) each photon has an energy equal to hf joules, then the optical power generated internally by the LED, P_{int}, is:

$$P_{int} = \eta_{int} \frac{i}{e} hf \quad (W) \tag{7.10}$$

Using Eq. (6.22) to express the internally generated power in terms of wavelength rather than frequency gives:

$$P_{int} = \eta_{int} \frac{hci}{e\lambda} \quad (W) \tag{7.11}$$

It is interesting to note that Eqs (7.10) and (7.11) display a linear relationship between the optical power generated in the LED and the drive current into the device (see Section 7.4.1). Similar relationships may be obtained for the optical power emitted from an LED but in this case the constant of proportionality η_{int} must be multiplied by a factor representing the external quantum efficiency* η_{ext} to provide an overall quantum efficiency for the device.

For the exponential decay of excess carriers depicted by Eq. (7.1) the radiative minority carrier lifetime is $\tau_r = \Delta n / r_r$ and the nonradiative minority carrier lifetime is $\tau_{nr} = \Delta n / r_{nr}$. Therefore, from Eq. (7.7) the internal quantum efficiency is:

$$\eta_{int} = \frac{1}{1 + (r_{nr}/r_r)} = \frac{1}{1 + (\tau_r/\tau_{nr})} \tag{7.12}$$

Furthermore, the total recombination lifetime τ can be written as $\tau = \Delta n / r_t$ which, using Eq. (7.5), gives:

$$\frac{1}{\tau} = \frac{1}{\tau_r} + \frac{1}{\tau_{nr}} \tag{7.13}$$

---

* The external quantum efficiency may be defined as the ratio of the photons emitted from the device to the photons internally generated. However, it is sometimes defined as the ratio of the number of photons emitted to the total number of carrier recombinations (radiative and nonradiative).

Hence Eq. (7.12) becomes:

$$\eta_{int} = \frac{\tau}{\tau_r} \tag{7.14}$$

It should be noted that the same expression for the internal quantum efficiency could be obtained from Eq. (7.7).

Example 7.1

The radiative and nonradiative recombination lifetimes of the minority carriers in the active region of a double-heterojunction LED are 60 ns and 100 ns respectively. Determine the total carrier recombination lifetime and the power internally generated within the device when the peak emission wavelength is 0.87 μm at a drive current of 40 mA.

Solution: The total carrier recombination lifetime is given by Eq. (7.13) as:

$$\tau = \frac{\tau_r \tau_{nr}}{\tau_r + \tau_{nr}} = \frac{60 \times 100 \text{ ns}}{60 + 100} = 37.5 \text{ ns}$$

To calculate the power internally generated it is necessary to obtain the internal quantum efficiency of the device. Hence using Eq. (7.14):

$$\eta_{int} = \frac{\tau}{\tau_r} = \frac{37.5}{60} = 0.625$$

Thus from Eq. (7.11):

$$P_{int} = \eta_{int} \frac{hci}{e\lambda} = \frac{0.625 \times 6.626 \times 10^{-34} \times 2.998 \times 10^{8} \times 40 \times 10^{-3}}{1.602 \times 10^{-19} \times 0.87 \times 10^{-6}}$$

$$= 35.6 \text{ mW}$$

The LED which has an internal quantum efficiency of 62.5% generates 35.6 mW of optical power, internally. It should be noted, however, that this power level will not be readily emitted from the device.

Although the possible internal quantum efficiency can be relatively high, the radiation geometry for an LED which emits through a planar surface is essentially Lambertian in that the surface radiance (the power radiated from a unit area into a unit solid angle, given in W sr^{-1} m^{-2}) is constant in all directions. The Lambertian intensity distribution is illustrated in Figure 7.1 where the maximum intensity I_0 is perpendicular to the planar surface but is reduced on the sides in proportion to the cosine of the viewing angle θ as the apparent area varies with this angle. This reduces the external power efficiency to a few percent as most of the light generated within the device is trapped by total internal reflection (see

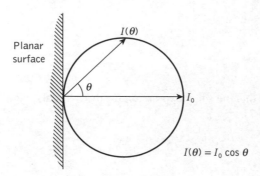

Figure 7.1 The Lambertian intensity distribution typical of a planar LED

Section 2.2.1) when it is radiated at greater than the critical angle for the crystal–air inter-face. As with the injection laser (see Section 6.4.1) the external power efficiency η_{ep} is defined as the ratio of the optical power emitted externally P_e to the electric power pro-vided to the device P or:

$$\eta_{ep} \simeq \frac{P_e}{P} \times 100\% \qquad (7.15)$$

Also, the optical power emitted P_e into a medium of low refractive index n from the face of a planar LED fabricated from a material of refractive index n_x is given approximately by [Ref. 6]:

$$P_e = \frac{P_{int}Fn^2}{4n_x^2} \qquad (7.16)$$

where P_{int} is the power generated internally and F is the transmission factor of the semiconductor–external interface. Hence it is possible to estimate the percentage of optical power emitted.

Example 7.2

A planar LED is fabricated from gallium arsenide which has a refractive index of 3.6.

(a) Calculate the optical power emitted into air as a percentage of the internal optical power for the device when the transmission factor at the crystal–air interface is 0.68.

(b) When the optical power generated internally is 50% of the electric power supplied, determine the external power efficiency.

Solution: (a) The optical power emitted is given by Eq. (7.16), in which the refractive index n for air is 1:

$$P_e \simeq \frac{P_{int}Fn^2}{4n_x^2} = \frac{P_{int}0.68 \times 1}{4(3.6)^2} = 0.013P_{int}$$

Hence the power emitted is only 1.3% of the optical power generated internally.

(b) The external power efficiency is given by Eq. (7.15), where:

$$\eta_{ep} = \frac{P_e}{P} \times 100 = 0.013 \frac{P_{int}}{P} \times 100$$

Also, the optical power generated internally $P_{int} = 0.5P$.

Hence:

$$\eta_{ep} = \frac{0.013P_{int}}{2P_{int}} \times 100 = 0.65\%$$

A further loss is encountered when coupling the light output into a fiber. Considerations of this coupling efficiency are very complex; however, it is possible to use an approximate simplified approach [Ref. 7]. If it is assumed for step index fibers that all the light incident on the exposed end of the core within the acceptance angle θ_a is coupled, then for a fiber in air, using Eq. (2.8):

$$\theta_a = \sin^{-1}(n_1^2 - n_2^2)^{\frac{1}{2}} = \sin^{-1}(NA) \tag{7.17}$$

Also, incident light at angles greater than θ_a will not be coupled. For a Lambertian source, the radiant intensity at an angle θ, $I(\theta)$, is given by (see Figure 7.1):

$$I(\theta) = I_0 \cos \theta \tag{7.18}$$

where I_0 is the radiant intensity along the line $\theta = 0$. Considering a source which is smaller than, and in close proximity to, the fiber core, and assuming cylindrical symmetry, the coupling efficiency η_c is given by:

$$\eta_c = \frac{\displaystyle\int_0^{\theta_a} I(\theta) \sin \theta \, d\theta}{\displaystyle\int_0^{\pi/2} I(\theta) \sin \theta \, d\theta} \tag{7.19}$$

Hence substituting from Eq. (7.18):

$$\eta_c = \frac{\displaystyle\int_0^{\theta_a} I_0 \cos \theta \sin \theta \, d\theta}{\displaystyle\int_0^{\pi/2} I_0 \cos \theta \sin \theta \, d\theta}$$

$$= \frac{\int_0^{\theta_a} I_0 \sin 2\theta \, d\theta}{\int_0^{\pi/2} I_0 \sin 2\theta \, d\theta}$$

$$= \frac{[-I_0 \cos 2\theta/2]_0^{\theta_a}}{[-I_0 \cos 2\theta/2]_0^{\pi/2}}$$

$$= \sin^2 \theta_a \tag{7.20}$$

Furthermore, from Eq. (7.17):

$$\eta_c = \sin^2 \theta_a = (NA)^2 \tag{7.21}$$

Equation (7.21) for the coupling efficiency allows estimates for the percentage of optical power coupled into the step index fiber relative to the amount of optical power emitted from the LED.

Example 7.3

The light output from the GaAs LED of Example 7.2 is coupled into a step index fiber with a numerical aperture of 0.2, a core refractive index of 1.4 and a diameter larger than the diameter of the device. Estimate:

(a) The coupling efficiency into the fiber when the LED is in close proximity to the fiber core.

(b) The optical loss in decibels, relative to the power emitted from the LED, when coupling the light output into the fiber.

(c) The loss relative to the internally generated optical power in the device when coupling the light output into the fiber when there is a small air gap between the LED and the fiber core.

Solution: (a) From Eq. (7.21), the coupling efficiency is given by:

$$\eta_c = (NA)^2 = (0.2)^2 = 0.04$$

Thus about 4% of the externally emitted optical power is coupled into the fiber.

(b) Let the optical power coupled into the fiber be P_c. Then the optical loss in decibels relative to P_e when coupling the light output into the fiber is:

$$\text{Loss} = -10 \log_{10} \frac{P_c}{P_e}$$

$$= -10 \log_{10} \eta_c$$

Hence:

$$\text{Loss} = -10 \log_{10} 0.04$$
$$= 14.0 \text{ dB}$$

(c) When the LED is emitting into air, from Example 7.2:

$$P_e = 0.013 P_{int}$$

Assuming a very small air gap (i.e. cylindrical symmetry unaffected), then from (a) the power coupled into the fiber is:

$$P_c = 0.04 P_e = 0.04 \times 0.013 P_{int}$$
$$= 5.2 \times 10^{-4} P_{int}$$

Hence in this case only about 0.05% of the internal optical power is coupled into the fiber.

The loss in decibels relative to P_{int} is:

$$\text{Loss} = -10 \log_{10} \frac{P_c}{P_{int}} = -10 \log_{10} 5.2 \times 10^{-4} = 32.8 \text{ dB}$$

If significant optical power is to be coupled from an incoherent LED into a low-NA fiber the device must exhibit very high radiance. This is especially the case when considering graded index fibers where the Lambertian coupling efficiency with the same NA (same refractive index difference) and $\alpha \simeq 2$ (see Section 2.4.4) is about half that into step index fibers [Ref. 8]. To obtain the necessary high radiance, direct bandgap semiconductors (see Section 6.3.3.1) must be used fabricated with DH structures which may be driven at high current densities. The principle of operation of such a device will now be considered prior to discussion of various LED structures.

7.2.1 The double-heterojunction LED

The principle of operation of the DH LED is illustrated in Figure 7.2. The device shown consists of a p-type GaAs layer sandwiched between a p-type AlGaAs and an n-type AlGaAs layer. When a forward bias is applied (as indicated in Figure 7.2(a)) electrons from the n-type layer are injected through the p–n junction into the p-type GaAs layer where they become minority carriers. These minority carriers diffuse away from the junction [Ref. 9], recombining with majority carriers (holes) as they do so. Photons are therefore produced with energy corresponding to the bandgap energy of the p-type GaAs layer. The injected electrons are inhibited from diffusing into the p-type AlGaAs layer because of the potential barrier presented by the p–p heterojunction (see Figure 7.2(b)). Hence, electroluminescence only occurs in the GaAs junction layer, providing both good internal quantum efficiency and high-radiance emission. Furthermore, light is emitted from the device without reabsorption because the bandgap energy in the AlGaAs layer is large in

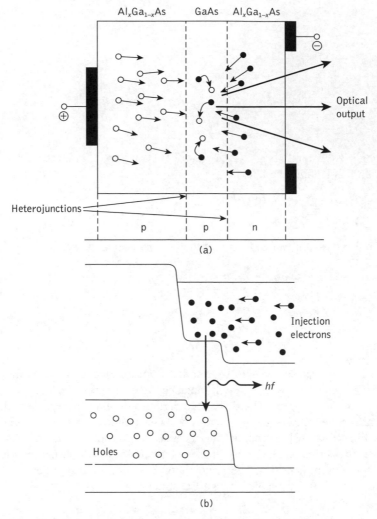

Figure 7.2 The double-heterojunction LED: (a) the layer structure, shown with an applied forward bias; (b) the corresponding energy band diagram

comparison with that in GaAs. The DH structure is therefore used to provide the most efficient incoherent sources for application within optical fiber communications. Nevertheless, these devices generally exhibit the previously discussed constraints in relation to coupling efficiency to optical fibers. This and other LED structures are considered in greater detail in the following section.

7.3 LED structures

There are six major types of LED structure and although only two have found extensive use in optical fiber communications, two others have become increasingly applied. These are the surface emitter, the edge emitter, the superluminescent and the resonant cavity LED respectively. The other two structures, the planar and dome LEDs, find more application as cheap plastic-encapsulated visible devices for use in such areas as intruder alarms, TV channel changers and industrial counting. However, infrared versions of these devices have been used in optical communications mainly with fiber bundles and it is therefore useful to consider them briefly before progressing to the high-radiance LED structures.

7.3.1 Planar LED

The planar LED is the simplest of the structures that are available and is fabricated by either liquid- or vapor-phase epitaxial processes over the whole surface of a GaAs substrate. This involves a p-type diffusion into the n-type substrate in order to create the junction illustrated in Figure 7.3. Forward current flow through the junction gives Lambertian spontaneous emission and the device emits light from all surfaces. However, only a limited amount of light escapes the structure due to total internal reflection, as discussed in Section 7.2, and therefore the radiance is low.

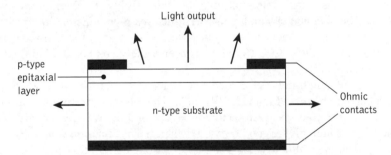

Figure 7.3 The structure of a planar LED showing the emission of light from all surfaces

7.3.2 Dome LED

The structure of a typical dome LED is shown in Figure 7.4. A hemisphere of n-type GaAs is formed around a diffused p-type region. The diameter of the dome is chosen to maximize the amount of internal emission reaching the surface within the critical angle of the GaAs–air interface. Hence this device has a higher external power efficiency than the planar LED. However, the geometry of the structure is such that the dome must be far larger than the active recombination area, which gives a greater effective emission area and thus reduces the radiance.

7.3.3 Surface emitter LEDs

A method for obtaining high radiance is to restrict the emission to a small active region within the device. The technique pioneered by Burrus and Dawson [Ref. 10] with homostructure devices was to use an etched well in a GaAs substrate in order to prevent heavy absorption of the emitted radiation, and physically to accommodate the fiber. These

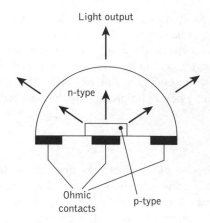

Figure 7.4 The structure of a dome LED

structures have a low thermal impedance in the active region allowing high current densities and giving high-radiance emission into the optical fiber. Furthermore, considerable advantage may be obtained by employing DH structures giving increased efficiency from electrical and optical confinement as well as less absorption of the emitted radiation. This type of surface emitter LED (SLED) has been widely employed within optical fiber communications.

The structure of a high-radiance etched well DH surface emitter* for the 0.8 to 0.9 μm wavelength band is shown in Figure 7.5 [Ref. 11]. The internal absorption in this device is very low due to the larger bandgap-confining layers, and the reflection coefficient at the back crystal face is high giving good forward radiance. The emission from the active layer is essentially isotropic, although the external emission distribution may be considered Lambertian with a beam width of 120° due to refraction from a high to a low refractive index at the GaAs–fiber interface. The power coupled P_c into a multimode step index fiber may be estimated from the relationship [Ref. 12]:

$$P_c = \pi (1 - r)AR_D(NA)^2 \tag{7.22}$$

where r is the Fresnel reflection coefficient at the fiber surface, A is the smaller of the fiber core cross-section or the emission area of the source and R_D is the radiance of the source. However, the power coupled into the fiber is also dependent on many other factors including the distance and alignment between the emission area and the fiber, the SLED emission pattern and the medium between the emitting area and the fiber. For instance, the addition of epoxy resin in the etched well tends to reduce the refractive index mismatch and increase the external power efficiency of the device. Hence, DH surface emitters often give more coupled optical power than predicted by Eq. (7.22). Nevertheless Eq. (7.22) may be used to gain an estimate of the power coupled, although accurate results may only be obtained through measurement.

* These devices are also known as Burrus-type LEDs

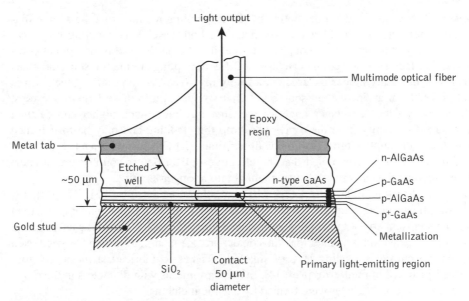

Figure 7.5 The structure of an AlGaAs DH surface-emitting LED (Burrus type). Reprinted from Ref. 11 with permission from Elsevier

Example 7.4

A DH surface emitter which has an emission area diameter of 50 μm is butt jointed to an 80 μm core step index fiber with a numerical aperture of 0.15. The device has a radiance of 30 W sr^{-1} cm^{-2} at a constant operating drive current. Estimate the optical power coupled into the fiber if it is assumed that the Fresnel reflection coefficient at the index matched fiber surface is 0.01.

Solution: Using Eq. (7.22), the optical power coupled into the fiber P_c is given by:

$$P_c = \pi(1 - r)AR_D(NA)^2$$

In this case A represents the emission area of the source.
    Hence:

$$A = \pi(25 \times 10^{-4})^2 = 1.96 \times 10^{-5} \text{ cm}^2$$

Thus:

$$P_c = \pi(1 - 0.01)1.96 \times 10^{-5} \times 30 \times (0.15)^2$$
$$= 41.1 \text{ μW}$$

In this example around 41 μW of optical power is coupled into the step index fiber.

However, for graded index fiber optimum direct coupling requires that the source diameter be about one-half the fiber core diameter. In both cases lens coupling may give increased levels of optical power coupled into the fiber but at the cost of additional complexity. Other factors which complicate the LED fiber coupling are the transmission characteristics of the leaky modes or large angle skew rays (see Section 2.4.1). Much of the optical power from an incoherent source is initially coupled into these large-angle rays, which fall within the acceptance angle of the fiber but have much higher energy than meridional rays. Energy from these rays goes into the cladding and may be lost. Hence much of the light coupled into a multimode fiber from an LED is lost within a few hundred meters. It must therefore be noted that the effective optical power coupled into a short length of fiber significantly exceeds that coupled into a longer length.

The planar structure of the Burrus-type LED and other nonetched well SLEDs [Ref. 13] allows significant lateral current spreading, particularly for contact diameters less than 25 µm. This current spreading results in a reduced current density as well as an effective emission area substantially greater than the contact area. A technique which has been used to reduce the current spreading in very small devices is to fabricate a mesa structure SLED, as illustrated in Figure 7.6 [Ref. 14]. In this case mesas with diameters in the range 20 to 25 µm at the active layer were formed by chemical etching.

These InGaAsP/InP devices which emitted at a wavelength of 1.3 µm had an integral lens formed at the exit face of the InP substrate in order to improve the coupling efficiency, particularly to single-mode fiber. Such monolithic lens structures provide a common strategy for improving the power coupled into fiber from LEDs, and alternative lens coupling techniques are discussed in Section 7.3.7. Moreover, there is increasing interest in coupling LEDs to single-mode fiber for shorter haul applications which, in the case of SLEDs, necessitates efficient lens coupling to obtain acceptable launch powers. For example, the

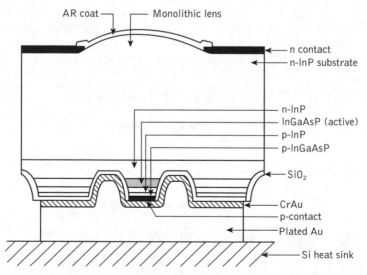

Figure 7.6 Small-area InGaAsP mesa-etched surface-emitting LED structure [Ref. 14]

LED illustrated in Figure 7.6 with a drive current of 50 mA was found to couple only around 2 µW of optical power into single-mode fiber [Ref. 14].

7.3.4 Edge emitter LEDs

Another basic high-radiance structure currently used in optical communications is the stripe geometry DH edge emitter LED (ELED). This device has a similar geometry to a conventional contact stripe injection laser, as shown in Figure 7.7. It takes advantage of transparent guiding layers with a very thin active layer (50 to 100 µm) in order that the light produced in the active layer spreads into the transparent guiding layers, reducing self-absorption in the active layer. The consequent waveguiding narrows the beam divergence to a half-power width of around 30° in the plane perpendicular to the junction. However, the lack of waveguiding in the plane of the junction gives a Lambertian output with a half-power width of around 120°, as illustrated in Figure 7.7.

Most of the propagating light is emitted at one end face only due to a reflector on the other end face and an antireflection coating on the emitting end face. The effective radiance at the emitting end face can be very high giving an increased coupling efficiency into small-*NA* fiber compared with the surface emitter. However, surface emitters generally radiate more power into air (2.5 to 3 times) than edge emitters since the emitted light is less affected by reabsorption and interfacial recombination. Comparisons [Refs 15–17] have shown that edge emitters couple more optical power into low *NA* (less than 0.3) than surface emitters, whereas the opposite is true for large *NA* (greater than 0.3).

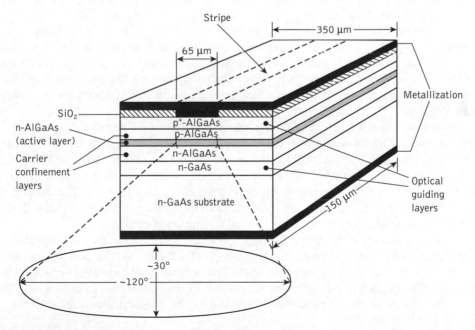

Figure 7.7 Schematic illustration of the structure of a stripe geometry DH AlGaAs edge-emitting LED

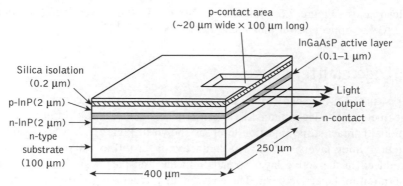

Figure 7.8 Truncated stripe InGaAsP edge-emitting LED [Ref. 20]

The enhanced waveguiding of the edge emitter enables it in theory [Ref. 16] to couple 7.5 times more power into low-*NA* fiber than a comparable surface emitter. However, in practice the increased coupling efficiency has been found to be slightly less than this (3.5 to 6 times) [Refs 16, 17]. Similar coupling efficiencies may be achieved into low-*NA* fiber with surface emitters by the use of a lens. Furthermore, it has been found that lens coupling with edge emitters may increase the coupling efficiencies by comparable factors (around five times).

The stripe geometry of the edge emitter allows very high carrier injection densities for given drive currents. Thus it is possible to couple approaching a milliwatt of optical power into low-*NA* (0.14) multimode step index fiber with edge-emitting LEDs operating at high drive currents (500 mA) [Ref. 18].

Edge emitters have also been found to have a substantially better modulation bandwidth of the order of hundreds of megahertz than comparable surface-emitting structures with the same drive level [Ref. 17]. In general it is possible to construct edge-emitting LEDs with a narrower linewidth than surface emitters, but there are manufacturing problems with the more complicated structure (including difficult heat-sinking geometry) which moderate the benefits of these devices.

Nevertheless, a number of ELED structures have been developed using the InGaAsP/ InP material system for operation at a wavelength of 1.3 μm. A common device geometry which has also been utilized for AlGaAs/GaAs ELEDs [Ref. 19] is shown in Figure 7.8 [Ref. 20]. This DH edge-emitting device is realized in the form of a restricted length, stripe geometry p-contact arrangement. Such devices are also referred to as truncated-stripe ELEDs. The short stripe structure (around 100 μm long) improves the external efficiency of the ELED by reducing its internal absorption of carriers.

It was mentioned in Section 7.1 that a particular impetus for the development of high-performance LEDs operating at a wavelength of 1.3 μm was their potential application in the future optical fiber access network. In this context the capacity to provide both high-speed transmission and significant launch powers into single-mode fiber are of prime concern. Aspects of these attributes are displayed by the two device structures shown in Figure 7.9.

The ELED illustrated in Figure 7.9(a) [Ref. 21] comprises a mesa structure with a width of 8 μm and a length of 150 μm for current confinement. The tilted back facet of the

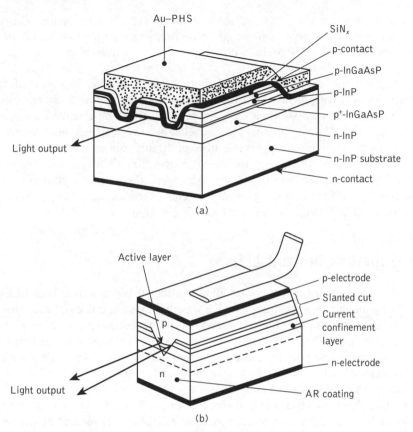

Au–PHS

SiN$_x$

p-contact

p-InGaAsP

p-InP

p$^+$-InGaAsP

n-InP

Light output

n-InP substrate

n-contact

(a)

Active layer

p-electrode

Slanted cut

Current
confinement
layer

p

n-electrode

n

Light output

AR coating

(b)

Figure 7.9 High speed InGaAsP edge-emitting LEDs: (a) mesa structure ELED [Ref. 21]; (b) V-grooved substrate BH ELED [Ref. 23]

device was formed by chemical etching in order to suppress laser oscillation. It should be noted that such ELED structures, being very similar to injection laser structures, could lase unless this mechanism is specifically avoided by removing the potential Fabry–Pérot cavity. This point is discussed in further detail in Section 7.3.5.

The ELED active layer was heavily doped with Zn to reduce the minority carrier lifetime and thus improve the device modulation bandwidth. In this way a 3 dB modulation bandwidth of 600 MHz was obtained [Ref. 21]. When operating at a speed of 600 Mbit s^{-1} the device, with lens coupling (see Section 7.3.7), launched an average optical power of approximately 4 µW into single-mode fiber at a peak drive current of 100 mA. An increase in the peak drive current to 240 mA provided an improvement in the coupled power to slightly over 6 µW. By contrast a BH ELED has been reported which couples 7 µW of optical power into single-mode fiber with a drive current of only 20 mA [Ref. 22]. This short-cavity device (100 µm) had a spectral width (FWHP) of 70 nm in comparison with a linewidth of 90 nm for the high-speed ELED shown in Figure 7.9(a).

Figure 7.9(b) displays another advanced InGaAsP ELED which was fabricated as a V-grooved substrate BH device [Ref. 23]. In this case the front facet was antireflection coated and the rear facet was also etched at a slat to prevent laser action. This device, which again emitted at a center wavelength of 1.3 µm, was reported to have a 3 dB modulation bandwidth around 350 MHz, with the possibility of launching 30 µW of optical power into single-mode fiber [Ref. 23].

Very high coupled optical power levels into single-mode fiber in excess of 100 µW have been obtained with InGaAsP ELEDs at drive currents as low as 50 mA [Ref. 24]. This device structure was based on the configuration of the p-substrate buried crescent injection laser [Ref. 25] with the rear facet beveled by chemical etching to suppress laser oscillation. Butt coupling to single-mode fiber of 10 µm core diameter provided launch powers of only 12 µW which were increased to over 200 µW using lens coupling (see Section 7.3.6) and drive currents of 100 mA. Moreover, the spectral widths of the ELEDs were as narrow as 50 nm which gave device characteristics approaching those of the superluminescent LEDs dealt with in the following section.

7.3.5 Superluminescent LEDs

Another device geometry which is providing significant benefits over both SLEDs and ELEDs for communication applications is the superluminescent diode or SLD. This device type offers advantages of: (a) a high output power; (b) a directional output beam; and (c) a narrow spectral linewidth – all of which prove useful for coupling significant optical power levels into optical fiber (in particular to single-mode fiber [Ref. 22]). Furthermore, the superradiant emission process within the SLD tends to increase the device modulation bandwidth over that of more conventional LEDs.

Figure 7.10 shows two forms of construction for the SLD. It may be observed that the structures in both cases are very similar to those of ELEDs or, for that matter, injection lasers. In effect, the SLD has optical properties that are bounded by the ELED and the injection laser. Similar to this latter device the SLD structure requires a *p–n* junction in the form of a long rectangular stripe (Figure 7.10(a) [Ref. 26]), a ridge waveguide [Ref. 27] or a BH (Figure 7.10(b) [Refs 22, 28]). However, one end of the device is made optically lossy to prevent reflections and thus suppress lasing, the output being from the opposite end.

For operation the injected current is increased until stimulated emission, and hence amplification, occurs (i.e. the initial step towards laser action), but because there is high loss at one end of the device, no optical feedback takes place. Therefore, although there is amplification of the spontaneous emission, no laser oscillation builds up. However, operation in the current region for stimulated emission provides gain causing the device output to increase rapidly with increases in drive current due to what is effectively single-pass amplification. High optical output power can therefore be obtained, together with a narrowing of the spectral width which also results from the stimulated emission.

An early SLD is shown in Figure 7.10(a) which employs a contact stripe together with an absorbing region at one end to suppress laser action. Such devices have provided peak output power of 60 mW at a wavelength of 0.87 µm in pulsed mode [Ref. 26]. Antireflection (AR) coatings can be applied to the cleaved facets of SLDs in order to suppress Fabry–Pérot resonance [Refs 22, 27, 29]. Such devices have launched 550 µW of optical power in multimode graded index fiber of 50 µm diameter at drive currents of 250 mA

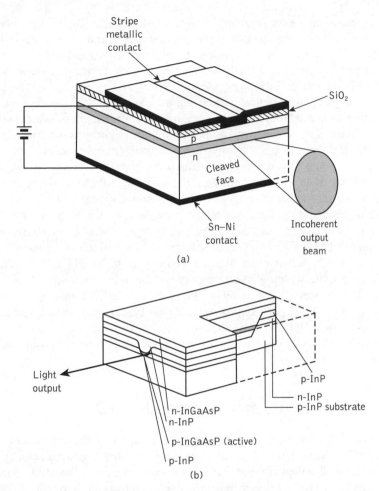

Figure 7.10 Superluminescent LED structures: (a) AlGaAs contact stripe SLD [Ref. 26]; (b) high output power InGaAsP SLD [Ref. 28]

[Ref. 22] and 250 µW into single-mode fiber using drive currents of 100 mA [Ref. 29]. In both cases the device linewidths were in the range 30 to 40 nm rather than the 60 to 90 nm spectral widths associated with conventional ELEDs.

The structure of an InGaAsP/InP SLD is illustrated in Figure 7.10(b) [Ref. 28]. The device which emits at 1.3 µm comprises a buried active layer within a V-shaped groove on the p-type InP substrate. This technique provides an appropriate structure for high-power operation because of its low leakage current. Unlike the aforementioned SLD structures which incorporate AR coatings on both end facets to prevent feedback, a light diffusion surface is placed within this device. The surface, which is applied diagonally on the active layer of length 350 µm, serves to scatter the backward light emitted from the active layer and thus decreases feedback into this layer. In addition, an AR coating is provided on the output facet. As it is not possible to achieve a perfect AR coating, the above structure is therefore not left totally dependent on this feedback suppression mechanism. The coupling of 1 mW of optical power into the spherically lensed end of a single-mode fiber (10 μm core diameter) has been demonstrated with this device operating at a drive current of 150 mA [Ref. 28]. Moreover, the spectral distribution from the SLD was observed to be a smooth envelope with an FWHP of 30 nm, while the device modulation bandwidth reached 350 MHz at the −1.5 dB point (see Section 7.4.3).

Although the incoherent optical power output from SLDs can approach that of the coherent output from injection lasers, the required current density is substantially higher (by around a factor of three times), necessitating high drive currents due to the long device active lengths (i.e. large areas). Improvements, however, in injection laser structures (see Sections 6.5 and 6.6) providing lower threshold currents for specific output powers have also made the SLD a more practical proposition. Nevertheless, other potential drawbacks associated with the SLD in comparison with conventional LEDs are the nonlinear output characteristic and the increased temperature dependence of the output power (see Section 7.4.1). It should be noted that the output of the SLD is spectrally broad (i.e. 20 to 150 nm) and therefore when these devices exhibit sufficient output signal power they can be used as broadband optical power sources [Refs 30, 31]. Commercially available SLDs can operate within a broad range of wavelengths from either 1.16 to 1.33 μm or 1.52 to 1.57 μm [Ref. 32]. In addition, these devices exhibit an output signal power around four to five times higher than a conventional ELED with optical output power of 8 mW being reported [Ref. 33]. Therefore such powerful broadband sources are suitable for the provision of spectrally sliced channels within wavelength division multiplexed systems (see Section 12.9.4).

7.3.6 Resonant cavity and quantum-dot LEDs

The resonant cavity light-emitting diode (RC-LED) is based on planar technology containing a Fabry–Pérot active resonant cavity between distributed Bragg reflector (DBR) mirrors. A quantum well is then embedded in this active cavity. Since the cavity is confined to a micrometer size, the RC-LED is therefore also referred to as a microcavity light-emitting diode [Ref. 34]. The basic structure for an RC-LED is shown in Figure 7.11(a) where an active region consisting of InGaAsP multiquantum wells is positioned in the optical resonant cavity which is located between two DBR mirrors, one each at the bottom and the top of the active cavity.

Current confinement is obtained through the ion implantation technique (see Section 6.6.2) in the top mirror while the RC-LED structure constitutes a Fabry–Pérot resonator where the optical cavity mode is in resonance amplifying the spontaneous emission from the active layer. The reflectivity of the bottom DBR mirror is kept to a maximum (i.e. higher than 90%) by incorporating a large number of gratings (i.e. more than 40) whereas the surface DBR mirror is made semitransparent by introducing fewer gratings (i.e. about 15) creating low facet reflectivity (i.e. 40 to 60%) to allow the optical signal to exit through this mirror. Since these devices incorporate DBR mirrors they may also be referred to as grating-assisted RC-LEDs [Refs 35, 36]. In this context the structure is similar to that of a vertical cavity surface-emitting laser (VCSEL) (see Section 6.6.3) excepting that the emitting side of the DBR mirror of the resonant cavity is semitransparent. Therefore light is emitted as a result of resonantly amplified spontaneous emission and stimulated emission

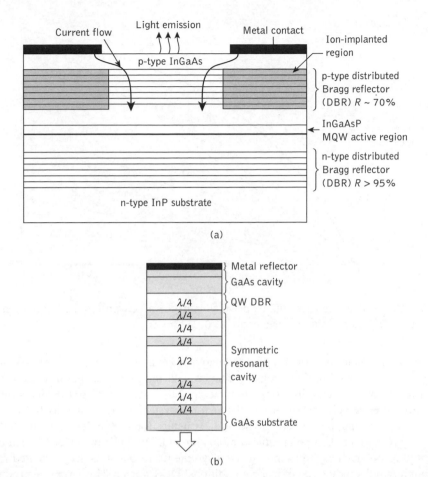

Figure 7.11 Structures of resonant cavity light-emitting diodes: (a) surface emitting using DBR mirrors; (b) bottom-emitting RC²LED using DBR mirror and resonant cavity reflector (RCR)

does not occur. The operation of the device is, however, similar to the VCSEL exhibiting low facet reflectivity on the top mirror and without threshold limitations.

Based on the cavity design, RC-LEDs can be constructed to emit from either the bottom or the surface of the device structure [Ref. 37]. Although they can be fabricated for longer wavelength operation at both 1.3 μm and 1.55 μm [Refs 38–40], RC-LEDs are generally fabricated for operation over a range of wavelengths between 0.85 and 0.88 μm and also at 0.65 μm for use with plastic optical fiber [Ref. 41]. Although the growth process for RC-LEDs is more complex than for conventional devices, their enhanced features, such as the highly directional circular output beam and improved fiber coupling efficiency, make overcoming the fabrication problems worthwhile. External quantum efficiency of the RC-LED is, however, reduced to around 6 to 10% when operating at a wavelength of 1.3 μm or 1.55 μm due to the increased linewidth broadening at these longer wavelengths [Ref. 42]. Nevertheless, even this value of external quantum efficiency proves sufficient to provide

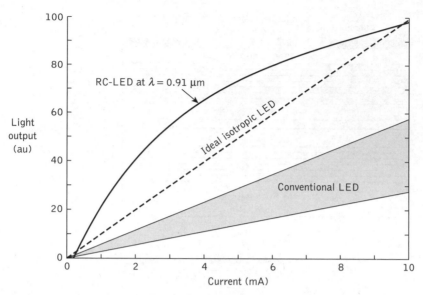

Figure 7.12 Optical intensity against drive current for a conventional LED and an RC-LED for an output signal wavelength of 0.91 μm

satisfactory data transmission at high modulation rates above 1 Gbit s^{-1} [Refs 43, 44]. In addition, device coupling limitations associated with the planar technology used for the RC-LED can be overcome by using photonic crystal-assisted RC-LED device structures [Ref. 36]. Alternatively, an advanced resonant cavity technique known as RC^{2}LED can improve the coupling performance. The basic structure for such a device is illustrated in Figure 7.11(b) which incorporates the combination of a DBR mirror and a resonant cavity to form a resonant cavity reflector. Hence a symmetric resonant cavity is created for the out coupling reflector instead of using a traditional DBR mirror. This structure produces a narrow radiation pattern and therefore it exhibits higher output signal power which is more than 50% greater than that provided by conventional RC-LED designs [Ref. 44].

Figure 7.12 provides a comparison of the output optical powers for a conventional LED and that of an RC-LED with increasing operating current at an output signal wavelength of 0.91 μm. The dashed line in Figure 7.12 represents the characteristics of an ideal isotropic emitting device possessing zero facet reflectivity which is assumed to emit light isotropically with 100% quantum efficiency for all wavelengths emitted from the active cavity. It can be seen in Figure 7.12 that the RC-LED provides a higher optical intensity than a conventional LED and it also surpasses the theoretical limit of an isotropic optical emitter [Ref. 45]. This higher optical intensity is due to the resonant amplified spontaneous emission output from the surface of the device. Hence the combination of both high efficiency and high radiance makes the RC-LED an ideal optical source for multimode fiber coupling in a range of applications.

Although fabrication of an RC-LED is similar to that of conventional planar LEDs, a quantum-dot structure used for the lasers (see Section 6.5.4), however, can also be applied to RC-LEDs and the resultant device is referred to as a quantum-dot or QD-LED. A single-mirror structure QD-LED is shown in Figure 7.13. In this structure an active layer

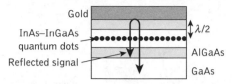

Figure 7.13 Structure of a quantum-dot LED

comprising a layer of InAs quantum dots covered by InGaAs is positioned at a distance from a gold-coated mirror on the device surface. The active region comprises a single layer of quantum dots while a AlGaAs layer is grown between the GaAs substrate and the active region in order to confine the injected carriers. To enhance output signal power, the quantum-dot layer is positioned at half the emission wavelength distance from the surface mirror. The optical signal reflected by the mirror therefore constructively interferes with the radiation emitted downwards from the active layer resulting in a fourfold increase in optical signal power being collected from the substrate side [Ref. 46]. QD-LEDs with 10 mW output power when operating at wavelengths of 1.30 μm and 1.55 μm have also been successfully fabricated [Refs 40, 47, 48].

QD-LEDs based on a resonant cavity with external quantum efficiency of greater than 20% have also been demonstrated [Refs 47, 49]. For nonresonant cavity QD-LEDs increased quantum efficiency can be obtained by introducing thin active layers at the surface of the LED. Such devices are referred to as surface-textured thin-film LEDs [Ref. 50]. In this structure an optical signal which suffers total internal reflection is scattered internally by the textured top surface and hence changes its angle of propagation. After reflection from the back reflector the optical signal can be coupled to the output of the LED. An external quantum efficiency of 29% at higher transmission rates of 1 Gbit s^{-1} has been demonstrated with this structure and it was further improved to 40% when incorporating an optical lens on the top of the device [Ref. 50].

7.3.7 Lens coupling to fiber

It is apparent that much of the light emitted from LEDs is not coupled into the generally narrow acceptance angle of the fiber. Even with the etched well surface emitter, where the low-*NA* fiber is butted directly into the emitting aperture of the device, coupling efficiencies are poor (of the order of 1 to 2%). However, it has been found that greater coupling efficiency may be obtained if lenses are used to collimate the emission from the LED, particularly when the fiber core diameter is significantly larger than the width of the emission region. There are several lens coupling configurations which include spherically polished structures not unlike the dome LED, spherical-ended or tapered fiber coupling, truncated spherical microlenses, GRIN-rod lenses and integral lens structures.

A GaAs/AlGaAs spherical-ended fiber-coupled LED is illustrated in Figure 7.14 [Ref. 51]. It consists of a planar surface-emitting structure with the spherical-ended fiber attached to the cap by epoxy resin. An emitting diameter of 35 μm was fabricated into the device and the light was coupled into fibers with core diameters of 75 and 110 μm. The geometry of the situation is such that it is essential that the active diameter of the device be substantially less (factor of 2) than the fiber core diameter if increased coupling efficiency

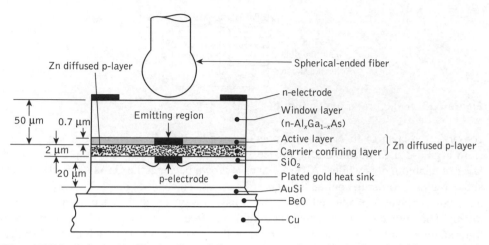

Figure 7.14 Schematic illustration of the structure of a spherical-ended fiber-coupled AlGaAs LED [Ref. 51]

is to be obtained. In this case good performance was obtained with coupling efficiencies around 6%. This is in agreement with theoretical [Ref. 52] and other experimental [Ref. 53] results which suggest an increased coupling efficiency of 2 to 5 times through the spherical fiber lens.

Another common lens coupling technique employs a truncated spherical microlens. This configuration is shown in Figure 7.15 for an etched well InGaAsP/InP DH surface emitter [Ref. 54] operating at a wavelength of 1.3 μm. Again, a requirement for efficient coupling is that the emission region diameter is much smaller than the core diameter of the fiber. In this case the best results were obtained with a 14 μm active diameter and an 85 μm core diameter for a step index fiber with a numerical aperture of 0.16. The coupling efficiency was increased by a factor of 13, again supported by theory [Ref. 52] which suggests possible increases of up to 30 times.

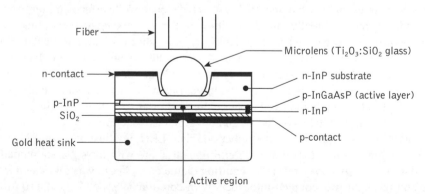

Figure 7.15 The use of a truncated spherical microlens for coupling the emission from an InGaAsP surface-emitting LED to the fiber [Ref. 54]

However, the overall power conversion efficiency η_{pc} which is defined as the ratio of the optical power coupled into the fiber P_c to the electric power applied at the terminals of the device P and is therefore given by:

$$\eta_{pc} = \frac{P_c}{P} \tag{7.23}$$

is still quite low. Even with the increased coupling efficiency η_{pc} was found to be around 0.4%.

Example 7.5

A lens-coupled surface-emitting LED launches 190 µW of optical power into a multimode step index fiber when a forward current of 25 mA is flowing through the device. Determine the overall power conversion efficiency when the corresponding forward voltage across the diode is 1.5 V.

Solution: The overall power conversion efficiency may be obtained from Eq. (7.23) where:

$$\eta_{pc} = \frac{P_c}{P} = \frac{190 \times 10^{-6}}{25 \times 10^{-3} \times 1.5} = 5.1 \times 10^{-3}$$

Hence the overall power conversion efficiency is 0.5%.

The integral lens structure shown in Figure 7.6 has become a favored power coupling strategy for use with surface emitters. In this technique a low-absorption lens is formed at the exit face of the substrate material instead of being fabricated in glass and attached to a planar SLED with epoxy. The method benefits from the elimination of the semiconductor–epoxy–lens interface which can limit the maximum lens gains of the SLEDs discussed above. An early example gave an improved coupling efficiency of around three times that of a planar SLED [Ref. 53], but for optimized devices it is predicted that coupling efficiencies should exceed 15% [Ref. 54].

It was mentioned in Section 7.3.4 that lens coupling can also be usefully employed with edge-emitting devices. In practice, lenses attached to the fiber ends or tapered fiber lenses are widely utilized to increase the coupling efficiency [Refs 13, 19]. An example of the former technique is illustrated in Figure 7.16(a) in which a hemispherical lens is epoxied onto the fiber end and positioned adjacent to the ELED emission region. The coupling efficiency has been increased by a factor of three to four times using this strategy [Refs 13, 19]. Alternatively, a truncated spherical lens glued onto the emitting facet of a superradiant ELED has given a coupling gain of a factor of five or 7 dB [Ref. 55].

Tapered fiber lenses have been extensively used to couple power from ELEDs into single-mode fiber. Butt coupling of optical power from LEDs into single-mode fiber is substantially reduced in comparison with that obtained into multimode fiber. It ranges from between 0.5 and 2 µW for a standard SLED up to around 10 to 12 µW for an ELED. The small core diameter of single-mode fiber does not allow significant lens coupling gain

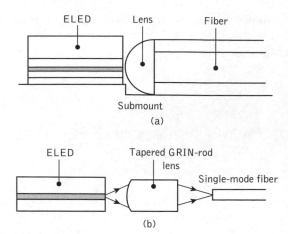

Figure 7.16 Lens coupling with edge-emitting LEDs: (a) lens-ended fiber coupling; (b) tapered (plano-convex) GRIN-rod lens coupling to single-mode fiber

to be achieved with SLEDs. For edge emitters, however, a coupling gain of around 5 dB may be realized using tapered fiber [Ref. 56].

An alternative strategy to improve the coupling efficiency from an ELED into single-mode fiber is depicted in Figure 7.16(b) [Ref. 24]. In this case a tapered GRIN-rod lens (see Section 5.5.1) was positioned between the high-power ELED and the fiber. A coupling efficiency defined as the ratio of the coupled power to the total emitted power of around 15% was obtained [Ref. 24]. The coupling efficiency can also be improved when microlenses with micrometer dimensions are integrated with the specific optical components (i.e. the LED or optical fiber). Using such microlenses the coupling of an SLED to fiber provided increased output power by a factor of 1.6 [Ref. 57]. Moreover, in comparison with a typical flat-end or arc-lensed fiber, the microlensed fiber gave an improvement in coupling efficiency of 40% and 18%, respectively [Ref. 58].

7.4 LED characteristics

7.4.1 Optical output power

The ideal light output power against current characteristic for an LED (depicted for an isotropic device in Figure 7.12) is shown in Figure 7.17. It is linear corresponding to the linear part of the injection laser optical power output characteristic before lasing occurs. Intrinsically the LED is a very linear device in comparison with the majority of injection lasers and hence it tends to be more suitable for analog transmission where severe constraints are put on the linearity of the optical source. However, in practice LEDs do exhibit significant nonlinearities which depend upon the configuration utilized. It is therefore often necessary to use some form of linearizing circuit technique (e.g. predistortion linearization or negative feedback) in order to ensure the linear performance of the device to allow its

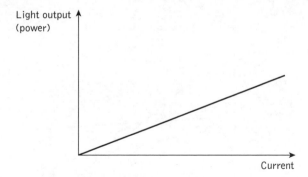

Figure 7.17 An ideal light output against current characteristic for an LED

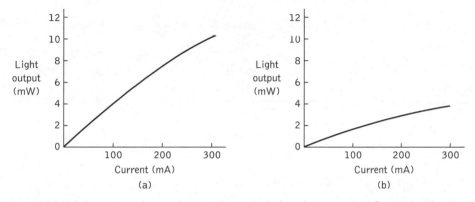

Figure 7.18 Light output (power) into air against d.c. drive current for typically good LEDs [Ref. 17]: (a) an AlGaAs surface emitter with a 50 μm diameter dot contact; (b) an AlGaAs edge emitter with a 65 μm wide stripe and 100 μm length

use in high-quality analog transmission systems [Ref. 59]. Figure 7.18(a) and (b) show the light output against current characteristics for typically good surface and edge emitters respectively [Ref. 17]. It may be noted that the surface emitter radiates significantly more optical power into air than the edge emitter, and that both devices are reasonably linear at moderate drive currents.

In a similar manner to the injection laser, the internal quantum efficiency of LEDs decreases exponentially with increasing temperature (see Section 6.7.1). Hence the light emitted from these devices decreases as the *p–n* junction temperature increases. The light output power against temperature characteristics for three important LED structures operating at a wavelength of 1.3 μm are shown, for comparison, in Figure 7.19 [Ref. 13]. It may be observed that the edge-emitting device exhibits a greater temperature dependence than the surface emitter and that the output of the SLD with its stimulated emission is *strongly* dependent on the junction temperature. This latter factor is further emphasized in the light output against current characteristics for a superluminescent LED displayed in Figure 7.20 [Ref. 27]. These characteristics show the variation in output power at a specific drive current over the temperature range 0 to 40 °C for a ridge waveguide device providing lateral current confinement. The nonlinear nature of the output characteristic

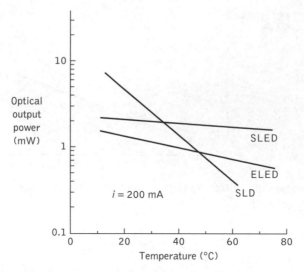

Figure 7.19 Light output temperature dependence for three important LED structures emitting at a wavelength of 1.3 μm [Ref. 13; © Elsevier]

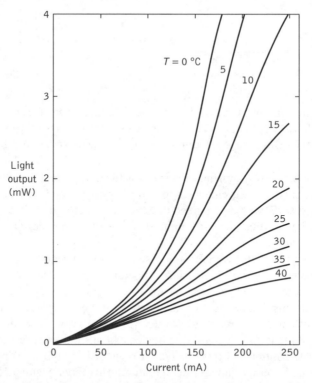

Figure 7.20 Light output against current characteristic at various ambient temperatures for an InGaAsP ridge waveguide SLD. Reprinted with permission from I. P. Kaminow, G. Eisenstein, L. W. Stulz and A. G. Dentai 'Lateral confinement InGaAsP superluminescent diode at 1.3 μm', *IEEE J. Quantum Electron.*, QE19, p. 78, 1983. Copyright ©1983, IEEE

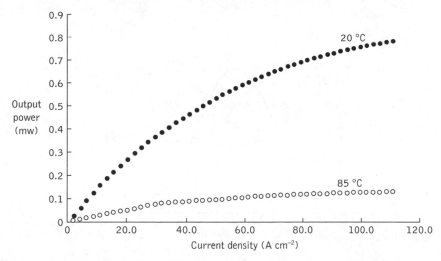

Figure 7.21 Output power of an AlGaInP resonant cavity LED at two different operating temperatures [Ref. 61]

typical of SLDs can also be observed with a knee becoming apparent at an operating temperature around 20 °C. Hence to utilize the high-power potential of such devices at elevated temperatures, the use of thermoelectric coolers may be necessary [Ref. 13].

It should also be noted that resonant cavity LEDs have shown a similar reduction in output power when operated at higher temperatures. For example, the maximum quantum efficiency at 15 °C for a 200 μm long RC-LED emitting at a signal wavelength of 0.66 μm was observed to be 2.8% which declined to 1.0% as the temperature was increased to 75 °C [Refs 60, 61]. Figure 7.21 displays the temperature dependency for the AlGaInP-based RC-LED operating at a signal wavelength of 0.66 μm designed for use with plastic optical fiber (see Section 4.5.5) [Ref. 61]. It shows optical output power against current density for the device operating at 20 °C and 85 °C. When operating at room temperature, however, RC-LEDs can provide high levels of optical output power. For example, another AlGaInP device fabricated on a GaAs substrate has demonstrated high external quantum efficiency of up to 23% [Ref. 62]. In particular it emitted 3.4 mW of output optical power at a wavelength of 0.65 μm with a drive current of 10 mA.

7.4.2 Output spectrum

The spectral linewidth of an LED operating at room temperature in the 0.8 to 0.9 μm wavelength band is usually between 25 and 40 nm at the half maximum intensity points (full width at half power (FWHP) points). For materials with smaller bandgap energies operating in the 1.1 to 1.7 μm wavelength region the linewidth tends to increase to around 50 to 160 nm. Examples of these two output spectra are shown in Figure 7.22 [Refs 7, 63]. Also illustrated in Figure 7.22(b) are the increases in linewidth due to increased doping levels and the formation of bandtail states (see Section 6.3.4). This becomes apparent in the differences in the output spectra between surface- and edge-emitting LEDs where the devices have generally heavily doped and lightly doped (or undoped) active layers

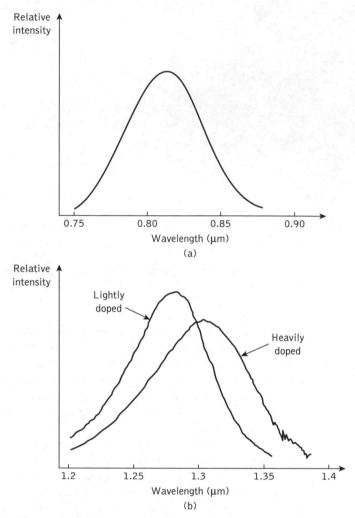

Figure 7.22 LED output spectra: (a) output spectrum for an AlGaAs surface emitter with doped active region [Ref. 7]; (b) output spectra for an InGaAsP surface emitter showing both the lightly doped and heavily doped cases. Reproduced with permission from A. C. Carter, *Radio Electron. Eng.*, 51, p. 41, 1981

respectively. It may also be noted that there is a shift to lower peak emission wavelength (i.e. higher energy) through reduction in doping in Figure 7.22(b), and hence the active layer composition must be adjusted if the same center wavelength is to be maintained.

The differences in the output spectra between InGaAsP SLEDs and ELEDs caused by self-absorption along the active layer of the devices are displayed in Figure 7.23. It may be observed that the FWHP points are around 1.6 times smaller for the ELED than the SLED [Ref. 13]. In addition, the spectra of the ELED may be further narrowed by the superluminescent operation due to the onset of stimulated gain and in this case the linewidth can be far smaller (e.g. 30 nm) than that obtained with the SLED.

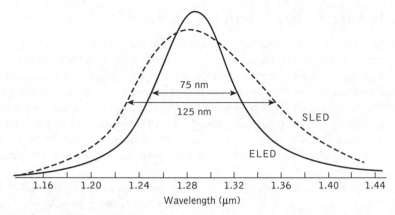

Figure 7.23 Typical spectral output characteristics for InGaAsP surface- and edge-emitting LEDs operating in the 1.3 μm wavelength region [Ref. 13; © Elsevier]

The output spectra also tend to broaden at a rate of between 0.1 and 0.3 nm °C^{-1} with increase in temperature due to the greater energy spread in carrier distributions at higher temperatures. Increases in temperature of the junction affect the peak emission wavelength as well, and it is shifted by +0.3 to 0.4 nm °C^{-1} for AlGaAs devices [Ref. 11] and by +0.6 nm°C^{-1} for InGaAsP devices [Ref. 64]. The combined effects on the output spectrum from a typical AlGaAs surface emitter are illustrated in Figure 7.24. It is clear that it may therefore be necessary to utilize heat sinks with LEDs for certain optical fiber communication applications, although this is far less critical (normally insignificant compared with the device linewidth) than the cooling requirements for injection lasers.

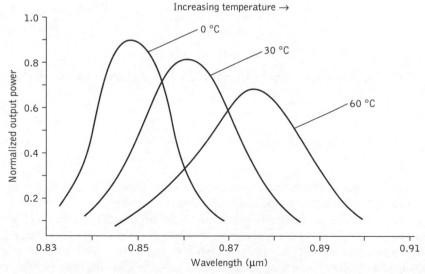

Figure 7.24 Typical spectral variation of the output characteristic with temperature for an AlGaAs surface-emitting LED

7.4.3 Modulation bandwidth

The modulation bandwidth in optical communications may be defined in either electrical or optical terms. However, it is often more useful when considering the associated electrical circuitry in an optical fiber communication system to use the electrical definition where the electrical signal power has dropped to half its constant value due to the modulated portion of the optical signal. This corresponds to the electrical 3 dB point or the frequency at which the output electric power is reduced by 3 dB with respect to the input electric power. As optical sources operate down to d.c. level we only consider the high-frequency 3 dB point, the modulation bandwidth being the frequency range between zero and this high-frequency 3 dB point.

Alternatively, if the 3 dB bandwidth of the modulated optical carrier (optical bandwidth) is considered, we obtain an increased value for the modulation bandwidth. The reason for this inflated modulation bandwidth is illustrated in Example 7.6 and Figure 7.25. In considerations of bandwidth within the text the electrical modulation bandwidth will be assumed unless otherwise stated, following current practice.

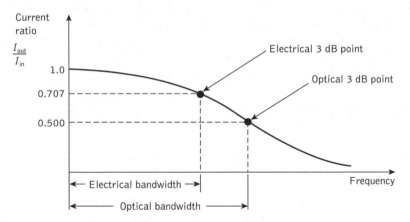

Figure 7.25 The frequency response for an optical fiber system showing the electrical and optical bandwidths

Example 7.6

Compare the electrical and optical bandwidths for an optical fiber communication system and develop a relationship between them.

Solution: In order to obtain a simple relationship between the two bandwidths it is necessary to compare the electric current through the system. Current rather than voltage (which is generally used in electrical systems) is compared as both the optical source and optical detector (see Section 8.6) may be considered to have a linear relationship between light and current.

Electrical bandwidth: The ratio of the electric output power to the electric input power in decibels RE_{dB} is given by:

$$RE_{dB} = 10 \log_{10} \frac{\text{electric power out (at the detector)}}{\text{electric power in (at the source)}}$$

$$= 10 \log_{10} \frac{I_{out}^2/R_{out}}{I_{in}^2/R_{in}}$$

$$\propto 10 \log_{10} \left[\frac{I_{out}}{I_{in}} \right]^2$$

The electrical 3 dB points occur when the ratio of electric powers shown above is $\frac{1}{2}$. Hence it follows that this must occur when:

$$\left[\frac{I_{out}}{I_{in}} \right]^2 = \frac{1}{2} \quad \text{or} \quad \frac{I_{out}}{I_{in}} = \frac{1}{\sqrt{2}}$$

Thus in the electrical regime the bandwidth may be defined by the frequency when the output current has dropped to $1/\sqrt{2}$ or 0.707 of the input current to the system.

Optical bandwidth: The ratio of the optical output power to the optical input power in decibels RO_{dB} is given by:

$$RO_{dB} = 10 \log_{10} \frac{\text{optical power out (received at detector)}}{\text{optical power in (transmitted at source)}}$$

$$\propto 10 \log_{10} \frac{I_{out}}{I_{in}}$$

(due to the linear light/current relationships of the source and detector). Hence the optical 3 dB points occur when the ratio of the currents is equal to $\frac{1}{2}$, and:

$$\frac{I_{out}}{I_{in}} = \frac{1}{2}$$

Therefore in the optical regime the bandwidth is defined by the frequencies at which the output current has dropped to $\frac{1}{2}$ or 0.5 of the input current to the system. This corresponds to an electric power attenuation of 6 dB.

The comparison between the two bandwidths is illustrated in Figure 7.25 where it may be noted that the optical bandwidth is significantly greater than the electrical bandwidth. The difference between them (in frequency terms) depends on the shape of the frequency response for the system. However, if the system response is assumed to be Gaussian, then the optical bandwidth is a factor of $\sqrt{2}$ greater than the electrical bandwidth [Ref. 65].

The modulation bandwidth of LEDs is generally determined by three mechanisms. These are:

(a) the doping level in the active layer;

(b) the reduction in radiative lifetime due to the injected carriers;

(c) the parasitic capacitance of the device.

Assuming negligible parasitic capacitance, the speed at which an LED can be directly current modulated is fundamentally limited by the recombination lifetime of the carriers, where the optical output power $P_e(\omega)$ of the device (with constant peak current) and angular modulation frequency ω is given by [Ref. 66]:

$$\frac{P_e(\omega)}{P_{dc}} = \frac{1}{[1 + (\omega\tau_i)^2]^{\frac{1}{2}}} \tag{7.24}$$

where τ_i is the injected (minority) carrier lifetime in the recombination region and P_{dc} is the d.c. optical output power for the same drive current.

Example 7.7

The minority carrier recombination lifetime for an LED is 5 ns. When a constant d.c. drive current is applied to the device the optical output power is 300 μW. Determine the optical output power when the device is modulated with an rms drive current corresponding to the d.c. drive current at frequencies of (a) 20 MHz; (b) 100 MHz.

It may be assumed that parasitic capacitance is negligible. Further, determine the 3 dB optical bandwidth for the device and estimate the 3 dB electrical bandwidth assuming a Gaussian response.

Solution: (a) From Eq. (7.24), the optical output power at 20 MHz is:

$$P_e(20\ \text{MHz}) = \frac{P_{dc}}{[1 + (\omega\tau_i)^2]^{\frac{1}{2}}}$$

$$= \frac{300 \times 10^{-6}}{[1 + (2\pi \times 20 \times 10^6 \times 5 \times 10^{-9})^2]^{\frac{1}{2}}}$$

$$= \frac{300 \times 10^{-6}}{[1.39]^{\frac{1}{2}}}$$

$$= 254.2\ \mu\text{W}$$

(b) Again using Eq. (7.24):

$$P_e(100\ \text{MHz}) = \frac{300 \times 10^{-6}}{[1 + (2\pi \times 100 \times 10^6 \times 5 \times 10^{-9})^2]^{\frac{1}{2}}}$$

$$= \frac{300 \times 10^{-6}}{[10.87]^{\frac{1}{2}}}$$

$$= 90.9\ \mu\text{W}$$

This example illustrates the reduction in the LED optical output power as the device is driven at higher modulating frequencies. It is therefore apparent that there is a somewhat limited bandwidth over which the device may be usefully utilized.

To determine the optical 3 dB bandwidth, the high-frequency 3 dB point occurs when $P_e(\omega)/P_{dc} = \frac{1}{2}$ Hence, using Eq. (7.24):

$$\frac{1}{[1 + (\omega\tau_i)^2]^{\frac{1}{2}}} = \frac{1}{2}$$

and $1 + (\omega\tau_i)^2 = 4$. Therefore $\omega\tau_i = \sqrt{3}$, and:

$$f = \frac{\sqrt{3}}{2\pi\tau} = \frac{\sqrt{3}}{\pi \times 10^{-8}} = 55.1 \text{ MHz}$$

Thus the 3 dB optical bandwidth B_{opt} is 55.1 MHz as the device, similar to all LEDs, operates down to d.c. level.

Assuming a Gaussian frequency response, the 3 dB electrical bandwidth B will be:

$$B = \frac{55.1}{\sqrt{2}} = 39.0 \text{ MHz}$$

Thus the corresponding electrical bandwidth is 39 MHz. However, it must be remembered that parasitic capacitance may reduce the modulation bandwidth below this value.

The carrier lifetime is dependent on the doping concentration, the number of injected carriers into the active region, the surface recombination velocity and the thickness of the active layer. All these parameters tend to be interdependent and are adjustable within limits in present-day technology. In general, the carrier lifetime may be shortened by either increasing the active layer doping or by decreasing the thickness of the active layer. However, in surface emitters this can reduce the external power efficiency of the device due to the creation of an increased number of nonradiative recombination centers.

Edge-emitting LEDs have a very thin, virtually undoped active layer and the carrier lifetime is controlled only by the injected carrier density. At high current densities the carrier lifetime decreases with injection level because of a bimolecular recombination process [Ref. 66]. This bimolecular recombination process allows edge-emitting LEDs with narrow recombination regions to have short recombination times, and therefore relatively high modulation capabilities at reasonable operating current densities [Ref. 67]. For instance, edge-emitting devices with electrical modulation bandwidths of 145 MHz have been achieved with moderate doping and extremely thin (approximately 50 nm) active layers [Ref. 68]

However, LEDs tend to be slower devices with significantly lower output powers than injection lasers because of the longer lifetime of electrons in their donor regions resulting

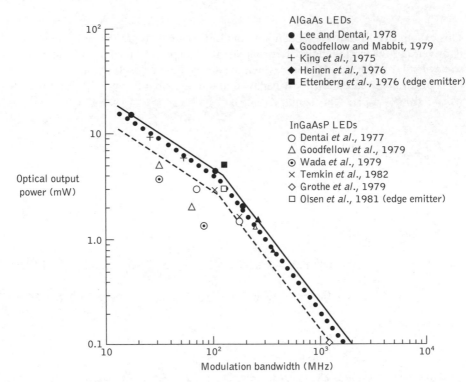

Figure 7.26 Reported optical output power against bandwidth for both AlGaAs and InGaAsP LEDs [Refs 69, 70]. Best results for AlGaAs devices (solid line). These AlGaAs LED results shifted to a wavelength of 1.3 μm (dotted line). Best results for InGaAsP devices (dashed line)

from spontaneous recombination rather than stimulated emission,* coupled with the increased numbers of nonradiative centers at higher doping levels. Thus at high modulation bandwidths the optical output power from conventional LED structures decreases as illustrated in Example 7.7 and also as shown in Figure 7.26.

The reciprocal relationship between modulation bandwidth and output power may be observed in Figure 7.26 which illustrates experimental results obtained with both AlGaAs and InGaAsP LEDs [Refs 69, 70]. The solid line gives an indication of the best results for AlGaAs LEDs, whereas, for comparison, the dotted line represents these AlGaAs results shifted by the ratio of the photon energy at 0.85 μm to that at 1.3 μm. Finally, the dashed line provides a contour of the best reported results for InGaAsP LEDs. It may be observed that the output power from AlGaAs LEDs is a factor of 2 higher than that of InGaAsP devices at all bandwidths, which partly results from the photon energy at the 1.3 μm wavelength being smaller (by a factor of 1.53) than that at 0.85 μm. Hence the center dotted line displays the adjustment of the AlGaAs LEDs for this factor showing that the best

* The superluminescent and resonant cavity LEDs are an exception in this respect and is therefore capable of high output power at relatively high modulation bandwidths (see Sections 7.3.5 and 7.3.6).

performance of InGaAsP LEDs is not far below that of AlGaAs LEDs. Moreover, the difference is probably due to the more advanced technology which is available for the latter devices combined with the enhanced wavelength saturation in the longer wavelength material [Ref. 13].

For surface-emitting AlGaAs LEDs a high output power of 15 mW has been obtained at modest bandwidths (17 MHz) [Ref. 71] whereas the very large bandwidth of 1.1 GHz was only achieved at the far lower output power of 0.2 mW [Ref. 72]. In general terms, to maximize the output power from SLEDs exhibiting low modulation bandwidths in the range 20 to 50 MHz, a thick active layer (2 to 2.5 µm) with low doping levels (less than 5×10^{17} cm^{-3}) can be employed. Thinner active layers (1 to 1.5 µm) and higher doping levels (0.5 to 1.0×10^{-18} cm^{-3}) are required for devices operating in the 50 to 100 MHz bandwidth region. In order to increase the modulation bandwidth into and beyond the 100 to 200 MHz range, however, very high doping levels in excess of 5×10^{18} cm^{-3} are necessary in combination with thin active layers.

Longer wavelength LEDs fabricated from the InGaAsP/InP material system for operation at a wavelength around 1.3 µm are widely commercially available. Such devices with undoped (i.e. with a residual n-type concentration between 1×10^{-17} and 5×10^{17} cm^{-3}) active layers provide modulation bandwidths in the range 50 to 100 MHz [Ref. 64]. Moreover, with higher doping densities (5×10^{18} cm^{-3}) and relatively thin active layers (400 nm), bandwidths of 690 MHz have been obtained [Ref. 73]. Modulation rates in the range 600 Mbit s^{-1} to 1.2 Gbit s^{-1} have also been achieved using high levels of Zn doping (1×10^{-19} to 1.3×10^{-19} cm^{-3}) in InGaAsP devices [Refs 21, 74]

7.4.4 Reliability

LEDs are not generally affected by the catastrophic degradation mechanisms which can severely affect injection lasers (see Section 6.7.6). Early or infant failures do, however, occur as a result of random and not always preventable fabricational defects. Such failures can usually be removed from the LED batch population over an initial burn-in operational period [Ref. 75]. In addition, LEDs do exhibit gradual degradation which may take the form of a rapid degradation mode* or a slow degradation mode.

Rapid degradation in LEDs is similar to that in injection lasers, and is due to both the growth of dislocations and precipitate-type defects in the active region giving rise to dark line defects (DLDs) and dark spot defects (DSDs), respectively, under device aging [Ref. 69]. DLDs tend to be the dominant cause of rapid degradation in GaAs-based LEDs. The growth of these defects does not depend upon substrate orientation but on the injection current density, the temperature and the impurity concentration in the active layer.

Good GaAs substrates have dislocation densities around 5×10^{-4} cm^{-2}. Hence, there is less probability of dislocations in devices with small active regions. DSDs, and the glide of existing misfit dislocations, however, predominate as the cause of rapid degradation in InP-based LEDs.

---

* LEDs which display rapid degradation are sometimes referred to as freak failures [Ref. 75] because they pass the burn-in period but fail earlier in operational life than the main device population.

LEDs may be fabricated which are largely free from these defects and are therefore subject to a slower long-term degradation process. This homogeneous degradation is thought to be due to recombination enhanced point defect generation (i.e. vacancies and interstitials), or the migration of impurities into the active region [Ref. 76]. The optical output power $P_e(t)$ may be expressed as a function of the operating time t, and is given by [Ref. 76]:

$$P_e(t) = P_{out} \exp(-\beta_r t) \tag{7.25}$$

where P_{out} is the initial output power and β_r is the degradation rate. The degradation rate is characterized by the activation energy of homogeneous degradation E_a and is a function of temperature. It is given by:

$$\beta_r = \beta_0 \exp(-E_a/KT) \tag{7.26}$$

where β_0 is a proportionality constant, K is Boltzmann's constant and T is the absolute temperature of the emitting region. The activation energy E_a is a variable which is dependent on the material system and the structure of the device. The value of E_a is in the range 0.56 to 0.65 eV, and 0.9 to 1.0 eV for surface-emitting GaAs/AlGaAs and InGaAsP/InP LEDs respectively [Ref. 9]. These values suggest 10^6 to 10^7 hours (100 to 1000 years) CW operation at room temperature for AlGaAs devices, and in excess of 10^9 hours for surface-emitting InGaAsP LEDs.

Example 7.8

An InGaAsP surface emitter has an activation energy of 1 eV with a constant of proportionality (β_0) of 1.84×10^7 h^{-1}. Estimate the CW operating lifetime for the LED with a constant junction temperature of 17°C, if it is assumed that the device is no longer useful when its optical output power has diminished to 0.67 of its original value.

Solution: Initially, it is necessary to obtain the degradation rate β_r. Thus from Eq. (7.26):

$$\beta_r = \beta_0 \exp(-E_a/KT)$$

$$= 1.84 \times 10^7 \exp\left(\frac{-1 \times 1.602 \times 10^{-19}}{1.38 \times 10^{-23} \times 290}\right)$$

$$= 1.84 \times 10^7 \exp(-40)$$

$$= 7.82 \times 10^{-11} \text{ h}^{-1}$$

Now, using Eq. (7.25):

$$\frac{P_e(t)}{P_{out}} = \exp(-\beta_r t) = 0.67$$

Therefore:

$$\beta_r t = -\ln 0.67$$

and:

$$t = \frac{\ln 0.67}{7.82 \times 10^{-11}} = \frac{0.40}{7.82 \times 10^{-11}}$$

$$= 5.1 \times 10^9 \text{ h}$$

Hence the estimated lifetime of the device under the specified conditions in Example 7.8 is 5.1×10^9 hours. It must be noted that the junction temperature, even for a device operating at room temperature, is likely to be well in excess of room temperature when substantial drive currents are passed. Also the diminished level of optical output in the example is purely arbitrary and for many applications this reduced level may be unacceptable.

Nevertheless it is quite common for the device lifetime or median life to be determined for a 50% drop in light output power from the device [Ref. 75]. It is clear, however, that with the long-term LED degradation process there is no absolute end-of-life power level and therefore to a large extent it is system dependent such that a trade-off can be made between the required system end-of-life power margin and the device reliability [Ref. 70]. Hence the allocated drop to end-of-life power can be substantially reduced to, say, 20% which will provide for an enhanced system power margin (e.g. increased link length) at the expense of the device median life. Overall, even with these more rigorous conditions, the anticipated median life for such LEDs is excellent and it is unlikely to cause problems in most optical fiber communication system applications.

Extrapolated accelerated lifetime tests are also in broad agreement with the theoretical estimates [Refs 73, 75–79] for the less sophisticated device structures. For example, a planar GaAs/AlGaAs DH LED exhibited a median life for a 50% output power reduction of 9×10^7 hours at a temperature of 25°C [Ref. 78]. By comparison, extrapolated half-power lifetimes in excess of 10^8 hours at a temperature of 60°C have been obtained with higher speed (greater than 200 Mbit s^{-1}) InGaAsP/InP LEDs [Ref. 79].

7.5 Modulation

In order to transmit information via an optical fiber communication system it is necessary to modulate a property of the light with the information signal. This property may be intensity, frequency, phase or polarization (direction) with either digital or analog signals. The choices are indicated by the characteristics of the optical fiber, the available optical sources and detectors, and considerations of the overall system.

However, at present in optical fiber communications considerations of the above for practical systems tend to dictate some form of intensity modulation of the source. Although much effort has been expended and considerable success has been achieved in the area of coherent optical communications (see Chapter 13), the widespread deployment of such systems will still take some further time. Therefore intensity modulation (IM) of the optical source and envelope or direct detection (DD) at the optical receiver is likely to remain the major modulation strategy* in the immediate future.

Intensity modulation is easy to implement with the electroluminescent sources available at present (LEDs and injection lasers). These devices can be directly modulated simply by variation of their drive currents at rates up to many gigahertz. Thus direct modulation of the optical source is satisfactory for many of the modulation bandwidths currently in use. However, there is increasing interest in integrated photonic devices (see Chapter 11) where external optical modulators [Refs 80, 81] are used in order to achieve greater bandwidths and to allow the use of optical amplifiers (see Sections 10.3 and 10.4) and nonsemiconductor sources (e.g. Nd : YAG laser) which cannot be directly modulated at high frequency (see Section 6.9.1). External optical modulators are active devices which tend to be used primarily to modulate the frequency or phase of the light, but may also be used for time division multiplexing and switching of optical signals. However, modulation considerations within this text (excepting Chapter 13) will mainly be concerned with the direct modulation of the intensity of the optical source.

Intensity modulation may be utilized with both digital and analog signals [Refs 82, 83]. Analog intensity modulation is usually easier to apply but requires comparatively large signal-to-noise ratios (see Section 9.2.5) and therefore it tends to be limited to relatively narrow-bandwidth, short-distance applications. Alternatively, digital intensity modulation gives improved noise immunity but requires wider bandwidths, although these may be small in comparison with the available bandwidth. It is therefore ideally suited to optical fiber transmission where the available bandwidth is large. Hence at present most fiber systems in the medium- to long-distance range use digital intensity modulation.

Problems

7.1   Describe with the aid of suitable diagrams the mechanism giving the emission of light from an LED. Discuss the effects of this mechanism on the properties of the LED in relation to its use as an optical source for communications.

7.2   Briefly outline the advantages and drawbacks of the LED in comparison with the injection laser for use as a source in optical fiber communications.

7.3   The power generated internally within a double-heterojunction LED is 28.4 mW at a drive current of 60 mA. Determine the peak emission wavelength from the device when the radiative and nonradiative recombination lifetimes of the minority carriers in the active region are equal.

7.4   The diffusion length L_D or the average distance moved by charge carriers before recombination in the active region of an LED is given by:

$$L_D = (D\tau)^{\frac{1}{2}}$$

* This strategy is often referred to as intensity modulation/direct detection, or IM/DD.

where D is the diffusion coefficient and τ is the total carrier recombination lifetime. Calculate the diffusion coefficient in gallium arsenide when the diffusion length is 21 µm and the radiative and nonradiative carrier recombination lifetimes are equal at 90 ns.

7.5   Estimate the external power efficiency of a GaAs planar LED when the transmission factor of the GaAs–air interface is 0.68 and the internally generated optical power is 30% of the electric power supplied. The refractive index of GaAs may be taken as 3.6.

7.6   The external power efficiency of an InGaAsP/InP planar LED is 0.75% when the internally generated optical power is 30 mW. Determine the transmission factor for the InP–air interface if the drive current is 37 mA and the potential difference across the device is 1.6 V. The refractive index of InP may be taken as 3.46.

7.7   A GaAs planar LED emitting at a wavelength of 0.85 µm has an internal quantum efficiency of 60% when passing a forward current of 20 mA s^{-1}. Estimate the optical power emitted by the device into air, and hence determine the external power efficiency if the potential difference across the device is 1 V. It may be assumed that the transmission factor at the GaAs–air interface is 0.68 and that the refractive index of GaAs is 3.6. Comment on any assumptions made.

7.8   The external power efficiency of a planar GaAs LED is 1.5% when the forward current is 50 mA and the potential difference across its terminals is 2 V. Estimate the optical power generated within the device if the transmission factor at the coated GaAs–air interface is 0.8.

7.9   Outline the common LED structures for optical fiber communications, discussing their relative merits and drawbacks. In particular, compare surface- and edge-emitting devices. Comment on the distinction between multimode and single-mode devices.

7.10   Derive an expression for the coupling efficiency of a surface-emitting LED into a step index fiber, assuming the device to have a Lambertian output. Determine the optical loss in decibels when coupling the optical power emitted from the device into a step index fiber with an acceptance angle of 14°. It may be assumed that the LED is smaller than the fiber core and that the two are in close proximity.

7.11   Considering the LED of Problem 7.5, calculate:
   (a) the coupling efficiency and optical loss in decibels of coupling the emitted light into a step index fiber with an NA of 0.15, when the device is in close proximity to the fiber and is smaller than the fiber core;
   (b) the optical loss relative to the optical power generated internally if the device emits into a thin air gap before light is coupled into the fiber.

7.12   Estimate the optical power coupled into a step index fiber of 50 µm core diameter with an NA of 0.18 from a DH surface emitter with an emission area diameter of 75 µm and a radiance of 60 W sr^{-1} cm^{-2}. The Fresnel reflection at the index-matched semiconductor–fiber interface may be considered negligible.

   Further, determine the optical loss when coupling light into the fiber relative to the power emitted by the device into air if the Fresnel reflection at the semiconductor–air interface is 30%.

7.13   Comment on the differences in the performance characteristics between the conventional LEDs used for optical fiber communications and superluminescent LEDs.

Describe, with the aid of a diagram, the structure of an SLD used for operation in the longer wavelength region and suggest potential application areas for such devices.

7.14   The Fresnel reflection coefficient at a fiber core of refractive index n_1 is given approximately from the classical Fresnel formulas by:

$$r = \left[\frac{n_1 - n}{n_1 + n}\right]^2$$

where n is the refractive index of the surrounding medium.
   (a) Estimate the optical loss due to Fresnel reflection at a fiber core from GaAs each of which have refractive indices of 1.5 and 3.6 respectively.
   (b) Calculate the optical power coupled into a step index fiber of 200 μm core diameter with an NA of 0.3 from a GaAs surface-emitting LED with an emission diameter of 90 μm and a radiance of 40 W sr^{-1} cm^{-2}. Comment on the result.
   (c) Estimate the optical power emitted into air for the device in (b).

7.15   Determine the overall power conversion efficiency for the LED in Problem 7.14 if it is operating with a drive current of 100 mA and a forward voltage of 1.9 V.

7.16   Describe what is meant by an isotropic light-emitting diode. Sketch the optical output power intensity against operating current for an isotropic light-emitting diode, comparing the characteristic with those exhibited by conventional LEDs and a resonant cavity light-emitting diode (RC-LED). Hence explain the performance attributes of the RC-LED.

7.17   Define the term resonance cavity in relation to LEDs and discuss the different structures for realizing resonant cavity light-emitting diodes. Explain how these device structures differ from those of conventional LEDs.

7.18   Explain the operation of a quantum-dot LED and describe a strategy to enhance the quantum efficiency of a nonresonant cavity quantum-dot LED.

7.19   Discuss lens coupling of LEDs to optical fibers and outline the various techniques employed.

7.20   Describe the relationship between the electrical and optical modulation bandwidths for an optical fiber communication system. Estimate the 3 dB optical bandwidth corresponding to a 3 dB electrical bandwidth of 50 MHz. A Gaussian frequency response may be assumed.

7.21   Determine the optical modulation bandwidth for the LED of Problem 7.14 if the device emits 840 μW of optical power into air when modulated at a frequency of 150 MHz.

7.22   Estimate the electrical modulation bandwidth for an LED with a carrier recombination lifetime of 8 ns. The frequency response of the device may be assumed to be Gaussian.

7.23 Discuss the reliability of LEDs in comparison with injection lasers.

Estimate the CW operating lifetime for an AlGaAs LED with an activation energy of 0.6 eV and a constant of proportionality (β_0) of 2.3×10^3 h^{-1} when the junction temperature of the device is constant at 50 °C. It may be assumed that the LED is no longer useful when its optical output power is 0.8 of its original value.

7.24 What is meant by the intensity modulation of an optical source? Give reasons for the major present use of direct intensity modulation of semiconductor optical sources and comment on possible alternatives.

Answers to numerical problems

7.3	1.31 μm	**7.12**	0.12 mW, 16.9 dB
7.4	9.8×10^{-3} m s^{-1}	**7.14**	(a) 0.81 dB
7.5	0.4%		(b) 600 μW
7.6	0.70		(c) 5.44 mW
7.7	230 μW, 1.15%	**7.15**	0.32%
7.8	97.2 mW	**7.20**	70.7 MHz
7.10	12.3 dB	**7.21**	40.6 MHz
7.11	(a) 16.7 dB	**7.22**	24.4 MHz
	(b) 35.2 dB	**7.23**	2.21×10^5 hours

References

[1] D. N. Christodoulides, L. A. Reith and M. A. Saifi, 'Theory of LED coupling to single-mode fibers', *J. Lightwave Technol.*, **LT-5**(11), pp. 1623–1629, 1987.

[2] L. Hafskjaer and A. S. V. Sudbo, 'Attenuation and bit-rate limitations in LED/single-mode fiber transmission systems', *J. Lightwave Technol.*, **6**(12), pp. 1793–1797, 1988.

[3] T. P. Lee and A. G. Dentai, 'Power and modulation bandwidth of GaAs–AlGaAs high radiance LEDs for optical communication systems', *IEEE J. Quantum Electron.*, **QE-14**(3), pp. 150–156, 1978.

[4] G. Keiser, *Optical Fiber Communications* (3rd edn), McGraw-Hill, 1999.

[5] H. Kressel, 'Electroluminescent sources for fiber systems', in M. K. Barnoski (Ed.), *Fundamentals of Optical Fiber Communications*, pp. 109–141, Academic Press, 1976.

[6] R. C. Goodfellow and R. Davis, 'Optical source devices', in M. J. Howes and D. V. Morgan (Eds), *Optical Fibre Communications*, pp. 27–106, Wiley, 1980.

[7] J. P. Wittke, M. Ettenburg and H. Kressel, 'High radiance LED for single fiber optical links', *RCA Rev.*, **37**(2), pp. 160–183, 1976.

[8] T. G. Giallorenzi, 'Optical communications research and technology: fiber optics', *Proc. IEEE*, **66**, pp. 744–780, 1978.

[9] A. A. Bergh and P. J. Dean, *Light-Emitting Diodes*, Oxford University Press, 1976.

[10] C. A. Burrus and R. W. Dawson, 'Small area high-current density GaAs electroluminescent diodes and a method of operation for improved degradation characteristics', *Appl. Phys. Lett.*, **17**(3), pp. 97–99, 1970.

[11] C. A. Burrus and B. I. Miller, 'Small-area double heterostructure aluminum-gallium arsenide electroluminescent diode sources for optical fiber transmission lines', *Opt. Commun.*, **4**, pp. 307–369, 1971.

[12] T. P. Lee, 'Recent developments in light emitting diodes for optical fiber communication systems', *Proc. SPIE*, **224**, pp. 92–101, 1980.

[13] T. P. Lee, C. A. Burrus Jr and R. H. Saul, 'Light-emitting diodes for telecommunications', in S. E. Miller and I. P. Kaminow (Eds), *Optical Fiber Telecommunications II*, pp. 467–507, Academic Press, 1988.

[14] T. Uji and J. Hayashi, 'High-power single-mode optical-fiber coupling to InGaAsP 1.3 µm mesa-structure surface emitting LEDs', *Electron. Lett.*, **21**(10), pp. 418–419, 1985.

[15] D. Gloge, 'LED design for fibre system', *Electron. Lett.*, **13**(4), pp. 399–400, 1977.

[16] D. Marcuse, 'LED fundamentals: comparison of front and edge emitting diodes', *IEEE J. Quantum Electron.*, **QE-13**(10), pp. 819–827, 1977.

[17] D. Botez and M. Ettenburg, 'Comparison of surface and edge emitting LEDs for use in fiber-optical communications', *IEEE Trans. Electron Devices*, **ED-26**(3), pp. 1230–1238, 1979.

[18] M. Ettenburg, H. Kressel and J. P. Wittke, 'Very high radiance edge-emitting LED', *IEEE J. Quantum Electron.*, **QE-12**(6), pp. 360–364, 1979.

[19] D. H. Newman, M. R. Matthews and I. Garrett, 'Sources for optical fiber communications', *Telecommun. J. (Eng. Ed.) Switzerland*, **48**(2), pp. 673–680, 1981.

[20] D. H. Newman and S. Ritchie, 'Sources and detectors for optical fiber communications applications: the first 20 years', *IEE Proc., Optoelectron.*, **133**(3), pp. 213–228, 1986.

[21] S. Fujita, J. Hayashi, Y. Isoda, T. Uji and M. Shikada, '2 Gbit/s and 600 Mbit/s single-mode fibre transmission experiments using a high speed Zn-doped 1.3 µm edge-emitting LED', *Electron. Lett.*, **13**(12), pp. 636–637, 1987.

[22] D. M. Fye, 'Low-current 1.3 µm edge-emitting LED for single-mode fiber subscriber loop applications', *J. Lightwave Technol.*, **LT-4**(10), pp, 1546–1551, 1986.

[23] T. Ohtsuka, N. Fujimoto, K. Yamaguchi, A. Taniguchi, N. Naitou and Y. Nabeshima, 'Gigabit single-mode fiber transmission using 1.3 µm edge-emitting LEDs for broadband subscriber loops', *J. Lightwave Technol.*, **LT-5**(10), pp. 1534–1541, 1987.

[24] S. Takahashi, K. Goto, T. Shiba, K. Yoshida, E. Omura, H. Namizaki and W. Susaki, 'High-coupled-power high-speed 1.3 µm edge-emitting LED with buried crescent structure on p-InP substrate', *Tech. Dig. Optical Fiber Communication Conf., OFC'88*, USA, paper WB5, January 1988.

[25] Y. Sakakibara, H. Higuchi, E. Oomura, Y. Nakajima, Y. Yamamoto, K. Goto, H. Namizaki, K. Ikeda and W. Susaki, 'High-power 1.3 µm InGaAsP p-substrate buried cresent lasers', *J. Lightwave Technol.*, **LT-3** (5), pp. 978–984, 1985.

[26] T. P. Lee, C. A. Burrus and B. I. Miller, 'A stripe-geometry double-heterostructure amplified-spontaneous-emission (superluminescent) diode', *IEEE J. Quantum Electron.*, **QE-9**, p. 820, 1973.

[27] I. P. Kaminow, G. E. Eisenstein, L. W. Stulz and A. G. Dentai, 'Lateral confinement InGaAsP superluminescent diode at 1.3 µm', *IEEE J. Quantum Electron.*, **QE-19**(1), pp. 78–82, 1983.

[28] Y. Kashima, M. Kobayashi and T. Takano, 'High output power GaInAsP/InP superluminescent diode at 1.3 µm', *Electron. Lett.*, **24**(24), pp. 1507–1508, 1988.

[29] G. Arnold, H. Gottsman, O. Krumpholz, E. Schlosser and E. A. Schurr, '1.3 µm. edge-emitting diodes launching 250 µW into single-mode fiber at 100 mA', *Electron. Lett.*, **21**(21), pp. 993–994, 1985.

[30] M. Rossetti, A. Markus, A. Fiore, L. Occhi and C. Velez, 'Quantum dot superluminescent diodes emitting at 1.3 µm', *IEEE Photonics Technol. Lett.*, **17**(3), pp. 540–542, 2005.

[31] L. Burrow, F. Causa and J. Sarma, '1.3-W ripple-free superluminescent diode', *IEEE Photonics Technol. Lett.*, **17**(10), pp. 2035–2037, 2005.

[32] DL-CS Series Superluminescent LEDs Catalog, http://www.denselight.com/SLED_DL_CS_catalogv1.htm, 29 February 2008.

[33] R. M. Measures, *Structural Monitoring with Fiber Optic Technology*, Elsevier, 2001.

[34] J. Potfajova, J. M. Sun, S. Winnerl, T. Dekorsy, W. Skorupa, B. Schmidt, M. Helm, S. Mantl and U. Breuer, 'Silicon-based electrically driven microcavity LED', *Electron. Lett.*, **40**(14), pp. 904–906, 2004.

[35] D. Danae, S. Carl, M. Ingrid, D. Peter Van and G. B. Roel, 'Electrically pumped grating-assisted resonant-cavity light-emitting diodes', *Proc. SPIE*, **4641**, pp. 42–49, 2002.

[36] D. Delbeke, R. Bockstaele, P. Bienstman, R. Baets and H. Benisty, 'High-efficiency semiconductor resonant-cavity light-emitting diodes: a review', *IEEE J. Sel. Top. in Quantum Electron.*, **8**(2), pp. 189–206, 2002.

[37] S. B. Constant, T. J. C. Hosea, L. Toikkanen, I. Hirvonen and M. V. Pessa, 'Accurate methods for study of light emission from quantum wells confined in a microcavity', *IEEE J. Quantum Electron.*, **38**(8), pp. 1031–1038, 2002.

[38] N. E. J. Hunt, E. F. Schubert, R. A. Logan and G. J. Zydzik, 'Extremely narrow spectral widths from resonant cavity light-emitting diodes (RCLEDs) suitable for wavelength-division multiplexing at 1.3 μm and 1.55 μm', *IEEE Int. Electron Devices Meet.*, San Francisco, USA, pp. 651–654, 13–16 December 1992.

[39] X. Jin, S.-Q. Yu, Y. Cao, D. Ding, J.-B. Wang, N. Samal, Y. Sadofyev, S. R. Johnson and Y.-H. Zhang, '1.3 μm GaAsSb resonant-cavity light-emitting diodes grown on GaAs substrate', *IEEE/LEOS Lasers Electro-Optics Society 16th Annu. Meet.*, USA, Vol. 1, pp. 69–70, 26–30 October 2003.

[40] L. Krestnikov, N. A. Maleev, A. V. Sakharov, A. R. Kovsh, A. E. Zhukov, A. F. Tsatsul'nikov, V. M. Ustinov, Z. I. Alferov, N. N. Ledentsov, D. Bimberg and J. A. Lott, '1.3 μm resonant-cavity InGaAs/GaAs quantum dot light-emitting devices', *J. Semicond. Sci. Technol.*, **16**(10), pp. 844–848, 2001.

[41] M. Pessa, M. Guina, M. Dumitrescu, I. Hirvonen, M. Saarinen, L. Toikkanen and N. Xiang, 'Resonant cavity light emitting diode for a polymer optical fibre system', *J. Semicond. Sci. Technol.*, **17**(6), pp. R1–R9, 2002.

[42] B. Depreter, I. Moerman, R. Baets, P. Van Daele and P. Demeester, 'InP-based 1300 nm microcavity LEDs with 9% quantum efficiency', *Electron. Lett.*, **36**(15), pp. 1303–1304, 2000.

[43] J. W. Gray, D. Childs, S. Malik, P. Siverns, C. Roberts, P. N. Stavrinou, M. Whitehead, R. Murray and G. Parry, 'Quantum dot resonant cavity light emitting diode operating near 1300 nm', *Electron. Lett.*, **35**(3), pp. 242–243, 1999.

[44] P. Bienstman and R. Baets, 'The RC²LED: a novel resonant-cavity LED design using a symmetric resonant cavity in the out coupling reflector', *IEEE J. Quantum Electron.*, **36**(6), pp. 669–673, 2000.

[45] E. F. Schubert, N. E. J. Hunt, R. J. Malik, M. Micovic and D. L. Miller, 'Temperature and modulation characteristics of resonant-cavity light-emitting diodes', *J. Lightwave Technol.*, **14**(7), pp. 1721–1729, 1996.

[46] A. Fiore, U. Oesterle, R. P. Stanley and M. Ilegems, 'High-efficiency light emitting diodes at 1.3 μm using InAs-InGaAs quantum dots', *IEEE Photonics Technol. Lett.*, **12**(12), pp. 1601–1603, 2000.

[47] M. Kicherer, A. Fiore, U. Oesterle, R. P. Stanley, M. Ilegems and R. Michalzik, 'Data transmission using GaAs-based InAs-InGaAs quantum dot LEDs emitting at 1.3 μm wavelength', *Electron. Lett.*, **38**(16), pp. 906–907, 2002.

[48] M. Francardi, A. Gerardino, L. Balet, N. Chauvin, D. Bitauld, C. Zinoni, L. H. Li, B. Alloing, N. Le Thomas, R. Houdré and A. Fiore, 'Towards a LED based on a photonic crystal nanocavity for single photon sources at telecom wavelength', *Microelectron. Eng.*, doi:/0.1016/j.mee.2007.12.063, 14 May, 2008.

[49] F. K. Yam and Z. Hassan, 'Innovative advances in LED technology', *Microelectron. J.*, **36**(2), pp. 129–137, 2005.

[50] R. Windisch, A. Knobloch, M. Kuijk, C. Rooman, B. Dutta, P. Kiesel, G. Borghs, G. H. Dohler and P. Heremans, 'Large-signal-modulation of high-efficiency light-emitting diodes for optical communication', *IEEE J. Quantum Electron.*, **36**(12), pp. 1445–1453, 2000.

[51] M. Abe, I. Umebu, O. Hasegawa, S. Yamakoshi, T. Yamaoka, T. Kotani, H. Okada and H. Takamashi, 'Highly efficient long lived GaAlAs LEDs for fiber-optical communications', *IEEE Trans. Electron Devices*, **ED-24**(7), pp. 990–994, 1977.

[52] R. A. Abram, R. W. Allen and R. C. Goodfellow, 'The coupling of light emitting diodes to optical fibres using sphere lenses', *J. Appl. Phys.*, **46**(8), pp. 3468–3474, 1975.

[53] O. Wada, S. Yamakoshi, A. Masayuki, Y. Nishitani and T. Sakurai, 'High, radiance InGaAsP/InP lensed LEDs for optical communication systems at 1.2–1.3 μm, *IEEE J. Quantum Electron.*, **QE-17**(2), pp. 174–178, 1981.

[54] R. C. Goodfellow, A. C. Carter, I. Griffith and R. R. Bradley, 'GaInAsP/InP fast, high radiance, 1.05–1.3 μm wavelength LEDs with efficient lens coupling to small numerical aperture silica optical fibers', *IEEE Trans. Electron Devices*, **ED-26**(8), pp. 1215–1220, 1979.

[55] J. Ure, A. C. Carter, R. C. Goodfellow and M. Harding, 'High power lens coupled 1.3 μm edge-emitting LED for long haul 140 Mb/s fiber optics systems', *IEEE Specialist Conf. on Light Emitting Diodes and Photodetectors*, Canada, paper 20, p. 204, 1982.

[56] R. H. Saul, W. C. King, N. A. Olsson, C. L. Zipfel, B. H. Chin, A. K. Chin, I. Camlibel and G. Minneci, '180 Mbit/s, 35 km transmission over single-mode fiber using 1.3 μm edge-emitting LEDs', *Electron. Lett.*, **21**(17), pp. 773–775, 1985.

[57] E. F. Schubert, *Light-Emitting Diodes*, Cambridge University Press, 2006.

[58] E.-H. Park, M.-J. Kim and Y.-S. Kwon, 'Microlens for efficient coupling between LED and optical fiber', *IEEE Photonics Technol. Lett.*, **11**(4), pp. 439–441, 1999.

[59] J. Straus, 'Linearized transmitters for analog fiber links', *Laser Focus*, **14**(10), pp. 54–61, 1978.

[60] K. Streubel, N. Linder, R. Wirth and A. Jaeger, 'High brightness AlGaInP light-emitting diodes', *IEEE J. Sel. Top. Quantum Electron.*, **8**(2), pp. 321–332, 2002.

[61] G. Blume, T. J. C. Hosea, S. J. Sweeney, P. de Mierry and D. Lancefield, 'AlGaInN resonant-cavity LED devices studied by electromodulated reflectance and carrier lifetime techniques', *IEE Proc., Optoelectron.*, **152**(2), pp. 118–124, 2005.

[62] R. Joray, M. Ilegems, R. P. Stanley, W. Schmid, R. Butendeich, R. Wirth, A. Jaeger and K. Streubel, 'Far-field radiation pattern of red emitting thin-film resonant cavity LEDs', *IEEE Photonics Technol. Lett.*, **18**(9), pp. 1052–1054, 2006.

[63] A. C. Carter, 'Light-emitting diodes for optical fibre systems', *Radio Electron. Eng. J. IERE*, **51**(7/8), pp. 341–348, 1981.

[64] H. Tempkin, C. L. Zipfel, M. A. DiGiuseppe, A. K. Chin, V. G. Keramides and 'R. H. Saul, 'InGaAsP LEDs for 1.3 μm optical transmission', *Bell Syst. Tech. J.*, **62**(1), pp. 1–24, 1983.

[65] I. Garrett and J. E. Midwinter, 'Optical communication systems', in M. J. Howes and D. V. Morgan (Eds), *Optical Fibre Communications*, pp. 251–300, Wiley, 1980.

[66] H. Kressel and J. K. Butler, *Semiconductor Lasers and Heterojunction LEDs*, Academic Press, 1977.

[67] G. F. Neumark, I. L. Kuskovsky and H. Jiang, *Wide Bandgap Light Emitting Materials and Devices*, Wiley-VCH, 2007.

[68] H. F. Lockwood, J. P. Wittke and M. Ettenburg, 'LED for high data rate, optical communications', *Opt. Commun.*, **16**, p. 193, 1976.

[69] T. P. Lee, 'Recent development in light emitting diodes (LEDs) for optical fiber communications systems', *Proc. SPIE*, **340**, pp. 22–31, 1982.

[70] R. H. Saul, 'Recent advances in the performance and reliability of InGaAsP LEDs for lightwave communication systems', *IEEE Trans. Electron Devices*, **ED-30**(4), pp. 285–295, 1983.

[71] T. P. Lee and A. G. Dentai, 'Power and modulation bandwidth of GaAs–AlGaAs high radiance LEDs for optical communication systems', *IEEE J. Quantum Electron.*, **QE-14**, pp. 150–159, 1978.

[72] J. Heinen, W. Huber and W. Harth, 'Light-emitting diodes with a modulation bandwidth of more than 1 GHz', *Electron. Lett.*, **12**, p. 533, 1976.

[73] W. C. King, B. H. Chin, I. Camlibel and E. L. Zipfel, 'High-speed high-power 1.3 µm InGaAsP/InP surface emitting LEDs for short-haul wide-bandwidth optical fiber communications', *IEEE Electron Device Lett.*, **EDL-6**, p. 335, 1985.

[74] A. Suzuki, Y. Inomoto, J. Hayashi, Y. Isoda, T. Uji and H. Nomura, 'Gbit/s modulation of heavily Zn-doped surface-emitting InGaAsP/InP DH LED', *Electron. Lett.*, 20, p. 274, 1984.

[75] N. K. Dutta and C. L. Zipfel, 'Reliability of lasers and LEDs', in S. E. Miller and I. P. Kaminow (Eds), *Optical Fiber Telecommunications II*, pp. 671–687, Academic Press, 1988.

[76] S. Yamakoshi, A. Masayuki, O. Wada, S. Komiya and T. Sakurai, 'Reliability of high radiance InGaAsP/InP LEDs operating in the 1.2–1.3 µm wavelength', *IEEE J. Quantum Electron.*, **QE-17**(2), pp. 167–173, 1981.

[77] S. Yamakoshi, T. Sugahara, O. Hasegawa, Y. Toyama and H. Takanashi, 'Growth mechanism of <100> dark-line defects in high radiance GaAlAs LEDs', *Int. Electronic Devices Mtg*, pp. 642–645, 1978.

[78] C. L. Zipfel, A. K. Chin, V. G. Keramidas and R. H. Saul, 'Reliability of DH $Ga_{1-x}Al_xAs$ LEDs for lightwave communications', *Proc. 19th Annu., IEEE Int. Reliability Physics Symp.*, pp. 124–129, 1981.

[79] A. Suzuki, T. Uji, Y. Inomoto, J. Hayashi, Y. Isoda and H. Nomura, 'InGaAsP/InP 1.3 µm wavelength surface-emitting LED's for high-speed short-haul optical communication systems', *J. Lightwave Technol.*, **LT-3**(6), pp. 1217–1222, 1985.

[80] A. Mahapatra and E. J. Murphy, 'Electrooptic modulators', in I. P. Kaminow and T. Li (Eds), *Optical Fiber Telecommunications IVA: Components*, pp. 258–294, Academic Press, 2002.

[81] M. Doi, M. Sugiyama, K. Tanaka and M. Kawai, 'Advanced $LiNbO_3$ optical modulators for broadband optical communications', *IEEE J. Sel. Top. Quantum Electron.*, **12**(4), pp. 745–750, 2006.

[82] P. J. Winzer and R. J. Essiambre, 'Advanced modulation formats for high-capacity optical transport networks', *J. Lightwave Technol.*, **24**(12), pp. 4711–4728, 2006.

[83] P. M. Krummrich, 'Advanced modulation formats for more robust optical transmission systems', *AEU-Int. J. Electron. Commun.*, **61**(3), pp. 141–146, 2007.

CHAPTER 8

Optical detectors

8.1 Introduction

We are concerned in this chapter with photodetectors currently in use and under investigation for optical fiber communications.

The detector is an essential component of an optical fiber communication system and is one of the crucial elements which dictate the overall system performance. Its function is to convert the received optical signal into an electrical signal, which is then amplified before further processing. Therefore when considering signal attenuation along the link, the system performance is determined at the detector. Improvement of detector characteristics and performance thus allows the installation of fewer repeater stations and lowers both the capital investment and maintenance costs.

The role the detector plays demands that it must satisfy very stringent requirements for performance and compatibility. The following criteria define the important performance and compatibility requirements for detectors which are generally similar to the requirements for sources.

1. *High sensitivity at the operating wavelengths.* The first-generation systems have wavelengths between 0.8 and 0.9 μm (compatible with AlGaAs laser and LED emission lines). However, considerable advantage may be gained at the detector from second-generation sources with operating wavelengths above 1.1 μm as both fiber attenuation and dispersion are reduced. There is much research activity at present in this longer wavelength region, especially concerning wavelengths around 1.3 μm where attenuation and material dispersion can be minimized. In this case semiconductor materials are currently under investigation (see Section 8.4.3) in order to achieve good sensitivity at normal operating temperatures (i.e. 300 K).

2. *High fidelity.* To reproduce the received signal waveform with fidelity, for analogy transmission the response of the photodetector must be linear with regard to the optical signal over a wide range.

3. *Large electrical response to the received optical signal.* The photodetector should produce a maximum electrical signal for a given amount of optical power; that is, the quantum efficiency should be high.

4. *Short response time to obtain a suitable bandwidth.* Current single-channel, single-mode fiber systems extend up to many tens of gigahertz. However, it is apparent that future wavelength division multiplexed systems (see Section 12.9.4) will operate in the multiple terahertz (10^{12} Hz) range, and possibly above.

5. *A minimum noise introduced by the detector.* Dark currents, leakage currents and shunt conductance must be low. Also the gain mechanism within either the detector or associated circuitry must be of low noise.

6. *Stability of performance characteristics.* Ideally, the performance characteristics of the detector should be independent of changes in ambient conditions. However, the detectors currently favored (photodiodes) have characteristics (sensitivity, noise, internal gain) which vary with temperature, and therefore compensation for temperature effects is often necessary.

7. *Small size.* The physical size of the detector must be small for efficient coupling to the fiber and to allow easy packaging with the following electronics.

8. *Low bias voltages.* Ideally the detector should not require excessive bias voltages or currents.

9. *High reliability*. The detector must be capable of continuous stable operation at room temperature for many years.

10. *Low cost*. Economic considerations are often of prime importance in any large-scale communication system application.

We continue the discussion in Section 8.2 by briefly indicating the various types of device which could be employed for optical detection. From this discussion it is clear that semiconductor photodiodes currently provide the best solution for detection in optical fiber communications. Therefore, in Sections 8.3 and 8.4 we consider the principles of operation of these devices, together with the characteristics of the semiconductor materials employed in their construction. Sections 8.5 to 8.7 then briefly outline the major operating parameters (quantum efficiency, responsivity, long-wavelength cutoff) of such photodiodes. Then, in Sections 8.8 and 8.9, we discuss the structure, operation and performance characteristics of the major device types (*p–n*, *p–i–n* and avalanche photodiodes) for optical detection over the wavelength range 0.8 to 1.6 μm. Then in Section 8.10 developments associated with photodiodes for mid-infrared and far-infrared detection including quantum-dot photodetectors are considered prior to discussion in Sections 8.11 and 8.12 of other semiconductor devices (heterojunction phototransistors and metal–semiconductor–metal photodetectors which are finding wider use as detectors for optical fiber communications.

8.2 Device types

To detect optical radiation (photons) in the near-infrared region of the spectrum, both external and internal photoemission of electrons may be utilized. External photoemission devices typified by photomultiplier tubes and vacuum photodiodes meet some of the performance criteria but are too bulky, and require high voltages for operation. However, internal photoemission devices, especially semiconductor photodiodes with or without internal (avalanche) gain, provide good performance and compatibility with relatively low cost. These photodiodes are made from semiconductors such as silicon, germanium and an increasing number of III–V alloys, all of which satisfy in various ways most of the detector requirements. They are therefore used in all major current optical fiber communication systems.

The internal photoemission process may take place in both intrinsic and extrinsic semiconductors. With intrinsic absorption, the received photons excite electrons from the valence to the conduction bands in the semiconductor, whereas extrinsic absorption involves impurity centers created within the material. However, for fast response coupled with efficient absorption of photons, the intrinsic absorption process is preferred and at present all detectors for optical fiber communications use intrinsic photodetection.

Silicon photodiodes [Refs 1–3] have high sensitivity over the 0.8–0.9 μm wavelength band with adequate speed (tens of gigahertz), negligible shunt conductance, low dark current and long-term stability. They are therefore widely used in first-generation systems and are currently commercially available. Their usefulness is limited to the first-generation wavelength region as silicon has an indirect bandgap energy (see Section 8.4.1) of 1.14 eV giving a loss in response above 1.09 μm. Thus for second-generation systems in the longer wavelength range 1.1 to 1.6 μm research is devoted to the investigation of semiconductor materials which have narrower bandgaps. Interest has focused on germanium and III–V alloys which give a good response at the longer wavelengths. Again, the performance characteristics of such devices have improved considerably over recent years and a wide selection of III–V alloy photodiodes as well as germanium photodiodes are now commercially available.

In addition to the development of advanced photodiode structures fabricated from III–V semiconductor alloys for operation at wavelengths of 1.3 and 1.55 μm, similar material systems are under investigation for use at the even longer wavelengths required for mid-infrared and far-infrared transmission (2 to 12 μm). Interest has also been maintained in other semiconductor detector types, namely the heterojunction phototransistor and the photoconductive detector, both of which can be usefully fabricated from III–V alloy material systems. In particular, the latter device type has more recently found favor as a potential detector over the 1.1 to 1.6 μm wavelength range. Nevertheless, at present the primary operating wavelength regions remain 0.8 to 0.9 μm, 1.3 μm and 1.55 μm, with the major device types being the p–i–n and avalanche photodiodes. We shall therefore consider these devices in greater detail before discussing mid-infrared photodiodes, phototransistors and photoconductive detectors.

8.3 Optical detection principles

The basic detection process in an intrinsic absorber is illustrated in Figure 8.1 which shows a p–n photodiode. This device is reverse biased and the electric field developed across the p–n junction sweeps mobile carriers (holes and electrons) to their respective majority sides (p- and n-type material). A depletion region or layer is therefore created on either side of the junction. This barrier has the effect of stopping the majority carriers crossing the junction in the opposite direction to the field. However, the field accelerates minority carriers from both sides to the opposite side of the junction, forming the reverse leakage current of the diode. Thus intrinsic conditions are created in the depletion region.

A photon incident in or near the depletion region of this device which has an energy greater than or equal to the bandgap energy E_g of the fabricating material (i.e. $hf \geq E_g$) will excite an electron from the valence band into the conduction band. This process leaves an empty hole in the valence band and is known as the photogeneration of an electron–hole (carrier) pair, as shown in Figure 8.1(a). Carrier pairs so generated near the junction are separated and swept (drift) under the influence of the electric field to produce a displacement by current in the external circuit in excess of any reverse leakage current (Figure 8.1(b)). Photogeneration and the separation of a carrier pair in the depletion region of this reverse-biased p–n junction is illustrated in Figure 8.1 (c).

The depletion region must be sufficiently thick to allow a large fraction of the incident light to be absorbed in order to achieve maximum carrier pair generation. However, since long carrier drift times in the depletion region restrict the speed of operation of the photodiode it is necessary to limit its width. Thus there is a trade-off between the number of photons absorbed (sensitivity) and the speed of response.

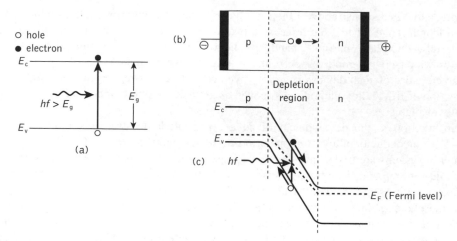

Figure 8.1 Operation of the $p–n$ photodiode: (a) photogeneration of an electron–hole pair in an intrinsic semiconductor; (b) the structure of the reverse-biased $p–n$ junction illustrating carrier drift in the depletion region; (c) the energy band diagram of the reverse-biased $p–n$ junction showing photogeneration and the subsequent separation of an electron–hole pair

8.4 Absorption

8.4.1 Absorption coefficient

The absorption of photons in a photodiode to produce carrier pairs and thus a photocurrent is dependent on the absorption coefficient α_0 of the light in the semiconductor used to fabricate the device. At a specific wavelength and assuming only bandgap transitions (i.e. intrinsic absorber) the photocurrent I_p produced by incident light of optical power P_0 is given by [Ref. 4]:

$$I_p = \frac{P_0 e(1-r)}{hf}\,[1 - \exp(-\alpha_0 d)] \tag{8.1}$$

where e is the charge on an electron, r is the Fresnel reflection coefficient at the semiconductor–air interface and d is the width of the absorption region.

The absorption coefficients of semiconductor materials are strongly dependent on wavelength. This is illustrated for some common semiconductors [Ref. 4] in Figure 8.2. It may be observed that there is a variation between the absorption curves for the materials shown and that they are each suitable for different wavelength applications. This results from their differing bandgaps energies, as shown in Table 8.1. However, it must be noted that the curves depicted in Figure 8.2 also vary with temperature.

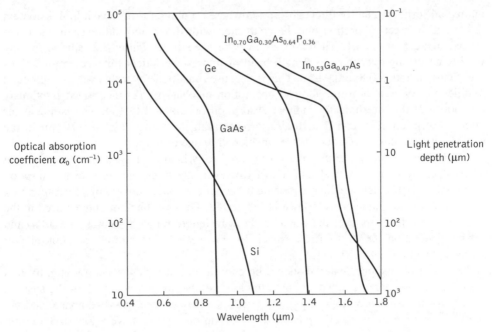

Figure 8.2 Optical absorption curves for some common semiconductor photodiode materials (silicon, germanium, gallium arsenide, indium gallium arsenide and indium gallium arsenide phosphide)

8.4.2 Direct and indirect absorption: silicon and germanium

Table 8.1 indicates that silicon and germanium absorb light by both direct and indirect optical transitions. Indirect absorption requires the assistance of a photon so that momentum as well as energy are conserved. This makes the transition probability less likely for

Table 8.1 Bandgaps for some semiconductor photodiode materials at 300 K

| | Bandgap (eV) at 300 K | |
	Indirect	*Direct*
Si	1.14	4.10
Ge	0.67	0.81
GaAs	–	1.43
InAs	–	0.35
InP	–	1.35
GaSb	–	0.73
$In_{0.53}Ga_{0.47}As$	–	0.75
$In_{0.14}Ga_{0.86}As$	–	1.15
$GaAs_{0.88}Sb_{0.12}$	–	1.15

indirect absorption than for direct absorption where no photon is involved. In this context direct and indirect absorption may be contrasted with direct and indirect emission discussed in Section 6.3.3.1. Therefore, as may be seen from Figure 8.2, silicon is only weakly absorbing over the wavelength band of interest in optical fiber communications (i.e. first-generation 0.8 to 0.9 μm). This is because transitions over this wavelength band in silicon are due only to the indirect absorption mechanism. As mentioned previously (Section 8.2) the threshold for indirect absorption occurs at 1.09 μm. The bandgap for direct absorption in silicon is 4.10 eV, corresponding to a threshold of 0.30 μm in the ultraviolet, and thus is well outside the wavelength range of interest.

Germanium is another semiconductor material for which the lowest energy absorption takes place by indirect optical transitions. However, the threshold for direct absorption occurs at 1.53 μm, below which germanium becomes strongly absorbing, corresponding to the kink in the characteristic shown in Figure 8.2. Thus germanium may be used in the fabrication of detectors over the whole of the wavelength range of interest (i.e. first- and second-generation 0.8 to 1.6 μm), especially considering that indirect absorption will occur up to a threshold of 1.85 μm.

Ideally, a photodiode material should be chosen with a bandgap energy slightly less than the photon energy corresponding to the longest operating wavelength of the system. This gives a sufficiently high absorption coefficient to ensure a good response, and yet limits the number of thermally generated carriers in order to achieve a low dark current (i.e. displacement current generated with no incident light (see Figure 8.5)). Germanium photodiodes have relatively large dark currents due to their narrow bandgaps in comparison with other semiconductor materials. This is a major disadvantage with the use of germanium photodiodes, especially at shorter wavelengths (below 1.1 μm).

8.4.3 III–V alloys

The drawback with germanium as a fabricating material for semiconductor photodiodes has led to increased investigation of direct bandgap III–V alloys for the longer wavelength region. These materials are potentially superior to germanium because their bandgaps can be tailored to the desired wavelength by changing the relative concentrations of their constituents, resulting in lower dark currents. They may also be fabricated in heterojunction structures (see Section 6.3.5) which enhances their high-speed operations.

Ternary alloys such as InGaAs and GaAlSb deposited on InP and GaSb substrates, respectively, have been used to fabricate photodiodes for the longer wavelength band. Although difficulties were experienced in the growth of these alloys, with lattice matching causing increased dark currents, these problems have now been reduced. In particular the alloy $In_{0.53}Ga_{0.47}As$ lattice matched to InP, which responds to wavelengths up to around 1.7 μm (see Figure 8.2), has been extensively utilized in the fabrication of photodiodes for operation at both 1.3 and 1.55 μm. Quaternary alloys can also be used for detection at these wavelengths. Both InGaAsP grown on InP and GaAlAsSb grown on GaSb have been studied, with the former material system finding significant application within advanced photodiode structures.

8.5 Quantum efficiency

The quantum efficiency η is defined as the fraction of incident photons which are absorbed by the photodetector and generate electrons which are collected at the detector terminals:

$$\eta = \frac{\text{number of electrons collected}}{\text{number of incident photons}} \tag{8.2}$$

Hence:

$$\eta = \frac{r_e}{r_p} \tag{8.3}$$

where r_p is the incident photon rate (photons per second) and r_e is the corresponding electron rate (electrons per second).

One of the major factors which determines the quantum efficiency is the absorption coefficient (see Section 8.4.1) of the semiconductor material used within the photodetector. The quantum efficiency is generally less than unity as not all of the incident photons are absorbed to create electron–hole pairs. Furthermore, it should be noted that it is often quoted as a percentage (e.g. a quantum efficiency of 75% is equivalent to 75 electrons collected per 100 incident photons). Finally, in common with the absorption coefficient, the quantum efficiency is a function of the photon wavelength and must therefore only be quoted for a specific wavelength.

8.6 Responsivity

The expression for quantum efficiency does not involve photon energy and therefore the responsivity R is often of more use when characterizing the performance of a photodetector. It is defined as:

$$R = \frac{I_p}{P_o} (\text{A W}^{-1}) \tag{8.4}$$

where I_p is the output photocurrent in amperes and P_o is the incident optical power in watts (i.e. output optical power from the fiber). The responsivity is a useful parameter as it gives the transfer characteristic of the detector (i.e. photocurrent per unit incident optical power).

The relationship for responsivity (Eq. (8.4)) may be developed to include quantum efficiency as follows. Considering Eq. (6.1) the energy of a photon $E = hf$. Thus the incident photon rate r_p may be written in terms of incident optical power and the photon energy as:

$$r_p = \frac{P_o}{hf} \tag{8.5}$$

In Eq. (8.3) the electron rate is given by:

$$r_e = \eta r_p \qquad (8.6)$$

Substituting from Eq. (8.5) we obtain:

$$r_e = \frac{\eta P_o}{hf} \qquad (8.7)$$

Therefore, the output photocurrent is:

$$I_p = \frac{\eta P_o e}{hf} \qquad (8.8)$$

where e is the charge on an electron. Thus from Eq. (8.4) the responsivity may be written as:

$$R = \frac{\eta e}{hf} \qquad (8.9)$$

Equation (8.9) is a useful relationship for responsivity which may be developed a stage further to include the wavelength of the incident light.

The frequency f of the incident photons is related to their wavelength λ and the velocity of light in air c, by:

$$f = \frac{c}{\lambda} \qquad (8.10)$$

Substituting into Eq. (8.9) a final expression for the responsivity is given by:

$$R = \frac{\eta e \lambda}{hc} \qquad (8.11)$$

It may be noted that the responsivity is directly proportional to the quantum efficiency at a particular wavelength.

The ideal responsivity against wavelength characteristic for a silicon photodiode with unit quantum efficiency is illustrated in Figure 8.3(a). Also shown is the typical responsivity of a practical silicon device. Figure 8.3(b), however, compares the responsivities and quantum efficiencies of the photodiodes based on silicon, germanium and the InGaAs ternary alloy. It shows the lower values of responsivity of 0.45 and 0.65 A W^{-1} at signal wavelengths of 0.90 μm and 1.30 μm, respectively, for silicon and germanium photodiodes. High responsivity values of 0.9 and 1.0 A W^{-1} at signal wavelengths of 1.30 μm

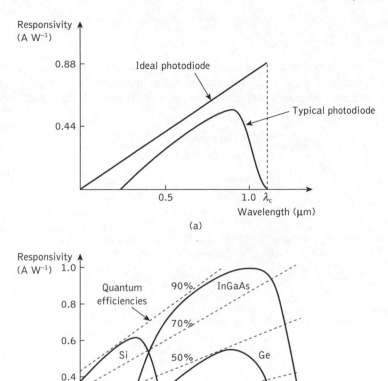

Figure 8.3 Responsivity against wavelength characteristics: (a) an ideal silicon photodiode with a typical device also shown; (b) silicon, germanium and InGaAs photodiodes with quantum efficiencies also shown. After Ref. 5. G. Keiser, *Optical Communications Essentials (Telecommunications)*, © 2003 with permission from The McGraw Hill Companies.

and 1.55 μm, respectively, for the photodiode fabricated from InGaAs alloy can also be observed. Moreover nearly 90% quantum efficiencies can be obtained for both the InGaAs and silicon photodiodes [Ref. 5]. It should also be noted that the responsivity drops rapidly at the cutoff wavelength for each of the photodiode materials. This factor is in accordance with Eq. (8.11) which provides the quantum efficiency as a function of signal wavelength which is critically dependent on the photodiode material bandgap energy. For a particular material, as the wavelength of the incident photon becomes longer the photon energy eventually is less than the energy required to excite an electron from the valance band to the conduction band and at this point the responsivity falls to zero.

Example 8.1

When 3×10^{11} photons each with a wavelength of 0.85 μm are incident on a photodiode, on average 1.2×10^{11} electrons are collected at the terminals of the device. Determine the quantum efficiency and the responsivity of the photodiode at 0.85 μm.

Solution: From Eq. (8.2):

$$\text{Quantum efficiency} = \frac{\text{number of electrons collected}}{\text{number of incident photons}}$$

$$= \frac{1.2 \times 10^{11}}{3 \times 10^{11}}$$

$$= 0.4$$

The quantum efficiency of the photodiode at 0.85 μm is 40%.

From Eq. (8.11):

$$\text{Responsivity } R = \frac{\eta e \lambda}{hc}$$

$$= \frac{0.4 \times 1.602 \times 10^{-19} \times 0.85 \times 10^{-6}}{6.626 \times 10^{-34} \times 2.998 \times 10^{8}}$$

$$= 0.274 \text{ A W}^{-1}$$

The responsivity of the photodiode at 0.85 μm is 0.27 A W^{-1}.

Example 8.2

A photodiode has a quantum efficiency of 65% when photons of energy 1.5×10^{-19} J are incident upon it.

(a) At what wavelength is the photodiode operating?

(b) Calculate the incident optical power required to obtain a photocurrent of 2.5 μA when the photodiode is operating as described above.

Solution: (a) From Eq. (6.1), the photon energy $E = hf = hc/\lambda$. Therefore:

$$\lambda = \frac{hc}{E} = \frac{6.626 \times 10^{-34} \times 2.998 \times 10^{8}}{1.5 \times 10^{-19}}$$

$$= 1.32 \text{ μm}$$

The photodiode is operating at a wavelength of 1.32 μm.

(b) From Eq. (8.9):

$$\text{Responsivity } R = \frac{\eta e}{hf} = \frac{0.65 \times 1.602 \times 10^{-19}}{1.5 \times 10^{-19}}$$

$$= 0.694 \text{ A W}^{-1}$$

Also from Eq. (8.4):

$$R = \frac{I_p}{P_o}$$

Therefore:

$$P_o = \frac{25 \times 10^{-6}}{0.694} = 3.60 \ \mu\text{W}$$

The incident optical power required is $3.60 \ \mu\text{W}$.

8.7 Long-wavelength cutoff

It is essential when considering the intrinsic absorption process that the energy of incident photons be greater than or equal to the bandgap energy E_g of the material used to fabricate the photodetector. Therefore, the photon energy:

$$\frac{hc}{\lambda} \geq E_g \qquad\qquad (8.12)$$

giving:

$$\lambda \leq \frac{hc}{E_g} \qquad\qquad (8.13)$$

Thus the threshold for detection, commonly known as the long-wavelength cutoff point λ_c, is:

$$\lambda_c = \frac{hc}{E_g} \qquad\qquad (8.14)$$

The expression given in Eq. (8.14) allows the calculation of the longest wavelength of light to give photodetection for the various semiconductor materials used in the fabrication of detectors.

It is important to note that the above criterion is only applicable to intrinsic photo-detectors. Extrinsic photodetectors violate the expression given in Eq. (8.12), but are not currently used in optical fiber communications.

Example 8.3

GaAs has a bandgap energy of 1.43 eV at 300 K. Determine the wavelength above which an intrinsic photodetector fabricated from this material will cease to operate.

Solution: From Eq. (8.14), the long wavelength cutoff:

$$\lambda_c = \frac{hc}{E_g} = \frac{6.626 \times 10^{-34} \times 2.998 \times 10^8}{1.43 \times 1.602 \times 10^{-19}}$$

$$= 0.867 \ \mu m$$

The GaAs photodetector will cease to operate above 0.87 μm.

8.8 Semiconductor photodiodes without internal gain

Semiconductor photodiodes without internal gain generate a single electron–hole pair per absorbed photon. This mechanism was outlined in Section 8.3, and in order to understand the development of this type of photodiode it is now necessary to elaborate upon it.

8.8.1 The *p–n* photodiode

Figure 8.4 shows a reverse-biased *p–n* photodiode with both the depletion and diffusion regions. The depletion region is formed by immobile positively charged donor atoms in the *n*-type semiconductor material and immobile negatively charged acceptor atoms in the *p*-type material, when the mobile carriers are swept to their majority sides under the influence of the electric field. The width of the depletion region is therefore dependent upon the doping concentrations for a given applied reverse bias (i.e. the lower the doping, the wider the depletion region). For the interested reader, expressions for the depletion layer width are given in Ref. 6.

Photons may be absorbed in both the depletion and diffusion regions, as indicated by the absorption region in Figure 8.4. The absorption region's position and width depend upon the energy of the incident photons and on the material from which the photodiode is fabricated. Thus in the case of the weak absorption of photons, the absorption region may extend completely throughout the device. Electron–hole pairs are therefore generated in both the depletion and diffusion regions. In the depletion region the carrier pairs separate and drift under the influence of the electric field, whereas outside this region the hole diffuses towards the depletion region in order to be collected. The diffusion process is very slow compared with drift and thus limits the response of the photodiode (see Section 8.8.3).

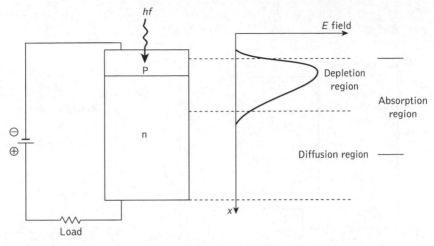

Figure 8.4 The *p–n* photodiode showing depletion and diffusion regions

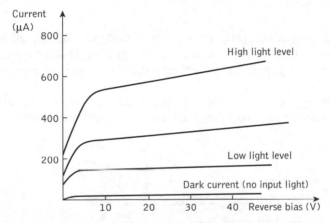

Figure 8.5 Typical *p–n* photodiode output characteristics

It is therefore important that the photons are absorbed in the depletion region. Thus it is made as long as possible by decreasing the doping in the *n*-type material. The depletion region width in a *p–n* photodiode is normally 1 to 3 μm and is optimized for the efficient detection of light at a given wavelength. For silicon devices this is in the visible spectrum (0.4 to 0.7 μm) and for germanium in the near infrared (0.7 to 0.9 μm).

Typical output characteristics for the reverse-biased *p–n* photodiode are illustrated in Figure 8.5. The different operating conditions may be noted moving from no light input to a high light level.

8.8.2 The *p–i–n* photodiode

In order to allow operation at longer wavelengths where the light penetrates more deeply into the semiconductor material, a wider depletion region is necessary. To achieve this the

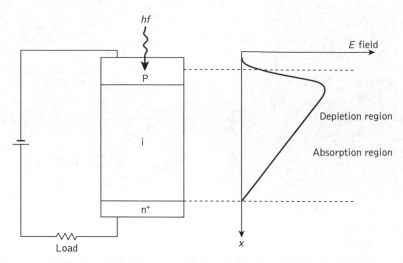

Figure 8.6 The *p–i–n* photodiode showing the combined absorption and depletion region

n-type material is doped so lightly that it can be considered intrinsic, and to make a low-resistance contact a highly doped *n*-type (n$^+$) layer is added. This creates a *p–i–n* (or PIN) structure, as may be seen in Figure 8.6 where all the absorption takes place in the depletion region.

Figure 8.7 shows the structures of two types of silicon *p–i–n* photodiode for operation in the shorter wavelength band below 1.09 μm. The front-illuminated photodiode, when

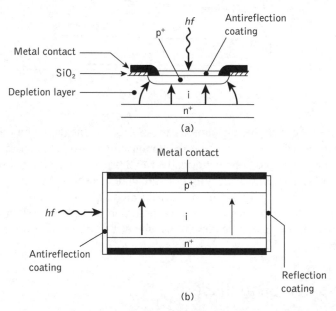

Figure 8.7 (a) Structure of a front-illuminated silicon *p–i–n* photodiode. (b) Structure of a side-illuminated (parallel to junction) *p–i–n* photodiode

operating in the 0.8 to 0.9 μm band (Figure 8.7(a)), requires a depletion region of between 20 and 50 μm in order to attain high quantum efficiency (typically 85%) together with fast response (less than 1 ns) and low dark current (1 nA). Dark current arises from surface leakage currents as well as generation–recombination currents in the depletion region in the absence of illumination. The side-illuminated structure (Figure 8.7(b)), where light is injected parallel to the junction plane, exhibits a large absorption width (≈500 μm) and hence is particularly sensitive at wavelengths close to the bandgap limit (1.09 μm) where the absorption coefficient is relatively small.

Germanium *p–i–n* photodiodes which span the entire wavelength range of interest are also commercially available, but as mentioned previously (Section 8.4.2) the relatively high dark currents are a problem (typically 100 nA at 20 °C increasing to 1 μA at 40 °C). However, as outlined in Section 8.4.3, III–V semiconductor alloys have been employed in the fabrication of longer wavelength region detectors. The favored material is the lattice-matched $In_{0.53}Ga_{0.47}As/InP$ system [Ref. 7] which can detect at wavelengths up to 1.67 μm. A typical planar device structure is shown in Figure 8.8(a) [Ref. 8] which requires epitaxial growth of several layers on an *n*-type InP substrate. The incident light is absorbed in the low-doped *n*-type InGaAs layer generating carriers, as illustrated in the energy band diagram Figure 8.8(b) [Ref. 9]. The discontinuity due to the homojunction

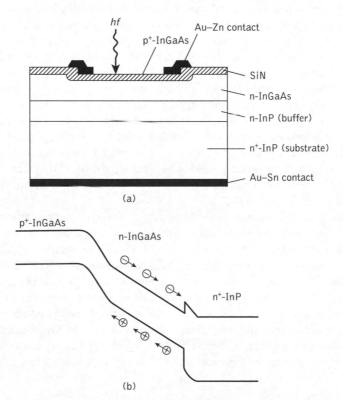

Figure 8.8 Planar InGaAs *p–i–n* photodiode: (a) structure; (b) energy band diagram showing homojunction associated with the conventional *p–i–n* structure

between the n$^+$-InP substrate and the n-InGaAs absorption region may be noted. This can be reduced by the incorporation of an *n*-type InP buffer layer.

The top entry* device shown in Figure 8.8(a) is the simplest structure, with the light being introduced through the upper p$^+$-layer. However, a drawback with this structure is a quantum efficiency penalty which results from optical absorption in the undepleted p$^+$-region. In addition, there is a limit to how small such a device can be fabricated as both light access and metallic contact are required on the top. To enable smaller devices with lower capacitances to be made, a substrate entry technique is employed. In this case light enters through a transparent InP substrate and the device area can be fabricated as small as may be practical for bonding.

Conventional growth techniques for III–V semiconductors can be employed to fabricate these devices, although liquid-phase epitaxy (LPE) tends to be preferred because of the relative ease in obtaining the low doping levels needed (around 10^5 cm^{-3}) to obtain low capacitance (less than 0.2 pF). However, LPE does not easily allow low-impurity-level concentrations and it is necessary to use long baking procedures over several days to purify the source material. High-quality devices have been produced using metal oxide vapor-phase epitaxy (MOVPE) [Ref. 10], a technique which appears much more appropriate for large-scale production of such devices.

A substrate entry† *p–i–n* photodiode is shown in Figure 8.9(a). This device incorporates a p$^+$-InGaAsP layer to provide a heterojunction structure (Schottky barrier) which improves quantum efficiency. Moreover, it is fabricated as a mesa structure which reduces parasitic capacitances [Ref. 11]. Unfortunately, charge trapping can occur at the n^-p^+-InGaAs/InGaAsP interface which may be observed in the energy band diagram of Figure 8.9(b). This may cause limitations in the response time of the device [Ref. 9]. However, small-area substrate entry devices can be produced with extremely low capacitance (less than 0.1 pF), quantum efficiency between 75% and 100% and dark currents less than 1 nA.

In both device types a depleted InGaAs layer of around 3 µm is used which provides high quantum efficiency and bandwidth. Furthermore, low doping permits full depletion of the InGaAs layer at low voltage (5 V). The short transit times in the relatively narrow depletion layers give a theoretical bandwidth of approximately 15 GHz. However, the bandwidth of commercially available packaged detectors is usually between 1 and 2 GHz due to limitations of the packaging.

A photodiode containing a waveguide structure, known as a mushroom waveguide, can, however, be used to overcome the bandwidth–quantum efficiency trade-off between the device capacitance and contact resistance [Ref. 12]. This structure, which is illustrated in Figure 8.10, comprises a thin layer of InGaAs (thickness of 0.20 µm) used as the absorption material which is lattice matched to an InP substrate thus providing operation at a wavelength of 1.55 µm. It may be observed that two graded layers of InGaAsP material, each having a thickness of 0.80 µm, are also employed above and below the absorption layer to avoid charge trapping. Since the device is side illuminated its quantum efficiency is therefore a function of the length of the absorption layer and also the thickness of this layer determines the amount of electron drift time. Thus a long and thin

---

* Top entry is also referred to as front illumination.
† Substrate entry is also referred to as back illumination.

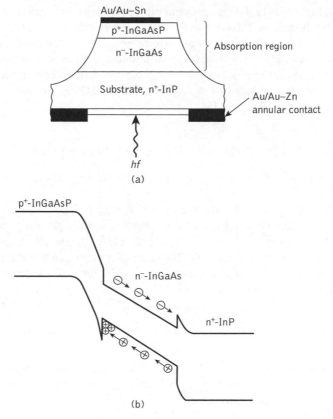

Figure 8.9 Substrate entry InGaAs *p–i–n* photodiode: (a) structure; (b) energy band diagram illustrating the heterojunction and charge trapping

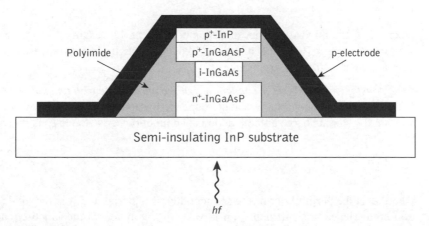

Figure 8.10 Structure of a mushroom waveguide photodiode

absorption layer provides both high quantum efficiency and fast response times [Ref. 12]. High-speed operation up to 110 GHz with 50% quantum efficiency using such structures been demonstrated [Refs 13, 14]. It should also be noted that in the mushroom waveguide structure the light and the carriers travel in different directions and therefore the device bandwidth and the quantum efficiency are not too dependent on each other. Hence quantum efficiencies of greater than 80% at a bandwidth of 10 GHz have been obtained using this waveguide structure [Refs 15, 16].

8.8.3 Speed of response and traveling-wave photodiodes

Three main factors limit the speed of response of a photodiode. These are [Ref. 16]:

1. *Drift time of carriers through the depletion region.* The speed of response of a photodiode is fundamentally limited by the time it takes photogenerated carriers to drift across the depletion region. When the field in the depletion region exceeds a saturation value, the carriers may be assumed to travel at a constant (maximum) drift velocity v_d. The longest transit time, t_{drift}, is for carriers which must traverse the full depletion layer width w and is given by:

$$t_{drift} = \frac{w}{v_d} \tag{8.15}$$

A field strength above 2×10^4 V cm^{-1} in silicon gives maximum (saturated) carrier velocities of approximately 10^7 cm s^{-1}. Thus the transit time through a depletion layer width of 10 μm is around 0.1 ns.

2. *Diffusion time of carriers generated outside the depletion region.* Carrier diffusion is a comparatively slow process where the time taken, t_{diff}, for carriers to diffuse a distance d may be written as:

$$t_{diff} = \frac{d^2}{2D_c} \tag{8.16}$$

where D_c is the minority carrier diffusion coefficient. For example, the hole diffusion time through 10 μm of silicon is 40 ns whereas the electron diffusion time over a similar distance is around 8 ns.

3. *Time constant incurred by the capacitance of the photodiode with its load.* A reverse-biased photodiode exhibits a voltage-dependent capacitance caused by the variation in the stored charge at the junction. The junction capacitance C_j is given by:

$$C_j = \frac{\varepsilon_s A}{w} \tag{8.17}$$

where ε_s is the permittivity of the semiconductor material and A is the diode junction area. Hence, a small depletion layer width w increases the junction capacitance. The capacitance of the photodiode C_d is that of the junction together with the capacitance of the leads and packaging. This capacitance must be minimized in order to reduce the *RC* time constant which also limits the detector response time (see Section 9.3.2).

Example 8.4

A silicon *p–i–n* photodiode has an intrinsic region with a width of 20 μm and a diameter of 500 μm in which the drift velocity of electrons is 10^5 m s^{-1}. When the permittivity of the device material is 10.5×10^{-13} F cm^{-1}, calculate: (a) the drift time of the carriers across the depletion region; (b) the junction capacitance of the photodiode.

Solution: (a) The drift time for the carriers across the depletion region for the photodiode can be obtained using Eq. (8.15) as:

$$t_{\text{drift}} = \frac{w}{v_\text{d}}$$

$$= \frac{20 \times 10^{-6}}{1 \times 10^5}$$

$$= 2 \times 10^{-10} \text{ s}$$

The drift time for the carriers across the depletion region is therefore 200 ps.
(b) The junction capacitance is given by Eq. (8.17) as:

$$C_\text{j} = \frac{\varepsilon_\text{s} A}{w}$$

where the area $A = \pi \times r^2 = 3.14 \times (500 \times 10^{-6})^2 = 0.79 \times 10^{-6}$ m^2. Therefore:

$$C_\text{j} = \frac{10.5 \times 10^{-13} \times 0.79 \times 10^{-6}}{20 \times 10^{-6}}$$

$$= 0.41 \times 10^{-13}$$

The photodiode has a junction capacitance of 4 pF.

Although all the above factors affect the response time of the photodiode, the ultimate bandwidth of the device is limited by the drift time of carriers through the depletion region t_{drift}. In this case, when assuming no carriers are generated outside the depletion region and that there is negligible junction capacitance, the maximum photodiode 3 dB bandwidth B_m is given by [Ref. 17]:

$$B_\text{m} = \frac{1}{2\pi t_{\text{drift}}} = \frac{v_\text{d}}{2\pi w} \tag{8.18}$$

Moreover, when there is no gain mechanism present within the device structure, the maximum possible quantum efficiency is 100%. Hence the value for the bandwidth given by Eq. (8.18) is also equivalent to the ultimate gain–bandwidth product for the photodiode.

Example 8.5

The carrier velocity in a silicon p–i–n photodiode with a 25 μm depletion layer width is 3×10^4 m s^{-1}. Determine the maximum response time for the device.

Solution: The maximum 3 dB bandwidth for the photodiode may be obtained from Eq. (8.18) where:

$$B_{\mathrm{m}} = \frac{\upsilon_{\mathrm{d}}}{2\pi w} = \frac{3 \times 10^4}{2\pi \times 25 \times 10^{-6}} = 1.91 \times 10^8 \text{ Hz}$$

The maximum response time for the device is therefore:

$$\text{Max. response time} = \frac{1}{B_{\mathrm{m}}} = 5.2 \text{ ns}$$

It must be noted, however, that the above response time takes no account of the diffusion of carriers in the photodiode or the capacitance associated with the device junction and the external connections.

The response of a photodiode to a rectangular optical input pulse for various device parameters is illustrated in Figure 8.11. Ideally, to obtain a high quantum efficiency for the photodiode the width of the depletion layer must be far greater than the reciprocal of the absorption coefficient (i.e. $1/\alpha_0$) for the material used to fabricate the detector so that most of the incident light will be absorbed. Hence the response to a rectangular input pulse of a low-capacitance photodiode meeting this condition, and exhibiting negligible diffusion outside the depletion region, is shown in Figure 8.11(a). It may be observed in this case that the rising and falling edges of the photodiode output follow the input pulse quite well. When the detector capacitance is larger, however, the speed of response becomes limited by the RC time constant of this capacitance and the load resistor associated with the receiver circuit (see Section 9.3.2), and thus the output pulse appears as illustrated in Figure 8.11(b).

Furthermore, when there is significant diffusion of carriers outside the depletion region, as is the case when the depletion layer is too narrow ($w \leq 1/\alpha_0$) and carriers are therefore created by absorption outside this region, then the output pulse displays a long tail caused by the diffusion component to the input optical pulse, as shown in Figure 8.11(c). Thus devices with very thin depletion layers have a tendency to exhibit distinctive fast response and slow response components to their output pulses, as may be observed in Figure 8.11(c), the former response resulting from absorption in the thin depletion layer.

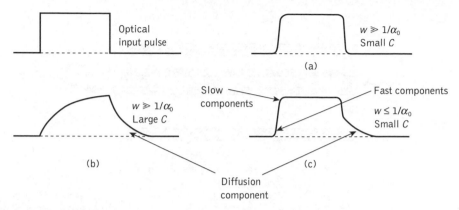

Figure 8.11 Photodiode responses to rectangular optical input pulses for various detector parameters

A recent approach to reduce the RC time constant limitation is to use a traveling-wave (TW) photodiode structure in which the absorption and carrier drift regions are positioned orthogonally to each other. Such a p–i–n photodiode is illustrated in Figure 8.12(a) where the photogenerated carriers are controlled by the electrical transmission lines and the absorption occurs in an optical waveguide that collects the photogenerated carriers [Ref. 18]. This approach distributes the capacitance along the electrical transmission lines which can be terminated with a matching impedance thus rendering the bandwidth independent of capacitance. However, a shortcoming of the structure is that both electrical and optical signals do not arrive at the same time due to their mismatched velocities.

Figure 8.12(b) represents a scheme to match the electrical and optical wave velocities. It consists of TW photodiodes and electrical transmission lines coupled to an optical waveguide. The waveguide geometry and its material composition determine the distribution of the incident power along the device. The group velocity of the optical traveling wave is fixed and therefore the only way to ensure that the velocity matching between optical and electrical waves can be achieved is by tuning the radio-frequency phase velocity (i.e. by varying the electrode dimensions). The same principle is implemented in the photodetector illustrated in Figure 8.12(c) [Ref. 18] where several photodiodes at regular intervals are produced above the waveguide, underneath the electrical transmission line. In this case the absorption occurs in a series of discrete photodiodes (i.e. instead of a single absorption layer) positioned periodically along the optical waveguide. This structure distributes the optical signal power into each high-speed photodiode which is then collected by bringing together the photocurrent from each photodiode on a low-loss electrical transmission line to reduce the microwave loss and to improve the output signal power. It is therefore possible to closely match the velocities of the optical and electrical waves using this generic structure [Refs 19, 20]. In addition, high-bandwidth performance up to 190 GHz has been demonstrated using a metal–semiconductor–metal traveling-wave (see Section 8.12) photodetector [Ref. 21].

A further TW p–i–n photodiode device, known as unitraveling carrier (UTC) structure, is illustrated in Figure 8.13. Although the operation of a UTC photodiode is similar to a conventional p–i–n photodiode, absorption occurs in a thin p-type layer instead of the

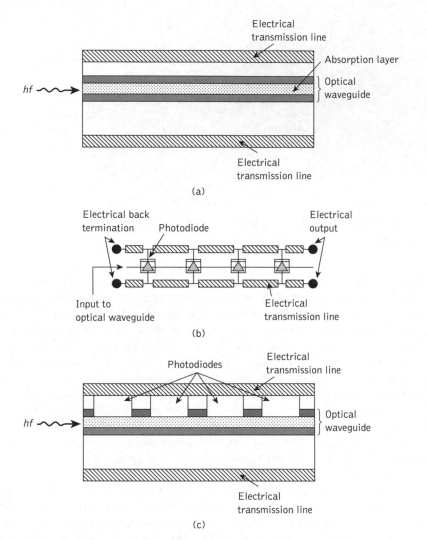

Figure 8.12 Traveling-wave photodiodes: (a) basic structure; (b) transmission line velocity matching scheme; (c) periodic traveling-wave photodiode

intrinsic i-region of the *p–i–n* photodiode. In the structure shown, based on the InGaAs/ InP material system, the photocarriers (i.e. electrons and holes pairs) are generated in the *p*-type absorption layer. However, when photons are absorbed to form the electron–hole pairs, the holes join these existing holes (instead of traveling) thus increasing the majority hole population. Hole carriers are replaced by electrons, which drift across the depletion region, and thus it is the electrons that generate the photocurrent. The bandwidth of the UTC photodiode is therefore determined by the electron diffusion time in the *p*-type absorption layer. When the absorption layer is thin the electrons can drift across it faster resulting in a higher bandwidth for the photodiode. Furthermore, using a small-gradient conduction band in the absorption layer can also speed up the diffusion time. Hence a

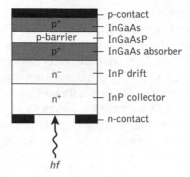

Figure 8.13 Unitraveling carrier (UTC) photodiode

photodiode structure with a 0.30 μm thin absorption layer has provided a bandwidth as large as 310 GHz [Ref. 22]. Similar photodiodes have demonstrated high transmission rates of 100 Gbit s^{-1} and 160 Gbit s^{-1} with an output peak voltage of 0.8 V [Ref. 23]. Moreover, such wideband photodiodes have also been produced for optical wireless communication and a monolithic UTC photodiode operating at wavelength of 1.55 μm was utilized for the purpose of generating a photonic CW signal transmitting over a bandwidth up to 1.5 THz [Ref. 24].

In order to improve the absorption efficiency of high-speed photodiodes, an optical resonance cavity similar to the Fabry–Pérot cavity can also be employed. When resonance occurs in this structure the incoming photons are reflected back and forth between the two mirror face reflectors. If a thin absorption layer is placed within this mirror cavity the absorption efficiency is enhanced due to multiple passes of photons and the resultant device is known as a resonant cavity enhanced (RCE) photodiode [Ref. 25]. A simple RCE photodiode in which the absorption layer is placed in between two reflector mirrors of InGaAs n- and p-type material is shown in Figure 8.14. Various approaches can be used to produce the top and bottom distributed Bragg reflector (DBR) mirrors comprising several alternating layers of low-index and high-index material. For example, InGaAs/InP [Ref. 26], AlGaAs/GaAs [Ref. 27], InAlGaAs/InAlAs or InGaAsP/InAlAs [Ref. 28]

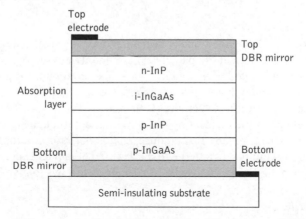

Figure 8.14 Schematic cross-section of a resonant cavity enhanced photodiode

material systems can be employed to construct DBR mirrors operating at the wavelengths of 1.30 and 1.55 μm [Ref. 29]. In addition, silicon-on-insulator technology (see Section 11.2) can also be used to fabricate DBR mirrors using germanium on silicon substrates for an RCE photodiode to operate at long wavelength. For example, RCE p–i–n photodiodes using germanium material on a double silicon-on-insulator substrate operating at the wavelength of 1.55 μm have exhibited high quantum efficiencies of 59% [Ref. 30]. These devices, which also provide a 3 dB bandwidth of around 13 GHz, are considered appropriate for reception at transmission rates of 10 Gbit s^{-1}. Furthermore, the small area of these devices (i.e. 10 to 70 μm) could prove useful for photonic integration (see Section 10.6).

8.8.4 Noise

The overall sensitivity of a photodiode results from the random current and voltage fluctuations which occur at the device output terminals in both the presence and absence of an incident optical signal. Although the factors that determine the sensitivity of the optical receiver are dealt with in Chapter 9, it is appropriate at this stage to consider the sources of noise that arise within photodiodes, which do not have an internal gain mechanism. The photodiode dark current mentioned in Section 8.8.2 corresponds to the level of the output photocurrent when there is no intended optical signal present. However, there may be some photogenerated current present due to background radiation entering the device.

The inherent dark current can be minimized through the use of high-quality, defect-free material which reduces the number of carriers generated in the depletion region as well as those which diffuse into this layer from the p^-- and n^+-regions. Moreover, the surface currents can be minimized by careful fabrication and surface passivation such that the surface state and impurity ion concentrations are reduced. Nevertheless, it is the case that the detector average current $\bar{I}$ always exhibits a random fluctuation about its mean value as a result of the statistical nature of the quantum detection process (see Section 9.2.3). This fluctuation is exhibited as shot noise [Ref. 31] where the mean square current variation $\overline{i_s^2}$ is proportional to $\bar{I}$ and the photodiode received bandwidth B. Thus the rms value of this shot noise current is:

$$(\overline{i_s^2})^{\frac{1}{2}} = (2eB\bar{I})^{\frac{1}{2}} \tag{8.19}$$

Various figures of merit have traditionally been employed to assess the noise performance of optical detectors. Although these parameters are not always appropriate for the evaluation of the high-speed photodiodes used in optical fiber communications, it is instructive to define those most commonly utilized. These are: the noise equivalent power (*NEP*); the detectivity (*D*); and the specific detectivity (*D**).

The *NEP* is defined as the incident optical power, at a particular wavelength or with a specified spectral content, required to produce a photodetector current equal to the rms noise current within a unit bandwidth (i.e. $B = 1$ Hz). To obtain an expression for the *NEP* at a specific wavelength, Eq. (8.8) must be rearranged as follows to give:

$$P_o = \frac{I_p hf}{\eta e} = \frac{I_p hc}{\eta e \lambda} \tag{8.20}$$

Then putting the photocurrent I_p equal to the rms shot noise current in Eq. (8.19) gives:

$$I_p = (2e\bar{I}B)^{\frac{1}{2}} \tag{8.21}$$

Moreover, the photodiode average current $\bar{I}$ may be represented by $(I_p + I_d)$ where I_d is the dark current within the device. Hence:

$$I_p = [2e(I_p + I_d)B]^{\frac{1}{2}} \tag{8.22}$$

When $I_p \gg I_d$, then:

$$I_p \simeq 2eB \tag{8.23}$$

Substituting Eq. (8.23) into Eq. (8.20) and putting $B = 1$ Hz gives the *NEP* as:

$$NEP = P_o \simeq \frac{2hc}{\eta\lambda} \tag{8.24}$$

It should be noted that the *NEP* for an ideal photodetector is given by Eq. (8.24) when the quantum efficiency $\eta = 1$.

When $I_p \ll I_d$, then from Eq. (8.22) the photocurrent becomes:

$$I_p \simeq [2eI_dB]^{\frac{1}{2}} \tag{8.25}$$

Hence for a photodiode in which the dark current noise is dominant, the use of Eq. (8.20) with $B = 1$ Hz gives an expression for the *NEP* of:

$$NEP = P_o \simeq \frac{hc(2eI_d)^{\frac{1}{2}}}{\eta e\lambda} \tag{8.26}$$

The detectivity D is defined as the inverse of the *NEP*. Thus:

$$D = \frac{1}{NEP} \tag{8.27}$$

Considering a photodiode receiving monochromatic radiation with the dark current as its dominant noise source, then from Eqs (8.26) and (8.27):

$$D = D_\lambda = \frac{\eta e\lambda}{hc(2eI_d)^{\frac{1}{2}}} \tag{8.28}$$

The specific detectivity D^* is a parameter which incorporates the area of the photodetector A in order to take account of the effect of this factor on the amplitude of the device dark current. This proves necessary when background radiation and thermal generation rather than surface conduction are the major causes of dark current. Therefore the specific detectivity is given by:

$$D^* = DA^{\frac{1}{2}} = \frac{\eta e \lambda}{hc(2eI_\mathrm{d}/A)^{\frac{1}{2}}} \tag{8.29}$$

It should be noted, however, that the above definition for D^* assumes a bandwidth of 1 Hz. Hence the specific detectivity over a bandwidth B would be equal to $D(AB)^{1/2}$.

Example 8.6

A germanium p–i–n photodiode with active dimensions of 100×50 μm has a quantum efficiency of 55% when operating at a wavelength of 1.3 μm. The measured dark current at this wavelength is 8 nA. Calculate the noise equivalent power and specific detectivity for the device. It may be assumed that dark current is the dominant noise source.

Solution: The noise equivalent power is given by Eq. (8.26) as:

$$NEP \simeq \frac{hc(2eI_\mathrm{d})^{\frac{1}{2}}}{\eta e \lambda}$$

$$= \frac{6.626 \times 10^{-34} \times 2.998 \times 10^{8}(2 \times 1.602 \times 10^{-19} \times 8 \times 10^{-9})^{\frac{1}{2}}}{0.55 \times 1.602 \times 10^{-19} \times 1.3 \times 10^{-6}}$$

$$= 8.78 \times 10^{-14}\ \mathrm{W}$$

Substituting for the detectivity D in Eq. (8.29) from Eq. (8.27) allows the specific detectivity to be written as:

$$D^* = \frac{A^{\frac{1}{2}}}{NEP} = \frac{(100 \times 10^{-6} \times 50 \times 10^{-6})^{\frac{1}{2}}}{8.78 \times 10^{-14}}$$

$$= 8.1 \times 10^{8}\ \mathrm{m\ Hz^{\frac{1}{2}}\ W^{-1}}$$

The above parameters are solely concerned with the noise performance of the photodiodes used within optical fiber communications. However, it is the noise associated with the optical receiver which also incorporates a load resistance and a preamplifier that is the major concern. This more general issue is dealt with in Chapter 9.

8.9 Semiconductor photodiodes with internal gain

8.9.1 Avalanche photodiodes

The second major type of optical communications detector is the avalanche photodiode (APD). This has a more sophisticated structure than the p–i–n photodiode in order to create

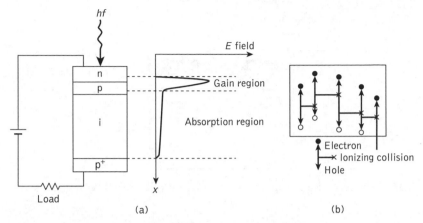

Figure 8.15 (a) Avalanche photodiode showing high electric field (gain) region. (b) Carrier pair multiplication in the gain region of an avalanche photodiode

an extremely high electric field region (approximately 3×10^5 V cm^{-1}), as may be seen in Figure 8.15(a). Therefore, as well as the depletion region where most of the photons are absorbed and the primary carrier pairs generated, there is a high-field region in which holes and electrons can acquire sufficient energy to excite new electron–hole pairs. This process is known as impact ionization and is the phenomenon that leads to avalanche breakdown in ordinary reverse-biased diodes. It often requires high reverse bias voltages (50 to 400 V) in order that the new carriers created by impact ionization can themselves produce additional carriers by the same mechanism as shown in Figure 8.15(b). More recently, however, it should be noted that devices which will operate at much lower bias voltages (15 to 25 V) have become available.

Carrier multiplication factors as great as 10^4 may be obtained using defect-free materials to ensure uniformity of carrier multiplication over the entire photosensitive area. However, other factors affect the achievement of high gain within the device. Microplasmas, which are small areas with lower breakdown voltages than the remainder of the junction, must be reduced through the selection of defect-free materials together with careful device processing and fabrication [Ref. 32]. In addition, excessive leakage at the junction edges can be eliminated by the use of a guard ring structure as shown in Figure 8.16. At present silicon, germanium and InGaAs APDs are generally available.

Operation of these devices at high speed requires full depletion in the absorption region. As indicated in Section 8.8.1, when carriers are generated in undepleted material, they are collected somewhat slowly by the diffusion process. This has the effect of producing a long 'diffusion tail' on a short optical pulse. When the APD is fully depleted by employing electric fields in excess of 10^4 V m^{-1}, all the carriers drift at saturation-limited velocities. In this case the response time for the device is limited by three factors. These are:

(a) the transit time of the carriers across the absorption region (i.e. the depletion width);

(b) the time taken by the carriers to perform the avalanche multiplication process; and

(c) the *RC* time constant incurred by the junction capacitance of the diode and its load.

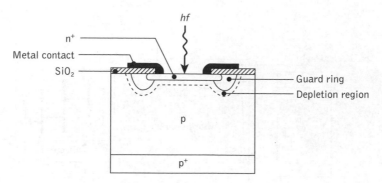

Figure 8.16 Structure of a silicon avalanche photodiode (APD) with guard ring

At low gain the transit time and *RC* effects dominate giving a definitive response time and hence constant bandwidth for the device. However, at high gain the avalanche build up time dominates and therefore the device bandwidth decreases proportionately with increasing gain. Such APD operation is distinguished by a constant gain–bandwidth product.

Often an asymmetric pulse shape is obtained from the APD which results from a relatively fast rise time as the electrons are collected and a fall time dictated by the transit time of the holes traveling at a slower speed. Hence, although the use of suitable materials and structures may give rise times between 150 and 200 ps, fall times of 1 ns or more are quite common and limit the overall response of the device.

8.9.2 Silicon reach through avalanche photodiodes

To ensure carrier multiplication without excess noise for a specific thickness of multiplication region within the APD it is necessary to reduce the ratio of the ionization coefficients for electrons and holes k (see Section 9.3.4). In silicon this ratio is a strong function of the electric field varying from around 0.1 at 3×10^5 V m^{-1} to 0.5 at 6×10^5 V m^{-1}. Hence for minimum noise, the electric field at avalanche breakdown must be as low as possible and the impact ionization should be initiated by electrons. To this end a 'reach through' structure has been implemented with the silicon APD. The silicon 'reach through' APD (RAPD) consists of p^+–π–p–n^+ layers as shown in Figure 8.17(a). As may be seen from the corresponding field plot in Figure 8.17(b), the high-field region where the avalanche multiplication takes place is relatively narrow and centered on the p–n^+ junction. Thus under low reverse bias most of the voltage is dropped across the p–n^+ junction.

When the reverse bias voltage is increased the depletion layer widens across the p-region until it 'reaches through' to the nearly intrinsic (lightly doped) π-region. Since the π-region is much wider than the p-region the field in the π-region is much lower than that at the p–n^+ junction (see Figure 8.17(b)). This has the effect of removing some of the excess applied voltage from the multiplication region to the π-region giving a relatively slow increase in multiplication factor with applied voltage. Although the field in the π-region is lower than in the multiplication region it is high enough (2×10^4 V cm^{-1}) when the photodiode is operating to sweep the carriers through to the multiplication region at their scattering limited velocity (10^7 cm s^{-1}). This limits the transit time and ensures a fast response (as short as 0.5 ns).

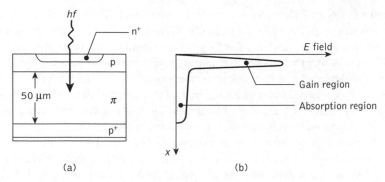

(a)   (b)

Figure 8.17 (a) Structure of a silicon RAPD. (b) The field distribution in the RAPD showing the gain region across the p–n^+ junction

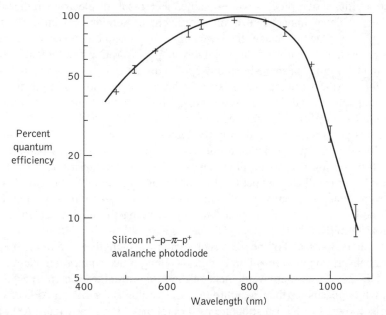

Figure 8.18 Measurement of quantum efficiency against wavelength for a silicon RAPD. After Ref. 33, from *The Bell System Technical Journal*. © 1978, AT&T

Measurements [Ref. 33] for a silicon RAPD for optical fiber communication applications at a wavelength of 0.825 μm have shown a quantum efficiency (without avalanche gain) of nearly 100% in the working region, as may be seen in Figure 8.18. The dark currents for this photodiode are also low and depend only slightly on bias voltage.

8.9.3 Germanium avalanche photodiodes

The elemental semiconductor germanium has been used to fabricate relatively sensitive and fast APDs that may be used over almost the entire wavelength range of primary interest at present (0.8 to 1.6 µm). However, it was clear from an early stage that higher dark currents together with larger excess noise factors (see Section 9.3.3) than those in silicon APDs were a problem with these devices. The large dark currents were associated with edge and surface effects resulting from difficulties in passivating germanium, and were also a direct consequence of the small energy bandgap as mentioned earlier in Section 8.4.2.

In the late 1970s when interest increased in the fabrication of detectors for longer wavelength operation (1.1 to 1.6 µm), germanium APDs using a conventional n^+p structure similar to the silicon APD shown in Figure 8.16 were produced [Ref. 34]. However, such devices exhibited dark currents near breakdown of between 100 nA and 300 nA which were very sensitive to temperature variations [Ref. 7]. Furthermore, unlike the situation with silicon APDs, these dark currents had significant components of both bulk (multiplied) and surface (unmultiplied) current. It was the multiplied component (typically 100 nA for the n^+p structure) which needed to be reduced (to around 1 nA) in order to provide low-noise operation. In addition, large excess noise factors associated with the avalanche multiplication process were obtained as a result of electrons rather than holes (which have a higher impact ionization coefficient in germanium) initiating the multiplication process. One advantage, however, of such germanium APDs over their silicon counterparts is that because of the relatively high absorption coefficient exhibited by germanium at 1.3 µm, avalanche breakdown voltages are quite low (typically 25 V).

Germanium APD structures have been fabricated to provide multiplication initiated by holes and thus to reduce the excess noise factor in the longer wavelength region. For example, an n^+np structure has been demonstrated [Ref. 35] which goes some way to achieving this performance by reducing the factor by some 30% on that obtained in n^+p devices. However, multiplied dark current around 1 µA was obtained when operating at a wavelength of 1.3 µm and a multiplication factor of 10. An alternative device providing similar results utilizes the p^+n structure [Ref. 36] shown in Figure 8.19(a). In this case dark currents were reduced to between 150 and 250 nA by using an ion-implanted technology [Ref. 37] and subsequently to around 5 nA by reducing the device sensitive area from 100 µm to 30 µm [Ref. 38].

Unfortunately, the speed of the p^+n structure at a wavelength of 1.5 µm is poor because most of the absorption in germanium at this wavelength takes place outside the depletion region.* This has led to the development of the p^+nn^- structure shown in Figure 8.19(b) [Ref. 39] which resembles the reach through structure used for silicon APDs (see Section 8.9.2). It is known as a Hi–Lo structure as it combines high bandwidth (700 MHz) with low multiplied dark current (33 nA) and good excess noise performance. However, the breakdown voltage is higher at +85 V and the unmultiplied dark currents are around 1 µA. Nevertheless, these Hi–Lo devices appear to be among the highest performance germanium APDs for longer wavelength operation and are only eclipsed by the emerging III–V alloy APDs which do not exhibit quite the same fundamental material limitations.

8.9.4 III–V alloy avalanche photodiodes

Due to the drawbacks with germanium APDs for longer wavelength operation, much effort has been expended in the study of III–V semiconductor alloys for the fabrication of

---

* The absorption length in germanium at a wavelength of 1.5 µm is 10 µm.

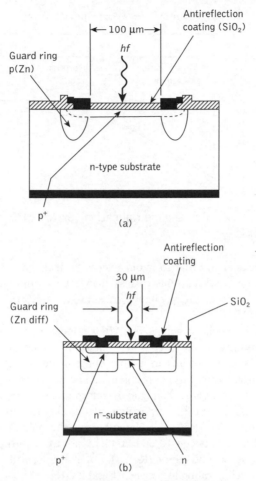

Figure 8.19 Germanium APDs: (a) p^+n structure. Reprinted with permission from Ref. 36 © IEEE 1980; (b) Hi-Lo (p^+nn^-) structure [Ref. 39]

APDs. In particular, the ternary InGaAs/InP and quaternary InGaAsP/InP material systems have been successfully employed. In common with the silicon reach through APD (see Section 8.9.2) separate absorption and multiplication regions are provided, as illustrated in Figure 8.20. This defines the so-called SAM APD which is a heterostructure device designed so that the multiplication takes place in the InP p–n junction [Ref. 40]. The performance of such long-wavelength APDs is limited, however, by the fundamental properties of the material systems.

A first limitation is related to the large tunneling currents associated with the narrow bandgap required for longer wavelength optical absorption. The band-to-band or defect-assisted tunneling currents become large before the electric field is high enough to obtain significant avalanche gain. This problem is substantially reduced using a separate absorption and multiplication region with the gain occurring at the InP p–n junction where the tunneling is much less [Ref. 7]. However, control over the doping and thickness of the

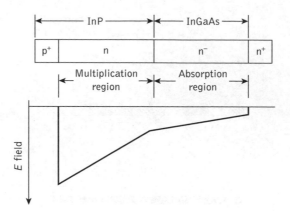

Figure 8.20 Separate absorption and multiplication (SAM) APD layer composition and electric field profile

n-type InP layer is critical in order to avoid excessive leakage current. Nevertheless, it is possible to obtain low dark currents of less than 10 nA (unmultiplied) together with quantum efficiencies of 80%, a capacitance of approximately 0.5 pF and an operating voltage of around 100 V.

A second limitation associated with SAM APDs concerns the trapping of holes in the valence band discontinuity at the InGaAs/InP heterointerface, as illustrated in Figure 8.21(a) [Ref. 41]. This factor results in a slow component of the photoresponse which causes a speed limitation. However, the problem can be alleviated by incorporating a thin grading layer of InGaAsP (whose bandgap is intermediate between InGaAs and InP) between these two layers (Figure 8.21(b)) to smooth out the discontinuity and thus provide improved speed performance [Ref. 42]. Nevertheless, the gain–bandwidth products for such devices are still only between 10 and 20 GHz, not quite sufficient for high bit rate systems in the gigabit per second region [Ref. 7]. For example, with a gain around 10 such devices will only provide operation to between 1 and 2 GHz.

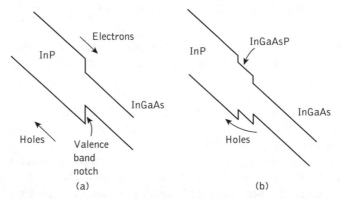

Figure 8.21 Energy band diagrams for SAM APDs: (a) InGaAs/InP heterojunction illustrating the notch in which holes may be trapped; (b) similar heterojunction to (a) with InGaAsP layer to reduce the effect of the notch

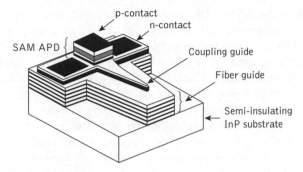

Figure 8.22 Three-dimensional schematic view of an asymmetric twin-waveguide SAM APD

A high-performance planar SAM APD based on an asymmetrical twin-waveguide technique (see Section 11.3) has demonstrated a high gain–bandwidth product greater than 150 GHz [Ref. 43]. The three-dimensional schematic view of this InGaAs/InAlAs-based device fabricated onto a semi-insulating InP substrate is depicted in Figure 8.22. It may be observed that the SAM APD structure incorporates a tapered fiber coupler which transfers the light from the fiber guide to the coupling guide sections. In this structure a layer of InAlAs material is used as the gain region in order to achieve high-bandwidth operation. The device is capable of handling a transmission rate of 40 Gbit s^{-1}. Moreover, it displayed an external quantum efficiency of around 48% at a signal wavelength of 1.55 μm while the measured value of responsivity at unity gain for the symmetrical twin-waveguide SAM APD structure remained 0.6 A W^{-1} [Ref. 43].

An improved technique for increasing the speed of response of the device is to provide several (two or three) InGaAsP buffer layers to create compositional grading at the heterojunction interface [Refs 17, 40]. This may be achieved by interposing a thin multiquantum-well (MQW) structure between the narrow and wideband gap layer. The configuration of a recent top-illuminated planar structure separate absorption, grading, charge and multiplication (SAGCM) InGaAs APD is shown in Figure 8.23. Although this structure also

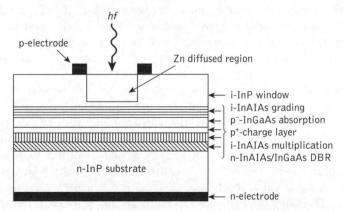

Figure 8.23 Separate absorption, grading, charge and multiplication (SAGCM) APD structure

has similarities with the SAM ADP, an additional charge layer has been incorporated between the absorption and multiplication layers in the active cavity [Ref. 44]. Figure 8.23 also contains a distributed Brogg reflector (DBR) formed from the InAlAs/InGaAs material system constituting a resonant cavity enhanced (RCE) structure in a similar manner to the RCE p–i–n photodiode (see Section 8.4). The charge layer contains an electric field to provide isolation between the absorption region and the multiplication layer. Since the charge layer separates the absorption and multiplication layers in the active cavity region, the photon absorption and the carrier multiplication processes are independent and can be optimized individually to improve both the device noise and speed performance. This device type has displayed a gain-bandwidth product of up to 120 GHz [Ref. 44] thus allowing operations at bandwidths of 10 GHz or higher. Similar RCE APD structures have demonstrated high gain–bandwidth products of around 300 GHz allowing operation at bandwidths of 24 GHz while exhibiting external quantum efficiencies around 70% [Refs 45, 46].

Overall, advanced photodiode developments are targeted at devices with improved sensitivities for operation at very high bandwidths, together with the fabrication of structures with improved functionality at low cost [Ref. 17]. Initial efforts to improve sensitivity focused upon semiconductor superlattices in the form of MQW structures [Ref. 47] and staircase APDs [Ref. 48]. Both of these APD structures had gain regions comprising MQWs formed by alternately growing thin layers of wide- and narrow-bandgap materials such as AlGaAs and GaAs respectively. By using materials exhibiting these properties the conduction and valence band discontinuities differed significantly, resulting in different ionization coefficients for electrons and holes. This factor should therefore give improvements in the noise performance of such III–V alloy APDs by reducing the ratio of the ionization coefficients for electrons and holes (k value, see Section 9.3.4) because the ionization coefficients of the two carrier types are normally approximately equal. In addition, the other major advantage in using MQW APDs results from their improved bandwidth capabilities caused by the reduction in avalanche buildup time provided by the multilayer structures.

The step-like MQW energy band structure where the discontinuity in the conduction band is greater than that in the valence band is shown in Figure 8.24. The structure, which is illustrated both unbiased and biased, could comprise about 100 layers of alternate wide- and narrow-bandgap semiconductors. Although such devices have been fabricated using the AlGaAs/GaAs material system by molecular beam epitaxy (MBE), the structure does not provide the same favorable k value reduction when using InP-based alloys for longer wavelength operation. In this case problems with tunneling in the high-field regions containing the narrow-bandgap layers tend to destroy the sensitivity improvement provided by the MQWs [Ref. 17].

The energy band structure of a more complex scheme known as a staircase APD is shown in Figure 8.25. In this technique a narrow-bandgap region is compositionally graded over a distance of 10 to 20 nm into a material with a minimum of twice the bandgap at the narrow end of the step. Again, the composition is abruptly changed to obtain the narrow bandgap as a second step is formed. The primary advantage of this staircase structure is that carrier multiplication caused by carrier transitions from the wide- to the narrow-bandgap material can occur at much lower electric field densities than that required with MQW devices. The majority of photodiodes fabricated using MQW bandgap engineering technology [Refs 49–51], however, have been designed for operation in the mid-infrared wavelength region (see Section 8.10).

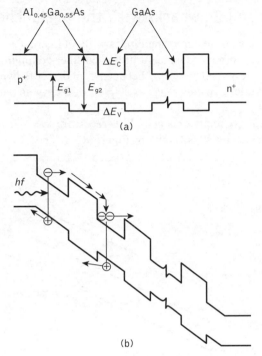

Figure 8.24 Energy band diagrams for MQW superlattice APD structure:
(a) unbiased, showing alternate layers of wide- and narrow-bandgap semiconductors;
(b) the biased device [Ref. 47]

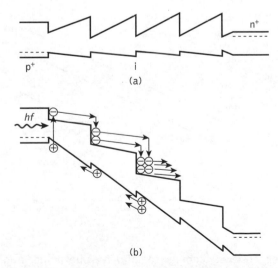

Figure 8.25 Energy band diagrams for the staircase APD: (a) the unbiased device;
(b) the biased device under normal operation [Ref. 48]

8.9.5 Benefits and drawbacks with the avalanche photodiode

APDs have a distinct advantage over photodiodes without internal gain for the detection of the very low light levels often encountered in optical fiber communications. They generally provide an increase in sensitivity of between 5 and 15 dB over *p–i–n* photodiodes while often giving a wider dynamic range as a result of their gain variation with response time and reverse bias.

The optimum sensitivity improvement of APD receivers over *p–i–n* photodiode devices is illustrated in the characteristics shown in Figure 8.26. The characteristics display the

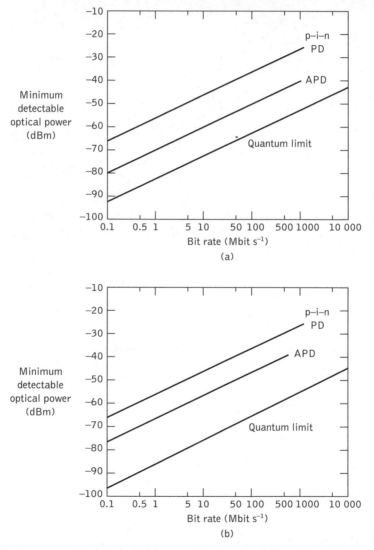

Figure 8.26 Receiver sensitivity comparison of *p–i–n* photodiode and APD devices at a bit-error-rate of 10^{-9}: (a) using silicon detectors operating at a wavelength of 0.82 μm; (b) using InGaAs detectors operating at a wavelength of 1.55 μm

minimum detectable optical power for direct detection (see Section 9.2.4) against the transmitted bit rate in order to maintain a bit-error-rate (BER) of 10^{-9} (see Section 12.6.3) in the shorter and longer wavelength regions. Figure 8.26(a) compares silicon photodiodes operating at a wavelength of 0.82 μm where the APD is able to approach within 10 to 13 dB of the quantum limit. In addition, it may be observed that the *p–i–n* photodiode receiver has a sensitivity around 15 dB below this level. InGaAs photodiodes operating at a wavelength of 1.55 μm are compared in Figure 8.26(b). In this case the APD requires around 20 dB more power than the quantum limit, whereas the *p–i–n* photodiode receiver is some 10 to 12 dB less sensitive than the APD. Finally, it should be noted that as a consequence of the rapid emergence of quantum crytography as a field within optical fiber communications, APDs for operation as single-photon-counting avalanche photodetectors (SPADs) have been receiving significant attention worldwide [Ref. 52].

Avalanche photodiodes, however, also have several drawbacks which include:

(a) fabrication difficulties due to their more complex structure and hence increased cost;

(b) the random nature of the gain mechanism which gives an additional noise contribution (see Section 9.3.3);

(c) the high bias voltages required particularly for silicon devices (150 to 400 V) which although lower for germanium and InGaAs APDs (20 to 40 V) are similarly wavelength dependent;

(d) the variation of the gain (multiplication factor) with temperature as shown in Figure 8.27 for a silicon RAPD [Ref. 33]; thus temperature compensation is necessary to stabilize the operation of the device.

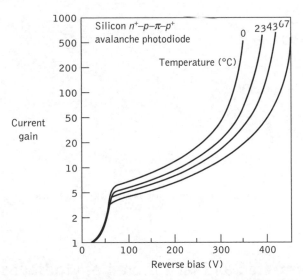

Figure 8.27 Current gain against reverse bias for a silicon RAPD operating at a wavelength of 0.825 μm. After Ref. 33, from *The Bell System Technical Journal*. © 1978, AT&T

8.9.6 Multiplication factor

The multiplication factor M is a measure of the internal gain provided by the APD. It is defined as:

$$M = \frac{I}{I_p} \tag{8.30}$$

where I is the total output current at the operating voltage (i.e. where carrier multiplication occurs) and I_p is the initial or primary photocurrent (i.e. before carrier multiplication occurs).

Example 8.7

The quantum efficiency of a particular silicon RAPD is 80% for the detection of radiation at a wavelength of 0.9 μm. When the incident optical power is 0.5 μW, the output current from the device (after avalanche gain) is 11 μA. Determine the multiplication factor of the photodiode under these conditions.

Solution: From Eq. (8.11), the responsivity:

$$R = \frac{\eta e \lambda}{hc} = \frac{0.8 \times 1.602 \times 10^{-19} \times 0.9 \times 10^{-6}}{6.626 \times 10^{-34} \times 2.998 \times 10^{8}}$$

$$= 0.581 \text{ A W}^{-1}$$

Also, from Eq. (8.4), the photocurrent:

$$I_p = P_o R$$
$$= 0.5 \times 10^{-6} \times 0.581$$
$$= 0.291 \text{ μA}$$

Finally, using Eq. (8. 30):

$$M = \frac{I}{I_p} = \frac{11 \times 10^{-6}}{0.291 \times 10^{-6}}$$

$$= 37.8$$

The multiplication factor of the photodiode is approximately 38.

8.10 Mid-infrared and far-infrared photodiodes

Developments of photodiodes for mid-infrared and far-infrared transmission systems have progressed alongside the activities concerned with fibers and sources in these wavelength regions (2 to 12 μm). Obtaining suitable lattice matching for III–V alloy materials has been a problem, however, when operating at wavelengths greater than 2 μm. For example, a lattice-matched InGaAsSb/GaSb material system was utilized in a p–i–n photodiode for high-speed operation at wavelengths up to 2.3 μm [Ref. 53].

An alternative approach which achieved some success was the use of indium alloys that, due to the high indium content for operation above 2 μm, were mismatched with respect to the InP substrate causing inherent problems of dislocation-induced junction leakage and low quantum efficiency [Ref. 54]. However, these problems were reduced by utilizing a compositionally graded buffer layer to accommodate the lattice mismatch. One technique involved the replacement of the conventional p–i–n homojunction with an InGaAs/AlInAs heterojunction in which a wider bandgap p-type AlInAs layer acted as a transparent window at long wavelengths to ensure that optical absorption occurred in a lightly doped n-type region of the device. This device, which exhibited a useful response out to a wavelength of 2.4 μm, displayed a quantum efficiency as high as 95% over the wavelength range 1.3 to 2.25 μm with dark currents as low as 35 nA [Ref. 54]. A similar approach was demonstrated with the ternary alloys $In_xGa_{1-x}As/InAs_yP_{1-y}$ to produce mesa structure photodiodes which operate at a wavelength of 2.55 μm [Ref. 55]. A compositionally graded region of InGaAs or InAsP accommodates the lattice mismatch between the ternary layers and the InP substrate. These devices, which have $In_{0.85}Ga_{0.15}As$ absorbing layers and lattice-matched $InAs_{0.68}P_{0.32}$ capping layers, displayed a quantum efficiency of 52% at a wavelength of 2.55 μm but with dark currents of between 10 and 20 μA. It has been suggested [Ref. 55] that these high dark currents were a result of electrically active defects associated with misfit dislocations in the active regions of the diode which could be reduced by improved materials growth.

In common with injection lasers, the HgCdTe material system has been utilized to fabricate long-wavelength photodiodes (see Section 6.11). $Hg_{1-x}Cd_xTe$ ternary alloys form a continuous family of semiconductors whose bandgap energy variation with x enables optical detection from 0. 8 μm to the far-infrared. Furthermore, with this material system the hole ionization coefficient exhibits a resonant characteristic which is a function of the alloy composition. This phenomenon is a band structure effect which is also displayed by the $Ga_{1-x}Al_xSb$ alloys [Ref. 55]. Resonant impact ionization processes in such materials yield a high ionization coefficient ratio* providing enhanced sensitivity at particular operating wavelengths. However, for the HgCdTe material system this occurs for a composition in the vicinity of $x = 0.7$, which is suitable for detection at a wavelength of 1.3 μm. Hence both HgCdTe p–i–n photodiodes and APDs that exhibit low sensitivity and high-speed response (500 MHz) have been produced for operation at this wavelength [Ref. 56]. In addition, $Hg_{0.4}Cd_{0.6}Te$ APDs have been fabricated for detection at a wavelength of 1.55 μm [Refs 57–60].

APDs based on HgCdTe are attractive for array applications in both the near and far infrared since their material properties possess uniform avalanche gain across an array where the variation in gain as a function of electric field is lower in HgCdTe as compared with silicon or InGaAs [Ref. 59]. Additionally, heteroepitaxial growth of HgCdTe on large-area silicon substrates is also possible which provides for the production of low-cost, large-format infrared APD arrays using molecular beam epitaxy. Recently, considerable

---

* In this case the ratio of hole ionization coefficient to electron ionization coefficient.

progress in the development of multiwavelength HgCdTe APD arrays has been demonstrated for military applications and nontelecommunication applications [Ref. 60].

8.10.1 Quantum-dot photodetectors

The detection mechanism in the quantum-dot photodetector relies on photoexcitation of electrons from confined states in conduction band quantum wells or dots. The process of photoexcitation of electrons in photodetectors is similar to that in the quantum-dot semiconductor optical amplifier (see Section 10.3). Hence several layers of quantum dots constitute the active cavity region of the device and the resultant device is also known as the dots-in-well (DWELL) structure as depicted in Figure 8.28 [Ref. 62]. Several layers of

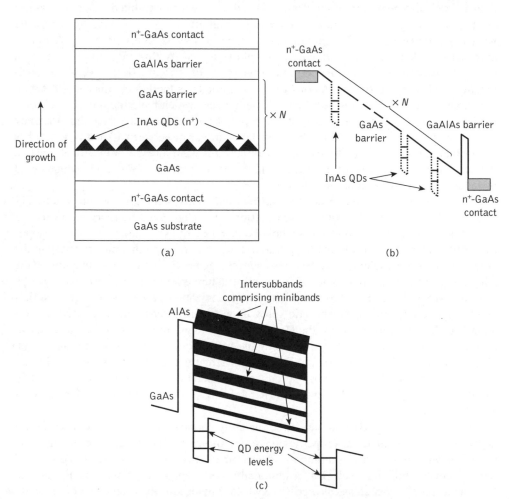

Figure 8.28 Quantum-dot photodetector: (a) the dots-in-well (DWELL) structure; (b) the conduction band profile under a bias condition; (c) the conduction band showing the formation of intersubbands

quantum dots (i.e. $N = 70$ in this particular case) are sandwiched between GaAs barrier layers to obtain an active cavity region for the purpose of photodetection as illustrated in Figure 8.28(a). These barriers isolate different layers and therefore increase the material absorption characteristics of the active cavity region. The barriers are usually kept wide enough to isolate quantum dots and therefore the quantum dots are not correlated between layers, which reduces any tunneling effects, thus helping to suppress dark current. Furthermore, although a large number of quantum dots are required to produce increased absorption, they should be aligned in the growth direction in order to maintain the optical coupling profile (i.e. edge- or surface-emitting operation). These structures can be produced using either GaAs or InGaAs and AlGaAs alloy material systems, whereas InAs or InAl may be used to fabricate quantum dots on GaAs or InP substrates, respectively [Refs 61–63]. Furthermore, the absorption spectrum of these material systems can be precisely tailored and hence the detection peak wavelength can be tuned from 3.0 to 25 μm and beyond, while the spectral width can be varied from half to several micrometers.

Figure 8.28(b) displays the conduction band profile of a DWELL photodetector. It constitutes some 70 layers of InAs quantum dots each followed by a GaAs barrier comprising a single quantum well until it is finally terminated with an GaAlAs barrier connecting with n^+-contacts of GaAs material. Figure 8.28(c) depicts the DWELL photodetector creating intersubband energy levels in which a stack of minibands are produced within a single energy band.* In the DWELL heterostructure photodetector shown an aluminum arsenide barrier layer is positioned beneath the quantum dots in addition to the GaAs barrier layer. Since aluminum possesses less mobility as compared with indium, it increases the optical absorption characteristics of the active cavity region and therefore it necessitates fewer layers of InAs quantum dots (in this case $N = 10$).

Quantum-dot photodetectors have also been shown to provide additional features of bias tunability for multiwavelength operation [Ref. 65]. A quantum-dot photodetector based on the InGaAs/InAs material system has demonstrated wavelength tunablity within the range of 5.5 to 5.9 μm with a responsivity of 3.5 to 3.0 mA W^{-1} at a temperature of 77 K [Ref. 66]. Germanium has also been used to fabricate quantum-dot photodetectors [Refs 67, 68]. Devices utilizing this material have demonstrated a response over the wavelength range from 2.2 to 3.1 μm [Ref. 69]. In addition, the dark current density of these devices at a temperature of 77 K remained around 6.4 mA cm^{-2} at a bias voltage of 1 V while the responsivity was maintained at around 3.0 mA W^{-1} [Ref. 70].

8.11 Phototransistors

The problems encountered with APDs for use in the longer wavelength region stimulated a renewed interest in bipolar phototransistors in the late 1970s. Hence, although these devices have been investigated for a number of years, they have yet to find use in major optical fiber communication systems. In common with the APD the phototransistor provides internal gain of the photocurrent. This is achieved through transistor action rather

---

* The detection mechanism in the DWELL photodetector may also include intersubband transitions (also known as type II transitions) which comprise minibands within a single energy band [Ref. 64].

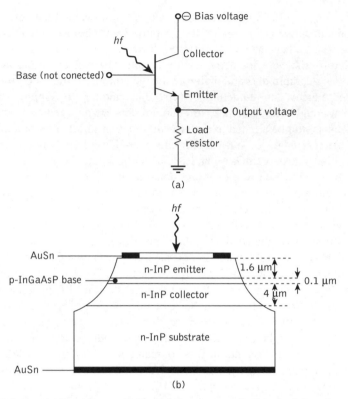

Figure 8.29 (a) Symbolic representation of the *n*–*p*–*n* phototransistor showing the external connections. (b) Cross-section of an *n*–*p*–*n* InGaAsP/InP heterojunction phototransistor. Reused with permission from Ref. 71 © 1980, American Institute of Physics

than avalanche multiplication. A symbolic representation of the *n*–*p*–*n* bipolar phototransistor is shown in Figure 8.29(a). It differs from the conventional bipolar transistor in that the base is unconnected, the base–collector junction being photosensitive to act as a light-gathering element. Thus absorbed light affects the base current giving multiplication of primary photocurrent through the device.

The structure of an *n*–*p*–*n* InGaAsP/InP heterojunction phototransistor is shown in Figure 8.29(b) [Ref. 71]. The three-layer heterostructure (see Section 6.3.5) is grown on an InP substrate using liquid-phase epitaxy (LPE). It consists of an *n*-type InP collector layer followed by a thin (0.1 μm) *p*-type InGaAsP base layer. The third layer is a wide-bandgap *n*-type InP emitter layer. Radiation incident on the device passes unattenuated through the wide-bandgap emitter and is absorbed in the base, base–collector depletion region and the collector. A large secondary photocurrent between the emitter and collector is obtained as the photogenerated holes are swept into the base, increasing the forward bias on the device. The use of the heterostructure permits low emitter–base and collector–base junction capacitances together with low base resistance. This is achieved through low emitter and collector doping levels coupled with heavy doping of the base, and allows large current gain. In addition the potential barrier created by the heterojunction at the emitter–base junction effectively eliminates hole injection from the base when the junction is forward biased. This gives good emitter base injection efficiency. The optical gain G_o of the device is given approximately by [Ref. 71].

$$G_o \simeq \eta h_{FE} = \frac{hf}{e} \frac{I_c}{P_o} \tag{8.31}$$

where η is the quantum efficiency of the base–collector photodiode, h_{FE} is the common emitter current gain, I_s is the collector current, P_o is the incident optical power, e is the electronic charge and hf is the photon energy. Moreover, the phototransistor shown in Figure 8.29(b) is capable of operating over the 0.9 to 1.3 µm wavelength band giving optical gains in excess of 100, as demonstratëd in Example 8.8.

Example 8.8

The phototransistor of Figure 8.29(b) has a collector current of 15 mA when the incident optical power at a wavelength of 1.26 µm is 125 µW. Estimate:

(a) the optical gain of the device under the above operating conditions;

(b) the common emitter current gain if the quantum efficiency of the base–collector photodiode at a wavelength of 1.26 µm is 40%.

Solution: (a) Using Eq. (8.31), the optical gain is given by:

$$G_o \simeq \frac{hf}{e} \frac{I_c}{P_o} = \frac{hc}{\lambda e} \frac{I_c}{P_o}$$

$$= \frac{6.626 \times 10^{-34} \times 2.998 \times 10^8 \times 15 \times 10^{-3}}{1.26 \times 10^{-6} \times 1.602 \times 10^{-19} \times 125 \times 10^{-6}}$$

$$= 118.1$$

(b) The common emitter current gain is:

$$h_{FE} = \frac{G_o}{\eta} = \frac{118.1}{0.4} = 295.3$$

In this example a common emitter current gain of 295 gives an optical gain of 118. It is therefore possible that this type of device will become an alternative to the APD for optical detection at wavelengths above 1.1 µm [Refs 72–74].

Phototransistors based on a heterojunction structure using the InGaAs/InAlAs material system have more recently been demonstrated to function as high-performance photo-detectors operating at a wavelength of 1.3 µm [Refs 75, 76]. These devices are generally

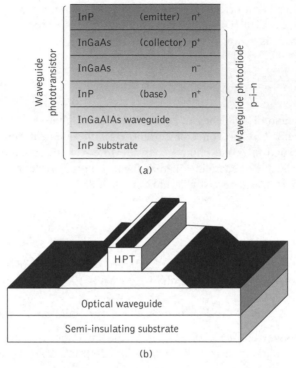

Figure 8.30 Waveguide phototransistor: (a) different waveguide sections also indicating the p–i–n photodiode; (b) three-dimensional view showing the three terminals in dark shading (HPT, heterojunction phototransistor)

known as waveguide phototransistors [Ref. 77] as they utilize a passive waveguide layer (in this case InGaAlAs onto an InP substrate) under the active transistor region as depicted in Figure 8.30(a), while a three-dimensional view for the same device structure is provided in Figure 8.30(b). It can be seen that the waveguide phototransistor structure also provides a waveguide p–i–n photodiode as identified on the right hand side of the Figure 8.30(a). Hence the p–i–n photodiode can be used as a separate photodetector by effectively removing the emitter layer of the heterojunction phototransistor.

The benefit of the waveguide structure, however, as compared with the conventional p–i–n heterojunction phototransistor, is that it facilitates an increased photocurrent which results in increased values of responsivity up to 29 A W^{-1} [Ref. 77]. Furthermore, higher values of responsivity can be obtained when the crystalline structure of a high-electron-mobility transistor (HEMT) (see Section 9.6) is incorporated in the design of the phototransistor [Ref. 78]. Such a phototransistor employing the InGaAlAs/GaAs material system has demonstrated a responsivity of 140 A W^{-1} at an operating wavelength in the far infrared around 6.0 μm and at a temperature of 23 K [Ref. 78]. The quantum efficiency, however, remained low (i.e. 16%) as the electrons that escaped by thermal emission from the quantum wells were also amplified, which contributed towards the high responsivity but reduced the signal-to-noise ratio.

8.12 Metal–semiconductor–metal photodetectors

Metal–semiconductor–metal (MSM) photodetectors are photoconductive detectors or photoconductors which provide what is conceptually the simplest form of semiconductor optical detection and have not until recently been considered as a serious contender for photodetection within optical fiber communications. Lately, however, there has been renewed interest in such devices, particularly for use in the longer wavelength region because of the suitability of III–V semiconductors for photoconductive detection applications [Refs 17, 79]. The basic detection process in a semiconductor discussed in Section 8.3 indicated that an electron may be raised from the valence band to the conduction band by the absorption of a photon provided that the photon energy was greater than the bandgap energy (i.e. $hf \geq E_g$). In this case, as long as the electron remains in the conduction band it will cause an increase in the conductivity of the semiconductor, a phenomenon referred to as photoconductivity. This forms the basic mechanism for the operation of MSM photodetectors.

A typical MSM device structure designed for operation in the longer wavelength region is shown in Figure 8.31 [Ref. 17]. It comprises two metallic electrodes on the top of the absorption layer. The conducting channel comprises a thin layer (1 to 2 μm) of n-type

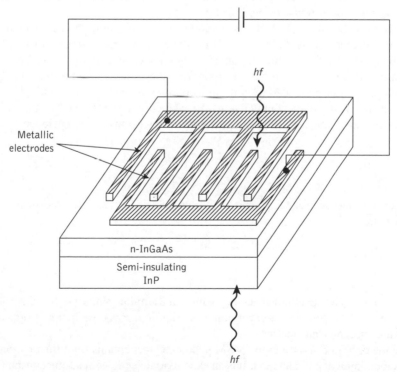

Figure 8.31 Metal–semiconductor–metal photodetector structure for operation in the 1.1 to 1.6 μm wavelength range

InCaAs which can absorb a significant amount of the incident light over the wavelength ranee 1.1 to 1.6 μm. In particular, good sensitivities at reasonably high bandwidths have been obtained with lightly doped (less than 5×10^{14} cm^{-3}) n-type In$_{0.53}$Ga$_{0.47}$As channel layers. Moreover, the composition of the InGaAs is arranged to be lattice matched to the semi-insulating InP substrate to avoid the formation of dislocations and other crystalline imperfections in the epitaxial layer. Low-resistance contacts are made to the conducting layer through the use of interdigital anodes and cathodes, as illustrated in Figure 8.31. In addition, these contacts are designed to maximize the coupling of light into the absorbing region by minimizing their obstruction of the active area while reducing the distance that photogenerated carriers have to travel prior to being collected at one of the electrodes. The optical coupling efficiency can also be improved by the application of an antireflection coating to the surface of the photoconductor facing the optical input.

In operation the incident light on the channel region is absorbed, thereby generating additional electron–hole pairs. These photogenerated carriers increase the channel conductivity, which results in an increased current in the external circuit. The optical receiver must therefore be sensitive to very small changes in resistance induced by the incident light. Furthermore, once the carriers have been generated the electrons will be swept by the applied electric field towards the anode while the holes move towards the cathode. In general, however, the mobility of the holes is considerably smaller than that of the electrons in III–V alloys such as GaAs and InGaAs. Thus the electrons, being the fastest charge carriers, provide the minimum time for detection and hence the limitation on the speed of response of the MSM photodetector.

In addition, while the fast electrons are collected at the anode, the corresponding holes are still proceeding across the channel. This creates an absence of electrons and hence a net positive charge in the channel region. However, the excess charge is immediately compensated by the injection of further electrons from the cathode into the channel. Thus further electrons may be generated from the absorption of a single photon. This factor creates what is known as the photoconductive gain G which may be defined as the ratio of the slow carrier transit time (or lifetime) t_s to the fast carrier transit time t_f. Hence:

$$G = \frac{t_s}{t_f} \tag{8.32}$$

Moreover, the photocurrent I_p produced by the MSM photodetector following Eq. (8.8) can be written as [Ref. 80]:

$$I_p = \frac{\eta P_o e}{hf} G \tag{8.33}$$

where η is the device quantum efficiency which in a similar manner to the absorption process (see Eq. (8.1)) follows an exponential distribution, P_o is the incident optical power and e is the charge on an electron.

Since the current response in the MSM photodetector remains for a time t_s after the end of an incident optical pulse and exhibits an exponential decay with a time constant equal to this slow carrier transit time (or lifetime), the maximum 3 dB bandwidth B_m for the device is given by:

$$B_{m} = \frac{1}{2\pi t_{s}}$$ (8.34)

Equations (8.32) and (8.34) can be combined to provide an expression for the gain–bandwidth product of a photoconductor as:

$$G \times B_{m} = \frac{1}{2\pi t_{f}}$$ (8.35)

It may be noted that an implication of Eq. (8.35) is that when t_{f} is fixed, photoconductive gain can only be obtained at the expense of the maximum bandwidth permitted by the device. There is therefore a trade-off between gain (and hence sensitivity) and speed of response.

Example 8.9

The electron transit time in an InGaAs MSM photodetector is 5 ps. Determine the maximum 3 dB bandwidth permitted by the device when its photoconductive gain is 70.

Solution: Using Eq. (8.35) the maximum 3 dB bandwidth provided by the MSM may be written as:

$$B_{m} = \frac{1}{2\pi t_{f}G} = \frac{1}{2\pi \times 5 \times 10^{-12} \times 70}$$

$$= 454.7 \text{ MHz}$$

The sensitivity of a MSM photodetector is limited by the noise generated within the device, even though the quantum efficiency may be quite high. For example, in an InGaAs photodetector with a 2 μm channel the quantum efficiency is around 88%. However, there are numerous sources contributing to the noise component within MSM photodetectors [Ref. 81]. In particular, noise arises from two sources: Johnson noise associated with the thermal noise from the bulk resistance of the photoconductor slab, and generation–recombination noise caused by fluctuations in the generation and recombination rates of the photogenerated carrier pairs. The former noise source, which is often dominant, results in a finite dark conductivity for the device which generates a randomly varying background dark current. It can be shown [Ref. 17] that the signal-to-noise ratio of a photoconductor receiver increases with increasing channel resistance and gain. Therefore, a method for increasing the sensitivity of the photodetector is to increase its photoconductive gain. Unfortunately, as indicated previously, this reduces the device response time.

Alternatively the MSM photodetector can be designed such that the dark current is reduced to a level such that the generation–recombination noise, which is fundamental, tends to dominate. Some success has been achieved in this direction using *p*-type substrates [Ref. 82]. In this case, however, the speed of response for the device was limited by the electrode spacing required to improve the sensitivity.

More recently, however, there have been several demonstrations proposing changes to the materials and the device geometry for producing ultrafast MSM photodetectors [Refs 83–86]. For example, an MSM photodetector structure with buried or recessed electrodes grown on GaAs has provided a significant improvement in the electric field distribution inside the photodetector [Ref. 86]. The MSM interdigitated electrode finger width of this device was 1.0 µm with finger spacing of 1.5 µm. It exhibited a reduced capacitance of 19×10^{-15} F when compared with a planar MSM photodetector with the same dimensions which had a capacitance of 69×10^{-15} F [Ref. 86]. This much lower capacitance reduced the RC time constant, thus improving the overall speed of the device. Commercially available MSM photodetectors based on GaAs are also specified with fast response times of 30 ps when operating at a wavelength of 1.50 µm while maintaining low dark currents of 100 pA at a temperature of 25 °C [Ref. 87]. Furthermore, an ultrahigh bandwidth traveling wave MSM photodetector exhibiting a risk time of just 0.8 ps and therefore a 570 GHz transform bandwidth has been experimentally reported [Ref. 88].

An inverted MSM photodetector based on the InGaAs/InP material system operating at the signal wavelength of 1.55 µm for vertical illumination has also been demonstrated. In the inverted MSM structure the substrate was removed, leaving the electrode fingers on the bottom of the device. This structure therefore eliminated the finger shadowing effect (i.e. the shadow cast by the metallic electrode fingers over the active area in a conventional MSM photodetector which restricts incident light from reaching this light-collecting region) and enhanced sensitivity such that the device exhibited a responsivity of 0.16 A W^{-1} with the dark current remaining under 1.2 nA [Ref. 89].

Metal–semiconductor–metal photodetectors based on InGaAs rather than GaAs suffer from reduced quantum efficiency. Introducing a thin layer, known as the barrier enhancement layer, containing InP or InAlAs, however, between the InGaAs layer and the metallic contacts improves the device performance. For example, a 20 nm layer of InAlAs produced a quantum efficiency of 92% for a back-illuminated MSM photodetector operating at a wavelength of 1.3 µm [Ref. 15].

Tunable MSM photodetectors which can potentially switch wavelengths have also been produced [Ref. 90]. Figure 8.32(a) displays the structure of a monolithic tunable MSM photodetector which comprises multiple electrode fingers with 1 µm finger spacing and width. The pattern in the center region of the device enables wavelength switching either to port 1 or 2 (i.e. on or off). The electrode fingers on the left hand side region are used to adjust the spectral response to obtain wavelength selection, while the right hand side region is not used but is included in the structure to provide device symmetry.

Device operation to achieve wavelength switching which is based on the principle of the optical delay interferometer (see Section 10.5.2) is illustrated in Figure 8.32(b) [Ref. 91]. The structure is divided into four different regions showing below the electrode fingers of the MSM photodetector with their corresponding interference patterns for positive and negative bias voltages. The alternate top fingers of the MSM device (which are actually connected to a common virtual ground electrode) serve to collect the net photocurrent. When a positive bias voltage is applied in any of the top fingers the photocurrent flows from that particular top finger to the bottom fingers while it flows in the reverse direction for a negative biasing condition as identified by patterns A and B, respectively. Therefore the appropriate interference pattern is used to select a specific wavelength signal.

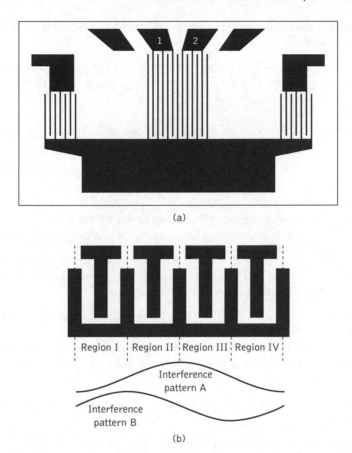

(a)

Region I ¦ Region II ¦Region III¦ Region IV

Interference
pattern A

Interference
pattern B

(b)

Figure 8.32 Tunable MSM photodetector: (a) monolithic integrated device; (b) four regions formed by applying alternative (positive and negative) bias voltages to the top and bottom fingers and the corresponding interference patterns (A and B) for two wavelengths are also shown. Reprinted with permission from Ref. 91 © IEEE 2005

Problems

8.1   Outline the reasons for the adoption of the materials and devices used for photodetection in optical fiber communications. Discuss in detail the p–i–n photodiode with regard to performance and compatibility requirements in photodetectors.

8.2   A p–i–n photodiode on average generates one electron–hole pair per three incident photons at a wavelength of 0.8 μm. Assuming all the electrons are collected calculate:
(a) the quantum efficiency of the device;
(b) its maximum possible bandgap energy;
(c) the mean output photocurrent when the received optical power is 10^{-7} W.

8.3 Explain the detection process in the p–n photodiode. Compare this device with the p–i–n photodiode.

8.4 Define the quantum efficiency and the responsivity of a photodetector.

Derive an expression for the responsivity of an intrinsic photodetector in terms of the quantum efficiency of the device and the wavelength of the incident radiation.

Determine the wavelength at which the quantum efficiency and the responsivity are equal.

8.5 A p–n photodiode has a quantum efficiency of 50% at a wavelength of 0.9 μm. Calculate:
(a) its responsivity at 0.9 μm;
(b) the received optical power if the mean photocurrent is 10^{-6} A;
(c) the corresponding number of received photons at this wavelength.

8.6 When 800 photons per second are incident on a p–i–n photodiode operating at a wavelength of 1.3 μm they generate on average 550 electrons per second which are collected. Calculate the responsivity of the device.

8.7 Explain what is meant by the long-wavelength cutoff point for an intrinsic photo-detector, deriving any relevant expressions.

Considering the bandgap energies given in Table 8.1, calculate the long-wavelength cutoff points for both direct and indirect optical transitions in silicon and germanium.

8.8 Describe the advantages and drawbacks associated with the mushroom-waveguide photodiode.

8.9 A p–i–n photodiode ceases to operate when photons with energy greater than 0.886 eV are incident upon it; of which material is it fabricated?

8.10 (a) The time taken for electrons to diffuse through a layer of p-type silicon is 28.8 ns. If the minority carrier diffusion coefficient is 3.4×10^{-3} m^2 s^{-1}, determine the thickness of the silicon layer.
(b) Assuming the depletion layer width in a silicon photodiode corresponds to the layer thickness obtained in part (a) and that the maximum response time of the photodiode is 877 ps, estimate the carrier (hole) drift velocity.

8.11 A silicon p–i–n photodiode with an area of 1.5 mm^2 is to be used in conjunction with a load resistor of 100 Ω. If the requirement for the device is a fast response time, estimate the thickness of the intrinsic region that should be provided. It may be assumed that the permittivity for silicon is 1.04×10^{-10} F m^{-1} and that the electron saturation velocity is 10^7 m s^{-1}.

8.12 Classify the different traveling-wave approaches to design a photodiode and identify the most useful method in practice.

8.13 Explain the structure and operation of a resonant cavity enhanced (RCE) photodiode. Discuss its applications and the reasons for the growing interest in its use within integrated photonics.

8.14 Define the noise equivalent power (*NEP*) for a photodetector. Commencing with Eq. (8.8) obtain an expression for the *NEP* of a photodiode in which the dark current noise dominates.

A silicon *p–i–n* photodiode with active dimensions 10 μm has a specific detectivity of 7×10^{10} m Hz$^{\frac{1}{2}}$ W^{-1} when operating at a wavelength of 0.85 μm. The device quantum efficiency at this wavelength is 64%. Assuming that it is the dominant noise source, calculate the dark current over a 1 Hz bandwidth in the device.

8.15 The specific detectivity of a wide area silicon photodiode at its operating wavelength is 10^{11} m Hz$^{\frac{1}{2}}$ W^{-1}. Estimate the smallest detectable signal power at this wavelength when the sensitive area of the device is 25 mm^2 and the signal bandwidth is 1 kHz.

8.16 Discuss the operation of the silicon RAPD, describing how it differs from the *p–i–n* photodiode.

Outline the advantages and drawbacks with the use of the RAPD as a detector for optical fiber communications.

8.17 Compare and contrast the structure and performance characteristics of germanium and III–V semiconductor alloy APDs for operation in the wavelength range 1.1 to 1.6 μm.

8.18 An APD with a multiplication factor of 20 operates at a wavelength of 1.5 μm. Calculate the quantum efficiency and the output photocurrent from the device if its responsivity at this wavelength is 0.6 A W^{-1} and 10^{10} photons of wavelength 1.5 μm are incident upon it per second.

8.19 Given that the following measurements were taken for an APD, calculate the multiplication factor for the device.

> Received optical power at 1.35 μm $= 0.2$ μW
> Corresponding output photocurrent $= 4.9$ μA
> (after avalanche gain)
> Quantum efficiency at 1.35 μm $= 40\%$

8.20 An APD has a quantum efficiency of 45% at 0.85 μm. When illuminated with radiation of this wavelength it produces an output photocurrent of 10 μA after avalanche gain with a multiplication factor of 250. Calculate the received optical power to the device. How many photons per second does this correspond to?

8.21 When 10^{11} photons per second each with an energy of 1.28×10^{-19} J are incident on an ideal photodiode, calculate:
(a) the wavelength of the incident radiation;
(b) the output photocurrent;
(c) the output photocurrent if the device is an APD with a multiplication factor of 18.

8.22 A silicon RAPD has a multiplication factor of 10^3 when operating at a wavelength of 0.82 μm. At this operating point the quantum efficiency of the device is 90% and the dark current is 1 nA.

Determine the number of photons per second of wavelength 0.82 μm required in order to register a light input to the device corresponding to an output current (after avalanche gain) which is greater than the level of the dark current (i.e. $I > 1$ nA).

8.23   Indicate the material systems under investigation and discuss their application in the fabrication of photodiodes for use in the mid-infrared wavelength region.

8.24   An InGaAsP heterojunction phototransistor has a common emitter current gain of 170 when operating at a wavelength of 1.3 μm with an incident optical power of 80 μW. The base–collector quantum efficiency at this wavelength is 65%. Estimate the collector current in the device.

8.25   Discuss the major features of a quantum-dot photodetector and with the aid of a diagram explain the DWELL structure for such a device.

8.26   Describe the construction of a waveguide phototransistor and outline its significant benefits over the conventional *p–i–n* phototransistor.

8.27   Describe the basic detection process in a photoconductive detector.

   The maximum 3 dB bandwidth allowed by an InGaAs photoconductive detector is 380 MHz when the electron transit time through the device is 7.6 ps. Calculate the photocurrent obtained from the device when 10 μW of optical power at a wavelength of 1.32 μm is incident upon it, and the device quantum efficiency is 75%.

8.28   Explain the term interdigitated finger width in relation to an MSM photodetector and describe its role in facilitating a tunable device.

Answers to numerical problems

8.2   (a) 33%; (b) 24.8×10^{-20} J;
(c) 21.3 nW

8.4   1.24 μm

8.5   (a) 0.36 A W^{-1}; (b) 2.78 μW;
(c) 1.26×10^{13} photon s^{-1}

8.6   0.72 A W^{-1}

8.7   0.3 μm, 1.09 μm, 1.53 μm, 1.85 μm

8.9 $In_{0.7}Ga_{0.3}As_{0.64}P_{0.36}$

8.10   (a) 14 μm; (b) 10^5 ms^{-1}

8.11   395 μm

8.14 1.23×10^{-14} A

8.15   1.58 pW

8.18   50%, 15.9 nA

8.19   24.1

8.20   77.8 nW, 3.33×10^{11} photon s^{-1}

8.21   (a) 1.55 μm; (b) 1.6 μA;
(c) 28.8 μA

8.22 6.94×10^6 photon s^{-1}

8.24   9.3 mA

8.27   0.44 mA

References

[1] J. C. Marshall, *Semiconductor Photodetectors*, SPIE, 2005.

[2] R. Sasa, A.-J. Annema and N. Bram, *High-Speed Photodiodes in Standard CMOS Technology*, Springer-Verlag, 2006.

[3] J. Brouckaert, G. Roelkens, D. Van Thourhout and R. Baets, 'Thin-film III–V photodetectors integrated on silicon-on-insulator photonic ICs', *J. Lightwave Technol.*, **25**(4), pp. 1053–1060, 2007.

[4] T. P. Lee and T. Li, 'Photodetectors', in S. E. Miller and A. G. Chynoweth (Eds), *Optical Fiber Telecommunications*, pp. 593–626, Academic Press, 1979.

[5] G. Keiser, *Optical Communications Essentials (Telecommunications)*, McGraw-Hill Professional, 2003.

[6] S. M. Sze and K. K. Ng, *Physics of Semiconductor Devices* (3nd edn), Wiley, 2006.

[7] D. H. Newman and S. Ritchie, 'Sources and detectors for optical fibre communications applications: the first 20 years', *IEE Proc., Optoelectron.*, **133**(3), pp. 213–229, 1986.

[8] R. W. Dixon and N. K. Dutta, 'Lightwave device technology', *AT&T Tech. J.*, **66**(1), pp. 73–83, 1987.

[9] J. E. Bowers and C. A. Burrus, 'Ultrawide-band long-wave *p–i–n* photodetectors', *J. Lightwave Technol.*, **LT–5**(10), pp. 1339–1350, 1987.

[10] A. W. Nelson, S. Wong, S. Ritchie and S. K. Sargood, 'GaInAs PIN photodiodes grown by atmospheric-pressure MOVPE', *Electron. Lett.*, **21**(19), pp. 838–840, 1985.

[11] S. Miura, H. Kuwatsuka, T. Mikawa and O. Wada, 'Planar embedded InP/GaInAs *p–i–n* photodiode for very high speed operation', *J. Lightwave Technol.*, **LT–5**(10), pp. 1371–1376, 1987.

[12] Y. M. El-Batawy and M. J. Deen, 'Analysis, circuit modeling, and optimization of mushroom waveguide photodetector (mushroom-WGPD)', *J. Lightwave Technol.*, **23**(1), pp. 423–431, 2005.

[13] K. Kato, A. Kozen, Y. Muramoto, Y. Itaya, T. Nagatsuma and M. Yaita, '110-GHz, 50%-efficiency mushroom-mesa waveguide *p–i–n* photodiode for a 1.55 μm wavelength', *IEEE Photonics Technol. Lett.*, **6**(6), pp. 719–721, 1994.

[14] K. Kato, 'Ultrawide-band/high-frequency photodetectors', *IEEE Trans. Microw. Theory Technol.*, **47**(7), pp. 1265–1281, 1999.

[15] G. P. Agrawal, *Lightwave Technology: Components and Devices*, Wiley–IEEE, 2004.

[16] G. Keiser, *Optical Fiber Communications* (3rd edn), McGraw-Hill, 1999.

[17] S. R. Forrest, 'Optical detectors for lightwave communication', in S. E. Miller and I. P. Kaminow (Eds), *Optical Fiber Telecommunications II*, pp. 569–599, Academic Press, 1988.

[18] B. L. Kasper, O. Mizuhara and Y. K. Chen, 'High bit-rate receivers, transmitters and electronics', in I. P. Kaminow and T. Li (Eds), *Optical Fiber Telecommunications IVA*, pp. 784–853, Academic Press, 2002.

[19] Y. Hirota, T. Ishibashi and H. Ito, '1.55 μm wavelength periodic traveling-wave photodetector fabricated using unitraveling-carrier photodiode structures', *J. Lightwave Technol.*, **19**(11), pp. 1751–1758, 2001.

[20] J.-W. Shi and C.-K. Sun, 'Theory and design of a tapered line distributed photodetector', *J. Lightwave Technol.*, **20**(11), pp. 1942–1950, 2002.

[21] D. Lasaosa, J.-W. Shi, D. Pasquariello, K.-G. Gan, M.-C. Tien, H.-H. Chang, S.-W. Chu, C.-K. Sun, Y.-J. Chiu and J. E. Bowers, 'Traveling-wave photodetectors with high power-bandwidth and gain-bandwidth product performance', *IEEE J. Sel. Top. Quantum Electron.*, **10**(4), pp. 728–741, 2004.

[22] H. Ito, T. Furuta, S. Kodama and T. Ishibashi, 'InP/InGaAs uni-travelling-carrier photodiode with 310 GHz bandwidth', *Electron. Lett.*, **36**(21), pp. 1809–1810, 2000.

[23] Y. Muramoto, K. Yoshino, S. Kodama, Y. Hirota, H. Ito and T. Ishibashi, '100 and 160 Gbit/s operation of uni-travelling-carrier photodiode module', *Electron. Lett.*, **40**(6), pp. 378–380, 2004.

[24] H. Ito, T. Furuta, F. Nakajima, K. Yoshino and T. Ishibashi, 'Photonic generation of continuous THz wave using uni-traveling-carrier photodiode', *J. Lightwave Technol.*, **23**(12), pp. 4016–4021, 2005.

[25] H. Huang, Y. Huang, X. Wang, Q. Wang and X. Ren, 'Long wavelength resonant cavity photodetector based on InP/air-gap Bragg reflectors', *IEEE Photonics Technol. Lett.*, **16**(1), pp. 245–247, 2004.

[26] C.-H. Chen, K. Tetz and Y. Fainman, 'Resonant-cavity-enhanced *p–i–n* photodiode with a broad quantum-efficiency spectrum by use of an anomalous-dispersion mirror', *J. Appl. Opt.*, **44**(29), pp. 6131–6140, 2005.

[27] X. Sun, J. Hsu, X. G. Zheng, J. C. Campbell and A. L. Holmes Jr, 'GaAsSb resonant-cavity-enhanced photodetector operating at 1.3 μm', *IEEE Photonics Technol. Lett.*, **14**(5), pp. 681–683, 2002.

[28] M. Kikuo, N. Takeshi, S. Kazuhiro and T. Takeshi, '40Gbps waveguide photodiodes', *NEC J. Adv. Technol.*, **2**(3), pp. 234–240, 2005.

[29] B. Butun, N. Biyikli, I. Kimukin, O. Aytur, E. Ozbay, P. A. Postigo, J. P. Silveira and A. R. Alija, 'High-speed 1.55 μm operation of low-temperature-grown GaAs-based resonant-cavity-enhanced *p–i–n* photodiodes', *Appl. Phys. Lett.*, **84**(21), pp. 4185–4187, 2004.

[30] O. I. Dosunmu, D. D. Cannon, M. K. Emsley, L. C. Kimerling and M. S. Unlu, 'High-speed resonant cavity enhanced Ge photodetectors on reflecting Si substrates for 1550-nm operation', *IEEE Photonics Technol. Lett.*, **17**(1), pp. 175–177, 2005.

[31] M. Schwartz, *Information Transmission, Modulation and Noise* (4th edn), McGraw-Hill, 1990.

[32] T. P. Lee, C. A. Burrus Jr and A. G. Dentai, 'InGaAsP/InP photodiodes microplasma-limited avalanche multiplication at 1–1.3 μm wavelength', *IEEE J. Quantum Electron.*, **QE-15**, pp. 30–35, 1979.

[33] A. R. Hartman, H. Melchior, D. P. Schinke and T. E. Seidel, 'Planar epitaxial silicon avalanche photodiode', *Bell Syst. Tech. J.*, **57**, pp. 1791–1807, 1978.

[34] H. Ando, H. Kanbe, T. Kimura, T. Yamaoka and T. Kaneda, 'Characteristics of germanium avalanche photodiodes in the wavelength region 1–1.6 μm, *IEEE J. Quantum Electron.*, **QE-14**(11), pp. 804–809, 1978.

[35] T. Mikawa, S. Kagawa, T. Kaneda, T. Sakwai, H. Ando and O. Mikami, 'A low-noise n⁺np germanium avalanche photodiode'. *IEEE J. Quantum Electron.*, **QE-17**(2), pp. 210–216, 1981.

[36] O. Mikami, H. Ando, H. Kanbe, T. Mikawa, T. Kaneda and Y. Toyama, 'Improved germanium avalanche photodiodes', *IEEE J. Quantum Electron.*, **QE-16**(9), pp. 1002–1007, 1980.

[37] S. Kagawa, T. Kaneda, I. Mikawa, Y. Banba, Y. Toyama and O. Mikami, 'Fully ion-implanted p⁺-n germanium avalanche photodiodes', *Appl. Phys. Lett.*, **38**(6), pp. 429–431, 1981.

[38] T. Mikawa, T. Kaneda, H. Nishimoto, M. Motegi and H. Okushima, 'Small-active-area germanium avalanche photodiode for single-mode fibre at 1.3 μm wavelength', *Electron. Lett.*, **19**(12), pp. 452–453, 1983.

[39] M. Niwa, Y. Tashiro, K. Minemura and H. Iwasaki, 'High-sensitivity Hi–Lo germanium avalanche photodiode for 1.5 μm wavelength optical communication', *Electron. Lett.*, **20**(13), pp. 552–553, 1984.

[40] G. E. Stillman, 'Detectors for optical-waveguide communications', in E. E. Basch (Ed.), *Optical-Fiber Transmission*, H. W. Sams & Co., pp. 335–374, 1987.

[41] S. R. Forrest, O. K. Kim and R. G. Smith, 'Optical response time in $In_{0.53}Ga_{0.47}$ As/InP avalanche photodiodes', *Appl. Phys. Lett.*, **41**, pp. 95–98, 1982.

[42] J. C. Campbell, A. G. Dentai, W. S. Holder and B. L. Kasper, 'High performance avalanche photodiode with separate absorption "grading" and multiplication regions', *Electron. Lett.*, **19**, pp. 818–820, 1983.

[43] J. Wei, F. Xia and S. R. Forrest, 'A high-responsivity high-bandwidth asymmetric twin-waveguide coupled InGaAs-InP-InAlAs avalanche photodiode', *IEEE Photonics Technol. Lett.*, **14**(11), pp. 1590–1592, 2002.

[44] E. Yagyu, E. Ishimura, M. Nakaji, T. Aoyagi and Y. Tokuda, 'Simple planar structure for high-performance AlInAs avalanche photodiodes', *IEEE Photonics Technol. Lett.*, **18**(1), pp. 76–78, 2006.

[45] C. Lenox, H. Nie, P. Yuan, G. Kinsey, A. L. Homles Jr, B. G. Streetman and J. C. Campbell, 'Resonant-cavity InGaAs-InAlAs avalanche photodiodes with gain-bandwidth product of 290 GHz', *IEEE Photonics Technol. Lett.*, **11**(9), pp. 1162–1164, 1999.

[46] O.-H. Kwon, M. M. Hayat, J. C. Campbell, B. E. A. Saleh and M. C. Teich, 'Gain-bandwidth product optimization of heterostructure avalanche photodiodes', *J. Lightwave Technol.*, **23**(5), pp. 1896–1906, 2005.

[47] R. Chin, N. Holonyak, G. E. Stillman, J. Y. Tang and K. Hess, 'Impact ionisation in multilayer heterojuction structures', *Electron. Lett.*, **16**(12), pp. 467–468, 1980.

[48] F. Capasso, 'Band-gap engineering via graded gap, superlattice and periodic doping structures: applications and novel photodetectors and other devices', *J. Vac. Sci. Technol. B*, **1**(2), pp. 457–461, 1983.

[49] R. Sidhu, N. Duan, J. C. Campbell and A. L. Holmes Jr, 'A long-wavelength photodiode on InP using lattice-matched GaInAs-GaAsSb type-II quantum wells', *IEEE Photonics Technol. Lett.*, **17**(12), pp. 2715–2717, 2005.

[50] R. Sidhu, L. Zhang, N. Tan, N. Duan, J. C. Campbell, A. L. Holmes, C.-F. Hsu and M. A. Itzler, '2.4 μm cutoff wavelength avalanche photodiode on InP substrate', *Electron. Lett.*, **42**(3), pp. 181–182, 2006.

[51] D. Pasquariello, E. S. Bjorlin, D. Lasaosa, Y.-J. Chiu, J. Piprek and J. E. Bowers, 'Selective undercut etching of InGaAs and InGaAsP quantum wells for improved performance of long-wavelength optoelectronic devices', *J. Lightwave Technol.*, **24**(3), pp. 1470–1477, 2006.

[52] J. C. Campbell, 'Advances in photodetectors', in I. P. Kaminow, T. Li and A. E. Wilner (Eds) *Optical Fiber Telecommunications*, VA, pp. 221–268, Elsevier/Academic Press, 2008.

[53] J. E. Bowers, A. K. Srivastava, C. A. Burrus, J. C. Dewinter, M. A. Pollack and J. L. Zyskind, 'High-speed GaInAsSb/GaSb PIN photodetectors for wavelengths to 2.3 μm', *Electron. Lett.*, **22**(3), pp. 137–138, 1986.

[54] A. J. Moseley, M. D. Scott, A. H. Moore and R. H. Wallis, 'High-efficiency, low-leakage MOCVD-grown GaInAs/AlInAs heterojunction photodiodes for detection to 2.4 μm', *Electron Lett.*, **22**(22), pp. 1206–1207, 1986.

[55] R. U. Martinelli, T. J. Zamerowski and P. A. Longeway, 'In$_x$G$_{1-x}$As/InAs$_y$P$_{1-y}$ lasers and photodiodes for 2.55 μm optical fiber communications', *Optical Fiber Communication Conf., OFC'88*, USA, paper TUC6, January 1988.

[56] G. Pichard, J. Meslage, T. Nguyen Duy and F. Raymond, '1.3 μm CdHgTe avalanche photodiodes for fibre-optic applications', *Proc. 9th ECOC'83*, Switzerland, pp. 479–482, 1983.

[57] H. Haupt and O. Hildebrand, 'Lasers and photodetectors in Europe', *IEEE J. Sel. Areas Commun.*, **SAC-4**(4), pp. 444–456, 1986.

[58] P. Capper and C. T. Elliott, *Infrared Detectors and Emitters: Materials and Devices*, Springer-Verlag, 2001.

[59] P. Norton, 'HgCdTe infrared detectors', *Optoelectron. Rev.*, **10**(3), pp. 159–174, 2002.

[60] A Rogalski, 'HgCdTe infrared detector material: history, status and outlook', *Rep. Prog. Phys.*, **68**, pp. 2267–2336, 2005.

[61] S. Chakrabarti, A. D. Stiff-Roberts, X. H. Su, P. Bhattacharya, G. Ariyawansa and A. G. U. Perera, 'High-performance mid-infrared quantum dot infrared photodetectors', *J. Phys. D: Appl. Phys.*, **38**(13), pp. 2135–2141, 2005.

[62] H. C. Liu, 'Quantum dot infrared photodetector', *Optoelectron. Rev.*, **11**(1), pp. 1–5, 2003.

[63] M. Böberl, T. Fromherz, T. Schwarzl, G. Springholz and W. Heiss, 'IV–VI resonant-cavity enhanced photodetectors for the mid-infrared', *J. Semicond. Sci. Technol.*, **19**(12), pp. L115–L117, 2004.

[64] V. Ryzhi, *Intersubband Infrared Photodetectors*, World Scientific, 2003.

[65] S. Krishna, 'Quantum dots-in-a-well infrared photodetectors', *J. Phys. D: Appl. Phys.*, **38**(13), pp. 2142–2150, 2005.

[66] Z. Ye, J. C. Campbell, Z. Chen, E.-T. Kim and A. Madhukar, 'Voltage-controllable multi-wavelength InAs quantum-dot infrared photodetectors for mid- and far-infrared detection', *J. Appl. Phys.*, **92**(7), pp. 4141–4143, 2002.

[67] A. D. Stiff, S. Krishna, P. Bhattacharya and S. W. Kennerly, 'Normal-incidence, high-temperature, mid-infrared, InAs-GaAs vertical quantum-dot infrared photodetector', *IEEE J. Quantum Electron.*, **37**(11), pp. 1412–1419, 2001.

[68] B. Kochman, A. D. Stiff-Roberts, S. Chakrabarti, J. D. Phillips, S. Krishna, J. Singh and P. Bhattacharya, 'Absorption, carrier lifetime, and gain in InAs-GaAs quantum-dot infrared photodetectors', *IEEE J. Quantum Electron.*, **39**(3), pp. 459–467, 2003.

[69] F. Liu, S. Tong, J. Liu and K. L. Wang, 'Normal-incidence mid-infrared Ge quantum-dot photo-detector', *J. Quantum Mater.*, **33**(8), pp. 846–850, 2004.

[70] S. Tong, F. Liu, A. Khitun, K. L. Wang and J. L. Liu, 'Tunable normal incidence Ge quantum dot midinfrared detectors', *J. Appl. Phys.*, **96**(1), pp. 773–776, 2004.

[71] P. D. Wright, R. J. Nelson and T. Cella, 'High gain InGaAsP–InP heterojunction phototransis-tors'. *Appl. Phys. Lett.*, **37**(2), pp. 192–194, 1980.

[72] R. A. Milano, P. D. Dapkus and G. E. Stillman, 'Heterojunction phototransistors for fiber-optic communications', *Proc. SPIE*, **272**, pp. 43–50, 1981.

[73] K. Tubatabaie-Alavi and C. G. Fonstad, 'Recent advances in InGaAs/InP phototransistors', *Proc. SPIE*, **272**, pp. 38–42, 1981.

[74] G. E. Stillman, L. W. Cook, G. E. Bulman, N. Tabatabaie, R. Chin and P. D. Dapkus, 'Long-wavelength (1.3 to 1.6 μm) detectors for fiber-optical communications', *IEEE Trans. Electron Devices*, **ED-29**(9), pp. 1355–1371, 1982.

[75] A. Leven, V. Houtsma, R. Kopf, Y. Baeyens and Y.-K. Chen, 'InP-based double-heterostructure phototransistors with 135 GHz optical-gain cutoff frequency', *Electron. Lett.*, **40**(13), pp. 833–834, 2004.

[76] C.-S. Choi, H.-S. Kang, W.-Y. Choi, D.-H. Kim and K.-S. Seo, 'Phototransistors based on InP HEMTs and their applications to millimeter-wave radio-on-fiber systems', *IEEE Trans. Microw. Theory Technol.*, **53**(1), pp. 256–263, 2005.

[77] W. K. Ng, C. H. Tan, P. A. Houston and A. Krysa, 'High current InP/InGaAs evanescently coupled waveguide phototransistor', *IEE Proc., Optoelectron.*, **152**(2), pp. 140–144, 2005.

[78] J. H. Oum, U. H. Lee, Y. H. Kang, J. R. Yang and S. Hong, 'Quantum-well infrared photo-transistor with pHEMT structure', *IEEE Electron Devices Lett.*, **26**(8), pp. 527–529, 2005.

[79] J. Brouckaet, G. Roelkens, D. van Thourhout and R. Baets, 'Compact InAlAs-InGaAs Metal–semiconductor–metal photodetectors integrated on silicon-on-insulator waveguides', *IEEE Photonics Technol. Lett.*, **19**(19), pp. 1484–1486.

[80] S. M. Sze, *Semiconductor Devices: Physics and Technology* (2nd edn), Wiley, 2002.

[81] S. R. Forrest, 'The sensitivity of photoconductor receivers for long-wavelength optical com-munications', *J. Lightwave Technol.*, **LT-3**(2), pp. 347–360, 1985.

[82] C. Y. Chen, A. G. Dentai, B. L. Kasper and P. A. Garbinski, 'High-speed junction-depleted $Ga_{0.4}$-$In_{0.53}$As photoconductive detectors', *Appl. Phys. Lett*, **46**, pp. 1164–1166, 1985.

[83] R. Chen, J. Fu, D. A. B. Miller and J. S. Harris, 'Spectral shaping of electrically controlled MSM-based tunable photodetectors', *IEEE Photonics Technol. Lett.*, **17**(10), pp. 2158–2160, 2005.

[84] S.-Y. Lo, Y.-L. Wei, R.-H. Yeh and J.-W. Hong, 'Suppressing dark-current in planar Si-based MSM photodetector with alternated i-a-Si:H/i-a-SiGe:H grade superlattice-like layers', *Electron. Lett.*, **41**(7), pp. 438–439, 2005.

[85] S. H. Hsu, Y. K. Su, S. J. Chang, W. C. Chen and H. L. Tsai, 'InGaAsN metal-semiconductor-metal photodetectors with modulation-doped heterostructures', *IEEE Photonics Technol. Lett.*, **18**(3), pp. 547–549, 2006.

[86] M. Mikulics, S. Wu, M. Marso, R. Adam, A. Forster, A. van der Hart, P. Kordos, H. Luth and R. Sobolewski, 'Ultrafast and highly sensitive photodetectors with recessed electrodes fabri-cated on low-temperature-grown GaAs', *IEEE Photonics Technol. Lett.*, **18**(7), pp. 820–822, 2006.

[87] MSM Photodetector G7096, http://sales.hamamatsu.com/assets/pdf/parts_G/G7096.pdf, 13 September 2007.

[88] J. W. Shi, K. G. Gan, Y. J. Chiu, Y. H. Chen, C. K. Sun, Y. J. Yang and J. E. Bowers, 'Metal–semiconductor–metal traveling-wave photodetectors, *IEEE Photonics Technol. Lett.*, **13**(6), pp. 623–625, 2001.

[89] S. Cho, S. Jokerst, N. M. Brown and A. S. Brooke, 'High-speed large-area inverted InGaAs thin-film metal-semiconductor-metal photodetectors', *IEEE J. Sel. Top. Quantum Electron.*, **10**(4), pp. 686–693, 2004.

[90] R. Chen, D. A. B. Miller, K. Ma and J. S. Harris Jr, 'Novel electrically controlled rapidly wavelength selective photodetection using MSMs', *IEEE J. Sel. Top. Quantum Electron.*, **11**(1), pp. 184–189, 2005.

[91] R. Chen, H. Chin, D. A. B. Miller, K. Ma and J. S. Harris Jr, 'MSM-based integrated CMOS wavelength-tunable optical receiver', *IEEE Photonics Technol. Lett.*, **17**(6), pp. 1271–1273, 2005.

CHAPTER 9

Direct detection receiver performance considerations

9.1 Introduction

The receiver in an intensity-modulated/direct detection (IM/DD) optical fiber communication system (see Section 7.5) essentially consists of the photodetector plus an amplifier with possibly additional signal processing circuits. Therefore the receiver initially converts the optical signal incident on the detector into an electrical signal, which is then amplified before further processing to extract the information originally carried by the optical signal.

The importance of the detector in the overall system performance was stressed in Chapter 8. However, it is necessary to consider the properties of this device in the context of the associated circuitry combined in the receiver. It is essential that the detector performs efficiently with the following amplifying and signal processing circuits. Inherent to this process is the separation of the information originally contained in the optical signal from the noise generated within the rest of the system and in the receiver itself, as well as any limitations on the detector response imposed by the associated circuits. These factors play a crucial role in determining the performance of the system.

In order to consider receiver design it is useful to regard the limit on the performance of the system set by the signal-to-noise ratio (SNR) at the receiver. It is therefore necessary to outline noise sources within optical fiber systems. The noise in these systems has different origins from that of copper-based systems. Both types of system have thermal noise generated in the receiver. However, although optical fiber systems exhibit little crosstalk the noise generated within the detector must be considered, as well as the noise properties associated with the electromagnetic carrier.

In Section 9.2 we therefore briefly review the major noise mechanisms which are present in direct detection optical fiber communication receivers prior to more detailed discussion of the limitations imposed by photon (or quantum) noise in both digital and analog transmission. This is followed in Section 9.3 with a more specific discussion of the noise associated with the two major receiver types (i.e. employing p–i–n and avalanche photodiode detectors). Expressions for the SNRs of these two receiver types are also developed in this section. Section 9.4 considers the noise and bandwidth performance of common preamplifier structures utilized in the design of optical fiber receivers. In Section 9.5 we present a brief account of low-noise field effect transistor (FET) preamplifiers which find wide use within optical fiber communication receivers. This discussion also includes consideration of p–i–n photodiode/FET (PIN–FET) hybrid receiver circuits which have been developed for optical fiber communications. Finally, major high-performance receiver design strategies to provide low-noise and high-bandwidth operation as well as wide dynamic range are described in Section 9.6.

9.2 Noise

Noise is a term generally used to refer to any spurious or undesired disturbances that mask the received signal in a communication system. In optical fiber communication systems we are generally concerned with noise due to spontaneous fluctuations rather than erratic disturbances which may be a feature of copper-based systems (due to electromagnetic interference etc.).

There are three main types of noise due to spontaneous fluctuations in optical fiber communication systems: thermal noise, dark current noise and quantum noise.

9.2.1 Thermal noise

This is the spontaneous fluctuation due to thermal interaction between, say, the free electrons and the vibrating ions in a conducting medium, and it is especially prevalent in resistors at room temperature.

The thermal noise current i_t in a resistor R may be expressed by its mean square value [Ref. 1] and is given by:

$$\overline{i_t^2} = \frac{4KTB}{R}$$

(9.1)

where K is Boltzmann's constant, T is the absolute temperature and B is the post-detection (electrical) bandwidth of the system (assuming the resistor is in the optical receiver).

9.2.2 Dark current noise

When there is no optical power incident on the photodetector a small reverse leakage current still flows from the device terminals. This dark current (see Section 8.4.2) contributes to the total system noise and gives random fluctuations about the average particle flow of the photocurrent. It therefore manifests itself as shot noise [Ref. 1] on the photocurrent. Thus the dark current noise $\overline{i_d^2}$ is given by:

$$\overline{i_d^2} = 2eBI_d \tag{9.2}$$

where e is the charge on an electron and I_d is the dark current. It may be reduced by careful design and fabrication of the detector.

9.2.3 Quantum noise

The quantum nature of light was discussed in Section 6.2.1 and the equation for the energy of this quantum or photon was stated as $E = hf$. The quantum behavior of electromagnetic radiation must be taken into account at optical frequencies since $hf > KT$ and quantum fluctuations dominate over thermal fluctuations.

The detection of light by a photodiode is a discrete process since the creation of an electron–hole pair results from the absorption of a photon, and the signal emerging from the detector is dictated by the statistics of photon arrivals. Hence the statistics for monochromatic coherent radiation arriving at a detector follow a discrete probability distribution which is independent of the number of photons previously detected.

It is found that the probability $P(z)$ of detecting z photons in time period τ when it is expected on average to detect z_m photons obeys the Poisson distribution [Ref. 2]:

$$P(z) = \frac{z_m^z \exp(-z_m)}{z!} \tag{9.3}$$

where z_m is equal to the variance of the probability distribution. This equality of the mean and the variance is typical of the Poisson distribution. From Eq. (8.7) the electron rate r_e generated by incident photons is $r_e = \eta P_o/hf$. The number of electrons generated in time τ is equal to the average number of photons detected over this time period z_m. Therefore:

$$z_m = \frac{\eta P_o \tau}{hf} \tag{9.4}$$

The Poisson distributions for $z_m = 10$ and $z_m = 1000$ are illustrated in Figure 9.1 and represent the detection process for monochromatic coherent light.

Incoherent light is emitted by independent atoms and therefore there is no phase relationship between the emitted photons. This property dictates an exponential intensity distribution for incoherent light which if averaged over the Poisson distribution [Ref. 2] gives:

$$P(z) = \frac{z_m^z}{(1 + z_m)^{z+1}} \tag{9.5}$$

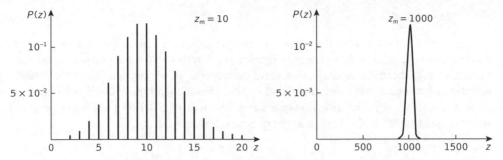

Figure 9.1 Poisson distributions for $z_m = 10$ and $z_m = 1000$

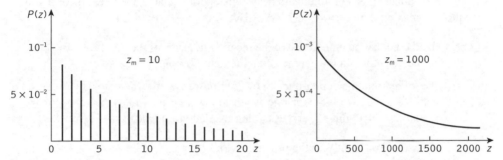

Figure 9.2 Probability distributions indicating the statistical fluctuations of incoherent light for $z_m = 10$ and $z_m = 1000$

Equation (9.5) is identical to the Bose–Einstein distribution [Ref. 3] which is used to describe the random statistics of light emitted in black body radiation (thermal light). The statistical fluctuations for incoherent light are illustrated by the probability distributions shown in Figure 9.2.

9.2.4 Digital signaling quantum noise

For digital optical fiber systems it is possible to calculate a fundamental lower limit to the energy that a pulse of light must contain in order to be detected with a given probability of error. The premise on which this analysis is based is that the ideal receiver has a sufficiently low amplifier noise to detect the displacement current of a single electron–hole pair generated within the detector (i.e. an individual photon may be detected). Thus in the absence of light, and neglecting dark current, no current will flow. Therefore the only way an error can occur is if a light pulse is present and no electron–hole pairs are generated. The probability of no pairs being generated when a light pulse is present may be obtained from Eq. (9.3) and is given by:

$$P(0|1) = \exp(-z_m) \tag{9.6}$$

Thus in the receiver described $P(0|1)$ represents the system error probability $P(e)$ and therefore:

$$P(e) = \exp(-z_{\mathrm{m}}) \tag{9.7}$$

However, it must be noted that the above analysis assumes that the photodetector emits no electron–hole pairs in the absence of illumination. In this sense it is considered perfect. Equation (9.7) therefore represents an absolute receiver sensitivity and allows the determination of a fundamental limit in digital optical communications. This is the minimum pulse energy E_{min} required to maintain a given bit-error-rate (BER) which any practical receiver must satisfy and is known as the quantum limit.

Example 9.1

A digital optical fiber communication system operating at a wavelength of 1 μm requires a maximum bit-error-rate of 10^{-9}. Determine:

(a) the theoretical quantum limit at the receiver in terms of the quantum efficiency of the detector and the energy of an incident photon;

(b) the minimum incident optical power required at the detector in order to achieve the above bit-error-rate when the system is employing ideal binary signaling at 10 Mbit s^{-1}, and assuming the detector is ideal.

Solution: (a) From Eq. (9.7) the probability of error:

$$P(e) = \exp(-z_{\mathrm{m}}) = 10^{-9}$$

and thus $z_{\mathrm{m}} = 20.7$.

z_{m} corresponds to an average number of photons detected in a time period τ for a BER of 10^{-9}.

From Eq. (9.4):

$$z_{\mathrm{m}} = \frac{\eta P_{\mathrm{o}} \tau}{hf} = 20.7$$

Hence the minimum pulse energy or quantum limit:

$$E_{\mathrm{min}} = P_{\mathrm{o}}\tau = \frac{20.7 hf}{\eta}$$

Thus the quantum limit at the receiver to maintain a maximum BER of 10^{-9} is:

$$\frac{20.7 hf}{\eta}$$

(b) From part (a) the minimum pulse energy:

$$P_{\mathrm{o}}\tau = \frac{20.7 hf}{\eta}$$

Therefore the average received optical power required to provide the minimum pulse energy is:

$$P_o = \frac{20.7hf}{\tau\eta}$$

However, for ideal binary signaling there are an equal number of ones and zeros (50% in the on state and 50% in the off state). Thus the average received optical power may be considered to arrive over two bit periods, and:

$$P_o(\text{binary}) = \frac{20.7hf}{2\tau\eta} = \frac{20.7hf\,B_T}{2\eta}$$

where B_T is the bit rate. At a wavelength of 1 μm, $f = 2.998 \times 10^{14}$ Hz, and assuming an ideal detector, $\eta = 1$.
   Hence:

$$P_o(\text{binary}) = \frac{20.7 \times 6.626 \times 10^{-34} \times 2.998 \times 10^{14} \times 10^7}{2}$$

$$= 20.6 \text{ pW}$$

In decibels (dB):

$$P_o \text{ in dB} = 10 \log_{10} \frac{P_o}{P_r}$$

where P_r is a reference power level.
   When the reference power level is 1 watt:

$$P_o = 10 \log_{10} P_o \quad \text{where } P_o \text{ is expressed in watts}$$
$$= 10 \log_{10} 2.06 \times 10^{-11}$$
$$= 3.14 - 110$$
$$= -106.9 \text{ dBW}$$

When the reference power level is 1 milliwatt:

$$P_o = 10 \log_{10} 2.06 \times 10^{-8}$$
$$= 3.14 - 80$$
$$= -76.9 \text{ dBm}$$

Therefore the minimum incident optical power required at the receiver to achieve an error rate of 10^{-9} with ideal binary signaling is 20.6 pW or -76.9 dBm.

The result of Example 9.1 is a theoretical limit and in practice receivers are generally found to be at least 10 dB less sensitive. Furthermore, although some 20.7 photons are required in order to detect a binary 1 with a BER of 10^{-9}, it is clear that these photons can arrive at the receiver over two bit periods if an equal number of transmitted ones and zeros are assumed (i.e. there are no photons transmitted in the zero-bit periods). Hence the 20.7 photons per pulse requirement can be considered as an average of around 10.4 photons per bit at the quantum limit.

9.2.5 Analog transmission quantum noise

In analog optical fiber systems quantum noise manifests itself as shot noise which also has Poisson statistics [Ref. 1]. The shot noise current i_s on the photocurrent I_p is given by:

$$\overline{i_s^2} = 2eBI_p \tag{9.8}$$

Neglecting other sources of noise the SNR at the receiver may be written as:

$$\frac{S}{N} = \frac{I_p^2}{i_s^2} \tag{9.9}$$

Substituting for $\overline{i_s^2}$ from Eq. (9.8) gives:

$$\frac{S}{N} = \frac{I_p}{2eB} \tag{9.10}$$

The expression for the photocurrent I_p given in Eq. (8.8) allows the SNR to be obtained in terms of the incident optical power P_o:

$$\frac{S}{N} = \frac{\eta P_o e}{hf \, 2eB} = \frac{\eta P_o}{2hfB} \tag{9.11}$$

Equation (9.11) allows calculation of the incident optical power required at the receiver in order to obtain a specified SNR when considering quantum noise in analog optical fiber systems.

Example 9.2

An analog optical fiber system operating at a wavelength of 1 μm has a post-detection bandwidth of 5 MHz. Assuming an ideal detector and considering only quantum noise on the signal, calculate the incident optical power necessary to achieve an SNR of 50 dB at the receiver.

Solution: From Eq. (9.11), the SNR is:

$$\frac{S}{N} = \frac{\eta P_o}{2hfB}$$

Hence:

$$P_o = \left(\frac{S}{N}\right)\frac{2hfB}{\eta}$$

For $S/N = 50$ dB, when considering signal and noise powers:

$$10 \log_{10} \frac{S}{N} = 50$$

and therefore $S/N = 10^5$

At 1 μm, $f = 2.998 \times 10^{14}$ Hz. For an ideal detector $\eta = 1$ and thus the incident optical power:

$$P_o = \frac{10^5 \times 2 \times 6.626 \times 10^{-34} \times 2.998 \times 10^{14} \times 5 \times 10^6}{1}$$

$$= 198.6 \text{ nW}$$

In dBm:

$$P_o = 10 \log_{10} 198.6 \times 10^{-6}$$
$$= -40 + 2.98$$
$$= -37.0 \text{ dBm}$$

Therefore the incident optical power required to achieve an SNR of 50 dB at the receiver is 198.6 nW which is equivalent to −37.0 dBm.

In practice, receivers are less sensitive than Example 9.2 suggests and thus in terms of the absolute optical power requirements analog transmission compares unfavorably with digital signaling.

However, it should be noted that there is a substantial difference in information transmission capacity between the digital and analog cases (over similar bandwidths) considered in Examples 9.1 and 9.2. For example, a 10 Mbit s^{-1} digital optical fiber communication system would provide only about 150 speech channels using standard baseband digital transmission techniques (see Section 12.5). In contrast a 5 MHz analog system, again operating in the baseband, could provide as many as 1250 similar bandwidth ($\approx$3.4 kHz) speech channels. A comparison of signal to quantum noise ratios between the two transmission methods, taking account of this information capacity aspect, yields less disparity although digital signaling still proves far superior. For instance, applying the figures quoted above within Examples 9.1 and 9.2, in order to compare two systems capable of transmitting the same number of speech channels (e.g. digital bandwidth of 10 Mbit s^{-1} and analog bandwidth of 600 kHz), gives a difference in absolute sensitivity in favor of digital transmission of approximately 31 dB. This indicates a reduction of around 9 dB on the 40 dB difference obtained by simply comparing the results over similar bandwidths. Nevertheless, it is clear that digital signaling techniques still provide a significant benefit in relation to quantum noise when employed within optical fiber communications.

9.3 Receiver noise

In order to investigate the optical receiver in greater detail it is necessary to consider the relative importance and interplay of the various types of noise mentioned in the preceding section. This is dependent on both the method of demodulation and the type of device used for detection.

The conditions for coherent detection are not met in IM/DD optical fiber systems for the reasons outlined in Section 7.5. Thus heterodyne and homodyne detection, which are very sensitive techniques and provide excellent rejection of adjacent channels, are not used, as the optical signal arriving at the receiver tends to be incoherent. In practice the vast majority of installed optical fiber communication systems use incoherent or direct detection in which the variation of the optical power level is monitored and no information is carried in the phase or frequency content of the signal. Therefore, the noise considerations in this section are based on a receiver employing direct detection of the modulated optical carrier which gives the same SNR as an unmodulated optical carrier. The significant developments in coherent optical fiber transmission, however, which have taken place over recent years are described in Chapter 13. Nevertheless, the major performance parameters associated with direct detection receivers which are discussed in this section and the following ones also apply to coherent optical receivers.

Figure 9.3 shows a block schematic of the front end of an optical receiver and the various noise sources associated with it. The majority of the noise sources shown apply to both

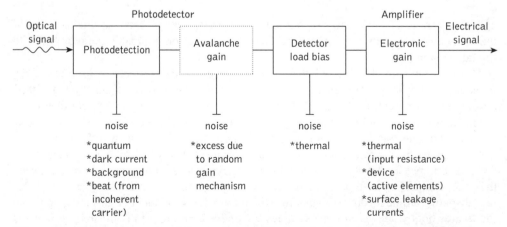

Figure 9.3 Block schematic of the front end of an optical receiver showing the various sources of noise

main types of optical detector (p–i–n and avalanche photodiode). The noise generated from background radiation, which is important in atmospheric propagation and some copper-based systems, is negligible in both types of optical fiber receiver, and thus is often ignored. Also the beat noise generated from the various spectral components of the incoherent optical carrier can be shown to be insignificant [Ref. 4] with multimode propagation and hence will not be considered. It is necessary, however, to take into account the other sources of noise shown in Figure 9.3.

The avalanche photodiode receiver is the most complex case as it includes noise resulting from the random nature of the internal gain mechanism (dotted in Figure 9.3). It is therefore useful to consider noise in optical fiber receivers employing photodiodes without internal gain, before avalanche photodiode receivers are discussed.

9.3.1 The p–n and p–i–n photodiode receiver

The two main sources of noise in photodiodes without internal gain are dark current noise and quantum noise, both of which may be regarded as shot noise on the photocurrent (i.e. effectively, analog quantum noise). When the expressions for these noise sources given in Eqs (9.2) and (9.4) are combined the total shot noise $\overline{i_{TS}^2}$ is given by:

$$\overline{i_{TS}^2} = 2eB(I_p + I_d) \tag{9.12}$$

If it is necessary to take the noise due to the background radiation into account then the expression given in Eq. (9.12) may be expanded to include the background-radiation-induced photocurrent I_b giving:

$$\overline{i_{TS}^2} = 2eB(I_p + I_d + I_b) \tag{9.13}$$

However, as I_b is usually negligible the expression given in Eq. (9.12) will be used in the further analysis.

When the photodiode is without internal avalanche gain, thermal noise from the detector load resistor and from active elements in the amplifier tends to dominate. This is especially the case for wideband systems operating in the 0.8 to 0.9 µm wavelength band because the dark currents in well-designed silicon photodiodes can be made very small. The thermal noise $\overline{i_t^2}$ due to the load resistance R_L may be obtained from Eq. (9. 1) and is given by:

$$\overline{i_t^2} = \frac{4KTB}{R_L} \tag{9.14}$$

The dominating effect of this thermal noise over the shot noise in photodiodes without internal gain may be observed in Example 9.3.

Example 9.3 does not include the noise sources within the amplifier, shown in Figure 9.3. These noise sources, associated with both the active and passive elements of the amplifier, can be represented by a series voltage noise source $\overline{v_a^2}$ and a shunt current noise source $\overline{i_a^2}$.

Example 9.3

A silicon p–i–n photodiode incorporated into an optical receiver has a quantum efficiency of 60% when operating at a wavelength of 0.9 μm. The dark current in the device at this operating point is 3 nA and the load resistance is 4 kΩ.

The incident optical power at this wavelength is 200 nW and the post-detection bandwidth of the receiver is 5 MHz. Compare the shot noise generated in the photodiode with the thermal noise in the load resistor at a temperature of 20 °C.

Solution: From Eq. (8.8) the photocurrent is given by:

$$I_\mathrm{p} = \frac{\eta P_\mathrm{o} e}{hf} = \frac{\eta P_\mathrm{o} e \lambda}{hc}$$

Therefore:

$$I_\mathrm{p} = \frac{0.6 \times 200 \times 10^{-9} \times 1.602 \times 10^{-19} \times 0.9 \times 10^{-6}}{6.626 \times 10^{-34} \times 2.998 \times 10^{8}}$$

$$= 87.1 \text{ nA}$$

From Eq. (9.12) the total shot noise is:

$$\overline{i_\mathrm{TS}^2} = 2eB(I_\mathrm{d} + I_\mathrm{p})$$
$$= 2 \times 1.602 \times 10^{-19} \times 5 \times 10^{6} \, [(3 + 87.1) \times 10^{-9}]$$
$$= 1.44 \times 10^{-19} \text{ A}^2$$

and the root mean square (rms) shot noise current is:

$$(\overline{i_\mathrm{TS}^2})^{\frac{1}{2}} = 3.79 \times 10^{-10} \text{ A}$$

The thermal noise in the load resistor is given by Eq. (9.14):

$$\overline{i_\mathrm{t}^2} = \frac{4KTB}{R_\mathrm{L}}$$

$$= \frac{4 \times 1.381 \times 10^{-23} \times 293 \times 5 \times 10^{6}}{4 \times 10^{3}}$$

$$= 2.02 \times 10^{-17} \text{ A}^2$$

($T = 20 \text{ °C} = 293 \text{ K}$).

Therefore the rms thermal noise current is:

$$(\overline{i_\mathrm{t}^2})^{\frac{1}{2}} = 4.49 \times 10^{-9} \text{ A}$$

In this example the rms thermal noise current is a factor of 12 greater than the total rms shot noise current.

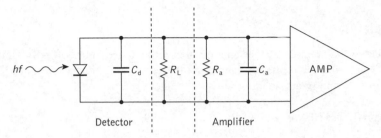

Figure 9.4 The equivalent circuit for the front end of an optical fiber receiver

Thus the total noise associated with the amplifier $\overline{i_{amp}^2}$ is given by:

$$\overline{i_{amp}^2} = \int_0^B (\overline{i_a^2} \times \overline{v_a^2} \,|\,Y\,|^2)\, df \tag{9.15}$$

where Y is the shunt admittance (combines the shunt capacitances and resistances) and f is the frequency. An equivalent circuit for the front end of the receiver, including the effective input capacitance C_a and resistance R_a of the amplifier, is shown in Figure 9.4. The capacitance of the detector C_d is also shown and the noise resulting from C_d is usually included in the expression for $\overline{i_{amp}^2}$ given in Eq. (9.15).

The SNR for the p–n or p–i–n photodiode receiver may be obtained by summing the noise contributions from Eqs (9.12), (9.14) and (9.15). It is given by:

$$\frac{S}{N} = \frac{I_p^2}{2eB(I_p + I_d) + \dfrac{4KTB}{R_L} + \overline{i_{amp}^2}} \tag{9.16}$$

The thermal noise contribution may be reduced by increasing the value of the load resistor R_L, although this reduction may be limited by bandwidth considerations which are discussed later. Also, the noise associated with the amplifier $\overline{i_{amp}^2}$ may be reduced with low detector and amplifier capacitance.

However, when the noise associated with the amplifier $\overline{i_{amp}^2}$ is referred to the load resistor R_L, the noise figure F_n [Ref. 1] for the amplifier may be obtained. This allows $\overline{i_{amp}^2}$ to be combined with the thermal noise from the load resistor $\overline{i_t^2}$ to give:

$$\overline{i_t^2} \times \overline{i_{amp}^2} = \frac{4KTBF_n}{R_L} \tag{9.17}$$

The expression for the SNR given in Eq. (9.16) can now be written in the form:

$$\frac{S}{N} = \frac{I_p^2}{2eB(I_p + I_d) + \dfrac{4KTBF_n}{R_L}} \tag{9.18}$$

Thus if the noise figure F_n for the amplifier is known, Eq. (9.18) allows the SNR to be determined.

Example 9.4

The receiver in Example 9.3 has an amplifier with a noise figure of 3 dB. Determine the SNR at the output of the receiver under the same conditions as Example 9.3.

Solution: From Example 9.3:

$$I_p = 87.1 \times 10^{-9} \text{ A}$$

$$\overline{i_{TS}^2} = 1.44 \times 10^{-19} \text{ A}^2$$

$$\overline{i_t^2} = 2.02 \times 10^{-17} \text{ A}^2$$

The amplifier noise figure:

$$F_n = 3 \text{ dB}$$
$$= 10 \log_{10} 2$$

Thus F_n may be considered as $\times 2$.

In Eq. (9.18) the SNR is given by:

$$\frac{S}{N} = \frac{I_p^2}{2eB(I_p + I_d) + \dfrac{4KTBF_n}{R_L}}$$

$$= \frac{I_p^2}{\overline{i_{TS}^2} + (\overline{i_t^2} \times F_n)}$$

$$= \frac{(87.1 \times 10^{-9})^2}{(1.44 \times 10^{-19}) + (2.02 \times 10^{-17} \times 2)}$$

$$= 1.87 \times 10^2$$

SNR in dB $= 10 \log_{10} 1.87 \times 10^2 = 22.72$ dB.

Alternatively it is possible to conduct the calculation in dB if we neglect the shot noise (say, $\overline{i_{TS}^2} = 0$).

In dB:

$$I_p = 9.40 - 80 = -70.60$$

Hence:

$$I_p^2 = -141.20 \text{ dB}$$

and:

$$\overline{i_t^2} = 3.05 - 170 = -166.95 \text{ dB}$$

The amplifier noise figure $F_n = 3$ dB.
  Therefore:

$$\frac{S}{N} = -141.20 + 166.95 - 3$$

$$= 22.75 \text{ dB}$$

A slight difference in the final answer may be noted. This is due to the neglected shot noise term.

A quantity discussed in Section 8.8.3 which is often used in the specification of optical detectors (or detector–amplifier combinations) is the noise equivalent power (*NEP*). It is defined as the amount of incident optical power P_o per unit bandwidth required to produce an output power equal to the detector (or detector–amplifier combination) output noise power. The *NEP* is therefore the value of P_o which gives an output SNR of unity. Thus the lower the *NEP* for a particular detector (or detector–amplifier combination), the less optical power is needed to obtain a particular SNR.

9.3.2 Receiver capacitance and bandwidth

Considering the equivalent circuit shown in Figure 9.4, the total capacitance for the front end of an optical receiver C_T is given by:

$$C_T = C_d + C_a \qquad\qquad (9.19)$$

where C_d is the detector capacitance and C_a is the amplifier input capacitance. It is important that this total capacitance is minimized not only from the noise considerations discussed previously, but also from the bandwidth penalty which is incurred due to the time constant of C_T and the load resistance R_L. We assume here that R_L is the total loading on the detector and therefore have neglected the amplifier input resistance R_a. However, in practical receiver configurations R_a may have to be taken into account (see Section 9.4.1). The reciprocal of the time constant $2\pi R_L C_T$ must be greater than, or equal to, the post-detection bandwidth B:

$$\frac{1}{2\pi R_L C_T} \geq B \qquad\qquad (9.20)$$

When the equality exists in Eq. (9.20) it defines the maximum possible value of B for the straightforward termination indicated in Figure 9.4.

Assuming that the total capacitance may be minimized, then the other parameter which affects B is the load resistance R_L. To increase B it is necessary to reduce R_L. However, this introduces a thermal noise penalty as may be seen from Eq. (9.14) where both the increase in B and decrease in R_L contribute to an increase in the thermal noise. A trade-off therefore exists between the maximum bandwidth and the level of thermal noise which may be tolerated. This is especially important in receivers which are dominated by thermal noise.

Example 9.5

A photodiode has a capacitance of 6 pF. Calculate the maximum load resistance which allows an 8 MHz post-detection bandwidth.

Determine the bandwidth penalty with the same load resistance when the following amplifier also has an input capacitance of 6 pF.

Solution: From Eq. (9.20) the maximum bandwidth is given by:

$$B = \frac{1}{2\pi R_L C_d}$$

Therefore the maximum load resistance:

$$R_L\,(\text{max}) = \frac{1}{2\pi C_d B} = \frac{1}{2\pi \times 6 \times 10^{-12} \times 8 \times 10^6}$$

$$= 3.32 \text{ k}\Omega$$

Thus for an 8 MHz bandwidth the maximum load resistance is 3.32 kΩ.

Also, considering the amplifier capacitance, the maximum bandwidth:

$$B = \frac{1}{2\pi R_L (C_d + C_a)} = \frac{1}{2\pi \times 3.32 \times 10^3 \times 12 \times 10^{-12}}$$

$$= 4 \text{ MHz}$$

As would be expected, the maximum post-detection bandwidth is halved.

9.3.3 Avalanche photodiode (APD) receiver

The internal gain mechanism in an APD increases the signal current into the amplifier and so improves the SNR because the load resistance and amplifier noise remain unaffected (i.e. the thermal noise and amplifier noise figure are unchanged). However, the dark current and quantum noise are increased by the multiplication process and may become a limiting factor. This is because the random gain mechanism introduces excess noise into the receiver in terms of increased shot noise above the level that would result from amplifying only the primary shot noise. Thus if the photocurrent is increased by a factor M (mean avalanche multiplication factor), then the shot noise is also increased by an excess noise factor M^x, such that the total shot noise $\overline{i_{SA}^2}$ is now given by:

$$\overline{i_{SA}^2} = 2eB(I_p + I_d)M^{2+x} \qquad (9.21)$$

where x is between 0.3 and 0.5 for silicon APDs and between 0.7 and 1.0 for germanium or III–V alloy APDs.

Equation (9.21) is often used as the total shot noise term in order to compute the SNR, although there is a small amount of shot noise current which is not multiplied through impact ionization. The shot noise current in the detector which is not multiplied is a device parameter and may be considered as an extra shot noise term. However, it tends to be insignificant in comparison with the multiplied shot noise and is therefore neglected in the further analysis (i.e. all shot noise is assumed to be multiplied).

The SNR for the APD may be obtained by summing the combined noise contribution from the load resistor and the amplifier given in Eq. (9.17), which remains unchanged, with the modified noise term given in Eq. (9.21). Hence the SNR for the APD is:

$$\frac{S}{N} = \frac{M^2 I_p^2}{2eB(I_p + I_d)M^{2+x} + \dfrac{4KTBF_n}{R_L}} \tag{9.22}$$

It is apparent from Eq. (9.22) that the relative significance of the combined thermal and amplifier noise term is reduced due to the avalanche multiplication of the shot noise term. When Eq. (9.22) is written in the form:

$$\frac{S}{N} = \frac{I_p^2}{2eB(I_p + I_d)M^x + \dfrac{4KTBF_n}{R_L} M^{-2}} \tag{9.23}$$

it may be seen that the first term in the denominator increases with increasing M whereas the second term decreases. For low M the combined thermal and amplifier noise term dominates and the total noise power is virtually unaffected when the signal level is increased, giving an improved SNR. However, when M is large, the thermal and amplifier noise term becomes insignificant and the SNR decreases with increasing M at the rate of M^x. An optimum value of the multiplication factor M_{op} therefore exists which maximizes the SNR. It is given by:

$$\frac{2eB(I_p + I_d)\, M_{op}^x}{(4KTBF_n/R_L)\, M_{op}^{-2}} = \frac{2}{x} \tag{9.24}$$

and therefore:

$$M_{op}^{2+x} = \frac{4KTF_n}{xeR_L(I_p + I_d)} \tag{9.25}$$

The variation in M_{op} for both silicon and germanium APDs is illustrated in Figure 9.5 [Ref. 5]. This shows a plot of Eq. (9.22) with F_n equal to unity and neglecting the dark current. For good silicon APDs where x is 0.3, the optimum multiplication factor covers a wide range. In the case illustrated in Figure 9.5, M_{op} commences at about 40 where the possible improvement in SNR above a photodiode without internal gain is in excess of

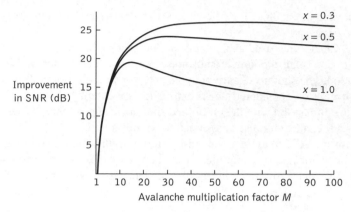

Figure 9.5 The improvement in SNR as a function of avalanche multiplication factor M for different excess noise factors M^x. Reproduced with permission from I. Garrett, *Radio Electron. Eng.*, **51**, p. 349, 1981

25 dB. However, for germanium and III–V alloy APDs where x may be equal to unity, it can be seen that less SNR improvement is possible (less than 19 dB). Moreover, the maximum is far sharper, occurring at a multiplication factor of about 12. Also it must be noted that Figure 9.5 demonstrates the variation of M_{op} with x for a specific case, and therefore only represents a general trend. It may be observed from Eq. (9.25) that M_{op} is dependent on a number of other variables apart from x.

Example 9.6

A good silicon APD ($x = 0.3$) has a capacitance of 5 pF, negligible dark current and is operating with a post-detection bandwidth of 50 MHz. When the photocurrent before gain is 10^{-7} A and the temperature is 18 °C, determine the maximum SNR improvement between $M = 1$ and $M = M_{op}$ assuming all operating conditions are maintained.

Solution: Determine the maximum value of the load resistor from Eq. (9.20):

$$R_L = \frac{1}{2\pi C_d B} = \frac{1}{2\pi \times 5 \times 10^{-12} \times 50 \times 10^6}$$

$$= 635.5\ \Omega$$

When $M = 1$, the SNR is given by Eq. (9.22):

$$\frac{S}{N} = \frac{I_p^2}{2eBI_p + \dfrac{4KTB}{R_L}}$$

where $I_d = 0$ and $F_n = 1$.

The shot noise is:

$$2eBI_p = 2 \times 1.602 \times 10^{-19} \times 50 \times 10^6 \times 10^{-7}$$
$$= 1.602 \times 10^{-18} \text{ A}^2$$

and the thermal noise is:

$$\frac{4KTB}{R_L} = \frac{4 \times 1.381 \times 10^{-23} \times 291 \times 50 \times 10^6}{636.5}$$
$$= 1.263 \times 10^{-15} \text{ A}^2$$

It may be noted that the thermal noise is dominating.
Therefore:

$$\frac{S}{N} = \frac{10^{-14}}{1.602 \times 10^{-18} \times 1.263 \times 10^{-15}} = 7.91$$

and the SNR in dB is:

$$\frac{S}{N} = 10 \log_{10} 7.91 = 8.98 \text{ dB}$$

Thus the SNR when $M = 1$ is 9.0 dB.
When $M = M_{op}$ and $x = 0.3$, from Eq. (9.25):

$$M_{op}^{2+x} = \frac{4KT}{xeR_1 I_p}$$

where $I_d = 0$ and $F_n = 1$. Hence:

$$M_{op}^{2.3} = \frac{4 \times 1.381 \times 10^{-23} \times 291}{0.3 \times 1.602 \times 10^{-19} \times 636.5 \times 10^{-7}}$$

and:

$$M_{op} = (5.255 \times 10^3)^{0.435}$$
$$= 41.54$$

The SNR at M_{op} may be obtained from Eq. (9.22):

$$\frac{S}{N} = \frac{M^2 I_p^2}{2eBI_p M^{2.3} + \dfrac{4KTB}{R_L}}$$

$$= \frac{(41.54)^2 \times 10^{-14}}{\{1.602 \times 10^{-18} \times (41.54)^{2.3}\} + 1.263 \times 10^{-15}}$$

$$= 1.78 \times 10^3$$

and the SNR in dB is:

$$\frac{S}{N} = 10 \log_{10} 1.78 \times 10^3 = 32.50 \text{ dB}$$

Therefore the SNR when $M = M_{op}$ is 32.5 dB and the SNR improvement over $M = 1$ is 23.5 dB.

Example 9.7

A germanium APD (with $x = 1$) is incorporated into an optical fiber receiver with a 10 kΩ load resistance. When operated at a temperature of 120 K, the minimum photocurrent required to give an SNR of 35 dB at the output of the receiver is found to be a factor of 10 greater than the dark current. If the noise figure of the following amplifier at this temperature is 1 dB and the post-detection bandwidth is 10 MHz, determine the optimum avalanche multiplication factor.

Solution: From Eq. (9.22) with $x = 1$ and $M = M_{op}$ (i.e. minimum photocurrent specifies that $M = M_{op}$) the SNR is:

$$\frac{S}{N} = \frac{M_{op}^2 I_p^2}{2eB(I_p + I_d)M_{op}^3 + \dfrac{4KTB}{R_L}}$$

Also from Eq. (9.25):

$$M_{op}^3 = \frac{4KTF_n}{eR_L(I_p + I_d)}$$

Therefore:

$$M_{op} = \left[\frac{4KTF_n}{eR_L(I_p + I_d)} \right]^{\frac{1}{3}}$$

Substituting into Eq. (9.22), this gives:

$$\frac{S}{N} = \frac{\left[\dfrac{4KTF_n}{eR_L(I_p + I_d)} \right]^{\frac{2}{3}} I_p^2}{\dfrac{8KTBF_n}{R_L} + \dfrac{4KTBF_n}{R_L}}$$

and as $I_d = 0.1I_p$ the SNR is:

$$\frac{S}{N} = \frac{\left(\frac{4KTF_n}{1.1eR_L}\right)^{\frac{2}{3}} I_p^{\frac{4}{3}}}{\frac{12KTBF_n}{R_L}}$$

Therefore the minimum photocurrent I_p:

$$I_p^{\frac{4}{3}} = \left(\frac{S}{N}\right) \frac{\frac{12KTBF_n}{R_L}}{\left(\frac{4KTF_n}{1.1eR_L}\right)^{\frac{2}{3}}}$$

where the SNR is:

$$\frac{S}{N} = 35 \text{ dB} = 3.16 \times 10^3$$

and as $F_n = 1$ dB which is equivalent to 1.26:

$$\frac{12KTBF_n}{R_L} = \frac{12 \times 1.381 \times 10^{-23} \times 120 \times 10^7 \times 1.26}{10^4}$$

$$= 2.51 \times 10^{-17}$$

Also:

$$\left(\frac{4KTF_n}{1.1eR_L}\right)^{\frac{1}{2}} = \left(\frac{4 \times 1.381 \times 10^{-23} \times 120 \times 1.26}{1.1 \times 1.602 \times 10^{-19} \times 10^4}\right)^{\frac{2}{3}}$$

$$= 2.82 \times 10^{-4}$$

Therefore:

$$I_p = \left(\frac{3.16 \times 10^3 \times 2.51 \times 10^{-17}}{2.82 \times 10^{-4}}\right)^{\frac{3}{4}}$$

$$= 6.87 \times 10^{-8} \text{ A}$$

To obtain the optimum avalanche multiplication factor we substitute back into Eq. (9.25), where:

$$M_{op} = \left(\frac{4 \times 1.381 \times 10^{-23} \times 120 \times 1.26}{1.602 \times 10^{-19} \times 10^3 \times 1.1 \times 6.87 \times 10^{-8}}\right)^{\frac{1}{3}}$$

$$= 8.84$$

In Example 9.7 the optimum multiplication factor for the germanium APD is found to be approximately 9. It shows the dependence of the optimum multiplication factor on the variables in Eq. (9.25), and although the example does not necessarily represent a practical receiver (some practical germanium APD receivers are cooled to reduce dark current), the optimum multiplication factor is influenced by device and system parameters as well as operating conditions.

9.3.4 Excess avalanche noise factor

The value of the excess avalanche noise factor is dependent upon the detector material, the shape of the electric field profile within the device and whether the avalanche is initiated by holes or electrons. It is often represented as $F(M)$ and in the preceding section we have considered one of the approximations for the excess noise factor, where:

$$F(M) = M^x \tag{9.26}$$

and the resulting noise is assumed to be white with a Gaussian distribution.

However, a second and more exact relationship is given by [Ref. 6]:

$$F(M) = M\left[1 - (1 - k)\left(\frac{M-1}{M}\right)^2 \right] \tag{9.27}$$

where the only carriers are injected electrons and k is the ratio of the ionization coefficients of holes and electrons. If the only carriers are injected holes:

$$F(M) = M\left[1 + \left(\frac{1-k}{k}\right)\left(\frac{M-1}{M}\right)^2 \right] \tag{9.28}$$

Figure 9.6 provides the excess avalanche noise factor as a function of multiplication factor M for several values of k which is obtained from Eq. (9.28). It should be noted that the hole injections for a value of $k = 1$, the excess noise factor, are simply equal to the multiplication factor. In the case of holes the smaller values of k produce high performance and therefore the best performance is achieved when k is small. Moreover, for silicon APDs k is between 0.02 and 0.10, whereas for germanium and III–V alloy APDs k is between 0.3 and 1.0.

With electron injection in silicon photodiodes, the smaller values of k obtained correspond to a larger ionization rate for the electrons than for the holes. As k departs from unity, only the carrier with the larger ionization rate contributes to the impact ionization and the excess avalanche noise factor is reduced. When the impact ionization is initiated by electrons this corresponds to fewer ionizing collisions involving the hole current which is flowing in the opposite direction (i.e. less feedback). In this case the amplified signal contains less excess noise. The carrier ionization rates in germanium photodiodes are often nearly equal and hence k approaches unity, giving a high level of excess noise.

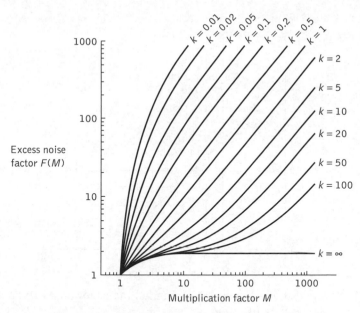

Figure 9.6 Excess noise factor against multiplication factor for different values of the ratio of ionization coefficients for holes and electrons k; hole injection based on Eq. (9.28).

9.3.5 Gain-bandwidth product

In addition to SNR a figure of merit for an APD receiver can also be expressed in terms of either the maximum 3 dB bandwidth or, alternatively, the gain–bandwidth product [Ref. 7]. The latter term is defined as gain multiplied by the bandwidth and determines a higher transmission rate limit related to the gain of the APD device. Since gain is a dimensionless quantity the gain–bandwidth product is therefore measured in the units of frequency (i.e. GHz).

Figure 9.7 shows the relationship between bandwidth and multiplication factor of an APD [Ref. 8]. The circles represent the 3 dB bandwidth values and constitute a nonlinear curve containing a linear part at higher values of the multiplication factor. The linearity occurs at a cut off frequency point after which the gain of the APD remains constant (i.e. in this particular case, at $M > 6$). When these point values displaying the linear part are connected as indicated in the Figure 9.7, the resultant slope line represents the theoretical value of the achievable gain–bandwidth product for an APD. For example, the 3 dB bandwidth at $M = 10$ remains 14 GHz which produces a gain–bandwidth product of 140 GHz. Furthermore, it maintains the same value of gain–bandwidth product for $M = 6$ with a 3 dB bandwidth of 23.3 GHz. This suggests a theoretical limit for the gain–bandwidth product of 140 GHz for the same device. The maximum 3 dB bandwidth for the APD, however, is shown as 40 GHz which is identified by the horizontal dashed line in Figure 9.7.

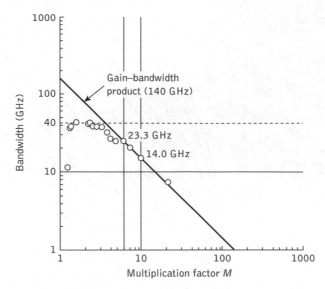

Figure 9.7 Bandwidth (3 dB) against multiplication factor for an APD; adapted with permission from Ref. 8 © IEEE 2004

Silicon APDs are recognized for their high gain–bandwidth products which result from the large asymmetry of electron and hole ionization coefficients leading to a ratio of ionization coefficients k of around 50 [Ref. 9]. However, these APDs do not operate at signal wavelengths between 1.3 and 1.6 μm and therefore they are not preferred for use in receivers operating at high transmission rates. Although InGaAs/InP-based APDs possess high quantum efficiency and can be utilized at signal wavelengths between 1.3 and 1.6 μm, they suffer from a lower gain–bandwidth product due to the small ratios of ionization coefficients in InP. Nevertheless, separate absorption, charge and multiplication (SACM) APD receivers (see Section 8.9.4) based on the InGaAs/InAlAs material system exhibit large gain–bandwidth products and a 290 GHz gain–bandwidth for such APDs has been achieved while maintaining the 3 dB bandwidth at 33 GHz [Ref. 10]. Hence such devices have found application in high-speed optical receivers where they provide greater sensitivity in comparison with *p–i–n* photodiodes.

9.4 Receiver structures

A full equivalent circuit for the digital optical fiber receiver, in which the optical detector is represented as a current source i_{det}, is shown in Figure 9.8. The noise sources (i_t, i_{TS} and i_{amp}) and the immediately following amplifier and equalizer are also shown. Equalization [Ref. 11] compensates for distortion of the signal due to the combined transmitter, medium and receiver characteristics. The equalizer is often a frequency-shaping filter which has a frequency response that is the inverse of the overall system frequency response. In wideband systems this will normally boost the high-frequency components to correct the overall

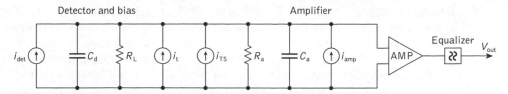

Figure 9.8 A full equivalent circuit for a digital optical fiber receiver including the various noise sources

amplitude of the frequency response. To acquire the desired spectral shape for digital systems (e.g. raised cosine, see Figure 12.39), in order to minimize intersymbol interference, it is important that the phase frequency response of the system is linear. Thus the equalizer may also apply selective phase shifts to particular frequency components.

However, the receiver structure immediately preceding the equalizer is the major concern of this section. In both digital and analog systems it is important to minimize the noise contributions from the sources shown in Figure 9.8 so as to maximize the receiver sensitivity while maintaining a suitable bandwidth. It is therefore useful to discuss various possible receiver structures with regard to these factors.

9.4.1 Low-impedance front-end

Three basic amplifier configurations are frequently used in optical fiber communication receivers. The simplest, and perhaps the most common, is the voltage amplifier with an effective input resistance R_a as shown in Figure 9.9. In order to make suitable design choices, it is necessary to consider both bandwidth and noise. The bandwidth considerations in Section 9.3.2 are treated solely with regard to a detector load resistance R_L. However, in most practical receivers the detector is loaded with a bias resistor R_b and an amplifier (see Figure 9.9). The bandwidth is determined by the passive impedance which appears across the detector terminals which is taken as R_L in the bandwidth relationship given in Eq. (9.20).

However, R_L may be modified to incorporate the parallel resistance of the detector bias resistor R_b and the amplifier input resistance R_a. The modified total load resistance R_{TL} is therefore given by:

$$R_{TL} = \frac{R_b R_a}{R_b + R_a} \tag{9.29}$$

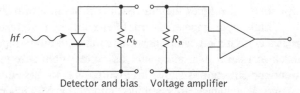

Detector and bias   Voltage amplifier

Figure 9.9 Low-impedance front-end optical fiber receiver with voltage amplifier

Considering the expressions given in Eqs (9.20) and (9.29), to achieve an optimum bandwidth both R_b and R_a must be minimized. This leads to a low-impedance front-end design for the receiver amplifier. Unfortunately this design allows thermal noise to dominate within the receiver (following Eq. (9.14)), which may severely limit its sensitivity. Therefore this structure demands a trade-off between bandwidth and sensitivity which tends to make it impractical for long-haul, wideband optical fiber communication systems.

9.4.2 High-impedance (integrating) front-end

The second configuration consists of a high input impedance amplifier together with a large detector bias resistor in order to reduce the effect of thermal noise. However, this structure tends to give a degraded frequency response as the bandwidth relationship given in Eq. (9.20) is not maintained for wideband operation. The detector output is effectively integrated over a large time constant and must be restored by differentiation. This may be performed by the correct equalization at a later stage [Ref. 12] as illustrated in Figure 9.10. Therefore the high-impedance (integrating) front-end structure gives a significant improvement in sensitivity over the low-impedance front-end design, but it creates a heavy demand for equalization and has problems of limited dynamic range (the ratio of maximum to minimum input signals).

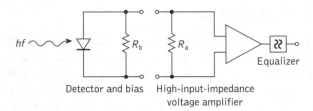

Detector and bias   High-input-impedance
                     voltage amplifier

Figure 9.10 High-impedance integrating front-end optical fiber receiver with equalized voltage amplifier

The limitations on dynamic range result from the attenuation of the low-frequency signal components by the equalization process which causes the amplifier to saturate at high signal levels. When the amplifier saturates before equalization has occurred, the signal is heavily distorted. Thus the reduction in dynamic range is dependent upon the amount of integration and subsequent equalization employed.

9.4.3 The transimpedance front-end

This configuration largely overcomes the drawbacks of the high-impedance front end by utilizing a low-noise, high-input-impedance amplifier with negative feedback. The device therefore operates as a current mode amplifier where the high input impedance is reduced by negative feedback. An equivalent circuit for an optical fiber receiver incorporating a transimpedance front-end structure is shown in Figure 9.11. In this equivalent circuit the parallel resistances and capacitances are combined into R_{TL} and C_T respectively. The open loop current to voltage transfer function $H_{OL}(\omega)$ for this transimpedance configuration

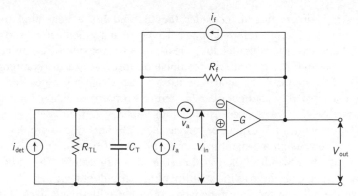

Figure 9.11 An equivalent circuit for the optical fiber receiver incorporating a transimpedance (current mode) preamplifier

corresponds to the transfer function for the two structures described previously which do not employ feedback (i.e. the low- and high-impedance front ends). It may be written as:

$$H_{OL}(\omega) = -G\frac{V_{in}}{i_{det}} = -G\frac{R_{TL}\dfrac{1}{j\omega C_T}}{R_{TL}+\dfrac{1}{j\omega C_T}} = \frac{-GR_{TL}}{1+j\omega R_{TL}C_T}\ (\text{V A}^{-1}) \qquad (9.30)$$

where G is the open loop voltage gain of the amplifier and ω is the angular frequency of the input signal. In this case the bandwidth (without equalization) is constrained by the time constant given in Eq. (9.20).*

When the feedback is applied, the closed loop current to voltage transfer function $H_{CL}(\omega)$ for the transimpedance configuration is given by (see Appendix E):

$$H_{CL}(\omega) \simeq \frac{-R_f}{1+(j\omega R_f C_T/G)}\ (\text{V A}^{-1}) \qquad (9.31)$$

where R_f is the value of the feedback resistor. In this case the permitted electrical bandwidth B (without equalization) may be written as:

$$B \leq \frac{G}{2\pi R_f C_T} \qquad (9.32)$$

Hence, comparing Eq. (9.32) with Eq. (9.20) it may be noted that the transimpedance (or feedback) amplifier provides a much greater bandwidth than do the amplifiers without feedback. This is particularly pronounced when G is large.

* The time constant can be obtained directly from Eq. (9.30) where the maximum bandwidth is defined by $\omega = 2\pi B = 1/R_{TL}C_T$.

Moreover, it is interesting to consider the thermal noise generated by the transimpedance front end. Using a referred impedance noise analysis it can be shown [Ref. 13] that to a good approximation the feedback resistance (or impedance) may be referred to the amplifier input in order to establish the noise performance of the configuration. Thus when $R_f \ll R_{TL}$, the major noise contribution is from thermal noise generated in R_f. The noise performance of this configuration is therefore improved when R_f is large, and it approaches the noise performance of the high-impedance front end when $R_f = R_{TL}$. Unfortunately, the value of R_f cannot be increased indefinitely due to problems of stability with the closed loop design. Furthermore, it may be observed from Eq. (9.32) that increasing R_f reduces the bandwidth of the transimpedance configuration. This problem may be alleviated by making G as large as the stability of the closed loop will allow. Nevertheless, it is clear that the noise in the transimpedance amplifier will always exceed that incurred by the high-impedance front-end structure.

Example 9.8

A high-input-impedance amplifier which is employed in an optical fiber receiver has an effective input resistance of 4 MΩ which is matched to a detector bias resistor of the same value. Determine:

(a) The maximum bandwidth that may be obtained without equalisation if the total capacitance C_T is 6 pF.

(b) The mean square thermal noise current per unit bandwidth generated by this high-input-impedance amplifier configuration when it is operating at a temperature of 300 K.

(c) Compare the values calculated in (a) and (b) with those obtained when the high-input-impedance amplifier is replaced by a transimpedance amplifier with a 100 kΩ feedback resistor and an open loop gain of 400. It may be assumed that $R_f \ll R_{TL}$, and that the total capacitance remains 6 pF.

Solution: (a) Using Eq. (9.29), the total effective load resistance:

$$R_{TL} = \frac{(4 \times 10^6)^2}{8 \times 10^6} = 2 \text{ M}\Omega$$

Hence from Eq. (9.20) the maximum bandwidth is given by:

$$B = \frac{1}{2\pi R_{TL} C_T} = \frac{1}{2\pi \times 2 \times 10^6 \times 6 \times 10^{-12}}$$

$$= 1.33 \times 10^4 \text{ Hz}$$

The maximum bandwidth that may be obtained without equalization is 13.3 kHz.

(b) The mean square thermal noise current per unit bandwidth for the high-impedance configuration following Eq. (9.14) is:

$$\overline{i_t^2} = \frac{4KT}{R_{TL}} = \frac{4 \times 1.381 \times 10^{-23} \times 300}{2 \times 10^6}$$

$$= 8.29 \times 10^{-27} \text{ A}^2 \text{ Hz}^{-1}$$

(c) The maximum bandwidth (without equalization) for the transimpedance configuration may be obtained using Eq. (9.32), where:

$$B = \frac{G}{2\pi R_f C_T} = \frac{400}{2\pi \times 10^5 \times 6 \times 10^{-12}}$$

$$= 1.06 \times 10^8 \text{ Hz}$$

Hence a bandwidth of 106 MHz is permitted by the transimpedance design.

Assuming $R_f \ll R_{TL}$, the mean square thermal noise current per unit bandwidth for the transimpedance configuration is given by:

$$\overline{i_t^2} = \frac{4KT}{R_f} = \frac{4 \times 1.381 \times 10^{-23} \times 300}{10^5}$$

$$= 1.66 \times 10^{-25} \text{ A}^2 \text{ Hz}^{-1}$$

The mean square thermal noise current in the transimpedance configuration is therefore a factor of 20 greater than that obtained with the high-input-impedance configuration.

The equivalent value in decibels of the ratio of these noise powers is:

$$\frac{\text{Noise power in the transimpedance configuration}}{\text{Noise power in the high-input-impedance configuration}} = 10 \log_{10} 20$$

$$= 13 \text{ dB}$$

Thus the transimpedance front-end in Example 9.8 provides a far greater bandwidth without equalization than the high-impedance front-end. However, this advantage is somewhat offset by the 13 dB noise penalty incurred with the transimpedance amplifier over that of the high-input-impedance configuration. Nevertheless it is apparent, even from this simple analysis, that transimpedance amplifiers may be optimized for noise performance, although this is usually obtained at the expense of bandwidth. This topic is pursued further in Ref. 14. However, wideband transimpedance designs generally give a significant improvement in noise performance over the low-impedance front-end structures using simple voltage amplifiers (see Problem 9.18). Finally it must be emphasized that the approach adopted in Example 9.8 is by no means rigorous and includes two important simplifications:

firstly, that the thermal noise in the high-impedance amplifier is assumed to be totally generated by the effective input resistance of the device; and secondly, that the thermal noise in the transimpedance configuration is assumed to be totally generated by the feedback resistor when it is referred to the amplifier input. Both these assumptions are approximations, the accuracy of which is largely dependent on the parameters of the particular amplifier. For example, another factor which tends to reduce the bandwidth of the transimpedance amplifier is the stray capacitance C_f generally associated with the feedback resistor R_f. When C_f is taken into account the closed loop response of Eq. (9.31) becomes:

$$H_{CL}(\Omega) \simeq \frac{-R_f}{1 + j\omega R_f(C_T/G + C_f)} \tag{9.33}$$

However, the effects of C_f may be cancelled by employing a suitable compensating network [Ref. 15].

The other major advantage which the transimpedance configuration has over the high-impedance front end is a greater dynamic range. This improvement in dynamic range obtained using the transimpedance amplifier is a result of the different attenuation mechanism for the low-frequency components of the signal. The attenuation is accomplished in the transimpedance amplifier through the negative feedback and therefore the low-frequency components are amplified by the closed loop rather than the open loop gain of the device. Hence for a particular amplifier the improvement in dynamic range is approximately equal to the ratio of the open loop to the closed loop gains. The transimpedance structure therefore overcomes some of the problems encountered with the other configurations and is often preferred for use in wideband optical fiber communication receivers [Ref. 16].

9.5 FET preamplifiers

The lowest noise amplifier device which is widely available is the silicon FET. Unlike the bipolar transistor, the FET operates by controlling the current flow with an electric field produced by an applied voltage on the gate of the device (see Figure 9.12) rather than with

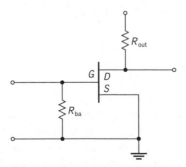

Figure 9.12 Grounded source FET configuration for the front end of an optical fiber receiver amplifier

a base current. Thus the gate draws virtually no current, except for leakage, giving the device an extremely high input impedance (can be greater than 10^{14} ohms). This, coupled with its low noise and capacitance (no greater than a few picofarads), makes the silicon FET appear an ideal choice for the front end of the optical fiber receiver amplifier. However, the superior properties of the FET over the bipolar transistor are limited by its comparatively low transconductance g_m (no better than 5 millisiemens in comparison with at least 40 millisiemens for the bipolar). It can be shown [Ref. 14] that a figure of merit with regard to the noise performance of the FET amplifier is g_m/C_T^2. Hence the advantage of high transconductance together with low total capacitance C_T is apparent. Moreover, as $C_T = C_d + C_a$, it should be noted that the figure of merit is optimized when $C_a = C_d$. This requires FETs to be specifically matched to particular detectors, a procedure which device availability does not generally permit in current optical fiber receiver design. As indicated above, the gain of the FET is restricted. This is especially the case for silicon FETs at frequencies above 25 MHz where the current gain drops to values near unity as the trans-conductance is fixed with a decreasing input impedance. Therefore at frequencies above 25 MHz, the bipolar transistor is a more useful amplifying device.*

Figure 9.12 shows the grounded source FET configuration which increases the device input impedance especially if the amplifier bias resistor R_{ba} is large. A large bias resistor has the effect of reducing the thermal noise but it will also increase the low-frequency impedance of the detector load which tends to integrate the signal (i.e. high-impedance integrating front-end). Thus compensation through equalization at a later stage is generally required.

9.5.1 Gallium arsenide MESFETs

Although silicon FETs have a limited useful bandwidth, much effort has been devoted to the development of high-performance microwave FETs since the mid-1970s. These FETs are fabricated from gallium arsenide and, being Schottky barrier devices [Refs 17–20], are called GaAs metal Schottky field effect transistors (MESFETs). They overcome the major disadvantage of silicon FETs in that they will operate with both low noise and high gain at microwave frequencies (GHz). Thus in optical fiber communication receiver design they present an alternative to bipolar transistors for wideband operation. These devices have therefore been incorporated into high-performance receiver designs using both p–i–n and detectors [Refs 21–32] . In particular, there has been much interest in hybrid integrated receiver circuits utilizing p–i–n photodiodes with GaAs MESFET amplifier front ends. The hybrid integration of a photodetector with a GaAs MESFET preamplifier having low leakage current, low capacitance (less than 0.5 pF) and high transconductance (greater than 30 millisiemens) provides a strategy for low-noise optical receiver design [Ref. 33].

---

* The figure of merit in relation to noise performance for the bipolar transistor amplifier may be shown [Ref. 14] to be $(h_{FE})^{\frac{1}{2}}/C_T$ where h_{FE} is the common emitter current gain of the device. Hence the noise performance of the bipolar amplifier may be optimized in a similar manner to that of the FET amplifier.

9.5.2 PIN–FET hybrid receivers

The p–i–n/FET, or PIN–FET, hybrid receiver utilizes a high-performance p–i–n photodiode followed by a low-noise preamplifier often based on a GaAs MESFET, the whole of which is fabricated using thick-film integrated circuit technology. This hybrid integration on a thick-film substrate reduces the stray capacitance to negligible levels giving a total input capacitance which is very low (e.g. 0.4 pF). The MESFETs employed have a transconductance of approximately 15 millisiemens at the bandwidths required (e.g. 40 Gbits^{-1} [Ref. 31]). Early work [Refs 22, 23] in the 0.8 to 0.9 µm wavelength band utilizing a silicon p–i–n detector showed the PIN–FET hybrid receiver to have a sensitivity of −45.8 dBm for a 10^{-9} BER which is only 4 dB worse than current silicon RAPD receivers (see Section 8.9.2).

The work was subsequently extended into the longer wavelength band (1.1 to 1.6 µm) utilizing III–V alloy p–i–n photodiode detectors. An example of a PIN–FET hybrid high-impedance (integrating) front-end receiver for operation at a wavelength of 1.3 µm using an InGaAs p–i–n photodiode is shown in Figure 9.13 [Refs 24–27]. This design, used by British Telecom, consists of a preamplifier with a GaAs MESFET and microwave bipolar transistor cascode followed by an emitter follower output buffer. The cascode circuit is chosen to ensure that sufficient gain is obtained from the first stage to give an overall gain of 18 dB. As the high-impedance front end effectively integrates the signal, the following digital equalizer is necessary. The pulse shaping and noise filtering circuits comprise two passive filter sections to ensure that the pulse waveform shape is optimized and the noise is minimized. Equalization for the integration (i.e. differentiation) is performed by monitoring the change in the integrated waveform over one period with a subminiature coaxial delay line followed by a high-speed, low-level comparator. The receiver is designed for use at a transmission rate of 140 Mbit s^{-1} where its performance is found to be comparable to germanium and III–V alloy APD receivers. For example, the receiver sensitivity at a BER of 10^{-9} is −44.2 dBm. Table 9.1 provides a comparison of typical sensitivities

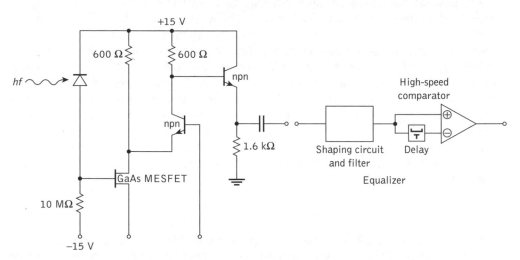

Figure 9.13 PIN–FET hybrid high-impedance integrating front-end receiver [Refs 24–27]

Table 9.1 Sensitivities for InGaAs PIN–FET and InAlAs APD receivers at the wavelength of 1.55 μm

Receiver type	Sensitivity (dBm)	Transmission rate (Gbit s⁻¹)	References
PIN–FET	−23.0	2.5	[Ref. 34]
APD	−34.0	2.5	[Ref. 34]
APD	−29.0	10	[Ref. 35]
APD	−27.1	10	[Ref. 36]
PIN–FET	−7.0	40	[Ref. 34]

obtained with an InGaAs hybrid PIN–FET receiver and an InAlAs APD receiver when both are operating at a wavelength of 1.55 μm. The hybrid PIN–FET receiver design displays a lower sensitivity than the APD receiver at a transmission rate of 2.5 Gbit s⁻¹ and although it can also function at the higher transmission rate of 40 Gbit s⁻¹, the PIN–FET receiver then exhibits a very poor sensitivity of only −7.0 dBm [Ref. 34]. It should be noted that it is difficult to achieve higher transmission rates using conventional PIN–FET or APD receivers due to limitations in their gain–bandwidth products. Furthermore, the trade-off between the multiplication factor requirement and the maximum 3 dB bandwidth also limits the performance of the conventional receivers. However, incorporating optical amplifiers into traveling-waveguide (TW) and unitraveling carrier (UTC) structures (see Section 8.9.4) facilitates the design of photoreceivers capable of operating at transmission rates higher than 40 Gbit s⁻¹ [Ref. 8] (also see Section 9.6).

When compared with the APD receiver the PIN–FET hybrid has both cost and operational advantages especially in the longer wavelength region. The low-voltage operation (e.g. +15 and −15 V supply rails) coupled with good sensitivity and ease of fabrication makes the incorporation of this receiver into wideband optical fiber communication systems commercially attractive. A major drawback with the PIN–FET receiver is the possible lack of dynamic range. However, the configuration shown in Figure 9.13 gave adequate dynamic range via a control circuit which maintained the mean voltage at the gate at 0 V by applying a negative voltage proportional to the mean photocurrent to the MESFET bias resistor. With a −15 V supply rail an optical dynamic range of some 20 dB was obtained. This was increased to 27 dB by reducing the value of the MESFET bias resistor from 10 to 2 MΩ which gave a slight noise penalty of 0.5 dB. These figures compare favorably with practical APD receivers.

Transimpedance front-end receivers have also been fabricated using the PIN–FET hybrid approach. An example of this type of circuit [Ref. 29] is shown in Figure 9.14. The amplifier consists of a GaAs MESFET followed by two complementary bipolar microwave transistors. A silicon *p–i–n* photodiode was utilized with the amplifier and the receiver was designed to accept data at a rate of 274 Mbits s⁻¹. In this case the effective input capacitance of the receiver was 4.5 pF giving a sensitivity around −35 dBm for a BER of 10⁻⁹.

These figures are somewhat worse than the high-impedance front-end design discussed previously. However, this design has the distinct advantage of a flat frequency response to a wider bandwidth which requires little, if any, equalization.

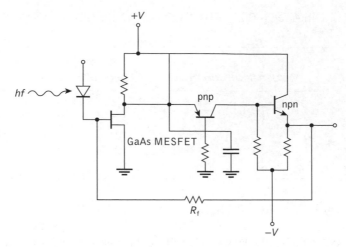

Figure 9.14 PIN–FET hybrid transimpedance front-end receiver [Ref. 29]

9.6 High-performance receivers

It is clear from the discussions in Sections 9.3 to 9.5 that noise performance is a major design consideration providing a limitation to the sensitivity which may be obtained with a particular receiver structure and component mix. However, two other important receiver performance criteria were also outlined in the aforementioned sections, namely bandwidth and dynamic range. Moreover, distinct trade-offs exist between these three performance attributes such that an optimized design for one criterion may display a degradation in relation to one or both of the other criteria. Nevertheless, although high-performance receiver design may seek to provide optimization for one particular attribute, attempts are generally made to minimize the degradations associated with the other performance parameters. In this section we describe further the strategies that have been adopted to produce high-performance receivers for optical fiber communications, together with some of the performance results which have been obtained over the last few years.

As mentioned in Section 9.5.2, low-noise performance combined with potential high-speed operation has been a major pursuit in the hybrid integration of p–i–n photodiodes with GaAs MESFETs. In this context it is useful to compare the noise performance of various transistor preamplifiers over a range of bandwidths. A theoretical state-of-the-art performance comparison for the silicon junction FET (JFET), the silicon metal oxide semiconductor FET (MOSFET) and the silicon bipolar transistor preamplifier with a GaAs MESFET device for transmission rates from 1 Mbit s^{-1} to 10 Gbit s^{-1} is shown in Figure 9.15 [Ref. 37]. It may be observed that at low speeds the three FET preamplifiers provide higher sensitivity than the Si bipolar device. In addition it is apparent that below 10 Mbit s^{-1} the Si MOSFET preamplifier provides a lower noise performance than the GaAs MESFET. Above 20 Mbit s^{-1}, however, the highest sensitivity is obtained with the GaAs MESFET device, even though at very high speeds the Si MOSFET and Si bipolar transistor preamplifiers exhibit a noise performance that is only slightly worse than the

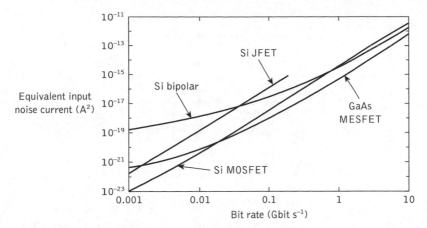

Figure 9.15 Noise characteristics for various optical receiver transistor preamplifiers. Reproduced with permission from B. L. Kasper, 'Receiver design', in S. E. Miller and I. P. Kaminow (Eds), *Optical Fiber Telecommunications II*, Academic Press Inc., p. 689, 1988, © Elsevier

aforementioned device. Furthermore, it is clear that, as indicated in Section 9.5, the Si bipolar transistor preamplifier displays a noise improvement over the Si JFET, in this case at speeds above 50 Mbit s^{-1}.

The optimization of PIN–FET receiver designs for sensitivity and high-speed operation has been investigated [Refs 38, 39]. Also a wideband (10 GHz), low-noise device using discrete commercial components has been reported [Ref. 40]. In addition, new high-speed, low-noise transistor types have been investigated for optical receiver preamplifiers. These devices include the heterojunction bipolar transistor (HBT) [Ref. 41] and high electron mobility transistor (HEMT) [Refs 42, 43]. The latter device type comprises a selectively doped heterojunction FET which has displayed 3 dB bandwidths up to 20 GHz within the three-stage optical preamplifier illustrated in Figure 9.16(a) [Ref. 42]. Each stage comprised a shunt feedback configuration containing a single HEMT with mutual conductance of 70 millisiemens and a gate to source capacitance of 0.36 pF (see Figure 9.16(b)). When operated with an InGaAs *p–i–n* photodiode, the preamplifier exhibited a 21.5 dB gain with an averaged input equivalent noise current density of 7.6 pA Hz$^{-\frac{1}{2}}$ over the range 100 MHz to 18 GHz.

Although the above discussion centered on *p–i–n* receiver preamplifier designs, high-speed APD optical receivers have also been investigated [Refs 44–46]. In particular, a high-sensitivity APD–FET receiver designed to operate at speeds up to 8 Gbit s^{-1} and at wavelengths in the range 1.3 to 1.5 μm is shown in Figure 9.17 [Ref. 45]. The receiver employed a 60 GHz gain–bandwidth product InGaAs/InGaAsP/InP APD followed by a hybrid GaAs MESFET high-impedance front end. Moreover, a receiver sensitivity of −25.8 dBm was obtained for a BER of 10^{-9}.

An additional strategy for the provision of wideband, low-noise receivers, especially using the *p–i–n* photodiode detector, involves the monolithic integration of this device type with III–V semiconductor alloy FETs or HBTs [Refs 47–51]. Such monolithic integrated receivers or optoelectronic integrated circuits (OEICs) are discussed further in

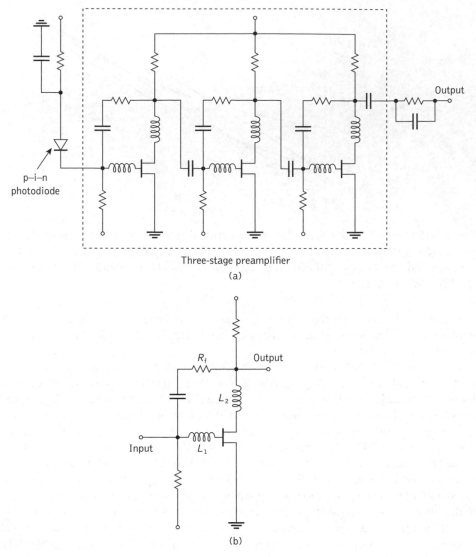

Figure 9.16 Circuit configuration for a high-speed optical receiver using an HEMT preamplifier [Ref. 42]: (a) *p–i–n*–HEMT optical receiver; (b) single-shunt feedback stage

Chapter 11. However, it should be noted that the major recent activities in this area have concerned devices for operation in the 1.1 to more 1.6 μm wavelength range. An example of the circuit configuration of a monolithic PIN–FET receiver is illustrated in Figure 9.18 [Ref. 51]. The design comprises a voltage variable FET feedback resistor which produces active feedback as an input shunt automatic gain control (AGC) circuit which extends the dynamic range by diverting excess photocurrent away from the input of the basic receiver. Furthermore, the shunt FET gives additional dynamic range extension through

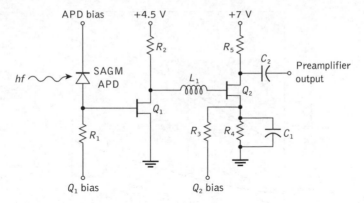

Figure 9.17 Circuit configuration for a high-sensitivity APD–FET optical receiver [Ref. 45]

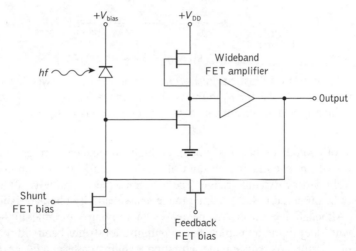

Figure 9.18 Monolithic PIN–FET optical receiver circuit configuration

the mechanism of active receiver bias compensation, which is discussed further in relation to Figure 9.20 following.

The receiver dynamic range is an important performance parameter as it provides a measure of the difference between the device sensitivity and its saturation or overload level. A receiver saturation or overload level is largely determined by the value of the photodiode bias resistor or, alternatively, the feedback resistor in the transimpedance configuration. Because the photodiode bias resistor has a small value in the low-impedance front-end design, the saturation level is high.* Similarly, the relatively low value of feedback resistor in the transimpedance configuration gives a high saturation level which, combined with a high sensitivity, provides a wide dynamic range, as indicated in Section 9.4.3. By contrast

---

* Unfortunately, the sensitivity of the low-impedance configuration is poor and hence the dynamic range is generally not large.

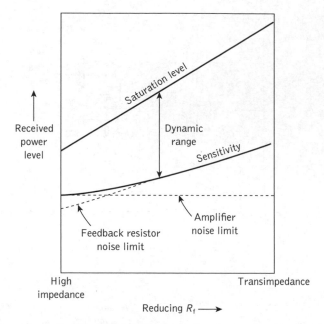

Figure 9.19 Characteristics illustrating the variation in received power level against the value of the feedback resistor R_f in the transimpedance front-end receiver structure. The high-impedance front-end receiver corresponds to $R_f = \infty$

the high value of photodiode bias resistor in the high-impedance front end causes a low saturation level which, even taking account of the high sensitivity of the configuration, gives a relatively narrow dynamic range. The difference between the two latter receiver structures may be observed in the dynamic range and sensitivity characteristics shown in Figure 9.19. Although the sensitivity decreases in moving from the high-impedance design (left hand side) to the transimpedance configuration (right hand side) as the value of the feedback resistor R_f is reduced, the saturation level increases at a faster rate, producing a significantly wider dynamic range for the transimpedance front-end receiver.

The significance of the receiver dynamic range becomes apparent when the reader considers the ideal multipurpose use of such a device for operation with a variety of optical source powers over different fiber lengths. Moreover, when a high-impedance receiver with a 1 MΩ bias resistor is utilized, the saturation level occurs at an input optical power of 0.5 μW or −33 dBm. Therefore, this device can only be employed in long-haul communication applications where the input power level is low. Corresponding figures for the transimpedance configuration (1 kΩ feedback resistor) and the low-impedance front end (200 Ω bias resistor) are 0.5 mW (−3 dBm) and 2.5 mW (+4 dBm). In all cases the saturation level can be substantially improved by using active receiver bias compensation, as illustrated schematically in Figure 9.20. Hence, as the d.c. voltage at the input to the amplifier increases with the incident optical power, the control loop applies an equal but opposite shift in the voltage to the other side of the bias resistor. In this way the voltage at the input to the preamplifier becomes independent of the detected power level. However, in practice the feedback voltage in the control loop cannot be unbounded and therefore the

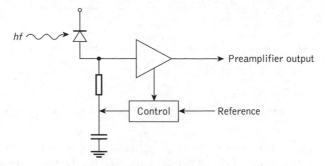

Figure 9.20 Active receiver bias compensation

technique has limitations. Nevertheless, saturation levels for high-impedance front-end receiver designs may be improved to around 20 µW, or 17 dBm, using this technique.

Even when using bias compensation with a high-impedance front-end receiver to improve the saturation level, the overall dynamic range tends to be poor. For such a receiver operating at a speed of 1 Gbit s^{-1} it is usually in the range 20 to 27 dB, whereas for a corresponding transimpedance receiver configuration without bias compensation, the dynamic range can be 30 to 39 dB.* Furthermore, in the latter case alternative design strategies have proved successful in increasing the receiver dynamic range. In particular the use of optically coupled feedback has demonstrated dynamic ranges of around 40 dB for p–i–n receivers operating at modest bit rates [Refs 52, 53].

The optical feedback technique, which is shown schematically in Figure 9.21, eliminates the thermal noise associated with the feedback resistor in the transimpedance front-end design. This strategy proves most useful at low transmission rates because in this case the feedback resistors employed are normally far smaller than the optimum value for low-noise performance so as to maintain the resistor at a practical size (e.g. 1 MΩ). Moreover, large values of feedback resistor limit the dynamic range of the conventional

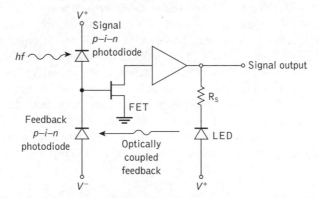

Figure 9.21 Schematic of optical feedback transimpedence receiver

* It should be noted that in both cases the bottom end of the range refers to p–i–n photodiode receivers while the top end of the range is only obtained with APD receivers.

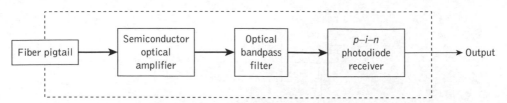

Figure 9.22 Block schematic of an SOA preamplified p–i–n photodiode receiver

transimpedance receiver structure, while also introducing parasitic shunt capacitance which can cause signal integration and hence restrict the bandwidth of the preamplifier. It may be observed from Figure 9.21 that the optical feedback signal is provided by an LED that is driven from the preamplifier output through a small resistor. This resistor acts as a load to generate an output voltage for the following amplifiers. The current feedback to the signal p–i–n photodetector is obtained from a second p–i–n photodiode which detects the optical feedback signal.

The removal of the feedback resistor in the optical feedback technique allows low-noise performance and hence high receiver sensitivity of the order of −64 dBm at transmission rates of 2 Mbit s^{-1} [Ref. 53]. In addition, as the feedback LED is a low-impedance device that can be driven with a low output voltage, the problem associated with amplifier saturation is much reduced. Therefore this factor, combined with the high sensitivity produced by the strategy, enables wide dynamic range. It should be noted, however, that some penalties occur when employing this technique in that there is an increase in receiver input capacitance and also an increase in detector dark current noise (resulting from the use of two photodiodes) in comparison with the conventional transimpedance preamplifier structure. Nevertheless, it is suggested that the optical feedback receiver component costs can be comparable with resistive feedback designs [Ref. 53] while providing a significant performance improvement.

An alternative strategy for the realization of high-sensitivity receivers, in this case for high-speed operation, is to employ preamplification using an optical amplifier prior to the receiver [Refs 54–58]. The two basic optical amplifier technological types, namely the semiconductor optical amplifier (SOA) and the fiber amplifier, are discussed in Sections 10.3 and 10.4 respectively. It is clear, however, that both device types may be utilized in this preamplification role which is illustrated schematically for an SOA device in Figure 9.22.* The SOA shown in Figure 9.22 operates as a near-traveling-wave amplifier and therefore the output emissions are predominantly spontaneous, creating a spectral bandwidth which is determined by the gain profile of the device. Because the typical spectral bandwidth is in the range 80 to 120 nm, a bandpass optical filter† is employed to reduce the intensity of the spontaneous emission reaching the optical detector. This has the effect of reducing the spontaneous noise products and thus improving the overall receiver sensitivity.

---

* It should be noted that the corresponding schematic showing a fiber amplifier fulfilling this role is provided in Figure 10.13(c).

† The optimum filter bandwidth is determined by a number of factors including the detector noise, the transmission rate, the transmitter chirp characteristics and the filter insertion loss but is typically in the range 0.5 to 3 nm.

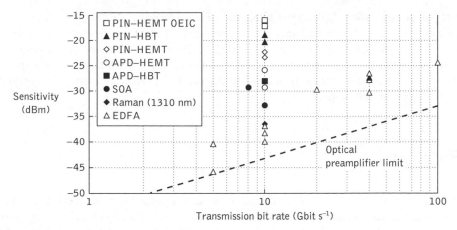

Figure 9.23 Characteristic showing receiver sensitivity against transmission bit rate for various receiver preamplifier types. Reprinted from Ref. 56 with permission from Elsevier

Although the sensitivity improvement introduced by the laser preamplifier is a function of the device internal gain, the coupling losses between the various elements and the bandwidth of the bandpass filter, it is typically in the range 10 to 15 dB when using an SOA. Moreover, it is interesting to observe from the sensitivity against transmission rate characteristics shown in Figure 9.23 [Ref. 56] that the SOA preamplifier $p–i–n$ photodiode configuration illustrated in Figure 9.22 displays a significant improvement over high-performance APD receivers, particularly at speeds of 10 Gbit s^{-1} and above. Figure 9.23 displays the sensitivity provided by different optical receiver preamplifier types in relation to increasing transmission rates up to 100 Gbit s^{-1} [Ref. 56]. It may be observed that the majority of the optical receivers shown operate at transmission rates from 10 to 40 Gbit s^{-1} [Refs 56–59] and that both the SOA and erbium-doped fiber amplifier (EDFA) are useful devices to provide for the optical preamplification. Furthermore, the latter device can attain a receiver sensitivity of −25 dBm with a noise figure value lying between 4 and 5 dB while enabling a high transmission rate of 40 Gbit s^{-1} [Ref. 56]. The SOA, however, exhibits a relatively high noise figure in the range 7 to 8 dB, but its small size, wider operating wavelength range and potential for monolithic integration make it an important device for optical preamplification (see Section 10.3). For example, an optically preamplified receiver using a vertical cavity SOA (see Section 10.3.3) operating at a signal wavelength of 1.55 μm exhibited a sensitivity of −28.5 dBm at higher transmission bit rates from 10 to 40 Gbit s^{-1}. An output signal power penalty of 4.7 dB was observed, however, when receiving the higher bit rates in the region from 20 to 40 Gbit s^{-1} [Ref. 59].

Finally, a germanium on silicon-on-insulator (Ge-on-SOI) technology (see Section 11.2) receiver comprising a $p–i–n$ photodiode paired with a high-gain CMOS amplifier has also been shown to operate at a transmission rate up to 15 Gbit s^{-1} with a sensitivity of −7.4 dBm [Re. 60]. Although the demonstrated sensitivity for this receiver was quite modest, it functioned with a single supply voltage of only 2.4 V benefiting from the integration of the Ge-on-SOI $p–i–n$ photodiode with the lower power silicon CMOS integrated circuit amplifier.

Although in this chapter we have focused on receiver performance and design techniques for intensity-modulated/direct detection optical fiber communication systems, many of the strategies discussed are also utilized within the generally more complex receiver structures required to enable coherent transmission. The various coherent demodulation schemes are discussed in some detail in Section 13.6 and the coherent receiver sensitivities are compared both with each other and with direct detection in Section 13.8. However, the specific preamplifier noise and technological considerations are not repeated as they apply equally to both detection techniques.

Problems

9.1   Briefly discuss the possible sources of noise in optical fiber receivers. Describe in detail what is meant by quantum noise. Consider this phenomenon with regard to:
(a) digital signaling,
(b) analog transmission,
giving any relevant mathematical formulas.

9.2   A silicon photodiode has a responsivity of 0.5 A W^{-1} at a wavelength of 0.85 μm. Determine the minimum incident optical power required at the photodiode at this wavelength in order to maintain a bit-error-rate of 10^{-7}, when utilizing ideal binary signaling at a rate of 35 Mbit s^{-1}.

9.3   An analog optical fiber communication system requires an SNR of 40 dB at the detector with a post-detection bandwidth of 30 MHz. Calculate the minimum optical power required at the detector if it is operating at a wavelength of 0.9 μm with a quantum efficiency of 70%. State any assumptions made.

9.4   A digital optical fiber link employing ideal binary signaling at a rate of 50 Mbit s^{-1} operates at a wavelength of 1.3 μm. The detector is a germanium photodiode which has a quantum efficiency of 45% at this wavelength. An alarm is activated at the receiver when the bit-error-rate drops below 10^{-5}. Calculate the theoretical minimum optical power required at the photodiode in order to keep the alarm inactivated. Comment briefly on the reasons why in practice the minimum incident optical power would need to be significantly greater than this value.

9.5   Discuss the implications of the load resistance on both thermal noise and post-detection bandwidth in optical fiber communication receivers.

9.6   A silicon p–i–n photodiode has a quantum efficiency of 65% at a wavelength of 0.8 μm. Determine:
(a) the mean photocurrent when the detector is illuminated at a wavelength of 0.8 μm with 5 μW of optical power;
(b) the rms quantum noise current in a post-detection bandwidth of 20 MHz;
(c) the SNR in dB, when the mean photocurrent is the signal.

9.7   The photodiode in Problem 9.6 has a capacitance of 8 pF. Calculate:
(a) the minimum load resistance corresponding to a post-detection bandwidth of 20 MHz;

(b) the rms thermal noise current in the above resistor at a temperature of 25 °C;

(c) the SNR in dB resulting from the illumination in Problem 9.6 when the dark current in the device is 1 nA.

9.8 The photodiode in Problems 9.6 and 9.7 is used in a receiver where it drives an amplifier with a noise figure of 2 dB and an input capacitance of 7 pF. Determine:

(a) the maximum amplifier input resistance to maintain a post-detection bandwidth of 20 MHz without equalization;

(b) the minimum incident optical power required to give an SNR of 50 dB.

9.9 A germanium photodiode incorporated into an optical fiber receiver working at a wavelength of 1.55 μm has a dark current of 500 nA at the operating temperature. When the incident optical power at this wavelength is 10^{-6} W and the responsivity of the device is 0.6 A W^{-1}, shot noise dominates in the receiver. Determine the SNR in dB at the receiver when the post-detection bandwidth is 100 MHz.

9.10 Discuss the expression for the SNR in an APD receiver given by:

$$\frac{S}{N} = \frac{M^2 I_{\mathrm{p}}^2}{2eB(I_{\mathrm{p}} + I_{\mathrm{d}})M^{2+x} + \dfrac{4KTBF_{\mathrm{n}}}{R_{\mathrm{L}}}}$$

with regard to the various sources of noise present in the receiver. How may this expression be modified to give the optimum avalanche multiplication factor?

9.11 A silicon RAPD has a quantum efficiency of 95% at a wavelength of 0.9 μm, an excess avalanche noise factor of $M^{0.3}$ and a capacitance of 2 pF. It may be assumed that the post-detection bandwidth (without equalization) is 25 MHz, and that the dark current in the device is negligible at the operating temperature of 290 K. Determine the minimum incident optical power which can yield an SNR of 23 dB.

9.12 With the device and conditions given in Problem 9.11, calculate:

(a) the SNR obtained when the avalanche multiplication factor for the RAPD falls to half the optimum value calculated;

(b) the increased optical power necessary to restore the SNR to 23 dB with $M = 0.5M_{\mathrm{op}}$.

9.13 What is meant by the excess avalanche noise factor $F(M)$? Give two possible ways of expressing this factor in analytical terms. Comment briefly on their relative merits.

9.14 Explain the gain–bandwidth product of an APD receiver in relation to the device multiplication factor. Briefly describe any two APD receivers with high gain–bandwidth products for operation in the longer wavelength (1.3 to 1.6 μm) region.

9.15 A germanium APD (with $x = 1.0$) operates at a wavelength of 1.35 μm where its responsivity is 0.45 A W^{-1}. The dark current is 200 nA at the operating temperature of 250 K and the device capacitance is 3 pF. Determine the maximum possible SNR when the incident optical power is 8×10^{-7} W and the post-detection bandwidth without equalization is 560 MHz.

9.16   The photodiode in Problem 9.15 drives an amplifier with a noise figure of 3 dB and an input capacitance of 3 pF. Determine the new maximum SNR when they are operated under the same conditions.

9.17   Discuss the three main amplifier configurations currently adopted for optical fiber communications. Comment on their relative merits and drawbacks.

A high-impedance integrating front-end amplifier is used in an optical fiber receiver in parallel with a detector bias resistor of 10 MΩ. The effective input resistance of the amplifier is 6 MΩ and the total capacitance (detector and amplifier) is 2 pF.

It is found that the detector bias resistor may be omitted when a transimpedance front-end amplifier design is used with a 270 kΩ feedback resistor and an open loop gain of 100.

Compare the bandwidth and thermal noise implications of these two cases, assuming an operating temperature of 290 K.

9.18   A p–i–n photodiode operating at a wavelength of 0.83 μm has a quantum efficiency of 50% and a dark current of 0.5 nA at a temperature of 295 K. The device is unbiased but loaded with a current mode amplifier with a 50 kΩ feedback resistor and an open loop gain of 32. The capacitance of the photodiode is 1 pF and the input capacitance of the amplifier is 6 pF.

Determine the incident optical power required to maintain an SNR of 55 dB when the post-detection bandwidth is 10 MHz. Is equalization necessary?

9.19   A voltage amplifier for an optical fiber receiver is designed with an effective input resistance of 200 Ω which is matched to the detector bias resistor of the same value. Determine:

(a) The maximum bandwidth that may be obtained without equalization if the total capacitance (C_T) is 10 pF.

(b) The rms thermal noise current generated in this configuration when it is operating over the bandwidth obtained in (a) and at a temperature of 290 K. The thermal noise generated by the voltage amplifier may be assumed to be from the effective input resistance to the device.

(c) Compare the values calculated in (a) and (b) with those obtained when the voltage amplifier is replaced by a transimpedance amplifier with a 10 kΩ feedback resistor and an open loop gain of 50. It may be assumed that the feedback resistor is also used to bias the detector, and the total capacitance remains 10 pF.

9.20   What is a PIN–FET hybrid receiver? Discuss in detail its merits and possible drawbacks in comparison with the APD receiver.

9.21   Identify the characteristics which are of greatest interest in the pursuit of high-performance receivers.

Discuss the major techniques which have been adopted in order to produce such high-performance receivers for use in long-haul optical fiber communications.

Answers to numerical problems

9.2 −70.4 dBm
9.3 −37.2 dBm
9.4 −70.1 dBm
9.6 (a) 2.01 μA; (b) 3.59 nA; (c) 55.0 dB
9.7 (a) 994.7 Ω; (b) 18.19 nA; (c) 39.3 dB
9.8 (a) 1.137 kΩ; (b) 19.58 μW
9.9 40.1 dB
9.11 −50.3 dBm

9.12 (a) 14.2 dB; (b) −49.6 dBm
9.15 23.9 dB
9.16 21.9 dB
9.17 High-impedance front-end: 21.22 kHz, 4.27×10^{-27} A² Hz⁻¹; Transimpedance front-end: 29.47 MHz, 5.93×10^{-26} A² Hz⁻¹
9.18 −23.1 dBm, equalization is unnecessary
9.19 (a) 159.13 MHz; (b) 160 nA; (c) 79.56 MHz, 11.3 nA, noise power 23 dB down

References

[1] (a) M. Schwartz, *Information Transmission, Modulation and Noise* (4th edn), McGraw-Hill, 1990. (b) M. P. Fitz, *Fundamentals of Communications Systems*, McGraw-Hill, 2007.

[2] P. Russer, 'Introduction to optical communications', in M. J. Howes and D. V. Morgan (Eds), *Optical Fibre Communications*, pp. 1–26, Wiley, 1980.

[3] M. Garbuny, *Optical Physics*, Academic Press, 1965.

[4] W. M. Hubbard, 'Efficient utilization of optical frequency carriers for low and moderate bit rate channels', *Bell Syst. Tech. J.*, 50, pp. 713–718, 1973.

[5] I. Garrett, 'Receivers for optical fibre communications', *Electron. Radio Eng.*, 51(7/8), pp. 349–361, 1981.

[6] C. Milorad, *Optical Transmission Systems Engineering*, Artech House, 2004.

[7] O.-H. Kwon, M. M. Hayat, J. C. Campbell, B. E. A. Saleh and M. C. Teich, 'Gain-bandwidth product optimization of heterostructure avalanche photodiodes', *J. Lightwave Technol.*, 23(5), pp. 1896–1906, 2005.

[8] H. Ito, 'Photoreceiver architectures beyond 40 Gbit/s', *IEEE Symp. on Compound Semiconductor Integrated Circuits*, Monterey, California, USA, pp. 85–88, 24–27 October 2004.

[9] T. Kamiya, H. Yajima and F. Saito, *Femtosecond Technology*, Springer-Verlag, 1999.

[10] C. Lenox, H. Nie, P. Yuan, G. Kinsey, A. L. Homles Jr, B. G. Streetman and J. C. Campbell, 'Resonant-cavity InGaAs-InAlAs avalanche photodiodes with gain-bandwidth product of 290 GHz', *IEEE Photonics Technol. Lett.*, 11(9), pp. 1162–1164, 1999.

[11] K. Azadet, E. F. Haratsch, H. Kim, F. Saibi, J. H. Saunders, M. Shaffer, L. Song and Y. Meng-Lin, 'Equalization and FEC techniques for optical transceivers', *IEEE J. Solid-State Circuits*, 37(3), pp. 317–327, 2002.

[12] S. D. Personick, 'Receiver design for digital fiber optic communication systems (Part I and II)', *Bell Syst. Tech. J.*, 52, pp. 843–886, 1973.

[13] J. L. Hullett and T. V. Muoi, 'Referred imepedance noise analysis for feedback amplifiers', *Electron. Lett.*, 13(13), pp. 387–389, 1977.

[14] R. G. Smith and S. D. Personick, 'Receiver design for optical fiber communication systems', in H. Kressel (Ed.), *Semiconductor Devices for Optical Communication* (2nd edn), Springer-Verlag, 1982.

[15] J. L. Hullett, 'Optical communication receivers', *Proc. IREE Australia*, 40(4), pp. 127–136, 1979.

[16] J. L. Hullett and T. V. Muoi, 'A feedback receiver amplifier for optical transmission systems', *Trans. IEEE*, **COM 24**, pp. 1180–1185, 1976.

[17] J. S. Barrera, 'Microwave transistor review, Part 1. GaAs field-effect transistors'. *Microwave J.* (USA), **19**(2), pp. 28–31, 1976.

[18] B. S. Hewitt, H. M. Cox, H. Fukui, J. V. Dilorenzo, W. O. Scholesser and D. E. Iglesias, 'Low noise GaAs MESFETs', *Electron. Lett.*, **12**(12), pp. 309–310, 1976.

[19] D. V. Morgan, F. H. Eisen and A. Ezis, 'Prospects for ion bombardment and ion implantation in GaAs and InP device fabrication', *IEE Proc.*, **128**(1–4), pp. 109–129, 1981.

[20] J. Mun, J. A. Phillips and B. E. Barry, 'High-yield process for GaAs enhancementmode MESFET integrated circuits', *IEE Proc.*, **128**(1–4), pp. 144–147, 1981.

[21] S. D. Personick, 'Design of receivers and transmitters for fiber systems', in M. K. Barnoski (Ed.), *Fundamentals of Optical Fiber Communications* (2nd edn), Academic Press, 1981.

[22] D. R. Smith, R. C. Hooper and I. Garrett, 'Receivers for optical communications: a comparison of avalanche photodiodes with PIN–FET hybrids', *Opt. Quantum Electron.*, **10**, pp. 293–300, 1978.

[23] R. C. Hooper and D. R. Smith, 'Hybrid optical receivers using PIN photodiodes', *IEE (London) Colloq. on Broadband High Frequency Amplifiers*, pp. 9/1–9/5, 1979.

[24] K, Ahmad and A. W. Mabbitt, 'Ga$_{1-x}$In$_x$As photodetectors for 1.3 micron PIN–FET receiver', *IEEE NY (USA) Int. Electron Devices Meet.*, Washington, DC, pp. 646–649, 1978.

[25] D. R. Smith, R. C. Hooper and R. P. Webb, 'High performance digital optical receivers with PIN photodiodes', *IEEE (NY) Proc. Int. Symp. on Circuits and Systems*, Tokyo, pp. 511–514, 1979.

[26] D. R. Smith, R. C. Hooper, K. Ahmad, D. Jenkins, A. W. Mabbitt and R. Nicklin, '*p–i–n*/FET hybrid optical receiver for longer wavelength optical communication systems', *Electron. Lett.*, **16**(2), pp. 69–71, 1980.

[27] R. C. Hooper, D. R. Smith and B. R. White, 'PIN–FET hybrids for digital optical receivers', *IEEE NY (USA) 30th Electronic Components Conf.*, San Francisco, pp. 258–260, 1980.

[28] S. Hata, Y. Sugeta, Y. Mizushima, K. Asatani and K. Nawata, 'Silicon *p–i–n* photodetectors with integrated transistor amplifiers', *IEEE Trans. Electron Devices*, **ED-26**(6), pp. 989–991, 1979.

[29] K. Ogawa and E. L. Chinnock, 'GaAs FET transimpedance front-end design for a wideband optical receiver', *Electron. Lett*, **15**(20), pp. 650–652, 1979.

[30] S. M. Abbott and W. M. Muska, 'Low noise optical detection of a 1.1 Gb/s optical data stream', *Electron. Lett.*, **15**(9), pp. 250–251, 1979.

[31] T. Otsuji, K. Murata, K. Nahara, K. Sano, E. Sano and K. Yamasaki, '20–40 Gbits/s class GaAs MESFET digital ICs for future optical fiber communication systems', in K.-C. Wang (Ed.), *High-Speed Circuits for Lightwave Communications*, pp. 87–124, World Scientific, 1999.

[32] E. Säckinger, *Broadband Circuits for Optical Fiber Communication*, Wiley, 2005.

[33] M. Brain and T. P. Lee, 'Optical receivers for lightwave communication systems', *J. Lightwave Technol.*, **LT-3**(6), pp. 1281–1300, 1985.

[34] R. Rajiv and S. Kumar, *Optical Networks: A Practical Perspective* (2nd edn), Morgan Kaufmann, 2001.

[35] B. F. Levine, J. A. Valdmanis, P. N. Sacks, M. Jazwiecki and J. H. Meier, '–29 dBm sensitivity, InAlAs APD-based receiver for 10Gb/s long-haul (LR-2) applications', *Proc. of Optical Fiber Communication, OFC/NFOEC'05*, Anaheim, California, USA, OFM5, pp. 1–3, 6–11 March 2005.

[36] J. M. Baek, H. S. Seo, B. O. Jeon, H. Y. Kang, D. Y. Rhee, S. K. Yang, M. K. Park, J. W. Burm and D. H. Jang, 'High sensitive 10-Gb/s APD optical receivers in low-cost TO-can-type packages', *IEEE Photonics Technol. Lett.*, **17**(1), pp. 181–183, 2005.

[37] B. L. Kasper, 'Receiver design', in S. E. Miller and I. P. Kaminow (Eds), *Optical Fiber Telecommunications II*, pp. 689–722, Academic Press, 1988.

[38] G. P. Vella-Coleiro, 'Optimization of optical sensitivity of *p–i–n* FET receivers', *IEEE Electron Device Lett.*, **9**(6), pp. 269–271, 1988.

[39] R. A. Minasian, 'Optimum design of 4-Gbit/s GaAs MESFET optical preamplifier', *J. Lightwave Technol.*, **LT-5**(3), pp. 373–379, 1987.

[40] M. A. R. Violas, D. J. T. Heatley, A. M. O. Duarte and D. M. Beddow, '10 GHz bandwidth low noise optical receiver using discrete commercial devices', *Electron. Lett.*, **26**(1), pp. 35–36, 1990.

[41] C. W. Farley, M. F. Chang, P. M. Asbeck, N. H. Sheng, R. Pierson, G. J. Sullivan, K. C. Wang and R. B. Nubling, 'High-speed (f_t = 78 GHz) AlInAs/GaInAs single heterojunction HBT', *Electron. Lett.*, **25**(13), pp. 846–847, 1989.

[42] N. Ohkawa, '20 GHz bandwidth low-noise HEMT preamplifier for optical receivers', *Electron. Lett.*, **24**(7), pp. 1061–1062, 1988.

[43] S. D. Walker, L. C. Blank, R. A. Garnham and J. M. Boggis, 'High electron mobility transistor lightwave receiver for broadband optical transmission system applications', *J. Lightwave Technol.*, **7**(3), pp. 454–458, 1989.

[44] B. L. Kasper and J. C. Campbell, 'Multigigabit-per-second avalanche photodiode lightwave receivers', *J. Lightwave Technol.*, **LT-5**(10), pp. 1351–1364, 1987.

[45] B. L. Kasper, J. C. Campbell, J. R. Talman, A. H. Gnauck, J. E. Bowers and W. S. Holden, 'An APD/FET optical receiver operating at 8 Gbit/s', *J. Lightwave Technol.*, **LT-5**(3), pp. 344–347, 1987.

[46] J. J. O'Reilly and R. S. Fyath, 'Performance of optical receivers employing ultralow noise avalanche photodiodes', *J. Opt. Commun.*, **9**(3), pp. 82–84, 1988.

[47] M. J. N. Sibley, R. T. Unwin, D. R. Smith, B. A. Boxall and R. J. Hawkins, 'A monolithic common collector front-end optical preamplifier', *J. Lightwave Technol.*, **LT-3**(1), pp. 13–15, 1985.

[48] Y. Archambault, D. Pavlidis and J. P. Guet, 'GaAs monolithic integrated optical preamplifier', *J. Lightwave Technol.*, **LT-5**(3), pp. 355–366, 1987.

[49] W. T. Colleran and A. A. Abidi, 'Wideband monolithic GaAs amplifier using cascodes', *Electron. Lett.*, **23**(18), pp. 951–952, 1987.

[50] S. Miura, T. Mikawa, T. Fujii and O. Wada, 'High-speed monolithic GaInAs pinFET', *Electron. Lett.*, **24**(7), pp. 394–395, 1988.

[51] G. F. Williams and H. P. Leblanc, 'Active feedback lightwave receivers', *J. Lightwave Technol.*, **LT-4**(10), pp. 1502–1508, 1986.

[52] B. L. Kasper, A. R. McCormick, C. A. Burrus Jr and J. R. Talman, 'An optical-feedback transimpedance receiver for high sensitivity and wide dynamic range at low bit rates', *J. Lightwave Technol.*, **6**(2), pp. 329–338, 1988.

[53] S. G. Methley, 'An optical feedback receiver, with high sensitivity', *Proc. SPIE*, **949**, pp. 51–55, April 1988.

[54] I. W. Marshall and M. J. O'Mahony, '10 GHz optical receiver using a travelling wave semiconductor laser preamplifier', *Electron. Lett.*, **23**(20), p. 1052, 1987.

[55] N. A. Olsson and M. G. Oberg, 'Ultra low reflectivity 1.5 μm semiconductor laser preamplifier', *Electron. Lett.*, **24**(9), pp. 569–570, 1988.

[56] B. L. Kasper, O. Mizuhara and Y. K. Chen, 'High bit-rates receivers, transmitters and electronics', in I. P. Kaminow and T. Li (Eds), *Optical Fiber Telecommunications IVA: Components*, Academic Press, pp. 784–852, 2002.

[57] S. Takashima, H. Nakagawa, S. Kim, F. Goto, M. Okayasu and H. Inoue, '40-Gbit/s receiver with −21 dBm sensitivity employing filterless semiconductor optical amplifier', *Proc. of Optical Fiber Communication, OFC'03*, Atlanta, Georgia, USA, pp. 471–472, 23–28 March 2003.

[58] P. Ossieur, D. Verhulst, Y. Martens, C. Wei, J. Bauwelinck, Q. Xing-Zhi and J. Vandewege, '1.25-gb/s burst-mode receiver for GPON applications', *IEEE J. Solid-State Circuits*, **40**(5), pp. 1180–1189, 2005.

[59] T. Kimura, S. Bjorlin, H.-F. Chou, Q. Chen, S. Wu and J. E. Bowers, 'Optically preamplified receiver at 10, 20, and 40 Gb/s using a 1550-nm vertical-cavity SOA', *IEEE Photonics Technol. Lett.*, **17**(2), pp. 456–458, 2005.

[60] S. J. Koester, C. L. Schow, L. Schares, G. Dehlinger, J. D. Schaub, F. E. Doany and R. A. John, 'Ge-on-SOI detector/Si-CMOS amplifier receivers for high performance optical communication applications', *J. Lightwave Technol.*, **25**(1), pp. 46–57, 2007.

CHAPTER | 10

Optical amplification, wavelength conversion and regeneration

10.1   Introduction

10.2   Optical amplifiers

10.3   Semiconductor optical amplifiers

10.4   Fiber and waveguide amplifiers

10.5   Wavelength conversion

10.6   Optical regeneration

       Problems

       References

10.1 Introduction

The preceding four chapters have been concerned with the devices employed to provide the electrical–optical interfaces within optical fiber communications. Optical sources were dealt with in Chapters 6 and 7 followed by optical detectors in Chapter 8 prior to consideration of optical recevier noise and its effect on receiver design in Chapter 9. These electrical–optical and optical–electrical conversion devices are crucial components for the realization of optical fiber communications, as may be observed in the following two chapters which discuss optical fiber systems. However, these devices are also, in a number of respects, a limiting factor within the implementation of optical fiber systems. The conversion of the information siganl from the electrical domain to the optical domain and vice versa often provides a bottleneck within optical fiber communications which may restrict both the operating bandwidth and the quality of the transmitted signal. Performing operations on signals in the optical domain combined with the pursuit of more efficient mechanisms to provide the electrical–optical interfaces has therefore assumed increasing significance within optical fiber communications and its associated application areas.

The above considerations have stimulated a growing activity in the area of active devices and components which allow optical signals to be manipulated without returning them back to the electrical regime where such operations have normally been carried out in the past. Potentially, such devices alleviate the possible bottleneck associated with the interfaces as well as providing more efficient, and hence cost-effective, methods for processing the optical signals. Moreover, in some cases the use of these devices and components may represent the only realistic solution for the implementation of particular optical fiber transmission techniques and systems.

This chapter therefore initiates the discussions of the area of active optical devices which may be utilized for a variety of functions within optical fiber communications. A major development which was stimulated by the massive effort to produce laser sources for optical fiber communications is that of optical amplification. The technology associated with such active optical devices is now well established and their use within optical fiber communications is widespread.

Section 10.2 introduces the concept of an optical amplifier and outlines the various generic types that are under investigation. Semiconductor optical amplifiers are then considered in some detail in Section 10.3. This is followed in Section 10.4 with a discussion of the various types of fiber and waveguide amplifier which have evolved more recently, assisted by the activities which have led to the realization of fiber lasers (see Section 6.9.2). The use of the optical amplifier and other device types to perform the important function of wavelength conversion is then described in Section 10.5. Finally, the growing application of optical amplification within optical regeneration for long-haul optical fiber transmission is dealt with in Section 10.6.

10.2 Optical amplifiers

Optical amplifiers, as their name implies, operate solely in the optical domain with no interconversion of photons to electrons. Therefore, instead of using regenerative repeaters which require optoelectronic devices for source and detector, together with substantial electronic circuitry for pulse slicing, retiming and shaping (see Section 12.6.1), optical amplifiers are placed at intervals along a fiber link to provide linear amplification of the transmitted optical signal. The optical amplifier, in principle, provides a much simpler solution in that it is a single in-line component which can be used for any kind of modulation at virtually any transmission rate. Moreover, such a device can be bidirectional and if it is sufficiently linear it may allow multiplex operation of several signals at different optical wavelengths (i.e. wavelength division multiplexing). In particular with single-mode fiber systems, the effects of signal dispersion can be small and hence the major limitation on repeater spacing becomes attenuation due to fiber losses. Such systems do not require full regeneration of the transmitted digital signal at each repeater, and optical amplification of the signal proves sufficient. Hence over recent years optical amplifiers have

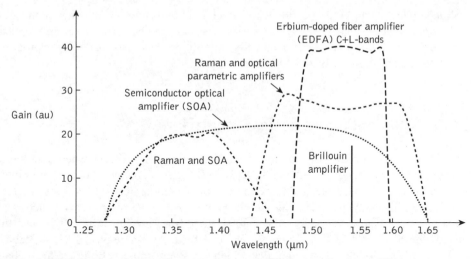

Figure 10.1 Gain–bandwidth characteristics of different optical amplifiers

emerged as promising network elements not just for use as linear repeaters but as optical gain blocks, wavelength converters, optical receiver preamplifiers and, when used in a non-linear mode, as optical gates, pulse shapers and routing switches [Ref. 1] (see Section 15.4).

The two main approaches to optical amplification to date have concentrated on semiconductor optical amplifiers which utilize stimulated emission from injected carriers and fiber amplifiers in which gain is provided by either stimulated Raman or Brillouin scattering* (see Sections 3.5 and 3.14), or by rare earth dopants (see Section 6.9.2). Both amplifier types (i.e. semiconductor and fiber; specifically rare earth and Raman) have the ability to provide high gain over wide spectral bandwidths, making them eminently suitable for optical fiber system applications. Semiconductor optical amplifiers, however, offer an advantage due to their smaller size and also because they can be integrated to produce subsystems which are an essential element of current optical communication systems and networks.

The typical gain profiles for various optical amplifier types based around the 1.3 and 1.5 μm wavelength regions are illustrated in Figure 10.1. It may be observed that the semiconductor optical amplifier (SOA), the erbium-doped fiber amplifier (EDFA) and the Raman fiber amplifier all provide wide spectral bandwidths. Hence these optical amplifier types lend themselves to applications involving wavelength division multiplexing [Ref. 3]. By contrast, the Brillouin fiber amplifier has a very narrow spectral bandwidth, possibly around 50 MHz, and therefore cannot be employed for wideband amplification. It could, however, be used for channel selection within a WDM system by allowing amplification of a particular channel without boosting other nearby channels.

Whereas SOAs exhibit low power consumption and their single-mode waveguide structures make them particularly appropriate for use with single-mode fiber, it is fiber

---

* Amplification from both stimulated Raman or Brillouin scattering can occur in undoped relatively long fiber lengths ($\approx$10 km) or doped short lengths ($\approx$10 m) of fiber [Ref. 2].

amplifiers which present fewer problems of compatibility for in-line interconnection within optical fiber links [Ref. 4]. At present, SOAs are the most developed optical amplifier generic type but research into fiber amplifiers has also made rapid progress towards commercial products over the last few years.

10.3 Semiconductor optical amplifiers

The SOA* is based on the conventional semiconductor laser structure where the output facet reflectivities are between 30 and 35% [Refs 4, 5]. Semiconductor optical amplifiers can be used in both nonlinear and linear modes of operation [Ref. 1]. Various types of SOA may be distinguished including the resonant or Fabry–Pérot amplifier which is an oscillator biased below oscillation threshold [Ref. 6], the traveling-wave (TW) and the near-traveling-wave (NTW) amplifiers which are effectively single-pass devices [Refs 1, 7] and the injection-locked laser amplifier, which is a laser oscillator designed to oscillate at the incident signal frequency [Ref. 8]. Such devices are capable of providing high internal gain (15 to 35 dB) with low power consumption and their single-mode waveguide structure makes them particularly suitable for use with single-mode fiber. Semiconductor optical amplifiers can, however, be classified into two main groups which are Fabry–Pérot amplifiers (FPAs) and traveling-wave amplifiers (TWAs) [Refs 1, 9], the difference between these groups being the facet reflectivities. A schematic diagram of an SOA is shown in Figure 10.2. It is based on the conventional semiconductor optical structure (gain- or index-guided) with an active region width w, thickness d and length L. When the input and output laser facet reflectivities denoted by R_1 and R_2 are each around 0.3, which depicts a normal semiconductor laser, then an FPA is obtained.[†] In this case, as the facet

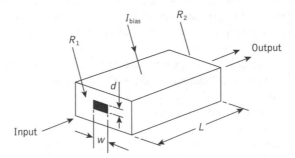

Figure 10.2 Schematic structure of the semiconductor optical amplifier

---

* These amplifiers were previously referred to as semiconductor laser amplifiers (SLAs) or semiconductor laser optical amplifiers (SLOAs). The SOA terminology was initially used for the traveling-wave device but has more recently been employed to also describe Fabry–Pérot and injection current distributed feedback laser amplifiers.
† An FPA may be defined as an amplifier with facet reflectivities in the order 0.01 to 0.3 [Ref. 1].

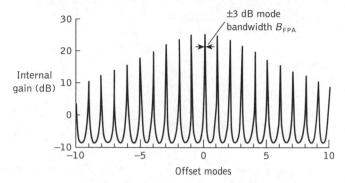

Figure 10.3 The Fabry–Pérot amplifier passband where mode 0 corresponds to the peak gain wavelength [Ref. 1]

reflectivity is large, a highly resonant amplifier is formed and the transmission characteristic comprises very narrow passbands, as displayed in Figure 10.3. The mode 0 corresponds to the peak gain wavelength and the mode spacing $\delta\lambda$ can be obtained from Eq. (6.16). For operation, the FPA is biased below the normal lasing threshold current, and light entering one facet appears amplified at the other facet together with inherent noise. In practice, the amplifier chip is bonded into a package with single-mode fiber pigtails which are used to guide light into and out of the amplifier. The inherent filtering of the FPA, although useful in certain applications, means the device is very sensitive to fluctuations in bias current, temperature and signal polarization. However, because of the resonant nature of FPAs, combined with their high internal fields, they are used within nonlinear applications – for example, to provide pulse shaping and bistable elements (see Section 11.7).

To form a traveling-wave SOA, antireflection coatings may be applied to the laser facets to reduce or eliminate the end reflectivities. This can be achieved by depositing a thin layer of silicon oxide, silicon nitride or titanium oxide on the end facets such that the reflectivities are reduced to 1×10^{-4} or less. Such a device becomes a TWA operating in the single-pass amplification mode in which the Fabry–Pérot resonance is suppressed by the reduction in facet reflectivity.* This has the effect of substantially increasing the amplifier spectral bandwidth and it makes the transmission characteristics less dependent upon fluctuations in bias current, temperature and input signal polarization. Hence the TWA proves superior to the FPA (particularly for linear applications) and also provides advantages in relation to both signal gain saturation and noise characteristics [Ref. 9]. Moreover, antireflection facet coatings have the effect of increasing the lasing current threshold, as illustrated in Figure 10.4, and so in practice such SOAs are operated at currents far beyond the normal lasing threshold current.

* In theory, a true TWA is the limiting case of a device with facets exhibiting zero reflectivity. However, in practice, even with the best antireflection coatings, some residual facet reflectivity remains (e.g. a low reflectivity of 1×10^{-5} has been obtained at a wavelength of 1.5 μm [Ref. 11]). Hence such devices are also referred to as near-traveling-wave amplifiers [Refs 1, 12].

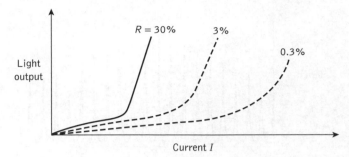

Figure 10.4 Light output against current characteristic for the semiconductor laser amplifier with different values of facet reflectivity R

10.3.1 Theory

The general equation for the cavity gain G of an SOA as a function of signal frequency f takes the form [Refs 1, 9, 10]:

$$G(f) = \frac{(1 - R_1)(1 - R_2)G_s}{(1 - \sqrt{R_1 R_2 G_s})^2 + 4\sqrt{R_1 R_2 G_s}\, \sin^2 \phi} \tag{10.1}$$

where R_1 and R_2 are the input and output facet reflectivities respectively, G_s is the single-pass gain and ϕ is the single-pass phase shift through the amplifier. It should be noted that Eq. (10.1) does not include coupling losses to and from the amplifier and that the phase shift ϕ may be written as [Ref. 9]:

$$\phi = \frac{\pi(f - f_o)}{\delta f} \tag{10.2}$$

where f_o is the Fabry–Pérot resonant frequency and δf is the free spectral range of the SOA.

The 3 dB spectral bandwidth of an FPA, or essentially the ± 3 dB single longitudinal mode bandwidth defined by the FWHP points B_{FPA}, is shown in Figure 10.3. It may be observed that using Eqs (10.1) and (10.2), B_{FPA} may be expressed as:

$$B_{FPA} = 2(f - f_o) = \frac{2\delta f}{\pi}\, \sin^{-1}\left[\frac{1 - \sqrt{R_1 R_2 G_s}}{2(\sqrt{R_1 R_2 G_s})^{\frac{1}{2}}}\right]$$

$$= \frac{c}{\pi n L}\, \sin^{-1}\left[\frac{1 - \sqrt{R_1 R_2 G_s}}{2(\sqrt{R_1 R_2 G_s})^{\frac{1}{2}}}\right] \tag{10.3}$$

where the mode separation frequency interval δf given by Eq. (6.14) combines the velocity of light c and the refractive index of the amplifier medium n with its length L. Alternatively the 3 dB spectral or optical bandwidth may be expressed as a function of the FPA cavity gain G following [Ref. 1]:

$$B_{FPA} = \frac{c}{\pi n L}\, \sin^{-1}\left[\frac{1}{2}\left(\frac{(1 - R_1)(1 - R_2)}{\sqrt{R_1 R_2 G}}\right)\right] \tag{10.4}$$

Example 10.1

An uncoated FPA has facet reflectivities of 30% and a single-pass gain of 4.8 dB. The amplifier has a 300 μm long active region, a mode spacing of 1 nm and a peak gain wavelength of 1.5 μm. Determine the refractive index of the active medium and the 3 dB spectral bandwidth of the device.

Solution: The refractive index of the active medium at the peak gain wavelength may be obtained by rearranging Eq. (6.16) such that:

$$n = \frac{\lambda^2}{2\delta\lambda L} = \frac{(1.5 \times 10^{-6})^2}{2 \times 1 \times 10^{-9} \times 300 \times 10^{-6}}$$

$$= 3.75$$

Using Eq. (10.3) the 3 dB spectral bandwidth is given by:

$$B_{\text{FPA}} = \frac{c}{\pi n L} \sin^{-1}\left[\frac{1 - \sqrt{R_1 R_2} G_s}{2(\sqrt{R_1 R_2} G_s)^{\frac{1}{2}}}\right]$$

$$= \frac{2.998 \times 10^8}{\pi \times 3.75 \times 300 \times 10^{-6}} \sin^{-1}\left[\frac{1 - \sqrt{0.09} \times 3.020}{2(\sqrt{0.09} \times 3.020)^{\frac{1}{2}}}\right]$$

$$= 8.482 \times 10^{10} \sin^{-1}\left[\frac{0.040}{1.904}\right]$$

$$= 8.482 \times 10^{10} \times 0.494 = 4.2 \text{ GHz}$$

The above result demonstrates the narrow spectral bandwidth obtained with an uncoated FPA.

The single-pass gain G_s defined in terms of the device parameters and the applied bias current following Eq. (6.18) for the semiconductor laser* is generally written in the form:

$$G_s = \exp(\bar{g}L) \tag{10.5}$$

where $\bar{g}$ is the net gain coefficient per unit length and L is the amplifier active length. However, the net gain per unit length $\bar{g}$ may be defined in terms of the material gain coefficient g_m, the optical confinement factor Γ and the effective loss coefficient per unit length $\bar{\alpha}$ as [Ref. 1]:

$$\bar{g} = \Gamma g_m - \bar{\alpha} \tag{10.6}$$

Furthermore, the material gain coefficient g_m is related to the signal intensity I following [Ref. 1]:

---

* The round trip gain of Eq. (6.18) includes a factor of 2 which is omitted for the single-pass gain.

$$g_{\mathrm{m}} = \frac{g_0}{1 + I/I_{\mathrm{s}}} \tag{10.7}$$

where g_0 is the unsaturated material gain coefficient in the absence of the input signal and I_{s} is the saturation intensity. Hence substitution of Eqs (10.6) and (10.7) in Eq. (10.5) for the single-pass gain gives:

$$G_{\mathrm{s}} = \exp[(\Gamma g_{\mathrm{m}} - \bar{\alpha})L] \tag{10.8}$$

$$= \exp\left[\left(\frac{\Gamma g_0}{1 + I/I_{\mathrm{s}}} - \bar{\alpha}\right)L\right] \tag{10.9}$$

It may be observed from Eqs (10.8) and (10.9) that the single-pass gain decreases with increasing intensity and that the material gain coefficient is reduced by a factor of 2 when the internal signal intensity I is equal to the saturation intensity I_{s}.

The phase shift ϕ_{s} associated with the single-pass amplifier includes the nominal phase shift ϕ_0 and an additional component resulting from the change in carrier density from the nominal density in the absence of a signal. Hence the total phase shift is given by [Ref. 13]:

$$\phi_{\mathrm{s}} = \phi_0 + \frac{g_0 bL}{2}\left(\frac{I}{1 + I_{\mathrm{s}}}\right) \tag{10.10}$$

where b is the linewidth broadening factor, and the nominal phase shift is:

$$\phi_0 = \frac{2\pi nL}{\lambda} \tag{10.11}$$

where n is the material refractive index. Thus Eqs (10.9) and (10.10) indicate that both single-pass gain and phase are functions of optical intensity. It is clear that for a constant signal intensity (i.e. with frequency modulation) there is no inherent signal distortion; however, with a time-varying intensity the gain and phase may also change with time, causing signal distortion. Furthermore, as G_{s} and ϕ_{s} are functions of the input signal intensity, then the SOA will exhibit nonlinear and bistable characteristics at high input powers.

It may be noted that Figure 10.3 shows the general form of the gain against wavelength for an SOA obtained from Eq. (10.1). Furthermore, Eqs (10.3) and (10.4) give the 3 dB spectral bandwidth for an FPA as the bandwidth of one longitudinal mode. The 3 dB spectral bandwidth, however, of a TWA is determined by the full gain width of the amplifier medium itself, as illustrated in Figure 10.5(a), rather than the Fabry–Pérot gain profile. Hence the 3 dB bandwidth of a TWA is three orders of magnitude larger than that of an FPA [Ref. 9]. Nevertheless, the passband comprises peaks and troughs whose relative amplitudes are determined by the facet reflectivities, the single-pass gain (and hence the applied bias current) and the input intensity. This gain undulation or peak–trough ratio of the passband ripple ΔG, which is defined as the difference between the resonant and non-resonant signal gain, may be observed in Figure 10.5(b). It is given by [Refs 1, 9]:

$$\Delta G = \left(\frac{1 + \sqrt{R_1 R_2}G_{\mathrm{s}}}{1 - \sqrt{R_1 R_2}G_{\mathrm{s}}}\right)^2 \tag{10.12}$$

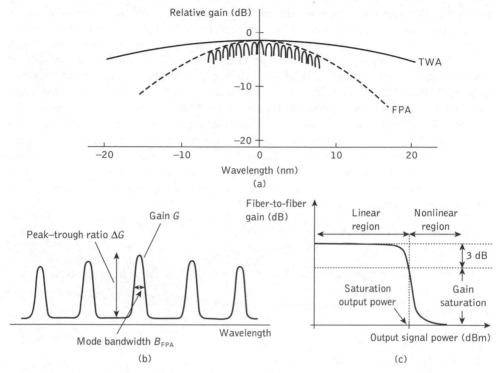

Figure 10.5 Characteristics for semiconductor optical amplifiers: (a) overall passband characteristics for both the traveling-wave and Fabry–Pérot amplifiers showing the large passband ripple in the latter case; (b) illustration of the peak–trough ratio of the passband ripple given In Eq. (10.12); (c) typical fiber-to-fiber gain against output signal power characteristic

For wideband operation the peak–trough ratio must be small and for convenience is normally considered to be less than 3 dB for TWAs over their signal–gain spectrum* [Refs 1, 9]. Hence an amplifier whose gain ripple significantly exceeds 3 dB is usually categorized as an FPA.

Most of the applications of the SOA use it as a basic optical gain block while the active cavity gain, G, of the device is defined as the ratio of output signal power to input signal power from the cavity [Ref. 14]. When the fiber coupling losses are included at both ends of the device, however, the gain is then referred to as the fiber-to-fiber gain of the device. This parameter is illustrated in Figure 10.5(c) which displays the fiber-to-fiber gain of an SOA plotted against the output signal power. This gain remains in the linear regime for small output signal power but it then saturates at higher output signal power levels. It is therefore called small signal gain[†] when the amplifier operates in this linear region. Further to this point the gain decreases with increasing output signal power in a nonlinear fashion. The output signal power at the point of intersection where the linear gain region meets the nonlinear gain region as indicated in Figure 10.5(c) is called the saturation output

* Sometimes this definition is said to apply to a near-traveling-wave amplifier as, in theory, a gain ripple of zero would correspond to a pure TWA.
† In this case the signal has negligible dependency on the gain coefficient of the optical amplifier.

power and after this point the gain will continue to decrease with increasing input signal power. The saturated output power of the amplifier, however, is defined at the point where the amplifier gain is reduced by 3 dB and this point is referred to as gain saturation. A high value of saturated output power is clearly preferred for the SOA and it depends on both the structure and the type of material used in the device. Hence typical values lie in the range of 5 to 20 dBm [Ref. 15]. Furthermore, the gain of the SOA depends not only on the intensity of the signal(s) present in the active cavity medium but also on the frequency (or wavelength) of the signal(s). Assuming for an input signal that the carrier concentration throughout the active cavity remains constant, then the gain of the amplifier is governed by the multiple reflections at the mirror facets. Finally, material systems the same as those employed for semiconductor lasers (see Section 6.3.6) are used to fabricate SOAs such as the direct bandgap III–V compounds which cover the full optical fiber communication wavelength range from 0.8 to 1.7 µm [Ref. 16].

Example 10.2

Derive an approximate expression for the cavity gain of an SOA in the limiting case of a 3 dB peak–trough ratio.

Solution: For a 3 dB peak–trough ratio Eq. (10.12) becomes:

$$\left(\frac{1 + \sqrt{R_1 R_2} G_s}{1 - \sqrt{R_1 R_2} G_s}\right)^2 = 0.5$$

Therefore:

$$1 + \sqrt{R_1 R_2} G_s = 0.707(1 - \sqrt{R_1 R_2} G_s)$$

$$\sqrt{R_1 R_2} G_s = \frac{0.293}{1.707} = 0.172$$

For a TWA, R_1, $R_2 \ll 1$ and, assuming a zero single-pass phase shift, Eq. (10.1) becomes:

$$G \simeq \frac{G_s}{(1 - \sqrt{R_1 R_2} G_s)^2}$$

Substituting for $\sqrt{R_1 R_2} G_s$ gives:

$$G \simeq \frac{G_s}{(1 - 0.172)^2} = \frac{0.172}{(1 - 0.172)^2 \sqrt{R_1 R_2}}$$

$$= \frac{0.25}{\sqrt{R_1 R_2}}$$

The approximate expression for the cavity gain for an SOA in the limiting case is $0.25/\sqrt{R_1 R_2}$. Thus for wide spectral bandwidth operation the available cavity gain is determined by the quality of the antireflection coatings on the device.

10.3.2 Performance characteristics

The wide spectral bandwidths that may be achieved using high-quality antireflection facet coatings on TWAs are in the region of 50 to 70 nm. However, in comparison with FPAs, such devices require significantly higher bias currents for operation as may be observed from Figure 10.4. In addition, whereas the narrow spectral bandwidth of FPAs provides inherent noise filtering, it is not obtained with TWAs and therefore they are subject to increased levels of noise.

The residual facet reflectivity in TWAs introduces a further problem when considering the use of such amplifiers within optical fiber communication systems. This problem results from the effect of backward gain within the devices. The gain of the backward-traveling signal G_b is defined as the ratio of the power in the backward-traveling signal P_b to the input signal power P_{in} into the amplifier. Hence the gain of the backward-traveling signal is given by [Ref. 17]:

$$G_b = \frac{P_b}{P_{in}} = \frac{(\sqrt{R_1} - \sqrt{R_2}G_s)^2 + 4\sqrt{R_1R_2}G_s \sin^2 \phi}{(1 - \sqrt{R_1R_2}G_s)^2 + 4\sqrt{R_1R_2}G_s \sin^2 \phi} \tag{10.13}$$

A graphical representation of Eq. (10.13) in which $R_1 = R_2$ for $G_s = 25$ dB is shown in Figure 10.6. It may be observed that the backward gain is approaching the potential forward gain at high facet reflectivity. Moreover, even at low facet reflectivity (0.01%) the backward gain is still very significant (10 dB). In systems with cascaded amplifiers, optical isolators may therefore be required to avoid the interaction of backward signals between the devices, unless the backward wave amplitude can be made sufficiently small [Ref. 1].

Structural configurations are also used to reduce the facet reflectivity and Figure 10.7 illustrates three different structures for the SOA. Figure 10.7(a) shows a top view of an angled facet SOA in which the active region is inclined away from the facet cleavage plane [Refs 14, 15]. The active region is tilted by θ_p with respect to the norm of the input signal. This angle depends on the Fresnel reflectivity (see Section 5.2) and is a key factor

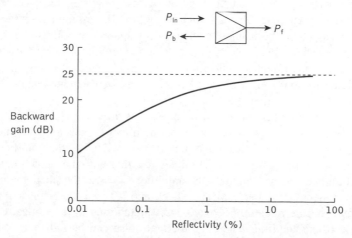

Figure 10.6 Backward gain against facet reflectivity for the traveling-wave amplifier determined from Eq. (10.13). P_f is the power in the forward-traveling signal

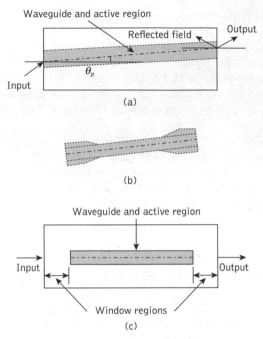

Figure 10.7 Semiconductor optical amplifier waveguide and active region structures to reduce the facet reflectivity: (a) angled facet; (b) angled facet–flared waveguide; (c) window-facet waveguide

in determining the coupling of the signal. A simple method to achieve an efficient coupling of the signal is to use a waveguide with a wider cross-sectional area that will decrease the relative reflectivity. This approach, however, will also produce higher order transverse modes [Ref. 18]. A straightforward way to preserve the single transverse mode is to broaden the end facets of the waveguide instead of using a wider waveguide as indicated in Figure 10.7(b). Such an SOA is known as an angled facet–flared waveguide amplifier [Ref. 15]. Alternatively, as shown in Figure 10.7(c), a window facet (or buried facet) structure in which the window contains a transparent region can be used to obtain reduced facet reflectivity [Refs 15, 19]. Increasing the length of the window region, however, will decrease the effective reflectivity of the SOA which consequently reduces its coupling efficiency and therefore there is a compromise between the length of the window region and the device effective reflectivity [Ref. 20].

Another very important characteristic of the SOA is the noise generated, since it largely determines the maximum number of devices which can be cascaded as linear repeaters within an optical fiber communication system [Re. 14]. During the amplification process the spontaneously emitted photons are amplified together with the signal photons and they are also accumulated at the output of the amplifier to cause the phenomenon known as amplified spontaneous emission (ASE) (see Section 6.3.2). It is the main source of noise in SOAs and as it occurs randomly it may cause fluctuations in the optical output signal. Unlike an electronic amplifier where the thermal noise can be reduced by lowering the temperature, the ASE noise in SOAs has no dependence on temperature. Since the spontaneously emitted photons interact directly with the optical signal at the output of the optical amplifier, the ASE is generally considered as additive noise with constant amplitude and hence it is referred to as white noise.

Since the ASE noise power depends on the amplifier gain then the noise power spectral density, P_{ASE}, can be written as [Refs 15, 21–23]:

$$P_{ASE} = mn_{sp}(G_s - 1)hf\,B \qquad (10.14)$$

where m, B and hf are the mode number, the optical bandwidth and the energy of photon, respectively. The parameter n_{sp} represents the spontaneous emission factor,* which is the fraction of spontaneous emission being emitted into the cavity mode (see Sections 6.2.3 and 6.10). It can be observed from Eq. (10.14) that along with the gain of the SOA, the ASE is dependent on the spontaneous emission factor which usually lies in the range of $1 \le n_{sp} \le 4.0$ where the maximum ASE noise will occur at values of n_{sp} around 4.0 [Ref. 22].

Another useful parameter to quantify optical amplifier noise is the noise figure F_n that is defined as the ratio of the input and output signal-to-noise ratios and is related to ASE noise following [Refs 15, 22]:

$$F_n = \frac{P_{ASE}}{G_s\,hf\,B} \qquad (10.15)$$

It can be observed from Eqs (10.14) and (10.15) that for an amplifier gain higher than unity with mode number $m = 2$ (i.e. when using a narrowband filter), the noise figure is reduced to a value $F_n = 2n_{sp}$. Furthermore, when using the lowest possible value of n_{sp} (i.e. unity), the 3 dB value of the noise figure becomes 2 which is considered as a suitable value of F_n to be used in the analysis of most optical amplification systems [Refs 21, 22].

Since the optical receiver will convert photons to electrons irrespective of their generation source, the ASE appearing at the output of the optical amplifier will give rise to noise in the electrical domain. The ASE noise spectrum also changes as a function of the bias current of the optical amplifier for a fixed level of input signal power as illustrated in Figure 10.8 [Ref. 23]. It may be observed that the ASE increases with increasing bias current since the population inversion, which causes the spontaneous emission, becomes higher. At the same time the maximum value of the ASE noise spectrum is shifted to shorter wavelengths with increasing values of the bias current.

The ASE noise is present in both polarization states (i.e. TE and TM) of the amplified signal and it beats with the signal within each polarization state generating signal–spontaneous noise for each of them. It also produces spontaneous–spontaneous noise as a result of reflections from the end facet causing self-beating. Therefore ASE noise has a significant impact on optical fiber systems in particular when a nonideal extinction ratio[†] is obtained at the receiving end. In this situation ASE noise may contribute towards optical power for a binary 0 signal and consequently the higher optical signal power can change it to a binary 1 signal level. The random nature of ASE can also produce the opposite effect by decreasing the optical signal power for a binary 1 to a binary 0.

---

* Also known as the population inversion factor.
† The ratio of the optical power used in transmitting a logic level 1 to the power used in transmitting a logic level 0.

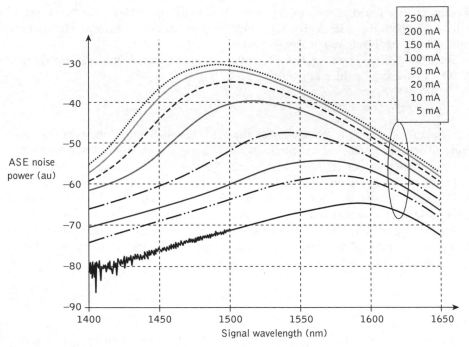

Figure 10.8 Spectral power variation in ASE noise with variation in bias current of a typical semiconductor optical amplifier for different signal wavelengths. Reprinted with permission from Ref. 23 © IEEE 2001

Example 10.3

An SOA operating at a signal wavelength of 1.55 μm produces a gain of 30 dB with an optical bandwidth of 1 THz. The device has a spontaneous emission factor of 4 and the mode number is equal to 2.2 when the net gain coefficient over the length of amplifier is 200. Determine: (a) the length of the device; (b) the ASE noise signal power at the output of the amplifier.

Solution: (a) The length of the amplifier can be calculated by rearranging Eq. (10.5) where:

$$G = \exp(\bar{g}L)$$

$$L = \frac{G_s \, (\text{dB})}{10 \times \bar{g} \times \log e}$$

$$= \frac{30}{10 \times 200 \times 0.434}$$

$$= 34.56 \times 10^{-3} \text{ m}$$

The length of the SOA is therefore 34.6 mm.

(b) The noise power spectral density P_{ASE} is given by Eq. (10.14) as:

$$P_{ASE} = mn_{sp}(G_s - 1)hfB$$
$$= 2.2 \times 4 \times (1000 - 1) \times 6.63 \times 10^{-34} \times 1.94 \times 10^{14} \times 1.0 \times 10^{12}$$
$$= 1.13 \text{ mW}$$

The ASE noise power generated within the SOA is a high level of 1.13 mW which is mainly caused by the large value of the spontaneous emission factor for the device.

It was mentioned previously that the gain of the FPA was very sensitive to changes in temperature and signal polarization. A dependence on these two parameters is also observed in TWAs. For example, at an operating wavelength of 1.5 µm the gain decreases by around 3 dB when the temperature of a TWA of length 500 µm is increased by 5 °C [Ref. 1]. Moreover, although a decrease in temperature increases the device gain, it also increases the passband ripple when there is residual reflectivity. With the FPA these effects are compounded by a shift in mode wavelength of approximately 10 GHz °C^{-1} caused by the variation in refractive index with temperature. Hence it is suggested that for high-gain FPAs the temperature must be controlled to within 0.1 °C [Ref. 1].

The dependence of the gain on the polarization of the input signal results from the difference in the single-pass gain for the TE and TM polarization modes. It is caused by a difference in their optical confinement factors (i.e. $\Gamma_{TE} \neq \Gamma_{TM}$). Furthermore, this effect is magnified in a resonant cavity because the mode propagation constants and the modal facet reflectivities are also polarization dependent. Hence the use of polarization controllers may be necessary when employing FPAs. The gain difference is minimized, however, with TWAS. For example, the gain difference for an FPA and a TWA both operating at a wavelength of 1.5 µm was found to be 10 dB and 2.5 dB respectively [Ref. 1].

10.3.3 Gain clamping

Gain clamping is a method used to maintain or clamp the carrier concentrations to a fixed level in the SOA active cavity medium. This technique is required to avoid the situation where the gain of the SOA can change due to variation of input signal power. For example, although the gain of the SOA saturates when the input power is increased, when two optical signals at different wavelengths are applied to the device, the gain produced by one optical signal can modify the response of the other due to the nonlinear effects of cross-gain modulation and four-wave mixing (see Section 3.14). These effects can, however, degrade the performance of the SOA. For smooth operation it is therefore necessary to maintain the active cavity gain of the device at a constant level.

Gain clamping may be achieved by incorporating lasing action into the amplifier, since the steady-state conditions for the laser can be obtained when the round trip gain is kept constant and equal to the round trip loss as explained in Section 6.2.5. In order to facilitate a gain-clamped SOA, mirrors are placed at either end of the device. This arrangement

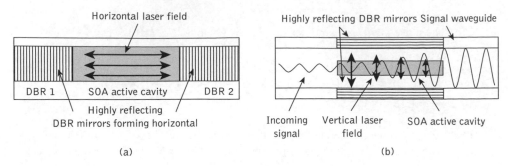

Figure 10.9 Gain clamped SOA with highly reflective distributed Bragg reflectors: (a) with horizontal cavity; (b) with vertical surface cavity

creates a resonant cavity, similar to the Fabry–Pérot laser, which is used to stabilize the gain of the optical amplifier as shown in Figure 10.9(a). It may be observed that the gain-clamped SOA has two distributed Bragg reflectors (DBR 1 and DBR 2) which are used as mirrors (see Section 6.6.2). The lasing field is longitudinal, however, and parallel to the direction of the optical signal. Hence highly reflective DBR mirrors are needed to stabilize the gain of the device. Either one or both of the DBR sections can be selected as the active region, which is accomplished by applying an injection current to it. A clear benefit of the use of an active DBR region rather than a passive DBR region is to provide compensation for the loss that occurs at the input to the amplifier [Ref. 24].

Another approach to fabricate a gain-clamped SOA is to employ a DBR mirror structure as depicted in Figure 10.9(b) where the laser field is both perpendicular to the signal and vertical within the amplifier structure. The optical signal to be amplified passes horizontally through the active medium, directly through the field from the laser which is pumping photons vertically in the active medium. Then the circulating optical power from the vertical cavity laser overlaps with the amplifier waveguide and creates optical feedback to maintain a constant local gain in the amplifier. Thus the vertical laser action linearizes the amplifier gain and provides ultrafast optical feedback in response to changes at the input of the SOA. As a result of this feature, the device is sometimes also referred to as a linear optical amplifier (LOA) [Ref. 25].

Since the gain of the amplifier remains stable for a gain-clamped SOA, it is therefore possible to modify the active cavity orientation to enable the SOA to emit vertically from the surface. Such a device is known as a vertical cavity SOA [Ref. 26]. Surface emission is an important feature in the fabrication of a vertical cavity SOA as most of the light emitted can more easily be coupled to other interfaced or integrated devices (see Section 11.6). Figure 10.10 shows a vertical cavity SOA in which the active cavity medium comprising multiquantum wells based on the InGaAsP/InP material system is set vertically between two highly reflective DBR mirrors fabricated from GaAs/AlGaAs, one each at the top and bottom located on the GaAs substrate. As the gain of the vertical cavity SOA medium is confined to an extremely small region, a large number of multiquantum wells become necessary to achieve high gain by allowing the signal to traverse through the gain region many times. The device can operate in either transmission or reflection mode but for simplicity reflection mode operation only is illustrated in Figure 10.10. In this case the input at

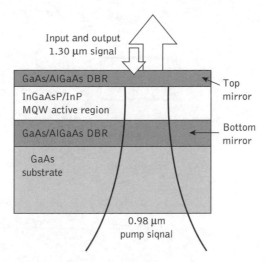

Input and output
1.30 μm signal

GaAs/AlGaAs DBR — Top mirror

InGaAsP/InP
MQW active region

GaAs/AlGaAs DBR ← Bottom mirror

GaAs
substrate

0.98 μm
pump siqnal

Figure 10.10 Reflection mode vertical cavity SOA

the signal wavelength of 1.30 μm enters into the amplifier from the side opposite the pump source emitting at a signal wavelength of 0.98 μm and the output amplified signal exits along the entrance path.

The vertical cavity SOA operates in a similar manner to that of the Fabry–Pérot laser and therefore it is sometimes referred to as a vertical cavity surface-emitting laser (VCSEL) [Refs 14, 26]. A bandpass filter is not needed for a vertical cavity SOA in order to minimize ASE noise as the spontaneous emission bandwidth is limited by the Fabry–Pérot cavity. Signal–spontaneous beat noise and shot noise increase with input signal power and at high signal power levels signal–spontaneous beat noise is the main contributor to the output noise. It should be noted, however, that the output ASE noise, and hence the signal–spontaneous beat noise, is greatly affected by the mirror reflectivity [Ref. 27].

10.3.4 Quantum dots

Quantum dots are crystalline forms of semiconductor materials composed of the periodic groups' II–VI, III–V or IV–VI material systems. Due to their small size (i.e. between 2 and 10 nanometers wide) they are also called nanocrystals. Semiconductor material systems at nanometer-scale dimensions possess variable bandgaps, unlike conventional bulk semiconductor materials. Variable bandgaps in nanometer size semiconductors result from discrete electronic energy levels as compared with continuous electronic energy levels in bulk semiconductors. In addition, a quantum-dot material can change its refractive index due to the discrete electronic energy levels so that these variable refractive index and bandgap properties enable emission over a range of wavelengths [Ref. 28]. Quantum dots occur naturally in quantum-well structures because of the monolayer fluctuations in the thickness of the quantum well. Hence self-assembled quantum dots can be fabricated simply by using molecular beam epitaxy when a material is grown on a substrate with a different lattice structure. The mis-matched lattices result in strain, producing strained islands on

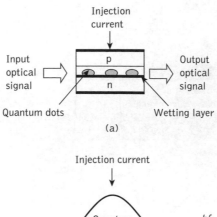

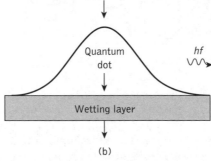

Figure 10.11 Quantum-dot semiconductor optical amplifier: (a) schematic illustration showing a layer of quantum dots situated on the wetting layer; (b) carrier injection current in the conduction band of a quantum dot

top of a two-dimensional layer, and the islands can be subsequently buried to form the quantum dots. The main limitations of this approach are the cost of fabrication and the controlled positioning of individual dots. However, individual quantum dots can also be grown employing electron beam lithography in which a pattern is etched onto a semiconductor chip and conducting metal is then deposited onto the pattern.

Although the conventional SOA based on multiquantum-well technology is already small in size as compared with the other types of optical amplifiers, a further reduction in size, with the additional tunable refractive index and bandgap features offering several advantages in high-speed photonic integration (see Section 11.6), can be obtained using quantum-dot (QD) SOA technology [Refs 29, 30]. The basic structure of a QD-SOA is shown in Figure 10.11(a) where *n*- and *p*-type AlGaAs material is used to sandwich the quantum dots fabricated from InAs grown on a GaAs substrate. A thin layer of material (in this case AlGaAs), the so-called wetting layer, is usually formed beneath the quantum dots. Each quantum dot is isolated such that each one exchanges carriers with the wetting layer. It is mainly through this layer that charge carriers are fed into the active states in the dots and therefore the properties of the wetting layer are important in determining the overall performance and emission wavelength(s) of the device. Figure 10.11(b) illustrates the carrier injection in the conduction band of a single quantum dot and the wetting layer. The carriers present in the wetting layer are launched into an excited state and finally drop back to the ground state. During this process some of the carriers recombine by means of a combination of phonon and Auger recombination processes, releasing a photon of energy $E = hf$.

Although conventional SOAs possess slow gain recovery time, QD-SOAs display an extremely fast speed gain recovery process. By employing quantum dots, gain saturation response time can be accelerated to picoseconds enabling QD-SOAs to perform high-speed switching operations requiring only about one-thousandth the time needed by conventional SOAs [Ref. 30]. Additionally, the QD-SOA also allows the amplification of a data-modulated pulse train without patterning effects in which optical signals do not contain different pulse power patterns for pseudorandom bit pulse sequences with increasing transmission rate [Ref. 31]. Moreover, a QD-SOA with the potential for operation at a transmission rate of 40 Gbit s^{-1} at a signal wavelength of 1.55 μm has been demonstrated [Ref. 32].

10.4 Fiber and waveguide amplifiers

The basic requirements for an optical amplifier include attributes such as high signal gain, high saturated output power, minimal noise generation (within the amplifier) and ultra-wide bandwidth. Fiber and waveguide amplifiers have become essential components for high-performance optical fiber communication systems as they are capable of fulfilling all the aforementioned requirements. Although the various fiber and waveguide amplifier types have significantly different performance characteristics, some of which do not match those obtained with SOAs, in all fiber and waveguide amplifier devices the spectral bandwidths and center wavelengths are largely defined by the atomic structure and not the mechanical geometry. Variations resulting from temperature changes, aging and pump power are therefore less significant in fiber and waveguide amplifiers than in SOAs.

A general representation of a fiber amplifier is shown in Figure 10.12. The gain medium normally comprises a length of single-mode fiber connected to a dichroic coupler (i.e. a wavelength division multiplexing coupler; see Section 5.6.3) which provides low insertion loss at both signal and pump wavelengths. Excitation occurs through optical pumping from a high-power solid-state or semiconductor laser which is combined with the optical input signal within the coupler. The amplified optical signal is therefore emitted from the other end of the gain medium.

The major options for implementing fiber amplifiers were mentioned in Section 10.2: namely, rare-earth-doped fiber amplifiers, Raman fiber amplifiers and Brillouin fiber amplifiers.* In particular, the former two device types, in common with traveling-wave

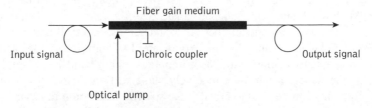

Figure 10.12 Schematic of a fiber amplifier

---

* In these devices the gain medium is often a standard single-mode fiber.

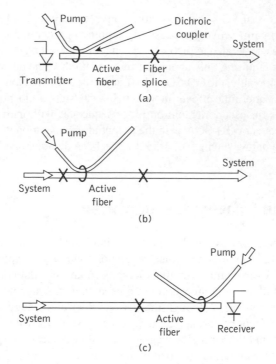

Figure 10.13 Some potential system applications for the fiber amplifier: (a) a power amplifier at the transmitter; (b) an optical repeater; (c) a preamplifier at the receiver

SOAs, are used across a broad range of system applications, some of which are illustrated in Figure 10.13. These include use as: a power amplifier at the transmitter, an in-line optical repeater amplifier and an optical preamplifier at the receiver.

10.4.1 Rare-earth-doped fiber amplifiers

Both neodymium- and erbium-doped fiber lasers were discussed in Section 6.9.2. To date, work on rare-earth-doped fiber amplifiers has concentrated on the erbium dopant, particularly in silica-based single-mode fibers. High gains of between 30 and 40 dB with low noise have been demonstrated [Refs 3, 33, 34] with optical pump powers in the range 50 to 100 mW. Such devices can be made to lase over the longer wavelength region (1.5 to 1.6 μm) of interest within optical fiber communications. Practical pump bands exist at wavelengths of 532 nm, 670 nm, 807 nm, 980 nm and 1480 nm [Refs 2, 35]. However, the latter three wavelengths comprise the most important pump bands. The amplification is dependent on the material gain of a relatively short section (1 to 100 m) of the fiber. Aluminum codoping can be used to broaden the spectral bandwidth to around 40 nm (see Figure 10.1). It should be noted, however, that the spectral dependence on gain is not always as constant as that illustrated in Figure 10.1 (e.g. see Figure 6.44) and hence the spectral bandwidth for erbium-doped silica fibers may be restricted to around 300 GHz (2.4 nm) [Ref. 33].

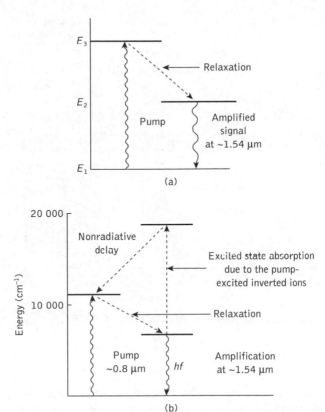

Figure 10.14 Energy-level diagrams for erbium-doped silica fiber laser: (a) the three-level lasing scheme provided by Er^{3+} doping; (b) an illustration of excited state absorption, which is a major limitation

A factor which limits the gain available from an erbium-doped fiber amplifier is a phenomenon known as excited state absorption (ESA). This process is illustrated in the energy-level diagrams for an erbium-doped fiber system shown in Figure 10.14. Erbium provides a three-level lasing scheme which is illustrated in Figure 10.14(a). However, in the erbium fiber amplifier photons at the pump wavelength tend to promote the electrons in the upper lasing level into a still higher state of excitation, as shown in Figure 10.14(b). These electrons then decay nonradiatively to intermediate levels, such as the pump bands, and then eventually back to the upper lasing level. Hence ESA reduces the pumping efficiency of the device and as a result it is necessary to pump at a higher power to obtain a specific gain.

The reduction of ESA in erbium-doped fiber amplifiers is therefore being pursued [Refs 36, 37]. This may be achieved by changing the location of the energy levels through codoping of the erbium–silica fiber amplifier with other compounds such as phosphorus pentoxide. Another technique is to pump the fiber amplifier at a wavelength which does not cause the population of an excited state. Unfortunately, significant ESA is present at

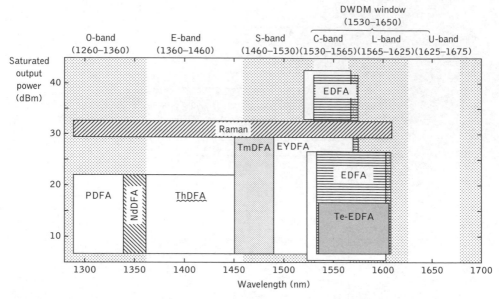

Figure 10.15 Optical amplification wavelength range for different fiber amplifiers

the favored 807 nm pump band. Nevertheless, improved efficiency is obtained with the 980 and 1480 nm pump wavelengths. In particular the 980 nm wavelength displays high efficiency (twice the dB W^{-1} gain figure of the 1480 nm wavelength) but pump sources are not readily available, whereas operation at 1480 nm can be facilitated by both semiconductor and solid-state laser sources.

An alternative solution to avoid ESA, however, is to change to another glass technology in place of silica. In this context the success that has been achieved in lasing with a fluorozirconate host glass (see Section 6.9.2) may provide a way forward. Signal amplification has already been obtained in an erbium-doped multimode fluorozirconate fiber using a 488 nm pump wavelength to provide gain at a wavelength of 1.525 μm [Ref. 38].

The bandwidth potential of rare-earth-doped fiber amplifiers can be utilized when the appropriate doping element is selected to enable the development of a fiber amplifier offering flatter gain and also increased bandwidth. Figure 10.15 provides an optical spectrum for fiber amplifiers displaying the main rare-earth-doped fiber and Raman fiber systems and their corresponding amplification bands. It displays the saturated output power for each type of fiber amplifier plotted as the function of the optical wavelength. The range of different optical communication wavelength bands (i.e. wavelength windows) is also provided at the top of Figure 10.15. Since each material possesses different absorption–emission properties to absorb energy either in a single step or multiple steps, and to emit light in one or more narrow spectral ranges, it is therefore not possible to construct a single rare-earth-doped fiber amplifier which can provide the amplification for all fiber bands. It should be noted, however, that the Raman fiber amplifier does give amplification across all fiber bands covering both the 1.3 μm and 1.5 μm windows (see Section 10.4.2) [Ref. 39].

Trivalent neodymium (Nd^{3+}) and erbium (Er^{3+}) are the most common rare earth elements that are used for the purpose of optical fiber amplification around the wavelength windows of 1300 nm and 1500 nm, respectively. The erbium–ytterbium-doped fiber amplifier (EYDFA) provides amplification over the wavelength range from 1535 to 1567 nm whereas the tellurium–erbium-doped fluoride fiber amplifier (Te-EDFA) can be used to obtain amplification for the range of wavelengths from 1530 to 1608 nm. It should be noted from Figure 10.15 that the praseodymium-doped fiber amplifier (PDFA) is the only rare-earth-doped fiber amplifier capable of amplification in the 1300 nm window [Refs 40, 41]. Moreover, the neodymium-doped fiber amplifier (NdDFA) may also provide amplification for the range of wavelengths between 1260 and 1360 nm but it provides more consistent performance at a signal wavelength of 1345 nm [Ref. 42]. Finally, the thorium-doped fiber amplifier (ThDFA) and thulium-doped fiber amplifier (TDFA) cover the amplification ranges in the E- and S-bands, respectively [Ref. 43].

Recently, an erbium-based microfiber amplifier (EMFA) has also been realized that uses erbium-doped glass to produce high optical gain over just a few centimeters of fiber, rather than over many meters, as with traditional EDFAs [Ref. 44]. The EMFA uses Er^{3+}-doped phosphate glass that supports the doping concentrations of the erbium ions at higher levels in comparison with conventional glasses. For example, a commercially available EMFA exhibits a high gain of 15 dB within a wavelength range from 1530 to 1565 nm while the device also displays greater than 12 dBm output signal power [Ref. 44].

10.4.2 Raman and Brillouin fiber amplifiers

Nonlinear effects within optical fiber may also be employed to provide optical amplification. Such amplification can be achieved by using stimulated Raman scattering, stimulated Brillouin scattering or stimulated four-photon mixing, giving parametric gain (see Sections 3.5 and 3.14) by injecting a high-power laser beam into undoped (or doped) optical fiber. Among these Raman amplification exhibits advantages of self-phase matching between the pump and signal together with a broad gain–bandwidth or high-speed response in comparison with the other nonlinear processes. In particular the broad gain–bandwidth associated with Raman amplification is attractive for current wavelength division multiplexed (WDM) systems since fiber Raman amplifiers in comparison with doped fiber amplifiers provide gain over the entire fiber band (i.e. 0.8 to 1.6 μm) [Ref. 39].

The pump signal optical wavelengths in Raman fiber amplifiers are typically 500 cm^{-1} higher in frequency than the signal to be amplified, and the pumping signal can propagate in either direction along the fiber. A schematic representation of both the forward and backward pumping capability of Raman fiber amplifiers is shown in Figure 10.16. Moreover, continuous-wave Raman gains exceeding 20 dB have been demonstrated experimentally in silica fiber [Ref. 45] which in principle exhibits a broad spectral bandwidth of up to 100 nm with suitable doping of the fiber [Ref. 2]. In addition, Raman gain in excess of 40 dB has been obtained using fluoride glass fiber in which the Raman shift is 590 cm^{-1} [Ref. 46]. Raman fiber amplifiers have been investigated for WDM system applications. For example, the simultaneous amplification using 60 mW pump power of three DFB laser diodes operating at wavelengths in the 1300 and 1500 nm windows provided each channel with 5 dB gain [Refs 47, 48]. Furthermore, the gain–bandwidth of

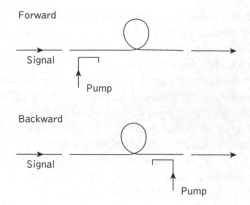

Figure 10.16 Illustrations of the forward and backward pumping capability associated with the fiber Raman amplifier

this fiber Raman amplifier was estimated to be in the range 90 to 300 nm, as indicated in Figure 10.15.

The Raman gain G_R is dependent on a number of factors including the fiber length, the fiber attenuation and the fiber core diameter.* It may be expressed as a function of the optical pump power P_p as [Ref. 49]:

$$G_R = \exp\left(\frac{g_R P_p L_{eff}}{A_{eff} k}\right) \tag{10.16}$$

where g_R is the power Raman gain coefficient, and A_{eff} and L_{eff} are the effective fiber core cross-sectional area and length, respectively, and k is a numerical factor that accounts for polarization scrambling between the optical pump and signal [Ref. 50]. The term $g_R/A_{eff}k$ represents the Raman gain efficiency measured in $W^{-1}\,km^{-1}$. It should be noted that for complete polarization scrambling, as in conventional single-mode fiber, $k = 2$. The effective fiber core area and length are given by:

$$A_{eff} = \pi r_{eff}^2 \tag{10.17}$$

$$L_{eff} = \frac{1 - \exp(-\alpha_p L)}{\alpha_p} \tag{10.18}$$

where r_{eff} is the effective core radius, α_p is the fiber transmission loss at the pump wavelength and L is the actual fiber length.

The theoretical Raman gain characteristics as a function of fiber length for standard 10 μm core single-mode fibers with a pump input power of 1.6 W are shown in Figure 10.17 [Ref. 51]. It may be observed that the Raman gain becomes larger as the fiber lengths increase up to around 50 km where asymptotically it reaches a constant value. Moreover,

---

* In standard single-mode fibers there are relatively low concentrations of germanium in the core, which increases the peak Raman gain in comparison with pure silica core fiber.

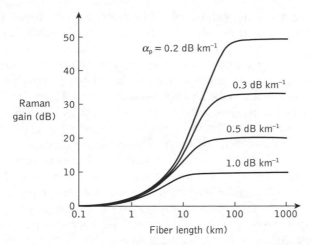

Figure 10.17 Raman gain dependence on fiber length and pump loss (α_p) for a pump input power of 1.6 W and a fiber core diameter of 10 μm. Reprinted with permission from Y. Aoki, 'Properties of fiber Raman amplifiers and their applicability to digital optical communication systems', *J. Lightwave Technol.*, **6**, p. 1225, 1988. Copyright @ 1988 IEEE

it is clear that higher Raman gains can be obtained with lower loss fibers. Although not apparent from Figure 10.17 it is also the case that the Raman gain is increased as the fiber core diameter is decreased (see Eq. (10.16)). Nevertheless, in general the optical pump power required for Raman amplification tends to be high.

Raman fiber amplifiers can be divided into two main categories, namely discrete and distributed. Discrete Raman amplifiers are also commonly referred as lump Raman amplifiers since these devices are used as a lumped element inserted into the transmission line to provide the gain. In this case all of the pump power is confined to the lumped element whereas it is distributed when the amplification takes place along several kilometers of fiber. Therefore a distributed Raman amplifier extends the pump power into the transmission line fiber [Ref. 52]. Both lumped and distributed Raman fiber amplifiers can be combined together for use in wideband applications where their combined amplification increases the overall amplified spectral bandwidth and such devices are referred to as hybrid Raman amplifiers (see Section 10.4.5).

In Raman amplifiers the ASE contributes most of the noise since the common sources of noise include beating of the signal with the ASE and also generation from nonlinearities caused by the medium such as four-wave mixing, self-phase modulation and cross-phase modulation (see Section 3.14). Moreover, the length of fiber can directly influence the noise within Raman amplifiers, for example, as a result of the double Rayleigh scattering reflections* traveling along the fiber which have a magnitude proportional to the length of

---

* Double Rayleigh scattering reflections are also known as multipath interference where the light signal reaches the receiver by more than one optical path; it occurs in fiber Raman amplifiers due to backward and forward propagation of the signal as a result of the nonuniform composition of the glass [Ref. 53].

the fiber. Hence the ASE noise will be reflected together with the signal thus causing it to be increased several times [Ref. 54]. The adverse effect of double Rayleigh scattering reflections, however, can be reduced if two or more stages of the amplification are employed instead of a single amplification stage over the full length of the fiber [Ref. 55]. In addition to ASE noise, the selection and number of pump signals can also directly influence the amplifier noise. In the fiber Raman amplifier (and particularly the distributed Raman amplifier) the pump signal and input signal interact for a longer time over several kilometers of the fiber and therefore any fluctuations in pump power (i.e. the pump noise) will be transferred to the transmitted signal. This pump noise is generally referred to as relative intensity noise (RIN) (see Section 6.7.4) and it becomes more serious when multi-pump signals are used in order to achieve wideband amplification [Ref. 55]. It can be minimized, however, by reducing the interaction time between the pump and the input signal. This can be achieved by using counterpropagation of the signals where the interaction time for the pump and the signal is very short (i.e. the backward pump method as illustrated in Figure 10.16).

By contrast, stimulated Brillouin scattering is a very efficient nonlinear amplification mechanism that can provide high gains at modest optical pump powers of around 1 mW [Ref. 35]. However, it results from the scattering process in which the pump wavelength is often only around 20 GHz distance from the frequency of the optical signal to be amplified. Moreover, it is a narrowband process and the gain–bandwidth may only be in the range 15 to 20 MHz in silica fiber at a wavelength of 1.5 μm [Ref. 56]. The limitation on the spectral bandwidth in a pure silica fiber is around 50 MHz [Ref. 2] which fundamentally restricts the use of Brillouin amplifiers to relatively low-speed communications. Although it is possible to extend the spectral bandwidth to 100 to 200 MHz with germanium doping of the fiber core, it does not significantly alleviate this problem.

Nevertheless, when the fiber is pumped with a CW laser at a power in the range 5 to 10 mW, gains in excess of 15 dB can be obtained [Ref. 2]. A very precise frequency difference of around 11 GHz, however, must be maintained between the optical pump and the signal to ensure that the Brillouin scattering phenomenon continues unabated. This fiber amplifier type is therefore perceived to have a rather restricted range of application. However, the narrowband process could be useful in the provision of tunable filters within WDM systems. It can provide channel selection by allowing amplification of a particular channel without boosting other nearby channels. For example, Brillouin amplification was investigated for channel selection in densely packed, single-mode fiber systems [Ref. 57]. In this case data transmitted at a rate of 45 Mbit s^{-1} was detected without errors with an interfering channel spaced only 140 MHz away at an operating wavelength of 1.5 μm.

A possible limitation of Brillouin amplification, however, for bidirectional WDM applications results from crosstalk due to Brillouin gain if the frequency difference between the counterpropagating waves coincides with the Brillouin shift of around 20 GHz [Ref. 58]. The power level at which significant crosstalk can occur is only of the order of 100 μW [Ref. 59]. Fortunately, since the Brillouin gain–bandwidth is particularly narrow such crosstalk can generally be avoided by a correct choice of signal wavelengths, without restricting the channel packing density. In addition this narrow bandwidth can prove useful for tunable filters to provide channel selection by allowing optical amplification of a particular channel without boosting other nearby channels [Refs 60, 61].

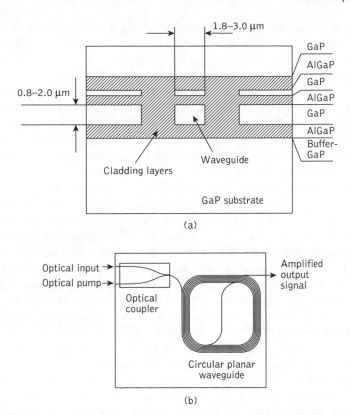

1.8–3.0 μm

0.8–2.0 μm

GaP
AlGaP
GaP
AlGaP
GaP
AlGaP
Buffer-
GaP

Cladding layers

Waveguide

GaP substrate

(a)

Optical input →
Optical pump →

Amplified
output
signal

Optical
coupler

Circular planar
waveguide

(b)

Figure 10.18 Raman amplifier: (a) structure of a Raman amplifier waveguide. Reprinted with permission from Ref. 63 © IEEE 2002; erbium-doped waveguide amplifier with circular planar waveguide

10.4.3 Waveguide amplifiers and fiber amplets

Recent advances in waveguide technology have produced semiconductor waveguide Raman lasers and amplifiers using a GaP core layer and AlGaP cladding layers [Ref. 62]. Figure 10.18(a) shows the cross-sectional view of a Raman amplifier waveguide formed on a GaP substrate. The core of the waveguide is fabricated from a GaP epitaxial layer and the cladding layers comprise AlGaP epitaxial layers. The thickness and the width of the core section within this device are in the range 0.8 to 2.0 μm and 1.8 to 3.0 μm, respectively, whereas the length of the waveguide is 5.0 to 10.0 mm. Raman gain for the semiconductor waveguide device is proportional to the pump power density and the waveguide length. These features of Raman waveguide amplifiers can be used to perform an optical gate function to select a single or desired number of channels from several channels extending over terahertz bandwidth rather than demultiplexing a WDM signal [Ref. 63].

Furthermore, for best performance the waveguide structure requires high crystalline qualities with only minor imperfections and irregularities in the optical waveguide [Ref. 64]. Such imperfections, however, reduce the gain of waveguide Raman amplifiers in particular where both sides of the waveguide are tapered for the purpose of efficient interconnection instead of using a straight waveguide structure. For instance, in an experimental demonstration a Raman fiber amplifier gave a maximum gain in excess of 23 dB using a straight waveguide, whereas it was reduced to just 4.2 dB when a tapered structure at both sides was employed [Ref. 65].

Another category of doped fiber amplifier that utilizes a planar waveguide is the erbium-doped waveguide amplifier (EDWA). This is similar to EDFA except that it provides a higher gain by using a short waveguide (i.e. 1 or 2 centimeters in length) rather than a coil of several meters of fiber as does the conventional EDFA [Ref. 66]. Additionally the EDWA employs a more complex planar circuit to perform optical functions including amplification, variable attenuation, splitting, filtering and pump sharing [Ref. 66]. The last function enables several amplifiers to be driven by a single-pump laser. This is important in particular for the subband and per-channel amplification since each amplifier requires only modest power to produce optical gain. Figure 10.18(b) shows the structure of the EDWA. It consists of input ports for the optical input and pump signals together with the amplifying medium containing phosphate glass. In order to reduce the size without reducing the gain, the planar waveguide is placed as a circular or elliptical spiral [Ref. 67]. It is also possible to increase erbium ion doping which provides higher optical gain per unit length thus reducing the length of the gain medium. This is achieved, however, at the cost of reduced efficiency since the addition of more erbium ions to the glass increases ion cluster formation, the ions in which also exchange energy in the excitation state and therefore not all these ions contribute to the gain mechanism.

EDWAs can be arranged in the form of an array as depicted in Figure 10.19 which comprises four EDWAs in an array sharing the same optical pump source. The pump signal is combined with the input signal via four coarse WDM couplers (see Section 5.6.3), followed by four EDWAs. The pump-sharing feature reduces the cost of amplification in situations where only moderate gain and power are required. Alternatively, an amplified splitter that contains an optical power splitter and a single optical amplifier reduces the number of amplifiers required. In addition, the amplified splitter may include couplers, multiplexer, optical filter and optical pump source. Amplified splitters utilizing EDWAs are commercially available which can splitter/combine 4 to 12 input signal wavelengths into a single output fiber [Ref. 68].

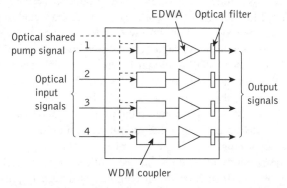

Figure 10.19 A 4 × 4 array comprising erbium-doped waveguide amplifiers sharing the same optical pump source; WDM couplers and filters are also shown as a part of the complete package

The cost-effective use of an optical amplifier depends not only on the size of the network but also on the selection of an appropriate type of amplifier. For example, a sub-banded dense WDM system application may require only 4 to 8 channels as compared with 32 to 40 channels for typical broadband dense WDM applications. In this case the total power requirements reduce by 9 to 10 dB and it requires a narrowband optical amplifier system providing a means to access few channels instead of all the channels within a multiplexed signal. Additionally, dynamic gain control at the add/drop multi-plexer may be required for a single or few channels only [Ref. 69]. In order to meet such requirements another category of optical amplifiers is emerging known as amplets. Amplets are small-sized optical amplifier architectures which comprise miniature taps, planar optical couplers, isolators, and filters to reduce the overall amplifier architecture dimension by around 30 to 40%. Therefore an amplet can be used for optical amplification of a single or few wavelength channels consuming lower power levels and occupying less space. Amplet architectures can be implemented using any optical amplifier technology (i.e. SOA or fiber amplifiers). The name amplet is used to distinguish them from their larger and more complex predecessors.

Figure 10.20 shows an architecture for an amplet based on two EDFA stages with built-in narrow-bandpass add/drop filters. The center taps between the two stages allow inser-tion of devices to achieve other networking functions such as dispersion compensation and dynamic gain control. In order to reduce the noise an ASE filter is also incorporated in the return path.

The physical dimensions of optical amplifiers with (and without) additional components required to achieve control operation are illustrated in Figure 10.21. An EDFA amplet can be configured based on the requirements of the particular amplification system. For example, with only a single add or drop tap (i.e. single port mini-amplet), with two add/drop ports (i.e. two-port amplet) or with multiple add/drop ports, while in the latter case the amplets can be cascaded using an array structure for expansion of capacity [Ref. 70].

Finally, Table 10.1 provides a comparison of the performance parameters for an EDFA amplet with typical EDWA and SOA devices [Ref. 69]. It can be seen that the EDFA amplet provides the highest output signal power (greater than 15 dBm) and that the gain of the amplet is nearly equal to that of the SOA. Further advantages of the EDFA amplet are

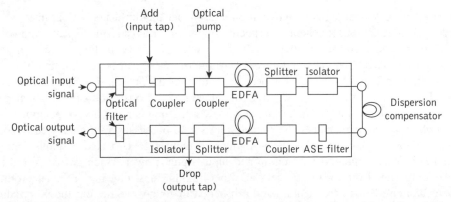

Figure 10.20 A two-port optical amplet architecture based on the erbium-doped fiber amplifier

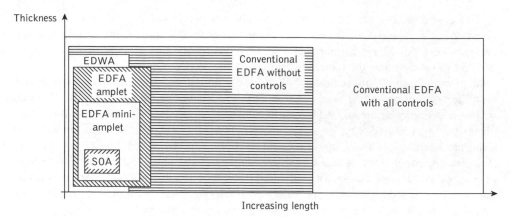

Figure 10.21 Device relative dimensions of optical amplifier technologies

Table 10.1 Comparison of EDFA amplet with EDWA and SOA [Ref. 69]

	EDFA amplet	EDWA	SOA
Output power (dBm)	>15	>12	>13
Gain (dB)	>24	>20	>25
Noise figure (dB)	<6	<6	<8
Max. gain flatness deviation (dB nm^{-1})	<0.75	<0.75	<0.1
Polarization mode disersion (dB)	<0.3	<0.5	<0.5
Power consumption (W)	<1	<1	<5

that it exhibits the lowest polarization mode dispersion and it also consumes substantially less power than the SOA with power consumption approximately equal to the EDWA.

10.4.4 Optical parametric amplifiers

When a strong optical signal is incident on a nonlinear medium (i.e. optical fiber, waveguide, photonic crystal) it generates a spectral region around itself due to scattering within the medium and therefore another small signal present in the medium may gain some signal power. This phenomenon, known as parametric gain, can be used to amplify weak optical signals. The parametric gain mechanism utilizes the principle of four-wave mixing (FWM) (see Section 3.14). Figure 10.22(a) illustrates the operation of an optical parametric amplifier when three signals, namely the pump, the idler and the weak input signal, are applied to a nonlinear fiber. An optical idler signal possesses a frequency equal to the difference of frequencies between the pump and input signal. It also contains a higher signal power as compared with pump and input signal power which is transferred to a weak input signal during the process of wave mixing. A strong pump signal produces Stokes and anti-Stokes frequencies and as a result of four-wave mixing the pump signal power is transferred to the optical input signal. Optical couplers are used to ensure four-wave mixing of the signals to produce the desired output signal since the parametric gain

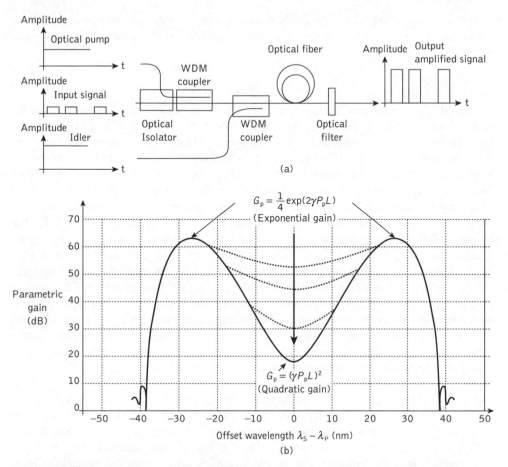

Figure 10.22 Optical parametric amplifier: (a) amplifier structure using a nonlinear fiber; (b) gain characteristic where the arrows identify the regions with exponential and quadratic gain

yields unwanted signals and therefore an optical filter is used to isolate these from the output amplified signal as shown in Figure 10.22(a). The peak parametric amplification or gain G_p can be written as an exponential gain following [Ref. 71]:

$$G_p = \frac{1}{4}\exp(2\gamma P_p L) \tag{10.19}$$

where γ is the fiber nonlinear coefficient, P_p is the pump signal power and L is the length of the fiber. The fiber nonlinear coefficient which depends on the refractive index of the fiber cladding n_2 and the signal wavelength λ is given by:

$$\gamma = \frac{2\pi n_2}{\lambda A_{\text{eff}}} \tag{10.20}$$

where A_{eff} is the effective core area of the fiber.

For an input signal frequency λ_s equal (or near) to the value of the pump signal frequency λ_p (i.e. $\lambda_s - \lambda_p = 0$) then the signal gain G_p in Eq. (10.19) reduces to the quadratic gain following [Ref. 71]:

$$G_p \approx (\gamma P_p L)^2 \tag{10.21}$$

Figure 10.23(b) shows a calculated parametric gain profile as a function of wavelength for a fiber of length 500 m. The horizontal axis represents the signal wavelength offset from the pump signal wavelength (i.e. $\lambda_s - \lambda_p$). The amplifier bandwidth is therefore determined from the gain profile at this offset wavelength which shows an exponential gain at the top leading to a quadratic parametric gain where the difference in the pump and the input signal frequency becomes negligible or zero. Furthermore, the use of short highly nonlinear fiber efficiently decreases the fiber length and thereby increases the nonlinear coefficient γ. This approach enables achievement of the maximum gain with increased bandwidth for the amplifier. Either one-pump or two-pump schemes can be used to provide parametric amplification; however, the latter scheme increases the bandwidth significantly with improved exponential gain from the parametric amplifier [Ref. 72]. The main advantages of the optical fiber parametric amplifier include the large signal gain of greater than 30 dB, the high saturated output signal power of more than 20 dBm and also a low-noise figure of less than 3 dB with a wide bandwidth of around 200 nm [Ref. 73]. In addition, the parametric amplifier can be used to provide various all-optical network functions such as wavelength conversion, signal sampling, pulse generation and all-optical regeneration [Ref. 71]. A high gain optical parametric amplifier has also been implemented within a LiNbO$_3$ waveguide and the resultant waveguide device was capable of amplification over the O- to C-band wavelength range (1.260 to 1.625 μm) with a signal gain of up to 30 dB [Ref. 74]. Although the operation of this parametric waveguide amplifier is similar to the one shown in Figure 10.22, the device implementation was realized monolithically by combining WDM couplers with a periodically polled LiNbO$_3$ waveguide (see Section 11.4.3) [Refs 74–76]. It should be noted, however, that in order to achieve higher gain and higher efficiency the waveguide amplifier was required to have long interaction length (i.e. necessitating a long waveguide) and hence a compact 160 mm long waveguide device achieved only a modest signal gain of 12 dB [Ref. 75].

It should be noted that Eqs (10.19) and (10.21) apply to the perfect phase-matching case in which all the signals are symmetrically positioned relative to each other. In order to obtain this situation, the wave propagation constant mismatch, $\Delta\beta$, must be reduced to zero (i.e. $\Delta\beta = \beta_s + \beta_i - 2\beta_p = 0$) where β_s, β_i and β_p are the propagation constants for the input (or source), the idler and the pump signals, respectively.

Example 10.4

A parametric optical amplifier is 500 m in length with an optical pump operating at a signal wavelength of 1.55 μm with signal power of 1.4 W. When the input signal wavelength is 1.56 μm and the parametric peak gain is 62.2 dB, calculate: (a) the fiber nonlinear coefficient; (b) the parametric gain in dB when it is reduced to quadratic gain.

Solution: (a) The fiber nonlinear coefficient γ can be obtained from Eq. (10.19) where the peak parametric gain is:

$$G_p = \frac{1}{4}\exp(2\gamma P_p L)$$

which can be rewritten in decibels as:

$$G_p(\text{dB}) = 10 \times \log_{10}\left[\frac{1}{4}\exp^{(2\gamma P_p L)}\right]$$

Therefore:

$$\gamma = \frac{G_p\,(\text{dB}) - 10 \times \log_{10}(0.25)}{P_p L} \times \frac{1}{10 \times \log_{10}(2.718\ 281\ 828\ 4)^2}$$

$$= \frac{62.2 + 6}{1.4 \times 500} \times \frac{1}{8.7}$$

$$= 11.19 \times 10^{-3}\ \text{W}^{-1}\ \text{km}^{-1}$$

Hence the fiber nonlinear coefficient of the parametric amplifier is $11.2\ \text{W}^{-1}\ \text{km}^{-1}$.

(b) The quadratic gain G_p for an optical parametric amplifier is given by Eq. (10.21) which can be rewritten to provide the decibel gain as follows:

$$G_p(\text{dB}) = 10 \times \log_{10}(\gamma P_p L)^2$$
$$= 10 \times \log_{10}(11.2 \times 10^{-3} \times 1.4 \times 500)^2$$
$$= 17.88\ \text{dB}$$

The gain of the parametric amplifier when it is reduced to the quadratic gain is therefore 17.9 dB.

10.4.5 Wideband fiber amplifiers

The combinations of fiber amplifiers which can be used to provide amplification extending over a wide wavelength range are often referred to as wideband, or hybrid, amplifiers. These combinations can be in serial, parallel or a combination of both configurations while the optical fiber amplifiers used in such combinations can be rare-earth-doped fiber, Raman fiber or the combination of both types of fiber amplifiers. Figure 10.23 illustrates these different ways to obtain wideband amplification using fiber amplifiers. A dual or serial configuration is shown in Figure 10.23(a) where two EDFAs are connected in series to accomplish wideband amplification, each providing the gain for both C- and L-bands. Figure 10.23(b) represents the parallel configuration with two EDFAs, each one placed in a branch to achieve wideband amplification for C- and L-bands, respectively. For example, an experimental setup has demonstrated that such an amplifier configuration with only two

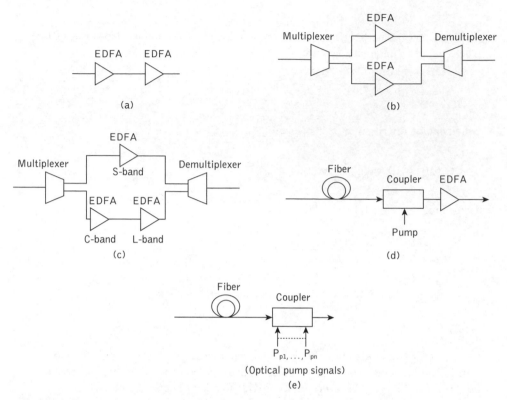

Figure 10.23 Various amplifier configurations for providing wideband amplification: (a) serial erbium-doped fiber amplifier (EDFA); (b) parallel EDFA; (c) combined parallel–serial, S-band (1460 to 1530 nm), C-band (1530 to 1565 nm), L-band (1565 to 1625 nm); (d) combination of fiber Raman amplifier and EDFA; (e) multiple pump fiber Raman amplifier

EDFAs in parallel can exhibit a flat gain of 15 dB for a wide bandwidth of 105 nm from 1515 to 1620 nm [Ref. 77].

Further expansion in amplification bandwidth can be achieved when the serial configuration is introduced within any one arm of the parallel configuration as indicated in Figure 10.23(c). This hybrid technique comprising two EDFAs in parallel and another in series can achieve a spectral bandwidth of 120 nm. A maximum gain of 33 dB was delivered using this hybrid configuration over the wavelength range from 1480 to 1600 nm, covering the S-, C- and L-bands [Ref. 78]. It is also possible to obtain wideband amplification using different types of fiber amplifiers where again the configuration can be arranged by using a serial, parallel or hybrid combination. For example, Figure 10.23(d) displays an EDFA in serial combination with a fiber Raman amplifier [Ref. 79].

It is also possible to achieve wideband amplification using only fiber Raman amplifiers. In this case, instead of using a single optical wavelength pump, multiple pump sources can be used to accomplish wideband amplification. Figure 10.23(e) illustrates the concept of a multiple-pump wavelength fiber Raman amplifier where P_p represents the pump signal power with additional subscript (i.e. 1, 2, 3, . . . , n) describing the number for each optical pump source. In this case complete amplification can be obtained using multiple stages (i.e. two or three) each for S-, C- and L-bands, respectively. This multistage technique, however, requires the wavelength multiplexing of the optical signal to be amplified at each stage. An experimental demonstration utilizing such multiple pump sources using fiber Raman amplifiers has provided an optical bandwidth greater than 100 nm [Ref. 80]. Another demonstration using a tellurite-based fiber Raman amplifier employing a 250 m fiber length and multiple-pump sources gave a 160 nm bandwidth with a gain of more than 10 dB over the wavelength range from 1490 to 1650 nm [Ref. 81].

10.5 Wavelength conversion

Wavelength conversion is defined as a process by which the wavelength of the transmitted signal is changed without altering the data carried by the signal. The device that performs this function is usually called a wavelength converter but it is also referred to as a wavelength (or frequency) changer, shifter or translator. It is termed an up-converter when the converted signal wavelength is longer than the original signal wavelength and it is called a down-converter if the converted signal wavelength is shorter than the original signal wavelength. A wavelength converter should be capable of receiving an incoming signal at any wavelength (i.e. a variable wavelength) at the input port and must produce the converted signal at a particular wavelength (i.e. a fixed wavelength) at the output port. Therefore the input/output (I/O) ports of the converter must possess the capability of a variable input–fixed output (VIFO) converter and the majority of the optical switching networks use this type of device (see Section 14.5).

Based on component enabling technologies the different device types can be described as either optoelectronic, or optical and all-optical wavelength converters. An optoelectronic wavelength converter comprises an optical receiver and transmitter that receives the incoming signal at one wavelength and then retransmits on another signal wavelength. The basic function of an optoelectronic wavelength converter is illustrated in Figure 10.24 where an optical intensity-modulated input signal at the wavelength λ_{in} is first detected in an optical-to-electrical (O/E) block using a photodiode and then the received data is processed in the electrical domain. The processor section may also contain an electronic regenerator circuit that reconstitutes the electrical signal from the received data thereby rectifying any errors induced by transmission through the fiber (see Section 12.6.1). The electrical data is then converted back to an optical signal in an electrical-to-optical (E/O) section by modulating an optical transmitter that emits a compliant wavelength at the output port (i.e. converted wavelength, λ_{conv}). This method can be modular and hence flexible, but the transmission rate tends to be restricted due to limitations of the O/E/O interfaces [Ref. 82].

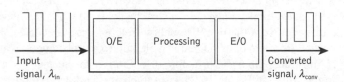

Figure 10.24 Optoelectronic wavelength converter

The main disadvantage of optoelectronic wavelength converters is that all wavelengths must be terminated and reprocessed when a single or only few wavelengths may be intended for that particular destination. In addition, these converters generally cannot process different signal modulation formats and therefore the information contained in the intensity, frequency, phase or polarization of the signal is required to be reprocessed for the purpose of wavelength conversion or it will be lost. Hence optoelectronic wavelength converters are not transparent to signal modulation formats. Nevertheless, currently such converters find some limited use in existing optical networks [Ref. 83]. For example, the SONET repeater provides optoelectronic wavelength conversion when the output port transmits a compliant wavelength [Ref. 84].

The other major wavelength converter type comprises optical and all-optical configurations where the signal remains entirely in the optical domain from the input to the output port. The control, however, can be provided electronically or optically and therefore these converters are referred to as either optical wavelength converters (OWCs) or all-optical wavelength converters (AOWCs), respectively. The strategy used for the implementation of optical wavelength conversion utilizes the nonlinearity of the optical medium which can be either active or passive, each producing different nonlinear effects (see Section 3.14). In this context devices can be further subdivided according to their operating principles into two broad categories which are coherent and cross-modulation wavelength converters [Ref. 84]. Coherent wavelength converters make use of a passive optical medium to exploit nonlinear effects such as four-wave mixing and difference frequency generation [Ref. 84]. The other category of optical wavelength converter is based on the cross-modulation properties of an active nonlinear medium in which an optical control signal experiences the changes produced due to the intensity variation of an intensity-modulated input signal present in the active cavity. The process of imposing the nonlinear response of the medium onto the control signal is known as a cross-modulation scheme and depending upon the properties of the nonlinear medium (i.e. optical gain and material absorption) and the modulation on the optical signal (i.e. intensity, phase or polarization), this scheme can be divided into four main approaches: cross-gain modulation (XGM); cross-phase modulation (XPM); cross-absorption modulation (XAM); and differential polarization modulation (DPM).

10.5.1 Cross-gain modulation wavelength converter

A cross-gain modulation (XGM) wavelength converter utilizes the nonlinear properties of a semiconductor optical amplifier to perform the conversion process. According to the cross-gain modulation principle, the intensity-modulated data on one signal wavelength, generally called the pump signal, produces variations in carrier density within the SOA which provides inverted gain modulation in the SOA medium (see Section 10.3). These gain modulations can be imprinted onto a continuous wave (CW) signal called a probe signal. Thus the probe signal acquires an inverse copy of the data and when the wavelength of the probe signal is different from the pump signal, wavelength conversion is obtained where data has been shifted from the pump signal wavelength to a probe signal wavelength.

Figure 10.25 illustrates the basic principle of operation for an XGM wavelength converter employing an SOA. In Figure 10.25(a) an intensity-modulated input signal at wavelength λ_{in} and CW probe signal at wavelength λ_{cw} are applied to the SOA and following the

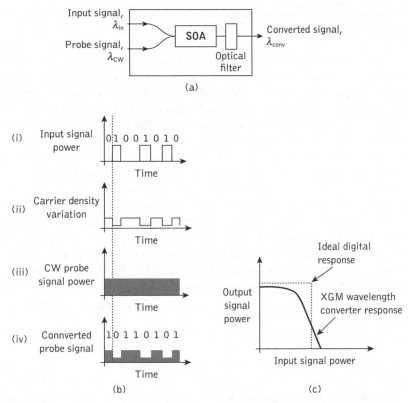

Figure 10.25 Cross-gain modulation (XGM) wavelength conversion with a semiconductor optical amplifier (SOA): (a) structure using copropagation of the input and CW probe signals; (b) signal time chart showing the operating principle of XGM wavelength conversion; (c) transfer function characteristic for the inverted output signal power

XGM the converted probe signal at wavelength λ_{conv} is obtained through an optical bandpass filter. This process of XGM wavelength conversion is also displayed in the signal time chart of Figure 10.25(b). In the first step an input signal containing intensity-modulated data 01001010 is applied to the SOA. The carrier density variations (ΔN) produce an inverted copy of this data as 10110101 in the SOA medium which is identified in the following step. A CW probe signal with a constant signal power is then applied to the SOA. The carrier-density-induced variations constituting the data modulate this CW probe signal power and thus the converted probe signal contains the data 10110101 as shown in the final step. Hence the converted probe signal now incorporates the original data from the intensity-modulated input signal. Figure 10.25(c) provides the transfer function characteristic for the XGM wavelength converter. It may be observed that the ideal transfer function indicates a step or rectangular (i.e. digital) response but as a result of the slow gain recovery of the SOA, the transfer function does not maintain a rectangular shape, rather its amplitude gradually decreases producing the response as shown in bold in Figure 10.26(c).

The XGM wavelength converter shown in Figure 10.25(a) is configured as a copropagating scheme where the input and CW probe signal are fed from the same side of the SOA and propagate in the same direction. It is also possible to achieve XGM wavelength conversion using a counterpropagating scheme where the input and CW probe signals are fed into the SOA from opposite sides. In either case the process of XGM conversion is not immune from the generation of noise due to the ASE noise within the SOA and therefore an optical filter is a necessary requirement at the output port of the device [Ref. 85].

The speed of operation of XGM wavelength conversion is determined by the carrier dynamics of the SOA which is restricted due to its relatively slow carrier recovery that, however, can be increased when the interaction time between the input and CW probe signal is increased. This situation can be achieved by increasing the length of the SOA or by cascading more SOAs in a copropagating scheme [Ref. 85]. In this manner XGM wavelength converters can generally be used for transmission rates of 40 Gbit s^{-1} [Ref. 86] and error-free conversion of an intensity-modulated signal at 100 Gbit s^{-1} based on grating-assisted XGM in a 2 mm long SOA has been demonstrated [Ref. 87]. Although the XGM wavelength converter has a relatively simple structure, the main drawback with the device concerns the degradation of extinction ratio (see Section 10.2) which causes severe limitations in relation to the cascadability of such devices in an optical network [Ref. 88].

10.5.2 Cross-phase modulation wavelength converter

During the process of XGM in an SOA the carrier density variation gives rise to a change in the refractive index of the nonlinear medium which is proportional to the amplitude of the input signal and inversely proportional to carrier density variations. These refractive index variations produce phase modulation when a CW probe signal is coupled to the SOA. When these phase modulations are converted into amplitude modulation using an interferometric arrangement, the resultant probe signal contains an exact copy of the intensity-modulated pattern and the converter is then referred to as a cross-phase modulation (XPM) wavelength converter. The basic structure of an XPM wavelength converter is illustrated in Figure 10.26. It is similar to the XGM wavelength converter but it incorporates an additional phase modulation–amplitude modulation (PM–AM) element with another path to it from the probe signal creating an interferometric structure. Hence the CW probe signal splits into two parts and each is applied to the SOA and the PM–AM element. The probe signal at wavelength λ_{CW} propagates through the optical filter acquiring the refractive-index-induced phase modulation in the active cavity of the SOA while the input signal at λ_{in} is blocked by this optical filter. Hence the CW probe signal combines with the phase-modulated probe signal in the PM–AM element when the interference is

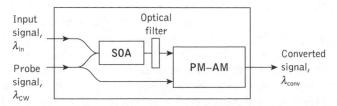

Figure 10.26 Structure of a cross-phase modulation wavelength converter

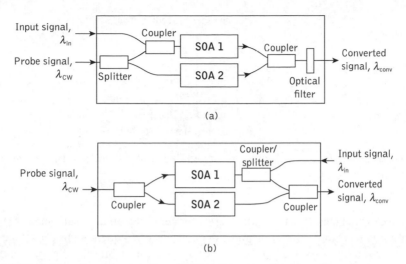

(a)

(b)

Figure 10.27 Cross-phase modulation (XPM) wavelength converter using semiconductor optical amplifiers (SOAs) in a Mach–Zehnder interferometer (MZI) arrangement: (a) copropagating scheme; (b) counterpropagating scheme

constructive. The output at the PM–AM element therefore incorporates a probe signal that contains a copy of the data of the input signal but it is now at a different wavelength, being that of the probe signal, thus giving XPM wavelength conversion.

There are several possiblities for realizing an XPM wavelength converter utilizing SOAs in an interferometeric configuration. Such a wavelength converter based on a Mach–Zehnder interferometer (MZZ) configuration is depicted in Figure 10.27 which comprises two SOAs located in the two paths or arms. The operation is similar to the basic XPM wavelength converter while the use of another SOA (i.e. SOA 2) in the lower path increases the gain of the converted probe signal. Phase to amplitude modulation can be obtained when a relative phase difference is introduced between the interferometer paths which can be provided by setting different path lengths [Ref. 89]. Alternatively, a phase–shifting element may be placed in the lower path of the interferometer before the optical filter at which the probe signal is recombined producing constructive interference [Ref. 90].

In the copropagating scheme shown in Figure 10.27(a) the CW probe and the input signals are launched into SOA 1 and both the CW probe and input signals travel in the same direction, whereas in the counterpropagation scheme displayed in Figure 10.27(b) the input signal at wavelength λ_{in} and CW probe signal at λ_{CW} are applied to SOA 1 from opposite directions. Although the operation of the wavelength converter is similar for both schemes, the counterpropagating scheme has the advantage of reduced noise as the CW probe signal does not interact and hence an optical filter may not be required [Ref. 84].

During the process of XGM and XPM another phenomenon, namely that of instantaneous frequency variation, occurs which is commonly referred to as frequency (or wavelength) chirp. It is defined as the deviation in the emission frequency with respect to time when a laser is driven by a time-varying current source (i.e. intensity-modulated digital signal). Since refractive index is related to frequency then a variation in the refractive index produces a variation in frequency at each instant. This chirp is commonly observed in the intensity-modulated converted probe signal affecting the leading and trailing edges

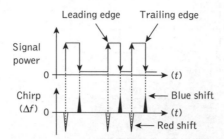

Figure 10.28 Frequency chirp: intensity-modulated signal (top) and the corresponding positive chirp (bottom)

of the pulses when the refractive index varies suddenly due to an instantaneous change in pulse amplitude. Figure 10.28 illustrates the concept of frequency chirp associated with the intensity-modulated signal shown in the top diagram, while the diagram underneath depicts the corresponding frequency chirp at the leading and trailing edges. When frequency chirp shifts the optical frequency towards the shorter wavelength it is known as blue shift, while a shift of optical frequency towards the longer wavelength is called red shift. The sign of the chirp is said to be positive when the leading edge of the pulse is red shifted in relation to the central wavelength and the trailing edge is blue shifted, and when the shifts are opposite to the foregoing then the sign of chirp will be negative. Hence Figure 10.28 displays a positive chirp condition.

The relationship between the frequency chirp Δf and the phase of the converted signal ϕ is given by [Refs 91, 92]:

$$\Delta f(t) = -\frac{1}{2\pi}\frac{\mathrm{d}\phi(t)}{\mathrm{d}t} \tag{10.22}$$

Although Eq. (10.22) shows the influence of phase variations on the frequency chirp, the phase of the converted probe signal is dependent upon the intensity-modulated signal power. If $P_{in}(t)$ is the input signal power then the phase variation can be written as [Ref. 93]:

$$\frac{\mathrm{d}\phi}{\mathrm{d}t} = \frac{\alpha}{2}\frac{1}{P_{in}(t)}\frac{\mathrm{d}P_{in}(t)}{\mathrm{d}t} \tag{10.23}$$

where the term α represents the linewidth enhancement factor which is generally referred as the α-parameter (see Section 6.2). In the context of the material properties of the nonlinear medium the α-parameter has a strong dependence on the signal wavelength and may be defined in terms of gain and refractive index variation as [Ref. 94]:

$$\alpha = \frac{4\pi}{\lambda}\frac{\mathrm{d}n_r/\mathrm{d}n}{\mathrm{d}g/\mathrm{d}n} \tag{10.24}$$

where λ is the signal wavelength, $\mathrm{d}n_r/\mathrm{d}n$ represents the differential refractive index showing the change in the real part of refractive index with respect to carrier density variation, $\mathrm{d}g/\mathrm{d}n$ is the differential gain which defines the change in gain with respect to change in carrier density.

Example 10.5

A semiconductor optical amplifier (SOA) operating at a signal wavelength of 1.55 μm produces an output signal power of 10 mW with an input signal power of 0.5 mW. The differential refractive index of the device is -1.2×10^{-26} m^3, the linewidth enhancement factor is -1.0 and the input signal power variation is 0.01 μm. Calculate: (a) the frequency chirp variation at the output signal; (b) the differential gain required in order for the device to operate at the same signal wavelength.

Solution: The frequency chirp Δf can be obtained by combining Eqs (10.22) and (10.23) where:

$$\Delta f(t) = -\frac{\alpha}{4\pi} \frac{1}{P_{in}(t)} \frac{dP_{in}(t)}{dt}$$

$$= -\frac{-1}{4 \times 3.14} \times \frac{1}{0.5 \times 10^{-3}} \times 0.01 \times 10^{-6}$$

$$= 1.59 \times 10^{-6}$$

$$= 1.6 \text{ MHz}$$

The frequency chirp variation at the output signal is 1.6 MHz and it has a positive sign indicating that it is blue-shifted chirp.

(b) The differential gain dg/dn can be obtained by rewriting Eq. (10.24) as:

$$dg/dn = \frac{4\pi}{\lambda} \frac{(dn_r/dn)}{\alpha}$$

$$= \frac{4 \times 3.14}{1.55 \times 10^{-6}} \times \frac{-1.2 \times 10^{-26}}{-1}$$

$$= 9.72 \times 10^{-20} \text{ m}^2$$

The differential gain required for the SOA to operate at a signal wavelength of 1.55 μm is 9.7×10^{-20} m^2.

It is also possible to implement a wavelength converter using a Mach–Zehnder interferometer (MZI) with only one SOA. Such a device based on a delay interferometer structure is shown in Figure 10.29(a). Although the operation of this device is similar to the MZI wavelength converter, this interferometer arrangement utilizes different path lengths for each arm. Furthermore, a phase shifter is inserted into the shorter arm to achieve the balance in relation to the longer arm that possesses a time delay Δt due to its relatively longer length. The phase shifter and the time delay Δt operate together as a phase modulation to amplitude modulation element (i.e. the same as described for Figure 10.26), and these two together provide the control to obtain the balance between the arms creating either destructive or constructive interference. For constructive interference, when an

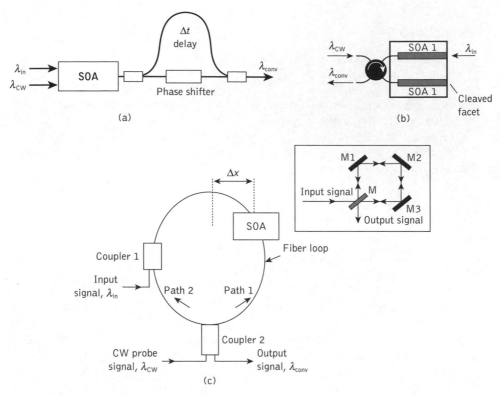

Figure 10.29 Wavelength converters based on cross-phase modulation (XPM) using semiconductor optical amplifiers (SOAs) in interferometer configurations: (a) delay interferometric wavelength conversion arrangement; (b) Michelson interferometer; (c) nonlinear optical loop mirrors showing a Sagnac interferometer in the inset

input signal at wavelength λ_{in} and the CW probe signal at wavelength λ_{CW} are injected into the SOA simultaneously, then the converted probe signal with wavelength λ_{conv} appears at the output port of the device, thus providing wavelength conversion. Such wavelength converters are preferred for their compact size and effective operational control for transmission rates higher than 40 Gbit s^{-1} [Ref. 95]. Moreover, there have been successful demonstrations of wavelength conversion using delay interferometer configurations with SOAs operating at a transmission rate of 160 Gbit s^{-1} [Refs 96–98].

The XPM wavelength converter can also be implemented in other interferometric arrangements such as the Michelson interferometer (MI) and the nonlinear optical loop mirrors (NOLM) configurations. Figure 10.29(b) and (c) show these configurations for XPM-based wavelength conversion utilizing the SOA as a nonlinear element. Although in these two interferometric arrangements the basic principle of operation of the XPM converter is similar to the MZI configuration where the phase-modulated probe signal is converted into amplitude modulation, their construction differs from it.

The MI arrangement depicted in Figure 10.29(b) can be seen as the half, or folded, version of the MZI (see Figure 10.27(b)) in the counterpropagation scheme. An SOA is placed in both the upper and lower arms of the interferometer and a CW probe signal at wavelength λ_{CW} is coupled to both these arms which splits equally between both paths and is then reflected by the cleaved facets. The probe signal then recombines either constructively or destructively in the same manner as that of the SOA–MZI wavelength converter shown in Figure 10.27(b). When an input signal at wavelength λ_{in} is coupled to SOA 1 only in the opposite direction to the CW probe signal, the refractive-index-induced phase variations in the medium of SOA 1 are modulated onto the CW probe signal, which then appears at the output terminal incorporating the phase modulation, and where a circulator (see Section 5.6) is used to recombine it with the reflected CW probe signal from SOA 2. It should be noted that the circulator output produces a converted probe signal at wavelength λ_{conv} when constructive interference between two interferometer paths is maintained. MI wavelength converters comprise a simple structure utilizing only one coupler and they are therefore both smaller in size and require less signal power than MZI converters. Although compact monolithically integrated MI wavelength converters using SOAs have also been demonstrated operating at a transmission rate of 40 Gbit s^{-1} [Ref. 99], these devices are not currently the preferred commercial implementation for wavelength conversion.

In addition to the MZI and MI arrangements, nonlinear optical loop mirrors (NOLMs) can also be used to provide the wavelength conversion function. Such a configuration is shown in Figure 10.29(c). The NOLM structure utilizes the basic principle of XPM in a SOA in a manner similar to that used in the classic Sagnac interferometer [Ref. 84] as illustrated by the inset of Figure 10.29(c). In this latter configuration two optical paths are configured using mirrors such that the input signal splits into two parts with one part traveling in a clockwise (i.e. via mirror M1) and the other in a counterclockwise (i.e. via mirror M3) direction. If the distance between both paths is different (i.e. path via M3 ≠ path via M1) then it causes a phase shift when both parts of the signal recombine at mirror M. In an ideal case, when equal lengths for each path are chosen, the signals recombine constructively (i.e. there is no phase shift). However, when one path length is made shorter than the other, the interferometer becomes asymmetrical inducing a phase shift on the output signal. The NOLM approach uses the technique described above where an SOA is embedded into an optical loop as depicted in the main diagram of Figure 10.29(c). In order to obtain asymmetry in the interferometer, the SOA is placed in the loop slightly away from the center point, which creates a different path length for CW probe signal path 1 and path 2 from the 3 dB coupler 1.

An intensity-modulated input at signal wavelength λ_{in} is coupled into the loop via coupler 1 in Figure 10.29(c). It propagates in a clockwise direction where it saturates the gain of the SOA, and at the same time it produces refractive index variations. These refractive index variations caused by the gain saturation effect in the SOA contain an inverted copy of the intensity-modulated data on the input signal. A CW probe signal at wavelength λ_{CW} is then injected into the loop using the 3 dB symmetrical coupler 2 which splits it into two halves, each half propagating in either a clockwise or a counterclockwise direction. The CW probe signal entering into the SOA from a clockwise direction experiences phase shifts due to the refractive index variations and thus the phase of the CW probe signal is modulated. Since the refractive index variations contain the data on the intensity-modulated input signal, this phase-modulated probe signal then incorporates the same information. When the phase-modulated probe signal and counterclockwise CW probe signal arrive at coupler 2, they interfere and create an intensity-modulated output signal. The output signal therefore emerges as a converted signal at wavelength λ_{conv} of the CW probe signal including the data from the input signal.

An interesting application of the NOLM wavelength conversion scheme is that the same structure can be used as an optical switch when the CW probe signal is replaced by an optical clock signal and in this case the structure is known as a terahertz optical asymmetrical demultiplexer (TOAD) [Ref. 100]. Although NOLM wavelength converters are useful devices, they must be driven by very short return to zero pulses in order to avoid crosstalk problems. They can, however, be efficiently employed for ultrafast wavelength conversion by incorporating additional fiber Bragg grating elements [Ref. 101].

10.5.3 Cross-absorption modulation wavelength converters

Cross-absorption modulation (XAM) exploits the nonlinear effect in an electroabsorption modulator (EAM) where absorption in the active cavity of an EAM due to an intensity-modulated input signal is transferred onto a CW probe signal [Ref. 102]. A block schematic for the electroabsorption modulator illustrating the basic principle of an XAM wavelength converter is depicted in Figure 10.30(a) in which the CW probe signal at wavelength λ_{CW} and the intensity-modulated input signal at wavelength λ_{in} are fed into the EAM and the converted output at signal wavelength λ_{conv} is obtained at the output of an optical filter. Figure 10.30(b) illustrates the operation of the EAM as a wavelength converter. A random bit sequence of binary pulses containing data 10101101 as an optical input signal together with a CW probe signal are shown fed into the EAM. The EAM absorbs signal power P_{in} from the input signal for a binary 1, whereas the CW probe signal power P_{CW} is not absorbed for this time slot and thus it appears at the output as binary 1 (i.e. input = 1, output = 1). The input signal with binary 0, however, does not offer enough signal power to the EAM and therefore it completely absorbs the CW probe signal power thus leaving no signal power in the probe signal (i.e. input = 0, output = 0). Hence the input signal pulse power pattern is replicated as an exact copy in the CW probe signal power pattern. When the CW probe contains the compliant wavelength, the XAM wavelength conversion is obtained giving P_{conv} at wavelength λ_{conv} at the output. As both P_{in} and P_{CW} signals are fed from the same direction in Figure 10.30(a) it constitutes a copropagating

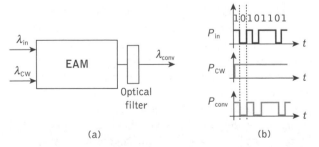

(a)   (b)

Figure 10.30 Cross-absorption modulation (XAM) wavelength conversion using an electroabsorption modulator (EAM): (a) block schematic of an implementation for a copropagating scheme; (b) operation of the converter

scheme, although a counterpropagation scheme for these signals is also possible [Ref. 103]. Since the EAM possesses faster absorption characteristics in comparison with the slow gain recovery of an SOA, XAM-based wavelength converters are faster than wavelength converters based on SOA technology and transmission rates of 100 Gbit s^{-1} have been successfully demonstrated using EAM wavelength conversion [Ref. 104].

10.5.4 Coherent wavelength converters

Coherent wave mixing relies on the principle of four-wave mixing (FWM) and difference frequency generation (DFG) in a nonlinear optical medium (see Section 3.14). In these mechanisms, when two or more optical signals interact with each other in a nonlinear medium they generate new signals. The intensity of the newly generated signals is dependent on the intensities of the interacting signals whereas the phase and the frequency of the newly generated signals are the linear combination of the signals present in the wave mixing. Therefore the information contained in any signal component (i.e. intensity, frequency, phase or polarization) is preserved. If the resultant signals contain a signal with a compliant signal frequency then wavelength conversion is achieved where the converted signal is obtained through an optical filter. Figure 10.31 shows a block diagram representing wave mixing of two signals in a nonlinear medium. It can be seen that the input and CW probe signals at angular frequencies ω_{in} and ω_{CW} interact in the passive nonlinear medium and when an output signal with angular frequency ω_{conv} emerges from the optical filter, the device has produced the wavelength conversion.

The newly generated signals possess angular frequencies that are dependent on the FWM process which is a nonlinear wave-mixing process arising from a third-order optical nonlinearity (see Section 3.14) [Refs 84, 105] as illustrated in Figure 10.32(a). Two optical signals, an input and the CW probe with angular frequencies ω_{in} and ω_{CW} respectively, interact in the nonlinear medium (i.e. the optical fiber). The nonlinear interaction of these optical signals creates sidebands for each thus producing two new signal components, namely the converted probe and the satellite signal. The former possesses an angular frequency, $\omega_{conv} = 2\omega_{CW} - \omega_{in}$ which can be seen on the right hand side of the CW probe signal, while the latter signal appears with angular frequency $\omega_{s} = 2\omega_{in} - \omega_{CW}$. Since the satellite signal is an unwanted component, an optical filter is used to remove it at the output port of the wavelength converter. It should be noted that the converted probe signal appears on the right hand side of the frequency spectrum exhibiting a phase difference of 180° in relation to the input signal. This property, which is known as spectral inversion (also referred to as phase conjugation), is a very useful characteristic of these wavelength converters as it assists them to overcome noise problems [Ref. 105].

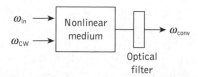

Figure 10.31 Basic principle of wavelength conversion using wave mixing in nonlinear medium

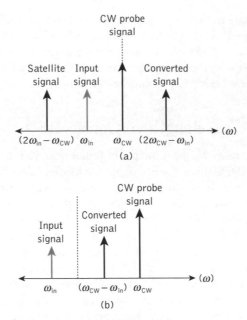

Figure 10.32 Spectral diagram of the wavelength conversion based on: (a) four-wave mixing (FWM); (b) difference frequency generation (DFG)

Since the FWM process generates optical signals whose properties are dependent on the frequencies and the amplitudes of the interacting signals, it is therefore possible to generate more than one desired output signal when more than one CW probe signal is used in the wavelength conversion process [Ref. 84]. In this case the same information contained in the input at one signal wavelength can be transferred to several wavelengths using simultaneous multiple-wavelength conversion. This attribute is considered to be a very useful property for use in optical switching networks where identical copies of information may be required for different destinations. In wavelength division multiplexed networks (see Section 15.2.3) it is desirable to use a bank of wavelength converters (see Section 15.4) for the simultaneous wavelength conversion of multiple-wavelength channels. In such cases the FWM wavelength converter is a suitable candidate for providing waveband conversion where different wavelength bands each comprising several channels are simultaneously converted to the desired output wavelengths [Refs 106, 107].

The efficiency of a wavelength converter is an important parameter that can be defined as the ability to convert an optical signal at any wavelength while providing the maximum output signal power. Although in comparison with other categories the FWM wavelength converter exhibits low conversion efficiency, it can be improved by increasing the interacting length of the optical signals (i.e. by increasing the length of the nonlinear fiber). Alternatively, employing an active medium (i.e. an SOA) for the FWM instead of optical fiber can also improve the conversion efficiency [Ref. 106].

In addition to FWM wavelength conversion there exists another process which employs the second-order nonlinearity of the optical medium (see Section 3.14). By contrast to FWM, this process produces only a single output signal because the frequency of the newly generated optical signal is determined on the basis of difference in the frequencies of the interacting optical signals. Therefore it is known as difference frequency generation (DFG) and it is possible to construct a wavelength converter using the principle of DFG. Figure 10.32(b) displays the spectral diagram for DFG wavelength conversion where the converted signal contains an angular frequency $\omega_{conv} = \omega_{CW} - \omega_{in}$ that is generated as a result of an interacting CW probe signal at an angular frequency ω_{CW} with the input signal at angular frequency ω_{in}. Since the DFG process produces only the single desired output signal and no satellite signal, it is therefore the preferred process over FWM wavelength conversion [Ref. 84]. Moreover, an important feature of the DFG process is that it enables simultaneous multiple-wavelength conversions to be performed without the generation of unwanted signals and thus the output signal from a DFG wavelength converter is relatively free from crosstalk as compared with FWM wavelength converters. Phase matching of the interacting waves is, however, a complicated issue that makes this type of wavelength converter difficult to realize practically [Ref. 108]. Although there have been several successful experimental demonstrations of DFG wavelength converters using periodical waveguide structures in lithium niobate, both the device length and interface requirements make these difficult to fabricate in a monolithic integrated optical structure [Ref. 108].

10.6 Optical regeneration

Optical regeneration is the name given to the technique used to reproduce the original signal while overcoming optical transmission losses. Since the development of optical amplifiers the letter R together with a number has been used to classify the type of regeneration according to its features and also its increasing complexity. These classifications are:

(a) 1R (Reamplification)

(b) 2R (1R+R, Reamplification + Reshaping)

(c) 3R (2R+R, Reamplification + Reshaping + Retiming)

(d) 4R (3R+R, Reamplification + Reshaping + Retiming + Reallocation of wavelengths)

The suffix 'Re' emphasizes the requirement for action to be taken to enhance the quality of the signal. Figure 10.33 illustrates the different stages of the amplification/regeneration process and the corresponding effect of those stages on the received optical pulse.

On the left-hand side of Figure 10.33 two pulses, each representing the original (transmitted) and noisy (received) signals, are shown. The received pulse when compared with the transmitted pulse displays an irregular rectangular shape, decreased amplitude and contains noise as indicated by the dark area on it. Straightforward amplification can increase the amplitude of the signal and noise as identified by 1R, or reamplification only. A further stage of improvement to restore the rectangular shape of the pulse can be provided by 2R (or reamplification and reshaping). The dark areas in the pulse shown by vertical columns represent the timing jitter which is caused by accumulation of noise due to

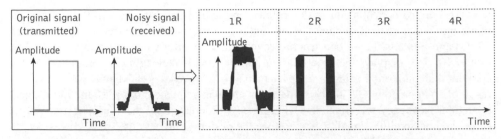

Figure 10.33 Various stages of the optical amplification/regeneration process

the random arrival of the digital optical signals as compared with their ideal position in time (see Section 12.6.1).

However, this time jitter can be corrected by using optical retiming circuits and is referred to as 3R (i.e. reamplification, reshaping and retiming). To enable the transmission of an optical signal through to its destination it may be required to change the wavelength of the signal by reassigning new wavelengths and this necessitates 4R (i.e. reallocation of wavelengths in addition to 3R). Also included in 4R is the improvement of the connectivity for a WDM optical network (see Section 15.2.3) by reallocating each destination with a different optical signal wavelength and thus avoiding any wavelength conflict that may arise when two or more nodes use identical signal wavelengths.

Implementation of the amplification/regeneration is shown in the block schematic shown in Figure 10.34 which illustrates the devices required to provide this functionality at different stages. The 1R regeneration is shown at termination point A in Figure 10.34 where straightforward amplification has been provided using an optical amplifier. It should be noted that such amplification cannot differentiate between an optical signal and the noise it contains and therefore both appearing at point A has been amplified, thereby degrading the quality of the optical signal. Hence employing several 1R stages for longer

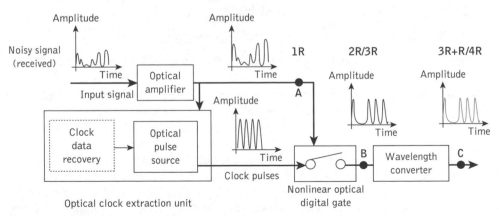

Figure 10.34 Optical amplification/regeneration block schematic: 1R regeneration at termination point A; 2R/3R regeneration at termination point B; 3R+R or 4R regeneration at termination point C

transmission distances causes additional problems since the accumulated distortion and noise on the fiber link can introduce errors. Furthermore, cascading several 1R stages utilizing optical amplifiers results in increased buildup of amplified spontaneous emission noise generated within the optical amplifiers (see Section 10.3.2).

To reduce errors it is useful to reshape the pulse to produce realignment for all the components contained in the optical signal. Therefore, in addition to reamplification, reshaping (i.e. 2R) is also required to obtain high performance over longer transmission distances. Reshaping of the digital signal can be accomplished using a decision circuit implemented through the use of a nonlinear optical digital gate (NODG). The 2R regenerated output is obtained at terminal point B in Figure 10.34 in which the main functional blocks for 2R are an optical pulse source (called the clock) and the NODG which can be realized using semiconductor optical amplifiers (see Section 10.5.2). An input signal is applied to the optical gate to control the flow of pulses generated by the clock source and the process functions as a digital AND gate in which with the presence of sufficient input pulse power (i.e. above the preset threshold level) produces a binary 1 at the output of the device, whereas when the input signal power is below the threshold level the optical gate remains closed resulting in a binary 0 at the output. The time period of the clock pulse is selected to be shorter than the input signal pulses and hence the output signal power appears consistent for binary 1 and 0, thus providing the required reshaping.

It is possible to extract the clock signal from the received input signal and use the extracted clock to generate clock pulses for the purpose of realignment to improve the pulse signal timing which is required to reposition the pulses of the received signal in order to maintain a regular spacing between successive pulses. This process, which is known as retiming, is also identified in Figure 10.34 at terminal point B where the 2R regeneration output is located. A difference lies in the method for the clock generation technique, however, where in 2R regeneration external clock pulses are used as an optical clock source whereas in 3R regeneration the clock is recovered from the incoming signal. Clock recovery from the incoming signal is achieved using a clock extraction technique through the incorporation of a clock data recovery unit [Ref. 109]. It should be noted that the recovered clock signal must possess the same frequency and also its phase should be synchronized with the data signal. Hence the clock data recovery unit extracts a low-jitter clock (see Section 12.6.1) from the noisy input signal and this clock signal subsequently drives the optical pulse source in order to generate the synchronized optical pulse stream. The data is then sampled at the clock frequency by the decision gate which operates in the same manner as for 2R regeneration.

Finally, 3R regeneration with an additional wavelength conversion function provides the wavelength reallocation for 4R regeneration and this is shown as located at terminal point C in Figure 10.34. The 4R regeneration capability can be combined within a 3R regenerator by using a tunable optical pulse source which is controlled externally for the purpose of wavelength reallocation [Ref. 110]. It should be noted that the reallocation of the signal wavelength improves the optical network connectivity while reducing the wavelength contention or collision that may otherwise arise when two signals at the same wavelengths are intended for the same destination. Wavelength reallocation or reassignment depends on the network type where several optical switching and routing techniques may be used to direct the signals to their anticipated destinations while avoiding any conflict among the wavelengths present in the network [Ref. 111] (see Sections 15.4 and 15.5).

Problems

10.1   Give the major reasons which have led to the development of optical amplifiers, outlining the attributes and application areas for these devices.

Describe the two main SOA types and indicate their distinguishing features.

10.2   An SOA has facet reflectivities of 23% and a single-pass gain of 6 dB. The device has an active region with a refractive index of 3.6, a peak gain wavelength of 1.55 μm with a spectral bandwidth of 5 GHz. Determine the length of the active region for the SOA and also its mode spacing.

10.3   The following parameter values apply to a semiconductor TWA operating at a wavelength of 1.3 μm:

material gain coefficient	1000 cm^{-1}
effective loss coefficient	22 cm^{-1}
active region length	200 μm
facet reflectivities	0.1%
optical confinement factor	0.3

Calculate in decibels both the minimum and the maximum optical gain that could be obtained from the device.

10.4   Determine the peak–trough ratio of the passband ripple for the TWA of Problem 10.3. Compare this value with that obtained using a device with the same specification as Problem 10.3 excepting that the facet reflectivities are reduced to 0.03%.

Estimate the cavity gain for the latter semiconductor TWA.

10.5   Describe the phenomenon of backward gain in a semiconductor TWA and suggest a way in which it might be limited in systems that employ cascaded amplifiers.

A semiconductor TWA has a maximum cavity gain of 17 dB with a peak–trough ratio of 3 dB. Estimate the backward gain exhibited by the device under maximum gain operation.

10.6   (a) Sketch the major elements of a fiber amplifier and describe the operation of the device. Indicate the benefits of fiber amplifier technology in comparison with that associated with SOAs.

(b) Using an energy band diagram, briefly discuss the mechanism for the provision of stimulated emission in the erbium-doped silica fiber amplifier. Name and describe a phenomenon occurring in this material system which creates a limitation to the optical gain that may be obtained from the device.

10.7   Explain the gain process in a Raman fiber amplifier and comment upon the flexibility associated with the pumping process in this fiber amplifier type.

The Raman gain coefficient for a silica-based fiber of 10 μm core diameter at a pump wavelength of 1.2 μm is $6.3 \times 10^{-14} \text{ m W}^{-1}$. Determine the Raman gain obtained in a 25 km length of the fiber when it is pumped at this wavelength with an input power of 1.4 W and when the transmission loss is 0.8 dB km^{-1}. It may be assumed that the effective core radius is 1.15 times as large as the actual core radius and that complete polarization scrambling occurs.

10.8   Discuss the need for the different types of optical fiber amplifiers. Sketch their amplification wavelength ranges when they are used in long-haul optical telecommunication systems.

10.9   Outline EDFA designs incorporating both a co- and counterpropagating pump. State the main advantages of each configuration.

10.10 Explain the term amplified spontaneous emission (ASE) noise and describe its impact on the optical output signal.

10.11 A semiconductor optical amplifier (SOA) operating at signal wavelength of 1.3 μm produces a gain of 20 dB with an optical bandwidth of 900 GHz. The device has a spontaneous emission factor of 1.5 and the mode number is equal to 2.0 with a net signal gain of 300. Determine: (a) the length of the SOA; (b) the ASE noise signal power at the output of the amplifier; (c) the noise figure of the amplifier.

10.12 Compare the SOA, EDWA and the EDFA amplet in relation to the provision of amplification within optical fiber communications.

10.13 Explain the concept of gain clamping in relation to the operation of an SOA. Indicate the ways in which it can be implemented.

10.14 Explain the nature of the quantum dot and describe with the aid of diagrams how it is employed within SOAs.

10.15 Outline five different optical amplifier configurations to achieve wideband optical amplification. Which is preferred practically and why?

10.16 Define the term parametric gain and describe its application within optical amplification.

10.17 A 450 m long optical parametric amplifier is fed with a 1.0 W optical pump signal operating at a wavelength of 1.3 μm. An input signal power of 5 mW at a wavelength of 1.31 μm is also present at the input port of the amplifier. When the fiber nonlinear coefficient is 10.6 mW^{-1} km^{-1}, calculate (a) the exponential parametric gain and (b) the quadratic gain; (c) comment on the results obtained for (a) and (b).

10.18 How does the frequency chirp affect the optical output signal? Explain its impact on the output signal of an optical wavelength converter employing: (a) a semiconductor optical amplifier; (b) an electroabsorption modulator.

10.19 A semiconductor optical amplifier is coupled with an input signal power of 0.75 mW operating at a signal wavelength of 1.3 μm and in response it produced an output signal power of 15 mW. When the differential refractive index of the device is -1.8×10^{-26} m^3, the linewidth enhancement factor is 2.5 and the optical signal power variation in input signal is 20 nW, determine: (a) the frequency chirp variation at the output signal; (b) the differential gain required in order for the device to operate at the same signal wavelength. Comment on the result obtained.

10.20 Describe the different approaches to implement an optical wavelength converter. Indicate which wavelength conversion schemes possess higher conversion efficiency.

10.21 Explain with the help of a sketch(es) the four optical regeneration stages for an optical signal and outline their applications in optical fiber transmission.

Answers to numerical problems

10.2 234 μm, 1.43 nm

10.3 22.1 dB, 26.8 dB

10.4 4.6 dB, 0.7 dB
29.2 dB

10.5 12.1 dB

10.7 31.8 dB

10.11 15.4 mm, 40 μm, 3.0

10.17 35.5 dB, 13.6 dB

10.19 5.3 MHz, 7×10^{-20} m^2

References

[1] M. J. O'Mahony, 'Semiconductor laser optical amplifiers for use in future fiber systems', *J. Lightwave Technol.*, **6**(4), pp. 531–544, 1988.

[2] T. Akiyama, M. Ekawa, M. Sugawara, K. Kawaguchi, S. Hisao, A. Kuramata, H. Ebe and Y. Arakawa, 'An ultrawide-band semiconductor optical amplifier having an extremely high penalty-free output power of 23 dBm achieved with quantum dots', *IEEE Photonics Technol. Lett.*, **17**(8), pp. 1614–1616, 2005.

[3] J. M. Senior and S. D. Cusworth, 'Devices for wavelength multiplexing and demultiplexing', *IEE Proc., Optoelectron.*, **136**(3), pp. 183–202, 1989.

[4] M. J. O'Mahony, 'Optical amplification techniques using semiconductors and fibres', *Proc. SPIE*, **1120**, pp. 43–44, 1989.

[5] R. Baker, 'Optical amplification', *Phys. World*, pp. 41–44, March 1990.

[6] J. Buus and R. Plastow, 'Theoretical and experimental investigations of 1.3 μm Fabry–Perot amplifiers', *IEEE J. Quantum Electron.*, **QE-21**(6), pp. 614–618, 1985.

[7] G. Eisenstein, B. L. Johnson and G. Raybon, 'Travelling-wave optical amplifier at 1.3 μm', *Electron. Lett.*, **23**(19), pp. 1020–1022, 1987.

[8] G. N. Brown, 'A study of the static locking properties of injection locked laser amplifiers', *Br. Telecom Technol. J.*, **4**(1), pp. 71–80, 1986.

[9] T. Saitoh and T. Mukai, 'Recent progress in semiconductor laser amplifiers', *J. Lightwave Technol.*, **6**(11), pp. 1656–1664, 1988.

[10] M. J. O'Mahony, I. W. Marshall and H. J. Westlake, 'Semiconductor laser amplifiers for optical communication systems', *Br. Telecom Technol. J.*, **5**(3), pp. 9–18, 1987.

[11] T. Saitoh, T. Mukai and O. Mikami, 'Theoretical analysis of antireflection coatings on laser diode facets', *J. Lightwave Technol.*, **LT-3**(2), pp. 288–293, 1985.

[12] G. Eisentein and R. M. Jopson, 'Measurements of the gain spectrum of near-travelling-wave and Fabry–Perot semiconductor optical amplifiers at 1.5 μm', *Int. J. Electron.*, **60**(1), pp. 113–121, 1986.

[13] M. J. Adams, H. J. Westlake, M. J. O'Mahony and I. D. Henning, 'A comparison of active and passive bistability in semiconductors', *IEEE J. Quantum Electron.*, **QE-21**(9), pp. 1498–1504, 1985.

[14] H. Ghafouri-Shiraz, *The Principles of Semiconductor Laser Diodes and Amplifiers: Analysis and Transmission Line Laser Modelling*, Imperial College Press, 2004.

[15] M. J. Connelly, *Semiconductor Optical Amplifiers*, Kluwer Academic, 2002.

[16] M. Razeghi, 'Optoelectronic devices based on III-V compound semiconductors which have made a major scientific and technological impact in the past 20 years', *IEEE J. Sel. Top. Quantum Electron.*, **6**(6), pp. 1344–1354, 2000.

[17] I. D. Henning, M. J. Adams and J. V. Collins, 'Performance predictions from a new optical amplifier model', *IEEE J. Quantum Electron.*, **QE-21**, pp. 609–613, 1985.

[18] J. Shim, J. Kim, D. Jang, Y. Eo and S. Arai, 'Facet reflectivity of a spot-size-converter integrated semiconductor optical amplifier', *IEEE J. Quantum Electron.*, **38**(6), pp. 665–673, 2002.

[19] B. Mason, J. Barton, G. A. Fish, L. A. Coldren and S. P. Denbaars, 'Design of sampled grating DBR lasers with integrated semiconductor optical amplifiers', *IEEE Photonics Technol. Lett.*, **2**(7), pp. 762–764, 2000.

[20] T. Yamatoya and F. Koyama, 'Optical preamplifier using antireflection-coating-free semiconductor optical amplifier with signal-inverted ASE', *IEEE Photonics Technol. Lett.*, **15**(8), pp. 1047–1049, 2003.

[21] L. Qiao and P. J. Vella, 'ASE analysis and correction for EDFA automatic control', *J. Lightwave Technol.*, **25**(3), pp. 771–778, 2007.

[22] T. Briant, P. Grangier, R. Tualle-Brouri, A. Bellemain, R. Brenot and B. Thedrez, 'Accurate determination of the noise figure of polarization-dependent optical amplifiers: theory and experiment', *J. Lightwave Technol.*, **24**(3), pp. 1499–1503, 2006.

[23] E. Udvary, 'Noise performance of semiconductor optical amplifiers', *Int. Conf. on Trends in Communication, EUROCON'01*, Bratislava, Slovakia, **1**, pp. 161–163, July 2001.

[24] K. Vyrsokinos, G. Toptchiyski and K. Petermann, 'Comparison of gain clamped and conventional semiconductor optical amplifiers for fast all-optical switching', *J. Lightwave Technol.*, **20**(10), pp. 1839–1846, 2002.

[25] C. Y. Jin, Y. Z. Huang, L. J. Yu and S. L. Deng, 'Numerical and theoretical analysis of the crosstalk in linear optical amplifiers', *IEEE J. Quantum Electron.*, **41**(5), pp. 636–641, 2005.

[26] P. Royo, R. Koda and L. A. Coldren, 'Vertical cavity semiconductor optical amplifiers: comparison of Fabry-Perot and rate equation approaches', *IEEE J. Quantum Electron.*, **38**(3), pp. 279–284, 2002.

[27] E. S. Bjorlin and J. E. Bowers, 'Noise figure of vertical-cavity semiconductor optical amplifiers', *IEEE J. Quantum Electron.*, **38**(1), pp. 61–66, 2002.

[28] T. Akiyama, H. Kuwatsuka, T. Simoyama, Y. Nakata, K. Mukai, M. Sugawara, O. Wada and H. Ishikawa, 'Nonlinear gain dynamics in quantum-dot optical amplifiers and its application to optical communication devices', *IEEE J. Quantum Electron.*, **37**(8), pp. 1059–1065, 2001.

[29] O. Qasaimeh, 'Optical gain and saturation characteristics of quantum-dot semiconductor optical amplifiers', *IEEE J. Quantum Electron.*, **39**(6), pp. 793–798, 2003.

[30] A. V. Uskov, E. P. O'Reilly, R. J. Manning, R. P. Webb, D. Cotter, M. Laemmlin, N. N. Ledentsov and D. Bimberg, 'On ultrafast optical switching based on quantum-dot semiconductor optical amplifiers in nonlinear interferometers', *IEEE Photonics Technol. Lett.*, **16**(5), pp. 1265–1267, 2004.

[31] A. V. Uskov, T. W. Berg and J. Mrk, 'Theory of pulse-train amplification without patterning effects in quantum-dot semiconductor optical amplifiers', *IEEE J. Quantum Electron.*, **40**(3), pp. 306–320, 2004.

[32] T. Aklyama, M. Ekawa, M. Sugawara, K. Kawaguchi, H. Sudo, H. Kuwatsuka, H. Ebe, A. Kuramata and Y. Arakawa, 'Quantum dots for semiconductor optical amplifiers', *Proc. Optical Fiber Communication, OFC'05*, Anaheim, California, USA, paper OWM2, 6–11 March 2005.

[33] R. J. Mears, L. Reekie, I. M. Jauncey and D. N. Payne, 'Low-noise erbium-doped fibre amplifier at 1.54 μm', *Electron Lett.*, **23**, pp. 1026–1028, 1987.

[34] E. Desurvire, J. R. Simpson and P. C. Parker, 'High-gain erbium-doped travelling-wave fibre amplifier', *Opt. Lett.*, **12**, pp. 888–890, 1987.

[35] D. N. Payne and L. Reekie, 'Rare-earth-doped fibre lasers and amplifiers', *14th Eur. Conf. on Optical Communications, ECOC'88*, pp. 49–53, September 1988.

[36] G. N. Brown and D. M. Spirit, 'Gain saturation and laser linewidth effects in a Brillouin fibre amplifier', *15th Eur. Conf. on Optical Communications, ECOC'89*, Gottenburg, pp. 70–73, September 1989.

[37] E. Desurvire, D. Bayart, B. Deshieux and S. Bigo, *Erbium-Doped Fiber Amplifiers*, Wiley, 2002.

[38] M. L. Dakss and P. Melman, 'Amplified stimulated Raman scattering and gain in fiber Raman amplifiers', *J. Lightwave Technol.*, **LT-3**, pp. 806–813, 1985.

[39] M. N. Islam, 'Raman amplifiers for telecommunications', *IEEE J. Sel. Top. Quantum Electron.*, **8**(3), pp. 548–559, 2002.

[40] Y. Nishida, M. Yamada, T. Kanamori, K. Kobayashi, J. Temmyo, S. Sudo and Y. Ohishi, 'Development of an efficient praseodymium-doped fiber amplifier', *IEEE J. Quantum Electron.*, **34**(8), pp. 1332–1339, 1998.

[41] Michel J. Digonnet and D. J. Digonnet, *Rare Earth Doped Fiber Lasers and Amplifiers*, CRC Press, 2001.

[42] K. Fujiura and S. Sudo, '1.3 μm fiber amplifiers', in M. J. F. Digonnet (Ed.), *Rare-earth-doped Fiber Lasers and Amplifiers*, pp. 681–755, CRC Press, 2001.

[43] T. Sakamoto, S. Aozasa, M. Yamada and M. Shimizu, 'High-gain hybrid amplifier consisting of cascaded fluoride-based TDFA and silica-based EDFA in 1458–1540 nm wavelength region', *Electron. Lett.*, **39**(7), pp. 597–599, 2003.

[44] NP Photonics, Erbium Micro Fiber Amplifier (EMFA), http://www.np-photonics.com, 13 September 2007.

[45] Y. Aoki, S. Kishida, H. Honomon, K. Washio and M. Sugimoto, 'Efficient backward and forward pumping CW Raman amplification for InGaAsP laser light in silica fibres', *Electron. Lett.*, **19**, pp 620–622, 1983.

[46] Y. Durteste, M. Monerie and P. Lamouler, 'Raman amplification in fluoride glass fibres', *Electron. Lett.*, **21**, p. 723, 1985.

[47] N. Edegawa, K. Mochizuki and Y. Imamoto, 'Simultaneous amplification of wavelength division multiplexed signals by highly efficient amplifier pumped by higher power semiconductor lasers', *Electron. Lett.*, **23**, pp. 556–557, 1987.

[48] J. Bromage, 'Raman amplification for fiber communications systems', *J. Lightwave Technol.*, **22**(1), pp. 79–93, 2004.

[49] C. A. Millar, M. C. Brierley and P. W. France, 'Optical amplification in an erbium-doped fluorozirconate fibre between 1480 nm and 1600 nm', *IEE Conf Publ.*, **292**, *Pt I*, pp. 66–69, 1988.

[50] R. J. Stolen, 'Polarization effects in fiber Raman and Brillouin lasers', *IEEE J. Quantum Electron.*, **QE-15**, pp. 1157–1160, 1979.

[51] Y. Aoki, 'Properties of fiber Raman amplifiers and their applicability to digital optical communication systems', *J. Lightwave Technol.*, **6**(7), pp. 1225–1239, 1988.

[52] S. Saito, T. Kimura, T. Tanabe, K. Suto, Y. Oyama and J. I. Nishizawa, 'Fabrication and characteristics of GaP-AlGaP tapered waveguide semiconductor Raman amplifiers', *J. Lightwave Technol.*, **21**(1), pp. 170–175, 2003.

[53] P. Parolari, L. Marazzi, L. Bernardini and M. Martinelli, 'Double Rayleigh scattering noise in lumped and distributed Raman amplifiers', *J. Lightwave Technol.*, **21**(10), pp. 2224–2228, 2003.

[54] S. A. E. Lewis, S. V. Chernikov and J. R. Taylor, 'Characterization of double Rayleigh scatter noise in Raman amplifiers', *IEEE Photonics Technol. Lett.*, **12**(5), pp. 528–530, 2000.

[55] M. D. Mermelstein, K. Brar and C. Headley, 'RIN transfer measurement and modeling in dual-order Raman fiber amplifiers', *J. Lightwave Technol.*, **21**(6), pp. 1518–1523, 2003.

[56] C. G. Atkins, D. Cotter, D. W. Smith and R. Wyatt, 'Application of Brillouin amplification in coherent optical transmission', *Electron. Lett.*, **22**, pp. 556–557, 1986.

[57] A. R. Charplyvy and R. W. Tkach, 'Narrow-band tunable optical filter for channel selection in densely packed WDM systems', *Electron. Lett.*, **22**, pp. 1084–1085, 1986.

[58] R. G. Waarts and R. P. Braun, 'Crosstalk due to stimulated Brillouin scattering in monomode fiber', *Electron. Lett.*, **21**, p. 1114, 1985.

[59] E. J. Bachus, R. P. Braun, W. Eutin, E. Grossman, H. Foisel, K. Heims and B. Strebel, 'Coherent optical fibre subscriber line', *Electron. Lett.*, **21**, p. 1203, 1985.

[60] S. J. Strutz and K. J. Williams, 'Low-noise hybrid erbium/Brillouin amplifier', *Electron. Lett.*, **36**(16), pp. 1359–1360, 2000.

[61] S. Sternklar and E. Granot, 'Narrow spectral response of a Brillouin amplifier.' *Opt. Lett.*, **28**(12), pp. 977–979, 2003.

[62] C. Headley and G. Agrawal, *Raman Amplification in Fiber Optical Communication Systems*, Academic Press, 2004.

[63] K. Suto, T. Saito, T. Kimura, J. I. Nishizawa and T. Tanabe, 'Semiconductor Raman amplifier for terahertz bandwidth optical communication', *J. Lightwave Technol.*, **20**(4), pp. 705–711, 2002.

[64] S. Saito, K. Suto, T. Kimura and J. I. Nishizawa, '80-ps and 4-ns pulse-pumped gains in a GaP-AlGaP semiconductor Raman amplifier', *IEEE Photonics Technol. Lett.*, **16**(2), pp. 395–397, 2004.

[65] S. Saito, J. I. Nishizawa, K. Suto and T. Kimura, 'The structure and maximal gain of CW-pumped GaP-AlGaP semiconductor Raman amplifier with tapers on both sides', *IEEE Photonics Technol. Lett.*, **16**(1), pp. 48–50, 2004.

[66] I. Mozjerin, A. A. Hardy and S. Ruschin, 'Effect of chip area limitation on gain and noise of erbium-doped waveguide amplifiers', *IEEE J. Sel. Top. Quantum Electron.*, **11**(1), pp. 204–210, 2005.

[67] D. Portch, R. R. A. Syms and W. Huang, 'Folded-spiral EDWAs with continuously varying curvature', *IEEE Photonics Technol. Lett.*, **16**(7), pp. 1634–1636, 2004.

[68] S. Demiguel, N. Sahri, M. Hartlaub, F. Blache, H. Gariah, S. Vuiye, D. Carpentier, D. Barbier and J. C. Campbell, 'Low-cost photoreceiver integrating an EDWA and waveguide PIN photodiode for 40 Gbit/s applications', *Electron. Lett.*, **43**(1), pp. 51–52, 2007.

[69] D. R. Zimmerman and L. H. Spiekman, 'Amplifiers for the masses: EDFA, EDWA, and SOA amplets for metro and access applications', *J. Lightwave Technol.*, **22**(1), pp. 63–70, 2004.

[70] J. Z. Kevin, L. H. Darlene, I. K. Koo and C. M. Brian, 'Erbium-doped fiber amplet and its application as a modular bandwidth management subsystem for dispersion compensation and power management', *J. Opt. Eng.*, **40**(7), pp. 1199–1203, 2001.

[71] J. Hansryd, P. A. Andrekson, M. Westlund, J. Li and P. O. Hedekvist, 'Fiber-based optical parametric amplifiers and their applications', *IEEE J. Sel. Top. Quantum Electron.*, **8**(3), pp. 506–520, 2002.

[72] P. Parolari, L. Marazzi, E. Rognoni and M. Martinelli, 'Influence of pump parameters on two-pump optical parametric amplification', *J. Lightwave Technol.*, **23**(8), pp. 2524–2530, 2005.

[73] M. C. Ho, K. Uesaka, M. Marhic, Y. Akasaka and L. Kazovsky, '200-nm-bandwidth fiber optical amplifier combining parametric and Raman gain', *J. Lightwave Technol.*, **19**(7), pp. 977–981, 2001.

[74] C. A. Chung, W. T. Dong, L. Y. Yin, L. C. Wai, C. Y. Hung, W. B. Cheng, H. Y. Chieh, S. J. Tsong, L. Y. Pin, C. Y. Fu and T. P. Hsi, 'Pulsed optical parametric generation, amplification, and oscillation in monolithic periodically poled lithium niobate crystals', *IEEE J. Quantum Electron.*, **40**(6), pp. 791–799, 2004.

[75] W. Sohler, W. Grundkotter, H. Herrmann, H. Hu, S. L. Jansen, J. H. Lee, Y. H. Min, V. Quiring, R. Ricken and S. Reza, 'All-optical signal processing devices with (periodically poled) lithium niobate waveguides', *Proc. Optical Fiber Communication and National Fiber Optic Engineers (OFC/NFOEC 2007)*, Anaheim, California, USA, OME3, pp. 1–3, March 2007.

[76] A. R. Pandey, P. E. Powers and J. W. Haus, 'Experimental performance of a two-stage periodically poled lithium niobate parametric amplifier', *IEEE J. Quantum Electron.*, **44**(3), pp. 203–208, 2008.

[77] Y. B. Lu, P. L. Chu, A. Alphones and P. Shum, 'A 105-nm ultrawideband gain-flattened amplifier combining C- and L-band dual-core EDFAs in a parallel configuration', *IEEE Photonics Technol. Lett.*, **16**(7), pp. 1640–1642, 2004.

[78] C. H. Yeh, C. C. Lee and S. Chi, '120-nm bandwidth erbium-doped fiber amplifier in parallel configuration', *IEEE Photonics Technol. Lett.*, **16**(7), pp. 1637–1639, 2004.

[79] J. H. Lee, Y. M. Chang, Y. G. Han, S. H. Kim, H. Chung and S. B. Lee, 'Performance comparison of various configurations of single-pump dispersion-compensating Raman/EDFA hybrid amplifiers', *IEEE Photonics Technol. Lett.*, **17**(4), pp. 765–767, 2005.

[80] M. Yamada and M. Shimizu, 'Ultra-wideband amplification technologies for optical fiber amplifiers', *NTT Tech. Rev. Lett.*, **1**(3), pp. 80–84, 2003.

[81] A. Mori, H. Masuda, K. Shikano and M. Shimizu, 'Ultra-wideband tellurite-based fiber Raman amplifier', *J. Lightwave Technol.*, **21**(5), pp. 1300–1306, 2003.

[82] Y. Suzuki and H. Toba, 'Recent research and development of all-optical wavelength conversion devices', *NTT Tech. Rev.*, **1**(1), pp. 26–31, 2003.

[83] M. Matsuura, N. Kishi and T. Miki, 'Ultrawideband wavelength conversion using cascaded SOA-based wavelength converters', *J. Lightwave Technol.*, **25**(1), pp. 38–45, 2007.

[84] S. J. B. Yoo, 'Wavelength conversion technologies for WDM network applications', *J. Lightwave Technol.*, **14**(6), pp. 955–966, 1996.

[85] M. M. de la Corte and J. M. H. Elmirghani, 'Accurate noise characterization of wavelength converters based on XGM in SOAs', *J. Lightwave Technol.*, **21**(1), pp. 182–197, 2003.

[86] M. L. Nielsen B. Lavigne and B. Dagens, 'Polarity-preserving SOA-based wavelength conversion at 40 Gbit/s using bandpass filtering', *Electron. Lett.*, **39**(18), pp. 1334–1335, 2003.

[87] A. D. Ellis, A. E. Kelly, D. Nesset, D. Pitcher, D. G. Moodie and R. Kashyap, 'Error free 100 Gbit/s wavelength conversion using grating assisted cross-gain modulation in 2 mm long semiconductor amplifier', *Electron. Lett.*, **34**(20), pp. 1958–1959, 1998.

[88] X. Zheng, F. Liu and A. Kloch, 'Experimental investigation of the cascadability of a cross-gain modulation wavelength converter', *IEEE Photonics Technol. Lett.*, **12**(3), pp. 272–274, 2000.

[89] J. Leuthold, C. H. Joyner, B. Mikkelsen, G. Raybon, J. L. Pleumeekers, B. I. Miller, K. Dreyer and C. A. Burrus, '100 Gbit/s all-optical wavelength conversion with integrated SOA delayed-interference configuration', *Electron. Lett.*, **36**(13), pp. 1129–1130, 2000.

[90] K. E. Stubkjaer, 'Semiconductor optical amplifier-based all-optical gates for high-speed optical processing', *IEEE J. Sel. Top. Quantum Electron.*, **6**(6), pp. 1428–1435, 2000.

[91] M. Saitoh, B. Ma and Y. Nakano, 'Static and dynamic characteristics analysis of all-optical wavelength conversion using directionally coupled semiconductor optical amplifiers', *IEEE J. Quantum Electron.*, **36**(8), pp. 984–990, 2000.

[92] S. C. Cao and J. C. Cartledge, 'Characterization of the chirp and intensity modulation properties of an SOA-MZI wavelength converter', *J. Lightwave Technol.*, **20**(4), pp. 689–695, 2002.

[93] M. Y. Jamro, J. M. Senior, M. S. Leeson and G. Murtaza, 'Modeling chirp and phase inversion in wavelength converters based on symmetrical MZI-SOAs for use in all-optical networks', *Photonics Netw. Commun.*, **5**(3), pp. 289–300, 2003.

[94] H. Wang, J. Wu and J. Lin, 'Spectral characteristics of optical pulse amplification in SOA under assist light injection', *J. Lightwave Technol.*, **23**(9), pp. 2761–2771, 2005.

[95] J. Leuthold, G. Raybon, Y. Su, R. J. Essiambre, S. Cabot, J. Jaques and M. Kauer, '40 Gbit/s transmission and cascaded all-optical wavelength conversion over 1 000 000 km', *Electron. Lett.*, **38**(16), pp. 890–892, 2002.

[96] J. Leuthold, L. Moller, J. Jaques, S. Cabot, L. Zhang, P. Bernasconi, M. Cappuzzo, L. Gomez, E. Laskowski, E. Chen, A. Wong-Foy and A. Griffin, '160 Gbit/s SOA all-optical wavelength converter and assessment of its regenerative properties', *Electron. Lett.*, **40**(9), pp. 554–555, 2004.

[97] M. Matsuura, N. Kishi and T. Miki, 'Widely pulsewidth-tunable multiwavelength synchronized pulse generation utilizing a single SOA-based delayed interferometric switch', *IEEE Photonics Technol. Lett.*, **17**(4), pp. 902–904, 2005.

[98] N. Y. Kim, X. Tang, J. C. Carfledge and A. K. Atich, 'Design and performance of an all-optical wavelength converter based on a semiconductor optical amplifier and delay interferometer', *J. Lightwave Technol.*, **25**(12), pp. 730–738, 2007.

[99] B. Mikkelsen *et al.*, '40 Gbit/s all-optical wavelength converter and RZ-to-NRZ format adapter realized by monolithic integrated active Michelson interferometer', *Electron. Lett.*, **33**(2), pp. 133–134, 1997.

[100] L. Xu, I. Glcsk, V. Baby and P. R. Prucnal, 'All optical wavelength conversion using SOA at nearly symmetric position in a fiber-based Sagnac interferometric loop', *IEEE Photonics Technol. Lett.*, **16**(2), pp. 539–541, 2004.

[101] S. L. Jansen, H. Chayet, E. Granot, S. B. Ezra, D. den van Borne, P. M. Krummrich, D. Chen, G. D. Khoe and H. De Waardt, 'Wavelength conversion of a 40-Gbit/s NRZ signal across the entire C-band by an asymmetric Sagnac loop', *IEEE Photonics Technol. Lett.*, **17**(10), pp. 2137–2139, 2005.

[102] M. Y. Jamro and J. M. Senior, 'Optimising negative chirp of an electroabsorption modulator for use in high-speed optical networks', *Eur. Trans. Telecommun.*, **18**(4), pp. 369–380, 2007.

[103] M. Hayashi, H. Tanaka, K. Ohara, T. Otani and M. Suzuki, 'OTDM transmitter using WDM-TDM conversion with an electroabsorption wavelength converter', *J. Lightwave Technol.*, **20**(2), pp. 236–242, 2002.

[104] K. Nishimura, R. Inohara, M. Usami and S. Akiba, 'All-optical wavelength conversion by electroabsorption modulator', *IEEE J. Sel. Top. Quantum Electron.*, **11**(1), pp. 278–284, 2005.

[105] H. Jang, S. Hur, Y. Kim and J. Jeong, 'Theoretical investigation of optical wavelength conversion techniques for DPSK modulation formats using FWM in SOAs and frequency comb in 10 Gb/s transmission systems', *J. Lightwave Technol.*, **23**(9), pp. 2638–2646, 2005.

[106] G. Contestabile, M. Presi and E. Ciaramella, 'Multiple wavelength conversion for WDM multicasting by FWM in an SOA', *IEEE Photonics Technol. Lett.*, **16**(7), pp. 1775–1777, 2004.

[107] C. Politi, D. Klonidis and M. J. O'Mahony, 'Waveband converters based on four-wave mixing in SOAs', *J. Lightwave Technol.*, **24**(3), pp. 1203–1217, 2006.

[108] S. Yu and W. Gu, 'Wavelength conversions in quasi-phase matched $LiNbO_3$ waveguide based on double-pass cascaded χ^2 SFG+DFG interactions', *IEEE J. Quantum Electron.*, **40**(11), pp. 1548–1554, 2004.

[109] Z. Hu, H. F. Chou, K. Nishimura, M. Usami, J. E. Bowers and D. J. Blumenthal, 'Optical clock recovery circuits using traveling-wave electroabsorption modulator-based ring oscillators for 3R regeneration', *IEEE J. Sel. Top. Quantum Electron.*, **11**(2), pp. 329–337, 2005.

[110] J. Martin, 'Benefits of partial reconfigurability in circuit-switched WDM networks', *J. High Speed Netw.*, **14**(3), pp. 201–213, 2005.

[111] C. C. Sue, 'Wavelength routing with spare reconfiguration for all-optical WDM networks', *J. Lightwave Technol.*, **23**(6), pp. 1991–2000, 2005.

CHAPTER | 11

Integrated optics and photonics

11.1 Introduction

The previous chapter introduced the concept of optical amplification and the device technologies associated with this process which also provide for wavelength conversion and optical signal regeneration. This chapter contains the discussion of both active and passive optical components and devices which are utilized within optical fiber communications to enable both signal manipulation and processing which facilitates the implementation of high-performance optical fiber communication systems and networks. The integration of such components and devices is also an important and integral aspect to give multiple optical processing functions which are carried out without recourse to conversion back to the electrical domain. Hence the development of integrated optics (IO) for the realization of optical and electro-optical elements integrated together and, more recently, integrated photonics (IP) in which device integration in large numbers can be provided on a single substrate are crucial to the further development of the optical fiber telecommunications network.

In particular, IP refers to the fabrication and integration of several or many components onto a single planar substrate. Such components include beam splitters, couplers, gratings, polarization controllers, interferometers, sources, detectors and optical amplifiers. These components when integrated with planar waveguides constitute the basic building blocks to fabricate more complex planar devices that perform not only optical signal guiding and coupling but also controlling functions such as switching, splitting, multiplexing and demultiplexing of optical signals. The basic concepts of the technologies associated with IO and IP are initially introduced in Section 11.2. Planar waveguides, however, are fundamental elements for both IO and IP device technologies. Section 11.3 therefore describes the characteristics and types of planar waveguides that can be used for the interconnection of various optical elements forming optical circuits which are also known as planar lightwave circuits (PLCs). This is followed by consideration of some integrated optical device types in Section 11.4. In particular this section concentrates on directional couplers and switches, together with modulator devices which provide an alternative to direct current modulation for optical sources. Planar lightwave circuits are further divided into two categories determined by the control of the optical signal flow from the input and the output, the first of which, optoelectronic integrated circuits, are dealt with in Section 11.5. The second category concerned with photonic integrated circuits is then discussed in Section 11.6. Optical bistability and digital optics are introduced in Section 11.7 to provide the reader with an insight into this important area, together with an understanding of how these phenomena may be utilized within optical fiber communications. Finally, in Section 11.8 developments in the field of optical computation are outlined. Although it is clear that these latter developments do not, at present, significantly influence optical fiber communications, it is likely that in the future there will be a requirement for the combination of optical communication, optical switching and optical computational technologies, within what has now become predominantly optical fiber telecommunication network.

11.2 Integrated optics and photonics technologies

Integrated technology for optical devices has developed within optical fiber communications so that it is now possible to fabricate a complete system onto a single chip. Integration for such devices has become a confluence of several optical or photonic disciplines. Both IO and IP technologies are referred to in the above where the control of the optical devices distinguishes one technology from the other. Electronic control of the optical devices determines the terminology of IO whereas photons control the operation of IP devices. In addition, IP does not involve any optoelectronic conversion of optical signals and hence this technology is also termed as 'all-optical'. Both IO and IP use planar waveguide technology to provide interconnections between optical components including the basic components for guiding and control of optical signals. IP technology, however, enables the fabrication of subsystems and systems to be realized onto a single substrate. Hence when both active and passive devices are monolithically integrated onto a single substrate in a multilayered integration then these devices are normally referred to as IP devices, while when both active and passive elements fabricated as individual devices are interconnected together they form larger IO devices or circuits. Thus IP can also be seen as a process for the miniaturization and integration of optical systems on a single substrate, and therefore it may be considered as a further enhancement of IO technology, not necessarily as an alternative technology. Both IO and IP seek to provide an alternative to the conversion of an optical signal back into the electrical regime prior to signal processing by allowing such processing to be performed on the optical signal. Hence thin transparent dielectric layers on planar substrates which act as optical waveguides are used in IO and IP to produce small-scale and miniature optical components and circuits.

The birth of IO may be traced back to basic ideas outlined by Anderson in 1966 [Ref. 1]. He suggested that a microfabrication technology could be developed for single-mode optical devices with semiconductor and dielectric materials in a similar manner to that which had taken place with electronic circuits. It was in 1969, however, after Miller [Ref. 2] had introduced the term integrated optics when discussing the long-term outlook in the area, that research began to gain momentum.

Developments in IO have passed the stage where both signal processing and logic functions can be physically realized. Furthermore, such devices may form the building blocks for future digital optical computers. Nevertheless, a number of these advances combine to be closely linked with developments in lightwave communication employing optical fibers.

A major factor in the development of IO is that it is essentially based on single-mode optical waveguides and therefore tends to be incompatible with multimode fiber systems. Hence IO did not make a significant contribution to early deployed optical fiber systems (see Section 15.1). The advent, however, of single-mode transmission technology further stimulated work in IO in order to provide devices* and circuits for these more advanced third-generation systems. Furthermore, the continued expansion of single-mode optical fiber communications has created a growing market for such IO components. It is also likely that new generations of optical fiber communication systems employing coherent and possibly soliton transmission will lean heavily on IO and IP techniques for their implementation.

The proposals for IO and IP devices and circuits which in many cases involve reinventions of electronic devices and circuits exhibit major advantages other than solely a compatibility with optical fiber communications. Electronic circuits have a practical limitation on speed of operation at a frequency of around 10^{10} Hz resulting from their use of metallic conductors to transport electronic charges and build up signals. The large transmission bandwidths (over 100 GHz) currently under investigation for optical fiber communications are already causing difficulties for electronic signal processing within the terminal equipment. The use of light with its property as an electromagnetic wave of extremely high frequency (10^{14} to 10^{15} Hz) offers the possibility of high-speed operation around 10^4 times faster than that conceivable employing electronic circuits. Interaction of light with materials

---

* This is especially the case in relation to the fabrication of single-mode injection lasers (see Section 6.6).

such as semiconductors or transparent dielectrics occurs at speeds in the range of 10^{-12} (pico) to approaching 10^{-15} (femto) seconds, thus providing a basis for subpicosecond optical switching.

The other major attribute provided by optical signals interacting within a responsive medium is the ability to utilize lightwaves of different frequencies (or wavelengths) within the same guided wave channel or device. Such frequency division multiplexing allows an information transfer capacity far superior to anything offered by electronics. Moreover, in signal processing terms it facilitates parallel access to information points within an optical system. This possibility for powerful parallel signal processing coupled with ultrahigh-speed operation offers tremendous potential for applications within both communications and computing.

The devices of interest in IO and IP are often the counterparts of microwave or bulk optical devices. These include junctions and directional couplers, switches and modulators, filters and wavelength multiplexers, lasers and amplifiers, detectors and bistable elements. It is envisaged that developments in this technology will provide the basis for the next generation of optical networks identified in Section 15.1 where full monolithic integration will be used.

The technology associated with the design and fabrication of IP circuits and devices depends upon different factors that mainly result from the characteristics of the substrate material on which the various devices are to be fabricated. The IP process may require serial, parallel or hybrid integration of independent devices. In serial integration of devices, different elements of the optical chip can be interconnected in a consecutive manner and therefore side-, or edge-emitting, or conducting optical devices can be readily integrated on the same substrate. In the parallel case the chip is constructed by developing columns of devices in which surface- or bottom-emitting devices can effectively be used whereas in hybrid integration IP technology the devices are fabricated using both serial and parallel integration on the same substrate. To gain control of the optical signals, however, additional elements can be developed separately or be directly attached to the IP circuit, or be interconnected to it by optical fibers [Ref. 3]. In addition, both active and passive devices may be required to be located on the same substrate and therefore hybrid IP integration demands multilayered IP circuits and components to be produced on a single substrate such that they must be compatible with three-dimensional structures of other IO or IP devices [Ref. 4].

The enabling technologies for IP mainly rely on silica-on-silicon (SOS) where the waveguide structures comprise three layers, namely the buffer, the core and the cladding [Ref. 5]. Due to its refractive index match to silica-based optical fiber, vertical light confinement in SOS can be achieved by increasing the refractive index of the core layer relative to the surrounding glass, whereas lateral confinement is obtained through structuring of the core layer. The real benefit of SOS, however, is the ability to apply wafer-scale, planar lithography and processing techniques to integrate substantial numbers of functions either as arrays of identical devices or in the form of customized circuit configurations on single or multiple chips. This integration capability offers an efficient platform for the implementation of typical fiber-based functionalities such as optical power splitters or combiners, couplers, wavelength-selective couplers, multiplexers/demultiplexers and optical gain elements. Furthermore, optical switches and controllable attenuators based on the thermo-optic effect can also be fabricated.

In addition, some devices can be fabricated using a silicon-on-liquid (SOL) gel process [Ref. 6] which cannot be implemented using SOS techniques. The SOL gel process is a versatile solution-based technique for making ceramic and glass materials which involves the transition of a system from a liquid (or solution) into a gel. Applying the SOL gel and SOS techniques, it is possible to fabricate thin-film coatings, ceramic fibers and waveguide-based optical amplifiers [Ref. 7]. Further to the above integration technologies used for IP devices, a silicon-on-insulator (SOI) approach is also used to produce microwaveguide bends and couplers at a reduced scale while maintaining compatibility with the standard silicon fabrication techniques [Ref. 8]. SOI based on CMOS technology that has revolutionized both electronic and optoelectronic integrated circuit technologies is now usefully applied as an IP technology. Using this integration technique, active components like optical sources, detectors and amplifiers can be coupled to other IP devices [Ref. 9]. Furthermore, the more recently developed photonic crystal waveguide technology (see Section 2.6) is also compatible with this integration technology [Refs 8, 10].

11.3 Planar waveguides

The use of circular dielectric waveguide structures for confining light is universally utilized within optical fiber communications. Both IO and IP involve an extension of this guided wave optical technology through the use of planar optical waveguides to confine and guide the light in guided wave devices and circuits. The mechanism of optical confinement in symmetrical planar waveguides was discussed in Section 2.3 prior to investigation of circular structures. In fact the simplest dielectric waveguide structure is the planar slab guide shown in Figure 11.1. It comprises a planar film of refractive index n_1 sandwiched between a substrate of refractive index n_2 and a cover layer of refractive index n_3 where $n_1 > n_2 \geq n_3$. Often the cover layer consists of air where $n_3 = n_0 = 1$, and it exhibits a substantially lower refractive index than the other two layers. In this case the film has layers of different refractive index above and below the guiding layer and hence performs as an asymmetric waveguide.

In the discussions of optical waveguides given in Chapter 2 we were solely concerned with symmetrical structures. When the dimensions of the guide are reduced so are the

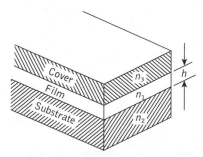

Figure 11.1 A planar slab waveguide. The film with high refractive index n_1 acts as the guiding layer and the cover layer is usually air where $n_3 = n_0 = 1$

number of propagating modes. Eventually the waveguide dimensions are such that only a single-mode propagates, and if the dimensions are reduced further this single-mode still continues to propagate. Hence there is no cutoff for the fundamental mode in a symmetric guide. This is not the case for an asymmetric guide where the dimensions may be reduced until the structure cannot support any modes and even the fundamental is cut off. If the thickness or height of the guide layer of a planar asymmetric guide is h (see Figure 11.1), then the guide can support a mode of order m with a wavelength λ when [Ref. 11]:

$$h \geq \frac{(m + \frac{1}{2})\lambda}{2(n_1^2 - n_2^2)^{\frac{1}{2}}} \tag{11.1}$$

Equation (11.1), which assumes $n_2 > n_3$, defines the limits of the single-mode region for h between values when $m = 0$ and $m = 1$. Hence for a typical thin-film glass guide with $n_1 = 1.6$ and $n_2 = 1.5$, single-mode operation is maintained only when the guide has a thickness in the range $0.45\lambda \leq h \leq 1.35\lambda$.

An additional consideration of equal importance is the degree of confinement of the light to the guiding layer. The light is not exclusively confined to the guiding region and evanescent fields penetrate into the substrate and cover. An effective guide layer thickness h_{eff} may be expressed as:

$$h_{\text{eff}} = h + x_2 + x_3 \tag{11.2}$$

where x_2 and x_3 are the evanescent field penetration depths for the substrate and cover regions respectively. Furthermore, we can define a normalized effective thickness H for an asymmetric slab guide as:

$$H = kh_{\text{eff}}(n_1^2 - n_2^2)^{\frac{1}{2}} \tag{11.3}$$

where k is the free space propagation constant equal to $2\pi/\lambda$. The normalized frequency (sometimes called the normalized film thickness) for the planar slab guide following Eq. (2.68) is given by:

$$V = kh(n_1^2 - n_2^2)^{\frac{1}{2}} \tag{11.4}$$

An indication of the degree of confinement for the asymmetric slab waveguide may be observed by plotting the normalized effective thickness against the normalized frequency for the TE modes. A series of such plots is shown in Figure 11.2 [Ref. 12] for various values of the parameter a which indicates the asymmetry of the guide, and is defined as:

$$a = \frac{n_2^2 - n_3^2}{n_1^2 - n_2^2} \tag{11.5}$$

It may be observed in Figure 11.2 that the confinement improves with decreasing film thickness only up to a point where $V \simeq 2.5$. For example, the minimum effective thickness

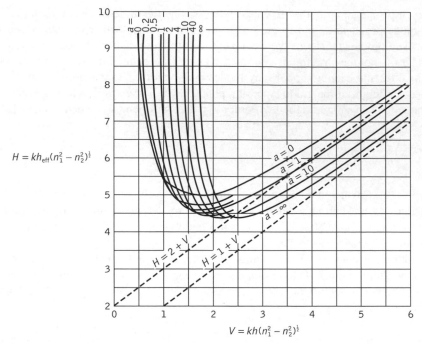

$$H = kh_{\text{eff}}(n_1^2 - n_2^2)^{\frac{1}{2}}$$

$$V = kh(n_1^2 - n_2^2)^{\frac{1}{2}}$$

Figure 11.2 The normalized effective thickness H as a function of the normalized frequency V for a slab waveguide with various degrees of asymmetry. Reproduced with permission from H. Kogelnik and V. Ramaswamy, *Appl. Opt.*, **13**, p. 1857, 1974

for a highly asymmetric guide ($a = \infty$) occurs when $H_{\text{min}} = 4.4$ at $V = 2.55$. Using Eq. (11.3) this gives a minimum effective thickness of:

$$(h_{\text{eff}})_{\text{min}} = \frac{4.4}{k}(n_1^2 - n_2^2)^{-\frac{1}{2}}$$

$$= 0.7\lambda(n_1^2 - n_2^2)^{-\frac{1}{2}} \tag{11.6}$$

Therefore considering a typical glass waveguide ($n_1 = 1.6$ and $n_2 = 1.5$), we obtain a minimum effective thickness of:

$$(h_{\text{eff}})_{\text{min}} = 1.26\lambda \tag{11.7}$$

Assuming a minimum operating wavelength to be 0.8 μm limits the effective thickness of the guide, and hence the confinement, to around 1 μm. Therefore it appears there is a limit to possible fabrication with IO which is not present in other technologies* [Ref. 14]. At present there is still ample scope but confinement must be considered along with packing density and the avoidance of crosstalk.

* The 1 μm barrier to confinement applies with all suitable waveguide materials. However, metal-clad waveguides are not so limited but are plagued by high losses [Ref. 13].

The planar waveguides for IO may be fabricated from glasses and other isotropic materials such as silicon dioxide and polymers. Although these materials are used to produce the simplest IO components, their properties cannot be controlled by external energy sources and hence they are of limited interest. In order to provide external control of the entrapped light to cause deflection, focusing, switching and modulation, active devices employing alternative materials must be utilized. A requirement for these materials is that they have the correct crystal symmetry to allow the local refractive index to be varied by the application of electrical, magnetic or acoustic energy.*

To date, interest has centered on the exploitation of the electro-optic effect due to the ease of controlling electric fields through the use of electrodes together with the generally superior performance of electro-optic devices. Acousto-optic devices have, however, found a lesser role, primarily in the area of beam deflection. Magneto-optic devices [Ref. 15] utilizing the Faraday effect are not widely used, as, in general, electric fields are easier to generate than magnetic fields.

A variety of electro-optic and acousto-optic materials have been employed in the fabrication of individual devices. Two basic groups can be distinguished by their refractive indices. These are materials with a refractive index near 2 ($LiNbO_3$, $LiTaO_3$, NbO_5, ZnS and ZnO) and materials with a refractive index greater than 3 ($GaAs$, InP and compounds of Ga and In with elements of Al, As and Sb).

Planar waveguide structures are produced using several different techniques which have in large part been derived from the microelectronics industry. For example, passive devices may be fabricated by radio-frequency sputtering to deposit thin films of glass onto glass substrates. Alternatively, active devices are often produced by titanium (Ti) diffusion into lithium niobate ($LiNbO_3$) or by ion implantation into gallium arsenide [Ref. 17].

The planar slab waveguide shown in Figure 11.1 confines light in only one direction, allowing it to spread across the guiding layer. In many instances it is useful to confine the light in two dimensions to a particular path on the surface of the substrate. This is achieved by defining the high-index guiding region as a thin strip (strip guide) where total internal reflection will prevent the spread of the light beam across the substrate. In addition the strips can be curved or branched as required. Examples of such strip waveguide structures are shown in Figure 11.3. They may be formed as either a ridge on the surface of the substrate or by diffusion to provide a region of higher refractive index below the substrate, or a rib of increased thickness within a thin planar slab. Techniques employed to obtain the strip pattern include electron and laser beam lithography as well as photolithography. The rectangular waveguide configurations illustrated in Figure 11.3 prove very suitable for use with electro-optic deflectors and modulators giving a reduction in the voltage required to achieve a particular field strength. In addition they allow a number of optical paths to be provided on a given substrate.

A trade-off also exists between the minimum radius of curvature which is required for high-density integration and the ease of fabrication. It is clear from Eq. 11.6 that the waveguide dimensions are dependent upon the refractive index change. When the change is large, the dimensions of the waveguide may be reduced, even though the scattering losses become larger. As the maximum confinement of the single-mode guide occurs when it is operated near to the cutoff of the second-order mode, then when the refractive

---

* Using the electro-optic, magneto-optic or acousto-optic effects [Ref. 16].

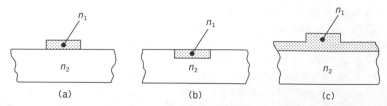

Figure 11.3 Cross-section of some strip waveguide structures: (a) ridge guide; (b) diffused channel (embedded strip) guide; (c) rib guide

index change is large, the radius of curvature of the waveguide can also be made very small. It is therefore necessary to find a compromise for the waveguide material used.

Titanium in diffusion of $LiNbO_3$ gives rise to refractive index increases in the order of 0.01 to 0.02 which dictates a bend radius of the order of a few centimeters for negligible losses. It is, however, possible to use a proton exchange technique to increase the refractive index change up to 0.15 [Ref. 18]. By contrast, semiconductor III–V alloy waveguides based on compositional modification of the crystal give an index change of around 0.1 or more [Ref. 19]. Therefore, bend radii of the order of 1 mm or less may be obtained using these compounds. Moreover, although the effects of interest in IO are usually exhibited over short distances of around one wavelength, efficient devices require relatively long interaction lengths, the effects being cumulative. Hence, typical device lengths range from 0.5 to 10 mm.

Optical connections to and from waveguide devices are normally made by optical fibers. The overall insertion loss for such devices therefore comprises a waveguide–fiber coupling loss as well as the waveguide optical propagation loss. Careful fabrication of Ti : $LiNbO_3$ waveguides with mode spot sizes well matched to that of typical single-mode fibers has yielded coupling losses in the range 0.5 to 1.0 dB per connection [Ref. 20]. In general, however, semiconductor waveguide devices exhibit larger fiber coupling losses because they operate with smaller spot sizes.

Propagation losses within both slab and strip waveguides are generally much greater than those obtained in single-mode optical fibers. However, the propagation losses for Ti : $LiNbO_3$ waveguides have gone below 0.2 dB cm^{-1}, with excess bend losses being maintained below 0.1 dB per bend [Ref. 21]. By contrast propagation losses in semiconductor waveguides around 1 dB cm^{-1} are obtained when operating at wavelengths corresponding to the bandgap energy. Much lower losses of approximately 0.2 dB cm^{-1}, however, have to be achieved at operating wavelengths far below the bandgap energy [Ref. 22]. Recently a new technique based on ion-implanted IO (known as I3O) technology has been demonstrated that enables miniaturization of components by either hybrid or monolithic integration [Ref. 10]. It utilizes titanium ion implantation in bulk silica to fabricate passive compact planar lightwave circuits. Using this technique various components can be obtained from an oxidized silicon platform using photolithography to pattern the waveguides, to perform ion implantation and also to provide the thermal annealing to control the temperature required in the process. In addition, the surface of components remains planar after I3O processing which provides for vertical integration of the waveguides and the components.

In order to achieve monolithic photonic integration the twin-waveguide structure has been proposed which can be achieved using two waveguides of symmetrical or asymmetrical

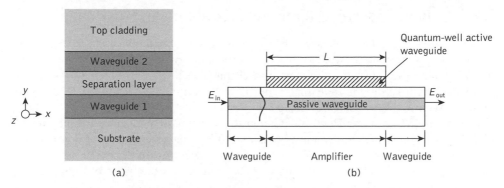

Figure 11.4 Twin-waveguide structures: (a) symmetric waveguides; (b) asymmetric waveguides incorporating an optical amplifier on the same substrate

refractive indices as shown in Figure 11.4 [Refs 23, 24]. The symmetrical waveguide structure which is depicted in Figure 11.4(a) comprises two waveguides (i.e. waveguide 1 and 2) in which the layers are stacked by epitaxial growth in the y direction and the guided waves travel along the x axis and therefore the refractive indices of both waveguides must be closely matched [Ref. 24]. This requirement does not permit the production of dissimilar elements (i.e. active and passive devices) on the same substrate, hence necessitating a separation layer to isolate them. Figure 11.4(b) displays an asymmetrical waveguide structure where both the active and/or passive devices can be formed in separate, vertically displaced waveguides on the same substrate. In this twin-waveguide structure the quantum-well active waveguide provides a gain section while the central section of the device of length L_a functions as an amplifier. This approach therefore introduces an asymmetry in waveguide refractive index profiles and hence the technique can be used to overcome the problems associated with monolithic integration of dissimilar active and/or passive devices on the same substrate [Ref. 23]. Based on this method various active/passive devices have been successfully demonstrated [Refs 25–29].

11.4 Some integrated optical devices

In this section some examples of various types of IO devices together with their salient features are considered. However, the numerous developments in this field exclude any attempt to provide other than general examples in the major areas of investigation which are pertinent to optical fiber communications. The requirement for multichannel communication within the various systems considered in Chapters 12 and 13 demands the combination of information from separate channels, transmission of the combined signals over a single optical fiber link, and separation of the individual channels at the receiver prior to routing to their individual destinations. Hence the application of IO in this area is to provide optical methods for multiplexing, modulation and routing. These various functions may be performed with a combination of optical beam splitters, switches, modulators, filters, sources and detectors.

11.4.1 Beam splitters, directional couplers and switches

Beam splitters are a basic element of many optical fiber communication systems often providing a Y-junction by which signals from separate sources can be combined, or the received power divided between two or more channels. A passive Y-junction beam splitter fabricated from $LiNbO_3$ is shown in Figure 11.5. Unfortunately, the power transmission through such a splitter decreases sharply with increasing half angle γ, the power being radiated into the substrate. Hence the total power transmission depends critically upon γ which, for the example chosen, must not exceed 0.5° if an acceptable insertion loss is to be achieved [Ref. 30]. In order to provide effective separation of the output arms so that access to each is possible, the junction must be many times the width of the guide. For example, around 3000 wavelengths are required to give a separation of about 30 μm between the output arms. Therefore, for practical reasons, the device is relatively long.

The passive Y-junction beam splitter finds application where equal power division of the incident beam is required. However, the Y-junction is of wider interest when it is fabricated from an electro-optic material, in which case it may be used as a switch. Such materials exhibit a change in refractive index δn which is directly proportional to an applied electric field* E following:

$$\delta n = \pm \tfrac{1}{2} n_1^3 r E \tag{11.8}$$

where n_1 is the original refractive index, and r is the electro-optic coefficient. Hence an active Y-junction may be fabricated from a single-crystal electro-optic material as illustrated in Figure 11.6. Lithium niobate is often utilized as it combines relatively low loss with large values of electro-optic coefficients[†] (as high as 30.8×10^{-12} mV^{-1}). Metal electrodes are attached so that when biasing is applied, one side of the waveguide structure exhibits an increased refractive index while the value of refractive index on the other side is reduced. The light beam is therefore deflected towards the region of higher refractive index causing it to follow the corresponding output arm. Furthermore, the field is maintained in the electrodes which extend beyond the junction ensuring continuation of the process. With switching voltages around 30 V, these devices prove to be quite efficient allowing for larger junction angles to be tolerated than those of the passive Y-junction

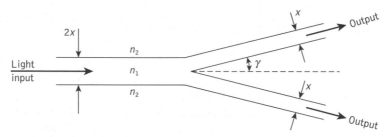

Figure 11.5 A passive Y-junction beam splitter

---

* The linear variation of refractive index with the electric field is known as the Pockels effect [Ref. 16].
† The change in refractive index is related by the applied field via the linear and quadratic electro-optic coefficients [Ref. 15].

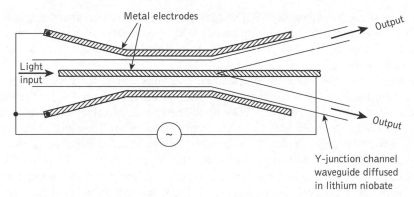

Figure 11.6 An electro-optic Y-junction switch

beam splitter. However, a physical length of several hundred wavelengths is still required for the switch. These devices therefore serve the function of optical signal routing. In addition, high-speed switches can be used to provide time division multiplexing of several lower bit rate channels onto a single-mode fiber link.

Switches may also be fabricated by placing two parallel strip waveguides in close proximity to each other as illustrated in Figure 11.7. The evanescent fields generated outside the guiding region allow transverse coupling between the guides. When the two waveguide modes have equal propagation constants β with amplitudes A and B (Figure 11.7), then the coupled mode equations may be written as [Ref. 31]:

$$\frac{dA}{dz} = j\beta A + jCB$$

$$\frac{dB}{dz} = j\beta B + jCA \tag{11.9}$$

where C is the coupling coefficient per unit length. In this case, assuming no losses, all the energy from waveguide X will be transferred to waveguide Y over a coupling length l_0.

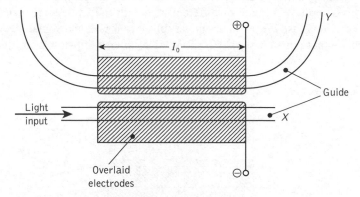

Figure 11.7 Electro-optically switched directional coupler. The COBRA configuration uses two electrodes. Reused with permission from Ref. 33 © 1975, American Institute of Physics

Furthermore it can be shown [Ref. 32] that for this complete energy transfer l_0 is given by $\pi/2C$. If the waveguide modes have different propagation constants, however, only part of the energy from guide X will be coupled into guide Y, and this energy will be subsequently recoupled back into X.

It is also noted that when the propagation constants differ, the coupling length l is reduced from the matched value l_0 and although less energy is transferred, the exchange occurs more rapidly. This property may be utilized to good effect in the formation of an optical switch. The mismatch in propagation constants can be adjusted such that the coupling length l is reduced to $l_0/2$. In this case, energy coupled from one guide into the other over a distance $l_0/2$ will be recoupled into the original guide over a similar distance. Hence two distinct cases exist for a switch of length l_0, namely the matched case whereby all the energy is transferred from one guide to the other and the mismatched case when $l = l_0/2$ where over a distance l_0 the energy is recoupled into the original guide.

Optical switches of the above type use electrodes placed on the top of each matched waveguide (Figure 11.7) so that the refractive indices of the guides are differentially altered to produce the differing propagation constants for the mismatched case. A widely used switch utilizing this technique is called the COBRA (*Commutateur Optique Binaire Rapide*) [Ref. 33] and is normally formed from titanium-diffused lithium niobate. Fabrication of the device, however, is critical in order to provide a coupling length which is exactly l_0 or an odd multiple of l_0. An electrode structure which avoids this problem by dividing the electrodes into halves with opposite polarities on each half is shown in Figure 11.8. With this device, which is called the stepped $\Delta\beta$ reversal coupler, it is always possible to obtain both the matched and mismatched cases described previously by applying suitable values of the reversed voltage. Hence the fabricated coupling length is no longer critical as the effective coupling length of the device may be adjusted electrically to achieve l_0.

A multimode interference (MMI) coupler similar to a fused fiber coupler (see Section 5.6) can be produced using planar waveguides. Multimode interference filter devices are based on the principle of self-imaging operation as shown in Figure 11.9(a) where an input field profile in a waveguide is reproduced at the output terminal and the device operates as a directional coupler. Single or multiple images (depending on the type of waveguide and number of input/output ports) at periodic intervals along the propagation direction of the waveguide, as a result of beating of different modes in the waveguide, can be obtained [Refs 34, 35]. In this case the signal is shown split into four equal signal levels appearing at each of the output ports. A simple 2×2 MMI coupler is illustrated in Figure 11.9(b).

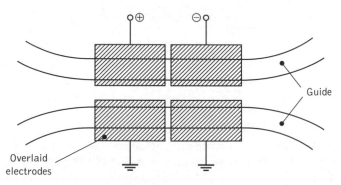

Figure 11.8 The stepped $\Delta\beta$ reversal coupler switch

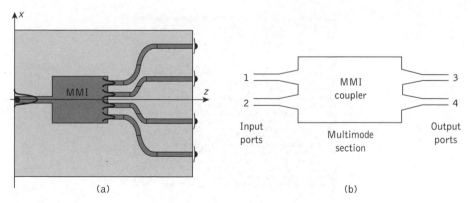

Figure 11.9 Multimode interference (MMI) coupler: (a) concept of reproducing self-images to produce a coupler; (b) schematic diagram of a 2 × 2 planar device

The coupler has a finite length because of the distance required to deliver constructive interference between the modes (see Section 5.6.1). Furthermore, this distance is wavelength dependent because the beat length depends on the difference in propagation constants between successive pairs of modes, which in turn varies with the wavelength of the input signal (see Section 5.6.3).

The most widely used 2 × 2 MMI coupler is the 3 dB coupler where power in each of the two inputs is split equally between the two output ports. The complete process of self-imaging relies on the interference patterns of the self-images at one interval leading to the formation of new self-images at the next interval and finally to the output images. It is also possible to change output images by modifying the refractive index of the MMI section and then the device operates as a tunable multimode interference coupler [Ref. 36]. The MMI coupler is also referred to as a cross-coupler (generally identified as X-coupler) when the input signal present at input port 1 in Figure 11.9(b) is obtained at output port 4, or similarly if the output is obtained from port 3 when there is an input present at port 2. Such cross-couplers are useful for the design of various types of multiport couplers for optical switches [Ref. 37]. Furthermore, although it is possible to fabricate an $N \times N$ array MMI coupler which can function as an optical add–drop multiplexer in WDM networks [Refs 38, 39], the output signal power is limited due to insertion losses (see Section 5.6).

Photonic crystal (see Section 2.6) based MMI couplers have also been demonstrated [Ref. 40]. Figure 11.10 displays a photonic crystal Y-junction coupler waveguide structure. A useful feature of these couplers is that they enable an optical signal to travel through a Y-junction making a wide angle of 60° (i.e. a 120° split), or even a 90° bend at a junction is feasible [Ref. 41]. These attributes associated with photonic crystals can also be used to produce WDM couplers. Figure 11.11 shows an example of an optical 2 × 2 coupler that can be used for wavelength division multiplexing and demultiplexing, which efficiently guides the optical signal to the desired output. It also illustrates the geometry of the coupler with two rows of dielectric pillars in the interaction region constituting the four-port device. The three-dimensional view of an enlarged area shown for a selected portion in Figure 11.11 displays the pillar formation creating a straight waveguide. Moreover, the dimensions (i.e. radius and the distance between each pillar) together with the refractive index of the pillar material determine the properties of the coupler to select/reject a particular optical wavelength or range of wavelengths [Ref. 42]. Thus the desired optical

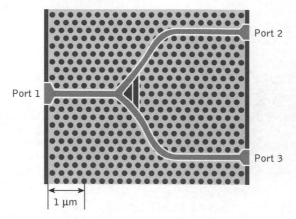

Figure 11.10 Y-junction coupler using a photonic crystal waveguide structure

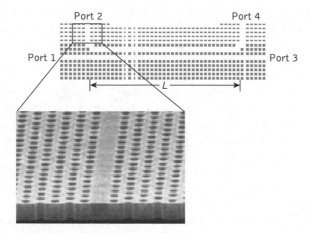

Figure 11.11 Photonic crystal-based 2 × 2 optical wavelength division multiplexed (WDM) waveguide coupler. The enlarged area shows a three-dimensional view of the pillar formation resulting in a straight waveguide structure

signal can be coupled to another waveguide over the interaction length L as indicated in Figure 11.11. Several wavelength division multiplexing/demultiplexing devices have been demonstrated based on MMI photonic crystal couplers operating at signal wavelengths of both 1.3 μm and 1.5 μm for transmission rates from 2.5 to 40 Gbit s^{-1} [Refs 42, 43].

The increasing deployment of optical fiber, particularly in the telecommunications network, has stimulated a great interest in optical or photonic switching in order to provide routing in a circuit-switched network [Refs 44–51]. The technology discussed in Sections 11.4 to 11.7 provides the basic building blocks for such optical switching systems. These switching systems can be classified in terms of their switching mechanism into space division switches, time division switches and wavelength or frequency* division switches [Ref. 44].

* In the optical domain these two terms are often used to indicate the same principle.

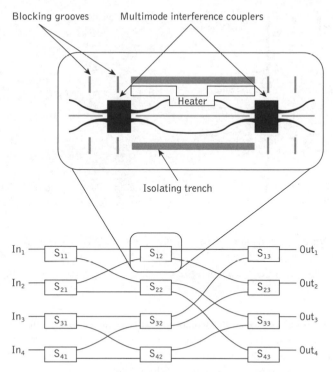

Figure 11.12 A 4×4 nonblocking optical space division switch matrix

Optical switching matrices based on the aforementioned switching mechanisms have been realised using IO technology [Refs 44–47]. Moreover, optical space division switches incorporating electro-optically controlled couplers have also been demonstrated. A non-blocking 4×4 optical space division switch is shown in Figure 11.12 which employed thermo-optical couplers and a Mach–Zehnder interferometer (MZI) arrangement was designed and fabricated in silicon-on-insulator wafer (see Section 11.2) [Ref. 44]. The switch matrix included 12 switching units and the total area covered was 0.1×4 cm^2. In each switching unit paired multimode interference couplers were used as a power splitter and a combiner in the MZI structure as identified in the inset of Figure 11.2. An optical signal at a wavelength of 1.5 μm from the single-mode fiber was coupled into the rib waveguides through the polished facet of the input waveguide. The device incurred an average insertion loss of 17 dB and had a switching time of 15 μs [Ref. 44]. Isolating trenches and blocking grooves were positioned between the neighboring arms in order to avoid mode coupling and hence they reduced the switching power consumption to 0.3 W. Although the average crosstalk was measured at 16.5 dB, it is indicated that this can be lowered by use of double MZI switches in the switching matrix [Ref. 45].

Simultaneous all-optical switching of several wavelength division multiplexed (WDM) channels has also been achieved employing a comb switch incorporating a micro-ring resonator [Refs 48–51]. For example, 20 continuous WDM channels were experimentally switched by the comb switch using a 104 GHz micro-ring resonator [Ref. 49]. The switching was achieved for a multiwavelength message cohesively by obtaining a small free-spectral

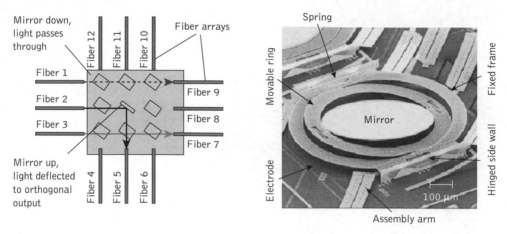

Figure 11.13 Optical micro-electromechanical system (OMEMS): (a) basic operation showing an optical cross-connect using a 3 × 3 matrix of micromirrors to route optical signals from arrays of input to output fibers; (b) a single mirror with supporting assembly [Ref. 55]

range of 0.8 nm and enabling several resonator modes to each switch one channel of the WDM signal simultaneously. It should be noted that the energy required to switch several channels is the same as that required to switch or select a single channel and therefore utilizing additional resonator modes enabled the switching of increased signal bandwidth without significant power penalty. Furthermore, at a transmission rate of 160 Gbit s^{-1} (i.e. 16×10 Gbit s^{-1}) the bit-error-rate measurements incurred negligible power penalty as the number of wavelength channels exiting the drop port were scaled from one to sixteen with a peak signal power of −6 dBm per channel. Such high-speed broadband switching and wavelength selective devices are considered a crucial element to provide for the deployment of interconnection networks based on silicon photonic integrated circuits [Refs 50, 51].

Optical micro-electromechanical systems (OMEMS)* are an assembly of very small electromechanical devices machined on silicon using a photolithographic technique. Figure 11.13(a) illustrates the basic operation involved for an OMEMS device to provide switching within an optical cross-connect. It shows a 3 × 3 matrix array of mirrors which form an optical switch interconnection of 12 optical fibers allowing an optical signal to be switched as required. The optical signal passes through imaging lenses to an array of mirrors that can tilt on two axes depending on the control signal (i.e. up or down). Three signals are shown entering into fibers 1, 2 and 3. When the mirror is set upward the signal simply passes through as in the cases of the signal traveling from fiber 1 to fiber 9 (shown by the dashed line) and fiber 3 to fiber 7 (identified by a straight line). However, when the mirror is rotated upwards it reflects the optical signal of the fiber 2 coupling it to the fiber 5 shown by the arrowed solid line. Therefore the use of reflecting mirror arrays (i.e. by using multiple reflections) can create any desired path between 12 fibers. A major problem, however, stems from the need to provide effective control of the operation for

---

* Also referred to as optical MEMS representing the technology and not the name of a specific device.

OMEMS which is difficult to realize since these systems use mirrors with thin membranes which are extremely sensitive to strain and deformation. Nevertheless, an additional assembly to hold the mirrors and to remove both strain and variation can be used to mitigate these adverse affects [Refs 52, 53]. Such an assembly is shown in Figure 11.13(b) where the mirror rotates with respect to the movable ring and the frame by elastically deforming the attaching springs. Furthermore the structure is designed to be sufficiently stiff in order to sustain mechanical vibrations.

Optical MEMS combined with other photonic devices such as vertical cavity semiconductor optical amplifiers (see Section 10.3) can also be used in optical signal switching and routing applications operating over a range of wavelengths from 1.3 to 1.6 μm [Ref. 54]. Silicon-on-insulator technology is preferred for the fabrication of OMEMS due to its simplicity and the small number of process steps involved in fabrication [Ref. 52]. For example, a 1024 × 1024 nonblocking optical cross-connect-based OMEMS has been implemented with a mean insertion loss of 1.33 dB and a maximum loss of 2.0 dB for all connections [Ref. 55]. In addition, the polarization-dependent loss of the structure, including the fiber pigtails and all other optical components, was shown to remain as low as 0.1 dB with the optical crosstalk being reduced below −40 dB.

11.4.2 Modulators

Although limitations imposed by direct current modulation of semiconductor injection lasers currently restricts the maximum achievable modulation, however more than 100 GH$_2$ have been demonstrated [Ref. 56]. Furthermore, with most injection lasers high-speed current modulation also creates undesirable wavelength modulation which imposes problems for systems employing WDM. Thus to extend the bandwidth capability of single-mode fiber systems there is a requirement for high-speed modulation which can be provided by IO waveguide intensity modulators. Simple on/off modulators may be based on the techniques utilized for the active beam splitters and switches described in Section 11.4.1. In addition a large variety of predominantly electro-optic modulators have been reported [Ref. 56] which exhibit good characteristics. For example, an important waveguide modulator is based upon a Y-branch interferometer which employs optical phase shifting produced by the electro-optic effect.

The change in refractive index exhibited by an electro-optic material with the application of an electric field given by Eq. (11.8) also provides a phase change for light propagating in the material. This phase change $\delta\phi$ is accumulative over a distance L within the material and is given by [Ref. 57]:

$$\delta\phi = \frac{2\pi}{\lambda}\,\delta nL \tag{11.10}$$

When the electric field is applied transversely to the direction of optical propagation we may substitute for δn from Eq. (11.8) giving:

$$\delta\phi = \frac{\pi}{\lambda}\,n_1^3 rEL \tag{11.11}$$

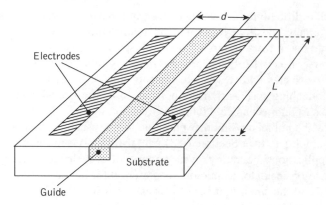

Figure 11.14 A simple strip waveguide phase modulator

Furthermore, taking E equal to V/d, where V is the applied voltage and d is the distance between electrodes, gives:

$$\delta\phi = \frac{\pi}{\lambda} n_1^3 r \frac{VL}{d} \qquad (11.12)$$

It may be noted from Eq. (11.12) that in order to reduce the applied voltage V required to provide a particular phase change, the ratio L/d must be made as large as possible.

A simple phase modulator may therefore be realized on a strip waveguide in which the ratio L/d is large as shown in Figure 11.14. These devices, when, for example, fabricated by diffusion of Nb into LiTaO$_3$, provide a phase change of π radians with an applied voltage in the range 5 to 10 V.

Example 11.1

A lithium niobate strip waveguide phase modulator designed for operation at a wavelength of 1.3 μm is 2 cm long with a distance between the electrodes of 25 μm. Determine the voltage required to provide a phase change of π radians given that the electro-optic coefficient for lithium niobate is 30.8×10^{-12} m V^{-1} and its refractive index is 2.1 at 1.3 μm.

Solution: When the phase change is π radians, using Eq. (11.12) we can write:

$$\delta\phi = \pi = \frac{\pi}{\lambda} n_1^3 r \frac{V_\pi L}{d}$$

Hence the voltage required to provide a π radian phase change is:

$$V_\pi = \frac{\lambda}{n_1^3 r} \frac{d}{L}$$

$$= \frac{1.3 \times 10^{-6} \times 25 \times 10^{-6}}{(2.1)^3 \times 30.8 \times 10^{-12} \times 2 \times 10^{-2}} = 5.7 \text{ V}$$

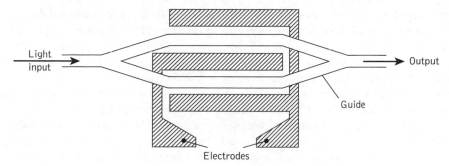

Figure 11.15 A Y-junction interferometric modulator based on the Mach–Zehnder interferometer

The result obtained in Example 11.1 has assumed the spatially uniform electric field of an ideal parallel plate capacitor. However, because the electro-optic refractive index change is small this is rarely the case and its effect on the optical phase velocity is dependent on the overlap integral of the electrical and optical fields. The consequence of these nonuniform fields can be incorporated into an overlap integral α, having a value between 0 and 1 which gives a measure of the overlap between the electrical and optical fields [Refs 18, 21]. The electro-optic refractive index change of Eq. (11.8) therefore becomes:

$$\delta n = \frac{\pm \alpha n_1^3 r}{2} \frac{V}{d} \qquad (11.13)$$

where the factor α represents the efficiency of the electro-optic interaction relative to an idealized parallel plate capacitor with the same distance between the electrodes.

As mentioned previously, the electro-optic property can be employed in an interferometric intensity modulator. Such a Mach–Zehnder interferometer is shown in Figure 11.15. The device comprises two Y-junctions which give an equal division of the input optical power. With no potential applied to the electrodes, the input optical power is split into the two arms at the first Y-junction and arrives at the second Y-junction in phase giving an intensity maximum at the waveguide output. This condition corresponds to the 'on' state. Alternatively, when a potential is applied to the electrodes, which operate in a push–pull mode on the two arms of the interferometer, a differential phase change is created between the signals in the two arms. The subsequent recombination of the signals gives rise to constructive or destructive interference in the output waveguide. Hence the process has the effect of converting the phase modulation into intensity modulation. A phase shift of π between the two arms gives the 'off' state for the device.

High-speed interferometric modulators have been demonstrated incorporating lithium niobate waveguides. A 100 GHz modulation bandwidth has been reported [Ref. 56] for an interferometer employing a less than 5 V on/off voltage. Similar devices incorporating electrodes on one arm only may be utilized as switches and are generally referred to as balanced bridge interferometric switches [Ref. 57]. An interferometric modulator based on planar waveguides has also demonstrated performance as an optical power attenuator [Ref. 58]. This device, referred to as a variable optical attenuator (VOA), is useful in wavelength division networks (see Section 15.2.3). In its simplest form the VOA attenuates optical signal power to a desired level which may be required for controlling optical power levels prior to optical amplifiers and receivers, or for channel equalization. A typical range of attenuation obtained from such a VOA is 0 to −20 dB while specific devices can provide higher attenuation up to −38 dB [Ref. 59]. A VOA providing this high level of attenuation can be used, for example, to block a WDM channel.

Useful modulators may also be obtained employing the acousto-optic effect. These devices, which deflect a light beam, are based on the diffraction of light produced by an acoustic wave traveling through a transparent medium. The acoustic wave produces a periodic variation in density (i.e. mechanical strain) along its path which, in turn, gives rise to corresponding changes in refractive index within the medium due to the photoelastic effect. Therefore, a moving optical phase diffraction grating is produced in the medium. Any light beam passing through the medium and crossing the path of the acoustic wave is diffracted by this phase grating from the zero-order into higher order modes.

Two regimes of operation are of interest: the Bragg regime and the Raman–Nath regime. The interaction, however, is of greatest magnitude in the Bragg regime where the zero-order mode is partially deflected into only one higher order (i.e. first-order) mode, rather than the multiplicity of higher order modes obtained in the Raman–Nath regime. Hence most acousto-optic modulators operate in the Bragg regime providing the highest modulation depth for a given acoustic power.

The Bragg regime is obtained by effecting a suitably long interaction length for the device so that it performs as a 'thick' diffraction grating. An IO acousto-optic Bragg deflection modulator is shown in Figure 11.16. It consists of a piezoelectric substrate (e.g. lithium niobate) onto the surface of which a thin-film optical waveguide is formed by, for example, titanium indiffusion or lithium outdiffusion. An acoustic wave is launched parallel to the surface of the waveguide forming a surface acoustic wave (SAW) [Ref. 60] in which most of the wave energy is concentrated within a depth of one acoustic wavelength. The wave is generated from an interdigital electrode system comprising parallel electrodes deposited on the substrate. A light beam guided by the thin-film waveguide interacts with the SAW giving beam deflection since both the light and the acoustic energy

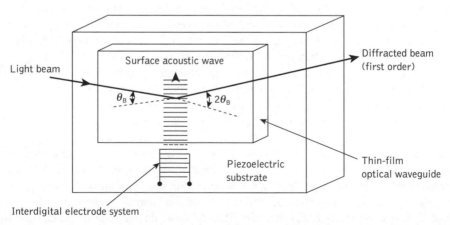

Figure 11.16 An acousto-optic waveguide modulator. The device gives deflection of a light beam due to Bragg diffraction by surface acoustic waves

are confined to the same surface layer. The conditions for Bragg diffraction between the zero- and first-order mode are met when [Ref. 17]:

$$\sin \theta_{\mathrm{B}} = \frac{\lambda_1}{2\Lambda} \qquad (11.14)$$

where θ_{B} is the angle between the light beam and the acoustic beam wavefronts, λ_1 is the wavelength of light in the thin-film waveguide and Λ is the acoustic wavelength. In this case the light is deflected by $2\theta_{\mathrm{B}}$ from its original path as illustrated in Figure 11.16.

The fraction of the light beam deflected depends upon the generation efficiency and the width of the SAW, the latter also defining the interaction length for the device. Although diffraction efficiencies are usually low (no more than 20%), the diffracted on/off ratio can be very high. Hence these devices provide effective switches as well as amplitude or frequency modulators.

11.4.3 Periodic structures for filters and injection lasers

Periodic structures may be incorporated into planar waveguides to form integrated optical filters and resonators. Light is scattered in such a guide in a similar manner to light scattered by a diffraction grating. A common example of a periodic waveguide structure is the corrugated slab waveguide shown in Figure 11.17. When light propagating in the guide impinges on the corrugation, some of the energy will be diffracted out of the guide into either the cover or the substrate. The device, however, acts as a one-dimensional Bragg diffraction grating, and light which satisfies the Bragg condition is reflected back along the guide at 180° to the original direction of propagation (Figure 11.17).

The Bragg condition for the case of 180° reflection can be obtained from Eq. (11.14) if we let the corrugation period D (Figure 11.16) equal the acoustic wavelength Λ and let λ_1 equal $\lambda_{\mathrm{B}}/n_{\mathrm{e}}$ where λ_{B} (the Bragg wavelength) is the optical wavelength in a vacuum and n_{e} is the effective refractive index of the guide. If we also assume that λ_{B} is equal to 90°, then Eq. (11.14) becomes:

$$D = \frac{l\lambda_{\mathrm{B}}}{2n_{\mathrm{e}}} \qquad (11.15)$$

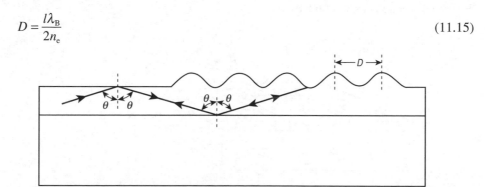

Figure 11.17 A slab waveguide with surface corrugation giving reflection back along the guide when the Bragg condition is met. Hence the structure performs as a one-dimensional Bragg diffraction grating

where $l = 1, 2, 3, \ldots, m$ is the order of the grating which was unity in Eq. (11.14) because diffraction took place between the zero- and first-order mode. The vacuum wavelength of light that will be reflected through 180° by such a grating is therefore:

$$\lambda_B = \frac{2n_e D}{l} \tag{11.16}$$

When the reflected light is incident at an angle (Figure 11.17) then [Ref. 15]:

$$n_e = n_1 \sin 2\theta \tag{11.17}$$

where n_1 is the refractive index of the guide. Hence, depending on the corrugation period of the structure, all the incident power at a particular wavelength will be reflected. Devices of this type therefore behave as frequency-selective rejection filters or mirrors. An example of such a reflection filter is shown in Figure 11.18. It comprises an InGaAsP/InP grating waveguide device in which the surface corrugation is typically written as a photoresist mask using two interfering ultraviolet beams before chemical or physical etching. The filter bandwidth could be quite small (6 Å 0 ⅞ 6×10^{-10} m) with modest interaction lengths using this technique [Ref. 61]. Although the low substrate-waveguide refractive index difference using lithium niobate devices combined with the inherent etching difficulties have limited the development of Ti : $LiNbO_3$ waveguide reflection filters, a 9 mm long Ti : $LiNbO_3$ waveguide grating filter with 2.6 nm bandwidth has been experimentally demonstrated [Ref. 62]. A periodically poled lithium niobate (PPLN) crystal was used where the controlled heating mechanism for the crystal provided the fine tuning for the phase-matching. Periodic poling is a process where the formation of layers takes place with an alternate orientation in the birefringent material (i.e. double refraction) and thus the domains are regularly spaced with a period which can be set within the desired operating wavelengths. The periods which typically range from 5 to 35 μm can be used to achieve the desired nonlinear functionality and therefore the length of period determines the properties of resultant device. For example, a short period is used for second harmonic generation whereas long periods are used for optical parametric oscillations (i.e. for wavelength conversion (see Section 10.5)).

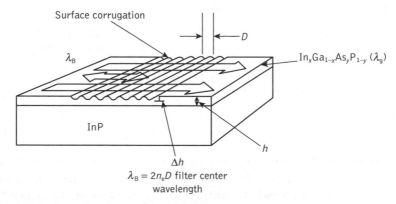

$\lambda_B = 2n_e D$ filter center wavelength

Figure 11.18 An InGaAsP/InP waveguide grating filter

For a waveguide grating filter which exhibits a large change in effective refractive index with a fine grating period, the 3 dB fractional bandwidth is given approximately by [Ref. 21]:

$$\frac{\delta\lambda}{\lambda} \simeq \frac{D}{L} \tag{11.18}$$

where L is the grating length. Hence Eq. (11.18) allows estimates of the filter 3 dB bandwidth, $\delta\lambda$, to be obtained.

Example 11.2

A 1 cm long InGaAsP/InP first-order grating filter is designed to operate at a center wavelength of 1.52 μm. The reflected light is incident at an angle of 1° and the refractive index of InGaAsP is 3.1. Determine the corrugation period and estimate the filter 3 dB bandwidth. A large change in effective refractive index may be assumed.

Solution: The effective refractive index of the waveguide is given by Eq. (11.17) as:

$$n_e = n_1 \sin 2\theta = 3.1 \sin 2°$$
$$= 0.11$$

The corrugation period for the first-order grating may be obtained from Eq. (11.15) as:

$$D = \frac{\lambda_B}{2n_e} = \frac{1.52 \times 10^{-6}}{2 \times 0.11} = 6.9 \ \mu m$$

Finally, the filter 3 dB bandwidth can be estimated from Eq. (11.18) where:

$$\delta\lambda \simeq \frac{D\lambda}{L} = \frac{6.9 \times 10^{-6} \times 1.52 \times 10^{-6}}{10^{-2}}$$

$$= 10.5 \ \text{Å} \ (\simeq 1 \ \text{nm})$$

It may be observed that a relatively narrow filter bandwidth is obtained in Example 11.2. Such devices could find use for wavelength demultiplexing of a larger number of channels. Alternatively, wide-bandwidth filters may be realized by forming gratings which exhibit a gradual change in the corrugation period. Such grating devices are said to have a chirped structure [Ref. 63]. It should also be noted that the corrugated gratings discussed above are also incorporated into advanced single-mode injection laser structures, namely the distributed feedback and the distributed Bragg reflector lasers (see Section 6.6.2).

Periodic optical filters used for WDM devices based on thin-film or arrayed-waveguide grating (AWG) (see Section 5.6.3) result in the loss of channels because of the need for a guard band in between adjacent channels. For example, an experimental demonstration of an eight-band filter with each band comprising four channels and leaving a single channel

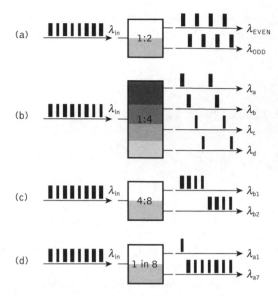

Figure 11.19 Interleaver waveband filter functions: (a) with even and odd channels being separated onto two different ports; (b) separation of channels out to 1:4 or higher; (c) banded Interleaver which separates even and odd bands of channels with a 4:8 combination; (d) asymmetric interleaver separating one channel in eight

between each band (i.e. 4-skip-1)* allowed a total of 32 channels, each spaced by 100 GHz, spanning an overall optical bandwidth of 4 THz [Ref. 64]. Clearly it is important to use the full bandwidth potential which can be achieved by employing periodic waveband filters, where instead of demultiplexing single-wavelength channels specified numbers of wavelength channels can be grouped together into bands thereby minimizing bandwidth usage. Such devices, known as interleaver waveband filters, can then be used to separate the required wavelength signals [Ref. 65].

Figure 11.19 illustrates the different functions of an interleaver waveband filter performing as a wavelength division demultiplexer. A demultiplexing scheme where even and odd channels are separated onto the two different output ports identified is shown in Figure 11.19(a). Further separation of specific wavelength channels can be obtained as indicated in Figure 11.19(b) where a multiplexed input signal is demultiplexed into four wavelength channels. It is also possible to achieve a bandpass filter operation while isolating the required number of channels in each band. This process is illustrated in Figure 11.19(c) where the banded interleaver splits into an even and odd band of channels with each band containing four discrete channels. In addition, a demultiplexing filter can be used to isolate

---

* M-skip-N, where M is the number of channels in each band whereas N is the number of channels excluded between any two bands utilizing this bandwidth for a guard band. For example, the ITU wavelength channel allocation for 32 channels employs a 4-skip-1 scheme utilizing channels in the C-band (1530 to 1560 nm), starting at channel 20 to channel 60 and excluding a single channel after every four channels, is given as eight bands: 21 to 24, 26 to 29, 31 to 34, 36 to 39, 41 to 44, 46 to 49, 51 to 54 and finally 56 to 59.

a single required channel from a WDM comb.* This situation is represented in Figure 11.19(d) where an asymmetric interleaver is shown to separate one channel from a WDM comb signal containing eight wavelength multiplexed signals. For example, a silica-based planar circuit utilizing a bandpass interleaver with a five-band 8-skip-0 waveband filter has demonstrated flexible waveband optical networking performance for 40 channels without guard bands at a transmission rate of 40 Gbit s^{-1} [Ref. 66].

Example 11.3

Design a wavelength channel plan for an 8-band, 32-channel dense WDM interleaver waveband filter with channel spacing of 100 GHz. Using the M-skip-N scheme with M equal to 4 and values for N of (i) 0, (ii) 1, (iii) 2, determine: (a) the total number of channels required for each interleaver band filter; and (b) the overall bandwidth of the filter in each case.

Solution: (a) In an M-skip-N scheme M represents the total number of channels in each band and N is the number of channels excluded between two consecutive wavebands to provide a guard band. Therefore, the overall number of channels C_{total} can be written as:

$$C_{\text{total}} = \text{number of bands} \times M + C_{\text{skip}}$$

where C_{skip} is the total number of channels skipped between two wavelength bands and is given by:

$$C_{\text{skip}} = (\text{number of bands} - 1) \times N$$

(i) 4-skip-0:

$$C_{\text{skip}} = (\text{number of bands} - 1) \times N$$
$$= (8 - 1) \times 0 - 0$$

$$C_{\text{total}} = (\text{number of bands} \times M) + C_{\text{skip}}$$
$$= (8 \times 4) + 0 = 32$$

As anticipated, the total number of wavelength channels to provide the 4-skip-0 scheme is just 32 as no channels are skipped between the wavebands.

(ii) 4-skip-1:

$$C_{\text{skip}} = (8 - 1) \times 1 = 7$$
$$C_{\text{total}} = (8 \times 4) + 7 = 39$$

Hence the total number of wavelength channels required for the 4-skip-1 scheme is 39 with a single channel being skipped between each wavelength band.

▶

---

* The term comb is used to describe equally spaced WDM signal channels.

(iii) 4-skip-2:

$$C_{skip} = (8 - 1) \times 2 = 14$$
$$C_{total} = (8 \times 4) + 14 = 46$$

Therefore the total number of wavelength channels needed for the 4-skip-2 scheme is 46 when 2 channels are skipped between each wavelength band.

(b) A channel spacing of 100 GHz corresponds to 0.8 nm based on the absolute reference of 193.1 THz (i.e. 1552.52 nm) (see Section 12.9.4). This information enables the production of Table 11.1 which provides the wavelength channel plan for 32 channels in 8 × 4 bands for 4-skip-0, 4-skip-1 and 4-skip-2 shown in the first, second and third columns, respectively. Furthermore, the values identified as (skipped) in the second and third columns are of those wavelength channels which are used as guard bands between two consecutive wavebands. The bandwidth for each of the three interleaver band filters identified in Table 11.1 can therefore be determined by subtracting the final wavelength channel from the initial one:

Table 11.1 Wavelength channel plan for 8-band, 32-channel dense WDM interleaver band filter for 4-skip-0, 4-skip-1 and 4-skip-2

4-skip-0		4-skip-1		4-skip-2	
Number of channels	Wavelength (nm)	Number of channels	Wavelength (nm)	Number of channels	Wavelength (nm)
1	1552.52	1	1552.52	1	1552.52
2	1553.32	2	1553.32	2	1553.32
3	1554.12	3	1554.12	3	1554.12
4	1554.92	4	1554.92	4	1554.92
5	1555.72	(skipped) 5	1555.72	(skipped) 5	1555.72
6	1556.52	6	1556.52	(skipped) 6	1556.52
7	1557.32	7	1557.32	7	1557.32
8	1558.12	8	1558.12	8	1558.12
9	1558.92	9	1558.92	9	1558.92
10	1559.72	(skipped) 10	1559.72	10	1559.72
11	1560.52	11	1560.52	(skipped) 11	1560.52
12	1561.32	12	1561.32	(skipped) 12	1561.32
13	1562.12	13	1562.12	13	1562.12
14	1562.92	14	1562.92	14	1562.92
15	1563.72	(skipped) 15	1563.72	15	1563.72
16	1564.52	16	1564.52	16	1564.52
17	1565.32	17	1565.32	(skipped) 17	1565.32
18	1566.12	18	1566.12	(skipped) 18	1566.12
19	1566.92	19	1566.92	19	1566.92
20	1567.72	(skipped) 20	1567.72	20	1567.72
21	1568.52	21	1568.52	21	1568.52

Table 11.1 (*continued*)

4-skip-0		4-skip-1		4-skip-2	
Number of channels	*Wavelength (nm)*	*Number of channels*	*Wavelength (nm)*	*Number of channels*	*Wavelength (nm)*
22	1569.32	22	1569.32	22	1569.32
23	1570.12	23	1570.12	(skipped) 23	1570.12
24	1570.92	24	1570.92	(skipped) 24	1570.92
25	1571.72	(skipped) 25	1571.72	25	1571.72
26	1572.52	26	1572.52	26	1572.52
27	1573.32	27	1573.32	27	1573.32
28	1574.12	28	1574.12	28	1574.12
29	1574.92	29	1574.92	(skipped) 29	1574.92
30	1575.72	(skipped) 30	1575.72	(skipped) 30	1575.72
31	1576.52	31	1576.52	31	1576.52
32	1577.32	32	1577.32	32	1577.32
		33	1578.12	33	1578.12
		34	1578.92	34	1578.92
		(skipped) 35	1579.72	(skipped) 35	1579.72
		36	1580.52	(skipped) 36	1580.52
		37	1581.32	37	1581.32
		38	1582.12	38	1582.12
		39	1582.92	39	1582.92
				40	1583.72
				(skipped) 41	1584.52
				(skipped) 42	1585.32
				43	1586.12
				44	1586.92
				45	1587.72
				46	1588.52

(i) For the 4-skip-0 the filter bandwidth is:

$$1577.32 - 1552.52 = 24.8 \text{ nm}$$

(ii) For the 4-skip-1 the filter bandwidth is:

$$1582.92 - 1552.52 = 30.4 \text{ nm}$$

(iii) For the 4-skip-2 the filter bandwidth is:

$$1588.52 - 1552.52 = 36.0 \text{ nm}$$

Hence the overall filter bandwidths for the 4-skip-0, 4-skip-1 and 4-skip-2 are 24.8, 30.4 and 36.0 nm, respectively.

11.4.4 Polarization transformers and wavelength converters

The polarization status of optical signals is an important factor in high-speed optical fiber communication system design as with increasing transmission rates optical devices become more sensitive to polarization-related impairments. In addition to polarization mode dispersion observed in optical fibers (see Section 3.13.2), the impact of the state of polarization (SOP) affects the performance of optical devices including losses incurred in optical elements (i.e. electro-optic modulators), the gain in optical amplifiers, the center wavelength in WDM filters and the responsivity of the receivers (see Section 12.4.2). Additionally, it may be required to modify the SOP in accordance with the requirements of the system to allow for the smooth propagation of optical signals. To facilitate this modification various techniques are used to transform or control polarization related impairments [Ref. 67]. Alternatively, the polarization is scrambled when the SOP of fully polarized light is made to vary randomly at a relatively low rate. This can be achieved by using polarization modulation methods incorporating a lithium niobate modulator [Ref. 68]. By contrast, a polarization descrambler can be used to keep the SOP constant using an automatic polarization stabilizer based on a polarization controller [Ref. 69].

The electro-optic effect typically in lithium niobate waveguide devices can be used to facilitate TE–TM mode conversion. However, to allow the transformation of an arbitrary input polarization, not just TE or TM, it is necessary to control the relative phase between the TE and TM components. Such polarization transformers which operate as TE–TM mode converters can be employed as elements within intensity modulators (when combined with a polarizer), optical filters or polarization controllers. A basic example of the latter device is shown in Figure 11.20 [Ref. 70]. It comprises two phase modulators and a single TE–TM mode converter on X-cut* lithium niobate.

The first phase modulator is required to adjust the phase difference between the incoming TE and TM modes to be $\pi/2$ so that the polarization controller can operate with all

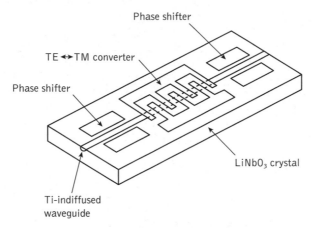

Figure 11.20 An IO polarization controller

---

* Conventional Y-cut lithium niobate is not normally used as the electro-optic coefficient is smaller, necessitating higher operating voltages.

incoming polarization states. When this condition is satisfied the central phase-matched mode converter is operated as a linear polarization rotator. Although a linear output polarization of either TE or TM is sufficient in some applications, for full polarization control a second phase shifter is required to adjust the output phase to a desired value of elliptical output polarization.

A number of electro-optic waveguide devices can be used to provide wavelength conversion of an optical signal (see Section 10.5) [Ref. 71]. A common technique is to employ a phase modulator in a serrodyne configuration to alter the optical frequency by a linearly increasing voltage applied to the device electrodes [Ref. 72]. In practice a continuously increasing ramp signal voltage cannot readily be produced and hence a sawtooth voltage waveform is used. However, sawtooth waveforms with instantaneous fall times can be generated and hence additional frequency components tend to be produced. This factor, combined with the need to vary the rate of change of the applied voltage to alter the extent of the frequency shift, limits the use of this device to applications where a small constant frequency shift is required. In applications where large frequency translations are necessary and where the device is used as a control element in a feedback loop (e.g. coherent optical receivers), alternative frequency translators have been utilized [Ref. 18].

Devices based on TE–TM mode conversion are also capable of generating frequency-translated or wavelength-converted optical signals [Ref. 73]. When the region where the mode conversion takes place is made to move relative to the direction of the optical wave, the source of the converted signal appears to be moving to a stationary observer and the light is therefore Doppler shifted. To generate the effect of a moving coupling grating in practice, a mode converter is divided into several sections and each is driven with a correctly phase-shifted sinusoidal signal which has a frequency equal to the desired up or down frequency translation. In principle, this technique should be highly efficient and generate no unwanted optical signals. However, significant unwanted sidebands have been observed with such devices which appear to arise from parasitic electrical fields [Ref. 18]. Careful device design is therefore necessary to maintain these signals at an acceptable level.

Mach–Zehnder interferometric Y-junction modulators (see Figure 11.15) can also be used to generate double-sideband wavelength-converted optical signals when they are modulated with an intensity-modulated waveform. In this case the optical wavelength shift is proportional to the wavelength of the modulating signal. However, in simple device structures the charging of the electrode capacitance limits the maximum modulation frequency and thus the magnitude of the frequency translation that can be obtained. To overcome this problem Mach–Zehnder interferometers (MZIs) with traveling-wave electrode structures have been designed which provide multigigahertz bandwidths. When a nonlinear medium (e.g. a semiconductor optical amplifier) is inserted into the arms of an MZI, however, the absorption characteristics of the nonlinear medium determine the refractive index variation with respect to change in the intensity of an input signal which produces phase variation in the output signal emerging from the output port of the MZI. This intensity–phase variation produces changes in the wavelength of an optical signal and therefore by controlling the optical signal power in each MZI arm, it is possible to produce any desired wavelength at the output of the wavelength converter (see Section 10.5) [Ref. 74].

Integration of the aforementioned electro-optic devices into a single lithium niobate substrate, for use in coherent optical fiber communication systems (see Chapter 13), is an area of interest to reduce both losses between individual devices as well as system cost.

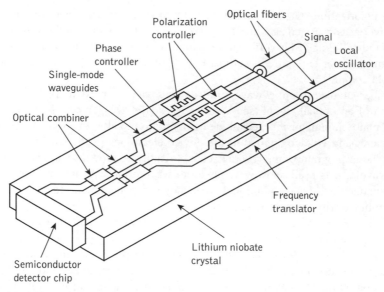

Figure 11.21 Coherent optical receiver device

For example, the configuration of a potential coherent optical receiver device is illustrated in Figure 11.21 [Refs 18, 75]. It was fabricated on Z-cut lithium niobate and comprises a polarization controller with output phase controller, and a frequency translator together with a directional coupler for mixing the two optical signals. Successful operation of this integrated device was demonstrated and a similar X-cut lithium niobate device requiring a lower operating voltage has been proposed [Ref. 75]. More recently, a coherent integrated optical receiver based on an optical phased locked loop implemented using InP technology has been demonstrated experimentally, which comprised a pair of balanced uni-traveling carrier photodetectors, a 2×2 multimode interference coupler and phase modulators located on a 1×2 mm substrate [Ref. 76].

11.5 Optoelectronic integration

The integration of interconnected optical and electronic devices is an important area of investigation for applications within optical fiber systems [Ref. 76]. Monolithic opto-electronic integrated circuits (OEICs) incorporating both optical sources and detectors have been successfully realized for a number of years. Monolithic integration for optical sources has been generally confined to the use of group III–V semiconductor compounds. These materials prove useful as they possess both optical and electronic properties which can be exploited to produce high-performance devices. Circuits have been fabricated from GaAs/AlGaAs for operation in the shorter wavelength region between 0.8 and 0.9 μm. Such a circuit is shown in Figure 11.22(a) where an injection laser is fabricated on a GaAs substrate with a MESFET (metal Schottky FET, see Section 9.5.1) which is used to bias and modulate the laser. Alternatively, Figure 11.22(b) demonstrates the integration of a

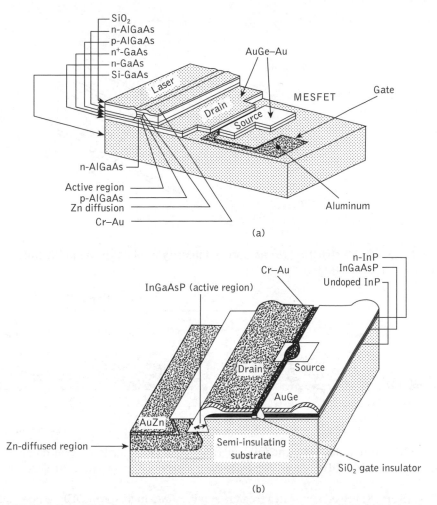

Figure 11.22 Monolithic integrated transmitter circuits: (a) GaAs/AlGaAs injection laser fabricated with a MESFET on a GaAs substrate; (b) InGaAsP/InP injection laser fabricated with a MISFET on a semi-insulating InP substrate

longer wavelength (1.1 to 1.6 µm) injection laser fabricated from InGaAsP/InP together with a MISFET (metal integrated-semiconductor FET) where the conventional *n*-type substrate is replaced by a semi-insulating InP substrate.

The realization of OEICs, however, lagged behind other developments in IO using dielectric materials such as lithium niobate. This situation was caused by the inherent difficulties in the fabrication of OEICs even when III–V compound semiconductors are employed [Ref. 77]. Compositional and structural differences between photonic devices and electronic circuits created problems in epitaxial crystal growth, planarization for lithography, electrical interconnections, thermal and chemical stability of materials, electrical matching between photonic and electrical devices together with heat dissipation. Nevertheless, the maturing of gallium arsenide technology for integrated circuits (as opposed to

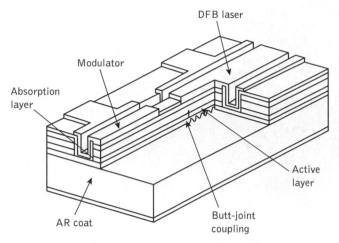

Figure 11.23 Device structure for an optical intensity modulator monolithically integrated with a DFB laser [Ref. 79]

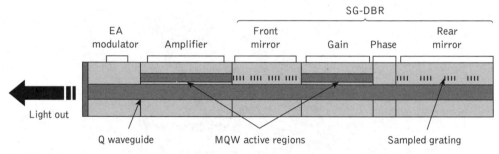

Figure 11.24 Integrated laser modulator (ILM) based on a sampled grating distributed Bragg reflector (SG-DBR) laser. Reprinted with permission from Ref. 80 © IEEE 2003

OEICs) [Ref. 78] helped stimulate research activities into high-speed OEICs. For example, the structure of a monolithically integrated DFB laser with an optical intensity modulator is shown in Figure 11.23 [Ref. 79]. The InGaAsP/InP devices, which were designed to avoid the large chirp associated with directly modulated semiconductor lasers, possess good dynamic characteristics at a modulation rate of 40 Gbit s^{-1} when operating at a wavelength of 1.55 μm. Figure 11.24 shows a monolithically integrated laser modulator (ILM) consisting of a sampled-grating distributed Bragg reflector (SG-DBR) laser integrated with a semiconductor optical amplifier [Ref. 80]. The operating wavelength for this ILM can be tuned within the range from 1520 to 1570 nm. Typical performance characteristics for CW mode operation for this device include a fiber-coupled output power of greater than 10 mW, a linewidth less than 2 MHz and a side-mode suppression ratio greater than 40 dB for 90 channels spectrally spaced at 50 GHz [Ref. 80].

Optoelectronic integrated circuits devices based on heterojunction bipolar transistor (HBT) and high electron mobility transistor (HEMT) technology (see Section 9.6) using GaAs and InP are capable of operating at transmission rates higher than 40 Gbit s^{-1} [Refs 81, 82]. Figure 11.25 illustrates a p–i–n photoreceiver based on an HEMT, comprising

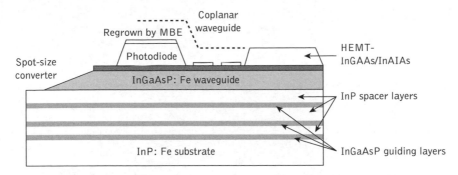

Figure 11.25 Monolithic integrated *p–i–n* HEMT photoreceiver. Reprinted with permission from Ref. 84 © IEEE 2005

a spot-size converter integrated with a photodiode. The spot-size converter increases the fiber alignment tolerances by one order of magnitude and thus enables the use of a cleaved instead of a lensed fiber for connection to the OEIC [Ref. 83]. It is integrated using InGaAsP guiding layers where more confinement is provided employing an InGaAsP : Fe waveguide, which feeds the photodiode while the latter is vertically tapered with a ramp for conversion of the spot size. The photodiode on the left side of Figure 11.25 is connected to the input of the traveling-wave amplifier via an air bridge. In addition, the HEMT is coupled to the coplanar transmission lines by microstrip lines formed by air bridges over a ground plane. Finally, this device, which was grown on an InP : Fe substrate, operated at transmission rates higher than 80 Gbit s^{-1} at a signal wavelength of 1.55 μm [Ref. 84].

An optical power coupler (or splitter) integrated with an optical waveguide amplifier is considered a useful solution for optical networks where the aim is to reduce the number of amplifiers in the system while also reaching the maximum number of the nodes [Ref. 85]. A simple amplified splitter configuration employing four erbium-doped waveguide amplifiers each sharing a single optical pump source for four incoming WDM signals was described in Figure 10.20 (see Section 10.4.3). Such planar waveguide amplified splitters with a splitting ratio of, for example, 1 : 8 are commercially available [Ref. 85].

Most of the applications for OEICs can be found within optical networks (see Chapter 15) where a large switching capacity is the desired feature which is required to support a large number of WDM channels. In this case the optical switches alone cannot provide the functionality and hence this necessitates the use of intelligent optical switches which offer control of both optical signal wavelength and signal power. Figure 11.26 shows such an intelligent 8 × 8 switching matrix constituting a wavelength interchange cross-connect (WIXC) (see Section 15.2.2) [Ref. 86]. The nonblocking switching matrix comprises several optical components on a single polymer-on-silica planar lightwave circuit (see Sections 11.1 and 11.2). It includes optical switches, photodiodes, variable optical attenuators (VOAs) and optical couplers acting as power taps. Moreover, the OEIC chip contained in an operational package is depicted in Figure 11.26(b). The single-mode fiber-to-fiber insertion loss for the matrix package operating over a range of signal wavelengths from 1.52 to 1.61 μm was 4 dB. The VOAs exhibited a dynamic range of 20 dB with an insertion loss of 0.1 to 0.3 dB. The chromatic dispersion (see Section 3.9) was 0.1 ps nm^{-1} and the polarization mode dispersion of the module was specified as 0.01 ps. Furthermore, the

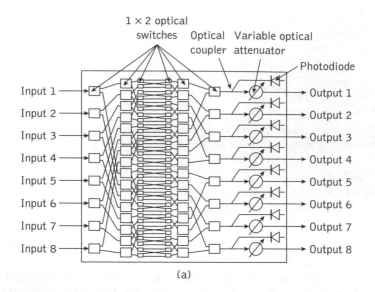

(a)

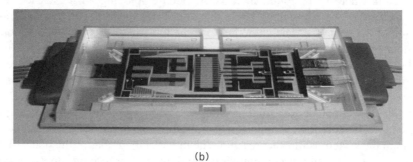

(b)

Figure 11.26 Chip architecture for an 8 × 8 intelligent optical switching matrix with single-mode fiber: (a) structure block schematic; (b) OEIC module package [Ref. 86]

switching matrix exhibited a crosstalk of 50 dB between any two ports [Ref. 86]. These performance parameters would therefore enable the implementation of a larger switching fabric providing an increased number of wavelength channels. In addition, using this 8 × 8 switching matrix an eight-port reconfigurable optical add/drop multiplexer (ROADM) (see Section 15.2.2) subsystem has also been demonstrated [Ref. 87]. This subsystem, which employed an array of integrated photodiodes combined onto a polymer platform to reduce the overall module size, successfully transmitted 40 wavelength channels.

More recently, optical replication technology has been introduced to produce OEICs on a larger scale where the integrated devices exhibit low optical coupling loss [Refs 88, 89]. Replication technology essentially employs hot embossing, molding and ultraviolet lithography techniques to produce high-quality optical components and devices. The development of wafer-scale replication technology using ultraviolet curable SOL gel (see Section 11.2) and polymer materials enables refractive and diffractive micro-optical elements to be replicated directly onto glass substrates or onto silicon wafers of material systems of the III–V group alloys [Ref. 89]. The SOL gel materials allow the combination of replication with lithography to leave selected areas material-free for sawing and bonding. This technology is therefore useful for the production of both planar micro-optical elements and stacked optical microsystems. Moreover, polymer-based devices such as optical micro-electromechanical systems (OMEMS) (see Section 11.4.1) can be replicated using this technology. The structure of polymer-based OMEMS with fiber interconnection, where the coupling loss remained 0.9 dB for single-mode fiber coupling at a signal wavelength of 1.5 μm, is shown in Figure 11.27(a) [Ref. 88].

The chip-to-chip interconnection of optical components illustrating a photodiode and a vertical cavity surface-emitting laser (VCSEL) is depicted in Figure 11.27(b). These devices are assembled on microtrenches where embedded electrodes are connected through the passive junction of the polymer waveguide on alignment pits, thus achieving optical interconnection between the optoelectronic devices. The averaged insertion loss in this case was 2.45 dB when coupling single-mode optical fiber with 8 μm core diameter at a wavelength of 1.55 μm while the estimated excessive coupling loss remained 0.92 dB per coupling interface [Ref. 89]. In addition, the replication of microlenses on active components can also be obtained with precisely controlled alignment and thickness. Moreover, wafer-scale replication of microlenses on VCSELs fabricated on GaAs wafers which reduces the beam divergence can be realized using this technology [Refs 89, 90]. For example, a section of a wafer comprising arrays of VCSELs with microlenses replicated onto the VCSEL devices is shown in Figure 11.27(c). It comprises a 10.2 cm wafer containing a large number of VCSELs arrays whereas in the magnified picture it displays only two VCSEL devices with diffractive microlenses grown on the top of each device. Finally, the area between replicated microlenses which is used for subsequent bonding and dicing of the devices is kept free of the SOL gel material.

In a further example referred to as Terabus, OEIC technologies have supported terabit per second rates for chip-to-chip data transfer [Ref. 91]. In this approach a chip-like optoelectronic packaging structure known as an optochip is assembled directly onto another optical chip (optocard) as indicated in Figure 11.28. In this case arrays of VCSELs and photodiodes were flip-chip bonded to the driver and receiver IC arrays. The optocard incorporated 48 parallel multimode optical waveguides each with a core diameter of 62.5 μm and the optochip comprised 4×12 arrays of VCSELs and p–i–n photodiodes operating at a signal wavelength of 0.98 μm. In order to provide optical coupling between the optochip and the optocard microlenses, arrays were etched onto the back side of the optoelectronic arrays and onto 45° mirrors in the waveguides. Transmitter and receiver operation was demonstrated to 10 Gbit s^{-1} per channel. Furthermore, the silicon carrier strip provided a platform to combine multilayer wiring and vias (i.e. stubs) for high-performance electrical interconnection, with the ability to integrate heterogeneous components including IC and optoelectronic devices using the flip-chip bonding technology. It is apparent that such systems would permit higher bandwidth transmission between processors or modules in future high-performance optical computation systems (see Section 11.8).

There is a demand for compact, low-cost optical devices within optical networks and, in particular, for passive optical networks (PONs) (see Section 15.6.3) which will operate at transmission rates in the range of 2.5 to 40 Gbit s^{-1} [Refs 92–94]. For example, optical transceivers for performing electrical/optical conversion for bidirectional transmission of a burst signal at wavelengths of 1.3 μm and 1.5 μm as depicted in Figure 11.29 are commercially available [Refs 94–100]. Such transceivers are designed for use with the gigabit PON (see Section 15.6.3) operating at transmission rates of up to 10 Gbit s^{-1}.

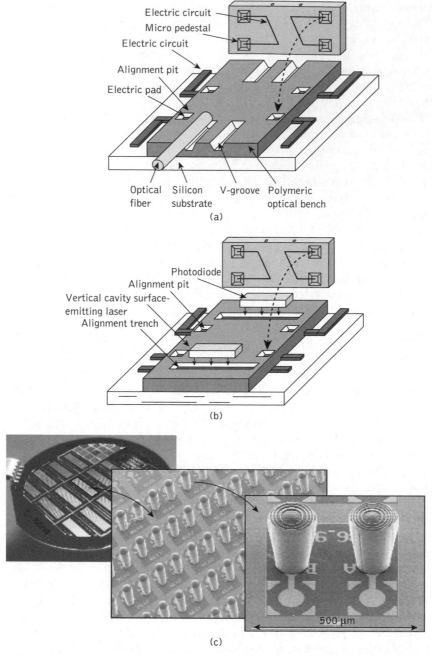

Figure 11.27 Optical replication technology: (a) polymer optical micro-electromechanical systems structure for single-mode optical fiber; (b) the chip-to-chip optical interconnection of electro-optic components [Ref. 88]; (c) (from left to right), replicated SOL-gel lensed arrays on a 10.2 cm diameter silicon wafer; a rectangular section containing VCSELs and finally a pair of mounted and bonded VCSEL devices with SOL-gel lenses incorporated on the top surface. Reprinted with permission from Ref. 90 © IEEE 2004

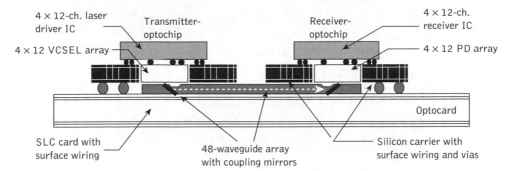

Figure 11.28 Structure of the Terabus transceiver comprising 4 × 12 arrays of vertical cavity surface-emitting lasers (VCSELs), p–i–n photodiodes and multimode fiber waveguides. The surface laminar circuitry (SLC) card accommodates the higher wiring density. Reprinted with permission from Ref. 91 © IEEE 2006

Figure 11.29 Gigabit passive optical network (GPON) transceiver [Ref. 94]

11.6 Photonic integrated circuits

A photonic integrated circuit (PIC) is the monolithic integration of two or more optical components onto a single substrate. The optical components can be either a passive or active element. Based on planar waveguide technology (see Section 11.3) it is possible to fabricate both active and passive components simultaneously onto the same substrate. Figure 11.30 presents the structural strategy to integrate several optical devices on a single indium phosphate substrate where the building blocks are divided into two columns. Each column separately accommodates active and passive functions in a layered structure

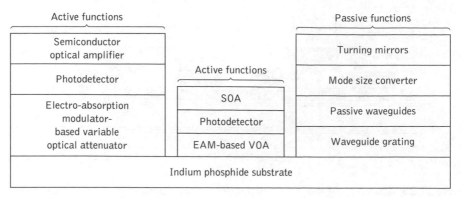

Figure 11.30 Structural strategy for the fabrication of photonic integrated circuits (PICs) onto an indium phosphide substrate

where similar devices types can be fabricated. Using this technique various devices can be incorporated onto a single chip providing flexibility while also achieving a reduction in the overall circuit size. Integrated devices include arrayed waveguide grating (AWG) (see Section 5.6.3) based optical cross-connects [Ref. 101], multiwavelength laser sources [Refs 102, 103] and tunable wavelength converters [Refs 104, 105]. However, the state of the art for PIC integration does not yet provide the capability to combine several dissimilar technology devices in the same column onto a single substrate.

The prime aim of a PIC is to reduce the overall size of optical functions, so the interconnection of several modules being grown on the same substrate is therefore important. Further reduction in the overall size of an integrated device can be achieved if multilevel or layered interconnections are incorporated [Ref. 4]. This reduction can be accomplished using fused AWGs arranged in a stack of layers forming a vertical coupler. Vertical fused coupling is obtained by repeating the fusion process for each layer containing the planar waveguides onto an indium phosphide wafer [Ref. 106]. Figure 11.31 illustrates this concept by providing the comparison between a conventional PIC and a multilevel fused AWG module [Refs 107–109]. A conventional PIC comprising an array of photonic crossbar switches interconnected by planar waveguides to switch the optical signal from one stage to another is shown in Figure 11.31(a). This arrangement therefore incorporates multistage switching of the optical signal to reach its destination.

By contrast an advanced single-level photonic switch module employing an AWG-based planar coupler (see Section 5.6.3) is displayed in Figure 11.31(b) which illustrates an $N \times N$ wavelength channel cross-connect fabricated using wafer-based vertical fused AWG couplers [Ref. 107] which can be viewed as an upgrade of the conventional AWG-based demultiplexer/multiplexer (see Section 5.6.3) where the number of ports has been increased on both the input and output sides. The upgraded configuration can function as a passive wavelength-selective, strictly nonblocking cross-connect. The AWG cross-connect comprises N input and N output waveguides, two planer regions functioning as multimode interference (MMI) couplers (see Section 11.3.1) and an M array of curved waveguides connecting the two planar regions. The operation of the device is similar to two cascaded passive star couplers (see Section 5.6.2) each comprising $N \times M$ and $M \times N$ combinations of input/output ports where M is greater than N [Ref. 107]. Such passive

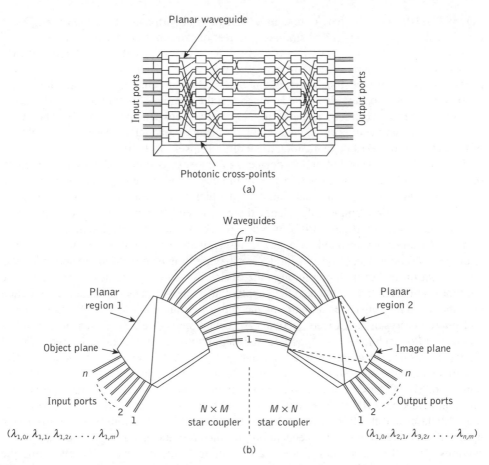

Figure 11.31 Interconnections for photonic integrated circuits (PICs): (a) a conventional planar PIC array of crossbar switches; (b) an $N \times N$ wavelength channel cross-connect based on an m-arrayed-waveguide grating (AWG)

cascaded couplers constitute an imaging system where the object and the image planes are formed at the opposite sides of two couplers (i.e. planar regions 1 and 2) which are connected by waveguides of unequal lengths.

The AWG-based coupler demultiplexes an incoming WDM signal comprising a number of wavelengths M (i.e. λ_1 to λ_m) on each input port in the planar region 1. Each of the M wavelength demultiplexed channels after traveling separately through different lengths of the curved waveguides are then multiplexed in the planar region 2. A specific wavelength channel from an input port is directed into the reconstructed WDM signal which appears at precisely one output port of the second planar region. The reconstituted spectrum of the WDM signal at any output port contains a different set of wavelength channels with at least one wavelength channel from each input port and therefore when the number of WDM channels M is present at N input ports then the output port 1 always produces a WDM signal containing a wavelength signal from each of the input ports. For example, in

Figure 11.31(b) the input port 1 contains a WDM signal with a number of wavelengths M (i.e. $\lambda_{1,0}$, $\lambda_{1,1}$, $\lambda_{1,2}$, ..., $\lambda_{1,m}$) and similarly the rest of the input ports from 2 to N contain $(\lambda_{2,0}, \lambda_{2,1}, \lambda_{2,2}, ..., \lambda_{2,m})$, ..., $(\lambda_{n,0}, \lambda_{n,1}, \lambda_{n,2}, ..., \lambda_{n,m})$. In this case the output port 1 always produces a WDM signal containing $\lambda_{1,0}$, $\lambda_{2,1}$, $\lambda_{3,2}$, ..., $\lambda_{n,m}$. The remaining output ports produce different WDM signals comprising one wavelength channel from each input port following the same pattern. It should be noted that the number of output ports is dependent on the number of channels and also the separation between the channels. Therefore, increasing the number of ports and narrowing the channel spacing increases the complexity of the device and makes its fabrication more difficult [Refs 107–109].

The current dimensions of optical devices (a number of components still have a length of several hundred micrometers to a few millimeters) and circuits generally remain too large to provide for very large-scale integration which is necessary to develop a complete photonic system on a single chip. Hence there is a need to reduce the size of photonic components and in particular PICs in order to achieve larger scale integration. Pursuit of this objective has resulted in a newer research field referred to as nanophotonics [Ref. 110]. Nanophotonics deals with devices having dimensions on the nanometer scale where light instead of traveling through the device is guided by the device and therefore the fundamentals of the optical near field apply. Furthermore, many of the nanodevices and circuits under investigation can be implemented using quantum-dot (see Section 10.3.4) and photonic crystal-based (see Section 2.6) technologies [Ref. 8]. Based on the afore-mentioned PIC technology it is possible to design photonic circuits to perform all-optical signal processing functions where both active and passive devices are monolithically integrated onto the same substrate. Figure 11.32 shows the structure of an integrated laser modulator (ILM) based on an asymmetrical twin-waveguide technique incorporating a distributed feedback (DFB) laser and an electroabsorption modulator [Ref. 23]. Moreover, the DFB laser and the ILM waveguide comprise InGaAlAs quantum wells in a p–n–i–p epitaxial structure allowing independent bias for the laser and electroabsorption modulator sections. The DFB laser can operate at a high optical output power of 3 mW while the

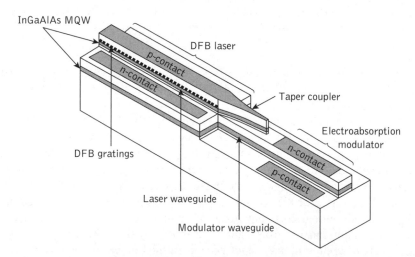

Figure 11.32 Integrated laser modulator (ILM) using a distributed feedback (DFB) laser and electroabsorption modulator (EAM) using a tapered coupler

modulator achieves a good extinction ratio of 6.5 dB for a voltage swing of 2.5 V and low insertion losses of 4 dB over a wide range of temperatures from 0 to 85 °C with no power penalty being incurred for transmission over 30 km of single-mode fiber [Ref. 23].

Three examples of the PICs produced onto an indium phosphide platform are depicted in Figure 11.33. Since these structures utilize phased AWGs they are also known as phased-array structures (PHASAR) [Ref. 111]. An experimental demonstration of the use of PICs in WDM transmission systems is shown in Figure 11.33(a). The 10-channel transmitter module comprises tunable DFB lasers, electroabsorption modulators (EAMs), variable optical attenuators (VOAs) and optical power monitors (OPMs). The AWG-based

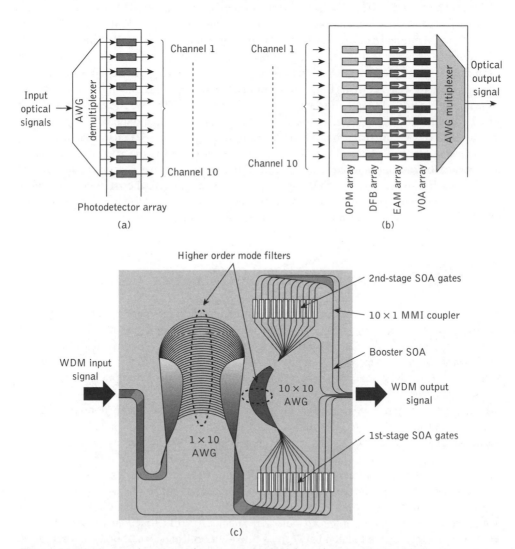

Figure 11.33 Arrayed-waveguide grating (AWG) based PICs: (a) 10-channel receiver section; (b) 10-channel transmitter section. Reprinted with permission from Ref. 113 © IEEE 2007; (c) 100-channel WDM channel selector. Reprinted with permission from Ref. 114 © IEEE 2004

optical multiplexer combines the channels to create an optical multiplexed signal emerging from the output of the module. The receiver architecture comprising a polarization-independent AWG demultiplexer is displayed in Figure 11.33(b). This is followed by an array of 10 high-speed, side-illuminated waveguide photodiodes. In each of the aforementioned modules, there are more than 50 components integrated monolithically onto an indium phosphide substrate enabling single-channel transmission at 40 Gbit s^{-1} leading to an overall transmission rate of 400 Gbit s^{-1} [Refs 112, 113].

A PIC comprising a 100-channel selector for WDM signals is illustrated in Figure 11.33(c) [Ref. 114]. In order to select any single channel from 100 multiplexed channels using a 10×10 array of semiconductor optical amplifiers (SOAs) would normally require 100 SOAs but this number was reduced in this case to just 20 devices. Therefore the PIC configuration consists of 10 AWG structures and two 10×10 gate arrays with 10 SOAs in each array. Both the AWG arrays and SOA gate arrays are combined using a multimode interference coupler. The operation of the channel selector relies on the gate arrays allowing each SOA to function as an open gate while the incoming 100-channel WDM signal is divided into 10 bands via an array of 1×10 AWGs (i.e. in which each band contains 10 channels). Each band is then fed into the first stage of the gate array where each SOA can permit only a single band to progress (i.e. 10 channels) when it is operating as an open switch. These 10 channels are forwarded to the 10×10 AWG array which is interconnected to the second stage of the gate array that comprises an array of SOAs. At this stage, when an SOA operates as an open switch, the desired wavelength channel is allowed to proceed. Thus any channel can be selected using the appropriate combination of first- and second-stage SOAs in the gate arrays.

11.7 Optical bistability and digital optics

Bistable optical devices have been under investigation for a number of years to provide a series of optical processing functions. These include optical logic and memory elements, power limiters and pulse shapers, differential amplifiers and A–D converters. Moreover, the bistable optical device (BOD) in providing for digital optical logic – namely, a family of logic gates whose response to light is nonlinear – gives the basis for optical computation.

In its simplest form the BOD comprises a Fabry–Pérot cavity containing a material in which variations in refractive index with optical intensity are nonlinear (nonlinear optical absorption also gives rise to bistability), as shown in Figure 11.34(a). In a similar manner to the laser such a cavity exhibits a sharp resonance to optical power passing into and through it when the optical path length in the nonlinear medium is an integer number of half wavelengths. By contrast with the laser the value of refractive index within the cavity controls the optical transmission giving high optical output on resonance and low optical output off resonance. The transfer characteristic for the device exhibits two-state hysteresis which results from tuning into and out of resonance, as illustrated in Figure 11.34(b). BODs are therefore able to latch between two distinct optical states (0 or 1) in response to an external signal to act as a memory or flip-flop. Furthermore, by careful adjustments of the device bias and input levels, the BOD can act as an AND gate, an OR gate, or a NOT gate, hence providing logic functions [Ref. 115].

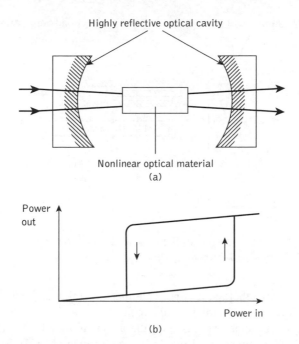

Figure 11.34 A generalized bistable optical device: (a) schematic structure;
(b) typical transfer characteristic

Although the switching speed of BODs is dependent on drive power, they offer the potential for very fast switching at low power levels. Investigations were therefore directed towards the possibility of picosecond switching using only picojoules of energy. A BOD exhibiting these properties would prove far superior to an electronic device which performs the same function. However, suitable nonlinear materials and device structures (e.g. nanotechnology) to give this performance are still under investigation. BODs may be separated into two basic classes: all-optical or intrinsic devices which utilize a nonlinear optical medium between a pair of partially reflecting mirrors forming a nonlinear etalon in which the feedback is provided optically; and hybrid devices where the feedback is provided electrically.

In some cases hybrid devices employ an artificial nonlinearity such as an electro-optic medium within the cavity to produce variations in refractive index via the electro-optic effect. In materials such as lithium niobate and gallium arsenide this produces strong artificial nonlinearity which can be combined with an electronic feedback loop. Such hybrid BODs had been fabricated in integrated optical form. A typical device is shown in Figure 11.35 [Ref. 116]. It consists of a titanium-diffused optical waveguide on a lithium niobate substrate with cleaved and silvered end faces to form the resonant optical cavity. The light emitted from the cavity is detected and amplified by an avalanche photodiode (APD). The electrical signal thus obtained is then fed back to the electrodes deposited on either side of the cavity in order to produce refractive index variations. Such a device therefore exhibits hysteresis and bistability. Although these hybrid BODs provided flexibility for experimental study, their switching speeds were ultimately limited by the use of

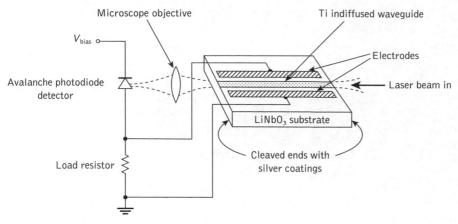

Figure 11.35 A hybrid integrated bistable optical device. Reused with permission from Ref. 116 © 1978 American Institute of Physics

electrical feedback. Nevertheless, it is possible that several such devices could be interconnected to provide a more complex logic circuit.

An alternative hybrid approach based on the use of inorganic superlattices had been pursued at AT&T Bell Laboratories [Ref. 117] and elsewhere [Ref. 118]. These materials were constructed by alternating thin films of two different semiconductor materials which exhibited nonlinear properties. Combinations used include gallium arsenide and gallium aluminum arsenide, mercury telluride and cadmium telluride, silicon, and indium phosphide. This work resulted in the development of the, so-called, self-electro-optic effect device (SEED) which exhibits hysteresis and bistate transmission. The device, a schematic of which is shown in Figure 11.36(a), comprises a single chip of alternating layers. Although the device is activated by light an electric field is required to 'prime' the material for switching. The switching results from wavelength-sensitive absorption within the superlattice structure which causes current flow, thus decreasing the bias voltage which in turn increases the absorption. Eventually a point is reached at which switching occurs and the device switches to high absorption as indicated in Figure 11.36(b).

Alternatively, all-optical or intrinsic BODs may employ an appropriate nonlinear optical medium. Investigations are centered around materials such as indium antimonide, zinc selenide, cadmium sulfide, gallium arsenide, indium arsenide, gallium aluminum arsenide and indium gallium arsenide phosphide in which optical absorption gives a change in refractive index. Unfortunately, these effects are generally weak and often require low temperatures to display themselves adequately. However, the possibility of low-power, low-energy, fast-switching integratable devices for use in real-time optical processing and digital optical computing is a proposition which has encouraged a concentrated activity in this area. Studies involved the use of indium antimonide (InSb) in the near infrared [Ref. 119] and also in the visible region, although operation is only achieved at low temperature (77 K) and at room temperatures using thermal nonlinearities in zinc selenide (ZnSe) interference filter configurations [Refs 120, 121]. This work was undertaken with visible light which provides the advantage that the switching and hysteresis effects can actually be seen.

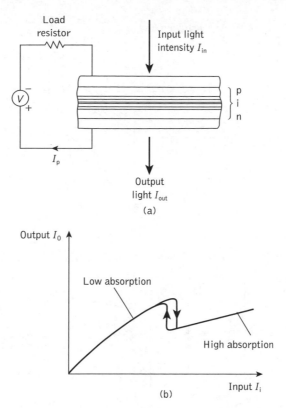

Figure 11.36 The self-electro-optic effect device (SEED): (a) schematic structure; (b) input/output response characteristic

Intrinsic optical bistability may also be obtained from large resonant nonlinearities available near the bandgaps of other semiconductor materials [Ref. 122]. Such bistability can be further distinguished as an active system which incorporates its own optical source, or a passive system which does not. The input to a passive bistable device is always optical while the input to an active device depends upon the method by which the source is to be excited. For example, in the former case room temperature bistability in bulk gallium arsenide at switching speeds of 30 ps was observed [Ref. 123]. In addition, nonlinear channel waveguide structures in GaAs/GaAlAs multiple-quantum-well material have shown optical bistability at relatively low power levels, but with slow switching speeds [Ref. 124].

The source of excitation for active bistability in semiconductors is normally provided by an injection current giving the configuration of a bistable laser diode [Ref. 125]. Semiconductor lasers exhibit optical bistability due to nonlinearities in absorption, gain, dispersion, waveguiding and the selection of the output polarization. One approach to laser diode bistability through nonlinear absorption is illustrated in Figure 11.37(a) [Ref. 126]. In this case the device was fabricated with a tandem electrode which provided two gain sections, with a loss region between them. The loss region acted as a saturable absorber creating the hysteresis characteristic displayed in Figure 11.37(b). Such devices

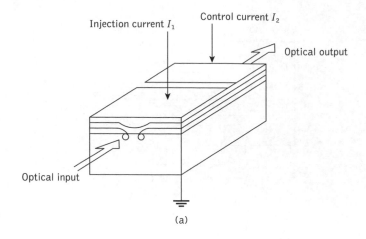

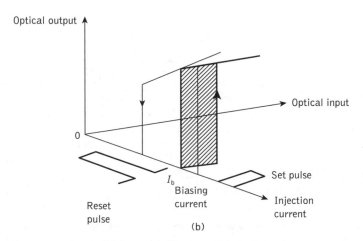

Figure 11.37 Bistable laser diode: (a) structure; (b) response characteristic

fabricated in GaAs/GaAlAs and InP/InGaAsP demonstrated nanosecond switching times with milliwatt power levels at room temperature [Ref. 122].

In conventional BODs with saturable absorbers it is difficult to implement the set/reset operation optically [Ref. 127]. However, using MMI couplers (see Section 11.4.1) it is possible to produce both set and reset optical functions on the same module. The structure of a device performing the basic operation of an all-optical flip-flop with set/rest functions is shown in Figure 11.38. It comprises a 2×2 MMI coupler with saturable absorbers and is capable of carrying out the all-optical digital flip-flop function. The length of the saturable absorber is 120 mm while the overall device length is around 1000 mm. Slow gain saturation, however, limits the device switching speed. Nevertheless, high transmission rates up to 40 Gbit s^{-1} have been demonstrated [Ref. 128].

In addition to conventional devices the concept of bistability can be applied to photonic crystal device technology and systems (see Section 2.6) where the inherent nonlinearity may be used for applications such as optical reshaping/retiming, optical switching and

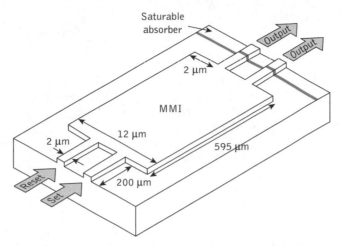

Figure 11.38 Structure of all-optical digital flip-flop based on a 2 × 2 multimode interference (MMI) coupler. Reprinted with permission from Ref. 128 © IEEE 2005

other logic functions [Ref. 129]. It has also been demonstrated experimentally that silicon-based photonic crystal nanocavities coupled to input and output waveguides can operate as on-chip optical bistable switching devices at quite low energies [Ref. 130]. Further development is still required, however, before such devices and systems become commercially available [Ref. 131].

The above BODs have been discussed primarily in relation to the provision of optical logic and memory elements. However, investigations of optical bistability have also included the other functions mentioned previously. Optical pulse shaping can be achieved using a BOD with a very narrow bistable loop. Such a device can be used to shape, clean up and amplify a noisy input pulse, as illustrated in Figure 11.39(a).

Nonlinear optical amplification can also be obtained with certain BODs. In particular, the Fabry–Pérot SOA can display dispersive bistability [Ref. 125] which, unlike its linear counterpart (see Section 10.3), provides a nonlinear gain characteristic, as shown in Figure 11.39(b). The optical amplification mechanism in this case can involve the interaction of at least two optical fields through the field-dependent dielectric constant of the non-linear material. The operation of such a BOD differential amplifier is also illustrated in Figure 11.39(b). The introduction of a weak second beam into the nonlinear optical cavity is used to control the resonance and transmission of the main beam through the additive effects of its own stored energy. Hence differential optical gain is provided by the device. With this configuration a weak beam can control an intense main beam producing the optical equivalent* of the electronic transistor.

Linear optical amplifiers exhibit the drawback of amplifying low-level noise signals together with the desired signal. Bistable amplifiers, however, are useful because of their signal regeneration capability [Ref. 132]. In the ideal case no amplification is provided for signals below a particular intensity level. Once an intensity threshold has been surpassed, the large gain can be determined by the slope of the curve in Figure 11.39(b). Moreover, a

---

* This two-beam optical transistor has been dubbed 'the transphasor' by the authors of Ref. 119.

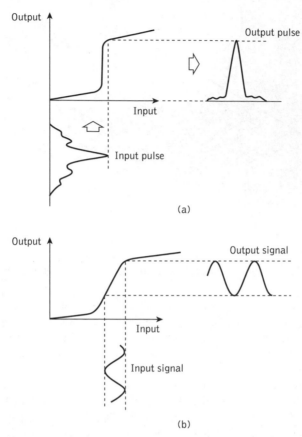

Figure 11.39 Illustration of two functions provided by the nonlinear characteristic of BODs: (a) optical pulse shaping; (b) optical amplification

saturated or maximum value of output intensity is also provided, displaying the power limiter function of the device.

Bistable optical devices are the main components for an optical gate performing digital logic functions which are considered to be the fundamental building blocks to provide for optical wavelength conversion and regeneration (see Sections 10.5 and 10.6). Mach–Zehnder interferometer (MZI) configurations using SOAs (see Section 10.3) and optical couplers are mainly employed to deliver the digital gate characteristics [Refs 133, 134]. Furthermore, it is possible to use SOAs in the MZI configuration with different combinations of clock pulses and input signals (i.e. clockwise and counterclockwise directions) to perform basic Boolean logic functions such as AND, NOT and OR operations. Three basic MZI configurations incorporating SOAs in each interferometer path to perform the nonlinear optical gate function are illustrated in Figure 11.40. For the AND gate operation the digital input signals A and B are applied to the interferometer in counterclockwise directions. In the presence of both digital signals the SOA in the upper path of the MZI produces a phase-modulated output as a result of cross-phase modulation (XPM)

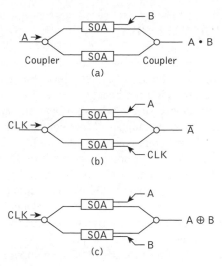

Figure 11.40 Digital optical gate using semiconductor optical amplifiers (SOAs) in an interferometric configuration to provide optical gate operation for: (a) an AND gate (A · B); (b) an inverter Ā; (c) an exclusive OR gate (A ⊕ B)

(see Section 3.14). This phase-modulated output signal is then converted into an amplitude-modulated signal at the output coupler after the digital signal A has traveled through the SOA in the lower path of the MZI.

Based on the same principle of the XPM in SOAs, the MZI arrangement in a counter-clockwise direction can produce different logic operations when the digital signals are applied from different positions. For example, an inverter gate operation (i.e. NOT gate) and an exclusive OR gate can be obtained as shown in Figure 11.40(b) and (c), respectively. Furthermore, it is possible to achieve all-optical signal processing and digital logical operations using different combinations of the digital inputs A and B and appropriate clock signals [Ref. 135]. Alternative interferometric configurations using SOAs can be employed to provide digital gate functions which include the Michelson interferometer [Ref. 136], the nonlinear optical loop mirror [Ref. 137] and the delayed interferometer [Ref. 138]. Apart from the SOA, the EAM can also perform the operation of an optical digital gate (see Section 10.5). Moreover, EAM devices have successfully provided optical digital gate and demultiplexing operations at transmission rates between 100 and 500 Gbit s^{-1} [Refs 139, 140].

All-optical logic gates using multibranch waveguide structures have also been demonstrated [Ref. 141]. In these structures only a small portion of the waveguide is configured to possess nonlinear characteristics instead of using an entire nonlinear waveguide. Thus the overall optical gate is confined to a smaller size as compared with conventional planar waveguides. In addition, photonic crystals have also been shown to exhibit useful characteristics to provide for all-optical logic switches and gates. Although the results obtained from experimental devices are promising, these structures based on a multibranch waveguide [Ref. 142] and photonic crystal waveguides (see Section 2.6) at present remain at a preliminary stage of development.

11.8 Optical computation

Although the maximum potential switching speeds of individual IO and IP logic devices have as yet to be accomplished, the use of parallel processing with optical signals mentioned in Section 11.1 can provide a net benefit over a similar serial electronic system, even at much slower optical device speeds. Conventional digital computers suffer from a bottleneck resulting from the limited number of interconnections which can be practically supported by an electronic-based communications technology. This restriction led to the classical von Neumann architecture for computing systems shown in Figure 11.41(a) in which the memory is addressed sequentially from the central processing unit (CPU). The CPU accesses the memory through a binary addressing unit and the memory contents are returned to the CPU via a small number of lines. This serial addressing of memory reduces the communications requirements and minimizes the number of lines, but this is achieved at the expense of overall computing speed. The problem, which is referred to as the von Neumann bottleneck, eventually limits the speed of the computer system.

With optical systems the situation is changed as they are capable of communicating many high-bandwidth channels in parallel without interference. Thus parallel communication can easily be provided within an optical computer system at relatively low cost. In theory this lends itself to the use of non-von Neumann architecture (see Figure 11.41(b)) in which all memory elements are accessible in parallel, thus removing the speed limitation caused by the bottleneck. The potential advantages offered by the digital optical computer are not therefore solely dependent upon the realization of subpicosecond optical switching devices.

For some time work in optical computation [Ref. 121] was directed towards particular requirements which were necessary to provide a practical optical computing system. These included:

(a) *High contrast*. Logic devices must exhibit a large change between logic 0 and logic 1 levels.

(b) *Steady-state bias*. To provide various different logic gates it is necessary that optical bias levels may be altered. For a BOD this implies that the device can be held

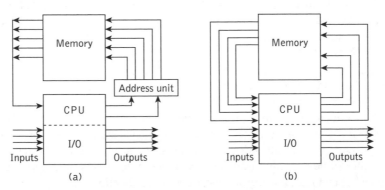

Figure 11.41 Computer architecture: (a) von Neumann; (b) non-von Neumann

indefinitely at any point on the characteristic with a CW laser beam. However, this holding beam necessitates a degree of thermal stability. Such stability was demonstrated with devices based on InSb at 77 K and on ZnSe at 300 K [Ref. 120].

(c) External addressing. The function of external addressing is to provide for separate external optical signals which can be combined with the holding beam to switch the device, thus giving logic functions. The switching energy can be derived from the holding beam which is then switched and propagates in transmission or reflection as the output beam to further devices in the optical circuit.

(d) Cascadability. The output from a particular device must be sufficient to switch at least one following device. This condition may be fulfilled by setting a holding beam near the switch point since the extra increment is then small in comparison with the change in output.

(e) Fan-out and fan-in. The advantage of parallel processing requires that a particular device has the ability to drive a large number of following devices. This could be achieved using free space propagation for addressing purposes. Furthermore, the summed effect of several elements could be focused onto one device to achieve fan-in.

(f) Gain. In order to maintain (d) and (e) above, there is a requirement for differential gain. This could possibly be achieved by the use of optical amplifier devices.

(g) Arrays. The easy construction of two dimensional (2-D) arrays within the technology must ideally be available.

(h) Speed and power. For 1-D circuits, subnanosecond or picosecond switching times are desirable, although this may be relaxed to microseconds for parallel arrays. Speed and power tend to be interchangeable but a low-power device is a necessity. The power requirements for a device should be in the milliwatt region or less.

Certain, although not all, of the above requirements are met by specific nonlinear devices described in Section 11.7. For example, it was suggested [Ref. 121] that the ZnSe interference devices, used as separate elements activated by external addresses, exhibit the possibilities of projection and display. Proposals for logic subsystems of this type are shown in Figure 11.42. In addition, Refs 121 and 143 indicate the possible arrangements for more complex optical logic subsystems, including a simple parallel processor, a serial to parallel converter, a shift register and a packet switch. Such proposals necessitate a solution which involves bulk optic, discrete elements together with possible monolithic IO and IP devices. Moreover, a hybrid approach for the implementation of high-speed switching matrices has been suggested [Ref. 144] which incorporates electronic logic elements with optical interconnections in order to exploit the best features of each technology.

In order to implement more complex optical logic subsystems, alternative nonlinear materials have been investigated including inorganic insulators and organic nonlinear compounds [Ref. 145]. Initial inorganic insulator materials such as lithium niobate, strontium barium niobate, bismuth silicon oxide and barium titanate, however, exhibited drawbacks in relation to poor thermal and mechanical properties, as well as slow response times (milliseconds). By contrast, the organic nonlinear materials listed in Table 11.2 have

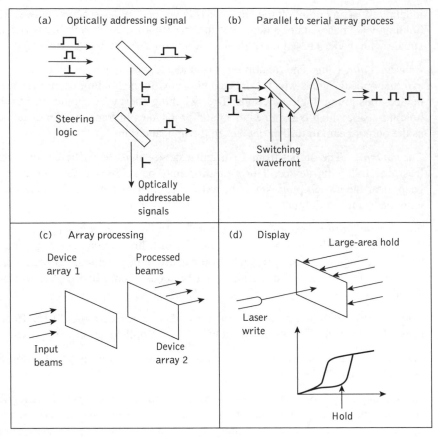

Figure 11.42 Proposals for logic subsystems based on the zinc selenide interference device, or similar [Ref. 121]

Table 11.2 Some organic nonlinear materials

Substituted and disubstituted acetylenes and diacetylenes
Anthracines and derivatives
Dyes
Macrocyclics
Polybenzimidazole
Polybenzimidazole and polybenzobisoxazole
Polyester and polyesteramids
Polyetherketone
Polyquinoxalines
Porphyrins and metal–porphyrin complexes
Metal complexes of TCNQ or TNAP
Urea

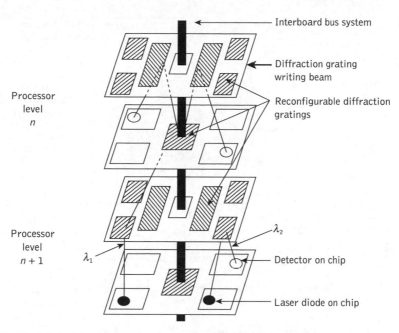

Figure 11.43 A hybrid optical/electronic multiprocessor architecture

displayed the potential for greater degrees of nonlinearity and much shorter response times. The major disadvantage with these materials is their relative environmental instability compared with inorganic materials (e.g. oxidation). However, work remains at a preliminary stage and more favorable results may be anticipated in the future [Ref. 146].

Success with such materials, together with further developments of the other optical devices mentioned in Section 11.7, could lead to the implementation of an all-optical computer. However, it is more likely that initially hybrid optical/electronic computational machines will evolve. For example, a multiprocessing system was developed, called the Connection Machine [Ref. 147], which comprised a large array of printed circuit boards, each containing 512 processing elements divided equally between 32 electronic chips. This particular concept is illustrated in Figure 11.43 where, for simplicity, only four chips per level are shown. It may be observed that each electronic board contained a frequency-selective filter (hologram) in addition to optoelectronic chips. These OEICs incorporated semiconductor lasers and photodiodes which enabled communications to be established between the chips. The switching operations for the interconnection process were performed by a planar array of reconfigurable diffraction gratings located on each board. Moreover, the architecture employed WDM to direct bit streams to the appropriate board as illustrated by λ_1 and λ_2 in Figure 11.43.

A possible all-optical digital multiprocessor architecture is shown in Figure 11.44 [Refs 145, 148]. The input to the machine is via either an array of independently address-able semiconductor lasers, or alternatively via a 2-D spatial light modulator (SLM). A laser array is capable of higher modulation speeds but necessitates more complex circuitry, especially when uniformity is required over the complete array. The gate array illustrated in Figure 11.44 comprises either another 2-D SLM with a nonlinear response

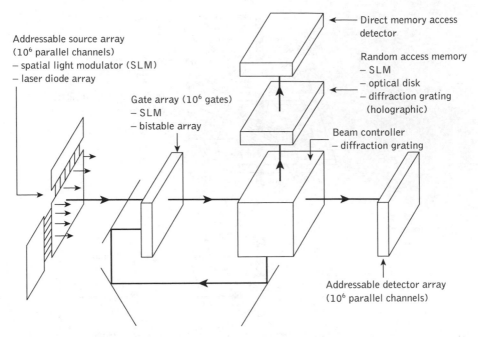

Figure 11.44 A possible all-optical multiprocessor architecture

or an array of BODs. In theory the BODs would provide much greater switching speed but exhibit the drawbacks mentioned in Section 11.6. The beam controller employs reconfigurable diffraction gratings in order to provide switching and interconnection. However, as a result of the large number of channels required in the all-optical computer it is likely that multiple planes of real-time hologram arrays would be utilized.

It may be observed in Figure 11.44 that all three computer interconnect systems (CPU–CPU, CPU–memory and CPU–I/O) are combined in the beam controller, although they could be implemented by three different components. Nevertheless, the ideal solution involves the beam controller directing any beam emerging from the gate array to any particular location on the detector array, the memory or the input of the gate array. Moreover, several logic elements can be interconnected via the beam controller to form a processing element. An example of a possible structure for a processing element, or node, in which individual elements in the gate array are designated to typical functions such as logic unit, clock, cache memory, etc., is illustrated in Figure 11.45. The example shown depicts a 5×5 rectangular array of logic elements, or gates, which gives 25 per processor. A practical multiprocessor might require 4×10^4 nodes giving 10^6 switching elements in the gate array.

A variant on the above architecture was developed by AT&T Laboratories which announced the first demonstration of a digital optical processor in 1990 [Ref. 149]. This processor, which operated at 10^6 cycles per second, is shown in schematic form in Figure 11.46. The hybrid bistable switching element was a GaAs/AlGaAs symmetric self-electro-optic effect device (S-SEED) which offered a potential of 10^9 operations per second with a switching energy of 1 pJ (see Section 11.6). The S-SEEDs which were 5 μm square

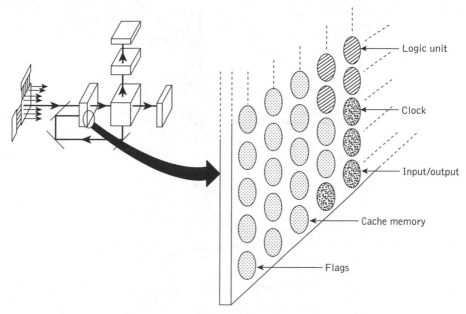

Figure 11.45 A processing element within the all-optical multiprocessor of Figure 11.44

and incorporated two mirrors with controllable reflectivity were formed into 32 device arrays. Each array also contained two 10 mW modulated injection laser diodes emitting at a wavelength of 0.85 μm as illustrated in Figure 11.46(a).

In the demonstration four S-SEED arrays were located within the processor, as may be obscrved in Figure 11.46(b). The injection lasers emitted many separate beams to provide communication between the arrays while each S-SEED drove two inputs. Interconnection between the four arrays was controlled by the lenses and masks, also shown in Figure 11.46. The masks comprised glass slides with patterns of transparent and opaque spots that allowed or impeded the transmission of light. Hence these patterns defined the conncctivity within the processor.

The processor logic was accomplished by each S-SEED operating as a NOR gate. Thus the output from each device array served as an input for the next array where the logic state of the S-SEEDs in the second array were determined by the state of the devices in the first array. Changing the on/off status of the switches in successive arrays allowed calculations to be performed. The memory resided in each S-SEED which did not change its state until the information represented by that state (i.e. a 0 or 1) was processed. In this way extensive pipelining of information was utilized within the processor (i.e. the output from one part of the machine served as the input for another part). Finally, the I/O was accomplished using both optical fibers and laser beams transmitted through free space.

Although research into optical computation has continued for several decades, its primary focus has been towards the development of nanophotonic devices and integrated circuits to provide computational functionality rather than the physical realization of an optical computer [Refs 150–153]. Therefore over the above period substantial progress

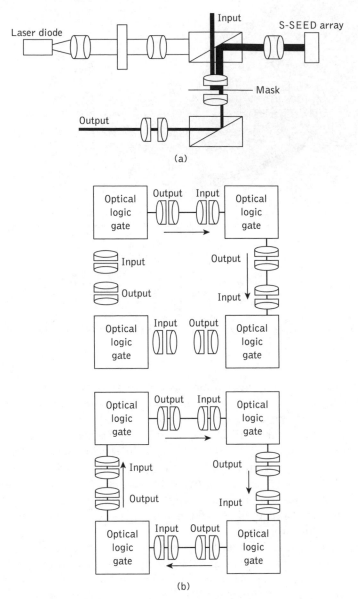

Figure 11.46 The AT&T digital optical processor: (a) structure of the processor S-SEED array module: (b) demonstration system using four array modules

has been made in relation to IO and IP device technology and the associated optical signal processing techniques. This progress has occurred as a result of the exploration of advanced materials including III–V semiconductors, polymers (see Section 11.2) and photonic crystal technology (see Section 2.6) [Refs 154, 155]. Moreover, the majority of the IO and IP devices using these new material technologies operate by employing the basic principles of nanophotonics and quantum physics. In this context optical computation has effectively become a branch of quantum computation [Ref. 156] where a number of IO and IP devices utilize quantum-dot, quantum-dash and quantum-well structures (see Section 10.4).

Quantum computation has progressed more quickly within nontelecommunication applications, however, where various devices and systems have been demonstrated, for example, in the fields of optics [Ref. 157], sensors [Refs 91, 158] and biotechnologies [Refs 159, 160]. Although it is clear that further developments in IO and IP (including digital optics) will have a direct impact on both the functionality and performance of future optical fiber communication systems and networks, it is less certain at what stage optical computers may become more fully embedded within telecommunication networks.

Problems

11.1 Briefly describe the waveguide structures employed for IO.

The normalized effective thickness H for an asymmetric slab guide is given by Eq. (11.3), obtain an expression for the minimum effective thickness in order to provide optical confinement.

Determine the minimum effective thickness for a lithium niobate IO waveguide structure which has film and substrate refractive indices of 2.1 and 2.0, respectively, when it is operated at a wavelength of 1.3 μm.

11.2 Calculate the minimum effective thickness to provide optical confinement in a III–V semiconductor compound IO waveguide operating at a wavelength of 1.3 μm which has film and substrate refractive indices of 3.5 and 2.7, respectively, at this wavelength.

Comment on the value obtained in comparison with that determined in Problem 11.1.

11.3 Outline the techniques that can be employed to provide directional coupling between waveguides with IO.

Commencing with Eq. (11.9), show that the power coupled from one waveguide to another when their propagation constants are equal is proportional to the factor $\sin^2(Cz)$. The boundary conditions $A(z = 0) = A(0)$ and $B(z = 0) = 0$ may be assumed.

11.4 Compare the voltages required to operate lithium niobate and III–V semiconductor compound strip waveguide phase modulators in order to produce a phase change of π radians when using 3 cm long devices with a distance between the electrodes of 30 μm. The electro-optic coefficients for lithium niobate and the III–V semiconductor compound may be taken as 30.8×10^{-12} mV^{-1} and 1.3×10^{-12} mV^{-1}, respectively, while the refractive indices are 2.1 and 3.1, respectively, at the operating wavelength of 1.3 μm.

Comment on the values for the voltages obtained.

11.5 Assuming that the waveguide size for the III–V semiconductor waveguide modulator of Problem 11.4 can be reduced by a factor of 5, determine the reduced voltage needed to obtain a phase shift of π radians.

Comment on the value calculated in comparison with that obtained for the lithium niobate phase modulator of Problem 11.4.

11.6   Explain the main features of implanted integrated optics technology. Discuss why it is called I3O and how it enables further miniaturization of optical components.

11.7   Describe with the aid of a sketch large-scale monolithic photonic integration using an asymmetrical twin-waveguide structure.

11.8   A lithium niobate phase modulator has $V_\pi L$ of 45 V mm. Determine the interaction length required so that an applied voltage of 10 V will produce a phase shift of 2π radians.

11.9   Explain the concept of a multimode interference coupler based on planar waveguides. Outline the use of photonic crystal technology in the construction of these devices.

11.10   A first-order InGaAsP/InP waveguide grating filter is required for operation at a wavelength of 1.56 µm. The reflected light is expected to be incident over an angle of 3°. Estimate the filter length required to provide a 3 dB bandwidth of 2.5 Å for the device.

11.11   With the aid of a sketch explain the construction and operating mechanism for OMEMS. Describe their applications in optical fiber telecommunication systems.

11.12   Discuss the function and operation of polarization transformers and wavelength converters with specific reference to coherent optical transmission.

11.13   Outline the importance of optoelectronic integration in relation to the future developments in optical fiber communications. Illustrate your answer with descriptions of OEICs which have been fabricated to provide optical transmitter, receiver and multiplexing functions.

11.14   Describe the generalized bistable optical device and mention the applications in which it is finding use. Indicate the primary reasons for the evolution of optical computational devices and discuss recent developments in this process which augur well for their future implementation.

11.15   Outline the main functions of an interleaver band filter and design the wavelength channel plan for a 128-channel, 16-band DWDM interleaver band filter for C-band (1528 to 1568 nm) operation with a channel spacing of 50 GHz for: (a) 8-skip-1 and (b) 8-skip-2; (c) calculate the bandwidth in each case and comment on the results obtained.

11.16   Describe the construction and operation of a monolithically integrated laser modulator based on a sampler grating distributed Bragg reflector.

11.17   Explain the significant aspects of a PIC and indicate using a sketch the active and passive building block strategies for its fabrication onto an indium phosphide platform.

11.18   Using appropriate diagrams outline the operation of digital optical gates employing semiconductor optical amplifiers in the Mach–Zehnder interferometric configuration for: (a) an AND gate; (b) an exclusive OR gate.

Answers to numerical problems

11.1   1.42 μm
11.2   0.41 μm
11.4   4.6 V, 33.6 V
11.5   6.7 V

11.8   9.0 mm
11.10   1.5 cm
11.15   (c) 56.8 nm and 62.8 nm

References

[1] D. B. Anderson, *Optical and Electrooptical Information Processing*, pp. 221–234, MII Press, 1965.

[2] S. E. Miller, 'Integrated optics: an introduction', *Bell Syst. Tech. J.*, **48**(7), pp. 2059–2069, 1969.

[3] H. V. Demir, V. A. Sabnis, O. Fidaner, J. F. Zheng, J. S. Harris Jr and D. A. B. Miller, 'Multifunctional integrated photonic switches', *IEEE J. Sel. Top. Quantum Electron.*, **11**(1), pp. 86–96, 2005.

[4] M. Raburn, B. Liu, K. Rauscher, Y. Okuno, N. Dagli and J. E. Bowers, '3-D photonic circuit technology', *IEEE J. Sel. Top. Quantum Electron.*, **8**(4), pp. 935–942, 2002.

[5] G. T. Reed and A. P. Knights, *Silicon Photonics: An Introduction*, Wiley, 2004.

[6] E. M. Yeatman, M. M. Ahmad, O. McCarthy, A. Martucci and M. Guglielmi, 'Sol-gel fabrication of rare-earth doped photonic components', *J. Sol-Gel Sci. Technol.*, **19**(1–3), pp. 231–236, 2000.

[7] W. Huang and R. R. A. Syms, 'Sol-gel silica-on-silicon buried-channel EDWAs', *J. Lightwave Technol.*, **21**(5), pp. 1339–1349, 2003.

[8] W. Bogaerts, R. Baets, P. Dumon, V. Wiaux, S. Beckx, D. Taillaert, B. Luyssaert, J. Van Campenhout, P. Bienstman and D. Van Thourhout, 'Nanophotonic waveguides in silicon-on-insulator fabricated with CMOS technology', *J. Lightwave Technol.*, **23**(1), pp. 401–412, 2005.

[9] S. M. Csutak, S. Dakshina Murthy and J. C. Campbell, 'CMOS-compatible planar silicon waveguide-grating-coupler photodetectors fabricated on silicon-on-insulator (SOI) substrates', *IEEE J. Quantum Electron.*, **38**(5), pp. 477–480, 2002.

[10] M. Loncar, T. Doll, J. Vuckovic and A. Scherer, 'Design and fabrication of silicon photonic crystal optical waveguides', *J. Lightwave Technol.*, **18**(10), pp. 1402–1411, 2000.

[11] L. Levi, *Applied Optics*, Vol. 2, Chapter 13, Wiley, 1980.

[12] H. Kogelnik and V. Ramaswamy, 'Scaling rules for thin-film optical waveguides', *Appl. Opt.*, **13**(8), pp. 1857–1862, 1974.

[13] A. Reisinger, 'Attenuation properties of optical waveguides with a metal boundary', *Appl. Phys. Lett.*, **23**(5), pp. 237–239, 1973.

[14] H. Kogelnik, 'Limits in integrated optics', *Proc. IEEE*, **69**(2), 232–238, 1981.

[15] T. Tamir (Ed.), *Integrated Optics* (2nd edn), Springer-Verlag, 1979.

[16] J. Wilson and J. F. B. Hawkes, *Optoelectronics: An Introduction* (2nd edn), Chapter 3, Prentice Hall, 1989.

[17] P. J. R. Laybourne and J. Lamb, 'Integrated optics: a tutorial review', *Radio Electron. Eng. (IERE J.)*, **51**(7/8), pp. 397–413, 1981.

[18] B. K. Nayar and R. C. Booth, 'An introduction to integrated optics', *Br. Telecom Technol. J.*, **4**(4), pp. 5–15, 1986.

[19] R. Th. Kersten, 'Integrated optics for sensors', in J. Dakin and B. Culshaw (Eds), *Optical Fiber Sensors: Principles and components*, Artech House, 1988.

[20] J. J. Veselka and S. K. Korotky, 'Optimization of Ti: $LiNbO_3$ optical waveguides and directional coupler switches for 1.5 μm wavelength, *IEEE J. Quantum Electron.*, **QE-22**, pp. 933–938, 1986.

[21] C. R. Doer, 'Planar lightwave devices for WDM', pp. 405–476; and A. Mahapatra and E. J. Murphy, 'Electrooptic modulators', in I. P. Kaminow and T. Li (Eds), *Optical Fiber Telecommunications IVA*, pp. 258–294, Academic Press, 2002.

[22] G. Bourdon, G. Alibert, A. Beguin, E. Guiot and R. Bellman, 'High-index contrast Ge-doped silica waveguide technology: optical performance and application to ultralow-loss ring resonators', *Proc. Optical Fiber Communications (OFC'03)*, Atlanta, Georgia, USA, **2**, ThD3, pp. 446–448, March 2003.

[23] V. M. Menon, F. Xia and S. R. Forrest, 'Photonic integration using asymmetric twin-waveguide (ATG) technology: part II – devices', *IEEE J. Sel. Top. Quantum Electron.*, **11**(1), pp. 30–42, 2005.

[24] F. Xia, V. M. Menon and S. R. Forrest, 'Photonic integration using asymmetric twin-waveguide (ATG) technology: part I – concepts and theory', *IEEE J. Sel. Top. Quantum Electron.*, **11**(1), pp. 17–29, 2005.

[25] W. Tong, V. M. Menon, F. Xia and S. R. Forrest, 'An asymmetric twin waveguide eight-channel polarization-independent arrayed waveguide grating with an integrated photodiode array', *IEEE Photonics Technol. Lett.*, **16**(4), pp. 1170–1172, 2004.

[26] G. B. Morrison, J. W. Raring, C. S. Wang, E. J. Skogen, Y. C. Chang, M. Sysak and L. A. Coldren, 'Electroabsorption modulator performance predicted from band-edge absorption spectra of bulk, quantum-well, and quantum-well-intermixed InGaAsP structures', *Solid State Electron.*, **51**(1), pp. 16–25, 2007.

[27] S. S. Agashe, K. T. Shiu and S. R. Forrest, 'Integratable high linearity compact waveguide coupled tapered InGaAsP photodetectors', *IEEE J. Quantum Electron.*, **43**(7), pp. 597–606, 2007.

[28] J. L. Zhang and T. F. Kuech, 'Fabrication and properties of an asymmetric waveguide containing nanoparticles', *J. Electron. Mater.*, **37**(2), pp. 135–144, 2008.

[29] K. C. Xin, K. S. Chiang and H. P. Chan, 'Broadband multiport dynamic optical power distributor based on thermooptic polymer waveguide vertical couplers', *IEEE Photonics Technol. Lett.*, **20**(4), pp. 273–275, 2008.

[30] H. Sasaki and I. Anderson, 'Theoretical and experimental studies on active Y-junctions in optical waveguides', *IEEE J. Quantum Electron.*, **QE-14**, pp. 883–892, 1978.

[31] D. Marcuse, 'The coupling of degenerate modes in two parallel dielectric waveguides', *Bell Syst. Tech. J.*, **50**(6), pp. 1791–1816, 1971.

[32] A. Yariv, 'Coupled mode theory for guided wave optics', *IEEE J. Quantum Electron.*, **QE-9**, pp. 919–933, 1973.

[33] M. Papuchon, Y. Combemale, X. Mathieu, D. B. Ostrowsky, L. Reiber, A. M. Roy, B. Sejourne and M. Werner, 'Electrically switched optical directional coupler: COBRA', *Appl. Phys. Lett.*, **27**(5), pp. 289–291, 1975.

[34] L. B. Soldano, F. B. Veerman, M. K. Smit, B. H. Verbeek, A. H. Dubost and E. C. M. Pennings, 'Planar monomode optical couplers based on multimode interference effects', *J. Lightwave Technol.*, **10**(12), pp. 1843–1850, 1992.

[35] J. M. Heaton and R. M. Jenkins, 'General matrix theory of self-imaging in multimode interference (MMI) couplers', *IEEE Photonics Technol. Lett.*, **11**(2), pp. 212–214, 1999.

[36] J. Leuthold and C. W. Joyner, 'Multimode interference couplers with tunable power splitting ratios', *J. Lightwave Technol.*, **19**(5), pp. 700–707, 2001.

[37] M. P. Earnshaw and D. W. E. Allsopp, 'Semiconductor space switches based on multimode interference couplers', *J. Lightwave Technol.*, **20**(4), pp. 643–650, 2002.

[38] M. T. Hill, X. J. M. Leijtens, G. D. Khoe and M. K. Smit, 'Optimizing imbalance and loss in 2×2 3-dB multimode interference couplers via access waveguide width', *J. Lightwave Technol.*, **21**(10), pp. 2305–2313, 2003.

[39] I. Molina-Fernandez, J. G. Wanguemert-Perez, A. Ortega-Monux, R. G. Bosisio and K. Wu, 'Planar lightwave circuit six-port technique for optical measurements and characterizations', *J. Lightwave Technol.*, **23**(6), pp. 2148–2157, 2005.

[40] T. Liu, A. R. Zakharian, M. Fallahi, J. V. Moloney and M. Mansuripur, 'Multimode interference-based photonic crystal waveguide power splitter', *J. Lightwave Technol.*, **22**(12), pp. 2842–2846, 2004.

[41] P. I. Borel, L. H. Frandsen, A. Harpoth, M. Kristensen, J. S. Jensen and O. Sigmund, 'Topology optimised broadband photonic crystal Y-splitter', *Electron. Lett.*, **41**(2), pp. 69–71, 2005.

[42] Y. Tanaka, H. Nakamura, Y. Sugimoto, N. Ikeda, K. Asakawa and K. Inoue, 'Coupling properties in a 2-D photonic crystal slab directional coupler with a triangular lattice of air holes', *IEEE J. Quantum Electron.*, **41**(1), pp. 76–84, 2005.

[43] M. Shirane, A. Gomyo, K. Miura, Y. Ohtera, H. Yamada and S. Kawakami, 'Optical add-drop multiplexers based on autocloned photonic crystals', *IEEE J. Sel. Areas Commun.*, **23**(7), pp. 1372–1377, 2005.

[44] D. Yang, Y. Li, F. Sun, S. Chen and J. Yu, 'Fabrication of a 4 × 4 strictly nonblocking SOI switch matrix', *Optics Commun.*, **250**(1–3), pp. 48–53, 2005.

[45] T. Goh, A. Himeno, M. Okuno, H. Takahashi and K. Hattori, 'High-extinction ratio and low-loss silica-based 8 × 8 strictly nonblocking thermooptic matrix switch', *J. Lightwave Technol.*, **17**(7), pp. 1192–1199, 1999.

[46] A. Chiba, T. Kawanishi, T. Sakamoto, K. Higuma and M. Izutsu, 'High-extinction-ratio (>55 dB) port selection by using a high-speed LiNbO₃ optical switch with intensity trimmers', *Proc. IEEE-LEOS Photon. Switch.*, San Francisco, USA, Tab1.4, pp. 51–52, August 2007.

[47] C. T. Lea, B. C. Lin and H. Mounir, 'An architecture for TSI-free nonblocking optical TDM switches', *J. Lightwave Technol.*, **25**(3), pp. 694–702, 2007.

[48] G. D. Kim and S. S. Lee, 'Photonic microwave channel selective filter incorporating a thermooptic switch based on tunable ring resonators', *IEEE Photon. Technol. Lett.*, **19**(13), pp. 1008–1010, 2007.

[49] N. Y. Han, M. R. Wang, L. Daqun, X. Wang, J. Martinez, R. R. Panepucci and K. Pathak, '1 × 4 Wavelength reconfigurable photonic switch using thermally tuned microring resonators fabricated on silicon substrate', *IEEE Photon. Technol. Lett.*, **19**(9), pp. 7704–7-6, 2007.

[50] B. G. Lee, A. Biberman, D. Po, M. Lipson and K. Bergman, 'All-Optical comb switch for multiwavelength message routing in silicon photonic networks', *IEEE Photon. Technol. Lett.*, **20**(10), pp. 767–769, 2008.

[51] J. Yang, N. K. Fontaine, Z. Pan, A. O. Karalar, S. S. Djordjevic, C. Yang, W. Chen, S. Chu, B. E. Little and S. J. B. Yoo, 'Continuously tunable, wavelength-selective buffering in optical packet switching networks', *IEEE Photon. Technol. Lett.*, **20**(12), pp. 1030–1032, 2008.

[52] J. L. Leclercq, M. Garrigues, X. Letartre, C. Seassal and P. Viktorovitch, 'InP-based MOEMS and related topics', *J. Micromech. Microeng.*, **10**(2), pp. 287–292, 2000.

[53] A. Amarendra, 'Analysis of out-of-plane thermal microactuators', *J. Micromech. Microeng.*, **16**(2), pp. 205–213, 2006.

[54] D. V. Plant, M. B. Venditti, E. Laprise, J. Faucher, K. Razavi, M. Chateauneuf, A. G. Kirk and J. S. Ahearn, '256-channel bidirectional optical interconnect using VCSELs and photodiodes on CMOS', *J. Lightwave Technol.*, **19**(8), pp. 1093–1103, 2001.

[55] V. A. Aksyuk *et al.*, 'Beam-steering micromirrors for large optical cross-connects', *J. Lightwave Technol.*, **21**(3), pp. 634–642, 2003.

[56] L. Thylen, U. Westergren, P. Holmstrom, R. Schatz and P. Janes, 'Recent development in high-speed optical modulators', in I. P. Kaminow, T. Li and A. E. Willner (Eds), *Optical Fiber Telecommunications VA*, pp. 183–220, Academic Press, 2008.

[57] D. B. Ostrowsky, 'Optical waveguide components', in M. J. Howes and D. V. Morgan (Eds), *Optical Fibre Communications*, pp. 165–188, John Wiley, 1980.

[58] S. M. Garner and S. Caracci, 'Variable optical attenuator for large-scale integration', *IEEE Photonics Technol. Lett.*, **14**(11), pp. 1560–1562, 2002.

[59] T. Hurvitz, S. Ruschin, D. Brooks, G. Hurvitz and E. Arad, 'Variable optical attenuator based on ion-exchange technology in glass', *J. Lightwave Technol.*, **23**(5), pp. 1918–1922, 2005.

[60] M. L. Maurício and V. S. Paulo, 'Modulation of photonic structures by surface acoustic waves', *Rep. Prog. Phys.*, **68**(7), pp. 1639–1701, 2005.

[61] R. C. Alferness, C. H. Joyner, M. D. Divino and L. L. Buhl, 'InGaAsP/InP, waveguide grating filters for $\lambda = 1.5$ μm', *Appl. Phys. Lett.*, **45**, pp. 1278–1280, 1984.

[62] C. Y. Huang, C. H. Lin, Y. H. Chen and Y. C. Huang, 'Electro-optic Ti:PPLN waveguide as efficient optical wavelength filter and polarization mode converter'. *Opt. Express.*, **15**(5), pp. 2548–2554, 2007.

[63] A. Katzir, A. C. Livanos, J. B. Shellan and A. Yariv, 'Chirped gratings in integrated optics', *IEEE J. Quantum Electron.*, **QE-13**(4), pp. 296–304, 1977.

[64] P. Peloso, M. Prunaire, L. Noirie and D. Penninckx, 'Applying optical transparency to a heterogeneous pan-European network', *Proc. Optical Fiber Communications (OFC'03)*, Atlanta, Georgia, USA, **3**, PD10, pp. 1–3, March, 2003.

[65] S. Cao, J. Chen, J. N. Damask, C. R. Doerr, L. Guiziou, G. Harvey, Y. Hibino, H. Li, S. Suzuki, K. Y. Wu and P. Xie, 'Interleaver technology: comparisons and applications requirements', *J. Lightwave Technol.*, **22**(1), pp. 281–289, 2004.

[66] S. Chandrasekhar, C. R. Doerr and L. L. Buhl, 'Flexible waveband optical networking without guard bands using novel 8-skip-0 banding filters', *IEEE Photonics Technol. Lett.*, **17**(3), pp. 579–581, 2005.

[67] A. E. Willner, S. M. R. M. Nezam, L. Yan, Z. Pan and M. C. Hauer, 'Monitoring and control of polarization-related impairments in optical fiber systems', *J. Lightwave Technol.*, **22**(1), pp. 106–125, 2004.

[68] Q. Yu, L. S. Yan, S. Lee, Y. Xie and A. E. Willner, 'Loop-synchronous polarization scrambling technique for simulating polarization effects using recirculating fiber loops', *J. Lightwave Technol.*, **21**(7), pp. 1593–1600, 2003.

[69] S. Kieckbusch, S. Ferber, H. Rosenfeldt, R. Ludwig, C. Boerner, A. Ehrhardt, E. Brinkmeyer and H. G. Weber, 'Automatic PMD compensator in a 160-Gb/s OTDM transmission over deployed fiber using RZ-DPSK modulation format', *J. Lightwave Technol.*, **23**(1), pp. 165–171, 2005.

[70] R. V. Alferness and L. L. Buhl, 'Waveguide electro-optic polarization transformer', *Appl. Phys. Lett.*, **38**(9), pp. 655–657, 1981.

[71] W. A. Stallard, D. J. T. Heatley, R. A. Lobbett, A. R. Beaumont, D. J. Hunkin, B. E. Daymond-John, R. C. Booth and G. R. Hill, 'Electro-optic frequency translators and their application in coherent optical fibre systems', *Br. Telecom Technol. J.*, **4**(4), pp. 16–22, 1986.

[72] K. K. Wong, R. De La Rue and S. Wright, 'Electro-optic waveguide frequency translator in LiNbO$_3$ fabricated by proton exchange', *Opt. Lett.*, **7**(11), pp. 546–548, 1982.

[73] F. Heismann and R. Ulrich, 'Integrated optical frequency translator with stripe waveguide', *Appl. Phys. Lett.*, **45**(5), pp. 490–492, 1984.

[74] F. Auracher and R. Keil, 'Method for measuring the RF modulating characteristics of Mach–Zehnder-type modulators', *Appl. Phys. Lett.*, **36**(8), pp. 626–629, 1980.

[75] M. Y. Jamro and J. M. Senior, 'Modelling chirp and phase inversion in wavelength converters based on symmetrical MZI-SOAs for use in all-optical networks', *Photonics Netw. Commun.*, **5**(3), pp. 289–300, 2003.

[76] A. Ramaswamy, L. A. Johansson, J. Klamkin, H. F. Chou, C. Sheldon, M. J. Rodwell, L. A. Coldren and J. E. Bowers, 'Integrated coherent receivers for high-linearity microwave photonic links', *J. Lightwave Technol.*, **26**(1), pp. 209–216, 2008.

[77] Y. Furukawa, H. Yonezu and A. Wakahara, 'Monolithic integration of III-V active devices into silicon platform for optoelectronic integrated circuits', *IEICE Trans. Electron.*, **E91**(2), pp. 145–149, 2008.

[78] Y. Zhang, C. S. Whelan, R. Leoni, P. F. Marsh, W. E. Hoke, J. B. Hunt, C. M. Laighton and T. E. Kazior, '40-Gbit/s OEIC on GaAs substrate through metamorphic buffer technology', *Electron. Lett.*, **24**(9), pp. 529–531, 2003.

[79] A. Klehr, J. Fricke, A. Knauer, G. Erbert, M. Walther, R. Wilk, M. Mikulics and M. Koch, 'High-power monolithic two-mode DFB laser diodes for the generation of THz radiation', *IEEE J. Sel. Top. Quantum Electron.*, **14**(2), pp. 289–294, 2008.

[80] L. A. Johansson, Y. A. Akulova, G. A. Fish and L. A. Coldren, 'Widely tunable EAM-integrated SGDBR laser transmitter for analog applications', *IEEE Photonics Technol. Lett.*, **15**(9), pp. 1285–1287, 2003.

[81] C. Meliani, M. Rudolph, J. Hilsenbeck and W. Heinrich, 'A 40 Gbps broadband amplifier for modulator-driver applications using a GaAs HBT technology', *Proc. Bipolar/BiCMOS Circuits and Technology Meet. Dig.*, Montreal, pp. 281–284, 2004.

[82] Y. K. Fukai, K. Kurishima, M. Ida, S. Yamahata and T. Enoki, 'Highly reliable InP-based HBTs with a ledge structure operating at high current density', *Electron. Commun. Jpn (Pt II: Electron.)*, **90**(4), pp. 1–8, 2007.

[83] M. Galarza, K. De Mesel, S. Verstuyft, C. Aramburu, M. Lopez-Amo, I. Moerman, P. Van Daele, and R. Baets, 'A new spot-size converter concept using fiber-matched antiresonant reflecting optical waveguides', *J. Lightwave Technol.*, **21**(1), pp. 269–274, 2003.

[84] G. G. Mekonnen, H. G. Bach, A. Beling, R. Kunkel, D. Schmidt and W. Schlaak, '80-Gb/s InP-based waveguide-integrated photoreceiver', *IEEE J. Sel. Top. Quantum Electron.*, **11**(2), pp. 356–360, 2005.

[85] EDWA: Amplified splitter series, Teem photonics®, http://www.teemphotonics.com/assets/files/PDF/1_4_1_8.pdf, 20 June 2008.

[86] http://www.photonics.dupont.com/downloads/ROADM_Overview.pdf, 17 October 2007.

[87] L. Eldada, J. Fujita, A. Radojevic, T. Izuhara, R. Gerhardt, J. Shi, D. Pant, F. Wang and A. Malek, '40-channel ultra-low-power compact PLC-based ROADM subsystem', *Conf. Proc. on Optical Fiber Communication and Natl Fiber Optic Engineering (OFC/NFOE'06)*, Anaheim, California, USA, pp. NThC4,1-9, March, 2006.

[88] M. T. Gale, C. Gimkiewicz, S. Obi, M. Schnieper, J. Söchtig, H. Thiele and S. Westenhöfer, 'Replication technology for optical microsystems', *Opt. Lasers Eng.*, **43**(3–5), pp. 373–386, 2005.

[89] J. T. Kim, J. J. Ju, S. Park and M. H. Lee, 'O/E integration of polymer waveguide devices by using replication technology', *IEEE J. Sel. Top. Quantum Electron.*, **13**(2), pp. 177–184, 2007.

[90] C. Gimkiewicz, C. Zschokke, S. Obi, C. Urban, J. S. Pedersen, M. T. Gale and M. Moser, 'Wafer-scale replication of optical components on VCSEL wafers', *Proc. Optical Fiber Communications (OFC'04)*, Los Angeles, USA, Vol. 1, 23–27 February 2004.

[91] L. Schares, J. A. Kash, F. E. Doany, C. L. Schow, C. Schuster, D. M. Kuchta, P. K. Pepeljugoski, J. M. Trewhella, C. W. Baks and R. A. John, 'Terabus: terabit/second-class card-level optical interconnect technologies', *IEEE J. Sel. Top. Quantum Electron.*, **12**(5), pp. 1032–1044, 2006.

[92] J. W. Raring and L. A. Coldren, '40-Gb/s widely tunable transceivers', *IEEE J. Sel. Top. Quantum Electron.*, **13**(1), pp. 3–14, 2007.

[93] H. Wei-Ping, L. Xun, X. Chang-Qing, H. Xiaobin, X. Chenglin and L. Wanguo, 'Optical transceivers for fiber-to-the-premises applications: system requirements and enabling technologies', *J. Lightwave Technol.*, **25**(1), pp. 11–27, 2007.

[94] Fujitsu® Optical transceiver for PON, http://www.fujitsu.com/global/services/telecom/optcompo/lineup/ompon/index.html, 20 June 2008.

[95] Intel® Optical transceivers (telecom), http://www.intel.com/design/network/products/optical/tel_transceivers.htm, 20 June 2008.

[96] Titan Photonics® G-PON SFF ONU transceiver, http://www.titanphotonics.com/files/PDF_074_GPON_SFF_ONU_1.244G_2.488G_Transceiver.pdf, 20 June 2008.

[97] M. Nogami and J. Nakagawa, 'Optical transceiver for optical access systems', *Adv. Tech. Rep. Mitsubishi Electric, Japan*, **6**, pp. 14–16, 2006.

[98] DenseLight Semiconductors® G.984 compliant GPON-ONU transceiver, http://www.denselight.com/products%20gpon%20overview.htm, 20 June 2008.

[99] SmartOptics® 4G and 10G Fibre channel transceivers, http://www.zycko.com/solutions/partners/s/Smartoptics/transceivers.asp, 20 June 2008.

[100] Opnext® Optical transceivers 40 Gbit/s SR transponder (SerDes Transceiver), http://www.msc-ge.com/frame/prodsearch/e/las/seli_optmod_TypeID/10.html, 20 June 2008.

[101] C. G. P. Herben, D. H. P. Maat, X. J. M. Leijtens, M. R. Leys Y. S. Oei and M. K. Smit, 'Polarization independent dilated WDM cross-connect on InP', *IEEE Photonics Technol. Lett.*, **11**(12), pp. 1599–1601, 1999.

[102] K. Sato, 'Semiconductor light sources for 40-Gb/s transmission systems', *J. Lightwave Technol.*, **20**(12), pp. 2035–2043, 2002.

[103] J. H. den Besten, R. G. Broeke, M. van Geemert, J. J. M. Binsma, F. Heinrichsdorff, T. van Dongen, E. A. J. M. Bente, X. J. M. Leijtens and M. K. Smit, 'An integrated 4×4-channel multiwavelength laser on InP', *IEEE Photonics Technol. Lett.*, **15**(3), pp. 368–370, 2003.

[104] M. L. Masanovic, V. Lal, J. A. Summers, J. S. Barton, E. J. Skogen, L. G. Rau, L. A. Coldren and D. J. Blumenthal, 'Widely tunable monolithically integrated all-optical wavelength converters in InP', *J. Lightwave Technol.*, **23**(3), pp. 1350–1362, 2005.

[105] P. Bernasconi, L. Zhang, W. Yang, N. Sauer, L. L. Buhl, J. H. Sinsky, I. Kang, S. Chandrasekhar and D. T. Neilson, 'Monolithically integrated 40-Gb/s switchable wavelength converter', *J. Lightwave Technol.*, **24**(1), pp. 71–75, 2006.

[106] S. C. Lee, R. Varrazza and S. Yu, 'Automatic per-packet dynamic power equalisation in a 4 × 4 active vertical coupler optical crosspoint switch matrix', *Proc. Eur. Conf. on Optical Communication (ECOC'05)*, Glasgow, UK, pp. 331–332, September 2005.

[107] P. Bernasconi, L. Stulz, J. Bailey, M. Cappuzzo, E. Chen, L. Gomez, E. Laskowski, R. Long and A. Wong-Foy, 'N × N arrayed waveguide gratings with improved frequency accuracy', *IEEE J. Sel. Top. Quantum Electron.*, **8**(6), pp. 1115–1121, 2002.

[108] F. Wang, K. Chen, W. Sun, H. Zhang, C. Ma, M. Yi, S. Liu and D. Zhang, '32-channel arrayed waveguide grating multiplexer using low loss fluorinated polymer operating around 1550 nm', *Opt. Commun.*, **259**(2), pp. 665–669, 2006.

[109] F. Xiao, G. Li and A. Xu, 'Modeling and design of irregularly arrayed waveguide gratings', *Opt. Express*, **15**(7), pp. 3888–3901, 2007.

[110] M. Ohtsu, K. Kobayashi, T. Kawazoe, S. Sangu and T. Yatsui, 'Nanophotonics: design, fabrication, and operation of nanometric devices using optical near fields', *IEEE J. Sel. Top. Quantum Electron.*, **8**(4), pp. 839–862, 2002.

[111] A. Yehia and D. Khalil, 'Design of a compact three-dimensional multimode interference phased array structures (3-D MMI PHASAR) for DWDM applications', *IEEE J. Sel. Top. Quantum Electron.*, **11**(2), pp. 444–451, 2005.

[112] R. Nagarajan *et al.*, '400 Gbit/s (10 channel × 40 Gbit/s) DWDM photonic integrated circuits', *Electron. Lett.*, **41**(6), pp. 347–349, 2005.

[113] D. F. Welch *et al.*, 'Large-scale InP photonic integrated circuits: enabling efficient scaling of optical transport networks', *IEEE J. Sel. Top. Quantum Electron.*, **13**(1), pp. 22–31, 2007.

[114] N. Kikuchi, Y. Shibata, H. Okamoto, Y. Kawaguchi, S. Oku, Y. Kondo and Y. Tohmori, 'Monolithically integrated 100-channel WDM channel selector employing low-crosstalk AWG', *IEEE Photonics Technol. Lett.*, **16**(11), pp. 2481–2483, 2004.

[115] W. K. Choi, D. G. Kim, D. G. Kim, Y. W. Choi, K. D. Choquette, S. Lee and D. H. Woo, 'Optical AND/OR gates based on monolithically integrated vertical cavity laser with depleted optical thyristor structure', *Opt. Express.*, **14**(24), pp. 11833–11838, 2006.

[116] P. W. Smith, I. P. Kaminow, P. J. Maloney and L. W. Stulz, 'Integrated bistable optical devices', *Appl. Phys. Lett.*, **33**(1), pp. 24–26, 1978.

[117] D. A. B. Miller, 'Multiple quantum well optical non-linearities: bistability from increasing absorption and the self-electro-optic effect', *Philos. Trans. R. Soc.*, **A313**, pp. 239–248, 1984.

[118] D. Jager, F. Forsman and H. C. Zhai, 'Hybrid optical bistability based on increasing absorption in depletion layer of an Si Schottky SEED device', *Electron. Lett.*, **23**(10), 490–491, 1987.

[119] A. C. Walker, F. A. P. Tooley, M. E. Prise, J. G. H. Mathew, A. K. Kar, M. R. Taghizadeh and S. D. Smith, 'InSb devices: transphasors with high gain, bistable switches and sequential logic gates', *Philos. Trans. R. Soc.*, **A313**, pp. 249–256, 1984.

[120] S. D. Smith, J. G. H. Mathew, M. R. Taghizadeh, A. C. Walker, B. S. Wherret and A. Henry, 'Room temperature, visible wavelength optical bistability in ZnSe interference filters', *Opt. Commun.*, **51**, pp. 357–362, 1984.

[121] S. D. Smith, 'Optical bistability, phononic logic and optical computation', *Appl. Opt.*, **25**, pp. 1550–1564, 1986.

[122] Yosia and P. Shum, 'Optical bistability in periodic media with third-, fifth-, and seventh-order nonlinearities', *J. Lightwave Technol.*, **25**, 875–882, 2007.

[123] N. Peyghambarian, 'Recent advances in optical bistability', *Fiber Integr. Opt.*, **6**(2), pp. 117–123, 1987.

[124] P. Li Kam Wa, J. W. Sitch, N. J. Mason, J. S. Roberts and P. N. Robson, 'All-optical multiple-quantum well waveguide switch', *Electron. Lett.*, **21**, pp. 26–27, 1985.

[125] J. G. McInerney and D. M. Heffernan, 'Optical bistability in semiconductor injection lasers', *IEE Proc., Optoelectron.*, **134**(1), pp. 41–50, 1987.

[126] Y. Odagiri, K. Komastu and S. Suzuki, 'Bistable laser diode memory for optical time-division switching applications', *Int. Conf. on Lasers and Electro-optics*, Anaheim, California, USA, paper ThJ3, 1984.

[127] Y. Hong, R. Ju, P. S. Spencer and K. A. Shore, 'Investigation of polarization bistability in vertical-cavity surface-emitting lasers subjected to optical feedback', *IEEE J. Quantum Electron.*, **41**(5), pp. 619–624, 2005.

[128] M. Takenaka, M. Raburn and Y. Nakano, 'All-optical flip-flop multimode interference bistable laser diode', *IEEE Photonics Technol. Lett.*, **17**(5), pp. 968–970, 2005.

[129] M. Soljacic, M. Ibanescu, S. G. Johnson, Y. Fink and J. D. Joannopoulos, 'Optimal bistable switching in nonlinear photonic crystals', *Phys. Rev. E*, **66**(5), pp. 055601–4, 2002.

[130] E. P. Barclay, K. Srinivasan and O. Painter, 'Nonlinear response of silicon photonic crystal microresonators excited via an integrated waveguide and fiber taper', *Opt. Express*, **1**(3), pp. 801–820, 2005.

[131] M. Notomi, A. Shinya, S. Mitsugi, G. Kira, E. Kuramochi and T. Tanabe, 'Optical bistable switching action of Si high-Q photonic-crystal nanocavities', *Opt. Express*, **13**(7), pp. 2678–2687, 2005.

[132] Y. Silberberg, 'All-optical repeater', *Opt. Lett.*, **11**(6), pp. 392–394, 1986.

[133] C. Bintjas, K. Vlachos, N. Pleros and H. Avramopoulos, 'Ultrafast nonlinear interferometer (UNI)-based digital optical circuits and their use in packet switching', *J. Lightwave Technol.*, **21**(11), pp. 2629–2637, 2003.

[134] H. J. S. Dorren, X. Yang, A. K. Mishra, Z. Li, H. Ju, H. de Waardt, G. D. Khoe, T. Simoyama, H. Ishikawa, H. Kawashima and T. Hasama, 'All-optical logic based on ultrafast gain and index dynamics in a semiconductor optical amplifier', *IEEE J. Sel. Top. Quantum Electron.*, **10**(5), pp. 1079–1092, 2004.

[135] T. Houbavlis *et al.*, 'All-optical signal processing and applications within the ESPRIT project DO_ALL', *J. Lightwave Technol.*, **23**(2), pp. 781–801, 2005.

[136] K. E. Stubkjaer, 'Semiconductor optical amplifier-based all-optical gates for high-speed optical processing', *IEEE J. Sel. Top. Quantum Electron.*, **6**(6), pp. 1428–1435, 2000.

[137] A. Bogoni, L. Poti, R. Proietti, G. Meloni, F. Ponzini and P. Ghelfi, 'Regenerative and reconfigurable all-optical logic gates for ultra-fast applications', *Electron. Lett.*, **41**(7), pp. 435–436, 2005.

[138] S. Randel, A. M. de Melo, K. Petermann, V. Marembert and C. Schubert, 'Novel scheme for ultrafast all-optical XOR operation', *J. Lightwave Technol.*, **22**(12), pp. 2808–2815, 2004.

[139] S. Kodama, T. Yoshimatsu and H. Ito, '500 Gbit/s optical gate monolithically integrating photodiode and electroabsorption modulator', *Electron. Lett.*, **40**(9), pp. 555–556, 2004.

[140] T. Yoshimatsu, S. Kodama, K. Yoshino and H. Ito, '100-gb/s error-free wavelength conversion with a monolithic optical gate integrating a photodiode and electroabsorption modulator', *IEEE Photonics Technol. Lett.*, **17**(11), pp. 2367–2369, 2005.

[141] Y. D. Wu, M. H. Chen and C. H. Chu, 'All-optical logic device using bent nonlinear tapered Y-junction waveguide structure', *J. Fiber Integr. Opt.*, **20**(5), pp. 517–524, 2001.

[142] Y. D. Wu, 'All-optical logic gates by using multibranch waveguide structure with localized optical nonlinearity', *IEEE J. Sel. Top. Quantum Electron.*, **11**(2), pp. 307–312, 2005.

[143] J. E. Midwinter, 'Light electronics, myth or reality', *IEE Proc., Optoelectron.*, **132**, pp. 371–383, 1985.

[144] J. E. Midwinter, 'Novel approach to the design of optically activated wide band switching matrices', *IEE Proc., Optoelectron.*, **134**(5), pp. 261–268, 1987.

[145] J. A. Neff, 'Major initiatives for optical computing', *Opt. Eng.*, **26**(1), pp. 2–9, 1987.

[146] N. S. Sariciftci and S. S. Sun, *Organic Photovoltaics: Mechanism, Materials, and Devices*, CRC Press, 2005.

[147] W. D. Hillis, *The Connection Machine*, MIT Press, 1985

[148] B. K. Jenkins, P. Chavel, R. Forchheimer, A. A. Sawchuk and T. C. Strand', Architectural implications of a digital optical processor', *Appl. Opt.*, **23**(19), pp. 3465–3474, 1984.

[149] 'Optical computer: is concept becoming reality', *OE Rep., SPIE*, **75**, pp. 1–2, March 1990.

[150] L. J. Irakliotis and P. A. Mitkas, 'Optics: a maturing technology for better computing', *Computer*, **31**(2), pp. 36–37, 1998.

[151] J. Shamir, 'Reconsidering the concepts of optical computing', *J. Opt. Memory Neural Netw.*, **10**(1), pp. 1–11, 2000.

[152] M. Ohtsu and K. Kobayashi, *Optical Near Fields: Introduction to Classical and Quantum Theories of Electromagnetic Phenomena at the Nanoscale*, Springer-Velag, 2004.

[153] H. M. A. Kelash, H. S. Soror and N. F. El-Halafawy, 'High-speed optical computing devices based on VCSEL diodes with feedback', *4th Workshop on Photonics and Its Application*, Giza, Egypt, pp. 36–43, May 2004.

[154] K. L. Y. Pan and S. Q. Zheng, *Parallel Computing Using Optical Interconnections*, Kluwer Academic, 1998.

[155] T. Tanabe, K. Nishiguchi, A. Shinya, E. Kuramochi, H. Inokawa, M. Notomi, K. Yamada, T. Tsuchizawa, T. Watanabe and H. Fukuda, 'Fast all-optical switching using ion-implanted silicon photonic crystal nanocavities', *Appl. Phys. Lett.*, **90**(1), pp. 031115, 1–3, 2007.

[156] L. A. B. Windover, J. N. Simon, S. A. Rosenau, K. S. Giboney, G. M. Flower, L. W. Mirkarimi, A. Grot, B. Law, C. K. Lin, A. Tandon, R. W. Gruhlke, H. Xia, G. Rankin, M. R. T. Tan and D. W. Dolfi, 'Parallel-optical interconnects >100 Gb/s', *J. Lightwave Technol.*, **22**(9), pp. 2055–2063, 2004.

[157] H. Rigneault, J. M. Lourtioz, C. Delalande and J. A. Levenson (Eds), *Nanophotonics*, ISTE Publishing, 2006.

[158] J. S. Wilson, *Sensor Technology Handbook*, Newnes, 2004.

[159] C. T. Leondes, *Computational Methods in Biophysics, Biomaterials, Biotechnology and Medical Systems*, Springer-Verlag, 2003.

[160] R. W. Waynant (Ed.), *Lasers in Medicine*, Springer-Verlag, 2006.

CHAPTER 12

Optical fiber systems 1: intensity modulation/direct detection

12.1 Introduction

The transfer of information in the form of light propagating within an optical fiber requires the successful implementation of an optical fiber communication system. This system, in common with all systems, is composed of a number of discrete components which are connected together in a manner that enables them to perform a desired task. Hence, to

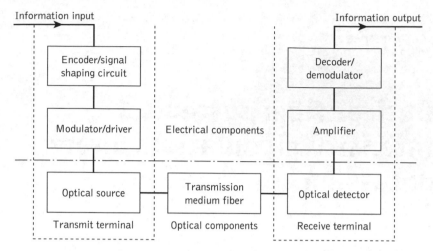

Figure 12.1 The principal components of an optical fiber communication system

achieve reliable and secure communication using optical fibers it is essential that all the components within the transmission system are compatible so that their individual performances, as far as possible, enhance rather than degrade the overall system performance.

The principal components of a general optical fiber communication system for either digital or analog transmission are shown in the system block schematic of Figure 12.1. The transmit terminal equipment consists of an information encoder or signal shaping circuit preceding a modulation or electronic driver stage which operates the optical source. Light emitted from the source is launched into an optical fiber incorporated within a cable which constitutes the transmission medium. The light emerging from the far end of the transmission medium is converted back into an electrical signal by an optical detector positioned at the input of the receive terminal equipment. This electrical signal is then amplified prior to decoding or demodulation in order to obtain the information originally transmitted.

The operation and characteristics of the optical components of this general system have been discussed in some detail within the preceding chapters. However, to enable the successful incorporation of these components into an optical fiber communication system it is necessary to consider the interaction of one component with another, and then to evaluate the overall performance of the system. Furthermore, to optimize the system performance for a given application it is often helpful to offset a particular component characteristic by trading it off against the performance of another component, in order to provide a net gain within the overall system. The electronic components play an important role in this context, allowing the system designer further choices which, depending on the optical components utilized, can improve the system performance.

The purpose of this chapter is to bring together the important performance characteristics of the individual system elements, and to consider their interaction within optical fiber communication systems. In particular we concentrate on the major current implementations of optical fiber communication systems which employ some form of intensity modulation (IM) of the optical source, together with simple direct detection (DD) of the modulated optical signal at the receiver (see Chapter 9). Such IM/DD optical fiber systems are in widespread use within many application areas and do not employ the more sophisticated coherent detection techniques which are discussed in Chapter 13.

It is intended that this chapter provide guidance in relation to the various possible component configurations which may be utilized for different IM/DD system applications, while also giving an insight into system design and optimization. Hence, the optical components and the associated electronic circuits will be discussed prior to consideration of general system design procedures. Although the treatment is by no means exhaustive, it will indicate the various problems involved in system design and provide a description of the basic techniques and practices which may be adopted to enable successful system implementation.

We commence in Section 12.2 with a discussion of the optical transmitter circuit. This includes consideration of the source limitations prior to description of various LED and laser drive circuits for both digital and analog transmission. In Section 12.3 we present a similar discussion for the optical receiver including examples of preamplifier and main amplifier circuits. General IM/DD system design considerations are then dealt with in Section 12.4. This is followed by a detailed discussion of digital systems, commencing with an outline of the operating principles of pulse code modulated (PCM) systems in Section 12.5, before continuing to consider the various aspects of digital IM/DD optical fiber systems in Section 12.6.

Analog IM/DD optical fiber systems are dealt with in Section 12.7 where the various possible analog modulation techniques are described and analyzed. Then, in Section 12.8, consideration is given to IM/DD optical fiber systems configured not simply as point-to-point links but as distribution systems. Next, in Section 12.9 the major multiplexing strategies (both digital and analog) which can be employed in IM/DD optical fiber systems are discussed in further detail to provide a greater understanding of the techniques which are being adopted to increase the information transmission capacity of both currently deployed and potential future systems. The deployment of optical amplifiers (see Chapter 10), particularly on long-haul IM/DD optical fiber links, is dealt with in Section 12.10 to complement the earlier consideration of multiplexing strategies and to enable the reader to appreciate the interaction of these important optical fiber communication techniques. Section 12.11 gives an insight into dispersion management, which has become an essential requirement for high-speed, long-haul optical fiber transmission systems. Finally, optical soliton systems are covered in Section 12.12 which provides an overview of both the system attributes and the issues for transmission of these ultrashort optical pulses.

12.2 The optical transmitter circuit

The unique properties and characteristics of the injection laser and the LED which make them attractive sources for optical fiber communications were discussed in Chapters 6 and 7. Although both device types exhibit a number of similarities in terms of their general performance and compatibility with optical fibers, striking differences exist between them in relation to both system application and transmitter design. It is useful to consider these differences, as well as the limitations of the two source types, prior to discussion of transmitter circuits for various applications.

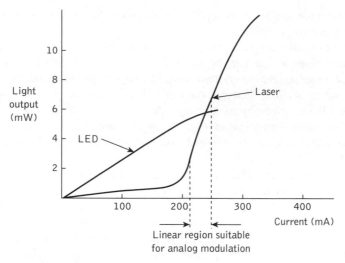

Figure 12.2 Light output (power) emitted into air as a function of d.c. drive current for a typical high-radiance LED and for a typical injection laser. The curves exhibit nonlinearity at high currents due to junction heating

12.2.1 Source limitations

12.2.1.1 Power

The electrical power required to operate both injection lasers and LEDs is generally similar, with typical current levels of between 20 and 300 mA (certain laser thresholds may be substantially higher than this – on the order of 1 to 2 A), and voltage drops across the terminals of 1.5 to 2.5 V. However, the optical output power against current characteristic for the two devices varies considerably, as indicated in Figure 12.2. The injection laser is a threshold device which must be operated in the region of stimulated emission (i.e. above the threshold) where continuous optical output power levels are typically in the range 1 to 10 mW.

Much of this light output may be coupled into an optical fiber because the isotropic distribution of the narrow-linewidth, coherent radiation is relatively directional. In addition, the spatial coherence of the laser emission allows it to be readily focused by appropriate lenses within the numerical aperture of the fiber. Coupling efficiencies near 30% may be obtained by placing a fiber close to a laser mirror, and these can approach 90% with a suitable lens and optical coupling arrangements [Refs 1, 2]. Therefore injection lasers are capable of launching between 0.5 and several milliwatts of optical power into a fiber.

LEDs are capable of similar optical output power levels to injection lasers depending on their structure and quantum efficiency, as indicated by the typical characteristic for a surface emitter shown in Figure 12.2. However, the spontaneous emission of radiation over a wide linewidth from the LED generally exhibits a Lambertian intensity distribution which gives poor coupling into optical fibers. Consequently only between 1% and up to 30% (using a good edge emitter and a resonant cavity LED [Ref. 3], respectively) of the emitted optical power from an LED may be launched into a multimode fiber, even with appropriate lens coupling (see Section 7.3.7). These considerations translate into optical power levels from a few to several hundred microwatts launched into individual multimode fibers. Thus the optical power coupled into a fiber from an LED can be 10 to 20 dB below that obtained with a typical injection laser. The power advantage gained with the injection laser is a major factor in the choice of source, especially when considering a long-haul optical fiber link.

12.2.1.2 Linearity

Linearity of the optical output power against current characteristic is an important consideration with both the injection laser and LED. It is especially pertinent to the design of analog optical fiber communication systems where source nonlinearities may cause severe distortion of the transmitted signal. At first sight the LED may appear to be ideally suited to analog transmission as its output is approximately proportional to the drive current. However, most LEDs display some degree of nonlinearity in their optical output against current characteristic because of junction heating effects which may either prohibit their use or necessitate the incorporation of a linearizing circuit within the optical transmitter. Certain LEDs (e.g. etched-well surface emitters) do display good linearity, with distortion products (harmonic and intermodulation) between 35 and 45 dB below the signal level [Refs 4, 5].

An alternative approach to obtaining a linear source characteristic is to operate an injection laser in the light-generating region above its threshold, as indicated in Figure 12.2. This may prove more suitable for analog transmission than would the use of certain LEDs. However, gross nonlinearities due to mode instabilities may occur in this region. These are exhibited as kinks in the laser output characteristic (see Section 6.5.1). Therefore, many of the multimode injection lasers have a limited use for analog transmission without additional linearizing circuits within the transmitter, although some of the single-mode structures have demonstrated linearity suitable for most analog applications. Alternatively, digital transmission, especially that utilizing a binary (two-level) format, is far less sensitive to source nonlinearities and is therefore often preferred when using both injection lasers and LEDs.

12.2.1.3 Thermal

The thermal behavior of both injection lasers and LEDs can limit their operation within the optical transmitter. However, as indicated in Section 6.7.1, the variation of injection laser threshold current with the device junction temperature can cause a major operating problem. Threshold currents of typical AlGaAs devices increase by approximately 1% per degree centigrade increase in junction temperature. Hence any significant increase in the junction temperature may cause loss of lasing and a subsequent dramatic reduction in the optical output power. This limitation cannot usually be overcome by simply cooling the device on a heat sink, but must be taken into account within the transmitter design, through the incorporation of optical feedback, in order to obtain a constant optical output power level from the device.

The optical output from an LED is also dependent on the device junction temperature, as indicated in Section 7.4.2. Most LEDs exhibit a decrease in optical output power following an increase in junction temperature, which is typically around −1% per degree centigrade. This thermal behavior, however, although significant, is not critical to the operation of the device due to its lack of threshold. Nevertheless, the temperature dependence can result in a variation in optical output power of several decibels over the temperature range 0 to 70 °C. It is therefore a factor within system design considerations which, if not tolerated, may be overcome by providing a circuit within the transmitter which adjusts the LED drive current with temperature.

12.2.1.4 Response

The speed of response of the two types of optical source is largely dictated by their respective radiative emission mechanisms. Spontaneous emission from the LED is dependent on the effective minority carrier lifetime in the semiconductor material (see Section 7.4.3). In heavily doped (10^{18} to 10^{19} cm^{-3}) gallium arsenide this is typically between 1 and 10 ns. However, the response of an optical fiber source to a current step input is often specified in terms of the 10–90% rise time, a parameter which is reciprocally related to the device frequency response (see Section 12.6.5). The rise time of the LED is at least twice the effective minority carrier lifetime, and often much longer because of junction and stray capacitance. Hence, the rise times for many conventional LEDs lie between 2 and 50 ns and give 3 dB bandwidths of around 7 to at best 175 MHz. Therefore, LEDs have tended to be restricted to lower bandwidth applications, although suitable drive circuits can maximize their bandwidth capabilities (i.e. reduce rise times).

Stimulated emission from injection lasers occurs over a much shorter period giving rise times of the order of 0.1 to 1 ns, thus allowing 3 dB bandwidths above 1 GHz. However, injection laser performance is limited by the device switch-on delay (see Section 6.7.2). To achieve the highest speeds it is therefore necessary to minimize the switch-on delay. Transmitter circuits, which prebias the laser to just below or just above threshold, in conjunction with high-speed drive currents which take the device well above threshold, prove useful in the reduction of this limitation. More recently, there has been tremendous progress in high-speed laser technology (see Section 6.6) leading to optical switching at speeds of terabits per second [Ref. 6]. The generation of ultrashort optical pulses from such laser sources with time periods reaching femto-seconds (i.e. order of 10^{-15}) has been realized [Refs 7–9]. These features therefore enable all-optical wavelength switching where optical signals each contained on a different wavelength are switched and sent to their destinations in real time (see Section 12.9.4).

12.2.1.5 Spectral width

The finite spectral width of the optical source causes pulse broadening due to material dispersion on an optical fiber communication link. This results in a limitation on the bandwidth–length product which may be obtained using a particular source and fiber. The incoherent emission from an LED usually displays a spectral linewidth of between 20 and 50 nm (full width at half power (FWHP) points) when operating in the 0.8 to 0.9 μm wavelength range. This limits the bandwidth–length product with a silica fiber to around 100 and 160 MHz km at wavelengths of 0.8 and 0.9 μm respectively. Hence the overall system bandwidth for an optical fiber link over several kilometers may be restricted by material dispersion rather than the response time of the source.

The problem may be alleviated by working at a longer wavelength where the material dispersion in high-silica fibers approaches zero (i.e. near 1.3 µm, see Section 3.9.1). In this region the source spectral width is far less critical and bandwidth–length products approaching 1 GHz km are feasible using LEDs.

Alternatively, an optical source with a narrow spectral linewidth may be utilized in place of the LED. The coherent emission from an injection laser generally has a linewidth of 1 nm or less (FWHP). Use of the injection laser greatly reduces the effect of material dispersion within the fiber, giving bandwidth–length products of 1 GHz km at 0.8 µm, and far higher at longer wavelengths. Hence, the requirement for a system operating at a particular bandwidth over a specific distance will influence both the choice of source and operating wavelength.

12.2.1.6 Nonzero extinction ratio

When the optical source is either intentionally prebiased-on during a 0 bit period, as indicated in Section 12.2.1.4, or simply not turned fully off, then some optical power will be emitted during the 0 pulse. This is particularly important with injection lasers when they are biased just below threshold and hence they launch spontaneous emissions into the fiber. In the case when optical power is incident on the photodetector during the 0 bit period, the system is said to exhibit a nonzero extinction ratio.

The extinction ratio ε is usually defined as the ratio of the optical energy emitted in the 0 bit period to that emitted during the 1 bit period. For an ideal system $\varepsilon = 0$, and the extinction ratio therefore varies between this value and unity. It should be noted, however, that in some cases the extinction ratio is defined as the reciprocal of the above, which implies that it takes up values between 1 and ∞.

Typical values for the extinction ratio are between 0.05 and 0.10 and such nonzero ratios give rise to a noise penalty (often called an extinction ratio penalty) within the optical fiber communication system. The extinction ratio penalty can be evaluated and in practice it is often found to be in the range 1 to 2 dB. Any dark current present in the photodetector will also appear to increase the extinction ratio as it adds to the signal current in both the 0 and 1 bit periods. The greatest penalty occurs, however, in the case of quantum noise limited detection.

12.2.2 LED drive circuits

Although the LED is somewhat restricted in its range of possible applications in comparison with the more powerful, higher speed injection laser, it is generally far easier to operate. Therefore in this section we consider some of the circuit configurations that may be used to convert the information voltage signal at the transmitter into a modulation current suitable for an LED source. In this context it is useful to discuss circuits for digital and analog transmission independently.

12.2.2.1 Digital transmission

The operation of the LED for binary digital transmission requires the switching on and off of a current in the range of several tens to several hundreds of milliamperes. This must be

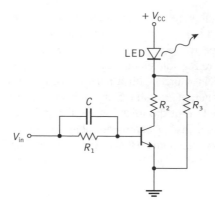

Figure 12.3 A simple drive circuit for binary digital transmission consisting of a common emitter saturating switch

performed at high speed in response to logic voltage levels at the driving circuit input. A common method of achieving this current switching operation for an LED is shown in Figure 12.3. The circuit illustrated uses a bipolar transistor switch operated in the common emitter mode. This single-stage circuit provides current gain as well as giving only a small voltage drop across the switch when the transmitter is in saturation (i.e. when the collector–base junction is forward biased, the emitter to collector voltage V_{CE} (sat) is around 0.3 V).

The maximum current flow through the LED is limited by the resistor R_2 while independent bias to the device may be provided by the incorporation of resistor R_3. However, the switching speed of the common emitter configuration is limited by space charge and diffusion capacitance; thus bandwidth is traded for current gain. This may, to a certain extent, be compensated by overdriving (pre-emphasizing) the base current during the switch-on period. In the circuit shown in Figure 12.3 pre-emphasis is accomplished by use of the speed-up capacitor C.

Increased switching speed may be obtained from an LED without a pulse shaping or speed-up element by use of a low-impedance driving circuit, whereby charging of the space charge and diffusion capacitance occurs as rapidly as possible. This may be achieved with the emitter follower drive circuit shown in Figure 12.4 [Ref. 10]. The use of this configuration with a compensating matching network (R_3C) provides fast direct modulation of LEDs with relatively low drive power. A circuit, with optimum values for the matching network, is capable of giving optical rise times of 2.5 ns for LEDs with capacitance of 180 pF, thus allowing 100 Mbit s^{-1} operation [Ref. 11].

Another type of low-impedance driver is the shunt configuration shown in Figure 12.5. The switching transistor in this circuit is placed in parallel with the LED, providing a low-impedance path for switching off the LED by shunting current around it. The switch-on performance of the circuit is determined by the combination of resistor R and the LED capacitance. Stored space charge may be removed by slightly reverse biasing the LED when the device is switched off. This may be achieved by placing the transistor emitter potential V_{EE} below ground. In this case a Schottky clamp (shown dotted) may be incorporated to limit the extent of the reverse bias without introducing any extra minority carrier stored charge into the circuit.

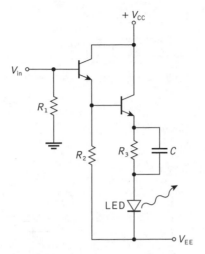

Figure 12.4 Low-impedance drive circuit consisting of an emitter follower with compensating matching network [Ref. 10]

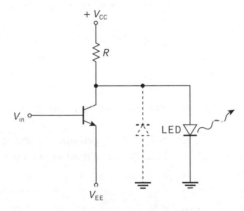

Figure 12.5 Low-impedance drive circuit consisting of a simple shunt configuration

A frequent requirement for digital transmission is the interfacing of the LED by a drive circuit with a common logic family, as illustrated in the block schematic of Figure 12.6(a). In this case the logic interface must be considered along with possible drive circuits. Compatibility with transistor–transistor logic (TTL) may be achieved by use of commercial integrated circuits, as shown in Figure 12.6(b) and (c). The configuration shown in Figure 12.6(b) uses a Texas Instruments' 74S140 line driver which provides a drive current of around 60 mA to the LED when R_1 is 50 Ω. Moreover, the package contains two sections which may be connected in parallel in order to obtain a drive current of 120 mA. The incorporation of a suitable speed-up capacitor (e.g. $C = 47$ pF) gives optical rise times of around 5 ns when using LEDs with between 150 and 200 pF capacitance [Ref. 12]. Figure 12.6(c) illustrates the shunt configuration using a standard TTL 75451 integrated circuit. The rise time of this shunt circuit may be improved through maintenance of charge

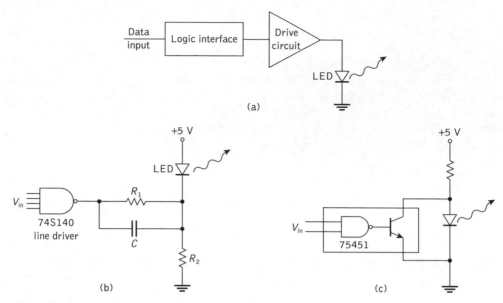

(a)

(b)

(c)

Figure 12.6 Logic interfacing for digital transmission: (a) block schematic showing the interfacing of the LED drive circuit with logic input levels; (b) a simple TTL-compatible LED drive circuit employing a Texas Instruments' 74S140 line driver [Ref. 12]; (c) a TTL shunt drive circuit using a commercially available integrated circuit [Ref. 12]

on the LED capacitance by placing a resistor between the shunt switch collector and the LED [Ref. 12].

An alternative important drive circuit configuration is the emitter-coupled circuit shown in Figure 12.7 [Ref. 12]. The LED acts as a load in one collector so that the circuit

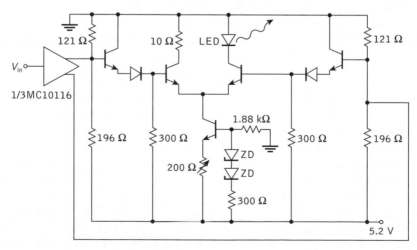

Figure 12.7 An emitter-coupled drive circuit which is compatible with ECL [Ref. 12]

provides current gain and hence a drive current for the device. Thus the circuit resembles a linear differential amplifier, but it is operated outside the linear range and in the switching mode. Fast switching speeds may be obtained due to the configuration's nonsaturating characteristic which avoids switch-off time degradations caused by stored charge accumulation on the transistor base region. The lack of saturation also minimizes the base drive requirements for the transistors, thus preserving their small signal current gain. The emitter-coupled driver configuration shown in Figure 12.7 is compatible with commercial emitter-coupled logic (ECL). However, to achieve this compatibility the circuit includes two level-shifting transistors which give ECL levels (high −0.8 V, low −1.8 V) when the positive terminal of the LED is at earth potential. The response of this circuit is specified [Ref. 12] at up to 50 Mbit s^{-1}, with a possible extension to 300 Mbit s^{-1} when using a faster ECL logic family and high-speed transistors. The emitter-coupled drive circuit configuration may also be interfaced with other logic families, and a TTL-compatible design is discussed in Ref. 13.

12.2.2.2 Analog transmission

For analog transmission the drive circuit must cause the light output from an LED source to follow accurately a time-varying input voltage waveform in both amplitude and phase. Therefore, as indicated previously, it is important that the LED output power responds linearly to the input voltage or current. Unfortunately, this is not always the case because of inherent nonlinearities within LEDs which create distortion products on the signal. Thus the LED itself tends to limit the performance of analog transmission systems unless suitable compensation is incorporated into the drive circuit. However, unless extremely low distortion levels are required, simple transistor drive circuits may be utilized.

Two possible high-speed drive circuit configurations are illustrated in Figure 12.8. Figure 12.8(a) shows a driver consisting of a common emitter transconductance amplifier which converts an input base voltage into a collector current. The circuit is biased for a class A mode of operation with the quiescent collector current about half the peak value. A similar transconductance configuration which utilizes a Darlington transistor pair in order

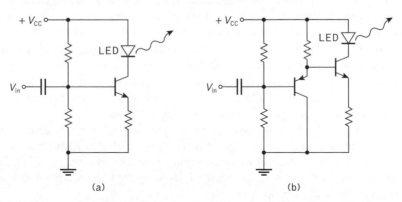

Figure 12.8 Transconductance drive circuits for analog transmission: (a) common emitter configuration; (b) Darlington transistor pair

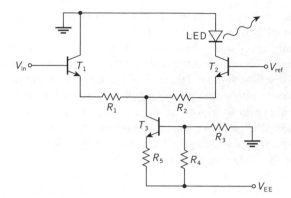

Figure 12.9 A differential amplifier drive circuit

to reduce the impedance of the source is shown in Figure 12.8(b). A circuit of this type has been used to drive high-radiance LEDs at frequencies of 70 MHz [Ref. 14].

Another simple drive circuit configuration is shown in Figure 12.9. It consists of a differential amplifier operated over its linear region which directly modulates the LED. The LED operating point is controlled by a reference voltage V_{ref} while the current generator provided by the transistor T_3 feeding the differential stage (T_1 and T_2) limits the maximum current through the device. The transimpedance of the driver is reduced through current series feedback provided by the two resistors R_1 and R_2 which are normally assigned equal values. Furthermore, variation between these feedback resistors can be used to compensate for the transfer function of both the drive circuit and the LED.

Although in many communication applications where a single analog signal is transmitted certain levels of amplitude and phase distortion can be tolerated, this is not the case in frequency multiplexed systems (see Section 12.4.2) where a high degree of linearity is required in order to minimize interference between individual channels caused by the generation of intermodulation products. Also, baseband video transmission of TV signals requires the maintenance of extremely low levels of amplitude and phase distortion. For such applications the simple drive circuits described previously are inadequate without some form of linearization to compensate for both LED and drive circuit nonlinearities. A number of techniques have been reported [Ref. 15], some of which are illustrated in Figure 12.10. Figure 12.10(a) shows the complementary distortion technique [Ref. 16] where additional nonlinear devices are included in the system. It may take the form of pre-distortion compensation (before the source drive circuit) or postdistortion compensation (after the receiver). This approach has been shown [Ref. 17] to reduce harmonic distortion by up to 20 dB over a limited range of modulation amplitudes.

In the negative feedback compensation technique shown in Figure 12.10(b), the LED is included in the linearization scheme. The optical output is detected and compared with the input waveform, the amount of compensation being dependent on the gain of the feedback loop. Although the technique is straightforward, large-bandwidth requirements (i.e. video) can cause problems at high frequencies [Ref. 18].

The technique shown in Figure 12.10(c) employs phase shift modulation for selective harmonic compensation using a pair of LEDs with similar characteristics [Ref. 19]. The

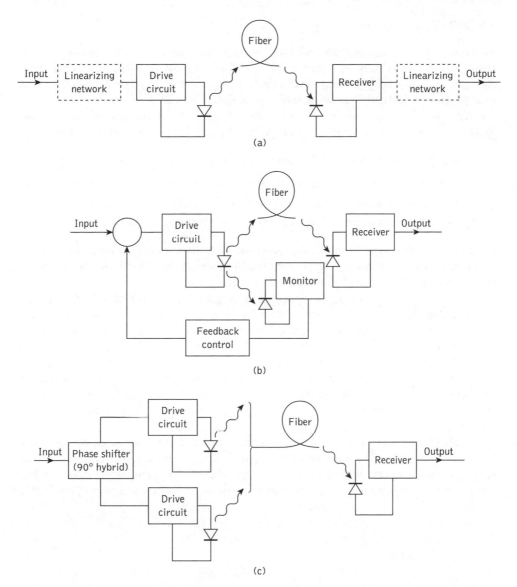

Figure 12.10 Block schematics of some linearization methods for LED drive circuits: (a) complementary distortion technique; (b) negative feedback compensation technique; (c) selective harmonic compensation technique

input signal is divided into equal parts which are phase shifted with respect to each other. These signals then modulate the two LEDs giving a cancellation of the second and third harmonic with a 90° and 60° phase shift respectively. However, although there is a high degree of distortion cancellation, both harmonics cannot be reduced simultaneously.

Other linearization techniques include cascade compensation [Ref. 20], feedforward compensation [Ref. 21] and quasi-feedforward compensation [Refs 22, 23].

12.2.3 Laser drive circuits

A number of configurations described for use as LED drive circuits for both digital and analog transmission may be adapted for injection laser applications with only minor changes. The laser, being a threshold device, has somewhat different drive current requirements from the LED. For instance, when digital transmission is considered, the laser is usually given a substantial applied bias, often referred to as prebias, in the off state. Reasons for biasing the laser near but below threshold in the off state are as follows:

1. It reduces the switch-on delay and minimizes any relaxation oscillations.

2. It allows easy compensation for changes in ambient temperature and device aging.

3. It reduces the junction heating caused by the digital drive current since the on and off currents are not widely different for most lasers.

Although biasing near threshold causes spontaneous emission of light in the off state, this is not normally a problem for digital transmission because the stimulated emission in the on state is generally greater by, at least, a factor of 10.

A simple laser drive circuit for digital transmission is shown in Figure 12.11. This circuit is a shunt driver utilizing a field effect transistor (FET) to provide high-speed laser operation. Sufficient voltage is maintained in series with the laser using the resistor R_2 and the compensating capacitor C such that the FET is biased into its active or pinch-off region. Hence for a particular input voltage V_{in} (i.e. V_{GS}) a specific amount of the total current flowing through R_1 is diverted around the laser leaving the balance of the current to flow through R_2 and provide the off state for the device. Using stable gallium arsenide MESFETs (see Section 9.5.1) the circuit shown in Figure 12.11 has modulated lasers at rates in excess of 1 Gbit s^{-1} [Ref. 24].

An alternative high-speed laser drive circuit employing bipolar transistors is shown in Figure 12.12 [Ref. 25]. This circuit configuration, again for digital transmission, consists of two differential amplifiers connected in parallel. The input stage, which is ECL compatible, exhibits a 50 Ω input impedance by use of an emitter follower T_1 and a 50 Ω

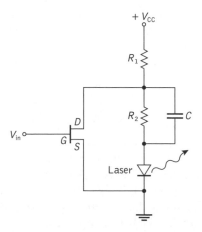

Figure 12.11 A shunt drive circuit for use with an injection laser

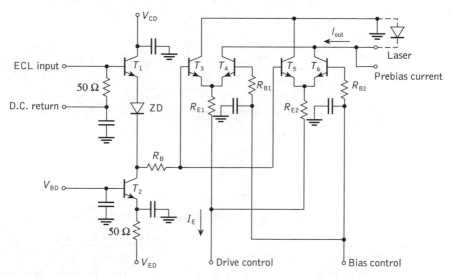

Figure 12.12 An ECL-compatible high-speed laser drive circuit. Reprinted with permission from Ref. 25 © IEEE 1978

resistor in parallel with the input. The transistor T_2 acts as a current source with the Zener diode ZD adjusting the signal level for ECL operation. The two differential amplifiers provide sufficient modulation current amplitude for the laser under the control of a d.c. control current I_E through the two emitter resistors R_{E1} and R_{E2}; I_E is provided by an optical feedback control circuit, to be discussed shortly. Finally, a prebias current is applied to the laser from a separate current source. This circuit when utilizing microwave transistors was operated with a return-to-zero digital format (see Section 3.8) at 1 Gbit s^{-1} [Ref. 25].

A major difference between the drive circuits of Figures 12.11 and 12.12 is the absence and use, respectively, of feedback control for adjustment of the laser output level. For this reason it is unlikely that the shunt drive circuit of Figure 12.11 would be used for a system application. Some form of feedback control is generally required to ensure continuous laser operation because the device lasing threshold is a sensitive function of temperature. Also, the threshold level tends to increase as the laser ages following an increase in internal device losses. Although lasers may be cooled to compensate for temperature variations, aging is not so easily accommodated by the same process. However, both problems may be overcome through control of the laser bias using a feedback technique. This may be achieved using low-speed feedback circuits which adjust the generally static bias current when necessary. For this purpose it is usually found necessary to monitor the light output from the laser in order to keep some aspect constant.

Several strategies of varying complexity are available to provide automatic output level control for the laser. The simplest and perhaps most common form of laser drive circuit incorporating optical feedback is the mean power control circuit shown in Figure 12.13(a). Often the monitor detector consists of a cheap, slow photodiode positioned next to the rear face of the laser package, as indicated in Figure 12.13(a). Alternatively, an optical coupler at the fiber input can be used to direct some of the radiation emitted from the laser into the monitor photodiode. The detected signal is integrated and compared with a reference by

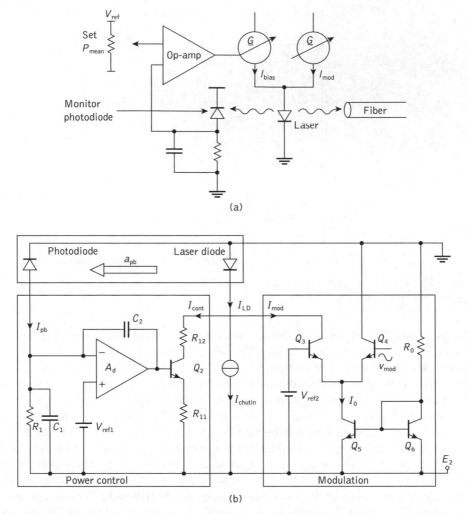

(a)

(b)

Figure 12.13 Laser drive circuits implementing mean power feedback control of the laser bias current: (a) using basic circuit; (b) using an operational amplifier as an inverter. Reprinted with permission from Ref. 26 © IEEE 2004

an operational amplifier which is used to servo-control the d.c. bias applied to the laser. Thus the mean optical power is maintained constant by varying the threshold current level. This technique is suitable for both digital and analog transmission. More recently, similar laser feedback control has been demonstrated which exhibits increased stability to control the average optical output power of a double-heterostructure laser [Ref. 26]. This laser drive circuit which uses an operational amplifier as an inverter is displayed in Figure 12.13(b). The drive circuit essentially comprises two sections that may be identified as providing power control and modulation. Again, monitoring for the mean transmitted optical signal power is realized through the photodiode located above the power control section which gives an input to an operational amplifier operating as an inverter.

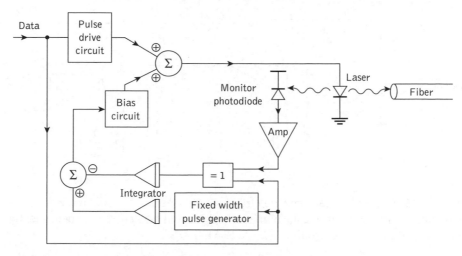

Figure 12.14 Switch-on delay feedback laser control circuit [Ref. 27]

An alternative control method for digital systems which offers accurate threshold tracking and very little device dependence is the switch-on delay technique illustrated in Figure 12.14 [Ref. 27]. This circuit monitors the switch-on delay of an optical pulse in order to control the laser bias current. The switch-on delay is measured for a zero level set below threshold and the feedback is set to a constant fixed delay to control it. Hence, the circuit provides a reference signal proportional to the delay period. This signal is used to control the bias level. The technique requires a fast monitor photodiode as well as a wideband amplifier to allow measurement of the small delay periods. It is also essential that the zero level is set below the lasing threshold because the feedback loop will only stabilize for a finite delay (i.e. the delay falls to zero at the threshold).

A major disadvantage, however, with just controlling the laser bias current is that it does not compensate for variations in the laser slope efficiency. The modulation current for the device is preset and does not take into account any slope changes with temperature and aging. In order to compensate for such changes, the a.c. and d.c. components of the monitored light output must be processed independently. This is especially important in the case of high-bit-rate digital systems where control of the on and off levels as well as the light level is required. A circuit which utilizes both a.c. and d.c. information in the laser output to control the device drive current and bias independently is shown in Figure 12.15 [Ref. 25]. The electrical output from the monitor photodiode is fed into a low-drift d.c. amplifier $A1$ and into a wideband amplifier $A2$. Therefore the mean value of the laser output power $P_e(\text{ave})$ is proportional to the output from $A1$ while the a.c. content of the monitoring signal is peak detected after the amplifier $A2$. The peak signals correspond to the maximum $P_e(\text{max})$ and the minimum $P_e(\text{min})$ laser output powers within a certain time interval. The difference signal proportional to $(P_e(\text{max}) - P_e(\text{min}))$ is acquired in $A3$ and compared with a drive reference voltage in order to control the current output from $A4$ and, consequently, the laser drive current. In this way the modulation amplitude of the laser is controlled. Control of the laser bias current is achieved from the difference between the output signal of $A1$ ($P_e(\text{ave})$) and $P_e(\text{min})$ which is acquired in $A5$. The output

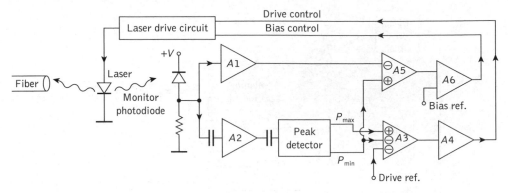

Figure 12.15 A laser feedback control circuit which uses a.c. and d.c. information in the monitored light output to control the laser drive and bias currents independently. Reprinted with permission from Ref. 25 © IEEE 1978

voltage of $A5$, which is proportional to $P_e(min)$, is compared with a bias reference voltage in $A6$ which supplies a current output to control the laser d.c. bias. This feedback control circuit was designed for use with the laser drive circuit shown in Figure 12.12 to give digital operation at bit rates in the gigahertz range.

12.3 The optical receiver circuit

The noise performance for optical fiber receivers incorporating both major detector types (the $p–i–n$ and avalanche photodiode) was discussed in Chapter 9. Receiver noise is of great importance within optical fiber communications as it is the factor which limits receiver sensitivity and therefore can dictate the overall system design. It was necessary within the analysis given in Chapter 9 to consider noise generated by electronic amplification (i.e. within the preamplifier) of the low-level signal as well as the noise sources associated with the optical detector. Also, the possible strategies for the configuration of the preamplifier were considered (see Section 9.4) as a guide to optimization of the receiver noise performance for a particular application. In this section we extend the discussion to consider different possible circuit arrangements which may be implemented to achieve low-noise preamplification, as well as further amplification (main amplification) and processing of the detected optical signal.

A block schematic of an optical fiber receiver is shown in Figure 12.16. Following the linear conversion of the received optical signal into an electric current at the detector, it is amplified to obtain a suitable signal level. Initial amplification is performed in the

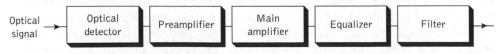

Figure 12.16 Block schematic showing the major elements of an optical fiber receiver

preamplifier circuit where it is essential that additional noise is kept to a minimum in order to avoid corruption of the received signal. As noise sources within the preamplifier may be dominant, its configuration and design are major factors in determining the receiver sensitivity. The main amplifier provides additional low-noise amplification of the signal to give an increased signal level for the following circuits.

Although optical detectors are very linear devices and do not themselves introduce significant distortion onto the signal, other components within the optical fiber communication system may exhibit nonlinear behavior. For instance, the received optical signal may be distorted due to the dispersive mechanisms within the optical fiber. Alternatively, the transfer function of the preamplifier–main amplifier combination may be such that the input signal becomes distorted (especially the case with the high-impedance front-end preamplifier). Hence, to compensate for this distortion and to provide a suitable signal shape for the filter, an equalizer is often included in the receiver. It may precede or follow the main amplifier, or may be incorporated in the functions of the amplifier and filter. In Figure 12.16 the equalizer is shown as a separate element following the amplifier and preceding the filter.

The function of the last element in the receiver, the filter, is to maximize the received signal-to-noise ratio while preserving the essential features of the signal. In digital systems the function of the filter is primarily to reduce intersymbol interference, whereas in analog systems it is generally required to hold the amplitude and phase response of the received signal within certain limits. The filter is also designed to reduce the noise bandwidth as well as inband noise levels. Finally, the general receiver consisting of the elements depicted in Figure 12.16 is often referred to as a linear channel because all operations on the received optical signal may be considered to be mathematically linear.

12.3.1 The preamplifier

The choice of circuit configuration for the preamplifier is largely dependent upon the system application. Bipolar or field effect transistors (FETs) can be operated in three useful connections. These are the common emitter or source, the common base or gate, and the emitter or source follower for the bipolar and field effect transistors respectively. Each connection has characteristics which will contribute to a particular preamplifier configuration. It is therefore useful to discuss the three basic preamplifier structures (low-impedance, high-impedance and transimpedance front-end) and indicate possible choices of transistor connection. In this context the discussion is independent of the type of optical detector utilized. However, it must be noted that there are a number of significant differences in the performance characteristics between the p–i–n and avalanche photodiode (see Chapter 8) which must be considered within the overall design of the receiver.

The simplest preamplifier structure is the low-input-impedance voltage amplifier. This design is usually implemented using a bipolar transistor configuration because of the high input impedance of FETs. The common emitter and the grounded emitter (without an emitter resistor) amplifier shown in Figure 12.17 are favored connections, as they may be designed with reasonably low input impedance and therefore give operation over a moderate bandwidth without the need for equalization. However, this is achieved at the expense of increased thermal noise due to the low effective load resistance presented to the detector.

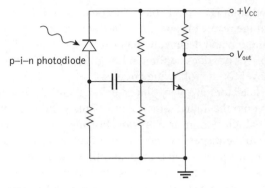

Figure 12.17 A *p–i–n* photodiode with a grounded emitter, low-input-impedance voltage preamplifier

Nevertheless, it is possible to reduce the thermal noise contribution of this preamplifier by choosing a transistor with characteristics which give a high current gain at a low emitter current in order to maintain the bandwidth of the stage. Also, an inductance may be inserted at the collector to provide partial equalization for any integration performed by the stage. The alternative connection giving very low input impedance is the common base circuit. Unfortunately, this configuration has an input impedance which gives insufficient power gain when connected to the high impedance of the optical detector.

The preferred preamplifier configurations for low-noise operation use either a high-impedance integrating front-end or a transimpedance amplifier (see Sections 9.4.2 and 9.4.3). Careful design employing these circuit structures can facilitate high gain coupled with low-noise performance and therefore enhanced receiver sensitivity. Although the bipolar transistor incorporated in the emitter follower circuit may be used to realize a high-impedance front-end amplifier, the FET is generally employed for this purpose because of its low-noise operation. It was indicated in Section 9.5 that the grounded source FET connection was a useful circuit to provide a high-impedance front-end amplifier. The same configuration with a source resistor (common source connection) shown in Figure 12.18

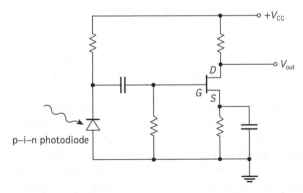

Figure 12.18 An FET common source preamplifier configuration which provides high-input impedance for the *p–i–n* photodiode

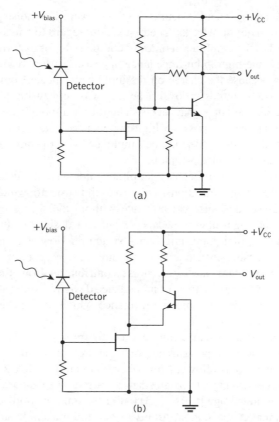

(a)

(b)

Figure 12.19 High-input-impedance preamplifier configurations: (a) grounded source FET followed by common emitter connection with shunt feedback; (b) cascode connection. The separate bias voltage indicates the use of either a *p–i–n* or avalanche photodiode

provides a similar high-input impedance and may also be used (often both configurations are referred to as the common source connection). When operating in this mode the FET power gain and output impedance are both high, which tends to minimize any noise contributions from the following stages. It is especially the case when the voltage gain of the common source stage is minimized in order to reduce the Miller capacitance [Ref. 28] associated with the gate to drain capacitance of the FET. This may be achieved by following the common source stage with a stage having a low-input impedance.

Two configurations which provide a low-input-impedance stage are shown in Figure 12.19. Figure 12.19(a) shows the grounded source FET followed by a bipolar transistor in the common emitter connection with shunt feedback over the stage. Another favored configuration to reduce Miller capacitance in the first-stage FET is shown in Figure 12.19(b). In this case the second stage consists of a bipolar transistor in the common base configuration which, with the initial grounded source FET, forms the cascode configuration.

The high-impedance front-end structure provides a very low-noise preamplifier design but suffers from two major drawbacks. The first is with regard to equalization, which must generally be tailored to the amplifier in order to compensate for distortion introduced onto the signal. Secondly, the high-input-impedance approach suffers from a lack of dynamic range which occurs because the charge on the input capacitance from the low-frequency components in the signal builds up over a period of time, causing premature saturation of the amplifier at high input signal levels. Therefore, although the circuits shown in Figure 12.19 are examples of possible high-impedance integrating front-end amplifier configurations, similar connections may be employed with overall feedback (to the first stage) to obtain a transimpedance preamplifier.

The transimpedance or shunt feedback amplifier finds wide application in preamplifier design for optical fiber communications. This front-end structure which acts as a current–voltage converter gives low-noise performance without the severe limitations on bandwidth imposed by the high-input-impedance front-end design. It also provides greater dynamic range than the high-input-impedance structure. However, in practice the noise performance of the transimpedance amplifier is not quite as good as that achieved with the high-impedance structure due to the noise contribution from the feedback resistor (see Section 9.4.3). Nevertheless, the transimpedance design incorporating a large value of feedback resistor can achieve a noise performance which approaches that of the high-impedance front-end.

Two examples of transimpedance front-end configurations are shown in Figure 12.20. Figure 12.20(a) illustrates a bipolar transistor structure consisting of a common emitter stage followed by an emitter follower [Ref. 29] with overall feedback through resistor R_f. The output signal level from this transimpedance pair may be increased by the addition of a second common emitter stage [Refs 30, 31] after the emitter follower. This stage is not usually included in the feedback loop. An FET front-end transimpedance design is shown in Figure 12.20(b) [Ref. 32].

The circuit consists of a grounded source configuration followed by a bipolar transistor cascade with feedback over the three stages. In this configuration the bias currents for the bipolar stages and the feedback resistance may be chosen to give good open loop bandwidth while making the noise contribution from these stages negligible.

Finally, for lower bandwidth, shorter haul applications an FET operational amplifier front-end is often adequate [Ref. 33]. Such a transimpedance preamplifier circuit, which is generally used with a $p–i–n$ photodiode, is shown in Figure 12.21. The choice of the operational amplifier is dependent on the gain–bandwidth product for the device. In a simple digital receiver design all that may be required in addition to the circuit shown in Figure 12.21 is a logic (e.g. TTL) interface stage following the amplifier.

12.3.2 Automatic gain control

It may be noted from the preceding section that the receiver circuit must provide a steady reverse bias voltage for the optical detector. With a $p–i–n$ photodiode this is not critical and a voltage of between 5 and 80 V supplying an extremely low current is sufficient. The avalanche photodiode requires a much larger bias voltage of between 100 and 400 V which defines the multiplication factor for the device. An optimum multiplication factor is usually chosen so that the receiver signal-to-noise ratio is maximized (see Section 9.3.3).

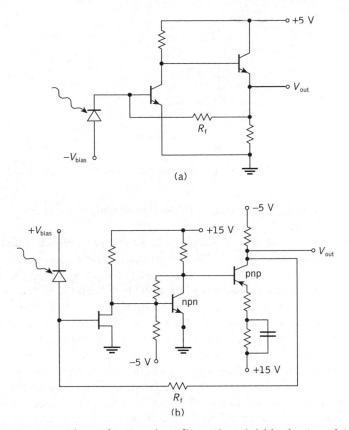

Figure 12.20 Transimpedance front-end configuration: (a) bipolar transistor design [Ref. 29]; (b) FET front-end and bipolar transistor cascade structure [Ref. 32]

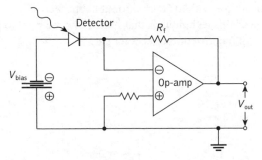

Figure 12.21 A typical circuit for an operational amplifier transimpedance front-end [Ref. 33]

The multiplication factor for the APD varies with the device temperature (see Section 8.9.5) making provision of fine control for the bias voltage necessary in order to maintain the optimum multiplication factor. However, the multiplication factor can be held constant by some form of automatic gain control (AGC). An additional advantage in the use of

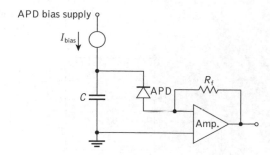

Figure 12.22 Bias of an APD with a constant current source to provide simple AGC

AGC is that it reduces the dynamic range of the signals applied to the preamplifier giving increased optical dynamic range at the receiver input.

One method of providing AGC is simply to bias the APD with a constant d.c. current source I_{bias}, as illustrated in Figure 12.22. The constant current source is decoupled by a capacitor C at all signal frequencies to prevent gain modulation. When the mean optical input power is known, the mean current to the APD is defined by the bias which gives a constant multiplication factor (gain) at all temperatures. Any variation in the multiplication factor will produce a variation in the charge on C, thus adjusting the biasing of the APD back to the required multiplication factor. Therefore, the output current from the photodetector is only defined by the input current from the constant current source, giving full AGC. However, this simple AGC technique is dependent on a constant, mean optical input power level, and takes no account of dark current generated within the detector.

A more widely used method which allows for the effect of variations in the detector dark current while providing critical AGC is to peak-detect the a.c. coupled signal after suitable low-noise amplification, as shown in Figure 12.23. The signal from the final stage of the main amplifier is compared with a preset reference level and fed back to adjust the high-voltage bias supply in order to maintain a constant signal level. This effectively creates a constant current source with the dark current subtracted.

A further advantage of this technique is that it may also be used to provide AGC for the main amplifier giving full control of the receiver gain.

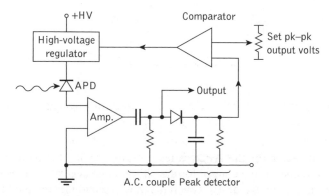

Figure 12.23 Bias of an APD by peak detection and feedback to provide AGC

12.3.3 Equalization

The linear channel provided by the optical fiber receiver is often required to perform equalization as well as amplification of the detected optical signal. In order to discuss the function of the equalizer it is useful to assume the light falling on the detector to consist of a series of pulses given by:

$$P_o(t) = \sum_{k=-\infty}^{+\infty} a_k h_p(t - k\tau) \tag{12.1}$$

where $h_p(t)$ is the received pulse shape, $a_k = 0$ or 1 corresponding to the binary information transmitted and τ is the pulse repetition time or pulse spacing. In digital transmission τ corresponds to the bit period, although the pulse length does not necessarily fill the entire time period τ. For a typical optical fiber link, the received pulse shape is dictated by the transmitted pulse shape $h_t(t)$ and the fiber impulse response $h_f(t)$ following:

$$h_p(t) = h_t(t) * h_f(t) \tag{12.2}$$

where * denotes convolution. Hence determination of the received pulse shape requires knowledge of the fiber impulse response which is generally difficult to characterize. However, it can be shown [Ref. 34] for fiber which exhibits mode coupling that the impulse response is close to a Gaussian shape in both the time and frequency domains.

It is likely that the pulses given by Eq. (12.1) will overlap due to pulse broadening caused by dispersion on the link giving intersymbol interference (ISI). Following detection and amplification Eq. (12.1) may be written in terms of a voltage $v_A(t)$ as:

$$v_A(t) = \sum_{k=-\infty}^{+\infty} a_k h_A(t - k\tau) \tag{12.3}$$

where the response $h_A(t)$ includes any equalization required to compensate for distortion (e.g. integration) introduced by the amplifier. Therefore, although there is equalization for degradations caused by the amplifier, distortion caused by the channel and the resulting intersymbol interference is still included in $h_A(t - k\tau)$. The pulse overlap causing this intersymbol interference may be reduced through the incorporation of a suitable equalizer with a frequency response $H_{eq}(\omega)$ such that:

$$H_{eq}(\omega) = \frac{\mathscr{F}\{h_{out}(t)\}}{\mathscr{F}\{h_A(t)\}} = \frac{H_{out}(\omega)}{H_A(\omega)} \tag{12.4}$$

where $h_{out}(t)$ is the desired output pulse shape and $\mathscr{F}$ indicates Fourier transformation. A block diagram indicating the pulse shapes in the time and frequency domains at the various points in an optical fiber system is shown in Figure 12.24.

An equalizer characterized by Eq. (12.4) will provide high-frequency enhancement in the linear channel to compensate for high-frequency roll-off in the received pulses, thus giving the desired pulse shape. However, in order to construct such an equalizer we require knowledge of $h_A(t)$ and therefore $h_p(t)$. In turn, this necessitates information on the fiber impulse response $h_f(t)$ which may not be easily obtained. Nevertheless, the

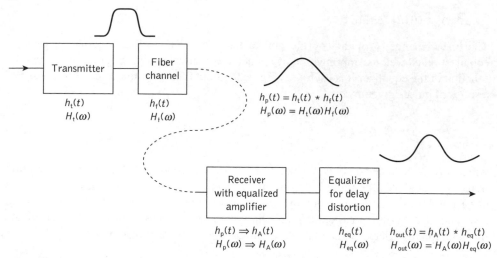

Figure 12.24 Block schematic of an optical fiber system illustrating the transmitted and received optical pulse shapes, together with electrical pulse shape, at the linear channel output

conventional transversal equalizer shown in Figure 12.25 may be incorporated into the linear channel to keep ISI at tolerable levels, even if it is difficult to design a circuit which gives the optimum system response indicated in Eq. (12.4).

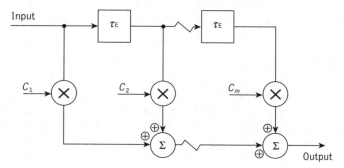

Figure 12.25 The transversal equalizer employing a tapped delay line

The transversal equalizer consists of a delay line tapped at τ_E second intervals. Each tap is connected through a variable gain device with tap coefficients c_i to a summing amplifier. ISI is reduced by filtering the input signal and by computing the values for the tap coefficients which minimize the peak ISI. It is likely that further reduction in ISI will be accomplished using adaptive equalization. ISI, however, can be compensated by the use of a dynamic gain equalizer (DGE).* In such an equalization technique the gain is required to automatically compensate for the variations in optical signal power levels. Devices operating on this principle generally utilize both variable optical attenuators (see Section 11.5) and channel monitoring mechanisms. These mechanisms incorporate a

---

* Also referred to as an adaptive gain equalizer.

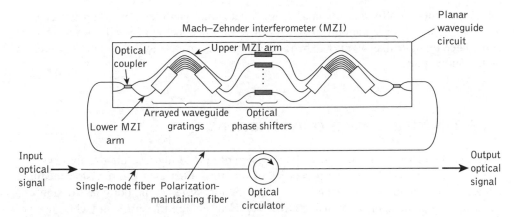

Figure 12.26 Planar dynamic gain equalizer

feedback approach to adjust the required amount of attenuation in order to equalize the gain over a range of wavelengths thus producing uniform signal power in each channel.

Both integrated optics and photonics technologies (see Sections 11.2, 11.5 and 11.6) can be used to provide dynamic gain equalization [Refs 35–38] and the majority of these device structures employ single or multiple cascaded stages of the Mach–Zehnder interferometer (MZI) configuration. Furthermore, MZI arrangements may incorporate passive devices including arrayed waveguide gratings (AWGs), fiber Bragg gratings and delay lines to obtain variable optical attenuation for dynamic gain equalization of each channel. A block schematic of a typical DGE for an optical wavelength division multiplexed system based on a planar waveguide circuit (see Section 11.3) is shown in Figure 12.26. It employs an AWG and phase shifters in the MZI paths to attenuate the optical signal power. The input signals are coupled to a planar waveguide circuit through an optical circulator (see Section 5.7) and polarization-maintaining fiber to avoid any polarization mode dispersion-related ISI noise. Since AWGs are temperature–wavelength-dependent devices (see Section 5.6.3) then the control of the MZI (i.e. its phase shift) can be achieved through electrical heating using thermo-optic couplers. Finally, all the wavelength channels in the output signal which have the same optical signal power are multiplexed in the second stage of the AWG and then coupled to the single-mode fiber via an optical circulator.

Dynamic gain equalizers can be categorized as either single-channel or multichannel devices providing operation using single- or multiple-signal wavelengths, respectively. Furthermore, the latter can be distinguished as either band or harmonic gain equalizers. In a band DGE a limited number of channels (i.e. generally eight channels) can be simultaneously provided with gain equalization and therefore all the channels are attenuated as a group. Harmonic DGEs, however, act on the entire transmission window by applying a sinusoidal transmission function across the whole range of wavelengths in order to reshape the gain curve of any optical amplifiers in the transmission link. By contrast, single-channel DGEs attenuate each wavelength channel individually and such devices are particularly useful at system/network nodes where a number of wavelength channels may be continually added and dropped off. DGEs are therefore sometimes known as dynamic channel or spectral equalizers. In addition, they may also be referred to as wavelength blockers [Refs 39, 40] since dynamic gain equalization can block a certain wavelength channel(s). In this case the optical signal power of a channel at a specific wavelength is attenuated by more than 40 dB and hence that particular wavelength channel is not present at the output port.

12.4 System design considerations

Many of the problems associated with the design of optical fiber communication systems occur as a result of the unique properties of the glass fiber as a transmission medium. However, in common with metallic line transmission systems, the dominant design criteria for a specific application using either digital or analog transmission techniques are the required transmission distance and the rate of information transfer.

Within optical fiber communications these criteria are directly related to the major transmission characteristics of the fiber, namely optical attenuation and dispersion. Unlike metallic conductors where the attenuation (which tends to be the dominant mechanism) can be adjusted by simply changing the conductor size, entirely different factors limit the information transfer capability of optical fibers (see Chapter 3). Nevertheless, it is mainly these factors, together with the associated constraints within the terminal equipment, which finally limit the maximum distance that may be tolerated between the optical fiber transmitter and receiver. Where the terminal equipment is more widely spaced than this maximum distance, as in long-haul telecommunication applications, it is necessary to insert repeaters at regular intervals, as shown in Figure 12.27. The repeater incorporates a line receiver in order to convert the optical signal back into the electrical regime where, in the case of analog transmission, it is amplified and equalized (see Section 12.3.3) before it is retransmitted as an optical signal via a line transmitter. When digital transmission techniques are used the repeater also regenerates the original digital signal in the electrical regime (a regenerative repeater which is often simply called a regenerator) before it is retransmitted as a digital optical signal. In this case the repeater may additionally provide alarm, supervision and engineering order wire facilities. However, optical or all-optical regeneration techniques now provide signal amplification with additional reshaping and retiming capabilities (i.e. 3R regeneration, see Section 10.6) and they can be used to achieve longer transmission distances using optical amplifiers in comparison with optoelectronic repeaters.

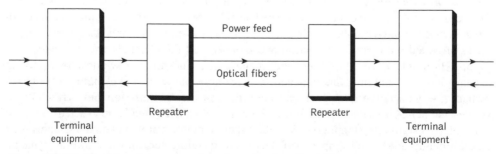

Figure 12.27 The use of repeaters in a long-haul optical fiber communication system

The installation of repeaters substantially increases the cost and complexity of any line communication system. Hence a major design consideration for long-haul telecommunication systems is the maximum distance of unrepeatered transmission so that the number of intermediate amplification or regeneration stages may be reduced to a minimum. In this respect optical fiber systems display a marked improvement over alternative line transmission systems using metallic conductors. However, this major advantage of optical fiber communications is somewhat reduced when there is a requirement for electronic signal processing at the repeater. This necessitates the supply of electric power to the intermediate repeaters via metallic conductors, as may be observed in Figure 12.27.

Before any system design procedures can be initiated it is essential that certain basic system requirements are specified. These specifications include:

(a) transmission type: digital or analog;

(b) acceptable system fidelity, generally specified in terms of the received BER for digital systems or the received SNR and signal distortion for analog systems;

(c) required transmission bandwidth;

(d) acceptable spacing between the terminal equipment or intermediate repeaters;

(e) cost;

(f) reliability.

However, the exclusive use of the above specifications inherently assumes that system components are available which will allow any system, once specified, to be designed and implemented. Unfortunately, this is not always the case, especially when the desired result is a wideband, long-haul system. In this instance it may be necessary to make choices by considering factors such as availability, reliability, cost and ease of installation and operation, before specifications (a) to (d) can be fully determined. A similar approach must be adopted in lower bandwidth, shorter haul applications where there is a requirement for the use of specific components which may restrict the system performance. Hence it is likely that the system designer will find it necessary to consider the possible component choices in conjunction with the basic system requirements.

12.4.1 Component choice

The system designer has many choices when selecting components for an optical fiber communication system. In order to exclude certain components at the outset it is useful if the operating wavelength of the system is established (i.e. shorter wavelength region 0.8 to 0.9 μm or longer wavelength region 1.1 to 1.7 μm). This decision will largely be dictated by the overall requirements for the system performance, the ready availability of suitable reliable components, and cost. Hence the major component choices are:

1. *Optical fiber type and parameters.* Multimode or single-mode; size, refractive index profile, attenuation, dispersion, mode coupling, strength, cabling, jointing, etc.

2. *Source type and characteristics.* Laser or LED; optical power launched into the fiber, rise and fall time, stability, etc.

3. *Transmitter configuration.* Design for digital or analog transmission; input impedance, supply voltage, dynamic range, optical feedback, etc.

4. *Detector type and characteristics.* p–n, p–i–n, or avalanche photodiode; responsivity, response time, active diameter, bias voltage, dark current, etc.

6. *Receiver configuration.* Preamplifier design (low impedance, high impedance or transimpedance front-end), BER or SNR, dynamic range, etc.

7. *Modulation and coding.* Source intensity modulation; using pulse modulation techniques for either digital (e.g. pulse code modulation, adaptive delta modulation) or analog (pulse amplitude modulation, pulse frequency modulation, pulse width modulation, pulse position modulation) transmission. Also, encoding schemes for digital transmission such as biphase (Manchester) and delay modulation (Miller) codes [Ref. 12]. Alternatively analog transmission using direct intensity modulation or frequency modulation of the electrical subcarrier (subcarrier FM). In the latter technique the frequency of an electrical subcarrier is modulated rather than the frequency of the optical source, as would be the case with direct frequency modulation. The electrical subcarrier, in turn, intensity modulates the optical source (see Section 12.7.5).

    Specific digital modulation techniques which require coherent detection are a focus of interest and system components which will permit these modulation methods to be utilized are increasingly available (see Chapter 13).

Decisions in the above areas are interdependent and may be directly related to the basic system requirements. The potential choices provide a wide variety of economic optical fiber communication systems. However, it is necessary that the choices are made in order to optimize the system performance for a particular application.

12.4.2 Multiplexing

In order to maximize the information transfer over an optical fiber communication link it is usual to multiplex several signals onto a single fiber. It is possible to convey these multichannel signals by multiplexing in the electrical time or frequency domain, as with conventional electrical line or radio communication, prior to intensity modulation of the optical source. Hence, digital pulse modulation schemes may be extended to multichannel operation by time division multiplexing (TDM) narrow pulses from multiple modulators under the control of a common clock. Pulses from the individual channels are interleaved and transmitted sequentially, thus enhancing the bandwidth utilization of a single-fiber link.

Alternatively, a number of baseband channels may be combined by frequency division multiplexing (FDM). In FDM the optical channel bandwidth is divided into a number of nonoverlapping frequency bands and each signal is assigned one of these bands of frequencies. The individual signals can be extracted from the combined FDM signal by appropriate electrical filtering at the receive terminal. Hence, FDM in an IM/DD system is generally performed electrically at the transmit terminal prior to intensity modulation of a single optical source. However, it is possible to utilize a number of optical sources, each operating at a different wavelength on the single-fiber link. In this technique, often referred to as wavelength division multiplexing (WDM), the separation and extraction of the multiplexed signals (i.e. wavelength separation) is performed with optical filters (e.g. interference filters, diffraction grating filters, or prism filters) [Refs 41, 42].

Finally, a multiplexing technique which does not involve the application of several message signals onto a single fiber is known as space division multiplexing (SDM). In SDM, each signal channel is carried on a separate fiber within a fiber bundle or multifiber cable form. The good optical isolation offered by fibers means that cross-coupling between channels can be made negligible. However, this technique necessitates an increase in the number of optical components required (e.g. fiber, connectors, sources, detectors) within a particular system and therefore has not been widely used to date.

12.5 Digital systems

Most of the future expansion of the telecommunication network has been planned around digital telephone exchanges linked by digital transmission systems. The shift towards digitizing the network followed the introduction of digital circuit techniques and, especially, integrated circuit technology which made the transmission of discrete time signals both advantageous and economic. Digital transmission systems generally give superior performance over their analog counterparts, as well as providing an ideal channel for data communications and compatibility with digital computing and storage techniques.

Optical fiber communication is well suited to baseband digital transmission in several important ways. For instance, it offers a tremendous advantage with regard to the acceptable SNR at the optical fiber receiver over analog transmission by some 20 to 30 dB (for practical systems), as indicated in the noise considerations of Section 9.2. Also, the use of baseband digital signaling reduces problems involved with optical source (and sometimes detector) nonlinearities and temperature dependence which may severely affect analog transmission. Therefore, most high-capacity optical fiber communication systems convey digital information in the baseband using intensity modulation of the optical source.

In common with electrical transmission systems, analog signals (e.g. speech) may be digitized for transmission utilizing pulse code modulation (PCM). Encoding the analog signal into a digital bit pattern is performed by initially sampling the analog signal at a frequency in excess of the Nyquist rate (i.e. greater than twice the maximum signal frequency). Within the European telecommunication network where the 3 dB telephone bandwidth is defined as 3.4 kHz, the sampling rate is 8 kHz. Hence, the amplitude of the constant width sampling pulses varies in proportion to the sample values of the analog signal giving a discrete signal known as pulse amplitude modulation (PAM), as indicated in Figure 12.28. The sampled analog signal is then quantized into a number of discrete levels, each of which is designated by a binary code which provides the PCM signal. This process is also illustrated in Figure 12.28 using a linear quantizer with eight levels (or seven steps) so that each PAM sample is encoded into three binary bits. The analog signal is thus digitized and may be transmitted as a baseband signal or, alternatively, be modulated by amplitude, frequency or phase shift keying [Ref. 43]. However, in practical PCM

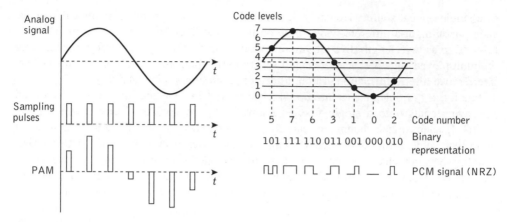

Figure 12.28 The quantization and encoding of an analog signal into PCM using a linear quantizer with eight levels

systems for speech transmission, nonlinear encoding (*A* law in Europe and μ law in North America) is generally employed over 256 levels (2^8), giving eight binary bits per sample (seven bits for code levels plus one polarity bit). Hence, the bandwidth requirement for PCM transmission is substantially greater (in this case by a factor of approximately 16) than the corresponding baseband analog transmission. This is not generally a problem with optical fiber communications because of the wideband nature of the optical channel.

Nonlinear encoding may be implemented via a mechanism known as companding where the input signal is compressed before transmission to give a nonlinear encoding characteristic and expanded again at the receive terminal after decoding. A typical nonlinear input–output characteristic giving compression is shown in Figure 12.29. Companding is used to reduce the quantization error on small-amplitude analog signal levels when they are encoded from PAM to PCM. The quantization error (i.e. the rounding off to the nearest discrete level) is exhibited as distortion or noise on the signal (often called quantization noise). Companding tapers the step size, thus reducing the distance between levels for small-amplitude signals while increasing the distance between levels for higher amplitude signals. This substantially reduces the quantization noise on small-amplitude signals at the expense of slightly increased quantization noise, in terms of signal amplitude, for the larger signal

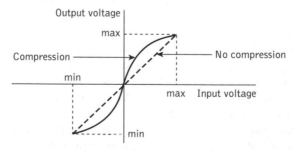

Figure 12.29 A typical nonlinear input–output characteristic which provides compression

levels. The corresponding SNR improvement for small-amplitude signals significantly reduces the overall signal degradation of the system due to the quantization process.

A block schematic of a simplex (one direction only) baseband PCM system is shown in Figure 12.30(a). The optical interface is not shown but reference may be made to Figure 12.1 which illustrates the general optical fiber communication system. It may be noted from Figure 12.30(a) that the received PCM waveform is decoded back to PAM via the reverse process to encoding, and then simply passed through a low-pass filter to recover the original analog signal.

The conversion of a continuous analog waveform into a discrete PCM signal allows a number of analog channels to be time division multiplexed for simultaneous transmission down one optical fiber link, as illustrated in Figure 12.30(b). The encoded samples from the different channels are interleaved within the multiplexer to give a single composite signal consisting of all the interleaved pulses. This signal is then transmitted over the optical channel. At the receive terminal the interleaved samples are separated by a synchronous switch or demultiplexer before each analog signal is reconstructed from the appropriate set of samples. Time division multiplexing a number of channels onto a single link can be used with any form of digital transmission and is frequently employed in the transmission of data as well as with the transmission of digitized analog signals. However, the telecommunication network is primarily designed for the transmission of analog speech signals, although the compatibility of PCM with data signals has encouraged the adoption of digital transmission systems.

A current European standard for speech transmisson using PCM on metallic conductors (i.e. coaxial line) is the 30-channel system. In this system the PAM samples from each channel are encoded into eight binary bits which are incorporated into a single time slot.

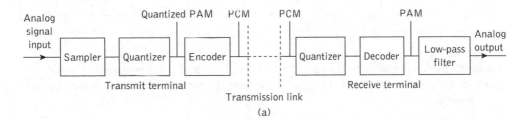

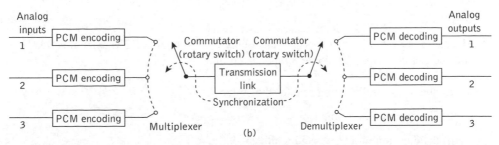

Figure 12.30 PCM transmission: (a) block schematic of a baseband PCM transmission system for single-channel transmission; (b) time division multiplexing of three PCM channels onto a single transmission link and subsequent demultiplexing at the link output

Time slots from respective channels are interleaved (multiplexed) into a frame consisting of 32 time slots. The 2 additional time slots do not carry encoded speech but signaling and synchronization information. Finally, 16 frames are incorporated into a multiframe which is a self-contained timing unit. The timing for this line signaling structure is shown in Figure 12.31 and calculated in Example 12.1.

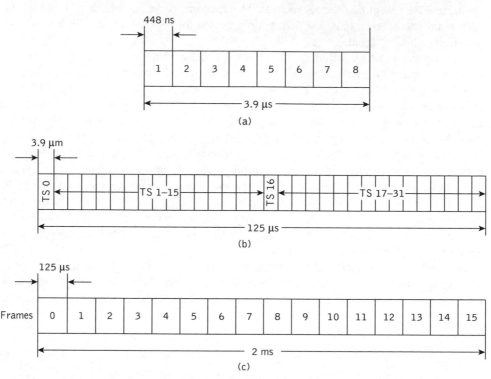

Figure 12.31 The timing for the line signalling structure of the European standard thirty channel PCM system: (a) bits per time slot; (b) time slots per frame; (c) frames per multiframe

Example 12.1

The sampling rate for each speech channel on the 30-channel PCM system is 8 kHz and each sample is encoded into 8 bits. Determine:

(a) the transmission or bit rate for the system:

(b) the duration of a time slot;

(c) the duration of a frame and multiframe.

Solution: (a) The 30-channel PCM system has 32 time slots each 8 bits wide which make up a frame. Therefore:

Number of bits in a frame $= 32 \times 8 = 256$ bits

This frame must be transmitted within the sampling period and thus 8×10^3 frames are transmitted per second. Hence, the transmission rate for the system is:

$$8 \times 10^3 \times 256 = 2.048 \text{ Mbit s}^{-1}$$

(b) The bit duration is simply:

$$\frac{1}{2.048 \times 10^6} = 488 \text{ ns}$$

Therefore, the duration of a time slot is:

$$8 \times 488 \text{ ns} = 3.9 \text{ μs}$$

(c) The duration of a frame is thus:

$$32 \times 3.9 \text{ μs} = 125 \text{ μs}$$

and the duration of a multiframe is:

$$16 \times 125 \text{ μs} = 2 \text{ ms}$$

The signaling structure shown in Figure 12.31 applies to 30-channel PCM systems which were originally designed to transmit over metallic conductors using a high-density bipolar line code (HDB 3). The increased bandwidth with optical fiber communications allows transmission rates far in excess of 2.048 Mbit s^{-1}. Therefore an increased number of telephone channels may be sampled, encoded, multiplexed and transmitted on an optical fiber link. In Europe the increased bit rates were chosen as multiples of the 30-channel system, whereas in North America and Japan they tend to be multiples of a 24-channel system. These bit rates and the corresponding number of transmitted telephone channels are specified in Table 12.1.

It must be noted that a bipolar code with a zero mean level (i.e. with positive- and negative-going pulses in the electrical regime) such as HDB 3 cannot be transmitted directly over an optical fiber link unless the mean level is raised to allow both positive- and negative-going pulses to be transmitted by the IM optical source. The resultant ternary (three-level) optical transmission is not always suitable for telecommunication applications and therefore binary coding after appropriate scrambling, biphase (Manchester encoding), delay modulation (Miller encoding), etc., is often employed. This involves additional complexity at the transmit and receive terminals as well as necessitating extra redundancy (i.e. bits which do not contain the transmitted information, thus giving a reduction in the information per transmitted symbol) in the line code. This topic is considered in greater detail in Section 12.6.7.

Table 12.1 Digital bit rates for multichannel PCM transmission in Europe, North America and Japan [Refs 44, 45]

Hierarchy	Telephone channels	Bit rates (Mbit s⁻¹)	Europe	North America	Japan
First level	24	1.544	T-1 (DS-1)*	–	J-1
	30	2.048	–	E-1	–
Intermediate level	48	3.152	T-1C (DS-1C)**	–	J-1C
Second level	96	6.312	T-2 (DS-2)	–	J2
	120	8.448	–	E-2	–
	480	32.064	–	–	J-3
	480	34.368	–	E-3	–
Third level					
	672	44.736	T-3 (DS-3)	–	–
	1440	97.728	–	–	J-3C
	1920	139.268	–	E-4	–
Fourth level					
	4032	274.176	T-4 (DS-4)	–	–
	5760	397.200	–	–	J-4
Fifth level	5760	400.352	T-5 (DS-5)	–	–
	7680	565.148	–	E-5	–

* T stands for trunk, DS for digital signal.
** The letter C represents the carrier which identifies the extended hierarchy level providing an increased number of channels. For example, two T-1 channels are added to produce T-1C (i.e. $2 \times 24 = 48$) and similarly three J-3 channels in the third level are added to produce the extended hierarchy level J-3C (i.e. $3 \times 480 = 1440$).

12.6 Digital system planning considerations

The majority of digital optical fiber communication systems for the telecommunication network or local data applications utilize binary intensity modulation of the optical source. Therefore, we choose to illustrate the planning considerations for digital transmission based on this modulation technique. Baseband PCM transmission using source intensity modulation is usually designated as PCM–IM.

12.6.1 The optoelectronic regenerative repeater

In the case of the long-haul, high-capacity digital systems, the most important overall system performance parameter is the spacing of the regenerative repeaters. It is therefore useful to consider the performance of the optoelectronic digital repeater, especially as it is usually designed with the same optical components as the terminal equipment. Figure 12.32 shows the functional parts of a typical optoelectronic regenerative repeater for optical fiber communications. The attenuated and dispersed optical pulse train is detected and amplified in the receiver unit. This consists of a photodiode followed by a low-noise preamplifier. The electrical signal thus acquired is given a further increase in power level

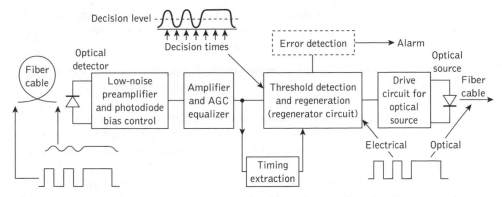

Figure 12.32 Block schematic showing a typical optoelectronic regenerative repeater for digital optical fiber communications

in a main amplifier prior to reshaping in order to compensate for the transfer characteristic of the optical fiber (and the amplifier) using an equalizer. Depending on the photodiode utilized, automatic gain control may be provided at this stage for both the photodiode bias current and the main amplifier (see Section 12.3.2).

Accurate timing (clock) information is then obtained from the amplified and equalized waveform using a timing extraction circuit such as a ringing circuit or phase-locked loop. This enables precise operation of the following regenerator circuit within the bit intervals of the original pulse train. The function of the regenerator circuit is to reconstitute the originally transmitted pulse train, ideally without error. This can be achieved by setting a threshold above which a binary 1 is registered, and below which a binary 0 is recorded, as indicated in Figure 12.32. The regenerator circuit makes these decisions at times corresponding to the center of the bit intervals based on the clock information provided by the timing circuit.

Hence the decision times are usually set at the mid-points between the decision-level crossings of the pulse train. The pulse train is sampled at a regular frequency equal to the bit rate, and at each sample instant a decision is made of the most probable symbol being transmitted. The symbols are then regenerated in their original form (either a binary 1 or 0) before retransmission as an optical signal using a source operated by an electronic drive circuit. Hence the possible regeneration of an exact replica of the originally transmitted waveform is a major advantage of digital transmission over corresponding analog systems. Optoelectronic repeaters in analog systems filter, equalize and amplify the received waveform, but are unable to reconstitute the originally transmitted waveform entirely free from distortion and noise. Signal degradation in long-haul analog systems is therefore accumulative, being a direct function of the number of repeater stages. In contrast the signal degradation encountered in PCM systems is purely a function of the quantization process and the system bit error rate.

Errors may occur in the regeneration process in the following situations:

1. The signal-to-noise ratio at the decision instant is insufficient for an accurate decision to be made. For instance, with high noise levels, the binary 0 may occur above the threshold and hence be registered as a binary 1.

2. There is intersymbol interference due to dispersion on the optical fiber link. This may be reduced by equalization which forces the transmitted binary 1 to pass through zero at all neighboring decision times.

3. There is a variation in the clock rate and phase degradations (jitter) such as distortion of the zero crossings and static decision time misalignment.

It should be noted that all-optical regeneration can provide simultaneous optical amplification and signal regeneration without any electrical to optical signal conversion (see Section 10.6). Hence such all-optical regeneration also performs optical signal reshaping and retiming.

A method which is often used to obtain a qualitative indication of the performance of a regenerative repeater or a PCM system is the examination of the received waveform on an oscilloscope using a sweep rate which is a fraction of the bit rate. The display obtained over two bit intervals' duration, which is the result of superimposing all possible pulse sequences, is called an eye diagram or pattern. An illustration of an eye diagram for a binary system with little distortion and no additive noise is shown in Figure 12.33(a). It may be observed that the diagram has the shape of a human eye which is open and that the decision time corresponds to the center of the opening. To regenerate the pulse sequence without error the eye must be open thereby indicating that a decision area exists, and the decision crosshair (provided by the decision time and the decision threshold) must be within this open area. The effect of practical degradations on the pulses (i.e. intersymbol interference and noise) is to reduce the size of, or close, the eye, as shown in Figure 12.33(b). Hence for reliable transmission it is essential that the eye is kept open, the margin against an error occurring being the minimum distance between the decision crosshair and the edge of the eye.

In practice, a low bit error rate in the region 10^{-7} to 10^{-10} may be tolerated with PCM transmission. However, with data transmission (e.g. computer communications) any error can cause severe problems, and it is necessary to incorporate error detecting and possibly correcting circuits into the regenerator. This invariably requires the insertion of a small amount of redundancy into the transmitted pulse train (see Section 12.6.7).

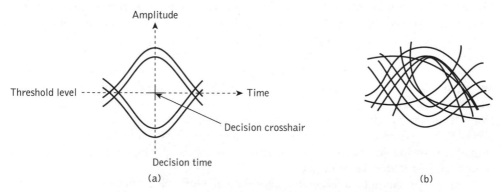

Figure 12.33 Eye diagrams in binary digital transmission: (a) the diagram obtained with a bandwidth limitation but no additive noise (open eye); (b) the diagram obtained with a bandwidth limitation and additive noise (partially closed eye)

Calculation of the possible repeater spacing must take account of the following system component performances:

(a) the average optical power launched into the fiber based on the end-of-life transmitter performance;

(b) the receiver input power required to achieve an acceptably low BER (e.g. 10^{-9}), taking into account component deterioration during the system's lifetime;

(c) the installed fiber cable loss including jointing and coupling (to source and detector) losses as well as the effects of aging and from anticipated environmental changes;

(d) the temporal response of the system including the effects of pulse dispersion on the channel; this becomes an important consideration with high-bit-rate multimode fiber systems which may be dispersion limited.

These considerations are discussed in detail in the following sections.

12.6.2 The optical transmitter and modulation formats

The average optical power launched into the fiber from the transmitter depends upon the type of source used and the required system bit rate, as indicated in Section 12.2.1. Typically, the laser launches around 1 mW, whereas usually the LED is limited to about 100 µW [Ref. 46]. It may also be noted that both device types emit less optical power at higher bit rates. However, the LED gives reduced output at modulation bandwidths in excess of 50 MHz, whereas laser output is unaffected below 40 GHz. Also, the fact that generally the optical power which may be launched into a fiber from an LED even at low bit rates is 10 to 15 dB down on that available from a laser is an important consideration, especially when receiver noise is a limiting factor within the system.

The signal generated by an optical source is required to be modulated in the transmitter prior to transmission over the optical fiber link. There are two major modulation formats which can be used in IM/DD-based digital optical communication systems, namely nonreturn-to-zero (NRZ) and return-to-zero (RZ) (see Section 3.8). RZ pulses can be produced using either two intensity modulators, or an intensity and phase modulator cascade. A conventional RZ signal format transmitter which comprises a CW laser source, two Mach–Zehnder modulators (i.e. MZM 1 and MZM 2 used as an intensity modulator and a pulse carver, respectively) together with a data encoder is shown in Figure 12.34. Initially RZ pulses are constructed in the intensity modulator which utilizes a clocked MZM to modulate the NRZ optical input signal. In order to generate the appropriate RZ pulses with constant amplitude, the drive signal for the pulse carver must be synchronized with the MZM clock signal. MZM 2 then produces a constant amplitude optical RZ pulse train at the output [Ref. 47]. Using this technique RZ transmitters operating at a bit rate of 40 Gbit s^{-1} have been successfully demonstrated [Refs 47, 48].

The RZ signal format displays considerable performance advantages over that of NRZ [Ref. 49]. For instance, it exhibits higher peak power together with greater noise immunity and therefore has a better BER performance. Moreover, it is less subject to fiber-related nonlinear effects [Refs 50, 51]. Eye diagrams for RZ and NRZ optical transmission are

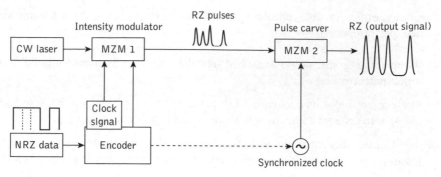

Figure 12.34 A return-to-zero (RZ) signal format transmitter

depicted in Figure 12.35(a). It can be observed that the eye diagram for the RZ format displays a larger vertical eye opening in comparison with that for NRZ. Furthermore, it should be noted that the narrow vertical eye opening (i.e. eye closure) determines the presence of intersymbol interference which arises as a consequence of fiber nonlinear effects (i.e. four-wave mixing and self/cross-phase modulation (see Section 3.14)) whereas the horizontal eye opening is related to deterministic jitter including pulse-width distortion [Ref. 53]. The wider vertical eye opening for the RZ signal format is indicative of a higher

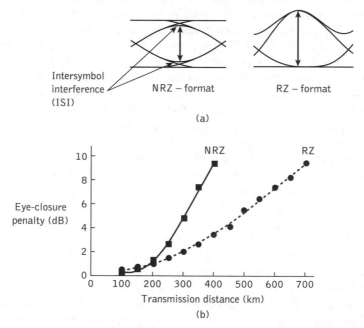

Figure 12.35 Performance comparison of the nonreturn to zero (NRZ) and the return to zero (RZ) modulation formats: (a) equivalent transmission rate eye diagrams; (b) eye-closure penalty against distance for a 16-channel WDM system operating at a transmission rate of 40 Gbit s^{-1}. Reprinted with permission from Ref. 52 © IEEE 1999

SNR at the receiver (i.e. by typically 1 to 3 dB) and hence the RZ format also exhibits greater tolerance to ISI for wavelength division multiplexed signals.

Figure 12.35(b) provides a comparison of eye-closure penalty for optical NRZ and RZ modulation formats when used in a 16-channel WDM system (see Section 12.9.4) operating at a transmission rate of 40 Gbit s^{-1} [Ref. 52]. It may be noted that a maximum eye-closure penalty of 9.6 dB was obtained for a 700 km single-mode fiber link when using the RZ signal format whereas the same value of eye-closure penalty was produced with the much shorter distance of only 400 km using an NRZ format.

Other characteristics of the RZ signal can also be manipulated to obtain more robust features for high-speed transmission. For example, prechirping* of the pulse with the sign of chirp opposite to that introduced by the fiber dispersion can be employed and the resultant signal format is referred to as chirped return-to-zero (CRZ) [Ref. 52]. This feature can enhance the performance of the optical transmission system due to the pulse compression effect introduced through the prechirping which will combat dispersion on the fiber link. In another RZ scheme an alternate bit phase inversion process removes or suppresses the carrier component from the power spectral density of the RZ signal and it is therefore known as a carrier-suppressed return-to-zero (CSRZ) [Ref. 53]. In particular, longer transmission distances can be realized using CSRZ in comparison with conventional RZ [Ref. 54].

For optical fiber multiplexed transmission, specifically when a WDM system is utilized, the signal modulation format becomes of increasing importance as higher spectral efficiency is required through reducing the channel spacing, for instance, to produce dense WDM transmission. The spectral efficiency which can be defined as the ratio of average channel capacity to the average channel spacing determines the overall density of a WDM system. For example, in a practical WDM system operating at 40 Gbit s^{-1} with a channel spacing of 100 GHz, the spectral efficiency for a conventional binary signal will be 0.4 bit s^{-1} Hz^{-1}. Clearly, decreasing the channel spacing increases the spectral efficiency. Higher values of ISI caused by fiber nonlinear effects may, however, be present on the resultant data signal. It should be noted that as the information carried on each channel occupies a particular spectral band, then if the channel spacing is decreased beyond a specific limit it leads to overlapping of adjacent channel information and hence to degradation of the data signals.

In order to decrease the optical spectral band occupied by a channel without decreasing the amount of information or data carried requires the use of an even more efficient modulation format such as alternate mark inversion (AMI) or duobinary (DB) which are translated from the same general class of partial response (PR) or correlative coding format signaling [Refs 50, 51]. In a PR or a correlative coding scheme the output response of a decoder is obtained by determining the correlation between the input data and an n bit delayed sequence of itself. For example, when transmitting a binary input data sequence of 100110 with a 1 bit delayed sequence of itself (i.e. 010011)† then the encoder can transmit a correlative output data sequence using a ternary scheme 11021. In this case the three states of the ternary scheme (i.e. binary 0, 1 and 2) may be identified by assigning three distinct electrical voltages values of 0, +V and −V, respectively producing an AMI signal

* Prechirping is a process which adds a chirp (see Section 6.7.3) to an optical signal before the transmission of the pulse.
† Assuming a binary 0 value for the initial bit in 1 bit delayed data as identified by underlined zero (i.e. 0).

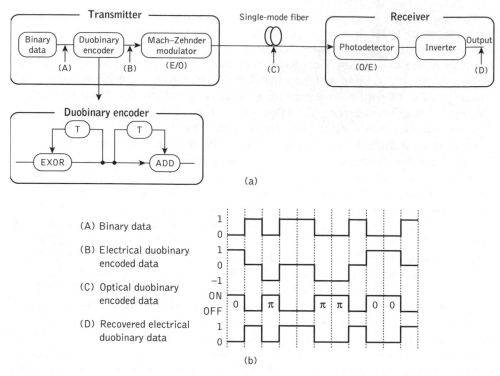

Figure 12.36 Duobinary transmitter and receiver: (a) block diagram of an optical duobinary transmission system; (b) signal waveforms obtained at points identified in the transmission system [Ref. 55]

format. These three voltage levels can be translated for use in an optical signal using a phase change to represent the $-V$ or $+V$ levels around the zero-phase point (i.e. 0, π). The MZI and delay interferometer (see Section 10.5.2) can be used to perform the phase shift operation on the optical signal.

A block schematic of a single-mode fiber system employing a duobinary transmission is illustrated in Figure 12.36(a) [Ref. 55]. The transmitter comprises an electrical duobinary encoder followed by a Mach–Zehnder modulator to provide the optical signal. It may be observed that the duobinary encoder consists of an exclusive OR gate, an adder and 1 bit delay circuit. At the receiving end a photodetector converts the optical signal back into the electrical domain. Figure 12.36(b) provides the corresponding waveforms obtained at the output of each section of the transmission system. Different combinations of logic circuits are employed in the duobinary encoder to produce the electrical duobinary data which forms a ternary or three-level coding scheme as displayed in waveform B.

The electrical duobinary data can be converted to an optical signal by using both ON/OFF and 0 or π phase values corresponding to a three-level electrical duobinary signal. This conversion process is identified at stage C in Figure 12.36(b) where an on state optical signal with 0 phase represents a binary 1 and an on state with a π phase indicates the minus one level corresponding to the electrical duobinary signal. The zero level of the electrical duobinary is simply produced by not transmitting an optical signal (i.e. OFF). Comparison of the data sequences at stages A and C indicates that the binary data can be recovered by simply inverting the optical intensity modulated signal as shown at stage D. This constitutes an important feature of optical duobinary systems in that the electrical signal can be recovered by direct detection at the photodiode followed by an electrical signal inversion process without the need to determine or recover the phase of the optical signal. It is possible, however, to use the optical phase change information if it is recovered by employing a coherent receiver (see Section 13.2) rather than a direct detection receiver [Ref. 56].

The optical duobinary signal format exhibits more tolerance to chromatic dispersion (see Section 3.9) than conventional binary signaling formats. This improvement occurs because the optical duobinary signal occupies only around half the bandwidth of an optical NRZ signal. Since fiber dispersion is dependent on signal bandwidth, optical duobinary therefore typically provides twice the dispersion tolerance to chromatic dispersion. Moreover, the narrow bandwidth of optical duobinary signaling also enables reduced channel spacings (e.g. 25 GHz) when combined with WDM. Therefore this format is employed with dense WDM over long-distance single-mode fiber links using conventional direct detection optical receivers operating at transmission rates of 10 Gbit s^{-1} [Refs 51, 55].

Another intensity modulated-based CSRZ format utilizes vestigial sideband (VSB) transmission which can be employed to achieve increased spectral efficiency [Refs 51, 54]. Hence VSB-CSRZ is an efficient technique because the complete information of a VSB channel is contained in only half of its spectrum while the other half of the spectrum is redundant information which can either be ignored or be reproduced from the half of the spectrum which has been retained [Ref. 57]. VSB signals are usually obtained by partially removing one of the sideband spectra from the conventional double-sideband spectrum (see Section 12.7.4) using an optical filter. The benefit of this process is that it therefore decreases the required channel spectrum and hence it enables channel spacing to be reduced.

12.6.3 The optical receiver

The input optical power required at the receiver is a function of the detector combined with the electrical components within the receiver structure. It is strongly dependent upon the noise (i.e. quantum, dark current and thermal) associated with the optical fiber receiver. The theoretical minimum pulse energy or quantum limit required to maintain a given BER was discussed in Section 9.2.4.

It was predicted that approximately 21 incident photons were necessary at an ideal photo-detector in order to register a binary 1 with a BER of 10^{-9}. However, this is a fundamental limit which cannot be achieved in practice and therefore it is essential that estimates of the minimum required optical input power are made in relation to practical devices and components.

Although the statistics of quantum noise follow a Poisson distribution, other important sources of noise within practical receivers (e.g. thermal) are characterized by a Gaussian probability distribution. Hence estimates of the required SNR to maintain particular BERs may be obtained using the procedure adopted for error performance of electrical digital systems where the noise distribution is considered to be white Gaussian. This Gaussian approximation [Ref. 58] is sufficiently accurate for design purposes and is far easier to evaluate than the more exact probability distribution within the receiver [Ref. 59]. The

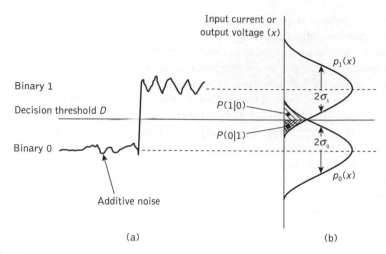

Figure 12.37 Binary transmission: (a) the binary signal with additive noise; (b) probability density functions for the binary signal showing the decision case. $P(0|1)$ is the probability of falsely identifying a binary 1 and $P(1|0)$ is the probability of falsely identifying a binary 0

receiver sensitivities calculated by using the Gaussian approximation are generally within 1 dB of those calculated by other methods [Ref. 30].

Although the transmitted signal consists of two well-defined light levels, in the presence of noise the signal at the receiver is not as well defined. This situation is shown in Figure 12.37(a) which illustrates a binary signal in the presence of noise. The signal plus the additive noise at the detector may be defined in terms of the probability density functions (PDFs) shown in Figure 12.37(b). These PDFs describe the probability that the input current (or output voltage) has a value i (or v) within the incremental range di (or dv). The expected values of the signal in the two transmitted states, namely 0 and 1, are indicated by $p_0(x)$ and $p_1(x)$ respectively. When the additive noise is assumed to have a Gaussian distribution, the PDFs of the two states will also be Gaussian. The Gaussian PDF which is continuous is defined by:

$$p(x) = \frac{1}{\sigma\sqrt{(2\pi)}} \exp\{-[(x-m)^2/2\sigma^2]\} \tag{12.5}$$

where m is the mean value and σ the standard deviation of the distribution. When $p(x)$ describes the probability of detecting a noise current or voltage, σ corresponds to the rms value of that current or voltage.

If a decision threshold D is set between the two signal states, as indicated in Figure 12.37, signals greater than D are registered as a one and those less than D as a zero. However, when the noise current (or voltage) is sufficiently large it can either decrease a binary 1 to a 0 or increase a binary 0 to a 1. These error probabilities are given by the integral of the signal probabilities outside the decision region. Hence the probability that a signal transmitted as a 1 is received as a 0, $P(0|1)$, is proportional to the shaded area indicated in Figure 12.37(b). The probability that a signal transmitted as a 0 is received as a 1, $P(1|0)$, is similarly proportional to the other shaded area shown in the diagram. If $P(1)$ and $P(0)$ are the probabilities of transmission for binary ones and zeros, respectively, then the total probability of error $P(\text{e})$ may be defined as:

$$P(\text{e}) = P(1)P(0|1) + P(0)P(1|0) \tag{12.6}$$

Now let us consider a signal current i_{sig} together with an additive noise current i_{N} and a decision threshold set at $D = i_{\text{D}}$. If at any time when a binary 1 is transmitted the noise current is negative such that:

$$i_{\text{N}} < -(i_{\text{sig}} - i_{\text{D}}) \tag{12.7}$$

then the resulting current $i_{\text{sig}} + i_{\text{N}}$ will be less than i_{D} and an error will occur. The corresponding probability of the transmitted 1 being received as a 0 may be written as:

$$P(0|1) = \int_{-\infty}^{i_{\text{D}}} p(i, i_{\text{sig}}) \, di \tag{12.8}$$

and following Eq. (12.5):

$$p_1(x) = p(i, i_{\text{sig}}) = \frac{1}{(\overline{i_{\text{N}}^2})^{\frac{1}{2}}\sqrt{(2\pi)}} = \exp\left\{-\left[\frac{(i - i_{\text{sig}})^2}{2(\overline{i_{\text{N}}^2})}\right]\right\} \tag{12.9}$$

$$= \text{Gsn} \left[i, i_{\text{sig}}, (\overline{i_{\text{N}}^2})^{\frac{1}{2}}\right] \tag{12.10}$$

where i is the actual current, i_{sig} is the peak signal current during a binary 1 (this corresponds to the peak photocurrent I_{p} when only a signal component is present), and $\overline{i_{\text{N}}^2}$ is the mean square noise current. Substituting Eq. (12.10) into Eq. (12.8) gives:

$$P(0|1) = \int_{-\infty}^{i_{\text{D}}} \text{Gsn} \left[i, i_{\text{sig}}, (\overline{i_{\text{N}}^2})^{\frac{1}{2}}\right] \, di \tag{12.11}$$

Similarly, the probability that a binary 1 will be received when a 0 is transmitted is the probability that the received current will be greater than i_{D} at some time during the 0 bit interval. It is given by:

$$P(1|0) = \int_{i_{\text{D}}}^{\infty} p(i, 0) \tag{12.12}$$

Assuming the mean square noise current in the zero state is equal to the mean square noise current in the one state $(\overline{i_{\text{N}}^2})$ (this is an approximation if shot noise is dominant), and that for a zero bit $i_{\text{sig}} = 0$, then following Eq. (12.5):

$$p_0(x) = p(i, 0) = \frac{1}{(\overline{i_{\text{N}}^2})^{\frac{1}{2}}\sqrt{(2\pi)}} \exp\left\{-\left[\frac{(i - 0)^2}{2(\overline{i_{\text{N}}^2})}\right]\right\} \tag{12.13}$$

$$= \text{Gsn} \left[i, 0, (\overline{i_{\text{N}}^2})^{\frac{1}{2}}\right] \tag{12.14}$$

Hence substituting Eq. (12.14) into Eq. (12.12) gives:

$$P(1|0) = \int_{i_D}^{\infty} \mathrm{Gsn}\, [i, 0, (\overline{i_N^2})^{\frac{1}{2}}]\, di \tag{12.15}$$

The integrals of Eqs (12.11) and (12.15) are not readily evaluated but may be written in terms of the error function (erf)* where:

$$\mathrm{erf}(u) = \frac{2}{\sqrt{\pi}} \int_0^u \exp(-z^2)\, dz \tag{12.16}$$

and the complementary error function is:

$$\mathrm{erfc}(u) = 1 - \mathrm{erf}(u) = \frac{2}{\sqrt{\pi}} \int_u^{\infty} \exp(-z^2)\, dz \tag{12.17}$$

Hence:

$$P(0|1) = \frac{1}{2}\left[1 - \mathrm{erf}\left(\frac{|i_{\mathrm{sig}} - i_D|}{(\overline{i_N^2})^{\frac{1}{2}}\sqrt{2}}\right)\right]$$

$$= \frac{1}{2}\,\mathrm{erfc}\left(\frac{|i_{\mathrm{sig}} - i_D|}{(\overline{i_N^2})^{\frac{1}{2}}\sqrt{2}}\right) \tag{12.18}$$

and:

$$P(1|0) = \frac{1}{2}\,\mathrm{erfc}\left(\frac{|0 - i_D|}{(\overline{i_N^2})^{\frac{1}{2}}\sqrt{2}}\right) = \frac{1}{2}\,\mathrm{erfc}\left(\frac{|-i_D|}{(\overline{i_N^2})^{\frac{1}{2}}\sqrt{2}}\right) \tag{12.19}$$

If we assume that a binary code is chosen such that the number of transmitted ones and zeros is equal, then $P(0) = P(1) = \frac{1}{2}$, and the net probability of error is one-half the sum of the shaded areas in Figure 12.37(b). Therefore Eq. (12.6) becomes:

$$P(e) = \tfrac{1}{2}[P(0|1) + P(1|0)] \tag{12.20}$$

and substituting for $P(0|1)$ and $P(1|0)$ from Eqs (12.18) and (12.19) gives:

$$P(e) = \frac{1}{2}\left[\frac{1}{2}\,\mathrm{erfc}\left(\frac{|i_{\mathrm{sig}} - i_D|}{(\overline{i_N^2})^{\frac{1}{2}}\sqrt{2}}\right) + \frac{1}{2}\,\mathrm{erfc}\left(\frac{|-i_D|}{(\overline{i_N^2})^{\frac{1}{2}}\sqrt{2}}\right)\right] \tag{12.21}$$

Equation (12.21) may be simplified by setting the threshold decision level at the mid-point between zero current and the peak signal current such that $i_D = i_{\mathrm{sig}}/2$. In electrical systems this situation corresponds to an equal minimum probability of error in both states due to the symmetrical nature of the PDFs. It must be noted that for optical fiber systems

---

* Another form of the error function denoted by Erf is defined in Problem 12.10.

this is not generally the case since the noise in each signal state contains shot noise contributions proportional to the signal level. Nevertheless, assuming a Gaussian distribution for the noise and substituting $i_D = i_{sig}/2$ into Eq. (12.21) we obtain:

$$P(e) = \frac{1}{2}\left[\frac{1}{2}\text{erfc}\left(\frac{|i_{sig}/2|}{(\overline{i_N^2})^{\frac{1}{2}}\sqrt{2}}\right) + \frac{1}{2}\text{erfc}\left(\frac{|-i_{sig}/2|}{(\overline{i_N^2})^{\frac{1}{2}}\sqrt{2}}\right)\right]$$

$$= \frac{1}{2}\text{erfc}\left(\frac{i_{sig}}{2(\overline{i_N^2})^{\frac{1}{2}}\sqrt{2}}\right) \tag{12.22}$$

The electrical SNR at the detector may be written in terms of the peak signal power to rms noise power (mean square noise current) as:

$$\frac{S}{N} = \frac{i_{sig}^2}{\overline{i_N^2}} \tag{12.23}$$

Comparison of Eq. (12.23) with Eq. (12.22) allows the probability of error to be expressed in terms of the analog SNR as:

$$P(e) = \frac{1}{2}\text{erfc}\left(\frac{(S/N)^{\frac{1}{2}}}{2\sqrt{2}}\right) \tag{12.24}$$

Estimates of the required SNR to maintain a given error rate may be obtained using the standard table for the complementary error function. A plot of $P(e)$ against $\frac{1}{2}\text{erfc}(u)$ is shown in Figure 12.38(a). This may be transposed into the characteristic illustrated in Figure 12.38(b) where the BER which is equivalent to the error probability $P(e)$ is shown as a function of the SNR following Eq. (12.24).

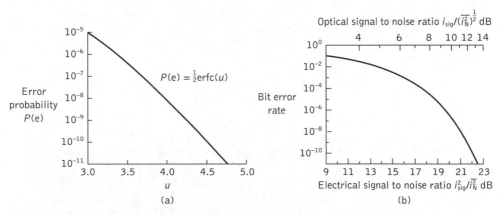

Figure 12.38 (a) A plot of the probability of error $\frac{1}{2}\text{erfc}(u)$ against the argument of the error function u. (b) The BER as a function of both the ratio of peak signal power to rms noise power (electrical SNR) and the ratio of peak signal current to rms noise current (optical SNR) for binary transmission

Example 12.2

Using the Gaussian approximation determine the required SNRs (optical and electrical) to maintain a BER of 10^{-9} on a baseband binary digital optical fiber link. It may be assumed that the decision threshold is set midway between the one and the zero level and that $2 \times 10^{-9} \simeq \text{erfc } 4.24$.

Solution: Under the above conditions, the probability of error is given by Eq. (12.24) where:

$$P(e) = \frac{1}{2}\text{erfc}\left(\frac{(S/N)^{\frac{1}{2}}}{2\sqrt{2}}\right) = 10^{-9}$$

Hence:

$$\text{erfc}\left(\frac{(S/N)^{\frac{1}{2}}}{2\sqrt{2}}\right) = 2 \times 10^{-9}$$

and:

$$\frac{(S/N)^{\frac{1}{2}}}{2\sqrt{2}} = 4.24$$

giving:

$$(S/N)^{\frac{1}{2}} = 4.24 \times 2\sqrt{2} \simeq 12$$

The optical SNR may be defined in terms of the peak signal current and rms noise current as $i_{\text{sig}}/(\overline{i_{\text{N}}^2})^{\frac{1}{2}}$. Therefore using Eq. (12.23):

$$\frac{i_{\text{sig}}}{(\overline{i_{\text{N}}^2})^{\frac{1}{2}}} = \left(\frac{S}{N}\right)^{\frac{1}{2}} = 12 \quad \text{or} \quad 10.8 \text{ dB}$$

The electrical SNR is defined by Eq. (12.23) as:

$$\frac{\overline{i_{\text{sig}}^2}}{\overline{i_{\text{N}}^2}} = \frac{S}{N} = 144 \quad \text{or} \quad 21.6 \text{ dB}$$

These results for the SNRs may be seen to correspond to a BER of 10^{-9} on the curve shown in Figure 12.38(b).

However, the plot shown in Figure 12.38(b) does not reflect the best possible results, or those which may be obtained with an optimized receiver design. In this case, if the system is to be designed with a particular BER, the appropriate value of the error function is established prior to adjustment of the parameter values (signal levels, decision threshold level, avalanche gain, component values, etc.) in order to obtain this BER [Ref. 60]. It is therefore necessary to use the generalized forms of Eqs (12.18) and (12.19) where:

$$P(0|1) = \frac{1}{2}\text{erfc}\left(\frac{|i_{\text{sig}\,1} - i_{\text{D}}|}{(\overline{i_{\text{N}}^2})^{\frac{1}{2}}\sqrt{2}}\right) \tag{12.25}$$

$$P(0|1) = \frac{1}{2}\text{erfc}\left(\frac{|i_{\text{D}} - i_{\text{sig}\,0}|}{(\overline{i_{\text{N0}}^2})^{\frac{1}{2}}\sqrt{2}}\right) \tag{12.26}$$

where $i_{\text{sig}\,\frac{1}{2}}$ and $i_{\text{sig}\,0}$ are the signal currents, in the 1 and 0 states, respectively, and $\overline{i_{\text{N1}}^2}$ and $\overline{i_{\text{N0}}^2}$ are the corresponding mean square noise currents which may include both shot and thermal noise terms. Equations (12.25) and (12.26) allow a more exact evaluation of the error performance of the digital optical fiber system under the Gaussian approximation [Ref. 60]. Unfortunately, this approach does not give as simple a direct relationship between the BER and the analog SNR as the one shown in Eq. (12.24). Thus for estimates of SNR within this text we will make use of the slightly poorer approximation given by Eq. (12.24). Although this approximation does not give the correct decision threshold level or optimum avalanche gain it is reasonably successful at predicting BER as a function of signal power and hence provides realistic estimates of the number of photons required at a practical detector in order to maintain given BERs.

For instance, let us consider a good avalanche photodiode receiver which we assume to be quantum noise limited. Hence we ignore the shot noise contribution from the dark current within the APD, as well as the thermal noise generated by the electronic amplifier. In practice, this assumption holds when the multiplication factor M is chosen to be sufficiently high to ensure that the SNR is determined by photon noise rather than by electronic amplifier noise, and the APD used has a low dark current. To determine the SNR for this ideal APD receiver it is useful to define the quantum noise on the primary photocurrent I_{p} within the device in terms of shot noise following Eq. (9.8). Therefore, the mean square shot noise current is given by:

$$\overline{i_{\text{s}}^2} = 2eBI_{\text{p}}M^2 \tag{12.27}$$

where e is the electronic charge and B is the post-detection or effective noise bandwidth. It may be observed that the mean square shot noise current $\overline{i_{\text{s}}^2}$ given in Eq. (12.27) is increased by a factor M^2 due to avalanche gain in the APD. However, Eq. (12.27) does not give the total noise current at the output of the APD as there is an additional noise contribution from the random gain mechanism. The excess avalanche noise factor $F(M)$ incurred was discussed in Section 9.3.4 and defined by Eqs (9.27) and (9.28). Equation (9.27) may be simplified [Ref. 61] to give an expression for electron injection in the low-frequency limit of:

$$F(M) = kM + \left(2 - \frac{1}{M}\right)(1 - k) \tag{12.28}$$

where k is the ratio of the carrier ionization rates. Hence, the excess avalanche noise factor may be combined into Eq. (12.27) to give a total mean square shot noise current $\overline{i_{\text{n}}^2}$ as:

$$\overline{i_n^2} = 2eBI_pM^2F(M) \tag{12.29}$$

Furthermore, the avalanche multiplication mechanism raises the signal current to MI_p and therefore the SNR in terms of the peak signal power to rms noise power may be written as:

$$\frac{S}{N} = \frac{(MI_p)^2}{2eBI_pM^2F(M)} = \frac{I_p}{2eBF(M)} \tag{12.30}$$

Now, if we let z_{md} correspond to the average number of photons detected in a time period of duration τ, then:

$$I_p = \frac{z_{md}e}{\tau} = \frac{z_me\eta}{\tau} \tag{12.31}$$

where z_m is the average number of photons incident on the APD and η is the quantum efficiency of the device. Substituting for I_p in Eq. (12.30) we have:

$$\frac{S}{N} = \frac{z_m\eta}{2B\tau F(M)} \tag{12.32}$$

Rearranging Eq. (12.32) gives an expression for the average number of photons required within the signaling interval τ to detect a binary 1 in terms of the received SNR for the good APD receiver as:

$$z_m = \frac{2B\tau F(M)}{\eta}\left(\frac{S}{N}\right) \tag{12.33}$$

A reasonable pulse shape obtained at the receiver in order to reduce intersymbol interference has the raised cosine spectrum shown in Figure 12.39. The raised cosine spectrum for the received pulse gives a pulse response resulting in a binary pulse train passing through either full or zero amplitude at the centers of the pulse intervals and with transitions passing through half amplitude at points which are midway in time between pulse centers. For raised cosine pulse shaping and full τ signaling $B\tau$ is around 0.6. Hence the

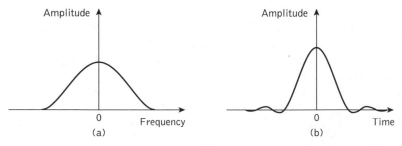

Figure 12.39 (a) Raised cosine spectrum. (b) Output of a system with a raised cosine output spectrum for a single input pulse

average number of photons required to detect a binary 1 using a good APD receiver at a specified BER may be estimated using Eq. (12.33) in conjunction with Eq. (12.24).

Example 12.3

A good APD is used as a detector in an optical fiber PCM receiver designed for base-band binary transmission with a decision threshold set midway between the zero and one signal levels. The APD has a quantum efficiency of 80%, a ratio of carrier ionization rates of 0.02 and is operated with a multiplication factor of 100. Assuming a raised cosine signal spectrum at the receiver, estimate the average number of photons which must be incident on the APD to register a binary 1 with a BER of 10^{-9}.

Solution: The electrical SNR required to obtain a BER of 10^{-9} at the receiver is given by the curve shown in Figure 12.38(b), or the solution to Example 12.2 as 21.6 dB or 144. Also, the excess avalanche noise factor $F(M)$ may be determined using Eq. (12.28) where:

$$F(M) = kM + \left(2 - \frac{1}{M}\right)(1 - k)$$

$$= 2 + (2 - 0.01)(1 - 0.02)$$
$$= 3.95 \simeq 4$$

The average number of photons which must be incident at the receiver in order to maintain the BER can be estimated using Eq. (12.33) (assuming $B\tau = 0.6$ for the raised cosine pulse spectrum) as:

$$z_m = \frac{2B\tau F(M)}{\eta}\left(\frac{S}{N}\right)$$

$$= \frac{2 \times 0.6 \times 4 \times 144}{0.8}$$

$$= 864 \text{ photons}$$

The estimate in Example 12.3 gives a more realistic value for the average number of incident photons required at a good APD receiver in order to register a binary 1 with a BER of 10^{-9} than the quantum limit of 21 photons determined for an ideal photodetector in Example 9.1. However, it must be emphasized that the estimate in Example 12.3 applies to a good silicon APD receiver (with high sensitivity and low dark current) which is quantum noise limited, and that no account has been taken of the effects of either dark current within the APD or thermal noise generated within the preamplifier. It is therefore likely that at least 1000 incident photons are required at a good APD receiver to register a binary 1 and provide a BER of 10^{-9} [Ref. 62]. Nevertheless somewhat lower values may be achieved by setting the decision threshold below the half amplitude level because the shot noise on the zero level is lower than the shot noise on the one level.

The optical power required at the receiver P_o is simply the optical energy divided by the time interval over which it is incident. The optical energy E_o may be obtained directly from the average number of photons required at the receiver in order to maintain a particular BER following:

$$E_o = z_m hf \tag{12.34}$$

where hf is the energy associated with a single photon which is given by Eq. (6.1). In order that a binary 1 is registered at the receiver, the optical energy E_o must be incident over the bit interval τ. For system calculations we can assume a zero disparity code which has an equal density of ones and zeros. In this case the optical power required to register a binary 1 may be considered to be incident over two bit intervals giving:

$$P_o = \frac{E_o}{2\tau} \tag{12.35}$$

Substituting for E_o from Eq. (12.34) we obtain:

$$P_o = \frac{z_m hf}{2\tau} \tag{12.36}$$

Also as the bit rate B_T for the channel is the reciprocal of the bit interval τ, Eq. (12.36) may be written as:

$$P_o = \frac{z_m hf B_T}{2} \tag{12.37}$$

Equation (12.37) allows estimates of the incident optical power required at a good APD receiver in order to maintain a particular BER, based on the average number of incident photons. In system calculations these optical power levels are usually expressed in dBm. It may also be observed that the required incident optical power is directly proportional to the bit rate B_T which typifies a shot-noise-limited receiver.

Example 12.4

The receiver of Example 12.3 operates at a wavelength of 1 μm. Assuming a zero disparity binary code, estimate the incident optical power required at the receiver to register a binary 1 with a BER of 10^{-9} at bit rates of 10 Mbit s^{-1} and 140 Mbit s^{-1}.

Solution: Under the above conditions, the required incident optical power may be obtained using Eq. (12.37) where:

$$P_o = \frac{z_m hf B_T}{2} = \frac{z_m hc B_T}{2\lambda}$$

At 10 Mbit s^{-1}:

$$P_o = \frac{864 \times 6.626 \times 10^{-34} \times 2.998 \times 10^8 \times 10^7}{2 \times 1 \times 10^{-6}}$$

$$= 858.2 \text{ pW}$$
$$= -60.7 \text{ dBm}$$

At 140 Mbit s^{-1}:

$$P_o = \frac{864 \times 6.626 \times 10^{-34} \times 2.998 \times 10^8 \times 14 \times 10^7}{2 \times 1 \times 10^{-14}}$$

$$= 12.015 \text{ nW}$$
$$= -49.2 \text{ dBm}$$

Example 12.4 illustrates the effect of direct proportionality between the optical power required at the receiver and the system bit rate. In the case considered, the required incident optical power at the receiver to give a BER of 10^{-9} must be increased by around 11.5 dB (factor of 14) when the bit rate is increased from 10 to 140 Mbit s^{-1}. Also, comparison with Example 9.1 where a similar calculation was performed for an ideal photodetector operating at 10 Mbit s^{-1} emphasizes the necessity of performing the estimate for a practical photodiode. The good APD receiver considered in Example 12.4 exhibits around 16 dB less sensitivity than the ideal photodetector (i.e. quantum limit).

The assumptions made in the evaluation of Examples 12.3 and 12.4 are not generally valid when considering p–i–n photodiode receivers because these devices are seldom quantum noise limited due to the absence of internal gain within the photodetector. In this case thermal noise generated within the electronic amplifier is usually the dominating noise contribution and is typically 1×10^3 to 3×10^3 times larger than the peak response produced by the displacement current of a single electron–hole pair liberated in the detector. Hence, for reliable performance with a BER of 10^{-9}, between 1 and 3×10^4 photons must be detected when a binary 1 is incident on the receiver [Ref. 63]. This translates into sensitivities which are about 30 dB or more, less than the quantum limit.

Finally, for a thermal-noise-limited receiver the input optical power is proportional to the square root of both the post-detection or effective noise bandwidth and the SNR (i.e. $P_o \propto |(S/N)B|^{\frac{1}{2}}$). However, this result is best obtained from purely analog SNR considerations and therefore is dealt with in Section 12.7.1.

12.6.4 Channel losses

Another important factor when estimating the permissible separation between regenerative repeaters or the overall link length is the total loss encountered between the transmitter(s) and receiver(s) within the system. Assuming there are no dispersion penalties on the link, the total channel loss may be obtained by simply summing in decibels the installed fiber cable loss, the fiber–fiber jointing losses and the coupling losses of the optical source and detector. The fiber cable loss in decibels per kilometer α_{fc} is normally specified by the manufacturer, or alternatively it may be obtained by measurement (see Sections 14.2 and 14.10). It must be noted that the cabled fiber loss is likely to be greater than the uncabled fiber loss usually measured in the laboratory due to possible microbending of the fiber within the cabling process (see Section 4.7.1).

Loss due to joints (generally splices) on the link may also, for simplicity, be specified in terms of an equivalent loss in decibels per kilometer α_j. In fact, it is more realistic to regard α_j as a distributed loss since the optical attenuation resulting from the disturbed mode distribution at a joint does not only occur in the vicinity of the joint. Finally, the loss contribution attributed to the connectors α_{cr} (in decibels) used for coupling the optical source and detector to the fiber must be included in the overall channel loss. Hence the total channel loss C_L (in decibels) may be written as:

$$C_L = (\alpha_{fc} + \alpha_j)L + \alpha_{cr} \tag{12.38}$$

where L is the length in kilometers of the fiber cable either between regenerative repeaters or between the transmit and receive terminals for a link without repeaters.

Example 12.5

An optical fiber link of length 4 km comprises a fiber cable with an attenuation of $5\ \text{dB km}^{-1}$. The splice losses for the link are estimated at $2\ \text{dB km}^{-1}$, and the connector losses at the source and detector are 3.5 and 2.5 dB respectively. Ignoring the effects of dispersion on the link determine the total channel loss.

Solution: The total channel loss may be simply obtained using Eq. (12.38) where:

$$
\begin{aligned}
C_L &= (\alpha_{fc} + \alpha_j)L + \alpha_{cr} \\
&= (5 + 2)4 + 3.5 + 2.5 \\
&= 34\ \text{dB}
\end{aligned}
$$

12.6.5 Temporal response

The system design considerations must also take into account the temporal response of the system components. This is especially the case with regard to pulse dispersion on the optical fiber channel. The formula given in Eq. (12.38) allows determination of the overall channel loss in the absence of any pulse broadening due to the dispersion mechanisms within the transmission medium. However, the finite bandwidth of the optical system may result in overlapping of the received pulses or ISI, giving a reduction in sensitivity at the optical receiver. Therefore, either a worse BER must be tolerated, or the ISI must be compensated by equalization within the receiver (see Section 12.3.3). The latter necessitates an increase in optical power at the receiver which may be considered as an additional loss penalty. This additional loss contribution is usually called the dispersion–equalization or

ISI penalty. The dispersion–equalization penalty D_L becomes especially significant in high-bit-rate multimode fiber systems and has been determined analytically for Gaussian-shaped pulses [Ref. 60]. In this case it is given by:

$$D_L = \left(\frac{\tau_e}{\tau}\right)^4 \text{ dB} \qquad (12.39)$$

where τ_e is the $1/e$ full width pulse broadening due to dispersion on the link and τ is the bit interval or period. For Gaussian-shaped pulses, τ_e may be written in terms of the rms pulse width σ as (see Appendix B):

$$\tau_e = 2\sigma\sqrt{2} \qquad (12.40)$$

Hence, substituting into Eq. (12.39) for τ_e and writing the bit rate B_T as the reciprocal of the bit interval τ gives:

$$D_L = 2(2\sigma B_T\sqrt{2})^4 \text{ dB} \qquad (12.41)$$

Since the dispersion–equalization penalty as defined by Eq. (12.41) is measured in decibels, it may be included in the formula for the overall channel loss given by Eq. (12.38). Therefore, the total channel loss including the dispersion–equalization penalty C_{LD} is given by:

$$C_{LD} = (\alpha_{fc} + \alpha_j)L + \alpha_{cr} + D_L \text{ dB} \qquad (12.42)$$

The dispersion–equalization penalty is usually only significant in wideband multimode fiber systems which exhibit intermodal as well as chromatic dispersion. Single-mode fiber systems which are increasingly being utilized for wideband long-haul applications are not generally limited by pulse broadening on the channel because of the absence of intermodal dispersion. However, it is often the case that intermodal dispersion is the dominant mechanism within multimode fibers. In Section 3.10.1 intermodal pulse broadening was considered to be a linear function of the fiber length L. Furthermore, it was indicated that the presence of mode coupling within the fiber made the pulse broadening increase at a slower rate proportional to $L^{\frac{1}{2}}$. Hence it is useful to consider the dispersion–equalization penalty in relation to fibers without and with mode coupling operating at various bit rates.

Example 12.6

The rms pulse broadening resulting from intermodal dispersion within a multimode optical fiber is 0.6 ns km^{-1}. Assuming this to be the dominant dispersion mechanism, estimate the dispersion–equalization penalty over an unrepeatered fiber link of length 8 km at bit rates of (a) 25 Mbit s^{-1} and (b) 150 Mbit s^{-1}. In both cases evaluate the penalty without and with mode coupling. The pulses may be assumed to have a Gaussian shape.

▶

Solution: (a) *Without mode coupling.* The total rms pulse broadening over 8 km is given by:

$$\sigma_T = \sigma \times L = 0.6 \times 8 = 4.8 \text{ ns}$$

The dispersion–equalization penalty is given by Eq. (12.41) where:

$$D_L = 2(2\sigma_T B_T \sqrt{2})^4 = 2(2 \times 4.8 \times 10^{-9} \times 25 \times 10^6 \sqrt{2})^4$$
$$= 0.03 \text{ dB}$$

With mode coupling. The total rms pulse broadening is:

$$\sigma_T \simeq \sigma\sqrt{L} = 0.6 \times \sqrt{8} = 1.7 \text{ ns}$$

Hence the dispersion–equalization penalty is:

$$D_L = 2(2 \times 1.7 \times 10^{-9} \times 25 \times 10^6 \sqrt{2})^4$$
$$= 4.2 \times 10^{-4} \text{ dB (i.e. negligible)}$$

(b) *Without mode coupling.* We have:

$$\sigma_T = 4.8 \text{ ns}$$
$$D_L = 2(2 \times 4.8 \times 10^{-9} \times 150 \times 10^6 \sqrt{2})^4 = 34.38 \text{ dB}$$

With mode coupling. We have:

$$\sigma_T = 1.7 \text{ ns}$$
$$D_L = 2(2 \times 1.7 = 10^{-9} \times 150 \times 10^6 \sqrt{2})^4 = 0.54 \text{ dB}$$

Example 12.6(a) demonstrates that at low bit rates the dispersion–equalization penalty is very small if not negligible. In this case the slight advantage of the effect of mode coupling on the penalty is generally outweighed by increased attenuation on the link because of the mode coupling, which may be of the order of 1 dB km^{-1}. Example 12.6(b) indicates that at higher bit rates with no mode coupling the dispersion-as equalization penalty dominates to the extent that it would be necessary to reduce the repeater spacing to between 4 and 5 km. However, it may be observed that encouragement of mode coupling on the link greatly reduces this penalty and outweighs any additional attenuation incurred through mode coupling within the fiber. In summary, it is clear that the dispersion equalization penalty need only be applied when considering wideband systems. Moreover, it is frequently the case that lower bit rate systems may be up graded at a later date to a higher capacity without incurring a penalty which might necessitate a reduction in repeater spacing.

An alternative approach involving the calculation of the system rise time can be employed to determine the possible limitation on the system bandwidth resulting from the temporal response of the system components. Therefore, if there is not a pressing need to obtain the maximum possible bit rate over the maximum possible distance, it is sufficient within the system design to establish that the total temporal response of the system is adequate for the desired system bandwidth. Nevertheless this approach does allow for a certain amount of optimization of the system components, but at the exclusion of considerations regarding equalization and the associated penalty.

The total system rise time may be determined from the rise times of the individual system components which include the source (or transmitter), the fiber cable and the detector (receiver). These times are defined in terms of a Gaussian response as the 10–90% rise (or fall) times of the individual components. The fiber cable 10–90% rise time may be separated into rise times arising from intermodal T_n and chromatic or intramodal dispersion T_c. The total system rise time is given by [Ref. 64]:

$$T_{syst} = 1.1(T_S^2 + T_n^2 + T_c^2 + T_D^2)^{\frac{1}{2}} \qquad (12.43)$$

where T_S and T_D are the source and detector 10–90% rise times, respectively, and all the rise times are measured in nanoseconds. Comparison of the rise time edge with the overall pulse dispersion results in the weighting factor of 1.1.

The maximum system bit rate $B_T(max)$ is usually defined in terms of T_{syst} by consideration of the rise time of the simple RC filter circuit shown in Figure 12.40(a). For a voltage step input of amplitude V, the output voltage waveform $v_{out}(t)$ as a function of time t is:

$$v_{out}(t) = V[1 - \exp(-t/RC)] \qquad (12.44)$$

Hence the 10–90% rise time t_r for the circuit is given by:

$$t_r = 2.2RC \qquad (12.45)$$

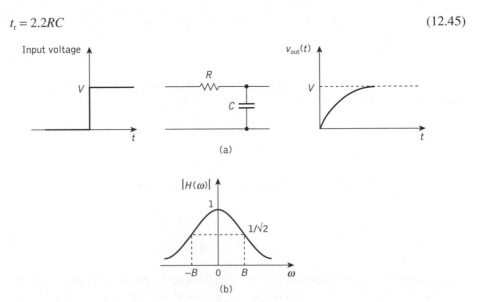

Figure 12.40 (a) The response of a low-pass RC filter circuit to a voltage step input. (b) The transfer function $H(\omega)$ for the circuit in (a)

The transfer function for this circuit is shown in Figure 12.38(b) and is given by:

$$|H(\omega)| = \frac{1}{(1 + \omega^2 C^2 R^2)^{\frac{1}{2}}} \tag{12.46}$$

Therefore the 3 dB bandwidth for the circuit is:

$$B = \frac{1}{2\pi RC} \tag{12.47}$$

Combining Eqs (12.45) and (12.47) gives:

$$t_r = \frac{2.2}{2\pi B} = \frac{0.35}{B} \tag{12.48}$$

The result for the 10–90% rise time indicated in Eq. (12.48) is of general validity, but a different constant term may be obtained with different filter circuits. However, for rise time calculations involving optical fiber systems the constant 0.35 is often utilized and hence in Eq. (12.48), $t_r = T_{\text{syst}}$. Alternatively, if an ideal (unrealizable) filter with an arbitrarily sharp cutoff is considered, the constant in Eq. (12.48) becomes 0.44. However, although this value for the constant is frequently employed when calculating the bandwidth of fiber from pulse dispersion measurements (see Section 14.3.1), the more conservative estimate obtained using a constant term of 0.35 is generally favored for use in system rise time calculations [Refs 64, 65]. Also, in both cases it is usually accepted [Ref. 43] that to conserve the shape of a pulse with a reasonable fidelity through the RC circuit then the 3 dB bandwidth must be at least large enough to satisfy the condition $B\tau = 1$, where τ is the pulse duration. Combining this relation with Eq. (12.48) gives:

$$T_{\text{syst}} = t_r = 0.35\tau \tag{12.49}$$

For an RZ pulse format, the bit rate $B_T = B = 1/\tau$ (see Section 3.8) and hence substituting into Eq. (12.49) gives:

$$B_T(\text{max}) = \frac{0.35}{T_{\text{syst}}} \tag{12.50}$$

Alternatively, for an NRZ pulse format $B_T = B/2 = 1/2\tau$ and therefore the maximum bit rate is given by:

$$B_T(\text{max}) = \frac{0.7}{T_{\text{syst}}} \tag{12.51}$$

Thus the upper limit on T_{syst} should be less than 35% of the bit interval for an RZ pulse format and less than 70% of the bit interval for an NRZ pulse format.

The effects of mode coupling are usually neglected in calculations involving system rise time, and hence the pulse dispersion is assumed to be a linear function of the fiber length. This results in a pessimistic estimate for the system rise time and therefore provides a conservative value for the maximum possible bit rate.

Example 12.7

An optical fiber system is to be designed to operate over an 8 km length without repeaters. The rise times of the chosen components are:

Source (LED)	8 ns
Fiber: intermodal	5 ns km^{-1}
(pulse broadening) intramodal	1 ns km^{-1}
Detector (p–i–n photodiode)	6 ns

From system rise time considerations, estimate the maximum bit rate that may be achieved on the link when using an NRZ format.

Solution: The total system rise time is given by Eq. (12.43) as:

$$T_{\text{syst}} = 1.1(T_{\text{S}}^2 + T_{\text{n}}^2 + T_{\text{c}}^2 + T_{\text{D}}^2)^{\frac{1}{2}}$$
$$= 1.1(8^2 + (8 \times 5)^2 + (8 \times 1)^2 + 6^2)^{\frac{1}{2}}$$
$$= 46.2 \text{ ns}$$

Hence the maximum bit rate for the link using an NRZ format is given by Eq. (12.51) where:

$$B_{\text{T}}(\text{max}) = \frac{0.7}{T_{\text{syst}}} = \frac{0.7}{46.2 \times 10^{-9}} \simeq 15.2 \text{ Mbit s}^{-1}$$

The rise time calculations indicate that this will support a maximum bit rate of 15.2 Mbit s^{-1} which for an NRZ format is equivalent to a 3 dB optical bandwidth of 7.6 MHz (i.e. the NRZ format has two bit intervals per wavelength).

Once it is established that pulse dispersion is not a limiting factor, the major design exercise is the optical power budget for the system.

12.6.6 Optical power budgeting

Power budgeting for a digital optical fiber communication system is performed in a similar way to power budgeting within any communication system. When the transmitter characteristics, fiber cable losses and receiver sensitivity are known, the relatively simple process of power budgeting allows the repeater spacing or the maximum transmission distance for the system to be evaluated. However, it is necessary to incorporate a system margin into the optical power budget so that small variations in the system operating parameters do not lead to an unacceptable decrease in system performance. The operating margin is often included in a safety margin M_{a} which also takes into account possible source and modal noise, together with receiver impairments such as equalization error, noise degradations and eye-opening impairments. The safety margin depends to a large extent on the system components as well as the system design procedure and is typically in the range 5 to 10 dB. Systems using an injection laser transmitter generally require a larger safety margin (e.g. 8 dB) than those using an LED source (e.g. 6 dB) because the temperature variation and aging of the LED are less pronounced.

The optical power budget for a system is given by the following expression:

$$P_i = P_o + C_L + M_a \text{ dB} \tag{12.52}$$

where P_i is the mean input optional power launched into the fiber, P_o is the mean incident optical power required at the receiver and C_L (or C_{LD} when there is a dispersion–equalization penalty) is the total channel loss given by Eq. (12.38) (or Eq. (12.42)). Therefore the expression given in Eq. (12.52) may be written as:

$$P_i = P_o + (\alpha_{fc} + \alpha_j)L + \alpha_{cr} + M_a \text{ dB} \tag{12.53}$$

Alternatively, when a dispersion–equalization penalty is included Eq. (12.52) becomes:

$$P_i = P_o + (\alpha_{fc} + \alpha_j)L + \alpha_{cr} + D_L + M_a \text{ dB} \tag{12.54}$$

Equations (12.53) and (12.54) allow the maximum link length without repeaters to be determined, as demonstrated in Example 12.8.

Example 12.8

The following parameters are established for a long-haul single-mode optical fiber system operating at a wavelength of 1.3 μm:

Mean power launched from the laser transmitter	−3 dBm
Cabled fiber loss	0.4 dB km^{-1}
Splice loss	0.1 dB km^{-1}
Connector losses at the transmitter and receiver	1 dB each
Mean power required at the APD receiver:	
when operating at 35 Mbit s^{-1} (BER 10^{-9})	−55 dBm
when operating at 400 Mbit s^{-1} (BER 10^{-9})	−44 dBm
Required safety margin	7 dB

Estimate:

(a) The maximum possible link length without repeaters when operating at 35 Mbit s^{-1} (BER 10^{-9}). It may be assumed that there is no dispersion–equalization penalty at this bit rate.

(b) The maximum possible link length without repeaters when operating at 400 Mbit s^{-9} (BER 10^{-9}) and assuming no dispersion–equalization penalty.

(c) The reduction in the maximum possible link length without repeaters of (b) when there is a dispersion–equalization penalty of 1.5 dB. It may be assumed for the purposes of this estimate that the reduced link length has the 1.5 dB penalty.

Solution: (a) When the system is operating at 35 Mbit s^{-1} an optical power budget may be performed using Eq. (12.53), where:

$$P_i - P_o = (\alpha_{fc} + \alpha_j)L + \alpha_{cr} + M_a \text{ dB}$$
$$-3 \text{ dBm} - (-55 \text{ dBm}) = (\alpha_{fc} + \alpha_j)L + \alpha_{cr} + M_a$$

Hence:

$$(\alpha_{fc} + \alpha_j)L = 52 - \alpha_{cr} - M_a$$
$$0.5L = 52 - 2 - 7$$

$$L = \frac{43}{0.5} = 86 \text{ km}$$

(b) Again using Eq. (12.53) when the system is operating at 400 Mbit s^{-1}:

$$-3 \text{ dBm} - (-44 \text{ dBm}) = (\alpha_{fc} + \alpha_j)L + \alpha_{cr} + M_a$$
$$(\alpha_{fc} + \alpha_j)L = 41 - 2 - 7$$

$$L = \frac{32}{0.5} = 64 \text{ km}$$

(c) Performing the optical power budget using Eq. (12.54) gives:

$$P_i - P_o = (\alpha_{fc} + \alpha_j)L + \alpha_{cr} + D_L + M_a$$

Hence:

$$0.5L = 41 - 2 - 1.5 - 7$$

and:

$$L = \frac{30.5}{0.5} = 61 \text{ km}$$

Thus there is a reduction of 3 km in the maximum possible link length without repeaters.

Although in Example 12.8 we have demonstrated the use of the optical power budget to determine the maximum link length without repeaters, it is also frequently used to aid decisions in relation to the combination of components required for a particular optical fiber communication system. In this case the maximum transmission distance and the required bandwidth may already be known. Therefore, the optical power budget is used to provide a basis for optimization in the choice of the system components, while also establishing that a particular component configuration meets the system requirements.

Example 12.9

Components are chosen for a digital optical fiber link of overall length 7 km and operating at 20 Mbit s^{-1} using an RZ code. It is decided that an LED emitting at 0.85 μm with graded index fiber to a *p–i–n* photodiode is a suitable choice for the system components, giving no dispersion–equalization penalty. An LED which is capable of launching an average of 100 μW of optical power (including the connector loss) into a graded index fiber of 50 μm core diameter is chosen. The proposed fiber cable has an attenuation of 2.6 dB km^{-1} and requires splicing every kilometer with a loss of 0.5 dB per splice. There is also a connector loss at the receiver of 1.5 dB. The receiver requires mean incident optical power of −41 dBm in order to give the necessary BER of 10^{-10}, and it is predicted that a safety margin of 6 dB will be required.

Write down the optical power budget for the system and hence determine its viability.

Solution:

Mean optical power launched into the fiber from the transmitter (100 μm)	−10 dBm
Receiver sensitivity at 20 Mbit s^{-1} (BER 10^{-10})	−41 dBm
Total system margin	31 dB
Cabled fiber loss (7 × 2.6 dB km^{-1})	18.2 dB
Splice losses (6 × 0.5 dB)	3.0 dB
Connector loss (1 × 1.5 dB)	1.5 dB
Safety margin	6.0 dB
Total system loss	28.7 dB
Excess power margin	2.3 dB

Based on the figures given, the system is viable and provides a 2.3 dB excess power margin. This could give an extra safety margin to allow for possible future splices if these were not taken into account within the original safety margin.

12.6.7 Line coding and forward error correction

The preceding discussions of digital system design have assumed that only information bits are transmitted, and that the 0 and 1 symbols are equally likely. However, within digital line transmission there is a requirement for redundancy in the line coding to provide efficient timing recovery and synchronization (frame alignment) as well as possible error detection and correction at the receiver. Line coding also provides suitable shaping of the transmitted signal power spectral density. Hence the choice of line code is an important consideration within digital optical fiber system design.

Binary line codes are generally preferred because of the large bandwidth available in optical fiber communications. In addition, these codes are less susceptible to any temperature dependence of optical sources and detectors. Under these conditions two-level codes are more suitable than codes which utilize an increased number of levels (multilevel codes) [Ref. 66]. Nevertheless, these factors do not entirely exclude the use of multilevel codes, and it is likely that ternary codes (three levels 0, 1, 2) which give increased information transmission per symbol over binary codes will be considered for some system applications. The corresponding symbol transmission rate (i.e. bit rate) for a ternary code may be reduced by a factor of 1.58 ($\log_2 3$), while still providing the same information transmission rate as a similar system using a binary code. It must be noted that this gain in information capacity for a particular bit rate is obtained at the expense of the dynamic range between adjacent levels as there are three levels inserted in place of two. This is exhibited as a 3 dB SNR penalty at the receiver when compared with a binary system at a given BER. Therefore ternary codes (and higher multilevel codes) are not attractive for long-haul systems.

For the reasons described above most digital optical fiber communication systems currently in use employ binary codes. In practice, binary codes are designed to insert extra symbols into the information data stream on a regular and logical basis to minimize the number of consecutive identical received symbols, and to facilitate efficient timing extraction at the receiver by producing a high density of decision level crossings. The reduction in consecutive identical symbols also helps to minimize the variation in the mean signal level which provides a reduction in the low-frequency response requirement of the receiver. This shapes the transmitted signal spectrum by reducing the d.c. component. However, this factor is less important for optical fiber systems where a.c. coupling is performed with capacitors, unlike metallic cable systems where transformers are often used, and the avoidance of d.c. components is critical. A further advantage is apparent within the optical receiver with a line code which is free from long identical symbol sequences, and where the continuous presence of 0 and 1 levels aids decision level control and avoids gain instability effects.

Two-level block codes of the nBmB type fulfill the above requirements through the addition of a limited amount of redundancy. These codes convert blocks of n bits into blocks of m bits where $m > n$ so that the difference between the number of transmitted ones and zeros is on average zero. A simple code of this type is the 1B2B code in which a 0 may be transmitted as 01, and a 1 as 10. This encoding format is shown in Figure 12.41(b) and is commonly referred to as biphase or Manchester encoding. It may be observed that with this code there are never more than two consecutive identical symbols, and that two symbols must be transmitted for one information bit, giving 50% redundancy. Thus twice the transmission bandwidth is required for the 1B2B code which restricts its use to systems where pulse dispersion is not a limiting factor. Another example of a 1B2B code which is illustrated in Figure 12.41(c) is the coded mark inversion (CMI) code. In this code a digit 0 is transmitted as 01 and the digit 1 alternately as 00 or 11.

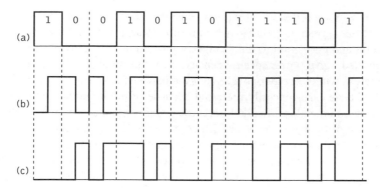

Figure 12.41 Examples of binary 1B2B codes used in optical fiber communications: (a) unencoded NRZ data; (b) biphase or Manchester encoding; (c) coded mark inversion (CMI) encoding

Timing information is obtained from the frequent positive to negative transitions, but, once again, the code is highly redundant requiring twice as many transmitted bits as input information bits.

Several optical line coding schemes have been proposed and implemented which include modified duobinary (see Section 12.6.2) and phased amplitude shift signaling (PASS) line codes [Refs 67–70]. The PASS codes employ duobinary codes with multilevels in which the phase shift is applied to each symbol during the transmission for spectral shaping, but they do not require phase detection and therefore a director detection optical receiver can be employed [Ref. 67]. For high-speed transmission the RZ signal format is preferred to facilitate reduced crosstalk between adjacent channels which is caused by four-wave mixing and cross-phase modulation (see Section 3.14). Line codes using both RZ and carrier suppressed RZ modulation formats (see section 12.6.2) have increasingly found use in experimental long-haul systems [Ref. 71]. More efficient codes of this type requiring less redundancy exist such as the 3B4B, 5B6B and the 7B8B codes. There is a trade-off within this class of code between the complexity of balancing the number of zeros and ones, and the added redundancy. The increase in line symbol rate (bit rate) and the corresponding power penalty over encoded binary transmission is given by the ratio $m : n$. Hence, considering the 5B6B code, the symbol rate is increased by a factor of 1.2 while the power penalty is also equal to 1.2 or about 0.8 dB. It is therefore necessary to take into account the increased bandwidth requirement and the power penalty resulting from coding within the optical fiber system design.

Simple error monitoring may be provided with block codes, at the expense of a small amount of additional redundancy, by parity checking. Each block of N bits can be made to have an even (even parity) or odd (odd parity) number of ones so that any single error in a block can be identified. More extensive error detection and error correction may be provided with increased redundancy and equipment complexity. Alternatively, error monitoring when using block codes may be performed by measuring the variation in disparity between the numbers of ones and zeros within the received bit pattern. Any variation in the accumulated disparity above an upper limit or below a lower limit allowed by a particular code is indicated as an error. Further discussion of error correction with relation to disparity may be found in Ref. 72. Moreover, variations on the above block codes to provide efficient high-speed digital transmission have been devised (e.g. [Ref. 73]). Such line coding schemes possess a good balance of ones and zeros together with jitter suppression and the capability to provide a simple error monitoring function.

The strategy of incorporating an error monitoring capability by adding redundant bits within the line code has more recently become termed as forward error correction (FEC). Following advances in high-speed electronics and optoelectronics, FEC is now a standard feature of both 10 Gbit s^{-1} and 40 Gbit s^{-1} commercial optical fiber transmission systems [Ref. 74]. In FEC operation the transmitter adds the error monitoring data, for instance in the form of block codes, into the transmitted bit stream enabling both the detection and correction of errors at the receiver without the need for any additional information from the transmitter.

Although a number of coding schemes have been used to achieve FEC in long-haul optical fiber systems, two main FEC code types predominate. These are the Bose, Chowdhry and Hocquenghem (BCH) [Ref. 75] and Reed–Solomon (RS) codes [Ref. 76] which are specified in ITU-T Recommendation G.975 [Ref. 77] for use in high-transmission-rate dense WDM submarine systems. Both these error detection and correction coding scheme types are well suited for the correction of random errors in compressed video data. Reed–Solomon codes are a nonbinary subset of BCH codes which can correct more bits in comparison with conventional BCH codes. Moreover, both code types are capable of correcting errors in burst data. For example, a 16-way interleaved RS(255, 239) code can correct errors in a data burst of up to 1024 bits in length. However, a drawback with this burst correcting capability is that if the burst is spread over several symbols then the RS code is not capable of correcting it, whereas the BCH coding scheme can correct such a burst type.

Forward error corrections electronic integrated circuits are often located in the first stage of digital signal processing and are therefore present as an integral part of the analog-to-digital conversion process. The use of FEC is standard practice in submarine optical cable systems in order to counter performance limitations resulting from the poor SNRs of the received signals. At transmission rates of 10 Gbit s^{-1} and beyond the implementation of FEC, however, is a challenging task due to the electronic circuit complexity and its consequent speed limitations especially since it requires more effective (and hence longer) FEC codes exhibiting a larger coding gain [Ref. 78]. Coding gain is the measure identifying the difference between the SNR of an error detection and correction coded system and an uncoded system which is required to attain the same BER [Ref. 78]. Another significant parameter related to coding gain is that of a code rate which determines the effectiveness of a coding scheme. It describes the amount of remaining nonredundant information (i.e. one minus the redundancy fraction) in the coded sequence and is always measured in bits per symbol. Ideally, code rates between 0.5 to 0.8 bits per symbol are preferred to provide an effective FEC coding scheme for the use in optical fiber communication systems [Ref. 79]. Nevertheless, recent advances in optoelectronic integrated circuits (see Section 11.5) have enabled the production of components for WDM systems operating at transmission rates from 10 Gbit s^{-1} to 40 Gbit s^{-1} and beyond while exhibiting very low code rates and also creating high coding gains [Ref. 80].

The component enabling technologies for FEC have evolved over a long period and can be divided into typically three generations, with each generation displaying improved

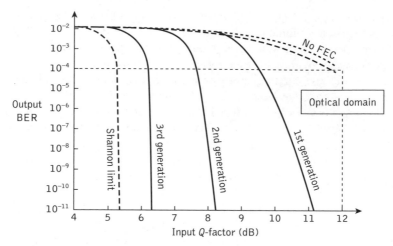

Figure 12.42 Typical error correction performances for the three generations of forward error correction displaying output bit-error-rate against input Q-factor [Ref. 79]

performance attributes but also exhibiting increased levels of complexity [Ref. 81]. For instance, the first generation of FEC which was based on linear codes RS(255, 239) employed around 7% redundancy (i.e. at a code rate of 0.93 bits per symbol) and provided a relatively low coding gain of only 5.8 dB. The second-generation FEC which is based on concatenated RS and/or BCH codes incorporates 7 to 21% redundancy and gives a coding gain of up to 8 dB. The third generation of FEC, however, utilizes soft decision decoding employing block turbo codes and/or low-density parity check codes (LDPC) [Ref. 82] which require up to 53% redundancy and deliver a coding gain greater than 10 dB. In soft decision coding both the demodulation and decoding functions are combined, and the output of a channel is passed directly to the decoder for error correction. The decoder accesses all the information in relation to the transmitted data and each bit is assigned a confidence level (i.e. quantized level) to decide if it will be assigned as binary 0 or 1.

The performances of the three generations of FEC are compared graphically in Figure 12.42 which depicts output BER as a function of input Q-factor.* It can be observed that each generation of FEC provides an improved BER performance with the ultimate goal of bringing it closer to the Shannon coding limit† with a narrow code rate of 0.4 bits per symbol at a BER of 1×10^{-13} for a soft decision decoder [Refs 74, 80, 84]. It should be

---

* The Q-factor or function, which represents the area under the tail of a zero mean, unit variance Gaussian probability function, is defined as $\int_u^\infty (1/\sqrt{2\pi}\exp[(e^{-z^2/2})])$. It is often used as an alternative to the complementary error function erfc(u) in the formulation of the probability of error (see Eq. (12.17)) and is related to it following $Q(u) = \frac{1}{2}\text{erfc}z/\sqrt{2}$.

† The Shannon coding theorem (also known as channel capacity theorem) states that there exists an error correcting code when the code rate is smaller than the capacity of the digital channel. If the system operates at a code rate greater than channel capacity then the system has a high probability of error, regardless of the choice of the coding scheme or type of detector [Ref. 83].

noted that at smaller values of code rate for the error detection and correction a high output BER from 10^{-9} to 10^{-15} is obtained in the optical fiber communication system in comparison with the lower range of BER of 10^{-3} to 10^{-6} delivered in a radio or satellite system. The first generation of FEC can perform detection and correction to give an output BER of 1×10^{-11} for an input BER of 1.4×10^{-4} (i.e. at input Q-factor of 9.5 dB). Second and third generations, however, display much improved performance and in particular the third generation which approaches very near to the theoretical limit by achieving output BER above 1×10^{-11} at an input Q-factor of 6.25 dB, which in this case is only 0.9 dB away from the Shannon limit. Although the coding gain and output BER performance for the third generation of FEC make it a suitable candidate for optical fiber communications, the transmission bit rate is restricted to 10 Gbit s^{-1} as a consequence of the limitations in the speed of the digital signal processing to produce the FEC. Nevertheless it is anticipated that third-generation FEC will push the output BER to the region around 10^{-13} while also further narrowing the gap between the input Q-factor and the Shannon limit [Ref. 85].

12.7 Analog systems

In Section 12.5 we indicated that the vast majority of optical fiber communication systems are designed to convey digital information (e.g. analog speech encoded as PCM). However, in a few areas of the telecommunication network or for particular non-telecommunication applications, information transfer in analog form is still likely to remain for some time to come, or be advantageous. Therefore, analog optical fiber transmission will undoubtedly have a part to play in future communication networks, especially in situations where the optical fiber link is part of a larger analog network (e.g. microwave relay network). Use of analog transmission in these areas avoids the cost and complexity of digital terminal equipment, as well as degradation due to quantization noise. This is especially the case with the transmission of video signals over short distances where the cost of high-speed A–D and D–A converters is not generally justified. Hence, there are many applications such as direct cable television and common antenna television (CATV) where analog optical fiber systems may be utilized.

There are limitations, however, inherent to analog optical fiber transmission, some of which have been mentioned previously. For instance, the unique requirements of analog transmission over digital are for high SNRs at the receiver output, which necessitates high optical input power (see Section 9.2.5), and high end-to-end linearity to avoid distortion and prevent crosstalk between different channels of a multiplexed signal (see Section 12.4.2). Furthermore, it is instructive to compare the SNR constraints for typical analog optical fiber and coaxial cable systems.

In a coaxial cable system the fundamental limiting noise is $4KTB$, where K is Boltzmann's constant, T is the absolute temperature, and B is the effective noise bandwidth for the channel. If we assume for simplicity that the coaxial cable loss is constant and independent of frequency, the SNR for a coaxial system is:

$$\left(\frac{S}{N}\right)_{\text{coax}} = \frac{V^2 \exp(-\alpha_N)}{Z_0 4KTB}$$

(12.55)

where α_N is the attenuation in nepers between the transmitter and receiver, V is the peak output voltage, and Z_0 is the impedance of the coaxial cable.

The SNR for an analog optical fiber system may be obtained by referring to Eq. (9.11) where:

$$\left(\frac{S}{N}\right)_{fiber} = \frac{\eta P_o}{2hfB} \tag{12.56}$$

The expression given in Eq. (12.56) includes the fundamental limiting noise for optical fiber systems which is $2hfB$. Although Eq. (12.56) is sufficiently accurate for the purpose of comparison, it applies to an unmodulated optical carrier. A more accurate expression would take into account the depth of modulation for the analog optical fiber system which cannot be unity [Ref. 63].* The average received optical power P_o may be expressed in terms of the average input (transmitted) optical power P_i as:

$$P_o = P_i \exp(-\alpha_N) \tag{12.57}$$

Substituting for P_o into Eq. (12.56) gives:

$$\left(\frac{S}{N}\right)_{fiber} = \frac{\eta P_i \exp(-\alpha_N)}{2hfB} \tag{12.58}$$

Equations (12.55) and (12.58) allow a simple comparison to be made of available SNR (or CNR) between analog coaxial and optical fiber systems, as demonstrated in Example 12.10.

Example 12.10

A coaxial cable system operating at a temperature of 17°C has a transmitter peak output voltage of 5 V with a cable impedance of 100 Ω. An analog optical fiber system uses an injection laser source emitting at 0.85 μm and launches an average of 1 mW of optical power into the fiber cable. The optical receiver comprises a photodiode with a quantum efficiency of 70%. Assuming the effective noise bandwidth and the attenuation between the transmitter and receiver for the two systems are identical, estimate in decibels the ratio of the SNR of the coaxial system to the SNR of the fiber system.

Solution: Using Eqs (12.55) and (12.58) for the SNRs of the coaxial and fiber systems respectively:

* Strictly speaking, Eq. (12.56) depicts the optical carrier-to-noise ratio (CNR).

$$\text{Ratio} = \frac{\left(\dfrac{S}{N}\right)_{\text{coax}}}{\left(\dfrac{S}{N}\right)_{\text{fiber}}} = \frac{\dfrac{V^2 \exp(-\alpha_{\text{N}})}{Z_0 4KTB}}{\dfrac{\eta P_i \exp(-\alpha_{\text{N}})}{2hfB}} = \frac{V^2 hf}{2KTZ_0 \eta P_i}$$

$$= \frac{V^2 hc}{2KTZ_0 \eta P_i \lambda}$$

Hence:

$$\text{Ratio} = \frac{25 \times 6.626 \times 10^{-34} \times 2.998 \times 10^8}{2 \times 1.385 \times 10^{-23} \times 290 \times 100 \times 0.7 \times 1 \times 10^{-3} \times 0.85 \times 10^{-6}}$$

$$= 1.04 \times 10^4 \simeq 40 \text{ dB}$$

The optical fiber channel in Example 12.10 has around 40 dB less SNR available than the alternative coaxial channel exhibiting similar channel losses. This results both from $2hfB$ being larger than $4KTB$ and from the far smaller transmitted power within the optical system. Furthermore, it must be noted that the comparison was made using an injection laser transmitter. If an LED transmitter with 10 to 20 dB less optical output power were compared, the coaxial system would display an advantage in the region 50 to 60 dB. For this reason it is difficult to match with fiber systems the SNR requirements of some analog coaxial links, even though the fiber cable attenuation may be substantially lower than that of the coaxial cable.

The analog signal can be transmitted within an optical fiber communication system using one of several modulation techniques. The simplest form of analog modulation for optical fiber communications is direct intensity modulation (D–IM) of the optical source. In this technique the optical output from the source is modulated simply by varying the current flowing in the device around a suitable bias or mean level in proportion to the message. Hence the information signal is transmitted directly in the baseband.

Alternatively, the baseband signal can be translated onto an electrical subcarrier by means of amplitude, phase or frequency modulation using standard techniques, prior to intensity modulation of the optical source. Pulse analog techniques where a sequence of pulses is used for the carrier may also be utilized. In this case a suitable parameter such as the pulse amplitude, pulse width, pulse position or pulse frequency is electrically modulated by the baseband signal. Again, the modulated electrical carrier is transmitted optically by intensity modulation of the optical source.

Direct modulation of the optical source in frequency, phase or polarization rather than by intensity requires these parameters to be well defined throughout the optical fiber system. There is much interest in this area and optical component technology has been developed which will allow practical system implementation. These techniques concerned with coherent optical transmission are discussed in Chapter 13.

12.7.1 Direct intensity modulation (D–IM)

A block schematic for an analog optical fiber system which uses direct modulation of the optical source intensity with the baseband signal is shown in Figure 12.43(a). Obviously, no electrical modulation or demodulation is required with this technique, making it both inexpensive and easy to implement.

The transmitted optical power waveform as a function of time $P_{opt}(t)$, an example of which is illustrated in Figure 12.40(b) may be written as:

$$P_{opt}(t) = P_i(1 + m(t)) \tag{12.59}$$

where P_i is the average transmitted optical power (i.e. the unmodulated carrier power) and $m(t)$ is the intensity modulating signal which is proportional to the source message $a(t)$. For a cosinusoidal modulating signal:

$$m(t) = m_a \cos \omega_m t \tag{12.60}$$

where m_a is the modulation index or the ratio of the peak excursion from the average to the average power as shown in Figure 12.43(b) and ω_m is the angular frequency of the modulating signal. Combining Eqs (12.59) and (12.60) we get:

$$P_{opt}(t) = P_i(1 + m_a \cos \omega_m t) \tag{12.61}$$

Furthermore, assuming the transmission medium has zero dispersion, the received optical power will be of the same form as Eq. (12.61), but with an average received optical

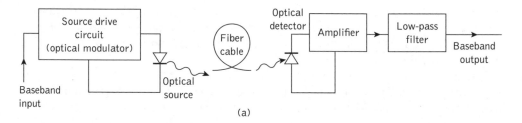

(a)

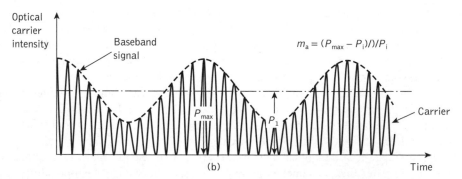

(b)

Figure 12.43 (a) Analog optical fiber system employing direct intensity modulation. (b) Time domain representation showing direct intensity modulation of the optical carrier with a baseband analog signal

power P_o. Hence the secondary photocurrent $I(t)$ generated at an APD receiver with a multiplication factor M is given by:

$$I(t) = I_p M (1 + m_a \cos \omega_m t) \tag{12.62}$$

where the primary photocurrent obtained with an unmodulated carrier I_p is given by Eq. (8.8) as:

$$I_p = \frac{\eta e}{hf} P_o \tag{12.63}$$

The mean square signal current $\overline{i_{\mathrm{sig}}^2}$ which is obtained from Eq. (12.62) is given by:

$$\overline{i_{\mathrm{sig}}^2} = \tfrac{1}{2}(m_a M I_p)^2 \tag{12.64}$$

The total average noise in the system is composed of quantum, dark current and thermal (circuit) noise components. The noise contribution from quantum effects and detector dark current may be expressed as the mean square total shot noise current for the APD receiver $\overline{i_{\mathrm{SA}}^2}$ given by Eq. (9.21) where the excess avalanche noise factor is written following Eq. (9.26) as $F(M)$ such that:

$$\overline{i_{\mathrm{SA}}^2} = 2eB(I_p + I_d)M^2 F(M) \tag{12.65}$$

where B is the effective noise or post-detection bandwidth.

Thermal noise generated by the load resistance R_L and the electronic amplifier noise can be expressed in terms of the amplifier noise figure F_n referred to R_L as given by Eq. (9.17). Thus the total mean square noise current $\overline{i_N^2}$ may be written as:

$$\overline{i_N^2} = 2eB(I_p + I_d)M^2 F(M) + \frac{4KTBF_n}{R_L} \tag{12.66}$$

The SNR defined in terms of the ratio of the mean square signal current to the mean square noise current (rms signal power to rms noise power) for the APD receiver is therefore given by:

$$\left(\frac{S}{N}\right)_{\mathrm{rms}} = \frac{\overline{i_{\mathrm{sig}}^2}}{\overline{i_N^2}} = \frac{\tfrac{1}{2}(m_a M I_p)^2}{2eB(I_p + I_d)M^2 F(M) + (4KTBF_n/R_L)} \quad \text{(APD)} \tag{12.67}$$

It must be emphasized that the SNR given in Eq. (12.67) is defined in terms of rms signal power rather than peak signal power used previously. When a unity gain photodetector is utilized in the receiver (i.e. p–i–n photodiode) Eq. (12.67) reduces to:

$$\left(\frac{S}{N}\right)_{\mathrm{rms}} = \frac{\tfrac{1}{2}(m_a I_p)^2}{2eB(I_p + I_d) + (4KTBF_n/R_L)} \quad (p\text{–}i\text{–}n) \tag{12.68}$$

Moreover, the SNR for video transmission is often defined in terms of the peak-to-peak picture signal power to the rms noise power and may include the ratio of luminance to composite video b. Using this definition in the case of the unity gain detector gives:

$$\left(\frac{S}{N}\right)_{\text{p-p}} = \frac{(2m_aI_pb)^2}{2eB(I_p + I_d) + (4KTBF_n/R_L)} \quad (p\text{--}i\text{--}n) \tag{12.69}$$

It may be observed that, excluding b, the SNR defined in terms of the peak-to-peak signal power given in Eq. (12.69) is a factor of 8 (or 9 dB) greater than that defined in Eq. (12.68).

Example 12.11

A single TV channel is transmitted over an analog optical fiber link using direct intensity modulation. The video signal which has a bandwidth of 5 MHz and a ratio of luminance to composite video of 0.7 is transmitted with a modulation index of 0.8. The receiver contains a p--i--n photodiode with a responsivity of 0.5 A W^{-1} and a preamplifier with an effective input impedance of 1 MΩ together with a noise figure of 1.5 dB. Assuming the receiver is operating at a temperature of 20 °C and neglecting the dark current in the photodiode, determine the average incident optical power required at the receiver (i.e. receiver sensitivity) in order to maintain a peak-to-peak signal power to rms noise power ratio of 55 dB.

Solution: Neglecting the photodiode dark current, the peak-to-peak signal rms noise power ratio is given following Eq. (12.69) as:

$$\left(\frac{S}{N}\right)_{\text{p-p}} = \frac{(2m_aI_pb)^2}{2eBI_p + (4KTBF_n/R_L)}$$

The photocurrent I_p may be expressed in terms of the average incident optical power at the receiver P_o using Eq. (8.4) as:

$$I_p = RP_o$$

where R is the responsivity of the photodiode. Hence:

$$\left(\frac{S}{N}\right)_{\text{p-p}} = \frac{(2m_aRP_ob)^2}{2eBRP_o + (4KTBF_n/R_L)}$$

and:

$$\left(\frac{S}{N}\right)_{\text{p-p}}\left(2eBRP_o = \frac{4KTBF_n}{R_L}\right) = (2m_aRP_ob)^2$$

Rearranging:

$$(2m_aRb)^2P_o^2 - \left(\frac{S}{N}\right)_{\text{p-p}} 2eBRP_o - \left(\frac{S}{N}\right)_{\text{p-p}} \frac{4KTBF_n}{R_L} = 0$$

where:

$$(2m_aRb)^2 = 4 \times 0.64 \times 0.25 \times 0.49$$

$$= 0.314$$

$$\left(\frac{S}{N}\right)_{\text{p-p}} 2eBR = 3.162 \times 10^5 \times 2 \times 1.602 \times 10^{-19} \times 5 \times 10^6 \times 0.5$$

$$= 2.533 \times 10^{-7}$$

$$\left(\frac{S}{N}\right)_{\text{p-p}} \frac{4KTBF_n}{R_L} = \frac{3.162 \times 10^5 \times 4 \times 1.381 \times 10^{-23} \times 293 \times 5 \times 10^6 \times 1.413}{10^6}$$

$$= 3.616 \times 10^{-14}$$

Therefore:

$$0.314P_o^2 - 2.533 \times 10^{-7} P_o - 3.616 \times 10^{-14} = 0$$

and:

$$P_o = \frac{2.533 \times 10^{-7} \pm \sqrt{[(2.533 \times 10^{-7})^2 - (-4 \times 0.314 \times 3.616 \times 10^{-14})]}}{0.628}$$

$$= 0.93 \ \mu\text{W}$$
$$= -30.3 \ \text{dBm}$$

It must be noted that the low-noise preamplification depicted in Example 12.11 may not always be obtained, and that higher thermal noise levels will adversely affect the receiver sensitivity for a given SNR. This is especially the case with lower SNRs, as illustrated in the peak-to-peak signal power to rms noise power ratio against average received optical power characteristics for a video system shown in Figure 12.44 [Ref. 86]. The performance of the system for various values of mean square thermal noise current $\overline{i_t^2} = 4KTBF_n/R_L$, where $\overline{i_t^2}$ is expressed as a spectral density in $A^2 \, Hz^{-1}$, is indicated. The value for the receiver sensitivity obtained in Example 12.11 is approaching the quantum limit, also illustrated in Figure 12.44, which is the best that could possibly be achieved with a noiseless amplifier.

The quantum or shot noise (when ignoring the photodetector dark current) limit occurs with large values of signal current (i.e. primary photocurrent) at the receiver. Considering a p–i–n photodiode receiver, this limiting case which corresponds to large SNR is given by Eq. (12.68) when neglecting the device dark current as:

$$\left(\frac{S}{N}\right)_{\text{rms}} \simeq \frac{m_a^2 I_p}{4eB} \quad \text{(quantum noise limit)} \tag{12.70}$$

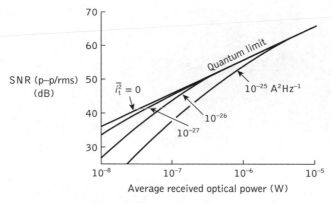

Figure 12.44 Peak-to-peak signal power to rms noise power ratio against the average received optical power for a direct intensity modulated video system and various levels of thermal noise given by $\overline{i_t^2}$. Reproduced with permission from G. G. Windus, *Marconi Rev.*, **XLI**, p. 77. 1981

Using the relationship between the average received optical power P_o and the primary photocurrent given in Eq. (12.63) allows Eq. (12.70) to be expressed as:

$$P_o \simeq \frac{4hf}{m_a^2 \eta} \left(\frac{S}{N}\right)_{\text{rms}} B \qquad (12.71)$$

Equation (12.71) indicates that for a quantum-noise-limited analog receiver, the optical input power is directly proportional to the effective noise or post-detection bandwidth B. A similar result was obtained in Eq. (12.37) for the digital receiver.

Alternatively, at low SNRs thermal noise is dominant, and the thermal noise limit when I_p is small, which may also be obtained from Eq. (12.68), is given by:

$$\left(\frac{S}{N}\right)_{\text{rms}} \simeq \frac{(m_a I_p)^2 R_L}{8KTBF_n} \qquad \text{(thermal noise limit)} \qquad (12.72)$$

Again substituting for I_p from Eq. (12.63) gives:

$$P_o \simeq \frac{hf}{e\eta m_a^2} \left(\frac{8KTF_n}{R_L}\right)^{\frac{1}{2}} \left(\frac{S}{N}\right)_{\text{rms}}^{\frac{1}{2}} B^{\frac{1}{2}} \qquad (12.73)$$

Therefore it may be observed from Eq. (12.73) that in the thermal noise limit the average incident optical power is directly proportional to $B^{1/2}$ instead of the direct dependence on B shown in Eq. (12.71) for the quantum noise limit. The dependence expressed in Eq. (12.73) is typical of the p–i–n photodiode receiver operating at low optical input power levels. Thus Eq. (12.73) may be used to estimate the required input optical power to achieve a particular SNR for a p–i–n photodiode receiver which is dominated by thermal noise.

Example 12.12

An analog optical fiber link employing D–IM has a p–i–n photodiode receiver in which thermal noise is dominant. The system components have the following characteristics and operating conditions:

p–i–n photodiode quantum efficiency	60%
Effective load impedance for the photodiode	50 kΩ
Preamplifier noise figure	6 dB
Operating wavelength	1 μm
Operating temperature	300 K
Receiver post-detection bandwidth	10 MHz
Modulation index	0.5

Estimate the required average incident optical power at the receiver in order to maintain an SNR, defined in terms of the mean square signal current to mean square noise current, of 45 dB.

Solution: The average incident optical power for a thermal-noise-limited p–i–n photodiode receiver may be estimated using Eq. (12.73) where:

$$P_{\mathrm{o}} \simeq \frac{hf}{e\eta m_{\mathrm{a}}^2}\left(\frac{8KTF_{\mathrm{n}}}{R_{\mathrm{L}}}\right)^{\frac{1}{2}}\left(\frac{S}{N}\right)_{\mathrm{rms}}^{\frac{1}{2}} B^{\frac{1}{2}}$$

and:

$$\frac{hf}{e\eta m_{\mathrm{a}}^2} = \frac{hc}{e\eta m_{\mathrm{a}}^2 \lambda} = \frac{6.626 \times 10^{-34} \times 2.998 \times 10^8}{1.602 \times 10^{-19} \times 0.6 \times 0.25 \times 1 \times 10^{-6}}$$

$$= 8.267$$

$$\left(\frac{8KTF_{\mathrm{n}}}{R_{\mathrm{L}}}\right)^{\frac{1}{2}} = \left(\frac{8 \times 1.381 \times 10^{23} \times 300 \times 4}{50 \times 10^3}\right)^{\frac{1}{2}}$$

$$= 1.628 \times 10^{-12}$$

$$\left(\frac{S}{N}\right)_{\mathrm{rms}}^{\frac{1}{2}} B^{\frac{1}{2}} = (3.162 \times 10^4 \times 10^7)^{\frac{1}{2}}$$

$$= 5.623 \times 10^5$$

Hence:

$$P_{\mathrm{o}} \simeq 8.267 \times 1.628 \times 10^{-12} \times 5.623 \times 10^5$$
$$= 7.57 \ \mu W$$
$$= -21.2 \ dBm$$

Therefore, as anticipated, the receiver sensitivity in the thermal noise limit is low.

12.7.2 System planning

Many of the general planning considerations for optical fiber systems outlined in Section 12.4 may be applied to analog transmission. However, extra care must be taken to ensure that the optical source and, to a lesser extent, the detector have linear input–output characteristics, in order to avoid distortion of the transmitted optical signal. Furthermore, careful optical power budgeting is often necessary with analog systems because of the generally high SNRs required at the optical receiver (40 to 60 dB) in comparison with digital systems (20 to 25 dB), to obtain a similar fidelity. Therefore, although analog system optical power budgeting may be carried out in a similar manner to digital systems (see Section 12.6.6), it is common for the system margin, or the difference between the optical power launched into the fiber and the required optical power at the receiver, for analog systems to be quite small (perhaps only 10 to 20 dB when using an LED source to p–i–n photodiode receiver). Consequently, analog systems employing D–IM of the optical source tend to have a limited transmission distance without repeaters which generally prohibits their use for long-haul applications.

Example 12.13

A D–IM analog optical fiber link of length 2 km employs an LED which launches mean optical power of -10 dBm into a multimode optical fiber. The fiber cable exhibits a loss of 3.5 dB km^{-1} with splice losses calculated at 0.7 dB km^{-1}. In addition there is a connector loss at the receiver of 1.6 dB. The p–i–n photodiode receiver has a sensitivity of -25 dBm for an SNR (i_{sig}^2/i_N^2) of 50 dB and with a modulation index of 0.5. It is estimated that a safety margin of 4 dB is required. Assuming there is no dispersion–equalization penalty:

(a) Perform an optical power budget for the system operating under the above conditions and ascertain its viability.

(b) Estimate any possible increase in link length which may be achieved using an injection laser source which launches mean optical power of 0 dBm into the fiber cable. In this case the safety margin must be increased to 7 dB.

Solution: (a) Optical power budget:

Mean power launched into the fiber cable from the LED transmitter	-10 dBm
Mean optical power required at the p–i–n photodiode receiver for SNR of 50 dB and a modulation index of 0.5	-25 dBm
Total system margin	15 dB
Fiber cable loss (2×3.5)	7.0 dB
Splice losses (2×0.7)	1.4 dB
Connector loss at the receiver	1.6 dB
Safety margin	4.0 dB
Total system loss	14.0 dB
Excess power margin	1.0 dB

Hence the system is viable, providing a small excess power margin.

(b) In order to calculate any possible increase in link length when using the injection laser source we refer to Eq. (12.53), where:

$$P_i - P_o = (\alpha_{fc} + \alpha_j)L + \alpha_{cr} + M_a \text{ dB}$$

Therefore:

$$0 \text{ dBm} - (-25 \text{ dBm}) = (3.5 + 0.7)L + 1.6 + 7.0$$

and:

$$4.2L = 25 - 8.6 = 16.4 \text{ dB}$$

giving:

$$L = \frac{16.4}{4.2} = 3.9 \text{ km}$$

Hence the use of the injection laser gives a possible increase in the link length of 1.9 km or almost a factor of 2. It must be noted that in this case the excess power margin has been reduced to zero.

The transmission distance without repeaters for the analog link of Example 12.13 could be extended further by utilizing an APD receiver which has increased sensitivity. This could facilitate an increase in the maximum link length to around 7 km, assuming no additional power penalties or excess power margin. Although this is quite a reasonable transmission distance, it must be noted that a comparable digital system could give in the region of 13 km transmission without repeaters.

The temporal response of analog systems may be determined from system rise time calculations in a similar manner to digital systems (see Section 12.6.5). The maximum permitted 3 dB optical bandwidth for analog systems in order to avoid dispersion penalties follows from Eq. (12.49) and is given by:

$$B_{opt}(\text{max}) = \frac{0.35}{T_{syst}} \tag{12.74}$$

Hence calculation of the total system 10–90% rise time T_{syst} allows the maximum system bandwidth to be estimated. Often this calculation is performed in order to establish that the desired system bandwidth may be achieved using a particular combination of system components.

Example 12.14

The 10–90% rise times for possible components to be used in a D–IM analog optical fiber link are specified below:

Source (LED)	10 ns
Fiber cable: intermodal	9 ns km^{-1}
chromatic	2 ns km^{-1}
Detector (APD)	3 ns

The desired link length without repeaters is 5 km and the required optical bandwidth is 6 MHz. Determine whether the above combination of components gives an adequate temporal response.

Solution: Equation (12.74) may be used to calculate the maximum permitted system rise time which gives the desired bandwidth where:

$$T_{syst}(max) = \frac{0.35}{B_{opt}} = \frac{0.35}{6 \times 10^6} = 58.3 \text{ ns}$$

The total system rise time using the specified components can be estimated using Eq. (12.43) as:

$$T_{syst} = 1.1(T_S^2 + T_n^2 + T_c^2 + T_D^2)^{\frac{1}{2}}$$
$$= 1.1(10^2 + (9 \times 5)^2 + (2 \times 5)^2 + 3^2)^{\frac{1}{2}}$$
$$\simeq 52 \text{ ns}$$

Therefore the specified components give a system rise time which is adequate for the bandwidth and distance requirements of the optical fiber link. However, there is little leeway for upgrading the system in terms of bandwidth or distance without replacing one or more of the system components.

12.7.3 Subcarrier intensity modulation

Direct intensity modulation of the optical source is suitable for the transmission of a baseband analog signal. However, if the wideband nature of the optical fiber medium is to be fully utilized it is essential that a number of baseband channels are multiplexed onto a single fiber link. This may be achieved with analog transmission through frequency division multiplexing of the individual baseband channels. Initially, the baseband channels must be translated onto carriers of different frequency by amplitude modulation (AM), frequency modulation (FM) or phase modulation (PM) prior to being simultaneously transmitted as an frequency division multiplexing signal. The frequency translation may be performed in the electrical regime where the baseband analog signals modulate electrical subcarriers and are then frequency division multiplexed to form a composite electrical signal prior to intensity modulation of the optical source.

A block schematic of an analog system employing this technique, which is known as subcarrier intensity modulation, is shown in Figure 12.45. The baseband signals are

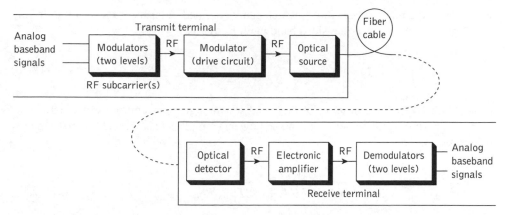

Figure 12.45 Subcarrier intensity modulation system for analog optical fiber transmission

modulated onto radio-frequency (RF) subcarriers by either AM, FM or PM and multiplexed before being applied to the optical source drive circuit.* Hence an intensity modulated (IM) optical signal is obtained which may be AM–IM, FM–IM or PM–IM. In practice, however, system output SNR considerations dictate that generally only the latter two modulation formats are used. Nevertheless, systems may incorporate two levels of electrical modulation whereby the baseband channels are initially amplitude modulated prior to FM or PM [Ref. 87]. The FM or PM signal thus obtained is then used to intensity modulate the optical source. At the receive terminal the transmitted optical signal is detected prior to electrical demodulation and demultiplexing (filtering) to obtain the originally transmitted baseband signals.

A further major advantage of subcarrier intensity modulation is the possible improvement in SNR that may be obtained during subcarrier demodulation. In order to investigate this process it is necessary to obtain a general expression for the SNR of the IM optical carrier which may then be applied to the subcarrier intensity modulation formats. Therefore, as with D–IM, considered in the preceding section, an electrical signal $m(t)$ modulates the source intensity. The transmitted optical power waveform is of the same form as Eq. (12.59), where:

$$P_{\text{opt}}(t) = P_{\text{i}}(1 + m(t)) \tag{12.75}$$

Also the secondary photon $I(t)$ generated at an APD receiver following Eq. (12.62) is given by:

$$I(t) = I_{\text{p}}M(1 + m(t)) \tag{12.76}$$

The mean square signal current $\overline{i_{\text{sig}}^2}$ may be written as [Ref. 84]:

$$\overline{i_{\text{sig}}^2} = (I_{\text{p}}M)^2 P_{\text{m}} \tag{12.77}$$

---

* When microwave frequency rather than radiofrequency subcarriers are employed the strategy is usually referred to as subcarrier multiplexing or SCM (see Section 12.9.2).

where P_m is the total power of $m(t)$, which can be defined in terms of the spectral density $S_m(\omega)$ of $m(t)$ occupying a one-sided bandwidth B_m Hz as:

$$P_m = \frac{1}{2\pi} \int_{-2\pi B_m}^{2\pi B_m} S_m(\omega)\, d\omega \tag{12.78}$$

Hence the SNR defined in terms of the mean square signal current to mean square noise current (i.e. rms signal power to rms noise power) using Eqs (12.77) and (12.66) can now be written as:

$$\left(\frac{S}{N}\right)_{rms} = \frac{\overline{i_{sig}^2}}{\overline{i_N^2}} = \frac{(I_p M)^2 P_m}{2eB_m(I_p + I_d)M^2 F(M) + (4KTBF_n/R_L)}$$

$$= \frac{I_p^2 P_m}{2B_m e(I_p + I_d)F(M) + (4KTBF_n/M^2 R_L)}$$

$$= \frac{(RP_o)^2 P_m}{2B_m N_o} \quad \text{(D–IM)} \tag{12.79}$$

where we substitute for I_p from Eq. (8.4) and for notational simplicity write:

$$N_o = e(I_p + I_d)F(M) + \frac{4KTBF_n}{M^2 R_L} \tag{12.80}$$

The result obtained in Eq. (12.79) gives the SNR for a direct intensity modulated optical source where the total modulating signal power is P_m. In this context Eq. (12.79) is simply a more general form of Eq. (12.67). However, we are now in a position to examine the signal-to-noise performance of various subcarrier intensity modulation formats.

12.7.4 Subcarrier double-sideband modulation (DSB–IM)

A simple way to translate the spectrum of the baseband message signal $a(t)$ is by direct multiplication with the subcarrier waveform $A_c \cos \omega_c t$ giving the modulated waveform $m(t)$ as:

$$m(t) = A_c a(t) \cos \omega_c t \tag{12.81}$$

where A_c is the amplitude and ω_c the angular frequency of the subcarrier waveform. For a cosinusoidal modulating signal ($\cos \omega_m t$) the subcarrier electric field $E_m(t)$ becomes:

$$E_m(t) = \frac{A_c}{2} \cos(\omega_c + \omega_m)t + \cos(\omega_c - \omega_m)t \tag{12.82}$$

giving the upper and lower sidebands. The time and frequency domain representations of the modulated waveform are shown in Figure 12.46. It may be observed from the frequency domain representation that only the two sideband components are present as

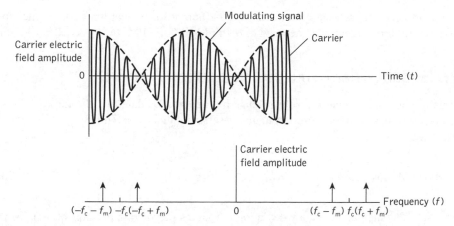

Figure 12.46 Time and frequency domain representations of double-sideband modulation

indicated in Eq. (12.82). This modulation technique is known as double-sideband modulation (DSB) or double-sideband suppressed carrier (DSBSC) AM. It provides a more efficient method of translating the spectrum of the baseband message signal than conventional full AM where a large carrier component is also present in the modulated waveform.

The DSB signal shown in Figure 12.46 intensity modulates the optical source. Therefore the transmitted optical power waveform is obtained by combining Eqs (12.75) and (12.81) where for simplicity we set the carrier amplitude A_c to unity, giving:

$$P_{opt}(t) = P_i(1 + a(t) \cos \omega_c t) \qquad (12.83)$$

Furthermore, in order to prevent overmodulation, the value of the message signal is normalized such that $|a(t) \leq 1|$ with power $P_a \leq 1$. The DSB modulated electrical subcarrier occupies a bandwidth $B_m = 2B_a$, and with a carrier amplitude of unity, $P_m = P_a/2$. Hence, the ratio of rms signal power to rms noise power obtained within the subcarrier bandwidth at the input to the DSB demodulator is given by Eq. (12.79) where:

$$\left(\frac{S}{N}\right)_{rms} \text{input DSB} = \frac{(RP_o)^2 P_a/2}{2 \times 2B_a N_o} = \frac{(RP_o)^2 P_a}{8B_a N_o} \qquad (12.84)$$

However, an ideal DSB demodulator gives a detection gain of 2 or 3 dB improvement in SNR [Ref. 87]. This yields an output SNR of:

$$\left(\frac{S}{N}\right)_{rms} \text{output DSB} = 2\left(\frac{S}{N}\right)_{rms} \text{input DSB} = \frac{(RP_o)^2 P_a}{4B_a N_o} \qquad (12.85)$$

Comparison of the result obtained in Eq. (12.85) with that using D–IM of the baseband signal given by Eq. (12.79) shows a 3 dB degradation in SNR when employing DSB–IM under the same conditions of bandwidth (i.e. $B_m = B_a$), modulating signal power (i.e. $P_m = P_a$), detector photocurrent and noise. For this reason DSB–IM systems (and also AM–IM systems in general) are usually not considered efficient for optical fiber communications. Therefore far more attention is devoted to both FM–IM and PM–IM systems.

12.7.5 Subcarrier frequency modulation (FM–IM)

In this modulation format, the subcarrier is frequency modulated by the message signal. The conventional form for representing the baseband signal which intensity modulates the optical source is [Ref. 87].

$$ m(t) = A_c \cos \left[\omega_c t + k_f \int_0^t a(\tau) \, d\tau \right] \tag{12.86} $$

where k_f is the angular frequency deviation in radians per second per unit of $a(t)$. To prevent intensity overmodulation, the carrier amplitude $A_c \leq 1$. The generally accepted expression for the bandwidth, which is referred to as Carson's rule, is given by:

$$ B_m \simeq 2(D_f + 1)B_a \tag{12.87} $$

where D_f is the frequency deviation ratio defined by:

$$ D_f = \frac{\text{peak frequency deviation}}{\text{bandwidth of } a(t)} = \frac{f_d}{B_a} \tag{12.88} $$

The peak frequency deviation in the subcarrier FM signal f_d is given by:

$$ f_d = k_f \max | a(t) | \tag{12.89} $$

Hence the SNR at the input to the subcarrier FM demodulator is:

$$ \left(\frac{S}{N} \right)_{rms} \text{input FM} = \frac{(RP_o)^2 (A_c^2/2)}{2 B_m N_o} \tag{12.90} $$

The subcarrier demodulator operating above threshold yields an output SNR [Ref. 84]:

$$ \left(\frac{S}{N} \right)_{rms} \text{output FM} = 6 D_f^2 (D_f^2 + 1) \frac{P_a (RP_o)^2 (A_c^2/2)}{2 B_m N_o} \tag{12.91} $$

Substituting for B_m from Eq. (12.87) gives:

$$ \left(\frac{S}{N} \right)_{rms} \text{output FM} = \frac{3 D_f^2 P_a (RP_o)^2 (A_c^2/2)}{2 B_a N_o} \tag{12.92} $$

The result obtained in Eq. (12.92) indicates that a significant improvement in the post-detection SNR may be achieved by using wideband FM–IM as demonstrated in the following example.

Example 12.15

(a) A D–IM and an FM–IM optical fiber communication system are operated under the same conditions of modulating signal power and bandwidth, detector photocurrent and noise. Furthermore, in order to maximize the SNR in the FM–IM system, the amplitude of the subcarrier is set to unity. Derive an expression for the improvement in post-detection SNR of the FM–IM system over the D–IM system. It may be assumed that the SNR is defined in terms of the rms signal power to rms noise power.

(b) The FM–IM system described in (a) has an 80 MHz subcarrier which is modulated by a baseband signal with a bandwidth of 4 kHz such that the peak frequency deviation is 400 kHz. Use the result obtained in (a) to determine the improvement in post-detection SNR (in decibels) over the D–IM system operating under the same conditions. Also estimate the bandwidth of the FM signal.

Solution: (a) The output SNR for the D–IM system is given by Eq. (12.79) where we can write $P_m = P_a$ and $B_m = B_a$. Hence:

$$\left(\frac{S}{N}\right)_{rms} \text{output D–IM} = \frac{(RP_o)^2 P_a}{2B_a N_o}$$

The corresponding output SNR for the FM–IM system is given by Eq. (12.92) where setting A_c to unity gives:

$$\left(\frac{S}{N}\right)_{rms} \text{output FM} = \frac{3D_f^2 P_a (RP_o)^2}{4B_a N_o}$$

Therefore the improvement in SNR of the FM–IM system over the D–IM system is given by:

$$\text{SNR improvement} = \frac{[3D_f^2 P_a (RP_o)^2]/(4B_a N_o)}{[(RP_o)^2 P_a]/(2B_a N_o)}$$

$$= \frac{3D_f^2}{2}$$

and:

$$\text{SNR improvement in decibels} = 10 \log_{10} \frac{3}{2} D_f^2$$

$$= 1.76 + 20 \log_{10} D_f$$

(b) The frequency deviation ratio is given by Eq. (12.88) where:

$$D_f = \frac{f_d}{B_a} = \frac{400 \times 10^3}{4 \times 10^3} = 100$$

Therefore the SNR improvement is:

SNR improvement $= 1.76 + 20 \log_{10} 100$
$$= 41.76 \text{ dB}$$

The bandwidth of the FM–IM signal may be estimated using Eq. (12.87) where:

$B_m \simeq 2(D_f + 1)B_a = 2(100 + 1)4 \times 10^3$
$$= 808 \text{ kHz}$$

This result indicates that the system is operating as a wideband FM–IM system.

Example 12.15 illustrates that a substantial improvement in the post-detection SNR over D–IM may be obtained using FM–IM. However, it must be noted that this is at the expense of a tremendous increase in the bandwidth required (808 kHz) for transmission of the 4 kHz baseband channel.

12.7.6 Subcarrier phase modulation (PM–IM)

With this modulation technique the instantaneous phase of the subcarrier is set proportional to the modulating signal. Hence in a PM–IM system the modulating signal $m(t)$ may be written as [Ref. 87]:

$$m(t) = A_c \cos(\omega_c t + k_p a(t)) \tag{12.93}$$

where k_p is the phase deviation constant in radians per unit of $a(t)$. Again the carrier amplitude $A_c \leq 1$ to prevent intensity overmodulation. Moreover, the bandwidth of the PM–IM signal is given by Carson's rule as:

$$B_m \simeq 2(D_p + 1)B_a \tag{12.94}$$

where D_p is the frequency deviation ratio for the PM–IM system. In common with subcarrier frequency modulation, the frequency deviation ratio is defined as:

$$D_p = \frac{f_d}{B_a} \tag{12.95}$$

where f_d is the peak frequency deviation of the subcarrier PM signal, which is given by:

$$f_d = k_p \max \left| \frac{da(t)}{dt} \right| \tag{12.96}$$

The SNR at the input to the subcarrier PM modulator is:

$$\left(\frac{S}{N} \right)_{\text{rms}} \text{ input PM} = \frac{(RP_o)^2 A_c^2/2}{2B_m N_o} \tag{12.97}$$

The output SNR from an ideal subcarrier PM demodulator operating above threshold is [Ref. 84]:

$$\left(\frac{S}{N}\right)_{\text{rms}} \text{ output PM} = \frac{D_p^2 P_a (RP_o)^2 A_c^2 / 2}{2 B_a N_o} \tag{12.98}$$

The result given in Eq. (12.98) suggests that an improvement in SNR over D–IM may be obtained using PM–IM, especially when the SNR is maximized with $A_c = 1$. However, comparison of PM–IM with FM–IM indicates that the latter modulation format gives the greatest improvement.

Example 12.16

A PM–IM and an FM–IM optical fiber communication system are operated under the same conditions of bandwidth, baseband signal power, subcarrier amplitude, frequency deviation, detector photocurrent and noise. Assuming the demodulators for both systems are ideal, determine the ratio (in decibels) of the output SNR from the FM–IM system.

Solution: The output SNR from the FM–IM system is given by Eq. (12.92) where:

$$\left(\frac{S}{N}\right)_{\text{rms}} \text{ output FM} = \frac{3 D_f^2 P_a (RP_o)^2 A_c^2 / 2}{2 B_a N_o}$$

Substituting for D_f from Eq. (12.88) gives:

$$\left(\frac{S}{N}\right)_{\text{rms}} \text{ output FM} = \frac{3 f_d^2 P_a (RP_o)^2 A_c^2 / 2}{2 B_a^3 N_o}$$

The output SNR for the PM–IM system is given by Eq. (12.98) where:

$$\left(\frac{S}{N}\right)_{\text{rms}} \text{ output PM} = \frac{D_p^2 P_a (RP_o)^2 A_c^2 / 2}{2 B_a N_o}$$

Substituting for D_p from Eq. (12.95) gives:

$$\left(\frac{S}{N}\right)_{\text{rms}} \text{ output PM} = \frac{f_d^2 P_a (RP_o)^2 A_c^2 / 2}{2 B_a^3 N_o}$$

The ratio of the output SNRs from the FM–IM and the PM–IM system is:

$$\text{Ratio} = \frac{[3 f_d^2 P_a (RP_o)^2 A_c^2 / 2] / (2 B_a^3 N_o)}{[f_d^2 P_a (RP_o)^2 A_c^2 / 2] / (2 B_a^3 N_o)}$$

$$= 3$$
$$= 4.77 \text{ dB}$$

Example 12.6 shows that the FM–IM system has a superior output SNR by some 4.77 dB over the corresponding PM–IM system. Nevertheless, this does not prohibit the use of PM–IM systems for analog optical fiber communications as they still exhibit a substantial improvement in output SNR over D–IM systems, as well as allowing frequency division multiplexing. It should be noted, however, that a similar bandwidth penalty to FM–IM is incurred using this modulation format.

12.7.7 Pulse analog techniques

Pulse modulation techniques for analog transmission, rather than encoding the analog waveform into PCM, were mentioned within the system design considerations of Section 12.4. The most common techniques are pulse amplitude modulation (PAM), pulse width modulation (PWM), pulse position modulation (PPM) and pulse frequency modulation (PFM). All the pulse analog techniques employ pulse modulation in the electrical regime prior to intensity modulation of the optical source. However, PAM–IM is affected by source non-linearities and is less efficient than D–IM, and therefore is usually discounted. PWM–IM is also inefficient since a large part of the transmitted energy conveys no information as only variations of the pulse width about a nominal value are of interest [Ref. 88]. Alternatively, PPM–IM and PFM–IM offer distinct advantages since the modulation affects the timing of the pulses, thus allowing the transmission of very narrow pulses. Hence, PPM–IM and PFM–IM provide similar signal-to-noise performance to subcarrier phase and frequency modulation while avoiding problems involved with source linearity. These techniques therefore prove advantageous for longer haul analog fiber links. Although PPM–IM is slightly more efficient, it provides less SNR improvement over D–IM than that gained with PFM–IM, where wideband FM gain may be obtained. Furthermore, the terminal equipment required for PFM–IM is less complex and therefore it is generally the preferred pulse analog technique [Refs 86, 89–94]. For these reasons the system aspects of pulse analog transmission will be considered in relation to PFM–IM.

A block schematic of a PFM–IM optical fiber system is shown in Figure 12.47. PFM in which the pulse repetition rate is varied in sympathy with the modulating signal is performed in the PFM modulator which consists of a voltage-controlled oscillator (VCO). This in turn operates the optical source by means of either a fixed pulse width or a fixed

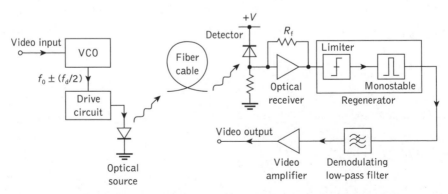

Figure 12.47 A PFM–IM optical fiber system employing regenerative baseband recovery

duty cycle (e.g. 50%). Demodulation in the system shown in Figure 12.47 is by regenerative baseband recovery, whereby the individual pulses are detected in a wideband receiver before they are regenerated with a limiter and monostable. This provides the desired modulating signal as a baseband component which is recovered through a low-pass filter.

Regenerative baseband recovery gives the best SNR at the system output. A simpler PFM demodulation technique for fixed width pulse transmission is direct baseband recovery. In this case, because a baseband component is generated at the transmit terminal, detection may be performed with a low-bandwidth receiver and the modulating signal obtained directly from a low-pass filter. However, this technique gives a reduced SNR for a given optical power and therefore does not find wide application.

The SNR in terms of the peak-to-peak signal power to rms noise power of a PFM–IM system using regenerative baseband recovery is given by [Ref. 86]:

$$\left(\frac{S}{N}\right)_{p-p} = \frac{3(T_0 f_D M R P_{po})^2}{(2\pi T_R B)^2 \overline{i_N^2}}$$ (12.99)

where T_0 is the nominal pulse period which is equivalent to the reciprocal of the pulse rate f_o, f_D is the peak-to-peak frequency deviation, R is the photodiode responsivity, M is the photodiode multiplication factor, P_{po} is the peak received optical power, T_R is the pulse rise time at the regenerator circuit input, B is the post-detection or effective baseband noise bandwidth and $\overline{i_N^2}$ is the receiver mean square noise current. It may be noted that improved SNRs are obtained with short rise time detected pulses. Moreover, the pulse rise time at the regenerator circuit input is dictated by the overall 10–90% system rise time T_{syst}, so there is no advantage in using a wideband receiver with a better pulse rise time than this. In fact, such a receiver would degrade the system performance by passing increased front-end noise. Therefore, in an optimized PFM regenerative receiver design, $T_R = T_{syst}$ and following Eq. (12.43)

$$T_R \simeq 1.1(T_S^2 + T_n^2 + T_c^2 + T_D^2)^{\frac{1}{2}}$$ (12.100)

where T_S, T_n, T_c and T_D are the rise times of the source (or transmitter), the fiber (intermodal and chromatic) and the detector (or receiver) respectively.

Example 12.17

An optical fiber PFM–IM system for video transmission employs regenerative baseband recovery. The system uses multimode graded index fiber and an APD detector and has the following operational parameters:

Nominal pulse rate	20 MHz
Peak-to-peak frequency deviation	5 MHz
APD responsivity	0.7
APD multiplication factor	60
Total system 10–90% rise time	12 ns
Baseband noise bandwidth	6 MHz
Receiver mean square noise current	$1 \times 10^{-17}\,\text{A}^2$

▶

Calculate: (a) the optimum receiver bandwidth; (b) the peak-to-peak signal power to rms noise power ratio obtained when the peak input optical power to the receiver is −40 dBm.

Solution: (a) For an optimized design the pulse rise time at the regenerator circuit is equal to the total system rise time, hence $T_R = 12$ ns. The optimum receiver bandwidth is simply obtained by taking the reciprocal of T_R giving 83.3 MHz.

(b) The nominal pulse period $T_0(= 1/f_0)$ is 5×10^{-8} s and the peak optical power at the receiver is 1×10^{-7} W. Therefore, the peak-to-peak signal to rms noise ratio may be obtained using Eq. (12.99), where:

$$\left(\frac{S}{N}\right)_{\text{p-p}} = \frac{3(T_0 f_D MRP_{\text{po}})^2}{(2\pi T_R B)^2 \overline{i_N^2}}$$

$$= \frac{3(5 \times 10^{-8} \times 5 \times 10^6 \times 60 \times 0.7 \times 10^{-7})^2}{(2\pi \times 12 \times 10^{-9} \times 6 \times 10^6)^2 \times 10^{-17}}$$

$$= 1.62 \times 10^6$$

$$= 62.1 \text{ dB}$$

The result of Example 12.17(b) illustrates the possibility of acquiring high SNRs at the output to a PFM–IM system using a regenerative receiver with achievable receiver noise levels and with moderate input optical signal power to the receiver.

12.8 Distribution systems

Thus far, the considerations in this chapter have effectively concerned only point-to-point and primarily unidirectional optical fiber communication systems. A strategy for obtaining bidirectional optical transmission on the same fiber link is described in Section 12.9.3, while in this section we discuss the implementation aspects of a growing area of activity within optical fiber communications, namely that of multiterminal distribution systems. For example, two major areas of application for such multiterminal distribution systems or networks which are dealt with in Chapter 15 are the telecommunication local access network (Section 15.6.3) and local area networks (Section 15.6.4).

Although many variants or hybrid topologies have been explored, the three basic multiterminal system architectures comprise the ring, bus and star configurations. The first topology, which has largely found implementation as a closed path or loop where consecutive nodes or terminals are connected by a series of point-to-point fiber links, is discussed in relation to the Fiber Distributed Data Interface covered in Section 15.6.4. With the latter two topologies, however, substantial progress has been made into the realization of multiterminal distribution systems and networks which do not simply comprise a series of point-to-point fiber links. In particular, they make use of the basic passive coupling devices described in Section 5.6.

It is instructive to form a comparison between the topological implementations of the bus and star distribution systems when each employ passive optical couplers to direct the signals to particular nodes. Block schematics for these two configurations are shown in Figure 12.48 where, for the purposes of the comparison, the linear nature of the bus is replicated in the star network through the positioning of the nodes in a linear manner. It is clear, however, that the star network configurations shown in Figures 15.4(c) and 15.28(d) are more representative of the use of the star topology to provide a widely distributed multiterminal network. Moreover the star–bus network implementation displayed in Figure 12.48(b) is not very economic in its use of fiber cable in comparison with the bus topology (Figure 12.48(a)) for a linear ordering of the network nodes.

The bus configuration illustrated in Figure 12.48(a) utilizes three port fiber couplers (see Section 5.6.1) to act as both beam splitter/combiner devices for the transmit and receive paths at each node, as well as passive fiber access couplers or taps along the bus link. However, whereas in the former case the split ratio is around 50%, in the latter tapping application the split ratio is often reduced to between 5 and 10% for the tap fiber so that the throughput optical power is a factor of 9 to 18 times greater than the optical power tapped off. Such an arrangement enables a larger optical power level to be transmitted down the bus and thus ensures adequate power at nodes distant from the transmit terminal.

Let us consider the total loss between node 1 and node $N - 1$. It should be noted in the configuration shown in Figure 12.48(a) that the path between nodes 1 and $N - 1$ exhibits the maximum loss because the final fiber tap couples only 10% of the incident optical power into the beam splitter of node $N - 1$. By contrast, the path to node N obtains a factor of 9 times this power level. Clearly, this situation could be modified by using a fiber beam splitter in place of the fiber tap in order to connect these two final nodes onto the bus.

Notwithstanding the above point we now consider the optical power budget for the worst case node interconnection (nodes 1 to $N - 1$) for the multiterminal bus system of Figure 12.48(a). It is apparent that to obtain the total channel loss $C_L(1, N - 1)$ between these two nodes the losses through each of the components must be summed. Let us commence at the transmit terminal, node 1. Then designating the connector losses in decibels as α_{cr} and assuming no excess loss in combining the transmitted signal onto the bus, a loss of $2\alpha_{cr}$ is obtained after the first beam splitter. The loss per kilometer exhibited by the fiber cable α_{fc} enables the total fiber cable loss between the two terminals to be written as $(N - 1)\alpha_{fc}L_{bu}$ where L_{bu} is equal to the fiber length between each of the access couplers. Furthermore, the total loss incurred by the signal in passing through the access couplers or taps between nodes 1 and $N - 1$ (excepting the final access coupler at which the signal to node $N - 1$ is tapped off) is given by $(2\alpha_{cr} + L_{ac}) (N - 3)$ where L_{ac} is the insertion loss of the access coupler. At the final access coupler before node $N - 1$ the loss obtained is $(2\alpha_{cr} + L_{tr})$ where L_{tr} is the loss due to the tap ratio of the device. Finally, a splitting loss L_{sp} occurs at the beam splitter together with a further connector loss α_{cr} at the optical receiver of node $N - 1$. The total channel loss between nodes 1 and $N - 1$ can therefore be written as:

$$C_L(1, N - 1) = 2\alpha_{cr} + (N - 1)\alpha_{fc}L_{bu} + (2\alpha_{cr} + L_{ac})(N - 3)$$
$$+ (2\alpha_{cr} + L_{tr}) + L_{sp} + \alpha_{cr} \tag{12.101}$$

To incorporate the overall channel losses into an optical power budget for the multiterminal bus distribution system, the mean power obtained at the optical transmitter P_t

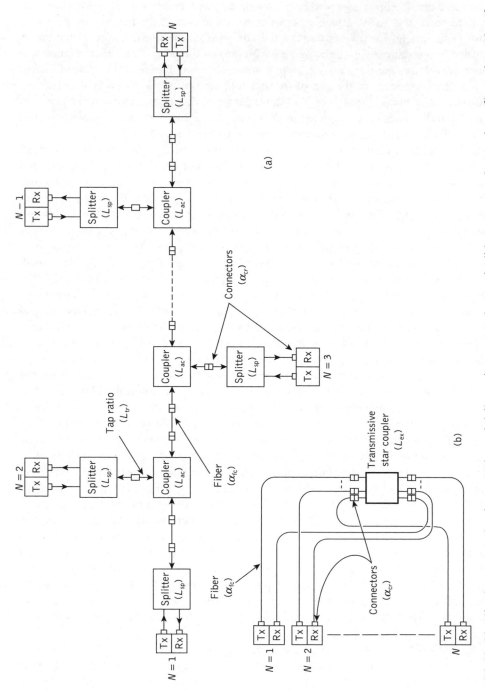

Figure 12.48 Distribution system implementations: (a) linear bus system/network; (b) star system/network configured as a bus for comparative purposes

at node 1, together with the mean incident optical power at the receiver, P_o of node $N-1$, must be included. Hence the optical power budget may be written as:

$$P_t = P_o + 2\alpha_{cr} + (N-1)\alpha_{fc}L_{bu} + (2\alpha_{cr} + L_{ac})(N-3)$$
$$+ (2\alpha_{cr} + L_{cr}) + L_{sp} + \alpha_{cr} + M_a \text{ dB} \tag{12.102}$$

where M_a is the system safety margin (see Section 12.6.6).

The star distribution system configuration displayed in Figure 12.48(b) employs a passive transmissive star coupler which provides two fibers to each node terminal (see Section 5.6.2). Hence an $N \times N$ star coupler allows the interconnection of N terminals. Assuming that the fiber cable lengths to each node are equal, then the same system loss is incurred for transmission between any two nodes. In this case the total system loss comprises the four connector losses at the transmitter, the receiver and the input and output ports of the star coupler $4\alpha_{cr}$; the total fiber cable loss $\alpha_{fc}L_{st}$ where L_{st} is the total fiber length in both arms of the star; the star splitting loss given by Eq. (5.18) as $10 \log_{10} N$ and the star excess loss L_{ex} provided by Eq. (5.19). In the case of equal fiber lengths the total channel loss between any two nodes is given by:

$$C_L(\text{star}) = 4\alpha_{cr} + \alpha_{fc}L_{st} + 10 \log_{10} N + L_{ex} \tag{12.103}$$

Again, to incorporate the overall channel losses into an optical power budget for the multi-terminal star distribution system, we designate the mean power obtained at the output of the optical transmitter P_t and the mean optical power incident at the receiver P_o so that:

$$P_t = P_o + 4\alpha_{cr} + \alpha_{fc}L_{st} + 10 \log_{10} N + L_{ex} + M_a \text{ dB} \tag{12.104}$$

where M_a is the system safety margin. A comparison of the optical power efficiencies of the two distribution systems is illustrated in the following example.

Example 12.18

Form a graphical comparison showing total channel loss against number of nodes for the bus and star distribution systems which incorporate components with the following performance parameters.

Connector loss:	1 dB
Access coupler insertion loss:	1 dB
Fiber cable loss:	5 dB km^{-1}
Access coupler tap ratio:	10 dB
Splitter loss:	3 dB
Star coupler excess loss:	0 dB

The distance between nodes on the bus system should be taken as 100 m and the worst case channel loss should be considered. It can be assumed that the total fiber cable length between all nodes on the star system is equal to 100 m.

▶

Solution: Bus distribution system. Using Eq. (12.101) the total channel loss is:

$$C_L(1, N-1) = 2 \times 1 + (N-1)5 \times 0.1 + (2 \times 1 + 1)(N-3) + (2 \times 1 + 10) + 3 + 1$$
$$= 0.5(N-1) + 3(N-3) + 18$$
$$= 3.5N + 8.5 \text{ dB}$$

Star distribution system. The total loss is given by Eq. (11.103) as:

$$C_L(\text{star}) = 4 \times 1 + 5 \times 0.1 + 10 \log_{10} N + 0$$
$$= 4.5 + 10 \log_{10} N \text{ dB}$$

The two expressions above for the bus and star distribution systems are plotted in Figure 12.49. It may be observed that the star configuration provides substantially greater efficiency in the utilization of optical power than the bus topology, particularly when the number of nodes becomes larger. It must be noted, however, that no excess loss for the star coupler has been included in the calculation and therefore it is anticipated that the total losses for the two distribution systems would be a little closer. Nevertheless, this factor would only make the optical power budgetary performance of the two configurations become similar when less than five terminals are interconnected.

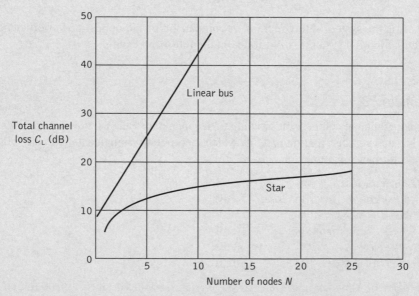

Figure 12.49 Characteristics showing the total channel loss against the number of nodes for the two distribution systems specified in Example 12.18

12.9 Multiplexing strategies

The basic multiplexing techniques which can be employed with IM/DD optical fiber systems were outlined in Section 12.4.2. Furthermore, the major baseband digital strategy, namely electrical time division multiplexing (ETDM), was discussed in some detail in Section 12.5. Although high speed ETDM plays an important role [Ref. 95], other significant multiplexing techniques are discussed in greater detail with particular emphasis on those strategies which allow greater exploitation of the available fiber bandwidth. We commence by further consideration of the multiplexing of digital signals prior to the more detailed description of techniques that may be employed for multiplexing either digital or analog intensity modulated signals, or a combination of both signal types.

12.9.1 Optical time division multiplexing

It was indicated in Section 11.6 that the practical limitations of the speed of electronic circuits have been pushed towards operational frequencies around 100 GHz. Therefore, although more recently the feasibility of 100 Gbit s^{-1} direct intensity modulation and transmission over substantial distances (480 km) has been demonstrated (e.g. [Ref. 96]), electronic multiplexing at such speeds is not straightforward and continues to present a restriction on the bandwidth utilization of a single-mode fiber link. Optoelectronic devices operating at transmission rates of 10 to 40 Gbit s^{-1} and above can be obtained and optical transmission systems using ETDM at 40 Gbit s^{-1} have become commercially available [Ref. 97]. An alternative strategy for increasing the bit rate of digital optical fiber systems beyond the bandwidth capabilities of the drive electronics is known as optical time division multiplexing (OTDM) [Ref. 98]. A block schematic of an OTDM system which has demonstrated 160 Gbit s^{-1} transmission over 100 km is shown in Figure 12.50 [Refs 99–101]. The principle of this technique is to extend ETDM by optically combining a number of lower speed electronic baseband digital channels. In the case illustrated in Figure 12.50, the optical multiplexing and demultiplexing ratio is 1 : 4, with a baseband channel rate of 40 Gbit s^{-1}. Hence the system can be referred to as a four-channel OTDM system.

The four optical transmitters in Figure 12.50 were driven by a common 40 GHz clock using quarter bit period time delays. Mode-locked semiconductor laser sources which produced short optical pulses (around 2 ps [Ref. 98] long) were utilized at the transmitters to provide low duty cycle pulse streams for subsequent time multiplexing. Data was encoded onto these pulse streams using integrated optical intensity modulators (see Section 11.6) which gave return-to-zero transmitter outputs at 40 Gbit s^{-1}. These IO devices were employed to eliminate the laser chirp (see Section 6.7.3) which would result in dispersion of the transmitted pulses as they propagated within the single-mode fiber, thus limiting the achievable transmission distance.

The four 40 Gbit s^{-1} data signals were combined using an OTDM multiplexer. Although four optical sources were employed, they all emitted at the same optical wavelength and the 40 Gbit s^{-1} data streams were bit interleaved to produce the 160 Gbit s^{-1} signal. At the receive terminal the incoming signal was decomposed into the 40 Gbit s^{-1} baseband components in a demultiplexer. Hence single-wavelength 160 Gbit s^{-1} optical transmission was obtained with electronics which only required a maximum bandwidth of about 40 GHz,

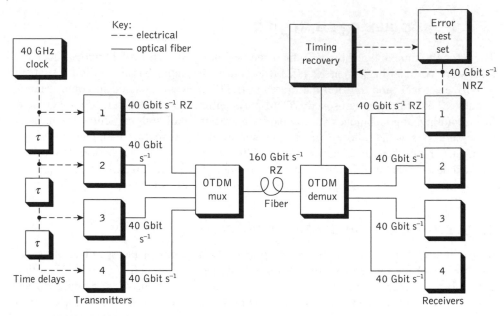

Figure 12.50 Four-channel OTDM fiber system

as return-to-zero pulses were employed. The transmitter and receiver sections shown in Figure 12.50 employed electroabsorption modulators (see Section 11.4.2) to provide for operation at the high transmission rate and furthermore negative dispersion fibers (see Section 3.12.3) were also incorporated to compensate for the positive dispersion of the standard single-mode fiber (SSMF). Moreover, a field trial employing such transmitters and receivers at a transmission rate of 160 Gbit s^{-1} over deployed SSMF has been successfully carried out [Ref. 101].

12.9.2 Subcarrier multiplexing

The use of RF subcarriers modulated by analog signals prior to intensity modulation of an optical source was discussed in Section 12.7.3. Moreover, the utilization of substantially higher frequency microwave subcarriers multiplexed in the frequency domain before being applied to intensity-modulate a high-speed injection laser source has generated significant interest [Refs 102–106]. Subcarrier multiplexing (SCM) is sometimes also referred to as optical SCM (OSCM) where the microwave frequency or RF electrical subcarriers are modulated with an optical carrier and then are transmitted using a single-wavelength signal [Refs 107–109].

Microwave subcarrier multiplexing enables multiple broadband signals to be transmitted over single-mode fiber and can be particularly attractive for video distribution systems [Refs 105, 106, 110]. In addition, with SCM, conventional microwave techniques can be employed to subdivide the available intensity modulation bandwidth in a convenient way. The result is a useful multiplexing technique which does not require sophisticated optics

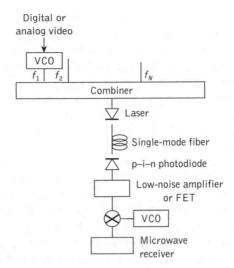

Figure 12.51 Basic subcarrier multiplexing (SCM) fiber system

or source wavelength specification (see Section 12.9.4). Either digital or analog modulation of the subcarriers can be utilized by upconverting to a narrowband channel at high frequency employing amplitude, frequency or phase shift keying (i.e. ASK, FSK or PSK), and amplitude, frequency or phase modulation (i.e. AM, FM or PM) respectively. For digital signals, FSK has the advantage of being simple to implement, at both the modulator and demodulator, whereas for analog video signals the modulation of the high-frequency carrier (upconversion) is often carried out using either AM–VSB (vestigial sideband) or FM techniques. In both cases, the multicarrier signal is formed by frequency division multiplexing of the modulated microwave subcarriers in the electrical domain prior to conversion to an intensity modulated optical signal.

A block schematic of a basic SCM system is shown in Figure 12.51 [Ref. 105]. The modulated microwave subcarrier signals are obtained by frequency upconversion from the baseband using voltage-controlled oscillators (VCOs). These subcarrier signals f_i are then summed in a microwave power combiner prior to the application of the composite signal to an injection laser which is d.c. biased at around 5 mW in order to produce the desired intensity modulation. The IM optical signal is then transmitted over single-mode fiber and directly detected using a wideband photodiode before demultiplexing and demodulation using a conventional microwave receiver.

Although relatively straightforward to implement using available components, SCM does exhibit some disadvantages, the most important of which is the problem associated with source nonlinearity [Ref. 106]. Distortion caused by this phenomenon can be particularly noticeable when several subcarriers are transmitted from a single optical source. Moreover, despite the fact that the receivers require narrow-bandwidth, SCM systems, with the exception of those employing AM–VSB modulation, must operate at high frequency, often in the gigahertz range. In addition, for digital systems SCM requires more bandwidth per channel than a TDM system. The upconversion results in the bandwidth expansion so that a 50 Mbit s^{-1} channel may require some 80 MHz of bandwidth

[Ref. 102]. Any reduction of this bandwidth overhead necessitates the adoption of more complex and less robust modulation techniques. For example, AM–VSB systems transmitting a standard cable television (CATV) multichannel spectrum tend to minimize the required bandwidth, but the signal must be received with a carrier-to-noise ratio of between 45 and 55 dB to avoid degradation of picture quality [Ref. 106].

The transmission of multiple CATV channels over substantial unrepeatered distances with good-quality reception has, however, been demonstrated with SCM using FM (i.e. FM–FDM). For example, 34 multiple sub-Nyquist-sampling encoding (MUSE) high-definition television (HDTV) channels, each requiring an FM bandwidth of 27 MHz, have been transmitted over an unrepeatered distance of 42 km [Ref. 110]. This transmission system, which operated at a wavelength of 1.3 μm, provided a carrier-to-noise ratio of 17.5 dB at the receive terminal. Furthermore, an unrepeatered transmission distance in excess of 100 km has also been demonstrated with SCM when operating at a wavelength of 1.54 μm [Ref. 111]. In this case some eight baseband video channels, each of which was frequency modulated to occupy around 30 MHz of bandwidth, were then frequency multiplexed over a range 840 to 1160 MHz before directly modulating a distributed feedback laser.

Apart from the possibility of combining digital and analog SCM signals into a composite signal, an alternative attractive strategy is the so-called hybrid SCM system which combines a baseband digital signal with a high-frequency composite microwave signal [Ref. 105]. In this case the receiver shown in Figure 12.51 cannot be narrowband but must have a bandwidth from d.c. level to beyond the highest microwave signal frequency employed. In such systems only a single channel needs to be selected for demodulation. Hence a tunable local oscillator, mixer and narrowband filter can be utilized at the receive terminals (Figure 12.51) to simultaneously select the desired SCM channel and down-convert it to a more convenient intermediate-frequency (IF) signal. Finally, the IF signal can be input to an appropriate demodulator to recover the baseband video signal.

An experimental 78-channel optical SCM CATV system operating at a signal wavelength of 1.55 μm has been successfully demonstrated [Ref. 112]. The overall transmission distance using an SSMF link without employing dispersion compensation was 740 km. Three erbium-doped fiber optical amplifiers, however, were incorporated in the fiber link to provide booster and in-line amplification. Various SCM-based optical fiber systems operating at transmission rates from 10 to 40 Gbit s^{-1} have also been demonstrated [Refs 113–115]. In order to obtain high-bandwidth capability, however, more recent SCM system developments have tended to employ it in combination with another multiplexing technique (see Section 12.9.6) [Ref. 107].

12.9.3 Orthogonal frequency division multiplexing

Orthogonal frequency division multiplexing (OFDM) is a multicarrier transmission technique which is based on frequency division multiplexing (FDM) (see Section 12.4.2). In conventional FDM multiple-frequency signals are transmitted simultaneously in parallel where the data contained in each signal is modulated onto subcarriers and therefore the subcarrier multiplexed signal typically contains a wide range of frequencies. Each subcarrier is separated by a guard band to avoid signal overlapping. The subcarriers are then demodulated at the receiver by using filters to separate the frequency bands. By contrast OFDM

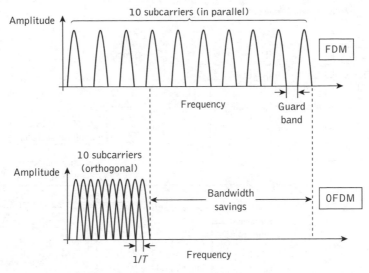

Figure 12.52 Orthogonal frequency division multiplexing (OFDM) compared with conventional frequency division multiplexing (FDM)

employs several subcarrier frequencies orthogonal to each other (i.e. perpendicular) and therefore they do not overlap. Hence this technique can squeeze multiple modulated carriers tightly together at a reduced bandwidth without the requirement for guard bands while at the same time keeping the modulated signals orthogonal so that they do not interfere with each other, as illustrated in Figure 12.52. In the upper spectral diagram 10 nonoverlapping subcarrier frequency signals arranged in parallel depicting conventional FDM are shown, each being separated by a finite guard band. OFDM is displayed in the bottom spectral diagram where the peak of one signal coincides with the trough of another signal. Each subcarrier, however, must maintain the Nyquist criterion separation with the minimum time period of T (i.e. a frequency spread of $1/T$) for each subcarrier.

OFDM uses the inverse fast Fourier transform (IFFT) for the purpose of modulation and the fast Fourier transform (FFT) for demodulation. Moreover, this is a consequence of the FFT operation by which subcarriers are positioned perpendicularly and hence the reason why the technique is referred to as orthogonal FDM. It may be observed that a large bandwidth saving in comparison with conventional FDM is identified in Figure 12.52 resulting from the orthogonal placement of the subcarriers. Since the orthogonal feature allows high spectral efficiency near the Nyquist rate where efficient bandwidth use can be obtained, OFDM generally exhibits a nearly white frequency spectrum (i.e. without electromagnetic interference between the adjacent channels). OFDM, also being tolerant to signal dispersion, thus enables high-speed data transmission across a dispersive channel and it has been widely used in high-bit-rate cable and wireless communication systems [Refs 116–119].

For applications within optical fiber communications it is necessary to incorporate an optical source to convert the OFDM signals into an optical signal format before coupling onto an optical fiber, while at the receiving end the intensity modulated signal can be recovered

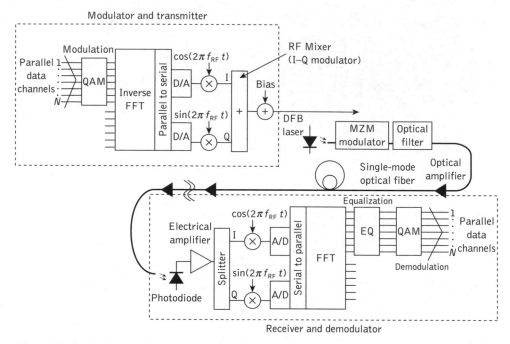

Figure 12.53 Optical orthogonal frequency division multiplexed system [Ref. 119]

using a direct detection receiver. To distinguish it from conventional OFDM it is referred to as optical OFDM (OOFDM). Although the multiplexing approach is similar to optical SCM (see Section 12.9.2), the orthogonal nature of the subcarriers is unique to OOFDM.

A block schematic of an experimental OOFDM system employing IM/DD is shown in Figure 15.53 [Ref. 119]. The input data comprising N channels is modulated onto N equally spaced electrical subcarriers using quadrature amplitude modulation (QAM) (see Section 13.5.3).* Using QAM the amplitudes of two carriers (usually sinusoids) which are 90° out of phase with each other are modulated. Since the orthogonal electrical carriers occupy the same frequency band but differ by a 90° phase shift, each can be modulated independently enabling transmission within the same frequency band.

Each QAM data channel in the modulator of Figure 12.53 is then sent for computation of the IFFT [Refs 51, 120]. The parallel-to-serial converter and digital-to-analog (D–A) converters produce a complex electrical signal waveform containing a superposition of all of the subcarriers. This waveform is modulated onto an RF carrier f_{RF}, using an in-phase and quadrature phase (I–Q) modulator, producing a real-valued OFDM waveform. A d.c. component (via a bias) is added to the modulated signal which enables recovery of the QAM symbols by direct detection at the receiving end. The OFDM signal is then input to a Mach–Zehnder modulator (MZM) which provides intensity modulation of the signal from the DFB laser source thus providing the OOFDM signal. The output from the MZM

---

* It should be noted that QAM is performed in the electrical domain prior to intensity modulation of the distributed feedback (DFB) laser source, whereas the systems described in Section 13.5.3 use PM or QAM of the optical signals.

is filtered to remove all unwanted frequencies leaving a suppressed single-sideband optical carrier signal which is then coupled into a single-mode fiber for transmission.

After propagation through single-mode optical fiber, direct detection by the photodiode at the receiver produces an electrical waveform which is converted back to the in-phase and quadrature components by mixing with the zero and $\pi/2$ phase differences of electrical local oscillators operating at frequency f_{RF}. Analog-to-digital (A–D) conversion then takes place prior to a serial-to-parallel conversion to establish a parallel digital bit stream. Hence serial I and Q waveforms are converted to parallel OFDM subcarriers and then the FFT is computed to enable recovery of the QAM signals. When the IFFT (at the transmitter) and FFT (at the receiver) operations are synchronized in time, each FFT window at the receiving end functions as a set of closely spaced narrowband filters. In the frequency domain, each of these channels is equalized to compensate for phase and amplitude distortions due to the optical and electrical paths. Finally, each channel is demodulated using a QAM demodulator to produce the original N parallel data channels.

Using the above strategy a number of OOFDM systems have been demonstrated [Refs 119–123]. For example, a system comprising a 400 km long SSMF link operating at a signal wavelength of 1.55 μm has been successfully operated at a transmission bit rate of 12 Gbit s^{-1} using commercially available electro-optic devices and optical amplifiers [Refs 54, 123]. An even higher transmission rate of 20 Gbit s^{-1} was obtained but in this case the length of the fiber link was reduced to 320 km. Finally, it should be noted that the system performance was mainly limited by the amplified spontaneous emission noise generated within the optical amplifiers.

12.9.4 Wavelength division multiplexing

Wavelength division multiplexing (WDM) involves the transmission of a number of different peak wavelength optical signals in parallel on a single optical fiber. Although in spectral terms optical WDM is analogous to electrical FDM, it has the distinction that each WDM channel effectively has access to the entire intensity modulation fiber bandwidth which with current technology is of the order of several gigahertz. The technique is illustrated in Figure 12.54 where a conventional (i.e. single nominal wavelength) optical fiber communication system is shown together with a duplex (i.e. two different nominal wavelength optical signals traveling in opposite directions providing bidirectional transmission), and also a multiplex (i.e. two or more different nominal wavelength optical signals transmitted in the same direction) fiber communication system. It is the latter WDM operation which has generated particular interest within telecommunications. For example, two-channel WDM is very attractive for a simple system enhancement such as piggybacking a 565 Mbit s^{-1} system onto an installed 140 Mbit s^{-1} link, or for doubling the capacity of a 565 Mbit s^{-1} link [Ref. 124]. Moreover, this multiplexing strategy overcomes certain power budgetary restrictions associated with electrical TDM. When the transmission rate over a particular optical link is doubled using TDM, a further 3 to 6 dB of optical power is generally required at the receiver (see Section 12.6.3). In the case of WDM, however, additional losses are also incurred from the incorporation of wavelength multiplexers and demultiplexers (see Section 5.6.3).

Wavelength division multiplexing in IM/DD optical fiber systems can be implemented using either LED or injection laser sources with either multimode or single-mode fiber.

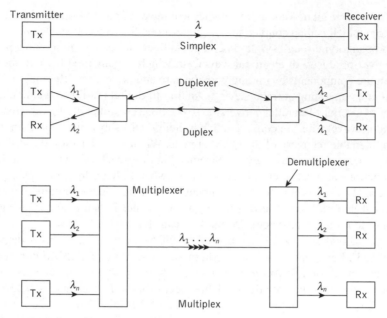

Figure 12.54 Optical fiber system operating modes illustrating wavelength division multiplexing (WDM)

However, the widespread deployment of single-mode fiber has encouraged the investigation of WDM on this transmission medium. In particular, developments concerned with single-mode fiber WDM transmission can be distinguished into two broad categories, namely coarse WDM (CDWM) and dense WDM (DWDM). Although both categories use the same concept of multiple-wavelength channels on a single fiber, they differ in the channel spacing they employ. CWDM as implied by the terminology uses wider channel spacing and hence provides significantly fewer channels than DWDM.

Coarse WDM is specified in ITU-T Recommendation G.694.2 [Ref. 125] which defines a wavelength grid with 20 nm channel spacings and includes 18 wavelengths between 1271 and 1611 nm as depicted in Figure 12.55. Moreover, both the unidirectional and bidirectional CDWM are provided in ITU-T Recommendation G.695 [Ref. 126]. In addition, Recommendation G.694.2 provides optical interface specifications for multichannel CWDM systems on target distances of 40 km and 80 km. Figure 12.55 also displays an attenuation characteristic for standard single-mode fiber shown by the dashed line which indicates that five of the CWDM wavelength channels fall within the E-band that cannot be used due to the water peak. However, as low-water-peak fiber (LWPF) can be employed in the E-band wavelength region (see Section 3.3.2) with an attenuation characteristic shown by the bold line in Figure 12.55, then 16-channel bidirectional CWDM modular systems have become commercially available [Refs 127–130]. For example, the performance of a 16-channel CWDM system with each channel operating at 2.5 Gbit s^{-1} over a distance of 75 km has been reported [Ref. 131]. The system demonstrated an aggregate bandwidth of 40 Gbit s^{-1} employing LWPF. It should also be noted that the reach of such systems can be extended to more than 200 km when using optical amplification [Refs 132, 133].

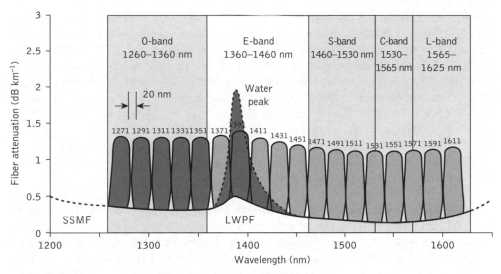

Figure 12.55 Optical wavelength channel allocation for coarse wavelength division multiplexed systems as specified by ITU-T Recommendation G.694.2

Dense WDM was originally concerned with optical signals multiplexed in the 1.55 μm wavelength region using the capabilities of erbium-doped fiber amplifiers (EDFAs) (see Section 10.4) to increase system capacity and therefore to reduce system cost. Figure 12.56 shows a block schematic for a DWDM system where a large number of channels N, each utilizing a single wavelength (i.e. from λ_1 to λ_N), are multiplexed onto a single-fiber transmission medium. Both the deployment of EDFAs and dispersion compensation are required for long-haul DWDM systems to offset any optical signal power losses caused by optical wavelength multiplexers and other passive optical devices [Ref. 134]. Finally, a wavelength demultiplexer distributes each channel to the corresponding receiver.

Dense WDM systems use narrow channel spacings and can therefore accommodate several hundred wavelength channels on a single optical fiber.* The three possible channel

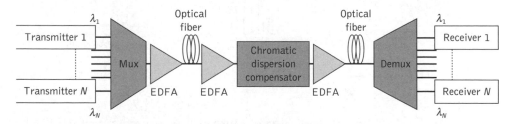

Figure 12.56 Block schematic of a dense wavelength division multiplexed system

* Since the majority of the commercially available CWDM systems use eight wavelength channels in S-, C- and L-bands, systems accommodating more than eight wavelengths are sometimes also referred to as DWDM systems. This is, however, incorrect with DWDM requiring more than 18 channels and with the typical lowest channel count for a DWDM system being 20.

spacings specified for DWDM systems are 1.6 nm (200 GHz), 0.8 nm (100 GHz) and 0.4 nm (50 GHz) [Refs 135, 136] while an even smaller channel spacing of 0.1 nm (12.5 GHz) is feasible in which case the system may also be referred to as super-DWDM [Ref. 137]. Table 12.2 displays a sample portion of the ITU-T frequency–wavelength grid allocating 100 GHz and 50 GHz channel spacings within each of the S-, C- and L-bands. For example, ITU-T Recommendation G.698.1 specifies DWDM applications at 2.5 and 10 Gbit s^{-1} with 0.8 nm (100 GHz) channel spacings [Refs 135, 138]. Moreover, a field trial for such a DWDM system demonstrated an aggregated transmission rate of 1 Tbit s^{-1} (i.e. 1×10^{12} bit s^{-1}) [Ref. 139]. In this system 25 DWDM channels each operating at a rate of 40 Gbit s^{-1} were successfully transmitted over a 6250 km link comprising 15 EDFAs together with nonzero-dispersion-shifted fibers.

Unlike CWDM transmission, DWDM systems use narrowband optical filters in the demultiplexing section due to the narrow channel spacing requirement. Furthermore, DWDM transmitters require temperature-controlled laser sources to stabilize the emitted signal wavelengths from each transmitter and also the large number of channels consumes a much higher power level. For example, a 16-channel CWDM system consumes as little as 4 W of power whereas a conventional DWDM system requires about 80 W to transmit the same number of channels [Ref. 140]. Recent DWDM system developments, however, include pluggable and software-tunable transceiver modules which are capable of transmitting 40 or 80 channels [Ref. 141]. Such transceivers can be plugged in as required at any wavelength and therefore they can handle a full range of wavelengths signals while reducing the overall DWDM system cost. Furthermore, the power consumption of these transceivers remains between 1 and 4 W when operating over a temperature range of −5 to 70 °C [Refs 142, 143].

Another WDM strategy which has been investigated for both telecommunication and nontelecommunication applications is illustrated in Figure 12.57 [Ref. 144]. In this case, in place of narrow linewidth injection laser sources, wide spectral width (63 nm) edge-emitting LEDs were utilized to provide the multiwavelength optical carrier signals which

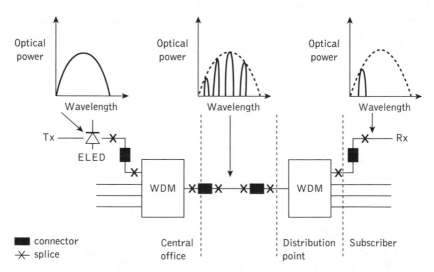

Figure 12.57 Spectral slicing of LED outputs to form several WDM channels

Table 12.2 Sample of the ITU-T frequency–wavelength grid for dense wavelength division multiplexed systems with 100 GHz (0.8 nm) and 50 GHz (0.4 nm) channel spacings for L-, C- and S-bands [Refs 135, 138]. The emboldened frequency at 193.1 THz corresponding to a wavelength of 1552.52 nm is the ITU-T reference frequency which centrally locates the range of grid frequencies [Ref. 101]

L-band				C-band				S-band			
100 GHz		50 GHz		100 GHz		50 GHz		100 GHz		50 GHz	
THz	nm	THz	nm	THz	nm	THz	nm	THz	nm	THz	nm
188.00	1594.64	188.05	1594.22	193.00	1553.33	193.05	1552.93	198.00	1514.10	198.05	1513.72
188.10	1593.79	188.15	1592.52	**193.10**	1552.52	193.15	1551.12	198.10	1513.34	198.15	1512.96
188.20	1592.95	188.25	1592.52	193.20	1551.72	193.25	1551.32	198.20	1512.58	198.25	1512.19
188.30	1592.10	188.35	1591.68	193.30	1550.92	193.35	1550.52	198.30	1511.81	198.35	1511.43
188.40	1591.26	188.45	1590.83	193.40	1550.12	193.45	1549.72	198.40	1511.05	198.45	1510.67
188.50	1590.41	188.55	1589.99	193.50	1549.32	193.55	1548.91	198.50	1510.29	198.55	1509.91
188.60	1589.57	188.65	1589.15	193.60	1548.51	193.65	1548.11	198.60	1509.63	198.65	1509.15
188.70	1588.73	188.75	1588.30	193.70	1547.72	193.75	1547.32	198.70	1508.77	198.75	1508.39
188.80	1587.88	188.85	1587.46	193.80	1546.92	193.85	1546.52	198.80	1508.01	198.85	1507.63
188.90	1587.04	188.95	1586.62	193.90	1546.12	193.95	1545.72	198.90	1507.25	198.95	1506.87
189.00	1586.20	189.05	1585.78	194.00	1545.32	194.05	1544.92	199.00	1506.49	199.05	1506.12

were transmitted on single-mode optical fiber. The full spectral output from each ELED was not, however, transmitted for each wavelength channel. Instead, a relatively narrow spectral slice (3.65 nm) for each separate channel was obtained using the diffraction grating WDM multiplexer device, as shown in Figure 12.57 prior to transmission down the optical link. This technique, which is known as spectral slicing, could enable LEDs with the same overall spectral output to be employed while still providing the distinctive wavelength channels for transmission between each subscriber terminal. In this case a WDM demultiplexer device is located at a distribution point in order to separate and distribute the different wavelength optical channels to the appropriate subscriber receive terminals. A similar strategy has been demonstrated for 16 channels using superluminescent LEDs, again transmitting on single-mode fiber [Ref. 145]. Moreover, the technique has also been employed in nontelecommunication areas to provide multiple wavelength channels from single-LED sources, usually on multimode fiber, in order to service, for example, a multiple optical sensor system in which each wavelength channel supplies a signal to a different optical sensor device [Refs 146, 147].

Although much of the earlier work on spectrum slicing focused on LEDs [Refs 148, 149], optical amplifiers (see Sections 10.3 and 10.4) can also be used for this purpose [Refs 150–152]. In this case the broad amplified spontaneous emission output characteristics of these devices can be used to produce multiwavelength sources. For example, a spectrum-sliced optical amplifier can provide 140 multiwavelength channels each with a spacing of just 0.57 nm covering a wavelength range over both the C- and L-bands [Ref. 153].

Semiconductor laser sources can also be used for the generation of optical spectrally sliced signals [Ref. 153]. Figure 12.58 shows a block schematic of a subsystem used to generate a number of spectrum-sliced channels using a femtosecond laser and EDFAs. It also incorporates optical modulators and delay lines to operate on the multiple signals which are amplified before the splitting stage and after combining stage. Based on this scheme, a 110-channel DWDM system with each channel operating at a transmission rate of 2.35 Gbit s^{-1} has been demonstrated [Ref. 154]. Moreover, 40-channel spectrum slicing with each channel spaced at 0.57 nm covering the C-band and L-band providing an overall transmission rate of 1.36 Tbit s^{-1} has also been reported [Ref. 153]. In addition, a number of other experimental demonstrations have also been shown to support terabit per second

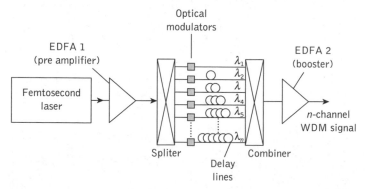

Figure 12.58 Block schematic of a spectrum slicing technique using a femtosecond laser source and erbium-doped fiber amplifiers (EDFAs). Reprinted with permission from Ref. 154 © IEEE 1999

transmission rates [Refs 155–157]. These systems, however, incorporated arrayed waveguide gratings (see Section 11.6) and used OTDM (see Section 12.9.1) to produce the large number of high-speed channels.

12.9.5 Optical code division multiplexing

Optical code division multiplexing (OCDM), sometimes termed optical code division multiple access (OCDMA)* is a digital technique where, instead of each channel occupying a given wavelength, frequency or time slot, the information is transmitted using a coded sequence of pulses. Each channel employs a specific code to transmit and recover the original signal. It utilizes the basic principle of spread spectrum transmission where all users share the fiber channel bandwidth simultaneously. The basic OCDM technique is illustrated in Figure 12.59 where multiplexing of three channels is accomplished by transmitting a unique time-dependent series of short pulses. Each bit to be transmitted is subdivided into a number of small intervals n known as chips (e.g. $n = 64$ or 128). Each user is then assigned a unique chip sequence of the n bit code. Many coding schemes exist for the generation of the chip sequences to encode/decode OCDM channels, with the overall premise that the greater the number of unique sequences needed (i.e. number of users), the larger the code sequence required [Refs 158–161]. A decoder is then used at the receiving end to recover the particular channel employing autocorrelation with the original chip sequence. As each data bit is converted in many chips, however, OCDM is not a bandwidth-efficient multiplexing technique.

In order to enable an increasing number of OCDM channels to be transmitted on a single fiber, however, ultrashort pulses (i.e. 10^{-15} second pulse duration) are required to be used in OCDM systems. Furthermore, the larger the number of channels, the longer the code sequences needed to provide a unique channel code. Nevertheless an optical OCDM multiplexer has been reported which provided bidirectional data transmission over 100 km of single-mode fiber [Ref. 162]. In addition, an experimental test bed has successfully demonstrated both 160 and 320 Gbit s^{-1} OCDM transmission systems [Ref. 163]. These systems using signal wavelengths centered at 1.55 µm supported 16 and 32 channels, respectively, with each channel operating at a transmission rate of 10 Gbit s^{-1}.

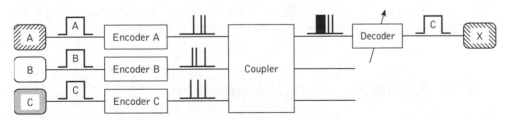

Figure 12.59 Block schematic showing three channels of an optical fiber code division multiplexing system

---

* The multiple-access terminology indicates that the technique requires a common communication medium to be shared by all channels.

12.9.6 Hybrid multiplexing

When two (or more) different multiplexing techniques are combined to allow optical signal multiplexing for several optical signals, the resultant is referred to as hybrid multiplexing. It should be noted that different multiplexing strategies (see Sections 12.9.1 to 12.9.5) exhibit their own advantages and drawbacks and therefore the combination of different multiplexing techniques can be used to overcome the problems associated with a specific technique. A hybrid multiplexing system can comprise either optical or electrical domain multiplexing, or combination of both signal types. Common examples of optical hybrid multiplexing are WDM being combined with OTDM, OCDM or SCM.

Hybrid WDM/OTDM systems can support terabit per second transmission rates when several WDM channels are combined with OTDM technology. In such a hybrid WDM/OTDM system each WDM channel can, in principle, operate at a transmission rate of 10, 40 or 160 Gbit s^{-1}. For example, a hybrid WDM/OTDM system, where six OTDM channels each operating at 170.6 Gbit s^{-1} were wavelength division multiplexed, supported an overall transmission rates of 1 Tbit s^{-1} [Ref. 164]. Another hybrid strategy employed OCDM with WDM [Refs 165–167]. Since OCDM uses coded signals it therefore reduced the requirements on both time and frequency management when distributing optical multiplexed signals at dropping nodes.

Another example is an experimental two-stage hybrid system utilizing OCDM/OTDM multiplexing in the first stage and then OTDM/WDM in the second stage of hybrid multiplexing. This two-stage hybrid multiplexed system produced a large-capacity signal to support a transmission rate of 6.4 Tbit s^{-1} [Ref. 168]. The hybrid OCDM/OTDM/WDM signal converter incorporated arrayed waveguide grating-based spectrum slicing to generate a large number of optical channels. Furthermore, the system demonstrated the feasibility of this hybrid technique for supporting transmission rates up to 40 Tbit s^{-1} [Ref. 168].

Finally, when SCM is combined with WDM the resultant hybrid multiplexing strategy can benefit from readily available commercial components [Refs 169–171]. As compared with conventional WDM systems, SCM can employ much narrower channel spacing and it exhibits better dispersion compensation tolerances. Therefore SCM can improve the optical transmission system when it is used in combination with WDM. A WDM/SCM long-haul optical fiber system operating at a transmission rate of 10 Gbit s^{-1} was accomplished by combining four SCM data streams each transmitting at a speed of 2.5 Gbit s^{-1} [Ref. 172]. Moreover, hybrid SCM/WDM multiplexing has been demonstrated for use with a passive optical network (see Section 15.6.3) [Ref. 173].

12.10 Application of optical amplifiers

The use of electronics-based regenerative repeaters in long-haul optical fiber communications was discussed in Sections 12.4 and 12.6.1. It is clear, however, that such devices not only increase the cost and complexity of the optical communication system, but also act as a bottleneck by restricting the system operational bandwidth. Hence the recent developments in optical amplifier technology described in Sections 10.2 to 10.4 have started to provide an additional, welcome flexibility in the design and implementation of IM/DD optical fiber systems.

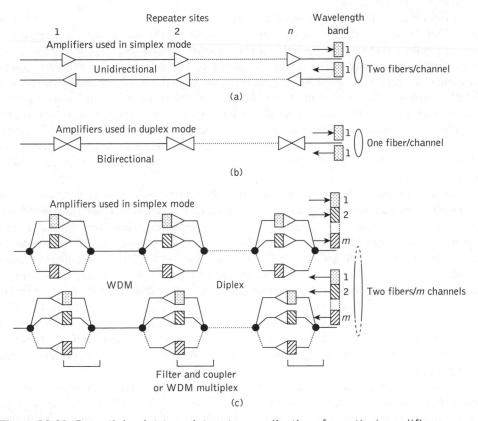

Figure 12.60 Potential point-to-point system applications for optical amplifiers: (a) simplex mode; (b) duplex mode; (c) multiamplifier configuration for wavelength division multiplexing (WDM) operation

The above flexibility stems from the ability for the transmitted optical signal to remain in the optical domain over the entire length of a long-haul link. Optical amplifiers (both semiconductor and fiber devices) therefore exhibit interesting features which assist in the system design, as illustrated in Figure 12.60. It may be observed from Figure 12.60(a) that, in a similar manner to electronic repeaters (see Figure 12.27), optical amplifiers may be employed in a simplex mode where each transmitted optical signal is carried on a separate fiber link. However, optical amplifiers have the ability to operate simultaneously in both directions at the same carrier wavelength,* as shown in Figure 12.60(b). Moreover, in this bidirectional mode they offer an added degree of reliability in that a single fiber break would only disable one-half of communication capacity per fiber pair rather than causing a complete system failure, as is the case with the present unidirectional systems (i.e. one transmission path between all user pairs is disabled).

---

* It is obviously necessary to intensity-modulate the optical carriers at different speeds to avoid signal interference.

A further range of flexibility associated with optical amplifiers concerns the ability of particular devices to simultaneously amplify multiple WDM optical signals (see Section 12.9.4). Both semiconductor optical and fiber amplifiers with spectral bandwidths in the range 50 to 100 nm can be realized (see Sections 10.3 and 10.4) which will allow single amplifiers to support more than 100 intensity modulated WDM channels (see Section 15.3). Moreover, the parallel multiamplifier configuration illustrated in Figure 12.60(c) could be envisaged which would enable contiguously spectrally aligned amplifiers to span a complete wavelength widow (say around 1.55 μm). Such configurations could increase system reliability in the event of an individual amplifier failure, while also relaxing the linearity and overload characteristics for amplifiers operating with densely packed WDM hierarchies [Ref. 174].

It was mentioned in Section 10.4 that optical amplifiers could be used in a broad range of system applications: namely, as power amplifiers at the optical transmitter; as in-line repeater amplifiers; and as preamplifiers at optical receivers. Moreover, the last system application was dealt with in Section 9.6. Therefore, in this section we concentrate on the utilization of optical amplifiers as in-line repeaters within IM/DD optical fiber systems. It must be remembered, however, that in contrast to regenerative repeaters, optical amplifiers simply act as gain blocks on an optical fiber link and hence they do not reconstitute a transmitted digital optical signal. A drawback with this operation is that both noise and signal distortions are continuously amplified as the optical signal passes down a link which uses cascaded amplifiers. A benefit, however, is that optical amplifiers are transparent to any type of signal modulation (i.e. digital or analog) and to any modulation bandwidth.

It was mentioned in Section 10.3.2 that noise in traveling-wave SOAs may determine the number of devices that can be cascaded as linear repeaters. Since the mean noise power resulting from spontaneous emission in optical amplifiers accumulates in proportion to the number of repeaters, gain saturation occurs when the total noise power becomes equal to the signal power. A simple model for the noise behavior of such devices which is valid under conditions of high signal-to-noise ratio and high gain is shown in Figure 12.61. It comprises an ideal noiseless amplifier with gain G preceded by an additive noise source of spectral density $S(f)$ given by [Ref. 175]:

$$S(f) = Khf \tag{12.105}$$

where K, which is dependent on the population inversion and the cavity loss, provides a measure of the amplifier quality and hf constitutes photon energy. A minimum theoretical

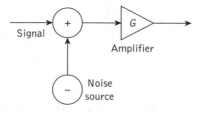

Figure 12.61 Noise model for traveling-wave optical amplifier.

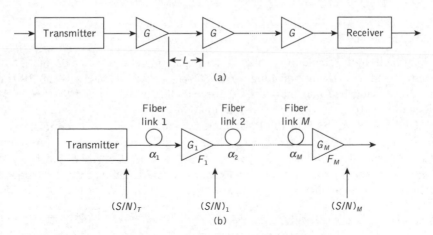

Figure 12.62 Cascaded optical amplifiers: (a) fiber system with cascaded optical amplifiers; (b) signal-to-noise ratios in a cascaded amplifier chain

value for K is unity, which would only occur for the ideal case of complete inversion and no cavity loss. However, in this case $S = hf$, which indicates that it is theoretically impossible for the traveling-wave SOA to be noiseless. Nevertheless, in practice $K < 2$ has been obtained, demonstrating that amplifiers with less than 3 dB more noise than the theoretical limit can be achieved [Ref. 176].

The cascading of optical amplifiers in a long-haul communication system is illustrated in Figure 12.62(a). Following each section of fiber cable length L there is an optical amplifier with gain G which just compensates for the fiber cable loss such that:

$$G = 10^{-(\alpha_{fc} + \alpha_j)L/10} \qquad (12.106)$$

where α_{fc} and α_j are the fiber cable losses and joint losses respectively, both in dB km^{-1}. Furthermore, as the optical signal travels through the amplifier cascade the noise levels increase because the additive noise from each device is cumulative. Hence, using Eqs (12.105) and (12.106) the SNR at the end of a cascaded link may be written as:

$$\frac{S}{N} \simeq \frac{P_i 10^{-(\alpha_{fc} + \alpha_j)L/10}}{(L_{to}/L)KhfB} \qquad (12.107)$$

where P_i is the power launched into the link at the transmitter, L_{to} the total system length and hence (L_{to}/L) is approximately equal to the total number of amplifier repeaters,* and the noise bandwidth equals the system bandwidth B. Equation (12.107) therefore enables the maximum transmission distance for a system using cascaded traveling-wave SOAs to be deduced.

---

* The total number of amplifier repeaters is actually $(L_{to}/L - 1)$; however, for a long link and a large number of repeaters this approximates to (L_{to}/L).

Example 12.19

A long-haul digital single-mode fiber system operating at a wavelength of 1.55 μm is envisaged employing traveling-wave SOAs spaced at intervals of 100 km. The power launched into the link at the transmitter is 0 dBm and the fiber cable attenuation is 0.22 dB km^{-1}. In addition, there are splice losses which average out at 0.03 dB km^{-1}. An SNR of 17 dB is required at the system receive terminal to provide an acceptable BER at the operating transmission rate of 1.2 Gbit s^{-1}. Assuming that the system bandwidth is equal to the transmission bit rate and that K for the amplifiers is equal to 4, estimate the maximum system length such that satisfactory performance is maintained.

Solution: Using Eq. (12.107), then:

$$(L_{to}/L) \simeq \frac{P_i\lambda 10^{-(\alpha_{fc} + \alpha_j)L/10}}{KhcB}\left(\frac{S}{N}\right)^{-1}$$

Hence for a link with a large number of cascaded amplifiers:

$$L_{to} \simeq \left(\frac{P_i\lambda 10^{-(\alpha_{fc} + \alpha_j)L/10}}{KhcB}\right)\left(\frac{S}{N}\right)^{-1} L$$

$$= \frac{(10^{-3} \times 1.55 \times 10^{-6} \times 10^{-2.5})\, 100 \times 10^3}{4 \times 6.626 \times 10^{-34} \times 2.998 \times 10^8 \times 1.2 \times 10^9 \times 50}$$

$$\simeq 1 \times 10^4 \text{ km}$$

Thus the maximum system length obtained in Example 12.19 is very large and would allow the interconnection of most points on the earth using a chain of optical amplifiers. However, the calculation does not take account of the nonregenerative nature of the amplifier repeaters in which pulse spreading as well as noise down the link is accumulated. Fiber dispersion therefore imposes serious limitations on the system performance, as discussed in Section 12.6.5, and it will restrict both the maximum system span as well as the maximum transmission rate.

Another parameter which is often specified in relation to the noise performance of optical amplifiers is the noise figure of the devices. The noise figure F for an optical amplifier is defined in a similar manner to an electrical amplifier as the signal-to-noise degradation between the device input and the device output:

$$F = \frac{(S/N)_{in}}{(S/N)_{out}} \tag{12.108}$$

Again, it is governed by factors including the population inversion, the number of transverse modes in the amplifier cavity, the number of incident photons on the amplifier and the optical bandwidth of the amplified spontaneous emissions. Typical noise figures range from 7 to 11 dB, with SOAs generally towards the bottom end of the range and fiber amplifiers towards the top end [Ref. 177].

We now consider a system with M cascaded optical amplifiers, as illustrated in Figure 12.62(b). In this case the link attenuation (both fiber cable and joint losses) in front of the kth amplifier is denoted by α_k, while the amplifier has a signal gain of G_k and a noise figure F_k. The input and output SNRs for such a cascaded link can be defined at the transmitter output T and the Mth amplifier output, respectively, so that in a similar manner to electrical amplifiers the total noise figure for the system F_{to} is:

$$F_{to} = \frac{(S/N)_T}{(S/N)_M}$$

$$= \frac{F_1}{\alpha_1} + \frac{F_2}{\alpha_1 G_1 \alpha_2} + \frac{F_3}{\alpha_1 G_1 \alpha_2 G_2 \alpha_3} + \ldots + \frac{F_M}{\alpha_M \sum\limits_{k=1}^{M-1}(\alpha_k G_k)} \tag{12.109}$$

The above expression can therefore enable determination of the total noise figure for the amplifier cascade.

Example 12.20

An optical fiber system is configured with a series of M optical amplifiers in cascade. The fiber cable and joint losses on each span between amplifiers on the link are compensated by the following amplifier gain. Obtain an expression for the total noise figure for the system and determine its value when all the amplifiers are identical.

Solution: As the amplifier gain compensates for the losses, then $\alpha_k G_k = 1$. Hence using Eq. (12.109) the total noise figure is given by:

$$F_{to} = \frac{F_1 G_1}{\alpha_1 G_1} + \frac{F_2 G_2}{\alpha_1 G_1 \alpha_2 G_2} + \ldots + \frac{F_M G_M}{\alpha_M \sum\limits_{k=1}^{M-1}(\alpha_k G_k)}$$

$$= F_1 G_1 + F_2 G_2 + \ldots + F_M G_M$$

$$= \sum_{k=1}^{M} F_k G_k$$

When all the repeaters are identical then $F_1 G_1, F_2 G_2, \ldots, F_M G_M$ are equal to FG. Therefore, the total noise figure becomes:

$$F_{to} = MFG$$

At the output from the first amplifier repeater a degradation in SNR of FG occurs followed by a decrease of $1/M$.

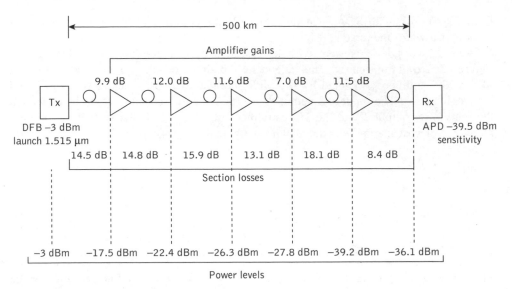

Figure 12.63 Experimental system incorporating five cascaded traveling-wave semiconductor optical amplifiers [Refs 178, 179]

An early experimental system configuration employing five traveling-wave SOAs and operating over a distance of some 500 km at a transmission rate of 565 Mbit s⁻¹ is shown in Figure 12.63 [Refs 178, 179]. The system was designed to give a BER better than 10^{-9}. In addition, as discussed above, the noise in SOAs can largely determine the number of devices that may be cascaded as linear repeaters. However, the spontaneous emission noise profile is relatively broadband, typically occupying around 30 nm, and therefore optical filtering can be used as a method of reducing overall noise levels. It was suggested that filters with bandwidths in the range 5 to 10 nm would be required for systems spanning greater than 500 km [Ref. 179]. Nevertheless, a higher speed IM/DD optical fiber system operating at 2.4 Gbit s⁻¹ over a distance of 516 km was subsequently demonstrated by cascading ten traveling-wave SOAs [Ref. 180].

Commercially available SOAs have been shown to exhibit optical gain in the range 25 to 30 dB with polarization sensitivity of 1 dB. More importantly, they can be used for amplification with wide flexibility in the choice of the gain peak wavelength from wavelengths between 1.30 and 1.65 μm when using InP-based devices (see Sections 10.3 and 11.6). However, when SOAs are considered for in-line amplification of WDM transmission systems they present several limitations due to their high noise figure of 7 to 11 dB [Ref. 177]. They also suffer from spontaneous emission noise and they cannot therefore operate at high transmission rates (i.e. 20 Gbit s⁻¹ or higher) over typical standard span lengths of 80 to 100 km [Refs 181–183]. The SOA therefore finds application for optical regeneration whereas erbium-doped fiber amplifiers are the preferred solution to provide amplification in long-haul optical fiber communication systems.

It should be noted that distance is not considered as a limitation when using optical regeneration in long-haul optical transmission systems. For example, an experimental system has successfully demonstrated optical transmission operating at a rate of 10 Gbit s⁻¹

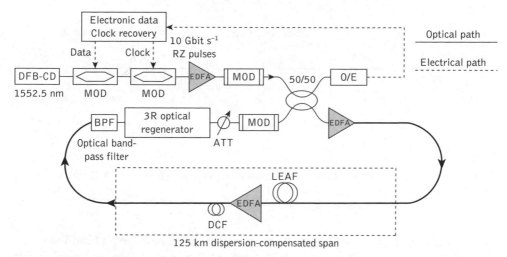

Figure 12.64 Experimental system setup to provide a 1.25×10^6 km transmission distance employing both fiber optical amplifiers and 3R regenerators. Reprinted with permission from Ref. 184 © IEEE 2006

over a distance of 1.25×10^6 km employing SOA-based optical regenerators and also a chain of EDFAs [Ref. 184]. Figure 12.64 shows the experimental system setup incorporating an optical fiber recirculating loop arrangement where EDFAs were included to provide optical amplification at 60 km intervals together with optical 3R regenerators employing SOAs at 125 km intervals enabling the provision of 10 000 stages. In addition, each 125 km long fiber span consisted of 65 km of large effective area single-mode fiber combined with 60 km of dispersion-compensating single-mode fiber (see Section 3.12.3) fiber in order to facilitate dispersion management.

Although the technology associated with SOAs is at present more established than that of fiber amplifiers, there appear to be significant drawbacks with these former devices in relation to their active nature, mechanical structure, reliability and yield performance which may inhibit their application in future systems. By contrast, in fiber amplifiers these parameters are largely defined by the atomic structure and hence they exhibit greater stability. Furthermore, although the signal gain–bandwidth of SOAs is generally more appropriate to WDM applications, they are also prone to crosstalk problems, which tends to limit their suitability in this area [Ref. 185].

When considering fiber amplifiers for use in WDM systems the Raman device (see Section 10.4.2) offers a gain bandwidth similar to the SOAs (i.e. 50 to 80 nm). However, in practice there are a number of difficulties associated with the use of these fiber amplifiers. In particular, very high pump source powers of the order of 300 mW into the fiber are required to provide gains of around 25 to 30 dB [Ref. 186].

The feasibility for the use of Raman fiber amplifiers for long-haul optical fiber communication has also been reported [Refs 187, 188]. Raman fiber amplifiers offer extremely low noise compared with conventional EDFAs and therefore the combined use of EDFA and Raman fiber amplifiers is considered important in the provision of long-haul optical fiber transmission. Figure 12.65 depicts the experimental setup for a 32-channel WDM transmission

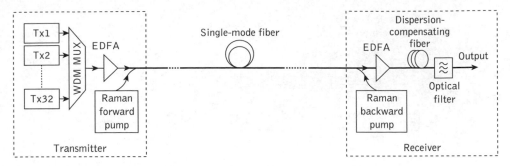

Figure 12.65 Experimental setup for a 23-channel WDM hybrid EDFA/Raman fiber amplified transmission system [Ref. 189]

system with 66 GHz channel spacing covering a wavelength range from 1547.2 to 1563.9 nm [Ref. 189]. Each channel was successfully transmitted over a 410 km long single-mode fiber link at a rate of 12.3 Gbit s^{-1}. In the transmitter section an optical signal was amplified using an EDFA as a booster amplifier and then a copropagating Raman forward pump was applied to create fiber Raman distributed amplification. The transmission link also included counter Raman pumping (i.e. backward pump) as identified in the receiver section of Figure 12.65. This device served as a preamplifier to the EDFA and improved the overall signal-to-noise ratio of the received signal. Furthermore, the receiving end the system also incorporated dispersion-compensating fiber to provide dispersion management over the link. Finally, the wavelength channels were then demultiplexed using an optical filter.

Commercially available hybrid EDFA/Raman fiber amplifiers are now available to provide optimized EDFA gain as an in-line amplifier with a Raman section to provide distributed amplification over a range of wavelengths within the C- and L-bands for operation at transmission rates from 10 to 40 Gbit s^{-1} [Refs 190–193]. Furthermore, a high transmission rate of 1.28 Tbit s^{-1} has also been demonstrated in a field trial employing distributed Raman fiber amplification in a commercially operated network [Ref. 188]. In this case the link constituted a repeaterless eight-channel WDM transmission system with each channel operating at 170 Gbit s^{-1} over a distance of 140 km.

12.11 Dispersion management

Dispersion management refers to the approaches to circumvent the transmission degradations caused by fiber dispersion (see Sections 3.8 to 3.12) using different types of single-mode optical fiber and other nonlinear passive optical devices. Hence multiple sections of constant dispersion single-mode fiber and dispersion-compensating elements whose lengths and group velocity dispersion are chosen to optimize the overall transmission performance of an optical fiber communication system are usually employed. It should be noted that single-mode fiber dispersion (see Section 3.11.2) tends to create limits for the generation, propagation and application of ultrashort pulses. In addition, optical amplifiers which are used in long-haul optical fiber systems also cause dispersion and thus restrict

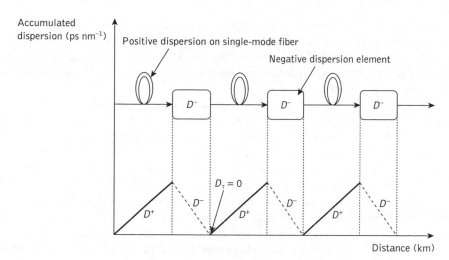

Figure 12.66 Example of a dispersion management scheme together with the appropriate dispersion management map on a single-mode fiber link

overall transmission distances. It is therefore necessary to control and manage the dispersion on a single-mode fiber link to constrain its effect on the optical fiber system.

A common method for managing dispersion is to combine two or more types of single-mode fiber to produce the desired dispersion over the entire link span. The total dispersion can be set at virtually any value as the contributions from different components may have opposite signs (i.e. either positive or negative) and hence they can partially, or completely, cancel each other. Dispersion-compensating fibers can be either placed at one location or distributed along the length of the fiber link [Ref. 194]. In addition, lumped dispersion-compensating devices, such as fiber gratings, can also be incorporated. Typically, dispersion management must consider single-mode fiber chromatic dispersion over a range of wavelengths. In particular, setting the total chromatic dispersion slightly above zero at all wavelengths helps to avoid detrimental four-wave-mixing noise [Ref. 195].

Figure 12.66 shows a dispersion management scheme for a single-mode fiber link. It incorporates a dispersion management map to compensate positive dispersion (i.e. identified as D^+) on the fiber with the negative dispersion* (i.e. identified as D^-) such that the chromatic or total first-order dispersion D_T (see Section 3.11.2) goes to zero.

Negative dispersion can be achieved by using dispersion-compensating fiber (DCF) (see Section 3.12.3) or lumped elements (e.g. fiber Bragg gratings). It should be noted, however, that single-mode fiber dispersion varies with wavelength and this property is referred to as the dispersion slope (see Section 3.11.2). Dispersion compensation involves altering the local dispersion between a positive and a negative group velocity dispersion (see Section 3.15) to produce a low overall (ideally zero) group velocity dispersion for the link. A zero-dispersion slope on a fiber link (i.e. comprising standard single-mode fiber and DCF at a specific wavelength) can be achieved when the DCF cancels the dispersion

---

* Negative dispersion is sometimes referred to as anomalous dispersion, while positive dispersion is generally regarded as normal or standard fiber dispersion.

effects of the SSMF. Dispersion management maps are two-dimensional plots of the accumulated dispersion against the fiber link distance. The same pattern can be mapped repeatedly extending the total transmission distance. Moreover, dispersion management is said to be periodic when the system repeatedly employs the same dispersion compensation over long transmission distances.

In practice, dispersion management is required in long-haul transmission systems where the length of the single-mode fiber may differ from one span to another depending on the particular optical amplifiers and other nonlinear components employed. In such cases it will not be possible to use periodic dispersion management and consideration of the difference in length of fiber spans is required. Therefore as a general criterion a dispersion management map is described for two different fiber lengths [Ref. 196]. Figure 12.67 depicts a typical dispersion management map period with two different single-mode fibers of different lengths L_1 and L_2 exhibiting negative D^- and positive dispersion D^+, respectively. The dispersion management map period (i.e. $L_{\text{map}} = L_1 + L_2$) should be selected in a manner so that the overall dispersion on both segments of the fiber link cancel one another and thus the mean value of the chromatic dispersion $\bar{D}_{\text{T}}$ becomes zero, which is given by:

$$\bar{D}_{\text{T}} = \frac{D^+ L_1 + D^- L_2}{L_{\text{map}}} \tag{12.110}$$

where the generic term DL is also known as the dispersion management map strength. It should be noted that the second-order fiber dispersion coefficient β_2 given in Eq. (3.69) (see Section 3.15) can be used to describe chromatic dispersion such that the dispersion management map strength can also be written as $-\beta_2 L$.

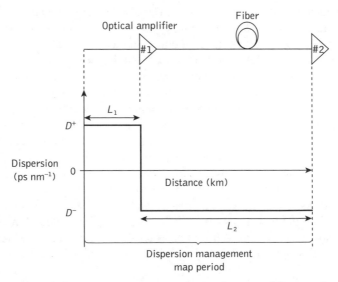

Figure 12.67 A typical dispersion management map for two different single-mode fiber types and lengths

For a periodic dispersion management map the same strategy can be repeated and the mean chromatic dispersion is calculated over the average path length of the single-mode fiber; hence $\bar{D}_\mathrm{T}$ is sometimes referred to as the path average dispersion [Refs 197–199] and Eq. (12.110) can be rewritten as:

$$\bar{D}_\mathrm{T} L_\mathrm{map} = -\beta_{21} L_1 - \beta_{22} L_2 \tag{12.111}$$

where the terms β_{21} and β_{22} represent the second-order dispersion coefficients for the two different fiber path lengths L_1 and L_2, respectively.

For perfect dispersion compensation the dispersion map strengths for the fiber path lengths L_1 and L_2 should cancel the negative and positive dispersion effects on each path and therefore $\bar{D}_\mathrm{T} L_\mathrm{map}$ must be equal to zero; hence Eq. (12.111) becomes:

$$\beta_{21} L_1 + \beta_{22} L_2 = 0 \tag{12.112}$$

Dispersion management for a multiwavelength channel is illustrated in Figure 12.68 where three wavelength channels are indicated as the short, the central or the long wavelengths which are designated as channels 1, 2 and 3, respectively. A typical DCF can be used to compensate for the positive dispersion in each channel on each fiber span. In the case of a WDM system with a large number of wavelength signals, the dispersion slope becomes longer and a periodic dispersion map is usually deemed to be sufficient if the dispersion compensation is mapped for a central wavelength channel only. Furthermore, at transmission rates between 2.5 and 10 Gbit s^{-1} periodic dispersion management can be used repeatedly in order to provide for longer transmission distances.

In a multiwavelength channel transmission system, if the dispersion is assumed to vary linearly with wavelength, then some dispersion always remains on the fiber link even after dispersion compensation. This remaining dispersion is known as residual dispersion. For multiwavelength channel dispersion compensation the residual dispersion depends on the

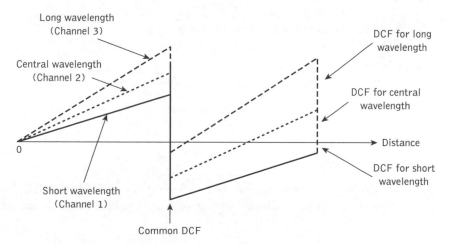

Figure 12.68 Dispersion management diagram employing dispersion-compensating fiber (DCF) for three different wavelength channels

variation in channel wavelength which can be calculated using the dispersion slopes given in Eq. (3.52) (see Section 3.11.2). The ratio of dispersion slope and the chromatic dispersion (i.e. S/D_T) is known as the relative dispersion slope (RDS) which determines the amount of dispersion compensation required to produce a zero dispersion slope for a dispersion management map period* [Ref. 194].

In multiwavelength transmission if the value of the RDS remains the same for the two single-mode fiber spans then the same dispersion management map can be repeated for the entire length of the optical link. Furthermore, when the RDS for the central wavelength remains the same as the RDS for the first wavelength, the same periodic dispersion management map can also serve for the other higher order wavelength channels present in the multiwavelength signal. However, the RDS must satisfy the dispersion slope condition for the central wavelength, λ_c, such that [Refs 197, 198]:

$$S_i(\lambda_c - \lambda_n) = D_{Ti}(\lambda_n - \lambda_c) \tag{12.113}$$

where the terms S_i and D_{Ti}, with $(i = 1, 2, \ldots, m)$, represent the dispersion slope and the amount of chromatic dispersion required to provide compensation on the transmission fiber at a particular wavelength, respectively. Equation (12.113) assumes that an optical fiber link is divided into a number of segments m where for the provision of periodic dispersion management for the entire fiber link each segment is further divided into two fiber lengths L_1 and L_2. The right hand side of Eq. (12.113) is related to dispersion slope for the central wavelength channel present in a multiwavelength channel transmission whereas the left-hand side represents the overall chromatic dispersion on the entire fiber link. Considering the three-wavelength transmission system shown in Figure 12.68, if the dispersion slopes for the first and central wavelength signals are given by S_1 and S_2 then by using Eqs (12.112) and (12.113) the dispersion management map period for the central wavelength must satisfy the condition [Ref. 197]:

$$S_1 L_1 + S_2 L_2 = 0 \tag{12.114}$$

Therefore the dispersion slope for the central wavelength is:

$$S_2 = -S_1 \left(\frac{L_1}{L_2} \right) \tag{12.115}$$

Using Eqs (12.112) and (12.114), S_2 can be rewritten as:

$$S_2 = -S_1 \left(\frac{L_1}{L_2} \right) = S_1 \left(\frac{\beta_{22}}{\beta_{21}} \right) \tag{12.116}$$

Equation (12.116) plays an important role in determining the value of the RDS for the central wavelength of a multiwavelength transmission system in order to determine the

---

* Typical SSMFs have dispersion of 17 ps nm^{-1} km^{-1} and a dispersion slope of 0.055 ps nm^{-2} km^{-1} at a wavelength of 1.55 μm which produces a relative dispersion slope of 0.0032 nm^{-1}.

periodic dispersion management map for multiwavelength channel operation. For example, when the dispersion slope for the central wavelength given in Eq. (12.116) is satisfied for both conditions (i.e. the fiber length and the second-order dispersion coefficient), the same dispersion management map period can be used for the other channels present in the multiwavelength signal.

Example 12.21

A dispersion management map strategy to provide dispersion compensation for a DWDM single-mode fiber system operating in the wavelength region around 1.55 μm is displayed in Figure 12.69. The two path lengths L_1 and L_2 are 160 km and 20 km, respectively. Furthermore, the second-order dispersion coefficient for the latter path L_2 is 17 ps nm^{-1} km^{-1}.

(a) Calculate the second-order dispersion coefficient for the first path length L_1 in order to achieve zero mean chromatic dispersion.

(b) If the dispersion slope for first fiber path L_1 is 0.075 ps nm^{-2} km^{-1} then determine the dispersion slope for the second fiber path L_2.

(c) Verify that the periodic dispersion management map will provide sufficient confidence to facilitate reliable DWDM transmission.

Solution. (a) The second-order dispersion coefficient in order for the first path β_{21} to achieve zero mean chromatic dispersion can be calculated using Eq. (12.112) as:

$$\beta_{21} = \frac{-\beta_{22}L_2}{L_1}$$

$$\beta_{21} = \frac{-17 \times 20}{160}$$

$$= -2.125 \text{ ps nm}^{-1} \text{ km}^{-1}$$

Therefore the second-order dispersion coefficient for first path is -2.125 ps nm^{-1} km^{-1}.

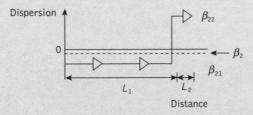

Figure 12.69 Dispersion management map for the DWDM optical fiber system in Example 12.21

(b) The dispersion slope for the second fiber path is given by Eq. (12.115) as:

$$S_2 = -S_1\left(\frac{L_1}{L_2}\right)$$
$$= -\frac{0.075 \times 160}{20}$$
$$= -0.6 \text{ ps nm}^{-2} \text{ km}^{-1}$$

Hence the chromatic dispersion slope for the second fiber path is -0.6 ps nm^{-2} km^{-1}.

(c) For multiwavelength channel operation the periodic dispersion management map in Figure 12.69 is considered to provide confidence for the other wavelengths present in the DWDM signal if the relative dispersion slope (RDS) remains the same for the first and the central wavelengths and therefore that Eq. (12.116) is satisfied following:

$$-S_1\left(\frac{L_1}{L_2}\right) = S_1\left(\frac{\beta_{22}}{\beta_{21}}\right)$$

Hence:

$$S_1\left(\frac{L_1}{L_2}\right) = S_1\left(\frac{\beta_{22}}{\beta_{21}}\right) = 0$$
$$= 0.075 \times \left(\frac{160}{20}\right) + 0.075 \times \left(\frac{17 \times 0.075}{-2.125}\right)$$
$$= 0.6 - 0.6$$
$$= 0$$

This outcome indicates that the RDS remains the same for the dispersion management map for the first and the central wavelengths present in the multiwavelength channel. Therefore the same dispersion management map can be repeated with confidence periodically over the entire optical fiber link to provide reliable dispersion management for the other wavelengths present in the DWDM signal.

12.12 Soliton systems

Solitons are nonlinear optical pulses (see Section 3.15) which have the potential to support very high optical transmission rates of many terabits per second over long distances. Figure 12.70 shows a block schematic for an optical fiber soliton transmission system. The major element in the transmitter section is a return-to-zero pulse generator. A simple

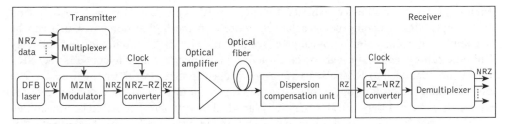

Figure 12.70 Block schematic of an optical fiber soliton transmission system

approach to generate RZ pulses is to employ an optical modulator and an NRZ-to-RZ converter which is driven by a DFB laser source. In this case a Mach–Zehnder modulator is used to modulate the NRZ data at the desired transmission rate (i.e. in the range 2.5 to 40 Gbit s^{-1}). Instead of using a single NRZ data stream, however, it is useful to modulate an optical NRZ signal incorporating several multiplexed NRZ data streams before the conversion into RZ pulses takes place. At the receiving end the incoming signal requires conversion back from RZ to NRZ and then finally a demultiplexer separates the specific NRZ data for each channel.

The generation of optical soliton pulses is crucial to achieve soliton transmission where the transmitter is required to produce ultrafast RZ pulses. Ideally, an RZ pulse source can be realized using an RZ laser source such as a mode-locked fiber ring laser [Refs 54, 199]. In practice such laser sources are difficult to realize since they are required to maintain a fixed frequency during the generation of ultrafast RZ pulses. An alternative practical RZ pulse generation scheme utilizing a similar concept to the above is illustrated in Figure 12.71 [Ref. 200]. A CW laser source generates an optical signal which is passed through the phase modulator driven by the encoded NRZ data signal as depicted in

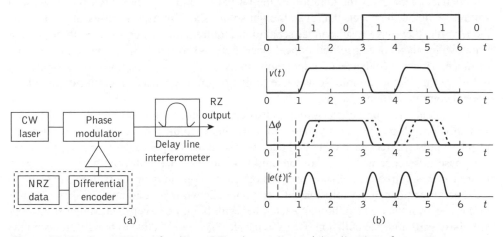

Figure 12.71 Generation of soliton RZ pulses using a delay line interferometer: (a) block schematic showing the NRZ-to-RZ conversion; (b) timing diagram for the four stages to produce the RZ pulses

Figure 12.71(a). The device exhibits a change in its logic level for each logical 1 bit to be sent, NRZ encoding being achieved entirely in the digital domain which can be accomplished using a digital optical gate (see Section 11.6.2). The following optical delay line interferometer is adjusted for constructive interference where it converts the NRZ phase modulation into RZ pulses with a width corresponding to the optical delay. The corresponding main four operational stages involved in this scheme are displayed in the timing diagram of Figure 12.71(b). Initially, an ideal binary on/off signal to be transmitted in RZ format is shown as the topmost waveform in Figure 12.71(b). Underneath this waveform are the encoded version of NRZ signal and the phases of the two interfering signals at the output of the interferometer indicated with solid and dashed lines, respectively. Finally, the bottom waveform depicts the phase difference $\Delta\phi$ between the two interferometer arms giving rise to an RZ pulse with a complex envelope of amplitude $|e(t)|^2$.

In theory, when a single soliton pulse is propagating over a long transmission distance, the pulse amplitude and timing are maintained since pulse propagation takes place within a bit pattern without any changes in its parameters. Furthermore, a single pulse travels without emitting any energy due to the exact balance between Kerr nonlinearity and the second-order nonlinear dispersion coefficient of the fiber (see Section 3.15). In practice, however, certain physical processes (i.e. amplifier noise and chromatic dispersion) can remove this balance resulting in deterioration of the bit pattern when several soliton pulses are propagating closely following each other. In such cases it is important to maintain a safe distance between two consecutive soliton pulses to avoid any destructive interaction which may cause one pulse to shed its energy to another pulse producing destructive interference and the phenomenon known as soliton interaction (see Section 3.15).

The distance between two interacting soliton pulses is commonly referred to as the strength of interaction [Ref. 201]. Figure 12.72 depicts the propagation of soliton pulses where they are shown both maintaining a safe distance and interacting with each other. The case where each soliton pulse occupies a small fraction of the bit period in order to maintain a safe distance between neighboring pulses is shown in Figure 12.72(a). The parameter T_0 represents the duration of the bit period and τ is the soliton pulse width. It should be noted, however, that even though solitons are ultrashort pulses of picosecond pulsewidth, they have a substantial separation requirement (which may be three or four times wider than for nonsoliton optical fiber systems) to maintain a noninteracting safe distance. Hence avoidance of interaction necessitates a large-bandwidth capacity and therefore this factor can impose a severe limitation on the transmission bit rate of soliton optical communication systems.

In an ideal situation optical soliton pulses collide with each other and their interaction strength depends on the relative phase difference between the interacting solitons such that in-phase pulses attract each other and out-of-phase pulses repel each other, whereas when the phase difference is $\pi/2$ the soliton pulses do not interact with each other [Ref. 201]. Moreover, a small-angle collision between optical solitons in their interaction can result in a deflection similar to the collisions of two pool balls while a large-angle collision results in the two solitons passing through each other with both, however, incurring a small spatial phase shift after the collision [Ref. 202]. This phase shift is due to the difference in the velocities of interacting soliton pulses. Figure 12.72(b) illustrates a collision leading to an interaction between two 180° out-of-phase optical soliton pulses where the two pulse

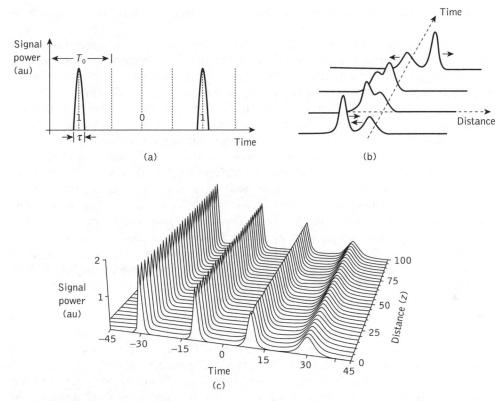

Figure 12.72 Propagation of return-to-zero optical soliton pulses: (a) soliton bit stream; (b) collision of two solitons; (c) four stable solitons at safe separation distances

envelopes overlap and separate from each other (i.e. passing through) with the solitons retaining their original shapes and velocities as prior to the collision.

The pulse width, however, of a single soliton remains unaffected by fiber dispersion even if it travels longer distances. The same is true for a train of soliton pulses provided that each pulse is well separated from the adjacent pulses as shown in Figure 12.72(c) where four in-phase pulses are propagating and each pulse is separated by a distance of around four times its pulse width. In this situation it can be observed that the solitons maintain their different relative amplitudes (i.e. 2.0, 1.5, 1.0 and 0.5) without any destructive interaction or collision [Ref. 203].

Optical fiber soliton transmission systems can be single-wavelength or multiwavelength channel. In a single-wavelength channel system only one transmitter is used to launch the RZ pulses onto the optical fiber. Multiwavelength-channel optical soliton systems, however, employ several transmitters simultaneously where the data is multiplexed using WDM (see Section 12.9.4). Although a tunable RZ laser source could provide a multiwavelength-channel optical soliton system, it is difficult to realize such a source to operate at the ultra-high frequencies needed to maintain the necessary narrow pulse widths.

It should be noted that long-haul optical soliton systems using a signal wavelength of 1.55 μm operating at transmission rates of 2.5 to 10 Gbit s^{-1} require repetition rates of 2.5 to 10.0 GHz necessitating pulse widths of 20 to 80 ps. Moreover, femtosecond pulse sources are required for the operation at transmission rates around 40 Gbit s^{-1} where the smallest frequency variation will incur a change in the shape of the pulse and the pulse width [Ref. 204]. In practice, however, it is difficult to realize such ideal optical sources since semiconductor optical laser devices are temperature dependent and their wavelength drifts slightly from its center wavelength with a variation in temperature [Ref. 199]. Furthermore, the ability to generate accurate ultrashort pulses declines with increasing transmission rate. For these reasons, at high transmission rates over long distances optical soliton pulses suffer from fiber attenuation and dispersive effects creating a low SNR which therefore necessitates both optical amplification and dispersion compensation. For example, in an experimental demonstration of an optical fiber soliton system operating at a transmission rate of 40 Gbit s^{-1} a series of optical amplifiers were used to support a transmission distance of more than 10 000 km [Ref. 205]. Furthermore, dispersion management (see Section 12.11) was also required to facilitate the operation of this system.

Optical fiber soliton systems which employ schemes to manage dispersion on the fiber between equal (or unequal) amplifier spacing (i.e. distance between two optical amplifiers) are known as dispersion-managed soliton (DMS) systems [Ref. 206]. As indicated previously, at higher transmission rates (i.e. greater than 20 Gbit s^{-1}) the soliton pulse-width requirement becomes very small (i.e. 5 to 10 ps) and increasing fiber transmission distance introduces attenuation and dispersive effects which will also be magnified by the amplifier. If the distance between two amplifiers is large, then it is necessary to take account of amplifier spacing in the dispersion management scheme. Such an approach is referred to as dense-dispersion management, which is depicted by a small dispersion management map period, and has been widely used in dispersion management schemes for experimental long-haul DWDM soliton transmission systems [Ref. 205].

The transmission bit rate of a soliton communication system is dependent on mainly two factors: namely, the soliton pulse width τ and the duration of the bit period T_0. Hence considering Figure 12.72(a) the maximum allowable transmission bit rate B_T for soliton pulses* can be written as [Ref. 197]:

$$B_T = \frac{1}{T_0} = \frac{1}{2q_0\tau} \tag{12.117}$$

where the term:

---

* It should be noted that the transmission bit rate provided in Eq. (12.117) is applicable to only soliton pulse propagation where the pulse width τ and thus the bit period T_0 remain the same even after traveling long distances; hence, $B_T = 1/T_0$, whereas the transmission bit rate for the ordinary non-soliton pulse propagation is given by Eq. (3.10) which allows for chromatic dispersion causing pulse-width broadening in single-mode fiber (see Section 3.8).

$$q_0 = \frac{1}{2}\left(\frac{T_0}{\tau}\right)$$

is a dimensionless parameter (usually stated in normalized units) that determines the pulse separation and provides the value of the separation required between adjacent soliton pulses to avoid interaction. Therefore the ratio T_0/τ determines the nature of the nonlinear propagation for soliton pulses. For example, higher interaction strength between soliton pulses is obtained when its value falls within the region of $0 < T_0/\tau < 1$ and the interaction reaches its maximum value for $T_0/\tau = 1$. The interaction between two neighboring solitons, however, decreases when $T_0/\tau \gg 1$. Hence, as a rule of thumb a value for T_0/τ in the range 6 to 8 is considered satisfactory to ensure a safe distance between neighboring soliton pulses [Ref. 197].

The pulse width of a soliton is characterized by the second-order dispersion coefficient β_2 in relation to the dispersion length L_D which is the length of single-mode optical fiber over which the pulse width of the soliton is broadened significantly due to dispersion effects* [Refs 198, 207, 208] following:

$$L_D = \frac{\tau^2}{|\beta_2|} \tag{12.118}$$

Therefore:

$$\tau = \sqrt{|\beta_2|L_D} \tag{12.119}$$

Assuming a soliton propagation condition where optical amplifiers are placed at a separation distance of 40 to 50 km such that the optical amplifier spacing L_A is smaller than the overall dispersion length (i.e. $L_A \ll L_D$), then Eq. (12.119) can be expressed as:

$$\tau \gg \sqrt{|\beta_2|L_A} \tag{12.120}$$

Using Eqs (12.117) and (12.120), the transmission bit rate for soliton propagation when the amplifier spacing is smaller than the dispersion length (i.e. $L_A \ll L_D$) is:

$$B_T \ll \frac{1}{2q_0}\frac{1}{\sqrt{|\beta_2|L_A}} \tag{12.121}$$

This expression indicates that when the amplifier spacing is smaller than the dispersion length it imposes severe limitations on both the transmission bit rate and the amplifier spacing in relation to large dispersion length (i.e. $L_D \gg L_A$). Therefore, in the situation where the typical amplifier spacing is between 40 and 50 km, the maximum achievable transmission bit rates for the optical fiber soliton system can be 20 Gbit s^{-1}.

---

* The dispersion length is a characteristic fiber length at which a pulse broadens by a factor of the square root of 2.

Example 12.22

An optical fiber soliton transmission system has an amplifier spacing of 50 km that is significantly smaller than the fiber dispersion length. In addition, the RZ pulse width is 6 ps with a bit period of 70 ps. If the second-order dispersion coefficient is $-0.5 \text{ ps}^2 \text{ km}^{-1}$ determine: (a) the separation for the soliton pulses to avoid interaction; (b) the transmission bit rate of the optical soliton communication system.

Solution: (a) The separation of the soliton pulses q_0 over a bit period length can be calculated from Eq. (12.117) as:

$$q_0 = \frac{1}{2}\left(\frac{T_0}{\tau}\right)$$

$$= \frac{1}{2}\left(\frac{70 \times 10^{-12}}{6 \times 10^{-12}}\right)$$

$$= 5.8$$

Therefore a separation of 5.8 is required to avoid interaction of the soliton pulses.

(b) The transmission bit rate of the optical soliton communication system where the distance between optical amplifiers is smaller than the dispersion length ($L_A \ll L_D$) is given by expression (12.121) as:

$$B_T \ll \frac{1}{2q_0} \frac{1}{\sqrt{|\beta_2| L_A}}$$

Therefore:

$$B_T \ll \frac{1}{2 \times 5.8} \frac{1}{\sqrt{(-50 \times 10^{-12} \times 10^{-12} \times 10^{-3})(50 \times 10^3)}}$$

$$\ll 17.24 \times 10^9$$

Hence the maximum bit rate will be much less than 17.2 Gbit s^{-1} for soliton transmission with the optical amplifiers placed at an interval of 50 km. In practice a suitable transmission rate would be around 10 Gbit s^{-1}.

Although optical solitons can be considered stable pulses, their signal power and pulse width do not remain the same when traveling at high transmission bit rates [Ref. 197]. In such cases, when the optical amplifier spacing is greater than the dispersion length (i.e. $L_A \gg L_D$) the main factor affecting the pulse width and signal power is the attenuation in the fiber. Therefore, combining fiber attenuation coefficient α (see Section 10.2) with the dispersion length (i.e. to give αL_D) describes the propagation of such soliton pulses more accurately. In this case each soliton pulse preserves its soliton nature after traveling a long distance maintaining its peak power and pulse width. There are three possible conditions for the propagation of optical pulse: $\alpha L_D = 1$, $\alpha L_D \gg 1$ and $\alpha L_D \ll 1$. When the values of αL_D are equal to or greater than unity, it indicates large fiber attenuation and the optical pulses cannot maintain their soliton nature. However, when αL_D is less than unity they do preserve their soliton nature [Ref. 198]. Hence for the condition $\alpha L_D \ll 1$, Eq. (12.118) can be rewritten as:

$$\alpha L_D \ll \frac{\alpha \tau^2}{|\beta_2|} \ll 1 \tag{12.122}$$

Therefore:

$$\tau \ll \frac{\sqrt{|\beta_2|}}{\sqrt{\alpha}} \tag{12.123}$$

Combining Eqs (12.11) and (12.123), the maximum achievable transmission bit rate is given by:

$$B_T \gg \frac{1}{2q_0} \sqrt{\frac{\alpha}{|\beta_2|}} \tag{12.124}$$

This expression implies that high transmission rates (e.g. greater than 40 Gbit s^{-1}) for ultrashort optical solitons can be supported subject to the condition $L_A \gg L_D$ which thus requires the deployment of optical amplifiers at separation distances typically greater than 100 km.

Example 12.23

An ultrashort pulse optical fiber soliton system for long-haul transmission has a dispersion length much smaller than the optical amplifier spacing. The second-order dispersion coefficient is −0.125 ps^2 km^{-1} and the attenuation coefficient is 0.2 dB km^{-1}. When the soliton pulse width is 4 ps with a bit period of 40 ps, determine the maximum transmission bit rate for the system.

Solution: The bit rate of an ultrashort optical soliton communication system where the dispersion length is much smaller than the amplifier spacing (i.e. $L_D \ll L_A$) is given by expression (12.124) as:

$$B_T \gg \frac{1}{2q_0} \sqrt{\frac{\alpha}{|\beta_2|}}$$

where the value of the dimensionless parameter q_0 can be obtained from Eq. (12.117) following:

▶

$$q_0 = \frac{1}{2}\left(\frac{T_0}{\tau}\right)$$

$$= \frac{1}{2}\left(\frac{40 \times 10^{-12}}{4 \times 10^{-12}}\right)$$

$$= 5.0$$

Therefore:

$$B_T \gg \frac{1}{2 \times 5.0}\sqrt{\frac{0.2 \times 10^{-3}}{-1.25(10^{-12} \times 10^{-12} \times 10^{-3})}}$$

$$\gg 4 \times 10^{10} \text{ bit s}^{-1}$$

Therefore the maximum bit rate of the ultrashort pulse optical soliton system can be significantly greater than 40 Gbit s^{-1}.

Optical amplifiers in soliton systems also contribute towards the amplitude fluctuation in the form of amplified spontaneous emission (ASE) noise (see Section 10.3.2). Furthermore, ASE noise variations accompanied by dispersion on the fiber gives rise to jitter in the pulse arrival times known as Gordon–Haus jitter (see Section 3.15) [Ref. 207]. ASE noise and Gordon–Haus jitter are two significant phenomena that also impose limitations on the maximum permissible transmission distance for an optical soliton transmission system. The limitations of Gordon–Haus jitter in relation to ASE noise with increasing optical signal power contained in transmitted soliton pulses are indicated in Figure 12.73.

As a result of the additive nature of the ASE noise from the amplifiers, a large SNR is required to differentiate binary zeros from ones necessitating an increasing optical signal power for longer transmission distances. However, at such transmission distances the Gordon–Haus effect dominates, which may cause bit errors due to random displacements of pulses ending up in neighboring time slots. A balance between the effect of ASE noise and Gordon–Haus jitter is therefore required to enable successful optical soliton system operation and this aspect is identified as the shaded area in Figure 12.73 [Ref. 208]. This shaded area depicts the acceptable range of values for the ASE noise and the Gordon–Haus jitter with increasing optical signal power which will permit low bit-error-rate transmission. In WDM soliton systems the above noise sources result in interchannel nonlinear crosstalk caused by both the four-wave-mixing (FWM) and cross-phase modulation (XPM) processes (see Section 3.14). Although adverse effects of FWM which generally leads to the amplitude fluctuation of the optical soliton pulses can be reduced by dispersion management, XPM sets the limit on the overall transmission distance and also the minimum channel spacing for WDM soliton systems. The use of in-line modulation or

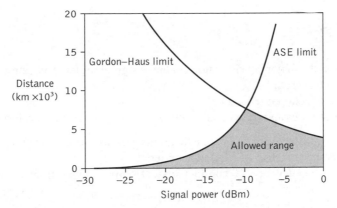

Figure 12.73 Optical soliton system transmission distance against input signal power at bit-error-rate of 10^{-9} [Ref. 208]

synchronous modulation schemes [Ref. 209] can, however, reduce these unwanted noise sources and very long transmission distances (i.e. greater than 10 000 km) have been successfully demonstrated employing such optical modulation techniques [Ref. 184].

As the optical soliton interaction depends on both the dispersion and the polarization properties of the fiber, however, these factors impose serious limitations with increasing transmission rates and distances. Almost all the transmission demonstrations at higher rates have been required to overcome polarization mode dispersion effects (see Section 3.13.2) [Refs 210, 211]. Furthermore, as the number of channels increases in WDM soliton systems and hence the channel spacing reduces (see Section 12.9.4), the inter-channel interaction becomes stronger and consequently degrades the system performance. Nevertheless, frequency guiding and sliding filters have been used to overcome these drawbacks [Refs 201, 212]. In these filter types the central frequency is gradually shifted along the length of the transmission fiber. For example, in a series of optical filters the transmission peak of each guiding filter is gradually shifted in frequency with respect to the peak frequency of the previous filter such that the center frequency slides gradually relative to the transmission distance [Ref. 212]. Furthermore, low bit-error-rate transmission at 10 Gbit s^{-1} over 40 000 km and 20 Gbit s^{-1} over 14 000 km has been successfully demonstrated using this frequency guiding and sliding filter approach [Ref. 201].

For long-haul DWDM soliton systems operating at higher transmission rates (i.e. above 40 Gbit s^{-1}) more complex dispersion management schemes are usually required which involve a rigorous computation to determine the required amount of dispersion compensation for unequal amplifying spans. Moreover, there have been several successful field trials of such transmission systems [Refs 196, 214–216]. Multiwavelength-channel optical soliton systems have also been demonstrated to provide high overall transmission rates [Refs 216–218]. For example, a 16-channel DWDM soliton system with each channel operating at a bit rate of 40 Gbit s^{-1} and a separation of 100 GHz successfully transmitted a total capacity of 640 Gbit s^{-1} over a distance of 1000 km [Ref. 217].

Problems

12.1   Discuss the major considerations in the design of digital drive circuits for:
(a) an LED source;
(b) an injection laser source.
Illustrate your answer with an example of a drive circuit for each source.

12.2   Outline, with the aid of suitable diagrams, possible techniques for:
(a) the linearization of LED transmitters;
(b) the maintenance of constant optical output power from an injection laser transmitter.

12.3   Discuss, with the aid of a block diagram, the function of the major elements of an optical fiber receiver. In addition, describe possible techniques for automatic gain control in APD receivers.

12.4   Equalization within an optical receiver may be provided using the simple frequency 'rollup' circuit shown in Figure 12.74(a). The normalized frequency response for this circuit is illustrated in Figure 12.74(b).

The amplifier indicated in Figure 12.74(a) presents a load of 5 kΩ to the photodetector and together with the photodetector gives a total capacitance of 5 pF. However, the desired response from the amplifier–equalizer configuration has an upper 3 dB point or corner frequency at 30 MHz. Assuming R_2 is fixed at 100 Ω determine the required values for C_1 and R_1 in order to obtain such a response.

12.5   Describe the conversion of an analog signal into a pulse code modulated waveform for transmission on a digital optical fiber link. Furthermore, indicate how several signals may be multiplexed onto a single fiber link.

A speech signal is sampled at 8 kHz and encoded using a 256-level binary code. What is the minimum transmission rate for this single pulse code modulated speech signal? Comment on the result.

12.6   A 1.5 MHz information signal with a dynamic range of 64 mV is sampled, quantized and encoded using a direct binary code. The quantization is linear with 512 levels. Determine:
(a) the maximum possible bit duration;
(b) the amplitude of one quantization level.

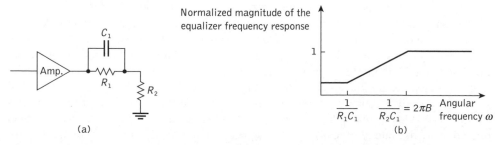

Figure 12.74 The equalizer of Problem 12.4: (a) the frequency 'rollup' circuit; (b) the spectral transfer characteristic for the circuit

12.7   Describe, with the aid of a suitable block diagram, the operation of an optical fiber optoelectronic regenerative repeater. Indicate reasons for the occurrence of bit errors in the regeneration process and outline a technique for establishing the quality of the channel.

12.8   Twenty-four 4 kHz speech channels are sampled, quantized, encoded and then time division multiplexed for transmission as binary PCM on a digital optical fiber link. The quantizer is linear with 0.5 mV steps over a dynamic range of 2.048 V.

Calculate:
(a) the frame length of the PCM transmission, assuming an additional channel time slot is used for signaling and synchronization;
(b) the required channel bandwidth assuming NRZ pulses.

12.9   Develop a relationship between the error probability and the received SNR (peak signal power to rms noise power ratio) for a baseband binary optical fiber system. It may be assumed that the numbers of ones and zeros are equiprobable and that the decision threshold is set midway between the one and zero level.

The electrical SNR (defined as above) at the digital optical receiver is 20.4 dB. Determine:
(a) the optical SNR;
(b) the BER.
It may be assumed that $\mathrm{erfc}(3.71) \simeq 1.7 \times 10^{-7}$.

12.10   The error function (erf) is defined in the text by Eq. (12.16). However, an error function also used in communications is defined as:

$$\mathrm{Erf}(u) = \frac{1}{\sqrt{(2\pi)}} \int_{-\infty}^{u} \exp(-x^2/2)\, \mathrm{d}x$$

where a capital E is used to denote this form of the error function. The corresponding complementary error function is:

$$\mathrm{Erfc}(u) = 1 - \mathrm{Erf}(u) = \frac{1}{\sqrt{(2\pi)}} \int_{u}^{\infty} \exp(-x^2/2)\, \mathrm{d}x$$

This complementary error function is also designated as $Q(u)$ in certain texts. Use of $\mathrm{Erfc}(u)$ or $Q(u)$ is sometimes considered more convenient within communication systems.

Develop a relationship for $\mathrm{erfc}(u)$ in terms of $\mathrm{Erfc}(u)$. Hence, obtain an expression for the error probability $P(\mathrm{e})$ as a function of the Erfc for a binary digital optical fiber system where the decision threshold is set midway between the one and zero levels and the numbers of transmitted ones and zeros are equiprobable. In addition, given that $\mathrm{Erfc}(4.75) \simeq 1 \times 10^{-6}$, estimate the required peak signal power to rms noise power ratios (both optical and electrical) at the receiver of such a system in order to maintain a BER of 10^{-6}.

12.11   Show that Eq. (9.27) reduces to Eq. (12.28). Hence determine $F(M)$ when $k = 0.3$ and $M = 20$.

12.12 A silicon APD detector is utilized in a baseband binary PCM receiver where the decision threshold is set midway between the one and zero signal level. The device has a quantum efficiency of 70% and a ratio of carrier ionization rates of 0.05. In operation the APD has a multiplication factor of 65. Assuming a raised cosine signal spectrum and a zero disparity code, and given that erfc(4.47) $\simeq 2 \times 10^{-10}$:
  (a) estimate the number of photons required at the receiver to register a binary 1 with a BER of 10^{-10};
  (b) calculate the required incident optical power at the receiver when the system is operating at a wavelength of 0.9 μm and a transmission rate of 34 Mbit s^{-1};
  (c) indicate how the value obtained in (b) should be modified to compensate for a 3B4B line code.

12.13 A p–i–n photodiode receiver requires 2×10^4 incident photons in order to register a binary 1 with a BER of 10^{-9}. The device has a quantum efficiency of 65%. Estimate in decibels the additional signal level required in excess of the quantum limit for this photodiode to maintain a BER of 10^{-9}.

12.14 An optical fiber system employs an LED transmitter which launches an average of 300 μW of optical power at a wavelength of 0.8 μm into the optical fiber cable. The cable has an overall attenuation (including joints) of 4 dB km^{-1}. The APD receiver requires 1200 incident photons in order to register a binary 1 with a BER of 10^{-10}. Determine the maximum transmission distance (without repeaters) provided by the system when the transmission rate is 1 Mbit s^{-1} and 1 Gbit s^{-1} such that a BER of 10^{-10} is maintained.
  Hence sketch a graph showing the attenuation limit on transmission distance against the transmission rate for the system.

12.15 An optical fiber system uses fiber cable which exhibits a loss of 7 dB km^{-1}. Average splice losses for the system are 1.5 dB km^{-1}, and connector losses at the source and detector are 4 dB each. After safety margins have been allowed, the total permitted channel loss is 37 dB. Assuming the link to be attenuation limited, determine the maximum possible transmission distance without a repeater.

12.16 Assuming a linear increase in pulse broadening with fiber length, show that the transmission rate $B_T(DL)$ at which a digital optical fiber system becomes dispersion limited is given by:

$$B_T(DL) \simeq \frac{(\alpha_{dB} + \alpha_j)}{5\sigma_T(km)} \frac{1}{10 \log_{10}(P_i/P_o)}$$

where $\sigma_T(km)$ is the total rms pulse broadening per kilometer on the link (hint: refer to Eqs (3.3) and (3.11)).
  (a) A digital optical fiber system using an injection laser source displays rms pulse broadening of 1 ns km^{-1}. The fiber cable has an attenuation of 3.5 dB km^{-1} and joint losses average out to 1 dB km^{-1}. Estimate the transmission rate at the dispersion limit when the difference in optical power levels between the input and output is 40 dB.

(b) Calculate the dispersion-limited transmission distance for the system described in (a) when the transmission rates are 1 Mbit s^{-1} and 1 Gbit s^{-1}. Hence sketch a graph showing the dispersion limit on transmission distance against the transmission rate for the system.

12.17 The digital optical fiber system described in Problem 12.16(a) has a transmission rate of 50 Mbit s^{-1} and operates over a distance of 12 km without repeaters. Assuming Gaussian-shaped pulses, calculate the dispersion–equalization penalty exhibited by the system for the cases when:
(a) there is no mode coupling; and
(b) there is mode coupling.

12.18 A digital optical fiber system uses an RZ pulse format. Show that the maximum bit rate for the system $B_T(\text{max})$ may be estimated using the expression:

$$B_T(\text{max}) = \frac{0.35}{T_{\text{syst}}}$$

where T_{syst} is the total system rise time. Comment on the possible use of the factor 0.44 in place of 0.35 in the above relationship.

An optical fiber link is required to operate over a distance of 10 km without repeaters. The fiber available exhibits a rise time due to intermodal dispersion of 0.7 ns km^{-1}, and a rise time due to intramodal dispersion of 0.2 ns km^{-1}. In addition, the APD detector has a rise time of 1 ns. Estimate the maximum rise time allowable for the source in order for the link to be successfully operated at a transmission rate of 40 Mbit s^{-1} using an RZ pulse format.

12.19 An edge-emitting LED operating at a wavelength of 1.3 μm launches −22 dBm of optical power into a single-mode fiber pigtail. The pigtail is connected to a single-mode fiber link which exhibits an attenuation of 0.4 dB km^{-1} at this wavelength. In addition, the splice losses on the link provide an average loss of 0.05 dB km^{-1}. The transmission rate of the system is 280 Mbit s^{-1} so that the sensitivity of the p–i–n photodiode receiver is −35 dBm. Penalties on the link require an allowance of 1.5 dB and a safety margin of 6 dB is also specified. If the connector losses at the LED transmitter and p–i–n photodiode receiver are each 1 dB, calculate the unrepeatered distance over which the link will operate.

12.20 A digital single-mode optical fiber system is designed for operation at a wavelength of 1.5 μm and a transmission rate of 560 Mbit s^{-1} over a distance of 50 km without repeaters. The single-mode injection laser is capable of launching a mean optical power of −13 dBm into the fiber cable which exhibits a loss of 0.25 dB km^{-1}. In addition, average splice losses are 0.1 dB at 1 km intervals. The connector loss at the receiver is 0.5 dB and the receiver sensitivity is −39 dBm. Finally, an extinction ratio penalty of 1 dB is predicted for the system. Perform an optical power budget for the system and determine the safety margin.

12.21 Briefly discuss the reasons for the use of block codes in digital optical fiber transmission. Indicate the advantages and drawbacks when a 5B6B code is employed.

12.22   A D–IM analog optical fiber system utilizes a p–i–n photodiode receiver. Derive an expression for the rms signal power to rms noise power ratio in the quantum limit for this system.

The p–i–n photodiode in the above system has a responsivity of 0.5 at the operating wavelength of 0.85 μm. Furthermore, the system has a modulation index of 0.4 and transmits over a bandwidth of 5 MHz. Sketch a graph of the quantum-limited receiver sensitivity against the received SNR (rms signal power to rms noise power) for the system over the range 30 to 60 dB. It may be assumed that the photodiode dark current is negligible.

12.23   In practice, the analog optical fiber receiver of Problem 12.22 is found to be thermal noise limited. The mean square thermal noise current for the receiver is 2×10^{-23} A^2 Hz^{-1}. Determine the peak-to-peak signal power to rms noise power ratio at the receiver when the average incident optical power is −17.5 dBm.

12.24   An analog optical fiber system has a modulation bandwidth of 40 MHz and a modulation index of 0.6. The system utilizes an APD receiver with a responsivity of 0.7 and is quantum noise limited. An SNR (rms signal power to rms noise power) of 35 dB is obtained when the incident optical power at the receiver is −30 dBm. Assuming the detector dark current may be neglected, determine the excess avalanche noise factor at the receiver.

12.25   A simple analog optical fiber link operates over a distance of 15 m. The transmitter comprises an LED source which emits an average of 1 mW of optical power into air when the drive current is 40 mA. Plastic fiber cable with an attenuation of 500 dB km^{-1} at the transmission wavelength is utilized. The minimum optical power level required at the receiver for satisfactory operation of the system is 5 μW. The coupling losses at the transmitter and receiver are 8 and 2 dB respectively. In addition, a safety margin of 4 dB is necessary. Calculate the minimum LED drive current required to maintain satisfactory system operation.

12.26   An analog optical fiber system employs an LED which emits 3 dBm mean optical power into air. However, a coupling loss of 17.5 dB is encountered when launching into a fiber cable. The fiber cable which extends for 6 km without repeaters exhibits a loss of 5 dB km^{-1}. It is spliced every 1.5 km with an average loss of 1.1 dB per splice. In addition, there is a connector loss at the receiver of 0.8 dB. The PIN–FET receiver has a sensitivity of −54 dBm at the operating bandwidth of the system. Assuming there is no dispersion–equalization penalty, perform an optical power budget for the system and establish a safety margin.

12.27   Indicate the techniques which may be used for analog optical fiber transmission where an electrical subcarrier is employed. Illustrate your answer with a system block diagram showing the multiplexing of several signals onto a single analog optical fiber link.

12.28   Subcarrier amplitude modulation (AM–IM) is employed on the RF carriers of an analog optical fiber system. When a large number of the RF subcarriers, each with random phases, are frequency division multiplexed, they add on a power basis so that the optical modulation index m is related to the per-channel modulation index m_k by:

$$m = \left(\sum_{k=1}^{N} m_k^2 \right)^{\frac{1}{2}}$$

where N equals the number of channels. An FDM signal incorporates 80 AM sub-carriers. When 40 of these signals have a per-channel modulation index of 2%, 20 signals have a 3% and the other 20 signals a 4% per-channel modulation index, calculate the optical modulation index of the transmitter.

12.29 A narrowband FM–IM optical signal has a maximum frequency deviation of 120 kHz when the frequency deviation ratio is 0.2. Compare the post-detection SNR of this signal with that of a DSB–IM optical signal having the same modulating signal power, bandwidth, detector photocurrent and noise. Also estimate the bandwidth of the FM–IM signal. Comment on both results.

12.30 A frequency division multiplexed optical fiber system uses FM–IM. It has 50 equal amplitude voice channels each bandlimited to 3.5 kHz. A 1 kHz guard band is provided between the channels and below the first channel. The peak frequency deviation for the system is 1.35 MHz. Determine the transmission bandwidth for this FDM system.

12.31 An FM–IM system utilizes pre-emphasis and de-emphasis to enhance its performance in noise [Ref. 84]. The de-emphasis filter is a first-order RC low-pass filter placed at the demodulator to reduce the total noise power. This filter may be assumed to have an amplitude response $H_{de}(\omega)$ given by:

$$| H_{de}(\omega) | = \frac{1}{1 + (\omega/\omega_c)^2}.$$

where $\omega_c = 2\pi f_c = 1/RC$.

The SNR improvement over FM–IM without pre-emphasis and de-emphasis is given by:

$$(S/N)_{de} \text{ improvement} = \frac{1}{3} \left(\frac{B_a}{f_c} \right)^2$$

where B_a is the bandwidth of the baseband signal and $f_c \ll B_a$.
   (a) Write down an expression for the amplitude response of the pre-emphasis filter so that there is no overall signal distortion.
   (b) Deduce an expression for the post-detection SNR improvement in decibels for the FM–IM system with pre-emphasis and de-emphasis over a D–IM system operating under the same conditions of modulating signal power and bandwidth, photocurrent and noise. It may be assumed that $f_c \ll B_a$.
   (c) A baseband signal with a bandwidth of 300 kHz is transmitted using the FM–IM system with pre-emphasis and de-emphasis. The maximum frequency deviation for the system is 4 MHz. In addition the de-emphasis filter comprises a 500 Ω resistor and a 0.1 µF capacitor. Determine the post-detection SNR improvement for this system over a D–IM system operating

under the same conditions of modulating signal power and bandwidth, photo-current and noise.

12.32 A PM–IM optical fiber system operating above threshold has a frequency devia-tion ratio of 15 and a transmission bandwidth of 640 kHz.
(a) Estimate the bandwidth of the baseband message signal.
(b) Compute the post-detection SNR improvement for the system over a D–IM system operating with the same modulating signal power and bandwidth, detector photocurrent and noise.

12.33 Discuss the advantages and drawbacks of the various pulse analog techniques for optical fiber transmission. Describe the operation of a PFM–IM optical fiber sys-tem employing regenerative baseband recovery.

12.34 An optical fiber PFM–IM system uses regenerative baseband recovery. The opti-cal receiver which incorporates a $p–i–n$ photodiode has an optimized bandwidth of 125 MHz. The other system parameters are:

Nominal pulse rate	35 MHz
Peak-to-peak frequency deviation	8 MHz
$p–i–n$ photodiode responsivity	0.6 A W^{-1}
Baseband noise bandwidth	10 MHz
Receiver mean square noise current	3×10^{-25} A^2 Hz^{-1}

(a) Calculate the peak level of incident optical power necessary at the receiver to maintain a peak-to-peak signal power to rms noise power ratio of 60 dB.
(b) The source and detector have rise times of 3.5 and 5.0 ns respectively. Estimate the maximum permissible total rise time for the fiber cable utilized in the system such that satisfactory operation is maintained. Comment on the value obtained.

12.35 Considering the bus and star distribution systems of Example 12.18, compare the losses associated with the addition of an extra node when the original number of nodes is 12. How does this alter if the connector losses increase to 1.5 dB and the distance between the nodes on the bus/combined length of two arms from the star hub increases to 400 m? Comment on the results.

12.36 An optical fiber data bus is to be implemented to interconnect nine stations, each separated by 50 m. Multimode fiber cable with an attenuation of 3 dB km^{-1} is to be used along with LED transmitters which launch 200 μW of optical power into their fiber pigtails. The PIN–FET hybrid receivers have a sensitivity of −50 dBm at the desired BER, while the connector losses are 1.1 dB each. The access cou-plers to be used have a tap ratio or power tap-off factor of 8% together with an insertion loss of 0.9 dB. Finally, the beam splitter can be assumed to have a loss of 3 dB. Determine the safety margin for the system when considering the highest loss terminal interconnection path.

12.37 Repeat Problem 12.36 when considering a 30-terminal star distribution network where the combined distance in the two fiber cable arms is 200 m and the excess loss of the star coupler hub is 3.4 dB. Comment on the result.

12.38 Compare and contrast the merits and drawbacks associated with the major multi-plexing techniques discussed in Section 12.9. Discuss the strategies for combining two of the techniques in order to provide increased information transfer.

12.39 Describe, with the aid of simple sketches, the ways in which optical amplifiers may be configured on long-haul telecommunication links in order to provide both bidirectional and multichannel optical transmission.

Suggest how optical amplifiers might be incorporated into optical fiber distri-bution systems to facilitate the interconnection of a larger number of nodes.

12.40 It is desired to obtain the maximum signal-to-noise ratio on a single-mode fiber communication link incorporating cascaded traveling-wave SOAs. If only noise considerations are taken into account, determine the optimum length between the amplifier repeaters when the system is operating at:
 (a) a wavelength of 1.3 μm where the fiber cable and joint losses are 0.41 dB km^{-1} and 0.04 dB km^{-1} respectively; and
 (b) a wavelength of 1.55 μm where the fiber cable and joint losses are 0.20 dB km^{-1} and 0.02 dB km^{-1} respectively.
 Comment on the values obtained.

12.41 The following specifications are envisaged for a single-mode fiber communica-tion system employing cascaded traveling-wave SOAs:

Operating wavelength: 1.30 μm
Power launched at transmitter: 2 dBm
Fiber cable loss: 0.40 dB km^{-1}
Fiber splice losses: 0.02 dB km^{-1}
K (amplifiers): 6
Amplifier separation: 50 km
Received SNR: 20 dB

Assuming amplifier noise to be the limiting factor, estimate the maximum trans-mission rate allowed for the system when the total system length is 8000 km.

12.42 The SNR at the transmitter output for a single-mode fiber system employing cas-caded traveling-wave SLAs is 48.6 dB. The final SLA is positioned as a pream-plifier to the optical receiver on the link and the output SNR from this device is 15 dB. In addition the amplifiers each have a noise figure of 4.5 dB and a fiber-to-fiber gain of 9.7 dB, and the attenuation of the link prior to each amplifier is 5.2 dB. Assuming the SNR position to remain constant, how many amplifiers can be cascaded on the link?

12.43 Discuss the need for dispersion management and explain its significance in long-haul optical fiber transmission.

12.44 Explain what is a dispersion management map. Describe its role in a multiwave-length-channel optical fiber transmission system.

12.45 Find the relative dispersion slopes for the single-mode fibers corresponding to the values of chromatic dispersion and dispersion slope provided in Table 12.3.

Table 12.3 Single-mode optical fiber dispersion and dispersion slope values

No.	Fiber type	Chromatic dispersion $(ps\ nm^{-1}\ km^{-1})$	Dispersion slope $(ps\ nm^{-2}\ km^{-1})$
i	Standard single-mode fiber	16.5	0.058
ii	Pure silica core fiber	18.5	0.055
iii	Nonzero dispersion shifted fiber	4.0	0.045

12.46 The dispersion management map for a DWDM optical fiber system is shown in Figure 12.75. The system operates at the wavelength of 1.55 μm and the two fiber spans L_1 and L_2 are 200 km and 30 km, respectively. In addition, the second-order dispersion coefficient for the first fiber span is -1.7 ps nm^{-1} km^{-1}. Determine the second-order dispersion coefficient for the second fiber span in order to achieve zero mean dispersion.

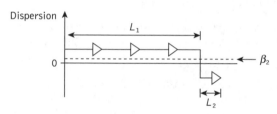

Figure 12.75 Dispersion management map for the DWDM optical fiber system in Problem 12.46

12.47 Identify and discuss the major property of a soliton pulse which enables low bit-error-rate propagation over longer optical fiber transmission distances.

12.48 Explain the term soliton interaction and discuss its role in a multiwavelength-channel soliton fiber system. What effect does it produce in optical fiber soliton systems at transmission rates of 10 Gbit s^{-1}, 40 Gbit s^{-1} and higher?

12.49 An optical soliton is transmitted over a distance of 10 000 km. The second-order dispersion coefficient is equal to -0.5 ps^2 km^{-1} and the attenuation coefficient is 0.22 dB km^{-1} while the separation between two soliton pulses is 4.0. Calculate the pulse width of the soliton when the dispersion length is: (a) 200 km; (b) 25 km. Comment on the significance of the results.

12.50 In Problem 12.49 determine the transmission bit rate when the second-order dispersion coefficient is -0.98 ps^2 km^{-1}.

12.51 Define the terms amplifier spacing and dispersion length. Briefly discuss why these both must be considered in order to achieve high transmission rates in an optical fiber soliton system.

Answers to numerical problems

12.4	53.0 pF, 472 Ω	**12.28**	25.7%
12.5	64 kbit s^{-1}	**12.29**	Ratio of output SNRs (rms signal power
12.6	(a) 37 ns; (b) 125 μV		to rms noise power) for, FM–IM to
12.8	(a) 300 bits; (b) 1.2 MHz		DSB–IM is −9.21 dB, 1200 kHz
12.9	(a) 10.2 dB; (b) 3.4 × 10^{-7}	**12.30**	3.15 MHz
12.10	$P(e) = \text{Erfc} \mid (S/N)^{\frac{1}{2}}/2 \mid$, 9.8 dB, 19.6 dB	**12.31**	(a) $1 + (\omega/\omega_c)^2$;
12.11	7.4		(b) $-3 + 20 \log_{10}(D_f B_a/f_c)$;
12.12	(a) 1400; (b) −52.8 dBm; (c) −51.8 dBm		(c) 59.0 dB
12.13	27.9 dB	**12.32**	(a) 20 kHz; (b) 20.5 dB
12.14	15.76 km, 8.26 km	**12.34**	(a) −24.4 dBm; (b) 4.6 ns
12.15	3.41 km	**12.35**	3.5 dB, 0.35 dB, 6.0 dB, 0.35 dB
12.16	(a) 22.5 Mbit s^{-1}; (b) 200 km, 0.2 km	**12.36**	3.7 dB
12.17	(a) 16.6 dB; (b) 0.1 dB.	**12.37**	19.8 dB
12.18	3.04 ns	**12.40**	(a) 9.6 km; (b) 19.7 km
12.19	7.8 km	**12.41**	858 Mbit s^{-1} (RZ)
12.20	2.1 dB	**12.42**	87
12.22	40 nW, 40 μW	**12.45**	(i) 0.0035 nm^{-1}; (ii) 0.0030 nm^{-1};
12.23	52.1 dB		(iii) 0.113 nm^{-1}
12.24	3.1	**12.46**	11.3 ps nm^{-1} km^{-1}
12.25	35.5 mA	**12.49**	(a) 10.0 ps; (b) 1.5 ps
12.26	5.4 dB	**12.50**	(a) 8.9 Gbit s^{-1}; (b) 59.2 Gbit s^{-1}

References

[1] T. Kitatani, K. Shinoda, T. Tsuchiya, H. Sato, K. Ouchi, H. Uchiyama, S. Tsuji and M. Aoki, 'Evaluation of the optical-coupling efficiency of InGaAlAs-InGaAsP butt joint using a novel multiple butt-jointed laser', *IEEE Photonics Technol. Lett.*, **17**(6), pp. 1148–1150, 2005.

[2] M. Maeda, I. Ikushima, K. Nagano, M. Tanaka, H. Naskshima and R. Itoh, 'Hybrid laser to fiber coupler with a cylindrical lens', *Appl. Opt.*, **16**(7), pp. 1966–1970, 1977.

[3] D. Delbeke, R. Bockstaele, P. Bienstman, R. Baets and H. Benisty, 'High-efficiency semiconductor resonant-cavity light-emitting diodes: a review', *IEEE J. Sel. Top. Quantum Electron.*, **8**(2), pp. 189–206, 2002.

[4] R. W. Dawson, 'Frequency and bias dependence of video distortion in Burrus-type homostructure and heterostructure LED's', *IEEE Trans. Electron Devices*, **ED-25**(5), pp. 550–553, 1978.

[5] J. Strauss, 'The nonlinearity of high-radiance light-emitting diodes', *IEEE J. Quantum Electron.*, **QE-14**(11), pp. 813–819, 1978.

[6] O. Wada, 'Femtosecond all-optical devices for ultrafast communication and signal processing', *New J. Phys.*, **6**(183), pp. 1–35, 2004.

[7] G. Steinmeyer, 'A review of ultrafast optics and optoelectronics', *J. Opt. A: Pure Appl. Opt.*, **5**(1), pp. R1–R15, 2003.

[8] K. A. Williams, M. G. Thompson and I. H. White, 'Long-wavelength monolithic mode-locked diode lasers', *New J. Phys.*, **6**(179), pp. 1–30, 2004.

[9] J. E. Simsarian, M. C. Larson, H. E. Garrett, X. Hong and T. A. Strand, 'Less than 5-ns wavelength switching with an SG-DBR laser', *IEEE Photonics Technol. Lett.*, **18**(4), pp. 565–567, 2006.

[10] K. Asatani and T. Kimura, 'Non-linear phase distortion and its compensation in LED direct modulation', *Electron. Lett.*, **13**(6), pp. 162–163, 1977.

[11] G. White and C. A. Burrus, 'Efficient 100 Mb/s driver for electroluminescent diodes', *Int. J. Electron.*, **35**(6), pp. 751–754, 1973.

[12] P. W. Shumate Jr and M. DiDomenico Jr, 'Lightwave transmitters', in H. Kressel (Ed.), *Semiconductor Devices for Optical Communications*, Topics in Applied Physics, Vol. 39, pp. 161–200, Springer-Verlag, 1982.

[13] L. Foltzer, 'Low-cost transmitters, receivers serve well in fibre-optic links', *EDN*, pp. 141–146, 20 October 1980.

[14] A. Albanese and H. F. Lenzing, 'Video transmission tests, performed on intermediate-frequency light wave entrance links', *J. SMPTE*, **87**(12), pp. 821–824, 1978.

[15] J. Strauss, 'Linearized transmitters for analog fiber links', *Laser Focus*, **14**(10), pp. 54–61, 1978.

[16] A. Prochazka, P. Lancaster and R. Neumann, 'Amplifier linearization by complementary pre or post distortion', *IEEE Trans. Cable Telev.*, **CATV-1**(1), pp. 31–39, 1976.

[17] K. Asatani and T. Kimura, 'Nonlinear distortions and their compensations in light emitting diodes', *Proc. Int. Conf. on Integrated Optics and Optical Fiber Communications*, p. 105, 1977.

[18] K. Asatani and T. Kimura, 'Linearization of LED nonlinearity by predistortions', *IEEE J. Solid State Circuits*, **SC-13**(1), pp. 133–138, 1978.

[19] J. Strauss, A. J. Springthorpe and O. I. Szentesi, 'Phase shift modulation technique for the linearisation of analogue optical transmitters', *Electron. Lett.*, **13**(5), pp. 149–151, 1977.

[20] J. Strauss and D. Frank, 'Linearisation of a cascaded system of analogue optical links', *Electron. Lett.*, **14**(14), 436–437, 1978.

[21] H. S. Black, US Patent 1686792, issued October 9, 1929.

[22] B. S. Kawasaki and K. O. Hill, 'Low-loss access coupler for multimode optical fiber distribution network', *Appl. Opt.*, **16**(7), p. 1794, 1977.

[23] J. Strauss and O. I. Szentesi, 'Linearisation of optical transmitters by a quasifeedforward compensation technique', *Electron. Lett.*, **13**(6), pp. 158–159, 1977.

[24] S. M. Abbott, W. M. Muska, T. P. Lee, A. G. Dentai and C. A. Burrus, '1.1 Gb/s pseudorandom pulse-code modulation of 1.27 μm wavelength CW InGaAsP/InP DH lasers', *Electron. Lett.*, **14**(11), pp. 349–350, 1978.

[25] J. Gruber, P. Marten, R. Petschacher and P. Russer, 'Electronic circuits for high bit rate digital fiber optic communication systems', *IEEE Trans. Commun.*, **COM-26**(7), pp. 1088–1098, 1978.

[26] P. Zivojinovic, M. Lescure and H. Tap-Beteille, 'Design and stability analysis of a CMOS feedback laser driver', *IEEE Trans. Instrum. Meas.*, **53**(1), pp. 102–108, 2004.

[27] S. R. Salter, D. R. Smith, B. R. White and R. P. Webb, 'Laser automatic level control for optical communications systems', *Third Eur. Conf. on Optical Communications*, Munich, Germany, September 1977, VDE-Verlag, 1977.

[28] S. D. Personick, 'Design of receivers and transmitters for fibre systems', in M. K. Barnoski (Ed.), *Fundamentals of Optical Fiber Communications* (2nd edn), pp. 295–328, Academic Press, 1981.

[29] R. G. Smith and S. D. Personick, 'Receiver design for optical fibre communication systems', in H. Kressel (Ed.), *Semiconductor Devices for Optical Communications*, Topics in Advanced Physics, Vol. 39, pp. 88–160, Springer-Verlag, 1982.

[30] R. G. Smith, C. A. Brackett and H. W. Reinbold, 'Atlanta fiber system experiment, optical detector package', *Bell Syst. Tech. J.*, **57**(6), pp. 1809–1822, 1978.

[31] J. L. Hullett and T. V. Muoi, 'A feedback amplifier for optical transmission systems', *IEEE Trans. Commun.*, **COM-24**, pp. 1180–1185, 1976.

[32] J. L. Hullett, 'Optical communication receivers', *Proc. IREE Australia*, pp. 127–134, September 1979.

[33] N. J. Bradley, 'Fibre optic systems design', *Electron. Eng.*, pp. 98–101, mid April 1980.

[34] S. D. Personick, 'Time dispersion in dielectric waveguides', *Bell Syst. Tech. J.*, **50**(3), pp. 843–859, 1971.

[35] B. J. Offrein, F. Horst, G. L. Bona, R. Germann, H. W. M. Salemink and R. Beyeler, 'Adaptive gain equalizer in high-index-contrast SiON technology', *IEEE Photonics Technol. Lett.*, **12**(5), pp. 504–506, 2000.

[36] K. Suzuki, T. Kitoh, S. Suzuki, Y. Inoue, Y. Hibino, T. Shibata, A. Mori and M. Shimizu, 'PLC-based dynamic gain equaliser consisting of integrated Mach-Zehnder interferometers with C- and L-band equalising range', *Electron. Lett.*, **38**(18), pp. 1030–1031, 2002.

[37] M. Barge, D. Battarel and J. L. B. de la Tocnaye, 'A polymer-dispersed liquid crystal-based dynamic gain equalizer', *J. Lightwave Technol.*, **23**(8), pp. 2531–2541, 2005.

[38] K. Maru, T. Chiba, K. Tanaka, S. Himi and H. Uetsuka, 'Dynamic gain equalizer using hybrid integrated silica-based planar lightwave circuits with LiNbO₃ phase shifter array', *J. Lightwave Technol.*, **24**(1), pp. 495–503, 2006.

[39] Dynamic gain equalizer, DGE1020, lightconnect, http://www.lightconnect.com/pdf/DGEO0202.pdf , 8 November 2007.

[40] Dynamic gain equalizer, Optogone, http://www.optogone.com/products/datasheet_dge.pdf, 8 November 2007.

[41] M. Rocks and R. Kerstein, 'Increase in fiber bandwidth for digital systems by means of multiplexing', *ICC'80 1980 Int. Conf. on Communications*, Seattle, WA, USA, Part 28, 5/1–5, June 1980.

[42] W. Koester and F. Mohr, 'Bidirectional optical link', *Electr. Commun.*, **55**(4), pp. 342–349, 1980.

[43] I. A. Glover and P. M. Grant, *Digital Communications* (2nd edn), Pearson, 2004.

[44] European Telecommunications Standards Institute (ETSI), http://www.etsi.org, 8 November 2007.

[45] ITU-T, '*ITU-T Recommendations*', International Telecommunication Union, http://www.itu.int/itu-t/publications/index.html, 8 November 2007.

[46] E. Garmire, 'Sources, modulators, and detectors for fiber-optic communication systems', in *Fiber Optic Handbook*, M. Bass and E. W. Van Stryland (Eds), pp. 4.1–4.80, McGraw-Hill, 2002.

[47] I. Kang, L. Mollenauer, B. Greene and A. Grant, 'A novel method for synchronizing the pulse carver and electroabsorption data modulator for ultralong-haul DWDM transmission', *IEEE Photonics Technol. Lett.*, **14**(9), pp. 1357–1359, 2002.

[48] S. Yiyai, G. Raybon, Z. Zheng, S. Chandrasekhar, R. Ryf and L. Moller, '40-Gb/s RZ signal transmission in a transparent network based on wavelength-selective optical cross connect', *IEEE Photonics Technol. Lett.*, **15**(10), pp. 1467–1469, 2003.

[49] R. M. Mu, T. Yu, V. S. Grigoryan and C. R. Menyuk, 'Dynamics of the chirped return-to-zero modulation format', *J. Lightwave Technol.*, **20**(1), pp. 47–57, 2002.

[50] P. J. Winzer and R. J. Essiambre, 'Advanced modulation formats for high-capacity optical transport networks', *J. Lightwave Technol.*, **24**(12), pp. 4711–4728, 2006.

[51] A. V. Ramprasad and M. Meenakshi, 'Performance of NRZ, RZ, CSRZ, and VSB-RZ modulation formats in the presence of four-wave mixing effect in DWDM optical systems', *J. Opt. Netw.*, **6**(2), pp. 146–156, 2007.

[52] M. I. Hayee, A. E. Willner, T. S. Syst and N. J. Eacontown, 'NRZ versus RZ in 10-40-Gb/s dispersion-managed WDM transmission systems', *IEEE Photonics Technol. Lett.*, **11**(8), pp. 991–993, 1999.

[53] E. Sackinger, *Broadband Circuits for Optical Fiber Communication*, Wiley, 2005.

[54] J. Lee, S. Kim, Y. Kim, Y. Oh, S. Hwang and J. Jeong, 'Optically preamplified receiver performance due to VSB filtering for 40-Gb/s optical signals modulated with various formats', *J. Lightwave Technol.*, **21**(2), pp. 521–527, 2003.

[55] K. Yonenaga and S. Kuwano, 'Dispersion-tolerant optical transmission system using duobinary transmitter and binary receiver', *J. Lightwave Technol.*, **15**(8), pp. 1530–1537, 1997.

[56] G. Goldfarb, K. Cheolhwan and L. Guifang, 'Improved chromatic dispersion tolerance for optical duobinary transmission using coherent detection', *IEEE Photonics Technol. Lett.*, **18**(3), pp. 517–519, 2006.

[57] H. Chen, M. Chen and S. Xie, 'PolSK label over VSB-CSRZ payload scheme in AOLS network', *J. Lightwave Technol.*, **25**(6), pp. 1348–1355, 2007.

[58] G. E. Stillman, 'Design consideration, for fibre optic detectors', Proc. *SPIE*, **239**, pp. 42–52, 1980.

[59] P. P. Webb, R. J. McIntyre and J. Conradi, 'Properties of avalanche photodiodes', *RCA Rev.*, **35**, pp. 234–278, 1974.

[60] J. E. Midwinter, *Optical Fibers for Transmission*, Wiley, 1979.

[61] R. J. McIntyre and J. Conradi, 'The distribution of gains in uniformly multiplying avalanche photodiodes', *IEEE Trans. Electron Devices*, **ED-19**, pp. 713–718, 1972.

[62] K. Mouthaan, 'Teleconmunications via glass–fibre cables', *Philips Telecommun. Rev.*, **37**(4), pp. 201–214, 1979.

[63] S. D. Personick, N. L. Rhodes, D. C. Hanson and K. H. Chan, 'Contrasting fiber-optic-component-design requirements in telecommunications, analog, and local data communications applications', *Proc. IEEE*, **68**(10), pp. 1254–1262, 1980.

[64] C. Kleekamp and B. Metcalf, *Designer's Guide to Fiber Optics*, Cahners Publishing, 1978.

[65] G. R. Elion and H. A. Elion, *Fiber Optics in Communications Systems*, Marcel Dekker, 1978.

[66] J. P. Fonseka, J. Liu and N. Goel, 'Multi-interval line coding technique for high speed transmissions', *IEE Proc., Commun.*, **153**(5), pp. 619–625, 2006.

[67] J. B. Stark, J. E. Mazo and R. Laroia, 'Phased amplitude-shift signaling (PASS) codes: increasing the spectral efficiency of DWDM transmission', *Eur. Conf. on Optical Communication (ECOC'98)*, Madrid, Spain, **1**, pp. 373–374, 20–24 September 1998.

[68] E. Forestieri and G. Prati, 'Novel optical line codes tolerant to fiber chromatic dispersion', *J. Lightwave Technol.*, **19**(11), pp. 1675–1684, 2001.

[69] I. B. Djordjevic and B. Vasic, 'Multilevel coding in M-ary DPSK/differential QAM high-speed optical transmission with direct detection', *J. Lightwave Technol.*, **24**(1), p. 420, 2006.

[70] C. C. Chien and I. Lyubomirsky, 'Comparison of RZ versus NRZ pulse shapes for optical duobinary transmission', *J. Lightwave Technol.*, **25**(10), p. 2953, 2007.

[71] I. B. Djordjevic, B. Vasic and V. S. Rao, 'Rate 2/3 modulation code for suppression of intra-channel nonlinear effects in high-speed optical transmission', *IEE Proc., Optoelectron.*, **153**(2), pp. 87–92, 2006.

[72] A. Luvison, 'Topics in optical fibre communication theory', *Optical Fibre Communications*, by Technical Staff of CSELT, pp. 647–721, McGraw-Hill, 1981.

[73] S. Kawanishi, N. Voshikai, J.-I. Yamada and K. Nakagawa, 'DmBIM code and its performance, in a very high-speed optical transmission system', *IEEE Trans. Commun.*, **36**(8), pp. 951–956, 1988.

[74] I. B. Djordjevic and B. Vasic, 'Constrained coding techniques for the suppression of intra-channel nonlinear effects in high-speed optical transmission', *J. Lightwave Technol.*, **24**(1), pp. 411–419, 2006.

[75] T. Mizuochi, Y. Miyata, T. Kobayashi, K. Ouchi, K. Kuno, K. Kubo, K. Shimizu, H. Tagami, H. Yoshida and H. Fujita, 'Forward error correction based on block turbo code with 3-bit soft decision for 10-Gb/s optical communication systems', *IEEE J. Sel. Top. in Quantum Electron.*, **10**(2), pp. 376–386, 2004.

[76] J. Yan, M. Chen, S. Xie and B. Zhou, 'Performance comparison of standard FEC in 40 Gbit/s optical transmission systems with NRZ, RZ and CS-RZ modulation formats', *Opt. Commun.*, **231**(1–6), pp. 175–180, 2004.

[77] ITU-T Recommendation, G.975.1, 'Forward error correction for high bit-rate DWDM submarine system', February 2004.

[78] K. Azadet, E. F. Haratsch, H. Kim, F. Saibi, J. H. Saunders, M. Shaffer, L. Song and Y. Meng-Lin, 'Equalization and FEC techniques for optical transceivers', *IEEE J. Solid-State Circuits*, **37**(3), pp. 317–327, 2002.

[79] E. Forestieri, *Optical Communication Theory and Techniques*, Springer-Verlag, 2006.

[80] J. Yuan, W. Ye, Z. Jiang, Y. Mao and W. Wang, 'A novel super-FEC code based on concatenated code for high-speed long-haul optical communication systems', *Opt. Commun.*, **273**(2), pp. 421–427, 2007.

[81] T. Mizuochi, 'Recent progress in forward error correction for optical communication systems', *IEICE Trans. Commun.*, **88**(5), pp. 1934–1946, 2005.

[82] J. Xu, L. Chen, I. Djurdjevic, S. Lin and K. Abdel-Ghaffar, 'Construction of regular and irregular LDPC codes: geometry decomposition and masking', *IEEE Trans. Inf. Theory*, **53**(1), pp. 121–134, 2007.

[83] S. R. Irving and C. Xuemin, *Error-Control Coding for Data Networks*, Springer-Verlag, 1999.

[84] H. Tagami, T. Kobayashi, Y. Miyata, K. Ouchi, K. Sawada, H. Kubo, K. Kuno, H. Yoshida, K. Shimizu, T. Mizuochi and K. Motoshima, 'A 3-bit soft decision IC for powerful forward error correction in 10-Gb/s optical communication systems', *IEEE J. of Solid State Circuits*, **40**(8), pp. 1695–1705, 2005.

[85] S. Kaneko, J. Kani, K. Iwatsuki, A. Ohki, M. Sugo and S. Kamei, 'Scalability of spectrum-sliced DWDM transmission and its expansion using forward error correction', *J. Lightwave Technol.*, **24**(3), pp. 1295–1301, 2006.

[86] G. G. Windus, 'Fibre optic systems for analogue transmission', *Marconi Rev.*, **XLIV**(221), pp. 78–100, 1981.

[87] K. Sam Shanmugam, *Digital and Analog Communication Systems*, Wiley, 1979.

[88] S. Y. Suh, 'Pulse width modulation for analog fiber-optic communications', *J. Lightwave Technol.*, **LT-5**(1), pp. 102–112, 1987.

[89] C. C. Timmerman, 'Signal-to-noise ratio of a video signal transmitted by a fiber-optic system using pulse-frequency modulation', *IEEE Trans. Broadcast.*, **BC-23**(1), pp. 12–16, 1976.

[90] C. C. Timmerman, 'A fiber optical system using pulse frequency modulation', *NTZ*, **30**(6), pp. 507–508, 1977.

[91] D. J. Brace and D. J. Heatley, 'The application of pulse modulation schemes for wideband distribution to customers (integrated optical fibre systems)', *Sixth Eur. Conf. on Optical Communications*, York, UK, 16–19 September 1980, pp. 446–449, 1980.

[92] E. Yoneda, T. Kanada and K. Hakoda, 'Design and performance of optical fibre transmission systems for color television signals', *Rev. Electr. Commun. Lab.*, **29**(11–12), pp. 1107–1117, 1981.

[93] T. Kanada, K. Hakoda and E. Yoneda, 'SNR fluctuation and nonlinear distortion in PFM optical NTSC video transmission systems', *IEEE Trans. Commun.*, **COM-30**(8), pp. 1868–1875, 1982.

[94] S. F. Heker, G. J. Herskowitz, H. Grebel and H. Wichansky, 'Video transmission in optical fiber communication systems using pulse frequency modulation', *IEEE Trans. Commun.*, **36**(2), pp. 191–194, 1988.

[95] K. Schuh and E. Lach, 'High-bit-rate ETDM transmission', in I. P. Kaminow, T. Li and A. E. Willner (Eds), *Optical Fiber Telecommunications VB*, pp. 179–200, Academic Press, 2008.

[96] R. H. Derksen, G. Lehmann, C. J. Weiske, C. Schubert, R. Ludwig, S. Ferber, C. Schmidt-Langhorst, M. Moller and J. Lutz, 'Integrated 100 Gbit/s ETDM receiver in a transmission

experiment over 480 km DMF', *Proc. OFC/Fiber Optic Enginecring Conf. (NFOEC'06)*, Anaheim, CA, USA, pp. 1–3, 5–10 March 2006.

[97] F. Density, 'Terabit in a thimble-optoelectronic revolution', *III–Vs Rev.*, **16**(3), pp. 44–47, 2003.

[98] R. S. Tucker, G. Eisenstein, S. K. Korotky, U. Koren, G. Raybon, J. J. Veselka, L. L. Buhl, B. L. Kasper and R. C. Alferness, 'Optical time-division multiplexing in a multigigabit/second fibre transmission system', *Electron. Lett.*, **23**(5), pp. 208–209, 1987.

[99] J. P. Turkiewicz, E. Tangdiongga, G. Lehmann, H. Rohde, W. Schairer, Y. R. Zhou, E. S. R. Sikora, A. Lord, D. B. Payne, G.-D. Khoe and H. de Waardt, '160 Gb/s OTDM networking using deployed fiber', *J. Lightwave Technol.*, **23**(1), pp. 225–235, 2005.

[100] T. Morioka, 'Ultrafast optical technologies for large-capacity TDM/WDM photonic networks', *J. Opt. Fiber Commun. Rep.*, **4**(1), pp. 14–40, 2007.

[101] H.-G. Weber and R. Ludwig, 'Ultra-high-speed OTOM transmission technology', in I. P. Kaminow, T. Li and A. E. Willner (Eds), *Optical Fiber Telecommunications VB*, pp. 201–232, Academic Press, 2008.

[102] T. E. Darcie, M. E. Dixon, B. L. Kasper and C. A. Burrus, 'Lightwave system using microwave subcarrier multiplexing', *Electron. Lett.*, **22**(15), pp. 774–775, 1986.

[103] R. Olshansky and V. A. Lanziera, '60 channel FM video subcarrier multiplexed optical communications system', *Electron. Lett.*, **23**, pp. 1196–1197, 1987.

[104] T. E. Darcie, P. P. Iannone, B. L. Kasper, J. R. Talman, C. A. Burrus and T. A. Baker, 'Wideband lightwave distribution system using subcarrier multiplexing', *J. Lightwave Technol.*, **7**(6), pp. 997–1005, 1989.

[105] R. Olshansky and V. A. Lanziera, 'Subcarrier multiplexed lightwave systems for broad-band distribution', *J. Lightwave Technol.*, **7**(9), pp. 1329–1342, 1989.

[106] T. E. Darcie, 'Subcarrier multiplexing for lightwave networks and video distribution systems', *IEEE J. Sel. Areas Commun.*, **8**(7), pp. 1240–1248, 1990.

[107] H. Rongqing, Z. Benyuan, H. Renxiang, C. T. Allen, K. R. Demarest and D. Richards, 'Subcarrier multiplexing for high-speed optical transmission', *J. Lightwave Technol.*, **20**(3), pp. 417–427, 2002.

[108] Z. Zuqing, V. J. Hernandez, Y. J. Min, C. Jing, P. Zhong and S. J. B. Yoo, 'RF photonics signal processing in subcarrier multiplexed optical-label switching communication systems', *J. Lightwave Technol.*, **21**(12), pp. 3155–3166, 2003.

[109] J. Chen, F. Saibi, E. Sackinger, J. Othmer, M. Yu, T. Huang, T. P. Liu and K. Azadet, 'A 40-Gb/s SCM optical communication system based on an integrated CMOS transceiver', *Microw. Opt. Technol. Lett.*, **49**(6), pp. 1272–1274, 2007.

[110] M. Maeda and M. Yamamoto, 'FM–FDM optical CATV transmission experiment and system design for MUSE HDTV signals', *IEEE J. Sel. Areas Commun.*, **8**(7), pp. 1257–1267, 1990.

[111] P. A. Rosher and S. C. Fenning, 'Multichannel video transmission over 100 km of step index single mode fibre using a directly modulated distributed feedback laser', *Electron. Lett.*, **26**(8), pp. 534–536, 1990.

[112] M. C. Wu, J. K. Wong, K. T. Tsai, Y. L. Chen and W. I. Way, '740-km transmission of 78-channel 64-QAM signals (2.34 Gb/s) without dispersion compensation using a recirculating loop', *IEEE Photonics Technol. Lett.*, **12**(9), pp. 1255–1257, 2000.

[113] T. O. C. View, 'Subcarrier multiplexing for high-speed optical transmission', *J. Lightwave Technol.*, **20**(3), pp. 417–427, 2002.

[114] A. M. E.-A. Diab, J. D. Ingham, R. V. Penty and I. H. White, 'Comprehensive statistical investigation of SCM-based transmission over 300 m of FDDI-grade multimode fiber', *Proc. Conf. an Lasers and Electro-Optics (CLEO)*, Baltimore, MD, USA, vol. 2, pp. 1351–1353, May 2005.

[115] G. Puerto, B. Ortega, A. Martinez, D. Pastor, M. D. Manzanedo, and J. Capmany, 'Scalability of 10 Gbit/s SCM optical label swapping networks featuring 2R multistage intra-node regeneration', *Electron. Lett.*, **42**(12), pp. 712–714, 2006.

[116] P. Ramjee, *OFDM for Wireless Communications Systems*, Artech House, 2004.

[117] J. Armstrong and A. J. Lowery, 'Power efficient optical OFDM', *Electron. Lett.*, **42**(6), pp. 370–372, 2006.

[118] I. B. Djordjevic and B. Vasic, 'Orthogonal frequency division multiplexing for high-speed optical transmission', *Opt. Express*, **14**(9), pp. 3767–3775, 2006.

[119] A. J. Lowery, L. B. Du and J. Armstrong, 'Performance of optical OFDM in ultralong-haul WDM lightwave systems', *J. Lightwave Technol.*, **25**(1), pp. 131–138, 2007.

[120] T. Li and J. Jean, *Fundamentals of Analog and Digital Signal Processing*, AuthorHouse, 2007.

[121] M. Mayrock and H. Hausdtein, 'PMD tolerant direct detection optical OFDM system', *Eur. Conf. on Optical Communication (ECOC'07)*, Berlin, Germany, **2**, pp. 217–218, 18 September 2007.

[122] A. J. Lowery, 'Fiber nonlinearity mitigation in optical links that use OFDM for dispersion compensation', *IEEE Photonics Technol. Lett.*, **19**(19), pp. 1556–1558, 2007.

[123] B. J. C. Schmidt, A. J. Lowery and J. Armstrong, 'Experimental demonstrations of 20 Gbit/s direct-detection optical OFDM and 12 Gbit/s with a colorless transmitter', *Optical Fiber Communications* (Postdeadline paper), Anaheim, CA, USA, pdp18, pp. 1–3, March 2007.

[124] A. C. Carter, 'Wavelength multiplexing for enhanced fibre-optic performance', *Telecommunications*, pp. 30–36, October 1986.

[125] ITU-T Recommendation G.694.2, 'Spectral grids for WDM applications: CWDM wavelength grid', December 2003.

[126] ITU-T Recommendation G.695, 'Optical interfaces for coarse wavelength division multiplexing applications', January 2005.

[127] P. J. Winzer, F. Fidler, M. J. Matthews, L. E. Nelson, H. J. Thiele, J. H. Sinsky, S. Chandrasekhar, M. Winter, D. Castagnozzi, L. W. Stulz and L. L. Buhl, '10-Gb/s upgrade of bidirectional CWDM systems using electronic equalization and FEC', *J. Lightwave Technol.*, **23**(1), pp. 203–210, 2005.

[128] S. K. Das, S. M. Mysore, R. A. Villa, J. J. Thomas, H. J. Thiele, L. Jiao and L. E. Nelson, '40 Gb/s (16 × 2.5 Gb/s) full spectrum coarse WDM transmission over 75 km low water peak fiber for low-cost metro and cable TV applications', *Conf. Proc. Natl. Fiber Optics Engineering Conf. (NFOEC)*, Dallas, TX, USA, pp. 881–887, 15–19 September 2002.

[129] Broaddata® Communications Inc., 'Fiber optic coarse wavelength division multiplexing (CWDM) system', 6540E, http://www.broadatacom.com/site/PDF/6540e.pdf, 11 September 2007.

[130] ITU-T Recommendation J.186, 'Transmission equipment for multi-channel television signals over optical access networks by sub-carrier multiplexing (SCM)', February 2002.

[131] C. M. B Lopes and E. A De Souza, 'Performance evaluation of a CWDM system operating in the O and E bands over standard fiber', *Microw. Opt. Technol. Lett.*, **48**(8), pp. 1540–1544, 2006.

[132] P. P. Iannone, K. C. Reichmann, X. Zhou and N. J. Frigo, '200 km CWDM transmission using a hybrid amplifier', *Proc. OFC/Fiber Optics Engineering Conf. (NFOEC)'05*, Anaheim, CA, USA, pp. 3, 6–11 March 2005.

[133] K. C. Reichmann, P. P. Iannone, Z. Xiang, N. J. Frigo and B. R. Hemenway, '240-km CWDM transmission using cascaded SOA Raman hybrid amplifiers with 70-nm bandwidth', *IEEE Photonics Technol. Lett.*, **18**(2), pp. 328–330, 2006.

[134] C.-H. Cheng, 'Signal processing for optical communication', *IEEE Signal Process. Mag.*, **23**(1), pp. 88–96, 2006.

[135] ITU-T Recommendation G.694.1, 'Spectral grids for WDM applications: DWDM frequency grid', June 2000.

[136] L. Sang-Mook, K. Min-Hwan and L. Chang-Hee, 'Demonstration of a bidirectional 80-km-reach DWDM-PON with 8-Gb/s capacity', *IEEE Photonics Technol. Lett.*, **19**(6), pp. 405–407, 2007.

[137] H. Suzuki, M. Fujiwara and K. Iwatsuki, 'Application of super-DWDM technologies to terrestrial terabit transmission systems', *J. Lightwave Technol.*, **24**(5), pp. 1998–2005, 2006.

[138] ITU-T Recommendation G.698.1, 'Multichannel DWDM applications with single channel optical interfaces', June 2005, http://www.itu.int/rec/T-REC-G.698.1/en, 8 November 2007.

[139] C. Jin-Xing *et al.*, 'Transmission of 40-Gb/s WDM signals over transoceanic distance using conventional NZ-DSF with receiver dispersion slope compensation', *J. Lightwave Technol.*, **24**(1), pp. 191–200, 2006.

[140] Network Innovation Laboratories Photonic Transport Network Laboratory, NTT® Japan, 'Optical soliton transmission system: new wave for long-distance, high-speed optical transmission', http://www.onlab.ntt.co.jp/en/pt/soliton/, 8 November 2007.

[141] DWDM Pluggable Transceiver, http://www.hotplugdwdm.org, 8 November 2007.

[142] M. Ichino, S. Yoshikawa, H. Oomori, Y. Maeda, N. Nishiyama, T. Takayama, T. Mizue, I. Tounai and M. Nishie, 'Small form factor pluggable optical transceiver module with extremely low power consumption for dense wavelength division multiplexing applications', *Proc. 55th Electronic Components and Technology Conf.*, Lake Buena Vista, USA, **1**, pp. 1044–1049, 31 May–3 June 2005.

[143] Luminent® Inc, DWDM SFP, http://www.luminentoic.com/products/datasheets/DSDWDMBr.pdf, 8 November 2007.

[144] A. Hunwicks, L. Bickers, M. H. Reeve and S. Hornung, 'An optical transmission system for single-mode local-loop applications using a sliced spectrum technique', *Twelfth Int. Fiber Optic Communications and Local Area Networks Exposition, FOC/LAN'88*, USA, pp. 237–240, September 1988.

[145] S. S. Wagner and T. E. Chapuran, 'Broadband high-density WDM transmission using superluminescent diodes', *Electron. Lett.*, **26**(11), pp. 696–697, 1990.

[146] J. M. Senior and S. D. Cusworth, 'Wavelength division multiplexing in optical sensor systems and networks: a review', *Opt. Laser Technol.*, **22**(2), pp. 113–126, 1990.

[147] K. H. Han, E. S. Son, H. Y. Choi, K. W. Lim and Y. C. Chung, 'Bidirectional WDM PON using light-emitting diodes spectrum-sliced with cyclic arrayed-waveguide grating', *IEEE Photonics Technol. Lett.*, **16**(10), pp. 2380–2382, 2004.

[148] G. J. Pendock and D. D. Sampson, 'Transmission performance of high bit rate spectrum-sliced WDM systems', *J. Lightwave Technol.*, **14**(10), pp. 2141–2148, 1996.

[149] G. Murtaza and J. M. Senior, 'WDM crosstalk analysis for systems employing spectrally-sliced LED sources', *IEEE Photonics Technol. Lett.*, **8**(3), pp. 440–442, 1996.

[150] A. D. McCoy, P. Horak, B. C. Thomsen, M. Ibsen and D. J. Richardson, 'Noise suppression of incoherent light using a gain-saturated SOA: implications for spectrum-sliced WDM systems', *J. Lightwave Technol.*, **23**(8), pp. 2399–2409, 2005.

[151] F. Koyama, T. Yamatoya and K. Iga, 'Highly gain-saturated GaInAsP/InP SOA modulator for incoherent spectrum-sliced light source', *Proc. Indium Phospide and Related Materials Conf.*, Williamsburg, VA, USA, pp. 439–442, 14–18 May 2000.

[152] S. Kaneko, J.-I. Kani, K. Iwatsuki, A. Ohki, M. Sugo and S. Kamei, 'Scalability of spectrum-sliced DWDM transmission and its expansion using forward error correction', *J. Lightwave Technol.*, **24**(3), pp. 1295–1301, 2006.

[153] W. Xiulin, M. Hai and H. Wencai, 'Spectrum sliced C+L-band multi-wavelength fiber source', *Proc. SPIE*, **6025**, pp. 340–346, 13 Jan, 2006.

[154] L. Boivin, M. Wegmueller, M. C. Nuss and W. H. Knox, '110 channels × 2.35 Gb/s from a single femtosecond laser', *IEEE Photonics Technol. Lett.*, **11**(4), pp. 466–468, 1999.

[155] K. Akimoto, J. Kani, M. Teshima and K. Iwatsuki, 'Super-dense WDM transmission of spectrum-sliced incoherent light for wide-area access network', *J. Lightwave Technol.*, **21**(11), pp. 2715–2722, 2003.

[156] E. H. Lee, Y. C. Bang, J. K. Kang, Y. C. Keh, D. J. Shin, J. S. Lee, S. S. Park, I. Kim, J. K. Lee, Y. K. Oh and D. H. Jang, 'Uncooled C-band wide-band gain lasers with 32-channel coverage and -20–20 dBm ASE injection for WDM-PON', *IEEE Photonics Technol. Lett.*, **18**(5), pp. 667–669, 2006.

[157] J. Briand, F. Payoux, P. Chanclou and M. Joindot, 'Forward error correction in WDM PON using spectrum slicing', *Opt. Switching and Netw.*, **4**(2), pp. 131–136, 2007.

[158] D. Sadot, U. Mahlab and V. B. Natan, 'New method for developing optical code division multiplexed access sequences using genetic algorithm', *Opt. Eng.*, **38**(1), pp. 151–156, 1999.

[159] R. M. H. Yim, L. R. Chen and J. Bajcsy, 'Design and performance of 2-D codes for wavelength-time optical CDMA', *IEEE Photonics Technol. Lett.*, **14**(5), pp. 714–716, 2002.

[160] A. R. Forouzan, J. A. Salehi and M. Nasiri-Kenari, 'Frame time-hopping fiber-optic code-division multiple access using generalized optical orthogonal codes', *IEEE Trans. Commun.*, **50**(12), pp. 1971–1983, 2002.

[161] T. Miyazawa and I. Sasase, 'Enhancement of tolerance to MAIs by the synergistic effect between M-ary PAM and the chip-level receiver for optical CDMA systems', *J. Lightwave Technol.*, **24**(2), pp. 658–666, 2006.

[162] G. C. Gupta, M. Kashima, H. Iwamura, H. Tamai, T. Ushikubo and T. Kamijoh, 'Over 100 km bidirectional, multi-channels COF-PON without optical amplifier', *Proc. OFC/Fiber Optics Engineering Conf. (NFOEC)'06*, Anaheim, CA, USA, pp. 1–3, 5–10 March 2006.

[163] W. Cong, C. Yang, R. P. Scott, V. J. Hernandez, N. K. Fontaine, B. H. Kolner, J. P. Heritage and S. J. B. Yoo, 'Demonstration of 160- and 320-Gb/s SPECTS O-CDMA network testbeds', *IEEE Photonics Technol. Lett.*, **18**(15), pp. 1567–1569, 2006.

[164] A. H. Gnauck, G. Raybon, P. G. Bernasconi, J. Leuthold, C. R. Doerr and L. W. Stulz, '1-Tb/s (6 × 170.6 Gb/s) transmission over 2000-km NZDF using OTDM and RZ-DPSK format', *IEEE Photonics Technol. Lett.*, **15**(11), pp. 1618–1620, 2003.

[165] P. R. Prucna, *Optical Code Division Multiple Access: Fundamentals and Applications*, Taylor & Francis, 2005.

[166] T. Pfeiffer, J. Kissing, J.-P. Elbers, B. Deppisch, M. Witte, H. Schmuck and E. Voges, 'Coarse WDM/CDM/TDM concept for optical packet transmission in metropolitan and access networks supporting 400 channels at 2.5 Gb/s peak rate', *J. Lightwave Technol.*, **18**(12) pp. 1928–1938, 2000.

[167] H. Sotobayashi, W. Chujo and K. Kitayama, 'Photonic gateway: multiplexing format conversions of OCDM-to-WDM and WDM-to-OCDM at 40 Gb/s (4 × 10 Gb/s)', *J. Lightwave Technol.*, **20**(12), pp. 2022–2028, 2002.

[168] H. Sotobayashi, W. Chujo, and K. Kitayama, '1.6-b/s/Hz 6.4-Tb/s QPSK-OCDM/WDM (4 OCDM × 40 WDM × 40 Gb/s) transmission experiment using optical hard thresholding', *IEEE Photonics Technol. Lett.*, **14**(4), pp. 555–557, 2002.

[169] S. L. Woodward and M. R. Phillips, 'Optimizing subcarrier-multiplexed WDM transmission links', *J. Lightwave Technol.*, **22**(3), pp. 773–778, 2004.

[170] J.-M. Kang and S.-K. Han, 'A novel hybrid WDM/SCM-PON sharing wavelength for up- and down-link using reflective semiconductor optical amplifier', *IEEE Photonics Technol. Lett.*, **18**(3), pp. 502–504, 2006.

[171] A. Murakoshi, K. Tsukamoto and S. Komaki, 'High-performance RF signals transmission in SCM/WDMA radio-on-fiber bus link using optical FM method in presence of optical beat interference', *IEEE Trans. Microw. Theory Tech.*, **54**(2), pp. 967–972, 2006.

[172] R. Hui, B. Zhu, R. Huang, C. Allen, K. Demarest and D. Richards, '10-Gb/s SCM fiber system using optical SSB modulation', *IEEE Photonics Technol. Lett.*, **13**(8), pp. 896–898, 2001.

[173] Y. Y. Won, H. C. Kwon, S. K. Han, E. S. Jung and B. W. Kim, 'OBI noise reduction using gain saturated SOA in reflective SOA based WDM/SCM-PON optical links', *Electron. Lett.*, **42**(17), pp. 992–993, 2006.

[174] P. Cockrane, 'Future directions in undersea fibre optic system technology', *Proc. IOOC'89*, Kobe, Japan, paper 21B1–2, July 1989.

[175] P. S. Henry, R. A. Linke and A. H. Gnauck, 'Introduction to lightwave systems', in S. E. Miller and I. P. Kaminow (Eds), *Optical Fiber Telecommunications II*, pp. 781–831, Academic Press, 1988.

[176] C. H. Henry, 'Theory of spontaneous emission noise in optical resonators and its application to lasers and optical amplifiers', *J. Lightwave Technol.*, **LT-4**, pp. 288–297, 1986.

[177] G. Keiser, *Optical Communications Essentials (Telecommunications)*, McGraw-Hill Professional, 2003.

[178] W. A. Stallard, J. D. Cox, D. J. Malyon, A. E. Ellis and K. H. Cameron, 'Long span high capacity optical transmission system employing laser amplifier repeaters', *Proc. Int. Conf. on Integrated Optics and Fibre Communications, IOOC'89*, Kobe, Japan, paper 21B2–2, July 1989.

[179] P. Cockrane, 'Future directions in long haul fibre optic systems', *Br. Telecom Technol. J.*, **8**(2), pp. 5–17, 1990.

[180] S. Yamamoto, H. Taga, N. Edagawa, K. Mochizuki anbd H. Wakabayashi, '516 km, 2.4 Gbit/s optical fiber transmission experiment using 10 semiconductor laser amplifiers and measurement of jitter accumulation', *Proc. Int. Conf. on Integrated Optics and Fibre Communications, IOOC'89*, Kobe, Japan, paper 20PDA-9. July 1989.

[181] H. K. Kim, S. Chandrasekhar, A. Srivastava, C. A. Burrus and L. Buhl, '10 Gbit/s based WDM signal transmission over 500 km of NZDSF using semiconductor optical amplifier as the in-line amplifier', *Electron. Lett.*, **37**(3), pp. 185–187, 2001.

[182] L. Zhihong, D. Yi, M. Jinyu, W. Yixin and L. Chao, '1050-km WDM transmission of 8×10.709 Gb/s DPSK signal using cascaded in-line semiconductor optical amplifier', *IEEE Photonics Technol. Lett.*, **16**(7), pp. 1760–1762, 2004.

[183] Z. Xiang and M. Birk, 'Performance comparison of an 80-km-per-span EDFA system and a 160-km hut-skipped all-Raman system over standard single-mode fiber', *J. Lightwave Technol.*, **24**(3), pp. 1218–1225, 2006.

[184] Z. Zuqing, M. Funabashi, P. Zhong, L. Paraschis and S. J. B. Yoo, '10 000-hop cascaded in-line all-optical 3R regeneration to achieve 1 250 000-km 10-Gb/s transmission', *IEEE Photonics Technol. Lett.*, **18**(5), pp. 718–720, 2006.

[185] N. H. Taylor and A. Hadjitotiou, 'Optical amplification and its applications', *Proc. SPIE*, **1314**, pp. 64–67, 1990.

[186] M. J. O'Mahony, 'Progress in optical amplifiers', *Int. Workshop on Digital Commun*, Session 6, paper 3, Tirrenia, Italy, September 1989.

[187] T. Kotanigawa, T. Matsuda and T. Kataoka, 'Applicable wavelength range of U-band signals in in-line Raman amplifier WDM systems', *Electron. Lett.*, **39**(13), pp. 999–1000, 2003.

[188] M. Schneiders, S. Vorbeck, R. Leppla, E. Lach, M. Schmidt, S. B. Papernyi and K. Sanapi, 'Field transmission of 8×170 Gb/s over high-loss SSMF link using third-order distributed Raman amplification', *J. Lightwave Technol.*, **24**(1), pp. 175–182, 2006.

[189] B. Bakhshi, L. Richardson, E. A. Golovchenko, G. Mohs and M. Manna, 'An experimental analysis of performance fluctuations in high-capacity repeaterless WDM systems', *Proc. OFC/Fiber Optics Engineering Conf. (NFOEC)'06*, Anaheim, CA, USA, pp. 3, 5–10 March 2006.

[190] Hybrid Raman–EDFA module, http://www.amonics.com/PDFs/EDFA_Raman.pdf, 8 November 2007.

[191] D. Z. Chen, T. J. Xia, G. Wellbrock, P. Mamyshev, S. Penticost, G. Grosso, A. Puc, P. Perrier and H. Fevrier, 'New field trial distance record of 3040 km on wide reach WDM with 10 and 40 Gb/s transmission including OC-768 traffic without regeneration', *J. of Lightwave Technol.*, **25**(1), pp. 28–37, 2007.

[192] U. Tiwari, K. Rajan and K. Thyagarajan, 'Multi-channel gain and noise figure evaluation of Raman/EDFA hybrid amplifiers', *Optics Communications*, **281**(6), pp. 1593–1597, 2008.

[193] S. Melle, R. Dodd, S. Grubb, C. Liou, V. Vusirikala and D. Welch, 'Bandwidth visualization enables long-haul WDM transport of 40 Gb/s and 100 Gb/s services', *IEEE Communications Magazine*, **46**(2), pp. 522–529, 2008.

[194] M. Nishimura, 'Optical fibers and fiber dispersion compensators for high-speed optical communication', *J. Opt. Fiber Commun. Rep.*, **2**(2), pp. 115–139, 2005.

[195] E. Poutrina and G. P. Agrawal, 'Impact of dispersion fluctuations on 40-Gb/s dispersion-managed lightwave systems', *J. Lightwave Technol.*, **21**(4), pp. 990–996, 2003.

[196] L. F. Mollenauer and J. P. Gordon, *Solitons in Optical Fibers: Fundamentals and Applications*, Academic Press, 2006.

[197] G. P. Agrawal, *Lightwave Technology, Telecommunication Systems*, Wiley Interscience, 2005.

[198] E. Iannone, F. Matera, A. Mecozzi and M. Settembre, *Nonlinear Optical Communication Networks*, Wiley, 1998.

[199] J. P. Dakin, *Handbook of Optoelectronics*, CRC Press, 2006.

[200] P. J. Winzer and J. Leuthold, 'Return-to-zero modulator using a single NRZ drive signal and an optical delay interferometer', *IEEE Photonics Technol. Lett.*, **13**(12), pp. 1298–1300, 2001.

[201] M. Bass and E. W. Van Strylan, *Handbook of Optics: Volume IV*, McGraw-Hill Professional, 2000.

[202] W. Krolikowski, M. Saffman, B. Luther-Davies and C. Denz, 'Anomalous interaction of spatial solitons in photorefractive media', *Phys. Rev. Lett.*, **80**(15), pp. 3240–3243, 1998.

[203] M. F. Ferreira, M. V. Facao, S. V. Latas and M. H. Sousa, 'Optical solitons in fibers for communication systems', *J. Fiber Integr. Opt.*, **24**(3/4), pp. 287–313, 2005.

[204] F. Gérôme, K. Cook, A. George, W. Wads worth and J. Kinglit, 'Delivery of sub-100fs pulses through 8 m of bellow-core fiber using soliton compression', *Optics Express*, **15**(12), pp. 7126–7131, 2007.

[205] M. Matsumoto, A. Matsumoto and A. Hasegawa, *Optical Solitons in Fibers*, Springer-Verlag, 2002.

[206] R. Rajiv and S. Kumar, *Optical Networks: A Practical Perspective* (2nd edn), Morgan Kaufmann, 2002.

[207] L.-S. Yan, S. M. R. M. Nezam, A. B. Sahin, J. E. McGeehan, T. Luo, Q. Yu and A. E. Willner, 'Performance optimization of RZ data format in WDM systems using tunable pulse-width management at the transmitter', *J. Lightwave Technol.*, **23**(3), pp. 1063–1067, 2005.

[208] A. Hasegawa, *Massive WDM and TDM Soliton Transmission Systems*, Springer-Verlag, 2002.

[209] M. Nakazawa, K. Suzuki and H. Kubota, 'Single-channel 80 Gbit/s soliton transmission over 10 000 km using in-line synchronous modulation', *Electron. Lett.*, **35**(2), pp. 162–164, 1999.

[210] R. Ludwig, U. Feiste, S. Diet, C. Schubert, C. Schmidt, H. J. Ehrke and H. G. Weber, 'Unrepeatered 160 Gbit/s RZ single-channel transmission over 160 km of standard fibre at 1.55 μm with hybrid MZI optical demultiplexer', *Electron. Lett.*, **36**(16), pp. 1405–1406, 2000.

[211] L. Yan, X. S. Yao, M. C. Hauer and A. E. Willner, 'Practical solutions to polarization-mode-dispersion emulation and compensation', *J. Lightwave Technol.*, **24**(11), pp. 3992–4005, 2006.

[212] L. A. Coldren and B. J. Thibeault, 'Vertical cavity surface emitting lasers', in I. P. Kaminow and T. L. Koch (Eds), *Optical Fiber Telecommunications IIIB*, pp. 200–266, Academic Press, 1997.

[213] K. Suzuki, H. Kubota and M. Nakazawa, 'Tb/s (40 Gb/s × 25 channel) DWDM quasi-DM soliton transmission over 1500 km using dispersion-managed single-mode fiber and conventional C-band EDFAs', *Proc. Conf. on Optical Fiber Communications, OFC'01*, Anaheim, CA, USA, vol. 2, pp. TuN7-1–TuN7-3, March 2001.

[214] G. E. Tudury, R. Salem, G. M. Carter and T. E. Murphy, 'Transmission of 80 Gbit/s over 840 km in standard fibre without polarisation control', *Electron. Lett.*, **41**(25), pp. 1394–1396, 2005.

[215] T. Otani, M. Hayashi, M. Daikoku, K. Ogaki, Y. Nagao, K. Nishijima and M. Suzuki, 'Field trial of 63 channels 40 Gbit/s dispersion-managed soliton WDM signal transmission over 320 km NZ-DSFs', *Proc. Conf. on Optical Fiber Communications, OFC'02*, Anaheim, CA, USA, pp. FC9-1–FC9-3, March, 2002.

[216] A. Labruyre, P. Tchofo Dinda and K. Nakkeeran, 'Transmission performance of $N \times$ 160 Gbit/s densely dispersion-managed optical fibre systems', *J. Mod. Opt.*, **52**(18), pp. 2731–2742, 2005.

[217] T. Masashi, N. Hirokazu, K. Suzuki, S. Kazunori, K. Hirokazu and N. Masataka, '640 Gbit/s (40 Gbit/s $\times$ 16 channel) dispersion-managed DWDM soliton transmission over 1000 km with spectral efficiency of 0.4 bit/s/Hz', *Tech. Rep. IEICE, OCS*, **1**(131), pp. 19–24, 2006.

[218] X. Liu, D. M. Gill and S. Chandrasekhar, 'Optical technologies and techniques for high bit rate fiber transmission', *Bell Labs Tech. J.*, **11**(2), pp. 83–104, 2006.

CHAPTER 13

Optical fiber systems 2: coherent and phase-modulated

13.1 Introduction

The direct detection of an intensity-modulated optical carrier is basically a photon counting process where each detected photon is converted into an electron–hole pair (or, in the case of the APD, a number of pairs due to avalanche gain). It was indicated in Section 7.5 that this process which ignores the phase and polarization of the electromagnetic carrier may be readily implemented with currently available optical components. Thus all the preceding discussion in Chapter 12 involving both digital and analog systems concerned only an intensity-modulated optical carrier being transmitted to a direct detection optical receiver or IM/DD optical fiber systems (see Section 12.1).

Conventional direct detection receivers, however, are generally limited by noise generated in the detector and preamplifier (see Chapter 9) except at very high signal-to-noise ratios (SNRs). The sensitivity of such square-law detection systems is therefore reduced below the fundamental quantum noise limit by at least 10 to 20 dB. This is particularly the case at longer wavelengths (i.e. 1.3 to 1.6 µm) and at higher transmission rates since the electronic preamplifier usually has a rising input optical power with frequency requirement* (see Eq. (12.37)). For a good APD receiver operating in the wavelength range 1.3 to 1.6 µm this corresponds to between 700 and 1000 photons per bit required to maintain a bit-error-rate (BER) of 10^{-9} [Ref. 1].

Improvements in receiver sensitivity, together with wavelength selectivity, may be obtained using the well-known coherent detection techniques (i.e. heterodyne and homodyne detection) for the optical signal [Ref. 2]. Unlike direct detection in which the optical signal is converted directly into a demodulated electrical output, such coherent optical receivers first add to the incoming optical signal from a locally generated optical wave prior to detecting the sum.[†] The resulting photocurrent is a replica of the original signal which is translated down in frequency from the optical domain (around 10^5 GHz) to the radio domain (up to several GHz) and where conventional electronic techniques can be used for further signal processing and demodulation. Hence an ideal coherent receiver operating in the 1.3 to 1.6 µm wavelength region requires a signal energy of only 10 to 20 photons per bit to achieve a BER of 10^{-9}. Hence coherent detection potentially provides a substantial benefit for high-speed systems operating at longer wavelengths.

A possible improvement in receiver sensitivity using conventional coherent detection of up to 20 dB can be obtained over direct detection [Ref. 4]. Furthermore, such enhanced receiver sensitivity could translate into increases in repeater spacings of the order of 100 km or more when using low-loss fiber at a wavelength of 1.55 µm. Hence, the improved sensitivity of 5 to 20 dB which results from the photomixing gain in the coherent receiver could provide:

(a) increased repeater spacings for both inland and undersea transmission systems;

(b) higher transmission rates over existing routes without reducing repeater spacings;

(c) increased power budgets to compensate for losses associated with couplers and optical multiplexer/demultiplexer devices (see Section 5.6) in distribution networks;

(d) improved sensitivity to optical test equipment such as optical time domain reflectometers (see Section 14.10.1).

---

* This requirement corresponds to a rising noise against frequency characteristic.
† Common usage in optical fiber communications is that the term 'coherent' refers to any system or technique which employs nonlinear mixing between two optical waves. Typically, one of these is the information-carrying signal and the other a locally generated signal (by a local oscillator or, alternatively, from the incoming signal itself which is termed self-coherent [Ref. 3]) at the receiver. The result of this process is a new signal (for heterodyne detection, the intermediate frequency) which appears at a microwave frequency given by the difference between the frequencies of the incoming signal and the local oscillator.

Although possible increases in the transmission distance between repeaters created the initial impetus for the pursuit of coherent transmission within optical fiber communications, by 1990, before the widespread deployment of optical amplifiers, it was also perceived that coherent techniques would enable a further massive step to be taken in the exploitation of the transmission capacity of optical fiber systems. This improvement would be facilitated by the inherent wavelength selectivity afforded by the coherent receiver allowing efficient access to the vast optical bandwidth available in single-mode fibers. For example, it was suggested for the low-loss window between 1.3 and 1.6 µm being over 50 000 GHz that coherent transmission would permit wavelength division multiplexing of huge channel numbers with channel spacings of only a few hundred megahertz [Ref. 5]. The successful introduction into the optical telecommunication network of erbium-doped fiber amplifiers in the early 1990s, however, caused the pursuit of coherent transmission to be virtually discontinued as both improved direct detection receiver sensitivity through amplification and hence increased transmission distance together with dense wavelength division multiplexing with channel spacings of 50 GHz (0.4 nm) or even 12.5 GHz (0.1 nm) could be facilitated (see Section 12.9.4).

The modulation formats that may be employed within coherent optical fiber communications are essentially the same as those used in coherent electrical line and radio communications. Modulation formats of this type were discussed in Sections 12.7.3 to 12.7.6 in a slightly different context, namely the generation of subcarriers for electrical frequency division multiplexing prior to intensity modulation of the optical source. In these cases direct detection of the optical signal is carried out at the receiver with subsequent electrical demodulation for the subcarriers. Such systems only provide improvements in the SNR over baseband IM/DD systems at the expense of a substantial bandwidth penalty. When a narrow-linewidth injection laser (less than 1 MHz) is used in an optical fiber communication system, however, it is possible to directly modulate the coherent optical carrier in amplitude (direct AM), frequency (direct FM) and phase (direct PM) prior to demodulation using a coherent optical receiver. In the case of digital transmission this implies amplitude, frequency or phase shift keying (i.e. ASK, FSK or PSK) modulation techniques [Ref. 2].

Direct modulation for coherent optical fiber transmission is provided using an external modulator which is fed by a semiconductor laser source operating in continuous wave mode. In electrical digital communications, coherent demodulation requires recovery of the carrier frequency signal through either heterodyne or homodyne detection [Ref. 2]. Coherent optical communications, however, has had a wider usage of the terminology. An optical fiber communication system was referred to as coherent when there was optical signal mixing even though carrier recovery may not occur. Hence an optical coherent system could employ a demodulator that did not perform carrier recovery but used noncoherent or envelope detection. In this context differential phase shift keying (DPSK) is considered a noncoherent electrical modulation technique but it has generally been referred to as a coherent optical communication modulation format. Moreover, a coherent optical receiver is referred to as synchronous or asynchronous (i.e. nonsynchronous) depending upon whether it operates with or without phase tracking, respectively. The latter asynchronous receiver normally employs power or envelope detection.

Although, as indicated above, research in coherent optical fiber communications effectively ceased in the early 1990s as a result of the successful introduction of fiber amplifiers, DPSK using, in particular, asynchronous detection started to receive renewed interest following specific experimental demonstrations in 2002 [Refs 6, 7]. In the context of the renewed focus these high-performance, long-haul systems with return-to-zero DPSK transmission, where a phase-to-intensity conversion takes place at the receiver prior to a direct detection process, have subsequently been widely demonstrated. As optical mixing with an independent optical signal does not necessarily take place at the receiver to achieve carrier recovery, such systems tend to be referred to as being optically phase-modulated [Ref. 8] or phase shift keyed [Ref. 9], or when appropriate, as self-coherent [Ref. 3]. The term coherent is then often utilized for the case of coherent detection in which the phase reference at the receiver is typically provided by a local oscillator laser that beats with the received optical signal to produce constructive and destructive interference (see Section 13.2). There is an increasing adoption of this stricter approach to the use of the terminology and therefore the chapter title reflects this trend by incorporating both the 'coherent' and 'phase-modulated' system terms in order to encompass systems using advanced modulation formats with direct detection, and also to emphasize the growth of phase modulation as the currently preferred advanced optical modulation format in competition with intensity modulation [Ref. 10].

While prior to the 1990s a major factor in the pursuit of coherent and phase-modulated optical fiber systems had been the perceived improvement in receiver sensitivities to enable transmission over longer distances, a main focus of the renewed interest since 2002 has been to increase spectral efficiency. Moreover, direct detection phase-modulated optical fiber systems do also provide for a sensitivity gain over IM/DD systems, albeit at more modest levels than when using synchronous detection. In particular, it should be noted that the theoretical sensitivity gains which can be determined for both PSK and DPSK over IM/DD are substantially reduced when optical amplifiers are deployed within the latter systems [Ref. 8].

The discussion is continued in Section 13.2 through a brief historical review of the development of coherent and phase-modulated optical transmission prior to the description of the conventional coherent optical fiber communication system, together with its important features. This leads into the consideration of the fundamental detection principles associated with the coherent optical receiver (i.e. heterodyne and homodyne detection) in Section 13.3. There are, however, a number of practical constraints which have in the past inhibited the development of coherent optical fiber systems, and they create certain limitations on the choice of system components. These issues are therefore dealt with in Section 13.4. This is followed in Section 13.5 by discussion of the various modulation formats that may be employed for coherent optical transmission with an emphasis on PSK prior to the description of the numerous demodulation schemes which have been applied within the detection process in Section 13.6. Recent developments concerned with DPSK as a modulation technique of increasing importance are then outlined in some detail in Section 13.7 with a specific focus on the return-to-zero signal format and direct detection at the receiver. A comparison of the various major modulation and detection techniques in relation to receiver sensitivity is then provided in Section 13.8. Finally, in Section 13.9 advanced multicarrier transmission strategies and systems are described including both polarization multiplexing and polarization interleaving together with discussion of experimental coherent quadrature phase shift keyed and orthogonal frequency division multiplexed systems in order to provide an indication of current research directions in the field.

13.2 Basic coherent system

Since the invention of the laser in 1960, research efforts have focused on techniques by which the coherent properties of laser light could be utilized for coherent optical communications. Improved SNRs over direct detection of an intensity modulated signal were demonstrated in free space optical communication systems using gas lasers in the late 1960s [Ref. 11]. In addition, the concept of optical frequency division multiplexing using coherent detection schemes was proposed in 1970 [Ref. 12]. However, it was only in the latter half of the 1970s, when single-mode transmission from a narrow-linewidth AlGaAs semiconductor laser was demonstrated [Ref. 13], that the proposals for coherent optical fiber transmission began taking shape. Nevertheless, it was appreciated that the polarization stability of the transmission medium was crucial for successful coherent detection. Ideally, for coherent transmission the fiber would be required to maintain a single linear polarization state throughout its length. This factor, in part, led to the investigations on polarization-maintaining (PM) fibers in the early 1980s (see Section 3.13.3). Moreover, for a brief period the use of PM fibers was a favored potential approach to the problem [Ref. 14]. For example, the use of optical adaptors (e.g. birefringent plates) at the transmit and receive terminals was suggested in order to allow only a single polarization state of the fundamental mode to be launched into, and received from, the PM fiber [Ref. 15].

It was clear, however, that if conventional circularly symmetric single-mode fiber, which did not maintain a single polarization state over its length (see Section 13.3.1), were to be employed, then some form of polarization matching of the incoming optical signal with the locally generated optical signal would be necessary. Although the first successful demonstration of optical frequency shift keyed heterodyne detection using a semiconductor laser source and local oscillator was reported in 1989 [Ref. 16], a period elapsed before the polarization stability measurements on installed conventional single-mode fiber indicated the real possibility of its use within coherent optical transmission systems [Ref. 17].

It was indicated in Section 13.1 that worldwide research activities in coherent optical communications had largely ceased by the early 1990s as a consequence of the improved performance facilitated by optically amplified IM/DD systems. This loss of interest in coherent optical transmission lasted for a decade until the renewal of research activity following the reporting of two interferometric direct detection DPSK experimental system demonstrations in 2002 [Refs 6, 7]. Although a significant range of coherent optical communication systems research and development has therefore been re-established, the major focus in this more recent period has been associated with phase-modulated systems [Ref. 8].

A block schematic of a generalized coherent optical fiber communication system is illustrated in Figure 13.1. The dashed lines enclose the main elements which distinguish the coherent system from its IM/DD equivalent. At the transmitter a CW narrow-linewidth semiconductor laser is shown which acts as an optical frequency oscillator. An external optical modulator usually provides amplitude, frequency or phase shift keying of the optical carrier by the information signal. At present external modulators are generally waveguide devices fabricated from lithium niobate or the group III–V compound semiconductors (see Section 11.4.2). Internal modulation of the injection laser drive current may, however, also be utilized to produce either ASK or FSK [Refs 18, 19].

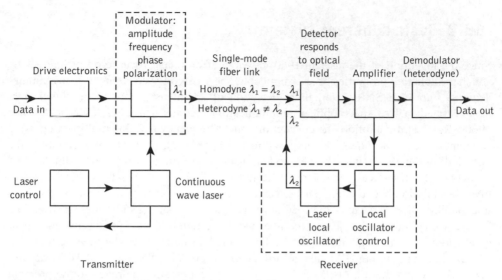

Figure 13.1 The generalized coherent optical fiber system

Modulated carrier waveforms for the three standard modulation techniques with binary data are illustrated in Figure 13.2. It may be observed from Figure 13.2(a) why binary ASK is often referred to as on–off keying (OOK). Figure 13.2(b) shows FSK in which the binary 1 is transmitted at a higher optical frequency than the binary 0 bit. The 180° phase shift between the binary 1 and 0 bits displayed in Figure 13.2(c) depicts PSK. Furthermore,

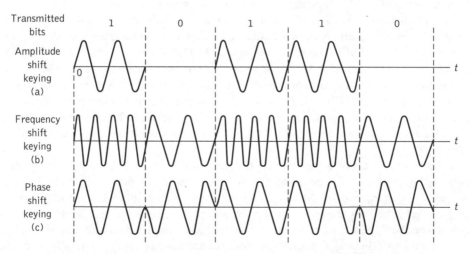

Figure 13.2 Modulated carrier waveforms used for binary data transmission:
(a) amplitude shift keying (ASK) or on-off keying (OOK); (b) frequency shift keying
(FSK); (c) phase shift keying (PSK)

it should be noted that whereas with ASK the amplitude of the carrier waveform is effectively switched on and off, the amplitude of the optical carrier remains constant in the other two modulation schemes shown in Figure 13.2. Variants on these standard modulation techniques exist, such as continuous phase FSK and differential PSK, have also been applied in experimental coherent optical fiber transmission systems. Moreover, an alternative digital scheme based on the modulation of the polarization properties of the optical signal has more recently come under investigation. This strategy, known as polarization shift keying (PolSK), is discussed further along with the other modulation formats in Section 13.5.4.

Referring now to the receiver shown in Figure 13.1, the incoming signal is combined (or mixed) with the optical output from a semiconductor laser local oscillator. This function can be provided by a single-mode fiber fused biconical coupler (see Section 5.6.1), a device which gives excellent wavefront matching of the two optical signals. However, integrated optical waveguide couplers (see Section 11.4.1) may also be utilized. The combined signal is then fed to a photodetector for direct detection in the conventional square-law device. Nevertheless, to permit satisfactory optical coherent detection, the optical coupler device must combine the polarized optical information-bearing signal field with the similarly polarized local oscillator signal field in the most efficient manner.

When the optical frequencies (or wavelengths) of the incoming signal and the local oscillator laser output are identical, then the receiver operates in a homodyne mode and the electrical signal is recovered directly in the baseband. For heterodyne detection, however, the local oscillator frequency is offset from the incoming signal frequency and therefore the electrical spectrum from the output of the detector is centered on an intermediate frequency (IF) which is dependent on the offset and is chosen according to the information transmission rate and the modulation characteristics. This IF, which is a difference signal (or difference frequency), contains the information signal and can be demodulated using standard electrical techniques [Ref. 2].

The electrical demodulator block shown in Figure 13.1 is required in particular for an optical heterodyne detection system which can utilize either synchronous or asynchronous/ nonsynchronous electrical detection. Synchronous or coherent* demodulation implies an estimation of phase of the IF signal in transferring it to the baseband. Such an approach requires the use of phase-locking techniques in order to follow phase fluctuations in the incoming and local oscillator signals. Alternatively, asynchronous or noncoherent (envelope) IF demodulation schemes may be employed which are less demanding but generally produce a lower, performance than synchronous detection techniques [Ref. 4]. Optical homodyne detection is by definition, however, a synchronous demodulation scheme and as the detected signal is brought directly into the baseband, then optical phase estimation is required. These issues are discussed further in Section 13.6.

---

* It is a little confusing but reference is made in the literature to (a) heterodyne, coherent detection and (b) to heterodyne, noncoherent or incoherent detection, both of which are coherent optical detection schemes. The former terminology means an optical heterodyne receiver using a synchronous electrical demodulation technique, whereas the latter corresponds to a heterodyne receiver with an asynchronous demodulation scheme.

13.3 Coherent detection principles

A simple coherent receiver model for ASK is displayed in Figure 13.3. The low-level incoming signal field e_S is combined with a second much larger signal field e_L derived from the local oscillator laser. It is assumed that the electromagnetic fields obtained from the two lasers (i.e. the incoming signal and local oscillator devices) can be represented by cosine functions and that the angle $\phi = \phi_S - \phi_L$ represents the phase relationship between the incoming signal phase ϕ_S and the local oscillator signal phase ϕ_L defined at some arbitrary point in time. Hence, as depicted in Figure 13.3, the two fields may be written as [Ref. 20]:

$$e_S = E_S \cos(\omega_S t + \phi) \tag{13.1}$$

and:

$$e_L = E_L \cos(\omega_L t) \tag{13.2}$$

where E_S is the peak incoming signal field and ω_S is its angular frequency, and E_L is the peak local oscillator field and ω_L is its angular frequency. The angle $\phi(t)$ representing the phase relationship between the two fields contains the transmitted information in the case of FSK or PSK. However, with ASK $\phi(t)$ is constant and hence it is simply written as ϕ in Eq. (13.1), the information being contained in the variation of E_S for ASK as may be observed in Figure 13.2(a).

For heterodyne detection, the local oscillator frequency ω_L is offset from the incoming signal frequency ω_S by an intermediate frequency such that:

$$\omega_S = \omega_L + \omega_{IF} \tag{13.3}$$

where ω_{IF} is the angular frequency of the IF. As mentioned in Section 13.1, the IF is usually in the radio-frequency region and may be a few tens or hundreds of megahertz. By contrast, within homodyne detection there is no offset between ω_S and ω_L and hence $\omega_{IF} = 0$. In this case the combined signal is therefore recovered in the baseband.

The two wavefronts from the incoming signal and the local oscillator laser must be perfectly matched at the surface of the photodetector for ideal coherent detection. This factor

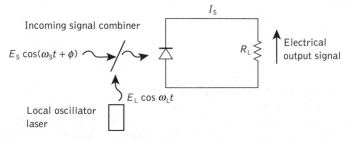

Figure 13.3 Basic coherent receiver model

creates the normal requirement for polarization control of the incoming optical signal which is discussed further in Section 13.4.

In the case of both heterodyne and homodyne detection the optical detector produces a signal photocurrent I_p which is proportional to the optical intensity (i.e. the square of the total field for the square-law photodetection process) so that:

$$I_p \propto (e_S + e_L)^2 \tag{13.4}$$

Substitution in the expression (13.4) from Eqs (13.1) and (13.2) gives:

$$I_p \propto [E_S (\cos \omega_S t + \phi) + E_L \cos \omega_L t]^2 \tag{13.5}$$

Assuming perfect optical mixing expansion of the right hand side of the expression shown in Eq. (13.5) gives:

$$\begin{aligned} &[E_S^2 \cos^2(\omega_S t + \phi) + E_L^2 \cos^2 \omega_L t + 2E_S E_L \cos(\omega_S t + \phi)\cos \omega_L t] \\ &= [\tfrac{1}{2}E_S^2 + \tfrac{1}{2}E_S \cos(2\omega_S t + \phi) + \tfrac{1}{2}E_L^2 + \tfrac{1}{2}\cos 2\omega_L t \\ &\quad + E_S E_L(\cos \omega_S t + \phi - \omega_L t) + E_S E_L \cos(\omega_S t + \phi + \omega_L t)] \end{aligned}$$

Removing the higher frequency terms oscillating near the frequencies of $2\omega_S$ and $2\omega_L$ which are beyond the response of the detector and therefore do not appear in its output, we have:

$$I_p \propto \tfrac{1}{2}E_S^2 + \tfrac{1}{2}E_L^2 + 2E_S E_L \cos(\omega_S t - \omega_L t + \phi) \tag{13.6}$$

Then recalling that the optical power contained within a signal is proportional to the square of its electrical field strength, expression (13.6) may be written as:

$$I_p \propto P_S + P_L + 2\sqrt{P_S P_L} \cos(\omega_S t - \omega_L t + \phi) \tag{13.7}$$

where P_S and P_L are the optical powers in the incoming signal and local oscillator signal respectively.

Furthermore, a relationship was obtained between the output photocurrent from an optical detector and the incident optical power P_o in Eq. (8.8) of the form $I_p = \eta e P_o / hf$. Hence the expression in (13.7) becomes:

$$I_p = \frac{\eta e}{hf} [P_S + P_L + 2\sqrt{P_S P_L} \cos(\omega_S t - \omega_L t + \phi)] \tag{13.8}$$

where η is the quantum efficiency of the photodetector, e is the charge on an electron, h is Planck's constant and f is optical frequency. When the local oscillator signal is much larger than the incoming signal, then the third a.c. term in Eq. (13.8) may be distinguished from the first two d.c. terms and I_p can be replaced by the approximation I_S where [Ref. 4]:

$$I_S = \frac{\eta e}{hf} [2\sqrt{P_S P_L} \cos(\omega_S t - \omega_L t + \phi)] \tag{13.9}$$

Equation (13.9) allows the two coherent detection strategies to be considered. For heterodyne detection $\omega_S \neq \omega_L$ and substituting from Eq. (13.3) gives:

$$I_S = \frac{2\eta e}{hf} \sqrt{P_S P_L} \cos(\omega_{IF} t + \phi) \tag{13.10}$$

indicating that the output from the photodetector is centered on an IF. This IF is stabilized by incorporating the local oscillator laser in a frequency control loop. Temperature stability for the signal and local oscillator lasers is also a factor which must be considered (see Example 13.1). The stabilized IF current is usually separated from the direct current by filtering prior to electrical amplification and demodulation.

For the special case of homodyne detection, however, $\omega_S = \omega_L$ and therefore Eq. (13.9) reduces to:

$$I_S = \frac{2\eta e}{hf} \sqrt{P_S P_L} \cos \phi \tag{13.11}$$

or:

$$I_S = 2R \sqrt{P_S P_L} \cos \phi \tag{13.12}$$

where R is the responsivity of the optical detector. In this case the output from the photodiode is in the baseband and the local oscillator laser needs to be phase locked to the incoming optical signal.

Example 13.1

A semiconductor laser used to provide the local oscillator signal in an ASK optical heterodyne receiver exhibits an output frequency change of $19\,\text{GHz}\,°\text{C}^{-1}$. If the receiver has a nominal IF of 1.5 GHz and assuming that there is no other form of laser frequency control, estimate the maximum temperature change that could be allowed for the local oscillator laser in order that satisfactory detection could take place.

Solution: Initially it is necessary to estimate the maximum frequency excursion allowed for the IF signal such that detection can still be facilitated. This must be no greater than 10% of the frequency of the IF. Hence the maximum allowed frequency change to the local oscillator laser output is around 150 MHz.

The maximum temperature change allowed for the local oscillator laser is therefore:

$$\text{Max. temp. change} = \frac{150 \times 10^6}{19 \times 10^9}$$

$$\simeq 8 \times 10^{-3}\,°\text{C}\ (0.008\,°\text{C})$$

Very small temperature changes can therefore adversely affect the detection process if the IF is not otherwise stabilized.

It may be observed from the expressions given in Eqs (13.10) and (13.11) that the signal photocurrent is proportional to $\sqrt{P_S}$, rather than P_S as in the case of direct detection (Eq. (8.8)). Moreover, the signal photocurrent is effectively amplified by a factor $\sqrt{P_L}$ proportional to the local oscillator field. This local oscillator gain factor has the effect of increasing the optical signal level without affecting the receiver preamplifier thermal noise or the photodetector dark current noise (see Sections 9.2 and 9.3); hence the reason why coherent detection provides improved receiver sensitivities over direct detection.

The requirement for coherence between the incoming and local oscillator signals in order to obtain coherent detection was mentioned in Section 13.2 and is discussed further in Section 13.4.2. Hence for successful mixing to occur, some correlation must exist between the two signals shown in Figure 13.3. Care must therefore be taken to ensure that this is the case when two separate laser sources are employed to provide the signal and local oscillator beams. It may be noted that this problem is reduced when a single laser source is used with an appropriate path length difference as, for example, when taking measurements by interferometric techniques (see Section 14.4.1).

When the local oscillator signal power is much greater than the incoming signal power then the dominant noise source in coherent detection schemes becomes the local oscillator quantum noise. In this limit the quantum noise may be expressed as shot noise following Eq. (9.8) where the mean square shot noise current from the local oscillator is given by:

$$\overline{i^2_{SL}} = 2eBI_{pL} \tag{13.13}$$

Substituting for I_{pL} from Eq. (8.8), where the photocurrent generated by the local oscillator signal is assumed to be by far the major contribution to the photocurrent, gives:

$$\overline{i^2_{SL}} = \frac{2e^2\eta P_L B}{hf} \tag{13.14}$$

The detected signal power S, being the square of the average signal photocurrent,* is given by Eq. (13.9) as:

$$S = \left(\frac{\eta e}{hf}\right)^2 P_S P_L \tag{13.15}$$

Hence the SNR for the ideal heterodyne detection receiver when the local oscillator power is large (ignoring the electronic preamplifier thermal noise and photodetector dark current noise terms) may be obtained from Eqs (13.14) and (13.15) as:

$$\left(\frac{S}{N}\right)_{\text{het-lim}} = \left(\frac{\eta e}{hf}\right)^2 P_S P_L \bigg/ \frac{2e^2\eta P_L B}{hf} = \frac{\eta P_S}{hf\,2B} = \frac{\eta P_S}{hf\,B_{IF}} \tag{13.16}$$

Equation (13.16) provides the so-called shot noise limit for optical heterodyne detection in which the IF amplifier bandwidth B_{IF} is assumed to be equal to $2B^\dagger$ (i.e. $B_{IF} = 2B$)

---

* It is implicit from Eq. (13.9) that the photodetector is a unity gain device (e.g. p–i–n photodiode) and not an APD. In the latter case the effect of the multiplication factor M on both the signal and noise powers must be taken into account (see Sections 9.3.3 and 9.3.4).
† This constitutes the minimum bandwidth requirement for optical heterodyne detection.

[Ref. 21]. It is also interesting to note that this heterodyne shot noise limit corresponds to the quantum noise limit for analog direct detection derived in Eq. (9.11). However, it is clear that optical heterodyne detection allows a much closer approach to this limit than does direct detection.

The shot noise SNR limit for optical homodyne detection can be deduced from Eq. (13.16) by reducing the receiver bandwidth requirement from B_{IF} to B as the output signal from the photodetector appears in the baseband when using the homodyne scheme. Hence the SNR limit for optical homodyne detection is:

$$\left(\frac{S}{N}\right)_{\text{hom-lim}} = \frac{\eta P_{s}}{hfB} \tag{13.17}$$

It should be remembered that the expressions given in Eqs (13.16) and (13.17) are based on simple on–off keying (i.e. OOK) and have effectively been derived in terms of carrier-to-noise ratio. Nevertheless, they display the potential 3 dB improvement in SNR when using optical homodyne detection over heterodyne detection. The improvement occurs as a direct result of the reduction in the receiver bandwidth provided by the former technique. Therefore, homodyne detection displays the twin advantages over heterodyne detection of increased sensitivity coupled with a reduced receiver bandwidth requirement. The latter factor implies that a higher maximum transmission rate should be facilitated by coherent optical fiber systems employing homodyne detection as they will be less restricted by the speed of response of the photodetector.

Example 13.2

The incoming signal power to an optical homodyne receiver operating at a wavelength of 1.54 μm, and at its shot noise limit, is −55.45 dBm. When the photodetector in the receiver has a quantum efficiency of 86% at this wavelength and the received SNR is 12 dB, determine the operating bandwidth of the receiver.

Solution: The incoming signal power P_{S} is given by:

$$-85.45 = 10 \log_{10} P_{S}$$

Hence:

$$P_{S} = 10^{0.455} \times 10^{-9} = 2.851 \text{ nW}$$

The operating bandwidth B of the homodyne receiver may be obtained from Eq. (12.17) as:

$$B = \frac{\eta P_{S}}{hf}\left(\frac{S}{N}\right)_{\text{hom-lim}}^{-1} = \frac{\eta P_{S}\lambda}{hf}\left(\frac{S}{N}\right)_{\text{hom-lim}}^{-1} = \frac{0.86 \times 2.851 \times 10^{-9} \times 1.54 \times 10^{-6} \times 10^{-1.2}}{6.626 \times 10^{-34} \times 2.998 \times 10^{8}}$$

$$= 1.2 \text{ GHz}$$

The above analysis of SNR for optical heterodyne and homodyne detection applies only to ASK and is not appropriate for FSK and PSK. The final SNR at the signal decision point is therefore dependent on the modulation scheme utilized, and in the case of heterodyne detection on the type of IF demodulator and baseband filter employed. More detailed considerations of the SNR for different modulation schemes are dealt with in Section 13.8.

13.4 Practical constraints

It was indicated in Section 13.1 that various practical constraints had inhibited the development of coherent optical fiber communications. These constraints are largely derived from factors associated with the elements of the coherent optical fiber communication system shown in Figure 13.1, and they are exacerbated by the stringent demands of coherent transmission. Substantial developments, however, in the component technology associated with optical fiber communications have allowed the initial difficulties experienced with coherent optical fiber transmission to be largely overcome. Nevertheless, practical constraints still exist and they still dictate the performance characteristics required from components and devices which are to be utilized in coherent optical fiber systems. It is therefore important to consider the major constraints and their effect on the choice of system elements. We start by discussing the aspects which determine specific requirements for the achievement of coherent optical transmission at both the transmit and receive terminals prior to outlining certain limitations of the fiber transmission medium which may affect the performance of future coherent optical communication systems.

13.4.1 Injection laser linewidth

Coherent optical transmission is severely degraded by the phase noise associated with both the transmitter and local oscillator lasers. A crucial parameter that determines both the level of phase noise and the long-term phase stability is determined by the laser linewidth with reduced phase noise being obtained using narrow-linewidth devices. Laser linewidth reduction therefore improves the spectral purity of the device output and thus reduces its noise content. Hence single-mode laser linewidths less than 1 MHz are preferred to avoid unnecessary receiver sensitivity degradation in coherent optical communications. In addition, another major reason for the use of narrow-linewidth lasers within the coherent detection process is the phase-locking requirement (for synchronous detection) as well as the minimum frequency-locking requirement for asynchronous detection.

Moreover, injection laser phase or frequency noise (see Section 6.7.4) can affect the coherent system performance as it is the principal cause of linewidth broadening in such devices. Randomly occurring spontaneous emission events, which are an inevitable aspect of injection laser operation, lead to sudden shifts (of random magnitude and sign) in the phase of the electromagnetic field generated by the laser causing the broadening effect. Hence, phase noise together with other linewidth-broadening factors [Ref. 22] must be minimized in devices which are to be employed for coherent optical transmission. Nevertheless, phase-locking techniques may be employed within the coherent receiver (see

Section 13.6.1). It was not until the latter half of the 1970s that semiconductor device technology evolved to a point where injection lasers, with good reproducibility which could operate in a single longitudinal mode, could be fabricated. However, the spectral linewidths associated with the most sophisticated of these devices in the 1980s such as the distributed feedback laser (see Section 6.6.2) were of the order of 5 to 50 MHz which was too broad for most of the coherent techniques [Ref. 19].

Several approaches to the solution of this laser linewidth problem subsequently evolved. They included the narrowing of injection laser linewidths through the use of an external resonator cavity in the long external cavity (LEC) laser (see Section 6.10.1), together with the deployment of integrated external cavity lasers in the form of advanced DFB and DBR structures (see Section 6.10.2). Narrow-linewidth devices have also been obtained employing the multiquantum-well design (see Section 6.5.3) within a single section of a DFB laser providing values around 100 kHz (see Section 6.10.2). LEC lasers are also capable of providing linewidths in the range of 10 to 100 kHz depending on their cavity lengths [Refs 18, 23]. Furthermore, a coherent transmitter has been recently demonstrated which utilizes a frequency-stabilized fiber ring laser (Section 6.10.3) exhibiting a linewidth of just 4 kHz [Ref. 24].

Overall, the laser linewidth requirements are critically dependent on the modulation format employed (i.e. ASK, FSK or PSK), the coherent detection mechanism (i.e. heterodyne or homodyne) and the electrical demodulation technique (i.e. synchronous or asynchronous or other). Although these issues are discussed in greater detail in Section 13.6, the linewidth tolerance is considerably wider for heterodyne receivers, particularly when employing asynchronous ASK or FSK demodulation as these schemes use envelope detection and are not reliant on the phase information. Indeed, using the same modulation format, improved sensitivity is obtained at the cost of more stringent linewidth requirements while it is indicated that the laser linewidth is no longer a bottleneck problem for most coherent techniques as optical communication bit rates increase within the multigigabit per second region [Ref. 25]. Nevertheless it should be noted that the more sensitive coherent transmission techniques (e.g. PSK homodyne detection, see Section 13.8.4) are most affected by phase noise and hence are subject to the narrowest laser linewidth requirements. Finally, an alternative approach which overcomes the phase noise problem is to use a specially configured reception technique called phase diversity reception (see Section 13.6.15). Such receivers use two or more optical detectors whose outputs are combined to produce a signal that is independent of the phase difference and hence the phase noise.

Another important factor concerning the favored narrow-linewidth injection lasers for coherent optical transmission is their inherent tunability. This aspect, which is discussed in more detail in Sections 6.10.1 and 6.10.2, provides the ability to tune the frequency of the local oscillator laser to that of the incoming optical signal for homodyne detection, or alternatively to tune the appropriate frequency difference to maintain the correct IF signal for heterodyne detection.

13.4.2 State of polarization

To enable either heterodyne or homodyne detection the polarization states of the incoming optical signal and the local oscillator laser output must be well matched in order to provide efficient mixing of the two signals within the coupling element shown in Figure 13.3. Conventional circularly symmetric single-mode fiber allows two orthogonally polarized fundamental modes to propagate. Within a perfectly formed fiber both modes would travel together, but in practice the fiber contains random manufacturing irregularities which produce geometric and strain-related anisotropic effects. This results in a progressive spatial separation between the two polarization modes as they propagate along the fiber, an effect which is usually referred to as fiber birefringence (see Section 3.13.1). Hence at any particular point along the fiber the state of polarization (SOP) can be linear, elliptical or circular. Several countermeasures have been investigated to overcome this fluctuation in the SOP with coherent transmission. They are:

(a) the use of polarization-maintaining (PM) single-mode fiber;

(b) the use of an SOP control device at the coherent optical receiver;

(c) the use of a polarization diversity receiver, or a polarization scrambling transmitter.

As mentioned in Section 13.2, early studies focused on the polarization stability of the transmission medium which was considered sufficiently important to necessitate the use of specially fabricated PM fiber (see Section 3.13.3). Such fibers, however, generally exhibit higher losses and are more expensive to fabricate than conventional single-mode fiber. Furthermore, much circularly symmetric standard single-mode fiber has already been installed and therefore coherent transmission techniques which utilize this medium are desirable.

Measurements of polarization stability for light propagating in circularly symmetric single-mode fiber over a 96-hour period are shown in Figure 13.4 [Ref. 17]. A single-frequency linearly polarized helium–neon gas laser emitting at 1523 nm together with a receiver which contained a polarization-dependent beam splitter to isolate the components were used in these measurements. It may be observed from Figure 13.4 that, as expected, the polarization state of the optical signal was not temporarily constant. Nevertheless, although polarization changes did occur, it was over periods of minutes or hours. These observations of the relatively slow changes in the polarization state of the transmitted signal, which were also verified on long cable links [Ref. 26], provided the potential for polarization matching at the coherent optical receiver. Polarization control devices with achievable response times could therefore be located at the receiver to provide polarization correction and hence matching of the SOP of the incoming signal and the local oscillator signal.

Active polarization-state control can be accomplished using mechanical, electro-optic or magneto-optic techniques. A polarization error signal may be generated and fed back to the polarization control device in all cases. In general, the SOP is described by the amplitude ratio of the x and y components of the electric field vector and their relative phase difference (see Section 3.13.1). As the two parameters which vary randomly at the coherent receiver are the ellipticity of the SOP and its orientation, the error signal must correct for both of these factors such that the incoming signal and the local oscillator output have identical SOPs.

A number of mechanisms have been developed for polarization-state control within coherent optical fiber communications. Initial implementations of such devices which

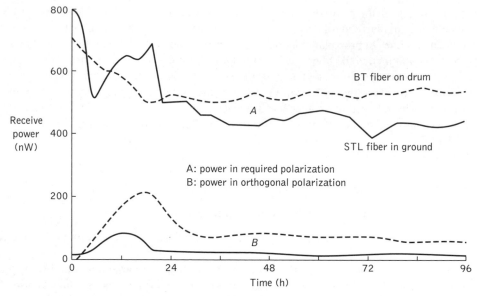

Figure 13.4 Characteristics displaying polarization stability in single-mode optical fiber. From D. W. Smith, R. A. Harmon and T. G. Hodgkinson, *Br. Telecom Technol. J.*, **1** (2), p. 12, 1983

were based on birefringent elements (i.e. elements which induce fiber birefringence in order to correct the SOP) included fiber squeezers [Ref. 27], electro-optic crystals [Ref. 28], rotatable fiber coils [Ref. 29], rotatable phase plates [Ref. 30] and rotatable fiber cranks [Ref. 31]. In addition, the possibility of using the Faraday effect to rotate the SOP of the incoming optical signal has also been demonstrated [Ref. 32] . Developments have also encompassed integrated optical electro-optic polarization control devices. Such controllers have been incorporated into integrated optical coherent receiver devices (see Section 11.4.4). At least two compensator devices are required to provide full polarization-state control. They can be placed in either the incoming signal path or the local oscillator output path; however, the latter position is preferable if the device introduces significant signal attenuation.

A major concern was the insufficient range exhibited by these polarization control schemes in tracking the continuously varying SOP which can change unpredictably over virtually unlimited range. Polarization-state control schemes with infinite ranges of adjustments were, however, demonstrated [Refs 33–35] in the late 1980s. In particular, a control technique using four fiber squeezers, which is illustrated in Figure 13.5(a), was found to provide an infinite range of adjustment or so-called endless polarization control [Ref. 33]. Stress was applied to the fiber in the local oscillator path using the squeezers which were angled at 45° to each other. The SOP could therefore be manipulated to the appropriate matching point. Moreover, the polarization control system was automated using a control algorithm whereby a dither signal was applied to the bias on each squeezer in a defined order, and the variation of the received demodulated signal was used to identify the optimum operating point. This strategy removed the need to optically sense the SOP.

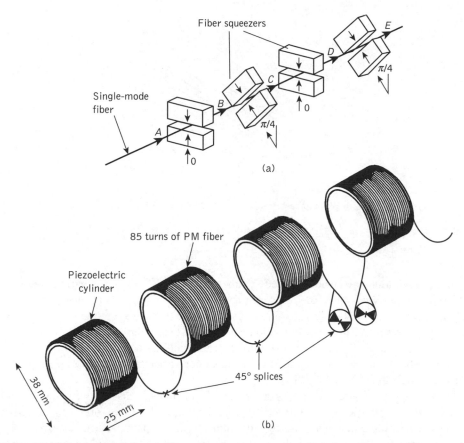

Figure 13.5 Techniques for endless polarization control: (a) four fiber squeezers; (b) polarization-maintaining (PM) fiber controller

Although the squeezers are simple to configure, a drawback with the technique is that they tend to damage the fiber. Moreover, it is questionable as to whether they could be engineered into reliable transducers for practical systems. A more rugged and reliable controller is shown in Figure 13.5(b) [Ref. 34]. It comprised four piezoelectric cylinders each wound around with 85 turns of PM fiber. The PM fibers on each cylinder were spliced together with the principal axis of the fibers mutually aligned at 45°, as illustrated in Figure 13.5(b). As the PM fiber used was highly birefringent with a beat length of only a few millimeters, the SOP changed many times along each element. The application of a voltage caused the piezoelectric cylinders to expand slightly and stretch the fiber, thus modifying the fiber birefringence. Hence, the overall effect was a variable retardation which gave polarization control. Again a control algorithm was devised to provide automatic operation.

An analogous control technique has also been demonstrated with an integrated optical, electro-optic polarization control device [Refs 34, 35]. This lithium niobate waveguide structure which comprised two elements is shown in Figure 13.6. Each element consisted

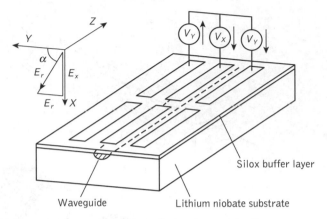

Figure 13.6 Lithium niobate waveguide polarization controller

of three longitudinal electrodes placed symmetrically over the Z-propagating waveguide diffused into an X-cut substrate. Voltages applied to the electrodes produced an electric field that could be orientated in any direction transverse to the waveguide to provide a virtually infinite range of polarization-state control. This technique appears to offer a robust mechanically stable method of polarization control which has been demonstrated in both a laboratory-based and a field-installed coherent optical fiber system, with no measurable sensitivity penalties [Refs 34, 35].

More recently, the need for polarization-state control has focused, in particular, on the exploitation of polarization multiplexing (see Section 13.9.1) in order to effectively double the spectral efficiency and hence the transmission capacity of the coherent optical fiber system [Ref. 36]. For example, a novel control scheme to provide endless polarization stabilization was proposed and described in Ref. 37. This new double-stage polarization stabilization strategy was successfully demonstrated by employing magneto-optic variable polarization rotators applied to an experimental polarization division multiplexed system [Ref. 36]. The magneto-optic polarization stabilizer operated by monitoring the SOP of a polarization channel identified by a pilot tone at 1 MHz and the system produced a measured bit-error-rate of less than 10^{-9} for the polarization demultiplexed channel at a transmission rate of 10 Gbit s^{-1}.

Alternative approaches which avoid the requirement for polarization-state control devices, but which also allow the use of circularly symmetric standard single-mode fiber, are polarization diversity reception [Refs 38–42] and polarization scrambling or spreading transmission [Refs 43, 44]. Both of these techniques are described in Section 13.6.6.

13.4.3 Local oscillator power

In a practical coherent receiver the theoretical performance may not be attained for the reason already outlined in Sections 13.4.1 and 13.4.2. In addition, there may be insufficient local oscillator power to achieve the shot noise detection limits discussed in Section 13.3. This factor, which highlights the need to ensure a low-loss signal path, can

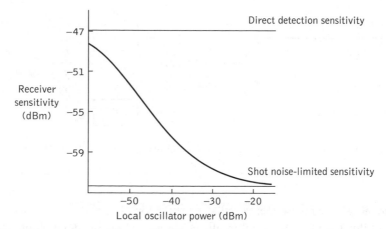

Figure 13.7 Theoretical characteristic showing the effect of local oscillator power on homodyne receiver sensitivity for a PIN–FET receiver operating at 140 Mbit s^{-1}. Reprinted with permission from D. W. Smith, 'Techniques for multigigabit coherent optical transmission', *J. Lightwave Technol.*, **LT-5**, p. 1466, 1987. Copyright © 1987 IEEE

be facilitated by an appropriate choice of an incoming signal/local oscillator combiner which has high coupling efficiency. In particular, it is clear that when the basic coherent receiver shown in Figure 13.3 is considered, the optical combiner or coupler has only one output port utilized, whereas in reality such devices have two output ports (see Section 5.6.1). There is, therefore, an optical loss associated with the power which is coupled to the other output port. Although the combiner or coupler can be designed so that the majority of the incoming signal power is coupled into the optical detector, a consequence of this process is that there will be a reduction in the power from the local oscillator laser coupled into the detector. It therefore becomes more difficult to maintain a high local oscillator signal power and thus to obtain shot noise-limited receiver performance.

The dramatic effect of the local oscillator power on an optical homodyne receiver sensitivity is illustrated by the theoretical characteristic shown in Figure 13.7 which corresponds to a PIN–FET receiver operating at 140 Mbit s^{-1} [Ref. 44]. One method to overcome limited local oscillator power is by the use of a low-noise photodiode/preamplifier combination such as the PIN–FET hybrid configuration (see Section 9.5.2) at the front end of the coherent receiver. Near shot-noise-limited detection has been obtained with just 1 μW of local oscillator power at a transmission rate of 140 Mbit s^{-1} when employing this strategy [Ref. 45]. In addition a heterodyne PIN–FET receiver with 8 GHz bandwidth and 10 pA Hz$^{-\frac{1}{2}}$ equivalent circuit noise current at an IF of 4.6 GHz has been demonstrated [Ref. 46].

An alternative approach which compensates for the losses due to coupling optics and also suppresses excess noise in the local oscillator signal is the use of a balanced receiver.* This scheme, which has often been employed for heterodyne detection in microwave

---

* It is also referred to as the balanced-mixer receiver [Ref. 47].

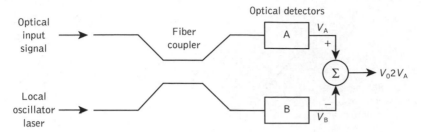

Figure 13.8 Schematic of a balanced optical receiver using two matched optical detectors

communications to suppress local oscillator fluctuations [Refs 48, 49], is shown in Figure 13.8. In this technique the local oscillator output and the incoming optical signal are usually combined using a four-port (i.e. 3 dB) single-mode fiber coupler.* The signal in one fiber in this device suffers a $\pi/2$ phase shift upon transfer to the other fiber. In effect complete coupling is only possible because this phase shift in the coupled signal is $\pi/2$ out of phase with the throughput signal.† Hence the throughput and coupled signals can be represented as a sine wave and cosine wave respectively. Considering Figure 13.8 the inputs to the optical detectors A and B can therefore be written as $E_S \sin \omega_S t + E_L \cos \omega_L t$ and $E_S \cos \omega_S t + E_L \sin \omega_L t$ respectively.

The two detector output voltages are thus given by:

$$V_A = E_S E_L \sin(\omega_S - \omega_L)t \tag{13.18}$$

$$V_B = E_S E_L \sin(\omega_L - \omega_S)t \tag{13.19}$$

It may be observed that these output voltages are similar but of opposite sign in that $V_A = -V_B$.

The two output voltages are operated upon by the combiner function (Σ) depicted in Figure 13.8 and as one is a positive input and the other is a negative input, then the output from the combiner function V_o will form the difference between the two inputs such that:

$$V_o = V_A - V_B = 2V_A \tag{13.20}$$

Equation (13.20) indicates that twice the voltage, or four times the power, is provided in comparison with the single optical detector scheme (Figure 13.3). This technique therefore gives a 6 dB improvement over the single optical detector. Furthermore, as the two photocurrents are effectively subtracted, the process results in both a cancelling out of the large d.c. term produced by the local oscillator signal, together with any local oscillator excess noise. It is particularly useful in reducing the excess AM noise generated by the

* Other devices which may be utilized are bulk optic or waveguide beam splitters.
† This signal becomes out of phase if it is coupled back to the throughput fiber where it would interfere destructively with the throughput signal.

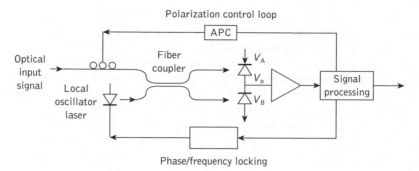

Figure 13.9 Dual optical detector balanced receiver for phase-modulated optical communications

local oscillator [Ref. 19]. Moreover, as a result of the efficient use of the local oscillator and incoming signal powers, the constraints imposed by the earlier requirement for a widespread ultra-low-noise preamplifier are relaxed. Close matching of the two arms of the balanced receiver is essential, however, if good excess noise cancellation is to be obtained. A dual detector balanced receiver for phase-modulated optical communications is shown in Figure 13.9. It may be noted that the necessary polarization alignment is achieved using automatic polarization control (APC) while phase or frequency locking is also required. Moreover the polarization control algorithm should be able to provide endless polarization without the need to reset (see Section 13.4.2).

13.4.4 Transmission medium limitations

Although, in common with IM/DD systems, the fiber loss is the major limitation on the performance of single-carrier coherent optical systems that can be ascribed to the transmission medium, there are nevertheless other factors that may well affect the operation of future coherent systems. These include intramodal or chromatic dispersion, polarization dispersion and the nonlinear scattering effects [Ref. 43].

The chromatic dispersion in standard single-mode fiber which has a dispersion zero at a wavelength of 1.31 μm is around 17 ps km^{-1} nm^{-1} when the fiber is operated at a wavelength of 1.55 μm. This factor can lead to significant dispersion penalties (i.e. receiver sensitivity degradations) for IM/DD systems even at modest transmission rates and distances. It results from the transmitted spectrum usually being far wider than the information spectrum due to laser frequency chirp (see Section 6.7.3) caused by the direct amplitude modulation of the semiconductor laser. By contrast, coherent optical transmission systems have the advantage of a compact spectrum even if the injection laser is directly modulated but particularly when an external modulator is employed. Receiver sensitivity degradation due to chromatic dispersion has, however, been observed in FSK transmission experiments at transmission rates greater than 4 Gbit s^{-1} [Ref. 43]. Furthermore, as both the transmission rates and distances for coherent transmission are increased, then greater chromatic dispersion penalties will be incurred. For example, calculations have indicated the maximum transmission rates to incur a 2 dB penalty after a 100 km distance on a standard single-mode fiber when operating at a wavelength of 1.55 μm to be in the range 5 to 9 Gbit s^{-1} depending on the modulation format used [Ref. 50].

As in the case of IM/DD the chromatic dispersion problem can be reduced through the use of dispersion-shifted fiber which exhibits a dispersion zero in the 1.55 μm wavelength window (see Section 3.12.1). However, this solution is only partially satisfactory as there is a massive base of installed standard single-mode fiber. An alternative strategy is to compensate for the chromatic dispersion in either the optical domain [Ref. 43] or the electrical domain [Ref. 51]. The latter method in particular has shown some limited success using stripline delay equalizers which have demonstrated the potential to compensate for dispersion-induced distortion up to transmission rates of 10 Gbit s^{-1}.

Polarization mode dispersion results from birefringence in the single-mode fiber and it corresponds to the difference in the propagation time associated with the two principal orthogonal polarization states (see Section 3.13.2). Whereas in IM/DD systems polarization mode dispersion simply results in pulse broadening due to the different spectral components arriving at different times, in coherent systems these components can also arrive with different polarizations. Moreover, both of these effects may be detrimental to coherent system performance. However, the differential propagation time is dependent upon the amount of mode mixing which takes place in the fiber, an effect which results from internal and external fiber perturbations. Assuming some mode mixing, the effects of polarization mode dispersion are therefore not expected to become important in a single-carrier system with a fiber distance of 100 km until transmission rates exceed 10 Gbit s^{-1} [Ref. 43]. In multicarrier systems (see Section 13.9) the problem associated with receiving optical carriers at different wavelengths in different polarization states, however, may be avoided by using PM fiber, polarization diversity reception or polarization scrambling transmission (see Section 13.6.6).

The nonlinear phenomena which may be of importance within coherent optical transmission include stimulated Raman scattering (SRS), stimulated Brillouin scattering (SBS), cross-phase modulation and four-wave mixing [Ref. 43]. ASK systems prove particularly susceptible to such nonlinear effects which result from optical power-level changes. Moreover, a potential advantage of the FSK and PSK modulation formats over ASK is that they produce a constant amplitude signal which provides some immunity to certain nonlinear effects. Although SRS should not be a consideration in single-carrier coherent optical systems operating at power levels below 1 W, Raman induced crosstalk may be a concern in multicarrier systems [Ref. 52].

Stimulated Brillouin scattering may be a problem at lower light levels than SRS as its threshold power level is significantly smaller (see Section 3.5). However, unlike SRS, SBS is a narrowband process with a bandwidth of only around 20 MHz at a wavelength of 1.55 μm (see Section 3.14.1). The maximum SBS gain which also is maximized in the reverse direction will therefore occur for lasers with linewidths less than 20 MHz. Moreover, as a result of information broadening of the linewidth, SBS will be greatly reduced when using modulation formats which do not contain a residual carrier component (i.e. PSK and narrow-deviation FSK, see Section 13.5).

Self-phase modulation is a phenomenon which occurs in single-carrier systems due to small refractive index changes induced by optical power fluctuations which affect the phase of the transmitted signal (see Section 3.14.2). For digitally modulated coherent optical systems this effect is perceived to be negligibly small at launched power levels up to a few hundred milliwatts [Ref. 43]. With multicarrier systems, however, a cross-phase modulation phenomenon occurs which can cause high levels of phase noise in long fiber lengths. In this case it has been shown that the power of each carrier should be restricted in order to limit the degradation caused by this phase crosstalk [Ref. 52]. Nevertheless, this limitation on transmitter power in a multicarrier system may not be as severe as the one imposed by the four-wave-mixing nonlinear phenomenon. It is suggested that this latter process will be present in all frequency division multiplexed systems with channel separations less than 10 GHz and that the crosstalk will restrict the maximum power per carrier to around 0.1 mW when the fiber lengths exceed 10 km [Refs 43, 52].

13.5 Modulation formats

13.5.1 Amplitude shift keying

Several techniques may be employed to amplitude-modulate an optical signal. Digital intensity modulation used in DD systems is essentially a crude form of ASK* in which the received signal is simply detected using the photodetector as a square-law device (see Figure 13.2(a)). It is apparent, therefore, that the simplest approach to ASK is by direct modulation of the laser drive current. A problem exists, however, with this approach because of the inability of semiconductor lasers to maintain a stable output frequency with changing drive current. The resulting frequency deviation, which can be of the order of 200 MHz mA^{-1}, broadens the linewidth of the modulated laser which creates difficulties for coherent optical detection (see Section 13.4.1).

Although direct modulation of the semiconductor laser in ASK coherent optical fiber systems has been demonstrated [Ref. 4], external modulation using active integrated optical devices, such as the directional coupler or the Mach–Zehnder interferometer (see Section 11.4.2), present attractive alternatives [Ref. 44]. It should be noted, however, that all external ASK modulators suffer the drawback that around half of the transmitter power is wasted. Nonsynchronous detection can also be employed with the ASK format which puts the least demands on the injection laser phase stability. In principle this modulation scheme can be used with laser sources exhibiting linewidths comparable with the bit transmission rate. In practice the linewidth in the range 10 to 50% of bit rate is normally specified for ASK heterodyne detection [Refs 44, 53], although some authors indicate 10 to 20% [Refs 4, 43]. Nevertheless, asynchronous heterodyne and phase diversity (see Section 13.6.5) detection receivers which use ASK can tolerate the linewidths of typical DFB lasers.

---

* It should be noted that ASK is also referred to as on–off keying (OOK).

13.5.2 Frequency shift keying

The frequency deviation property of a directly modulated semiconductor laser can be usefully employed with wideband FSK coherent optical fiber systems. Hence optical FSK (see Figure 13.2(b)) in common with ASK has the advantage that it does not necessarily require an external modulator, thus allowing higher launch powers and a more compact transmitter configuration, as illustrated in Figure 13.10(a). The direct frequency modulation characteristics of the laser are determined by changes in the device carrier density in the high-modulation-frequency region, and by the temperature modulation effect in the low-frequency region. Thus at frequencies above 1 MHz where the carrier modulation effect occurs, the frequency deviation is typically 100 to 500 MHz mA^{-1}, whereas below 1 MHz it is around 1 GHz mA^{-1} due to the predominant temperature effect [Ref. 44]. Although the response under frequency modulation of the semiconductor laser is therefore not uniform, a frequency shift of between 100 MHz to 1 GHz is readily obtained from the device without serious intensity modulation effects [Ref. 19].

The laser linewidth requirements for wide frequency deviation FSK heterodyne nonsynchronous detection are similar to those of ASK with nonsynchronous heterodyne detection and are in the range 10 to 50% of the transmission bit rate [Refs 44, 52]. Therefore, the use of FSK with broad-linewidth injection lasers has been relatively successful, particularly at low bit rates. For example, the direct modulation of the laser drive

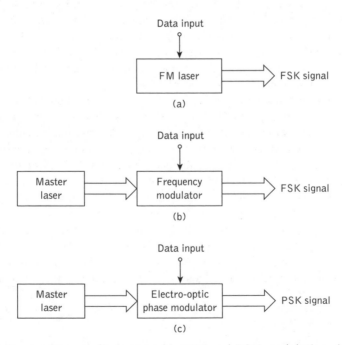

Figure 13.10 Transmitter configurations for FSK and PSK modulation: (a) FSK by direct modulation of an FM semiconductor laser injection current; (b) FSK using an external modulator; (c) PSK using an external electro-optic modulator

current causing variations in the lasing wavelength has provided FSK transmission over 300 km at a rate of 34 Mbit s^{-1} with a receiver sensitivity of 165 photons per bit [Ref. 54]. However, due to the presence of large thermal effects at linewidths less than 10 MHz, the FM response of semiconductor lasers is typically nonuniform and hence electronic equalization may be required [Ref. 19]. Although it has been observed that the use of a split electrode on the injection laser largely eliminates this effect [Ref. 55], alternative strategies have also been utilized. These include the use of bipolar optical FSK transmission [Ref. 56] and alternate mark inversion encoding [Ref. 57]. The former technique provided a transmission rate of 1 Gbit s^{-1} over 121 km whereas the latter operated at a rate of 565 Mbit s^{-1} using commercial DFB lasers.

When a single oscillator is switched between two frequencies, as is often the case with a semiconductor laser source, the phase of the signal is a continuous function of time and the modulation is known as continuous phase FSK (CPFSK) [Ref. 43]. This modulation scheme has been successfully demonstrated [Ref. 58] using integrated external cavity lasers (see Section 6.10.2). In this experiment transmission at 2 Gbit s^{-1} over 197 km of single-mode fiber was achieved. Continuous phase FSK is attractive because it allows d.c. modulation of the injection laser while also providing high receiver sensitivity. Furthermore, it is suitable for high-speed transmission since it creates no laser chirping degradation (see Section 6.7.3), as experienced in IM/DD systems. The multichannel properties of CPFSK with a small frequency deviation have also indicated that it is a potential technique for wavelength division multiplexing [Ref. 59].

External modulation techniques for FSK, shown schematically in Figure 13.10(b), include both acousto-optic and electro-optic approaches. Using bulk optic devices, FSK may be accomplished using a Bragg cell which employs traveling acoustic waves in a crystal to simultaneously diffract and frequency-shift the optical signal. Alternatively, an equivalent effect can be obtained by using surface acoustic waves on an integrated optical waveguide device (see Section 11.4.2). Frequency shift keying modulation can also be provided by a Mach–Zehnder interferometer with sinusoidal modulation applied to one of its branches. Such devices have been operated at modulation frequencies in excess of 1 GHz.

Finally, multilevel FSK (MFSK) offers the potential for improving the coherent optical receiver sensitivity by increasing the choice of signaling frequencies [Ref. 60]. In principle this M-ary scheme provides the best receiver performance in the limit of large channel spacing. Thus eight-level FSK yields an equivalent sensitivity to binary PSK but at the expense of a greater receiver bandwidth requirement.

13.5.3 Phase shift keying

Optical phase modulation can be achieved by d.c. modulation of a semiconductor laser into which external coherent laser light is injected [Ref. 61]. When the injected laser frequency is exactly tuned to the modulating signal frequency, the output signal phase relative to the modulating signal phase is zero. A relative phase change of $\pi/2$ is obtained when the injected laser frequency is detuned away from the modulated light frequency to the injection-locking limit. Hence the cutoff modulation frequency is determined by the injection-locking bandwidth. Furthermore, this technique has the effect of reducing the linewidth of the injection-locked laser to that of the injected signal device [Ref. 19].

Table 13.1 Laser linewidth requirements for various modulation formats as a percentage of the bit rate

| Modulation format | Homodyne | Heterodyne | |
		Synchronous	Asynchronous
ASK	0.005–0.1%	0.05–0.1%	10–50%
FSK (wide deviation)	No	0.05–0.1%	10–50%
FSK (narrow deviation)	No		0.3–2.0%
PSK	0.005–0.01%	0.1–0.5%	No
DPSK	No	0.3–0.5%	No

External modulation for PSK is relatively straightforward and therefore normally utilized to provide the modulation format (see Figure 13.10(c)) which allows the most sensitive coherent detection mechanism within the binary modulation schemes (see Section 13.8). Simple integrated optical phase modulators fabricated from electro-optic materials such as lithium niobate or III–V compound semiconductors may be employed to give the appropriate shift with the application of an electric field (see Section 11.4.2). Such devices, which exhibit a fiber-to-fiber insertion loss of 2 to 5 dB, require around 5 V drive to produce a phase shift of π radians. Moreover, modulation bandwidths in excess of 10 GHz have been obtained from traveling-wave structures.

The phase detection process for PSK, however, necessitates synchronous detection with the requirement for corresponding narrow laser linewidths. These very narrow-linewidth requirements for both PSK heterodyne and homodyne detection may be observed in Table 13.1 which presents the laser linewidths as a percentage of the transmission bit rate for the major modulation formats considered in Section 13.5. Moreover, it may be noted that the most stringent laser linewidth requirement is for homodyne detection with binary PSK where for efficient detection linewidths on the order of 0.01% of the transmission rate are required [Ref. 62].

By contrast, two level differential PSK (DPSK), which is also referred to as differential binary PSK (DBPSK) and is also indicated in Table 13.1 by the accepted term DPSK, is a less demanding form of PSK since information is encoded as a change (or the absence of a change) in the optical phase on a bit-by-bit basis. The relationship between DPSK and PSK is illustrated in Figure 13.11 where it may be observed that with DPSK the incoming bit is delayed in order that its phase can be compared with the next received bit. Hence the technique does not require phase comparisons over more than two bit intervals. Moreover, the SNR performance of DPSK is only a fraction of a decibel less than that of heterodyne (synchronous) PSK [Ref. 1]. As laser linewidths of the order of 0.3 to 0.5% of the transmission rate can be tolerated with DPSK, experimental systems operating at 1.2 Gbit s^{-1} were demonstrated using both an integrated external cavity DFB laser [Ref. 63] and an external fiber cavity DFB laser [Ref. 64] in the late 1980s. In addition DPSK is technically straightforward to implement at high transmission rates because the phase fluctuation between the two signal bits is reduced [Ref. 21].

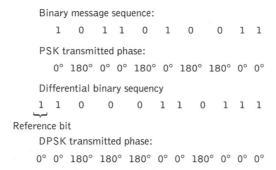

Binary message sequence:

1 0 1 1 0 1 0   0 1 1

PSK transmitted phase:

0° 180° 0° 0° 180° 0° 180° 180° 0° 0°

Differential binary sequency

1 1 0   0   0 1 1 0 1 1 1

Reference bit

DPSK transmitted phase:

0° 0° 180° 180° 180° 0° 0° 180° 0° 0° 0°

Figure 13.11 Comparison of a one-bit-at-a-time DPSK scheme with binary PSK. The differential binary sequence is obtained by repeating the preceding bit in the sequence if the message bit is a 1 or by changing to the opposite bit if the message bit is a 0

Unlike multilevel FSK which can provide improved receiver sensitivity by spectral expansion, M-ary PSK (and also, for that matter, M-ary ASK) can potentially provide spectral conservation through the use of multilevel signaling. This approach can be generalized for M-ary schemes where the spectral efficiency is increased by a factor of $\log_2 M$ (i.e. for M-level signaling) with respect to binary signaling. Moreover M-ary PSK, M-ary ASK and their combinations such as quadrature amplitude modulation (QAM) [Refs 2, 65] can avoid noise degradation in the electronic preamplifier by increasing the utilization of the IF band frequency within optical heterodyne detection [Ref. 66].

Following the widespread deployment of optical amplifiers, a major benefit associated with coherent transmission has become the improvement of the system spectral efficiency which can be further enhanced using multilevel modulation [Ref. 8]. The signal space or signal constellation diagrams for binary and quaternary, or four-level, ASK, or on–off keying (OOK) are shown in Figure 13.12(a). It may be observed that, although the optical carrier has both in- and quadrature phase represented in the two-dimensional constellation diagram, OOK uses only the positive axis in a single dimension to carry information. Therefore M-ary ASK as typified by 4-OOK employs four different amplitude levels each containing two information bits (i.e. dibits). By contrast, in the quadrature PSK (QPSK) or 4-QAM scheme depicted in Figure 13.12(b) all aspects of the signal space are employed (i.e. positive and negative sides of both dimensions) to carry information [Ref. 2].

For QPSK or other M-ary PSK modulation schemes, however, the signals are distinguished from each other in phase but are all of the same amplitude. In order to obtain greater spectral efficiency it is useful to combine M-ary PSK with M-ary ASK to produce QAM as illustrated in Figure 13.12(c). The 16-QAM constellation diagram demonstrates how four information bits can be contained in each transmitted symbol. It should also be noted that the error probability is primarily determined by the minimum Euclidean distance, d (in Figure 13.12), and therefore 16-QAM exhibits a lower error rate when subjected to noise than, for example, 16-level M-ary PSK [Ref. 67]. The receiver sensitivities associated with the above multilevel transmission techniques are discussed further in Section 13.8.7.

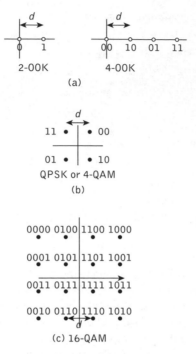

Figure 13.12 Signal space or constellation diagram for: (a) binary and quaternary ASK or on–off keying; (b) quadrature phase shift keying (QPSK) or four-point quadrature amplitude modulation (4-QAM); (c) 16-point quadrature amplitude modulation (16-QAM)

13.5.4 Polarization shift keying

An additional modulation format which has been investigated within coherent optical fiber communications involves use of the polarization characteristics of the transmitted optical signal. The digital transmission implementation of such polarization modulation is known as polarization shift keying (PolSK). As a single-mode optical fiber can support two polarizations, they can be used alternately to carry either a zero or a one bit. Unlike PSK or FSK, however, the PolSK data modulation format has received comparatively little attention, primarily as a consequence of the need for active polarization management at the receiver due to the random polarization changes in standard single-mode fiber [Ref. 10]. Hence PolSK requires the additional receiver complexity associated with the polarization control requirement without providing a significant sensitivity improvement over intensity modulation. Nevertheless, polarization is finding utilization to increase spectral efficiency by either transmitting two different signals at the same wavelength in two orthogonal polarizations to produce polarization multiplexing (see Section 13.9.1) or for the transmission of adjacent wavelength division multiplexed channels using alternating polarizations to reduce nonlinear interactions and coherent crosstalk between the channels giving polarization interleaving (see Section 13.9.2).

An early realization of coherent optical transmission using heterodyne detection with PolSK was obtained through external modulation by a lithium niobate phase modulator [Ref. 68]. This device produced a phase shift of π radians between the TE and TM modes, which rotated the signal polarization by 90°. These orthogonal polarization states were then maintained during transmission within a single-mode fiber. In this context a prerequisite for the fiber was that no coupling occurred between the two orthogonal polarization modes. The system was, however, successfully operated at a transmission rate of 560 Mbit s^{-1} and proved between 2 and 3 dB more sensitive than ASK modulation with heterodyne detection.

The differential variant of PolSK (DPolSK) has also been demonstrated [Ref. 69]. This modulation format eliminates the ambiguity involved in deciding whether a particular polarization represents a binary 0 or a 1. Furthermore, the DPolSK scheme can lead to the removal of the phase noise associated with both the laser source and the fluctuations from the transmission medium [Ref. 69]. This factor provides an improvement over PolSK where only the phase jitter of the laser may be cancelled.

The above binary schemes have been concerned with the two polarization modes of a single-mode fiber. Multilevel PolSK is also possible in which the transmitted symbols are each associated with different polarization states within the fiber [Refs 70, 71]. Moreover, such a modulation format can provide a performance improvement over the more traditional multilevel systems outlined in Sections 13.5.2 and 13.5.3 [Ref. 71].

13.6 Demodulation schemes

Basic receiver configurations for optical heterodyne and homodyne detection are shown in Figure 13.13. In both cases it has been assumed that some form of polarization control is required to match the incoming signal SOP to that of the local oscillator signal (see Section 13.4.2). This factor therefore implies the use of circularly symmetric standard single-mode fiber. For heterodyne detection (Figure 13.13(a)), a beat-note signal between the incoming optical signal and the local oscillator signal produces the IF signal which is obtained using the square-law optical detector (see Section 13.3). The IF signal, which generally has a frequency of between three and four times the transmission rate, is then demodulated into the baseband using either a synchronous or asynchronous detection technique.* An optical receiver bandwidth several times greater than that of a direct detection receiver is therefore required for a specific transmission rate. Moreover, as IF frequency fluctuation degrades the heterodyne receiver performance, then frequency stabilization may be achieved by feeding back from the demodulator through an automatic frequency control (AFC) circuit to the local oscillator drive circuit.

---

* A brief explanation of the terminology was provided in Section 13.2. It should be noted that asynchronous heterodyne detection does not strictly require phase matching between the incoming signal and the local oscillator. Spatial coherence between the two signals is, however, required when they are combined so that asynchronous detection schemes do fall within the broad heading of coherent optical fiber systems.

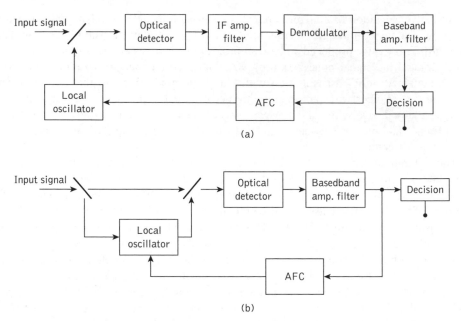

Figure 13.13 Basic coherent receiver configurations: (a) optical heterodyne receiver; (b) optical homodyne receiver illustrating the phase locking between the local oscillator and incoming signals

In the case of homodyne detection in which the phase of the local oscillator signal is locked to the incoming signal, then, by definition, a synchronous detection scheme must be employed. Moreover, the result of the mixing process in the optical detector produces an information signal which is in the baseband (see Section 13.3) and thus requires no further demodulation. An AFC loop is also shown within the homodyne receiver configuration of Figure 13.13(b) to provide the necessary frequency stabilization between the two signals. Hence any variant detection schemes based on homodyne detection, but in which the local oscillator laser is not phase locked to the incoming signal such as phase diversity or multiport detection, could be considered as a form of heterodyne rather than homodyne detection [Ref. 4]. However, this technique is dealt with separately in Section 13.6.4.

In both optical heterodyne and homodyne detection, where the incoming signal is demodulated using a local oscillator laser, FM noise in this device together with that resulting from the source laser causes SNR degradation in the receivers through FM to AM, or PM to AM, conversion which generally determines the lower limit of bit-error-rate performance [Ref. 72]. Frequency modulation noise which basically results from the spontaneous emission coupled to the lasing mode is, in the semiconductor laser, enhanced by AM noise caused by photon number fluctuation which is generated through the same mechanism [Ref. 73]. Moreover, excess AM noise within the local oscillator laser due to its resonance characteristics also deteriorates the SNR performance and hence degrades the receiver sensitivity. To reduce the effect of local oscillator FM noise a semiconductor laser with a narrowed or suppressed spectral linewidth must be used [see Section 13.4.1].

The excess AM noise in the semiconductor laser decreases with an increase in the bias level so that high-bias operation is effective in suppressing this mechanism [Ref. 73]. Furthermore, excess AM noise associated with the local oscillator can be suppressed by employing the balanced receiver configuration described in Section 13.4.3.

13.6.1 Heterodyne synchronous detection

Optical heterodyne synchronous detection necessitates an estimation of the phase of the IF signal in translating it to the baseband. Such an approach generally requires the use of phase-locking techniques at the receiver in order to track phase fluctuations in the incoming and local oscillator signals. Since the information signal is to be processed on an IF carrier, then electrical phase estimation may be employed. Hence the phase-locked loop (PLL) techniques and configurations appropriate to radio-frequency and microwave communications can be utilized [Ref. 1]. Such techniques have been investigated primarily for PSK demodulation where an estimation of the phase of the signal is required [Ref. 4]. Furthermore, synchronous PSK demodulation is the most sensitive of the heterodyne detection techniques (see Section 13.8.7). In order to achieve a measurement of the phase of a fully modulated PSK signal, it is necessary to obtain a phase reference from the phase of the average incoming optical signal within a particular time interval. Therefore the purpose of the PLL is to provide that reference where, in general, the time average is defined by the bandwidth of the loop.

An examination of the spectrum of a PSK signal reveals that no signal energy is present at the carrier frequency when the phase shift from the binary 1 to 0 states is a full 180°.* The introduction of a nonlinear element within the PLL is therefore necessary to enable efficient carrier recovery. A squaring loop technique illustrated in Figure 13.14(a) is particularly applicable to binary PSK, phase-noise-sensitive coherent optical fiber systems. By squaring the PSK signal frequency this method produces a carrier at twice the original frequency which can be filtered out and then used for phase estimation. A similar result may be obtained with the statistically equivalent Costas loop shown in Figure 13.14(b) [Ref. 74].

An alternative approach to carrier recovery is to reduce slightly the depth of the phase modulation so that a small component of the transmitted energy lies at the carrier frequency. This pilot carrier signal can then be amplified and recovered as a phase reference at the receiver. In this case of what is essentially a weak carrier, a much reduced loop bandwidth (i.e. reduced integration time) is required in order adequately to recover the carrier. Furthermore, while providing a stable reference, long integration times increase the sensitivity of the receiver to carrier phase noise.

A variant on the pilot carrier technique for PSK synchronous demodulation is shown in Figure 13.15 in which carrier recovery takes place at the IF stage [Ref. 75]. In this case the detected IF signal is divided into two routes, one being the signal route and the other being the carrier recovery route. Following the carrier recovery route the signal is doubled by a frequency doubler (FD) to remove the $(0, \pi)$ phase modulation component. Twice the IF frequency of the resultant signal is then divided by a frequency halver (FH) to recover the

---

* This situation typifies a suppressed carrier modulation type [Ref. 2].

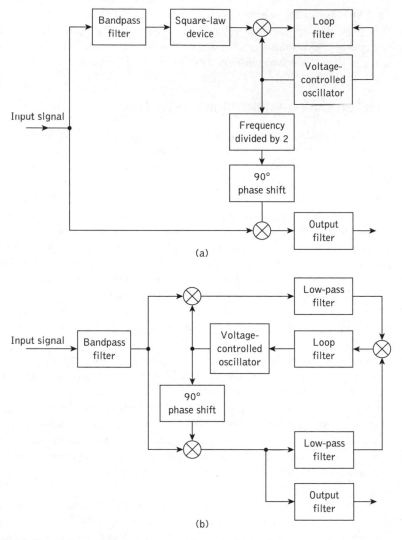

(a)

(b)

Figure 13.14 Techniques for carrier recovery used in coherent optical PSK receivers: (a) squaring loop; (b) Costas loop

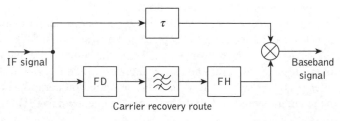

Carrier recovery route

Figure 13.15 A carrier recovery synchronous demodulator. Reprinted with permission from Ref. 78 © IEEE 1983

reference carrier signal. Finally, the recovered carrier and signal are mixed to give the demodulator output. This technique, which provides for suppression of phase noise, has, for example, been demonstrated in a coherent optical PSK system. [Refs 75, 76].

The above synchronous demodulation schemes can also be used within ASK and FSK heterodyne optical fiber systems but they do not provide the same potential receiver sensitivity performance as that of PSK. Moreover, alternative asynchronous techniques for ASK and FSK are often more reliable in the presence of phase noise and provide receiver sensitivities only slightly less than the corresponding synchronous methods. These asynchronous demodulation schemes are therefore discussed in Section 13.6.2.

13.6.2 Heterodyne asynchronous detection

It was indicated in Section 13.5 that both ASK and FSK may be demodulated using asynchronous detection techniques, which puts the least demands on laser linewidth and phase stability. Such demodulation schemes, which include ASK envelope detection as well as FSK single- and dual-filter detection, do not therefore require the extremely narrow laser linewidths associated with synchronous binary PSK demodulation. Heterodyne envelope detection of an ASK signal may be achieved using an IF bandpass filter followed by a peak detector to recover the baseband signal as shown in Figure 13.16(a). Such a scheme, however, incurs a receiver sensitivity penalty as a result of nonlinear filtering of the Gaussian-distributed noise in the peak detection process combined with the phase noise broadening of the signal spectrum such that a significant proportion of the signal energy can be translated outside the IF signal band. An optimum receiver bandwidth balances these factors and, for combined source and local oscillator linewidths of 10% of the transmitted data rate, the receiver penalty is around 3 dB [Ref. 43].

By employing parallel filters with channels centered on the two transmitted frequencies it is possible to use envelope detection on each channel of a binary FSK signal. The configuration for this dual-filter demodulation technique is provided in Figure 13.16(b) [Ref. 46]. At the output it produces a differential ASK signal with a receiver sensitivity which is slightly better than asynchronous ASK demodulation in the presence of phase noise (i.e. a 2 dB penalty when source and local oscillator laser linewidths are 10% of the transmitted data rate [Ref. 41]). This improvement results from the complementary behavior of the dual-filter approach where, for large spacing between the two FSK signal channels, it is possible to have a significant spectral broadening of the signal but with insufficient energy in the complementary channel to register an error.

As mentioned earlier, asynchronous detection cannot be employed in the case of the PSK modulation format primarily because the phases of the received optical signal and the local oscillator laser are not locked and they can therefore drift over time. The use of the DPSK (see Sections 13.5.3 and 13.7) format, however, does allow asynchronous demodulation as there is no similar requirement for phase locking. An asynchronous DPSK receiver is depicted in Figure 13.16(c) which utilizes a 1 bit delay scheme. Hence the received bit stream is multiplied by its replica that has been delayed by 1 bit period duration. This phase comparison demodulation reduces the synchronization problems associated with PSK as phase stability is only required over a few bit periods and it can therefore be implemented using distributed feedback lasers.

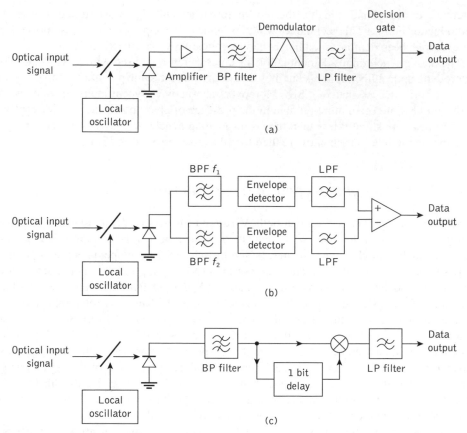

Figure 13.16 Asynchronous heterodyne detection: (a) ASK single envelope detector receiver; (b) FSK dual-filter receiver; (c) asynchronous DPSK receiver

13.6.3 Homodyne detection

The attraction of optical homodyne detection is not just the potential 3 dB improvement in receiver sensitivity (see Section 13.3) but also that it can ease the receiver bandwidth requirement considerably. This factor is illustrated in Figure 13.17 which compares the spectra at the output of the detector for PSK homodyne and PSK heterodyne detection. It may be observed that homodyne detection requires only the normal direct detection receiver bandwidth whereas heterodyne detection requires at least twice this bandwidth and often a factor of 3 or 4 times it. Unfortunately, optical homodyne detection using independent source and local oscillator lasers (i.e. not self-homodyne, see Section 13.7) has proved quite difficult to achieve because of the problems associated with remotely optical locking a local oscillator laser to a low-level modulated signal [Ref. 44]. Such phase locking is essential because the phase difference ϕ in Eq. (13.12) must be held near zero for high-sensitivity reception. Furthermore, if ϕ drifts to $\pi/2$, then the output signal current I_S will become zero and the detection process will cease.

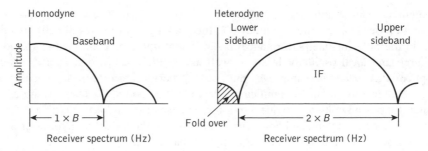

Figure 13.17 Comparison of the electrical spectra at the optical detector output for homodyne and heterodyne detection of a PSK signal

Two homodyne demodulation strategies have, however, proved successful in demonstration for coherent optical fiber reception. They are the use of either a pilot carrier or a decision-driven optical phase locked loop [Refs 1, 19, 44]. Unlike the electrical PLL techniques described in Section 13.6.1 which comprised an electrical phase detector, loop filter and voltage-controlled oscillator (VCO), in the optical PLL, the photodetector and laser local oscillator act as the phase detector and VCO respectively. Hence the optical phase difference between the incoming signal and the local oscillator signal, or the phase error signal, is detected by the photodetector prior to being fed back to correct the local oscillator frequency and phase. Although homodyne detection using an optical PLL has been demonstrated [Ref. 77], it is somewhat difficult to realize and puts stringent demands on the laser linewidths [Ref. 62].

The optical PLL configuration shown in Figure 13.18 [Ref. 77] employs a pilot carrier strategy for PSK homodyne detection. In common with other pilot carrier techniques (see Section 13.6.1) this carrier is generated by using incomplete (less than 180°) phase modulation. The pilot carrier signal, together with the incoming signal, are combined in a 3 dB fiber directional coupler and then detected using a balanced receiver (see Section 13.4.3). The output signal from the difference amplifier (Figure 13.18) is therefore a function of the phase error which may be used for phase locking through the loop filter to the optical local oscillator which performs as the VCO. It should be noted, however, that any carrier power used in this phase-locking process directly reduces the receiver sensitivity by an

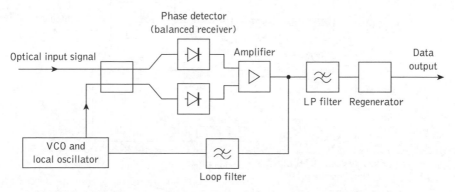

Figure 13.18 Pilot carrier optical phase locked loop receiver [Ref. 77]

equivalent amount. Furthermore, the signal power required to track the phase of the incoming carrier to a specified accuracy (i.e. the tracking error relates directly to a degradation in the bit-error-rate performance) is dependent upon the combined phase noise of the source and local oscillator lasers as well as the PLL bandwidth. Hence, extending the optical PLL bandwidth improves the tracking performance until a point is reached where the increased shot noise significantly degrades the loop SNR. There is, therefore, an optimum loop bandwidth to provide a minimum phase error and it is possible to improve the performance of the optical homodyne receiver when the local oscillator laser has substantial phase noise by simply increasing the PLL bandwidth.

The pilot carrier optical PLL approach using external cavity semiconductor lasers and a balanced transimpedance receiver has demonstrated a sensitivity of 72 photons per bit at 4 Gbit s^{-1} [Ref. 25]. Furthermore, there is a similar balanced optical PLL receiver configuration where part of the transmitted power must be used for unmodulated carrier transmission so that the local oscillator laser can lock in quadrature to the residual carrier which is therefore in phase with the data signal. This homodyne receiver has the benefit of suppressing the excess intensity noise of the lasers employed but it does impose very stringent requirements on the laser linewidths (see Section 13.4.1). Nevertheless, experimental demonstration of this approach has achieved receiver sensitivities of 25 and 332 photons per bit at transmission rates of 140 Mbit s^{-1} and 2 Gbit s^{-1} respectively.

The basic principle of the decision-driven optical PLL homodyne receiver mixes the received data with the local oscillator using a 90° optical hybrid as illustrated in Figure 13.19. This device, which typically incorporates a 3 dB fiber coupler and two polarization beam-splitting elements, ensures a 90° phase difference between the output signals from the two balanced receivers (see Section 13.6.6).

The 90° hybrid device can be realized using the phase shift properties of polarized light [Refs 79, 80]. A 90° phase shift can be obtained by combining a circularly polarized local oscillator signal with a linearly polarized incoming signal and then by resolving the combined signal into two orthogonal components with a polarization beam splitter. The linearly polarized incoming signal must, however, be aligned at 45° to the beam splitter plane.

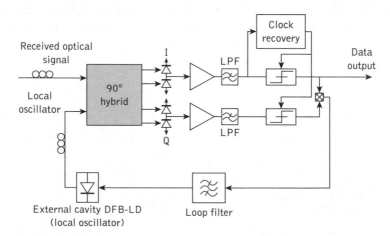

Figure 13.19 Decision-driven optical phase locked loop receiver

The combined local oscillator and data signals are in phase on the upper output of the 90% hybrid in Figure 13.19 which is input to the data collection branch, while the two signals are in quadrature on the lower output which is sent to the optical PLL. Clearly, using this approach there is no transmitted power wastage from a pilot carrier or residual component. Furthermore, a decision-driven optical PLL receiver with the configuration shown in Figure 13.19 has demonstrated a sensitivity of 297 photons per bit at a transmission rate of 10 Gbit s^{-1} [Ref. 81].

13.6.4 Intradyne detection

An alternative coherent optical detection concept to the common heterodyne or homodyne detection processes is that of intradyning. In the intradyne receiver the incoming signal is not precisely shifted to the baseband as in homodyne detection but shifted to a frequency much lower than the data transmission rate. Hence the electronic filtering can be undertaken using a baseband filter which is only slightly wider than that utilized for a homodyne receiver. The requirements on the automatic frequency control, however, for the local oscillator laser are similar to those of a heterodyne receiver. Use of an PLL can therefore be avoided with intradyne detection as there is a low IF generated and the IF is not zero as in homodyne receivers. A spectral diagram of the intradyne process for PSK comparing it with both heterodyne and homodyne PSK detection is provided in Figure 13.20(a). It may

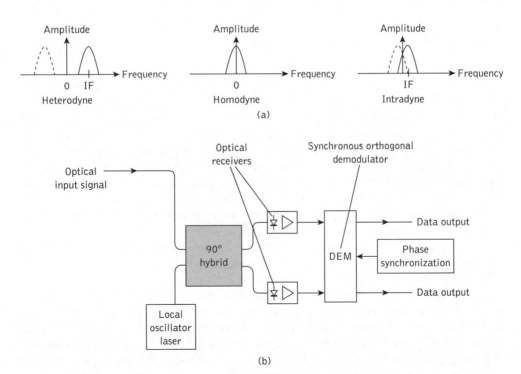

Figure 13.20 Optical intradyne detection: (a) spectral comparison with heterodyne and homodyne detection; (b) two-phase QPSK intradyne receiver

be observed that the intradyne spectrum is downconverted to produce an IF near the base-band as a consequence of the local oscillator laser only exhibiting approximately the same optical frequency as the incoming signal. Moreover, the spectra for heterodyne and homo-dyne detection can be related to those shown in Figure 13.17.

The intradyne detection process was initially implemented in an experimental quadra-dature PSK receiver which employed purely electronic digital carrier recovery [Ref. 82]. Figure 13.20(b) depicts a two-phase intradyne receiver structure. There is no optical PLL shown as the receiver exhibits nonzero, albeit very low, IF. A minimum of two phases, those being in-phase and quadrature (I and Q), is needed to avoid loss of information. This synchronous intradyne receiver incorporates a 90° hybrid (see Section 13.6.3) and two balanced receivers (see Section 13.4.3) using four identical photodiodes. The carrier recovery delivers the electronic carrier required to provide an input to the electronic PLL which facilitates the IF downconversion to convert the detected I and Q signals from the small IF signal into two data streams. This following synchronous demodulation is per-formed by an orthogonal demodulator [Ref. 82].

Although the intradyne approach removes the need for an optical PLL while exhibiting the nearly similar baseband, direct detection receiver bandwidth of homodyne detection, it can necessitate laser linewidth restrictions similar to those when an optical PLL is utilized. A feedforward carrier recovery QPSK/BPSK intradyne receiver strategy has, however, been explored which is predicted to be extremely tolerant to laser linewidths, thus enabling the use of distributed feedback lasers [Ref. 83].

13.6.5 Phase diversity reception

An additional demodulation scheme employed in microwave systems which has been applied to coherent optical systems is phase diversity reception, or multiport detection [Refs 84, 85]. In these techniques the local oscillator laser is operated at a frequency com-parable with the frequency of the incoming signal but the two signals are not phase locked. The phase diversity receiver does, however, convert the incoming signal directly to the baseband and therefore has the bandwidth advantage of homodyne detection. As the optical mixing is not phase synchronized, the demodulation strategy avoids this major problem associated with homodyne detection, but at best the receiver sensitivity is equi-valent to that of heterodyne detection. Hence, from the viewpoint of receiver bandwidth requirements, phase diversity reception behaves like an optical homodyne receiver but it is essentially asynchronous with the sensitivity performance of, or worse than, a heterodyne receiver. It cannot therefore be regarded strictly as a homodyne technique* and certain authors have suggested it is more appropriate to classify such multiport detection schemes with heterodyne receivers [Ref. 4].

A number of optical phase diversity reception schemes have been investigated operat-ing with two or more ports and two or more matched receivers. The technique utilizes a fixed phase relationship between the ports of a multiport coupler to provide the direct demodulation to the baseband without the requirement for an optical PLL. One variant of the optical phase diversity receiver known as the in-phase and quadrature (I and Q) receiver

---

* Many authors have, however, referred to it as a homodyne strategy [Refs 86–88].

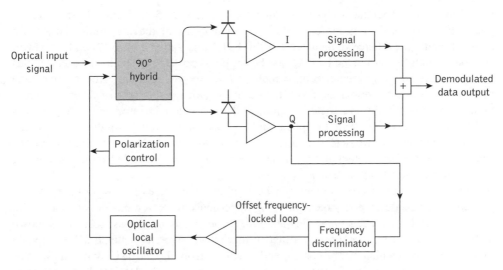

Figure 13.21 In-phase and quadrature phase diversity (two-phase) receiver

is shown in Figure 13.21. In this two-phase [Refs 85–89] scheme the incoming and local oscillator signals are combined in a 90° optical hybrid similar to the one described in Section 13.6.3. The 90° hybrid is connected to two detectors, the outputs of which are amplified and then passed through square-law devices prior to electrical recombination. The output signals from each receiver path prior to recombination can be written in terms of their voltages V_1 and V_2 as:

$$V_1^2 = k_1^2 m^2(t) \sin^2 \delta\phi \tag{13.21}$$

$$V_2^2 = k_2^2 m^2(t) \cos^2 \delta\phi \tag{13.22}$$

where k_1 and k_2 are constants, $m(t)$ is the modulation and $\delta\phi$ is the phase error. Hence the output signal from the receiver is:

$$V_1^2 + V_2^2 = k^2 [m(t)]^2 \tag{13.23}$$

where we assume $k = k_1 = k_2$. It may be observed from Eq. (13.23) that for ASK the demodulated signal is constant irrespective of the relative phase between the incoming and local oscillator signals. Therefore, with ASK modulation the laser linewidth requirements are comparable with heterodyne detection with asynchronous IF demodulation (see Section 13.5.1). For example, the experimental demonstration of such two-phase ASK demodulation was achieved using a commercial 1.5 μm DFB laser with a linewidth of 38 MHz at 150 Mbit s^{-1} providing a sensitivity of −55 dBm [Ref. 88].

For PSK modulation, however, the signal may be differentially demodulated by the inclusion of a 1 bit delay in one of the inputs to the square-law mixer. Hence the change in phase during a single bit period only is of concern and any longer term phase drift is removed. This DPSK I and Q demodulation is expected to have similar laser linewidth requirements to DPSK heterodyne detection (see Section 13.5.3) [Ref. 44]. The I and Q phase diversity receiver is more sensitive to fluctuations of the incoming SOP than a conventional optical heterodyne receiver [Ref. 89]. Polarization control should, however, be more straightforward in the I and Q case because it is possible to obtain electrical signals directly from the two receiver arms which provide exact information on the received polarization state. Nevertheless, problems do exist with two-port phase diversity reception, as in practice the electrical square-law demodulation is imperfect and produces additional terms which tend to appear in the baseband along with the demodulated signal. Moreover, the two detected currents must be 90° out of phase, which may only be achieved at the expense of additional signal processing [Ref. 79], and also the two arms of the receiver must be well matched.

Other phase diversity techniques can be used as alternatives to I & Q detection. In particular, the phase diversity receiver using three-phase reception has proved successful [Refs 85, 86, 91, 92]. A schematic diagram of the generalized three-phase receiver is shown in Figure 13.22. It may be observed that in this phase diversity scheme a 120° optical hybrid is required which can be conveniently realized from the intrinsic symmetry associated with the construction of a three-fiber fused biconical coupler [Ref. 93]. Furthermore, this strategy avoids the polarization sensitivity of the two-phase arrangement. However, the receiver sensitivity of this approach is poorer than the conventional optical heterodyne strategy as the additional port for the third detector introduces an extra 1.8 dB degradation [Ref. 84].

Demonstrations of three-phase schemes have taken place using helium–neon lasers operating at 650 Mbit s^{-1} [Ref. 91] and DFB lasers at 140 Mbit s^{-1} [Ref. 92]. Furthermore, demodulation of direct FSK signals generated from a DFB laser operating at 5 Gbit s^{-1} was demonstrated using a three-port single-filter phase diversity receiver [Ref. 87]. The single-filter detection was achieved by setting the local oscillator frequency equal to the center frequency of the stronger FSK sideband [Ref. 93]. Receiver sensitivities for a

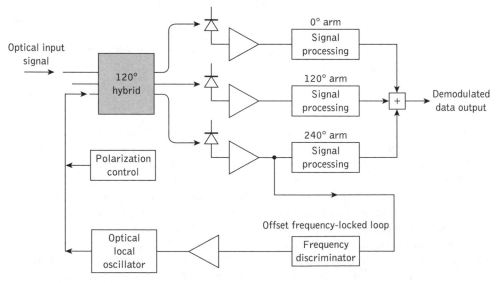

Figure 13.22 Phase diversity receiver using three-phase detection

bit-error-rate of 10^{-9} of -30.5 dBm and -27.0 dBm were obtained at transmission rates of 4 and 5 Gbit s^{-1} respectively [Ref. 87].

It has also been suggested [Ref. 44] that multiport detection could be extended to provide both phase and polarization diversity reception (see Section 13.6.6). Furthermore, greater numbers of ports than three for phase diversity receivers were envisaged (i.e. four, six or eight) combined with a balanced receiver approach (see Section 13.4.3) in order to reduce excess noise from the local oscillator laser [Refs 44, 84, 90, 94].

13.6.6 Polarization diversity reception and polarization scrambling

The polarization matching requirement of coherent optical receivers can be removed by using either polarization diversity reception or polarization scrambling [Refs 38–42] transmission. A block schematic of a generic polarization diversity receiver is shown in Figure 13.23. This scheme, which is essentially polarization insensitive, employs separate heterodyne or homodyne detection for the two orthogonal polarization states of the optical signal.

The polarization diversity receiver illustrated in Figure 13.23 exhibits an optical front end incorporating an optical 3 dB coupler and two polarization beam splitters which constitute a 90° optical hybrid (see Section 13.6.3). However, unlike coherent receivers requiring SOP control, the received optical signal is not linearly polarized but is generally elliptically polarized and uncontrolled. The local oscillator laser is linearly polarized at 45° with respect to the received signal and mixing takes place in the coupler prior to the two separate polarization beam splitters. Detection of the outputs from the polarization beam splitters takes place in the two balanced receivers (see Section 13.4.3) providing the in-phase and quadrature outputs. Finally, although being applicable to most modulation formats [Ref. 8] as indicated in Figure 13.23, the polarization diversity receiver has found utilization primarily for PM and FM systems.

Polarization-insensitive heterodyne detection has also been demonstrated using polarization scrambling or spreading transmission. In this technique the polarization state of

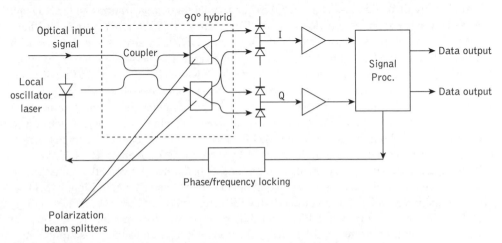

Figure 13.23 Generic polarization diversity receiver structure

the optical signal is deliberately changed at the coherent system transmitter so that all possible polarization states are propagated during a single bit period. Although this method significantly reduces the receiver complexity in comparison with polarization diversity reception, it incurs a reduction in sensitivity at the coherent optical receiver. While polarization diversity reception can be used with polarization multiplexed coherent optical systems (see Section 13.9.1), this is not the case for polarization scrambling transmission, which therefore is a significant limitation for the technique.

13.7 Differential phase shift keying

The initial measurements of bit-error-rate on experimental coherent optical fiber systems were reported in the early 1980s, followed by the first field demonstration of a DPSK system in 1988 [Ref. 95]. This system, which employed heterodyne detection, was installed using an 18-fiber cable in an underground duct over 176 km between Cambridge and Bedford in the UK, and successfully transmitted at a wavelength of 1.534 μm and a rate of 565 Mbit s^{-1} with a long-term measured bit-error-rate (BER) of 5×10^{-9} exhibiting a receiver sensitivity of −47.6 dBm (276 photons per bit). Error-free operation with no error floor at BER levels above 10^{-13} was also observed using the same system arrangement but operating over the 150 km of installed fiber cable. Moreover, subsequent improvements to the system included a hybrid balanced receiver using a GaAs–FET IC preamplifier to enhance the sensitivity to −52 dBm as well as the incorporation of an injection-laser-pumped, erbium-doped fiber amplifier repeater [Ref. 96].

As indicated in Section 13.1, the resurgence of interest in coherent and phase-modulated optical communications which has taken place since 2002 has been centered on phase-modulated systems and, in particular, differential binary PSK (DBPSK, which is often simply indicated as DPSK) and differential quadrature PSK (DQPSK) systems. These research activities resulted in excess of 20 long-haul experimental demonstrations of DPSK transmission at rates of between 10 and 170.6 Gbit s^{-1} before 2005 [Ref. 8]. In addition, more than 10 long-distance DQPSK experimental demonstrations have also been reported over the same period [Refs 8, 25]. Therefore in this section we focus on these more recent developments associated with DPSK and DQPSK transmission systems with specific reference to the return-to-zero (RZ) signal format. It is helpful, however, to adopt a now commonly used classification for optical PSK detection schemes that divides them into two broad categories which are referred to as coherent (where a local oscillator laser is employed) and interferometric (where a Mach–Zehnder delay interferometer in conjunction with differential direct detection is utilized). The latter category, which is also referred to as self-homodyning, is often used for DBPSK and is generalized to any differentially coherent signals including M-ary PSK by the term self-coherent [Refs 3, 25].

Optical DPSK signals carry the data in the phase difference between two consecutive symbols on the optical carrier as explained in Section 13.5.3 and shown in Figure 13.11. A DPSK transmitter, which is very similar to the PSK transmitter of Figure 13.10(c), is illustrated in Figure 13.24(a). The DPSK transmitter, however, also requires a precoder to enable the generation of the DPSK signal prior to its application to the phase modulator. Although a straight line phase modulator is displayed in Figure 13.24(a), a Mach–Zehnder

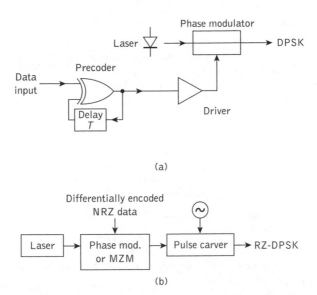

Figure 13.24 Differential phase shift keying transmitters: (a) DPSK transmitter; (b) RZ-DPSK transmitter

modulator can be used in its place to produce the phase modulation [Ref. 10]. Since the renewal of activity concerned with phase-modulated optical systems stimulated by the initial experimental demonstrations in 2002 [Refs 6, 7], the RZ pulse format has proved of most interest for the provision of both spectral efficiency and improved tolerance to fiber nonlinearities, enabling it to be utilized for long-haul transmission with optical amplifiers boosting the optical power levels [Ref. 8].

An optical RZ-DPSK transmitter is shown in Figure 13.24(b). In common with the DPSK transmitter in Figure 13.24(a), the nonreturn-to-zero (NRZ) data signal is differentially encoded using a precoder before it is applied to the phase modulator or the Mach–Zehnder modulator (MZM). The phase modulator only modulates the phase of the optical field resulting in a constant amplitude optical field, but it also introduces chirp across the bit transitions as a consequence of its noninstantaneous operation. By contrast an MZM always produces precise π phase transitions at the expense, however, of residual optical intensity dips at the phase transition locations. Since exact phase modulation is more important for both DPSK and RZ-DPSK than constant optical intensity, such transmitters are more frequently implemented using an MZM to perform the phase modulation. Finally, a pulse carver, usually comprising an MZM controlled by a synchronized sinusoidal drive with the same amplitudes and of opposite phase, provides the RZ-DPSK output [Ref. 9].

Two types of DPSK receiver are shown in Figure 13.25. In both cases a balanced optical receiver is employed (see Figure 13.9) incorporating a fiber coupler and two matched photodetectors. The heterodyne receiver depicted in Figure 13.25(a) then utilizes an electrical delay of bit period T in the phase comparison system together with the undelayed signal which then combine in a multiplier circuit to provide the differential phase. The output from the multiplexer is then integrated to eliminate residual noise, a positive voltage being obtained if the phases of the combined signals are the same. Alternatively, when the two

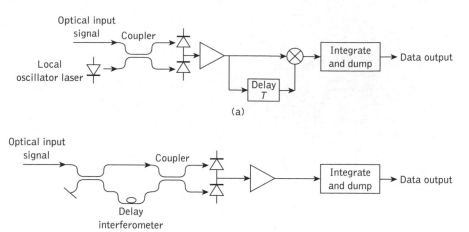

Figure 13.25 Differential phase shift keying receivers: (a) heterodyne DPSK receiver; (b) direct detection DPSK receiver

phases differ by 180°, a negative voltage is obtained at the data output. It should be noted that while frequency locking may be required, phase locking is not necessary for optical heterodyne DPSK demodulation.

Although the direct detection DPSK receiver shown in Figure 3.25(b) operates in the same manner as the heterodyne demodulator, there is no local oscillator laser signal and hence an asymmetric Mach–Zehnder interferometer, often termed a delay interferometer (DI), splits the incoming signal into two paths before recombining them in the fiber coupler after they have propagated with a path difference equivalent to the bit period T. Therefore in this interferometric detection scheme the balanced receiver follows the delay interferometer and acts as a mixer replacing the electrical multiplier circuit in the heterodyne demodulation process. Moreover, when using optical amplification before the receiver, then the direct detection DPSK receiver performance has been shown to be approximately the same as that of the heterodyne receiver [Ref. 97]. Furthermore, a matched filter (i.e. integrate and dump) is often utilized at the direct detection DPSK receiver output to reduce noise [Ref. 8]. Although using a DI to detect the DPSK signal creates more complexity in comparison with an IM/DD receiver, it is far more straightforward than deploying a heterodyne DPSK receiver with its requirement for a local oscillator laser. Hence interferometric direct detection for RZ-DPSK modulation has become an important area of renewed activity to facilitate long-haul optical transmission [Refs 98, 99]. It should be noted, however, that the theoretical sensitivity gain of DPSK over IM/DD indicated in Section 13.8.6 of up to 20 dB does not take into consideration the use of optical amplifiers within an IM/DD system. Both heterodyne and DI receivers systems exhibit much more limited sensitivity gain over optically amplified IM/DD systems such that the DPSK signal has around a 3 dB sensitivity gain, while the PSK signal exhibits only around a 3.5 dB sensitivity benefit [Ref. 8].

Direct detection (interferometric) DPSK transmission, using RZ pulses for the superior receiver sensitivity in comparison with on–off keying and the higher tolerance to fiber

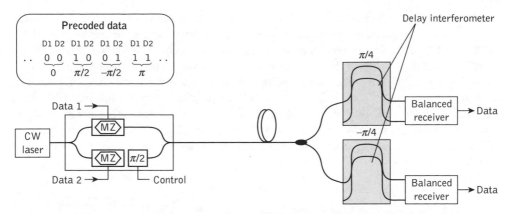

Figure 13.26 Typical DQPSK system implementation [Ref. 9]

nonlinearities [Ref. 100], its SOP-independent operation and its capability to provide high spectral efficiency, has attracted considerable attention since 2002 [Ref. 9]. For example, a sensitivity of just 8.7 photons per bit at a BER of 10^{-9} which is significantly below the 22 photons per bit quantum noise limit for direct detection RZ-DPSK has been demonstrated over a standard single-mode fiber transmission distance of 77 km at a speed of 40 Gbit s^{-1} when employing forward error correction [Ref. 101].

As indicated in Section 13.5.3, the spectral efficiency of PSK can be increased through the use of multilevel signaling. In particular, DQPSK has recently received much attention [Ref. 10]. The most commonly employed implementation of a DQPSK system is depicted in Figure 13.26 [Ref. 9]. It comprises a transmitter employing two parallel DPSK modulators (MZMs) that are integrated together in order to provide stability (it is possible to use a serial configuration which has also been experimentally demonstrated). The output from the CW laser source is divided by the splitter into two equal intensity paths to the two MZMs while an optical $\pi/2$ phase shifter in one of the paths and a combiner produce a single output signal with four phase shifts (i.e. 0, $\pi/2$, $-\pi/2$, π). Finally, a pulse carver (see Figure 13.24(b)) can be added to the transmitter structure to produce RZ-DQPSK.

The receiver effectively consists of two DPSK balanced receivers in which the phase difference in the arms of the two delay interferometers is set to $+\pi/4$ and $-\pi/4$. Hence the incoming DQPSK signal is first split into two equal parts and the differently biased DI receivers are employed in parallel to simultaneously demodulate the two binary data streams contained in the DQPSK signal. It should be noted that the DI delay must equal the symbol duration for the DQPSK demodulation which is twice the bit duration. Hence the benefit of DQPSK, being that for the same data rate the symbol rate is reduced by a factor of 2, is apparent. As a consequence the spectral occupancy is therefore reduced, the transmitter and receiver bandwidth requirements are lowered and the chromatic dispersion and polarization mode dispersion limitations are extended. In comparison with DPSK, however, the optical SNR needed to obtain a specific BER is increased in the range 1 to 2 dB [Ref. 9]. Furthermore, the frequency offset tolerance between the laser and the DI is around six times less than for DPSK, which causes more complexity in the DI design and stabilization.

Although the DQPSK receiver of Figure 13.26 is interferometric rather than coherent following the categorization provided earlier, coherent detection can be employed for both DPSK and DQPSK. In particular, homodyne optical PLL receiver structures (e.g. balanced and decision driven) [Refs 25, 102] together with phase diversity reception [Ref. 103] have been a focus of interest. Moreover, several techniques which are referred to as self-coherent have been utilized, including the decision-feedback-directed receiver [Ref. 104] and the sampled self-coherent optical receiver in which a digital representation of the received signal phase waveform is obtained [Ref. 105]. Both these approaches seek to implement a coherent receiver without the physical presence of a local oscillator laser source.

13.8 Receiver sensitivities

The basic detection principles for the ASK coherent optical receiver were discussed in Section 13.3. In addition, the 3 dB SNR improvement for the ASK homodyne receiver in comparison with the corresponding heterodyne receiver in the shot noise limit was demonstrated (Eqs (13.16) and (13.17)). Although a synchronous detection process was assumed in Section 13.3 for ASK with heterodyne detection, ASK with asynchronous detection can achieve approximately the same SNR limit [Ref. 106]. It is now important to consider the receiver sensitivities for the other major modulation schemes and detection processes so that the choices regarding the implementation of coherent optical fiber systems can be understood.

In Section 13.3 comparison between ASK heterodyne and homodyne reception was undertaken from determination of their respective shot- or quantum-noise-limited SNRs. As in this section we propose to extend this comparison to receiver sensitivities of other digital modulation schemes, it is useful to transfer from considerations of SNR to those of BER. In addition we will continue to consider minimum receiver sensitivities in the presence of quantum noise only, neglecting thermal and other noise sources in the electronic preamplifier discussed in Section 9.3, as well as excess noise sources in the local oscillator laser. Although these other noise sources are normally present, comparison of the modulation formats under quantum-noise-limited detection assists in the deliberations regarding the desirability of specific schemes. Furthermore, near-quantum-noise-limited reception is more readily achieved using heterodyne or homodyne detection than by employing direct detection. We concentrate on synchronous demodulation for the major modulation formats (i.e. ASK, FSK, PSK) prior to consideration of homodyne detection for the ASK and PSK modulation schemes.

13.8.1 ASK heterodyne detection

The ASK or OOK modulation format has similarities to digital transmission in an IM/DD optical fiber system, a BER analysis for the latter system being provided in Section 12.6.3. In a heterodyne receiver, however, the analyses of signal and noise phenomena are more complicated than in the IM–DD case because the optical detector output appears as an IF signal and not as a baseband signal. Hence the IF output current from the photodetector $I_S(t)$ which corresponds to the input current to the preamplifier from Eq. (13.9) can be written as:

$$I_S(t) = \begin{cases} I_{SH} \cos(\omega_{IF}t + \phi) & \text{for a 1 bit} \\ 0 & \text{for a 0 bit} \end{cases} \tag{13.24}$$

where:

$$I_{SH} = \frac{2\eta e}{hf} \sqrt{P_s P_L} \tag{13.25}$$

To obtain the IF noise current, two assumptions can be made. Firstly, it is assumed that the local oscillator signal power is much larger than the incoming signal power so that the total noise current is approximately equal to i_{SL}^2, given by Eq. (13.14), and this applies for both the 1 and the 0 bit. Secondly, it is assumed that this IF noise current $N(t)$ can be considered as narrowband noise which can be expressed as [Ref. 2]:

$$N(t) = x(t) \cos \omega_{IF}t + y(t) \sin \omega_{IF}t \tag{13.26}$$

where $x(t)$ and $y(t)$ are functions of time which vary at a much slower rate than the IF signal. It should be noted that the first and second terms in Eq. (13.26) represent the I and Q components respectively. Hence the mean square values of $x(t)$ and $y(t)$ may be written as:

$$\overline{x^2}(t) = \overline{y^2}(t) = \overline{i_{SL}^2} \tag{13.27}$$

For heterodyne synchronous detection, the IF amplifier is followed by a demodulation circuit which has a phase synchronized reference signal proportional to $\cos \omega_{IF}t$. Therefore, the detector output $V_d(t)$ is given by:

$$V_d(t) = k[I_S(t) + x(t)] \tag{13.28}$$

The probability density functions of $V_d(t)$ for the ASK signal $I_S(t)$ represented by Eq. (13.24) are illustrated in Figure 13.27. Moreover, it may be observed that these probability density functions are similar to those for digital IM–DD reception shown in Figure 12.37. Assuming this to be the case, the optimum decision threshold level D is set midway between the zero current (0 state) and the peak signal current (1 state) such that:

$$I_D \simeq \frac{I_{SH}}{2} = \frac{\eta e}{hf} \sqrt{P_s P_L} \tag{13.29}$$

The optical detector output given by Eq. (13.28) can now be considered as a baseband signal and noise contribution. Hence the analysis for BER can follow the method utilized for IM/DD in Section 12.6.3. We are therefore in a position to move straight to Eq. (12.21) and to substitute in the appropriate values from this derivation. Thus the probability of error $P(e)$ for ASK heterodyne synchronous detection can be written as:

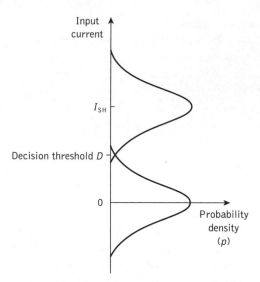

Figure 13.27 Probability density functions for ASK heterodyne synchronous detection

$$P(e) = \frac{1}{2}\left[\frac{1}{2}\text{erfc}\left(\frac{|I_{SH} - I_D|}{(\overline{i_{SL}^2})^{\frac{1}{2}}\sqrt{2}} \right) + \frac{1}{2}\text{erfc}\left(\frac{|-I_D|}{(\overline{i_{SL}^2})^{\frac{1}{2}}\sqrt{2}} \right) \right]$$ (13.30)

Substituting for I_D from Eq. (13.29) gives:

$$P(e) = \frac{1}{2}\left[\frac{1}{2}\text{erfc}\left(\frac{|I_{SH}/2|}{(\overline{i_{SL}^2})^{\frac{1}{2}}\sqrt{2}} \right) + \frac{1}{2}\text{erfc}\left(\frac{|-I_{SH}/2|}{(\overline{i_{SL}^2})^{\frac{1}{2}}\sqrt{2}} \right) \right]$$

$$= \frac{1}{2}\text{erfc}\left(\frac{I_{SH}}{2(\overline{i_{SL}})^{\frac{1}{2}}\sqrt{2}} \right)$$ (13.31)

Finally, substituting for I_{SH} using Eq. (13.25) and for $\overline{i_{SL}^2}$ from Eq. (13.14), where we replace B with the IF signal bandwidth B_{IF} because $I_S(t)$ was originally an IF signal exhibiting this bandwidth, then Eq. (13.31) can be written as:

$$P(e) = \frac{1}{2}\text{erfc}\left[\frac{2\eta e\sqrt{P_S P_L}}{hf} \middle/ 2\sqrt{2}\left(\frac{2e^2\eta P_L B_{IF}}{hf} \right)^{\frac{1}{2}} \right]$$

$$= \frac{1}{2}\text{erfc}\left(\frac{\eta P_S}{4hf B_{IF}} \right)^{\frac{1}{2}}$$ (13.32)

It is more appropriate to specify the probability of error in terms of the transmission bit rate B_T rather than the receiver bandwidth and, therefore, assuming transmission at a rate equivalent to twice the baseband bandwidth (see Eq. (3.12)), then $B_T = 2B \simeq B_{IF}$. Hence Eq. (13.32) becomes:

$$P(e) = \frac{1}{2}\text{erfc}\left(\frac{\eta P_S}{4hf B_T} \right)^{\frac{1}{2}}$$ (13.33)

This expression allows the shot-noise-limited performance of ASK heterodyne synchronous detection to be compared with alternative detection schemes.

For ASK heterodyne asynchronous or envelope detection it can be shown that the probability of error in the shot noise limit, under similar assumptions to those used above for synchronous detection, is given by the approximate expression [Refs 106, 107]:

$$P(e) \simeq \tfrac{1}{2}\exp\left[-\left(\frac{I_{SH}^2}{8(\overline{i_{SL}^2})}\right)\right] \tag{13.34}$$

Again substituting for I_{SH} and $\overline{i_{SL}^2}$ from Eqs (13.25) and (13.14), respectively, gives:

$$P(e) \simeq \tfrac{1}{2}\exp\left[-\left(\frac{\eta P_S}{4hfB_T}\right)\right] \tag{13.35}$$

Moreover, as indicated in Section 13.8.7, these formulas give approximately equivalent BER results to those provided by Eqs (13.31) and (13.33) for ASK heterodyne synchronous detection. The reason for this situation is that the approximation $erfc(u) \simeq \exp[-(u)^2]$ holds for large u and hence for a low BER [Ref. 107].

13.8.2 FSK heterodyne detection

We commence by considering FSK heterodyne synchronous detection in the shot or quantum noise limit. The two angular frequencies for the transmitted 1 and 0 bits are assumed to be ω_1 and ω_2 so that:

$$I_S(t) = \begin{cases} I_{SH}\cos(\omega_1 + \phi) & \text{for a 1 bit} \\ I_{SH}\cos(\omega_2 + \phi) & \text{for a 0 bit} \end{cases} \tag{13.36}$$

where I_{SH} is defined in Eq. (13.25) and ϕ, which is a function of time, represents the phase noise associated with the semiconductor laser. This is neglected, as before in Section 13.8.1, because we are concerned with shot-noise-limited detection.

It is assumed that the signal $I_S(t)$ is received using two receivers tuned to ω_1 and ω_2 and that the output voltages from receivers 1 and 2 are V_1 and V_2 respectively. Furthermore it is assumed that the two receivers exhibit ideal frequency selectivity such that there is no crosstalk between ω_1 and ω_2 and therefore any additional voltages are generated by shot noise effects only. It is possible to just consider the time slot when a 1 bit (ω_1) is transmitted without losing generality. Hence the PDF of the output from receiver 1 is given by [Ref. 106]:

$$p_1(V) = \frac{1}{(\overline{i_{SL}^2})^{\frac{1}{2}}\sqrt{2\pi}} \exp\left\{-\left[\frac{(I_{SH} - V_1)^2}{2(\overline{i_{SL}^2})}\right]\right\} \tag{13.37}$$

Again we assume the local oscillator output power to be much higher than that of the incoming signal so that the total noise current is approximately equal to $\overline{i_{SL}^2}$ provided by Eq. (13.14). The noise output from receiver 2 can therefore be written as:

$$p_2(V) = \frac{1}{(i_{SL}^2)^{\frac{1}{2}}\sqrt{2\pi}} \exp\left\{-\left[\frac{V_2^2}{2(i_{SL}^2)}\right]\right\}$$

(13.38)

As an error occurs when $V_2 > V_1$, then the probability of error $P(e)$ is equivalent to the probability that $V_1 - V_2 < 0$. Hence:

$$P(e) = \int_{-\infty}^{0} \frac{1}{[2\pi 2(i_{SL}^2)]^{\frac{1}{2}}} \exp\left\{-\left[\frac{(w - I_{SH})^2}{4(i_{SL}^2)}\right]\right\}dw$$

(13.39)

Changing the limits of the integration gives:

$$P(e) = \int_{I_{SH}/(i_{SL}^2)}^{\infty} \frac{1}{\sqrt{\pi}} \exp(-z^2)dz$$

(13.40)

Comparison with the definition for the complementary error function given in Eq. (12.17) allows Eq. (13.40) to be written as:

$$P(e) = \tfrac{1}{2}\mathrm{erfc}\left(\frac{I_{SH}}{2(i_{SL}^2)^{\frac{1}{2}}}\right)$$

(13.41)

Finally, substituting from Eq. (13.25) for I_{SH} and from Eq. (13.14) for $\overline{i_{SL}^2}$ where the bandwidth, which in this heterodyne detection case is B_{IF}, is again (see Section 13.8.1) written in terms of the transmission bit rate B_T, then the probability of error becomes:

$$P(e) = \tfrac{1}{2}\mathrm{erfc}\left(\frac{\eta P_S}{2hfB_T}\right)^{\frac{1}{2}}$$

(13.42)

The comparison of Eq. (13.42) with Eq. (13.33) indicates that FSK heterodyne synchronous detection has a receiver sensitivity which in the shot noise limit is 3 dB higher than that of ASK heterodyne synchronous detection. This improvement in sensitivity for FSK modulation may be attributed to the use of two frequencies (and hence dimensions) rather than only the one in the case of ASK. It should be noted, however, that a similar BER is obtained with the two modulation schemes when the same average power is transmitted. Whereas with ASK, zero signal power is transmitted for a binary 0 bit, in FSK a similar signal power is continuously transmitted [Ref. 108]. Nevertheless, there are advantages associated with the use of FSK over ASK, even when two systems with the same average signal power are considered. In particular, the optimization of the decision level proves easier and the spectrum broadening as a result of switching between a one and a zero state in practice is much reduced on that obtained with ASK.

Considering FSK heterodyne asynchronous or envelope detection, it can be shown that the probability of error in the shot noise limit under similar assumptions to those above for synchronous detection is given by the expression [Refs 106, 107]:

$$P(e) = \tfrac{1}{2}\exp\left[-\left(\frac{I_{SH}^2}{4(i_{SL}^2)}\right)\right]$$

(13.43)

Substituting for I_{SH} and $\overline{i_{SL}^2}$ from Eqs (13.25) and (13.14), respectively, gives:

$$P(e) = \tfrac{1}{2}\exp\left[-\left(\frac{\eta P_s}{2hfB_T}\right)\right]$$

(13.44)

This result for FSK asynchronous detection is approximately equivalent to the one obtained for synchronous detection (Eq. (13.42)) and shows a 3 dB improvement over ASK asynchronous detection (Eq. (13.35)).

13.8.3 PSK heterodyne detection

In this modulation format the information is transmitted by a carrier of one phase for a binary 1 and a different phase for a binary 0. The phase shift employed is normally π radians so that:

$$I_S(t) = \begin{cases} I_{SH}\cos(\omega_{IF}t + \phi) & \text{for a 1 bit} \\ I_{SH}\cos(\omega_{IF}t + \pi + \phi) & \\ \text{or} & \text{for a 0 bit} \\ -I_{SH}\cos(\omega_{IF}t + \phi) & \end{cases}$$

(13.45)

Therefore the synchronously detected signal $I_S(t)$ is positive for the 1 bits and negative for the 0 bits. In this case the optimum decision-level current is given as $I_D = 0$ instead of that obtained in Eq. (13.29) for ASK synchronous detection. Nevertheless, a similar method to obtain the probability of error $P(e)$ for the PSK heterodyne synchronous detection to that used in Section 13.8.1 for ASK detection may be employed. Hence, assuming that the output voltage from the receiver for a binary 1 is V_1 and for a binary 0 it is V_2, then:

$$P(e) = \frac{1}{2}\int_{-\infty}^{0} \frac{1}{(i_{SL}^2)^{\frac{1}{2}}\sqrt{2\pi}} \exp\left[-\frac{(I_{SH}-V_1)^2}{2(i_{SL}^2)}\right] dV_1$$

$$+ \frac{1}{2}\int_{0}^{\infty} \frac{1}{(i_{SL}^2)^{\frac{1}{2}}\sqrt{2\pi}} \exp\left[-\frac{(-I_{SH}-V_2)^2}{2(i_{SL}^2)}\right] dV_2$$

$$= \frac{1}{2}\,\text{erfc}\left(\frac{I_{SH}}{(i_{SL}^2)^{\frac{1}{2}}\sqrt{2}}\right)$$

(13.46)

Substituting from Eq. (13.25) for I_{SH} and from Eq. (13.14) for $\overline{i_{SL}^2}$ where the bandwidth, which in this heterodyne detection case is B_{IF}, and following Section 13.8.1, it is written in terms of the transmission bit rate B_T, then the probability of error becomes:

$$P(e) = \tfrac{1}{2}\text{erfc}\left(\frac{\eta P_s}{hfB_T}\right)^{\frac{1}{2}}$$

(13.47)

It may be noted that in the shot noise limit PSK heterodyne synchronous detection exhibits 3 dB and 6 dB more sensitivity than the FSK and ASK heterodyne synchronous detection schemes respectively (Eqs (13.42) and (13.33)). In practice, however, very small levels of phase fluctuation at the transmitter can significantly deteriorate the potential low BER of the PSK system.

Although asynchronous PSK detection is not strictly realizable, a more relaxed synchronous detection process is afforded by differential PSK in which the transmitted information is contained by the phase difference between two consecutive bit periods (see Sections 13.5.3 and 13.7). The probability of error in the detection of this modulation format at the shot noise limit and under the previous assumption is given by [Refs 106, 107]:

$$P(e) = \tfrac{1}{2}\exp\left[-\left(\frac{I_{SH}^2}{2(\overline{i_{SL}^2})}\right)\right] \tag{13.48}$$

Furthermore, substituting for I_{SH} and $\overline{i_{SL}^2}$ from Eqs (13.25) and (13.14) gives:

$$P(e) = \tfrac{1}{2}\exp\left[-\left(\frac{\eta P_S}{hf B_T}\right)\right] \tag{13.49}$$

The above expression for DPSK indicates a roughly equivalent probability of error for this scheme to that obtained with PSK synchronous detection (Eq. (13.47)). Moreover, it demonstrates a potential 3 dB and 6 dB improvement over the FSK and ASK asynchronous detection schemes respectively.

13.8.4 ASK and PSK homodyne detection

From the three basic modulation formats, ASK and PSK signals can be demodulated using a homodyne detection scheme, provided that both the frequency and the phase of the local oscillator output signal can be synchronized to the incoming carrier signal (see Section 13.6.3). It should be noted that FSK modulation can only be detected using a homodyne-type receiver when the device has two phase-controlled local oscillators. The only exception to this occurs when phase diversity reception (see Section 13.6.5) is employed, but these techniques cannot be regarded as true homodyne detection schemes.

It was shown in Section 13.3 that the reduction in the bandwidth requirement for homodyne detection produced a sensitivity improvement of 3 dB over the corresponding ASK heterodyne detection scheme. The probability of error for ASK homodyne detection can be derived from a slight modification to the result obtained in Eq. (13.31) for ASK heterodyne synchronous detection. This modification involves the noise power term $(\overline{i_{SL}^2})$ which in the homodyne case is reduced by a half because of the factor of two bandwidth reduction. Hence the probability of error for ASK homodyne detection is given by [Ref. 107].

$$P(e) = \tfrac{1}{2}\mathrm{erfc}\left(\frac{I_{SH}}{2(\overline{i_{SL}^2}/2)^{\frac{1}{2}}\sqrt{2}}\right)$$

$$= \tfrac{1}{2}\mathrm{erfc}\left(\frac{I_{SH}}{2(\overline{i_{SL}^2})^{\frac{1}{2}}}\right) \tag{13.50}$$

It should be noted, however, that the signal power in Eq. (13.50) remains the same as in the heterodyne case which may be observed to be correct by comparing Eqs (13.10) and (13.11). Substitution from Eq. (13.25) for I_{SH} and from Eq. (13.14) for $\overline{i_{SL}^2}$ where in this case the bit rate B_T is set equal to the baseband bandwidth B gives:

$$P(e) = \tfrac{1}{2}\mathrm{erfc}\left(\frac{\eta P_S}{2hfB_T}\right)^{\frac{1}{2}} \tag{13.51}$$

Considering now the PSK homodyne detection scheme, Eq. (13.46) for PSK synchronous detection can be modified in a similar manner to the above so that the probability of error is given by:

$$P(e) = \tfrac{1}{2}\mathrm{erfc}\left(\frac{I_{SH}}{(\overline{i_{SL}^2})^{\frac{1}{2}}}\right) \tag{13.52}$$

Again substituting from Eqs (13.25) and (13.14) gives the probability of error for PSK homodyne detection in the shot noise limit as:

$$P(e) = \tfrac{1}{2}\mathrm{erfc}\left(\frac{2\eta P_S}{hfB_T}\right)^{\frac{1}{2}} \tag{13.53}$$

The result obtained in Eq. (13.53) represents the lowest error probability and hence the highest receiver sensitivity of all the coherent detection schemes. As anticipated, it displays a 3 dB improvement over PSK heterodyne synchronous detection.

13.8.5 Dual-filter direct detection FSK

The dual-filter direct detection FSK receiver is illustrated in Figure 13.28(a). An example of dual optical filters is provided in Figure 13.28(b) which depicts two optical multilayer dielectric filters. It can be observed in Figure 13.28(a) that a balanced receiver configuration (see Figure 13.8) is utilized with one photodiode connected to the output of each optical filter. Bragg gratings in place of multilayer dielectric filters can also be employed to implement the optical filters.

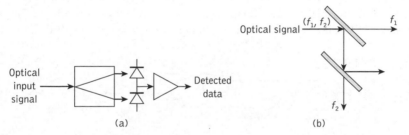

Figure 13.28 Dual-filter direct detection FSK receiver: (a) receiver structure; (b) dual filters using multilayered dielectric filters

When the two optical filters in Figure 13.28 are matched filters and the two received FSK signals are orthogonal to each other, the probability of error $P(e)$ for the receiver can be derived as [Ref. 8]:

$$P(e) = \tfrac{1}{2} \exp\left[-\left(\frac{\eta P_s}{2hf B_T}\right)\right] \left(1 + \frac{\eta P_s}{8hf B_T}\right) \tag{13.54}$$

Comparison of this expression with that for asynchronous FSK of Eq. (13.44) shows that the error probability is increased by a factor of $(1 + \eta P_s/8hf B_T)$ for direct detection as a consequence of the amplified noise from the orthogonal polarization. For an error probability of 10^{-9}, FSK synchronous detection is around 0.45 dB more sensitive than asynchronous heterodyne detection which has about 0.40 dB more sensitivity than a dual-filter direct detection FSK receiver.

13.8.6 Interferometric direct detection DPSK

A direct detection receiver for DPSK employing a delay interferometer was shown in Figure 13.25(b). The receiver incorporating some additional elements is provided again in Figure 13.29. At the receiver input there is an optical filter which is assumed to be an optical matched filter for the transmitted signal. Following this element the signal is split into two paths within the interferometer and then combined again after a path difference equivalent to a one bit period delay of T. In practice the path difference of T must be chosen such that $\exp(j\omega T)$ is equal to unity where ω is the angular frequency of the signal. A balanced receiver (see Figure 13.8) is then used to obtain the photocurrent while an electrical low-pass filter is deployed to reduce the receiver noise. It should be noted it is important that the lower pass filter has a wide bandwidth and hence does not distort the received signal.

With the assumption of the use of an optical matched filter the probability of error $P(e)$ for the receiver depicted in Figure 13.29 can be shown to be [Ref. 8]:

$$P(e) = \tfrac{1}{2} \exp\left[-\left(\frac{\eta P_s}{hf B_T}\right)\right] \left(1 + \frac{\eta P_s}{4hf B_T}\right) \tag{13.55}$$

Comparison of this expression with that for the asynchronous DPSK receiver of Eq. (13.49) indicates that the error probability is increased by a factor of $(1 + \eta P_s/4hf B_T)$ for direct detection. Moreover, this increase in the probability of error is attributed to the amplified noise from the orthogonal polarization at the receiver. Hence for an error probability of 10^{-9}, asynchronous DPSK heterodyne detection is marginally worse by 0.45 dB than synchronous

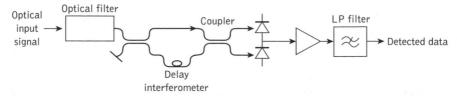

Figure 13.29 Interferometric direct detection DPSK receiver

PSK detection while direct detection DPSK is around 0.40 dB less sensitive than asynchronous DPSK. This direct detection DPSK receiver has, however, 3 dB more sensitivity than the direct detection FSK receiver of Section 13.8.5, and also outperforms ASK with asynchronous detection (see Section 13.8.1) by around 2.7 dB at a BER of 10^{-9} [Ref. 3].

13.8.7 Comparison of sensitivities

A comparison of the analytical results for the major modulation formats and detection schemes obtained in Sections 13.8.1 to 13.8.6 is provided in Table 13.2. Moreover, to allow a comparison to be made with IM-DD, an additional column records the details determined from Eq. (9.7) and Example 9.1 for ASK and the results from Sections 13.8.5 and 13.8.6 for FSK and PSK, respectively. It should be noted, however, that the average number of photons per bit required to maintain a BER of 10^{-9} assumes in the case of ASK that photons arrive over two bit periods because no light is transmitted for a binary 0. Hence the values shown must be doubled if the actual number of photons required to register a binary 1 with a BER of 10^{-9} is to be recorded (i.e. not the average over the two bit periods). This factor can lead to some confusion in the approaches adopted by different authors (e.g. Refs 8, 67, 107, 109). However, no such difficulties occur with the FSK and PSK modulation formats as a constant amplitude carrier signal is transmitted for both the binary 1 and 0 bits.

The average number of received photons per bit at a particular BER as given in Table 13.2 may be determined from the expressions which are provided for the respective error probabilities in the table. An analytical definition for the number of received photons per bit, however, is needed. This may be written down by considering Eq. (8.5) which simply provides the incident photon rate at an optical detector. To convert this into the photon number per bit N_p the expression must be simply multiplied by the signaling interval τ so that:

$$N_p = \frac{P_s \tau}{hf} = \frac{P_s}{hf B_T} \tag{13.56}$$

where we have written the peak signal power P_S in place of P_o in Eq. (8.5) for consistency with the notation in this chapter, and B_T is the transmission bit rate. Furthermore, it is useful to note that Eqs (13.16) and (13.17) which provide the SNRs for heterodyne and homodyne detection, respectively, in their shot noise limits can also be expressed in terms of the number of received photons per bit by taking $B_T = 2B$ as [Ref. 107]:

$$\left(\frac{S}{N}\right)_{\text{het-lim}} = \frac{\eta P_s}{hf B_T} = \eta N_p \tag{13.57}$$

and:

$$\left(\frac{S}{N}\right)_{\text{het-lim}} = \frac{\eta P_s}{hf B_T/2} = 2\eta N_p \tag{13.58}$$

At this stage, however, we are concerned with the determination of the average numbers of photons per bit for the various digital modulation schemes listed in Table 13.2. An explanation of how these values are obtained is provided in the following example.

Table 13.2 Comparison of optical receiver sensitivities in the quantum or shot noise limit. The upper entry in each detection technique provides the possibility of error determined for the different schemes, while the lower entry represents the average number of photons per bit required by an ideal binary receiver ($\eta = 1$) to achieve a BER of 10^{-9}

Modulation	Homodyne detection	Heterodyne Synchronous detection	Asynchronous detection	Direct detection
ASK or OOK	$\frac{1}{2}\mathrm{erfc}\left(\frac{\eta N_\mathrm{p}}{2}\right)^{1/2}$	$\frac{1}{2}\mathrm{erfc}\left(\frac{\eta N_\mathrm{p}}{4}\right)^{1/2}$	$\frac{1}{2}\exp\left[-\left(\frac{\eta N_\mathrm{p}}{4}\right)\right]$	$\frac{1}{2}\exp[-(\eta N_\mathrm{p})]$
Av. no. photons per bit*	18	36	40	10.4
FSK	No (only very special receiver)	$\frac{1}{2}\mathrm{erfc}\left(\frac{\eta N_\mathrm{p}}{2}\right)^{1/2}$	$\frac{1}{2}\exp\left[-\left(\frac{\eta N_\mathrm{p}}{2}\right)\right]$	$\frac{1}{2}\exp\left[-\left(\frac{\eta N_\mathrm{p}}{2}\right)\right]\left(1+\frac{\eta N_\mathrm{p}}{8}\right)$
Av. no. photons per bit	–	36	40	43.8
PSK	$\frac{1}{2}\mathrm{erfc}(2\eta N_\mathrm{p})^{1/2}$	$\frac{1}{2}\mathrm{erfc}(\eta N_\mathrm{p})^{1/2}$	DPSK† $\frac{1}{2}\exp[-(\eta N_\mathrm{p})]$	DPSK $\frac{1}{2}\exp\left[-(\eta N_\mathrm{p})\right]\left(1+\frac{\eta N_\mathrm{p}}{4}\right)$
Av. no. photons per bit	9	18	20	21.9

* Values provided assume that the photons arrive over two bit periods.
† Strictly speaking, there is not a heterodyne asynchronous demodulation scheme for PSK. Differential PSK (DPSK), however, exhibits a less stringent synchronous detection technique than conventional PSK when using a receiver local oscillator and is therefore included in the asynchronous column for convenience.

Example 13.3

Calculate the number of received photons per bit in order to maintain a BER of 10^{-9} for:

  (a) ASK heterodyne synchronous detection;

  (b) ASK heterodyne asynchronous detection;

  (c) PSK homodyne detection.

An ideal binary receiver may be assumed in all cases.

Solution: (a) Substituting N_p from Eq. (13.56) in Eq. (13.33) gives the probability of error for ASK heterodyne detection as:

$$P(e) = \tfrac{1}{2}\text{erfc}\left(\frac{\eta N_p}{4}\right)^{\frac{1}{2}}$$

To maintain a BER of 10^{-9} and with an ideal binary receiver ($\eta = 1$):

$$10^{-9} = \tfrac{1}{2}\text{erfc}\left(\frac{N_p}{4}\right)^{\frac{1}{2}}$$

and:

$$\left(\frac{N_p}{4}\right)^{\frac{1}{2}} = 4.24$$

Hence:

$$\frac{N_p}{4} \simeq 18 \quad \text{and} \quad N_p = 72$$

However, for ASK the 72 photons can arrive over two bit periods, assuming an equal number of ones and zeros. Hence the average number of photons per bit required is 36.

(b) Substituting N_p from Eq. (13.56) into Eq. (13.35) gives:

$$P(e) = \tfrac{1}{2}\exp\left[-\left(\frac{\eta N_p}{4}\right)\right]$$

In this case:

$$\exp\left[-\left(\frac{N_p}{4}\right)\right] = 2 \times 10^{-9}$$

and therefore:

$$\frac{N_p}{4} \simeq 20 \quad \text{and} \quad N_p = 80$$

Again we are considering ASK modulation so that the average number of received photons per bit is 40. It may be noted that this result is very approximately equal to the number obtained in (a).

(c) For PSK homodyne detection, N_p from Eq. (13.56) may be substituted in Eq. (13.47) to give:

$$P(e) = \tfrac{1}{2}\text{erfc}(2N_p)^{\frac{1}{2}}$$

Hence:

$$(2N_p)^{\frac{1}{2}} = 4.24$$

and:

$$N_p \simeq \frac{18}{2} = 9$$

In this case N_p is equal to the average number of photons per bit as 9 photons must be received for the zero bit as well as the one bit in order to achieve a BER of 10^{-9}.

As mentioned in Section 13.8.1, the approximation $\text{erfc}(u) \simeq \exp[-(u)^2]$ is used by some authors [Refs 106, 107] to indicate the rough equivalence between the error probabilities for the synchronous and asynchronous detection cases. However, it may be observed from Table 13.2 and Example 13.3 that the asynchronous detection in reality requires slightly more photons per bit (slightly higher incoming optical power) than does synchronous detection in the shot noise limit. It should be noted that a more accurate approximation relating the complementary error function to the negative exponential of $\text{erfc}(u) \simeq \exp[-(u)^2/x\sqrt{\pi}]$ for $x > 3$ is used by some authors [Ref. 109].

Nevertheless, aside from these approximations it is possible to write down a generalized expression for the error probabilities of the various modulation formats with synchronous detection schemes recorded in Table 13.2. For fully synchronous heterodyne and homodyne detection the generalized expression therefore takes the form:*

$$P(e) = \tfrac{1}{2}\text{erfc}\left(\frac{KZ\eta N_P}{4}\right)^{\frac{1}{2}} \tag{13.59}$$

where K is a constant which is equal to 1 for heterodyne detection and 2 for homodyne detection. The constant Z is determined by the modulation scheme as follows: for ASK and FSK, $Z = 1$; and for PSK, $Z = 4$. As with the expressions recorded in Table 13.2, the generalized relationship of Eq. (13.59) can also be used to determine the minimum detectable power level to maintain a particular BER with a specific modulation scheme and synchronous receiver.

---

* The exception to this is DPSK which, although in reality is synchronous when using a receiver local oscillator does not fulfill the necessary continuous phase-matching condition.

Example 13.4

Determine the minimum incoming optical power level required to detect a 400 Mbit s^{-1} FSK signal at a BER of 10^{-9} using an ideal heterodyne synchronous receiver operating at a wavelength of 1.55 μm.

Solution: For FSK heterodyne synchronous detection, $K = 1$ and $Z = 1$ and substituting Eq. (13.56) into Eq. (13.59) gives:

$$10^{-9} = \tfrac{1}{2}\operatorname{erfc}\left(\frac{\eta P_{\mathrm{S}}}{2hfB_{\mathrm{T}}}\right)^{\frac{1}{2}}$$

Hence for the ideal receiver:

$$\frac{P_{\mathrm{S}}}{2hfB_{\mathrm{T}}} = 18$$

and:

$$P_{\mathrm{S}} \simeq \frac{36hc\,B_{\mathrm{T}}}{\lambda} = \frac{36 \times 6.63 \times 10^{-34} \times 3 \times 10^{8} \times 400 \times 10^{6}}{1.55 \times 10^{-6}}$$

$$= 1.8 \text{ nW} \quad \text{or} \quad -57.4 \text{ dBm}$$

Therefore, the minimum incoming peak power level to maintain a BER of 10^{-9} is −57.4 dBm. Although this level may appear at first sight rather high, it must be remembered that the bit rate of 400 Mbit s^{-1} is also a factor which directly contributes to the sensitivity of the receiver (i.e. higher bit rates give lower receiver sensitivity).

The relative sensitivities of the different modulation formats and detection schemes can be readily deduced from consideration of the average numbers of photons per bit indicated for each detection technique in Table 13.2. However, a more immediate comparison of the relative receiver performances may be obtained from Figure 13.30. It may be observed that the sensitivity improvement of ASK heterodyne coherent detection over IM/DD is 10 to 20 dB in Figure 13.30, whereas in Table 13.2 IM-DD would appear to be around 6 dB more sensitive than ASK heterodyne detection. This second statement is true in theory but in practice practical direct detection receivers require in the order of 400 to 4000 photons per bit to maintain a BER of 10^{-9}. By contrast, practical ASK heterodyne coherent receivers are far more likely to be operated near the quantum or shot noise limit and thus provide the sensitivity indicated in Table 13.2. In addition, it can be observed from Figure 13.30 that PSK heterodyne and homodyne detection provide for the highest receiver sensitivity.

The probability of error as a function of the SNR for phase-modulated transmission is displayed in Figure 13.31. The three error probabilities obtained from Table 13.2 for heterodyne synchronous detection, heterodyne differential detection and interferometric direct detection are compared. Clearly synchronous detection exhibits the highest sensitivity

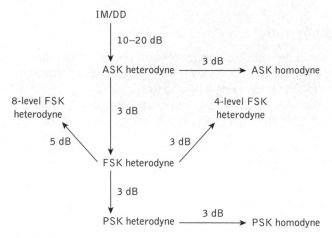

Figure 13.30 Receiver sensitivity improvements using various coherent modulation and demodulation schemes

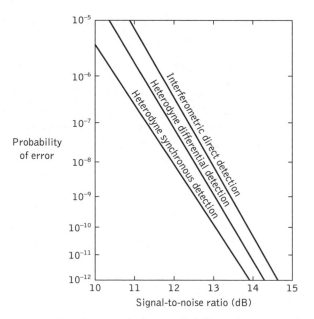

Figure 13.31 Phase-modulated transmission probability of error against the signal-to-noise ratio at the receiver

which translates into long potential repeater/amplifier spacing requirements for coherent optical fiber transmission. Moreover, a reduction in receiver sensitivity is caused by increases in the transmission bit rate as mentioned in Example 13.3. Hence both the choice of coherent transmission scheme and the bit rate will impact on the ultimate repeater/amplifier spacing for the link [Ref. 21].

Example 13.5

Calculate the absolute maximum repeater spacing that could be provided to maintain a BER of 10^{-9} within a coherent optical fiber system operating at a wavelength of 1.55 μm when the fiber and splice/connector losses average out at 0.2 dB km^{-1}, the optical power launched into the fiber link is 2.5 mW and the transmission rates are 50 Mbit s^{-1} and 1 Gbit s^{-1}. For both bit rates consider the following ideal receiver types:

(a) ASK heterodyne synchronous detection;

(b) PSK homodyne detection.

Solution: (a) Considering the 50 Mbit s^{-1} transmission rate and using the result obtained for ideal ASK heterodyne synchronous detection in Example 13.3 (average photons per bit required 36) from Eq. (13.56):

$$N_{\mathrm{p}} = \frac{P_{\mathrm{S}}}{hfB_{\mathrm{T}}} = \frac{P_{\mathrm{S}}}{hfB_{\mathrm{T}}} \simeq 36$$

Hence:

$$P_{\mathrm{S}} \simeq 36hc\, B_{\mathrm{T}} = \frac{36 \times 6.63 \times 10^{-34} \times 3 \times 10^{8} \times 50 \times 10^{6}}{1.55 \times 10^{-6}}$$

$$= 0.23 \text{ nW} \quad \text{or} \quad -66.4 \text{ dBm}$$

The maximum system margin with no overheads is therefore:

Max. system margin = 4 dBm − (−66.4) dBm = 70.4 dB

Moreover, the absolute maximum repeater spacing is:

$$\text{Max. repeater spacing} = \frac{70.4}{0.2} = 352 \text{ km}$$

For 1 Gbit s^{-1} we have:

$$P_{\mathrm{S}} \simeq \frac{36 \times 6.63 \times 10^{-34} \times 3 \times 10^{8} \times 10^{9}}{1.55 \times 10^{-6}}$$

$$= 4.6 \text{ nW} \quad \text{or} \quad -53.4 \text{ dBm}$$

Therefore the maximum system margin is 57.4 dB and:

$$\text{Max. repeater spacing} = \frac{57.4}{0.2} = 287 \text{ km}$$

Again using the result for PSK homodyne detection obtained in Example 13.3 and considering first the 50 Mbit s^{-1} rate, then from Eq. (13.56):

▶

$$N_p = \frac{P_S}{hf\,B_T} = \frac{P_S}{hf\,B_T} \simeq 9$$

and:

$$P_S \simeq 9hc\,B_T = \frac{9 \times 6.63 \times 10^{-34} \times 3 \times 10^8 \times 50 \times 10^6}{1.55 \times 10^{-6}}$$

$$= 58\ \text{pW} \quad \text{or} \quad 72.4\ \text{dBm}$$

The maximum system margin is now 76.4 dB so that:

$$\text{Max. repeater spacing} = \frac{76.4}{0.2} = 382\ \text{km}$$

For 1 Gbit s^{-1}:

$$P_S \simeq \frac{9 \times 6.63 \times 10^{-34} \times 3 \times 10^8 \times 10^9}{1.55 \times 10^{-6}}$$

$$= 1.15\ \text{nW} \quad \text{or} \quad -59.4\ \text{dBm}$$

Hence the maximum system margin is 63.4 dB and:

$$\text{Max. repeater spacing} = \frac{63.4}{0.2} = 317\ \text{km}$$

Thus the range of repeater spacings indicated by this example is between 287 and 382 km.

The relative receiver sensitivities of two, multilevel FSK modulation schemes are indicated in Figure 13.32. Hence, as anticipated, these four- and eight-level FSK heterodyne detection schemes display a reduction in sensitivity of 3 dB and 5 dB, respectively, over binary FSK heterodyne detection. Receiver sensitivity characteristics for other M-ary modulation schemes are shown in Figure 13.32(a) [Ref. 66]. The receiver sensitivities for binary PSK (BPSK), quaternary or 4-level PSK (QPSK) and 16-level quadrature amplitude modulation (QAM) against their information transmission capacities are displayed. Furthermore, these characteristics are a function of the receiver thermal noise parameters which are also shown in Figure 13.32(a). It may be observed that at higher transmission rates the receiver sensitivities are degraded by increases in thermal noise, as demonstrated by the corresponding characteristics given in Figure 13.32(b) [Ref. 66].

However, QPSK provides an M-ary modulation format which will enable the achievement of high-capacity coherent optical transmission with no sensitivity penalty in comparison with binary PSK. It may be noted that in Figure 13.32(a) the sensitivity of QPSK is at least equivalent to (or marginally better than) that of BPSK. Moreover, a feasibility study concerned with coherent optical QPSK transmission also supported this observation

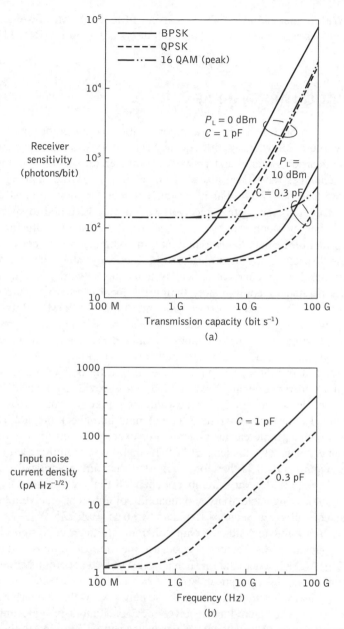

Figure 13.32 (a) Receiver sensitivities for *M*-ary modulation schemes. (b) Equivalent input noise current density of preamplifier used to determine characteristics provided in (a) where *C* is the total input capacitance, the transconductance is 50 mS and the numerical factor is 1.1. Reproduced with permission from K. Nosu and K. Iwashita, 'A consideration of factors affecting future coherent lightwave communication systems', *J. Lightwave Technol.*, **6**, p. 686, 1988

[Ref. 110]. With higher numbers of levels using PSK or QAM, however, the receiver sensitivity degrades and hence narrow linewidth lasers are required [Refs 111, 112].

13.9 Multicarrier systems

It was mentioned in Section 13.1 that a major attribute of coherent optical transmission was its ability to provide wavelength/frequency selectivity with narrow channel spacings for future multicarrier systems and networks. Wavelength division multiplexing technology and techniques have been discussed in relation to IM/DD optical fiber systems (see Sections 5.6.3 and 12.9.4) but it is apparent that far more optical carriers could be employed using coherent optical receivers which may be tuned to specific incoming carrier signals. For such coherent systems, because the channel widths are very narrow (often on the order of the data transmission rate) in comparison with conventional WDM, then the channel spacings are often measured in frequency rather than wavelength units. Consequently, multicarrier coherent optical systems and networks in which the different channels are separated by appropriately tuning the local oscillator of the receiver were previously referred to as frequency division multiplexed (FDM) [Refs 5, 113–116]. Alternatively, when the different optical channels are selected at the receiver by the optical front-end using, for example, a Fabry–Pérot interferometer, or another optical filter, the coherent system is said to be wavelength division multiplexed. In addition, more recently the WDM terminology has started to be utilized for both of the above cases.

It is clear that coherent optical WDM/FDM systems provide a powerful strategy for the utilization of the enormous optical bandwidth potential of fibers (over 50 000 GHz between 1.3 and 1.6 μm) while avoiding the potential bottleneck created by the speed of the electronics within single-carrier systems. For example, even at spacings of 10 GHz several thousand frequency/wavelength channels could be accommodated over the 1.3 to 1.6 μm wavelength band. It is therefore suggested that future applications for coherent systems could be to provide a large number of channels that are dynamically allocated and reused employing the high selectivity and tunability of coherent receivers in metropolitan area or access networks (see Sections 15.6.2 and 15.6.3) [Ref. 25].

Previously, a favored technique within coherent multicarrier systems was to use a passive star coupler to distribute or broadcast the optical signals over the network [Refs 113–115]. A block schematic for such an WDM/FDM coherent star network is illustrated in Figure 13.33. All the optical carriers (shown as λ_i, λ_j, λ_k, etc.) are generated and modulated individually for transmission over the network. At the network output, tunable, highly selective, optical heterodyne receiver local oscillators provide multiple-channel access. Alternatively, fixed wavelengths/frequencies can be assigned to the receivers and tunable lasers can be employed at the transmitters. The former strategy has been demonstrated for a small number of channels [Ref. 115]. In the latter demonstration four FSK modulated channels operating each at 200 Mbit s^{-1} and spaced 2.45 GHz apart were combined in a 16×16 passive star coupler. Demultiplexing was achieved using a heterodyne FM receiver with an IF of 850 MHz. Moreover, the minimum received optical power required to maintain a BER of 10^{-9} was −55.5 dBm (109 photons per bit) and the degradation due to adjacent channel interference was found to be negligible.

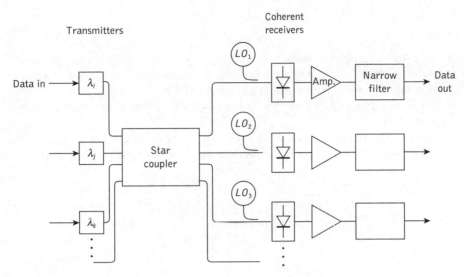

Figure 13.33 Wavelength (or frequency) division multiplexed coherent star network

Considering an $N \times N$ coherent passive star network of the form shown in Figure 13.33, the minimum power required at each receiver P_S can be obtained from Eq. (13.57) or Eq. (13.58) as:

$$P_S \geq N_p \, hf \, B_T \tag{13.60}$$

Then taking account of the star coupler distribution loss and assuming no excess loss through the device, the power required to be launched from each transmitter P_{tx} is:

$$P_{tx} \geq N_p \, hf \, B_T \, N \tag{13.61}$$

where N is the number of ports on the coupler.

The relationship given in expression (13.61) may be seen to be correct because the division of power on the network is inversely related to N (see Section 5.6.2), as is the portion of the total fiber optical bandwidth available to each terminal. However, since for an optical WDM network the total information throughput capacity $N \times B_T$ is limited by the fiber bandwidth B_{fib}, then the transmission capacity (bit rate) available to each terminal is:

$$B_T \leq \frac{B_{fib}}{N} \tag{13.62}$$

Substituting expression (13.62) into (13.61) gives:

$$P_{tx} \geq N_P \, hf \, B_{fib} \tag{13.63}$$

This is an interesting result which indicates that the transmitter power requirement is independent of the number of terminals on the network [Ref. 43].

Example 13.6

Estimate the minimum transmitter power requirement for an optical coherent WDM passive star network using heterodyne synchronous receivers which need an average of 150 photons per bit for reception at the desired BER. It may be assumed that the network is operating from a shortest wavelength of 1.3 μm with an optical bandwidth of 20 THz.

Solution: The minimum required transmitter power P_{tx} may be obtained directly from expression (13.63) as the worst case occurs at the shortest wavelength. Hence:

$$P_{tx} \simeq N_p \, hf B_{fib} = \frac{N_p \, hc \, B_{fib}}{\lambda}$$

$$= \frac{150 \times 6.63 \times 10^{-34} \times 3 \times 10^8 \times 20 \times 10^{12}}{1.3 \times 10^{-6}}$$

$$= 0.5 \text{ mW} \quad \text{or} \quad -3 \text{ dBm}$$

The result obtained in Example 13.6 indicates that relatively modest transmitter power will, in principle, facilitate the operation of optical coherent WDM star networks with arbitrarily large numbers of terminals. It must be emphasized, however, that significant losses on the network (coupler excess loss, fiber losses, etc.) have not been taken into account and that the restriction on the number of channels will generally be substantially greater than that dictated by expression (13.62) in which the channel bandwidth requirement was taken to be approximately equal to the transmission bit rate. In many cases the modulated spectrum can be much wider than this amount (e.g. for wide-deviation FSK) and, in addition, guard bands are necessary between channels.

More recently, as indicated in Section 13.7, interest in applying DPSK to long-haul, and particularly WDM, transmission has been revitalized following the observation that the RZ signal format may reduce nonlinear phase noise [Ref. 117]. Resulting from the conversion of amplified spontaneous emission to phase noise through self-phase modulation and cross-phase modulation, nonlinear phase noise can cause severe performance degradation* in such long-haul systems which employ a PSK signaling format [Ref. 118]. Schemes to mitigate the penalty induced by the nonlinear phase noise have been proposed which include overlapped pulse transmission [Ref. 119] and electronic compensation [Ref. 120]. Moreover, even taking into consideration the potential susceptibility of RZ-DPSK transmission to nonlinear phase jitter, it is indicated that the expectations for high-capacity RZ-DPSK transoceanic WDM transmission are great [Ref. 117].

Following on from the above, the first field trial using installed nondispersion slope matched submarine fibers successfully demonstrated RZ-DPSK WDM transmission [Ref. 121] over a distance of 13 100 km. The measurement configuration for the field trial is shown in Figure 13.34 which displays block diagrams of the RZ-DPSK transmitter

---

* The performance degradation due to nonlinear phase noise is also referred to as the Gordon–Mollenauer effect [Ref. 25].

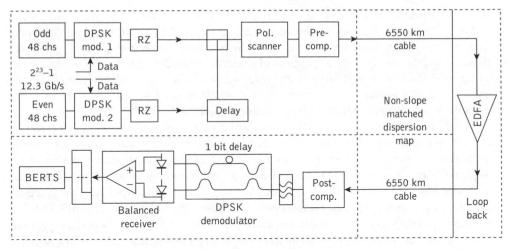

Figure 13.34 Measurement configuration for the RZ-DPSK submarine system field trial [Ref. 121]

and receiver together with the fiber submarine cable link comprising 2×6550 km. Single-stage erbium-doped fiber amplifier (EDFA) technology with 13 dBm output power provided the undersea repeaters, the average repeater spacing and span loss being 345 km and 10 dB respectively. In addition, the submarine fibers were conventional nondispersion slope matched dispersion-shifted fibers giving a system dispersion slope of around 0.08 ps nm^{-1} km^{-1} [Ref. 121].

Ninety-six 33 GHz spaced WDM channels were RZ-DPSK modulated employing two modulation paths (see Figure 13.34). A $2^{23} - 1$ pseudorandom bit sequence pattern at a transmitted bit rate of 12.3 Gbit s^{-1} (10 Gbit s^{-1} data plus the forward error correction (FEC) overhead (see Section 12.6.7)) with data for the even channels, and delayed and inverted data for the odd channels. Finally, no DPSK precoding was required due to the nature of the pseudorandom bit sequence patterns.

In the receiver channel demodulation took place using a 12.3 Gbit s^{-1} DPSK interferometric demodulator following dispersion postcompensation and optical filtering. Then the two optical outputs were detected utilizing a balanced receiver configuration prior to being sent to a bit-error-rate test set (BERTS). Although the accumulated dispersion varied over a range of $\pm 13\,000$ ps nm^{-1} across the signal bandwidth, all channels exhibited a Q-factor (see Section 12.6.2) performance better than 11 dB, therefore giving more than 3 dB forward error correction margin [Ref. 121].

13.9.1 Polarization multiplexing

Polarization division multiplexing (PDM or PolDM), often referred to simply as polarization multiplexing (POLMUX), provides a different approach to multilevel modulation. It facilitates the doubling of the feasible spectral efficiency through the transmission of independent information in each of the two orthogonal polarizations. POLMUX does, however, require polarization control at the receiver to separate the two polarization modes as a consequence of the random birefringence in optical fibers. A main benefit of POLMUX is that with an optical link that is not dominated by polarization mode dispersion or polarization-dependent loss, it proves to be a transparent multiplexing technique. Hence it displays many of the advantages of a multilevel modulation format in comparison with non-POLMUX signals for the same data rate. For example, it exhibits a higher chromatic dispersion tolerance, a narrower spectral width together with a better optical SNR sensitivity and tolerance to self-phase modulation than RZ-DQPSK.

The main drawback with POLMUX is the polarization-sensitive detection necessary at the receiver. Although this adds to the receiver complexity, reduces polarization mode dispersion tolerance and also increases the effects of cross-polarization nonlinearities, particularly in the case of WDM transmission, it can essentially be implemented without difficulty when employing a coherent receiver with, for instance, polarization diversity reception (see Section 13.6.5). Furthermore, when used in combination with RZ-DQPSK it results in 4 bits per symbol modulation which provides for a very narrow optical spectrum with a long symbol period in comparison with binary modulation formats. The spectral efficiency relates to the multiplexed modulation format and is measured in units of bit s^{-1} Hz^{-1}. For example, experimental demonstrations of POLMUX RZ-DQPSK transmission have exhibited spectral efficiency of 1.49 bit s^{-1} Hz^{-1} over 324 km in 2005 [Ref. 122] while, more recently, 1.6 bit s^{-1} Hz^{-1} over 1700 km has been obtained [Ref. 123].

Although interferometric direct detection can be utilized with POLMUX, it is significantly more sensitive to linear impairments in comparison with a system without POLMUX, whereas coherent detection is much more tolerant to such impairments [Ref. 124]. An experimental 2.8 Gbit s^{-1} polarization multiplexed QPSK system employing a digital coherent polarization diversity receiver is shown in Figure 13.35 [Ref. 125]. The system which is designed to improve spectral efficiency of existing fiber links employs synchronous QPSK transmission providing polarization-dependent loss tolerance. Moreover, it may be noted that the system uses distributed feedback lasers at both the transmitter and receiver local oscillator.

The signal from the DFB laser in Figure 13.35 was modulated with 2.8 Gbit s^{-1} precoded data utilizing two QPSK modulators to provide the two orthogonal polarization signals which were then combined in a polarization beam splitter generating the POLMUX QPSK signal. After transmission over 120 km of standard single-mode fiber the signals were received in the polarization diversity optical receiver using two integrated optical 90° hybrids prior to balanced detection. Following an analog-to-digital conversion, electronic manipulation within the digital signal processing (DSP) was undertaken to separate the two polarizations. Finally, a feedforward scheme [Ref. 126] recovered the optical carrier and the four data streams were synchronously demodulated.

13.9.2 High-capacity transmission

Although polarization interleaving transmission where adjacent WDM channels are transmitted in alternating polarizations to reduce crosstalk or nonlinear interactions between the channels can be implemented using interferometric reception [Ref. 127], coherent detection potentially enables ultra-dense WDM channels to be achieved. For instance, polarization-interleaved WDM transmission of 20 Gbit s^{-1} QPSK signals with a 12.5 GHz channel spacing using a digital coherent receiver has recently been demonstrated [Ref. 128].

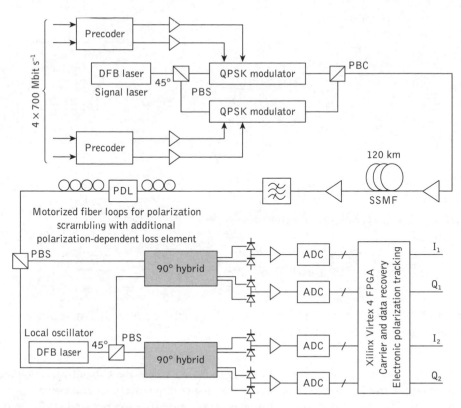

Figure 13.35 Experimental configuration for 2.8 Gbit s^{-1} polarization multiplexed synchronous QPSK transmission [Ref. 125]

In this experimental system even and odd WDM channels were orthogonally polarized employing polarization controllers and then interleaved on a 12.5 GHz spaced grid with a 2 × 2 fiber coupler. The transmission link comprised a total distance of 1074 km in 26 spans incorporating erbium doped fiber amplifiers. A phase diversity optical homodyne receiver based on carrier phase estimation [Ref. 103] was positioned after a polarizer, the latter to eliminate the adjacent orthogonally polarized channels. Finally, the restored in-phase and quadrature data were sampled and digitized using analog-to-digital converters.

Another multiwavelength strategy for which significant benefits have been indicated is referred to as coherent WDM. This approach seeks to increase the spectral density of NRZ binary format signals in a single polarization from 0.4 to 1 bit s^{-1} Hz^{-1} without the use of prefilters in the transmitter [Ref. 129]. Recently, successful coherent WDM signal transmission at 280 Gbit s^{-1} over 1200 km of standard single-mode fiber has been demonstrated [Ref. 130]. Hence it is suggested that the technique which exhibits high spectral density together with low implementation cost is an attractive option for use in long-haul transmission systems.

A more conventional WDM approach incorporating polarization multiplexing (see Section 13.9.1) together with coherent detection has also recently been demonstrated. The experimental test bed which is depicted in Figure 13.36 included 160 distributed feedback lasers with 50 GHz spacing which were separated into two spectrally interleaved sets

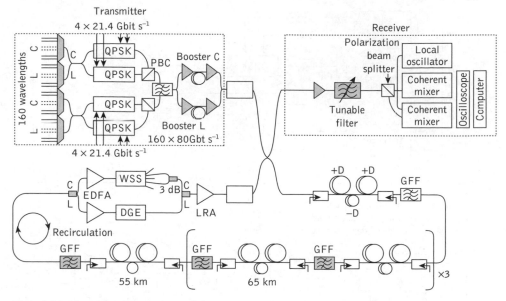

Figure 13.36 Experimental setup for 160 WDM channel polarization multiplexed transmission over 2550 km [Ref. 131]

[Ref. 131]. In each set the output from lasers spanning across the C-band was combined with the output from lasers spanning across the L-band through a polarization-maintaining multiplexer. Then the two wavelength signal combs produced were passed on to the two pairs of QPSK modulators. Polarization multiplexing was achieved by the polarization beam combiner (PBC) multiplexing the output from the first modulator of each pair with the output from the second one. Odd and even WDM channels were then spectrally inter-leaved using a 50 GHz interleaver. The resulting multiplexed signal was amplified and injected into the recirculating loop of the test bed comprising seven 65 km long fiber spans plus a 55 km span separated by Raman amplifiers incorporating gain-flattened filters (GFFs). At the end of the loop light was split into a separate C-band channel and an L-band channel path. Following the C-band path the multiplexed signal was passed through an erbium-doped fiber amplifier (EDFA) and a reconfigurable optical add/drop multi-plexer (ROADM) was emulated using a wavelength-selective switch (WSS). Along the L-band path the multiplexed signal was also amplified by an EDFA and spectrally equalized employing a dynamic gain equalizer (DGE). Both paths were then recombined in a C/L multiplexer. The 160 WDM channels each at a QPSK data rate of 40 Gbit s^{-1} giving a PDM–QPSK rate of 80 Gbit s^{-1} were transmitted over an overall distance of 2550 km of dispersion-compensated fiber prior to the receiver where a coherent mixer combined the signal with a narrow-linewidth local oscillator laser output for each of the two polarization axes. Finally, the corresponding waveforms were digitized and stored. In summary, a high data flow of 12.8 Tbit s^{-1} over this long distance was successfully obtained.

A multicarrier modulation format in which there has been growing interest to compen-sate for impairments in optical fiber transmission systems is orthogonal frequency division multiplexing (OFDM) (see Section 12.9.3). Recently, a coherent system approach, termed

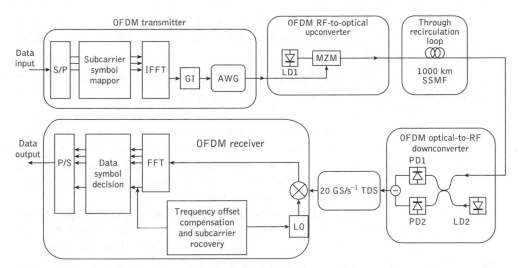

Figure 13.37 Coherent optical OFDM system: GI, guard time insertion; AWG, arrayed waveguide grating; LO, local oscillator laser; GS/s TDS, giga samples per second time domain sampling oscilloscope [Ref. 132]

coherent optical OFDM (CO-OFDM), has been proposed and demonstrated [Refs 132–135]. In particular, CO-OFDM has been shown to combat both fiber chromatic dispersion and polarization mode dispersion.

A generic CO-OFDM system with a direct conversion architecture is provided in Figure 13.37 [Ref. 132]. In the transmit block an RF OFDM transmitter generates a baseband OFDM signal which is then upconverted to the optical domain using an optical in-phase and quadrature (I and Q) modulator based on two Mach–Zehnder modulators (MZMs). For reception an optical-to-RF downconverter comprising a 90° optical hybrid and two balanced receivers is employed to downconvert the optical OFDM signal back to the baseband (i.e. homodyne detection). Finally, the RF OFDM receiver undertakes OFDM baseband processing and demodulation to recover the transmitted data.

The initial proof-of-concept experimental demonstration of CO-OFDM successfully transmitted 128 OFDM subcarriers with a nominal data rate of 8 Gbit s^{-1} over 1000 km of standard single-mode fiber without using dispersion compensation [Ref. 133]. Furthermore, the experimental CO-OFDM transmission of modulation formats with higher spectral efficiency has also been achieved. Both 16-QAM and 64-QAM transmission (see Section 13.5.3) at bit rates of 10.59 and 15.89 Gbit s^{-1} exhibiting spectral efficiencies of 2.8 and 4.2 bit s^{-1} Hz^{-1} respectively were successfully demonstrated [Ref. 135]. Finally, the concept of combining WDM with CO-OFDM has also been recently explored [Ref. 136].

In summary, coherent and, in particular, phase-modulated multicarrier transmission have become the focus of significant research activities worldwide for the provision of future ultra-high-capacity long-haul systems. Hence, although IM/DD WDM systems operating at channel rates of 10 Gbit s^{-1} and more recently 40 Gbit s^{-1} are commercially available with the implementation of 100 Gbit s^{-1} systems planned in the relatively near term, the overall benefits that can be provided in terms of both spectral efficiency tolerance in relation to signal degradations [Ref. 137] and improved receiver sensitivity are a strong indication that research and development of coherent and more specifically phase-modulated systems is set to continue, and potentially increase in pace [Refs 3, 25, 138, 139].

Problems

13.1　(a) Outline the initial rationale behind the pursuit of coherent optical systems. Indicate the major problems encountered in the realization of coherent optical transmission and briefly describe the developments that have taken place since 2002 which have stimulated a new momentum in the pursuit of coherent and particularly phase-modulated systems.

　　　(b) Discuss, with the aid of a suitable block diagram, a coherent optical fiber communication system. Comment on the differing system requirements to facilitate heterodyne detection in comparison with homodyne detection.

13.2　The frequency stability requirement for a local oscillator laser in an ASK optical heterodyne detection system is 10 MHz. When the laser is emitting at a center frequency of 1.55 μm and exhibits an output frequency change with temperature of 14 GHz °C^{-1}, estimate:

　　　(a) the fractional stability necessary for the device;

　　　(b) the maximum temperature change that could be permitted for the device when there is no other form of laser frequency control;

　　　(c) the maximum transmission bandwidth that would be allowed by the laser frequency stability.

13.3　(a) Obtain from first principles the theoretical SNR improvement in the shot noise limit for optical homodyne detection over heterodyne detection. Indicate the primary reason for this improvement.

　　　(b) Briefly discuss the strategies which have been adopted to provide optical homodyne detection with specific reference to PSK modulation.

13.4　A homodyne OOK receiver has a bandwidth of 250 MHz and utilizes a photodiode with a responsivity of 0.6 A W^{-1} at the operating wavelength. The device is shot noise limited and a received SNR of 11 dB is required to provide an acceptable BER. Compute the receiver sensitivity and the photocurrent obtained when the local oscillator laser output power is −3 dBm and the phase difference between this signal and the incoming one is 12°.

13.5　The incoming signal power to an ASK optical heterodyne receiver operating at its shot noise limit is 1.28 nW for a received SNR of 9 dB. Determine the transmission wavelength of the ASK system if the quantum efficiency of the photodetector is 75% at this wavelength and the transmission bandwidth is 400 MHz.

13.6　Outline the major practical constraints associated with coherent optical transmission and discuss the techniques which have been adopted to overcome them.

13.7　Explain the concept of intradyne detection in coherent optical communications, and, with the aid of a block schematic, outline the structure and operation of a QPSK intradyne receiver.

13.8 To allow asynchronous ASK heterodyne detection the linewidths of the signal and local oscillator lasers should be less than 50% of the transmitted bit rate. Estimate the maximum permitted linewidths in nanometers for such ASK system sources:

(a) emitting at a wavelength around $1.30\,\mu m$ when the transmission rate is 140 Mbit s^{-1};

(b) emitting at a wavelength around $1.55\,\mu m$ when the transmission rate is 2.4 Gbit s^{-1}.

13.9 Compare and contrast the attributes and drawbacks associated with direct modulation of the laser signal source and indirect modulation of the source in both ASK and FSK coherent optical fiber communication systems.

13.10 (a) Describe what is understood by continuous phase frequency shift keying (CPFSK) modulation within coherent optical transmission. Indicate the benefits of this modulation technique in comparison with FSK.

(b) Discuss the advantages and suggest a drawback associated with coherent optical differential phase shift keying (DPSK) in comparison with synchronous PSK heterodyne detection.

13.11 A coherent PSK optical fiber communication system employing synchronous heterodyne detection requires a minimum input optical power level of -58.2 dBm in order to receive with a BER of 10^{-9}. The system is operated at a transmission rate of 600 Mbit s^{-1} and the quantum efficiency of the receiver photodetector is 80%. Assuming shot-noise-limited operation at the receiver, obtain the transmission wavelength for the system.

13.12 Outline the major techniques employed to achieve asynchronous optical ASK and FSK heterodyne detection. Indicate the benefits of these schemes over the corresponding synchronous demodulation schemes.

13.13 Describe what is meant by phase diversity reception for coherent optical fiber communication systems. Discuss with the aid of a suitable block diagram the salient features of the in-phase and quadrature receiver when used for optical ASK demodulation.

13.14 Verify that in order to obtain a BER of 10^{-9}:

(a) an average of 18 photons per bit is required within an ideal ASK homodyne detection system;

(b) an average of 40 photons per bit is necessary for ideal asynchronous FSK heterodyne detection.

13.15 Determine the minimum detectable peak optical power levels for both of the detection schemes in Problem 13.14 when the transmission wavelength and bit rate are $1.31\,\mu m$ and 100 Mbit s^{-1} respectively.

13.16 A coherent DPSK system operating at a wavelength of $1.54\,\mu m$ uses a photodetector with a quantum efficiency of 83% at this wavelength. In shot-noise-limited performance a BER of 0.94×10^{-12} is obtained at the coherent optical receiver for a minimum detectable optical power level of 2.1 nW. Calculate both the average number of photons per bit required to maintain the BER and the transmission bit rate of the system under these circumstances.

13.17 An OOK coherent optical fiber system using asynchronous heterodyne detection has a transmission wavelength of 1.55 μm. Estimate the number of photons required for a one bit to provide a BER of 10^{-10} when there is shot-noise-limited detection and the responsitivity of the system photodetector at the operating wavelength is 0.7.

13.18 An FSK coherent optical fiber system employing synchronous heterodyne detection has a transmission wavelength of 1.3 μm where the fiber and splice/connector losses average out at 0.4 dB km^{-1}. If 2 mW of optical power is launched into the system link and an ideal photodetector is assumed, determine the absolute maximum repeater spacing to maintain a BER of 10^{-9} at transmission rates of: (a) 140 Mbit s^{-1}; (b) 2.4 Gbit s^{-1}.

13.19 Briefly describe a common method to facilitate direct detection for both optical FSK and PSK. In each case indicate the sensitivity performance in comparison with both asynchronous and synchronous heterodyne detection at a BER of 10^{-9}.

13.20 Suggest the reasons why DPSK transmission with return-to-zero pulses (i.e. RZ-DPSK) employing an interferometric direct detection receiver has recently emerged as a favored technique in order to demonstrate the future viability of commercial long-haul optical fiber communication systems.

13.21 A DPSK coherent optical fiber system operating at a transmission wavelength and bit rate of 1.55 μm and 250 Mbit s^{-1}, respectively, has a repeater spacing of 300 km. Assuming a launch power of 0 dBm with shot-noise-limited detection and average overall transmission losses of 0.2 dB km^{-1} at the operation wavelength, compute the minimum quantum efficiency required for the photodetector to enable the system to function with a BER of 10^{-10}.

13.22 Describe, with the aid of a diagram, a typical DQPSK system implementation with the particular focus on the use of two balanced receivers to provide interferometric reception.

13.23 Contrast the earlier understanding of FDM with WDM in relation to coherent optical transmission. Describe, with the aid of a suitable diagram, the possible implementation of a coherent multicarrier distribution system based on a passive star coupler.

  Estimate the number of photons per bit obtained with an optical coherent WDM passive star network, which is operating from the shortest wavelength of 1.50 μm with an optical bandwidth of 100 nm, when the transmitter powers are 0 dBm. Comment on the result.

13.24 The two spectral transmission regions for coherent multicarrier systems may be considered to be the O-band and the combined S, C and L bands. Determine the number of WDM channels that could be accommodated in each region when coherent optical PSK transmission at 10 Gbit s^{-1} is to be utilized on each channel. A 20% guard band frequency for filter roll-off should be assumed.

13.25 Outline the concept of polarization multiplexing while explaining its benefits in relation to increasing the capacity of coherent optical transmission systems.

13.26 Describe, giving experimental system examples, the use of multiwavelength strategies for the provision of potential future high-capacity optical fiber phase-modulated systems.

13.27 Briefly explain what is meant by coherent optical orthogonal frequency division multiplexing (CO-OFDM). Using a block schematic of a CO-OFDM system describe the operation of the direct conversion architecture approach.

Answers to numerical problems

13.2 (a) 1.93 in 10^7
     (b) 7×10^{-4} °C
     (c) 50 MHz
13.4 −60.8 dBm; 0.76 μA
13.5 1.32 μm
13.7 (a) 4×10^{-4} nm
     (b) 1×10^{-2} nm
13.11 1.57 μm

13.15 (a) 273 pW
      (b) 607 pW
13.16 27 photons per bit, 500 MHz
13.17 159 photons
13.18 (a) 160.3 km
      (b) 129.5 km
13.21 71.4%
13.23 603 photons per bit
13.24 5590; 6068

References

[1] J. Saltz, 'Modulation and detection for coherent lightwave communications', *IEEE Commun. Mag.*, **24**(6), pp. 38–49, 1986.

[2] I. A. Glover and P. M. Grant, *Digital Communications* (2nd edn), Pearson Education, 2004.

[3] X. Liu, S. Chandrasethar and A. Leven, 'Self-coherent optical transport systems', in I. P. Kaminow, T. Li and A. E. Willner (Eds), *Optical Fiber Telecommunications VB*, pp. 131–178, Elsevier/Academic Press, 2008.

[4] T. G. Hodgkinson, D. W. Smith, R. Wyatt and D. J. Malyon, 'Coherent optical fiber transmission systems', *Br. Telecom Technol. J.*, **3**(3), pp. 5–18, 1985.

[5] B. S. Glance, J. Stone, K. J. Pollock, P. J. Fitzgerald, C. A. Burrus Jr, B. L. Kasper and L. W. Stulz, 'Densely spaced FDM coherent star network with optical signals confined to equally spaced frequencies', *J. Lightwave Technol.*, **6**(11), pp. 1170–1181, 1988.

[6] A. H. Gnauck *et al.*, '2.5 Tb/s (64 × 42.7 Gb/s) transmission over 40 × 100 km NZDSF using RZ-DPSK format and all-Raman-amplified spans', *Optical Fiber Communication Conf. (OFC)*, Postdeadline paper FC2, 2002.

[7] R. A. Griffin, R. I. Johnstone, R. G. Walker, J. Hall, S. D. Wadsworth, K. Berry, A. C. Carter, M. J. Wale, P. A. Jerram and N. J. Parsons, '10 Gb/s optical differential quadrature phase shift key (DQPSK) transmission using GaAs/AlGaAs integration', *Optical Fiber Communication Conf. (OFC)*, Postdeadline paper FD6, 2002.

[8] K.-P. Ho, *Phase-Modulated Optical Communication Systems*, Springer Science & Business Media, 2005.

[9] A. H. Gnauck and P. J. Winzer, 'Optical phase-shift-keyed transmission', *J. Lightwave Technol.*, **23**(1), pp. 115–130, 2005.

[10] P. J. Winzer and R.-J. Essiambre, 'Advanced optical modulation formats', *IEEE Proc.*, **94**(5), pp. 952–985, 2006.

[11] M. C. Teich, 'Homodyne detection of infrared radiation from a moving diffuse target', *Proc. IEEE*, **57**(5), pp. 789–792, 1969.

[12] O. E. DeLange, 'Wide-band optical communication systems: part II – frequency division multiplexing', *Proc. IRE*, **58**(10), pp. 1683–1690, 1970.

[13] S. Machida, A. Kawana, K. Ishihara and H. Tsuchiya, 'Interference of a AlGaAs laser diode using 4.15 km single-mode fiber cable', *IEEE J. Quantum Electron.*, **QE-15**(7), pp. 535–537, 1979.

[14] T. Kimura and Y. Yamamoto, 'Progress of coherent optical fibre communication systems', *Opt. Quantum Electron.*, **15**, pp. 1–39, 1983.

[15] F. Favre, L. Jeunhomme, I. Joindot, M. Monerie and J. C. Simon, 'Progress towards heterodyne-type single-mode fibre communication systems', *IEEE J. Quantum Electron.*, **QE-17**(6), pp. 897–905, 1981.

[16] S. Saito, Y. Yamamoto and T. Kimura, 'Optical heterodyne detection of directly frequency modulated semiconductor laser signals', *Electron. Lett.*, **16**, pp. 826–827, 1980.

[17] D. W. Smith, R. A. Harmon and T. G. Hodgkinson, 'Polarisation stability requirements for coherent optical fibre transmission systems', *Br. Telecom Technol. J.*, **1**(2), pp. 12–16, 1983.

[18] I. W. Stanley, 'A tutorial review of techniques for coherent optical fibre transmission systems', *IEEE Commun. Mag.*, **23**(8), pp. 37–53, 1985.

[19] T. Kimura, 'Coherent optical fiber transmission', *J. Lightwave Technol.*, **LT-5**(4), pp. 414–428, 1987.

[20] D. W. Smith, 'Coherent fiberoptic communications', *Laser Focus*, pp. 92–106, November 1985.

[21] T. Okoshi, 'Ultimate performance of heterodyne/coherent optical fiber communications', *J. Lightwave Technol.*, **LT-4**(10), pp. 1556–1562, 1986.

[22] M. Osinski and J. Buus, 'Linewidth broadening factor in semiconductor lasers – an overview', *IEEE J. Quantum Electron.*, **QE-23**(1), pp. 9–29, 1987.

[23] J. Mellor, S. Al-Chalabi, K. H. Cameron, R. Wyatt, J. C. Regnault, V. W. Devlin and M. C. Brain, 'Performance characteristics of miniature external cavity semiconductor lasers', *Proc. CLEO'89*, Baltimore, MD, USA, paper FP1, April 1989.

[24] K. Kasai, J. Hongo, M. Yoshida and M. Nakazawa, 'Optical phase-locked loop for coherent transmission over 500 km using heterodyne detection with fiber lasers', *IEICE Electron. Express*, **4**(3), pp. 77–81, 2007.

[25] L. G. Kazovsky, G. Kalogerakis and W.-T. Shaw, 'Homodyne phase-shift-keying systems: past challenges and future opportunities', *J. Lightwave Technol.*, **24**(12), pp. 4876–4884, 2006.

[26] T. Myogadani, S. Tanaka and Y. Suetsugu, 'Polarization fluctuation in single mode fiber cables', *IOOC–ECOC'85*, **1**, pp. 151–154, 1985.

[27] R. Ulrich, 'Polarization stabilization on single-mode fiber', *Appl. Phys. Lett.*, **35**(11), pp. 840–842, 1979.

[28] M. Kubota, T. Oohara, K. Furuya and Y. Suematsu, 'Electrooptical polarization control on single-mode fibres', *Electron. Lett.*, **16**(15), p. 573, 1980.

[29] H. C. Lefevre, 'Single-mode fiber fractional wave devices and polarization controllers', *Electron. Lett.*, **16**(20), pp. 778–780, 1980.

[30] T. Imai, K. Nosu and H. Yamaguchi, 'Optical polarization control utilising an optical heterodyne detection scheme', *Electron. Lett.*, **21**(2), pp. 52–53, 1985.

[31] T. Okoshi, N. Fukaya and K. Kikuchi, 'A new polarization-state control device: rotatable fibre cranks', *Electron. Lett.*, **21**(20), pp. 895–896, 1985.

[32] T. Okoshi, Y. Cheng and K. Kikuchi, 'A new polarization-control scheme for optical heterodyne receivers', *Electron. Lett.*, **21**(18), pp. 787–788, 1985.

[33] M. J. Creaner, R. C. Steele, G. R. Walker and N. G. Walker, '565 Mbit/s optical PSK transmission system with endless polarization control', *Electron. Lett.*, **24**(5), pp. 270–271, 1988.

[34] N. G. Walker and G. R. Walker, 'Polarization control for coherent communications', *J. Lightwave Technol.*, **8**(3), pp. 438–458, 1990.

[35] N. G. Walker, G. R. Walker and J. Davidson, 'Endless polarization control using an integrated optic lithium niobate device', *Electron. Lett.*, **24**(5), pp. 266–268, 1988.

[36] J. P. Martelli, P. Boffi, M. Ferrario, L. Marazzi, P. Perolari, S. M. Pietralunga, R. Siano, A. Righetti and M. Martinelli, 'Polarization stabilizer for polarization-division multiplexed optical systems', *Proc. Eur. Conf. on Optical Communications*, Berlin, Germany, **3**, pp. 123–124, September 2007.

[37] M. Martinelli, P. Martinelli and S. M. Pietralunga, 'Polarization stabilization in optical communication systems', *J. Lightwave Technol.*, **24**(11), pp. 4172–4183, 2006.

[38] T. E. Darcie, B. Glance, K. Gayliard, J. R. Talman, B. L. Kasper and C. A. Burrus, 'Polarization-insensitive operation of coherent FSK transmission system using polarization diversity', *Electron Lett.*, **23**(25), pp. 1382–1384, 1987.

[39] T. G. Hodgkinson, R. A. Harmon and D. W. Smith, 'Performance comparison of ASK polarization diversity and standard coherent heterodyne receivers', *Electron. Lett.*, **24**(1), pp. 58–59, 1988.

[40] A. D. Kersey, M. J. Marrone and A. Dandridge, 'Adaptive polarisation diversity receiver configuration for coherent optical fibre communications', *Electron. Lett.*, **25**(4), pp. 275–277, 1989.

[41] I. Garrett and G. Jacobsen, 'Theoretical analysis of heterodyne optical receivers for transmission systems using (semiconductor) lasers with non-negligible linewidth', *J. Lightwave Technol.*, **LT-4**(3), pp. 323–334, 1986.

[42] T. G. Hodgkinson, R. A. Harmon and D. W. Smith, 'Polarisation insensitive heterodyne detection using polarisation scrambling', *Electron. Lett.*, **23**(10), pp. 513–514, 1987.

[43] R. A. Linke and A. H. Gnauck, 'High capacity coherent lightwave systems', *J. Lightwave Technol.*, **6**(11), pp. 1750–1769, 1988.

[44] D. W. Smith, 'Techniques for multigigabit coherent optical transmission', *J. Lightwave Technol.*, **LT-5**(10), pp. 1466–1478, 1987.

[45] T. G. Hodgkinson, R. Wyatt and D. W. Smith, 'Experimental assessment of a 140 Mbit/s coherent optical receiver at 1.52 microns', *Electron. Lett.*, **18**, pp. 523–525, 1982.

[46] J. L. Gimlett, 'Low noise 8 GHz *p–i–n*/FET optical receiver', *Electron. Lett.*, **23**, pp. 281–283, 1987.

[47] S. B. Alexander, 'Design of wide-band optical heterodyne balanced mixer receivers', *J. Lightwave Technol.*, **LT-5**(4), pp. 523–537, 1987.

[48] H. P. Yuen and V. W. S. Chan, 'Noise in homodyne and heterodyne detection', *Opt. Lett.*, **8**(3), pp. 177–179, 1983.

[49] G. L. Abbas, V. W. Chan and T. K. Yee, 'Dual detector optical heterodyne receiver for local oscillator noise suppression', *J. Lightwave Technol.*, **LT-3**(5), pp. 1110–1122, 1985.

[50] A. F. Elrefaie, R. E. Wagner, D. A. Atlas and D. G. Daut, 'Chromatic dispersion limitations in coherent lightwave transmission systems', *J. Lightwave Technol.*, **6**(5), pp. 704–709, 1988.

[51] K. Iwashita and N. Takachio, 'Chromatic dispersion compensation in coherent optical communications', *J. Lightwave Technol.*, **8**(3), pp. 367–375, 1990.

[52] A. R. Chraplyvy, 'Limitations on lightwave communications imposed by optical fiber non-linearities', *Optical Fiber Communication Conf. (OFC'88)*, New Orleans, USA, paper TUD3, January 1988.

[53] I. Garrett and G. Jacobson, 'The effect of laser linewidth on coherent optical receivers', *J. Lightwave Technol.*, **LT-5**(4), pp. 551–560, 1987.

[54] K. Emura, S. Yamazaki, S. Fujita, M. Shikada, I. Mito and K. Minemura, 'Over 300 km transmission experiment on an optical FSK heterodyne dual filter detection system', *Electron. Lett.*, **22**(21), pp. 1096–1097, 1986.

[55] K. Emura, S. Yamazaki, M. Shikada, S. Fujita, M. Yamaguchi, I. Mito and K. Minemura, 'System design and long-span transmission experiments on an optical FSK heterodyne single filter detection system', *J. Lightwave Technol.*, **LT-5**(4), pp. 469–477, 1987.

[56] R. S. Vodhanel, '1 Gbit/s bipolar optical FSK transmission experiment over 121 km of fibre', *Electron. Lett.*, **24**(3), pp. 163–164, 1988.

[57] R. C. Steele and M. Creaner, '565 Mbit/s AMI FSK coherent system using commercial DFB lasers', *Electron. Lett.*, **25**(11), pp. 732–734, 1989.

[58] K. Iwashita and N. Takachio, '2 Gbit/s optical CPFSK heterodyne transmission through 200 km single-mode fibre', *Electron. Lett.*, **23**(7), pp. 341–342, 1987.

[59] L. G. Kazovsky and G. Jacobsen, 'Multichannel CPFSK coherent optical communications system', *J. Lightwave Technol.*, **7**(6), pp. 972–982, 1989.

[60] L. L. Jeromin and V. W. S. Chan, 'M-ary FSK performance for coherent optical communication systems using semiconductor lasers', *IEEE Trans. Commun.*, **COM-34**(4), pp. 375–381, 1986.

[61] S. Kobayashi and T. Kimura, 'Optical phase modulation in an injection locked AlGaAs semiconductor laser', *IEEE J. Quantum Electron.*, **QE-18**(10), pp. 1662–1669, 1982.

[62] B. Glance, 'Performance of homodyne detection of binary PSK optical signals', *J. Lightwave Technol.*, **LT-4**(2), pp. 228–235, 1986.

[63] S. Yamazaki, S. Murata, K. Komatsu, Y. Koizumi, S. Fujita and K. Emura, '1.2 Gbit/s optical DPSK heterodyne detection transmission system using monolithic external-cavity DFB LDs', *Electron. Lett.*, **23**(16), pp. 860–862, 1987.

[64] T. Chikama, T. Naitou, H. Onaka, T. Kiyonaga, S. Watanabe, M. Suyama, M. Seino and H. Kuwahara, '1.2 Gbit/s, 201 km optical DPSK heterodyne transmission experiment using a compact, stable external fibre DFB laser module', *Electron. Lett.*, **24**(10), pp. 636–637, 1988.

[65] J. Hongs, K. Kasai, M. Yoshida and M. Nakazawa, '1-G symbol/s 64-QAM coherent optical transmission over 150 km', *IEEE Photonics Technol. Lett.*, **19**(9), pp. 638–640, 2007.

[66] K. Nosu and K. Iwashita, 'A consideration of factors affecting future coherent lightwave communication systems', *J. Lightwave Technol.*, **6**(5), pp. 686–694, 1988.

[67] S. Betti, G. De Marchis and E. Iannone, *Coherent Optical Communications Systems*, Wiley, 1995.

[68] E. Dietrich, B. Enning, R. Gross and H. Knupke, 'Heterodyne transmission of a 560 Mbit/s optical signal by means of polarization shift keying', *Electron. Lett.*, **23**(8), pp. 421–422, 1987.

[69] Y. Imai, K. Iizuka and R. T. B. James, 'Phase-noise-free coherent optical communication system utilizing differential polarization shift keying (DPolSK)', *J. Lightwave Technol.*, **8**(5), pp. 691–698, 1990.

[70] S. Betti, F. Curti, G. De Marchis and E. Iannone, 'Multilevel coherent optical system based on Stokes parameters modulation', *J. Lightwave Technol.*, **8**(7), pp. 1127–1136, 1990.

[71] P. Benedelta and P. Poggliolini, 'Performance evaluation of multilevel polarization shift keying modulation schemes', *Electron. Lett.*, **26**(4), pp. 244–246, 1990.

[72] S. Saito, Y. Yamamoto and T. Kimura, 'S/N and error rate evaluation for an optical FSK–heterodyne detection system using semiconductor lasers', *IEEE J. Quantum Electron.*, **QE-19**(2), pp. 180–193, 1983.

[73] Y. Yamamoto, 'AM and FM quantum noise in semiconductor lasers – Part I and II', *IEEE J Quantum Electron.*, **QE-19**(1), pp. 34–58, 1983.

[74] T. G. Hodgkinson, 'Costas loop analysis for coherent optical receivers', *Electron. Lett.*, **22**(7), pp. 394–396, 1986.

[75] S. Watanabe, T. Chikama, T. Naito and H. Kuwahara, '560 Mb/s optical PSK heterodyne detection using carrier recovery', *Electron. Lett.*, **25**, pp. 588–590, 1989.

[76] T. Chirkawa, S. Watanabe, T. Naito, H. Onaka, T. Kiyonaga, Y. Onoda, H. Miyata, M. Suyama, M. Seino and H. Kuwahara, 'Modulation and demodulation techniques in optical heterodyne PSK transmission systems', *J. Lightwave Technol.*, **8**(3), pp. 309–322, 1990.

[77] D. J. Malyon, D. W. Smith and R. Wyatt, 'Semiconductor laser homodyne optical phase lock loop', *Electron. Lett.*, **22**, pp. 421–422, 1986.

[78] H. K. Phillip, A. L. Scholtz, E. Bonek and W. Leeb, 'Costas loop experiments for a 10.6 μm communications receiver', *IEEE Trans. Commun.*, **COM-31**(8), pp. 1000–1002, 1983.

[79] L. Kazovsky, L. Curtis, W. Young and N. Cheung, 'All fiber 90° optical hybrid for coherent communications', *Appl. Opt.*, **26**(3), pp. 437–439, 1987.

[80] M. Seimetz and C.-M. Weinert, 'Options, feasibility and availability of 2 × 4 90° hybrids for coherent optical systems', *J. Lightwave Technol.*, **24**(3), pp. 1317–1322, 2006.

[81] S. Norimatsu, '10 Gb/s optical PSK homodyne transmission experiments using external cavity DFB LDs', *Electron. Lett.*, **26**(10), pp. 648–649, 1990.

[82] F. Derr, 'Coherent optical QPSK intradyne system: concept and digital receiver realization', *J. Lightwave Technol.*, **10**(9), pp. 1290–1296, 1992.

[83] R. Noé, 'Phase noise-tolerant synchronous QPSK/BPSK baseband-type intradyne receiver concept with feedforward carrier recovery', *J. Lightwave Technol.*, **23**(2), pp. 802–808, 2005.

[84] N. G. Walker and J. E. Carroll, 'Simultaneous phase and amplitude measurements on optical signals using a multiport junction', *Electron. Lett.*, **20**, pp. 981–983, 1984.

[85] A. Davis, M. Pettitt, J. King and S. Wright, 'Phase diversity techniques for coherent optical receivers', *J. Lightwave Technol.*, **LT-5**(4), pp. 561–572, 1987.

[86] L. Kazovsky, P. Meissner and E. Patzak, 'ASK multiport optical homodyne receiver', *J. Lightwave Technol.*, **LT-5**(6), pp. 770–791, 1987.

[87] K. Emura, R. S. Vodhanel, R. Welter and W. B. Sessa, '5 Gbit/s optical phase diversity homodyne detection experiment', *Electron. Lett.*, **25**(6), pp. 400–401, 1989.

[88] R. Welter and L. G. Kazovsky, '150 Mbit s^{-1} phase diversity ASK homodyne receiver with a DFB laser', *Optical Fiber Communication Conf. (OFC'88)*, New Orleans, USA, paper TU1, January 1988.

[89] T. G. Hodgkinson, R. A. Harmon and D. W. Smith, 'Demodulation of optical DPSK using in-phase and quadrature detection', *Electron. Lett.*, **21**, pp. 867–868, 1985.

[90] L. G. Kazovsky, 'Phase and polarization-diversity coherent optical techniques', *J. Lightwave Technol.*, **7**(2), pp. 279–292, 1989.

[91] A. W. Davis, S. Wright, M. J. Pettitt, J. P. King and K. Richards, 'Coherent optical receiver for 680 Mbit/s using phase diversity', *Electron. Lett.*, **22**, pp. 9–11, 1986.

[92] M. J. Pettitt, D. Remodios, A. W. Davies, A. Hadjifotiou and S. Wright, 'A coherent transmission system using DFB lasers and phase diversity', *Proc. IEE Colloq.*, UK, pp. 9/1–9/5, 1987.

[93] E. Gottwald and J. Pietzsch, 'Measurement method for determination of optical phase shifts in 3 × 3 fibre couplers', *Electron. Lett.*, **24**, pp. 265–266, 1988.

[94] J. Siuzdak and W. van Etten, 'BER evaluation for phase and polarization diversity optical homodyne receivers using non-coherent ASK and DPSK demodulation', *J. Lightwave Technol.*, **7**(4), pp. 584–599, 1989.

[95] M. J. Creaner, R. C. Steele, I. Marshall, G. R. Walker, N. G. Walker, J. Mellis, S. Al-Chalabi, I. Sturgess, M. Rutherford, J. Davidson and M. Brain, 'Field demonstration of 565 Mbit/s DPSK coherent transmission system over 176 km of installed fiber', *Electron. Lett.*, **24**(22), pp. 1354–1356, 1988.

[96] M. C. Brain, M. J. Creaner, R. C. Steele, N. G. Walker, G. R. Walker, J. Mellis, S. Al-Chalabi, J. Davidson, M. Rutherford and I. C. Sturgess, 'Progress towards the field deployment of coherent optical fiber systems', *J. Lightwave Technol.*, **8**(3), pp. 423–437, 1990.

[97] O. K. Tonguz and R. E. Wagner, 'Equivalence between preamplifier direct detection and heterodyne receivers', *IEEE Photonics Technol. Lett.*, **15**(7), pp. 975–977, 1991.

[98] A. H. Gnauck, G. Raybon, S. Chandrasekhar, J. Leuthold, C. R. Doerr, L. W. Schulz and E. Burrows, '25 40-Gb/s copolarized DPSK transmission over 12100-km NZDF with 50-GHz channel spacing', *IEEE Photonics Technol. Lett.*, **15**(3), pp. 467–469, 2003.

[99] C. Rasmussen *et al.*, 'DWDM 40G transmission over trans-Pacific distance (10 000 km) using CSRZ-DPSK, enhanced FEC and all-Raman amplified 100 km Ultra-Wave™ fiber spans', *J. Lightwave Technol.*, **22**(1), pp. 203–207, 2004.

[100] C. Xu, X. Liu and X. Wei, 'Differential phase-shift keying for high spectral efficiency optical transmissions', *IEEE J. Sel. Top. Quantum Electron.*, **10**(2), pp. 281–293, 2004.

[101] N. W. Spellmeyer, J. C. Gottschalk, D. O. Caplan and M. L. Stevens, 'High-sensitivity 40 Gb/s RZ-DPSK with forward error correction', *IEEE Photonics Technol. Lett.*, **16**(6), pp. 1579–1581, 2004.

[102] S. Camatel and V. Ferrero, 'Homodyne coherent detection of ASK and PSK signals performed by a subcarrier optical phase-locked loop', *IEEE Photonics Technol. Lett.*, **18**(1), pp. 142–144, 2006.

[103] D. S. Ly-Gagnon, S. Tsukamoto, K. Katoh and K. Kituchi, 'Coherent detection of optical quadrature phase-shift-keying signals with carrier phase estimation', *J. Lightwave Technol.*, **24**(1), pp. 12–21, 2006.

[104] M. Nazarathy, X. Liu, L. Christen, Y. Lize and A. Willner, 'Self coherent, decision-feedback-directed 40-Gb/s DQPSK receiver', *IEEE Photonics Technol. Lett.*, **19**(11), pp. 828–830, 2007.

[105] X. Liu, 'DSP enhanced differential direct-detection for DQPSK and M-ary DPSK', *Proc. 33rd Eur. Conf. on Optical Communications*, Berlin, Germany, **3**, pp. 143–146, September 2007.

[106] T. Okoshi, K. Emura, K. Kikuchi and R. Th. Kersten, 'Computation of bit-error rate of various heterodyne and coherent-type optical communication schemes', *J. Opt. Commun.*, **2**(3), pp. 89–96, 1981.

[107] T. Okoshi and K. Kikuchi, *Coherent Optical Fiber Communications*, KTK Scientific Kluwer Academic, 1988.

[108] L. G. Kazovsky, 'Optical heterodyning versus optical homodyning: a comparison', *J. Opt. Commun.*, **6**(1), pp. 18–24, 1985.

[109] P. S. Henry, R. A. Linke and A. H. Gnauck, 'Introduction to lightwave systems', in S. E. Miller and I. P. Kaminow (Eds) *Optical Fiber Telecommunications II*, pp. 781–831, Academic Press, 1988.

[110] S. Yamazaki, T. Fujta and K. Emura, 'Feasibility study on optical QPSK heterodyne systems', *Optical Fiber Communication Conf. (OFC'88)*, New Orleans, USA, paper WC3, January 1988.

[111] S. Norimatsu and K. Iwashita, 'Linewidth requirements for optical synchronous detection systems with non-negligible loop delay time', *J. Lightwave Technol.*, **10**(3), pp. 341–349, 1992.

[112] S. Savory and A. Hadjifotiou, 'Laser linewidth requirements for optical DQPSK systems', *IEEE Photonics Technol. Lett.*, **13**(3), pp. 930–932, 2004.

[113] E. J. Bachus, R. P. Braun, C. Casper, H. M. Foisel, E. Grobmann, B. Strebel and F. J. Westphal, 'Coherent optical multicarrier systems', *J. Lightwave Technol.*, **7**(2), pp. 375–384, 1989.

[114] R. A. Linke, 'Frequency division multiplexed optical networks using heterodyne detection', *IEEE Netw.*, pp. 13–20, March 1989.

[115] B. Glance, T. L. Koch, O. Scaramucci, K. C. Reichmann, U. Koren and C. A. Burrus, 'Densely spaced FDM coherent optical star network using monolithic widely frequency-tunable lasers', *Electron. Lett.*, **25**(10), pp. 672–673, 1989.

[116] S. Yamazaki, M. Shibutani, N. Shimoska, S. Murata, T. Ono, M. Kitamura, K. Emura and M. Shikada, 'A coherent optical FDM CATV distribution system', *J. Lightwave Technol.*, **8**(3), pp. 396–405, 1990.

[117] T. Mizuochi, K. Ishida, T. Kobayashi, J. Abe, K. Kinjo, K. Motoshima and K. Kasahara, 'A comparative study of DPSK and OOK WDM transmission over transoceanic distances and their performance degradations due to nonlinear phase noise', *J. Lightwave Technol.*, **21**(9), pp. 1933–1943, 2003.

[118] J. P. Gordon and L. F. Mollenauer, 'Phase noise in photonic communications systems using linear amplifiers', *Opt. Lett.*, **15**(23), pp. 1351–1355, 1990.

[119] X. Liu, C. Xu and X. Wei, 'Nonlinear phase noise in pulse-overlapped transmission based on return-to-zero differential-phase-shift-keying', *Eur. Conf. on Optical Communications (ECOC)*, Copenhagen, Denmark, paper 9.6.5, September 2002.

[120] K.-P. Ho and J. M. Kahn, 'Electronic compensation technique to mitigate nonlinear phase noise', *J. Lightwave Technol.*, **22**(3), pp. 779–783, 2004.

[121] J.-X. Cai *et al.*, 'RZ-DPSK field trial over 13 100 km of installed non-slope matched submarine fibers', *J. Lightwave Technol.*, **23**(1), pp. 95–103, 2005.

[122] S. Bhandare, D. Sandel, B. Milivojevic, A. Hidayat, A. A. Fauzi, H. Zhang, S. K. Ibrahim, F. Wust and R. Noé, '5.94-Tb/s 1.49-b/s/Hz (40 × 2 × 2 × 40 Gb/s) RZ-DQPSK polarization-division multiplex C-band transmission over 324 km', *IEEE Photonics Technol. Lett.*, **17**(4), pp. 194–196, 2005.

[123] D. van den Borne, S. L. Jansen, E. Gottwald, P. M. Krummrich, G. D. Khoe and H. de Waardt, '1.6-b/s/Hz spectrally efficient transmission over 1700 km of SSMF using 40 × 85.6 Gb/s POLMUX-RZ-DQPSK', *J. Lightwave Technol.*, **25**(1), pp. 222–232, 2007.

[124] G. Charlet, M. Salsi, J. Renaudier, O. B. Parder, H. Mardoyan and S. Bigo, 'Performance comparison of singly polarized and polarization-multiplexed coherent transmission at 10 Gbauds under linear impairments', *Proc. 33rd Eur. Conf. on Optical Communications*, Berlin, Germany, **3**, pp. 147–148, September 2007.

[125] T. Pfau *et al.*, 'Polarization-multiplexed 2.8 Gbit/s synchronous QPSK transmission with real-time polarization tracking', *Proc. 33rd Eur. Conf. on Optical Communications*, Berlin, Germany, **3**, pp. 263–264, September 2007.

[126] E. Ip and J. M. Kahn, 'Feedforward carrier recovery for coherent optical communications', *J. Lightwave Technol.*, **25** (9), pp. 2675–2692, 2007.

[127] B. Zhu, E. Nelson, A. H. Gnauck, C. Doerr, J. Leuthold, L. Grüner-Nielsen, M. O. Pedersen, J. Kim and R. L. Lingle Jr, 'High spectral density long-haul 40-Gb/s transmission using CSRZ-DPSK format', *J. Lightwave Technol.*, **22** (1), pp. 208–214, 2004.

[128] S.-Y. Kim and K. Kikuchi, '1000-km polarization-interleaved WDM transmission of 20-Gbit/s QPSK signals on the frequency grid with 12.5-GHz channel spacing using digital coherent receiver', *Proc. 33rd Eur. Conf. on Optical Communications*, Berlin, Germany, **3**, pp. 361–262, September 2007.

[129] A. D. Ellis and F. C. G. Gunning, 'Spectral density enhancement using coherent WDM', *IEEE Photonics Technol. Lett.*, **17**(2), pp. 504–506, 2005.

[130] T. Healy, F. C. G. Gunning, E. Princemin, B. Cuenot and A. D. Ellis, '1200 km SMF (100 km spans) 280 Gbit/s coherent WDM transmission using hybrid Raman/EDFA amplification', *Proc. 33rd Eur. Conf. on Optical Communications*, Berlin, Germany, **1**, pp. 59–60, September 2007.

[131] G. Charlet, J. Renandier, H. Mardoyan, O. Betran Pardo, F. Cérou, P. Tran and S. Bigo, '12.8 Tbit/s transmission of 160 PDM-QPSK (160 × 2 × 40 Gbit/s) channels with coherent detection over 2,550 km', *Proc. 33rd Eur. Conf. on Optical Communications*, Berlin, Germany, PD1.6, September 2007.

[132] W. Shieh and C. Athaudage, 'Coherent optical orthogonal frequency division multiplexing', *Electron. Lett.*, **42**(10), pp. 587–588, 2006.

[133] W. Shieh, X. Yi and Y. Tang, 'Transmission experiment of multi-gigabit coherent optical OFDM systems over 1000 km SSMF fibre', *Electron. Lett.*, **43**(3), pp. 134–135, 2007.

[134] Y. Tang, W. Shieh, X. Yi and R. Evans, 'Optimum design for RF-to-optical up-converter in coherent optical OFDM systems', *IEEE Photonics Technol. Lett.*, **19**(7), pp. 483–485, 2007.

[135] X. Yi, W. Shieh and Y. Ma, 'Phase noise on coherent optical OFDM systems with 16-QAM and 64-QAM beyond 10 Gb/s', *Proc. 33rd Eur. Conf. on Optical Communications*, Berlin, Germany, **2**, pp. 213–214, September 2007.

[136] H. C. Bao and W. Shieh, 'Transmission of wavelength-division-multiplexed channels with coherent optical OFDM', *IEEE Photonics Technol. Lett.*, **19**(12), pp. 922–924, 2007.

[137] M. H. Taghavi, G. C. Papen and P. H. Siegel, 'On the multiuser capacity of WDM in a non-linear optical fiber: coherent communication', *IEEE Trans. Inf. Theory*, **52**(11), pp. 5008–5022, 2006.

[138] S. Bigo, 'Coherent detection: a key enabler for next generation optical transmission systems', *IEEE Proc. 9th Int. Conf. on Transparent Optical Networks*, Rome, Italy, Tu B4.5, pp. 332–335, July 2007.

[139] K. Kituchi, 'Coherent optical communication systems', in I. P. Kaminow, T. Li and A. E. Willner (Eds), *Optical Fiber Telecommunications VB*, pp. 95–129, Elsevier/Academic Press, 2008.

CHAPTER   14

Optical fiber measurements

14.1 Introduction

In this chapter we are primarily concerned with measurements on optical fibers which characterize the fiber. These may be split into three main areas:

(a) transmission characteristics;

(b) geometrical and optical characteristics;

(c) mechanical characteristics.

Data in these three areas is usually provided by the optical fiber manufacturer with regard to specific fibers. Hence fiber measurements are generally performed in the laboratory and techniques have been developed accordingly. This information is essential for the optical communication system designer in order that suitable choices of fibers, materials and devices may be made with regard to the system application. However, although the system designer and system user do not usually need to take fundamental measurements of the fiber characteristics, there is a requirement for field measurements in order to evaluate overall system performance, and for functions such as fault location. Therefore, we also include some discussion of field measurements which take into account the effects of cabled fiber, splice and connector losses, etc.

Several organizations have become involved in standardization issues relating to optical fiber measurements. The International Telecommunication Union (ITU) [Ref. 1], formerly known as the International Telephone and Telegraph Consultative Committee (CCITT), has recommended several standards for fiber transmission systems and fiber measurements. These are known as Fiber Optic Test Procedures (FOTPs) and Optical Fiber System Test Procedures (OFSTPs) [Ref. 2]. These standards are published by the Electronic Industries Alliance (EIA) [Ref. 3] and Telecommunication Industries Association (TIA) [Ref. 4] in the United States. Both the TIA and EIA jointly publish several standards where each standard is identified with a specific series. For example, the TIA/EIA-455 series represents FOTPs and similarly the TIA/EIA-526 series corresponds to OFSTPs. The particular recommendation related to a specific FOTP or OFSTP is labeled at the end of the series. Thus TIA/EIA-455-220 represents FOTP-220 while TIA/EIA 526-19 relates to OFSTP-19. A full list of these notations and their details can be obtained from the Telecommunications Industry Association website [Ref. 5]. The test methods are further divided into reference test methods (RTMs) and alternative test methods (ATMs). An RTM provides a measurement of a particular characteristic, strictly according to the definition which usually gives the highest degree of accuracy and reproducibility, whereas an ATM may be more suitable for practical use but can deviate from the strict definition; however, there must be a way to relate such results to those obtained from the RTM.

The fiber transmission characteristics of greatest interest are those of attenuation and dispersion. For multimode fibers the latter parameter enables the bandwidth to be determined, whereas with single-mode fibers it is the intramodal or chromatic dispersion which is generally provided by manufacturers. Furthermore, the important geometrical and optical characteristics for multimode fibers are size (core and cladding diameters), numerical aperture (see Section 2.2.3) and refractive index profile, but for single-mode fibers they are the effective cutoff wavelength of the second-order mode (see Section 2.5.1) and the mode-field diameter (see Section 2.5.2). Measurements of the mechanical characteristics such as tensile strength and durability were outlined in Section 4.6.1 and are therefore pursued no further in this chapter.

When attention is focused on the measurement of the transmission properties of multimode fibers, problems emerge regarding the large number of modes propagating in the fiber. The various modes show individual differences with regard to attenuation and dispersion within the fiber. Moreover, mode coupling occurs giving transfer of energy from one mode to another (see Section 2.4.2). The mode coupling which is associated with perturbations in the fiber composition or geometry, and external factors such as microbends or splices, is for instance responsible for the increased attenuation (due to radiation) of the higher order modes. These multimode propagation effects mean that both the fiber loss and bandwidth are not uniquely defined parameters but depend upon the fiber excitation conditions and environmental factors such as cabling, bending, etc. Also, these transmission parameters may vary along the fiber length (i.e. they are not necessarily linear functions) due to the multimode propagation effects, making extrapolation of measured data to different fiber lengths less than meaningful.

It is therefore important that transmission measurements on multimode fibers are performed in order to minimize these uncertainties. In the laboratory, measurements are usually taken on continuous lengths of uncabled fiber in order to reduce the influence of external factors on the readings (this applies to both multimode and single-mode fibers). However, this does mean that the system designer must be aware of the possible deterioration in the fiber transmission characteristics within the installed system. The multimode propagation effects associated with fiber perturbations may be accounted for by allowing or encouraging the mode distribution to reach a steady-state (equilibrium) distribution. This distribution occurs automatically after propagation has taken place over a certain fiber length (coupling length) depending upon the strength of the mode coupling within the particular fiber. At equilibrium the mode distribution propagates unchanged and hence the fiber attenuation and dispersion assume well-defined values. These values of the transmission characteristics are considered especially appropriate for the interpretation of measurements to long-haul links and do not depend on particular launch conditions.

The equilibrium mode distribution may be achieved by launching the optical signal through a long (dummy) fiber to the fiber under test. This technique has been used to good effect [Ref. 6] but may require a kilometer of dummy fiber and is therefore not suitable for dispersion measurements. Alternatively there are a number of methods of simulating the equilibrium mode distribution with a much shorter length of fiber. Mode equilibrium may be achieved using an optical source with a mode output which corresponds to the steady-state mode distribution of the fiber under test. This technique may be realized experimentally using an optical arrangement which allows the numerical aperture of the launched beam to be varied (using diaphragms) as well as the spot size of the source (using pinholes). In this case the input light beam is given an angular width which is equal to the equilibrium distribution numerical aperture of the fiber and the source spot size on the fiber input face is matched to the optical power distribution in a cross-section of the fiber at equilibrium.

Other techniques involve the application of strong mechanical perturbations on a short section of the fiber in order quickly to induce mode coupling and hence equilibrium mode distribution within 1 m. These devices which simulate mode equilibrium over a short length of fiber are known as mode scramblers or mode filters.

A mode scrambler increases the power in the higher order modes relative to the lower order ones, whereas a mode filter typically reduces the power in the higher order modes without seriously affecting the power in the lower order modes. Mode filters are usually required with LED sources because more higher order mode excitation than the equilibrium mode distribution is obtained in most multimode fibers. By contrast, a mode scrambler may be necessary for lens-less excitation with laser sources, as this technique produces more lower order modes than are contained in the equilibrium mode distribution in most multimode fibers. Hence both strategies are employed to achieve equilibrium mode power distributions in multimode fibers.

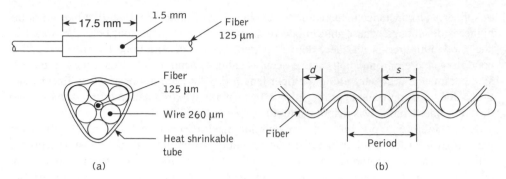

Figure 14.1 Mode scramblers: (a) heat shrinking technique [Ref. 8]; (b) bending technique [Ref. 9]

A simple mode scrambling method [Ref. 7] is to sandwich the fiber between two sheets of abrasive paper (i.e. sandpaper) placed on wooden blocks in order to provide a suitable pressure. Two slightly more sophisticated techniques are illustrated in Figure 14.1 [Refs 8, 9]. Figure 14.1(a) shows mechanical perturbations induced by enclosing the fiber with metal wires and applying pressure by use of a surrounding heat shrinkable tube. A method which allows adjustment and therefore an improved probability of repeatable results is shown in Figure 14.1(b). This technique involves inserting the fiber between a row of equally spaced pins, subjecting it to sinusoidal bends. Hence the variables are the number of pins giving the number of periods, the pin diameter d and the pin spacing s.

A common mode filtering technique uses a mandrel wrap applied before the test fiber as illustrated in Figure 14.2(a). In this method four or five turns of fiber are wrapped around a

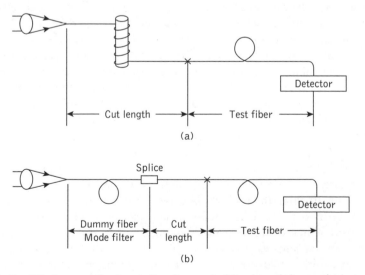

Figure 14.2 Equilibrium mode simulation by mode filtering: (a) mandrel wrap; (b) dummy fiber

20 to 30 mm diameter mandrel in order to simulate the equilibrium mode power distribution [Ref. 10]. The other popular mode filtering technique which was mentioned earlier is shown in Figure 14.2(b). This method employs a dummy fiber of length 0.5 to 1 km which is of a similar type to the test fiber. The dummy fiber is spliced before the test fiber such that an equilibrium mode distribution is established after the optical launch.

In order to test that a particular mode scrambler or filter gives an equilibrium mode distribution within the test fiber, it is necessary to check the insensitivity of the far-field radiation pattern (this is related to the mode distribution, see Section 2.4.1) from the fiber with regard to changes in the launch conditions. It is also useful to compare the far-field patterns from the device and a separate long length at the test fiber for coincidence [Ref. 6]. However, it must be noted that, at present, mode scramblers or filters tend to give only an approximate equilibrium mode distribution and their effects vary with different fiber types. Hence measurements involving the use of different mode scrambling methods can be subject to discrepancies. Nevertheless, the majority of laboratory measurement techniques to ascertain the transmission characteristics of multimode optical fibers use some form of equilibrium mode simulation in order to give values representative of long transmission lines. Moreover, the current standards agreements regarding equilibrium mode simulation are outlined in TIA/EIA-455-50 [Ref. 11].

We commence the discussion of optical fiber measurements in Section 14.2 by dealing with the major techniques employed in the measurement of fiber attenuation. These techniques include measurement of both total fiber attenuation and the attenuation resulting from individual mechanisms within the fiber (e.g. material absorption, scattering). In Section 14.3 fiber dispersion measurements in both the time and frequency domains are discussed. Various techniques for the measurement of the fiber refractive index profile are then considered in Section 14.4. The measurement of the fiber cutoff wavelength which has particular relevance to single-mode fibers is then dealt with in Section 14.5. In Section 14.6 we discuss simple methods for measuring fiber numerical aperture. Measurement of the fiber outer and core diameters are then described in Section 14.7.

A far more important parameter than the fiber numerical aperture and core diameter for single-mode fibers is the mode-field diameter. Hence the measurement of this single-mode fiber characteristic is discussed in Section 14.8. This is followed in Section 14.9 with a description of a measurement procedure for reflectance and optical return loss for either a fiber component or an optical link. Finally, field measurements which may be performed on optical fiber links, together with examples of measurement instruments, are discussed in Section 14.10. Particular attention is paid in this concluding section to optical time domain reflectometry (OTDR).

14.2 Fiber attenuation measurements

Fiber attenuation measurement techniques have been developed in order to determine the total fiber attenuation of the relative contributions to this total from both absorption losses and scattering losses. The overall fiber attenuation is of greatest interest to the system designer, but the relative magnitude of the different loss mechanisms is important in the development and fabrication of low-loss fibers. Measurement techniques to obtain the total fiber attenuation give either the spectral loss characteristic (see Figure 3.3) or the loss at a single wavelength (spot measurement).

14.2.1 Total fiber attenuation

A commonly used technique for determining the total fiber attenuation per unit length is the cut-back or differential method. Figure 14.3 shows a schematic diagram of the typical experimental setup for measurement of the spectral loss to obtain the overall attenuation spectrum for the fiber. It consists of a 'white' light source, usually a tungsten halogen or xenon are lamp. The focused light is mechanically chopped at a low frequency of a few hundred hertz. This enables the lock-in amplifier at the receiver to perform phase-sensitive detection. The chopped light is then fed through a monochromator which utilizes a prism or diffraction grating arrangement to select the required wavelength at which the attenuation is to be measured. Hence the light is filtered before being focused onto the fiber by means of a microscope objective lens. A beam splitter may be incorporated before the fiber to provide light for viewing optics and a reference signal used to compensate for output power fluctuations. As indicated in Section 14.1, when the measurement is performed on multimode fibers it is very dependent on the optical launch conditions. Therefore unless the launch optics are arranged to give the steady-state mode distribution at the fiber input, or a dummy fiber is used, then a mode scrambling device is attached to the fiber within the first meter. The fiber is also usually put through a cladding mode stripper, which

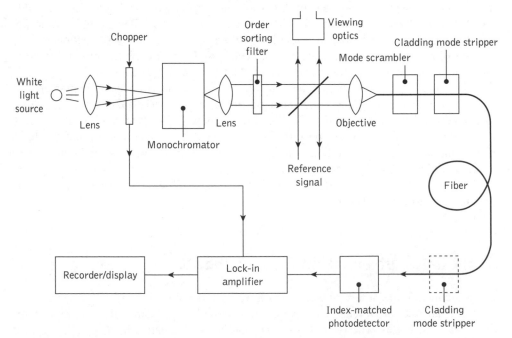

Figure 14.3 A typical experimental arrangement for the measurement of spectral loss in optical fibers using the cut-back technique

may consist of an S-shaped groove cut in the Teflon and filled with glycerine. This device removes light launched into the fiber cladding through radiation into the index-matched (or slightly higher refractive index) glycerine. A mode stripper can also be included at the fiber output end to remove any optical power which is scattered from the core into the cladding down the fiber length. This tends to be pronounced when the fiber cladding consists of a low-refractive-index silicone resin.

The optical power at the receiving end of the fiber is detected using a p–i–n or avalanche photodiode. In order to obtain reproducible results the photodetector surface is usually index matched to the fiber output end face using epoxy resin or an index-matching gell [Ref. 12]. Finally, the electrical output from the photodetector is fed to a lock-in amplifier, the output of which is recorded.

The cut-back method* involves taking a set of optical output power measurements over the required spectrum using a long length of fiber (usually at least a kilometer). This fiber is generally uncabled having only a primary protective coating. Increased losses due to cabling (see Section 4.7.1) do not tend to change the shape of the attenuation spectrum as they are entirely radiative, and for multimode fibers are almost wavelength independent. The fiber is then cut back to a point 2 m from the input end and, maintaining the same launch conditions, another set of power output measurements is taken. The following relationship for the optical attenuation per unit length α_{dB} for the fiber may be obtained from Eq. (3.3):

$$\alpha_{dB} = \frac{10}{L_1 - L_2} \log_{10} \frac{P_{02}}{P_{01}} \tag{14.1}$$

L_1 and L_2 are the original and cut-back fiber lengths respectively, and P_{01} and P_{02} are the corresponding output optical powers at a specific wavelength from the original and cut-back fiber lengths. Hence when L_1 and L_2 are measured in kilometers, α_{dB} has units of dB km^{-1}.

Furthermore Eq. (14.1) may be written in the form:

$$\alpha_{dB} = \frac{10}{L_1 - L_2} \log_{10} \frac{V_2}{V_1} \tag{14.2}$$

where V_1 and V_2 correspond to output voltage readings from the original fiber length and the cut-back fiber length respectively. The electrical voltages V_1 and V_2 may be directly substituted for the optical powers P_{01} and P_{02} of Eq. (14.1) as they are directly proportional to these optical powers (see Section 7.4.3). The accuracy of the results obtained for α_{dB} using this method is largely dependent on constant optical launch conditions and the achievement of the equilibrium mode distribution within the fiber. In this case only the fiber to detector power coupling changes between measurements and this variation can be made less than 0.01 dB [Ref. 10]. Hence the cut-back technique is regarded as the RTM for attenuation measurements by the EIA as well as the ITU.

* The cut-back method is outlined in TIA/EIA-455-46 and TIA/EIA-455-78 for multimode and single-mode fibers respectively [Refs 13, 14]. In addition, it is the ITU reference test method for fiber attenuation [Ref. 1].

Example 14.1

A 2 km length of multimode fiber is attached to apparatus for spectral loss measurement. The measured output voltage from the photoreceiver using the full 2 km fiber length is 2.1 V at a wavelength of 0.85 μm. When the fiber is then cut back to leave a 2 m length, the output voltage increases to 10.7 V. Determine the attenuation per kilometer for the fiber at a wavelength of 0.85 μm and estimate the accuracy of the result.

Solution: The attenuation per kilometer may be obtained from Eq. (14.2) where:

$$\alpha_{dB} = \frac{10}{L_1 - L_2} \log_{10} \frac{V_2}{V_1} = \frac{10}{1.998} \log_{10} \frac{10.7}{2.1}$$

$$= 3.5 \text{ dB km}^{-1}$$

The uncertainty in the measured attenuation may be estimated using:

$$\text{Uncertainty} = \frac{\pm 0.2}{L_1 - L_2} = \frac{\pm 0.2}{1.997} \approx \pm 0.1 \text{ dB}$$

The dynamic range of the measurements that may be taken depends upon the exact configuration of the apparatus utilized, the optical wavelength and the fiber core diameter. However, a typical dynamic range is in the region 30 to 40 dB when using a white light source at a wavelength of 0.85 μm and multimode fiber with a core diameter around 50 μm. This may be increased to around 60 dB by use of a laser source operating at the same wavelength. It must be noted that a laser source is only suitable for making a single-wavelength (spot) measurement as it does not emit across a broad band of spectral wavelengths.

Spot measurements may be performed on an experimental setup similar to that shown in Figure 14.3. However, interference filters are frequently used instead of the monochromator in order to obtain a measurement at a particular optical wavelength. These provide greater dynamic range (10 to 15 dB improvement) than the monochromator but are of limited use for spectral measurements due to the reduced number of wavelengths that are generally available for measurement. A typical optical configuration for spot attenuation measurements is shown in Figure 14.4. The interference filters are located on a wheel to allow measurement at a selection of different wavelengths. In the experimental arrangement shown in Figure 14.4 the source spot size is defined by a pinhole and the beam angular width is varied by using different diaphragms. However, the electronic equipment utilized with this setup is similar to that used for the spectral loss measurements illustrated in Figure 14.3. Therefore determination of the optical loss per unit length for the fiber at a particular wavelength is performed in exactly the same manner, using the cut-back method. Spot attenuation measurements are sometimes utilized after fiber cabling in order to obtain information on any degradation in the fiber attenuation resulting from the cabling process.

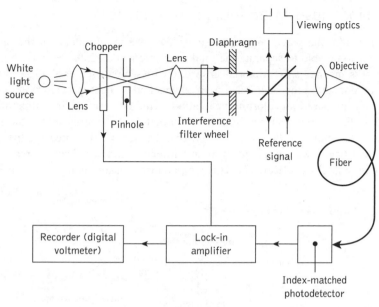

Figure 14.4 An experimental arrangement for making spot (single-wavelength) attenuation measurements using interference filters and employing the cut-back technique

Although widely used, the cut-back measurement method has the major drawback of being a destructive technique. Therefore, although suitable for laboratory measurement it is far from ideal for attenuation measurements in the field. Several nondestructive techniques exist which allow the fiber losses to be calculated through a single reading of the optical output power at the far end of the fiber after determination of the near-end power level. The simplest is the insertion or substitution* technique, which utilizes the same experimental configuration as the cut-back method. However, the fiber to be tested is spliced, or connected by means of a demountable connector to a fiber with a known optical output at the wavelength of interest. When all the optical power is completely coupled between the two fibers, or when the insertion loss of the splice or connector is known, then the measurement of the optical output power from the second fiber gives the loss resulting, from the insertion of this second fiber into the system. Hence the insertion loss due to the second fiber provides measurement of its attenuation per unit length. Unfortunately, the accuracy of this measurement method is dependent on the coupling between the two fibers and is therefore somewhat uncertain.

The most popular nondestructive attenuation measurement technique for both laboratory and field use only requires access to one end of the fiber. It is the backscatter measurement method which uses optical time domain reflectometry and also provides measurement of splice and connector losses as well as fault location. Optical time domain reflectometry finds major use in field measurements and is therefore discussed in detail in Section 14.10.1.

_____

* Description of the substitution method is provided in TIA/EIA-455-53 [Ref. 15].

14.2.2 Fiber absorption loss measurement

It was indicated in the preceding section that there is a requirement for the optical fiber manufacturer to be able to separate the total fiber attenuation into the contributions from the major loss mechanisms. Material absorption loss measurements allow the level of impurity content within the fiber material to be checked in the manufacturing process. The measurements are based on calorimetric methods which determine the temperature rise in the fiber or bulk material resulting from the absorbed optical energy within the structure.

The apparatus shown in Figure 14.5 [Ref. 16], which is used to measure the absorption loss in optical fibers, was modified from an earlier version which measured the absorption losses in bulk glasses [Ref. 17]. This temperature measurement technique, illustrated

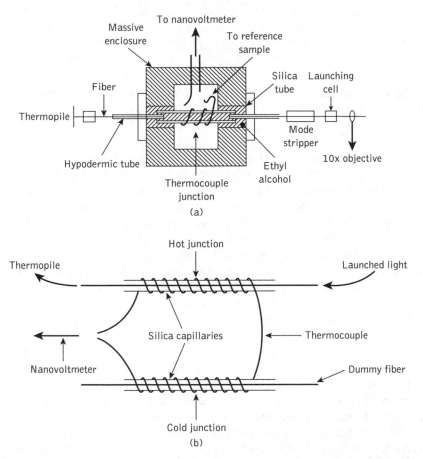

Figure 14.5 Calorimetric measurement of fiber absorption losses: (a) schematic diagram of a version of the apparatus [Ref. 16]; (b) the temperature measurement technique using a thermocouple

diagrammatically in Figure 14.5(b), has been widely adopted for absorption loss measurements. The two fiber samples shown in Figure 14.5(b) are mounted in capillary tubes surrounded by a low-refractive-index liquid (e.g. methanol) for good electrical contact, within the same enclosure of the apparatus shown in Figure 14.5(a). A thermocouple is wound around the fiber containing capillary tubes using one of them as a reference junction (dummy fiber). Light is launched from a laser source (Nd : YAG or krypton ion depending on the wavelength of interest) through the main fiber (not the dummy), and the temperature rise due to absorption is measured by the thermocouple and indicated on a nanovoltmeter. Electrical calibration may be achieved by replacing the optical fibers with thin resistance wires and by passing known electrical power through one. Independent measurements can then be made using the calorimetric technique and with electrical measurement instruments.

The calorimetric measurements provide the heating and cooling curve for the fiber sample used. A typical example of this curve is illustrated in Figure 14.6(a). The attenuation of the fiber due to absorption α_{abs} may be determined from this heating and cooling characteristic. A time constant t_c can be obtained from a plot of $(T_\infty - T_t)$ on a logarithmic scale against the time t, an example of which shown in Figure 14.6(c) was obtained from the heating characteristic displayed in Figure 14.6(b) [Ref. 17]. T_∞ corresponds to the maximum temperature rise of the fiber under test and T_t is the temperature rise at a time t. It may be observed from Figure 14.6(a) that T_∞ corresponds to a steady-state temperature for the fiber when the heat loss to the surroundings balances the heat generated in the fiber resulting from absorption at a particular optical power level. The time constant t_c may be obtained from the slope of the straight line plotted in Figure 14.6(c) as:

$$t_c = \frac{t_2 - t_1}{\ln(T_\infty - T'_{t_1}) - \ln(T_\infty - T_{t_2})} \tag{14.3}$$

where t_1 and t_2 indicate two points in time and t_c is a constant for the calorimeter which is inversely proportional to the rate of heat loss from the device.

From detailed theory it may be shown [Ref. 17] that the fiber attenuation due to absorption is given by:

$$\alpha_{abs} = \frac{CT_\infty}{P_{opt} t_c} \, \text{dB km}^{-1} \tag{14.4}$$

where C is proportional to the thermal capacity per unit length of the silica capillary and the low-refractive-index liquid surrounding the fiber, and P_{opt} is the optical power propagating in the fiber under test. The thermal capacity per unit length may be calculated, or determined by the electrical calibration utilizing the thin resistance wire. Usually the time constant for the calorimeter t_c is obtained using a high-absorption fiber which gives large temperature differences and greater accuracy. Once t_c is determined, the absorption losses of low-loss test fibers may be calculated from their maximum temperature rise T_∞, using Eq. (14.4). The temperatures are measured directly in terms of the thermocouple output (microvolts), and the optical input to the test fiber is obtained by use of a thermocouple or an optical power meter.

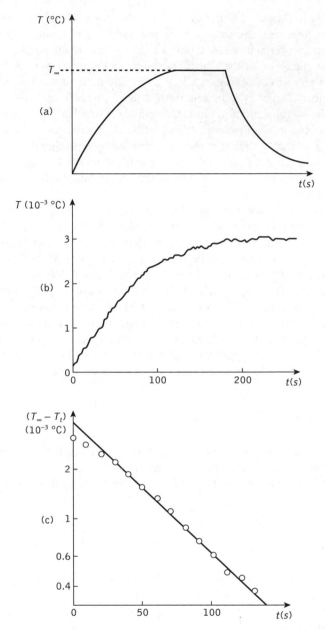

Figure 14.6 (a) A typical heating and cooling curve for a glass fiber sample. (b) A heating curve. (c) The corresponding plot of $(T_\infty - T_t)$ against time for a sample glass rod (bulk material measurement). Reproduced with permission from K. I. White and J. E. Midwinter. *Opto-electronics*, **5**, p. 323, 1973

Example 14.2

Measurements are made using a calorimeter and thermocouple experimental arrangement as shown in Figure 14.5 in order to determine the absorption loss of an optical fiber sample. Initially a high absorption fiber is utilized to obtain a plot of $(T_\infty - T_t)$ on a logarithmic scale against t. It is found from the plot that the readings of $(T_\infty - T_t)$ after 10 and 100 seconds are 0.525 and 0.021 μV respectively.

The test fiber is then inserted in the calorimeter and gives a maximum temperature rise of 4.3×10^{-4} °C with a constant measured optical power of 98 mW at a wavelength of 0.75 μm. The thermal capacity per kilometer of the silica capillary and fluid is calculated to be 1.64×10^4 J °C^{-1}.

Determine the absorption loss in dB km^{-1}, at a wavelength of 0.75 μm, for the fiber under test.

Solution: Initially, the time constant for the calorimeter is determined from the measurements taken on the high-absorption fiber using Eq. (14.3) where:

$$
\begin{aligned}
t_c &= \frac{t_2 - t_1}{\ln(T_\infty - T_{t_1}) - \ln(T_\infty - T_{t_2})} \\
&= \frac{100 - 10}{\ln(T_\infty - T_{10}) - \ln(T_\infty - T_{100})} \\
&= \frac{90}{\ln(0.525) - \ln(0.021)} \\
&= 28.0 \text{ s}
\end{aligned}
$$

Then the absorption loss of the test fiber may be obtained using Eq. (14.4) where:

$$
\begin{aligned}
\alpha_{abs} &= \frac{CT_\infty}{P_{opt} t_c} = \frac{1.64 \times 10^4 \times 4.3 \times 10^{-4}}{98 \times 10^{-3} \times 28.0} \\
&= 2.6 \text{ dB km}^{-1}
\end{aligned}
$$

Hence direct measurement of the contribution of absorption losses to the total fiber attenuation may be achieved. However, fiber absorption losses are often obtained indirectly from measurement of the fiber scattering losses (see the next section) by subtraction from the total fiber attenuation, measured by one of the techniques discussed in Section 14.2.1.

14.2.3 Fiber scattering loss measurement

The usual method of measuring the contribution of the losses due to scattering within the total fiber attenuation is to collect the light scattered from a short length of fiber and compare it with the total optical power propagating within the fiber. Light scattered from the fiber may be detected in a scattering cell as illustrated in the experimental arrangement shown in Figure 14.7. This may consist of a cube of six square solar cells (Tynes cell

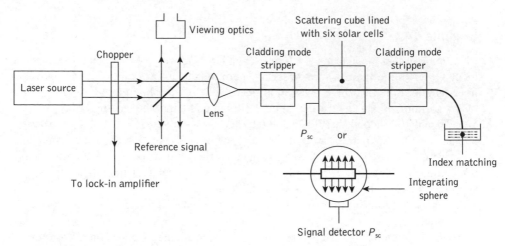

Figure 14.7 An experimental setup for measurement of fiber scattering loss illustrating both the solar cell cube and integrating sphere scattering cells

[Ref. 18]) or an integrating sphere and detector [Ref. 19]. The solar cell cube which contains index-matching fluid surrounding the fiber gives measurement of the scattered light, but careful balancing of the detectors is required in order to achieve a uniform response. This problem is overcome in the integrating sphere which again usually contains index-matching fluid but responds uniformly to different distributions of scattered light. However, the integrating sphere does exhibit high losses from internal reflections. Other variations of the scattering cell include the internally reflecting cell [Ref. 20] and the sandwiching of the fiber between two solar cells [Ref. 21].

A laser source (i.e. He–Ne, Nd : YAG, krypton ion) is utilized to provide sufficient optical power at a single wavelength together with a suitable instrument to measure the response from the detector. In order to avoid inaccuracies in the measurement resulting from scattered light which may be trapped in the fiber, cladding mode strippers (see Section 14.2.1) are placed before and after the scattering cell. These devices remove the light propagating in the cladding so that the measurements are taken only using the light guided by the fiber core. Also, to avoid reflections contributing to the optical signal within the cell, the output fiber end is index matched using either a fluid or suitable surface.

The loss due to scattering α_{sc} following Eq. (3.3) is given by:

$$\alpha_{sc} = \frac{10}{l(\text{km})} \log_{10} \left(\frac{P_{opt}}{P_{opt} - P_{sc}} \right) \text{dB km}^{-1} \tag{14.5}$$

where $l(\text{km})$ is the length of the fiber contained within the scattering cell, P_{opt} is the optical power propagating within the fiber at the cell and P_{sc} is the optical power scattered from the short length of fiber l within the cell. As $P_{opt} \gg P_{sc}$, then the logarithm in Eq. (14.5) may be expanded to give:

$$\alpha_{sc} = \frac{4.343}{l(\text{km})} \left(\frac{P_{sc}}{P_{opt}} \right) \text{dB km}^{-1} \tag{14.6}$$

Since the measurements of length are generally in centimeters and the optical power is normally registered in volts, Eq. (14.6) can be written as:

$$\alpha_{sc} = \frac{4.343 \times 10^5}{l(\text{cm})} \left(\frac{V_{sc}}{V_{opt}}\right) \text{dB km}^{-1} \tag{14.7}$$

where V_{sc} and V_{opt} are the voltage readings corresponding to the scattered optical power and the total optical power within the fiber at the cell. The relative experimental accuracy (i.e. repeatability) for scatter loss measurements is in the range ±0.2 dB using the solar cell cube and around 5% with the integrating sphere. However, it must be noted that the absolute accuracy of the measurements is somewhat poorer, being dependent on the calibration of the scattering cell and the mode distribution within a multimode fiber.

Example 14.3

An He–Ne laser operating at a wavelength of 0.63 μm was used with a solar cell cube to measure the scattering loss in a multimode fiber sample. With a constant optical output power the reading from the solar cell cube was 6.14 nV. The optical power measurement at the cube without scattering was 153.38 μV. The length of the fiber in the cube was 2.92 cm. Determine the loss due to scattering in dB km^{-1} for the fiber at a wavelength of 0.63 μm.

Solution: The scattering loss in the fiber at a wavelength of 0.63 μm may be obtained directly using Eq. (14.7) where:

$$\alpha_{sc} = \frac{4.343 \times 10^5}{l(\text{cm})} \left(\frac{V_{sc}}{V_{opt}}\right)$$

$$= \frac{4.343 \times 10^5}{2.92} \left(\frac{6.14 \times 10^{-9}}{153.38 \times 10^{-6}}\right)$$

$$= 6.0 \text{ dB km}^{-1}$$

14.3 Fiber dispersion measurements

Dispersion measurements give an indication of the distortion to optical signals as they propagate down optical fibers. The delay distortion which, for example, leads to the broadening of transmitted light pulses limits the information-carrying capacity of the fiber. Hence as shown in Section 3.8, the measurement of dispersion allows the bandwidth of the fiber to be determined. Therefore, besides attenuation, dispersion is the most important transmission characteristic of an optical fiber. As discussed in Section 3.8, there are three major mechanisms which produce dispersion in optical fibers (material dispersion, waveguide dispersion and intermodal dispersion). The importance of these different mechanisms to the total fiber dispersion is dictated by the fiber type.

For instance, in multimode fibers (especially step index), intermodal dispersion tends to be the dominant mechanism, whereas in single-mode fibers intermodal dispersion is nonexistent as only a single mode is allowed to propagate. In the single-mode case the dominant dispersion mechanism is chromatic (i.e. intramodal dispersion). The dominance of intermodal dispersion in multimode fibers makes it essential that dispersion measurements on these fibers are performed only when the equilibrium mode distribution has been established within the fiber, otherwise inconsistent results will be obtained. Therefore devices such as mode scramblers or filters must be utilized in order to simulate the steady-state mode distribution.

Dispersion effects may be characterized by taking measurements of the impulse response of the fiber in the time domain, or by measuring the baseband frequency response in the frequency domain. If it is assumed that the fiber response is linear with regard to power [Refs 22, 23], a mathematical description in the time domain for the optical output power $P_o(t)$ from the fiber may be obtained by convoluting the power impulse response $h(t)$ with the optical input power $P_i(t)$ as:

$$P_o(t) = h(t) * P_i(t) \tag{14.8}$$

where the asterisk * denotes convolution. The convolution of $h(t)$ with $P_i(t)$ shown in Eq. (14.8) may be evaluated using the convolution integral where:

$$P_o(t) = \int_{-\infty}^{\infty} P_i(t-x)h(x)\,\mathrm{d}x \tag{14.9}$$

In the frequency domain the power transfer function $H(\omega)$ is the Fourier transform of $h(t)$ and therefore by taking the Fourier transforms of all the functions in Eq. (14.8) we obtain:

$$\mathscr{P}_o(\omega) = H(\omega)\mathscr{P}_i(\omega) \tag{14.10}$$

where ω is the baseband angular frequency. The frequency domain representation given in Eq. (14.10) is the least mathematically complex, and by performing the Fourier transformation (or the inverse Fourier transformation) it is possible to switch between the time and frequency domains (or vice versa) by mathematical means. Hence, independent measurement of either $h(t)$ or $H(\omega)$ allows determination of the overall dispersive properties of the optical fiber. Thus fiber dispersion measurements can be made in either the time or frequency domains.

14.3.1 Time domain measurement

The most common method for time domain measurement of pulse dispersion in multimode optical fibers is illustrated in Figure 14.8 [Ref. 24]. Short optical pulses (100 to 400 ps) are launched into the fiber from a suitable source (e.g. AlGaAs injection laser) using fast driving electronics. The pulses travel down the length of fiber under test (around 1 km) and are broadened due to the various dispersion mechanisms. However, it is possible to take measurements of an isolated dispersion mechanism by, for example,

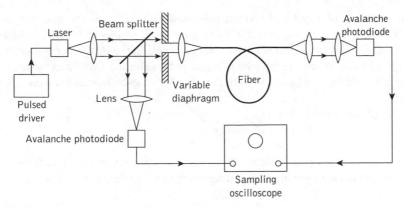

Figure 14.8 Experimental arrangement for making multimode fiber dispersion measurements in the time domain. Reprinted with permission from Ref. 24 © IEEE 1972

using a laser with a narrow spectral width when testing a multimode fiber. In this case the chromatic dispersion is negligible and the measurement thus reflects only intermodal dispersion. The pulses are received by a high-speed photodetector (i.e. avalanche photodiode) and are displayed on a fast sampling oscilloscope. A beam splitter is utilized for triggering the oscilloscope and for input pulse measurement.

After the initial measurement of output pulse width, the long fiber length may be cut back to a short length and the measurement repeated in order to obtain the effective input pulse width. The fiber is generally cut back to the lesser of 10 m or 1% of its original length [Ref. 10]. As an alternative to this cut-back technique, the insertion or substitution method similar to that used in fiber loss measurement (see Section 14.2.1) can be employed. This method has the benefit of being nondestructive and only slightly less accurate than the cut-back technique. These time domain measurement methods for multimode fiber are covered in TIA/EIA-455-51 [Ref. 25].

The fiber dispersion is obtained from the two pulse width measurements which are taken at any convenient fraction of their amplitude. However, unlike the considerations of dispersion in Sections 3.8 to 3.11 where rms pulse widths are used, dispersion measurements are normally made on pulses using the half maximum amplitude or 3 dB points. If $P_i(t)$ and $P_o(t)$ of Eq. (14.8) are assumed to have a Gaussian shape then Eq. (14.8) may be written in the form:

$$\tau_o^2(3\ dB) = \tau^2(3\ dB) + \tau_i^2(3\ dB) \tag{14.11}$$

where $\tau_i(3\ dB)$ and $\tau_o(3\ dB)$ are the 3 dB pulse widths at the fiber input and output, respectively, and $\tau(3\ dB)$ is the width of the fiber impulse response again measured at half the maximum amplitude. Hence the pulse dispersion in the fiber (commonly referred to as the pulse broadening when considering the 3 dB pulse width) in ns km^{-1} is given by:

$$\tau(3\ dB) = \frac{(\tau_o^2(3\ dB) - \tau_i^2(3\ dB))^{\frac{1}{2}}}{L}\ ns\ km^{-1} \tag{14.12}$$

where $\tau(3 \text{ dB})$, $\tau_i(3 \text{ dB})$ and $\tau_o(3 \text{ dB})$ are measured in ns and L is the fiber length in km. It must be noted that if a long length of fiber is cut back to a short length in order to take the input pulse width measurement, then L corresponds to the difference between the two fiber lengths in km. When the launched optical pulses and the fiber impulse response are Gaussian, the 3 dB optical bandwidth for the fiber B_{opt} may be calculated using [Ref. 26]:

$$B_{\text{opt}} \times \tau(3 \text{ dB}) = 0.44 \text{ GHz ns}$$
$$= 0.44 \text{ MHz ps} \tag{14.13}$$

Hence estimates of the optical bandwidth for the fiber may be obtained from the measurements of pulse broadening without resorting to rigorous mathematical analysis.

Example 14.4

Pulse dispersion measurements are taken over a 1.2 km length of partially graded multimode fiber. The 3 dB widths of the optical input pulses are 300 ps, and the corresponding 3 dB widths for the output pulses are found to be 12.6 ns. Assuming the pulse shapes and fiber impulse response are Gaussian calculate:

(a) the 3 dB pulse broadening for the fiber in ns km^{-1};

(b) the fiber bandwidth–length product.

Solution: (a) The 3 dB pulse broadening may be obtained using Eq. (14.12) where:

$$\tau(3 \text{ dB}) = \frac{(12.6^2 - 0.3^2)^{\frac{1}{2}}}{1.2} = \frac{(158.76 - 0.09)^{\frac{1}{2}}}{1.2}$$

$$= 10.5 \text{ ns km}^{-1}$$

(b) The optical bandwidth for the fiber is given by Eq. (14.13) as:

$$B_{\text{opt}} = \frac{0.44}{\tau(3 \text{ dB})} = \frac{0.44}{10.5} \text{ GHz km}$$

$$= 41.9 \text{ MHz km}$$

The value obtained for B_{opt} corresponds to the bandwidth–length product for the fiber because the pulse broadening in part (a) was calculated over a 1 km fiber length. Also it may be noted that in this case the narrow input pulse width makes little difference to the calculation of the pulse broadening. The input pulse width becomes significant when measurements are taken on low-dispersion fibers (e.g. single-mode).

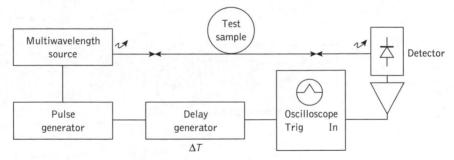

Figure 14.9 Experimental arrangement for the measurement of chromatic or intramodal dispersion by time delay

The above dispersion measurement techniques allow the total dispersion for multimode fibers to be determined. It is clear, however, that chromatic or intramodal dispersion is an important transmission parameter, particularly for single-mode fibers. Moreover, it can also be a significant distortion effect in multimode fibers even though intermodal dispersion is normally dominant. The time domain measurement of chromatic dispersion is outlined in TIA/EIA-455-168 [Ref. 27]. A typical experimental arrangement is shown in Figure 14.9. The pulse delay against optical wavelength is measured for both long and short fiber lengths. The source usually comprises multiple injection lasers possibly including wavelength-tunable devices (see Section 6.10). When $\Delta T(\lambda)$ is the delay difference for the length difference $L_1 - L_2$, then the specific group delay per unit length $\tau_g(\lambda)$ is given by [Ref. 10]:

$$\tau_g(\lambda) = \frac{\Delta T(\lambda)}{L_1 - L_2} \tag{14.14}$$

Differentiation of Eq. (14.14) provides the chromatic dispersion D_T following Eq. (3.46) where:

$$D_T(\lambda) = \frac{d\tau_g}{d\lambda} \text{ ps nm}^{-1} \text{ km}^{-1} \tag{14.15}$$

and the dispersion slope S from Eq. (3.52):

$$S(\lambda) = \frac{dD_T}{d\lambda} \text{ ps nm}^{-2} \text{ km}^{-1} \tag{14.16}$$

This pulse delay method is also one of the two RTMs to obtain chromatic dispersion in single-mode fibers which are recommended by the ITU-T [Ref. 1].

14.3.2 Frequency domain measurement

Frequency domain measurement is the preferred method for acquiring the bandwidth of multimode optical fibers. This is because the baseband frequency response $H(\omega)$ of the fiber

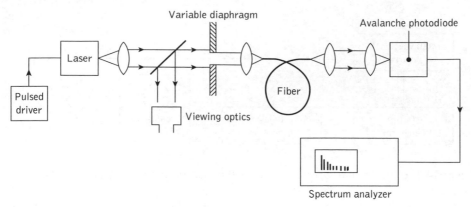

Figure 14.10 Experimental setup for making fiber dispersion measurements in the frequency domain using a pulsed laser source

may be obtained directly from these measurements using Eq. (14.10) without the need for any assumptions of Gaussian shape, or alternatively, the mathematically complex deconvolution of Eq. (14.8) which is necessary with measurements in the time domain. Thus the optical bandwidth of a multimode fiber is best obtained from frequency domain measurements.

One of two frequency domain measurement techniques is generally used. The first utilizes a similar pulsed source to that employed for the time domain measurements shown in Figure 14.8. However, the sampling oscilloscope is replaced by a spectrum analyzer which takes the Fourier transform of the pulse in the time domain and hence displays its constituent frequency components. The experimental arrangement is illustrated in Figure 14.10.

Comparison of the spectrum at the fiber output $\mathscr{P}_o(\omega)$ with the spectrum at the fiber input $\mathscr{P}_i(\omega)$ provides the baseband frequency response for the fiber under test following Eq. (5.10) where:

$$H(\omega) = \frac{\mathscr{P}_o(\omega)}{\mathscr{P}_i(\omega)} \qquad (14.17)$$

The second technique involves launching a sinusoidally modulated optical signal at different selected frequencies using a sweep oscillator. Therefore the signal energy is concentrated in a very narrow frequency band in the baseband region, unlike the pulse measurement method where the signal energy is spread over the entire baseband region. A possible experimental arrangement for this swept frequency measurement method is shown in Figure 14.11 [Ref. 28]. The optical source is usually an injection laser, which may be directly modulated (see Section 7.5) from the sweep oscillator. A spectrum analyzer may be used in order to obtain a continuous display of the swept frequency signal. Again, Eq. (14.17) is utilized to obtain the baseband frequency response, employing either the cut-back or substitution procedure in a similar manner to the time domain measurement (see Section 14.3.1). However, the spectrum analyzer provides no information on the phase of the received signal Therefore a vector voltmeter or ideally a network analyzer can be employed to give both the frequency and phase information. This multimode fiber frequency domain measurement method is described in TIA/EIA-455-30 [Ref. 29].

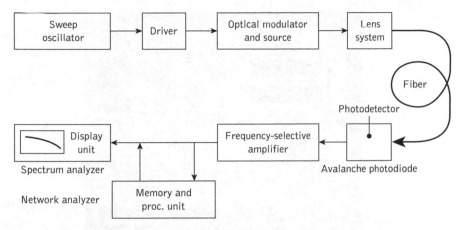

Figure 14.11 Block schematic showing an experimental arrangement for the swept frequency measurement method to provide fiber dispersion measurements in the frequency domain [Ref. 28]

The chromatic or intramodal dispersion for single-mode fibers may also be obtained using frequency domain measurement techniques. The second reference test method recommended by the ITU [Ref. 1] falls into this category and is known as the phase shift method. This technique is also covered in TIA/EIA-455-169 [Ref. 30]. To obtain the phase shift $\phi(\lambda)$ against wavelength, the pulse generator in Figure 14.9 (corresponding to the time domain measurement) is replaced by a high-frequency oscillator operating at a constant frequency and the delay generator and oscilloscope are replaced by a phase meter or vector voltmeter. Finally, an electrical or optical reference channel is connected between the oscillator and the meter.

When an optical signal, which is sinusoidally modulated in power with frequency f_m, is transmitted through a single-mode fiber of length L, then the modulation envelope is delayed in time by:

$$\frac{L}{v_g} = \tau_g L \tag{14.18}$$

where v_g is the group velocity which corresponds to the signal velocity. Since a delay of one modulation period T_m or $1/f_m$ corresponds to a phase shift of 2π, then the sinusoidal modulation is phase shifted in the fiber by an angle ϕ_m where:

$$\phi_m = \frac{2\pi\tau_g L}{T_m} = 2\pi f_m \tau_g L \tag{14.19}$$

Hence the specific group delay is given by:

$$\tau_g = \frac{\phi_m}{2\pi f_m L} \tag{14.20}$$

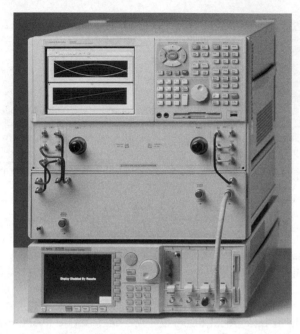

Figure 14.12 Optical fiber dispersion measurement and analyzer [Ref. 33].
© Agilent Technologies, Inc. 2005. Reproduced with permission, courtesy of Agilent
Technologies, Inc.

Again the chromatic dispersion and the dispersion slope can be obtained by differentiation
following Eqs (14.15) and (14.16) respectively.

   The widespread application of optical wavelength division multiplexed systems requires
the use of accurate wideband chromatic dispersion compensation, in particular where the
optical fiber networks comprise both earlier and new fiber types operating at signal wavelengths of 1.31 μm and 1.55 μm. Various commercial instruments [Refs 31–33] are available for the measurement of dispersion at transmission rates up to 40 Gbit s^{-1}. These
instruments are used in the frequency domain and the test and measurement procedures
are commonly referred to as phase shift methods. In these methods a phase difference is
obtained from the variation in timing between two periodic signals (i.e. original and a
delayed signal) [Ref. 34]. Furthermore, these instruments do not require an additional computation facility in order to provide simulation results and they can also display (or plot)
the measured data. An optical dispersion analyzer is shown in Figure 14.12 [Ref. 33].
Such dispersion measuring and analyzing instruments can perform dispersion and loss
measurements across the entire C- and L-band wavelength range (i.e. 1.49 to 1.64 μm) with
a single setup and connection, thus saving time and reducing the opportunities for error.

14.4 Fiber refractive index profile measurements

The refractive index profile of the fiber core plays an important role in characterizing the
properties of optical fibers. It allows determination of the fiber's numerical aperture and

the number of modes propagating within the fiber core, while largely defining any inter-modal and/or profile dispersion caused by the fiber. Hence a detailed knowledge of the refractive index profile enables the impulse response of the fiber to be predicted. Also, as the impulse response and consequently the information-carrying capacity of the fiber is strongly dependent on the refractive index profile, it is essential that the fiber manufacturer is able to produce particular profiles with great accuracy, especially in the case of graded index fibers (i.e. optimum profile). There is therefore a requirement for accurate measurement of the refractive index profile. These measurements may be performed using a number of different techniques each of which exhibit certain advantages and drawbacks. In this section we will discuss some of the more popular methods which may be relatively easily interpreted theoretically, without attempting to review all the possible techniques which have been developed.

14.4.1 Interferometric methods

Interference microscopes (e.g. Mach–Zehnder, Michelson) have been widely used to determine the refractive index profiles of optical fibers. The technique usually involves the preparation of a thin slice of fiber (slab method) which has both ends accurately polished to obtain square (to the fiber axes) and optically flat surfaces. The slab is often immersed in an index-matching fluid, and the assembly is examined with an interference microscope. Two major methods are then employed, using either a transmitted light interferometer (Mach–Zehnder [Ref. 35]) or a reflected light interferometer (Michelson [Ref. 36]). In both cases light from the microscope travels normal to the prepared fiber slice faces (parallel to the fiber axis), and differences in refractive index result in different optical path lengths [Ref. 37]. This situation is illustrated in the case of the Mach–Zehnder interferometer in Figure 14.13(a). When the phase of the incident light is compared with the phase of the emerging light, a field of parallel interference fringes is observed. A photograph of the fringe pattern may then be taken, an example of which is shown in Figure 14.13(a) [Ref. 38].

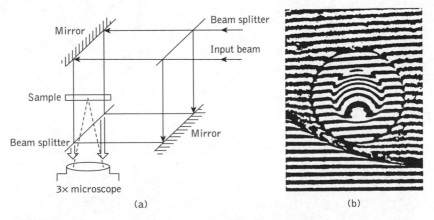

Figure 14.13 (a) The principle of the Mach–Zehnder interferometer [Ref. 35]. (b) The interference fringe pattern obtained with an interference microscope from a graded index fiber. Reused with permission from L. G. Cohen, P. Kaiser, J. M. MacChesney, P. N. O'Connor and H. M. Presby. *Appl. Phys. Lett.*, **26**, p. 472, 1975
© 1975 American Institute of Physics

The fringe displacements for the points within the fiber core are then measured using as reference the parallel fringes outside the fiber core (in the fiber cladding). The refractive index difference between a point in the fiber core (e.g. the core axis) and the cladding can be obtained from the fringe shift q, which corresponds to a number of fringe displacements. This difference in refractive index δn is given by [Ref. 10]:

$$\delta n = \frac{q\lambda}{x} \tag{14.21}$$

where x is the thickness of the fiber slab and λ is the incident optical wavelength. The slab method gives an accurate measurement of the refractive index profile, although computation of the individual points is somewhat tedious unless an automated technique is used. Figure 14.14 [Ref. 38] shows the refractive index profile obtained from the fringe pattern indicated in Figure 14.13(b).

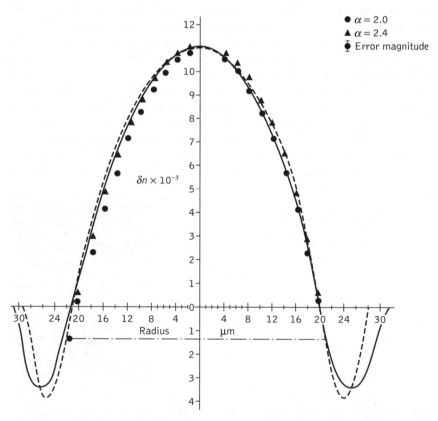

Figure 14.14 The fiber refractive index profile computed from the interference pattern shown in Figure 14.13(b). Reused with permission from L. G. Cohen, P. Kaiser, J. M. MacChesney, P. N. O'Connor and H. M. Presby. *Appl. Phys. Lett.*, **26**, p. 472, 1975 © 1975, American Institute of Physics

A limitation of this method is the time required to prepare the fiber slab [Ref. 39]. However, another interferometric technique has been developed [Ref. 40] which requires no sample preparation. In this method the light beam is incident to the fiber perpendicular to its axis; this is known as transverse shearing interferometry. Again fringes are observed from which the fiber refractive index profile may be obtained.

More recently, a further interferometric technique known as induced-grating auto-correlation (IGA) has been used to measure the nonlinear refractive index of the silica fiber [Refs 41, 42]. This method is based on the electro-optic effect where measuring the electric field autocorrelation function determines the refractive index of the optical fiber. The IGA function technique involves the determination of the self-phase modulation of an optical signal in an optical fiber (see Section 3.14). This is achieved by delaying one path of the optical signal. The IGA response is then produced by varying the time delay between the two optical signals which are then mixed together [Ref. 31].

Figure 14.15 shows the experimental setup used to observe an IGA response using a nonlinear optical loop mirror interferometer. It consists of a laser source and a combination of optical lenses and mirrors where a beam splitter separates the signal creating the delayed path. The two optical signals (i.e. original and delayed signals) combine at a point where a photorefractive crystal is placed which is the mixing element employed in this method. Several crystalline material systems, known as photorefractive crystals [Ref. 43], can be used to produce a diffraction grating in order to implement IGA. Photorefraction is, however, an electro-optic phenomenon in which the local index of refraction is modified by spatial variations of the light intensity.

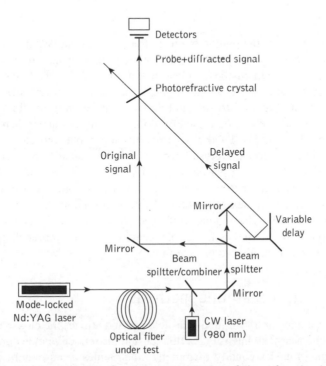

Figure 14.15 Experimental setup for the measurement of the refractive index of silica fiber using the induced-grating autocorrelation function technique

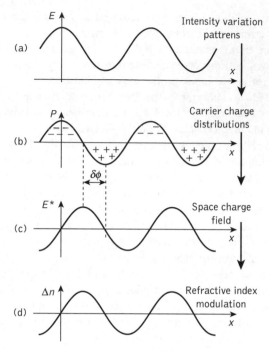

Figure 14.16 Photorefractive index effect

Figure 14.16 illustrates the photorefractive effect in a crystal. When two coherent light beams are superimposed on each other in a photorefractive crystal an interference pattern results with high and low intensities as identified in Figure 14.16(a). In the areas of higher intensity, charge carriers (i.e. electrons or holes) are excited where the gradient in the charge carrier density causes diffusion and thus the carriers migrate through the crystal. Eventually, they are trapped in the crystal resulting in a charge carrier distribution as indicated in Figure 14.16(b). The local charge distribution evokes an electric space charge field which in turn modifies the crystal refractive index causing the signal phase to be shifted by a quarter of a grating period (i.e. equivalent to a phase shift of $\delta\phi$ equal to $\pi/2$ with respect to the intensity pattern) as shown in Figure 14.16(c). Hence the electro-optic effect creates a refractive index distribution Δn proportional to the electric field as indicated in Figure 14.16(d). The required information on the nonlinear phase variation is gathered from two optical signals using this technique and it then enables the refractive index profile for an optical fiber to be determined.

14.4.2 Near-field scanning method

The near-field scanning or transmitted near-field method utilizes the close resemblance that exists between the near-field intensity distribution and the refractive index profile, for a fiber with all the guided modes equally illuminated. It provides a reasonably straightforward and rapid method for acquiring the refractive index profile. When a diffuse Lambertian

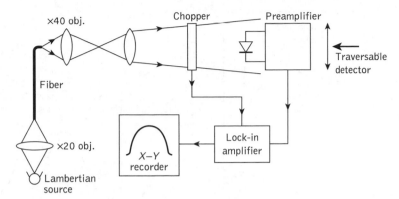

Figure 14.17 Experimental setup for the near-field scanning measurement of the refractive index profile. Reused with permission from Ref. 47 © 1976, American Institute of Physics

source (e.g. tungsten filament lamp or LED) is used to excite all the guided modes then the near-field optical power density at a radius r from the core axis $P_D(r)$ may be expressed as a fraction of the core axis near-field optical power density $P_D(0)$ following [Ref. 44]:

$$\frac{P_D(r)}{P_D(0)} = C(r, z)\left[\frac{n_1^2(r) - n_2^2}{n_1^2(0) - n_2^2}\right] \tag{14.22}$$

where $n_1(0)$ and $n_1(r)$ are the refractive indices at the core axis and at a distance r from the core axis respectively, n_2 is the cladding refractive index and $C(r, z)$ is a correction factor. The correction factor which is incorporated to compensate for any leaky modes present in the short test fiber may be determined analytically. A set of normalized correction curves is, for example, given in Ref. 45. For multimode fiber such a transmitted near-field method is described in TIA/EIA-455-43 [Ref. 46]. The transmitted near-field approach is, however, not similarly recommended for single-mode fiber.

An experimental configuration is shown in Figure 14.17. The output from a Lambertian source is focused onto the end of the fiber using a microscope objective lens. A magnified image of the fiber output end is displayed in the plane of a small active area photodetector (e.g. silicon *p–i–n* photodiode). The photodetector which scans the field transversely receives amplification from the phase-sensitive combination of the optical chopper and lock-in amplifier. Hence the profile may be plotted directly on an *X–Y* recorder. However, the profile must be corrected with regard to $C(r, z)$ as illustrated in Figure 14.18(a) which is very time consuming. Both the scanning and data acquisition can be automated with the inclusion of a minicomputer [Ref. 45].

The test fiber is generally 2 m in length to eliminate any differential mode attenuation and mode coupling. A typical refractive index profile for a practical step index fiber measured by the near-field scanning method is shown in Figure 14.18(b). It may be observed that the profile dips in the center at the fiber core axis. This dip was originally thought to result from the collapse of the fiber preform before the fiber is drawn in the manufacturing process but has been shown to be due to the layer structure inherent at the deposition stage [Ref. 48].

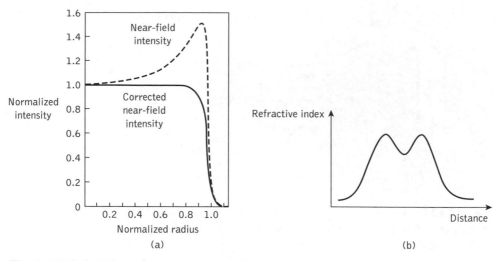

Figure 14.18 (a) The refractive index profile of a step index fiber measured using the near-field scanning method, showing the near-field intensity and the corrected near-field intensity. Reproduced with permission from F. E. M. Sladen, D. N. Payne and M. J. Adams, *Appl. Phys. Lett.*, **28**, p. 225, 1976 © 1976, American Institute of Physics. (b) The refractive index profile of a practical step index fiber measured by the near-field scanning method [Ref. 47]

14.4.3 Refracted near-field method

The refracted near-field (RNF) or refracted ray method is complementary to the transmitted near-field technique (see Section 14.4.2) but has the advantage that it does not require a leaky mode correction factor or equal mode excitation. Moreover, it provides the relative refractive index differences directly without recourse to external calibration or reference samples. The RNF method is the most commonly used technique for the determination of the fiber refractive index profile [Ref. 49] and is the EIA reference test method for both multimode and single-mode fibers. Details of the test procedure are provided in TIA/EIA-455-44 [Ref. 50].

A schematic of an experimental setup for the RNF method is shown in Figure 14.19. A short length of fiber is immersed in a cell containing a fluid of slightly higher refractive index. A small spot of light typically emitted from a 633 nm He–Ne laser for best resolution is scanned across the cross-sectional diameter of the fiber. The measurement technique utilizes that light which is not guided by the fiber but escapes from the core into the cladding. However, light escaping from the fiber core partly results from the power leakage from the leaky modes which is an unknown quantity. The effect of this radiated power reaching the detector is undesirable and therefore it is blocked using an opaque circular screen, as shown in Figure 14.19(a). The refracted ray trajectories are illustrated in Figure 14.19(b) where θ' is the angle of incidence in the fiber core, θ is the angle of refraction in the fiber core and θ'' constitutes the angle of the refracted inbound rays external to the fiber core. Any light leaving the fiber core below a minimum angle θ''_{min} is prevented from reaching the detector by the opaque screen (Figure 14.19(a)). Moreover, it may be observed from Figure 14.19(b) that this minimum angle corresponds to a minimum angle

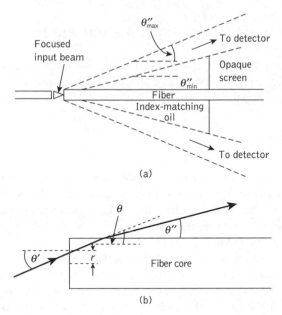

Figure 14.19 Refracted near-field method for the measurement of refractive index profile: (a) experimental arrangement; (b) illustration of the ray trajectories. After Ref. 51. A. H. Cherin, *An Introduction to Optical Fibers* © 1987 with permission from The McGraw Hill Companies

of incidence θ'_{min}. However, all light at an angle of incidence $\theta' > \theta'_{min}$ must be allowed to reach the detector. To ensure that this process occurs it is advisable that input apertures are used to limit the convergence angle of the input beam to a suitable maximum angle θ'_{max} corresponding to a refracted angle θ''_{max}. In addition, the immersion of the fiber in an index-matching fluid prevents reflection at the outer cladding boundary. Hence all the refracted light emitted from the fiber at angles over the range θ''_{min} to θ''_{max} may be detected.

The detected optical power as a function of the radial position of the input beam $P(r)$ is measured and a value $P(a)$ corresponding to the input beam being focused into the cladding is also obtained. The refractive index profile $n(r)$ for the fiber core is then given by [Ref. 52]:

$$n(r) = n_2 + n_2 \cos \theta''_{min}(\cos \theta''_{min} - \cos \theta'_{max}) \frac{P(a) - P(r)}{P(a)} \tag{14.23}$$

where n_2 is the cladding refractive index. Furthermore, Eq. (14.23) can be written as:

$$n(r) = k_1 - k_2 P(r) \tag{14.24}$$

It is clear that $n(r)$ is proportional to $P(r)$ and hence the measurement system can be calibrated to obtain the constants k_1 and k_2. For example, a calibration scheme in which the power that passes the opaque screen is monitored as it is translated along the optical axis provided an early strategy [Refs 53, 54]. Alternative calibration techniques which allow accurate RNF measurements are described in Ref. 49.

14.5 Fiber cutoff wavelength measurements

A multimode fiber has many cutoff wavelengths because the number of bound propagating modes is usually large. For example, considering a parabolic refractive index graded fiber, following Eq. (2.95) the number of guided modes M_g is:

$$M_g = \left(\frac{\pi a}{\lambda}\right)^2 (n_1^2 - n_2^2) \tag{14.25}$$

where a is the core radius and n_1 and n_2 are the core peak and cladding indices respectively. It may be observed from Eq. (14.25) that operation at longer wavelengths yields fewer guided modes. Therefore it is clear that as the wavelength is increased, a growing number of modes are cutoff where the cutoff wavelength of a LP_{lm} mode is the maximum wavelength for which the mode is guided by the fiber.

Usually the cutoff wavelength refers to the operation of single-mode fiber in that it is the cutoff wavelength of the LP_{11} mode (which has the longest cutoff wavelength) which makes the fiber single moded when the fiber diameter is reduced to 8 or 9 μm. Hence the cutoff wavelength of the LP_{11} is the shortest wavelength above which the fiber exhibits single-mode operation and it is therefore an important parameter to measure (see Section 2.5.1). The theoretical value of the cutoff wavelength can be determined from the fiber refractive index profile following Eq. (2.98). Because of the large attenuation of the LP_{11} mode near cutoff, however, the parameter which is experimentally determined is called the effective cutoff wavelength, which is always smaller than the theoretical cutoff wavelength by as much as 100 to 200 nm [Ref. 54]. It is this effective cutoff wavelength which limits the wavelength region for which the fiber is 'effectively' single-mode.

The effective cutoff wavelength is normally measured by increasing the signal wavelength in a fixed length of fiber until the LP_{11} mode is undetectable. Since the attenuation of the LP_{11} mode is dependent on the fiber length and its radius of curvature, the effective cutoff wavelength tends to vary with the method of measurement. Moreover, numerous methods of measurement have been investigated [Refs 10, 54] and because these techniques can give significantly different results, the measurement has caused some problems [Ref. 55]. Nevertheless, three methods were originally recommended by the ITU-T [Ref. 1], two of which, being transmitted power techniques, were recommended as RTMs. In addition, these two techniques correspond to the EIA standard test method TIA/EIA-455-80C [Ref. 56].

The effective cutoff wavelength has been defined by the ITU-T as the wavelength greater than which the ratio between the total power, including the launched higher order modes, and the fundamental mode power has decreased to less than 0.1 dB in a quasi-straight 2 m fiber length with one single loop of 140 mm radius.* Measurement configurations which enable the determination of fiber cutoff wavelength by the RTMs are shown in Figure 14.20. A single-turn configuration is illustrated in Figure 14.20(a), while the split mandrel configuration of Figure 14.20(b) proves convenient for fiber handling. The other test apparatus is the same as that employed for the measurement of fiber attenuation by

---

* It should be noted that the 2 m fiber length corresponds to the length specified in the cut-back attenuation measurements (Section 14.2.1).

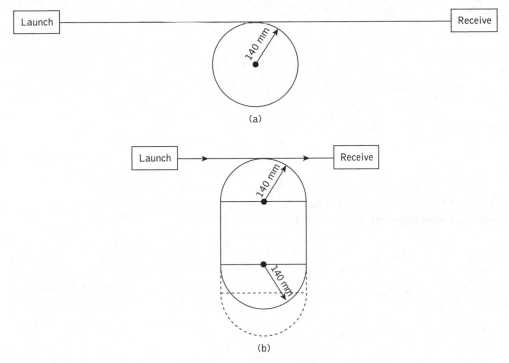

Figure 14.20 Configurations for the measurement of uncabled fiber cutoff wavelength: (a) single turn; (b) split mandrell [Refs 1, 56]

the cut-back method (Figure 14.2). However, the launch conditions used must be sufficient to excite both the fundamental and the LP_{11} modes, and it is important that cladding modes are stripped from the fiber.

In the bending-reference technique the power $P_s(\lambda)$ transmitted through the fiber sample in the configurations shown in Figure 14.20 is measured as a function of wavelength. Thus the quantity $P_s(\lambda)$ corresponds to the total power, including launched higher order modes, of the ITU-T definition for cutoff wavelength. Then keeping the launch conditions fixed, at least one additional loop of sufficiently small radius (60 mm or less) is introduced into the test sample to act as a mode filter to suppress the secondary LP_{11} mode without attenuating the fundamental mode at the effective cutoff wavelength. In this case the smaller transmitted spectral power $P_b(\lambda)$ is measured which corresponds to the fundamental mode power referred to in the definition. The bend attenuation $a_b(\lambda)$ comprising the level difference between the total power and the fundamental power is calculated as:

$$a_b(\lambda) = 10 \log_{10} \frac{P_s(\lambda)}{P_b(\lambda)}$$

(14.26)

The bend attenuation characteristic exhibits a peak in the wavelength region where the radiation losses resulting from the small loop are much higher for the LP_{11} mode than for the LP_{01} fundamental mode, as illustrated in Figure 14.21. It should be noted that the

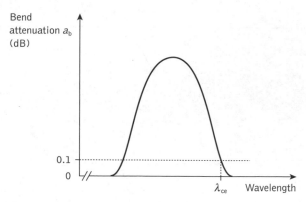

Figure 14.21 Bend attenuation against wavelength in the bending method for the measurement of cutoff wavelength λ_{ce}

shorter wavelength side of the attenuation maximum corresponds to the LP_{11} mode, being well confined in the fiber core, and hence negligible loss is induced by the 60 mm diameter loop, whereas on the longer wavelength side the LP_{11} mode is not guided in the fiber and therefore, assuming that the loop diameter is large enough to avoid any curvature loss to the fundamental mode, there is also no increase in loss. Using the ITU-T and EIA definition for the effective cutoff wavelength λ_{ce} it may be determined as the longest wavelength at which the bend attenuation or level difference $a_b(\lambda)$ equals 0.1 dB, as shown in Figure 14.21.

The other RTM is called the power step method [Ref. 57] or the multimode reference technique [Ref. 10]. Again, the fiber configurations shown in Figure 14.20 are employed with the test apparatus the same as that to measure fiber attenuation by the cut-back method. Furthermore, the launch conditions must again be sufficient to excite both the fundamental and LP_{11} modes and, as in the bending method, the transmitted power $P_s(\lambda)$ is measured as a function of wavelength. Next, however, the 2 m length of single-mode fiber is replaced by a short (1 to 2 m) length of multimode fiber and the spectral power $P_m(\lambda)$ emerging from the end of the multimode fiber is measured.

The relative attenuation $a_m(\lambda)$ or level difference between the powers launched into the multimode and single-mode fibers may be computed as:

$$a_m(\lambda) = 10 \log_{10} \frac{P_s(\lambda)}{P_m(\lambda)} \tag{14.27}$$

A typical characteristic showing the level difference as a function of wavelength is provided in Figure 14.22 in which the step reduction of level difference around cutoff may be observed. This results from the increase in power obtained at the output of the single-mode fiber from propagation of the LP_{11} mode, as well as the fundamental LP_{01} mode when going through the cutoff wavelength. To obtain the effective cutoff wavelength, the longest wavelength portion of the characteristic is fitted to a straight line and the intersection of the $a_m(\lambda)$ curve with another parallel straight line displaced by 0.1 dB produces the result. It should be noted, however, that accurate measurement requires an attenuation

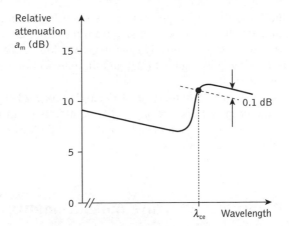

Figure 14.22 Relative attenuation against wavelength in the power step technique for the measurement of cutoff wavelength λ_{ce}

difference of not less than 2 dB [Ref. 10]. Such a difference may be readily obtained as there are two modes in the primary LP_{01} mode group and four in the secondary LP_{11} mode group. Hence with equal excitation of both groups the maximum attenuation difference is $10 \log_{10} (2 + 4)/2$, or 4.8 dB, when going through cutoff [Ref. 57]. Finally, this method and the bending reference technique have been shown to yield approximately the same values for the effective cutoff wavelength in a round robin test [Ref. 58].

A third method for determination of the effective cutoff wavelength which is recommended by the ITU-T as an alternative test method [Ref. 1] is the measurement of the change in spot size with wavelength [Ref. 59]. In this case the spot size is measured as a function of wavelength by the transverse offset method (see Section 14.8) using a 2 m length of fiber on each side of the joint with a single loop of radius 140 mm formed in each 2 m length. When the fiber is operating in the single-mode region, the spot size increases almost linearly with increasing wavelength [Ref. 54], as may be observed in Figure 14.23.

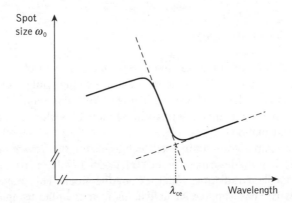

Figure 14.23 Wavelength dependence of the spot size in the spot size technique for the measurement of cutoff wavelength λ_{ce}

However, as the cutoff wavelength is approached, the contribution from the second-order mode creates a significant change in the spot size from the expected single-mode values. At this point two straight lines with a positive and negative slope can be fitted through the measured points, as illustrated in Figure 14.23, and the intersection point corresponds to the effective cutoff wavelength.

The effective cutoff wavelength for a cabled single-mode fiber will generally be smaller than that of the uncabled fiber because of bend effects (both micro- and macro-bending). A procedure for this measurement is outlined in TIA/EIA-455-170 [Ref. 60] similar to the transmitted power methods.

14.6 Fiber numerical aperture measurements

The numerical aperture is an important optical fiber parameter as it affects characteristics such as the light-gathering efficiency and the normalized frequency of the fiber (V). This in turn dictates the number of modes propagating within the fiber (also defining the single-mode region) which has consequent effects on both the fiber dispersion (i.e. intermodal) and, possibly, the fiber attenuation (i.e. differential attenuation of modes). The numerical aperture (NA) is defined for a step index fiber in air by Eq. (2.8) as:

$$NA = \sin \theta_a = (n_1^2 - n_2^2)^{\frac{1}{2}} \tag{14.28}$$

where θ_a is the maximum acceptance angle, n_1 is the core refractive index and n_2 is the cladding refractive index. It is assumed in Eq. (14.28) that the light is incident on the fiber end face from air with a refractive index (n_0) of unity. Although Eq. (14.28) may be employed with graded index fibers, the numerical aperture thus defined represents only the local NA of the fiber on its core axis (the numerical aperture for light incident at the fiber core axis). The graded profile creates a multitude of local NAs as the refractive index changes radially from the core axis. For the general case of a graded index fiber these local numerical apertures $NA(r)$ at different radial distances r from the core axis may be defined by:

$$NA(r) = \sin \theta_a(r) = (n_1^2(r) - n_2^2)^{\frac{1}{2}} \tag{14.29}$$

Therefore, calculations of numerical aperture from refractive index data are likely to be less accurate for graded index fibers than for step index fibers unless the complete refractive index profile is considered. However, if refractive index data is available on either fiber type from the measurements described in Section 14.4, the numerical aperture may be determined by calculation.

Alternatively, a simple commonly used technique for the determination of the fiber numerical aperture is now described by TIA/EIA-455-177 [Ref. 61] and involves measurement of the far-field radiation pattern from the fiber. This measurement may be performed by directly measuring the far-field angle from the fiber using a rotating stage, or by calculating the far-field angle using trigonometry. An example of an experimental

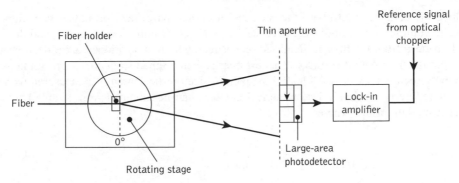

Figure 14.24 Fiber numerical aperture measurement using a scanning photodetector and a rotating stage

arrangement with a rotating stage is shown in Figure 14.24. A 2 m length of the graded index fiber has its faces prepared in order to ensure square smooth terminations. The fiber output end is then positioned on the rotating stage with its end face parallel to the plane of the photodetector input, and so that its output is perpendicular to the axis of rotation. Light at a wavelength of 0.85 μm is launched into the fiber at all possible angles (overfilling the fiber) using an optical system similar to that used in the spot attenuation measurements (Figure 14.4).

The photodetector, which may be either a small-area device or an apertured large-area device, is placed 10 to 20 cm from the fiber and positioned in order to obtain a maximum signal with no rotation (0°). Hence when the rotating stage is turned the limits of the far-field pattern may be recorded. The output power is monitored and plotted as a function of angle, the maximum acceptance angle being obtained when the power drops to 5% of the maximum intensity [Ref. 62]. Thus the numerical aperture of the fiber can be obtained from Eq. (14.28). This far-field scanning measurement may also be performed with the photodetector located on a rotational stage and the fiber positioned at the center of rotation. Moreover, TIA/EIA-455-177 also outlines a technique to obtain the numerical aperture from the refractive index profile of the fiber.

A less precise measurement of the numerical aperture can be obtained from the far-field pattern by trigonometric means. The experimental apparatus is shown in Figure 14.25

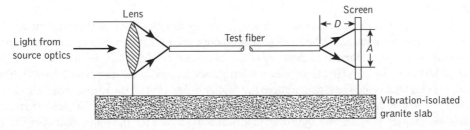

Figure 14.25 Apparatus for trigonometric fiber numerical aperture measurement

where the end prepared fiber is located on an optical base plate or slab. Again light is launched into the fiber under test over the full range of its numerical aperture, and the far-field pattern from the fiber is displayed on a screen which is positioned a known distance D from the fiber output end face. The test fiber is then aligned so that the optical intensity on the screen is maximized. Finally, the pattern size on the screen A is measured using a calibrated vernier caliper. The numerical aperture can be obtained from simple trigonometrical relationships where:

$$NA = \sin \theta_a = \frac{A/2}{[(A/2)^2 + D^2]^{\frac{1}{2}}} = \frac{A}{(A^2 + 4D^2)^{\frac{1}{2}}} \qquad (14.30)$$

Example 14.5

A trigonometrical measurement is performed in order to determine the numerical aperture of a step index fiber. The screen is positioned 10.0 cm from the fiber end face. When illuminated from a wide-angled visible source the measured output pattern size is 6.2 cm. Calculate the approximate numerical aperture of the fiber.

Solution: The numerical aperture may be determined directly, using Eq. (14.30) where:

$$NA = \frac{A}{(A^2 + 4D^2)^{\frac{1}{2}}} = \frac{6.2}{(38.44 + 400)^{\frac{1}{2}}} = 0.30$$

It must be noted that the accuracy of this measurement technique is dependent upon the visual assessment of the far-field pattern from the fiber.

The above measurement techniques are generally employed with multimode fibers only, as the far-field patterns from single-mode fibers are affected by diffraction phenomena. These are caused by the small core diameters of single-mode fibers which tend to invalidate simple geometric optics measurements. However, more detailed analysis of the far-field pattern allows determination of the normalized frequency and core radius for single-mode fibers, from which the numerical aperture may be calculated using Eq. (2.69) [Ref. 63].

Far-field pattern measurements with regard to multimode fibers are dependent on the length of the fiber tested. When the measurements are performed on short fiber lengths (around 1 m) the numerical aperture thus obtained corresponds to that defined by Eq. (14.28) or (14.29). However, when a long fiber length is utilized which gives mode coupling and the selective attenuation of the higher order modes, the measurement yields a lower value for the numerical aperture. It must also be noted that the far-field measurement techniques give an average (over the local NAs) value for the numerical aperture of graded index fibers. Hence alternative methods must be employed if accurate determination of the fiber's NA is required [Ref. 64].

14.7 Fiber diameter measurements

14.7.1 Outer diameter

It is essential during the fiber manufacturing process (at the fiber drawing stage) that the fiber outer diameter (cladding diameter) is maintained constant to within 1%. Any diameter variations may cause excessive radiation losses and make accurate fiber–fiber connection difficult. Hence on-line diameter measurement systems are required which provide accuracy better than 0.3% at a measurement rate greater than 100 Hz (i.e. a typical fiber drawing velocity is 1 m s^{-1}). Use is therefore made of noncontacting optical methods such as fiber image projection and scattering pattern analysis.

The most common on-line measurement technique uses fiber image projection (shadow method) and is illustrated in Figure 14.26 [Ref. 65]. In this method a laser beam is swept at a constant velocity transversely across the fiber and a measurement is made of the time interval during which the fiber intercepts the beam and casts a shadow on a photodetector. In the apparatus shown in Figure 14.26 the beam from a laser operating at a wavelength of 0.6328 μm is collimated using two lenses (G_1 and G_2). It is then reflected off two mirrors (M_1 and M_2), the second of which (M_2) is driven by a galvanometer which makes it rotate through a small angle at a constant angular velocity before returning to its original starting position. Therefore, the laser beam which is focused in the plane of the fiber by a lens (G_3) is swept across the fiber by the oscillating mirror, and is incident on the photodetector unless it is blocked by the fiber. The velocity $\mathrm{d}s/\mathrm{d}t$ of the fiber shadow thus created at the photodetector is directly proportional to the mirror velocity $\mathrm{d}\phi/\mathrm{d}t$ following:

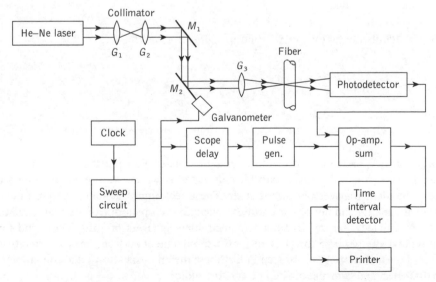

Figure 14.26 The shadow method for the on-line measurement of the fiber outer diameter. Reused with permission from Ref. 65 © 1973, American Institute of Physics

$$\frac{\mathrm{d}s}{\mathrm{d}t} = l\frac{\mathrm{d}\phi}{\mathrm{d}t} \qquad\qquad (14.31)$$

where l is the distance between the mirror and the photodetector.

Furthermore, the shadow is registered by the photodetector as an electrical pulse of width W_e which is related to the fiber outer diameter d_o as:

$$d_o = W_e\frac{\mathrm{d}s}{\mathrm{d}t} \qquad\qquad (14.32)$$

Thus the fiber outer diameter may be quickly determined and recorded on the printer. The measurement speed is largely dictated by the inertia of the mirror rotation and its accuracy by the rise time of the shadow pulse.

Example 14.6

The shadow method is used for the on-line measurement of the outer diameter of an optical fiber. The apparatus employs a rotating mirror with an angular velocity of 4 rad s^{-1} which is located 10 cm from the photodetector. At a particular instant in time a shadow pulse of width 300 μs is registered by the photodetector. Determine the outer diameter of the optical fiber in μm at this instant in time.

Solution: The shadow velocity may be obtained from Eq. (14.31) where:

$$\frac{\mathrm{d}s}{\mathrm{d}t} = l\frac{\mathrm{d}\phi}{\mathrm{d}t} = 0.1 \times 4 = 0.4 \text{ m s}^{-1}$$

$$= 0.4 \text{ μm μs}^{-1}$$

Hence the fiber outer diameter d_o in μm is given by Eq. (5.24):

$$d_o = W_e\frac{\mathrm{d}s}{\mathrm{d}t} = 300 \text{ μs} \times 0.4 \text{ μm μs}^{-1}$$

$$= 120 \text{ μm}$$

Other on-line measurement methods, enabling faster diameter measurements, involve the analysis of forward or backward far-field patterns which are produced when a plane wave is incident transversely on the fiber. These techniques generally require measurement of the maxima in the center portion of the scattered pattern from which the diameter can be calculated after detailed mathematical analysis [Refs 66–69]. They tend to give good accuracy (e.g. ±0.25 μm [Ref. 69]) even though the theory assumes a perfectly circular fiber cross-section. Also, for step index fibers the analysis allows determination of the core diameter, and core and cladding refractive indices.

Measurements of the fiber outer diameter after manufacture (off-line) may be performed using a micrometer or dial gage. These devices can give accuracies of the order of

±0.5 µm. Alternatively, off-line diameter measurements can be made with a microscope incorporating a suitable calibrated micrometer eyepiece.

14.7.2 Core diameter

The core diameter for step index fibers is defined by the step change in the refractive index profile at the core–cladding interface. Therefore the techniques employed for determining the refractive index profile (interferometric, near-field scanning, refracted ray, etc.) may be utilized to measure the core diameter. Graded index fibers present a more difficult problem as, in general, there is a continuous transition between the core and the cladding. In this case it is necessary to define the core as an area with a refractive index above a certain predetermined value if refractive index profile measurements are used to obtain the core diameter.

Core diameter measurement is also possible from the near-field pattern of a suitably illuminated (all guided modes excited) fiber. The measurements may be taken using a microscope equipped with a micrometer eyepiece similar to that employed for off-line outer diameter measurements. However, the core–cladding interface for graded index fibers is again difficult to identify due to fading of the light distribution towards the cladding, rather than the sharp boundary which is exhibited in the step index case. Nevertheless, details of the above measurement procedures are provided in TIA/EIA-455-58 [Ref. 70].

14.8 Mode-field diameter for single-mode fiber

It was indicated in Section 2.5.2 that for single-mode fiber the geometric distribution of light in the propagating mode rather than the core diameter or numerical aperture is what is important in predicting the operational properties such as waveguide dispersion, launching and jointing losses, and microbending loss. In particular, the mode-field diameter (MFD), which is a measure of the width of the distribution of the electric field intensity, is used to predict many of these properties. Alternatively, the spot size which is simply equal to half the MFD, or the mode-field radius, is utilized.

Since the field of the fundamental mode of a circularly symmetric fiber is bell shaped and exhibits circular symmetry (see Figure 2.31), not only is its extent readily described by the MFD, but it can be expressed in terms of both the near-field (i.e. the optical field distribution on the output face of the fiber) and the far-field (i.e. the radiation pattern at larger distances, typically a few millimeters, from the fiber end face) distributions [Ref. 71]. Hence direct measurement of the MFD may be obtained using either near-field or far-field scanning techniques. These basic methods are covered for standardization purposes in TIA/EIA-455-165 [Ref. 72] and TIA/EIA-455-164 [Ref. 73] respectively.

A typical experimental arrangement for measurement of the near-field intensity distribution using a scanning fiber is illustrated in Figure 14.27 [Ref. 71]. As would be expected, it has distinct similarities to the experimental setup for the near-field scanning of the refractive index profile shown in Figure 14.17. The arrangement utilizes a relatively

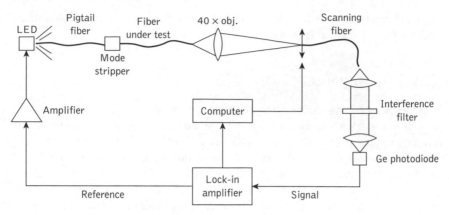

Figure 14.27 Experimental setup for near-field intensity distribution measurements (near-field scanning) to obtain mode-field diameter

intense light source (an LED or injection laser) operating at the desired wavelength to inject optical power into the fiber under test. A lens system is required to magnify the fiber output end, the image of which is scanned across a diameter using another fiber on a motor-driven translation stage pigtailed to a small-area photodiode. The near-field MFD, d_n, may be obtained using [Ref. 71]:

$$d_n = 2\sqrt{2} \left\{ \frac{\displaystyle\int_0^\infty E^2(r)r^3\,dr}{\displaystyle\int_0^\infty E^2(r)r\,dr} \right\}^{\frac{1}{2}} \tag{14.33}$$

where $|E(r)|^2$ is the local near-field intensity at radius r. Equation (14.33) assumes a non-Gaussian field distribution in which the near-field MFD is proportional to the rms width of the near-field distribution. The numerical integration of the local measured near-field intensities at intervals determined by the dynamic range of the setup thus allows d_n to be calculated. Although the near-field scanning technique provides a direct way to measure the MFD, the method suffers from inaccuracies resulting from lens distortion, difficulties in locating and stably holding the image plane at the detector, and a limited dynamic range with only a small portion of the optical power reaching the photodetector.

Another direct MFD measurement technique is obtained by scanning the far-field intensity distribution. This method is very straightforward to implement, as shown in Figure 14.28. The experimental arrangement required comprises a high-intensity light source (an injection laser is normally needed) and a photodetector mounted on a motor-driven rotational stage. It is necessary that the far-field intensity pattern be detected at a sufficiently large distance from the center of the fiber output end such that good angular resolution is achieved in detection. When using a pigtailed injection laser source, however, this distance may be as low as a few millimeters. Furthermore, the angular sector scanned in front of the fiber must be sufficiently wide (between ±20 and 25°) to completely include the main lobe of the radiation pattern. In particular, this aspect is critical when dispersion-modified fibers are scanned because they exhibit broad far-field distributions.

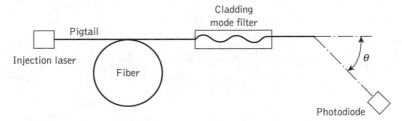

Figure 14.28 Experimental arrangement for far-field intensity distribution measurements (far-field scanning) to obtain mode-field diameter

The far-field MFD d_f can be obtained directly by inserting the measured far-field intensities into [Ref. 10]:

$$d_f = \frac{\sqrt{2}}{\pi} \left\{ \frac{F(\theta)\sin\theta\cos\theta\,d\theta}{F(\theta)\sin^3\theta\cos\theta\,d\theta} \right\}^{\frac{1}{2}} \tag{14.34}$$

where $F(\theta)$ corresponds to the measured data. Again, the integration can be performed numerically. It should be noted, however, that in this case the rms far-field, or Petermann II [Ref. 74], definition has been adopted in TIA/EIA-455-164. This definition applies to non-Gaussian measurements and is particularly appropriate for dispersion-modified fiber operating at a wavelength of 1.55 μm. Other integrative far-field methods also include various aperture techniques, two of which are reported in standards, namely the variable aperture method in TIA/EIA-455-167 [Ref. 75] and the knife-edge method in TIA/EIA-455-174 [Ref. 76].

Finally, an indirect method for the measurement of the MFD which has proved popular is the transverse offset technique [Refs 54, 71, 77, 78]. It overcomes some of the drawbacks associated with the near- and far-field methods by measurement of the power transmitted through a mechanical butt splice as one of the fibers is swept transversely through the alignment position. The experimental apparatus is shown in Figure 14.29 which employs the same single-mode fiber on either side of the joint. This technique makes use of the dependence of splice loss on spot size for Gaussian modes. Hence the variation of transmitted power with offset, $P(u)$, which is measured on a high-precision translation stage, can be fitted to the expected Gaussian dependence. For the case of identical fibers with an MFD of $2\omega_o$ this is given by [Ref. 78]:

$$P(u) = P_o \exp\left(\frac{-u^2}{2\omega_o^2}\right) \tag{14.35}$$

where u is the offset and P_o is the maximum transmitted power. The means of fit to Eq. (14.35) is very important as the pattern departs from the Gaussian distribution. Moreover, it has been found that an unweighted truncated fit with the truncation de-emphasizing the Gaussian tails gives good agreement with near-field and far-field techniques for circularly symmetric single-mode fiber [Ref. 78].

The transverse offset technique has several advantages; in particular it is efficient in its use of optical power since most of the light is intercepted and transmitted, in contrast to the near-field method. Furthermore, it is possible to use a tungsten lamp and

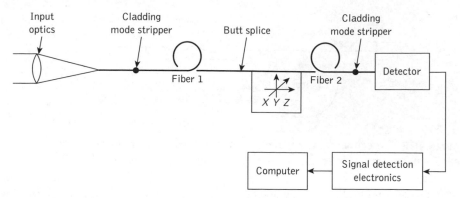

Figure 14.29 Experimental setup for the measurement of mode-field diameter by transverse offset technique

monochromatic combination to provide a tunable optical source which allows easy measurement of the MFD as a function of wavelength. The technique therefore lends itself to the determination of the cutoff wavelength as mentioned in Section 14.5. In addition, it is relatively rapid, quite accurate (with less than 2% error in spot size [Ref. 79]) and does not require complex mathematical evaluation. Finally, it is a technique which is described in ITU-T G.652 [Ref. 1].

14.9 Reflectance and optical return loss

It was indicated in Section 6.7.4 that reflections along a fiber link (i.e. optical feedback) can adversely affect injection laser stability. Furthermore, multiple reflections can contribute to the noise levels at the optical detector. Fresnel reflection r occurs at a fiber–air interface, as discussed in Section 5.2, giving a reflectance of around 4% or −14 dB. The optical return loss (ORL) is therefore defined as [Ref. 10]:

$$ORL = -10 \log_{10} r \tag{14.36}$$

It should be noted that the term reflectance is sometimes utilized when referring to single components whereas the optical return loss applies to a series of components, including the fiber, along a link.

Low values of reflectance can be obtained with fusion splicing and with carefully designed mechanical joints. For example, the use of index-matching gel can substantially reduce reflections. Nevertheless, certain mechanisms can cause larger values of reflectance. These include optical interference produced in the cavity between two fiber end faces as well as reflection from a high-index layer formed on the end face of a highly polished fiber. Ideally, the optical return loss needs to be maintained at levels above 40 dB to avoid detrimental effects on the performance of the fiber link [Ref. 10].

Optical return loss measurements can be performed using an optical continuous wave reflectometer (OCWR), as described in TIA/EIA-455-107 [Ref. 80]. In this arrangement, shown in Figure 14.30, a continuous wave LED or injection laser source is connected to

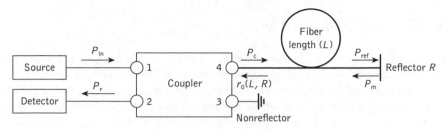

Figure 14.30 Optical return loss measurement using a four-port coupler

the input port 1 of a four-port coupler and a detector is connected to input port 2. Then a jumper cable with the reflecting components to be measured is spliced to output port 4 and output port 3 is made nonreflecting using an index-matching gel or a tight fiber loop. The optical power P_r at port 2, which results from reflections caused by the components and the coupler, is thus measured. Next the jumper cable is removed and replaced by a nonreflecting termination. This allows P_c due only to the coupler to be measured at port 2. The detector is then transferred to port 4 and the power incident upon the reflector P_{ref} is measured. Additionally, there is another method for termination in order to reduce the reflections to the optical source, which uses a mandrel wrap connecting a test jumper to the far end of the fiber or device under test (DUT) by wrapping it around a small-diameter (i.e. 5 to 15 mm) cylindrical rod [Ref. 81]. At a signal wavelength of 1.31 μm, eight or more wraps may be required whereas few turns may be sufficient for measurement at the operating wavelength of 1.55 μm.

Apart from the loss in transmission between port 4 and port 2, the fraction of the reflected power from the DUT is $(P_r - P_c)/P_{ref}$. To obtain the ports 4 to 2 loss, the source and detector are connected to ports 4 and 2, respectively, providing a measurement P_{out}. Finally, a power P_{in} is measured by connecting the source directly to the detector such that P_{out}/P_{in} is the fraction of optical power transmitted between port 4 and port 2. Hence the optical return loss is given by:

$$ORL = 10 \log_{10}\left(\frac{P_{out}P_{ref}}{P_{in}(P_r - P_c)}\right)$$ (14.37)

The OCWR is a d.c. instrument and only provides a measurement of the overall ORL for a component on a link; it does not allow information on the location of a number of reflecting components to be obtained. Accepted ORL values in the optical telecommunication industry vary according to the connector type; hence it is between 20 and 25 dB for a conventional physical contact connector whereas for an ultra-polished physical contact connector the range is from 35 to 55 dB. It can vary, however, from 55 to 70 dB for a typical physical contact connector depending on the angle of alignment [Refs 82, 83]. ORL instruments are commercially available in various sizes that are capable of measuring ORL as well as component/device reflectance for the purposes of quality control and inspection during product manufacturing, installation and system/component troubleshooting [Refs 82, 83]. Using such instruments it is possible to measure return loss at up to three wavelengths simultaneously in real time. Additionally, another device which can also be used for the measurement of ORL, albeit in a more complex manner, is the optical time domain reflectometer which is described in Section 14.10.1.

14.10 Field measurements

The measurements discussed in the preceding sections are primarily suited to the laboratory environment where quite sophisticated instrumentation may be used. However, there is a requirement for the measurement of the transmission characteristics of optical fibers when they are located in the field within an optical communication system. It is essential that optical fiber attenuation and dispersion measurements, connector and splice loss measurements and fault location be performed on optical fiber links in the field. Although information on fiber attenuation and dispersion is generally provided by the manufacturer, this is not directly applicable to cabled, installed fibers which are connected in series within an optical fiber system. Effects such as microbending (see Section 4.7.1) with the resultant mode coupling (see Section 2.4.2) affect both the fiber attenuation and dispersion. It is also found that the simple summation of the transmission parameters with regard to individual connected lengths of fiber cable does not accurately predict the overall characteristics of the link [Refs 82, 84]. Hence test equipment has been developed which allows these transmission measurements to be performed in the field.

In general, field test equipment differs from laboratory instrumentation in a number of aspects as it is required to meet the exacting demands of field measurement. Therefore the design criteria for field measurement equipment include:

1. Sturdy and compact encasement which must be portable.

2. The ready availability of electric power must be ensured by the incorporation of batteries or by connection to a generator. Hence the equipment should maintain accuracy under conditions of varying supply voltage and/or frequency.

3. In the event of battery operation, the equipment must have a low power consumption.

4. The equipment must give reliable and accurate measurements under extreme environmental conditions of temperature, humidity and mechanical load.

5. Complicated and involved fiber connection arrangements should be avoided. The equipment must be connected to the fiber in a simple manner without the need for fine or critical adjustment.

6. The equipment cannot usually make use of external triggering or regulating circuits between the transmitter and receiver due to their wide spacing on the majority of optical links.

Even if the above design criteria are met, it is likely that a certain amount of inaccuracy will have to be accepted with field test equipment. For example, it may not be possible to include adjustable launching conditions (i.e. variation in spot size and numerical aperture) in order to create the optimum. Also, because of the large dynamic range required to provide measurements over long fiber lengths, lossy devices such as mode scramblers may be omitted. Therefore measurement accuracy may be impaired through inadequate simulation of the equilibrium mode distribution.

A number of portable, battery-operated optical power meters are commercially available. Some of these instruments are of small dimension and therefore are designed to be

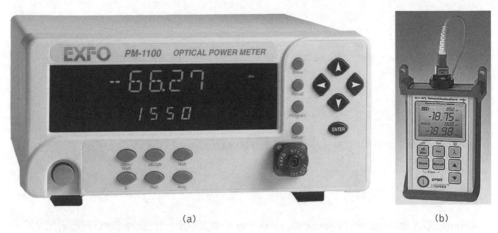

(a)   (b)

Figure 14.31 Optical power meter: (a) bench top or portable model [Ref. 83]; (b) handheld size [Ref. 85]

hand held, while others, which generally provide greater accuracy and stability, are slightly larger in size. A typical example of the latter type is shown in Figure 14.31(a) [Ref. 83]. Such devices usually measure optical power in dBm or dBμ (i.e. 0 dBm is equivalent to 1 mW and 0 dBμ is equivalent to 1 μW; see Example 14.7) over a specified range (e.g. 0.38 to 1.15 μm or 0.75 to 1.7 μm). In most cases the spectral range is altered by the incorporation of different sensors (i.e. wide-area photodiodes). For example, the optical power meter displayed in Figure 14.31(a) can be used with five different inbuilt sensors comprising either silicon, germanium or InGaAs photodiodes. The device is also specified to have a measurement range from −100 dBm (0.1 pW) to +3 dBm (2 mW) with an accuracy of ±1% when employing the latter sensor. Alternatively, handheld optical power meters are often preferred for their simple operation and compactness. Figure 14.31(b) displays a handheld optical power meter which can detect the fiber type and which switches automatically to perform optical power measurements. The device can be calibrated for various operating wavelengths from 0.85 to 1.6 μm based on the detector type (e.g. Si, Ge or InGaAs). It has a measurement accuracy of ±0.25 dB with a resolution of 0.01 dB and an operating temperature range between −10 and 50 °C [Ref. 85].

It must be noted, however, that although these instruments often take measurements over a certain spectral range, this simply implies that they may be adjusted to be compatible with the center emission frequency of particular optical sources so as to obtain the most accurate reading of optical power. Therefore, the devices do not generally give spectral attenuation measurements unless the source optical output frequency is controlled or filtered to achieve single-wavelength operation. Optical power meters may be used for measurement of the absolute optical attenuation on a fiber link by employing the cut-back technique. Other optical system parameters which may also be obtained using such instruments are the measurement of individual splice and connector losses, the determination of the absolute optical output power emitted from the source (see Sections 6.5 and 7.4.1) and the measurement of the responsivity or the absolute photocurrent of the photodetector in response to particular levels of input optical power (see Section 8.6).

Example 14.7

An optical power meter records optical signal power in either dBm or dBμ.

   (a) Convert the optical signal powers of 5 mW and 20 μW to dBm.

   (b) Convert optical signal powers of 0.3 mW and 80 nW to dBμ.

Solution: The optical signal power can be expressed in decibels using:

$$dB = 10 \log_{10}\left(\frac{P_o}{P_r}\right)$$

where P_o is the received optical signal power and P_r is a reference power level.
   (a) For a 1 mW reference power level:

$$dBm = 10 \log_{10}\left(\frac{P_o}{1 \text{ mW}}\right)$$

Hence an optical signal power of 5 mW is equivalent to:

   Optical signal power $= 10 \log_{10} 5 = 6.99$ dBm

and an optical power of 20 μW is equivalent to:

   Optical signal power $= 10 \log_{10} 0.02$
$$= -16.99 \text{ dBm}$$

   (b) For a 1 μW reference power level:

$$dB\mu = 10 \log_{10}\left(\frac{P_o}{1 \text{ μW}}\right)$$

Therefore an optical signal power of 0.3 mW is equivalent to:

   Optical signal power $= 10 \log_{10}\left(\frac{P_o}{1 \text{ μW}}\right) = 10 \log_{10} 30$
$$= 14.77 \text{ dBμ}$$

and an optical signal power of 800 nW is equivalent to:

   Optical signal power $= 10 \log_{10} 0.8$
$$= -0.97 \text{ dBμ}$$

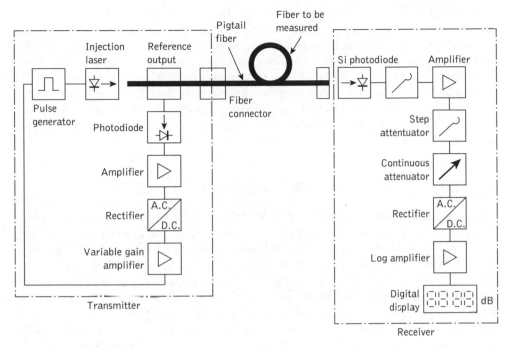

Figure 14.32 An optical attenuation meter [Refs 82, 84]

There are a number of portable measurement test sets specifically designed for fiber attenuation measurements which require access to both ends of the optical link. These devices tend to use the cut-back measurement technique unless correction is made for any difference in connector losses between the link and a short length of similar reference cable. A block schematic of an optical attenuation meter consisting of a transmitter and receiver unit is shown in Figure 14.32 [Refs 82, 84]. Reproducible readings may be obtained by keeping the launched optical power from the light source absolutely constant. A constant optical output power is achieved with the equipment illustrated in Figure 14.32 using an injection laser and a regulating circuit which is driven from a reference output of the source derived from a photodiode. Hence any variations in the laser output power are rectified by automatic adjustment of the modulating voltage, and therefore current, from the pulse generator. A large-area photodiode is utilized in the receiver to eliminate any effects from differing fiber end faces. It is generally found that when a measurement is made on multimode fiber a short cut-back reference length of a few meters is insufficient to obtain an equilibrium mode distribution. Hence unless a mode scrambling device together with a mode stripper are used, it is likely that a reference length of around 500 m or more will be required if reasonably accurate measurements are to be made. When measurements are made without a steady-state mode distribution in the reference fiber, a significantly higher loss value is obtained which may be as much as 1 dB km^{-1} above the steady-state attenuation [Refs 26, 86].

Several field test sets are available for making dispersion measurements on optical fiber links. These devices generally consist of transmitter and receiver units which take measurements in the time domain. Short light pulses ($\simeq$ 200 ns) are generated from an injection laser and are broadened by transmission down the optical link before being received by a fast response photodetector (i.e. avalanche photodiode) and displayed on a sampling oscilloscope. This is similar to the dispersion measurements in the time domain discussed in Section 14.3. If it is assumed that the pulses have a near-Gaussian shape, Eq. (14.13) may be utilized to determine the pulse broadening on the link, and hence the 3 dB optical bandwidth may be obtained.

14.10.1 Optical time domain reflectometry

A measurement technique which is far more sophisticated and which finds wide application in both the laboratory and the field is the use of optical time domain reflectometry (OTDR). This technique is often called the backscatter measurement method. It provides measurement of the attenuation on an optical link down its entire length giving information on the length dependence of the link loss. In this sense it is superior to the optical attenuation measurement methods discussed previously (Section 14.2) which only tend to provide an averaged loss over the whole length measured in dB km^{-1}. When the attenuation on the link varies with length, the averaged loss information is inadequate. OTDR also allows splice and connector losses to be evaluated as well as the rotation of any faults on the link. It relies upon the measurement and analysis of the fraction of light which is reflected back within the fiber's numerical aperture due to Rayleigh scattering (see Section 3.4.1). Hence the backscattering method, which was first described by Barnoski and Jensen [Ref. 87], has the advantages of being nondestructive (i.e. does not require the cutting back of the fiber) and of requiring access to one end of the optical link only.

The backscattered optical power as a function of time $P_{Ra}(t)$ may be obtained from the following relationship [Ref. 88]:

$$P_{Ra}(t) = \tfrac{1}{2} P_i S \gamma_R W_o v_g \exp(-\gamma v_g t) \tag{14.38}$$

where P_i is the optical power launched into the fiber, S is the fraction of captured optical power, γ_R is the Rayleigh scattering coefficient (backscatter loss per unit length), W_o is the input optical pulse width, v_g is the group velocity in the fiber and γ is the attenuation coefficient per unit length for the fiber. The fraction of captured optical power S is given by the ratio of the solid acceptance angle for the fiber to the total solid angle as:

$$S \simeq \frac{\pi (NA)^2}{4\pi n_1^2} = \frac{(NA)^2}{4 n_1^2} \tag{14.39}$$

It must be noted that the relationship given in Eq. (14.39) applies to step index fibers and the parameter S for a graded index fiber is generally a factor of 2/3 lower than for a step index fiber with the same numerical aperture [Refs 89, 90]. Hence using Eqs (14.38) and (14.39) it is possible to determine the backscattered optical power from a point along the link length in relation to the forward optical power at that point.

Example 14.8

An optical fiber link consists of multimode step index fiber which has a numerical aperture of 0.2 and a core refractive index of 1.5. The Rayleigh scattering coefficient for the fiber is 0.7 km^{-1}. When light pulses of 50 ns duration are launched into the fiber, calculate the ratio in decibels of the backscattered optical power to the forward optical power at the fiber input. The velocity of light in a vacuum is 2.998×10^8 m s^{-1}.

Solution: The backscattered optical power $P_{Ra}(t)$ is given by Eq. (14.39) where:

$$P_{Ra}(t) = \frac{1}{2} P_o S \gamma_R W_o v_g \exp(-\gamma v_g t)$$

At the fiber input $t = 0$; hence the power ratio is:

$$\frac{P_{Ra}(0)}{P_i} = \frac{1}{2} S \gamma_R W_o v_g$$

Substituting for S from Eq. (5.26) gives:

$$\frac{P_{Ra}(0)}{P_i} = \frac{1}{2} \left[\frac{(NA)^2 \gamma_R W_o v_g}{4 n_1^2} \right]$$

The group velocity in the fiber v_g is defined by Eq. (2.40) as:

$$v_g = \frac{c}{N_g} \simeq \frac{c}{n_1}$$

Therefore:

$$\frac{P_{Ra}(0)}{P_i} = \frac{1}{2} \left[\frac{NA^2 \gamma_R W_o c}{4 n_1^3} \right]$$

$$= \frac{1}{2} \left[\frac{(0.02)^2 0.7 \times 10^{-3} \times 50 \times 10^{-9} \times 2.998 \times 10^8}{4(1.5)^3} \right]$$

$$= 1.555 \times 10^{-5}$$

In decibels:

$$\frac{P_{Ra}(0)}{P_i} = 10 \log_{10} 1.555 \times 10^{-5}$$

$$= 48.1 \text{ dB}$$

Hence in Example 14.8 the backscattered optical power at the fiber input is 48.1 dB down on the forward optical power. The backscattered optical power should not be confused with any Fresnel reflection at the fiber input end face resulting from a refractive

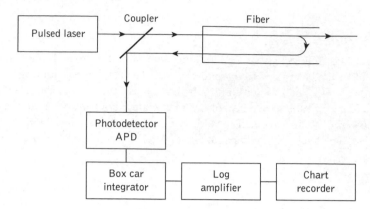

Figure 14.33 Optical time domain reflectometry or the backscatter measurement method

index mismatch. This could be considerably greater than the backscattered light from the fiber, presenting measurement problems with OTDR if it is allowed to fall onto the receiving photodetector of the equipment described below.

A block schematic of the backscatter measurement method is shown in Figure 14.33 [Ref. 91]. A light pulse is launched into the fiber in the forward direction from an injection laser using either a directional coupler or a system of external lenses with a beam splitter (usually only in the laboratory). The backscattered light is detected using an avalanche photodiode receiver which drives an integrator in order to improve the received signal-to-noise ratio by giving an arithmetic average over a number of measurements taken at one point within the fiber. This is necessary as the received optical signal power from a particular point along the fiber length is at a very low level compared with the forward power at that point by some 45 to 60 dB (see Example 14.8), and is also swamped with noise. The signal from the integrator is fed through a logarithmic amplifier and averaged measurements for successive points within the fiber are plotted on a chart recorder. This provides location-dependent attenuation values which give an overall picture of the optical loss down the link. A possible backscatter plot is illustrated in Figure 14.34 which shows the initial pulse caused by reflection and backscatter from the input coupler followed by a long tail caused by the distributed Rayleigh scattering from the input pulse as it travels down the link. Also shown in the plot is a pulse corresponding to the discrete reflection from a fiber joint, as well as a discontinuity due to excessive loss at a fiber imperfection or fault. The end of the fiber link is indicated by a pulse corresponding to the Fresnel reflection incurred at the output end face of the fiber. Such a plot yields the attenuation per unit length for the fiber by simply computing the slope of the curve over the length required. Also the location and insertion losses of joints and/or faults can be obtained from the power drop at their respective positions on the link. Finally the overall link length can be determined from the time difference between reflections from the fiber input and output end faces. Standard methods for these measurements are covered in TIA/EIA-455-59 to 61 [Refs 92–94] and they provide very powerful techniques for field measurements on optical fiber links. In addition, the measurement of splice or connector loss and the

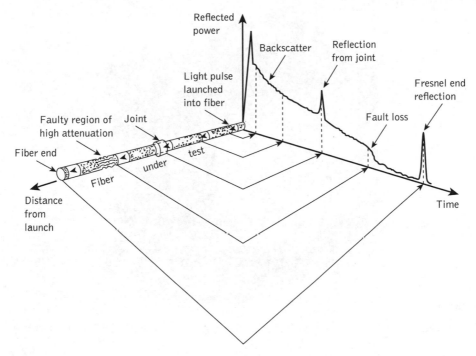

Figure 14.34 An illustration of a possible backscatter plot from a fiber under test [Ref. 92]

measurement of splice or connector return loss utilizing OTDR is contained in TIA/EIA-455-8 [Ref. 95].

A number of optical time domain reflectometers are commercially available for operation over the entire wavelength range. These instruments are capable of carrying out tests over single or dual wavelengths for multimode (i.e. 0.85/1.3 μm) and for single-mode optical fiber (i.e. 1.31/1.55 μm or 1.55/1.625 μm) links. Although the OTDR functionality is provided, these instruments are also often capable of performing a number of other optical system and network tests (e.g. optical loss, dispersion measurement etc.). Such instruments are usually referred as universal or optical network test systems rather than simply optical time domain reflectometers [Ref. 96].

An example of a portable optical test system demonstrating OTDR operation is shown in Figure 14.35. This flexible system can accommodate several different modules plugged into its mainframe. These modules can be configured in five arrangements each enabling a distinctive feature of OTDR measurement based on operating parameters (i.e. wavelength or band, data transmission rate etc.). The OTDR instrument has a dynamic range of 30–45 dB and the output optical signal power from the optical sources remain stable within a range from −5 to −10 dBm. Furthermore, it is capable of providing OTDR optical tests using short pulses from 5 to 260 ns over distances of 1.25–260 km with sampling resolution of 0.04–5.0 m from up to 52000 sampling points. It must be noted, however, that the accuracy of the distance measurement is around ±1 m. Dynamic range determines the total optical loss that the optical time domain reflectometer can analyze and hence the total

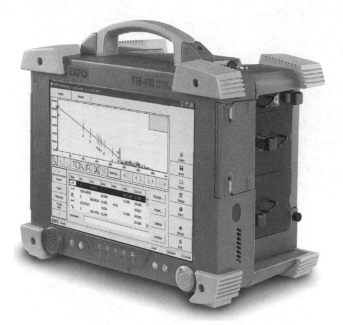

Figure 14.35 Optical time domain reflectometer, Model No. EXFO FTB-400 [Ref. 96]

length of the fiber link over which it will operate. OTDR instrument manufacturers tend to specify dynamic range differently, describing it in relation to pulse width, signal-to-noise ratio or averaging time. As a rule of thumb an optical time domain reflectometer is required to exhibit a dynamic range of 5 to 8 dB higher than the maximum loss incurred on the optical link. For example, a single-mode optical time domain reflectometer with a dynamic range of 35 dB has a usable dynamic range of around 30 dB.

Dead zones which are originated from reflective events within a specific fiber (i.e. connectors, mechanical splices, etc.) also affect the accuracy of the instrument. When a strong optical reflection from a reflective event reaches the optical time domain reflectometer, its detection circuit can become saturated for a specific time period (converted to distance in the instrument) until it recovers and continues to measure the backscattering accurately again. As a result of this saturation there is a certain portion of the fiber link following the reflective event that cannot be displayed by the instrument; hence the term dead zone. Analyzing the dead zone is important when specifying OTDR performance to ensure the entire link can be measured. In addition, current optical time domain reflectometers use appropriate software to enable fast manipulation of the measured data. This feature allows instant calculation of the optical power link budget and the generation of comprehensive reports. Furthermore, the information extracted from OTDR traces can also be used to determine ORL in an optical fiber network [Ref. 97].

A number of themes have been pursued in relation to OTDR performance. These include the enlargement of the instrument dynamic range, the enhancement of the instrument resolution, the reduction of noise levels intrinsic to single-mode fibers and the increase in the user friendliness of the equipment [Ref. 98]. Significant improvements have been obtained in the former two instrument performance characteristics with a range of strategies including the use of higher input optical power levels, decreasing the minimum detectable optical power and employing narrower pulse widths.

For example, one strategy which has proved successful is the use of a photon counting technique [Ref. 99] in which the backscattered photons are detected digitally. In this method the avalanche photodiode is operated in a Geiger tube breakdown mode [Ref. 100] by biasing the device above its normal operating voltage where it can detect a single photon. The photon counting technique has demonstrated significantly improved receiver sensitivity (i.e. −7 dB) than the best analog system at a wavelength of 1.3 μm. Moreover, a resolution of up to 1.5 cm with high sensitivity (3×10^{-10} W) has been reported when operating at a wavelength of 0.85 μm using a single-photon-detecting APD at room temperature [Ref. 101]. More recently, there have been several demonstrations of photon counting OTDR instruments operating at the wavelengths of 1.31 and 1.55 μm [Refs 102–104] where an optical spatial resolution of 5 cm has been achieved [Ref. 102].

Single-mode fiber optical time domain reflectometers exhibit an additional problem over multimode devices, namely polarization noise. In general, the state of polarization of the backscattered light differs from that of the laser pulse coupled into the fiber at the input end and is dependent on the distance of the backscattering fiber element from the input fiber end. This results in an amplitude fluctuation in the backscattered light known as polarization noise. Interestingly, this same phenomenon can be employed to measure the evolution of the polarization in the fiber with the so-called polarization optical time domain reflectometer (POTDR) [Ref. 54]. However, in a conventional single-mode OTDR instrument reduction of the polarization noise is necessary using a polarization-independent acousto-optic deflector (see Section 11.4.2) or, more usually, a polarization scrambler [Ref. 98].

In addition, a major limitation which is encountered with conventional optical time domain reflectometers during the test and measurement process is the presence of polarization mode dispersion (PMD) particularly in older fiber types [Refs 105, 106]. The POTDR, however, provides for the measurement of PMD on a fiber link. Figure 14.36(a) depicts a block schematic of a POTDR which comprises a conventional OTDR instrument, polarization controller, polarization analyzer and pulsed laser. The POTDR employs a narrow band external cavity laser or DFB laser (i.e. with pulse width of 10 ns or less) to give high spatial resolution with respect to the frequency-dependent distortions due to PMD. The fiber under test is connected to the POTDR using an optical circulator (see Section 5.7). Alternatively, an optical fiber coupler (see Section 5.6) can also be used to interface with the POTDR. The pulses are then sent through the optical circulator towards the fiber under test following which the instrument performs the measurements from the state of polarization (SOP) of the backscattered field while the polarization controller and polarization analyzer are used to determine both the SOP and degree of polarization (DOP) for the optical signals. The polarization analyzer then provides the data necessary to construct the OTDR traces.

Figure 14.36(b) displays an example of an output obtained from a commercial POTDR of the measurement on nonzero-dispersion-shifted fiber (NZ-DSF) [Ref. 106]. Several different segments of NZ-DSF fiber (provided by different suppliers) were spliced together to create an overall length of 23.5 km. The total link PMD was measured using a standard interferometric test and was recorded to be 8.82 ps. The PMD of the individual fiber segments, however, was initially assumed to be unknown. The trace in Figure 14.36(b) shows a POTDR output signal power characteristic also incorporating the

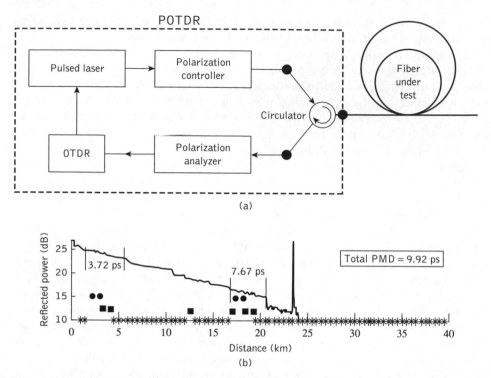

Figure 14.36 Polarization OTDR: (a) block schematic; (b) an output signal power characteristic showing a trace exhibiting different levels of PMD [Ref. 106]

values for high PMD (solid circles), medium PMD (solid boxes) and normal PMD (asterisks) that appeared at different positions on the fiber link. It should be noted that these PMD points were obtained from the analytical solution of the SOP and the DOP obtained from the POTDR trace [Refs 107–110]. The POTDR trace correctly identified the two fiber sections with high PMD (i.e. 3.72 ps and 7.67 ps) with each section being positioned at 2 to 4 km and 17 to 19 km, respectively.

Since the POTDR can identify the levels of PMD on the fiber link, these instruments are also used during the fiber manufacturing process before cabling in order to ensure uniform distribution of PMD along the fiber. Furthermore, they also find application in monitoring the levels of PMD when dispersion-compensating fibers are fabricated.

Problems

14.1   Describe what is meant by 'equilibrium mode distribution' and 'cladding mode stripping' with regard to transmission measurements in optical fibers. Briefly outline methods by which these conditions may be achieved when optical fiber measurements are performed.

14.2 Discuss with the aid of a suitable diagram the cut-back technique used for the measurement of the total attenuation in an optical fiber. Indicate the differences in the apparatus utilized for spectral loss and spot attenuation measurement.

A spot measurement of fiber attenuation is performed on a 1.5 km length of optical fiber at a wavelength of 1.1 μm. The measured optical output power from the 1.5 km length of fiber is 50.1 μW. When the fiber is cut back to a 2 m length, the measured optical output power is 385.4 μW. Determine the attenuation per kilometer for the fiber at a wavelength of 1.1 μm.

14.3 Briefly outline the principle behind the calorimetric methods used for the measurement of absorption loss in optical fibers.

A high-absorption optical fiber was used to obtain the plot of $(T_\infty - T_t)$ (on a logarithmic scale) against time shown in Figure 14.37(a). The measurements were achieved using a calorimeter and thermocouple experimental arrangement. Subsequently, a different test fiber was passed three times through the same calorimeter before further measurements were taken. Measurements on the test fiber produced the heating and cooling curve shown in Figure 14.37(b) when a constant 76 mW of optical power, at a wavelength of 1.06 μm, was passed through it. The constant C for the experimental arrangement was calculated to be 2.32×10^4 J °C^{-1}. Calculate the absorption loss in decibels per kilometer, at a wavelength of 1.06 μm, for the fiber under test.

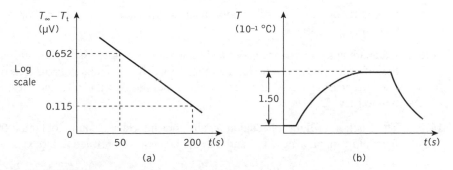

Figure 14.37 Fiber absorption for measurements for the Problem 14.3 plot of $(T_\infty - T_t)$ against time for a high-absorption fiber; (b) the heating and cooling curve for the test fiber

14.4 Discuss the measurement of fiber scattering loss by describing the use of two common scattering cells.

An Nd : YAG laser operating at a wavelength of 1.064 μm is used with an integrating sphere to measure the scattering loss in an optical fiber sample. The optical power propagating within the fiber at the sphere is 98.45 μW and 5.31 nW of optical power is scattered within the sphere. The length of fiber in the sphere is 5.99 cm. Determine the optical loss due to scattering for the fiber at a wavelength of 1.064 μm in decibels per kilometer.

14.5 Fiber scattering loss measurements are taken at a wavelength of 0.75 μm using a solar cell cube. The reading of the input optical power to the cube is 7.78 V with a gain setting of 10^5. The corresponding reading from the scattering cell which incorporates a 4.12 cm length of fiber is 1.56 V with a gain setting of 10^9. Previous measurements of the total fiber attenuation at a wavelength of 0.75 μm gave a value of 3.21 dB km^{-1}. Calculate the absorption loss for the fiber at a wavelength of 0.75 μm in decibels per kilometer.

14.6   Discuss with the aid of suitable diagrams the measurement of dispersion in optical fibers. Consider both time and frequency domain measurement techniques.

   Pulse dispersion measurements are taken on a multimode graded index fiber in the time domain. The 3 dB width of the optical output pulses from a 950 m fiber length is 827 ps. When the fiber is cut back to a 2 m length the 3 dB width of the optical output pulses becomes 234 ps. Determine the optical bandwidth for a kilometer length of the fiber assuming Gaussian pulse shapes.

14.7   Pulse dispersion measurements in the time domain are taken on a multimode and a single-mode step index fiber. The results recorded are:

	Input pulse width (3 dB)	Output pulse width (3 dB)	Fiber length (km)
(a) Multimode fiber	400 ps	31.20 ns	1.13
(b) Single-mode fiber	200 ps	425 ps	2.35

Calculate the optical bandwidth over 1 kilometer for each fiber assuming Gaussian pulse shapes.

14.8   Compare and contrast the major techniques employed to obtain a measurement of the refractive index profile for an optical fiber. In particular, suggest reasons why the refracted near-field method has been adopted as the reference test method by the EIA.

14.9   The fraction of light reflected at an air–fiber interface r can be obtained from the Fresnel formula of Eq. (5.1) and for small changes in refractive index:

$$\frac{\delta r}{r} = \left(\frac{4}{n_1^2 - 1}\right)\delta n_1$$

where n_1 is the fiber core refractive index at the point of reflection. Show that the fractional change in the core refractive index $\delta n_1/n_1$ may be expressed in terms of the fractional change in the reflection coefficient $\delta r/r$ following:

$$\frac{\delta n_1}{n_1} = \left(\frac{r^{\frac{1}{2}}}{1 - r}\right)\frac{\delta r}{r}$$

Hence, show that for a step index fiber with n_1 of 1.5, a 5% change in r corresponds to only a 1% change in n_1.

14.10   Describe, with the aid of suitable diagrams, the reference test methods which are utilized to determine the effective cutoff wavelength in single-mode fiber.

14.11   Compare and contrast two simple techniques used for the measurement of the numerical aperture of optical fibers.

   Numerical aperture measurements are performed on an optical fiber. The angular limit of the far-field pattern is found to be 26.1° when the fiber is rotated from a center zero point. The far-field pattern is then displayed on a screen where its size is measured as 16.7 cm. Determine the numerical aperture for the fiber and the distance of the fiber output end face from the screen.

14.12   Describe, with the aid of a suitable diagram, the shadow method used for the on-line measurement of the outer diameter of an optical fiber.

   The shadow method is used for the measurement of the outer diameter of an optical fiber. A fiber outer diameter of 347 μm generates a shadow pulse of 550 μs when the rotating mirror has an angular velocity of 3 rad s^{-1}. Calculate the distance between the rotating mirror and the optical fiber.

14.13   Define the mode-field diameter (MFD) in a single-mode fiber and indicate how this parameter relates to the spot size.

   Discuss the techniques which are commonly employed to measure the MFD by either direct or indirect methods. Comment on their relative attributes and drawbacks.

14.14   Outline the major design criteria of an optical fiber power meter for use in the field. Suggest any problems associated with field measurements using such a device.

   Convert the following optical power meter readings to numerical values of power: 25 dBm, −5.2 dBm, 3.8 dBμ.

14.15   Explain the optical return loss (ORL) measurement procedure based on the optical continuous wave reflectometer. Discuss the need for the use of an ORL test meter instead of using an OTDR.

14.16   Describe what is meant by optical time domain reflectometry. Discuss how the technique may be used to take field measurements on optical fibers. Indicate the advantages of this technique over other measurement methods to determine attenuation in optical fibers.

   A backscatter plot for an optical fiber link provided by OTDR is shown in Figure 14.38. Determine:
   (a) the attenuation of the optical link for the regions indicated A, B and C in decibels per kilometer.
   (b) the insertion loss of the joint at the point X.

14.17   Discuss the sensitivity of OTDR in relation to commercial reflectometers. Comment on an approach which may lead to an improvement in the sensitivity of this measurement technique.

   The Rayleigh scattering coefficient for a silica single-mode step index fiber at a wavelength of 0.80 μm is 0.46 km^{-1}. The fiber has a refractive index of 1.6 and a numerical aperture of 0.14. When a light pulse of 60 ns duration at a wavelength of 0.80 μm is launched into the fiber, calculate the level in decibels of the

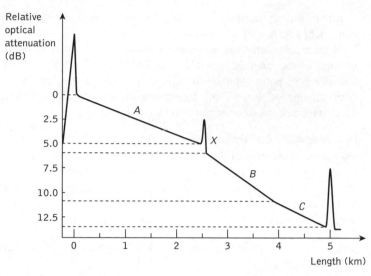

Figure 14.38 The backscatter plot for the optical link of Problem 14.16

backscattered light compared with the Fresnel reflection from a clean break in the fiber. It may be assumed that the fiber is surrounded by air.

14.18 Describe the structure and operation of a POTDR and compare its features with a conventional OTDR instrument.

Answers to numerical problems

14.2   5.92 dB km^{-1}
14.3   1.77 dB km^{-1}
14.4   3.91 dB km^{-1}
14.5   1.10 dB km^{-1}
14.6   525.9 MHz km
14.7   (a) 15.9 MHz km;
          (b) 7.3 GHz km

14.9   1.0%
14.11 0.44, 17.0 cm
14.12 21.0 cm
14.14 316.2 mW, 302 µW, 2.40 µW
14.16 (a) 2.0 dB km^{-1}, 3.0 dB km^{-1}, 2.5 dB km^{-1};
          (b) 1.0 dB
14.17 −37.3 dB

References

[1] ITU-T Recommendations, http://www.itu.int/itu-t/publications/index.html, 21 September 2007.
[2] The Fiber Optic Association Tech Topics, 'EIA-TIA fiber optic standards', http://thefoa.org/tech/standards.htm, 21 September 2007.
[3] Electronic Industries Alliance, EIA, http://www.eia.org/, 21 September 2007.
[4] Telecommunications Industry Association, TIA, http://www.tiaonline.org, 21 September 2007.
[5] IHS Inc, TIA – Telecommunications Industry Association collection, http://electronics.ihs.com/collections/tia/index.htm, 21 September 2007.

[6] M. Tateda, T. Horiguchi, M. Tokuda and N. Uchida, 'Optical loss measurement in graded index fiber using a dummy fiber', *Appl. Opt.*, **18**(19), pp. 3272–3275, 1979.

[7] M. Eve, A. M. Hill, D. J. Malyon, J. E. Midwinter, B. P. Nelson, J. R. Stern and J. V. Wright, 'Launching independent measurements of multimode fibres', *2nd Eur. Conf. on Optical Fiber Communications*, Paris, pp. 143–146, 1976.

[8] M. Ikeda, Y. Murakami and C. Kitayama. 'Mode scrambler for optical fibres', *Appl. Opt.*, **16**(4), pp. 1045–1049, 1977.

[9] S. Seikai, M. Tokuda, K. Yoshida and N. Uchida, 'Measurement of baseband frequency response of multimode fibre by using a new type of mode scrambler', *Electron. Lett.*, **13**(5), pp. 146–147, 1977.

[10] F. P. Kapron, 'Fiber-optic test methods', in F. C. Allard (Ed.), *Fiber Optics Handbook for Engineers and Scientists*, pp. 4.1–4.54, McGraw-Hill, 1990.

[11] TIA/EIA-455-50. Light launch conditions for long length graded-index optical fiber spectral attenuation measurements.

[12] J. P. Dakin, W. A. Gambling and D. N. Payne, 'Launching into glass–fibre waveguide', *Opt. Commun.*, **4**(5), pp. 354–357, 1972.

[13] TIA/EIA-455-46. Spectral attenuation measurement for long-length graded-index optical fibers.

[14] TIA/EIA-455-78. Spectral attenuation cutback measurement for single-mode fibers.

[15] TIA/EIA-455-53. Attenuation by substitution measurement – for multimode graded-index optical fibers or fiber assemblies used in long length communication systems.

[16] K. I. White, 'A calorimetric method for the measurement of low optical absorption losses in optical communication fibres', *Opt. Quantum Electron.*, **8**, pp. 73–75, 1976.

[17] K. I. White and J. E. Midwinter, 'An improved technique for the measurement of low optical absorption losses in bulk glass', *Opto-electronics*, **5**, pp. 323–334, 1973.

[18] A. R. Tynes, 'Integrating cube scattering detector', *Appl. Opt.*, **9**(12), pp. 2706–2710, 1970.

[19] F. W. Ostermayer and W. A. Benson, 'Integrating sphere for measuring scattering loss in optical fiber waveguides', *Appl. Opt.*, **13**(8), pp. 1900–1905, 1974.

[20] S. de Vito and B. Sordo, 'Misure do attenuazione e diffusione in fibre ottiche multimodo', *LXXV Riuniuone AEI*, Rome, 15–21 Sept. 1974.

[21] J. P. Dakin, 'A simplified photometer for rapid measurement of total scattering attenuation of fibre optical waveguides', *Opt. Commun.*, **12**(1), pp. 83–88, 1974.

[22] L. G. Cohen, P. Kaiser and C. Lin, 'Experimental techniques for evaluation of fiber transmission loss and dispersion', *Proc. IEEE*, **68**(10), pp. 1203–1208, 1980.

[23] S. D. Personick, 'Baseband linearity and equalization in fiber optic digital communication systems', *Bell Syst. Tech. J.*, **52**(7), pp. 1175–1194, 1973.

[24] D. Gloge, E. L. Chinnock and T. P. Lee, 'Self pulsing GaAs laser for fiber dispersion measurement', *IEEE J. Quantum Electron.*, **QE-8**, pp. 844–846, 1972.

[25] TIA/EIA 455-51. Pulse distortion measurement of multimode glass fiber information transmission capacity.

[26] F. Krahn, W. Meininghaus and D. Rittich, 'Measuring and test equipment for optical cable', *Philips Telecommun. Rev.*, **37**(4), pp. 241–249, 1979.

[27] TIA/EIA-455-168. Chromatic dispersion measurement of multimode graded-index and single-mode optical fibers by spectral group delay measurement in the time domain.

[28] I. Kokayashi, M. Koyama and K. Aoyama, 'Measurement of optical fibre transfer functions by swept frequency technique and discussion of fibre characteristics', *Electron. Commun. Jpn*, **60-C**(4), pp. 126–133, 1977.

[29] TIA/EIA-455-30. Frequency domain measurement of multimode optical fiber information transmission capacity.

[30] TIA/EIA-455-169. Chromatic dispersion measurement of single-mode fibers by phase-shift method.

[31]　K. Kuroda, *Progress in photorefractive nonlinear optics*, Taylor & Francis, 2002.

[32]　M. C. Michel, 'Application note 140: ORL measurements in field applications', EXFO, http://documents.exfo.com/appnotes/anote140-ang.pdf, 21 September 2007.

[33]　Agilent Inc., Optical fiber dispersion analyzer, http://cp.literature.agilent.com/litweb/pdf/5988-7200EN.pdf, 21 September 2007.

[34]　K. S. Abedin, M. Hyodo and N. Onodera, 'Measurement of the chromatic dispersion of an optical fiber by use of a Sagnac interferometer employing asymmetric modulation', *Opt. Lett.*, **25**(5), pp. 299–301, 2000.

[35]　W. E. Martin, 'Refractive index profile measurements of diffused optical waveguides', *Appl. Opt.*, **13**(9), pp. 2112–2116, 1974.

[36]　H. M. Presby, W. Mammel and R. M. Derosier, 'Refractive index profiling of graded index optical fibers', *Rev. Sci. Instrum.*, **47**(3), pp. 348–352, 1976.

[37]　B. Costa and G. De Marchis, 'Test methods (optical fibres)', *Telecommun. J. (Engl. Ed.) Switzerland*, **48**(11), pp. 666–673, 1981.

[38]　L. C. Cohen, P. Kaiser, J. B. MacChesney, P. B. O'Conner and H. M. Presby, 'Transmission properties of a low-loss near-parabolic-index fiber', *Appl. Phys. Lett.*, **26**(8), pp. 472–474, 1975.

[39]　L. C. Cohen, P. Kaiser, P. D. Lazay and H. M. Presby, 'Fiber characterization', in S. E. 'Miller and A. G. Chynoweth (Eds), *Optical Fiber Telecommunications*, pp. 343–400, Academic Press, 1979.

[40]　M. E. Marhic, P. S. Ho and M. Epstein, 'Nondestructive refractive index profile measurement of clad optical fibers', *Appl. Phys. Lett.*, **26**(10), pp. 574–575, 1975.

[41]　H. Garcia, A. M. Johnson, F. A. Oguama and T. Sudhir, 'New approach to the measurement of the nonlinear refractive index of short (>25 m) lengths of silica and erbium-doped fibers', *Opt. Lett.*, **28**(19), pp. 1796–1798, 2003.

[42]　F. A. Oguama, H. Garcia and A. M. Johnson, 'Simultaneous measurement of the Raman gain coefficient and the nonlinear refractive index of optical fibers: theory and experiment', *J. Opt. Soc. Am. B.*, **22**(2), pp. 426–436, 2005.

[43]　M. de Angelis, S. De Nicola, A. Finizio, G. Pierattini, P. Ferraro, S. Pelli, G. Righini and S. Sebastiani, 'Digital-holography refractive-index-profile measurement of phase gratings', *Appl. Phys. Lett.*, **88**(3), pp. 111–114, 2006.

[44]　P. L. Chu, 'Measurements in optical fibres', *Proc. IEEE Australia*, **40**(4), pp. 102–114, 1979.

[45]　M. J. Adams, D. N. Payne and F. M. E. Sladen, 'Correction factors for determination of optical fibre refractive-index profiles by near-field scanning techniques', *Electron. Lett.*, **12**(11), pp. 281–283, 1976.

[46]　TIA/EIA-455-43. Output near-field radiation pattern measurement of optical waveguide fibers.

[47]　F. E. M. Sladen, D. N. Payne and M. J. Adams, 'Determination of optical fibre refractive index profile by near field scanning technique', *Appl. Phys. Lett.*, **28**(5), pp. 255–258, 1976.

[48]　A. J. Ritger, 'Bandwidth improvement in MCVD multimode fibers by fluorine etching to reduce centre dip', *Eleventh Eur. Conf. on Optical Communications*, pp. 913–916, 1985.

[49]　K. W. Raine, J. G. N. Baines and D. E. Putland, 'Refractive index profiling – state of the art', *J. Lightwave Technol.*, **7**(8), pp. 1162–1169, 1989.

[50]　TIA/EIA-455-44. Refractive index profile, refracted ray method.

[51]　A. H. Cherin, *An Introduction to Optical Fibers*, McGraw-Hill, 1987.

[52]　K. I. White 'Practical application of the refracted near-field technique for the measurement of optical fiber refractive index profiles', *Opt. Quantum Electron.*, **11**(2), pp. 185–196, 1979.

[53]　W. J. Stewart, 'Optical fiber and preform profiling technology', *J. Quantum Electron.*, **QE-18**(10), pp. 1451–1466, 1982.

[54]　E.-G. Neuman, *Single-Mode Fibers: Fundamentals*, Springer-Verlag, 1988.

[55]　D. B. Payne, M. H. Reeve, C. A. Millar and C. J. Todd, 'Single-mode fiber specification and system performance', *Symp. on Optical Fiber Measurement, NBS spec. publ.*, **683**, pp. 1–5, 1984.

[56]　TIA/EIA-455-80. Cutoff wavelength of uncabled single-mode fiber by transmitted power.

[57] R. Srivastava and D. L. Franzen, 'Single-mode optical fiber characterization', *NBS Rep.*, pp. 1–101, July 1985.

[58] D. L. Franzen, 'Determining the effective cutoff wavelength of single-mode fibers: an interlaboratory comparison', *J. Lightwave Technol.*, **LT-3**, pp. 128–134, 1985.

[59] C. A. Millar, 'Comment: fundamental mode spot-size measurement in single-mode optical fibers', *Electron. Lett.*, **18**, pp. 395–396, 1982.

[60] TIA/EIA-455-170. Cable cutoff wavelength of single-mode fiber by transmitted power.

[61] TIA/EIA-455-177. Numerical aperture of graded-index optical fiber.

[62] D. L. Franzen, M. Young, A. H. Cherin, E. D. Head, M. J. Hackert, K. W. Raine and J. G. N. Baines, 'Numerical aperture of multimode fibers by several methods: resolving differences', *J. Lightwave Technol.*, **7**(6), pp. 896–901, 1989.

[63] W. A. Gambling, D. N. Payne and H. Matsumura, 'Propagation studies on single-mode phosphosilicate fibres', *2nd Eur. Conf. on Optical Fiber Communications*, Paris, pp. 95–100, 1976.

[64] F. T. Stone, 'Rapid optical fibre delta measurement by refractive index tuning', *Appl. Opt.*, **16**(10), pp. 2738–2742, 1977.

[65] L. G. Cohen and P. Glynn, 'Dynamic measurement of optical fibre diameter', *Rev. Sci. Instrum.*, **44**(12), pp. 1745–1752, 1973.

[66] H. M. Presby, 'Refractive index and diameter measurements of unclad optical fibres', *J. Opt. Soc. Am.*, **64**(3), pp. 280–284, 1974.

[67] P. L. Chu, 'Determination of diameters and refractive indices of step-index optical fibres', *Electron. Lett.*, **12**(7), pp. 150–157, 1976.

[68] H. M. Presby and D. Marcuse, 'Refractive index and diameter determinations of step index optical fibers and preforms', *Appl. Opt.*, **13**(12), pp. 2882–2885, 1974.

[69] D. Smithgall, L. S. Wakins and R. E. Frazee, 'High-speed noncontact fibre-diameter measurement using forward light scattering', *Appl. Opt.*, **16**(9), pp. 2395–2402, 1977.

[70] TIA/EIA-455-58. Core diameter measurement of graded-index optical fibers.

[71] M. Artiglia, G. Coppa, P. DiVita, M. Potenza and A. Sharma, 'Mode field diameter measurements in single-mode optical fibers', *J. Lightwave Technol.*, **7**(8), pp. 1139–1152, 1989.

[72] TIA/EIA-455-165. Single-mode fiber, measurement of mode field diameter by near field scanning.

[73] TIA/EIA-455-164. Single-mode fiber, measurement of mode field diameter by far-field scanning.

[74] K. Petermann, 'Constraints for the fundamental mode spot size for broadband dispersion-compensated single-mode fibres', *Electron. Lett.*, **19**, pp. 712–714, 1983.

[75] TIA/EIA-455-167. Mode field diameter measurement-variable aperture in the far field.

[76] TIA/EIA-455-174. Mode field diameter of single-mode optical fiber by knife-edge scanning in the far field.

[77] J. Streckert, 'New method for measuring the spot size of single-mode fibers', *Opt. Lett.*, **5**, pp. 505–506, 1980.

[78] M. L. Dakss, 'Optical-fiber measurements', in E. E. Basch (Ed.), *Optical-Fiber Transmission*, H. W. Sams, pp. 133–178, 1987.

[79] J. Streckert, 'A new fundamental mode spot size definition usable for non-Gaussian and non-circular field distributions', *J. Lightwave Technol.*, **LT-3**, pp. 328–331, 1985.

[80] TIA/EIA-455-107. Return loss.

[81] Z. F. Wang, W. Cao and Z. Lu, 'MOEMS: packaging and testing', *Microsyst. Technol.*, **12**(1), pp. 52–58, 2005.

[82] Acterna Inc., http://www.acterna.com, 21 September 2007.

[83] EXFO, http://documents.exfo.com/appnotes/anote044-ang.pdf, 21 September 2007; http://documents.exfo.com/specsheets/PM-1100-angHR.pdf, 21 September 2007.

[84] F. Krahn, W. Meininghaus and D. Rittich, 'Field and test measurement equipment for optical cables', *Acta Electron.*, **23**(3), pp. 269–275, 1979.

[85] Afltelecommunications Inc., Optical power meter and test equipments, http://www.afltelecommunications.com, 21 September 2007.

[86] R. Olshansky, M. G. Blankenship and D. B. Keck, 'Length-dependent attenuation measurements in graded-index fibres', *Proc. 2nd Eur. Conf. on Optical Communications*, Paris, pp. 111–113, 1976.

[87] M. K. Barnoski and S. M. Jensen, 'Fiber waveguides: a novel technique for investigating attenuation characteristics', *Appl. Opt.*, **15**(9), pp. 2112–2115, 1976.

[88] S. D. Personick, 'Photon probe, an optical fibre time-domain reflectometer', *Bell Syst. Tech. J.*, **56**(3), pp. 355–366, 1977.

[89] E. G. Newman, 'Optical time domain reflectometer: comment', *Appl. Opt.*, **17**(11), p. 1675, 1978.

[90] M. Ohashi, 'Novel OTDR technique for measuring relative-index differences of fiber links', *IEEE Photonics Technol. Lett.*, **18**(24), pp. 2584–2586, 2006.

[91] M. K. Barnoski and S. D. Personick, 'Measurements in fiber optics', *Proc. IEEE*, **66**(4), pp. 429–440, 1978.

[92] TIA/EIA-455-59. Measurement of fiber point defects using an OTDR.

[93] TIA/EIA-455-60. Measurement of fiber or cable length using an OTDR.

[94] TIA/EIA-455-61. Measurement of fiber or cable attenuation using an OTDR.

[95] TIA/EIA-455-8. Measurement of splice or connector loss and reflectance using an OTDR.

[96] EXFO OTDR FTB-7000B; http://otdrstore.com/, 21 September 2007.

[97] P. C. Noutsios, 'Optical return loss measurements and simulation of an arbitrary array of concatenated reflective elements on field-installed optical links', *J. Lightwave Technol.*, **24**(4), pp. 1697–1702, 2006.

[98] M. Tateda and T. Huriguchi, 'Advances in optical time-domain reflectometry', *J. Lightwave Technol.*, **7**(8), pp. 1217–1224, 1989.

[99] P. Healey, 'Optical time domain reflectometry by photon counting', *6th Eur. Conf. on Optical Communications*, UK, pp. 156–159, 1980.

[100] P. P. Webb *et al.*, 'Single photon detection with avalanche photodiodes', *Bull. Am. Phys. Soc. II*, **15**, p. 813, 1970.

[101] C. G. Bethea, B. F. Levine, S. Cova and G. Ripamonti, 'High-resolution, high sensitivity optical time-domain reflectometer', *Opt. Lett.*, **13**(3), pp. 233–235, 1988.

[102] M. Wegmuller, F. Scholder and N. Gisin, 'Photon-counting OTDR for local birefringence and fault analysis in the metro environment', *J. Lightwave Technol.*, **22**(2), pp. 390–400, 2004.

[103] E. Diamanti, C. Langrock, M. M. Fejer and Y. Yamamoto, '1.5 μm photon-counting optical time domain reflectometry with a single-photon detector using up-conversion in a PPLN waveguide', *Conf. on Lasers and Electronic Optics, CLEO'05*, Baltimore, MD, USA, pp. 1079–1081, 22–27 May 2005.

[104] Y. Inoue, T. Aiba and N. Shibata, 'Measuring modal birefringence along a holey fiber by photon-counting OTDR', *IEEE Photonics Technol. Lett.*, **17**(6), pp. 1238–1240, 2005.

[105] T. Ozeki, K. Shinotsuka and K. Nakayama, 'PMD distribution measurement by an OTDR with polarimetry assuming Rayleigh backscattering by randomly oriented non-spherical particles', *Conf. Proc. IEEE/LEOS on Fiber Optic Passive Components*, Mondello, Italy, pp. 181–186, 22–24 June 2005.

[106] EXFO P-OTDR, http://documents.exfo.com/appnotes/anote087-ang.pdf, 21 September 2007.

[107] A. Galtarossa and L. Palmieri, 'Reflectometric measurements of polarization properties in optical-fiber links', *J. Opt. Fiber Commun. Rep.*, **1**(2), pp. 150–179, 2004.

[108] T. Ozeki, S. Seki and K. Iwasaki, 'PMD distribution measurement by an OTDR with polarimetry considering depolarization of backscattered waves', *J. Lightwave Technol.*, **24**(11), pp. 3882–3888, 2006.

[109] R. Goto, S. Tanigawa, S. Matsuo and K. Himeno, 'On-spool PMD estimation method for low-PMD fibers with high repeatability by local-DGD measurement using POTDR', *J. Lightwave Technol.*, **24**(11), pp. 3914–3919, 2006.

[110] M. Wuilpart, C. Crunelle and P. Mégret, 'High dynamic polarization-OTDR for the PMD mapping in optical fiber links', *Opt. Commun.*, **269**(2), pp. 315–321, 2007.

CHAPTER 15

Optical networks

15.1 Introduction

An optical fiber network commonly referred to as an optical network is a telecommunications network with optical fiber as the primary transmission medium which is designed in such a way that it makes full use of the unique attributes of optical fibers. Over the last three decades, optical fiber has become the preferred medium for provision of the major infrastructure for voice, video and data transmission, because it offers far greater bandwidth and is less bulky than copper cables. In the latter part of this period the telecommunications industry has undergone unprecedented technological change due to the rapid growth of the Internet and the World Wide Web. With the ongoing implementation of more bandwidth-intensive communication applications the requirement for increasingly higher capacity networking capability continues apace. Optical networking technology

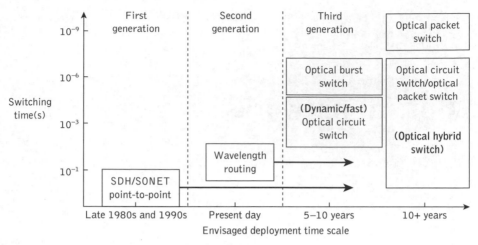

Figure 15.1 Optical fiber network evolution [Ref. 3]

and techniques have therefore evolved in order to meet these growing demands for efficient, cost-effective, reliable, high service level, worldwide communications.

Optical network evolution identifying the generations of the network development is illustrated in Figure 15.1. It may be noted that the synchronous digital hierarchy (SDH) or synchronous optical network (SONET) based point-to-point approaches of the 1990s are in the process of being upgraded with wavelength routed networking technologies. The main contributing factors leading to this evolution have been the network structure or configuration, the switching speed and the optical device enabling technologies. Based on improvements in these aspects, optical fiber networks can be currently divided into the three deployment stages or generations which are identified in Figure 15.1.

Present-day optical fiber networks that have developed from the largely point-to-point optical fiber infrastructure deployed over the last two decades can be viewed as second generation which heavily utilize wavelength routing techniques. These optical networks, however, are currently static in the sense that the allocated resources carrying the traffic cannot be reassigned automatically once the transmission has failed to reach a destination. Hence the existing network structures are largely not reconfigurable. Nevertheless, network infrastructures with individual wavelength links operating at transmission rates up to 40 Gbit s^{-1} are being deployed utilizing dense wavelength division multiplexing technology [Refs 1, 2].

The next or third generation of optical fiber networks is, however, expected to exhibit fast and reconfigurable network features by overcoming the existing static network architectures. This can be achieved using enhanced optical switching techniques. The three modes of optical switching which can be used to increase the operational speed and produce reconfigurable networks are circuit switching, packet switching and burst switching. In optical circuit switching (OCS) a path is set up between the source and destination before the transmission can take place and then after successful transmission of the entire message the path is removed. Alternatively, with optical packet switching (OPS), instead of achieving complete transmission in a single step the message is broken into small units or packets and each packet is sent on an established path. The packet therefore also contains control and addressing information regarding the destination.

Optical burst switching (OBS) is a technique which lies between optical circuit switching and optical packet switching. In this case a message is transmitted in data bursts on an established path while separate bursts of information are sent containing the control data. The switching setup time for the OBS mode is significantly higher than that of OCS and it can be assigned in microseconds. Moreover, a very fast switching speed of the order of nanoseconds can be obtained with OPS. In addition, OPS offers higher utilization of resources as compared with the moderate and poor utilization of resources obtained with OBS and OCS, respectively. Higher transmission rates up to 160 Gbit s^{-1} are also achievable using OPS but network architectures are more complex and the optical component enabling technology for such packet switching networks has not yet reached commercial deployment viability [Refs 4–6].

This chapter therefore considers both the basics and the current position together with potential future developments associated with optical fiber networks. In particular, substantial discussion is centered on wavelength routed and switched optical networks which have become the focus of advanced optical fiber networks. In order to cover all the main aspects the chapter is organized as follows. Concepts including routing, switching, connection setup and terminology used in the optical fiber networking are dealt with in Section 15.2. The fundamentals associated with optical transmission modes, layers and protocols are described in Section 15.3. Details of different network techniques for both synchronous and asynchronous transmission are described, followed by discussion of the generic Open Systems Interconnection model, the optical transport layer and the Internet Protocol.

The major topics of optical routing networks and optical switching networks are then dealt within Sections 15.4 and 15.5, respectively. Wavelength routing assignment and strategies are described in Section 15.4 while circuit and packet switching, including multiprotocol label switching and optical burst switching, are addressed in Section 15.5. This is followed in Section 15.6 by coverage of the continuing development of optical networks deployed both in the public telecommunications network and for on-site or on-premises applications. It includes long-haul, metropolitan area and access networks together with local area networks. Section 15.7 focuses on optical Ethernet which has evolved from the conventional copper cable Ethernet local area network to provide a networking capability which can be utilized in both metropolitan and access networks.

Finally, Section 15.8 addresses the important issues of optical network protection and survivability with, for example, the provision of alternative routes in the case of fiber link breakdown or network failures. Such features are essential for an optical network in order to maintain the appropriate quality of service and also to support a sustainable infrastructure when deploying new fiber within the network.

15.2 Optical network concepts

A network utilizing optical fiber as a transmission medium provides a connection between many users to enable them to communicate with each other by transporting information from a source to a destination. It may also require an intermediate stage to process the data

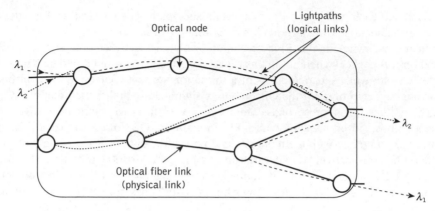

Figure 15.2 Optical network structure

for control operation. Thus each stage of information transfer is required to follow the fundamentals of optical networking. These fundamentals essentially involve the methodology for the interconnection of various optical devices together with the operational procedures for successful transportation of optical signals between the source and the destination nodes.

Figure 15.2 shows the structure of a simple optical network. It consists of optical nodes interconnected with optical fiber links. An optical node is a multifunctional element which basically acts as a transceiver unit capable of receiving, transmitting and processing (if required) the optical signal. Optical fibers provide point-to-point physical connections between network nodes. The point-to-point fiber links can be used to establish logical links where the destination node can be reached by traveling through one (or more intermediate nodes) in a single or multiple hops. Ideally, for an uninterrupted optical signal to reach a destination node using multihops each channel is assigned a specific signal wavelength from source to destination. A signal carried on a dedicated wavelength from source to a destination node is known as a lightpath (i.e. λ_1 and λ_2 are two lightpaths in this particular case). The data can be sent over these lightpaths once the connections are set up. In addition, a controlling mechanism is also required to provide for data flow during its transmission in order to authenticate the entire data transportation between each transmitting and receiving node. Based upon this simple optical network scenario, the fundamentals can be divided into three broad areas: namely, optical networking terminology, the functions and types of optical network node and switching elements, and wavelength division multiplexed optical networks. These aspects are further explained in the following sections prior to a brief overview of the public telecommunications networks.

15.2.1 Optical networking terminology

Networks employing optical fiber can transmit and receive using either unidirectional or bidirectional transmission over the same single optical fiber. Despite the differences in optical transmission techniques, the fiber networking fundamentals remain the same. In order to send and receive messages across an optical network a transmission path must be established either to switch or route the messages to their final destinations.

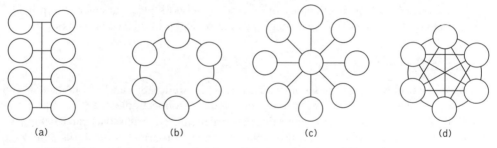

Figure 15.3 Network topologies: (a) bus; (b) ring; (c) star; (d) mesh

15.2.1.1 Network topology

The network structure formed due to the interconnectivity patterns is known as a topology, which can take a form of a bus, ring, star or a mesh structure. Figure 15.3 depicts these topologies where the circles represent the interconnected nodes. In the bus topology the data generally circulates bidirectionally. Data is input and removed from the optical fiber bus via four-port couplers located at the nodes (see Section 5.6.1). In the ring configuration the data usually circulates unidirectionally, being looped through the nodes at each coupling point, and hence it is repeatedly regenerated in phase and amplitude. By contrast an optical star coupler (see Section 5.6.2) forms a central hub to the network which may be either active or passive. In passive operation the star coupler at the hub splits the data in terms of power so that the signal is transmitted to all nodes. Overall losses are therefore primarily determined by the number of splits. With an active hub the data is split in an optoelectronic star coupler which carries out both O/E and E/O conversion and therefore this process does not contribute to the overall network losses.

The ring and star topologies are combined in the mesh configuration (Figure 15.3(d)) to provide interconnections among the network nodes. It is referred to as full-mesh when each network node is interconnected with all nodes in the network. Therefore the full-mesh configuration provides more interconnectivity than the other three topologies in Figure 15.3. Finally, optical networks can be implemented using either a single or a combination of topologies. The full-mesh topology, however, although complex, is often preferred for the provision of either a logical or virtual topology due to its high flexibility and interconnectivity features.

15.2.1.2 Network architecture

This provides for the implementation of networking functionalities in different layers of a reference model (see Section 15.3.3). The partitioning of functions following a logical methodology defines communication tasks into a set of hierarchical layers within a standardized network architecture. Each layer in the network architecture performs a related set of functions in order that a user system (normally a heterogeneous computer) will be able to communicate with another user system. The ultimate purpose of this approach is to produce a network that can be physically and operationally interconnected, and that offers large network capacity while also ensuring network protection/survivability (see Section

15.8). Therefore a major functionality of the optical network physical layer is to provide appropriate routing and switching strategies within optical multiplexed networks.

15.2.1.3 Networking modes

There are two networking modes in which a transmission path can be established and these are referred to as either connection oriented, or connectionless. In a connection-oriented network, an end-to-end connection setup, generally referred to as handshaking, is performed before the transmission takes place. Connection-oriented networks employ a bidirectional communication environment to initiate a connection so that both source and destination can communicate with each other. Once the path is determined all the subsequent information and the data are sent to and from the destination using the established connection.

By contrast, there is no dedicated end-to-end connection in the connectionless networking mode and therefore no explicit connection setup is performed before the actual data transmission. Hence the transmission is simply launched onto a common fiber channel such as the bus or ring which interconnects the network nodes. The connectionless networking communication mode is often achieved by transmitting information in one direction (i.e. from the source to destination) without confirming the presence or readiness of the destination to receive the information. In such networks both the addressing information and the data are organized in small blocks, known as packets, where each packet is assigned a unique number and sent in a sequence such that the destination node can reassemble them to construct the complete message.

15.2.1.4 Network switching modes

As indicated in Section 15.1, switching can be achieved in different modes. Two primary modes are circuit switching, and packet switching or cell switching. An end-to-end circuit is required to be set up before establishing a connection in the circuit-switched mode. All the specific network resources necessary for the switching are dedicated for the particular transmission during which no other transmission can access these resources. Furthermore, such transmissions are continuous and hence arrive in real time. Once the transmission is complete the circuit is terminated and resources become available to other users. Therefore the circuit switching mode can be characterized as continuous, selective and temporary. Circuit switching is utilized in connection-oriented networks.

The packet or cell switching mode implies a store-and-forward strategy where incoming messages (in the form of blocks) are forwarded to their corresponding destinations. Sending or receiving a complete message in a single block requires a buffering stage where information can be gathered before it is sent to the destination. Alternatively, messages can be sent in small blocks (i.e. packets or even smaller blocks called cells) which do not require such significant buffering stages. Hence the message is divided into packets which comprise variable length blocks of data with a tight upper bound limit or a cell which incorporates a fixed length block of data. Packets or cells from many different sources are statistically multiplexed [Ref. 7] and sent on to their destinations. In statistical multiplexing, a fixed bandwidth communication channel is divided into several variable channels and the desired or requested bandwidth is allocated to each end user. This contrasts with simple time or wavelength division multiplexing where a fixed bandwidth is assigned. Finally, both packet and cell switching are applied in connection-oriented networks. Packets, however, are also often employed in connectionless networks.

15.2.1.5 Virtual circuits

Unlike a physical circuit, which terminates the physical medium on specific physical ports, a virtual circuit is a series of logical connections between the sending and receiving devices. A virtual circuit is a connection between source and destination nodes which may consist of several different routes. These routes can change at any time, and the incoming return route does not have to mirror the outgoing route. When the connections are set up in advance, based on expected traffic patterns, these are termed provisional virtual circuits. In switched networks, however, where connections can be set up on demand and released after the data exchange is complete, such connections are known as switched virtual circuits.

15.2.1.6 Network routing

Routing refers to the process whereby a node finds one or more paths to possible destinations in a network. In this process control and data processing functions are performed to identify the route and to handle the data during the journey from source to destination. A simple routing process known as the control plane comprises three stages which are neighbor discovery, topology discovery and path selection. Neighbor discovery is the first step whereby the nodes in a network discover the identity of their immediate neighbors and their connections to the neighboring nodes. Topology discovery is the process by which a node discovers all other nodes in the network and how they are connected to each other.

Once the network topology is known, paths from a source to a destination can be determined using path computation algorithms or routing protocols. A protocol is the set of standard rules for data representation, signaling, authentication and error detection required to send information over a communication channel. The control plane therefore acquires and then provides the reachability information throughout the network. This reachability or routing information disseminated by the routing protocol is consolidated in a forwarding table at each node which contains the identity of the next node and the outgoing interface to forward packets addressed to each destination. In order to handle the data more efficiently it requires an additional procedure known as the data plane which allows an optical node to forward the incoming packets towards their destinations by using the information contained in a forwarding table. This operation involves examining the destination address of the packets, searching the forwarding table to determine the next hop and the corresponding outgoing interface, and then forwarding the packets on this next hop.

15.2.1.7 Modularity and scalability

Modularity defines the characteristics of a network which allows the addition or reduction of networking nodes in a modular fashion. Such modular networks therefore permit the use of wavelengths (i.e. wavelength reuse) in different sections of a large optical network without causing wavelength conflict.

Scalability is the property of a network which enables it to progressively accommodate a large number of nodes and end user systems without incurring excessive overheads. It is one of the most important design aspects of any network or protocol. For a network with a high number of nodes, scalability becomes difficult to realize as a large number of alternative routes are available. Scalability can be achieved in several different ways such as by employing a routing hierarchy, or by performing route aggregation [Ref. 8]. Both these techniques help contain the size of the routing databases. Another common technique for achieving scalability is to reduce the number of routing updates using threshold schemes [Ref. 9].

15.2.2 Optical network node and switching elements

An optical node is considered as a multifunctional element that performs several tasks depending upon its type and the network requirements. Essentially it sends, receives and resends or redirects optical signals to its neighboring connected nodes. The resending or redirecting of an optical signal to the desired networking nodes requires the node to perform either a routing or switching function. It should be noted that when several optical signals travel in a multiplexed form, the networking node becomes transmission system dependent since different optical multiplexing techniques can be used such as time division or wavelength division multiplexing. An optical node can also function as a router directing an input signal wavelength to a specified output port. A router essentially comprises optical couplers (see Section 5.6) where the desired signal from an incoming multiplexed signal can be isolated (i.e. demultiplexed) and then directed or routed to the output port. For this reason it is commonly referred to as a wavelength router. Moreover, it is also possible to change the wavelength of the signal before sending it to the specified output port. In this case the router will be called a wavelength converting router and since the device is used to switch the wavelength by changing it to the compliant signal wavelength it is also known as a wavelength routing switch or simply an optical switch [Ref. 10]. It should be noted that unlike semiconductor switches an optical switch performs wavelength operational functions (i.e. multiplexing/demultiplexing and switching) and the physical medium remains fixed with only the information contained in the messages being changed to send (or route/reroute) it to the corresponding destination node through the assigned route. Such wavelength switches can be constructed using integrated optical or optoelectronic devices (see Sections 10.3 to 10.6 and 11.2 to 11.6).

The four different functions of an optical router are depicted in Figure 15.4. A 1×2 wavelength demultiplexer is shown in Figure 15.4(a) which illustrates the splitting of an optical signal present at input port 1 containing two signal wavelengths (i.e. λ_1 and λ_2) and routing them to ports 2 and 3, respectively. A three-port wavelength multiplexer combining two wavelength signals is indicated in Figure 15.4(b). An optical add/drop multiplexer (OADM) which also comprises a wavelength add/drop device (WADD) is shown in Figure 15.4(c) where an incoming multiplexed signal comprising three wavelengths (i.e. λ_1, λ_2 and λ_3) is partially demultiplexed by dropping the λ_2 signal at an intermediate port 2. The OADM further adds another wavelength signal λ_4 at intermediate port 3 which is them multiplexed with the transmitted signal wavelengths so that the combined signal leaving port 4 contains wavelengths λ_1, λ_3 and λ_4. Devices which are configured to drop a particular wavelength (in this case λ_2) and add another specific wavelength (in this case λ_4) are also known as fixed OADM.

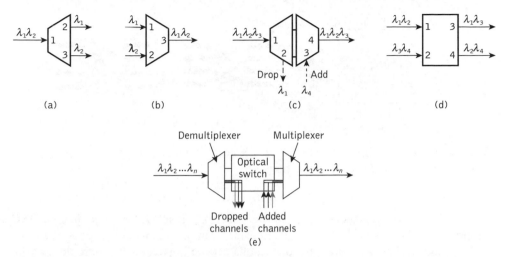

(a) (b) (c) (d)

(e)

Figure 15.4 Optical networking node elements: (a) wavelength demultiplexer; (b) wavelength multiplexer; (c) optical add/drop multiplexer; (d) 2 × 2 optical switch; (e) reconfigurable optical add/drop multiplexer

Figure 15.4(d) shows a simple 2 × 2 optical switch with two input and two output ports. The two optically multiplexed signals comprising wavelengths $\lambda_1\lambda_2$ and $\lambda_3\lambda_4$ are present at input ports 1 and at 2, respectively. At the output ports the wavelengths are required to be switched and multiplexed as $\lambda_1\lambda_3$ and $\lambda_2\lambda_4$ emerging at the output ports 3 and 4, respectively. Using this approach it is possible to produce an optical switch with a greater number of ports (i.e. an $N \times N$ optical switch, where N is a large number). A combination of an OADM and an optical switch producing a reconfigurable optical add/drop multiplexer (ROADM) is illustrated in Figure 15.4(e). This device can drop one or a desired number of wavelength channels after demultiplexing a wavelength multiplexed signal and, similarly, it can also add a new single or more wavelength channels through an optical switch. Finally, the multiplexer unit in Figure 15.4(e) brings together all the wavelength channels to produce a combined wavelength multiplexed output signal. A passive ROADM with this functionality can be constructed using a fiber Bragg grating and an optical circulator (see Section 5.7).

Although ROADMs prove useful in providing for simple optical network topologies, they cannot facilitate complex mesh topologies which incorporate a large number of nodes. Such networks therefore require an additional element to provide a means of cross-network interconnection with the added features of an ROADM. This additional element is an optical cross-connect (OXC) which has the capability of switching the connection between two interfaced points.

A block schematic of different OXC structures is provided in Figure 15.5. A large number of wavelength signals (i.e. $\lambda_1, \lambda_2, \ldots, \lambda_N$) can be demultiplexed at the input ports of an OXC as shown in Figure 15.5(a) and then these are internally connected to the desired output ports where the different wavelength signals are multiplexed for onward transmission. Furthermore, as indicated the OXC makes it possible to provide an interface between

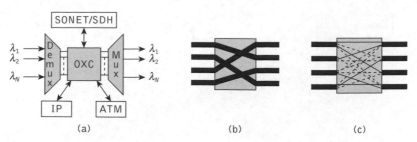

Figure 15.5 Optical cross-connect (OXC): (a) block schematic; (b) fiber cross-connections; (c) logical cross-connections

different transmission methodologies (see Section 15.3). It should be noted that the OXC comprises an $N \times N$ optical switching fabric which can provide interconnections between network nodes. Cross-network interconnections between optical and electronic network nodes, however, require optoelectronic conversion. Moreover, wavelength conversion may also be needed to establish a connection that is otherwise not possible using a single lightpath (see Section 15.4). An OXC possessing such wavelength conversion features can switch (or interchange) different wavelength signals and therefore is known as a wavelength interchangeable cross-connect (WIXC) [Refs 11, 12]. In addition, an OXC facilitating optoelectronic conversions is often referred to as an opaque OXC whereas one incorporating all-optical/optical wavelength converters is known as a transparent OXC. Furthermore, the $N \times N$ optical switching fabric in an OXC facilitates either physical or logical interconnections. Physical interconnection is provided by the fiber switching capability within the OXC as illustrated in Figure 15.5(b). In this case optical fibers are switched and connected to the desired network node. The OXC can also facilitate the logical wavelength interconnection features as indicated in Figure 15.5(c) for multiple-wavelength signals which can be routed to their destination either individually or in a group of different wavelength signals (i.e. wavebands). In the latter case several desired wavelength signals can be multiplexed and routed together to their destination nodes [Ref. 12].

The various types of logical interconnection (i.e. single wavelength, multiple wavelengths, wavebands) and physical interconnections (i.e. fiber cross-connect) can also be combined together in a multilayer structure. In such multilayered OXCs each layer, in addition to providing the specific switching capability, can be connected to another layer, thus offering increased flexibility. A multilayered OXC is sometimes referred to as a multigranular optical cross-connect (MG-OXC) [Ref. 13].

15.2.3 Wavelength division multiplexed networks

Optical fiber networks using wavelength division multiplexing (WDM) techniques (see Section 12.9.4) can be classified as either broadcast-and-select networks or wavelength routing networks. A broadcast-and-select network strategy based on a star coupler (see Section 5.6.2) is shown in Figure 15.6. The optical transmission is broadcast to all other nodes using fixed transmitters and a tunable receiver at the destination node extracts the desired signal from the entire group of wavelength multiplexed transmitted signals (i.e. λ_1,

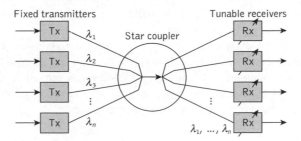

Figure 15.6 Broadcast-and-select network

$\lambda_2, \lambda_3, \ldots, \lambda_n$). It should be noted that all transmissions are broadcast to all network nodes and hence most of the transmitted power is depleted on the receivers which do not use it. Consequently, as the number of nodes increases, each station receives a small fraction of the overall transmitted power. Alternatively, a wavelength routing network can be used to avoid this wastage of transmitted power where each node within the network is provided with restricted connection(s) to the receiver(s). In wavelength routing, instead of distributing the message over the entire network, the signal is routed to the specific destination through either a single node or using multiple nodes.

The concept of wavelength routing is illustrated in Figure 15.7 where the physical bidirectional interconnections between five nodes (i.e. A, B, C, D and E) are shown in Figure 15.7(a). Using three wavelengths (i.e. λ_1, λ_2 and λ_3) any network node can transmit or receive a signal from another node within the network. This strategy of wavelength implementation or path selection shown in Figure 15.7(b) is known as routing and wavelength assignment (RWA). For example, node A can transmit to node B using wavelength λ_1 and it can simultaneously receive from node B using wavelength λ_2 only when the signal is routed through node E. To simplify such routing and wavelength assignments a virtual topology is generally used to describe only the enabled wavelength paths. The virtual topology is indicated Figure 15.7(c) where only wavelength signals identify the possible interconnections between nodes as given by the RWA. It can be observed by comparing Figure 15.7(a) and (c) that there exists no physical connection between nodes C and E but both nodes can communicate via the virtual connection set up using wavelengths λ_1 and λ_3.

Both broadcast-and-select and wavelength routing networks can be further classified into single-hop or multihop. Single-hop networks allow direct communication between

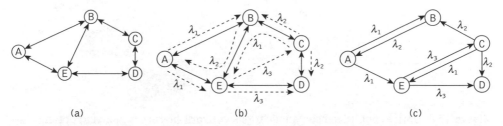

Figure 15.7 Wavelength routing: (a) physical connection; (b) wavelength assignments; (c) virtual topology

any two nodes and the data remains entirely in the optical domain (i.e. without optoelectrical conversion) until it reaches its destination.

In a multihop network a transmission may take place through intermediate nodes before reaching its destination. At each intermediate node the data can be switched electronically to the next possible node and it is then retransmitted as an optical signal. Although this conversion process is inefficient, it is necessary if there is no common wavelength path between two nodes in order to establish a direct connection.

15.2.4 Public telecommunications network overview

The telecommunications network providing services in the public domain is known as the public telecommunications network where the service providers (or carriers) offer a variety of services for the provision of voice, data and video transmission. A simple block hierarchy for the optical public telecommunications network is illustrated in Figure 15.8, which is divided into three tiers: long-haul, metropolitan and access networks (see Section 15.6). The long-haul network, also known as the core or sometimes the backbone

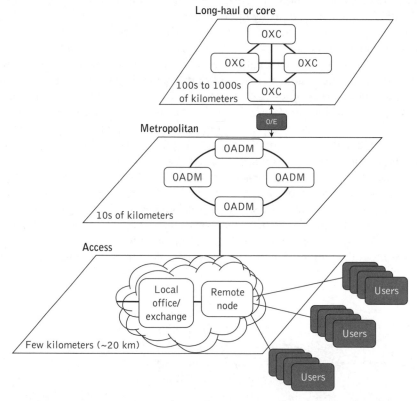

Figure 15.8 Optical public telecommunications network hierarchy showing optical cross-connects (OXCs) in the long-haul, optical add/drop multiplexers (OADMs) in the metropolitan and an optical fiber access network

network, provides national or global coverage with a reach of thousands of kilometers. These networks connect many metropolitan networks within a country or interconnect together several country-wide long-haul networks. Interconnection between optical nodes is generally accomplished by means of optoelectrical conversion and/or optical switches employing OXCs.

At the next lower hierarchal level resides the metropolitan area network (MAN), often called the metro, or sometimes the back-haul network. These networks offer a multiservice platform and may be confined to a region spreading to tens of kilometers. At present they are largely implemented using the ring topology. The interconnections between optical nodes in metro networks are also achieved using OADMs while the larger metro networks can also incorporate OXCs.

The lowest tier in the hierarchy is the local access network which may be extended from a few hundreds of meters to 20 kilometers or so. Hence the access network provides the initial interface to the telecommunications network for residential and business customers. Although these networks can be configured based on the bus, star or ring topologies (see Section 15.2.1.1), the fiber access network configuration illustrated in Figure 15.8 is currently gaining favor (see Section 15.6.3). The users are connected to a branching node known as a remote node (RN) which interconnects the users with the local office/telephone exchange. In order to provide larger interconnectivity, several local offices/exchanges are connected using a metropolitan area network.

15.3 Optical network transmission modes, layers and protocols

Although optical fiber networks imply an entirely optical framework, these network structures usually incorporate complex combinations of both optical and electronic infrastructure. The optical infrastructure present in such networks constitutes a transparent optical network utilizing optical signal transmission where electronic devices can be used for signal control, or to provide the method of interconnection to a number of other networks. Therefore a practical scenario dictates the use of a combination of the optical and electronic domains where different devices are required to communicate with each other in order to establish a connection for transmission between different segments of the optical fiber network. The end points of the optical network usually comprise network nodes and network stations. The nodes connect the fibers within the optical network while the stations connect the optical network to the nonoptical systems in the electronic domain. The nodes also provide functions that control the optical signals whereas the stations provide the terminating points for the optical signal. The stations and nodes therefore comprise both optoelectronic and photonic components (e.g. lasers, photodetectors, couplers/splitters, switches, amplifiers, regenerators, wavelength converters, etc.). Therefore this practical scenario dictates the use of a combination of both optical and electronic domains where various devices are required to communicate with each other in order to establish a connection for transmission between different segments of the optical network.

In order to establish useful communication among different network elements of the same or other networks it is necessary to employ certain physical network structures, transmission types, rules and protocols which make networking implementation more straightforward and also allow the optical network to be upgraded with new developments and to follow improved service trends.

15.3.1 Synchronous networks

The pre-existing multichannel PCM transmission hierarchies for the telecommunications network in Europe, North America and Japan were provided in Table 12.1 (Section 12.5), together with discussion of the time division multiplexing strategy. A schematic of the way in which the European hierarchy is multiplexed up to the 140 Mbit s^{-1} rate from the constituent 2 Mbit s^{-1} (30-channel) signals is shown in Figure 15.9(a). Difficulties arise, however, with this multiplexing strategy, which is currently adopted throughout the world, in that each 2 Mbit s^{-1} transmission circuit (taking the European example) has its own independent clock to provide for timing and synchronization. This results in slightly different frequencies occurring throughout a network and is referred to as pleisochronous* transmission [Ref. 14]. Although this strategy is well suited to the transport of bits it suffers a major drawback in that in order to multiplex the different levels (i.e. 2 to 8 to 34 to 140 Mbit s^{-1}) extra bits need to be inserted (bit stuffing) at each intermediate level so as to maintain pleisochronous operation.

The presence of bit stuffing in the existing pleisochronous digital hierarchy (PDH) makes it virtually impossible to identify and extract an individual channel from within a high-bit-rate transmission link [Ref. 15]. Thus to obtain an individual channel the whole demultiplexing procedure through the various levels (Figure 15.9(a)) must be carried out. This process is both complex and uneconomic, particularly when considering future telecommunication networking requirements such as drop and insert where individual channels are extracted or inserted at particular stages. Moreover, a substantial saving in electronic hardware together with increased reliability could be achieved by having a straight, say 2 to 140 Mbit s^{-1}, multiplexing/demultiplexing capability as illustrated in Figure 15.9(b) [Ref. 16].

It was therefore clear that a new fully synchronous digital hierarchy was required to enable the international telecommunications network to evolve in the optical fiber era. In particular, this would facilitate the add/drop of lower transmission rate channel groups from much larger higher speed groups without the need for banks of multiplexers and large, unreliable distribution frames.

Furthermore, by the mid-1980s the lack of standards for optical networks had led to a proliferation of proprietary interfaces where transmission systems produced by one manufacturer would not necessarily interconnect with those from any other manufacturer such that the ability to mix and match different equipment was restricted. Hence standardization towards a synchronous optical network termed SONET commenced in the United States in 1985 [Ref. 17]. However, two key areas resulted in some modification to the original proposals. These were to make the standard operational in a pleisochronous environment and still retain its synchronous nature and to develop it into an international

---

* Corresponding signals are defined as pleisochronous if their significant instants occur at nominally the same rate, any variation being constrained within specific limits.

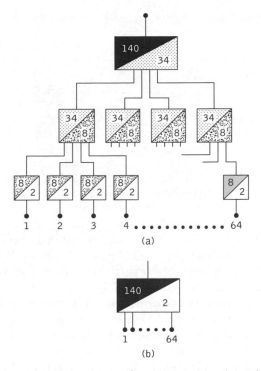

Figure 15.9 European multiplexing hierarchy: (a) existing pleisochronous structure; (b) synchronous multiplexing

transmission standard in which the incompatibilities between the existing European and North American signal hierarchies could be resolved. In this latter context the ITU-T (formerly CCITT) began deliberation of the SONET concepts in 1986 which resulted in basic recommendations for a new synchronous digital hierarchy (SDH) in November 1988. These recommendations were published [Refs 18–20]. Prior to these recommendations the American National Standards Institute (ANSI) had issued draft standards relating to SONET [Ref. 21] but as a result of the extensive discussions between the two standards authorities, the two hierarchies are effectively the same. Hence the synchronous optical network recommendations tend to be referred to as SONET in North America and SDH in Europe.

The SONET standard as ultimately developed by ANSI defines a digital hierarchy with a base rate of 51.840 Mbit s^{-1}, as shown in Table 15.1. The OC notation refers to the optical carrier level signal. Hence the base rate signal is OC-1. The STS level in brackets refers to a corresponding synchronous transport signal from which the optical carrier signal is obtained after scrambling (to avoid a long string of ones or zeros and hence enable clock recovery at receivers) and electrical to optical conversion.* Thus STS-1 is the basic building block of the SONET signal hierarchy. Higher level signals in the hierarchy are

---

* STS-1 corresponds to OC-1 which is the lowest level optical signal used at the SONET equipment and network interface.

Table 15.1 Levels of the SONET signal hierarchy

level		Line rate (Mbit s^{-1})
OC-1	(STS-1)	51.840
OC-3	(STS-3)	155.520
OC-9	(STS-9)	466.560
OC-12	(STS-12)	622.080
OC-24	(STS-18)	1 244.160
OC-36	(STS-36)	1 866.240
OC-48	(STS-48)	2 488.320
OC-196	(STS-192)	9 953.28
OC-768	(STS-768)	39 813.12

obtained by byte interleaving (where a byte is 8 bits) an appropriate number of STS-1 signals in a similar manner to that described for the European standard PCM system described in Section 12.5. This differs from the bit interleaving approach utilized in the existing North American digital hierarchy (see Table 12.1). The STS-1 frame structure shown in Figure 15.10 is precisely 125 µs and hence there are 8000 frames per second. This structure enables digital voice signal transport at 64 kbit s^{-1} (1 byte per 125 µs) and the North American DS1-24 channel (1.544 Mbit s^{-1}), as well as the European 30-channel (2.048 Mbit s^{-1}) signals (see Table 12.1) to be accommodated. Other signals in the two hierarchies can also be accommodated. The basic STS-1 frame structure illustrated in Figure 15.10 comprises nine rows, each of 90 bytes, which therefore provide a total of 810 bytes or 6480 bits per 125 µs frame. This results in the 51.840 Mbit s^{-1} base rate mentioned above.

The first 3 bytes in each row of the STS-1 frame contain transport overhead bytes, leaving the remaining 783 bytes to be designated as the synchronous payload envelope (SPE). Apart from the first column (9 bytes) which is used for the path overhead, the remaining 774 bytes in the SPE constitute the SONET data payload. The transport overhead bytes are utilized for functions such as framing, scrambling, error monitoring, synchronization and multiplexing while the path overhead within the SPE is used to provide end-to-end communication between systems carrying digital voice, video and other signals which are

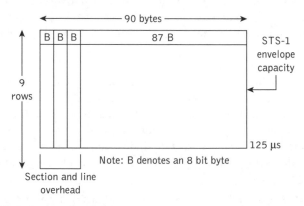

Figure 15.10 STS-1 frame structure

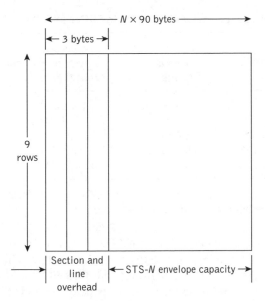

Figure 15.11 STS-N frame structure

to be multiplexed onto the STS-1 signal. In the latter case a path is defined to end at a point at which the STS-1 signal is created or taken apart (i.e. demultiplexed) into its lower bit rate signals.

The STS-1 SPE does not have to be contained within a single frame; it may commence in one frame and end in another. A 'payload pointer' within the transport overhead is employed to designate the beginning of the SPE within that frame. This provides the flexibility required in order to accommodate different bit rates and a variety of services. Moreover, to accommodate sub-STS-1 signal rates a virtual tributary (VT) structure is defined comprising four rates: 1.728 Mbit s^{-1} (VT 1.5); 2.304 Mbit s^{-1} (VT 2); 3.456 Mbit s^{-1} (VT 3); and 6.912 Mbit s^{-1} (VT 6). For example, it may be observed that the 1.544 Mbit s^{-1} and 2.048 Mbit s^{-1} signal streams can each be mapped into a VT 1.5 and VT 2 respectively.

Finally, the higher order multiplexing of a number of STS-1 signals is obviously important in order to achieve the higher bit rates required for wideband services. The format of the STS-N signal frame is shown in Figure 15.11 which, as mentioned previously, is obtained by byte interleaving N STS-1 signals. In this case the transport overhead bytes of each STS-1 (i.e. the first three single-byte columns of each STS-1 signal shown in Figure 15.10) are frame aligned to create the $3N$ bytes of transport overhead, which is illustrated in Figure 15.11. However, the SPEs do not require alignment since the service payload pointers within the associated transport overhead bytes provide the location for the appropriate SPEs.

The synchronous digital hierarchy, as defined by the ITU-T [Refs 18–20], operates in the same manner as described above but differs in some of its terminology [Ref. 22]. In this case the 125 μs frame structure is referred to as a synchronous transport module (STM) and the base rate STM-1 is 155.520 Mbit s^{-1} which corresponds to OC-3 (STS-3), as may be observed from Table 15.2. Hence the European 140 (i.e. 139.264) Mbit s^{-1} pleisochronous signal can be mapped within an STM-1 signal when including a suitable overhead.

Table 15.2 Corresponding levels and bit rates for SDH and SONET

SDH level	SONET level	Line rate (Mbit s⁻¹)	Synchronous payload envelope rate (Mbit s⁻¹)	Transport overhead rate (Mbit s⁻¹)
STM-1	OC-3*	155.520	150.3366	5.184
STM-4	OC-12	622.080	601.344	20.736
STM-16	OC-48	2 488.320	2 405.376	84.672
STM-64	OC 196	9 953.28	9 621.504	331.776
STM-256	OC 768	39 813.120	38 486.016	1327.104

* Rates such as OC-9, OC-18, OC-24, OC-36 and OC-96 are referenced in some of the standards documents but are not widely implemented.

As would be expected, the higher level STM signals also correspond to SONET optical carrier rates as well as providing a match to appropriate multiples in the European pleiso-chronous hierarchy. Again, these higher levels are formed by simple byte interleaving. Tributaries are used to incorporate the signal rates below the STM-1 rate into the frame format where they may be located by means of pointers. Hence a 155.520 Mbit s⁻¹ channel can be readily identified within a 40 Gbit s⁻¹ (i.e. 39 813.2 Mbit s⁻¹) stream. As with SONET this allows a network of high-capacity cross-connects to be established at nodes throughout the transmission network. Moreover, this facility also enables efficient management of an optical network to route and to distribute traffic between nodes, dropping off capacity to exchanges for traffic switching (see Sections 15.4 and 15.5) [Ref. 23]. As a further development in STM the available bandwidth can be reallocated to the users according to their requirements. This is known as dynamic synchronous transfer mode (DSTM), which employs circuit switching since it is the combination of optical switching and transport technology as recommended by the European Telecommunications Standards Institute (ESTI) [Ref. 24]. Dynamic synchronous transfer mode is a time division multiplexing strategy (see Section 12.4.2) where the time slots are divided into control and data slots. The control slots are used for signaling purposes and are statically allocated to a node, whereas the data slots are assigned dynamically for the user of the data transmission. Every node is allocated a few control slots and several data slots and if necessary the control slots can be converted into data slots (or vice versa) depending on traffic demand [Ref. 25]. Furthermore, DSTM can support multicast where a slot can be accessed by several nodes on a bus or ring topology (see Section 15.2.1.1) [Ref. 26].

When DSTM employs a SONET/SDH framing scheme it further extends the scheme with a dynamic reallocation mechanism that can redistribute bandwidth not being used by one user to another user who has a demand for it. This is possible since DSTM supports multirate bandwidth allocation in the slots of the order of 512 kbit s⁻¹. For example, a total 19 500 slots each comprising 512 kbit s⁻¹ constitute a capacity of 10 Gbit s⁻¹. Therefore, if DSTM is incorporated in a wavelength division multiplexed system then the large bandwidth can be segmented into 512 kbit s⁻¹ chunks which can be reallocated dynamically. Dynamic synchronous transfer mode technology includes both switching and also data transmission features and therefore it can be considered as a next-generation of SONET/SDH which incorporates switching and provides configurable, on-demand bandwidth allocation.

15.3.2 Asynchronous transfer mode

Asynchronous transfer mode (ATM) is a packetized multiplexing and switching technique which seeks to combine the benefits of packet switching and circuit switching. Asynchronous transfer mode transfers information in fixed size units called cells where each cell contains the information identifying the source of the transmission but which generally contain less data than packets. Unlike the fixed time division multiplexed technique (see Section 12.4.2) where each user waits to send in the allocated time slot, ATM is asynchronous and therefore the time slots are made available on demand. To enable correct segmentation and assembly of different cells at the destination, each cell contains significant information in addition to data. An ATM cell comprises a header and payload data as shown in Figure 15.12. It contains 48 bytes of data with 5 bytes of header information. Each single byte in the header field includes different information to identify destination, path, channel and the error control bits. Before sending ATM cells carrying user data, a virtual connection between source and destination has to be established. All connections follow the same path within the network. During the connection setup each control bit (1 or 0) generates an entry in the virtual path identifier (or virtual channel identifier) translation table to inform the destination to receive the incoming packet.

Header (5 bytes)	Payload (48 bytes)

Figure 15.12 Format of an ATM cell

15.3.3 Open Systems Interconnection reference model

Open Systems Interconnection (OSI) describes a standard architecture to be used for designing networks. The OSI model was specified jointly by the International Organization for Standardization (ISO) [Ref. 27] and ITU-T [Ref. 28]. Figure 15.13 shows

	Layers	Function	Data unit	
7	Application	Network process to application		
6	Presentation	Data representation and encryption	Data	Host layers
5	Session	Interhost communication		
4	Transport	End-to-end connections and reliability	Segments	
3	Network	Path determination and logical addressing	Packets	
2	Data link	Physical addressing	Frames	Media layers
1	Physical	Media, signal and binary transmission	Bits	

Figure 15.13 Description of the layers in the OSI reference model

the structure of the OSI network reference model and identifies the functions of each of the seven levels or layers. The application layer which sits at the top of the hierarchy as the seventh layer provides a means for a user to access information on or utilize the network by receiving a service (e.g. database management, network management). This layer is the main interface for the users to interact through applications with the network. The sixth level is the presentation layer which transforms data to provide a standard interface for the application layer. The session or fifth layer controls the dialogs (sessions) between intelligent devices. Furthermore, it establishes, manages and terminates the connections between the local and remote application. It also provides for either duplex or half-duplex operation and establishes check pointing, adjournment, termination and restart procedures. The top three layers (i.e. 5, 6 and 7) are therefore data-driven levels.

The fourth level is the transport layer which provides transparent transfer of data between end users, thus relieving the upper layers from any transport concern while providing reliable data transfer. The transport layer controls the reliability of a specific link through flow control, segmentation/desegmentation and error control. Some protocols are both state and connection oriented (see Section 15.2). This means that the transport layer can keep track of the packets and retransmit those that fail. The best known example of a layer four protocol is the Transmission Control Protocol (TCP) which is a connection-oriented protocol that provides basic data transfer between nodes offering interconnection, flow control, multiplexing with reliable priority and security [Refs 29, 30].

The network layer at the third level provides the functional and procedural method for transferring variable length data sequences from a source to a destination via one or more networks while maintaining the quality of service requested by the transport layer. The network layer performs network routing functions, and might also perform segmentation/desegmentation and report on delivery errors. Routers operate at this layer for sending data throughout the extended network, thus making Internet transmission possible.

The remaining two layers are the data link and physical media. The data link layer provides the functional and procedural mechanisms to enable transfer of data between network entities and to detect, and potentially correct, errors that may occur in the physical layer. It is the layer at which network bridges and switches operate and that connectivity is provided among locally attached network nodes forming layer 2 domains for unicast or broadcast forwarding. Other protocols can be implemented on the data frames where each frame is logically addressed (see Section 15.3.5). The best known example of such a protocol is carrier sense multiple access with collision detection (CSMA/CD) or the Ethernet protocol (see Section 15.6.4). Other examples of data link layer protocols are High-level Data Link Control (HDLC) (see Section 15.6.5) and the Advanced Data Communications Control Procedure (ADCCP) for point-to-point or packet-switched networks [Refs 31, 32].

The physical layer (PHY) is located at the bottom of the OSI reference model hierarchy and it defines all the electrical, optical and physical media specifications for devices (e.g. the fiber type, cable specifications, hubs, repeaters, network adaptors). It both establishes and terminates a connection to the communications medium. Furthermore, it also ensures that communication resources are effectively shared among multiple users. For example, it implements contention resolution and flow control while also carrying out data conversion between the equipment and the signal requirements of the communications channel.

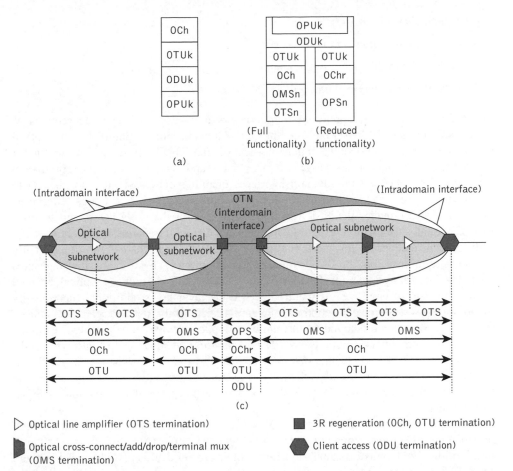

Figure 15.14 Optical transport network layers: (a) a typical hierarchy; (b) hierarchy as specified in ITU-T Recommendation 709 [Ref. 35] reproduced with the kind permission of ITU; (c) functional block diagram showing OTN layers

15.3.4 Optical transport network

An optical transport network (OTN) is regarded as the set of facilities using optical fiber interconnections to carry data between network elements that switch or route the data from different customers into the network [Ref. 33]. It is specified in ITU-T Recommendation G.873.1 [Ref. 34] and it provides for the transport, multiplexing, routing, supervision and survivability for client signals. The OTN has been proposed as the successor to SDH/ SONET which is the current standard for voice and data traffic in the public switched telecommunications network (see Section 15.3.1). Hence it combines the benefits of SDH/SONET with dense WDM (DWDM) technology (see Section 12.9.4).

ITU-T Recommendation G.709 [Refs 20, 34, 35] provides information about the hierarchical structure and interface functionalities of the OTN by describing the four layers shown in Figure 15.14(a). These are:

- optical channel (OCh);*

- optical channel transport unit (OTU);†

- optical channel data unit (ODU); and

- optical channel payload unit (OPU).

The OTN hierarchy to support these four layers may also include three sections as indicated in Figure 15.14(b), which are: the optical transmission section (OTS); the optical multiplexing section (OMS); the optical physical section (OPS). These sections are further divided into a number of subsections *n*. The layers and sections in an OTN produce a confluence of inter- and intradomain interfaces enabling optical signal transmission to reach from one end to the other using electronic, optoelectronic or all-optical interfaces. Figure 15.14(c) provides the basic functional structure for the OTN layers and sections which facilitates the method for interfacing between different transmission domains.

In order to get access for end-to-end networking of optical channels to transparently convey user information, the OCh layer is employed in the OTN structure. It also ensures connection rearrangement for flexible routing. Either full or reduced optical channel availability can be employed as required. The OTU layer gives additional functionality by adding forward error correction (see Section 12.6.7) to the network elements and by allowing carrier operators to reduce the number of optical devices and switches used in the network (i.e. amplifiers, multiplexers, 3R regenerators). It also encapsulates two additional layers, the optical channel data unit (ODU) and the optical channel payload unit (OPU). The ODU layer provides client-independent connectivity, connection protection and monitoring whereas the OPU enables access to the payload information of SDH/SONET signals.

Different physical sections of the OTN are interfaced based on the information obtained from OTS, OMS and OPS, respectively. For example, OTS provides the required information about optical transmission functionality for optical signals to travel through optical fibers and the necessary switching operation at fiber end points. The optical multiplexing section (OMS) gives the means of networking for a multiwavelength optical signal, whereas the OPS describes the optical characteristics of the physical section in order to provide reduced functionality excluding those sections which are not required to establish an interface between two domains. For example, a reduced functionality can be achieved by direct adaptation of the OPS into OChr functions since both OMS and OTS are not necessary in the interdomain interfaces as identified in Figure 15.14(b) and (c).

The control plane of the OTN refers to a set of protocols related to signaling and routing for distributed routing control, connection management and resource discovery when a connection is required to be established between two network nodes. Depending on the provision of fast, efficient and reliable network connection, the OTN can be made reconfigurable. Such an advanced type of reconfigurable OTN is referred to as an automatically switched optical network (ASON). This architecture is capable of switching the optical channels automatically when requested. Automatically switched optical networks are specified in ITUT-T Recommendation G.8080 [Ref. 36]. Unlike SDH/SONET, the OTN is

* OChr – the letter r indicates a reduced functionality.
† OTUk – the letter k is used to describe a sublayer.

a transport layer that can carry 10 gigabit Ethernet (GbE) (see Section 15.7) transmission from IP/Ethernet switches and routers at full bandwidth [Refs 37, 38]. Hence as a consequence of the ongoing migration towards IP/Ethernet-based infrastructure the OTN is becoming the transport layer of choice for network operators [Refs 39, 40]. Moreover, it also provides for legacy optical fiber networks to be combined onto a common network infrastructure.

15.3.5 Internet Protocol

The Internet Protocol (IP) is a network layer (i.e. layer 3 protocol in the OSI model) (see Section 15.3.3) that contains both addressing and control information to enable packets (or datagrams) to be routed within a network. The Internet can be characterized as a logical architecture (independent of any particular network) which can permit multiple different networks to be interconnected enabling each network node to communicate without the need to know which network it is using or how to route information between them [Ref. 41]. As indicated in Figure 15.15, the IP provides the means of communication between the link and transport layers. A virtual connection is established between nodes requiring communication when IP is combined with a specific higher level protocol such as the Transmission Control Protocol (TCP) or the User Datagram Protocol (UDP) [Ref. 42]. For this role TCP/IP is preferred, since UDP/IP does not guarantee reliable delivery of data in comparison with TCP/IP, which generally encapsulates data from the link layer protocols such as Ethernet (see Section 15.7). The IP provides protocols for both the functions of signaling and routing required to carry the entire signal operation necessary to transmit and receive from optical nodes. The signaling protocols include Multiprotocol Label Switching (MPLS) and Generalized Multiprotocol Label Switching (GMPLS) (see Section 15.5.3) while the routing protocols include the Open Shortest-Path First (OSPF), the Intermediate-System-to-Intermediate System (IS–IS) and the Border Gateway Protocol (BGP) [Refs 43, 44].

Three generic stages of deployment for optical IP networks are shown in Figure 15.16. The first generation used ATM to carry IP packets (i.e. IP over ATM), which has proved

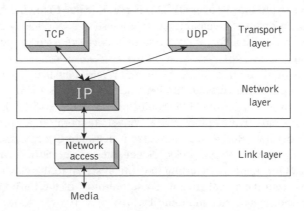

Figure 15.15 Internet Protocol for optical networks

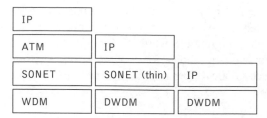

Figure 15.16 Different implementation schemes for IP over WDM/DWDM

not to be an efficient technique as compared with the direct use of SDH/SONET. This is because of the overhead requirements of the ATM cell structure together with the inherent overlap of IP packets which map significantly outside the ATM cell size. Therefore service providers are considering running IP content directly over SONET (i.e. IP over SONET). In order to carry IP packets, however, mapping of the IP packets directly to the SDH/SONET frames is required and the resultant technique is referred to as packet over SONET (PoS). This mapping of IP frames can be accomplished in three stages. In the first stage the data is segmented into IP packets which are then encapsulated via the Point-to-Point Protocol (PPP). The framing information is then added in the second stage and then finally, in the third stage, the resultant data is mapped synchronously onto the SDH/SONET frame.

The second stage and column in Figure 15.16 shows the three tiers comprising IP, a thin layer of SONET and dense WDM. The ongoing IP revolution suggests that SDH/SONET is not bandwidth efficient for IP packet transport [Ref. 4]. For example, SDH/SONET uses synchronous time division multiplexing where each time slot must be occupied otherwise there is a waste of capacity when there is no data traffic flowing on specific channel bandwidth. This situation can be improved if statistical multiplexing techniques (see Section 15.2.1.4) are employed where bandwidth is not assigned if the user has not requested [Ref. 45]. Thus a thin SONET layer is still present in the second stage with the switch over from WDM to DWDM at the physical layer. The latter case, however, is considered more useful if IP can be directly mapped onto DWDM. The removal of the ATM and SONET layers to produce IP over DWDM is possible due to the increasing sophistication of the optical and photonic enabling technologies (see Sections 11.2 to 11.6). Hence the third stage of deployment for IP optical networks is expected to incorporate optical/photonic switching [Refs 4, 46].

The realization of IP running directly on a DWDM physical layer has become an important goal in optical networking. Implementation schemes for IP over DWDM and the further development of IP over OTN are illustrated in Figure 15.17. IP over a point-to-point WDM network employing optoelectrical conversion at the edges for the IP routing is shown in Figure 15.17(a) [Ref. 47]. In this case routers are directly interconnected with high-capacity point-to-point WDM links. Hence the entire traffic is processed in the router, at the packet level, after converting the optical signal to the electronic domain and extracting packets from the signal stream. Therefore most of the traffic processed by the router is transit traffic, as indicated in Figure 15.17(b). The IP over WDM architecture utilizes the high capacity of the router requiring high-throughput and thus power-consuming

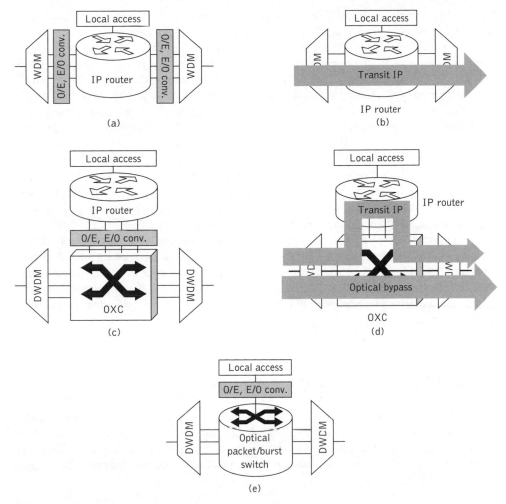

Figure 15.17 Internet protocol over physical layer evolution and traffic flow patterns: (a) IP over WDM; (b) IP over WDM traffic; (c) IP over OTN, (d) IP over OTN traffic; (e) IP over OPS/OBS

routers. Even though handling the transit traffic in a router at the packet level allows improved filling of the optical links due to statistical multiplexing, it increases the average delay and packet loss, which does not provide for an efficient network. In addition, it also introduces scalability problems since routers which are overloaded with transit traffic are not easy scaled in terms of increasing the number of high-speed ports.

Internet protocol over OTN where the traffic which passes through a node without termination can be either switched in an optical cross-connect (OXC) without optoelectrical conversion, or be forwarded by a router at the packet level after E/O conversion, is shown in Figure 15.17(c). Both IP over WDM and IP over OTN can be considered as nonreconfigurable since they rely heavily on the existing SDH/SONET, or ATM-based network hierarchies. Internet protocol over OTN does, however, allow the bypass of nontransit traffic by separately placing an IP router onto an OXC which then provides a wavelength switching OTN layer. This functionality is illustrated in Figure 15.17(d). In this case the traffic routing becomes independent of distance in the sense that traffic between any node pair can be routed in a single logical hop. Transit traffic can be therefore switched in the optical layer instead of being forwarded only in routers, as depicted by the optical bypass indicated in Figure 15.17(d). Such optical bypassing is especially important since it reduces the traffic load on individual routers, thus reducing packet delay and network congestion.

An advanced version of IP over optical switches where the packets are switched in the all-optical domain (i.e. without O/E conversion) using either optical packet switches or optical burst switches (OBS), has also been explored [Refs 4, 48, 49]. The E/O conversion, however, may still be required at the network edges for the control operation to enable access to the network. The successful implementation of such IP over optical switches depends on developments with photonic integrated circuits (see Section 11.6) which are not yet commercially available. Such networks in which wavelength paths (e.g. optical label switched paths (see Section 15.5.2)) can be dynamically routed in an integrated manner are potentially highly reconfigurable. Several aspects of this type of operation have been successfully demonstrated [Ref. 3] and it is therefore anticipated that the integration of IP and optical switching technology will evolve to provide the above performance in the near term [Ref. 50].

15.4 Wavelength routing networks

The optical layer is based on wavelength-dependent concepts when it lies directly above the physical layer. Hence the entire physical interconnected network provides wavelength signal service among the nodes using either single or multihop. This situation is illustrated in Figure 15.18. Three network nodes are interconnected using two wavelength channels

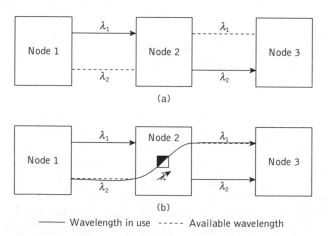

Figure 15.18 Wavelength-dependent interconnection: (a) fixed wavelength nodes; (b) wavelength convertible node (Node 2)

(i.e. λ_1 and λ_2) where the solid line connecting the nodes represents the available wavelength channel and the dashed line identifies that the wavelength channel is in use. If the network node 1 is required to connect with node 3 then as indicated in Figure 15.18(a) there is no single wavelength channel available to establish a lightpath between them. When a lightpath cannot be established on a link using a single wavelength channel it is referred to as a wavelength continuity constraint. A methodology to reduce this wavelength continuity constraint is to switch the wavelength channel at node 2 by converting the incoming wavelength λ_2 to λ_1 (which is available between nodes 2 and 3) to enable a link between node 2 and 3 to be established. This process is shown in Figure 15.18(b). Wavelength conversion (see Section 10.5) is required to convert from λ_2 to a compliant wavelength (i.e. λ_1) at the output port of network node 2 (which functions as an intermediate node) in order to provide a path. Hence the newly set up path uses two wavelength stages (i.e. two hops) to interconnect nodes 1 and 3. Such networks which employ wavelength conversion devices (or switches) are known as wavelength convertible networks.

Several network architectures can be employed to implement wavelength convertible networks. Three different WDM network architectures employing the wavelength conversion function are shown in Figure 15.19. Full wavelength conversion, where each network link utilizes a dedicated wavelength converter, is depicted in Figure 15.19(a). All the wavelength channels at the output port of the optical switch will be converted into their compliant wavelength channel by the appropriate wavelength converter (WC). For example, the topmost wavelength converter changes incoming λ_1 into λ_2 which is then connected to a multiplexer. There is no need, however, for wavelength conversion of the local add/drop channels.

It is not always required to provide the wavelength conversion function within every network node and it is more cost effective to implement networks with fewer and hence shared wavelength converters. So-called sparse wavelength convertible network architectures employing a number of wavelength converters as a wavelength converter bank (WCB) functioning on a shared basis per link and per node are shown in Figure 15.19(b) and (c), respectively. The arrangement of wavelength converters organized in a WCB is illustrated in the inset to Figure 15.19(b). This figure depicts a WCB servicing the optical fiber links where only the required wavelength channels are switched through the WCB (i.e. in Figure 15.19(b) the wavelength channel λ_2 is converted to wavelength channel λ_3). By contrast two optical switches are required to construct the shared per node wavelength convertible network architecture indicated in Figure 15.19(c). Optical switch 2 switches the converted wavelength channels to their designated nodes (i.e. in this case the wavelength channel λ_2 is converted to wavelength channel λ_3 via the shared WCB and is then switched through optical switch 2 to provide connection to the multiplexer).

A large number of wavelength channels on the network links, however, increase the complexity of switching nodes in ordinary OXCs which switch traffic only at the wavelength level. Moreover, the complexity worsens when multigranular OXCs (MG-OXCs) are used where the traffic is required to be accessed at multiple levels (i.e. at granularities such as the fiber, wavelength, digital cross-connects, etc.). In these cases the MG-OXC output traffic does not simply either terminate at or transparently pass through a node, but may also be required to transport from one layer to another via multiplexers/demultiplexers [Ref. 51]. This complexity can be reduced if more wavelength channels are grouped into one single waveband to be switched as a unique channel. Such waveband switching (WBS)

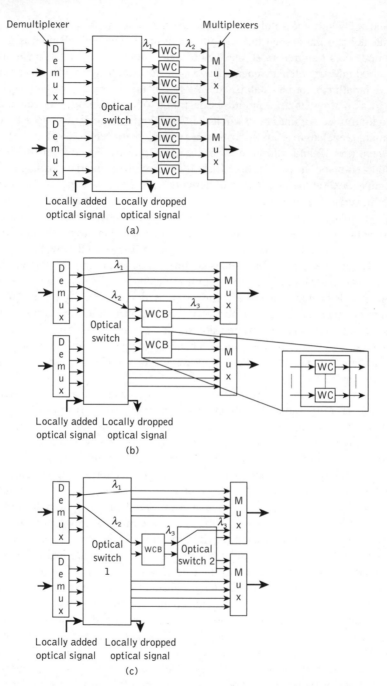

Figure 15.19 Wavelength convertible routing network architectures: (a) full or rededicated wavelength converters; (b) shared per link; (c) shared per node

networks have been proposed as a possible solution to ease the complexity of numerous wavelength-driven channels, especially in optical core networks [Refs 13, 52]. In comparison with a wavelength convertible network node a WBS node requires only a single input and single output port as the switching is independent of the number of wavelength channels incorporated inside the waveband.

An MG-OXC with three switching layers, including wavelength cross-connecting (WXC), waveband cross-connecting (BXC) and fiber cross-connecting (FXC) layers, is shown in Figure 15.20. Individual wavelength channels and wavebands are terminated (or switched) transparently through the WXC and BXC layers separately. The terminated waveband is demultiplexed into individual wavelength channels which are sent to the WXC layer as inputs. Furthermore, the output wavelength channels at the WXC layer can be multiplexed selectively into a waveband, which is then sent to the BXC layer as an input. The output wavelength channels from the WXC layer and the output wavebands from the BXC layer are grouped and transmitted along the output fiber link. A fiber may contain the same wavelength channels as individual channels and also be incorporated in wavebands, and therefore network control is required to determine which set of wavelength channels is grouped into a particular waveband [Ref. 53].

The advantage of WBS is that it supports a greater number of wavelength channels and it also reduces the number of switches or ports within the optical network. For example, considering Figure 15.20, if 10 individual fibers carry 1000 wavelength channels (i.e. 100 in each) which are further grouped in five bands then each group will comprise 20 wavelength channels. When a only single wavelength channel is intended to add or drop at any intermediate node then, instead of multiplexing/demultiplexing all 1000 wavelength

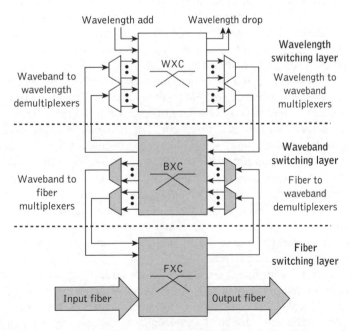

Figure 15.20 Waveband switching network architecture employing a multigranular optical cross-connect (MG-OXC)

channels, just a single band of 20 wavelength channels (from five groups) containing that destination wavelength channel will be multiplexed/demultiplexed without affecting traffic in the remaining nine fibers. In addition, four bands (out of five) in the selected fiber will also not be required to undergo multiplexing/demultiplexing to add/drop the single wavelength channel. Clearly, implementation of such a routing strategy requires the use of wavelength assignment and routing algorithms for these networks [Refs 54–57].

15.4.1 Routing and wavelength assignment

In dense WDM networks a lightpath is established by reserving a particular wavelength on the physical links between the source and destination edge nodes. It is a two-stage search-and-select process related to both routing (i.e. searching/selecting a suitable path) and wavelength assignment (i.e. searching/selecting or allocating an available wavelength for the connection). The overall process is often referred to as the routing and wavelength assignment (RWA) problem. Although there are various possible ways to determine the selection of path and the allocation of wavelengths, they fall into two basic categories of either sequential or combinational selections [Refs 58, 59]. Each of these categories fundamentally addresses the core issue of the wavelength continuity constraint (see Sections 10.5 and 15.4).

The implementation of RWA can be static or dynamic depending upon the traffic patterns in the network. Static RWA techniques are employed to provide a set of semipermanent connections, which remain active for a relatively longer time. The traffic patterns in this case are reasonably well known in advance and the variation in traffic pattern is not frequent. Therefore it is useful to optimize the way in which network resources (i.e. physical links and wavelengths) will be assigned to each connection. Moreover, the static RWA problem is often referred to as the virtual topology design problem (see Section 15.7).

Dynamic RWA deals with establishing the lightpath in frequently varying traffic patterns. In this case the traffic patterns are not known and therefore the connection requests are initiated in a random fashion, depending on the network state at the time of a request. The resources may (or may not) be sufficient to establish a lightpath between the corresponding source and destination edge node pair. The network state provides information about the physical path (i.e. route) and wavelength assignment for all active lightpaths. The state evolves randomly in time as new lightpaths are admitted and existing lightpaths are released. Thus, each time a request is made, an algorithm must be executed in real time to determine whether it is feasible to accommodate the request and, if so, to perform RWA. The algorithms are used to decide the shortest path available in a network to establish a connection between the source and destination nodes. In case of unavailability of the path on the same wavelength channel, then alternative routes including the possibility of wavelength conversion can be explored.

A five-node network with fixed connections where node 1 requested to establish a link with node 5 is illustrated in Figure 15.21. Although there is no direct physical connection or path available, there are four possibilities to establish the link between nodes 1 and 5, depending on the available or assigned wavelengths between each of the network nodes. These are: via node 2 using a single hop; nodes 4 and 2 comprising two hops; nodes 2 and 3 with two hops; and the longest possible route stretching over three hops via nodes 4, 2

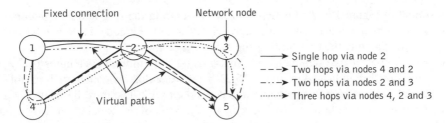

Figure 15.21 Wavelength routing and the selection of a path

and 3. Considering these four routes, the single hop remains the shortest path between nodes 1 and 5.

Depending upon whether the same wavelength is available between selected network nodes, the lightpath can be set up or, alternatively, wavelength conversion will need to be employed by changing it to a compliant wavelength. The second shortest path will be selected if wavelength conversion can be facilitated. In the case when there are no wavelength channels available, then no link can be set up between the desired nodes. Such unavailability of a network link between two nodes leads to blocking of the channel. Moreover, the probability of blocking increases if there are many network nodes with a small number of wavelength channels and fewer links between them [Ref. 60]. Thus an RWA algorithm plays an important role in determining the blocking probability of a network while also providing information about the availability of the path or link between the source and the destination. In addition, it also gives an indication regarding the location and the number of wavelength converters that may be required in an optical network [Refs 61, 62]. In order to develop a suitable RWA algorithm it is important to assign wavelengths within a network which offer the lowest or no blocking probability. Therefore wavelength assignment in RWA which is dependent on the network topology is also crucial.

An example of wavelength assignment for a ring network where the ring topology consists of four nodes, namely a, b, c and d, is shown in Figure 15.22. It may be observed that there exist four possibilities for establishing lightpaths between the network nodes as indicated by the two inner and the two outer half circles. Although four individual wavelengths (i.e. one wavelength per half circle) may be required, when wavelength reuse is introduced three wavelengths are sufficient to provide similar connectivity. This situation

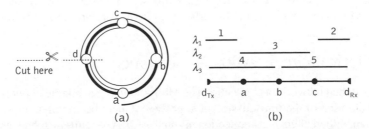

Figure 15.22 Wavelength assignment in a ring network topology: (a) lightpath arrangement for four network nodes; (b) wavelength assignments

is illustrated in Figure 15.22(b) which is drawn by cutting the ring and straightening the network wavelength connections. This approach splits network node d into two halves, d_{Tx} and d_{Rx} (i.e. separating the transmitting and receiving elements).

A total number of five lightpaths (i.e. links) using only three wavelengths (i.e. λ_1, λ_2 and λ_3) are therefore indicated in Figure 15.22(b). For example, wavelength λ_1 establishes two lightpaths, one each between nodes a to d_{Tx} and c to d_{Rx}. The third lightpath connecting nodes a to c via node b utilizes wavelength λ_2. Moreover, wavelength λ_3 is used to establish the fourth lightpath between nodes b to d_{Tx} via node a. Finally, the same wavelength λ_3 has been used to set up the fifth lightpath between network nodes b to d_{Rx} via node c. Such RWA also determines the length of a path in terms of the number of hops it has to travel to reach the destination node. Such patterns, which are referred to as graph coloring [Ref. 63], can therefore be used to locate the shortest path in relation to different network topologies (e.g. ring or mesh) [Refs 64–66].

15.5 Optical switching networks

An optical switch represents the single most dynamic element in an optical network which traditionally can switch data between different ports of a network. Broadly, as with electronic switching, optical switching can be classified into two categories which are circuit and packet switching. An optical switch performs various digital logic operations (see Section 11.4.1) allowing signals to switch from one state to another. Therefore larger arrays of optical switches can switch signals from one port to another. These arrays of switches form circuit switching fabrics known as optical cross-connects (OXCs) which incorporate switching connections or lightpaths. A router, on the other hand, routes data packets instead of providing circuit-switched connectivity. In this case the data is split into packets which are routed separately through the network and then are reassembled at the receiving terminal to reconstitute the original message. This process provides for more efficient use of the available bandwidth.

Both circuit and packet switching techniques are used in high-capacity networks. In circuit-switched networks the connection must be set up between the transmitter and receiver before initiating the transfer of data. When the network resources remain dedicated to the circuit connection during the entire transfer and the complete message follows the same path, the circuit is said to be static. SDH/SONET (see Section 15.3.1) is an example of a static circuit-switched network. However, if the network resources can be reallocated without being physically disturbed the circuit is referred to as being dynamic.

15.5.1 Optical circuit-switched networks

In circuit-switched networks a connection is established using available network resources for the full duration of the transmission of a message. Once the complete message is successfully transmitted then the connection is removed. A circuit-switched environment requires that an end-to-end circuit be set up before the actual transmission can take place. A fixed share of network resources is then reserved for the specific transmission which no

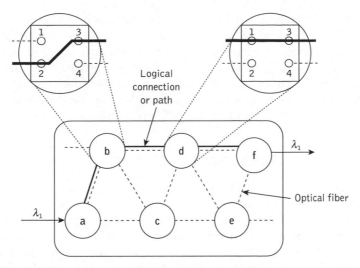

Figure 15.23 Optical circuit-switched network

other transmission can use. A request signal must, however, travel from the source to the destination and it should also be acknowledged before the transmission begins.

Figure 15.23 provides a block diagram illustrating an optical circuit-switched (OCS) network. In this configuration six optical nodes (i.e. a to f) are interconnected and a requested logical connection or path for optical signal wavelength, λ_1, is established producing a circuit path through network nodes a, b, d and f. Optical nodes of an OCS network contain optical switches where large multiport optical switches (i.e. a switching fabric) or an OXC (see Section 15.2.2) are used to establish connections between the desired input and output ports. Although different component enabling technologies (see Section 11.6) can be used to construct such optical switches, the basic optical switching function remains the same. Figure 15.23 illustrates this functionality at optical nodes b and d. For example, a 2×2 optical switch at node b enables cross-connection using ports 2 and 3, whereas node d employs a direct connection for ports 1 and 3. Hence a logical path or an optical circuit from node a to f is created for the lightpath using signal wavelength λ_1.

It should be noted that an optical switch can handle traffic from both SDH/SONET and IP networks (see Sections 15.3.1 and 15.3.5) with the limitation, however, that only in-service signals can use the available network resources. Furthermore, the wavelength continuity constraint (see Section 15.4) also limits the flexibility of OCS networks as lightpaths using the same wavelength can only carry the traffic from the source to the destination. Nevertheless this constraint can be overcome if the wavelength conversion function (see Section 10.5) is used at the output ports of the optical nodes to change the wavelength of the signal as required. In general OCS networks are suitable for implementation in public telecommunications networks (see Section 15.2.4) where a large volume of traffic is required to be switched in real time. OCS therefore has become an umbrella term to cover many network architectures based on two-way reservation [Ref. 67]. Various networks using the OCS concept have emerged to bring optical network services directly from the long-haul to business and residential customers [Refs 67–69].

A disadvantage of OCS is that it cannot efficiently handle bursty traffic (see Section 15.5.4). In such traffic conditions the data is sent in optical bursts of different lengths and therefore the resources cannot be readily assigned. For example, an active user (i.e. at peak times) needs large bandwidth when sending a burst as compared with relatively little bandwidth for a nonactive user (i.e. off-peak times). Bursty traffic is therefore character-ized by both peak and average bandwidth, and much of the bandwidth remains unused with the transmission of bursty traffic in as OCS network [Ref. 70].

15.5.2 Optical packet-switched networks

In an optical packet-switched (OPS) network data is transported entirely in the optical domain without intermediate optoelectrical conversions. An optical packet switch per-forms the four basic functions of routing, forwarding, switching and buffering. The rout-ing function provides network connectivity information often through preallocated routing tables, whereas forwarding defines the output for each incoming packet (i.e. based on a routing table). The switch directs each packet to the correct output (i.e. defined by the forwarding process) while buffering provides data storage for packets to resolve any con-tention problems which may occur during packet transmission.

Figure 15.24 shows the overall structure of a typical packet. It contains a header or label and the payload (i.e. data) and it requires a guard band to ensure the data is not overwritten. The label points to an entry in a lookup table that specifies to where the packet should be forwarded. Such a labeling technique is much faster than the traditional routing method where each packet is examined before a forwarding decision is made. At the receiving node these labels are required to be recognized from the lookup table and then the data is reassembled sequentially. Since labels define the routing criteria it is therefore sometimes referred to as label switching, which includes several functions involved in the labeling technique such as assignment (or writing) recognition (or reading), swapping (or exchanging) and forwarding etc. [Refs 71, 72].

The process of address lookup, however, can be replaced by address matching using optical correlators [Ref. 71], which are switching devices employing digital logic gates to perform optical pattern recognition similar to detecting an address from a lookup table. A set of optical correlators can therefore determine whether a packet is destined for a par-ticular switch by correlating the address encoded into the packet with the switch's own address. If there is a match, the payload is extracted and processed. Otherwise, the packet is forwarded to the next switch.

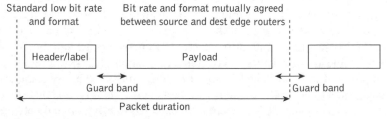

Figure 15.24 Optical packet-switched network packet format

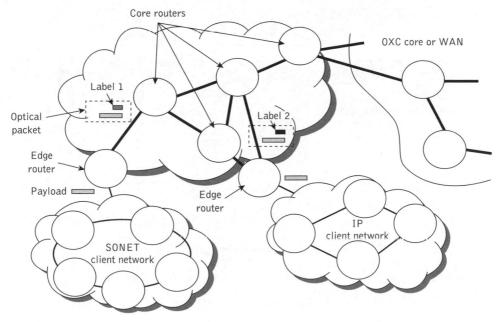

Figure 15.25 An optical packet-switched network. Reprinted with permission from Ref. 5 © IEEE 2006

The packets in an OPS network can carry different types of traffic (i.e. voice, data and video) and therefore such an optical packet switching scheme can integrate existing SDH/SONET and IP-based optical networks. An example of an OPS network is illustrated in Figure 15.25. It shows a long-haul or core network connected to both a SONET and an IP network. The edge routers (i.e. network nodes at the edges) perform the functions of attaching and detaching a label as identified for labels 1 and 2 attached to two different optical packets being transmitted to specific networks (i.e. to the IP and SONET client networks). Within the OPS network, however, the core router only processes the label where the content of the payload remains the same (i.e. the payload can carry ATM traffic or IP packets at different bit rates). This routing function typically involves the following four steps which are: (a) extraction of the label from the packet; (b) processing of the label to obtain routing information; (c) routing of the payload and contention resolution if necessary; and (d) rewriting of the label and recombining it with the payload. When an OPS network is implemented and only the label is processed electronically for routing purposes with the payload remaining in the optical domain, the OPS network is generally referred to as an optical label-switched (OLS) network.

Figure 15.26 depicts a generic OLS network configuration to route packets while also extracting and rewriting labels at input and output interfaces. In this particular implementation only the detached label is processed electronically and the payload (i.e. data) remains in optical buffers before being sent through the optical switch fabric to the desired output buffers. Finally, the routing processor and control section regenerate the label which is reattached with the data at the output interface.

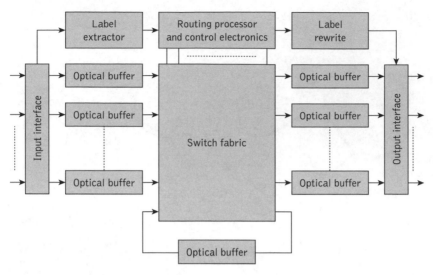

Figure 15.26 Generic optical label-switched network configuration. Reprinted with permission from Ref. 5 © IEEE 2006

15.5.3 Multiprotocol Label Switching

Multiprotocol Label Switching (MPLS) provides for the efficient designation, routing, forwarding and switching of traffic flows through an optical packet-switched network. It was originally proposed by Cisco Systems Inc. and was called tag switching when it was a Cisco proprietary proposal [Ref. 73], but it was later renamed label switching when the concept was adopted by the Internet Engineering Task Force for open standardization [Refs 74, 75]. Multiprotocol label switching uses labels to forward traffic across the various MPLS domains. When a labeled packet arrives at a label switching router (LSR), the incoming label will determine the path of this packet within the MPLS network. The MPLS label forwarding will then swap this label to the appropriate outgoing label and send the packet on the next network hop. These labels are assigned to packets based on grouping or forwarding equivalence classes. Packets belonging to the same forwarding equivalence class receive the same treatment. Hence the MPLS lookup and forwarding system allows explicit control of routing based on destination and source addresses. Since a packet's label class is assigned only once at the MPLS edge, and packets are switched based on the label information only, then MPLS is a fast and efficient protocol.

In practice, an LSR receives a packet from one interface, replaces the incoming label in the MPLS header by an outgoing label, and forwards the packet out to another interface. Thus MPLS defines a virtual circuit capability depending on the type of network and, in particular, the type of switching, by correlating the switching type with the label. Table 15.3 identifies the different traffic types and their related switching and implementations for MPLS. For example, in the time division multiplexed optical networks (see Section 15.3.1) a node switches data from time slots in an incoming interface to time slots in an outgoing interface. Therefore the label is essentially a time slot. For WDM networks (see Section 15.2.3), the nodes switch wavelengths and the label should identify the

Table 15.3 Summary of MPLS functions and implementations

Switching domain	Traffic type	Forwarding or implementation scheme
Time	TDM/SONET	Time slot in repeating cycle: implementation using digital cross-connect system (DXC) and ADM
Packet or cell	IP	IP router, ATM switch: implementation of label as a header
Wavelength	Transparent	Optical domain, DWDM switch: implementation using wavelength (i.e. lambdas)

wavelength to be switched from an incoming interface to an outgoing interface. Similarly, for IP or ATM there are labels or headers. These examples explain the implicit nature of the label for a specific transmission or switching technique. The labels in MPLS are, however, explicit and the same control plane mechanisms can be used for establishing label mappings in switches and also for setting up cross-connect tables.

In order to remain flexible in the current protocol landscape, MPLS is independent of both layer 2 and layer 3 in the OSI model (see Section 15.3.3), and it can support ATM, frame relay and Ethernet as a data link layer. One popular service provided by MPLS-based networks is IP virtual private networks (VPNs) deployed over SDH/SONET [Refs 76, 77]. These IP-VPNs allow business corporations to create a secure, dedicated wide area network (WAN) in order to connect their offices around the world, without the expense of private circuits or private data networks. Since MPLS is IP based it is an essential ingredient for future optical networking.

Data forwarding is not limited to the function of merely packet forwarding, and therefore there should be a general solution enabling the technique to retain the simplicity of forwarding using a label for a variety of devices that switch in time, wavelength or space (i.e. physical fiber ports). Furthermore, it is not possible for different networks to look into the contents of the received data and thereby extract a label. For instance, optical packet networks are able to parse the headers of the packets, check the label, and carry out decisions for the output interface (i.e. forwarding path) that they have to use. This is not the case for optical TDM or dense WDM networks. The equipment in these network types is not designed to have the ability to examine the content of the data that is fed into it. These aspects therefore necessitate generalizing the structure and switching of labels.

Generalized MPLS (GMPLS) therefore extends MPLS providing additional protocols for wavelength and physical space domains (i.e. optical fiber switching) [Refs 75, 78, 79]. It supports transparent traffic using DWDM and OXCs for the implementation of label switching. The former utilizes wavelengths as labels where the OXCs use fibers as labels. For devices that switch in any domain (i.e. packet, time, wavelength and fiber) GMPLS provides the control and management planes (i.e. routing, signaling and linking) [Ref. 80]. Generalized MPLS also incorporates the routing protocols for the automatic discovery of the network topology (i.e. backup route discovery) and announces resource availability to optical nodes (e.g. bandwidth or protection type (see Section 15.8)). Both these factors are significant when an optical link failure occurs.

The signaling protocols for the establishment of label-switched paths (LSPs) for the GMPLS therefore include the enhancements to MPLS as follows:

(a) label exchange to include nonpacket networks (i.e. generalized labels);

(b) establishment of bidirectional LSPs;

(c) signaling for the establishment of a backup path (protection information);

(d) expediting label assignment via a suggested label;*

(e) waveband switching support – a set of contiguous wavelengths switched together.

In addition, GMPLS incorporates a link management protocol which performs the following four functions:

• control of channel management which is established by negotiating the link parameters (e.g. frequency in sending keep-alive messages) and ensuring the health of a link (e.g. hello protocol);

• link connectivity verification which ensures the physical connectivity of the link between the neighboring nodes;

• link property correlation which identifies the link properties of the adjacent nodes (e.g. protection mechanism);

• fault isolation to enable a single or multiple faults to be isolated in the optical domain.

Recently, there have been several demonstrations of successful GMPLS-based optical network implementations [Refs 81, 82]. The issue of generalization for multiprotocol switching is, however, dependent on transmission and device enabling technologies and therefore, in particular, speed and efficiency performance improvements are anticipated in the near future [Refs 83, 84].

15.5.4 Optical burst switching networks

Combining important aspects of optical circuit switching (see Section 15.5.1) and optical packet switching (see Section 15.5.2) results in optical burst switching (OBS). Moreover, as OBS operates at the subwavelength level it therefore provides for rapid setup and teardown of optical network lightpaths. This hybrid switching and routing technology uses electronics to control routing decisions but keeps data in the optical domain as it passes through each optical node. Packets with a common destination are aggregated in edge routing nodes into larger transmission units called a burst or a data burst (DB), each of which is transmitted separately from the data control packet called the burst header cell

---

* An upstream optical node can suggest a label to its downstream node; otherwise the OXC is required to receive a label from the downstream node in order to configure its network switches. Thus suggested labels enhance the speed of the configuration process.

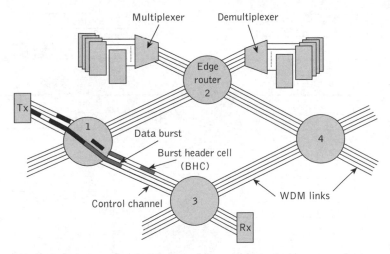

Figure 15.27 Concept of optical burst switching

(BHC)* containing necessary information (i.e. for switching and destination address). Figure 15.27 illustrates the concept of OBS where four edge routers of a large network are shown to establish links between data sources (Tx) and receivers (Rx) individually or by using multiplexers or demultiplexers, respectively.

Optical bursts containing both the data burst and the BHC travel on a control channel. An idle channel on the access link is selected when a data burst is required to be sent, whereas the BHC travels on the control channel ahead of its associated data burst in time and is processed electronically at every node along the path. The OBS edge router, on receiving the BHC, assigns the incoming burst to an available channel on the outgoing link leading towards the desired destination and establishes a path between the specified channel on the access link and the channel selected to carry the burst. It also forwards the BHC on the control channel of the selected link, after modifying the cell to specify the channel on which the burst is being forwarded. This process is repeated at every routing node along the path to the destination. The BHC also includes an offset field which contains the time between the transmission of the first bit of the BHC and the first bit of the burst, and a length field specifying the time duration of the burst. One or several channels on each link can be reserved for control information that is used to control the dynamic assignment of the remaining channels to user data bursts. It should be noted that the WDM transmission links shown in Figure 15.27 carry a number of wavelength channels and the user data bursts can be dynamically assigned to any of these channels by the OBS routers.

Figure 15.28 depicts the edge router's function in more detail providing burst assembly and disassembly operations at ingress and egress, respectively. In Figure 15.28(a) each of the users operating on different formats (i.e. IP, SONET, ATM, WDM/DWDM, etc.) sends different data to the edge router. The router disassembles the data and issues BHCs (i.e. C1 and C2 shown in Figure 15.28(a)) on the data control channel (DCC) in advance for Burst 1 and Burst 2, respectively. Each burst may contain different data: for example,

* Sometimes also referred to as burst header packet (BHP).

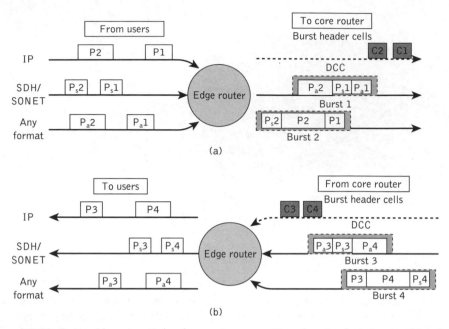

Figure 15.28 Optical burst switch edge router operation showing burst assembly at ingress and disassembly at egress: (a) from users to core router; (b) from core router to users

Burst 1 contains data P_a2, P_s1 and P_a2, where the subscripts a and s represent any format and SDH/SONET, respectively. In order to perform burst assembly as shown in Figure 15.28(b) the control channel provides BHCs (i.e. C3 and C4) for the data bursts Burst 3 and Burst 4, respectively. The users then receive their corresponding disassembled data that the edge router received from the core router.

It should be noted from Figure 15.28 that the disassembly at egress edges (i.e. sending to users) is simpler than the assembly (or transmission) of the burst (i.e. burstification) at ingress edges. In the latter case, since different signal formats may be used then various signaling and control protocols can be employed to enable optical burstification and OBS routing. Furthermore, the reservation and scheduling of resources for the burst switching in this process are important factors where the former considers end-to-end burst transmission and the latter focuses on assigning and managing resources for individual bursts within OBS nodes [Ref. 4].

Depending on the network dimension and granularity (i.e. burst size), either one- or two-way signaling protocols are used at the edge routers of an OBS network. With two-way signaling, sending back an acknowledgment signal to confirm the safe arrival of the signal is required, whereas no such feedback mechanism is available when using one-way signaling protocols. One-way protocols include tell-and-go (TAG) [Ref. 85] and just-enough-time (JET) [Ref. 86]. These are also referred to as one-pass reservation protocols [Refs 4, 87]. Examples of two-way signaling protocols are tell-and-wait (TAW) [Ref. 88] and just-in-time (JIT) [Ref. 89], which are predominantly used for the purpose of burst reservation and scheduling [Ref. 49]. In case of the JET or TAG protocols, the burst transmission does not wait for the acknowledgment of successful end-to-end path setup and the burst transmission is initiated immediately, or shortly after the burst has been assembled following the control packet being sent out. If the burst transmission with the TAG protocol is delayed with respect to the control header, then the delay is referred to as offset time and can be reduced by compensating the processing times. In this scheme less bandwidth is therefore wasted but the burst drop rate increases and it is not considered to be as reliable. Due to the submillisecond burst duration assumed in TAG burst management, this scheme is usually considered for application in metropolitan and access networks (see Sections 15.6.2 and 15.6.3) where distances are comparatively short [Ref. 4].

The JIT and TAW protocols utilize ATM delayed transmission and they wait for an acknowledgment before sending a burst. This process assumes conventional end-to-end transmission (i.e. virtual path) leading to a setup delay for the optical bursts (i.e. in the millisecond range for long-haul networks). If the intermediate switches are set in advance during the setup phase to avoid this delay, then the bandwidth wasted can be much higher than the bandwidth actually needed for burst transmission.

In addition to burst reservation, burst scheduling assigns and manages the resources for individual burst switching nodes. Burst scheduling schemes can be classified based on the duration for which resources are scheduled for a burst. The reserve-a-limited duration (RLD) and reserve-a-fixed duration (RFD) schemes are commonly adopted for burst reservation. The RLD requires the sender to signal the start and end of a burst and resources are explicitly reserved until the end of burst transmission. For each resource, the idle time (i.e. the duration when the resource is free or available) is recorded. The RFD scheme, however, considers the exact start and end time of bursts for resource scheduling. For example, the gaps (or voids) between already reserved bursts can be used for newly arriving bursts. Moreover, several designs have been used to optimize resource allocation of both the RLD and RFD schemes by improving wavelength selection or by minimizing voids [Refs 85, 86, 90].

Advanced techniques referred to as adaptive and autonomic OBS have also been proposed which can learn and adapt new routes after acquiring network information such as wavelength routing, wavelength selection, protection and the information related to the restoration mechanisms (see Section 15.8). Such OBS techniques use a feedback mechanism to optimize the selection of control and routing information and therefore they are capable of being both self-protecting and self-optimizing [Refs 91, 92].

15.6 Optical network deployment

Optical networks are dependent on the use of advanced fiber transmission systems, developments in optical component enabling technologies and evolution in the network topological strategies together with the associated protocols to provide efficient, high-capacity operation. Significant progress in all these aspects has been made in recent years and therefore current optical networks operate at higher speed with increased functionality and reliability. The major deployed optical fiber networks are in the area of the public telecommunications network (see Section 15.2.4) together with on-site, or on-premises, local area networks (see Section 15.6.5). An example of a modern, complex optical network is illustrated in Figure 15.29(a) which depicts a DWDM backbone incorporating add/drop

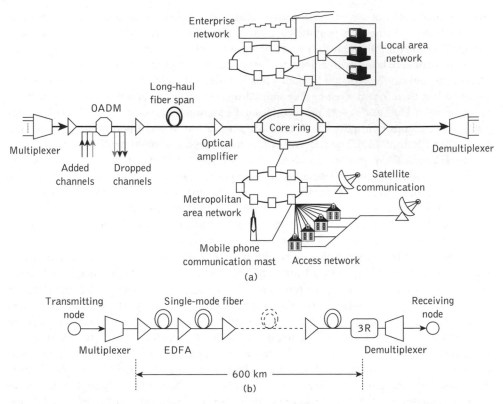

Figure 15.29 Optical network: (a) a modern, complex dense wavelength division multiplexed (DWDM) backbone network with add/drop channels, also interconnecting metropolitan, access, local area and enterprise networks; (b) structure of a point-to-point DWDM link for the long-haul network

channels together with a core ring feeding both metropolitan area and access networks together with enterprise and local area networks. Both the present and future deployment of optical networks, particularly in the constituent areas of the public telecommunications network and also in the crossover region to local area networks, are therefore further described in this section.

15.6.1 Long-haul networks

A long-haul network as the name implies is a network connecting several regional or national networks together. These networks are also referred to as core or backbone networks and they also interconnect other long-haul networks to extend global interconnectivity between national domains. A current long-haul optical network typically comprises point-to-point DWDM links with optical regenerators at end points and with erbium-doped fiber amplifiers (EDFAs) placed between the end terminals as shown in Figure 15.29(b). Moreover, an optical 3R regenerator (see Section 10.6) is often used at typically 600 km intervals to reduce overall signal degradation on the link. In order to achieve improved connectivity these point-to-point DWDM links can be interconnected in a mesh topology (see Section 15.2.1.1).

Using the above network structures it is possible to interconnect optical networking nodes situated in different cities around the globe [Ref. 93]. For example, a European network organization euNetworks [Ref. 94] carries a high capacity optical fiber long-haul network. It extends over a 5400 km route and interconnects 18 cities in five countries: namely Germany, France, Belgium, the Netherlands and the UK. Two submerged cables link Europe to the UK while long-haul optical fiber links interconnect the metropolitan area networks (see Section 15.6.2) of the 18 cities. Similarly, 30 major cities in the United States are also interconnected by a long-haul optical fiber network as illustrated in Figure 15.30.

Existing long-haul optical networks are in a constant state of evolution that is mainly driven by the changing traffic patterns, the improvements in optical device enabling technologies and the economic pressure from service providers. Long-haul optical fiber networks are now classified in relation to their maximum achievable distance without optical signal regeneration by designations of long-haul, extended long-haul (ELH) and ultra long-haul (ULH). The ranges of the transmission distances for these designations are [Refs 95, 96]:

- long-haul optical fiber networks from 600 to 1000 km;

- extended long-haul (ELH) from 1000 to 2000 km;

- ultra long-haul (ULH) from 2000 to 4000 km.

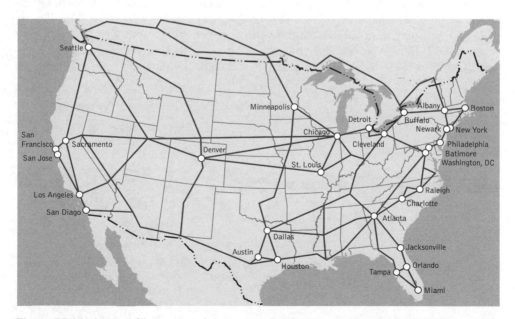

Figure 15.30 Optical fiber network interconnectivity among the major US cities [Ref. 93]

Improvements in transmission systems incorporating DWDM (see Section 12.9.4), advanced signal modulation formats [Refs 97, 98] and extended forward error control [Refs 99, 100] play a significant role in making the above longer haul distances feasible. Moreover, the majority of existing long-haul DWDM networks operate at channel rates of either 2.5 Gbit s^{-1} over 64 or more channels and 10 Gbit s^{-1} with up to 40 wavelength channels, but it is apparent that transmission rates from 40 to 160 Gbit s^{-1} per channel with more than 100 wavelength channels will be deployed in the near term [Refs 101–103].

Submerged or transoceanic optical fiber networks typically cover very long distances between the continents (i.e. in the 3000 to 10 000 km range) where most of the optical fiber cable lies in deep sea water. In order to construct such long-length ultra long-haul networks both the fiber attenuation and dispersion must be reduced so that they have minimum impact. Moreover, submerged networks primarily use repeaters that only reamplify the signal (i.e. 1R amplifiers are commonly referred to as repeaters in submerged deployments). Optical amplifiers also require electrical power which is fed to them through insulated copper cables that run down the length of the fiber cable. The power is transmitted from the land-based terminal landing site and is delivered in parallel down the length of the cable to the sealed underwater amplifiers. Submerged systems therefore demand the highest fiber transmission performance under the most stringent optical power budgets and the harshest environmental conditions.

The various elements that comprise a submerged cable system are shown in Figure 15.31 [Ref. 104]. This system basically consists of two sections with dry and wet plants residing on dry land (i.e. landing section) and in sea water (i.e. submerged section). The landing station houses the terminal equipment that interconnects the optical signal from the submerged cable and passes it on to a terrestrial system. The underwater cable includes repeaters, gain equalizers and branching elements to facilitate access to other landing stations. Some of the major deployed submerged optical cable networks are identified in Table 15.4. It may be noted that the transatlantic optical fiber cables (TAT) which have reached TAT-14 now provide interconnections between the United States and five European countries [Ref. 105]. TAT-14 employs a DWDM bidirectional ring configuration

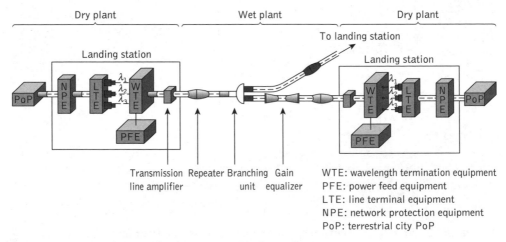

Figure 15.31 A submerged cable system [Ref. 104]

Table 15.4 Examples of deployed submerged optical cable networks [Ref. 108]
www.atlantic-cable.com

Name	Date(s) in service	Transmission capacity	Location/ownership
TAT-8	1988–2002	~20 Mbit s^{-1}	USA–France
TAT-9	1992–2004	~40 Mbit s^{-1}	USA–Spain
TAT-10	1992–2003	2 × 565 Mbit s^{-1}	USA–Germany
TAT-11	1993–2003	2 × 565 Mbit s^{-1}	USA–France
TAT-12/13	1996–	12 × 2.5 Gbit s^{-1}	USA–UK, France
TAT-14	2000–	2 × 16 × 10 Gbit s^{-1} (320 Gbit s^{-1})	USA–UK, France, the Netherlands, Germany, Denmark
CANTAT-3	1994–	2 × 2.5 Gbit s^{-1}	Nova Scotia–Europe
Apollo	2002–	10 Gbit s^{-1}	USA–UK, France (Cable & Wireless)
AC-1	1998–	10 Gbit s^{-1}	USA–UK, Germany and the Netherlands (Global Crossing Ltd)
AC-2/Yellow	2000–	10 to 40 Gbit s^{-1}	USA–UK (Level 3 Communication Ltd)
FLAG Atlantic	2000–	10 Gbit s^{-1}	Across the globe (FLAG Telecom Ltd, Reliance Communications Ltd)
SAT-3/WASC/ SAFE	2001–	10 to 40 Gbit s^{-1}	South Atlantic 3 (West Africa Submerged Cable Ltd)
ATLANTIS-2	1999–	10 to 40 Gbit s^{-1}	Argentina, Brazil, Senegal, Cape Verde, Canary Islands and Portugal
COLUMBUS-III	1999–	20 Gbit s^{-1}	USA, Portugal, Spain and Italy

comprising four optical fibers (a fiber pair for operation and another for protection (see Section 15.8)). A single fiber in TAT-14 can carry 16 wavelength channels where each channel can potentially allow a transmission rate of STM-64 (i.e. ~10 Gbit s^{-1}) and therefore it possesses an operational capacity of 320 Gbit s^{-1} (i.e. 2 × 16 × 10 Gbit s^{-1}). Finally, a comparative chart showing different worldwide submerged long-haul networks and their corresponding operational transmission capacities is provided in Figure 15.32 (p. 1012). Although submerged system requirements increase every year, the transatlantic capacity remains the largest element of the submerged cable network with an operational capacity in 2006 approaching 3 Tbit s^{-1} [Refs 106, 107].

15.6.2 Metropolitan area networks

By definition metropolitan area networks (MANs) or metro networks provide the regional interface interconnecting the access network end users (i.e. business or residential customers) with the long-haul networks. Metro networks are subject to specific requirements and traffic demands that are quite different from long-haul or access networks. For example, they are characterized by more rapidly changing traffic patterns requiring the networks to be fast, scalable, modular and also reconfigurable. In addition, the MAN must be cost effective in terms of both operation and maintenance. To achieve these features, optical MANs are usually structured as ring topologies [Ref. 109].

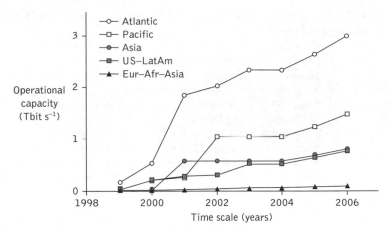

Figure 15.32 Operational capacities of submerged cable networks across the different regions of the world [Ref. 106]

Two metro ring networks interconnected by a digital cross-connect (DXC) which acts as a hub node between the two networks are depicted in Figure 15.33. Ring 1 constitutes a metro network providing the means of interconnection to the long-haul and other metro rings (i.e. ring 2) while ring 2 interconnects metro networking nodes with several access or enterprise networks. For this reason MANs tend to be divided into two segments: namely, the central ring with the neighboring ring interconnecting the access networks with the central ring. Since the central ring connects to the long-haul network it is sometimes referred to as a core ring (i.e. metro core) while the neighboring rings are called access rings (i.e. metro access) or collector rings since they collect traffic and forward it to the DXC which interconnects the two rings. Such cross-connecting nodes, providing interconnection between metro and long-haul networks, are also sometimes referred to as points-of-presence (POP) [Ref. 110].

An example of a MAN indicating both the distances and the number of channels provided for a typical large city is shown in Figure 15.34. The network includes a central metro-core ring and five metro-access rings which are connected to an additional twelve dedicated links from the premises [Ref. 109]. Furthermore, the core ring is interconnected to the access/collector rings though a DXC while the long-haul interconnection is provided by either a DXC or OXC depending upon the traffic type being carried through the cross-connect.

Since optical metro networks always provide the interconnectivity between long-haul and access networks, their infrastructure therefore employs a number of technologies

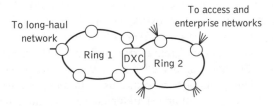

Figure 15.33 Structure for a metropolitan area network

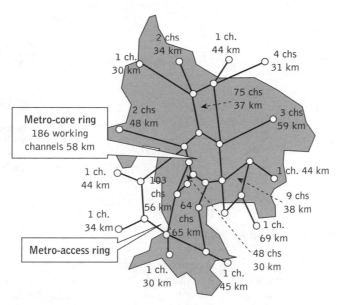

Figure 15.34 Metropolitan area network for a typical large city [Ref. 109]

which are layered upon each other. These technologies include SDH/SONET, DWDM, MPLS and Ethernet. The most common and currently lowest cost metro network solution is the SDH/SONET-based ring network [Refs 111, 112]. Since reconfigurable optical add/drop multiplexers (ROADMs) (see Section 15.2.2), however, enable dynamic and flexible node-to-node connection via lightpaths, these devices are now also considered as key components within metro ring network nodes. For example, a successful field trial of the flexible ROADM network with a wavelength- and packet-selective switch has recently been reported [Ref. 113]. Bit-error-rates of less than 10^{-12} for 16 wavelength channels were obtained over a transmission distance of 90 km at 10 Gbit s^{-1} for this network. Moreover, the combination of both metro and access networks (i.e. into a single network solution) using long-reach passive optical networks (see Section 15.6.3) is being explored [Refs 114, 115], which may well be an aspect of future network deployment.

15.6.3 Access networks

The access network is an element of a public telecommunications network that connects access nodes to either individual users (i.e. business, residential) or MANs. Therefore it can be considered as the last link in a network between the customer's premises* and the first point of connection to the network infrastructure (i.e. local exchange/switching center or local office). The transmission media choices providing strategies for the development of an access network are shown in Figure 15.35. Using these different media options the access network can be provided from a single point that can be located at the local exchange or a regional hub (metro) or at points-of-presence (POP) for a long-haul network

---

* Sometimes it is also referred to as the subscriber loop or last mile.

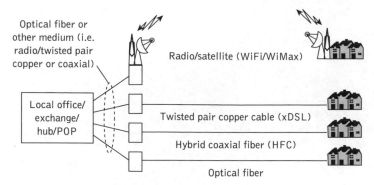

Figure 15.35 Access network transmission media deployment strategies

Table 15.5 Different types of access network

Type	Transmission medium	Transmission rate downstream/ upstream	Standards
Cable modems	Hybrid fiber coaxial (HFC)	160 /120 (Mbit s⁻¹) (with 4 Channel bonding†)	ANSI/SCTE-40, DOCSIS 3.0, ITU-T J.222 [Refs 116–120]
ADSL (asymmetric digital subscriber line)	Twisted pair copper (5.5 km)	7/0.8 (Mbit s⁻¹)	ITU-T G.992.1
ADSL2 : ADSL2+	(5.5 km)	8/1 : 24/1 (Mbit s⁻¹)	ITU-T G.992.3, ITU-T G.992.5
VDSL (very high-speed DSL)	(1.5 km)	55/15 (Mbit s⁻¹)	ITU-T G.993.1
VDSL2 : VDSL2+	(0.3 to 5 km)	55/30 :100/100 (Mbit s⁻¹)	ITU-T G.993.2 [Refs 117, 121–125]
Wifi WiMax	Wireless (radio/satellite)	24/1 (Mbit s⁻¹) 70/25 (Mbit s⁻¹)	IEEE 802.11b, 802.16e [Refs 24, 117, 126]
Fiber access network	Optical fiber	2.5/1.25 (Gbit s⁻¹) and 2.5/2.5 (Gbit s⁻¹)	ITU-T G.983.1, G.984.1, FSAN [Refs 117, 127–129]

† Channel bonding is a technique that combines two or more network resources (or interfaces) for redundancy or higher throughput. Bonding of multiple transmission lines (i.e. with similar data rates) into a single channel increases the overall bandwidth by the number of channels bonded.

connection to service end users. The different types of access network characterized by transmission media are also provided in Table 15.5.

Access networks based on hybrid fiber coaxial (HFC) are a combination of two technologies employing both optical fiber and coaxial cable as the media. Originally, HFC was a cable TV (CATV) concept to provide TV broadcasting and reception in rural areas [Refs 116–120]. A generic form of HFC network providing cable TV services to business or 500 to 2000 residential users is depicted in Figure 15.36. The optical nodes employ optoelectronic transceivers (see Sections 11.5 and 11.6) to connect the optical fiber and

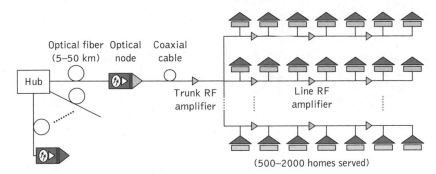

Figure 15.36 A generic hybrid fiber coaxial (HFC) cable TV network

coaxial cable infrastructure. In these networks RF amplifiers are used as trunk and line amplifiers to amplify signal power on the coaxial cables to enable a large number of customers' premises to be reached. In addition to coaxial cable, other media (i.e. twisted pair copper cable and radio) have also been used to carry voice, data and video signals and therefore the term HFX* is sometimes used to signify the combination of fiber with other media instead of just the coaxial cable implied by HFC [Refs 116–120].

Digital subscriber line (DSL) provides broadband access over, usually existing, twisted pair copper telephone lines. It utilises a discrete multitone modulation format which employs a large number of 4 kHz-spaced orthogonal subcarriers in the twisted copper pair bandwidth. The different variants of DSL enable transmission rates up to 100 Mbit s^{-1} (i.e. VDSL2+) but at this high-end speed is limited to a maximum transmission distance of only 300 m. By contrast, wireless access networks offer the advantage of mobility using wireless modems with IEEE standards for both WiFi and WiMax [Ref. 126]. Although this approach offers per-user bandwidth from 1.5 to 75 Mbit s^{-1} reaching maximum distances of up to 50 km [Ref. 117], there are future limitations on the bandwidth that can be made available. Access networks employing an optical fiber from the local exchange to users, however, can potentially support a growing and large per-user bandwidth from 10 Mbit s^{-1} to 2.5 Gbit s^{-1} [Refs 127–129].

An optical fiber access network can be divided into feeder and distribution sections as shown in Figure 15.37(a). The section of the network between the local exchange and the remote node (RN) is the feeder whereas the other section between the RN and network interface unit (NIU) is the distribution network. The NIU enables communication between devices that use different protocols by providing a common transmission protocol. To facilitate this situation an NIU may, however, require conversion of a specific device protocol to the common one and vice versa. A single RN can serve several NIUs located either in a building or outside at street/curb level serving a number of customers.

Optical fiber deployment strategies for the access network are illustrated in Figure 15.37(b). Three different approaches are shown depending upon the distance between end users and the local exchange/office. An optical network unit (ONU) is used to terminate the fiber connections. The location of ONUs and their distances from a local exchange and an NIU determine the specific configuration for the fiber access network. For example, it is known as fiber-to-the-cabinet (FTTCab) or fiber-to-the-curb (FTTC) when ONUs are located in

---

* Where X represents coaxial of twisted pair copper cable, or radio.

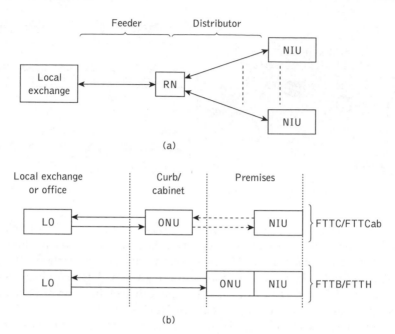

(a)

(b)

Figure 15.37 Optical fiber access network: (a) feeder and distribution sections; (b) different deployment strategies

a cabinet situated in main streets or near to the curb. When the ONU is placed within a major building the fiber access network is referred to as fiber-to-the-building (FTTB). FTTC and FTTB tend to be employed for both business and domestic end users. Moreover, when the ONU is located within a single house or home (i.e. very near to the NIU) such a fiber access network is referred to as fiber-to-the-home (FTTH) and data is transmitted over the fiber from the local exchange (or hub) directly into the end user premises.

An optical fiber access network primarily employing passive optical components and configured around a passive splitter/combiner is called a passive optical network (PON). It was initially referred to as passive since no components required electrical power, except at the fiber end points which may incorporate optoelectronic devices (i.e. in the ONU or NIU). A basic PON structure comprising essentially three elements connecting a service provider to an output device located near the customers' premises is shown in Figure 15.38.

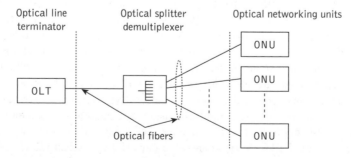

Figure 15.38 General structure of a passive optical network (PON)

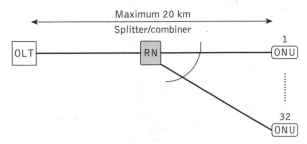

Figure 15.39 Asynchronous transfer mode passive optical network (APON)

It incorporates a passive optical splitter (see Section 5.6) between an optical line termination (OLT) at the provider end and an ONU situated at the customer end.

In a PON serving a significant number of customers, the output device is a $1 \times n$ optical power splitter.* Hence the cost of maintaining the feeder section of the access network is shared among many customers as both the capital expenditure (i.e. CAPEX) and operational costs (i.e. OPEX) are major issues when designing an optical access network.

Current PON-based access networks can be classified into three categories which are:

- ATM PON (APON) or broadband PON (BPON);

- Ethernet PON (EPON) and Gigabit Ethernet PON (GE-PON) (see Section 15.7);

- Gigabit PON (GPON).

Figure 15.39 shows an APON with a 1×32 optical splitter at the RN location covering an overall maximum distance of 20 km [Refs 1, 127]. APON, which utilizes asynchronous transfer mode (see Section 15.3.2) over PON strategy, has evolved into a network that facilitates voice/data communication which combines the capacity for supporting multiple services at ATM bit rates with the transparent transmission feature of the PON, and therefore it is also referred to as broadband PON (BPON) [Refs 127, 129].

Passive optical networks utilize separate wavelength channels in the regions of 1.5 and 1.3 μm to carry transmission from OLT to ONUs (i.e. downstream) and from ONU to OLT (upstream), respectively. Hence for an APON the ATM transmission is broadcast downstream from OLT to the ONUs while the transmission from each ONU is granted upstream bandwidth using a time-shared (i.e. TDM) approach.

Although the data rates used in APON (i.e. 155 Mbit s^{-1} upstream and 622 Mbit s^{-1} downstream) are equivalent to those used in SDH/SONET (i.e. OC-3 and OC-12), APON cannot readily carry the IP or Ethernet data packets (see Sections 15.3.5 and 5.7) which are required to be broken up and reassembled at the user terminal. This can be accomplished, however, using Ethernet PON (EPON) technology. It is quite straightforward to carry IP packets via Ethernet frames and hence data transmission which may occur in variable length packets as illustrated for the EPON in Figure 15.40. Each packet contains information on the destination and therefore all the packets arrive in a sequential order

---

* More recently, in order to provide for WDM operation the splitter has been replaced by a $1 \times n$ wavelength multiplexer/demultiplexer.

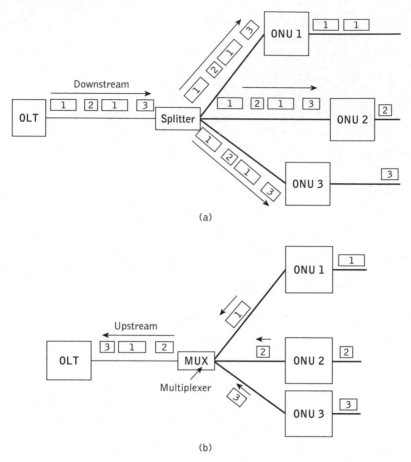

Figure 15.40 Ethernet PON (EPON): (a) downstream broadcast; (b) upstream multiplexing

enabling a complete message to be constructed. Figure 15.40(a) depicts the downstream transmission for EPON where the data is broadcast from the OLT to multiple ONUs in variable length packets to the three ONUs shown. At the splitter, the traffic is coupled to three separate fibers, each carrying all the same traffic. When the data reaches the ONU it accepts the packets that are intended for it and discards the remaining packets. For example, ONU 1 receives two packets identified as packet 1, and the packets 2 and 3 are discarded. Similarly, ONU 2 and 3 receive their intended packets 2 and 3, respectively.

The upstream traffic in EPON is managed by employing a TDM approach as illustrated in Figure 15.40(b). The specific transmission time slots are dedicated to each ONU in order not to interfere with each other once the data is coupled onto the common fiber. For example, ONU 3 transmits packet 3 in the first time slot, ONU 1 transmits packet 1 in a second nonoverlapping time slot, and finally ONU 2 transmits packet 2 in a third nonoverlapping time slot. The three packets each with variable packet length are TDM interleaved by the multiplexer unit and the resultant multiplexed signal is transmitted to the OLT.

Ethernet PON can be implemented using either two- or three-wavelength signals, one each for downstream and upstream transmissions, where the third wavelength region provides for the RF (i.e. for CATV) or for dense WDM transmission. For example, when the signal wavelengths of 1.49 µm and 1.31 µm are used for downstream and upstream transmission respectively, then the signal wavelength at 1.55 µm is allocated for the CATV overlay. Alternatively, the 1.55 µm window (i.e. 1.53 to 1.56 µm) can enable a DWDM wavelength band overlay to be incorporated on the PON if additional DWDM component technology is used [Refs 130, 131]. Such DWDM upgrades are designed to provide for future increased capacity demands.

Gigabit PON (GPON) is a high-capacity PON development which is designed to facilitate IP packet transmission rather than just ATM or Ethernet packets. A typical structure for a GPON is shown in Figure 15.41. As indicated in the ITU-T Recommendation and FSAN* G. 984.1 [Refs 128, 129, 132], it has a similar configuration to EPON comprising an OLT, ONUs and an optical splitter with a maximum split ratio of 1×128. Due to the fiber dispersion, however, a 1×64 split ratio is regarded to be most appropriate for longer network spans [Ref. 133] with the maximum specified transmission distance between the ONU and OLT normally being 20 km [Ref. 134]. Gigabit PON can also support single-fiber bidirectional transmission with transmission rates up to 2.4 Gbit s^{-1} for both up- and downstream transmission [Refs 126, 135].

Gigabit PON can carry both ATM and Ethernet as well as IP packets [Ref. 136] and therefore it allows different traffic mixes to be accommodated as depicted in Figure 15.42. Since different transmission modes employ different packet/frame structures, it is necessary to provide a common framing structure to incorporate the range of transmission structures. Figure 15.42(a) shows the strategy used where ATM, Ethernet and IP packets are mapped onto the GPON frame structure. ATM traffic is incorporated in an ATM frame while a generic frame procedure (GFP) is usually employed for both IP and Ethernet traffic [Ref. 137]. It should be noted that the GFP approach was introduced to map or encapsulate SDH/SONET frames and that the ITU-T has restructured the GFP for PON calling it the GPON encapsulation method (GEM), which can be regarded as GFP over PON [Ref. 138].

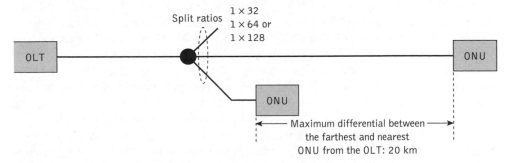

Figure 15.41 Gigabit passive optical network (GPON) [Ref. 126]

* Full Service Access Network (FSAN) group [Ref. 129] comprises major telecommunications services providers, independent test laboratories and equipment suppliers for implementations of PON-based technologies. Since 1995 this group has proposed a large number of PON structures and implementations which have resulted in various ITU-T recommendations.

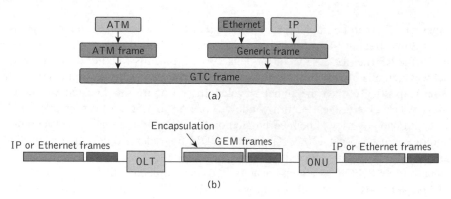

(a)

(b)

Figure 15.42 GPON frame transmission: (a) combined transmission of ATM, IP and Ethernet; (b) GPON encapsulation method

Hence, both GEM and the ATM frames are combined using the GPON transmission convergence (GTC) frame structure as shown in Figure 15.42(a) [Ref. 138]. Finally, the GEM frame structure for a GPON illustrating the encapsulation of the IP or Ethernet frames and the mapping of them onto a GPON frame is depicted in Figure 15.42(b) [Ref. 126].

It should be noted that GPON is not backwards compatible to APON (or BPON), but that instead it has been designed for optimal efficiency in terms of bandwidth utilization, maintenance, scalability and support for multiple services. Since ATM-based PONs are not cost effective and many service providers are now discarding the ATM layer, GPON may be used to carry just IP and/or Ethernet frames in future. Recognizing this potential development, the ITU-T has specified GPON-lite which is already configured to not carry ATM frames and will only incorporate IP and Ethernet packet structures [Ref. 139]. In a parallel development under the IEEE 802.3 Committee, a gigabit-capable Ethernet has been specified as a standardized PON which uses only Ethernet frames and is therefore referred to as Gigabit Ethernet PON (GE-PON) [Refs 135, 140, 141].

The overall bandwidth provided by a PON can be increased if WDM techniques (see Section 12.9.4) are employed. WDM can potentially facilitate large bandwidth with relatively moderate complexity [Refs 142–144]. A potential WDM-based PON implementation is shown in Figure 15.43. The OLT comprises both WDM source and receiving units which send and receive WDM signals. An optical bandpass filter (OBPF) is used to separate the optical signals between the transmitting and receiving units in the OLT. The OLT is connected to the RN by a single optical fiber that carries bidirectional signal transmission enabling two wavelength bands, one each for the downstream (i.e. $\lambda_1, \lambda_2, \ldots, \lambda_n$) and upstream (i.e. $\lambda_{n+1}, \lambda_{n+2}, \ldots, \lambda_{2n}$). The RN section incorporates optical multiplexer and demultiplexer units which usually comprise arrayed waveguide gratings (AWGs) (see Section 11.6) and these devices are used to route WDM signals from the RN to a particular ONU and vice versa. Such WDM-PONs are also referred to as wavelength-routed PONs (WRPONs) [Ref. 145]. Moreover, each ONU comprises essentially an optical filter to separate incoming and outgoing wavelength signals. For example, ONU 1 uses signal wavelengths λ_1 and λ_n for downstream and upstream transmission, respectively.

It is also possible to employ AWGs to route a band of wavelength signals to a specific group or a number of ONUs. This can be accomplished using a multistage configuration

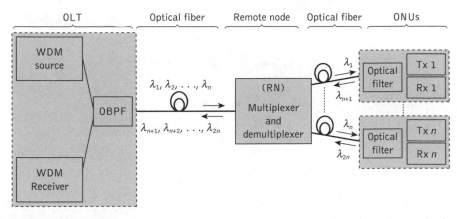

Figure 15.43 Block schematic of a wavelength division multiplexed passive optical network (WDM-PON)

for the RN incorporating several AWG devices where each RN stage provides access to a specific number of ONUs. Although not yet deployed, such multistage AWG-based DWDM PONs have been demonstrated [Refs 131, 146, 147]. A multistage AWG implementation of the RN in a DWDM-based PON is shown in Figure 15.44. The two-stage AWG configuration comprises a 2×8 device, followed by eight 1×4 AWGs in each stage. It incorporates a downstream grouping strategy for a total of 32 ONUs where a single 1×4 AWG serves an individual group of four ONUs. A similar approach utilizing an AWG has also been proposed to facilitate a multiple PON access network [Ref. 148].

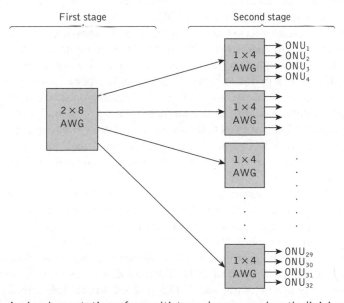

Figure 15.44 An implementation of a multistage dense wavelength division multiplexed passive optical network (DWDM-PON) showing two stages of arrayed-waveguide gratings (AWGs) for downstream access in a remote node

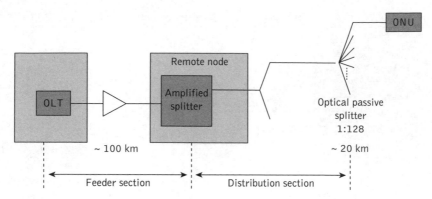

Figure 15.45 Long-reach PON (LR-PON) configuration potentially connecting the access network directly to the core network

It should be noted that current PONs can serve only a limited number of ONUs and they also operate over typical access network maximum spans of around 20 km. Longer transmission and distribution distances for a PON can be achieved when optical amplifiers are incorporated and such an access network was initially referred to as a superPON [Ref. 149]. These networks are now, however, more often called long-reach PONs (LR-PONs) [Refs 150, 151]. The concept behind the LR-PON is to provide for the expansion of the physical size of the access network to over 100 km which may in certain locations remove the requirement for an MAN tier in the telecommunications network hierarchy (see Section 15.2.4) [Ref. 134]. In addition, as LR-PONs can incorporate a significantly greater splitting ratio the number of ONUs served by such long-reach networks is also substantially increased over the standard GPON.

Figure 15.45 depicts a typical structure for an LR-PON showing the reach extended up to 120 km for the overall access network. In this PON configuration the feeder section incorporates an optical amplifier(s) while the remote node houses an optical splitter and an optical amplifier (i.e. amplified splitter) to increase the optical signal power to provide for a greater number of splits [Refs 152, 153]. The passive optical splitter configuration in the distribution section following the amplified splitter can have a splitting ratio of 1×128, 1×1024, 1×2048, or even more depending upon the network optical power budget. In addition, the LR-PON can be configured as a wavelength-routed PON potentially enabling end users to interconnect with the metropolitan, or directly with the long-haul or core, network [Ref. 151].

A hybrid DWDM/TDM approach which has also been demonstrated using an LR-PON structure is illustrated in Figure 15.46 [Ref. 150]. In this configuration the PON can be divided into four operational sections, namely the core exchange, the local exchange, the street cabinet and the customer ONU. The section at the core exchange or central office comprises a transceiver incorporating DWDM channels using a combination of AWGs and an EFDA. The DWDM signal has 100 GHz channel spacing splitting the C-band into two halves (i.e. 1529 to 1541.6 nm and 1547.2 to 1560.1 nm) each carrying both downstream and upstream channels.

The middle section which is located in the local exchange functions as a repeater employing EDFAs, an AWG and the centralized DWDM optical sources for each ONU.

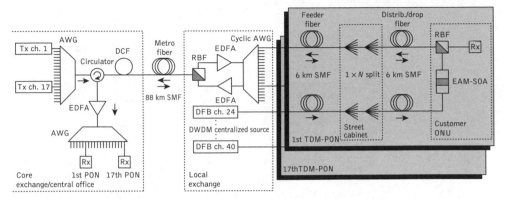

Figure 15.46 Structure of a dense wavelength division multiplexed long-reach PON (LR-PON) [Ref. 150]

The street cabinet section mainly houses $1 \times n$ splitters to interconnect the feeder and distribution network sections, each of which is 6 km long, using two separate fiber links (i.e. distribution and drop fiber). Finally, the customer ONU section incorporates an electroabsorption modulator (EAM) monolithically integrated with semiconductor optical amplifiers (SOAs) (see Section 10.2), providing sufficient overall gain and bandwidth to support large splitting ratios and upstream transmission rates of up to 10 Gbit s^{-1} [Ref. 144]. This hybrid LR-PON structure demonstrated a total reach of 100 km with the potential for supporting 17 TDM-PONs each operating at a different wavelength with up to 256 ONUs, giving an aggregate network ONU count of 4352.

15.6.4 Local area networks

A local area network (LAN), unlike the local telecommunications network, is an interconnection topology which is usually confined to either a single building or group of buildings contained entirely within a confined site or establishment (e.g. industrial, educational, military, etc.). The LAN is therefore operated and controlled by the owning body rather than by a common carrier.* Optical fiber communication technology has found application within LANs to meet the on-site communication requirements of large commercial organizations and to enable access to distributed or centralized computing resources.

Figure 15.47 shows the Open Systems Interconnection (OSI) seven-layer network model (see Section 15.3.3) which was originally developed for wide area networks (WANs). As there are fundamental differences between LANs and WANs it has been found

---

* Another possible definition of a LAN, based on speed and range of operation, is that a LAN typically operates at a transmission rate of between 10 Mbit s^{-1} and 10 Gbit s^{-1} over distances of 200 m to 5 km. Hence a LAN is intermediate between a short-range multiprocessor network (usually data bus) and a wide area network which provides data transmission over very long distances (i.e. possibly worldwide) using a range of communications technology. However, it must be noted that there are always exceptions to these general definitions which tend to increase in number as the technology advances.

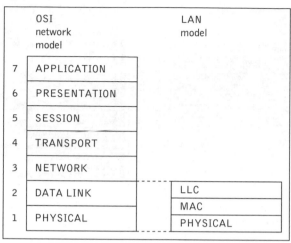

Figure 15.47 Network reference models: OSI and LAN modification

necessary to redefine the two bottom layers of the OSI model into three layers, as displayed in Figure 15.47. The physical layer, as in the WAN, is responsible for the physical transmission of information.

The functions of the data link layer, however, are separated into two layers: namely, the logical link control (LLC) layer which assembles/disassembles data frames or packets and provides the appropriate address and error checking fields; and the medium access control (MAC) layer which organizes communications over the link [Ref. 154]. The MAC layer embodies the set of logical rules or the access protocol which allow nodes to access the common communication channel, and several MAC options may therefore be provided for the same LLC layer. Furthermore, another major difference between WANs and LANs is that each node on the network can be directly connected to all others, thus eliminating the need for routing via the network layer. As a result of this high degree of connectivity, the bus and ring topologies have become commonplace [Ref. 155].

The basic optical fiber LAN topologies are the bus, ring and star (see Section 15.2.1.1). In the bus topology, data generally circulates bidirectionally. Data is input and removed from the bus via four-port couplers located at the nodes (see Section 5.6.1). In the ring, configuration data usually circulates unidirectionally, being looped through the nodes at each coupling point and hence repeatedly regenerated in phase and amplitude. The optical star forms a central hub to the network which may be either active or passive. In passive operation, the star coupler (see Section 5.6.2) at the hub splits the data in terms of power, so that it is available to all nodes. Line losses are therefore primarily determined by the degree of splitting. With an active hub the data is split electrically in the active star coupler and therefore the only causes of line loss are the fiber and splice/connector losses.

The design of the MAC layer is crucial in LANs because, in general, it is the efficiency of the access protocol which governs the availability of the bandwidth provided by the network for the dual functions of data transmission and channel arbitration. Three specific types of access protocol have gained a fair degree of acceptance, primarily because of their simplicity. These are: (1) random access protocols, the most popular example of which is carrier sense multiple access with collision detection (CSMA/CD) used on the

Ethernet LAN; (2) token passing protocols; and (3) time division multiple access (TDMA) protocols.

With CSMA/CD protocols, nodes are allowed to transmit their data as soon as the communication channel is found to be idle (i.e. carrier sense). If the transmissions from two or more nodes collide, the event is detected by the physical layer hardware and the nodes involved terminate transmission before attempting retransmission after a random time interval. Token passing protocols behave as distributed polling systems in which nodes sequentially obtain permission to use the channel. The channel arbitration is determined by the possession of a small distinctive bit packet, known as a token, which can only be held by one node (which is then permitted to use the channel) at any one time. When the data transmission from that node is completed it passes the token packet to the next node in the logical sequence. Finally, with basic (i.e. fixed assignment) TDMA protocols, each node on the network is assigned a fixed length time slot during which it may transmit data. In this way, the protocol operates in a similar manner to that of TDM on a point-to-point link (see Section 12.4.2).

An optical fiber regenerative repeater (see Section 12.6.1) can be used to restore the attenuated signal power in an optical fiber LAN. In addition, this device may be employed as an active tap to allow the interconnection of numerous nodes within ring or linear bus topologies. The use of such repeater interfaces allows similar topologies to be realized for optical fiber LANs as their metallic counterparts, but with certain cost penalties. In order to reduce the number of repeaters required, a range of passive tapping components are available (see Section 5.6). Commercial taps, however, incur an excess loss of between 0.1 and 1 dB. In addition to this excess loss the tap must also divert some of the optical signal power to the tapping receiver. Even if the tap only diverts 1% of the transmitted optical signal power, this places a limit of around 30 on the number of taps which may be provided unless active electrical components are included to allow regeneration of the transmitted optical signal.

One of the major problems associated with optical fiber buses which use the tapping methods discussed previously is that there is an uneven distribution of the transmitted optical power between the connected receivers. This has two distinct drawbacks, namely that the optical receivers must have a large dynamic range and that detection of the collision events required for CSMA/CD protocols becomes difficult [Refs 63, 156]. One solution to this problem is to use the passive star couplers discussed in Section 5.6.2. Each node in the star topology has a single input and a single output fiber connected to the hub of the star. Optical power within the input fibers can be distributed more evenly between the output fibers, thus achieving more uniform reception levels. Passive star couplers are currently available using up to 100×100 ports. Typical port-to-port insertion losses for 32×32 and 64×64 port transmissive passive stars employing multimode fiber are around 18 dB and 21 dB respectively. Variations on the passive star topology include the use of an active star hub [Ref. 157] and the use of cascaded passive stars [Ref. 158].

Optical fiber LANs are now well-established products provided by equipment suppliers to complement and extend the capacity of their copper- and wireless-based LANs. Furthermore, organizations such as the Telecommunications Industry Association (TIA) have developed computer-based interactive LAN cost models to help network planners and end users to compare more easily the cost of a network based on fiber (or copper/coaxial/radio etc.) covering different aspects of networking [Refs 159, 160]. Besides determining the cost and efficiency performance, these models also consider different standards, topologies, component mixes as well as operational issues. Regular updates are issued including new media choices, aggregate pricing that reflects current market conditions, and graphic comparisons of different network structures.

In the late 1980s, standardization within optical fiber local area networking was centered on the Fiber Distributed Data Interface (FDDI) which had been specified by the X3T9.5 working group of the American National Standards Institute (ANSI). Although originally based on the emerging IEEE 802.5 token ring operating at a data rate of 4 Mbit s^{-1}, FDDI was intended for use at the much higher data rate of 100 Mbit s^{-1} employing optical fiber transmission media [Refs 161, 162]. Hence the facilities supported by the FDDI could be summarized as follows:

(a) 100 Mbit s^{-1} data transfers with a 125 Mbaud transmission using a 4B5B line code;

(b) up to 100 km of dual-ring fiber;

(c) up to 500 network nodes;

(d) up to 2 km between nodes;

(e) timed token passing protocol.

The architecture for the FDDI was made compatible with the OSI model and the IEEE 802 structure, as illustrated in Figure 15.48. The various blocks shown in Figure 15.48 were each the subject of specific areas within the standardization proposals. Hence the physical layer was divided into two sublayers, these being the physical medium dependent (PMD) layer and the physical layer protocol (PHY). The former specified the required performance of the optical hardware including the optical fibers, connectors, transmitters, receivers and bypass switches [Ref. 163]. Interfacing between the PMD and the data link

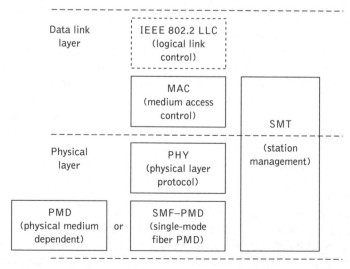

Figure 15.48 FDDI reference model and its relationship to the OSI model

layer was provided by the PHY together with such functions as encoding, clock synchronization and symbol alignment [Ref. 164]. Medium access control for the data link layer was provided by the MAC layer [Ref. 165]. Finally, station management (SMT) controlled processes within stations including configuration, fault tolerance and link quality [Ref. 166].

The FDDI network topology consisted of stations which were logically connected into a ring. However, two contrarotating rings were employed, each of which was capable of supporting a data rate of 100 Mbit s^{-1}, as shown in Figure 15.49(a). Furthermore, the FDDI could be physically installed as a dual ring incorporating concentrators (or star points) to form a dual ring of trees. This comprised a series of duplex, point-to-point optical fiber links.

The FDDI had inbuilt recovery mechanisms to enhance its robustness and was an example of a protected network (see Section 15.8). Network operation was normally designated to the primary ring (Figure 15.49(a)). However, if this ring was broken, then transmission was switched to the secondary ring. The reconfiguration of stations around a cable break, as illustrated in Figure 15.49(b), was controlled by the station management function. In addition, station failure did not affect the network operation due to the provision of optical bypass switches within all the stations. This ensured that the optical path through a station was maintained in the event of a station fault, or power down.

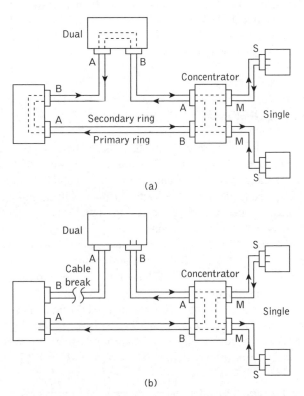

Figure 15.49 FDDI: (a) the dual-ring structure; (b) ring reconfiguration on fault detection

The token passing protocol aspects of the FDDI were somewhat similar to the IEEE 802.5 token passing ring. Each station in the scheme was allowed to transmit only when it is in possession of a token which was circulated sequentially around the stations on the ring. The basic FDDI protocol had provision for handling two priority classes of data traffic: synchronous and asynchronous. At each station, transmission under these classes was controlled by employing timers and by specifying the percentage of the ring bandwidth that could be used by synchronous traffic [Ref. 167].

In order to cater for periodic deterministic traffic, traffic with a single time reference (e.g. speech, possibly video), a circuit-switched mode of operation within the protocol was defined [Ref. 167]. This development, which was referred to as FDDI-II, allocated the 100 Mbit s^{-1} bandwidth of FDDI to circuit-switched data increments of 6.144 Mbit s^{-1} isochronous channels. The FDDI optical fiber links were specified to use InGaAsP LED sources emitting at a wavelength of 1.3 μm to take advantage of the zero (first-order) material dispersion point. In addition, the two primary fiber recommendations were 62.5/125 μm and 85/125 μm graded index exhibiting a bandwidth–distance product better than 400 MHz km^{-1} with losses of less than 2 dB km^{-1} at 1.3 μm. The power budget over the 2 km maximum span for multimode fiber was 11 dB. Moreover, a single-mode fiber version of the physical medium dependent (SMF–PMD) shown in Figure 15.48 could be operated over distances up to 60 km [Ref. 162]. Nevertheless, although both the circuit-switched capability [Ref. 167] and the single-mode fiber option were introduced, they were not widely adopted due to the higher speed, low cost and growing acceptance of optical Gigabit Ethernet (see Section 15.7) which has replaced FDDI-II from the late 1990s making it now become a legacy network [Refs 167–169].

15.7 Optical Ethernet

Optical Ethernet is similar to the conventional Ethernet LAN (see Section 15.6.4) with the exception of the physical layer. It is the fourth generation of the Ethernet family and unlike earlier generations (i.e. X.25 and Frame Relay) and ATM (see Section 15.3.2) [Ref. 170] it uses IP-based technology (see Section 15.3.5). Furthermore a Gigabit Ethernet (GbE) network was developed in 1998 by merging together two technologies (i.e. IEEE 802.3 Ethernet and ANSI X3T11 fiber channel [Ref. 135]) to enable it to operate at higher transmission rates (i.e. 100 Mbit s^{-1} to 10 Gbit s^{-1}) using optical fiber. However, GbE normally implies a transmission rate of 1 Gbit s^{-1} while the higher 10 Gbit s^{-1} speed is usually referred to as 10 GbE. The ITU-T and IEEE have both defined a physical layer for optical Ethernet [Refs 32, 122]. For example, ITU-T Recommendation G.985 [Ref. 171] specifies a 100 Mbit s^{-1} optical Ethernet* using a bidirectional WDM transmission system incorporating a single-mode optical fiber [Refs 172–175]. This does not preclude, however, the use of other fiber types (i.e. low-water-peak single-mode fiber) or even multimode fiber operating at 0.85 μm for short transmission distances [Refs 175, 176].

---

* It should be noted that since the earlier Ethernet generations were also capable of carrying large transmission data rates, a general consensus now suggests the higher transmission rates (i.e. 1 Gbit s^{-1} or 10 Gbit s^{-1}) define optical Ethernet. Nevertheless, optical Ethernet can also operate at the lower transmission rates of either 100 Mbit s^{-1} or 10 Mbit s^{-1}.

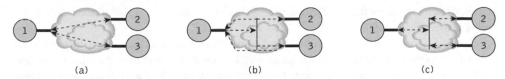

(a)   (b)   (c)

Figure 15.50 Different types of optical Ethernet connections: (a) point-to-point; (b) point-to-multipoint; (c) multipoint-to-multipoint

Three different types of optical Ethernet connection are depicted in Figure 15.50. A point-to-point connection is shown in Figure 15.50(a) where only a single network node is connected with another node. A point-to-multipoint connection which enables a single node interconnection with two or more network nodes is illustrated in Figure 15.50(b). For example, node 1 can transmit to nodes 2 and 3 simultaneously while it can also receive transmissions from both these nodes. This is not the case, however, for node 2 or 3 which can transmit simultaneously to other nodes only when using the multipoint-to-multipoint network connections as indicated in Figure 15.50(c). In this case the nodes have similar connectivity patterns where each can be connected to a common fiber channel. The increased network node connectivity patterns of the multipoint-to-multipoint configuration can resemble a bus, tree or a mesh topology (see Section 15.2.1.1). Such a multipoint-to-multipoint junction can be employed to work as a switching hub with the nonblocking switching features. Since it facilitates a switching function between different Ethernet users it is commonly referred to as Ethernet switch [Refs 177, 178]. In addition, optical Ethernet has also become an accepted technology for virtual connectivity. For example, the line services which are point-to-point in nature include services such as Ethernet virtual private lines [Refs 179, 180].

Optical Ethernet employs the standard Ethernet protocol which incorporates four basic aspects: the frame, MAC, signaling components and the physical medium. An Ethernet frame is a standardized set of specific fields each comprising a different number of bits which is used to carry data over the network as shown in Figure 15.51. The basic Ethernet frame format begins with a field called the *Preamble* which informs receiving nodes that a frame is being transmitted, alerting them to start receiving the data. Unique *Destination* and *Source address* fields allow the optical Ethernet equipment to avoid the problem of two or more Ethernet interfaces in a network having the same address. This feature also eliminates any need to locally administer and manage Ethernet addresses. Following in the Ethernet frame is the *Type* or *Length* field which is used to identify the type or the length of other network protocols being carried in the data field. The *Data* field comes next and this ensures that the frame signals must stay on the network long enough for every Ethernet node to detect the frame within the correct time limit. Finally, the *Frame check sequence* field provides a checking procedure for the integrity of the data in the entire

Preamble	Destination address	Source address	Type/ length	Data	Frame check sequence

Figure 15.51 Ethernet frame format

frame. This field allows the receiving optical Ethernet interface to verify that the bits in the frame have arrived without an error.

The MAC protocol (see Section 15.6.4) provides a set of rules embedded in each Ethernet interface to access the shared channel among the multiple networking nodes. Each network node primarily performs two tasks which are either to transmit or to receive data where the received data may also be required to be forwarded to other networking nodes. Therefore these nodes constitute a group of signaling components and may comprise standalone devices such as repeaters, switches, couplers or routers. Finally, the physical medium consists of the cables and other hardware used to carry the optical Ethernet signals between networking nodes.

When optical Ethernet uses Ethernet switches in a full-duplex operation then each node can be connected to the switch via a separate point-to-point dedicated link between the nodes. In this approach either the switch or the node can send and receive independently without collisions, thus removing the requirement for the conventional CSMA/CD protocol (see Section 15.6.4). Hence the CSMA/CD protocol is not used when the Ethernet connections are configured for a full-duplex operation. This situation may lead, however, to an increased frame dropping rate because without collision detection the sender cannot detect a dropped frame and resend it. Nevertheless switches drop frames only if the buffer is full; when this occurs the sender will not be able receive the lost frame data. In addition, error control is achieved by encapsulating control packets between data packets in an MAC frame arriving from the upper layer. Figure 15.52 shows the Ethernet sublayers in the OSI reference model (see Sections 15.3.3 and 15.6.4) which indicates the division of the data link layer where the two sublayers (MAC and physical signaling) below the LLC and MAC bridging sublayer are specific to an Ethernet LAN.

In addition to being a LAN, optical Ethernet also provides switching capabilities in both layer 2 and layer 3 (i.e. IP routing (see Section 15.3.5)) and it is therefore usually considered to be a layer 2/3 switched network rather than as a layer 2 network which typifies a conventional LAN [Refs 177, 181]. Figure 15.53 shows two configurations for an optical Ethernet network. A point-to-point optical Ethernet operating as an access network (see

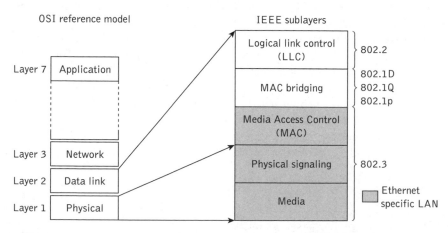

Figure 15.52 Ethernet sublayers as given by IEEE 802.3. Reprinted with permission from Ref. 32 © IEEE

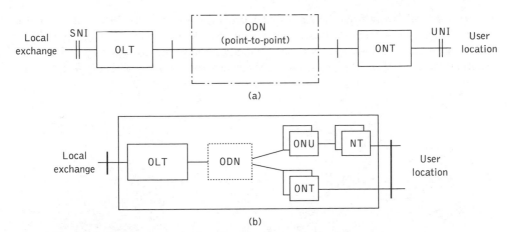

Figure 15.53 Optical Ethernet: (a) point-to-point access network [Ref. 171, ITU-T Recommendation 985]; (b) point-to-multipoint access network (Ethernet PON) [Ref. 182, ITU-T Recommendation 838] reproduced of with the kind permission of ITU

Section 15.6.3) is shown in Figure 15.53(a). The optical distribution node (ODN) provides point-to-point access on a bidirectional single-mode optical fiber. Two different wavelength regions of 1.26 to 1.36 µm and 1.48 to 1.58 µm are used for upstream and downstream transmission, respectively. The service network interface (SNI) is connected to the optical line termination (OLT) at the local exchange/office while the user network interface (UNI) is connected to the optical network termination (ONT) at the user locations.

A point-to-multipoint Ethernet PON (EPON) can also be configured based on the point-to-point optical Ethernet as illustrated in Figure 15.53(b). The network termination (NT) provides the user network interface line termination function and the ONT combines the functions of the ONU and NT. The bandwidth in the EPON can be assigned dynamically or statically and is chosen by the OLT. In the static assignment the bandwidth can be divided into upstream and downstream, whereas in the dynamic assignment procedure the OLT sets the upstream bandwidth for each ONU (or ONT) according to the request from ONUs (or ONTs).

Optical Ethernet has therefore migrated from being simply a LAN to be increasingly considered for deployment in the access network where the practical limits to the size of the network are not geographic but include bandwidth, node count and the overlying protocol capability (i.e. broadcast traffic, routing table size, etc.). Hence optical Ethernet could provide a solution not only in the access network but also in metropolitan and even potentially in long-haul networks (i.e. carrier Ethernet) [Refs 37, 183, 184]. This situation has occurred because Ethernet switches also support Multiprotocol Label Switching (see Section 15.5.3) which is the major desired feature for an MAN [Ref. 185]. The resultant structure of optical Ethernet when using Ethernet switches at high transmission rate (i.e. 10 Gbit s^{-1}) therefore extends the network capabilities (i.e. 10 GbE) into the areas of the telecommunications network [Refs 173, 181, 186]. Furthermore, full-duplex operation is utilized in 10 GbE networks which also possess backward compatibility for earlier deployed optical Ethernet transmission rates.

The ability of telecommunication organizations and business communities to better manage their corporate LANs by strategically choosing the location of their data center

and server farms away from their main sites or campuses using 10 GbE is illustrated in
Figure 15.54(a). For example, this approach can provide interconnection among multiple
site locations (i.e. locations A, B and C) typically within that 40 km range [Ref. 186].
Alternatively, a 10 GbE MAN is shown in Figure 15.54(b) where switching facilities (i.e.

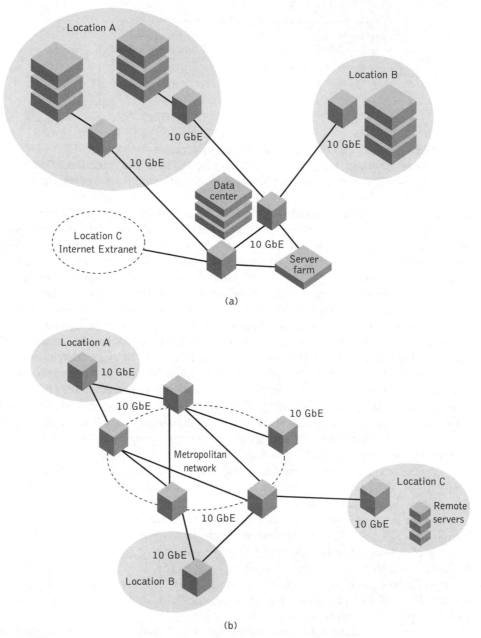

Figure 15.54 The 10 Gigabit Ethernet: (a) local area network; (b) metropolitan
area network

switch-to-switch or switch-to-server) can be provided within the data centers. Such metro Ethernet interfaces employing optical transceivers and multimode or single-mode optical fibers transmitting at three different signal wavelength bands (i.e. 0.85 μm, 1.31 μm and 1.55 μm) can also currently reach a span of 40 km [Refs 122, 186].

Recently, field trials of generalized MPLS (GMPLS) (see Section 15.5.3) based on lambda-on-demand were successfully carried out incorporating 10 GbE transmission at optimal cross-connect switches (see Section 15.2.2) [Refs 187, 188]. The WDM network based on a GMPLS control plane consisted of a single optimal cross-connect node and three optical add/drop multiplexing nodes interconnected in a ring topology through bidirectional single-mode optical fibers. Furthermore, higher transmission rates up to 100 Gbit s^{-1} have also been demonstrated using optical Ethernet technology [Refs 186, 189–191].

Finally, it is often convenient to be able to transmit Ethernet-configured data packets across an optical network which does not include optical Ethernet networks. A large portion of the existing optical network infrastructure utilizes SDH/SONET-based transmission (see Section 15.3.1) and in this case it is useful if the network also exhibits transparency to Ethernet. It should be noted that Ethernet traffic cannot be carried directly over the SDH/SONET network due to the different transmission methodologies and data rates. In order to provide an interface between Ethernet and SONET an interworking protocol is required to facilitate Ethernet frame transmission over SONET. Such an Ethernet over SONET (EoS) protocol enables the different networks to communicate with each other without losing data. EoS therefore encapsulates the Ethernet frames into the SONET frame structure [Refs 192, 193]. For example, virtual concatenation allows nonstandard SDH/SONET (i.e. Ethernet) transmission to be transported over SONET.

Virtual concatenation (VC) is a splitting or demultiplexing technique used to split SDH/SONET bandwidth into logical groups, which may be transported or routed independently [Refs 194, 195]. The basic principle behind VC is that a number of smaller containers are concatenated and assembled together to create a bigger container that carries more data per second. In SDH/SONET this technique works well across the existing infrastructure and it can also significantly increase network utilization by effectively spreading the load across the whole network. Therefore, it provides for both bandwidth efficiency and flexibility using the service provider's current SDH/SONET equipment to transport, for instance, two GbE links over a SONET STS-48 link while leaving capacity for six STS-1 channels for other traffic as illustrated in Figure 15.55.

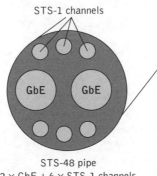

Figure 15.55 An example of virtual concatenation

15.8 Network protection, restoration and survivability

Network protection which not only provides the information when a breakdown in the network occurs but also offers the necessary protection to overcome the failure is an essential aspect of all network operations. In case of a link failure there should be efficient and immediate solutions to recover from such faults to enable the network to both sustain and maintain customer service. The breakdown of an optical network component such as a fiber splice can lead to the failure of all lightpaths that utilize that specific link. Moreover, in high-capacity optical fiber networks each lightpath is expected to operate at a rate of many gigabits per second and therefore a link failure can lead to a high data loss. Hence network survivability becomes a critical issue in network design and real-time operation in order to provide protection and fast restoration against any failures. Although higher protocol layers have procedures to recover from link failures, the recovery time to minimize data losses is still quite long (i.e. in seconds) as compared with the expected restoration time in the optical layer (i.e. less than a few milliseconds). Higher layer protection and restoration mechanisms therefore need to be fast and efficient [Refs 196, 197].

Survivability is the ability of a network to withstand and recover from network failures. It determines the capability to provide continuous service in the presence of network failures. The basic function is to restore and provide protection in case of an optical network component failure. The restoration or provisioning process is required to discover a dynamic backup route whereas the network protection process ensures the availability of network resources to offer that backup route. Network protection can be either preconfigured or provided as required during the failure. In preconfigured (or preplanned) protection the spare resources are allocated in advance when a lightpath is required to be set up and therefore it provides for fast recovery in real time [Refs 198, 199].

Network protection mechanisms are driven by algorithms which maintain services carried by a specific network topology. These can be divided into two categories which are path- and link-oriented protections. In path protection the source and destination nodes of each connection statically reserve the backup paths and wavelengths during call setup. The working (or primary) and protecting (or backup) paths for a connection need to utilize separate optical fibers as shown in Figure 15.56. All the connections on a failed link are rerouted to the backup route. The source nodes of the failed link in Figure 15.56(a) (i.e. primary or operational fiber) are informed about the failure via messages from the nodes adjacent to the failed link (i.e. in this case, network node 2 sends a link failure message to

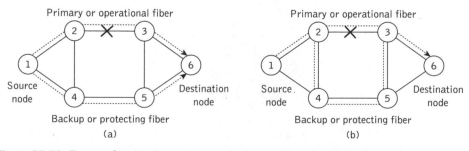

Figure 15.56 Types of optical network protection: (a) path; (b) link

node 1). The source immediately terminates the link with node 2 and sets up a new path to establish a connection via the backup fiber between nodes 1 and 6 through nodes 4 and 5. It should be noted that in providing link protection only the affected connections on the failed link are rerouted around that failed link. The source and destination nodes of the connections remain unaware of the link failure and the necessary rerouting. This situation is illustrated in Figure. 15.56(b) where a connection is set up dynamically using a protecting fiber employing network nodes 4 and 5 (i.e. interconnecting nodes 2, 4, 5 and 3) in order to restore connection between nodes 1 and 6.

Restoration or provisioning to provide network protection enables the network to restore and establish new connections replacing the faulty link by establishing a new path or the link. However, restoration offers increased flexibility over more rigid preconfigured protection. In preconfigured protection the resources are dedicated to establish a path or a link whereas restoration protection may require either shared or dedicated resources.

Protection is referred to as 1 + 1 protection when a single protecting or backup fiber is made available for a single working or primary fiber. It is called 1:1 or 1:N protection when the single protecting fiber provides backup for one or a number N of operational fibers [Refs 198, 200]. Furthermore, passive optical couplers/splitters (see Section 5.2) and optical switches (see Sections 10.5 and 10.6) can be used to share or switch over the connection. Network protection using an optical switch and an optical splitter is illustrated in Figure 15.57. A single protecting fiber facilitates backup for a single operational fiber

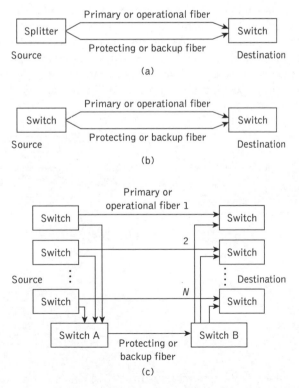

Figure 15.57 Network protection: (a) 1 + 1 protection; (b) 1:1 protection using optical switch; (c) 1:N protection using optical switches

with an optical splitter at the source end and an optical switch at the destination end. This demonstrates 1 + 1 protection where a source is connected via an optical splitter dividing the incoming signal and coupling it to both the operational and protecting fibers. Thus the optical switch at the destination end can directly establish a link to access the protecting fiber.

A scheme for 1:1 protection is depicted in Figure 15.57(b) where optical switches are used at both ends. To facilitate protection both optical switches must be used to switch over the connection. It is possible to extend the 1:1 to 1:N protection in which a single protecting fiber can provide protection for any single failed operational fibers. For example, 1:N protection with multiport optical switches A and B provides interconnection for the protecting fiber for any single broken link from N fibers as illustrated in Figure 15.57(c). Each operational fiber is connected to optical switches at both the source and destination ends. For normal operation both A and B switches remain in their OFF positions and when an operational fiber fails they are turned ON for that specific fiber, enabling access to the protecting fiber and thus restoring the connection between the source and destination. Furthermore, in order to reduce the switching time optical splitters can also be used in place of optical switches at the source and destination. In this case only optical switches A and B are required to provide access to the protection fiber but it should be noted that the splitter attenuates the optical signal power.

Either 1 + 1 or 1:N protection can be used in WDM networks depending on the network survivability requirements. Different network configurations (i.e. bus, ring or mesh) can be used to provide protection and to restore the optical network link. In a WDM ring topology 1 + 1 protection is commonly employed for its simplicity. Figure 15.58 shows two 1 + 1 protection schemes based on the ring topology for WDM networks with and without optical layer protection. WDM networks when using the optical layer enable all the networking resources (i.e. SONET and IP) to share the same wavelength, which in normal conditions carry specific traffic types on different wavelength signals, thus not

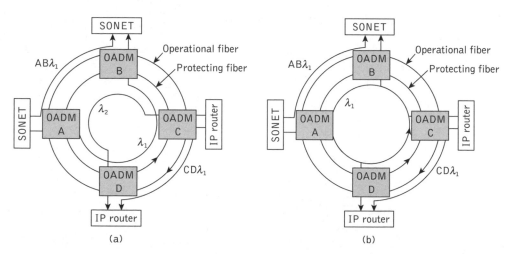

(a)   (b)

Figure 15.58 Two 1 + 1 protection schemes for WDM rings: (a) without the optical layer; (b) with the optical layer

allowing them to share the same lightpath. In both cases the ring network comprises a single operational fiber (i.e. outer ring) and a protecting fiber (i.e. inner ring) and four optical add/drop multiplexers (OADMs) at points A, B, C and D are also interconnected with two SONET line terminals and two IP routers. Such protected rings are referred to as self-healing since they can provide automatic protection in the case of link failure and also they can divert the traffic using a safe alternative route without loss of data. Assuming a unidirectional ring structure for each fiber, the operational fiber carries a clockwise transmission while the protecting fiber transmits in a counterclockwise direction. For this reason it is also known as a bidirectional line-switched ring (BLSR) [Refs 201, 202].

The wavelength continuity constraint (see Section 15.4) dictates that a particular lightpath can travel only if it maintains the same wavelength and direction (i.e. clockwise or counterclockwise) to avoid wavelength collision. Therefore two lightpaths AB and CD on the operational fiber both using wavelength λ_1 can be set up as illustrated in Figure 15.58(a). These lightpaths are physically separate and hence they do not interact with each other. In the case of an AB link failure, the protection fiber (i.e. inner ring) can use wavelength λ_1 via IP routers employing OADMs at C and D. In order to provide protection at the same time from C to D another wavelength (i.e. λ_2) will be required since λ_1 is already in use on the protecting fiber providing an alternate route for lightpath AB via C and D. Therefore in this case the SONET line terminals A, B and IP routers C, D cannot share the same wavelength λ_1 to provide 1 + 1 protection. Such 1 + 1 protection can be facilitated, however, using the optical layer protection [Ref. 109] where all the equipment (i.e. SONET line terminals, IP routers, multiplexers and demultiplexers, etc.) can share the same protection as identified in Figure 15.58(b). It can be seen that only a single protecting ring employing a single wavelength λ_1 facilitates protection simultaneously for both SONET line terminals and IP routers. For example, in case of a failure between A and D (or D and C) the protection can be made available using signal wavelength λ_1. It should be noted that using such optical layer protection employing a single wavelength provides only for a single failure protection as compared with a situation as identified in Figure 15.58(a) where more simultaneous failures could be provided with protection (i.e. in this case protection against two faults using two signal wavelengths). Optical layer protection offers, however, a capacity benefit when a large number of fiber pairs require protection.

In contrast to conventional SONET or IP layer protection, optical layer protection can employ optical switches and wavelength routing [Ref. 203]. This situation is illustrated in Figure 15.59 which compares optical protection of the IP layer with optical layer protection. Using the IP layer all the protection is handled by the routers as shown in Figure 15.59(a). Two WDM links (i.e. WDM link 1 and 2) each providing multiplexed ports for operation and protection are established using four optical multiplexers/demultiplexers. By comparison, when employing optical layer protection only a single protection fiber is used to share a common multiplexer as indicated in Figure 15.59(b). This latter solution is possible since three operational fibers and a single protection fiber are further multiplexed and the resultant WDM signal is divided using an optical splitter after the multiplexer providing 1 + 1 protection. The WDM signal is spilt between both the operational and protection fibers on the WDM link. Alternatively, an optical switch can be used to direct the incoming signal from the operational fiber to the protection fiber. In the case of the operational fiber link failure, the optical switch is turned on to restore connection via the protecting fiber as indicated in Figure 15.59(b). Thus optical layer protection quickly and

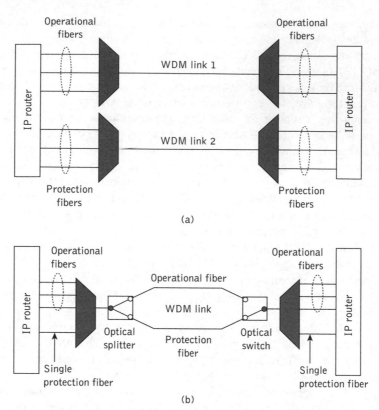

Figure 15.59 Comparison of conventional optical protection: (a) IP layer;
(b) optical layer

effectively restores all the channels by simply using a single optical switch. It should be
noted, however, that shared protection (i.e. 1:N protection) limits alternative routes as only
one fiber is available to facilitate protection for N operational fibers.

Problems

15.1   Briefly discuss the evolution of optical networks. Indicate the significant features
of the optical network generations.

15.2   Differentiate between connection-oriented and connection-less optical networks.
Describe a suitable topology in each case to provide large connectivity among the
network users.

15.3   Explain the network terms virtual circuit, virtual topology and broadcast-and-
select. Indicate their application and role in optical networks.

15.4   How does an optical router differ from an optical switch? With the aid of block diagrams illustrate the main networking node elements used in optical routing and switching to establish links or connections between different networking nodes.

15.5   Explain the modularity and scalability features of an optical network. Outline their roles in the development of flexible and physically expanding optical networks.

15.6   Describe the concept of an OXC and a ROADM. Outline how they are utilized in the development of large-scale wavelength division multiplexed networks.

15.7   Explain the distinguishing features of optical switching and optical wavelength routing. With the aid of block diagrams outline the optical network hierarchy for the public telecommunications network.

15.8   Explain synchronous transmission mode (STM) and discuss its hierarchical features in relation to SONET and SDH optical networks. Modify the frame structure of an STS-1 to be used for the STS-*N* frames. Indicate the purpose and the limitations of each field.

15.9   Define what is ATM and its application in optical networks. Compare the frame format of an ATM cell with a SONET fame.

15.10   Describe the purpose and the layered structure of the Open System Interconnection (OSI) reference model. Summarize the functions of each of the layers.

15.11   Outline the main features of the optical transport network (OTN) and describe its hierarchy specified by the ITU-T.

15.12   Explain the terms protocol and Internet Protocol (IP) and using the OSI reference model discuss the implementation aspects of the IP over: (a) ATM; (b) SONET; (c) DWDM.

15.13   Describe IP over OTN and IP over OPS/OBS. Outline their important characteristics and identify the future-proofing existing within the technique(s) for providing bandwidth-efficient transmission.

15.14   Classify the conventional architectures of a wavelength routing network employing wavelength converters. With the help of a block diagram discuss their implementation for the provision of cost-efficient optical networks.

15.15   Define the term waveband and its application in wavelength conversion with particular reference to a multigranular optical cross-connect.

15.16   Differentiate between static and dynamic routing and wavelength assignments explaining their implementations using a ring topology.

15.17   Describe the main features and drawbacks of optical circuit-switched networks.

15.18   Discuss the operation of optical packet-switched networks thereby explaining the frame format and also differentiating the functions of edge and core routers in these networks.

15.19 Indicate the purpose of optical labels in optical packet-switched networks. Sketch a generic optical label switching network structure and describe its implementation.

15.20 Briefly discuss the applications of Multiprotocol Label Switching for different traffic types in the domains of: (a) time; (b) packet or cell; (c) wavelength.

15.21 Explain the concept of optical burst switching networks. Draw the burst sending and assembling operations at a core router for SDH/SONET and IP.

15.22 Explain the operation of the protocols JIT and JET, and TAW and TAG. Hence identify their main features and applications in optical burst switching networks.

15.23 Describe the basic requirements of long-haul optical networks. Outline the main features of an optical long-haul network employing a dense wavelength division multiplexed transmission system.

15.24 Indicate the major elements of a submerged cable system and describe their role in providing a transoceanic optical fiber network.

15.25 A metropolitan area network (MAN) provides the link between long-haul and access networks. Discuss the basic requirements and functionality of an optical MAN.

15.26 Identify the different possible strategies for the implementation of an access network and outline the functional blocks for an optical fiber access network to provide FTTC and FTTH.

15.27 Classify the types of passive optical network (PON) and describe the PON features which provide the preferred network strategy for the implementations of FTTC/FTTCab.

15.28 Explain the term optical Ethernet and its application in the development of an Ethernet-PON. Sketch and outline the network configuration for up- and downstream broadcasting.

15.29 GPON and GE-PON are two different types of passive optical network. Comment on their operational differences.

15.30 Sketch a functional block diagram for a WDM-PON. Explain the significant role of an optical amplified splitter in designing a remote node of a WDM-PON.

15.31 Describe the architecture of the bidirectional long-reach PON and explain its main components with reference to both up- and downstream transmissions.

15.32 Define the term local area network (LAN). Describe three common media access protocols used in LANs.

15.33 Explain the FDDI and with the help of suitable diagram describe the FDDI network model. Give reasons for the decline in the use of FDDI for the provision of optical LANs.

15.34 Describe the development of optical Ethernet and its deployment in the telecommunications network. Comment on the use of GbE for provisioning from the local area through to the metropolitan area network.

15.35 Define virtual concatenation and describe its application in providing a bandwidth-efficient optical network.

15.36 Compare the two main types of optical network protection. Outline their salient features and discuss which type provides the more robust network protection when there are a large number of link failures.

15.37 Explain the concept of optical layer protection and give reasons why it is preferred over other conventional protection methods. Describe its applications in the design of self-healing optical rings.

References

[1] C. Fow Sen, X. Zhao, Y. Xiuqin, J. Lin, J. P. Zhang, Y. Gu, G. Ru, Z. Guansong, L. Longjun, X. Huiping, H. Hadimioglu and H. J. Chao, 'An optical packet switch based on WDM technologies', *J. Lightwave Technol.*, **23**(3), pp. 994–1014, 2005.

[2] R. Nagarajan *et al.*, '400 Gbit/s (10 channel × 40 Gbit/s) DWDM photonic integrated circuits', *Electron. Lett.*, **41**(6), pp. 347–349, 2005.

[3] M. J. O'Mahony, C. Politi, D. Klonidis, R. Nejabati and D. Simeonidou, 'Future optical networks', *J. Lightwave Technol.*, **24**(12), pp. 4684–4696, 2006.

[4] S. J. B. Yoo, 'Optical packet and burst switching technologies for the future photonic internet', *J. Lightwave Technol.*, **24**(12), pp. 4468–4492, 2006.

[5] C. Gee-Kung, Y. Jianjun, Y. Yong-Kee, A. Chowdhury and J. Zhensheng, 'Enabling technologies for next-generation optical packet-switching networks', *Proc. IEEE*, **94**(5), pp. 892–910, 2006.

[6] J. Wei, Z. Min and Y. Peida, 'All-optical-packet header and payload separation for unslotted optical-packet-switched networks', *J. Lightwave Technol.*, **25**(3), pp. 703–709, 2007.

[7] A. Stavdas, H. C. Leligou, K. Kanonakis, C. Linardakis and J. Angelopoulos, 'Scheme for performing statistical multiplexing in the optical layer', *J. Opt. Netw.*, **4**(5), pp. 237–247, 2005.

[8] A. S. Benjamin, S. Assaf and B. Keren, 'A modular, scalable, extensible, and transparent optical packet buffer', *J. Lightwave Technol.*, **25**(4), pp. 978–985, 2007.

[9] S. Lee, K. Sriram, H. Kim and J. Song, 'Contention-based limited deflection routing protocol in optical burst-switched networks', *IEEE J. Sel. Areas Commun.*, **23**(8), pp. 1596–1611, 2005.

[10] Y. Fukushima, H. Harai, S. Arakawa and M. Murata, 'Design of wavelength-convertible edge nodes in wavelength-routed networks', *J. Opt. Netw.*, **5**(3), pp. 196–209, 2006.

[11] Z. Lei, P. Ji, W. Ting, O. Matsuda and M. Cvijetic, 'Study on wavelength cross-connect realized with wavelength selective switches', *Proc. of Optical Fiber Communication and National Fiber Optic Engineering Conf. (OFC/NFOCE'06)*, Anaheim, CA, p. 7, March 2006.

[12] H. S. Hamza and J. S. Deogun, 'Wavelength-exchanging cross connects (WEX) – a new class of photonic cross-connect architectures', *J. Lightwave Technol.*, **24**(3), pp. 1101–1111, 2006.

[13] X. Cao, A. Vishal, J. Li and C. Xin, 'Waveband switching networks with limited wavelength conversion', *IEEE Commun. Lett.*, **9**(7), pp. 646–648, 2005.

[14] ITU-T Recommendation G.705, 'Characteristics of plesiochronous digital hierarchy (PDH) equipment functional blocks', October 2000.

[15] A. Ehrhardt, 'Next generation optical networks: an operator's point of view', *Proc. Int. Conf. on Transparent Optical Networks (ICTON'06)*, Nottingham, UK, pp. 93–97, 2006.

[16] P. Cochrane and M. Brain, 'Future optical fiber transmission technology and networks', *IEEE Commun. Mag.*, **26**(11), pp. 45–60, 1988.

[17] R. Ballart and Y. C. Ching, 'SONET: now it's the standard optical network', *IEEE Commun. Mag.*, **27**(3), pp. 8–15, March 1989.

[18] ITU-T Recommendation G.707, 'Network node interface for the synchronous digital hierarchy (SDH)', 1996.

[19] ITU-T Recommendation G.707/Y.1322, 'Network node interface for the synchronous digital hierarchy (SDH)', October 2000.

[20] ITU-T Recommendation G.709 (05/05), 'Interfaces for the Optical Transport Network (OTN)', May 2005.

[21] ANSI T1.105.07–1996 (R2001), 'Synchronous Optical Network (SONET) – Sub-STS-1 Interface Rates and Formats Specification', 1996.

[22] ITU-T Recommendation G.711, 'Pulse code modulation (PCM) of voice frequencies', February 2002.

[23] H. Liu and F. A. Tobagi, 'Traffic grooming in WDM SONET rings with multiple line speeds', *IEEE J. Sel. Areas Commun.*, **25**(3), pp. 68–81, 2007.

[24] European Telecommunications Standards Institute (ESTI), http://www.etsi.org, 17 October 2007.

[25] M. N. O. Sadiku, 'Dynamic synchronous transfer mode', in M. Ilyas and H. T. Mouftah (Eds), *The Handbook of Optical Communication Networks*, CRC Press, pp.103–110, 2003.

[26] C. Chih-Jen and A. A. Nilsson, 'Queuing networks modelling for a packet router architecture using the DTM technology', *Proc. IEEE Int. Conf. on Communications (ICC'2000)*, New Orleans, USA, pp. 186–191, July 2000.

[27] International Organization far Standardization (ISO), http://www.iso.org, 17 October 2007.

[28] International Telecommunication Union – Telecommunication Sector (ITU-T), http://www.itu.int/ITU-T/publications/index.html, 17 October 2007.

[29] F. Callegati, M. Casoni and C. Raffaelli, 'Packet optical networks for high-speed TCP-IP backbones', *IEEE Commun. Mag.*, **37**(1), pp. 124–129, 1999.

[30] W. Zhong-Zhen and C. Han-Chiang, 'Design and implementation of TCP/IP offload engine system over Gigabit Ethernet', *Proc. 15th Int. Conf. on Computer Communications and Networks*, Virginia, USA, pp. 245–250, October 2006.

[31] A. Rodriguez-Moral, P. Bonenfant, S. Baroni and R. Wu, 'Optical data networking: protocols, technologies, and architectures for next generation optical transport networks and optical internetworks', *J. Lightwave Technol.*, **18**(12), pp. 1855–1870, 2000.

[32] IEEE 802.3 CSMA/CD (ETHERNET), http://www.ieee802.org/3, 17 October 2007.

[33] ITU-T, 'Optical transport network: tutorial', http://www.itu.int/ITU-T/studygroups/com15/otn/OTNtutorial.pdf, 17 October 2007.

[34] ITU-T Recommendation, G.873.1, 'Optical Transport Network (OTN): Linear protection', March 2006.

[35] ITU-T Recommendation G.709/Y.1331, 'Interfaces for the Optical Transport Network (TON)', February 2003.

[36] ITU-T Recommendation G.8080/Y.1304, 'Architecture for the automatically switched optical network (ASON)', June 2006.

[37] C. Meirosu, P. Golonka, A. Hirstius, S. Stancu, B. Dobinson, E. Radius, A. Antony, F. Dijkstra, J. Blom and C. de Laat, 'Native 10 Gigabit Ethernet experiments over long distances', *Future Gener. Comput. Syst.*, **21**(4), pp. 457–468, 2005.

[38] O. Hideki, A. Nahoko and M. Toshio, 'A study for technology of controlling next-generation backbone network, *J. Natl Inst. Inf. Commun. Technol.*, **53**(2), pp. 127–132, 2006.

[39] W. Bigos, B. Cousin, S. Gosselin, M. Le Foll and H. Nakajima, 'Optimized design of survivable MPLS over optical transport networks', *J. Opt. Switch. Netw.*, **3**(3–4), pp. 202–218, 2006.

[40] D. F. Welch *et al.*, 'Large-scale InP photonic integrated circuits: enabling efficient scaling of optical transport networks', *IEEE J. Sel. Top. Quantum Electron.*, **13**(1), pp. 22–31, 2007.

[41] ITU-T, 'A handbook on Internet Protocol (IP)-based networks and related topics and issues', http://www.itu.int/ITU-D/e-strategy/publications-articles/pdf/IP%20Policy%20Handbook.pdf, 17 October 2007.

[42] J. He and S. H. G. Chan, 'TCP and UDP performance for Internet over optical packet-switched networks', *Comput. Netw.*, **45**(4), pp. 505–521, 2004.

[43] H. Uose, 'GEMnet2: NTT's new network testbed for global R&D', *Proc. Testbeds and Research Infrastructures for the Development of Networks and Communities (Tridentcom'05)*, Trento, Italy, pp. 232–241, February 2005.

[44] M. Tatipamula, F. Le Faucheur, T. Otani and H. Esaki, 'Implementation of IPv6 services over a GMPLS-based IP/optical network', *IEEE Commun. Mag.*, **43**(5), pp. 114–122, 2005.

[45] S. Bjornstad, D. R. Hjelme and N. Stol, 'A packet-switched hybrid optical network with service guarantees', *IEEE J. Sel. Areas Commun.*, **24**(8), pp. 97–107, 2006.

[46] J. Strand and A. Chiu, 'Realizing the advantages of optical reconfigurability and restoration with integrated optical cross-connects', *J. Lightwave Technol.*, **21**(11), pp. 2871–2882, 2003.

[47] D. M. M. I. Javier Aracil, 'Traffic management for IP-over-WDM networks', in D. Sudhir (Ed.), *IP over WDM: Building the Next-generation Optical Internet*, pp. 329–356, Willey-VCH, 2004.

[48] J. Phuritatkul, J. Yusheng and Z. Yongbing, 'Blocking probability of a preemption-based bandwidth-allocation scheme for service differentiation in OBS networks', *J. Lightwave Technol.*, **24**(8), pp. 2986–2993, 2006.

[49] Z. Rosberg, A. Zalesky, H. L. Vu and M. Zukerman, 'Analysis of OBS networks with limited wavelength conversion', *IEEE/ACM Trans. Network.*, **14**(5), pp. 1118–1127, 2006.

[50] W. Wei, Z. Qingji, O. Yong and L. David, 'High-performance hybrid-switching optical router for IP over WDM integration', *Photonics Netw. Commun.*, **9**(2), pp. 139–155, 2005.

[51] Y. Wang, Q. Zeng, C. He, L. Lu and Z. Sui, 'Multi-granular OXC node based on lambda-group technology', *IEICE Trans. Commun.*, **E89-B**(6), pp. 1883–1887, 2006.

[52] C. Politi, D. Klonidis and M. J. O'Mahony, 'Waveband converters based on four-wave mixing in SOAs', *J. Lightwave Technol.*, **24**(3), pp. 1203–1217, 2006.

[53] L. Mengke and R. Byrav, 'Dynamic waveband switching in WDM mesh networks based on a generic auxiliary graph model', *Photonics Netw. Commun.*, **10**(3), pp. 309–331, 2005.

[54] X. Masip-Bruin, S. Sanchez-Lopez and D. Colle, 'Routing and wavelength assignment under inaccurate routing information in networks with sparse and limited wavelength conversion', *Proc. Global Telecommunications (GLOBECOM'03)*, San Francisco, USA, pp. 2575–2579, December 2003.

[55] C. Xiaowen, L. Bo and I. Chlamtac, 'Wavelength converter placement under different RWA algorithms in wavelength-routed all-optical networks', *IEEE Trans. Commun.*, **51**(4), pp. 607–617, 2003.

[56] A. Todimala, R. Byrav and N. V. Vinodchandran, 'On computing disjoint paths with dependent cost structure in optical networks', *Proc. 2nd Int. Conf on Broadband Networks*, Boston, MA, USA, pp. 145–154, 2005.

[57] Z. Jing, Z. Keyao and B. Mukherjee, 'Backup reprovisioning to remedy the effect of multiple link failures in WDM mesh networks', *IEEE J. Sel. Areas Commun.*, **24**(8), pp. 57–67, 2006.

[58] H. Zang, J. P. Jue and B. Mukherjee, 'A review of routing and wavelength assignment approaches for wavelength-routed optical WDM networks', *Opt. Netw. Mag.*, **1**(1), pp. 47–60, 2000.

[59] P. Manohar, D. Manjunath and R. K. Shevgaonkar, 'Routing and wavelength assignment in optical networks from edge disjoint path algorithms', *IEEE Commun. Lett.*, **6**(5), pp. 211–213, 2002.

[60] A. Sridharan and K. N. Sivarajan, 'Blocking in all-optical networks', *IEEE/ACM Trans. Network.*, **12**(2), pp. 384–397, 2004.

[61] C. Xiaowen and L. Bo, 'Dynamic routing and wavelength assignment in the presence of wavelength conversion for all-optical networks', *IEEE/ACM Trans. Network.*, **13**(3), pp. 704–715, 2005.

[62] N. Skorin-Kapov, 'Heuristic algorithms for the routing and wavelength assignment of scheduled lightpath demands in optical networks', *IEEE J. Sel. Areas Commun.*, **24**(8), pp. 2–15, 2006.

[63] S. Huang, R. Dutta and G. N. Rouskas, 'Traffic grooming in path, star, and tree networks: complexity, bounds, and algorithms', *IEEE J. Sel. Areas Commun.*, **24**(4), pp. 66–82, 2006.

[64] I. Chlamtac, A. Ganz and G. Karmi, 'Lightnets: topologies for high-speed optical networks', *J. Lightwave Technol.*, **11**(5), pp. 951–961, 1993.

[65] R. Ramaswami and G. Sasaki, 'Multiwavelength optical networks with limited wavelength conversion', *IEEE/ACM Trans. Network.*, **6**(6), pp. 744–754, 1998.

[66] S. Janardhanan, A. Mahanti, D. Saha and S. K. Sadhukhan, 'A routing and wavelength assignment (RWA) technique to minimize the number of SONET ADMs in WDM rings', *Proc. 39th Int. Conf. on System Sciences*, Hawaii, USA, pp. 1–10, January 2006.

[67] Z. Rosberg, A. Zalesky and M. Zukerman, 'Packet delay in optical circuit-switched networks', *IEEE/ACM Trans. Network.*, **14**(2), pp. 341–354, 2006.

[68] M. Veeraraghavan and Z. Xuan, 'A reconfigurable Ethernet/SONET circuit-based metro network architecture', *IEEE J. Sel. Areas Commun.*, **22**(8), pp. 1406–1418, 2004.

[69] X. Zhu, X. Zheng, M. Veeraraghavan, Z. Li, Q. Song, I. Habib and N. S. V. Rao, 'Implementation of a GMPLS-based network with end host initiated signaling', *Proc. IEEE Conf. on Communications (ICC'06)*, Istanbul, Turkey, pp. 2710–2716, 2006.

[70] H. Luo, G. Hu and L. Li, 'Burstification queue management in optical burst switching networks', *Photonics Netw. Commun.*, **11**(1), pp. 87–97, 2006.

[71] J. M. Martinez, J. Herrera, F. Ramos and J. Marti, 'All-optical address recognition scheme for label-swapping networks', *IEEE Photonics Technol. Lett.*, **18**(1), pp. 151–153, 2006.

[72] A. Chowdhury, J. Yu and G. K. Chang, 'Same wavelength packet switching in optical label switched networks', *J. Lightwave Technol.*, **24**(12), pp. 4838–4849, 2006.

[73] Multiprotocol Label Switching (MPLS), http://www.cisco.com/en/US/products/ps6557/products_ios_technology_home.html, 17 October 2007.

[74] Internet Engineering Task Force (IETF), http://www.ietf.org, 17 October 2007.

[75] ITU-T Recommendation G.7713.3, 'Distributed call and connection management: signalling mechanism using GMPLS CR-LDP', March 2003.

[76] C. Metz, 'Multiprotocol label switching and IP. Part 2. Multicast virtual private networks', *IEEE Internet Comput.*, **10**(1), pp. 76–81, 2006.

[77] C. Li-Der and H. Mao Yuan, 'Design and implementation of two-level VPN service provisioning systems over MPLS networks', *Proc. Int. Symp. on Computer Networks*, Istanbul, Turkey, pp. 42–48, June 2006.

[78] S. Tanaka *et al.*, 'Field test of GMPLS all-optical path rerouting', *IEEE Photonics Technol. Lett.*, **17**(3), pp. 723–725, 2005.

[79] T. D. Nadeau and H. Rakotoranto, 'GMPLS operations and management: today's challenges and solutions for tomorrow', *IEEE Commun. Mag.*, **43**(7), pp. 68–74, 2005.

[80] I. W. Habib, S. Qiang, L. Zhaoming and N. S. V. Rao, 'Deployment of the GMPLS control plane for grid applications in experimental high-performance networks', *IEEE Commun. Mag.*, **44**(3), pp. 65–73, 2006.

[81] T. Lehman, J. Sobieski and B. Jabbari, 'DRAGON: a framework for service provisioning in heterogeneous grid networks', *IEEE Commun. Mag.*, **44**(3), pp. 84–90, 2006.

[82] S. Spadaro, A. D'Alessandro, A. Manzalini and J. Solé-Pareta, 'TRIDENT: an automated approach to traffic engineering in IP/MPLS over ASON/GMPLS networks', *Comput. Netw.*, **51**(1), pp. 207–223, 2007.

[83] J. M. Kim, O. H. Kang, J. I. Jung and S. U. Kim, 'Control mechanism for QoS guaranteed multicast service in OVPN over IP/GMPLS over DWDM', *J. Commun.*, **2**(1), pp. 45, 2007.

[84] W. Xiaodong, H. Weisheng and P. Yunfeng, 'Lightpath-based flooding for GMPLS-controlled all-optical networks', *IEEE Commun. Lett.*, **11**(1), pp. 91–93, 2007.

[85] Y. Xiang, L. Jikai, C. Xiaojun, C. Yang and Q. Chunming, 'Traffic statistics and performance evaluation in optical burst switched networks', *J. Lightwave Technol.*, **22**(12), pp. 2722–2738, 2004.

[86] L. Yanjun, Z. GuoQing and C. Lu, 'A novel parallel scheduling protocol based on JET in optical burst switching', *Proc. Optical Fiber Communication and National Fiber Optic Engineering Conf. (OFC/NFOEC'06)*, Anaheim, CA, USA, JThB49 pp. 1–3, March 2006.

[87] N. Barakat and E. H. Sargent, 'Separating resource reservations from service requests to improve the performance of optical burst-switching networks', *IEEE J. Sel. Areas Commun.*, **24**(4), pp. 95–107, 2006.

[88] P. Kirci and A. H. Zaim, 'Comparison of OBS protocols', *Proc. Int. Symp. on Computer Networks*, Istanbul, Turkey, pp. 158–161, 16–18 June 2006.

[89] J. Y. Wei and R. I. McFarland Jr, 'Just-in-time signaling for WDM optical burst switching networks', *J. Lightwave Technol.*, **18**(12), pp. 2019–2037, 2000.

[90] S. Yongmei, T. Hashiguchi, M. Vu Quang, X. Wang, H. Morikawa and T. Aoyama, 'Design and implementation of an optical burst-switched network testbed', *IEEE Commun. Mag.*, **43**(11), pp. S48–S55, 2005.

[91] Y. Li and G. N. Rouskas, 'Adaptive path selection in OBS networks', *J. Lightwave Technol.*, **24**(8), pp. 3002–3011, 2006.

[92] J. Praveen, B. Praveen, T. Venkatesh, Y. V. Kiran and C. Siva Ram Murthy, 'A first step toward autonomic optical burst switched networks', *IEEE J. Sel. Areas Commun.*, **24**(12), pp. 94–105, 2006.

[93] http://www.telegeography.com, 17 October 2007.

[94] euNetworks: European optical fiber link, http://www.eunetworks.com/GV_Maps/map.htm, 17 October 2007.

[95] A. A. M. Saleh and J. M. Simmons, 'Evolution toward the next-generation core optical network', *J. Lightwave Technol.*, **24**(9), pp. 3303–3321, 2006.

[96] S. Chatterjee and S. C. Yang, 'Managing the optical networking solution: is increased bandwidth enough to reduce delays?', *IT Prof.*, **7**(1), pp. 46–50, 2005.

[97] C. J. Chen Yong, Chen Ting and Jian Shuisheng, 'Advanced modulation formats for long-haul optical-transmission systems with dispersion compensation by chirped FBG', *Microw. Opt. Technol. Lett.*, **48**(2), pp. 344–347, 2006.

[98] P. J. Winzer and R. J. Essiambre, 'Advanced modulation formats for high-capacity optical transport networks', *J. Lightwave Technol.*, **24**(12), pp. 4711–4728, 2006.

[99] G. E. Tudury, R. Salem, G. M. Carter and T. E. Murphy, 'Transmission of 80 Gbit/s over 840 km in standard fibre without polarisation control', *Electron. Lett.*, **41**(25), pp. 1394–1396, 2005.

[100] D. A. Fishman, D. L. Correa, E. H. Goode, T. L. Downs, A. Y. Ho, A. Hale, P. Hofmann, B. Basch and S. Gringeri, 'LambdaXtreme® National Network Deployment', *Bell Labs Tech. J.*, **11**(2), pp. 55–63, 2006.

[101] S. L. Jansen, D. van den Borne, P. M. Krummrich, S. Spalter, G. D. Khoe and H. de Waardt, 'Long-haul DWDM transmission systems employing optical phase conjugation', *IEEE J. Sel. Top. Quantum Electron.*, **12**(4), pp. 505–520, 2006.

[102] H. Maeda, G. Funatsu and A. Naka, 'Ultra-long-span 500 km 16×10 Gbit/s WDM unrepeatered transmission using RZ-DPSK format', *Electron. Lett.*, **41**(1), pp. 34–35, 2005.

[103] E. Le Rouzic and S. Gosselin, '160-Gb/s optical networking: a prospective techno-economical analysis', *J. Lightwave Technol.*, **23**(10), pp. 3024–3033, 2005.

[104] Submarine cable system diagram, http://www.telegeography.com/ee/free_resources/ figures/ib-02.php, 17 October 2007.

[105] TAT-14 Transatlantic cable network, https://www.tat-14.com/tat14, 17 October 2007.

[106] Submarine cable map 2007, http://telegeography.com/products/map_cable/index.php, 17 October 2007.

[107] E. B. Desurvire, 'Capacity demand and technology challenges for lightwave systems in the next two decades', *J. Lightwave Technol.*, **24**(12), pp. 4697–4710, 2006.

[108] Deployment of submarine optical fiber cable and communication systems since 2001, http://www.atlantic-cable.com/Cables/CableTimeLine/index 2001.htm, 17 October 2007.

[109] L. Ming-Jun, M. J. Soulliere, D. J. Tebben, L. Nederlof, M. D. Vaughn and R. E. Wagner, 'Transparent optical protection ring architectures and applications', *J. Lightwave Technol.*, **23**(10), pp. 3388–3403, 2005.

[110] S. Chamberland, M. St-Hilaire and S. Pierre, 'A heuristic algorithm for the point of presence design problem in IP networks', *IEEE Commun. Lett.*, **7**(9), pp. 457–459, 2003.

[111] A. Tychopoulos, I. Papagiannakis, D. Klonidis, A. Tzanakaki, J. Kikidis, O. Koufopavlou and I. Tomkos, 'A low-cost inband FEC scheme for SONET/SDH optical metro networks', *IEEE Photonics Technol. Lett.*, **18**(24), pp. 2581–2583, 2006.

[112] N. Vanderhorn, S. Balasubramanian, M. Mina, R. J. Weber and A. K. Somani, 'Light-trail testbed for metro optical networks', *Proc. Testbeds and Research Infrastructures for the Development of Networks and Communities (TRIDENTCOM'06)*, Barcelona, Spain, pp. 1–6, March 2006.

[113] N. Kataoka, W. Naoya, K. Sone, Y. Aoki, H. Miyata, H. Onaka and K. Kitayama, 'Field trial of data-granularity-flexible reconfigurable OADM with wavelength-packet-selective switch', *J. Lightwave Technol.*, **24**(1), pp. 88–94, 2006.

[114] L. Sang-Mook, M. Sil-Gu, K. Min-Hwan and L. Chang-Hee, 'Demonstration of a long-reach DWDM-PON for consolidation of metro and access networks', *J. Lightwave Technol.*, **25**(1), pp. 271–276, 2007.

[115] P. S. Darren and E. M. John, 'A 10-Gb/s 1024-way-split 100-km long-reach optical-access network', *J. Lightwave Technol.*, **25**(3), pp. 685–693, 2007.

[116] J. C. Point, 'HFC and WIMAX evolutions for service convergence', *Proc. IET Int. Conf. on Access Technologies*, Cambridge, UK, pp. 17–20, June, 2006.

[117] L. Goleniewski and K. W. Jarrett, *Telecommunications Essentials: The Complete Global Source*, Addison Wesley, 2006.

[118] ANSI/SCTE-40, 'Digital cable network interface standard', 2004.

[119] Data over cable service interface specifications, DOCSIS 3.0, Physical layer specification, CM-SP-PHYv3.0-I05-070803; Cable Television Laboratories, Inc., 8 March 2007.

[120] ITU-T J.222.3, 'Interactive cable television services – IP cable modems', December 2007.

[121] ITU-T G.992.1, 'Asymmetric digital subscriber line (ADSL) transceivers', July 1999.

[122] ITU-T G.992.3, 'Asymmetric digital subscriber line transceivers 2 (ADSL2)', January 2005.

[123] ITU-T Recommendation G.992.5, 'Asymmetric digital subscriber line (ADSL) transceivers – extended bandwidth ADSL2 (ADSL2+)', January 2005.

[124] ITU-T Recommendation G.993.1, 'Very high speed digital subscriber line transceivers', June 2004.

[125] ITU-T G.993.2, 'Very high speed digital subscriber line transceivers 2 (VDSL2)', February 2008.

[126] http://standards.ieee.org/getieee802/drafts.html, 25 June 2008.

[127] ITU-T G.983.1, 'Broadband optical access systems based on passive optical networks (PON)', January 2005.

[128] ITU-T Recommendation G.984.1, 'Gigabit-capable passive optical networks (GPON): general characteristics', March 2008.

[129] FSAN: Full Service Access Network, http://www.fsanweb.org, 17 October 2007.

[130] D. Liu, B. Hu and H. Liu, 'Low-cost DWDM technology used for the Ethernet passive optical network (EPON)', *Proc. SPIE*, **4870**, pp. 465–470, 2003.

[131] A. Banerjee, Y. Park, F. Clarke, H. Song, S. Yang, G. Kramer, K. Kim and B. Mukherjee, 'Wavelength-division-multiplexed passive optical network (WDM-PON) technologies for broadband access: a review', *J. Opt. Netw.*, **4**(11), pp. 737–758, 2005.

[132] ITU-T Recommendation G.984.2, 'Gigabit-capable passive optical networks (G-PON): physical media dependent (PMD) layer specification, March 2008.

[133] T. Gilfedder, 'Deploying GPON technology for backhaul applications', *BT Technol. J.*, **24**(2), pp. 20–25, 2006.

[134] R. P. Davey, P. Healey, I. Hope, P. Watkinson, D. B. Payne, O. Marmur, J. Ruhmann and Y. Zuiderveld, 'DWDM reach extension of a GPON to 135 km', *J. Lightwave Technol.*, **24**(1), pp. 29–31, 2006.

[135] H. Frazier, 'The 802.3z Gigabit Ethernet Standard', *IEEE Netw.*, **12**(3), pp. 6–7, 1998.

[136] G. E. Keiser, *FTTX Concepts and Applications*, Wiley–IEEE Press, 2006.

[137] ITU-T Recommendation G.7041, 'Generic framing procedure (GFP)', August 2005.

[138] ITU-T Recommendation G.984.3, Amendment 1 (07/05), 'Gigabit-capable passive optical networks (G-PON): transmission convergence layer specification', March 2008.

[139] M. Abrams and A. Maislos, 'Insights on delivering an IP triple play over GE-PON and GPON', *Proc. Optical Fiber Communication and National Fiber Optic Engineering Conf., (OFC/NFOEC) 2006*, Anaheim, CA, USA, p. 8, March 2006.

[140] IEEE 802, http://www.ieee802.org/3, 17 October 2007.

[141] T. Nomura, H. Ueda, C. Itoh, H. Kurokawa, T. Tsuboi and H. Kasai, 'Design of optical switching module for gigabit Ethernet optical switched access network', *IEICE Trans. Commun.*, **89**(11), pp. 3021, 2006.

[142] W. Lee, M. Y. Park, S. H. Cho, J. Lee, C. Kim, G. Jeong and B. W. Kim, 'Bidirectional WDM-PON based on gain-saturated reflective semiconductor optical amplifiers', *IEEE Photonics Technol. Lett.*, **17**(11), pp. 2460–2462, 2005.

[143] M. Herzog, M. Maier and A. Wolisz, 'RINGOSTAR: an evolutionary AWG-based WDM upgrade of optical ring networks', *J. Lightwave Technol.*, **23**(4), pp. 1637–1651, 2005.

[144] R. Davey, J. Kani, F. Bourgart and K. McCammon, 'Options for future optical access networks', *IEEE Commun. Mag.*, **44**(10), pp. 50–56, 2006.

[145] S. Dong Jae *et al.*, 'Low-cost WDM-PON with colorless bidirectional transceivers', *J. Lightwave Technol.*, **24**(1), pp. 158–165, 2006.

[146] K. Lee, J. H. Song, H. K. Lee and W. V. Sorin, 'Multistage access network for bidirectional DWDM transmission using ASE-injected FP-LD', *IEEE Photonics Technol. Lett.*, **18**(6), pp. 761–763, 2006.

[147] C. H. Lee, S. M. Lee, K. M. Choi, J. H. Moon and S. G. Mun, 'WDM-PON experiences in Korea', *J. Opt. Netw.*, **6**(5), pp. 451–464, 2007.

[148] Y. Shachaf, C. H. Chang, P. Kourtessis and J. M. Senior, 'Multi-PON access network using a coarse AWG for smooth migration from TDM to WDM PON', *Opt. Express*, **15**(12), pp. 7840–7844, 2007.

[149] J. M. Senior, A. J. Phillips, M. S. Leeson, R. Johnson, M. O. Van Deventer, J. M. Peter and I. Van de Voorde, 'Upgrading SuperPON: next step for future broadband access networks', *Proc. SPIE*, **2919**, pp. 260–266, 1996.

[150] G. Talli and P. D. Townsend, 'Hybrid DWDM-TDM long-reach PON for next-generation optical access', *J. Lightwave Technol.*, **24**(7), pp. 2827–2834, 2006.

[151] R. P. Davey, D. Nesset, A. Rafel, D. B. Payne and A. Hill, 'Designing long reach optical access networks', *BT Technol. J.*, **24**(2), pp. 13–19, 2006.

[152] A. J. Phillips, J. M. Senior, R. Mercinelli, M. Valvo, P. J. Vetter, C. M. Martin, M. O. Van Deventer, P. Vaes and X. Z. Qiu, 'Redundancy strategies for a high splitting optically amplified passive optical network', *J. Lightwave Technol.*, **19**(2), pp. 137–149, 2001.

[153] I. T. Monroy, F. Ohman, K. Yvind, R. Kjaer, C. Peucheret, A. M. J. Koonen and P. Jeppesen, '85 km long reach PON system using a reflective SOA-EA modulator and distributed Raman fiber amplification', *Proc. IEEE/LEOS'06*, Montreal, Canada, pp. 705–706, October/ November 2006.

[154] J. Zheng and H. T. Mouftah, 'Media access control for Ethernet passive optical networks: an overview', *IEEE Commun. Mag.*, **43**(2), pp. 145–150, 2005.

[155] C. Guan and V. W. S. Chan, 'Topology design of OXC-switched WDM networks', *IEEE J. Sel. Areas Commun.*, **23**(8), pp. 1670–1686, 2005.

[156] E. Wong and C. Chang-Joon, 'CSMA/CD-based Ethernet passive optical network with optical internetworking capability among users', *IEEE Photonics Technol. Lett.*, **16**(9), pp. 2195–2197, 2004.

[157] M. Barranco, J. Proenza, G. Rodriguez-Navas and L. Almeida, 'An active star topology for improving fault confinement in CAN networks', *IEEE Trans. Ind. Inf.*, **2**(2), pp. 78–85, 2006.

[158] L. Wen-Piao, K. Ming-Seng and C. Sien, 'The modified star-ring architecture for high-capacity subcarrier multiplexed passive optical networks', *J. Lightwave Technol.*, **19**(1), pp. 32–39, 2001.

[159] Nemertes Research, Cost models and economics, http://www.nemertes.com/cost_models _economics, 17 October 2007.

[160] Fiber optics LAN section, http://www.fols.org, 17 October 2007.

[161] F. Ross, 'FDDI-a tutorial', *IEEE Commun. Mag.*, **24**(5), pp. 10–17, 1986.

[162] F. E. Ross, 'An overview of FDDI: the fiber distributed data interface', *IEEE J. Sel. Areas Commun.*, **7**(7), pp. 1043–1051, 1989.

[163] ANSI INCITS 166 Information Systems – Fiber Distributed Data Interface (FDDI) – Token Ring Physical Layer Medium Dependent (PMD), January 1990.

[164] ANSI INCITS 148 Information Systems – Fiber Distributed Data Interface (FDDI) – Token Ring Physical Layer Protocol (PHY), January 1988.

[165] ANSI INCITS 139 Information Systems – Fiber Distributed Data Interface (FDDI) – Token Ring Media Access Control (MAC), January 1987.

[166] ANSI INCITS 229 Information Systems – Fiber Distributed Data Interface (FDDI) – Station Management (SMT), January 1994.

[167] M. Teener and R. Gvozdanovic, 'FDDI-II operation and architectures', *Proc. Local Computer Networks Conf.*, Minneapolis, USA, pp. 49–61, March, 1989.

[168] J. P. Weem, P. Kirkpatrick and J. M. Verdiell, 'Electronic dispersion compensation for 10 Gigabit communication links over FDDI legacy multimode fiber', *Proc. Optical Fiber Communication and National Fiber Optic Engineering Conf. (OFC/NFOEC'05)*, California, USA, pp. 1–3, March 2005.

[169] A. M. E. A. Diab, J. D. Ingham, R. V. Penty and I. H. White, '10-Gb/s transmission on single-wavelength multichannel SCM-based FDDI-grade MMF links at lengths over 300 m: a statistical investigation', *J. Lightwave Technol.*, **25**(10), pp. 2976–2983, 2007.

[170] M. H. Sherif, 'Technology substitution and standardization in telecommunication services', *Proc. Standardization and Innovation in Information Technology Conf.*, pp. 241–252, October 2003.

[171] ITU-T Recommendation. G.985 (03/2003), '100 Mbit/s point-to-point Ethernet based optical access system', March 2003.

[172] M. P. McGarry, M. Reisslein and M. Maier, 'WDM Ethernet passive optical networks', *IEEE Commun. Mag.*, **44**(2), pp. 15–22, 2006.

[173] M. N. Petersen, M. H. Olesen and M. S. Berger, '10 Gb/s non-regenerated pure Ethernet field trials over dark fibers', *Proc. Optical Network Design and Modeling (ONDM'05) Conf.*, Milan, Italy, pp. 49–54, February 2005.

[174] C. F. Lam, D. Tsang and C. C. K. Chan, 'Optical Ethernet II: introduction to the feature issue', *J. Opt. Netw.*, **6**(2), pp. 121–122, 2007.

[175] ISO/IEC standard 11801, Cabling standards, http://www.iso.org, 17 October 2007.

[176] 10 Gigabit Ethernet Alliance (10GEA), http://www.ethernetalliance.org, 17 October 2007.

[177] M. V. Lau, S. Shieh, W. Pei-Feng, B. Smith, D. Lee, J. Chao, B. Shung and S. Cheng-Chung, 'Gigabit Ethernet switches using a shared buffer architecture', *IEEE Commun. Mag.*, **41**(12), pp. 76–84, 2003.

[178] P. Szegedi, 'VLAN-sensitive optical protection for scalable SLA definition of switched Ethernet services', *Proc. Int. Conf. on Transparent Optical Networks (TON'04)*, Wroclaw, Poland, pp. 255–258, July 2004.

[179] ITU-T Recommendation G.8011.2/Y.1307.2, 'Ethernet virtual private line (EVPL) service', September 2005.

[180] J. M. Jo, S. J. Lee, K. D. Hong, C. J. Lee, O. H. Kang and S. U. Kim, 'Virtual source-based minimum interference path multicast routing in optical virtual private networks', *Photonics Netw. Commun.*, **13**(1), pp. 19–30, 2007.

[181] L. Sang-Woo, J. Yong-Sung, K. Ki-Young and J. Jong-Soo, 'Implementation of 10Gb Ethernet switch hardware platform with a network processor and a 10Gb EMAC', *Proc. Int. Conf. on Advanced Communications Technology (ICACT'07)*, Korea, pp. 563–566, February 2007.

[182] ITU-T Recommendation Q.838.1, 'Requirements and analysis for the management interface of Ethernet passive optical networks (EPON)', October 2004.

[183] H. Nakamura, H. Suzuki, K. Jun-ichi and K. Iwatsuki, 'Reliable wide-area wavelength division multiplexing passive optical network accommodating gigabit Ethernet and 10-Gb Ethernet services', *J. Lightwave Technol.*, **24**(5), pp. 2045–2051, 2006.

[184] C. F. Lam and W. I. Way, 'Optical Ethernet: protocols, managment and 1–100 G technologies', in I. P. Kaminow, T. Li and A. E. Willner (Eds), *Optical Fiber Telecommunications VB*, pp. 345–400, Elsevier/Academic Press, 2008.

[185] M. Ali, G. Chiruvolu and A. Ge, 'Traffic engineering in metro Ethernet', *IEEE Netw.*, **19**(2), pp. 10–17, 2005.

[186] A. Zapata, M. Duser, J. Spencer, P. Bayvel, I. de Miguel, D. Breuer, N. Hanik and A. Gladisch, 'Next-generation 100-gigabit metro Ethernet (100 GbME) using multiwavelength optical rings', *J. Lightwave Technol.*, **22**(11), pp. 2420–2434, 2004.

[187] T. Yukio *et al.*, 'The first application-driven lambda-on-demand field trial over a US nation wide network', *Proc. Optical Fiber Communication and National Fiber Optic Engineering Conf. (OFC/NFOEC'06)*, Anaheim, CA, USA, post deadline publication paper PDP-48, March 2006.

[188] L. Zhou, T. Y. Chai, C. V. Saradhi, Y. Wang, V. Foo, Q. Qiang, J. Biswas, C. Lu, M. Gurusamy and T. H. Cheng, 'Development of a GMPLS-capable WDM optical network testbed and distributed storage application', *IEEE Commun. Mag.*, **44**(2), pp. S26–S32, 2006.

[189] P. J. Winzer, G. Raybon and M. Duelk, '107-Gb/s optical ETDM transmitter for 100G Ethernet transport', *Proc. Eur. Conf. on Optical Communications (ECOC'05)*, Glasgow, UK, **6**, pp. 1–2, 25–29 September 2005.

[190] M. Daikoku, I. Morita, H. Taga, H. Tanaka, T. Kawanishi, T. Sakamoto, T. Miyazaki and T. Fujita, '100 Gbit/s DQPSK transmission experiment without OTDM for l00G Ethernet transport', *Proc. Optical Fiber Communication and National Fiber Optic Engineering Conf. (OFC/NFOEC'06)*, Anaheim, CA, USA, pp. 1–3, 2–10 March, 2006.

[191] A. Schmid-Egger, A. Kirstädter and J. Eberspächer, 'Ethernet in the backbone: an approach to cost-efficient core networks', in E. M. Kern, H.-G. Hegering and B. Brügge (Eds), *Managing Development and Application of Digital Technologies*, Springer-Verlag, pp. 195–212, 2006.

[192] J. Mocerino, 'Carrier class Ethernet service delivery migrating SONET to IP & triple play offerings', *Proc. Optical Fiber Communication and National Fiber Optic Engineering Conf. (OFC/NFOEC'06)*, Anaheim, CA, USA, pp. 396–401, March 2006.

[193] P. Kloppenburg, 'Optical Ethernet is evolving as a delivery vehicle for metro retail services', *Proc. Optical Fiber Communication and National Fiber Optic Engineering Conf. (OFC/NFOEC'06)*, Anaheim, CA, USA, p. 10, March 2006.

[194] O. Canhui, L. H. Sahasrabuddhe, Z. Keyao, C. U. Martel and B. Mukherjee, 'Survivable virtual concatenation for data over SONET/SDH in optical transport networks', *IEEE/ACM Trans. Netw.*, **14**(1), pp. 218–231, 2006.

[195] G. Bernstein, D. Caviglia, R. Rabbat and H. Van Helvoort, 'VCAT-LCAS in a clamshell', *IEEE Commun. Mag.*, **44**(5), pp. 34–36, 2006.

[196] S. Hongsheng, X. Yunbin, G. Xuan, Z. Jie and G. Wanyi, 'Design and implementation of intelligent optical network management system', *Proc. Communications Technology Int. Conf. (ICCT'03)*, **1**, pp. 625–628, April 2003.

[197] L. Ruan and F. Tang, 'Survivable IP network realization in IP-over-WDM networks under overlay model', *Comput. Commun.*, **29**(10), pp. 1772–1779, 2006.

[198] M. Guido, P. Achille, P. De Simone and M. Mario, 'Optical network survivability: protection techniques in the WDM layer', *Photonics Netw. Commun.*, **4**(3), pp. 251–269, 2002.

[199] Y. Xi, S. Lu and B. Ramamurthy, 'Survivable lightpath provisioning in WDM mesh networks under shared path protection and signal quality constraints', *J. Lightwave Technol.*, **23**(4), pp. 1556–1567, 2005.

[200] H. Höller and S. Voss, 'A heuristic approach for combined equipment-planning and routing in multi-layer SDH/WDM networks', *Eur. J. Oper. Res.*, **171**(3), pp. 787–796, 2006.

[201] S. Ramamurthy, L. Sahasrabuddhe and B. Mukherjee, 'Survivable WDM mesh networks', *J. Lightwave Technol.*, **21**(4), pp. 870–883, 2003.

[202] W. D. Grover, *Mesh-based Survivable Networks: Options and Strategies for Optical, MPLS, SONET, and ATM Networking*, Prentice Hall PTR, 2004.

[203] L. Guo, J. Cao, H. Yu and L. Li, 'Path-based routing provisioning with mixed shared protection in WDM mesh networks', *J. Lightwave Technol.*, **24**(3), pp. 1129–1141, 2006.

Appendices

A. The field relations in a planar guide

Let us consider an electromagnetic wave having an angular frequency ω propagating in the z direction with propagation vector (phase constant) β. Then as indicated in Section 2.3.2, the electric and magnetic fields can be expressed as:

$$\mathbf{E} = \mathrm{Re}\{\mathbf{E}_0(x, y)\,\exp[\,\mathrm{j}(\omega t - \beta z)]\} \tag{A1}$$

$$\mathbf{H} = \mathrm{Re}\{\mathbf{H}_0(x, y)\,\exp[\,\mathrm{j}(\omega t - \beta z)]\} \tag{A2}$$

For the planar guide the Cartesian components of $\mathbf{E}_0$ and $\mathbf{H}_0$ become:

$$\frac{\partial E_z}{\partial y} + \mathrm{j}\beta E_y = -\mathrm{j}\mu_r\mu_0\omega H_x \tag{A3}$$

$$\mathrm{j}\beta E_x + \frac{\partial E_z}{\partial x} = \mathrm{j}\mu_r\mu_0\omega H_y \tag{A4}$$

$$\frac{\partial E_y}{\partial x} - \frac{\partial E_x}{\partial y} = -\mathrm{j}\mu_r\mu_0\omega H_z \tag{A5}$$

$$\frac{\partial H_z}{\partial y} + \mathrm{j}\beta H_y = \mathrm{j}\omega\varepsilon_r\varepsilon_0 E_x \tag{A6}$$

$$-\mathrm{j}\beta H_x - \frac{\partial H_z}{\partial x} = \mathrm{j}\omega\varepsilon_r\varepsilon_0 E_y \tag{A7}$$

$$\frac{\partial H_y}{\partial x} - \frac{\partial H_x}{\partial y} = \mathrm{j}\omega\varepsilon_r\varepsilon_0 E_z \tag{A8}$$

If we assume that the planar structure is an infinite film in the y–z plane, then for an infinite plane wave traveling in the z direction the partial derivative with respect to y is zero ($\partial/\partial y = 0$). Employing this assumption we can simplify the above equations to demonstrate fundamental relationships between the fields in such a structure. These are:

$$\mathrm{j}\beta E_y = -\mathrm{j}\mu_r\mu_0\omega H_x \quad \text{(TE mode)} \tag{A9}$$

$$j\beta E_x + \frac{\partial E_z}{\partial x} = j\mu_r\mu_0\omega H_y \quad \text{(TM mode)} \tag{A10}$$

$$\frac{\partial E_y}{\partial x} = -j\mu_r\mu_0\omega H_z \quad \text{(TE mode)} \tag{A11}$$

$$j\beta H_y = j\omega\varepsilon_r\varepsilon_0 E_x \quad \text{(TM mode)} \tag{A12}$$

$$-j\beta H_x - \frac{\partial H_z}{\partial x} = j\omega\varepsilon_r\varepsilon_0 E_y \quad \text{(TE mode)} \tag{A13}$$

$$\frac{\partial H_y}{\partial x} = j\omega\varepsilon_r\varepsilon_0 E_z \quad \text{(TM mode)} \tag{A14}$$

It may be noted that the fields separate into TE and TM modes corresponding to coupling between E_y, H_x, H_z ($E_z = 0$) and H_y, E_x, E_z ($H_z = 0$) respectively.

B. Gaussian pulse response

Many optical fibers, and in particular jointed fiber links, exhibit pulse outputs with a temporal variation that is closely approximated by a Gaussian distribution. Hence the variation in the optical output power with time may be described as:

$$P_o(t) = \frac{1}{\sqrt{(2\pi)}} \exp\left[-\left(\frac{t^2}{2\sigma^2}\right)\right] \tag{B1}$$

where σ and σ^2 are the standard deviation and the variance of the distribution respectively. If t_e represents the time at which $P_o(t_e)/P_o(0) = 1/e$ (i.e. $1/e$ pulse width), then from Eq. (B1) it follows that:

$$t_e = \sigma\sqrt{2}$$

Moreover, if the full width of the pulse at the $1/e$ points is denoted by τ_e then:

$$\tau_e = 2t_e = 2\sigma\sqrt{2}$$

In the case of the Gaussian response given by Eq. (B1) the standard deviation σ is equivalent to the rms pulse width.

The Fourier transform of Eq. (B1) is given by:

$$\mathscr{P}(\omega) = \frac{1}{\sqrt{(2\pi)}} \exp\left[-\left(\frac{\omega^2\sigma^2}{2}\right)\right] \tag{B2}$$

The 3 dB optical bandwidth B_{opt} is defined in Section 7.4.3 as the modulation frequency at which the received optical power has fallen to one-half of its constant value. Thus using Eq. (B2):

$$\frac{[\omega(3 \text{ dB opt})]^2}{2} \sigma^2 = 0.693$$

and:

$$\omega(3 \text{ dB opt}) = 2\pi B_{opt} = \frac{\sqrt{2} \times 0.8326}{\sigma}$$

Hence:

$$B_{opt} = \frac{\sqrt{2} \times 0.8326}{2\pi\sigma} = \frac{0.530}{\tau_e} = \frac{0.187}{\sigma} \text{ Hz}$$

When employing a return-to-zero pulse where the maximum bit rate $B_T(\text{max}) = B_{opt}$, then:

$$B_T(\text{max}) \simeq \frac{0.2}{\sigma} \text{ bit s}^{-1}$$

Alternatively, the 3 dB electrical bandwidth B occurs when the received optical power has dropped to $1/\sqrt{2}$ of the constant value (see Section 7.4.3) giving:

$$B = \frac{0.530}{\tau_e\sqrt{2}} = \frac{0.375}{\tau_e} = \frac{0.133}{\sigma} \text{ Hz}$$

C. Variance of a random variable

The statistical mean (or average) value of a discrete random variable X is the numerical average of the values which X can assume weighted by their probabilities of occurrence. For example, if we consider the possible numerical values of X to be $x_1, x_2, \ldots, x_i$, with probabilities of occurrence $P(x_1), P(x_2), \ldots, P(x_i)$, then as the number of measurements N of X goes to infinity, it would be expected that the outcome $X = x_1$ would occur $NP(x_1)$ times, the outcome $X = x_2$ would occur $NP(x_2)$ times, and so on. In this case the arithmetic sum of all N measurements is:

$$x_1 P(x_1)N + x_2 P(x_2)N + \ldots + x_i P(x_i)N = N \sum_i x_i P(x_i) \tag{C1}$$

The mean or average value of all these measurements which is equivalent to the mean value of the random variable may be calculated by dividing the sum in Eq. (C1) by the number of measurements N. Furthermore, the mean value for the random variable X which can be denoted as $\bar{X}$ (or m) is also called the expected value of X and may be represented by $E(X)$. Hence:

$$\bar{X} = m = E(X) = \sum_{i=1}^{N} x_i P(x_i) \tag{C2}$$

Moreover, Eq. (C2) also defines the first moment of X which we denote as M_1. In a similar manner the second moment M_2 is equal to the expected value of X^2 such that:

$$M_2 = \sum_{i=1}^{N} x_i^2 P(x_i) \tag{C3}$$

M_2 is also called the mean square value of X which may be denoted as $\overline{X^2}$.

For a continuous random variable, the summation of Eq. (C2) approaches an integration over the whole range of X so that the expected value of X:

$$M_1 = E(X) = \int_{-\infty}^{\infty} x p_X(x) \, dx \tag{C4}$$

where $p_X(x)$ is the probability density function of the continuous random variable X. Similarly, the expected value of X^2 is given by:

$$M_2 = E(X^2) = \int_{-\infty}^{\infty} x^2 p_X(x) \, dx \tag{C5}$$

It is often convenient to subtract the first moment $M_1 = m$ prior to computation of the second moment. This is analogous to moments in mechanics which are referred to the center of gravity rather than the origin of the coordinate system. Such a moment is generally referred to as a central moment. The second central moment represented by the symbol σ^2 is therefore defined as:

$$\sigma^2 = E[(X - m)^2] = \int_{-\infty}^{\infty} (x - m)^2 p_X(x) \, dx \tag{C6}$$

where σ^2 is called the variance of the random variable X. Moreover, the quantity σ which is known as the standard deviation is the root mean square (rms) value of $(X - m)$.

Expanding the squared term in Eq. (C6) and integrating term by term we find:

$$\begin{aligned}
\sigma^2 &= E[X^2 - 2mX + m^2] \\
&= E(X^2) - 2mE(X) + E(m^2) \\
&= E(X^2) - 2m^2 - m^2 \\
&= E(X^2) - m^2
\end{aligned} \tag{C7}$$

As $E(X^2) = M_2$ and $m = M_1$, the variance may be written as:

$$\sigma^2 = M_2 - (M_1)^2$$

D. Variance of the sum of independent random variables

If a random variable $W = g(X, Y)$ is a function of two random variables X and Y, then extending the definition in Eq. (C4) for expected values gives the expected value of W as:

$$E(W) = \int_{-\infty}^{\infty} \int_{-\infty}^{\infty} g(x, y) p_{XY}(x, y) \, dx \, dy \tag{D1}$$

where $p_{XY}(x, y)$ is the joint probability density function. Furthermore, the two random variables X and Y are statistically independent when:

$$p_{XY}(x, y) = p_X(x)p_Y(y) \tag{D2}$$

Now let X and Y be two statistically independent random variables with variances σ_X^2 and σ_Y^2 respectively. In addition we assume the sum of these random variables to be another random variable denoted by Z such that $Z = X + Y$, where Z has a variance σ_Z^2. If the mean values of X and Y are zero, employing the definition of variance given in Eq. (C6) together with the expected value for a function of two random variables (Eq. (D1)) we can write:

$$\sigma_Z^2 = \int_{-\infty}^{\infty} \int_{-\infty}^{\infty} (x + y)^2 p_{XY}(x, y) \, dx \, dy \tag{D3}$$

As X and Y are statistically independent we can utilize Eq. (D2) to obtain:

$$\sigma_Z^2 = \int_{-\infty}^{\infty} \int_{-\infty}^{\infty} (x + y)^2 p_X(x)p_Y(y) \, dx \, dy$$

$$= \int_{-\infty}^{\infty} x^2 p_X(x) \, dx + \int_{-\infty}^{\infty} y^2 p_Y(y) \, dy$$

$$+ 2 \int_{-\infty}^{\infty} x p_X(x) \, dx \int_{-\infty}^{\infty} y p_Y(y) \, dy \tag{D4}$$

The two factors in the last term of Eq. (C4) are equal to the mean values of the random variables (X and Y) and hence are zero. Thus:

$$\sigma_Z^2 = \sigma_X^2 + \sigma_Y^2$$

E. Closed loop transfer function for the transimpedance amplifier

The closed loop transfer function $H_{CL}(\omega)$ for the transimpedance amplifier shown in Figure 9.11 may be derived by summing the currents at the amplifier input, remembering that the amplifier input resistance is included in R_{TL}. Hence:

$$i_{det} + \frac{V_{out} - V_{in}}{R_f} = V_{in}\left(\frac{1}{R_{TL}} + j\omega C_T\right) \tag{E1}$$

As $V_{in} = -V_{out}/G$, then:

$$i_{det} = -V_{out}\left(\frac{1}{R_f} + \frac{1}{GR_f} + \frac{1}{GR_{TL}} + \frac{j\omega C_T}{G}\right) \tag{E2}$$

Therefore:

$$H_{CL}(\omega) = \frac{V_{out}}{i_{det}} = \frac{1}{1 + (1/G) + (R_f/GR_{TL}) + (j\omega C_T R_f/G)}$$

$$= \frac{-R_f/(1 + 1/G + R_f/GR_{TL})}{[1 + j\omega C_T R_f/(1 + R_f/R_{TL} + G)]} \tag{E3}$$

Since:

$$G \gg \left(1 + \frac{R_f}{R_{TL}}\right) \tag{E4}$$

then Eq. (E3) becomes:

$$H_{CL}(\omega) \simeq \frac{-R_f}{1 + (j\omega R_f C_T/G)} \text{ V A}^{-1}$$

Index